D1609484

Mammals of Mexico

MAMMALS OF MEXICO

Edited by
Gerardo Ceballos

Johns Hopkins University Press
Baltimore

Printed in China on acid-free paper
9 8 7 6 5 4 3 2 1

Johns Hopkins University Press
2715 North Charles Street
Baltimore, Maryland 21218-4363
www.press.jhu.edu

An earlier edition of this work appeared in Spanish as *Los Mamíferos Silvestres de México*, edited by Gerardo Ceballos and Gisselle Oliva (Comisión Nacional para el Conocimiento y Uso de la Biodiversidad Fondo de Cultura Económica, 2005)

Library of Congress Cataloging-in-Publication Data

Mammals of Mexico / edited by Gerardo Ceballos.
974 pages cm
Includes bibliographical references and index.
ISBN-13: 978-1-4214-0843-9 (hardcover : alk. paper)
ISBN-13: 978-1-4214-0879-8 (electronic)
ISBN-10: 1-4214-0843-0 (hardcover : alk. paper)
ISBN-10: 1-4214-0879-1 (electronic)
1. Mammals — Mexico. I. Ceballos, Gerardo.
QL722.M36 2013
599 — dc23 2012036684

A catalog record for this book is available from the British Library.

Introductory photographic plates:
Black-tailed prairie dog (*Cynomys ludovicianus*) in the grasslands of the Janos Biosphere Reserve, Chihuahua. Photo: Gerardo Ceballos.
Free tailed-bats (*Tadarida brasiliensis*) form colonies of millions of individuals, northern Mexico. Photo: Gerardo Ceballos.
Jaguar (*Panthera onca*) in a semi-evergreen forest in Calakmul Biosphere Reserve, Campeche. Photo: Gerardo Ceballos.
Page ii Ring-tailed cat (*Bassarisus astutus*) in an oak forest, Picachos, Nuevo Leon. Photo: Marcelo Sada.
Page vi North American porcupine (*Erethizon dorsatum*) in the Janos Biosphere Reserve, Chihuahua. Photo: Rurik List.
Page ix Elk (*Cervus canadensis*) were reintroduced to the Serranias del Burro region of Coahuila in the 1960s. Photo: Gerardo Ceballos.
Page x Gray whale (*Eschrichtius robustus*) in Laguna San Ignacio, Vizcaíno Biosphere Reserve, Baja Sur. Photo: Pieter Folkens.
Page xii Blue whale (*Balaenoptera musculus*) near Aqua Verde, Sea of Cortez, Baja. Photo: Michael Fishbach.

Publication was made possible, in part, by a grant from CONABIO

To Guadalupe, Pablo, and Regina
the light of my life and the joy of my existence

To my late dad, Oscar, and my mother, Leonor,
for all their love and support, that forged me as
a biologist

To Paul R. Ehrlich, the late Sir James Goldsmith, and José Sarukhan,
for their friendship, teachings, support, and guidance.
I have no words to express my gratitude to them.

To the Universidad Nacional Autonoma de Mexico,
the greatest educational institution in the history of Mexico

Contents

Preface

This book is the product of an idea that I had in the early 1990s, when I realized that there was a clear need for a volume on the mammals in Mexico. So I decided to edit such a book. I was completely unaware of the complexity of the task that I was going to face, which became one of the most important challenges of my professional life. To prepare the edition in Spanish took almost a decade of hard work. The road to complete the book was long and difficult, but also extremely interesting and stimulating. Once printed, the book was sold out in less than a year, indicating the need for such a volume. My coeditor, Gisselle Oliva, and I were extremely happy. A few years later, I had a conversation with Vincent Burke, the executive editor of Johns Hopkins University Press, and we decided to have the book translated and published in English.

The main objective of the book is to present information on the mammals of Mexico in a clear and simple way to a wide audience. The introductory material is divided into three sections. The first presents a synthesis of the biodiversity and conservation of mammals in Mexico. The second is devoted to a brief overview of the history of mammalogy in Mexico, featuring authors and works that we believe have had the greatest impact on the development of this science. The third presents a systematic list of the mammals of the country updated to 2011, with comments on the compilation of the list and the taxonomic decisions made about the species included in the book. This introductory material is followed by species accounts for all the species of mammals recorded in the country, including all marine ones. Appendixes A and B supplement the species accounts by providing the tracks of medium-sized and large species and photographs of the skulls of most genera.

In order to make the information more accessible to general readers, the use of specialized terms or jargon is minimized. The nomenclature of species is based on Wilson and Reeder (2005), with a number of modifications indicated in the introductory Systematic List of Species. The species accounts are arranged phylogenetically at the order and family levels, and alphabetically at the genera and species levels. Each account includes sections with the scientific name, common name, description, natural history and ecology, vegetational associations and elevation range, conservation status, subspecies in Mexico, external measures and weight, dental formula, and distribution. A map with the historical distribution of species in Mexico, based on published maps and an extensive database with more than 100,000 records of mammals in Mexico, was created as the first part of this project. In some cases where there is adequate information, such as the wolf and the antelope, the historical and present distribution is presented. A great effort was made to include photographs of most species.

The external measures and weight are from the literature, given the range (minimum and maximum) of adult animals. In some cases only one value is recorded. For certain species where there is no measures or weight information, the recorded measures of the most closely related species are presented.

Information on natural history and ecology is a synthesis of published information. It is clear that for some species the information available is very extensive and for others very limited. The vegetation types refer to major biomes, which mainly follows the classification of Rzedowski (1978). In general, we tried to standardize the common names of plants according to Martinez (1979).

The book includes an extensive bibliography, covering more than 200 years of publications on the mammals of Mexico. Most references are cited in the text, but we have included a wealth of sources that were not cited in the text, to present a more complete bibliography. Since the publication of the Spanish edition in 2005 there have been more than 500 publications on the mammals of Mexico. Not all of them are mentioned in this volume, but the most important ones have been included.

Although I made a great effort to ensure that the information in the book is complete, accurate, and objective, it is clearly incomplete, for many reasons such as the rapid publication of scientific studies and the breadth of the subject. I believe, however, that this volume presents the most synthetic view of mammals of my beloved country.

Acknowledgments

COMPILING A BOOK like this one would be literally impossible without the help of dozens of people. I am deeply grateful to my colleagues and friends who have been part of this long endeavor.

This book was published in Spanish in 2005, after almost 10 years of work. I am extremely grateful to my wife (Guadalupe) and my children (Pablo and Regina) for their love and support during the long hours that I spent working on the book.

I am particularly grateful to Gisselle Oliva, my coeditor of the Spanish edition, for all her hard work in bringing that volume to publication. The huge task of translating the book into English was done by my friend, student, and colleague Angelica Menchaca.

The National Commission of Biodiversity (Comisión Nacional para el Conocimiento y Uso de la Biodiversidad, CONABIO) provided most of the funds to publish the Spanish edition and sponsored the color illustrations in the present volume. Special thanks to Jorge Soberon, Jose Sarukhan, and Sebastian Ortiz.

For help with ideas, discussion of specific topics, illustrations, the design of the Spanish edition, or comments on specific sections in the book, I thank my friends Jesús Pacheco, Lourdes Martinez, Yolanda Domínguez, Rodrigo Medellín, Joaquín Arroyo, Rurik List, Eduardo Espinosa, Rosalba Becerra, Eric Mellink, Clementina Equihua, Heliot Zarza, Cuauhtemoc Chávez, Juan Cruzado, Anabel Bieler, and Marcelo Aranda.

The excellent photographs were kindly provided by Scott Altenbach, Miguel Álvarez del Toro, Héctor T. Arita, Joaquin Arroyo, Horacio Barcenas, Troy L. Best, David E. Brown, Susette Castañeda Rico, Cuauhtémoc Chávez, Claudio Contreras Koob, Juan Cruzado, Jonas A. Delgadillo V., Fulvio Eccardi, M.B. Fenton, Michael Fishbach, Pieter Folkens, Francois Gohier, Alberto Gonzalez, Carlos Guichard, David J. Hafner, Mark S. Hafner, Anna Horvath, Edmundo Huerta, Adriana Juan, Brian and Annika Keeley, Rurik List, Oscar Mendoza Mayorga, Rodrigo Medellín, Eric Mellink, Daniel Navarro, Jesús Pacheco, Ricardo Paredes León, B. S. Pash, Oliver Pergman, Sandra Pompa, Bernal Rodriguez, Patricio Robles Gil, Jens Rydell, Marcelo Sada, David J. Schmidly, Vinicio Sosa, Alejandro Torres, Marko Tschapka, Jorge Urbán, Bernardo Villa, Don Wilson, and Heliot Zarza. The Wildlife Conservation Society, the Detroit Public Library, and the Kansas State Historical Society kindly provided historic photographs. The illustrations of the marine mammals are by Pieter Folkens. The photographs of the skulls were taken by Ana Isabel Bieler. Marcelo Aranda made the track drawings.

The National Autonomous University of Mexico (UNAM) has provided me with support during all of my professional career, and for that I am most grateful.

Finally, the English-language edition was made possible only through the incredible support of my good friend Vince Burke, executive editor at Johns Hopkins University Press, and other Press staff, especially Jennifer Malat and Mary Lou Kenney, and copy editor Maria denBoer. My student Lourdes Martinez took on the strenuous work of helping me review the proofs, and Teresa Aguiar compiled the index. To all of them, my deepest gratitude.

Mammals of Mexico

Diversity and Conservation

Gerardo Ceballos, Joaquín Arroyo-Cabrales, Rodrigo A. Medellín, Luis Medrano González, and Gisselle Oliva

The freezing dawn surprised Edward W. Nelson, who, half awake, remained sheltered on his cot. The last ember burned in the fire. He remembered with nostalgia that, 7 years before, he had begun his almost nonstop trip across Mexico. He was collecting mammals for an ambitious project to which he would devote 14 years of his life as the official field agent of the Ornithology and Mammalogy Division of the U.S. Department of Agriculture. The trip to head back home seemed a long ways off. The date was June 1899.

Nelson had set up camp at the top of a wide plateau covered with enormous pines and firs. The landscape was breathtaking. How could he forget those endless mountains on the horizon with their beautiful ravines and the deep blue sky above? Getting there had been a long and arduous trip. Most of it he had ridden on a mule, but from Ciudad Juarez to Casas Grandes he traveled by wagon and then on horseback to Colonia Garcia, a small town at the outskirts of the mountains. A week later, he reached the northern foothills of the Sierra Madre Occidental, the extensive mountain range that skirts the western part of the country for more than 1,000 km.

He had been awakened the night before by the howling of a pack of wolves. Wildlife was everywhere in the region. He had seen white-tailed deer, black bears, cougars, wild turkeys, and imperial woodpeckers—the largest woodpeckers in the world. He collected a large number of different species of bats, mice, squirrels, rabbits, and carnivores as well as birds, reptiles, and amphibians. But he was most astonished by the grizzly bears, weighing more than 300 kg, that lived in these forests and that were later described as a full species native to these Sierras, *Ursus nelsoni.* That species is currently considered to be a synonym of the brown bear of North America (*U. arctos*).

Nelson got to Mexico by way of Manzanillo, Colima, in the company of his 18-year-old assistant, Edward A. Goldman, in January 1892. The efforts of Nelson and Goldman would more than repay the investment of the Department of Agriculture; their work culminated in the collection of more than 17,000 specimens of mammals and the description of 354 species and subspecies of vertebrates, mainly mammals and birds (Goldman, 1951; Sterling, 1991). Nelson's work represents the beginning of the modern study of Mexican mammals. It led to recognition of the enormous biodiversity of mammals in Mexico, of this biodiversity's importance at the global level, and the critical need for conservation of these mammals. The following description of the general patterns of distribution and conservation of Mexican mammals is an extension of the work begun by Nelson and Goldman. It is also a humble tribute to them and to the hundreds of scientists and amateurs, foreigner and Mexican, who have contributed their efforts and resources to expand our knowledge and promote the conservation of Mexican mammals.

Global Mammal Species Diversity

Mexico's biological diversity is widely recognized. Since pre-Columbian times, the great diversity of species and ecosystems has amazed observers. Baron Alexander von Humboldt considered the rugged Mexican territory a true biological paradise. The mammal species richness of Mexico became evident at the end of the nineteenth century, when Goldman and Nelson's research revealed that the country had hundreds of species and that the magnitude of its biodiversity was greater than that of the rest of North America combined (Goldman, 1951). This unique characteristic of the country came to the attention of specialists only a few decades ago, however, when the study and conservation of biodiversity became issues of worldwide relevance (Wilson and Feer, 2005).

The information gathered in recent decades about distribution patterns of mammals in the world has clearly established Mexico as one of the richest countries in mammal species (Ceballos and Brown, 1995; Mittermeier et al., 1997). Even though its territory represents about 1.6% of the continental surface of the planet (1,972,547 km^2), it supports about 11% of all mammal species.

Mexico, Indonesia, Brazil, and China have the highest number of mammal species, all of them with more than 500 (table 1; fig. 1). Other countries with high mammal diversity include Peru, the Democratic Republic of the Congo (formerly Zaire), and India. In general, countries with high mammal species richness are rich in other taxonomic groups as well; this is the case for Mexico, which occupies first place in the combined reptile and amphibian richness, fourth in vascular plants, and eleventh in birds (Groombridge and Jenkins, 2002; Mittermeier and Goettsch, 1992; Mittermeier et al., 1997; Toledo, 1988). The countries that support an exceptionally greater number of species than expected for their size are termed megadiverse, and recognition of their high species diversity has provided a strong impulse toward the worldwide importance of diversity (Ceballos and Brown, 1995; Mittermeier et al., 1997).

Table 1. Top 19 countries in mammal species diversity, with area, total species number, and endemic species (modified from Ceballos and Brown, 1995; Groombridge and Jenkins, 2002). The megadiverse countries include, among others, Indonesia, Brazil, Mexico, China, Colombia, Peru, Australia, Democratic Republic of Congo, and India. The number of species shown in the table reflects the information in the original publications.

Country	Area (km^2)	Species Number	Endemic Species
Indonesia	1,919,440	560	201
Mexico	1,972,547	544	170
Brazil	8,513,800	540	96
China	9,600,000	510	77
Colombia	1,139,155	456	28
Peru	1,285,220	452	45
United States	9,372,614	428	93
Democratic Republic of the Congo	2,336,889	415	28
India	3,288,800	350	44
Bolivia	1,098,585	316	17
Uganda	235,885	315	4
Kenya	582,646	309	21
Tanzania	945,090	306	14
Myanmar	417,964	300	8
Cameroon	477,279	297	0
Venezuela	912,050	288	11
Malaysia	329,750	286	27
Australia	7,686,848	282	210
Ecuador	284,655	280	25

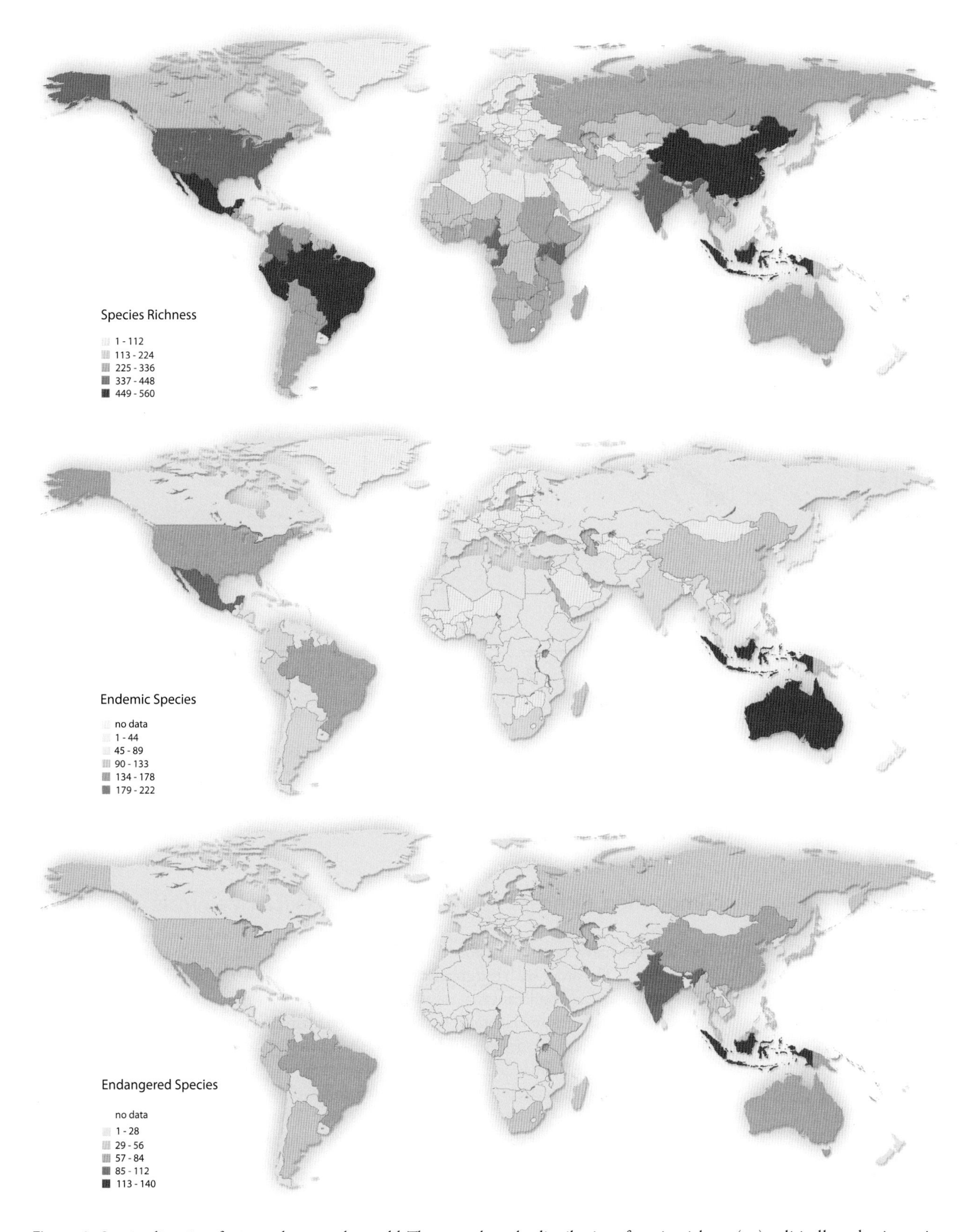

Figure 1. Species diversity of mammals across the world. The maps show the distribution of species richness (*top*), politically endemic species (species endemic to one country, sensu Ceballos and Ehrlich, 2002) (*middle*), and endangered species (*bottom*). Megadiverse countries that support a large proportion of all species include Mexico, Peru, Brazil, China, and Indonesia, among others.

Besides being species-rich, Mexico has a high percentage of endemic species (170, 30%), placing it third after Indonesia and Australia (see table 1; fig. 1). Other countries with a high number of endemic mammals are Brazil, China, the Philippines, Madagascar, and Papua New Guinea (Ceballos and Brown, 1995). The percentage of endemic species is greater in insular countries such as Australia, the Philippines, and Indonesia; however, Mexico occupies first place among continental countries. Mexico is remarkable as well because it supports many more endemic species than might be expected for its territory or its total number of species (fig. 2). This pattern is relatively generalized because countries with a high concentration of endemic mammals have a high concentration of endemic species in many other animal and plant groups. Overall Mexico places eighth in endemic birds, second in reptiles, and third in amphibians (Groombridge and Jenkins, 2002; Mittermeier et al., 1997).

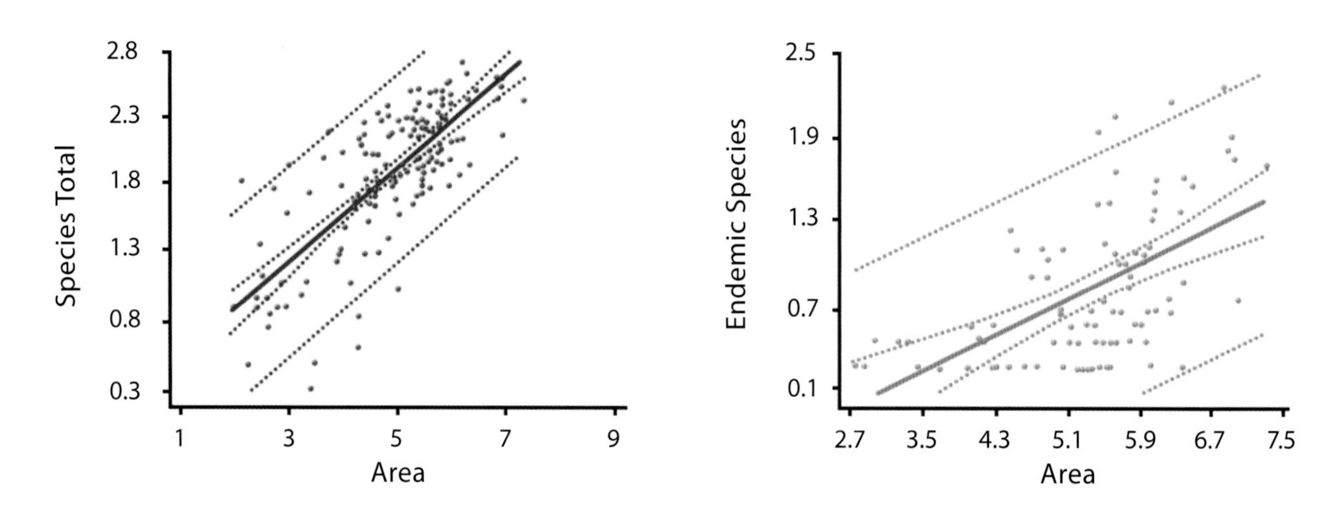

Figure 2. Correlation between mammal species number and the area (km^2) of countries in different continents. Note that the positive relationship indicates that larger countries have more species, as expected. Total species number (*left*) and endemic species (*right*) (modified from Ceballos and Brown, 1995).

Composition, Richness, and Endemism of Mexican Mammals

There are 544 native species recorded in Mexico, representing 202 genera, 46 families, and 13 orders (table 2). The order with the greatest diversity is Rodentia, with 243 species (44%), followed by bats, carnivores, and cetaceans that together make up 86% of all species. Other orders with great richness are insectivores and lagomorphs (fig. 3). Marine mammals are represented by 3 orders, 11 families, 33 genera, and 50 species, which comprise 9% of the country's mammals (Aurioles, 1993; Ceballos et al., 2010; Salinas and Ladrón de Guevara, 1993; Torres et al., 1995). Among marine mammals, the order Cetacea is the best represented, followed by Carnivora and Sirenia. Mexican marine mammal species make up 40% of the world's total (Pompa et al., 2011; Rice, 1998).

The average number of species per genus is low (approximately two), although there is a wide variation with many monotypic genera and some with large numbers of species such as *Peromyscus*, *Myotis*, *Chaetodipus*, *Neotoma*, *Reithrodontomys*, *Cryptotis*, and *Sorex*. The genera represented by only one species belong, in general, to orders with few genera or species in Mexico, such as perissodactyls, sirenians, primates, xenarthrans, and artiodactyls. In Mexico, the species-rich orders, such as rodents, insectivores, and lagomorphs, show a relatively high species number per genus.

Table 2. Composition, diversity, endemism, and conservation status of Mexican mammals. The number of species at risk of extinction includes all the categories of the official list of endangered species.

Order	Families	Genera	Species	Endemic Species	Endangered Species
Didelphimorphia	1	7	8	1	3
Sirenia	1	1	1	0	1
Cingulata	1	2	2	0	1
Pilosa	2	2	2	0	2
Primates	1	2	3	0	3
Lagomorpha	1	3	15	7	6
Rodentia	8	50	243	116	64
Soricomorpha	2	6	38	24	13
Carnivora	8	28	42	3	23
Perissodactyla	1	1	1	0	1
Artiodactyla	4	8	10	0	5
Cetacea	7	25	40	1	40
Chiroptera	9	67	139	18	36
Total	46	202	544	170	198

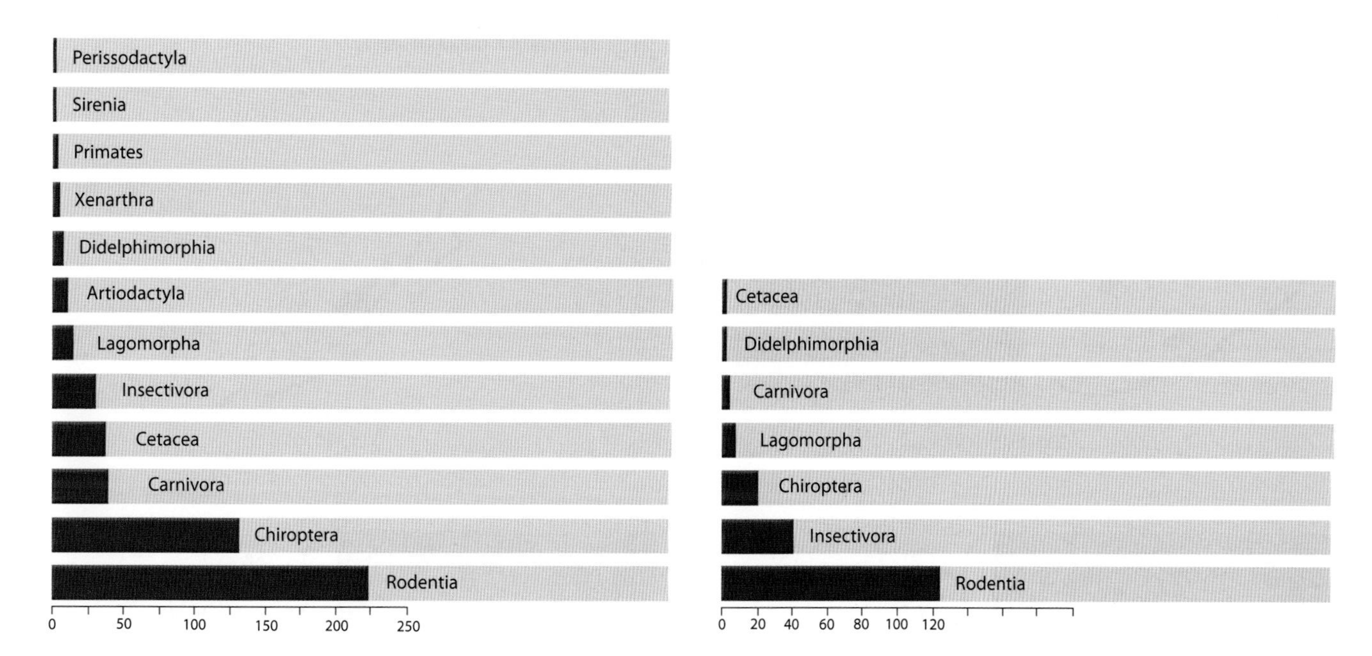

Figure 3. Species richness (*left*) and endemism (*right*) among the orders of mammals from Mexico. Most species belong to Rodentia (rodents) and Chiroptera (bats). The orders Sirenia and Perissodactyla have a single species each. In the case of endemic species, both rodents and bats are the orders with the highest number of species. There are six orders that have no endemic species in Mexico (Cingulata, Pilosa, Primates, Sirenia, Perissodactyla, and Artiodactyla).

Other orders, however, while showing a tendency to contain monotypic genera, are represented by highly diverse species. This is the case for chiropterans, carnivores, and cetaceans.

Approximately 31% of the species (170) and 4% of the genera (Soricomorpha: *Megasorex*; Didelphimorphia: *Tlacuatzin*; Chiroptera: *Musonycteris*; Rodentia: *Hodomys*, *Pappogeomys*, *Zygogeomys*, *Osgoodomys*, *Megadontomys*, *Nelsonia*, *Neotomodon*, *Xenomys*; Lagomorpha: *Romerolagus*) are endemic to Mexico (fig. 4). The majority of endemic species are rodents, which is not surprising given their high diversity and low mobility in comparison to other orders (see table 2; Ceballos and Rodríguez, 1993; Ramírez-Pulido and Müdespacher, 1987). The 116 endemic species of Mexican rodents include cricetids, heteromyids, pocket gophers, squirrels, and one agouti. The insectivores, roughly 60% endemic species, all of which are shrews except for one mole, represent the second-richest group of endemics. The rest of the endemic species belong to bats and four other orders. Mexico is one of five countries in the world with an endemic cetacean, the vaquita (*Phocoena sinus*) of the Gulf of California, which has the most restricted distribution of any marine mammal (Pompa et al., 2011).

Figure 4. (*Below and opposite*) Geographic distribution of genera of mammals endemic to Mexico, most of which are monotypic (i.e., have a single species). The genera belong to Didelphimorphia (genus *Tlacuatzin*, 1 species), Soricomorpha (*Megasorex*, 1 species), Chiroptera (*Musonycteris*, 1 species), Rodentia (*Hodomys*, 1 species; *Megadontomys*, 3 species; *Nelsonia*, 2 species; *Neotomodon*, 1 species; *Osgoodomys*, 1 species; *Pappogeomys*, 1 species; *Xenomys*, 1 species; *Zygogeomys*, 1 species), and Lagomorpha (*Romerolagus*, 1 species). All photos by Gerardo Ceballos, except for *Musonycteris*, by Marko Tschapka.

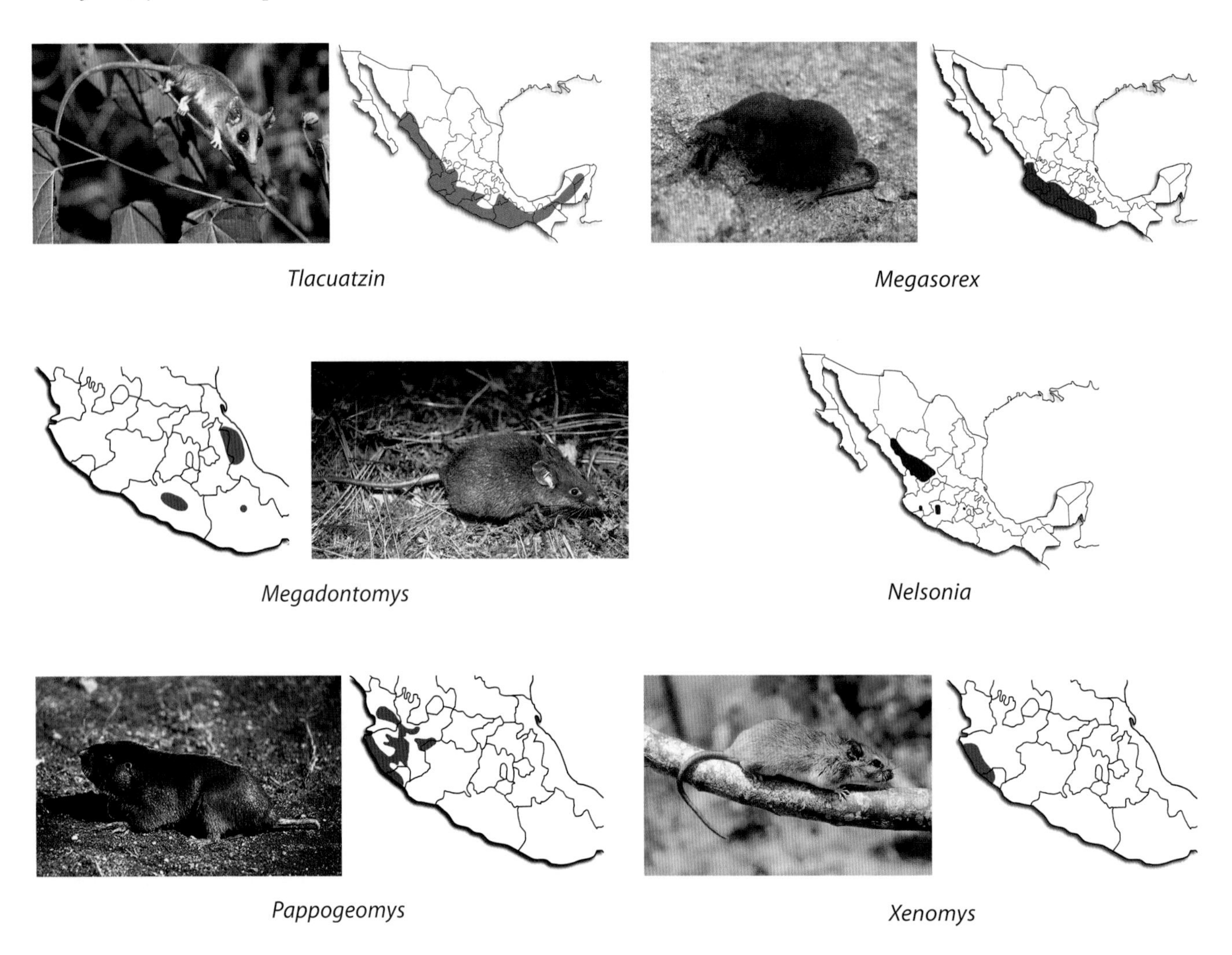

Endemic genera of marsupials (*Tlacuatzin*), insectivores (*Megasorex*), bats (*Musonycteris*), rodents (*Pappogeomys, Zygogeomys, Osgoodomys, Megadontomys, Nelsonia, Neotomodon, Xenomys, Hodomys*), and rabbits (*Romerolagus*) are distributed exclusively in the dry forests of the Pacific and the temperate mountains of the Sierra Madre Occidental, Eje Volcanico Transversal, and Sierra Madre del Sur (Ceballos et al., 1998). Why is there such richness of species and a high concentration of endemic species in Mexico? These are questions that have intrigued scientists for a long time. They have been explained as the result of a complex interaction of several factors, such as area, latitude, geologic history, climate patterns, complex topography, and diverse vegetation types (see Álvarez and Lachica, 1974; Arita and Rodríguez, 2002; Ceballos and Navarro, 1991; Fa and Morales, 1993; Goldman and Moore, 1946). Nonetheless, it is clear that we are still far from understanding precisely the role that each factor plays and the interaction among the factors. This is an interesting challenge because this understanding is fundamental to long-term conservation of mammals and other animals (Ceballos et al., 1998; Sanchez-Cordero et al., 2005).

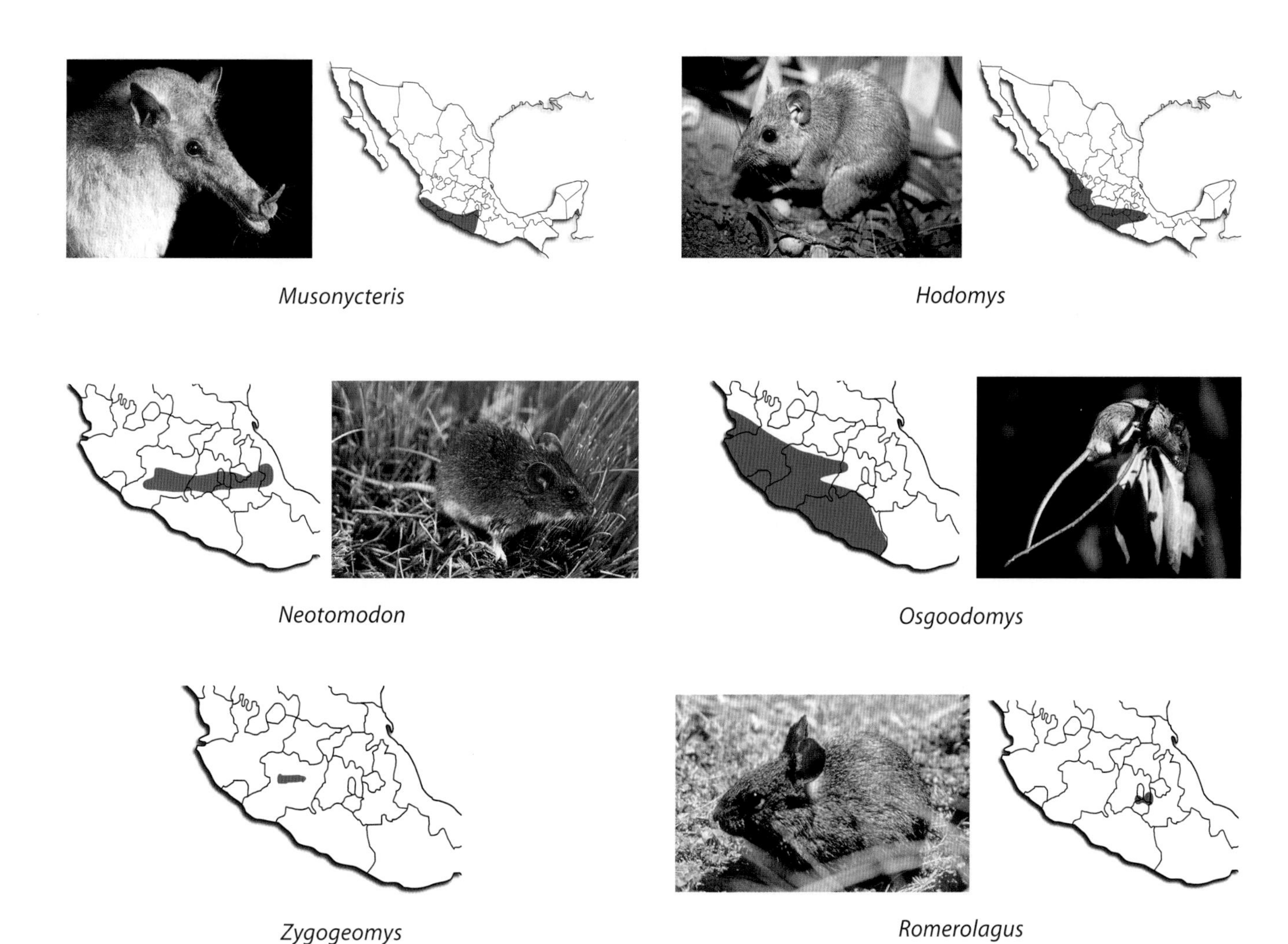

Distribution Patterns

Having a well-established general pattern of mammal species distribution reflects the positive relationship between increasing species richness and increasing area; in other words, larger areas contain more species per area (Brown and Lomolino, 1999). Indeed, area explains nearly 70% of the variation in species number among countries (Ceballos and Brown, 1995). Although the largest countries are generally the richest in species number, there are some remarkable exceptions, especially when we compare temperate and tropical countries, as tropical countries tend to contain more species (see fig. 2). Mexico is the fifteenth-largest country in the world, so it is not surprising that it possesses a high number of species. A statistical evaluation of the relation between species number and area reveals that, at a global scale, Mexico supports a greater number of species per unit area than expected (Ceballos and Brown, 1995). When comparing Mexico to other political entities in America, the same pattern is evident: it exceeds the expected species number according to the regression of species number and country area (Arita, 1993a, 1997). When comparing individual Mexican states with the same political units of the continent, however, they do not show remarkably high species numbers. In fact, although few data are available, it is estimated that, individually, Mexican localities are not particularly rich in species (Arita, 1993a, 1997).

Mexico's dynamic geologic history, which is in part related to its geographic position, is considered one of the main factors explaining its high biodiversity. Mexico is the only country that, within its frontiers, contains the entire length of the boundary between two biogeographical regions of the world, the Nearctic and the Neotropical (fig. 5). These regions interdigitate around the Tropic of Cancer, close to 24 degrees north, Mexico covers nearly 60% and 40% of the land for each region, respectively, with a transition area of variable size and complexity in the intermediate zone (Fa and Morales, 1993; Goldman and Moore, 1946; Sanchez-Cordero et al., 2005). The temperate and tropical fauna mixture is abrupt in some areas, while in others it is smooth. This mixture is responsible for the unique structure of many communities.

If we classify terrestrial species by their distribution depending on their Nearctic (temperate) or Neotropical (tropical) affinity, 7 families exhibit Nearctic affinity, 9 Neotropical, and 19 share both affinities (table 3; Arita and Ortega, 1999; Ceballos and Navarro, 1991). At the species level,

Figure 5. Limits between the Nearctic (*blue*) and Neotropical (*yellow*) biogeographic regions. Mexico is the only continental country where two biogeographic regions have complete intergradation. This is one of the likely reasons the country has such a rich mammalian fauna, with both tropical and temperate species.

Table 3. Zoogeographic affinities of the terrestrial mammal families of Mexico according to the current distribution of species (modified from Álvarez and Lachica, 1974; Ceballos and Navarro, 1991). Most of the families are shared, and are followed by the ones with a tropical origin, also called Neotropical. The families with a temperate confinement (Nearctic) are less numerous.

ORDER/Family	Nearctic	Neotropical	Shared
DIDELPHIMORPHIA			
Didelphidae			X
CINGULATA			
Dasypodidae			X
PILOSA			
Myrmecophagidae		X	
Cyclopedidae		X	
PRIMATES			
Atelidae		X	
LAGOMORPHA			
Leporidae			X
RODENTIA			
Sciuridae			X
Castoridae	X		
Heteromyidae			X
Geomyidae			X
Erethizontidae			X
Muridae			X
Dasyproctidae		X	
Cuniculidae		X	
SORICOMORPHA			
Soricidae	X		
Talpidae	X		
CARNIVORA			
Felidae			X
Canidae			X
Ursidae	X		
Mustelidae			X
Mephitidae			X
Procyonidae			X
PERISSODACTYLA			
Tapiridae	X		
ARCTIODACTYLA			
Tayassuidae			X
Cervidae			X
Antilocapridae	X		
Bovidae	X		
CHIROPTERA			
Emballonuridae		X	
Phyllostomidae			X
Mormoopidae			X
Noctilionidae		X	
Thyropteridae		X	
Natalidae		X	
Molossidae			X
Vespertilionidae			X
TOTAL	7	9	19

29% (153 species) are associated with the Nearctic region and 28% (147) with the Neotropical region, while 13% (65) share both affinities and 30% are endemic to the country (see the checklist of mammals of Mexico in the Systematic List of Species, beginning on page 49).

Mexico played an important role in what is known as the Great American Exchange. The emergence from the ocean of the Isthmus of Panama 3 million years ago activated a process of biological dispersal, extinction, and replacement of species, for both North and South America (Ferrusquía-Villafranca, 1977; Miller and Carranza-Castañeda, 2002; Stehli and Webb, 1985). Mexico was a corridor for the dispersal of all the species that migrated to the south, such as heteromyid rodents and sciurids, felids, canids and other carnivores, and tapirs and peccaries. It received many species that migrated to the north, such as armadillos, opossums and other marsupials, sigmodontine rodents, and monkeys. Many other groups of mammals, extinct today, followed the same patterns (Simpson, 1980; Stehli and Webb, 1985). This dispersal is another factor determining the current presence of many species, genera, and families in Mexico's northern or southern distribution limits. It explains the distributions of the order Primates (monkeys) and families such as Didelphidae (slender mouse opossums), Myrmecophagidae (anteaters), Emballonuridae and Moormopidae (bats), Tapiridae (tapirs), and Dasyproctidae (agoutis), whose northern limit is the tropical lands of Mexico. It also explains the distribution of families Talpidae (moles), Ursidae (bears), Antilocapridae (pronghorn antelopes), Bovidae (bison and bighorn sheep), and Castoridae (beavers), whose southern limit is the temperate forests and scrubs of the northern and central part of the country.

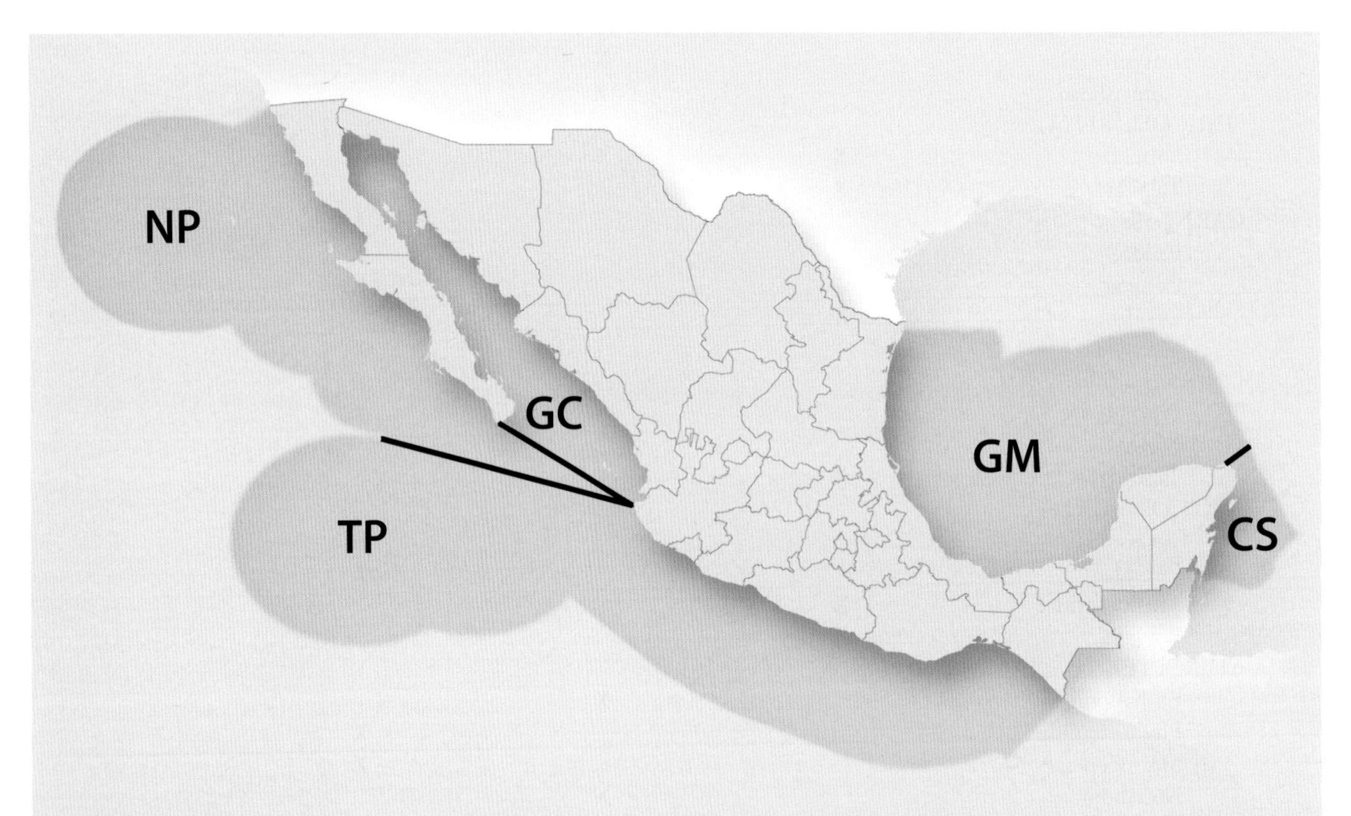

Figure 6. Oceanographic regions of the exclusive economic zone of Mexico (modified from Ayala-Castañares, 1982; De la Lanza Espino, 1991; Salinas and Ladrón de Guevara, 1993). The line that connects Cabo Corrientes with Cabo San Lucas limits the Gulf of California (GC) region. The delimitation of the Tropical Pacific (TP) and the Northern Pacific (NP) includes the Revillagigedo Islands, although this region actually is the transition area between the California Current and the eastern tropical Pacific. The delimitation between the Gulf of Mexico (GM) and the Caribbean Sea (CS) is an imaginary line between Cabo Catoche and Cuba.

An additional environmental factor that likely contributes significantly to the terrestrial mammal richness in Mexico is its multifaceted environmental heterogeneity, reflected in a complex geographic configuration. Mexico has two peninsulas, Baja California in the northwest and Yucatan in the southeast. The Gulf and Pacific coastal plains cover a vast territory along the country's coasts. The complex intergradation with the mountain ranges that limits the coastal plains creates favorable conditions for the formation of diverse biomes and climate patterns. The Mexican Altiplane is isolated from the coasts and the southern part of the country by mountain chains, and it combines an array of temperate and tropical ecosystems. Mountain chains, ranging from the Sierras Madre Occidental and Oriental to the Eje Volcanico Transversal and the Nudo Mixteco, with its high peaks that sometimes reach more than 5,000 m, contribute significant topographic changes. These changes are partly responsible for the different plant communities, which in turn give rise to the high species diversity. Environmental heterogeneity is reflected in a corresponding diversity in climate, rainfall, and vegetation (Arita, 1993a). This subject is clearly one of the most interesting for researchers, with important theoretical and practical implications.

Even though many regions show high alpha-diversity values, such as the tropical rainforest in Lacandona in Chiapas or Los Tuxtlas in Veracruz, the best explanation for the great species richness is the so-called beta diversity, communities that experience spatially changing environments. In addition, there is evidence showing that the turnover rate of species among regions or states depends on the environmental heterogeneity level (Arita, 1993a, 1997; Rodríguez et al., 2003).

The species richness of marine mammals is similar to that of the terrestrial groups, where communities of different biogeographic regions converge (Fa and Morales, 1998). In the Gulf of Mexico and the Caribbean Sea, one can find mammals primarily associated with tropical and subtropical water currents of the Gulf and the Antilles (fig. 6). In the Pacific, we find species of tropical origin associated with the North Equatorial and the Costa Rican Currents as well as the cold and temperate water of the Californian Current. The division into coastal and oceanic regions is also a consideration. The vast extension of the continental platform in the Gulf of Mexico and the Caribbean Sea only allows species of pelagic habits, such as physeterids and ziphiids, to be observed far from the coast. In contrast, the narrow extension of the continental platform in the Pacific makes it easy to observe pelagic species near the coast and even in the bays with deep areas such as Bahia de Banderas, where ziphiids and kogids are often seen. The Gulf of California provides specific conditions in all oceanographic aspects; the mammal communities also have unique characteristics such as a resident population of fin whales and even a species endemic to the region of the Upper Gulf, the vaquita (*Phocoena sinus*). Marine mammals move in these heterogeneous environments in a complex way in accordance with the direct effects of spatial and temporal oceanographic variations and food. The pygmy beaked whale (*Mesoplodon peruvianus*) and West Indian manatee (*Trichechus manatus*) are examples of species that require warm water, while the North Pacific right whale (*Eubalaena japonica*) and Dall's porpoise (*Phocoenoides dalli*) live in cold water. Humpback whales (*Megaptera novaeangliae*) and gray whales (*Eschrichtius robustus*), during their annual migratory cycle, experience a wide variation in environmental temperature, but their distribution in Mexico depends on specific conditions for breeding. Dolphins and sea lions are less restricted by oceanographic conditions and move in response to food fluctuations.

The number of species of terrestrial mammals in North America increases with decreasing latitude (Ceballos and Brown, 1995; Fleming, 1974; Simpson, 1964; Wilson, 1974). This continental pattern is also observed in Mexico, where the largest concentrations of species per unit area are located in the more equatorial latitudes (Ceballos and Navarro, 1991; Ceballos et al., 1998). Analyzing the distribution of mammals in squares of 1 to 0.5 degrees has been useful in assessing these patterns in detail. In a grid of 0.5 degrees it is clear that the number of species increases dramatically from north to south (fig. 7). The species gradients vary from about 22 to 158 species per cell, representing less than 2% of all terrestrial species to about 34%, respectively. Quadrants with the lowest number of species are located in the Baja California Peninsula and the Mexican Plateau (Ceballos and Navarro, 1991). The Yucatan Peninsula, the Eje Volcanico Transversal, and the Sierra Madre Occidental and Sierra Madre Oriental are regions with an average number of species. The regions with the greatest richness are located in eastern and southeastern Mexico, including Veracruz, Tabasco, Oaxaca, and Chiapas. This pattern is similar for most mammalian orders that show an increase of species with decreasing latitude (Santos del Prado, 1996). Artiodactyls, lagomorphs, and rodents, however, are exceptions because of their Nearctic affinities (see fig. 7). For example, rodents show a pattern of higher species richness in the Mexico–United States border in the Chihuahua and Sonora states in Mexico and Arizona and New Mexico in the United States. However, states with the highest number of species, which are also the most ecologically heterogeneous, are those located in the Nearctic-Neotropical transition: Veracruz, Michoacan, Guerrero, Oaxaca, and Chiapas (table 4; Ramírez-Pulido et al., 1983, 1986).

Endemic species-richness patterns are quite different from patterns that describe overall richness; most mammalian endemisms are found in the central and western part of the country, and small concentrations are found in the tropical southern part (fig. 8). The regions with the highest

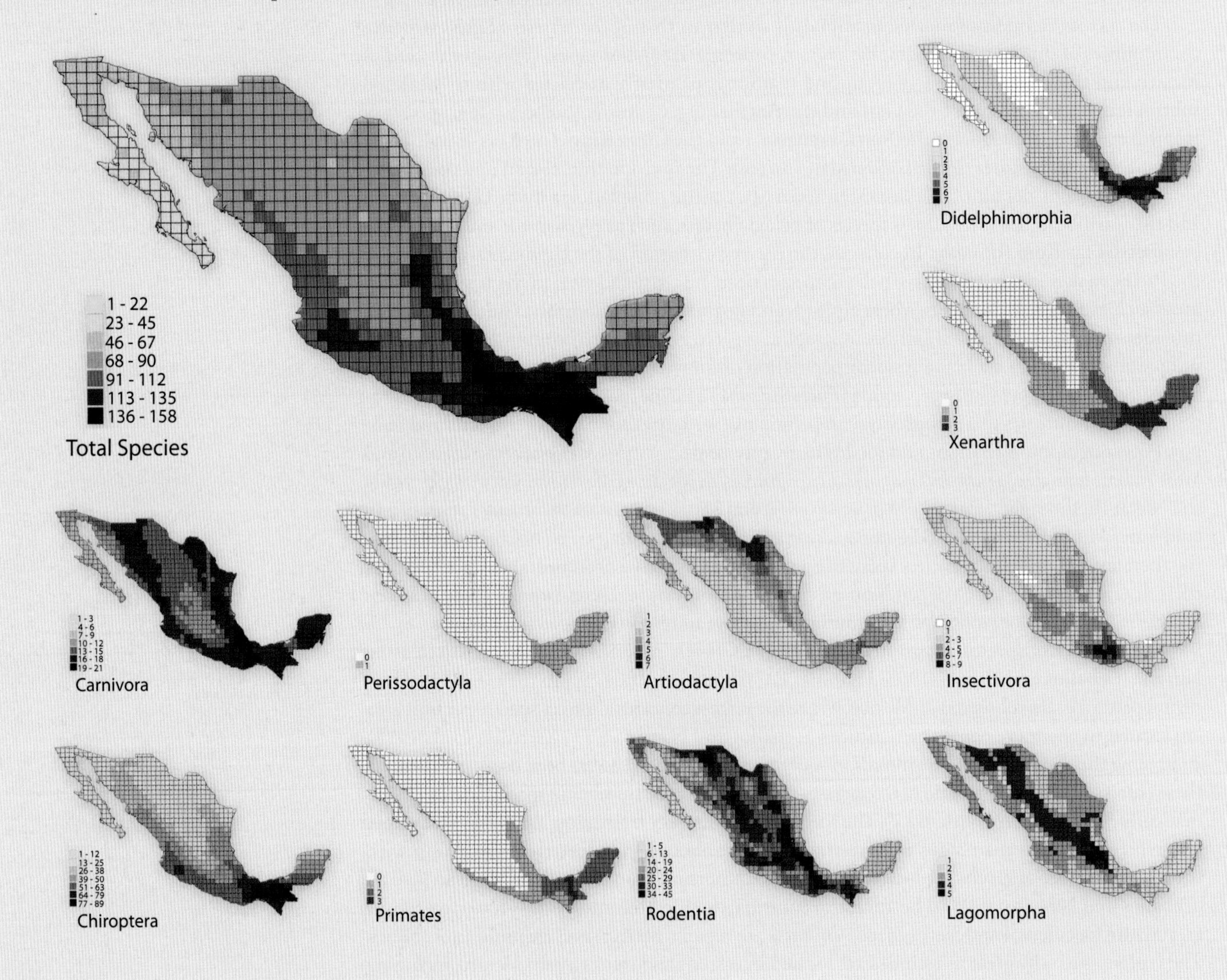

Figure 7. Latitudinal patterns of species richness of the mammals of Mexico, showed in squares of 0.5 degrees. Note that the number of species increases as latitude decreases, so regions in Oaxaca and Chiapas in southern Mexico have the highest number of species. The patterns among different orders are, however, quite variable, and a few orders such as Arctyodactyla have more species in northern latitudes. For comprehensiveness, we have merged the orders Cingulata and Pilosa as the superorder Xenarthra.

concentration of endemic species, up to 29 species per square of 0.5 degrees, are the temperate forests of the Eje Volcanico Transversal and northwestern Oaxaca, and the dry forests of the central part of the west coast (Ceballos and Rodríguez, 1993; Ceballos et al., 1998; Fa and Morales, 1993; Goldman and Moore, 1946; Harris et al., 2000; Sullivan et al., 2000). Altogether, these regions support approximately 75% of the Mexican mammalian endemisms; the rest of the endemics are found in the islands of the Gulf of California, the dry Sierras of the northeast and central-northern parts of the country as well as in other islands of the Pacific and the Caribbean. In general, in these regions, endemisms of different groups of vertebrates, such as birds and amphibians, also occur, highlighting the general nature of these patterns and the underlying causal processes (Escalante et al., 1993; Flores-Villela, 1993; Hanken and Wake, 1998).

Table 4. Mammal species richness in the federative entities of Mexico. The number of species was taken from several references, particularly from Arita (1993), Ramírez Pulido et al. (1983, 1986, 1993), and monographs of the mammals by state (see table 9 for references).

State	Area	Richness
Aguascalientes	5,471	50
Baja California	69,921	90
Baja California Sur	73,475	64
Campeche	50,812	71
Chiapas	74,211	166
Chihuahua	244,938	129
Coahuila	149,982	107
Colima	5,191	86
Distrito Federal	1,479	63
Durango	123,181	120
Guanajuato	30,141	65
Guerrero	64,281	115
Hidalgo	20,813	97
Jalisco	80,836	163
Mexico	21,355	119
Michoacan	59,928	128
Morelos	4,950	86
Nayarit	26,979	97
Nuevo Leon	64,924	91
Oaxaca	93,952	191
Puebla	33,902	110
Queretaro	11,449	67
Quintana Roo	50,212	82
San Luis Potosi	63,068	140
Sinaloa	58,328	108
Sonora	182,052	128
Tabasco	26,267	79
Tamaulipas	79,384	139
Tlaxcala	4,016	28
Veracruz	71,699	170
Yucatan	38,402	95
Zacatecas	73,252	115

Hotspots of endemic species vary among orders. Bats, for example, show the greatest concentration of endemic species in western Mexico, along the coast and neighboring mountains of southern Jalisco, Colima, and northern Michoacan (see fig. 8). Endemic insectivores are concentrated in the Sierra Madre del Sur and the eastern section of the Eje Volcanico Transversal, between Oaxaca and Veracruz. Finally, endemic rodents are most diverse along the Eje Volcanico Transversal.

The dry tropical forests of the Pacific support most of the genera (6) and endemic species (35), representing 21% of Mexican endemics. The endemic genera restricted to these forests are *Tlacuatzin*, *Megasorex*, *Musonycteris*, *Hodomys*, *Osgoodomys*, and *Xenomys* (see fig. 4). For mammals, however, endemic species represent six orders, dominated by bats (36%), including several species of *Rhogeesa*, and rodents (50%), with species such as *Peromyscus perfulvus* and *Xenomys nelsoni*. A large number of bats and the only endemic carnivore of the Mexican continental territory, the pygmy spotted skunk (*Spilogale pygmaea*), inhabit the dry tropical forest of western Mexico.

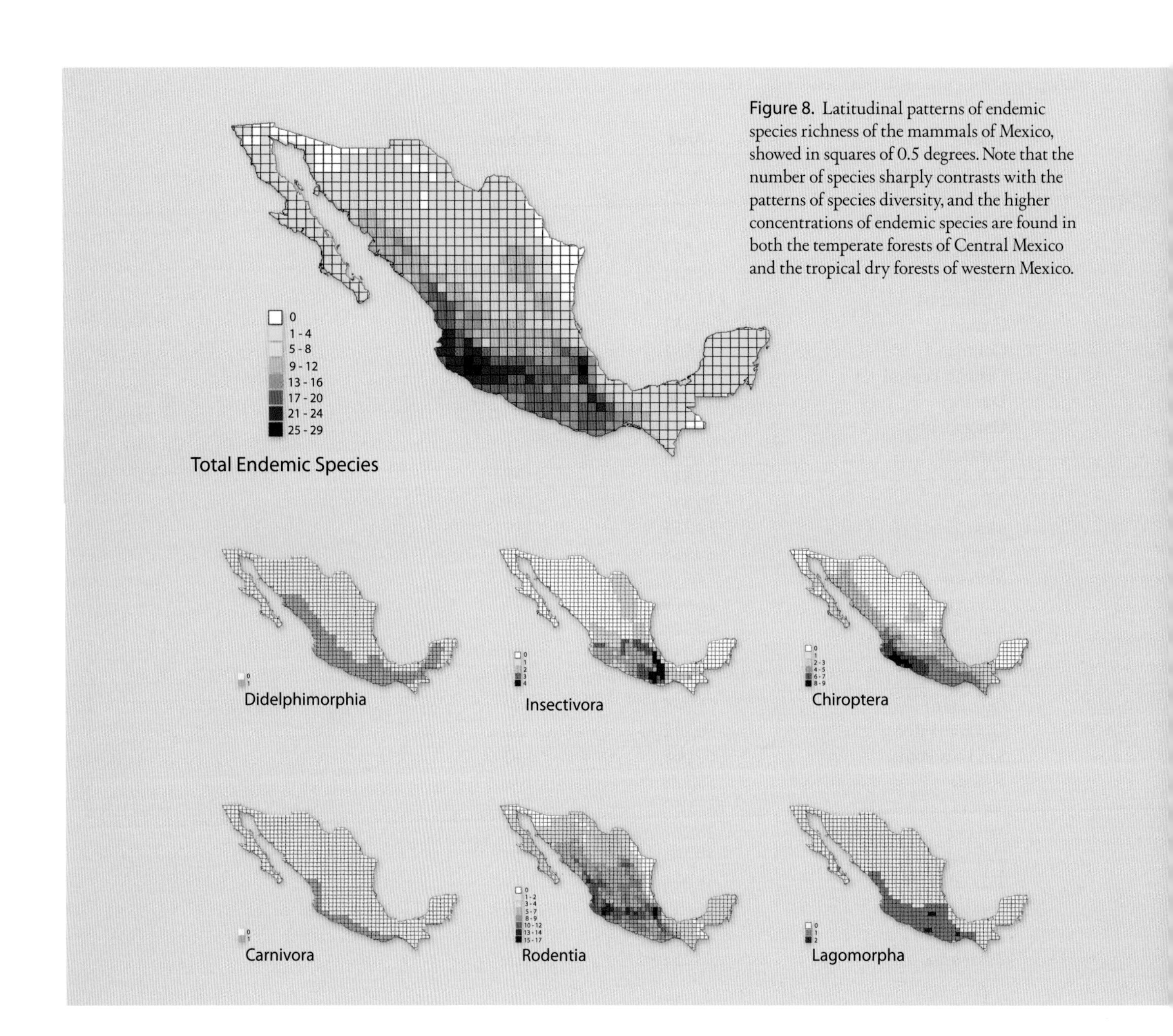

Figure 8. Latitudinal patterns of endemic species richness of the mammals of Mexico, showed in squares of 0.5 degrees. Note that the number of species sharply contrasts with the patterns of species diversity, and the higher concentrations of endemic species are found in both the temperate forests of Central Mexico and the tropical dry forests of western Mexico.

In the Eje Volcanico Transversal, there are 5 endemic genera, which include *Pappogeomys, Zygogeomys, Neotomodon, Megadontomys*, and *Romerolagus*, and 35 endemic species of mammals (21%), primarily rodents, several species of insectivores, and one lagomorph. This pattern is related to the Nearctic biogeographic affinities of these groups. In the Baja California Peninsula and on the islands of the Gulf of California, there are 22 (15% of the total) endemic species; 18 of them are, as expected, small rodents associated with arid areas, for example, *Ammospermophilus insularis* and *Peromyscus guardia*. Finally, the mountains of the Sierra Madre del Sur in Guerrero, Oaxaca, and Chiapas support an endemic genus (*Megadontomys*) and 22 species (14% of the total), which include insectivores such as *Sorex stizodon*, one lagomorph (*Sylvilagus insonus*), and rodents such as *Microtus oaxacensis*.

The rest of the endemic taxa (38%) represent one complex group that includes 42 rodents (70%), 11 insectivores (18%), and other groups. This arrangement indicates a Nearctic affinity given that most of the other groups are found in the temperate mountains of the northeastern and southern parts of Mexico, in regions where the Nearctic region breaks into small, isolated fragments with an altitude higher than 1,500 m, and where the vegetation corresponds to that of a temperate climate.

An analysis of the regional composition of marine mammals shows a fundamental difference between the communities of the Pacific and those of the Atlantic: the Pacific communities contain greater diversity (fig. 9; Pompa et al., 2011). Caution is warranted with regard to this finding as the diversity of marine mammals from the tropical regions is not yet completely described. It does appear that more species can be found in regions affected by cold water. The similarities and differences become important when comparing the tropical Pacific, the Gulf of California, and the northwestern Pacific because the emerging pattern suggests a difference in the historic flow of species between the north Pacific and tropical Pacific, and from these regions to the Gulf of California. The Gulf of Mexico and the Caribbean have a similar composition of species.

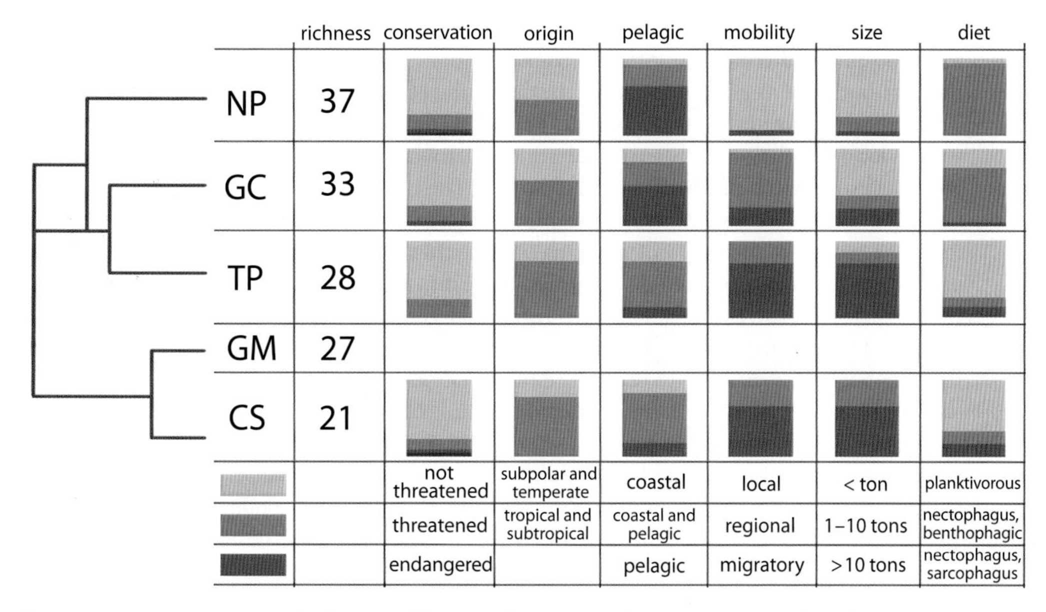

Figure 9. Species composition (richness and functional composition) of marine mammals in the different oceanographic regions in Mexico. With a data matrix of the species present in each zone, the distance between regions was determined, that is, the fraction of species whose occurrence or absence does not match (Hamming distance). Recent records of mammals in the Gulf of Mexico were updated by Ortega Ortiz (2000). Matrix distances are resolved in a dendrogram using an analysis of clusters with the UPGMA (Unweighted Pair Group Method with Arithmetic Mean) algorithm. Bars indicate the species composition of each region in terms of their conservation status, their geographic origin, and the functional composition of the apparently invasive species, species of tropical origin in the North Pacific (NP), every island in the Gulf of California (GC), and the cold and warm waters in the tropical Pacific (TP), Gulf of Mexico (GM), and Caribbean Sea (CS).

Species such as the pantropical spotted dolphin (*Stenella attenuata*), the bottlenosed dolphin (*Tursiops truncatus*), the pygmy beaked whale (*Mesoplodon peruvianus*), and even the vaquita (*Phocoena sinus*) in the Gulf of California crossed from the tropical Pacific to other regions. Fin whales (*Balaenoptera physalus*), northern elephant seals (*Mirounga angustirostris*), and harbor seals (*Phoca vitulina*) dispersed as well from the northeastern Pacific to other regions. Animals such as the North Pacific right whale (*Eubalaena japonica*), the northern right whale dolphin (*Lissodelphis borealis*), and Dall's porpoise (*Phocoenoides dalli*) are found only in the northern Pacific coast of Baja California associated with the California Current. The region that surrounds the Revillagigedo Islands is a transition area where marine mammals from the northeastern and tropical regions live in a pattern similar to that of the marine invertebrates (Bautista-Romero et al., 1994).

The composition of species in a region coincides with the oceanography (species' tropical or cold-water origin) but shows an obvious mix between regions. If we would like to know the process that has shaped the communities of marine mammals, we must consider unexpected (intruder) species in terms of their pelagic behavior, mobility, size, and diet. Because of these features, the composition of the communities is termed functional composition. In the north Pacific there is a greater proportion of tropical intruder species than there are cold-water intruder species in the tropical regions. Most of the intruder species of the north Pacific have pelagic habits, move seasonally but do not migrate, are smaller than one ton, and feed on nekton; that is, they are basically small cetaceans.

In Mexico, pinnipeds are restricted to the Gulf of California and the north Pacific because of their need for adequate sites for reproduction, for which the absence of terrestrial and aquatic predators and the presence of highly productive areas are essential. Even though pinnipeds can move long distances to find food, to the north and the tropical Pacific, the lack of sites for reproduction limits their distribution to the area between Baja California and the Galápagos Islands (Ceballos et al., 2010; Riedman, 1990). In tropical regions, intruder species have combined coastal and pelagic habits, are larger, and feed on plankton and flesh; that is, they are basically large cetaceans. Thus, there is a greater tendency for tropical species to break into cold-water areas, suggesting a boreal displacement of marine mammals parallel to climate change that will vary according to the magnitude of this change, its effects on marine ecosystems, and the ability of each species to make use of that alternate ecosystem. In tropical regions, cold-water species occur as remnants of the distribution of migratory species or the dispersion of pandemic species such as killer whales. The Gulf of California has characteristics that allow dissimilar assemblages of intruder species, such as local populations of baleen whales and killer whales and species of intermediate trophic levels such as small-toothed whales and pinnipeds. These findings show that a better understanding of the history of Mexican marine mammals requires a deeper study of these species' dynamics at the mouth of the Gulf of California because this area seems to be critical for dispersion processes related to oceanographic dynamics. These distribution patterns, with apparent contractions, expansions, and exchanges between regions, suggest that marine mammal communities in Mexico have been shaped by, and are currently undergoing modifications as a result of, changes in the distribution associated with oceanographic variations caused by climate change.

The similarity between marine mammal communities of the Gulf of Mexico and those in the Caribbean indicates that the effect of the warming of the sea on the distribution of marine mammals promotes the integration of these communities, perhaps as a result of the expansion of the Gulf Current and the currents from the Antilles. It is well known that conditions in the Caribbean and the Gulf of Mexico have remained relatively constant and that these areas were the least affected by the last glaciations, which froze much of the north Atlantic. The Mexican Pacific coast as a zone of interaction between bodies of cold and tropical waters is a dynamic region. The effects of the end of the last glaciation have mainly influenced the Gulf of California and the waters of the Pacific facing Baja California (Ruddiman, 1987; Williams et al., 1993), generating the exchange of marine mammals between the Pacific and the northeast tropical Pacific and the expansion of these communities into the Gulf of California. These movements can be observed at the end of cycles of El Niño and La Niña. This implies that the factors described above suggest that conservation of marine mammals in Mexico will require, among other things, minimizing the effects of global climate change that would otherwise threaten the abundance, distribution, and habits of these species.

Vegetation and Mammals

Mexico has between 10 and 60 different plant communities, depending on which classification system is used. In this discussion we use the vegetation classification of Rzedowski (1978). In general, the vegetation of Mexico can be divided into temperate and tropical and further into dry and wet (fig. 10). Temperate vegetation is found in the central and northern regions of the country and includes temperate forests, xeric scrubs, and grasslands (fig. 11). Grasslands and xeric scrubs cover approximately 50% of the territory and are located primarily in the Baja California Peninsula and the Mexican Plateau. Temperate forests represent around 21% of the territory and include pine (*Pinus*), fir (*Abies*), oak (*Quercus*), and juniper (*Juniperus*). They are located along the mountains of Sierra Madre Occidental, Sierra Madre Oriental, Sierra Madre del Sur, and Eje Volcanico Transversal (Rzedowski, 1978).

Tropical vegetation located along the coastal plains of the Gulf and the Pacific and in the southern part of the country include the subtropical dry forests (also known as deciduous forests), the tropical rainforest, mangrove forests, swamps, and wetlands (see fig. 11). The tropical dry forest covers 17% of the territory and is characterized by its deciduous vegetation. These forests are primarily located on the Pacific coast from southern Sonora to Chiapas, with isolated spots in southern Tamaulipas and northern Veracruz, and in the northern Yucatan Peninsula. The tropical rainforest has a more restricted distribution; it has 11% of the country's vegetation and is found along the Gulf coast from Chiapas to San Luis Potosí. The mangrove swamps and other types of wetland are located along the coastal plains and the coast of the Yucatan Peninsula.

The tropical rainforest stands out because of its high mammal species richness. It is the most threatened ecosystem in Mexico (Challenger, 1998) and is mainly found in the southern and southeastern parts of the country. The mammalian fauna is characterized by a strong Amazonic influence. This is not surprising because the vegetation type is shared not just with many countries of Central America in a continuous way but also with South America. Some characteristic species include the naked-tailed armadillo (*Cabassous centralis*), the silky anteater (*Cyclopes didactylus*), the howler monkeys (*Alouatta* spp.), the Central American spider monkey (*Ateles geoffroyi*), the Yucatan brown brocket (*Mazama pandora*), the tapir (*Tapirus bairdii*), the white-lipped peccary (*Tayassu pecari*), marsupials such as the water opossum (*Chironectes minimus*), the Central American woolly opossum (*Caluromys derbianus*), and the brown four-eyed opossum (*Metachirus nudicaudatus*), many Phyllostomid bats such as *Vampyrum spectrum*, *Trachops cirrhosus*, *Tonatia* spp., and *Mimon* spp., and rodents of different groups (*Peromyscus mexicanus*, *Tylomys nudicaudus*, *Otonyctomys hattii*, *Heteromys desmarestianus*, and *Orthogeomys grandis*).

The deciduous forest is seriously threatened by destruction and fragmentation and is only marginally protected by law (Ceballos and Garcia, 1995; Maass et al., 2005). It is primarily located in the Pacific slope from south Sonora to Chiapas. This vegetation type provides one of the most important sources of endemic species in Mexico, including the gray mouse opossum (*Tlacuatzin canescens*), the Mexican shrew (*Megasorex gigas*), several bats (*Musonycteris harrisoni*, *Glossophaga morenoi*, *Artibeus hirsutus*, *Rhogeessa alleni*, and others), the pygmy spotted skunk (*Spilogale pygmaea*), Collie's squirrel (*Sciurus colliaei*), the Magdalena rat (*Xenomys nelsoni*), and the Tehuantepec jackrabbit (*Lepus flavigularis*).

The temperate cloud forest is one of the most endangered types of vegetation in Mesoamerica (Challenger, 1998). In Mexico it is primarily found in the Sierra Madre de Chiapas and south Oaxaca and Guerrero, and in the Sierra Madre Oriental in Veracruz, Puebla, San Luis Potosí, and Tamaulipas. A few of the most characteristic species of the temperate cloud forest are shrews (*Cryptotis goodwini* and *Sorex stizodon*), bats (*Hylonycteris underwoodi* and *Eptesicus brasiliensis*), Goldman's pocket mouse (*Chaetodipus goldmani*), rodents (*Microtus umbrosus* and *Megadontomys thomasi*), and the Omilteme cottontail (*Sylvilagus insonus*). The distribution range of most species characteristic of this type of vegetation is restricted by the small, fragmented area occupied by the temperate cloud forest. Many of these species are endemic to Mexico.

In Mexico the pine-oak forest has a relatively broad distribution that extends to the United States. It has a mixture of species with an extensive distribution and endemic species. Examples of the most characteristic species include some shrews (*Sorex oreopolus* and *Sorex saussurei*), bats

(*Anoura geoffroyi* and *Myotis velifer*), the black bear (*Ursus americanus*), the southern flying squirrel (*Glaucomys volans*), several species of tree squirrels (genus *Sciurus*), rodents (*Cratogeomys merriami*, *Neotomodon alstoni*, and *Peromyscus melanotis*), the volcano rabbit (*Romerolagus diazi*), and the Mexican cottontail (*Sylvilagus cunicularius*).

The xerophytic scrub covers a great proportion of the northern deserts of Mexico, including the Baja California Peninsula and the islands of the Gulf of California. For this reason, its mammalian fauna is a mixture of species, some with a broad and extended distribution range and some endemic to certain areas. Sometimes the species are shared with the United States. Characteristic species of the xerophytic scrub include the shrews *Notiosorex crawfordi* and *Sorex arizonae*, bats (*Myotis vivesi*, *Euderma maculatum*, and *Antrozous pallidus*), the desert fox (*Vulpes*

Figure 10. Main ecosystems (i.e., plant communities) of Mexico according to Rzedowski (1978). The map indicates historic (*left*) and present (*right*) distributions. Note that large regions of the country have lost most of their natural vegetation due to agriculture, cattle-grazing, and human settlements. The tropical rainforest is the most affected plant community environment and is considered in critical danger of disappearing.

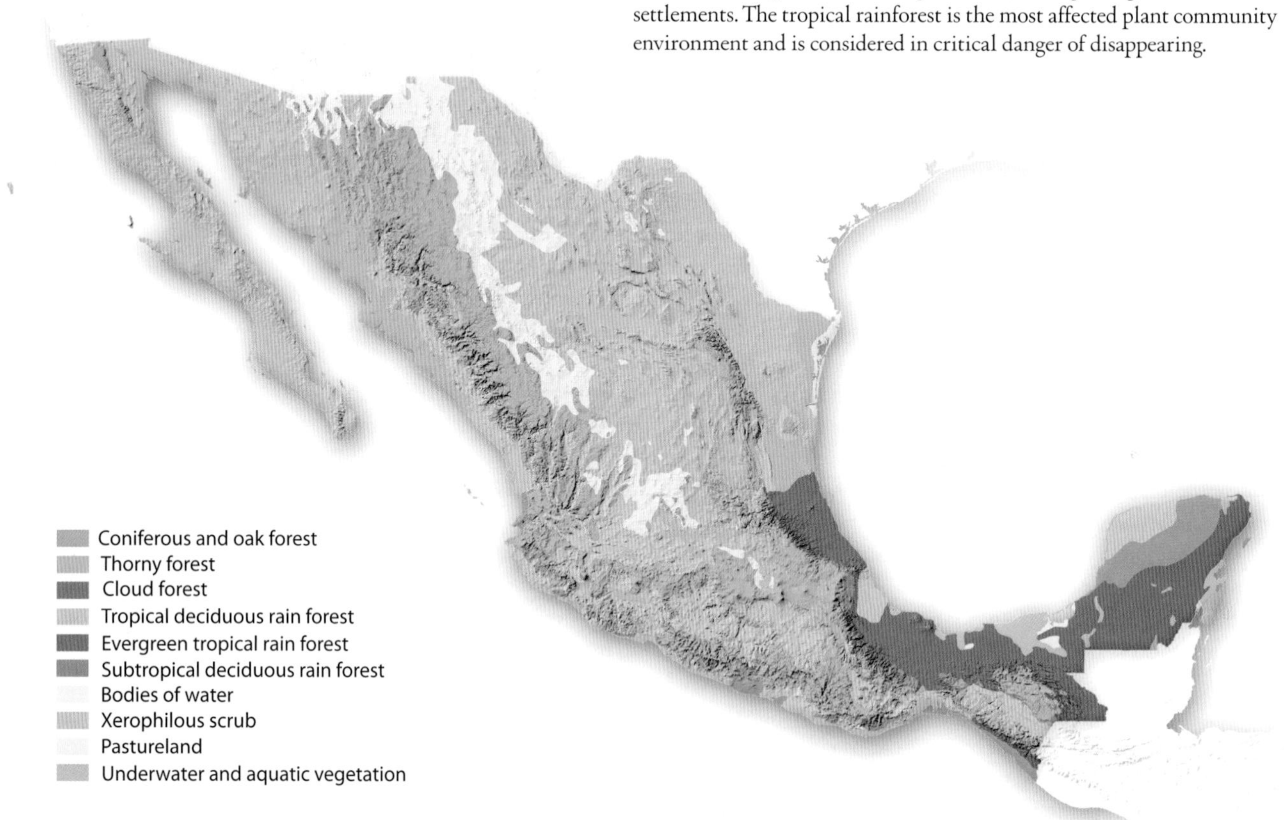

Coniferous and oak forest
Thorny forest
Cloud forest
Tropical deciduous rain forest
Evergreen tropical rain forest
Subtropical deciduous rain forest
Bodies of water
Xerophilous scrub
Pastureland
Underwater and aquatic vegetation

macrotis), the bighorn sheep (*Ovis canadensis*), the mule deer (*Odocoileus hemionus*), rodents such as *Xerospermophilus tereticaudus*, *Thomomys umbrinus*, *Neotoma leucodon*, and the kangaroo rats (*Dipodomys* spp.), the desert cottontail (*Sylvilagus audubonii*), and the black-eared jackrabbit (*Lepus californicus*).

Finally, the grasslands is another type of vegetation severely disturbed by human activities; its original expanse has been dramatically reduced (Ceballos et al., 2010). Today, the largest areas of grasslands are found in northern Chihuahua, Sonora, Durango, and Zacatecas. Some characteristic mammal species include the American badger (*Taxidea taxus*), the black-footed ferret (*Mustela nigripes*), the pronghorn (*Antilocapra americana*), the American bison (*Bison bison*), prairie dogs (*Cynomys mexicanus* and *C. ludovicianus*), and the white-sided jackrabbit (*Lepus callotis*).

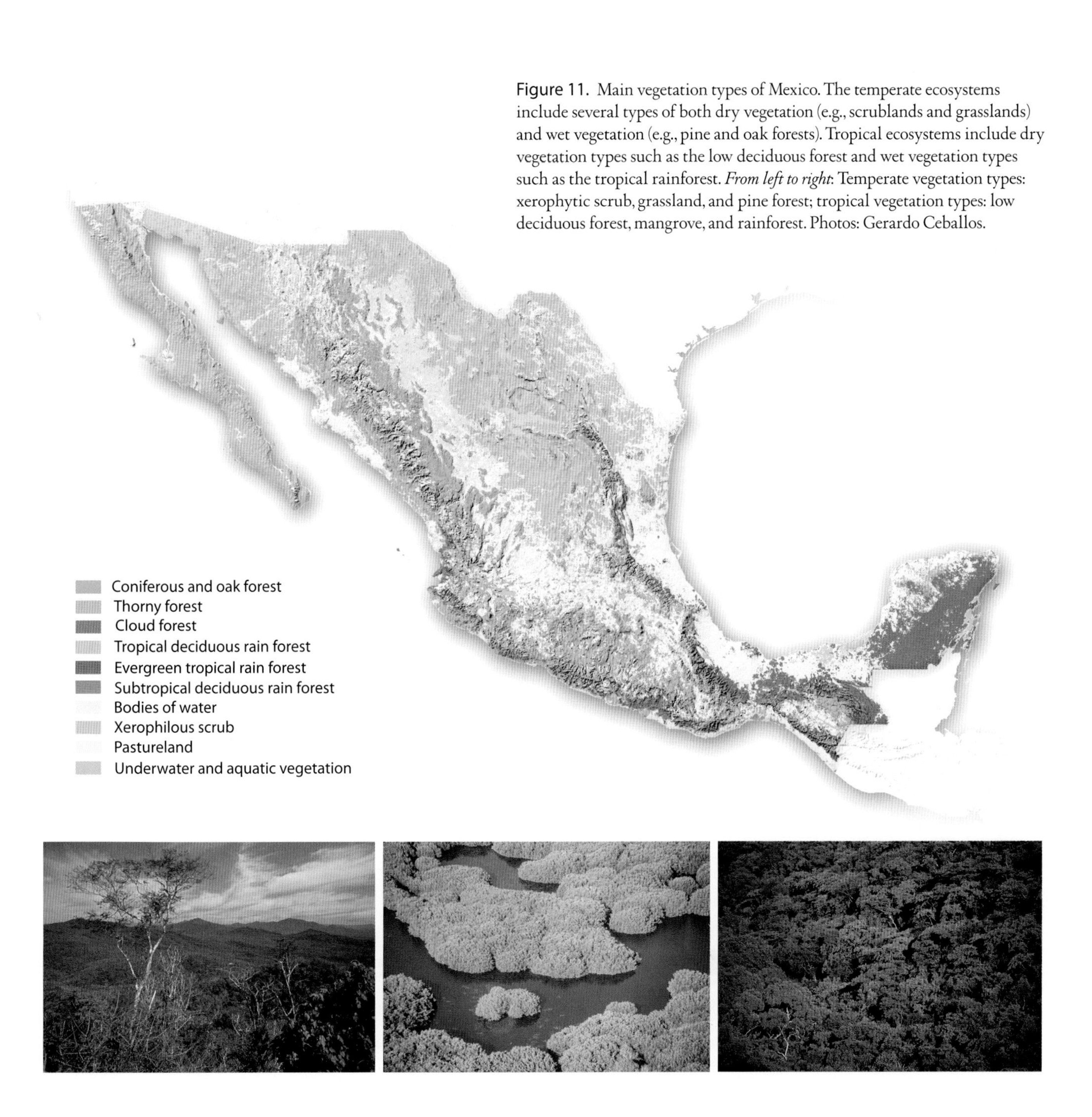

Figure 11. Main vegetation types of Mexico. The temperate ecosystems include several types of both dry vegetation (e.g., scrublands and grasslands) and wet vegetation (e.g., pine and oak forests). Tropical ecosystems include dry vegetation types such as the low deciduous forest and wet vegetation types such as the tropical rainforest. *From left to right*: Temperate vegetation types: xerophytic scrub, grassland, and pine forest; tropical vegetation types: low deciduous forest, mangrove, and rainforest. Photos: Gerardo Ceballos.

Body Masses and Geographic Ranges

Patterns related to body size of Mexican mammals have been thoroughly analyzed (Arita and Figueroa, 1999; Ceballos and Navarro, 1991; Ceballos et al., 1998; Torres et al., 1995). Most of the terrestrial species (66%) are small in size (up to 100 g), with a variation range from 3 g or 4 g in some bats and shrews, up to 450 kg in the bison (*B. bison*) and with an average of 9,400 g (fig. 12; table 5). About 27% of the species, represented by marsupials, primates, carnivores, armadillos, anteaters, rodents, and lagomorphs, are intermediate in size, between 101 g and 10 kg. Only 7% of the carnivore species, perissodactyls, and artiodactyls are large (> 10 kg). There are differences in the distribution of flying mammals (bats) and non-flying mammals because the latter vary by several orders of magnitude greater in comparison to bats (Arita and Figueroa, 1999).

In marine mammals, which on average are larger than terrestrial mammals, body size ranges from about 46 kg for the vaquita (*Phocoena sinus*) up to 170 tons for the blue whale (*Balaenoptera musculus*). The extremes of this interval correspond to the smallest and largest species in the world (Torres et al., 1995). The frequency distribution of body sizes of terrestrial mammals shows a wide variation at different geographical scales (Arita and Figueroa, 1999). Evaluated in quadrants of 4, 2, 1, and 0.5 degrees, the distribution tends to become more uniform at regional and local scales, as expected for the observed continental scales (Brown and Nicoletto, 1991).

Most endemic species are small (< 100 g; see table 5), with significant differences between the endemic and non-endemic species; the average sizes are 287 g and 9,393 g for each group of species, respectively (Ceballos and Rodríguez, 1993; Ceballos et al., 1998). The small-sized endemics include shrews, bats, and rodents. The medium-sized group includes pocket gophers, squirrels, and some rats, while the large-sized group includes only three carnivores (e.g., *Spilogale pygmaea*), seven lagomorphs (e.g., *Lepus flavigularis*), and a rodent (the black agouti, *Dasyprocta mexicana*). The only endemic marine mammal in this category is the vaquita (*Phocoena sinus*).

On average, terrestrial mammals have distribution ranges covering about 25% of the area of Mexico (see table 5), even though there is enormous variation, from species such as the puma (*Puma concolor*) with a range that covers the entire country to species such as the vole *Microtus pennsylvanicus* that inhabits an area smaller than 100 ha. There are interesting differences between groups. On average, endemic species occupy much smaller areas, equivalent to 3% of the territory. Species with a restricted distribution range include 161 endemic species, for example, the Mexican prairie dog *(Cynomys mexicanus)*, or species with a marginal distribution such as the bison (*Bison bison*) near the northern border or the bat *Eumops hansae* in the southern border. These species occupy very small ranges in Mexico, equivalent to 1% of the territory and they are vulnerable to extinction or extirpation (Arita et al., 1997; Ceballos et al., 1998). Species considered endangered tend to occupy distribution areas geographically smaller than species not at risk.

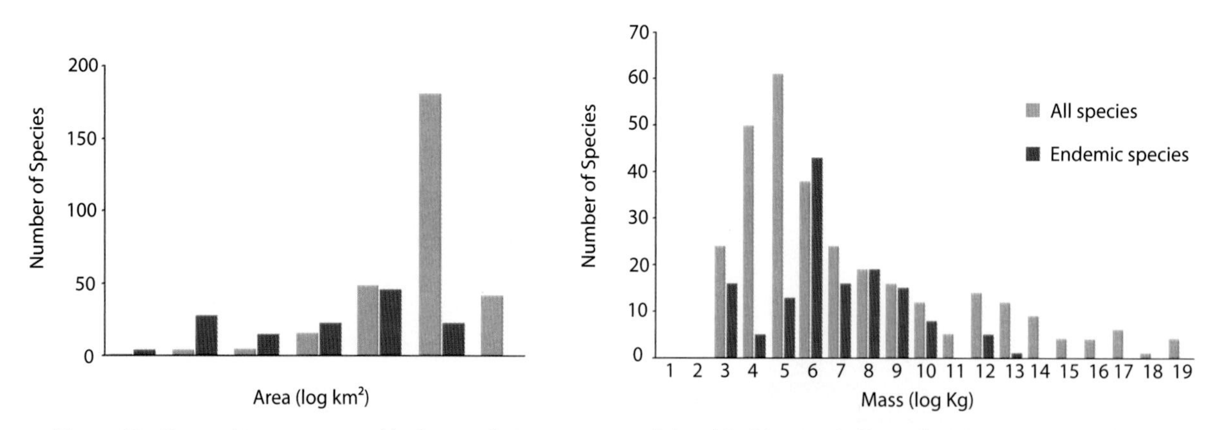

Figure 12. Geographic range size and body size of Mexican mammals (modified from Ceballos and Rodriguez, 1993; Ceballos et al., 1998). Note that most species are small and that the average geographic range is equivalent to 25% of the land area of Mexico.

Table 5. Body sizes and geographical distribution areas of the different groups of terrestrial mammals of Mexico (modified from Ceballos et al., 1998). In this analysis all the species are included, incorporating the ones described or registered after 1998.

Group of Species	N	Average Geographic Distribution (km^2)	Average Body Size (g)
1. All species	461	428,407	9,404
Endemic	147	64,561	287
Not endemic	314	427,183	9,393
2. Restricted species	161	10,090	624
Endemic	100	8,463	235
Not endemic	61	11,461	9,270
Species with a widespread distribution	300	474,732	11,149
Endemic	47	181,843	262
Not endemic	253	525,187	11,465
3. Endangered species	96	159,199	24,734
Endemic	49	9,495	411
Not endemic	47	335,365	32,701
Species with no risk	365	397,811	1,764
Endemic	98	117,659	220
Not endemic	267	485,199	2,256

Trophic Guilds

A simple way to evaluate the trophic relations of mammals is to classify them into guilds, or general feeding categories. In this study we use the seven general categories of Ceballos and Navarro (1991) because knowledge about diet is still scarce. In general, most species are herbivores (245, 51%). This category includes rabbits, hares, mice, some squirrels, pocket gophers, and large mammals such as deer, pronghorn, tapir, and bison (fig. 13). These are followed by insectivorous species (138, 29%), represented by shrews, moles, bats, anteaters, and some rodents. A smaller number of species constitutes each of the other trophic categories. Fruit eaters (38, 8%) include monkeys, some marsupials, bats, squirrels, agoutis, and pacas. The carnivores (27, 6%) include canids, felids, bears, weasels, and some bats. Finally, nectar eaters (10, 2%) and blood eaters (3, >1%) are represented only by bats.

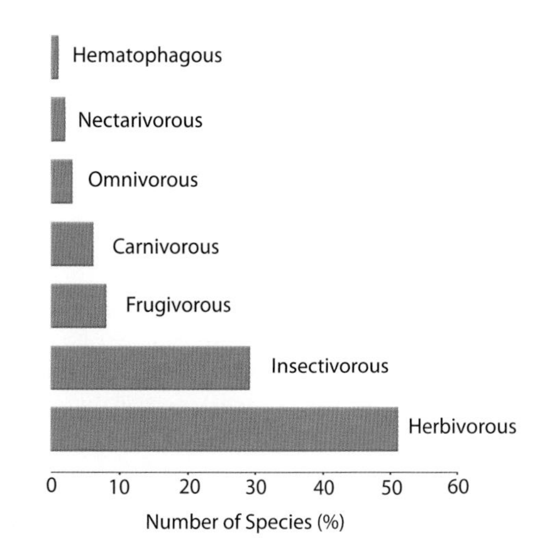

Figure 13. Trophic guilds of the terrestrial mammals of Mexico. In these simplified classifications, most species are either herbivorous or insectivorous.

Migrations

Some bats and marine mammals carry out annual migrations to find adequate places for reproduction, hibernation, and feeding (Eisenberg, 1981). Even though there are not many studies of the migration of species such as *Leptonycteris nivalis*, *L. yerbabuenae*, and *Tadarida brasiliensis*, it is known that the populations travel hundreds or thousands of kilometers in their annual migration (fig. 14). For example, one portion of the population of *L. yerbabuenae* travels every spring from the coast of Jalisco to the Baja California Peninsula, Sonora, and Arizona looking for caves in which to roost and reproduce. At the end of the summer and fall these animals fly back to the warm coasts of the Mexican Pacific (Ceballos et al., 1997). Although it was thought that the migratory corridors varied in *Tadarida brasiliensis*, supposedly moving from New Mexico and Texas through the Mexican Plateau to the center and south of the country (Villa-R., 1956), it has been shown that at least for this species, this is not true (Russell et al., 2005). Rather, the species behaves as a single, panmictic population (Russell et al., 2005).

The migratory behavior of whales is better known. Every year, gray whales (*Eschrichtius robustus*) make a long journey of more than 15,000 km, from the frozen waters of the Arctic, where they spend the summer and fall taking advantage of abundant food, to the coastal lagoons of Guerrero Negro, Bahia Magdalena, and San Ignacio on the Pacific coast of the Baja California Peninsula, where they reproduce. Similarly, humpback whales (*Megaptera novaeangliae*) travel from the coasts of Jalisco and Revillagigedo Islands to different regions of the north Pacific, from California to the Arctic (fig. 15; Urbán-R. et al., 2000).

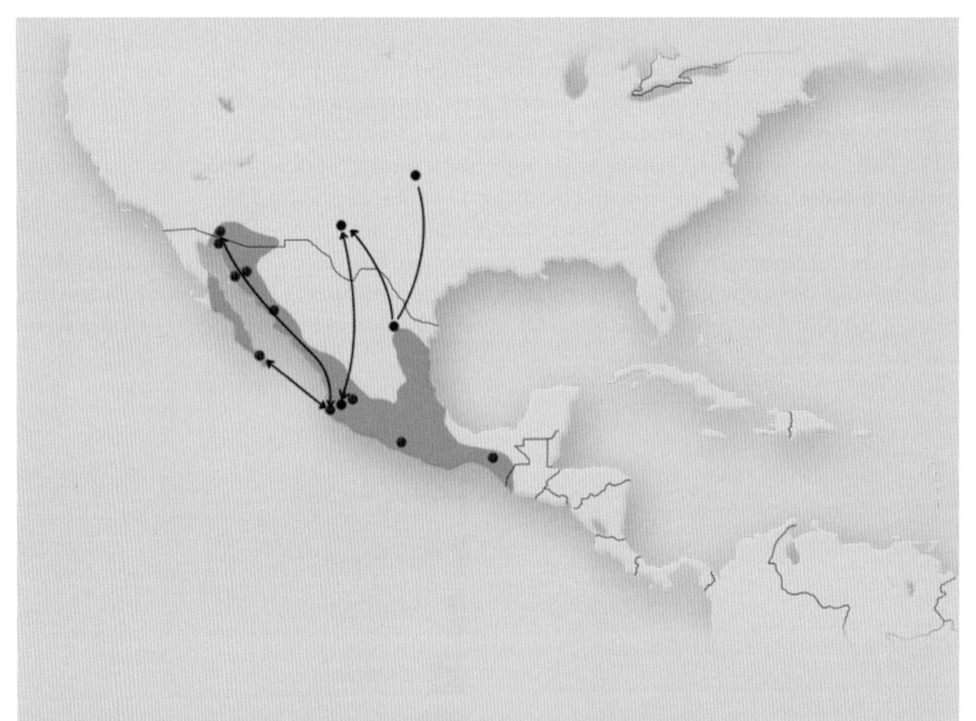

Figure 14. Migratory routes of some bat species in Mexico. Species such as *Leptonycteris yerbabuenae* and *Tadarida brasiliensis* travel extraordinary distances to reproduce. *L. yerbabuenae* is mainly distributed in central and western Mexico (in red) and migrates along the Pacific coast, from Jalisco to Sonora in northern Mexico and Arizona in the southern United States (map modified from Ceballos et al., 1997). In contrast, *T. brasiliensis* travels (in blue) along the continental land from the southern United States and northern Mexico to central and southern Mexico (map modified from Villa-R., 1956).

Pleistocene Effects on Mammals

The current distribution of Mexican mammals is, in large part, the result of a response to climate changes in the late Pleistocene and the early Holocene. But information on the intensity of climate fluctuations during the Pleistocene and the effect on the distribution patterns of Mexican mammals remains sketchy at best. Recently, Ceballos et al. (2010) analyzed this subject based on fossil evidence from the late Pleistocene and the Holocene, examining diversity, structural composition of communities, extinct species, and general patterns of individual responses. The following synthesis is based almost entirely on that analysis.

Environmental changes associated with glacial periods during the Pleistocene and the Holocene were intense. During the last maximum glacial period 18,000 years ago, an enormous ice layer covered most of the Northern Hemisphere. In North America, from the Arctic to Kansas, an ice layer almost 3 km deep covered practically the entire continent. The glacial period caused climate changes that in general made it colder and provoked eustatic changes in the level of the sea, which receded, exposing great extensions of the continental plateau. Specifically, there were variations in rainfall, atmospheric composition, and wind currents that created environments quite different from the present ones, all of these consequences of the size, height, and intense reflectivity of the ice layer (Polaco and Arroyo-Cabrales, 2001).

The effects of the environmental dynamics of the Pleistocene and the Holocene have been studied using the fossil record and current distribution patterns. Even though the fossil record is a limited tool, there are data on about 800 localities that represent 284 species, 145 genera, 42

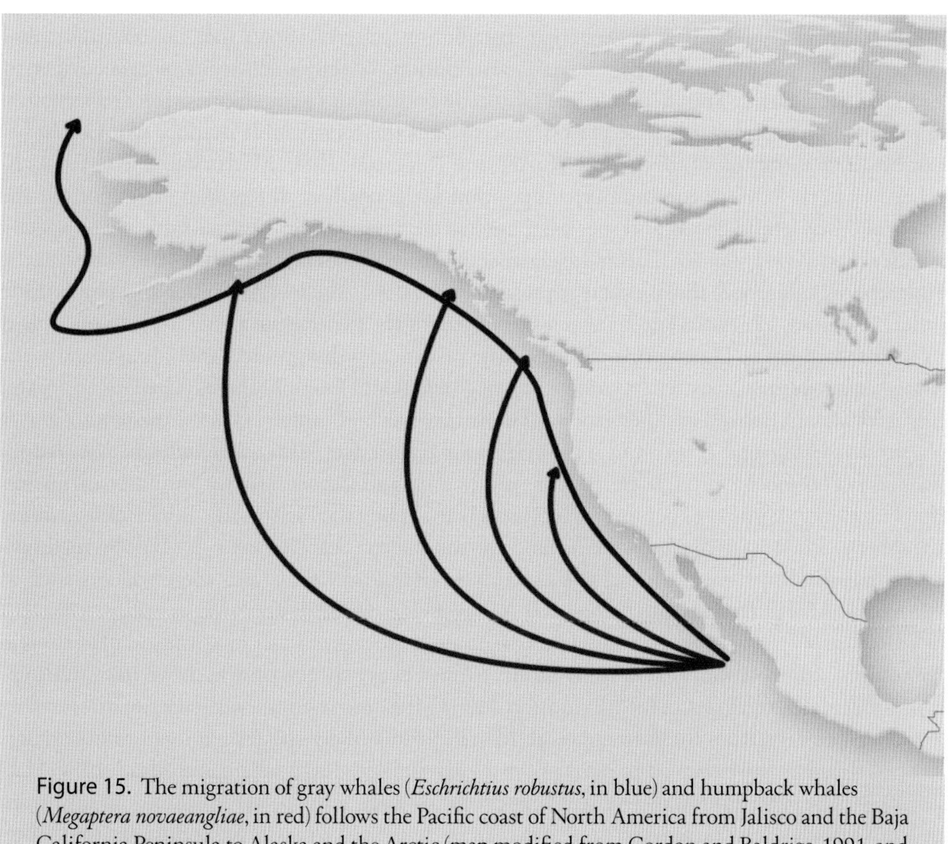

Figure 15. The migration of gray whales (*Eschrichtius robustus*, in blue) and humpback whales (*Megaptera novaeangliae*, in red) follows the Pacific coast of North America from Jalisco and the Baja California Peninsula to Alaska and the Arctic (map modified from Gordon and Baldrige, 1991, and Urbán et al., 2002).

Table 6. Species richness of the Mexican mammals in the late Pleistocene and the present. Note that the order Proboscidea became extinct in Mexico and North America at the end of the Pleistocene.

Order	Family	Genera	Species
Didelphimorphia	3/1	4/7	7/8
Cingulata	5/1	10/2	13/2
Pilosa	1/2	1/2	1/2
Primates	1/1	2/2	3/3
Lagomorpha	1/1	5/3	13/15
Rodentia	8/8	36/50	96/243
Soricomorpha	1/2	3/6	7/38
Carnivora	6/8	23/28	39/42
Artiodactyla	5/4	19/8	34/10
Perissodactyla	2/1	2/1	15/1
Proboscidea	3/0	5/0	6/0
Chiroptera	6/9	35/67	49/139
Total	42/38	145/176	283/503

families, and 11 orders of terrestrial mammals (table 6; Arroyo-Cabrales et al., 2002; Ceballos et al., 2010). The order with the largest number of fossil species is rodents (Rodentia), followed by bats (Chiroptera) and carnivores (Carnivora).

Analysis of this database and current distribution patterns shows that the environmental dynamics resulted in extinctions, vicariant distributions of the same species or related species, and speciation events. The effects of the Pleistocene are evident in the composition and structure of mammalian communities. The most drastic effect is the considerable number of taxa that became extinct at the end of this period (Kurtén and Anderson, 1980; Martin and Klein, 1984). In other words, the current community composition is the result of differential extinctions at the levels of order, family, genus, and species; these had a tremendous impact, particularly on species smaller than 100 kg (fig. 16). This indicates that the mammalian fauna of the Pleistocene and the Holocene were far more diverse than at present, especially at the levels of order and family. One order, 5 families, 27 genera, and 85 species recorded for the Pleistocene are no longer represented in the modern fauna. During this period all the proboscides disappeared, including mammoths (*Mammuthus columbi*), mastodons (*Mammut americanum*), and Gomphotheres (*Cuvieronius* sp. and *Stegomastodon* sp.) as well as other groups of herbivores such as camels (*Camelops hesternus*), horses (*Equus conversidens* and *E. mexicanus*), bison (*Bison antiquus*), deer (*Odocoileus halli* and *Navahoceros fricki*), and Pleistocenic pronghorns (*Stockoceros conklingi* and *Capromeryx* sp.). Similarly, carnivores that fed on large herbivores became extinct as well, including the Pleistocenic lion (*Panthera atrox*), the saber-toothed tiger (*Smilodon fatalis*), and the short-faced bear (*Arctodus* sp.). Finally, medium- and small-sized mammals such as the cacomixtle (*Bassariscus ticuli*) and some rodents such as *Microtus meadensis*, *Baiomys intermedius*, and *Neotoma magnodonta* disappeared as well.

Because of the vast climate changes during the Pleistocene, species distribution patterns and community composition were very dynamic. Almost every species on which we have information suffered deep changes in its distribution, leaving in some regions to follow climate corridors and adequate vegetation for their survival (Martin, 1960). Many species managed to find refuge in areas they had not occupied in northern or southern latitudes (Graham and Mead, 1987; Harris, 1974). The extra-limital distributions, as they have been termed, were common in taxa from order to species level. The affected organisms can be divided into two large groups. The first group includes the ones that disappeared from Mexico but survived in other countries, and the second, the ones that survived in Mexico but whose current distribution is different from that in the Pleistocene.

Figure 16. Mammoths were widely distributed in North America. Fossil deposits are plentiful. The isolines of the map connect regions with a similar number of records (map modified from Agenbroad, 1984). Mammoths were apparently more abundant in temperate regions in Mexico. Some fossil deposits, in the Mexican Basin, such as the one in Tocuila near Texcoco in the state of Mexico, are extraordinary for their number of individuals. There are three skulls in this photograph *(below)*. Drawing: courtesy of George "Rinaldino" Teichmann; Photo: Gerardo Ceballos.

The extra-limital distributions that represent extinctions in Mexico include at least one order (Proboscidea), four families (e.g., Camelidae), and eight genera (e.g., *Cuon*). Of the order Proboscidea, represented by extinct species of mammoths and mastodons, there are three elephant species in Africa and Asia. Among the extinct families we find the horses (Equidae) alive today in Africa and Asia, the capybaras (Hydrochoeridae) in South America, and the camelids (Camelidae) in South America and Asia. Among the species that disappeared we find the cuon (*Cuon alpinus*), the lion (*Panthera leo*) that survives in Africa and Asia, and the spectacled bear (genus *Tremarctos*) in South America. Other species, such as the giant anteater (*Myrmecophaga tridactyla*) recorded near Sonora and the southern bog lemming (*Synaptomys cooperi*) recorded in Nuevo León, are now found about 1,300 km south in Honduras and 1,600 km to the northeast in Arkansas, respectively (fig. 17).

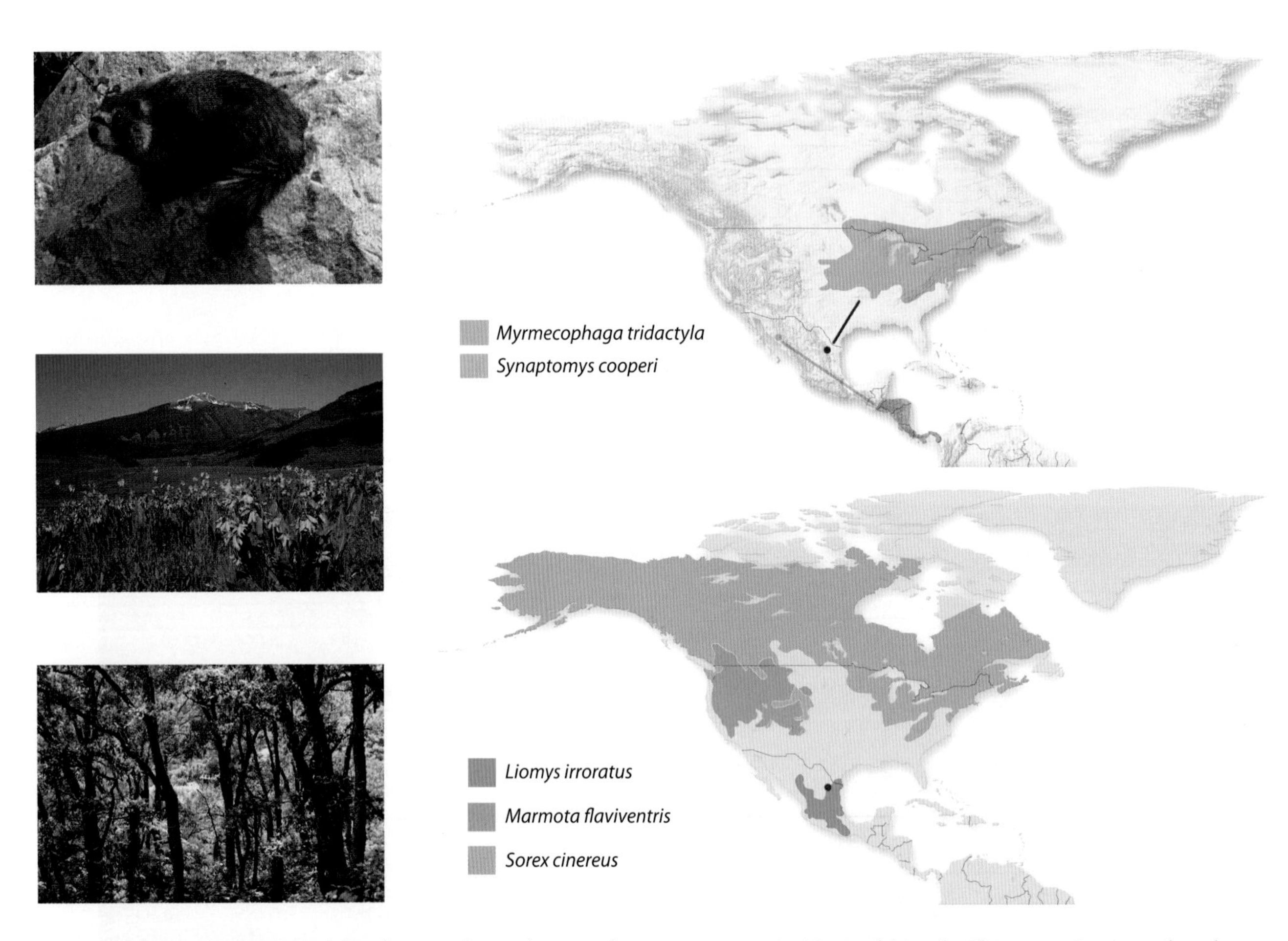

Figure 17. (*Top map*) Extra-limital distributions of several mammalian species present in Mexico during the Pleistocene. Species such as the lemming (*Synaptomys cooperi*) that is currently distributed in eastern Canada and the United States (yellow color) was found in Nuevo León. (orange circle) and the giant anteater (*Myrmecophaga tridactyla*), presently distributed from Central America to the south (green color), was found in Sonora (green circle). Both species are now thousands of kilometers from their Pleistocene sites in Mexico.

(*Bottom map*) Due to the fact that the mammal communities were quite different in the past, they are known as disharmonic faunas. A good example is the San Josecito region in the state of Nuevo León (red circle), where species such as *Marmota flaviventris*, *Liomys irroratus*, and *Sorex cinereus* coexisted at that time but now are geographically separate by long distances from each other. The habitat where *M. flaviventris* presently lives, for example, such as the mountains in Colorado is very different from the one in Nuevo León, Mexico.

(*Top photo*) *M. flaviventris*, Salt Lake City, Utah. Photo: Gerardo Ceballos. (*Middle photo*) Habitat of *M. flaviventris*, Crested Butte, Colorado. Photo: Gerardo Ceballos. (*Bottom photo*) Vegetation near San Josecito, Nuevo León. Photo: Joaquín Arroyo.

One consequence of extra-limital distributions in community compositions is disharmonic fauna. During the Pleistocene, species or groups of species apparently coexisted in a more diverse manner than the current compositions. The available evidence shows that the current species assemblages are relatively recent. For example, in San Josecito, Nuevo León, there are records showing sympatric species that presently are separated by hundreds or thousands of kilometers; this is the case, for example, for *Marmota flaviventris*, *Liomys irroratus*, *Synaptomys cooperi*, and *Sorex cinereus* (see fig. 17).

In other cases, species with continuous ranges were separated. This could explain the endemics, relictuals, and species with detached distributions in the southwestern United States and northern Mexico (Brown, 1971; Martin, 1955, 1960; Martin and Harrell, 1957; Patterson, 1980; Riddle et al., 2000a, b). Among the endemics in this region are *Cynomys mexicanus* (derived from *Cynomys ludovicianus*), *Callospermophilus madrensis* (from *C. lateralis*), *Tamiasciurus mearnsi* (from *T. douglasi*),

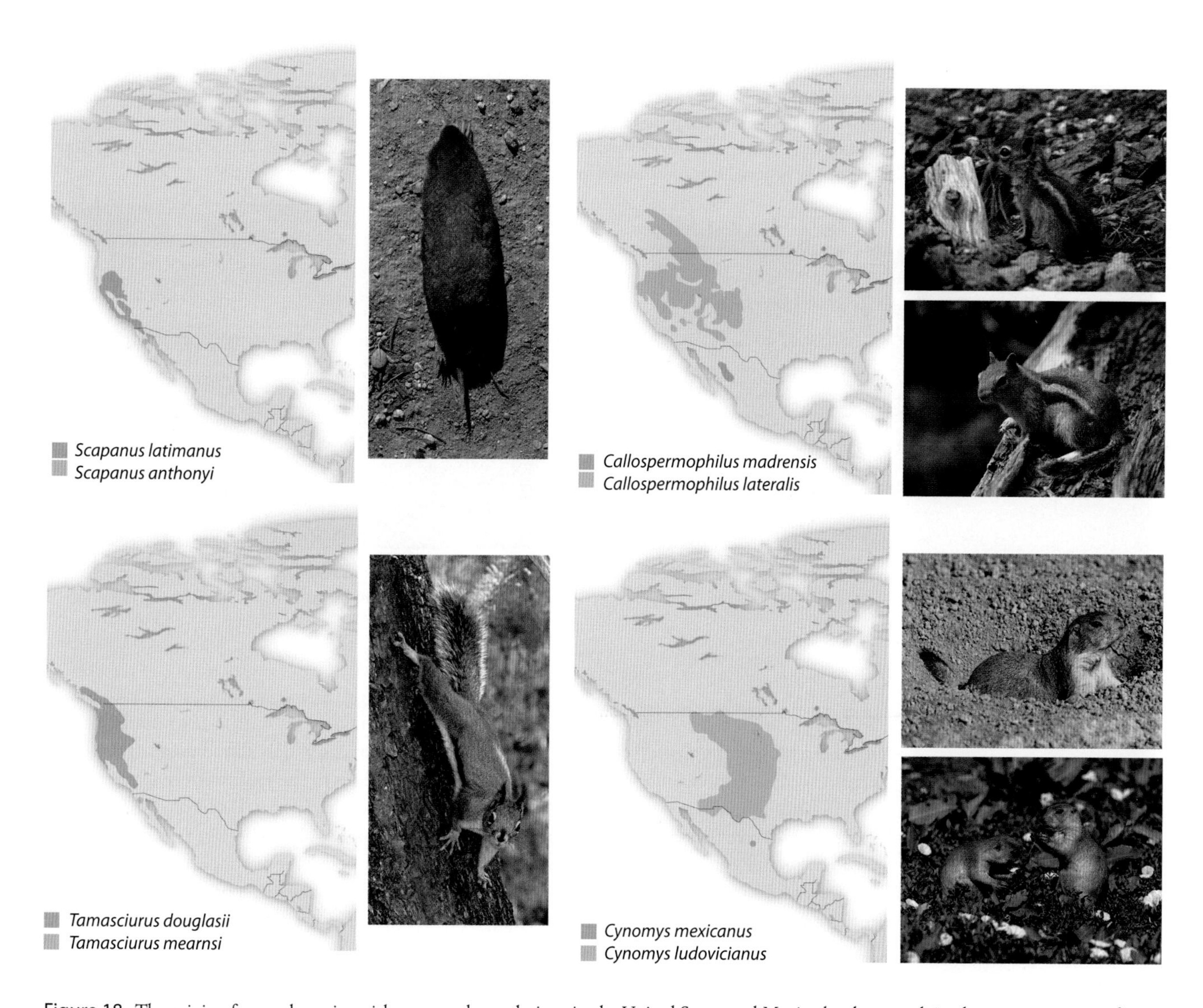

Figure 18. The origin of several species with separated populations in the United States and Mexico has been explained as a consequence of isolation caused by climate changes during the Pleistocene (see a review in Ceballos et al., 2010). Some examples include widely distributed species such as *Glaucomys volans* or Mexican endemics closely related to species from northern latitudes such as prairie dogs (*Cynomys ludovicianus* and *C. mexicanus*), ground squirrels (*Callospermophilus lateralis* and *C. madrensis*), tree squirrels (*Tamasciurus douglasii* and *T. mearnsi*), and moles (*Scapanus latimanus* and *Scapanus anthonyi*). All photos by G. Ceballos except *T. mearnsi*, by B. S. Pash, and *C. madrensis*, by R. List.

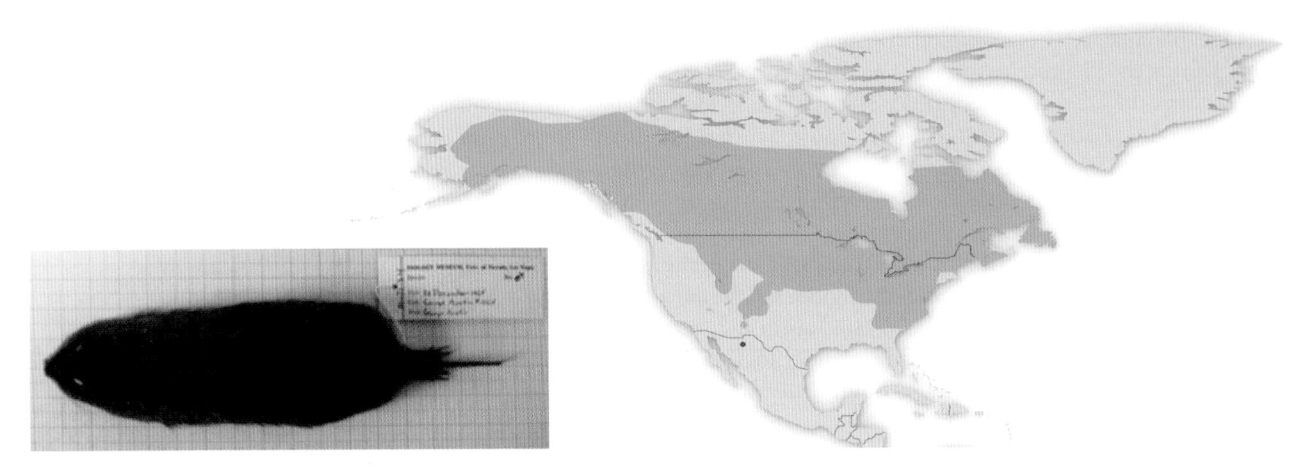

Figure 19. The meadow vole (*Microtus pennsylvanicus chihuahuensis*) became extinct in Mexico in the last decade (List et al., 2010) (inset photo: Oliver Pergman). Although the species is widely distributed in the United States and Canada (in green), its disappearance in Mexico is a clear example of the extinction of populations that in many cases is the prelude to species extinctions and the loss of genetic variability (Ceballos and Ehrlich, 2002). The species had a relictual distribution in Mexico (in red), isolated in a single locality known as Ojo de Galeana in the state of Chihuahua, where it inhabited a small marsh surrounded by xeric scrub vegetation. Very likely such distribution was the result of a range contraction after the environment became drier during the late Pleistocene (Ceballos et al., 2010 for a review). The species was not found from 2000 to 2004, and the spring dried up because of agricultural expansion and the meadow vole habitat disappeared in 2004.

Typical habitat of *M. pennsylvanicus* in Crested Butte, Colorado. Photo: Gerardo Ceballos.

Marsh vegetation in Ojo de Galeana, Chihuahua, before the water spring dried out in 2004 by overuse. Photo: Gerardo Ceballos.

and *Scapanus anthonyi* (from *S. latimanus*; fig. 18). Among the species with relict populations in northern Mexico the clearest examples are *Microtus pennsylvanicus*, widely distributed in the northern and central United States and Canada, and relict populations in the southwestern United States and in Chihuahua. The population in Chihuahua, probably extinct today (List et al., 2010), inhabited about 40 ha of flooded vegetation near Ojo de Galeana springs, located in an arid zone (fig. 19).

The presence of endemics, relictual species, and disjunct populations supports the hypothesis of biogeographic filters and of important speciation centers in isolated geographic ranges. These filters caused species to cross differentially, partially determining the current species composition. Examples of these filters are the eastern United States and the Sierra Madre Oriental, with species such as *Glaucomys volans*, *Microtus quasiater*, and *Scalopus aquaticus*. There is also the Rocky Mountains in the central-eastern United States and the Sierra Madre Occidental, with species such as *Sciurus aberti*, *Callospermophilus madrensi*, and *Microtus pennsylvanicus*; the central United States and northern Mexico, characterized by *Cynomys mexicanus*, *Peromyscus nasutus*, and *Onychomys torridus*; and the Eje Volcanico Transversal-Sierra Madre del Sur, with *Sorex sassurei*, *Reithrodontomys microdon*, *Peromyscus furvus*, and *G. volans* as characteristic species.

Conservation

Sooner or later, one has to take sides, if one is to remain human.
— Graham Greene, *The Quiet American*

"The apparently fragile and harmless kitty cat was abandoned one afternoon on Estanque Island. He sat on the sand feeling confused. For a while he could see the fishermen's boat drifting away until it finally disappeared on the horizon. He waited. Cold waves soaked his feet . . . but he did not move. But then in the dead of night a sudden movement caught his attention. He instinctively jumped and caught an endemic mouse (*Peromyscus guardia*) with his claws! The following months the cat managed to survive by eating these mice until finally in the summer of 1997 they were totally exterminated; the last specimen of this subspecies, not yet described, died at the claws of the cat" (Ceballos, 1999:1). The story of *Peromyscus guardia* is just one example of the many problems mammals have to face and shows how an insular species can be severely affected by an exotic species (Mellink et al., 2002; Vázquez-Domínguez et al., 2004).

During recent decades many species in Mexico and the world have become extinct. Their disappearance is just a part of what is considered to be one of the most severe environmental problems of the century: the loss of biodiversity. Because of human activities, hundreds of thousands of populations and species are at risk of extinction. The magnitude of this catastrophe is as huge as the extinctions that occurred at the end of several geological periods such as the Permian and the Cretaceous (Benton and Hasper, 1995). And even more alarming is the difference between the time in which those extinctions occurred compared to what is happening in the present: an extinction process that took millions of years is happening today in just decades and will certainly have severe consequences in the history of life on Earth.

At least 43 species of vertebrates have become extinct in Mexico, 15 of them mammals (Ceballos, 1993; Ehrlich and Ceballos, 1997). These extinctions mark the modern history of the country and have left a deep scar on its ecosystems. The survival probabilities of some species will decrease if conservation and management measures are not implemented quickly. Conservation of species and resources management should be a priority of every country.

Conservation of Mexican mammals should be based on a thorough analysis of social struggle, which in the words of Carlos Fuentes is the breeding ground of environmental problems. The twenty-first century is marked by a severe crisis facing both natural resources and wild mammals. The origin of this crisis is strongly linked to excessive population growth and social inequality. What can be done to reverse this gloomy scenario and to support Mexico's biodiversity? The main strategy has been to create legal protection for endangered species and to establish protected natural areas and nature reserves. In recent years, to promote conservation of a considerable part of the country's diversity, many economic activities, such as extensive agriculture, forestry, and mining, are being progressively conducted in a rational and sustainable manner (see, e.g., Daily et al., 2003).

Endangered and Extinct Species

Over the past several centuries, Mexico has demonstrated its leadership in the evaluation and classification of endangered species in an effort to protect and preserve them. This concern goes back to pre-Columbian times, when Moctezuma established zoos and banned the taking of particular species such as quetzals and jaguars. At the end of the nineteenth century, the disappearance of large mammals in the north of the country caught the attention of scientists and conservationists (Herrera, 1890; Leopold, 1959). Before 1990, however, there was no methodical effort to evaluate the situation of mammals and no qualitative method of assessment (Ceballos and

Navarro, 1991). The method that was eventually developed was based on the analysis of biological features associated with being vulnerable to extinction, such as geographic distribution, body size, and trophic guilds, and features associated with human activities, such as overexploitation, habitat loss, and use of pesticides (table 7); by then more than 120 species were endangered and at least 9 were extinct. In 2007, other authors devised a new protocol (Sánchez et al., 2007) that became the federal law under which species at risk would be identified, following a specified protocol (SEMARNAT, 2010). Today, other countries are adapting this protocol to evaluate the extinction risk.

Currently, the Mexican list of endangered species includes about 30% of the species that are threatened or at risk of extinction (table 8). If we consider the 230 subspecies, then 44% of the species and subspecies (see table 2) are threatened or at risk of extinction (SEMARNAT, 2010). This catalogue includes species that are globally classified at risk, for example, the volcano rabbit (*Romerolagus diazi*) and the Mexican prairie dog (*Cynomys mexicanus*), but it also emphasizes species that have large populations in other countries but are at risk in Mexico, for example, the pronghorn (*Antilocapra americana*) and the beaver (*Castor canadensis*).

Mexican mammals are under-represented in international lists of endangered species. The most recent list from the International Union for Conservation of Nature (IUCN) includes 133 species (IUCN, 2010). Most species not included are small mammals. In the Mexican list, 36 species of bats and 13 insectivores are included, but in the IUCN list only 14 bats and 11 insectivores are included. There are two reasons for this. First, the IUCN list includes only those species considered at global risk, so it omits those whose Mexican populations are endangered but are not at risk in other countries. Second, the information used by IUCN is not entirely up-to-date and does not include complete data on the Mexican regulation concerning endangered species and thus does not include several endemic species with conservation issues at a global scale.

As discussed earlier, the number of extinct species has grown in recent years (Baillie and Groombridge, 1996; Ceballos 1993; IUCN, 2010). In the first complete evaluation, which took place in 1988, 7 Mexican species were classified as extinct (Ceballos and Navarro, 1991). This number had increased to 11 almost 10 years later (Ehrlich and Ceballos, 1997) and then to 16 today, if *N. martinensis* and *N. bunkeri* are included; these species were recently relegated to either subspecies status or synonyms of other *Neotoma* species (Patton et al., 2007). In this decade the species listed have changed for two main reasons. First, paradoxical as it may seem, several spe-

Table 7. Criteria for the classification of endangered species (modified from Ceballos and Navarro, 1991). In this classification we consider criteria to determine the impact of human activities (Category I) and criteria to evaluate biological characteristics co-related to extinction vulnerability (Category II).

Category I: Evaluation of the impact of human activities

a. Overexploitation
   Commercial exploitation
   Subsistence hunting
   Considered as plague or predator
b. Habitat destruction
   Migratory species
   Tropical species
   Carnivore species

Category II: Biological features associated with extinction

c. Population size
d. Body size
e. Restricted distribution
f. Trophic guild

Table 8. Extinct or extirpated Mexican mammalian species. Status: EX = extinct; ET = extirpated; EX?, EC? = Probably extinct or extirpated. Causes: OE = overexploitation, HD = habitat or modified habitat; PO = pollution, IE = introduced species.

Order/species	Status	Causes
Carnivora		
Canis lupus	ET	OE, HD
Ursus arctos	ET	OE, HD
Monachus tropicalis	EX	OE
Enhydra lutris	ET	OE
Rodentia		
Dipodomys gravipes	EX?	HD
Dipodomys insularis	EX?	
Microtus pennsylvanicus	ET	HD, IE
Neotoma anthonyi	EX	IE
Neotoma bunkeri	EX	IE
Neotoma insularus	EX	IE
Neotoma turneri	EX?	IE
Oryzomys nelsoni	EX	HD, IE
Peromyscus guardia	EX?	IE
Peromyscus pembertoni	EX	?

cies that were considered extirpated from Mexico have recovered naturally or with management programs such as the American deer or elk (*Cervus canadensis*) and the black-footed ferret (*Mustela nigripes*); in other cases, species such as the bison (*Bison bison*) have been rediscovered in the country. Second, several species of small island rodents have been added to the list, and their possible extinction has been documented in recent years. Ten of the extinct species are small rodents. Most of these extinctions have been attributed to the actions of exotic predators such as domestic cats introduced to the islands (Mellink et al., 2002; Vázquez-Domínguez et al., 2004).

Among the species considered possibly extinct, 10 represent global extinctions, some because they are endemic species of Mexico or because they are globally extinct (for example, the Caribbean monk seal, *Monachus tropicalis*). The rest represent local extirpations because populations exist in other countries. The distribution of extinct species indicates that most of them were in the north of the country, in island and marine ecosystems (fig. 20). See table 3 for a synthesis of those species.

Canis lupus: The Mexican wolf disappeared from its natural habitat in the southwestern United States (New Mexico and Arizona) and in northern Mexico more than 20 years ago (fig. 21). The last record of its presence dates back to 1978 in Durango (McBride, 1980). The captive population has recovered and has already been reintroduced in Arizona, where the population numbers more than 50 individuals, but they are still facing many problems. Five individuals were reintroduced in the Sierra de San Luis in northern Chihuahua in the spring of 2012, but only one still survived after a month.

Dipodomys gravipes: The San Quentin kangaroo rat was abundant in the valley by the same name in Baja California. The development of intensive farming has almost completely destroyed their habitat. Coupled with this, the presence of dogs, cats, and rats certainly has impacted their populations. It is likely that the species is extinct (Mellink, 1992a).

Dipodomys insularis: This rat was endemic to San José Island in the Gulf of California. At the end of the 1980s, they were common on the island (Best and Thomas, 1991a). In recent years, however, they have not been found despite fieldwork (Álvarez-Castañeda and Ortega-Rubio, 2003).

Enhydra lutris: More than a century ago, the excessive fur trade undertaken by Spaniards,

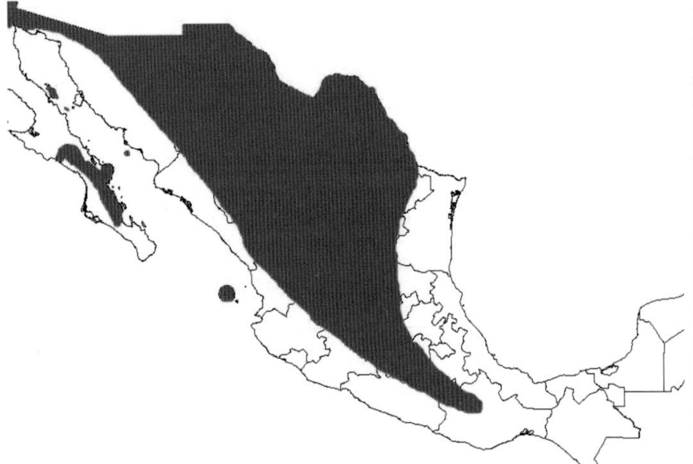

Figure 20. Regions where extinct mammal species were found in Mexico. Extinct species inhabited mostly the northern part of the country, islands, and marine ecosystems.

Figure 21. The Mexican wolf (*Canis lupus baileyi*) became extinct in the wild in Mexico around 1978. It still survives in captivity. Five wolves were released in northern Sonora in the spring of 2012. By autumn only a female was surviving. The specimen in the photograph was poisoned during a national campaign for eradication of wolves at the end of the 1950s. Photo: Bernardo Villa.

Russians, and Americans exterminated the sea otter (fig. 22). In recent years, however, it has recolonized the coast of Baja California, where two records of their populations have been made, at Cedros Island and Bahia Magdalena (Gallo-Reynoso and Rathbun, 1997; Rodríguez Jaramillo and Gendron, 1996). Even so, it should be considered as practically extinct. It is possible, however, that the species may recover naturally in the following decades.

Monachus tropicalis: This tropical seal was distributed along the coast of the Gulf of Mexico, with some sightings on several islands near the coast of the Yucatan Peninsula (fig. 23). Apparently, their extermination began during the time of the Spanish conquest of Mexico. They were hunted because of their high oil content and abundance of meat. The last specimens were captured in the early twentieth century (Timm et al., 1997).

Neotoma anthonyi: This rat was endemic to Todos Santos Island, but domestic cats probably caused its disappearance (Mellink, 1992b). Its biology is almost entirely unknown. Recently, the species was relegated to the subspecies status of *N. bryanti* (Patton et al., 2007).

Neotoma bunkeri: This species was endemic to Coronado Island; it probably became extinct because of domestic cats introduced by fishermen. In 1991, it was considered endangered (Ceballos and Navarro, 1991), but a couple of years later it was confirmed that it was extinct (Smith et al., 1993). Patton et al. 2007 treat this taxon as a synonym of *Neotoma bryanti*.

Neotoma insularis: This species was endemic to Angel de la Guarda Island and is probably extinct. It was considered a subspecies of *N. lepida* until recently (Patton et al., 2007).

Neotoma turneri: This rat was endemic to Turner Island in the Gulf of California near the coast of Sonora. Intense fieldwork indicates that it is probably extinct (B. Tershy, 2002, pers. comm.); the cause is probably the introduction of domestic cats.

Microtus pennsylvanicus: The Mexican population was restricted to an area of 40 ha of wetlands near a spring in the arid zone of Ojo Galeana, Chihuahua (Ceballos and Navarro, 1991). The last sighting was made in 1998 (see fig. 14; List et al., 2010).

Oryzomys nelsoni: This rat is known from four specimens that were used for its original description in 1899. It was endemic to the island of Maria Madre in the Marias archipelago. The causes of its extinction are unknown, but it is likely that the introduction of rats and goats destroyed its habitat and displaced the population. In an intensive search on the island of Maria

Figure 22. The sea otter (*Enhydra lutris*) became extinct in Mexico after a brutal extermination for its pelt in the nineteenth century by Spanish, American, Russian, and Japanese hunters. A single Russian ship obtained more than 13,900 sea otter pelts in the winter of 1806 (Ceballos and Navarro, 1991). In recent years there have been a few sightings and several births reported off the coast of Baja California (Gallo-Reynoso and Rathbun, 1997; Rodríguez Jaramillo and Gendron, 1996). These two sea otters were photographed in Monterey Bay, California, where one of the two largest populations of the species survives. Photo: Gerardo Ceballos.

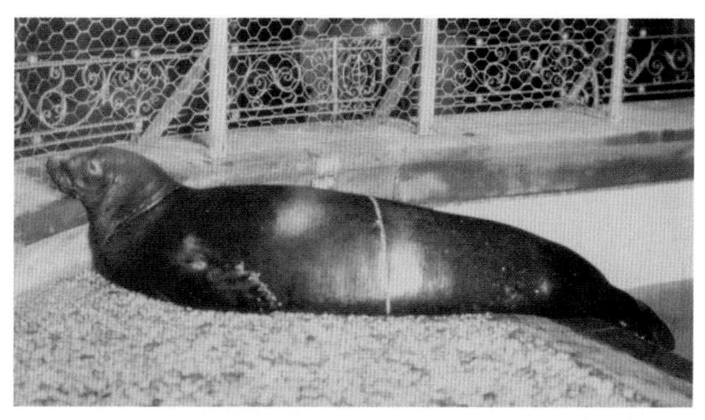

Figure 23. The tropical monk seal (*Monachus tropicalis*) was one of very few species of tropical seals. It became extinct around 1952 due to overharvesting. The specimen in the photo was kept in captivity in the New York Zoo. Photo: courtesy of the New York Zoological Society.

Madre, in 2002, only introduced rats were found (B. Tershie, 2002, pers. comm.). Further fieldwork by Rodrigo Medellín's field crew showed no additional sightings in 2008 and 2009.

Peromyscus guardia: This mouse was endemic to Angel de la Guarda Island, Mejia Island, Granito Island, and Estanque Island in the Gulf of California (fig. 24). It is probably extinct, but there are still hopes of finding some individuals on Angel de la Guarda Island (Mellink et al., 2002). The cause of its extinction is likely the introduction of domestic cats. The last known population, which lived in Estanque, was large until 1995, but a single cat exterminated it. In 1999, no more specimens were found. This is a clear example of the fragility of island species (Vázquez-Domínguez et al., 2004).

Peromyscus madrensis: This species is endemic to the Islas Marias and had been considered extinct. During a widespread search on the Island Maria Madre in 2002, only introduced rats were found (B. Tershie, 2002, pers. comm.). But fieldwork in 2010 showed that the species is still alive and probably relatively abundant (J. Cruzado, pers. comm.).

Figure 24. The Angel de la Guarda island deermouse (*Peromycus guardia*) was found in four islands off Baja California and is now probably extinct (Vázquez-Domínguez et al., 2004). There is a slim chance that the species still survives in Angel de la Guarda Island. Photo: Gerardo Ceballos.

Figure 25. The Mexican brown bear (*Ursus arctos horribilis*) is extinct. This huge specimen was one of the last individuals in Mexico; it was hunted in Sierra del Nido in Chihuahua in 1954. Photo: courtesy of David E. Brown.

Figure 26. The vaquita (*Phocoena sinus*) is one of Mexico's most critically endangered mammalian species. This individual was taken in a gill net. Photo: Jorge Urbán.

Peromyscus pembertoni: Only 11 specimens, captured in 1931, are known for this species. It was endemic to San Pedro Nolasco Island in the Gulf of California off the coast of Sonora (Burt, 1932a). The causes of its extinction are unknown (Ceballos and Navarro, 1991).

Ursus arctos: At the end of the nineteenth century, grizzly bears were quite common on the border between Mexico and the United States (fig. 25). But the indiscriminate hunting and destruction of habitat caused its extinction in Mexico and the southwestern United States (Brown, 1985). A small population survived in the Sierra del Nido, where the last specimen was killed in 1962. Until 1971, there were unconfirmed reports of its presence in the Sierra Madre Occidental within the limits of Sonora and Chihuahua (Tinker, 1978). But the skull of a bear killed in Sonora in 1976 apparently belonging to this species was reported (Gallo-Reynoso et al., 2008).

Most of the following species require intense management programs to preserve them for the long term, given their small population sizes.

Phocoena sinus: The vaquita is the smallest cetacean in the world and is the only one endemic to a single country (fig. 26). Distributed in the upper Gulf of California, its population is less than 400 individuals. The main threat to this species is the unauthorized use of long gill nets (D'Agrosa et al., 2000).

Bison bison: The bison was considered extinct, but in 1988 a population of approximately 100 individuals was found living in Janos-Casas Grandes in northeastern Chihuahua (List and Solis, 2008; Pacheco et al., 2000). It is now considered endangered; these animals move freely between New Mexico and Chihuahua (fig. 27). The expansion of small towns and poaching are the main threats to its long-term conservation. In November 2009, 23 genetically pure bison were reintroduced in Janos in an effort to increase the bison population in Mexico (List et al., 2010b).

Lontra canadensis: This river otter is common in the United States and Canada. Its distribution in Mexico is marginal, confined to the Colorado River and San Pedro River in Sonora and to the Rio Bravo in Tamaulipas (Ceballos, 1985; Leopold, 1965). It was considered extinct in Mexico (Ceballos and Navarro, 1991), but there are apparently new records in the Rio Colorado in Coahuila (Gallo-Reynoso, 1997). It is possible that some individuals are still living near the Amistad dam or in Laguna Madre in Tamaulipas.

Mustela nigripes: The black-footed ferret was not included in any list of species at risk in Mexico since it was only known by subfossil remains from Chihuahua. At the beginning of this century it was reintroduced to the region of Janos-Casas Grandes in Chihuahua, where an estimated population of less than 50 animals lives (Pacheco et al., 2002).

Figure 27. The bison (*Bison bison*) became nearly extinct due to excessive hunting in the mid-nineteenth century. In Mexico bison was considered extinct until a small herd was discovered in the Janos region in northwestern Chihuahua (Ceballos and Navarro, 1991). In the same region a herd of 23 genetically pure bison was reintroduced in 2009, and now several individuals have been born in Mexico (List et al., 2009). (*Top left*) A stack of thousands of skulls of the last great bison herd exterminated in two months between 1877 and 1878; more than 100,000 animals were hunted between December and January. Photo: courtesy of the Burton Historical Collection, Detroit Public Library. (*Top right*) Part of a shipment of more than 40,000 bison hides in Dodge City, Kansas, 1874, waiting to be transported to the tannery industry in the eastern United States. Photo: courtesy of the Kansas State Historical Society. (*Below*) Herd of bison on the border between Chihuahua (Mexico) and New Mexico (United States) in the spring of 2004. Photo: Rurik List.

Antilocapra americana: The pronghorn has been at critical risk for decades. Currently, about 1,000 individuals are distributed in a wide region in the northern part of the country, and some populations, such as the one in Baja California, are increasing. The species was reintroduced to the Colombia Valley of Coahuila (Medellín et al., 2005).

Sylvilagus insonus: This rabbit species is endemic to Omiltemi in the state of Guerrero. It is critically endangered, but the causes are unknown. Subsistence hunting is probably one of the main reasons for its rarity. It was recently found after a century without sightings (Cervantes et al., 2004).

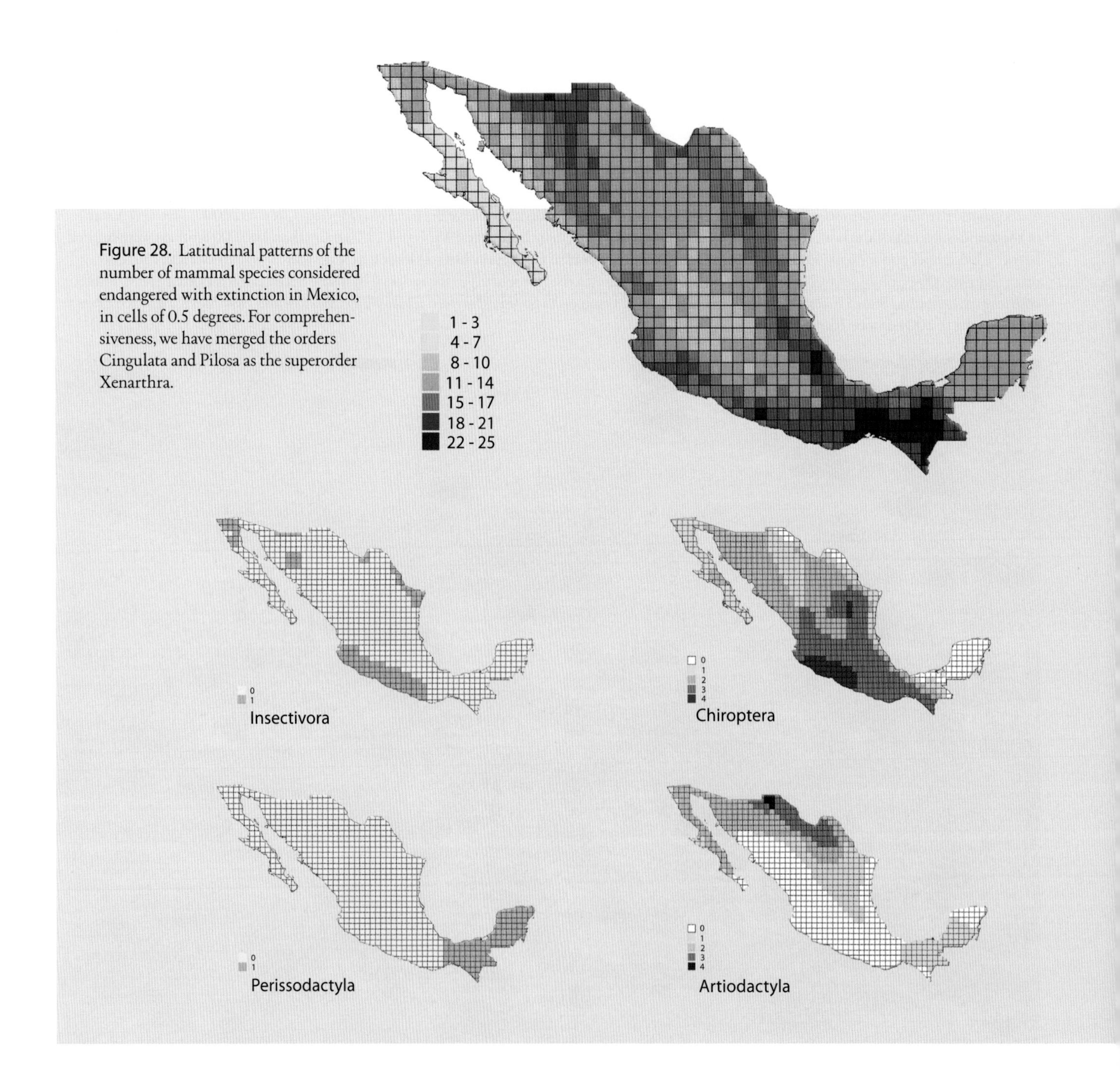

Figure 28. Latitudinal patterns of the number of mammal species considered endangered with extinction in Mexico, in cells of 0.5 degrees. For comprehensiveness, we have merged the orders Cingulata and Pilosa as the superorder Xenarthra.

The distribution of species at risk illustrates the historical patterns of land use and exploitation of species (fig. 28). In general, the greatest concentrations of at-risk species are located in the southern part of the country, where the rainforests have been devastated (Ceballos, 2007). Other regions such as the Sierra Madre Occidental and the Sierra Madre Oriental have large concentrations of species at risk as well, mainly because of hunting. Rodents at risk are located in the eastern part of the country, while artiodactyls are concentrated in the north and bats in the east. These differences are related to patterns of endemism and distribution of species in each order with regions that have suffered more negative impacts by human activities.

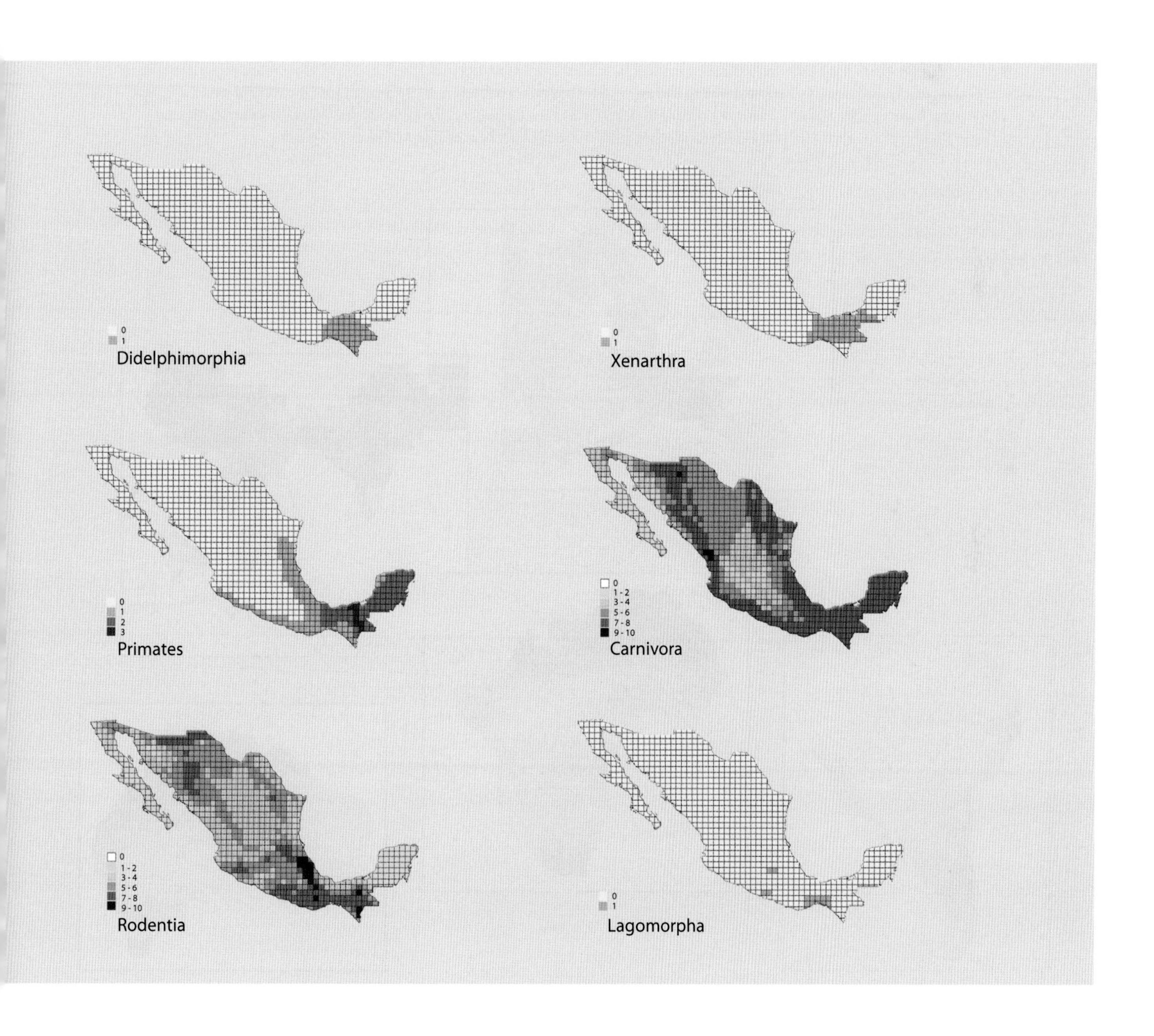

The magnitude of the extinction crisis is even more severe if one considers the extinction of populations (and subspecies). Most species, both those considered at risk and those relatively common, have lost many populations (Ceballos and Ehrlich, 2002). Habitat destruction, poaching, and illegal trade have caused the disappearance of numerous populations from many parts of the national territory (fig. 29). The prospect of survival for many species is limited if factors responsible for deterioration are not eliminated or reversed and if adequate management, with excellent recovery strategies, is not implemented.

Figure 29. The extinction of populations, even the most common ones, is a very serious problem in Mexico, which in the worst-case scenario causes the extinction of the species. This problem is illustrated at two different scales. On a global scale it is illustrated by the disappearance of mammal populations in various continents (*below*). These maps show the number of extinct populations of some mammal species, mostly large species. It is important to note that there are large regions where most populations have been lost (modified from Ceballos and Ehrlich, 2002). On a more regional scale (*right*), the problem is illustrated by the current and historical distribution of the pronghorn (*Antilocapra americana, top*), the bighorn sheep (*Ovis canadensis*), the black bear (*Ursus americanus*), and the tapir (*Tapirus bairdii, bottom*) in Mexico. Photos: pronghorn, Claudio Contreras Koob; bighorn sheep, black bear, and tapir, Gerardo Ceballos.

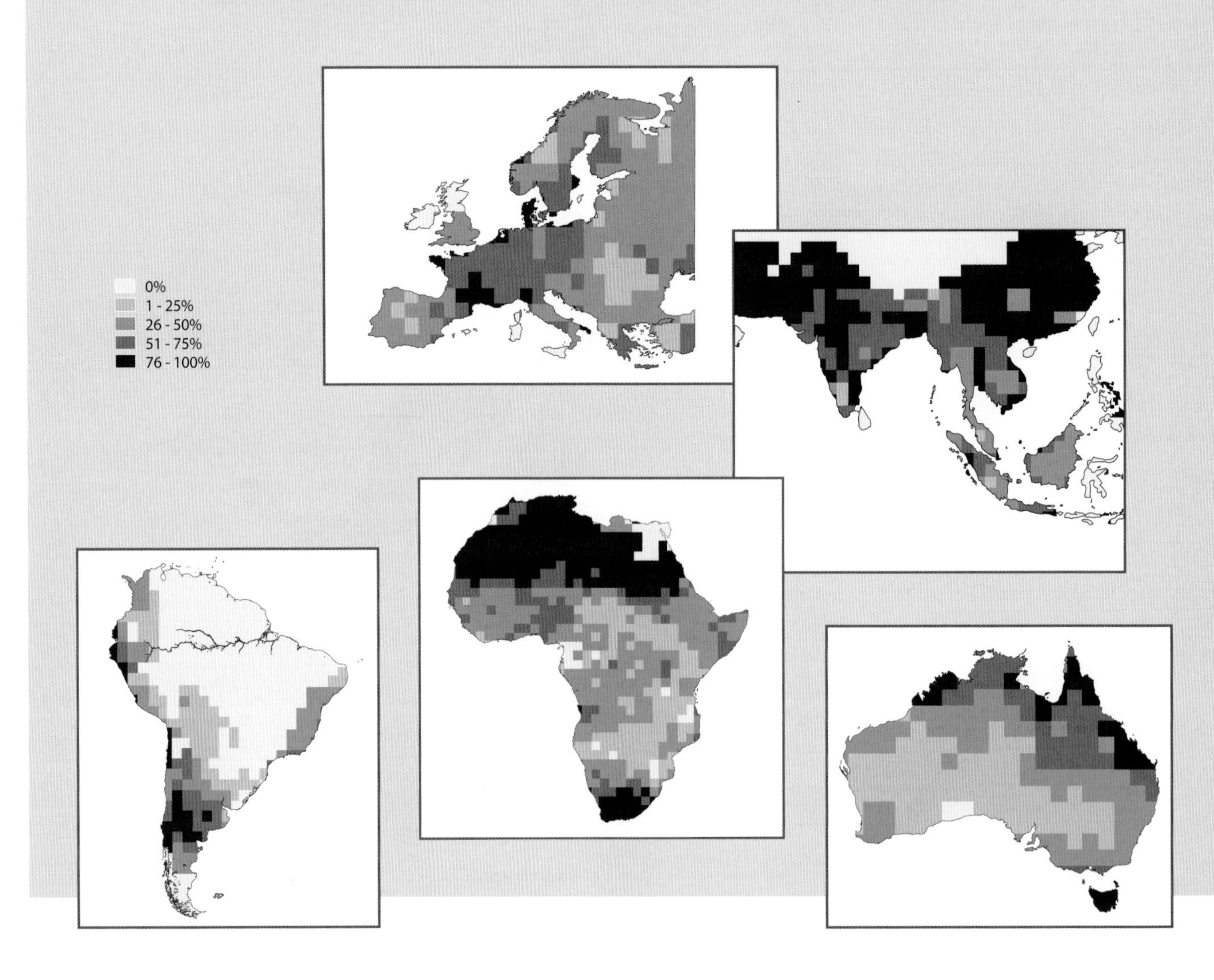

Historical distribution
Modern distribution

Causes of Extinctions

The causes of the extinction of mammal species in Mexico are diverse, but they can be attributed ultimately to the ongoing growth of human population and social inequality, factors that have generated an irrational exploitation of natural resources (Challenger, 1998; Ehrlich and Ceballos, 1997). Deforestation, overexploitation, and the introduction of exotic species have most severely affected mammals. It is clear that many other factors such as pollution play important roles in the disappearance of groups or individual species of mammals. Deforestation is, undoubtedly, the major cause of the disappearance of populations and species of mammals in Mexico. Unfortunately, the rates of deforestation in the country are among the highest in the world (see fig. 10). It has been estimated that about 600,000 ha of forests are lost annually, with the consequent loss of habitat (Velázquez et al., 2002). The most threatened ecosystems are rainforests, especially the tropical rainforests, which covered 22 million ha in the early twentieth century and currently have less than 600,000 ha remaining (Challenger, 1998). Other ecosystems severely affected are tropical deciduous forests, mangroves, and temperate cloud forests.

The negative effects of the destruction and fragmentation of natural vegetation are heightened by the overexploitation of species, from sport and subsistence hunting and illegal wildlife trade. The different impacts of these factors, including hunting and illegal trade, are considerable, but their exact magnitude is unknown (fig. 30). It is clear that, for species like the white-lipped peccary (*Tayassu pecari*) and the tapir (*Tapirus bairdii*), hunting is the main threat to their long-term conservation. There are dramatic examples of the impact of subsistence hunting on species such as Omiltemi cottontail (*Sylvilagus insonus*), which was recently recorded after a specimen was hunted in the mountains of Guerrero after more than a century without any reported sightings (Cervantes et al., 2004).

The introduction of species is an underevaluated problem, but surely this is having a negative impact on many native species. In Mexico there are well-established feral populations of domesticated species, including dogs, cats, donkeys, pigs, goats, and rabbits. There are at least 60 species of exotic mammals, 20 species of exotic birds, 10 exotic reptiles, and 5 exotic amphibians that are severely affecting many regions of Mexico; these species have had serious negative impacts on Mexican biological diversity (Álvarez et al., 2008). A dramatic example of the impact of introduced species is the probable extinction of the last known population of *Peromyscus guardia*, where, as noted earlier, a single cat exterminated it completely (Mellink et al., 2002; Vásquez-Domínguez et al., 2004).

Figure 30. Poaching and illegal trade of wildlife is a severe problem that affects many species of mammals considered at risk of extinction. For example, in 2000 more than 25 hides of illegally hunted jaguars were seized from a taxidermist in the city of Chetumal, Quintana Roo, in southern Mexico. Photo: Gerardo Ceballos.

Priority Areas for Conservation

A genuine concern of scientists and other groups is to reduce the rate of extinction. In most countries, conservation of ecosystems and species focuses on the establishment of nature reserves such as national parks, biosphere reserves, sanctuaries, and wildlife refuges. In the past decade, various proposals emerged using methods of complementarity (heuristics and optimization) to define priority areas for conservation that maximize the number of represented species in the lowest number of protected areas (e.g., Caldecott et al., 1996; Pressey et al., 1993; Rodríguez and Gaston, 2002). These methods are based on the idea that there will be greater effectiveness with limited resources in areas with higher concentrations of species in general or on those groups of species most threatened.

Mexico is one of the countries in which these analyses have led to considerable progress (Ceballos, 1999). The analysis of the 30 protected areas most important for their size, location, and ecological uniqueness has demonstrated interesting patterns, as described below (Ceballos, 1999, 2007). Seventy-five percent of the species are found in reserves. This could be taken as an indication of the effectiveness of the system of protected natural areas for conservation of mammals. A more detailed analysis indicates, however, that the groups most at risk, including the endemic and endangered species, are under-represented in these reserves, since only 50% and 67% of their species, respectively, have populations in a reserve. In other words, the system of protected areas does not represent properly the species that are most vulnerable to extinction, so it is not adequately fulfilling its role of protecting the country's biodiversity. Most of the species represented in reserves are found in only one or two reserves, suggesting that other reserves are required to protect additional populations by reducing the risk of losing some species by natural or anthropogenic factors.

The number of species protected in reserves varies from around 33 in the temperate forests of the Desierto de los Leones National Park, the oldest park in the country, to more than 120 species in the tropical forests of the Montes Azules Biosphere Reserve in Chiapas. A complementarity analysis of all reserves indicates that 24 additional reserves are necessary to protect all species (see fig. 30). The most important reserves, for their contribution of species, are located in major biomes of the country and include the tropical rainforest (Montes Azules Biosphere Reserve, Chiapas), the dry forests of the Pacific (Chamela-Cuixmala Biosphere Reserve, Jalisco), the shrublands and grasslands (Islas del Golfo Biosphere Reserve, Baja California, and the Janos Biosphere Reserve in Janos-Casas Grandes, Chihuahua), and the temperate forests (Ajusco National Park). Finally, there are 13 additional areas in which we would have 98% of mammal species represented if they were set aside as reserves (fig. 31); these areas are located in the north of the Baja California Peninsula, the Sierra Madre Occidental, southern Coahuila, the Eje Volcanico Transversal, the Sierra de Juarez in Oaxaca, and the east coast of Quintana Roo.

Conservation in Rural Areas

Agriculture, forestry, and cattle ranching are among the most important causes of the disappearance of species and ecosystems. Because of the enormous growth of the human population and its demand for food and other natural products, it is anticipated that the impact of these activities will increase in the next decades. That is, landscapes dominated by human activities will increase and become the dominant environments on the planet (Daily, 1997).

In the past ten years it has become clear that a considerable percentage of species can survive in environments dominated by human activities, so there is a chance to keep a fraction of the biological diversity in areas with some kind of management. This is possible in regions with high deforestation (Daily et al., 2003). It is the responsibility of academics and conservationists to assess the potential of these regions and to define rules for management to optimize the number of native species in these landscapes disturbed by anthropogenic activities (figs. 32 and 33).

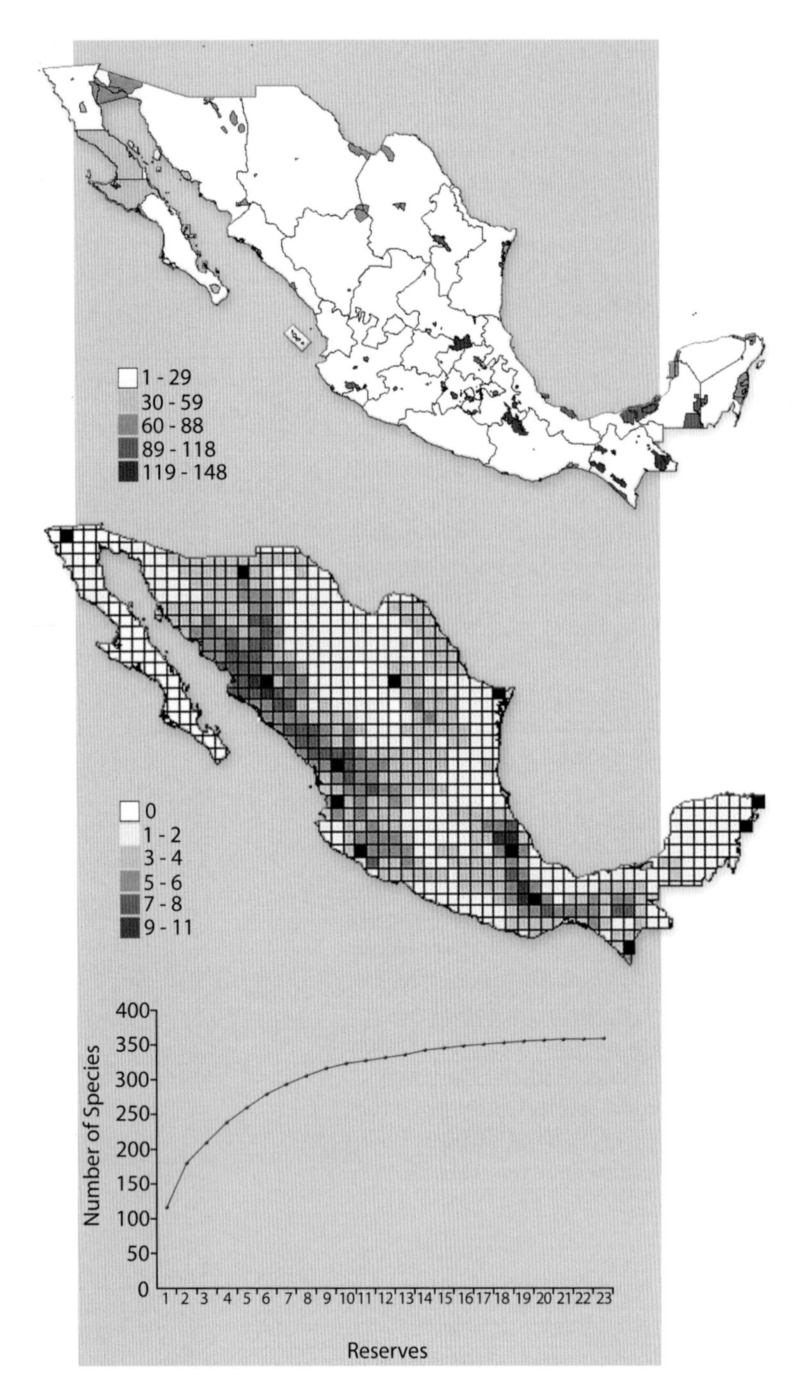

Figure 31. The federally protected areas cover more than 10% of the national territory (*top*). There are 117 species not represented in reserves. Overlapped geographic ranges indicate that they are concentrated in the Sierra Madre Occidental in western Mexico and in the mountains of Eje Volcanico Transversal (*middle*). The map indicates in black the cells where an additional 24 reserves are required to have all mammal species (75% of the species in the country). An optimization analysis indicates that the most important reserves in terms of their species richness and complementarity are the following ones (*bottom*): 1. Montes Azules; 2. Janos; 3. Gulf of California Islands; 4. El Triunfo; 5. Chamela-Cuixmala; 6. Sierra de San Pedro Martir; 7. Maderas del Carmen; 8. Calakmul; 9. Zempoala Lagoons; 10. Omiltemi; 11. The Tuxtlas; 12. La Encrucijada; 13. El Cielo; 14. El Pinacate; 15. Iztaccihuatl-Popocateptl; 16. Sierra de Manantlán; 17. Sierra de la Laguna; 18. La Sepultura; 19. Mapimí; 20. La Malinche; 21. Nevado de Toluca; 22. Sian Ka'an; 23. El Vizcaíno.

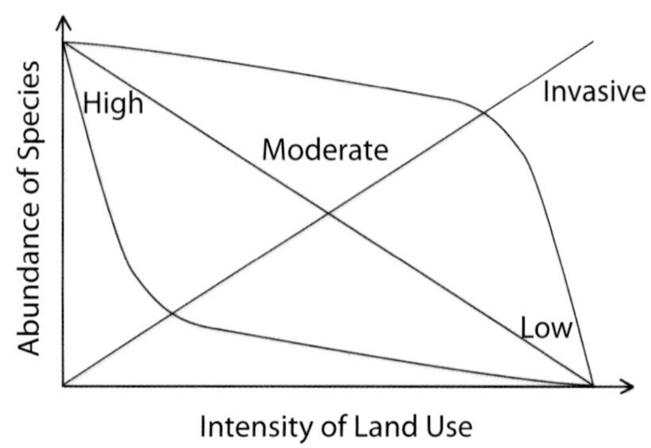

Figure 32. Schematic model of the effect of the intensity of land use on mammal species with different susceptibility. Species with high susceptibility can only withstand minor disturbances, while species with low or moderate susceptibility maintain populations in quite disturbed environments. The abundance of invasive species increases with the intensity of land use (modified from Daily et al., 2003).

Figure 33. Scientific research is essential to establish firm strategies for the conservation of species in the long term. Studies in Costa Rica, southern Campeche, and Quintana Roo have shown that there are many species that survive in environments with varying degrees of disturbance (Daily et al., 2003; G. Ceballos, pers. obs.). The species here were photographed with automatic cameras in the Forest Ejido Caobas, Quintana Roo, in areas with poaching and logging. The number of surviving species is impressive. Photos: Gerardo Ceballos and Cuauhtemoc Chávez.

Concluding Remarks

It is relevant to conclude this section of the book with comments we published some years ago:

Perhaps the most fundamental challenge of this generation of mammalogists, in Mexico and Latin America, is the conservation of biodiversity and the enormous implications this represents, in order to maintain the environmental services and the welfare of the population. An excellent indicator that we are working in the right direction is the increasing number of professionals dedicated to mammalogy and the increasing focus in solving problems with a firm scientific basis. However, because of the press of time and the severity of the problem this effort will be insufficient if it does not come together with answers of the proper magnitude from the government and civil society that are able to establish strategies that contemplate the biological diversity and the natural resources as a fundamental axis for development, alongside social, economic, and political conditions. Such responses must involve schemes that accommodate the huge international pressure of globalization with the national needs, which should favor the elimination of extreme poverty from the destruction of natural resources and the loss of biodiversity. (Ceballos et al., 2002:405)

History of Mammalogy in Mexico

Joaquín Arroyo-Cabrales, Luis Medrano González, and Gerardo Ceballos

The mammal fauna of Mexico is one of the most diverse in the world. So, it is important to know how mammalogy developed there. Until relatively recently there was no information available on this topic. José Ramírez-Pulido and his co-workers, however, have done an eloquent synthesis of progress in learning about Mexican mammals since 1830 (Ramírez-Pulido and Britton, 1981; Ramírez-Pulido and González-Ruiz, 2006; Ramírez-Pulido and Müdespacher, 1987a). We have complemented his work with additional information from León Paniagua (1994), Martín del Campo (1941), and Sanchez (1985). This section is based on our recent work (Ceballos et al., 2002). Knowledge prior to the synthesis of Ramírez-Pulido and collaborators was scant and the few contributions on the historical development of Mexican mammalogy were reviews of collections made during large expeditions (Goldman, 1951; Saussure, 1993) and chapters in some regional and state monographs (table 9), such as Chihuahua (Anderson, 1972), Coahuila (Baker, 1956), Chiapas (Álvarez del Toro, 1977), Durango (Baker and Greer, 1962), Guanajuato (Sanchez and Magaña-Cota, 2008), San Luis Potosí (Dalquest, 1953a), Mexico (Ceballos et al., 2009), Valle de Mexico (Ceballos and Galindo, 1984), and Veracruz (Hall and Dalquest, 1963).

We have divided the description of the research on the mammals of Mexico into five periods: (1) up to 1829; (2) 1830–1883; (3) 1884–1919; (4) 1920–1942; (5) 1943–present. This classification is a modification of the proposal by Ramírez-Pulido and Britton (1981). A summary of the 1,196 described species and subspecies known up until 1987 indicates that 24 authors described 75% of the taxa (895), while the remaining 144 described 301 (25%). Only 11 of these taxa were classified by national researchers or in collaboration with foreign researchers.

The first period, covering up to 1829, includes mainly stories gathered by religious missionaries from Native Americans, as well as the work of botanical and zoological travelers and explorers in New Spain. Such works are descriptive and illustrative compilations, not systematic compilations. Nevertheless, ordering and classifying of the species was made on the basis of utility, behavior, habitats, and general morphology.

The oldest information on Mexican mammals is from paintings in states such as Baja California, where one can find some of the most ancient cave paintings (2,000 to 8,000 years old) from the Sierras San Francisco and Guadalupe, depicting marine mammals (Esquivel, 1998; Lorenzo, 1986; Schobinger, 1997). With the Spanish conquest, knowledge of marine mammals grew hand in hand with their intended use as well as with the spread of scientific knowledge from Europe. There are some manuscripts by clerics and naturalists such as Francisco Hernández and Francisco Xavier Clavijero, and navigators such as Juan Loyola Salinas, almost entirely restricted to species that were subjected to exploitation, particularly the manatee (Durand, 1950). Descriptions of the mammals and other animals are often confusing or inaccurate. For example, the famous naturalist Francisco Hernández (1571–1576) described the manatee, but his description includes some details that suggest confusion between this species and the monk seal. Hernández said: "They live in either the ocean or fish lakes. Haitians call it manatee, beast almost shapeless like a calf, with rounded head and front arms as a goat, dingy, rare and covered with weird hair, ferocious but does not bite. Lives both at sea and on the beach (and sometimes they even deviate from the waters), feeds on grasses and coastal marine figs . . . it mates in a human manner, the female lies on the beach with its shoulders to the ground and the male lies on it relatively quickly."

In the late eighteenth century, there were great explorations of Mexico that contributed to the knowledge of its mammals. Some important naturalists at that time were Martin of Sessé,

Table 9. Examples of published studies about mammals in the states of Mexico, indicating if the study is a monograph including all the mammals, a checklist, or a monograph of a specific order.

State	Source	Group
Aguascalientes	De la Riva (1989)	Checklist
	Espinoza (1982)	Chiroptera's monograph
	Hesselbach and Pérez (2001)	Monograph
Baja California	Alvarez and Patton (1999, 2000)	Monograph
Campeche	See Yucatan	Monograph
	Sánchez-H. and Romero (1995)	Chiroptera's monograph
Chiapas	Álvarez del Toro (1977)	Monograph
	Álvarez-Castañeda and Álvarez (1991)	Chiroptera's monograph
	Aranda and March (1987)	Field guide
Chihuahua	Anderson (1972)	Monograph
Coahuila	Baker (1956)	Monograph
Colima	Kennedy et al. (1984)	Chiroptera's monograph
State of Mexico	Ceballos and Galindo (1984)	Monograph
	Chávez and Ceballos (1998)	Checklist
	Ceballos et al. (2009)	Monograph
	Ramírez-Pulido et al. (1995)	Checklist
Distrito Federal	Ceballos and Galindo (1984)	Monograph
Guanajuato	Sáncez and Magaña Cota (2008)	Chiroptera's monograph
Jalisco	Watkins et al. (1972)	Chiroptera's monograph
	Ceballos and Miranda (2000)	Checklist
	Guerrero and Cervantes (2003)	Checklist
Michoacan	Hall and Villa (1949, 1950)	Checklist
	Álvarez et al. (1987)	Coastal mammals except Chiroptera
	Álvarez and Sánchez-Casas (1997)	Chiroptera's monograph
		Chiroptera and Rodentia
	Polaco and Muñiz-Martínez	Chiroptera's monograph
	Wang et al. (2003)	Chiroptera's monograph
Morelos	Álvarez-Castañeda (1996)	Monograph
Nayarit	Carleton et al. (1982)	Rodentia's monograph
Nuevo Leon	Jiménez et al. (1999)	Monograph
Queretaro	Padilla García and Pineda López (1988)	Checklist
	Gutiérrez G. et al. (2007)	Monograph
Quintana Roo	See Yucatan	Monograph
San Luis Potosí	Dalquest (1953)	Monograph
Sinaloa	Armstrong and Jones (1972)	Monograph Marsupialia, Insectivora, Edentata, and Lagomorpha
	Jones et al. (1972)	Lagomorpha's monograph
	Armstrong et al. (1973)	Monograph Chiroptera, Carnivora, and Artiodactyla
Sonora	Burt (1938)	Monograph
	Caire (1997)	Monograph
Tamaulipas	Álvarez (1963)	Monograph
Tabasco	Sánchez-H. and Romero (1995)	Chiroptera's monograph
Veracruz	Hall and Dalquest (1963)	Monograph
	Estrada and Coates Estrada (1986)	Field guide
	Gaona et al. (2003)	Checklist
Yucatan	Gaumer (1917)	Monograph
	Genoways and Jones (1975)	Monograph
		Chiroptera's monograph
		Rodentia's monograph
	Jones et al. (1973, 1974 a, b)	Monograph

José Mariano Mociño, and Alejandro Malaspina. In his history of Mexico, Clavijero (1780) cites as common "fish" sea whales, dolphins, manatees, and sea lions. In that period, only 12 specific taxa from Mexico were described (Ramírez-Pulido and Britton, 1981). With the independence of Mexico in 1810, there was a nearly complete halt in mammal research, but from 1830 on, naturalists from institutions in other countries, mostly Germany, Great Britain, France, and the United States, appeared more frequently in the country.

The second period (1830–1883) corresponds to the great expansion of exploration activities in the western United States, a result of which was that the study and collection of biological material in the northern Mexican states began (Baker, 1991). This period was particularly important since the first work to have a profound impact on mammalogy research in America was published (Audubon and Bachman, 1845-1853). During the same period the Smithsonian Institution in Washington, D.C., was established; it would become one of the most important institutions supporting field explorations in Mexico. In this period 129 taxa were described. Swiss naturalist Henri de Saussure described 19 taxa, while Spencer Fullerton Baird, one of the greatest naturalists of the late nineteenth century, described 20 taxa (Baird, 1858).

During the nineteenth century, because of hunting by British and other Europeans and thereafter by Americans, copious information about the sea otter, common sea lions, harbor seals, whales, and sperm whales was gathered. Such compilations were concentrated in the Gulf of California or Mar de Cortes and the Pacific Ocean off Baja California. The exploitation of marine mammals was so intense that it caused the disappearance of some species, such as the sea otter from Baja California, and the near extinction of both the gray whale and the Guadalupe sea lion (Henderson, 1975). As formal knowledge, the natural history of Mexican marine mammals was developed almost exclusively by the work of foreign naturalists such as Scammon (1874).

The third period (1884–1919) was marked by the strong interest of museums and government agencies in the United States in exploring, researching, and collecting biological material in Mexico. The result was the creation of collections of mammals, among other groups, from different regions of the country that were deposited in several scientific institutions. The last part of this period coincided with both the Mexican revolution and the First World War, which resulted in a low number of taxa being described (Goldman, 1951). This period can be considered as "owned" by Clinton Hart Merriam, not only because of the high number of species and subspecies he described but also because under the command of the Office of Biological Exploration of the United States, he supported E.A. Goldman and E.W. Nelson, who played a leading role in biological studies in Mexico (Sterling, 1991). During this period, 605 taxa were described. Merriam classified 208 taxa (34%), while other important researchers were J.A. Allen (84 taxa), W. Osgood (47), O. Thomas (39), E.A. Mearns (37), and E.W. Goldman (30). In 1905, Jesús Díaz de León was the first Mexican to describe a species, the volcano rabbit (*Romerolagus diazi*).

In the first half of the twentieth century the sea lion was hunted intensively in California, and by 1952 the monk seal was no longer seen. Between 1920 and 1940, the Mexican government legislated protective actions for some species of marine mammals and signed an agreement with the International Whaling Commission, but no systematic efforts were made outside the research carried out by foreigners. At the same time, biology emerged as a professional career in Mexico, but only the manatee and the gray whale would receive some attention (Aurioles Gamboa et al., 1993)

The fourth period (1920–1942) was characterized by renewed interest in exploring Mexico: 219 taxa were described. It can be considered as the Nelson and Goldman era since, together, they described 114 taxa, 52% of the period's total (Sterling, 1991). Other researchers who were notable for the number of taxa described were W. Burt (25) and L. Huey (24). Bernardo Villa, who is considered the founder of mammalogy in Mexico, described one species in this period, *Tylomis gymnurus* Villa, 1941 (= *Tylomys nudicaudus gymnurus*).

The modern period (1943–present) is one of conceptual synthesis, which has shifted from simple typological descriptions to those that incorporate biological and phylogenetic analyses. In the past few decades, mammalogy in Mexico has experienced a huge boom, as the number of specialists has multiplied exponentially and a considerable number of collaborations with disciplines such as molecular genetics, physiology, ecology, evolution, paleontology, biogeography, and conservation have taken place (Ceballos and Ehrlich, 2002; Ceballos et al., 2005; Corona-M. et al., 2008; Guevara-Chumacero et al., 2001; López O. and Ramirez-Pulido, 1999). Up until 1987, 214 taxa were described, including 30 by L. Huey and 20 by G. Goodwin. In this period, Mexican researchers, alone or in collaboration with foreigners, described several species (Bradley et al., 2004; Carleton et al., 2002; Engstrom et al., 1992). The institutions having the most impact during this period are the National Autonomous University of Mexico, the National Polytechnic Institute, the Metropolitan Autonomous University, the Institute of Ecology, Xalapa, and the Autonomous University of Veracruz. The most important institutions abroad have been the University of Kansas, Michigan State University, the Smithsonian Institution (National Museum of Natural History), the American Museum of Natural History, Texas A&M University, and the Texas Technological University.

With regard to the development of marine mammalogy in Mexico, the second half of the twentieth century revealed the vast richness of aquatic mammals in the country and the existence of a cultural complex with which these animals interact. In 1958, K.S. Norris and W.N. McFarland discovered an endemic species of sea porpoise (*Phocoena sinus*) in the northern part of the Gulf of California; it is commonly known as cochinito or vaquita. But not until the 1970s was it determined that this species is in serious danger of extinction due to its association with the fishing of the totoaba (*Totoaba macdonaldi*), a fish itself at risk of extinction. Since then, the vaquita has been a source of growing attention by the scientific community and the Mexican government. Before 1980, Mexicans undertook only a few, isolated studies on marine mammals, some notable exceptions being those on the sea lion of California and the gray whale.

The capture of marine mammals for human consumption, their use as bait, and the interference of these animals with fishing activities explain why Mexican seafarers know these animals so well (Aguayo, 1989; Aurioles Gamboa et al., 1993; Zavala-González et al., 1994). Besides fishing activities, tourism involving marine mammals has grown in recent years (Ávila Foucat and Saad Alvarado, 1998). This tourism poses a major disturbance of coastal, marine, and other adjacent ecosystems, but it has also promoted cultural exchange among the coastal inhabitants.

The development of mammalogy in Mexico must grapple with the fact that marine mammals affect important economic activities of the poorest sectors of the human population. Therefore, while mammalogy in Mexico should not isolate itself from scientific progress in the world, it needs to focus on alternative socioeconomic development, enrichment of knowledge, and education. It must seek ways to encourage the harmonious interaction of man and nature. Because the Mexican mammalian fauna is part of a dynamic process affected by global climate change, anthropogenic disturbance can be a severe problem in the long-term conservation and evolutionary potential of these animals.

In the past decade, detailed analyses of various aspects have been published, for example, type localities listing of Mexican mammals (Álvarez et al., 1997), lists of specimens collected in Mexico and deposited in museums in the United States and Canada (López-Wilchis and López-Jardinez, 1995, 1998, 1999, 2000), a list of publications on Mexican mammals (Guevara-Chumancero et al., 2001), a synthesis of mammals from regions or states (Álvarez-Castañeda and Patton, 1999; Aranda and March, 1987; Ceballos and Miranda, 2000), and studies on systematic and filogeography of various species or groups (Bradley et al., 2004; Demastes et al., 2003; Harris et al., 2000; Riddle et al., 2000a, b; Ruedas, 1998; Schmidly et al., 1988; Tiemann-Boege et al., 2000).

Moreover, in the past twenty years, there has been a large number of studies on ecology, biogeography, and conservation. This has allowed us to have a better understanding of the patterns and processes of the structure and function of populations and communities of mammals of Mexico. A synthesis of some of this work is in books such as Avances en el estudio de los mamíferos de México (Medellín and Ceballos, 1993) and Bibliografia reciente de los mamiferos de México 1994-2000 (Ramírez-Pulido et al., 2000), and in 15 volumes of the journal of Mexican mammalogy (Revista Mexicana de Mastozoología), which is an important forum for the diffusion of research on Mexican mammals. An analysis of these publications shows that there are hundreds of recent references, which include topics such as taxonomy, biogeography, behavior, conservation, macroecology, and systematics (Arita et al., 1997; Castro-Campilla and Ramírez-Pulido, 2000; Ceballos et al., 1998, 2005; Conroy et al., 2001; Montellano-Ballesteros and Arroyo-Cabrales, 2002; Ortega and Arita, 2000; Peterson et al., 1999; Polaco et al., 2001; Ramírez-Pulido et al., 1994; Sánchez-Cordero, 2001; Sánchez-Hernández et al., 2001; Valenzuela and Ceballos, 2000).

Systematic List of Species

Gerardo Ceballos and Joaquín Arroyo-Cabrales

In this section we present the complete updated checklist of the mammals of Mexico up to June 2011. We include the primary affinities and precedence, as well as endemism, insularity, and conservation status of each species. Also, we incorporate a section discussing exotic species and their current populations in Mexico. In recent years, several authors have compiled lists of terrestrial and marine Mexican mammals (Arita and Ceballos, 1997; Aurioles, 1993; Cervantes et al., 1994; Ceballos et al., 2005; Ramírez-Pulido et al., 1983, 1986, 1996, 2005; Salinas and Ladrón de Guevara, 1993; Torres et al., 1995); however, there have been many changes in the taxonomic descriptions of new species and new records of species, which have significantly increased the inventory of mammals in Mexico (table 10).

The dynamics of these changes, at family, genera, and species levels, have been intense. For example, it was established that skunks belong to the family Mephitidae and species of several genera such as *Scapanus*, *Cryptotis*, *Sorex*, *Notiosorex*, *Cynomops*, *Micronycteris*, *Conepatus*, *Spilogale*, *Cratogeomys*, *Habromys*, *Megadontomys*, *Neotoma*, *Peromyscus*, *Oryzomys*, *Reithrodontomys*, and *Sylvilagus* were described, revalidated, or synonymized. This list is based on our recent compilations (Ceballos et al., 2002b, 2005). We excluded exotic murid rodents (*Mus musculus*, *Rattus norvergicus*, and *R. rattus*) and domestic species with wild populations, such as dogs, cats, goats, and donkeys (Álvarez Romero et al., 2008). The nomenclature used follows the one proposed by Wilson and Reeder (2005), with modifications listed below.

Patterns of distribution were determined using maps in Patterson et al. (2008), Hall (1981), and Medellín et al. (1997) as well as new information published in many other more specific publications. We classified the species according to their current geographical distribution, as follows: (1) Mexican species shared with other North American countries (NA); (2) Mexican species shared

Table 10. Systematic composition of the mammals of Mexico.

Order	Families	Genera	Species	Endemic Species
Didelphimorpha	1	7	8	1
Sirenia	1	1	1	0
Cingulata	1	2	2	0
Pilosa	2	2	2	0
Primates	1	2	3	0
Lagomorpha	1	3	15	7
Rodentia	8	50	243	116
Soricomorpha	2	6	38	24
Carnivora	8	28	42	3
Perissodactyla	1	1	1	0
Arctiodactyla	4	8	10	0
Cetacea	7	25	40	1
Chiroptera	9	67	139	18
TOTAL	46	202	544	170

with other South American countries (SA); (3) species with large areas of distribution that include both North and South America (AM); (4) species that are endemic to Central America: Mexico, Central America (MA); and (5) Mexican endemic species (MX). The islands' species list was compiled from Ceballos and Rodríguez (1993), Engstrom et al. (1989), Jones and Lawlor (1965), Lawlor (1983), Ramírez-Pulido and Müdespacher (1987a), Sánchez-Herrera (1986), and Wilson (1991). The species were described as completely insular (I); totally continental (C), that is, present on the continent; continental-insular (CI) for species that combine both patterns; and marine (M).

The conservation statuses of species were compiled from the list of Mexican species at risk (SEMARNAT, 2010), IUCN (2010; see also www.redlist.org/search/search-expert.php), and CITES (2010; see also www.cites.org, especially www.cites.org/eng/disc/text.shtml#II). CITES classifies species subject to international trade in three appendixes. Appendix I includes all species threatened with extinction that are or may be affected by trade. Trade in specimens of these species must be subject to particularly strict regulation and must be authorized only in exceptional circumstances. CITES prohibits international trade in species threatened with extinction "except when the purpose of the import is not commercial." Appendix II includes "(a) all species which although not necessarily now threatened with extinction may become so unless trade in specimens of such species is subject to strict regulation in order to avoid utilization incompatible with their survival." It also includes so-called look-alike species, that is, those "species which must be subject to regulation in order that trade of specimens referred to in sub-paragraph (a) . . . may be brought under effective control." Finally, appendix III includes "all species which any Party identifies for the purpose of preventing or restricting exploitation, and as needing the co-operation of other Parties in the control of trade."

The additions and modifications of the Mexican species lists, based on the list proposed by Wilson and Reeder (2005), are as follows:

1. We agree with the proposal of Voss and Jansa (2003) to recognize *Marmosa canescens* within a different genus, *Tlacuatzin*, endemic to Mexico.
2. We agree with the proposals of Woodman and Timm (1999, 2000) to reassign several species in the genus *Cryptotis*.
3. We recognize three species of *Notiosorex* following Carraway and Timm (2000).
4. Based on DNA sequencing of the mitochondrial cytochrome b and an intron of the nuclear beta fibrinogen, Baker et al. (2003) described a new species (*Notiosorex cockrumi*).
5. Carraway (2007) proposed two new *Sorex* species for Mexico, *S. ixtlanensis* and *S. mediopua*, as well as a new subspecies (*S. veraecrucis altoensis*). Two polytypic species (*S. monticolus* and *S. saussurei*) were reassigned as monotypic. See also Esteva et al. (2010).
6. We accept the proposal of Yates and Salazar-Bravo (2005) to recognize *Scapanus anthony* as a species.
7. We agree with the proposal of Simmons and Handley (1998), who determined that northern populations of bats of the genus *Centronycteris*, including the ones that reach Mexico, belong to a separate species (*C. centralis*), different from *C. maximiliani*.
8. We follow the correct way to cite the author and year of *Mormoops megalophylla* (Peters, 1864); in the original description he used the name of *Mormops*.
9. The arrangement of the subfamilies in the family Phyllostomidae is under discussion. Wetterer et al. (2000) presented a supposedly robust analysis based on "the total evidence," proposing the existence of seven subfamilies, but Baker et al. (2000) analyzed a set of different features (DNA sequencing of the gen-2 recombination-activated), and their results differ from some of the proposals by Wetterer et al. (2000). There are agreements, however, on the monophyletic groups such as Desmodontinae, Glossophaginae (for the known genera in Mexico), and Stenodermatinae (except *Carollia*). Other monophyletic groups included in the family Phyllostomidae are still being debated, as well as the inclusion of *Carollia* in Stenodermatinae. Here, we follow the classification of McKenna and Bell (1997), acknowledging the subfamily Phyllostominae without Tribe classifications and leaving *Carollia* in Stenodermatinae.

10. We consider more than one genus of *Micronycteris*, previously known as monotypic, following Simmons and Voss (1998) and Wetterer et al. (2000). Also, we agree with Simmons (1996) that the valid name for *M. megalotis* is *M. microtis*.
11. We include the bat *Trinycteris nicefori*, recently recorded in Mexico by Escobedo-Morales et al. (2006).
12. Simmon and Voss (1998) discussed the taxonomic status of *Mimon bennettii* and *M. cozumelae*, and proposed that both are valid, with *M. cozumelae* being distributed in Mexico and Central America. This proposal has been questioned by Gregorin et al. (2008).
13. We follow Simmons and Voss (1998) and Wetterer et al. (2000) in recognizing *Phylloderma* as a separate genus from *Phyllostomus*.
14. Lee et al. (2002) revised the taxonomic status of the species in the genus *Tonatia* and proposed that *T. brasiliensis* and *T. evotis* should be considered in a different taxonomic unit, *Lophostoma*. We follow their proposal.
15. Another controversy that is not yet resolved, even with detailed analyses (Baker et al., 2000; Van Den Bussche et al., 1998; Wetterer et al., 2000), is if *Dermanura* is a subgenus or sister genus of *Artibeus*. Recent findings suggest strong data to support the existence of two separated genera (Hoofer et al., 2006; Solari et al., 2011).
16. We follow the proposal of Koopman (1993, 1994), Simmons (2005), Redondo et al. (2008), and Hoofer et al. (2006) to consider *Artibeus intermedius* as a synonym of *A. lituratus* (intraspecific variation), contrary to what Davis (1984) and Guerrero et al. (2004) have mentioned, mostly due to the lack of definitive diagnostic characters that separate both taxa at species level.
17. We accept the proposal of Baker et al. (2000) to divide the populations of *Carollia brevicauda* into two species, with the northern population called *C. sowelli*.
18. Gardner (2008) and Sánchez-Hernández et al. (2005) have pointed out the actual presence of *Sturnira hondurensis* in Mexico, rather than *S. ludovici*. Such a proposal has been discussed for a long time, but it seems that Iudica (2000), in his unpublished comprehensive review of the genus, supports such a conclusion.
19. Lim et al. (2003) re-evaluated the taxonomic status of the populations of *Vampyressa pusilla*, and according to differences in the morphology, measurements, chromosomes, and mitochondrial analysis, they propose that *V. pusilla* is endemic to the forests of southeastern South America, while *V. thyone* is allopatric in northern South America, Central America, and Mexico.
20. We follow the recent comprehensive review of the genus *Natalus* by Tejedor (2011), and agree with the presence of two species in Mexico, *N. lanatus* and *N. mexicanus*.
21. We follow the proposal of Simmons (2005) to use subfamilies Vespertilioninae, Myotinae, and Antrozoinae.
22. We follow the conclusions of Piaggio et al. (2002) in recognizing *Myotis occultus* as a different species from *Myotis lucifugus* due to the fact that they have separate evolutionary and monophyletic lineages.
23. During the past decades, several authors (Baker and Patton, 1967; Hoofer and Van Den Bussche, 2003; Hoofer et al., 2006; Horáček and Hanák, 1985/1986; Menu, 1984), have pointed out the lack of a correct generic name for the taxa assigned to New World *Pipistrellus*. Based on their analyses, we recognize the genera *Parastrellus* and *Perimyotis* for the species *hesperus* and *subflavus*, respectively.
24. *Rhoogessa bickhami* is a species recently described from the same type locality as *R. genowaysi* in Chiapas. It is not included in the species accounts because it was described when the book was already in press.
25. We agree with Peters et al. (2002) in recognizing as a valid name *Cynomops mexicanus* instead of the previously known *Molossops greenhalli mexicanus*. This species is endemic to western Mexico.
26. We follow Eger (1977) in recognizing *Eumops nanus* to occur in Mexico, rather than *E. bonariensis*. Furthermore, we agree with McDonough et al. (2008) on the presence of *Eumops ferox* in Mexico, rather than *E. glaucinus*.

27. López-González and Presley (2001) studied medium-sized bats of the genus *Molossus*, and reidentified the only specimen of *M. bondae* in Quintana Roo, Mexico. They named it *M. aztecus*, meaning that *M. bondae* is not distributed in Mexico.
28. González-Ruiz et al. (2011) studied the specimens pertaining to *Molossus sinaloae* from Mexican collections, and found that there are two taxa mixed, the actual *M. sinaloae* that occurred on the Pacific versant and a new species, *Molossus alvarezi*, from the Yucatan Peninsula.
29. According to a large phylogenetic database of species in the order Primates, Groves (2001) proposed the division of the family Cebidae, with the genera *Allouata* and *Ateles*, the only ones naturally distributed in Mexico, inside the family Atelidae. The name for the subfamily of Alouatta was changed from Allouattinae to Mycetinae, derived from the synonym of Allouata, Mycetes.
30. Wozencraft (2005) recognized the genus *Herpailurus*, and we follow his recommendation.
31. Mercure et al. (1993) analyzed molecular data of populations of *Vulpes* that inhabited the southern United States, which supports the formal proposal that *V. macrotis* is a valid species, separated from *V. velox*, and the only one distributed in Mexico.
32. *Mustela nigripes* was reintroduced to the biological reserve at Janos-Casas Grandes in northeastern Chihuahua (Pacheco et al., 2002). In the summer of 2002, the first black-footed ferret, born in historic times in Mexico, was videotaped (J. Pacheco, 2005, pers. comm.).
33. We follow Dragoo and Honeycutt (1997) in recognizing the family Mephitidae.
34. Dragoo et al. (1993) analyzed allozymic data of *Spilogale* nucleotic sequences, proposing that two species are distributed in the United States and Mexico, *S. gracilis* and *S. putorius*, and being both similarly divergent from *Conepatus* species, they are recognized as separate species. Wozencraft (2005) agreed with separating *Spilogale angustifrons* for populations occurring in southern and southeastern Mexico, while *S. putorius* is restricted to the United States.
35. We agree with the conclusion of Dragoo et al. (2003), who evaluated the taxonomic relationships between *Conepatus mesoleucus* and *C. leuconotus*, finding that they represent the same species, with *C. leuconotus* being the valid name for the species since it has page priority at the original description.
36. *Mesoplodon peruvianus* was recorded in Mexico off La Paz, Baja California Sur, by Urbán-R. and Aurioles (1992).
37. An additional unidentified species of *Mesoplodon* was filmed in Mexico at the beginning of the 1990s (Salinas and Ladrón de Guevara, 1993). We think that this unidentified specimen could represent the newly described *M. perrini* from the San Diego coast in California (Dalebout et al., 2002).
38. The Galapagos fur seal (*Arctocephalus galapagoensis*) has been recorded off Baja California (Aurioles-Gamboa et al., 2004).
39. The Galapagos sea lion (*Zalophus wollebaeki*) was recently recorded in Chiapas (Ceballos et al., 2010).
40. Northern fur seals (*Callorhinus ursinus*) have been recorded in Mexico (Aurioles-Gamboa et al. 1993; Ceballos et al., 2010).
41. We accept the proposal to maintain the Cozumel coati (*Nasua nelsoni*) as a full species, critically endangered (Cuarón et al., 2009).
42. We follow the conclusions that the Tres Marias Islands raccoon (*Procyon lotor insularis*) is only a subspecies of the mainland raccoon (Helgen and Wilson, 2005).
43. Groves and Grubb (2011) comprehensively reviewed the taxonomic relationships of artiodactyls and perissodactyls based on recent molecular, cytogenetic, and morphological data, as well as the author's own observations, all of which supports new classifications. Specially, for Mexico some of the changes include the use of the genus name *Tapirella* for the tapir, two species of collared javelina (*Pecari angulatus* and *P. crassus*), and the use of the genus name *Bos* for the American bison. Such proposals have not been considered by American mammalogists yet, however, so we expect further discussion on the validity of these and so decide to stay with Wilson and Reeder's (2005) uses.

44. We follow Geist (1998) and Randi et al. (2001) in recognizing as distinct species *Cervus elaphus* from Europe and *Cervus canadensis* from North America.
45. The Yucatan brown brocket *Mazama pandora* was recognized at species level, separated from the red brocket *M. americana* (Medellín et al., 1998).
46. Helgen et al. (2009) reviewed in detail the North American specimens previously assigned to the genus *Spermophilus*. Based on several character data (morphologic, cytogenetic, ecologic, and behavioral), they recognized eight genera to which to assign those specimens: *Notocitellus* A.H. Howell, 1938; *Otospermophilus* Brandt, 1844; *Callospermophilus* Merriam, 1897; *Ictidomys* J.A. Allen, 1877; *Poliocitellus* A.H. Howell, 1938; *Xerospermophilus* Merriam, 1892; and *Urocitellus* Obolenskij, 1927. *Spermophilus* sensu stricto would be restricted to the northern Palearctic region.
47. Jameson (1999) and Piaggio and Spicer (2001) proposed that the three subgenera within the chipmunk genus *Tamias* should be considered at the species level. Following such an arrangement, the Mexican species would be allocated to the genus *Neotamias*. The same authors constrained the genera presence in Mexico to two species, *N. dorsalis* and *N. merriami*, but Banbury and Spicer (2007) added *N. obscurus* to those chipmunks living in Mexico, while Thorington and Hoffmann (2005) also recognized the validity of *N. bulleri* and *N. durangae*.
48. Demastes et al. (2003), based on the phylogenetic revision of *Pappogeomys alcorni* using 424 base pairs on the mitochondrial cytochrome b, and supported by a morphometric analysis of 101 *Pappogeomys* specimens, concluded that *P. alcorni* is a disjunct subspecies of *P. bulleri*, rather than a monotypical species.
49. The systematic revision of *C. merriami* by Hafner et al. (2005) divided *C. merriami* into three species, *C. merriami*, *C. fulvescens*, and *C. perotensis*.
50. Hafner et al. (2004) revised pocket gophers of the *C. gymnurus* species group (now known as the *C. fumosus* species group) and synonymized *C. gymnurus*, *C. neglectus*, *C. tylorhinus*, and *C. zinseri* under *C. fumosus*.
51. Hafner et al. (2008) provided a formal description of *C. goldmani*.
52. Álvarez-Castañeda (2010), based on genetic and molecular assays and through a phylogeographic analysis, proposed the splitting of the complex *Thomomys umbrinus-bottae* into eight species, but this proposal has not been accepted. We accept the recommendation of Hafner et al. (2011) that such arrangement needs further support.
53. Hafner et al. (2011) recently recognized the pocket gophers of northwestern Mexico, inhabiting the dry, thorny scrubland along the Pacific versant of the Sierra Madre Occidental from northern Sinaloa to western Durango, northwestern Jalisco, and western Nayarit as a different species (*Thomomys atrovarius*). We accept their recommendation but their paper was published when this book was already in press, so there is no species account.
54. Following comments by Engstrom et al. (1987), Rogers (1990), Rogers and Rogers (1992), and Williams et al. (1993a, b) for considering *Heteromys goldmani* as a subspecies within *H. desmarestianus*, the recognized Mexican species for *Heteromys* sensu stricto are *H. desmarestianus*, *H. gaumeri*, and *H. nelsoni*.
55. Based on molecular data (2,381 base pairs for the 3-gene data set), Hafner et al. (2007) found that *Liomys* is paraphyletic to *Heteromys*, and placed *Liomys* as its synonym. We will not follow such arrangement until it is widely accepted.
56. Following Hall (1981), *Chaetodipus anthonyi* and *C. dalquesti* are considered as valid species, different from *C. fallax* and *C. arenarius*, respectively, since the proposal for merging the subspecies pairs was not justified (Williams et al., 1993a, b).
57. *Chaetodipus eremicus* is considered distinct from *C. penicillatus*, based on the extensive analyses by Lee et al. (1996).
58. Populations of *Chaetodipus baileyi* west of the Colorado River from southern California of the California Peninsula were recently recognized as a distinct species, *C. rudinoris*, based on the mitochondrial DNA analysis by Riddle et al. (2000a). We follow such a recommendation.
59. Weskler (2006), based on a comprehensive review of morphological and molecular data, proposed a new phylogenetic hypothesis for the species within the tribe Oryzomyini and

proposed that within the recently described genus *Handleyomys* Voss et al., 2002, some species from *Oryzomys* should be reallocated, including *O. alfaroi*, *O. chapmani*, *O. melanotis*, *O. rhapdops*, *O. rostratus*, and *O. saturatior*. Such a proposal was also supported by Weksler et al. (2006).

60. Riddle et al. (2000b) recognized the populations of *Peromyscus eremicus* west from the Colorado River, from southern California to the Baja California Peninsula, as a distinct species (*P. fraterculus*).
61. *Peromyscus sagax* is recognized as a valid species, endemic to central Michoacán, based on Bradley et al. (1996), who used morphometrical, karyological, and molecular data for their analyses.
62. Based on genetic and molecular data, Hafner et al. (2001) evaluated the island populations of Baja Californian *Peromyscus*, proposing that the species *P. stephani*, *P. interparietalis*, *P. caniceps*, and *P. dickeyi* are actually subspecies within *P. boylii*, *P. eremicus*, *P. fraterculus*, and *P. merriami*, respectively. We do not accept such a proposal until further studies using morphometics, karyotypic, and allozymic data support those.
63. In accordance with Edwards et al. (2001), the populations of *Neotoma albigula* west from the Conchos River (Chihuahua) pertain to a different species, *Neotoma leucodon*.
64. Edwards and Bradley (2002) analyzed the cytochrome b mitochondrial DNA gene sequences for 15 individuals from 6 populations of *Neotoma mexicana*, and their results support their suggestion that, at least 2 southern Mexican populations (*N. m. isthmica* and *N. m. picta*) may represent sister taxa to *N. mexicana*.
65. Matocq (2002) analyzed external morphological, cranial morphology, and genetic data (ADN and nuclear microsatellites) and separated the southern populations of *Neotoma fuscipes* as *Neotoma macrotis*.
66. Carleton et al. (2002) described a new species of *Habromys*, *H. delicatulus*, from mixed temperate forests from the state of Mexico.
67. *Reithrodontomys bakeri* was described from the temperate forests of the Sierra Madre del Sur in Guerrero (Bradley et al., 2004).
68. Peppers et al. (2002) separated *Sigmodon toltecus* from *S. hispidus* on the basis of differences of cytochrome b.
69. Analyses by Carleton and Arroyo (2009) and Hanson et al. (2010) have defined the current status within the *Oryzomys palustris* complex in Mexico, recognizing six species: *O. albiventer*, *O. couesi*, *O. peninsulae*, *O. mexicanus*, *O. nelsoni*, and *O. texensis*.
70. We accept the conclusion of Musser and Carleton (2005) that *Sigmodon planifrons* and *S. zanjonensis* are different from *S. hispidus*.
71. We follow Musser and Carleton (2005) in their consideration that *Oryzomys palustris* is not found in Mexico.
72. We follow the recommendation of the International Commission on Zoological Nomenclature (ICZN, 1998) in recognizing the Cuniculidae family and the *Cuniculus* genus.
73. Ruedas and Salazar-Bravo (2007) based on a karyotypic study, excised *Sylvilagus brasiliensis gabbi* into two species, *S. gabbi* occurring from Panama to the north, and *S. brasiliensis* occurring to the south of Panama. The subspecies for Mexico is *S. g. truei*.
74. The populations of *Sylvilagus floridanus* of southern Texas and Maderas del Carmen in Coahuila are considered as a different species (*S. robustus*; Ruedas, 1998).

Checklist of Mammals of Mexico

In the following checklist we mention all the species by order, and within it by family, subfamily, and sometimes tribe (TAXA column). Orders and families are listed phylogenetically, and species within a family by alphabetical sequence. See p. 872 for key to abbreviations.

TAXA	DISTRIBUTION		CONSERVATION		
	Land/Marine	Continent	Mexico	CITES	IUCN
ORDER DIDELPHIMORPHIA					
FAMILY DIDELPHIDAE					
SUBFAMILY DIDELPHINAE					
Chironectes minimus (Zimmermann, 1780)	C	SA	E		
Didelphis marsupialis Linnaeus, 1758	IC	SA			
Didelphis virginiana Kerr, 1792	IC	AM			
Marmosa mexicana Merriam, 1897	C	MA			
Metachirus nudicaudatus (È. Geoffroy St.-Hilaire, 1803)	C	SA	T		
Philander opossum (Linnaeus, 1758)	C	SA			
Tlacuatzin canescens (J.A. Allen, 1893)	IC	MX			
SUBFAMILY CALUROMYINAE					
Caluromys derbianus (Waterhouse, 1841)	C	SA	T		
ORDER SIRENIA					
FAMILY TRICHECHIDAE					
Trichechus manatus Linnaeus, 1758	M	AM	E	I	VU
ORDER CINGULATA					
FAMILY DASYPODIDAE					
SUBFAMILY DASYPODINAE					
Dasypus novemcinctus Linnaeus, 1758	IC	AM			
SUBFAMILY TOLYPEUTINAE					
Cabassous centralis (Miller, 1899)	C	SA	E	III	
ORDER PILOSA					
FAMILY CYCLOPEDIDAE					
Cyclopes didactylus (Linnaeus, 1758)	C	SA	E		
FAMILY MYRMECOPHAGIDAE					
Tamandua mexicana (Saussure, 1860)	C	SA	E		
ORDER PRIMATES					
FAMILY ATELIDAE					
SUBFAMILY MYCETINAE					
Alouatta palliata (Gray, 1849)	C	SA	E		EN[1]
Alouatta pigra Lawrence, 1933	C	MA	E		EN
SUBFAMILY ATELINAE					
Ateles geoffroyi Kuhl, 1820	C	MA	E		EN[2]
ORDER LAGOMORPHA					
FAMILY LEPORIDAE					
SUBFAMILY LEPORINAE					
Lepus alleni Mearns, 1890	IC	NA	*		
Lepus californicus Gray, 1837	IC	NA	*		
Lepus callotis Wagler, 1830	C	NA			NT
Lepus flavigularis Wagner, 1844	C	MX	E		EN

TAXA	DISTRIBUTION		CONSERVATION		
	Land/Marine	Continent	Mexico	CITES	IUCN
Lepus insularis W. Bryant, 1891	I	MX	Pr		NT
Romerolagus diazi (Ferrari-Pérez, 1893)	C	MX	E	I	EN
Sylvilagus audubonii (Baird, 1858)	C	NA			
Sylvilagus bachmani (Waterhouse, 1839)	IC	NA	*		
Sylvilagus cunicularius (Waterhouse, 1848)	C	MX			
Sylvilagus floridanus (J.A. Allen, 1890)	C	AM			
Sylvilagus gabbi truei (J.A. Allen, 1890)	C	SA			
Sylvilagus graysoni (J.A. Allen, 1877)	I	MX	P		EN
Sylvilagus insonus (Nelson, 1904)	C	MX	P		EN
Sylvilagus mansuetus Nelson, 1907	I	MX	E		NT
Sylvilagus robustus (V. Bailey, 1905)	C	NA			EN
ORDER RODENTIA					
FAMILY SCIURIDAE					
SUBFAMILY SCIURINAE					
Ammospermophilus harrisii (Audubon & Bachman, 1854)	C	NA			
Ammospermophilus insularis Nelson & Goldman, 1909	I	MX	T		
Ammospermophilus interpres (Merriam, 1890)	C	NA			
Ammospermophilus leucurus (Merriam, 1889)	C	NA			
Callospermophilus madrensis Merriam, 1901	C	MX	Pr		NT
Cynomys ludovicianus (Ord, 1815)	C	NA	T		
Cynomys mexicanus Merriam, 1892	C	MX	E		EN
Glaucomys volans (Linnaeus, 1758)	C	NA	T		
Ictidomys mexicanus (Erxleben, 1777)	C	NA			
Ictidomys parvidens (Mearns, 1896)	C	NA			
Neotamias bulleri (J.A. Allen, 1889)	C	MX			
Neotamias dorsalis (Baird, 1855)	C	NA			
Neotamias durangae (J.A. Allen, 1903)	C	MX			
Neotamias merriami (J.A. Allen, 1889)	C	NA	Pr		
Neotamias obscurus (J.A. Allen, 1890)	C	NA			
Notocitellus adocetus (Merriam, 1903)	C	MX			
Notocitellus annulatus (Audubon & Bachman, 1842)	C	MX			
Otospermophilus atricapillus (W.E. Bryant, 1889)	C	MX			EN
Otospermophilus beecheyi (Richardson, 1829)	C	NA			
Otospermophilus variegatus (Erxleben, 1777)	IC	NA			
Sciurus aberti Woodhouse, 1853	C	NA	Pr		
Sciurus alleni Nelson, 1898	C	MX			
Sciurus arizonensis Coues, 1867	C	NA	T		
Sciurus aureogaster F. Cuvier, 1829	C	MA			
Sciurus colliaei Richardson, 1839	C	MX			
Sciurus deppei Peters, 1863	C	MA		III	
Sciurus griseus Ord, 1818	C	NA	T		
Sciurus nayaritensis J.A. Allen, 1890	C	NA			
Sciurus niger Linnaeus, 1758	C	NA			
Sciurus oculatus Peters, 1863	C	MX	Pr		
Sciurus variegatoides Ogilby, 1839	C	MA	Pr		
Sciurus yucatanensis J.A. Allen, 1877	C	MA			
Tamiasciurus mearnsi Townsend, 1897	C	MX	T		E
Xerospermophilus perotensis (Merriam, 1893)	C	MX	T		EN
Xerospermophilus spilosoma (Bennett, 1833)	C	NA			
Xerospermophilus tereticaudus (Baird, 1858)	IC	NA			

TAXA	DISTRIBUTION		CONSERVATION		
	Land/Marine	Continent	Mexico	CITES	IUCN
FAMILY CASTORIDAE					
Castor canadensis Kuhl, 1820	C	NA	E		
FAMILY HETEROMYIDAE					
SUBFAMILY DIPODOMYINAE					
Dipodomys compactus True, 1889	C	NA			
Dipodomys deserti Stephens, 1887	C	NA			
Dipodomys gravipes Huey, 1925	C	MX	E		CR
Dipodomys insularis Merriam, 1907	I	MX	E		CR
Dipodomys merriami Mearns, 1890	IC	NA	*		EN[3]
Dipodomys nelsoni Merriam, 1907	C	MX			
Dipodomys ordii Woodhouse, 1853	C	NA			
Dipodomys phillipsii Gray, 1841	C	MX	Pr		
Dipodomys simulans (Merriam, 1904)	C	NA			
Dipodomys spectabilis Merriam, 1890	C	NA			NT
SUBFAMILY HETEROMYINAE					
Heteromys desmarestianus Gray, 1868	C	SA			
Heteromys gaumeri J.A. Allen & Chapman, 1897	C	MA			
Heteromys nelsoni Merriam, 1902	C	MX	Pr		EN
Liomys irroratus (Gray, 1868)	C	NA			
Liomys pictus (Thomas, 1893)	C	MA			
Liomys salvini (Thomas, 1893)	C	MA			
Liomys spectabilis Genoways, 1971	C	MX	Pr		EN
SUBFAMILY PEROGNATHINAE					
Chaetodipus anthonyi (Osgood, 1900)	I	MX	T		
Chaetodipus arenarius (Merriam, 1894)	C	MX	*		
Chaetodipus artus (Osgood, 1900)	C	MX			
Chaetodipus baileyi (Merriam, 1894)	IC	NA	*		
Chaetodipus californicus (Merriam, 1889)	C	NA			
Chaetodipus dalquesti (Roth, 1976)	C	MX	Pr		VU
Chaetodipus eremicus (Mearns, 1898)	C	NA			
Chaetodipus fallax (Merriam, 1889)	C	NA			
Chaetodipus formosus (Merriam, 1889)	C	NA			
Chaetodipus goldmani (Osgood, 1900)	C	MX			NT
Chaetodipus hispidus (Baird, 1858)	C	NA			
Chaetodipus intermedius (Merriam, 1889)	IC	NA	*		
Chaetodipus lineatus (Dalquest, 1951)	C	MX			
Chaetodipus nelsoni (Merriam, 1894)	C	NA			
Chaetodipus penicillatus (Woodhouse, 1852)	IC	NA	*		
Chaetodipus pernix (J.A. Allen, 1898)	C	MX			
Chaetodipus rudinoris (Elliot, 1903)	IC	NA			
Chaetodipus spinatus (Merriam, 1889)	IC	NA	*		
Perognathus amplus Osgood, 1900	C	NA	*		
Perognathus flavescens Merriam, 1889	C	NA			
Perognathus flavus Baird, 1855	C	NA			
Perognathus longimembris (Coues, 1875)	C	NA			
Perognathus merriami J.A. Allen, 1892	C	NA			
FAMILY GEOMYIDAE					
Cratogeomys castanops (Baird, 1852)	C	NA			
Cratogeomys fulvescens Merriam, 1895	C	MX			
Cratogeomys fumosus (Merriam, 1892)	C	MX	T		
Cratogeomys goldmani Merriam, 1895	C	MX			

TAXA	DISTRIBUTION		CONSERVATION		
	Land/Marine	Continent	Mexico	CITES	IUCN
Cratogeomys merriami (Thomas, 1893)	C	MX			
Cratogeomys perotensis Merriam, 1895	C	MX			
Cratogeomys planiceps (Merriam, 1895)	C	MX			
Geomys arenarius Merriam, 1895	C	NA			NT
Geomys personatus True, 1889	C	NA	T		
Geomys tropicalis Goldman, 1915	C	MX	T		CR
Orthogeomys cuniculus Elliot, 1905	C	MX	T		
Orthogeomys grandis (Thomas, 1893)	C	MA			
Orthogeomys hispidus (Le Conte, 1852)	C	MA			
Orthogeomys lanius (Elliot, 1905)	C	MX	T		CR
Pappogeomys bulleri (Thomas, 1892)	C	MX	Pr		
Thomomys atrovarius J.A. Allen, 1898	C	MX			
Thomomys bottae (Eydoux & Gervais, 1836)	IC	NA			
Thomomys umbrinus (Richardson, 1829)	C	NA			
Zygogeomys trichopus Merriam, 1895	C	MX	E		EN
FAMILY CRICETIDAE					
SUBFAMILY ARVICOLINAE					
Microtus californicus (Peale, 1848)	C	NA	E		
Microtus guatemalensis Merriam, 1898	C	MA	T		NT
Microtus mexicanus (Saussure, 1861)	C	NA			
Microtus oaxacensis Goodwin, 1966	C	MX	T		EN
Microtus pennsylvanicus (Ord, 1815)	C	NA	E		
Microtus quasiater (Coues, 1874)	C	MX	Pr		NT
Microtus umbrosus Merriam, 1898	C	MX	Pr		EN
Ondatra zibethicus (Linnaeus, 1766)	C	NA	T		
SUBFAMILY NEOTOMINAE					
Baiomys musculus (Merriam, 1892)	C	MA			
Baiomys taylori (Thomas, 1887)	C	NA			
Habromys chinanteco (Robertson & Musser, 1976)	C	MX			CR
Habromys delicatulus Carleton et al., 2002	C	MX			CR
Habromys ixtlani (Goodwin, 1964)	C	MX			CR
Habromys lepturus (Merriam, 1898)	C	MX			CR
Habromys lophurus (Osgood, 1904)	C	MA			NT
Habromys schmidlyi Romo-Vázquez et al., 2005	C	MX			
Habromys simulatus (Osgood, 1904)	C	MX	Pr		EN
Hodomys alleni (Merriam, 1892)	C	MX			
Megadontomys cryophilus (Musser, 1964)	C	MX	T		EN
Megadontomys nelsoni (Merriam, 1898)	C	MX	T		EN
Megadontomys thomasi (Merriam, 1898)	C	MX	Pr		EN
Nelsonia goldmani Merriam, 1903	C	MX	Pr		EN
Nelsonia neotomodon Merriam, 1897	C	MX	Pr		NT
Neotoma albigula Hartley, 1894	IC	NA	*		
Neotoma angustapalata Baker, 1951	C	MX			EN
Neotoma bryanti Merriam, 1887	I	MX	T		EN
Neotoma devia Goldman, 1927	C	NA			
Neotoma goldmani Merriam, 1903	C	MX			
Neotoma insularis Townsend, 1912	I	MX	T		
Neotoma isthmica Goldman 1904	C	MX			
Neotoma lepida Thomas, 1893	IC	NA	*		
Neotoma leucodon Merriam, 1894	C	NA			
Neotoma macrotis Thomas, 1893	C	NA			

TAXA	DISTRIBUTION		CONSERVATION		
	Land/Marine	Continent	Mexico	CITES	IUCN
Neotoma mexicana Baird, 1855	C	NA			
Neotoma micropus Baird, 1855	C	NA			
Neotoma nelsoni Goldman, 1905	C	MX			CR
Neotoma palatina Goldman, 1905	C	MX			VU
Neotoma phenax (Merriam, 1903)	C	MX	Pr		NT
Neotoma picta Goldman 1904	C	MX			
Neotomodon alstoni Merriam, 1898	C	MX			
Nyctomys sumichrasti (Saussure, 1860)	C	MA			
Oligoryzomys fulvescens (Saussure, 1860)	C	SA			
Onychomys arenicola Mearns, 1896	C	NA			
Onychomys leucogaster (Wied-Neuwied, 1841)	C	NA			
Onychomys torridus (Coues, 1874)	C	NA			
Oryzomys albiventer Merriam, 1901	C	MX			
Oryzomys alfaroi (J.A. Allen, 1891)	C	SA			
Oryzomys chapmani Thomas, 1898	C	MX	Pr		
Oryzomys couesi (Alston, 1877)	IC	AM	*		
Oryzomys melanotis Thomas, 1893	C	MX			
Oryzomys mexicanus J.A. Allen, 1897	C	MX			
Oryzomys nelsoni Merriam, 1898	I	MX	E		EX
Oryzomys rhabdops Merriam, 1901	C	MA			
Oryzomys rostratus Merriam, 1901	C	MA			
Oryzomys saturatior Merriam, 1901	C	MA			
Oryzomys texensis J.A. Allen, 1894	C	NA			
Osgoodomys banderanus (J.A. Allen, 1897)	C	MX			
Otonyctomys hatti Anthony, 1932	C	MA	T		
Ototylomys phyllotis Merriam, 1901	C	MA			
Peromyscus aztecus (Saussure, 1860)	C	MA			
Peromyscus beatae Thomas, 1903	C	MX			
Peromyscus boylii (Baird, 1855)	IC	NA	*		
Peromyscus bullatus Osgood, 1904	C	MX	Pr		CR
Peromyscus californicus (Gambel, 1848)	C	NA			
Peromyscus caniceps Burt, 1932	I	MX	Pr		CR
Peromyscus crinitus (Merriam, 1891)	IC	NA	T		
Peromyscus dickeyi Burt, 1932	I	MX	Pr		CR
Peromyscus difficilis (J.A. Allen, 1891)	C	MX			
Peromyscus eremicus (Baird, 1858)	IC	NA	*		
Peromyscus eva Thomas, 1898	C	MX	T		
Peromyscus fraterculus (Miller, 1892)	C	NA			
Peromyscus furvus J.A. Allen & Chapman, 1897	C	MX			
Peromyscus gratus Merriam, 1898	C	NA			
Peromyscus guardia Townsend, 1912	I	MX	E		CR
Peromyscus guatemalensis Merriam, 1898	C	MA			
Peromyscus gymnotis Thomas, 1894	C	MA			
Peromyscus hooperi Lee & Schmidly, 1977	C	MX			
Peromyscus hylocetes Merriam, 1898	C	MX			
Peromyscus interparietalis Burt, 1932	I	MX	T		CR
Peromyscus leucopus (Rafinesque, 1818)	IC	NA	*		
Peromyscus levipes Merriam, 1898	C	MX			
Peromyscus madrensis Merriam, 1898	I	MX	T		EN
Peromyscus maniculatus (Wagner, 1845)	IC	NA	*		
Peromyscus megalops Merriam, 1898	C	MX			

TAXA	DISTRIBUTION		CONSERVATION		
	Land/Marine	Continent	Mexico	CITES	IUCN
Peromyscus mekisturus Merriam, 1898	C	MX	T		CR
Peromyscus melanocarpus Osgood, 1904	C	MX			EN
Peromyscus melanophrys (Coues, 1874)	C	MX			
Peromyscus melanotis J.A. Allen & Chapman, 1897	C	NA			
Peromyscus melanurus Osgood, 1909	C	MX			EN
Peromyscus merriami Mearns, 1896	C	NA			
Peromyscus mexicanus (Saussure, 1860)	C	MA			
Peromyscus nasutus (J.A. Allen, 1891)	C	NA			
Peromyscus oaxacensis Merriam, 1898	C	MA			
Peromyscus ochraventer Baker, 1951	C	MX			EN
Peromyscus pectoralis Osgood, 1904	C	NA			
Peromyscus pembertoni Burt, 1932	I	MX	E		EX
Peromyscus perfulvus Osgood, 1945	C	MX			
Peromyscus polius Osgood, 1904	C	MX			NT
Peromyscus pseudocrinitus Burt, 1932	I	MX	T		CR
Peromyscus sagax Elliot, 1903	C	MX			
Peromyscus schmidlyi Bradley et al., 2004	C	MX			
Peromyscus sejugis Burt, 1932	I	MX	T		EN
Peromyscus simulus Osgood, 1904	C	MX			VU
Peromyscus slevini Mailliard, 1924	I	MX	T		CR
Peromyscus spicilegus J.A. Allen, 1897	C	MX			
Peromyscus stephani Townsend, 1912	I	MX	T		CR
Peromyscus truei (Shufeldt, 1885)	C	NA			
Peromyscus winkelmanni Carleton, 1977	C	MX	Pr		EN
Peromyscus yucatanicus J.A. Allen & Chapman, 1897	C	MX			
Peromyscus zarhynchus Merriam, 1898	C	MX	Pr		VU
Reithrodontomys bakeri Bradley et al., 2004	C	MX			
Reithrodontomys burti Benson, 1939	C	MX			
Reithrodontomys chrysopsis Merriam, 1900	C	MX			
Reithrodontomys fulvescens J.A. Allen, 1894	C	NA			
Reithrodontomys gracilis J.A. Allen & Chapman, 1897	IC	MA	*		
Reithrodontomys hirsutus Merriam, 1901	C	MX			VU
Reithrodontomys megalotis (Baird, 1858)	C	NA			
Reithrodontomys mexicanus (Saussure, 1860)	C	SA			
Reithrodontomys microdon Merriam, 1901	C	MA	T		
Reithrodontomys montanus (Baird, 1855)	C	NA			
Reithrodontomys spectabilis Jones & Lawlor, 1965	I	MX	T		CR
Reithrodontomys sumichrasti (Saussure, 1861)	C	MA			
Reithrodontomys tenuirostris Merriam, 1901	C	MA			VU
Reithrodontomys zacatecae Merriam, 1901	C	MX			
Rheomys mexicanus Goodwin, 1959	C	MX	Pr		EN
Rheomys thomasi Dickey, 1928	C	MA	*		NT
Scotinomys teguina (Alston, 1877)	C	MA	Pr		
Sigmodon alleni Bailey, 1902	C	MX			VU
Sigmodon arizonae Mearns, 1890	C	NA			
Sigmodon fulviventer J.A. Allen, 1889	C	NA			
Sigmodon hirsutus (Burmeister, 1854)	C	SA			
Sigmodon hispidus Say & Ord, 1825	C	AM			
Sigmodon leucotis Bailey, 1902	C	MX			
Sigmodon mascotensis J.A. Allen, 1897	C	MX			
Sigmodon ochrognathus Bailey, 1902	C	NA			

TAXA	DISTRIBUTION		CONSERVATION		
	Land/Marine	Continent	Mexico	CITES	IUCN
Sigmodon planifrons Nelson & Goldman, 1933	C	MA			
Sigmodon toltecus (Saussure, 1860)	C	NA			
Sigmodon zanjonensis Goodwin, 1932	C				
Tylomys bullaris Merriam, 1901	C	MX	T		CR
Tylomys nudicaudus Peters, 1866	C	MA			
Tylomys tumbalensis Merriam, 1901	C	MX	Pr		CR
Xenomys nelsoni Merriam, 1892	C	MX	T		EN
FAMILY ERETHIZONTIDAE					
SUBFAMILY ERETHIZONTINAE					
Erethizon dorsatum (Linnaeus, 1758)	C	NA	E		
Sphiggurus mexicanus (Kerr, 1792)	C	MA	T	III	
FAMILY DASYPROCTIDAE					
Dasyprocta mexicana Saussure, 1860	C	MX			CR
Dasyprocta punctata Gray, 1842	IC	SA		III	
FAMILY CUNICULIDAE					
Cuniculus paca (Linnaeus, 1766)	IC	SA		III	
ORDER SORICOMORPHA					
FAMILY SORICIDAE					
SUBFAMILY SORICINAE					
Cryptotis alticola (Merriam, 1895)	C	MX			SP
Cryptotis goldmani (Merriam, 1895)	C	MX	*		
Cryptotis goodwini Jackson, 1933	C	MA			
Cryptotis griseoventris Jackson, 1933	C	MA			VU
Cryptotis magna (Merriam, 1895)	C	MX	Pr		VU
Cryptotis mayensis (Merriam, 1901)	C	MA	Pr		
Cryptotis merriami Choate, 1970	C	MA			
Cryptotis mexicana (Coues, 1877)	C	MX	*		
Cryptotis nelsoni (Merriam, 1895)	C	MX			CR
Cryptotis obscura (Merriam, 1895)	C	MX			VU
Cryptotis parva (Say, 1823)	C	AM	*		
Cryptotis peregrina (Merriam, 1895)	C	MX			
Cryptotis phillipsii (Schaldach, 1966)	C	MX			VU
Cryptotis tropicalis (Merriam, 1895)	C	MX			
Megasorex gigas (Merriam, 1897)	C	MX	T		
Notiosorex cockrumi Baker et al., 2003		NA			
Notiosorex crawfordi (Coues, 1877)	IC	NA	T		
Notiosorex evotis (Coues, 1877)	C	MX			
Notiosorex villai Carraway & Timm, 2000	C	MX			VU
Sorex arizonae Diersing & Hoffmeister, 1977	C	NA	E		
Sorex emarginatus Jackson, 1925	C	MX			
Sorex ixtlanensis Carraway, 2007	C	MX			
Sorex macrodon Merriam, 1895	C	MX	Pr		VU
Sorex mediopua Carraway, 2007	C	MX			
Sorex milleri Jackson, 1947	C	MX	Pr		VU
Sorex monticolus Merriam, 1890	C	NA			
Sorex oreopolus Merriam, 1892	C	MX			
Sorex orizabae Merriam 1895	C	MX			
Sorex ornatus Merriam, 1895	C	NA	*		
Sorex saussurei Merriam, 1892	C	MA	*		
Sorex sclateri Merriam, 1897	C	MX	Pr		CR

TAXA	DISTRIBUTION		CONSERVATION		
	Land/Marine	Continent	Mexico	CITES	IUCN
Sorex stizodon Merriam, 1895	C	MX	Pr		CR
Sorex ventralis Merriam, 1895	C	MX			
Sorex veraecrucis Jackson, 1925	C	MX			
Sorex veraepacis Alston, 1877	C	MA	Pr		
FAMILY TALPIDAE					
SUBFAMILY TALPINAE					
Scalopus aquaticus (Linnaeus, 1758)	C	NA	E		
Scapanus anthonyi J.A. Allen, 1893	C	MX	T		
Scapanus latimanus (Bachman, 1842)	C	NA	T		
ORDER CARNIVORA					
FAMILY FELIDAE					
SUBFAMILY FELINAE					
Herpailurus yagouaroundi (Lacépède, 1809)	C	AM	T	I	EN[4]
Leopardus pardalis (Linnaeus, 1758)	C	AM	E	I	EN[5]
Leopardus wiedii (Schinz, 1821)	C	AM	E	I	NT
Lynx rufus (Schreber, 1777)	C	NA		II	
Puma concolor (Linnaeus, 1771)	C	AM			
SUBFAMILY PANTHERINAE					
Panthera onca (Linnaeus, 1758)	C	AM	E		NT
FAMILY CANIDAE					
Canis latrans Say, 1823	IC	NA			
Canis lupus Linnaeus, 1758	C	NA	E	I	EW[6]
Urocyon cinereoargenteus (Schreber, 1775)	IC	AM			
Vulpes macrotis Merriam, 1888	C	NA	T		
FAMILY URSIDAE					
SUBFAMILY URSINAE					
Ursus americanus Pallas, 1780	C	NA	*		
Ursus arctos Linnaeus, 1758	C	NA	E	I	EX[7]
FAMILY OTARIIDAE					
Arctocephalus galapagoensis Heller, 1904	I	SA	I		VU
Arctocephalus townsendi Merriam, 1897	M	NA	E	I	NT
Callorhinus ursinus (Linnaeus, 1758)	M	AM			
Zalophus californianus (Lesson, 1828)	M	AM	Pr		
Zalophus wollebaeki (Sivertsen, 1953)	M	AM			
FAMILY PHOCIDAE					
Mirounga angustirostris (Gill, 1866)	M	NA	T		
Monachus tropicalis (Gray, 1850)	M	MA	E		EN
Phoca vitulina Linnaeus, 1758	M	NA	Pr		
FAMILY MUSTELIDAE					
SUBFAMILY LUTRINAE					
Enhydra lutris (Linnaeus, 1758)	M	NA	E	I	EN
Lontra canadensis (Schreber, 1777)	C	NA		II	
Lontra longicaudis (Olfers, 1818)	C	SA	T	IV	
SUBFAMILY MUSTELINAE					
Eira barbara (Linnaeus, 1758)	C	SA	E	III	EN[8]
Galictis vittata (Schreber, 1776)	C	SA	T	III	
Mustela frenata Lichtenstein, 1831	C	AM			
Mustela nigripes (Audubon & Bachman, 1851)	C	NA		I	EN
SUBFAMILY TAXIDIINAE					
Taxidea taxus (Schreber, 1777)	C	NA	T		

TAXA	DISTRIBUTION		CONSERVATION		
	Land/Marine	Continent	Mexico	CITES	IUCN
FAMILY MEPHITIDAE					
Conepatus leuconotus (Lichtenstein, 1832)	C	NA			
Conepatus semistriatus (Boddaert, 1785)	C	SA	*		
Mephitis macroura Lichtenstein, 1832	C	AM			
Mephitis mephitis (Schreber, 1776)	C	NA			
Spilogale angustifrons Howell, 1902					
Spilogale gracilis Merriam, 1890	C	NA			
Spilogale pygmaea Thomas, 1898	C	MX	T		VU
FAMILY PROCYONIDAE					
SUBFAMILY POTOSINAE					
Potos flavus (Schreber, 1774)	C	SA	E	III	
SUBFAMILY PROCYONINAE					
Bassariscus astutus (Lichtenstein, 1830)	IC	NA	*		
Bassariscus sumichrasti (Saussure, 1860)	C	MA	Pr	III	
Nasua narica (Linnaeus, 1766)	C	AM	*	III	
Nasua nelsoni (Merriam, 1901)	I	MX	E		CR[9]
Procyon lotor (Linnaeus, 1758)	C	AM			
Procyon pygmaeus Merriam, 1901	I	MX	E		CR
ORDER PERISSODACTYLA					
FAMILY TAPIRIDAE					
Tapirus bairdii (Gill, 1865)	C	SA	E		EN
ORDER ARTIODACTYLA					
FAMILY TAYASSUIDAE					
Pecari tajacu (Linnaeus, 1758)	IC	AM		II	
Tayassu pecari (Link, 1795)	C	SA	E	II	NT
FAMILY CERVIDAE					
SUBFAMILY CERVINAE					
Cervus canadensis Erxleben, 1777	C	NA	Ex		
SUBFAMILY ODOCOILEINAE					
Mazama americana (Erxleben, 1777)	C	SA	*		
Mazama pandora Merriam, 1901	C	MA			VU
Odocoileus hemionus (Rafinesque, 1817)	IC	NA	*		VU[10]
Odocoileus virginianus (Zimmermann, 1780)	IC	AM			
FAMILY ANTILOCAPRIDAE					
Antilocapra americana (Ord, 1815)	C	NA	E		EN/CR[11]
FAMILY BOVIDAE					
SUBFAMILY BOVINAE					
Bison bison (Linnaeus, 1758)	C	NA	E		NT
SUBFAMILY CAPRINAE					
Ovis canadensis Shaw, 1804	C	NA	Pr		IEN/CR[12]
ORDER CETACEA					
SUBORDER MYSTICETI					
FAMILY BALAENIDAE					
Eubalaena japonica (Lacépède, 1818)	M	NA	E		EN
FAMILY BALAENOPTERIDAE					
Balaenoptera acutorostrata Lacépède, 1804	M	AM	Pr	I	
Balaenoptera borealis Lesson, 1828	M	AM	Pr	I	EN
Balaenoptera edeni Anderson, 1879	M	AM	Pr	I	

TAXA	DISTRIBUTION		CONSERVATION		
	Land/Marine	Continent	Mexico	CITES	IUCN
Balaenoptera physalus (Linnaeus, 1758)	M	AM	Pr	I	EN
Balaenoptera musculus (Linnaeus, 1758)	M	AM	Pr	I	EN
Megaptera novaeangliae (Borowski, 1781)	M	AM	Pr	I	
FAMILY ESCHRICHTIDAE					
Eschrichtius robustus (Lilljeborg, 1861)	M	NA	Pr	I	
SUBORDER ODONTOCETI					
FAMILY PHYSETERIDAE					
Kogia breviceps (De Blainville, 1838)	M	AM	Pr	II	
Kogia sima (Owen, 1866)	M	AM	Pr	II	
Physeter macrocephalus Linnaeus, 1758	M	AM	Pr	I	VU
FAMILY ZIPHIIDAE					
Berardius bairdii Stejneger, 1883	M	NA	Pr	I	
Indopacetus pacificus (Longman, 1926)	M	AM	Pr	I	
Mesoplodon carlhubbsi Moore, 1963	M	NA	Pr		
Mesoplodon densirostris (De Blainville, 1817)	M	AM	Pr	II	
Mesoplodon europaeus (Gervais, 1855)	M	NA	Pr	II	
Mesoplodon ginkgodens Nishiwaki & Kamiya, 1958	M	NA	Pr	II	
Mesoplodon perrini Dalebout et al. 2002	M	NA	Pr		
Mesoplodon peruvianus Reyes et al., 1991	M	AM	Pr	II	
Ziphius cavirostris G. Cuvier, 1823	M	AM	Pr	II	
FAMILY DELPHINIDAE					
Delphinus capensis Gray, 1828	M	AM	Pr	II	
Delphinus delphis Linnaeus, 1758	M	AM	Pr	II	
Feresa attenuata Gray, 1875	M	AM	Pr	II	
Globicephala macrorhynchus Gray, 1846	M	AM	Pr	II	
Grampus griseus (G. Cuvier, 1812)	M	AM	Pr	II	
Lagenodelphis hosei Fraser, 1956	M	AM	Pr	II	
Lagenorhynchus obliquidens Gill, 1865	M	NA	Pr		
Lissodelphis borealis (Peale, 1848)	M	NA	Pr	II	
Orcinus orca (Linnaeus, 1758)	M	AM	Pr	II	
Peponocephala electra (Gray, 1846)	M	AM	Pr	II	
Pseudorca crassidens (Owen, 1846)	M	AM	Pr	II	
Stenella attenuata (Gray, 1846)	M	AM	Pr	II	
Stenella clymene (Gray, 1846)	M	AM	Pr	II	
Stenella coeruleoalba (Meyen, 1833)	M	AM	Pr	II	
Stenella frontalis (G. Cuvier, 1829)	M	AM	Pr	II	
Stenella longirostris (Gray, 1828)	M	AM	Pr	II	
Steno bredanensis (G. Cuvier in Lesson, 1828)	M	AM	Pr	II	
Tursiops truncatus (Montagu, 1821)	M	AM	Pr	II	
FAMILY PHOCOENIDAE					
Phocoena sinus Norris & McFarland, 1958	M	MX	Pr	I	CE
Phocoenoides dalli (True, 1885)	M	NA	Pr	II	
ORDER CHIROPTERA					
FAMILY EMBALLONURIDAE					
SUBFAMILY EMBALLONURINAE					
Balantiopteryx io Thomas, 1904	C	MA			VU
Balantiopteryx plicata Peters, 1867	IC	SA			
Centronycteris centralis Thomas, 1912	C	SA	Pr		
Diclidurus albus Wied-Neuwied, 1820	C	SA			
Peropteryx kappleri Peters, 1867	C	SA	Pr		

TAXA	DISTRIBUTION		CONSERVATION		
	Land/Marine	Continent	Mexico	CITES	IUCN
Peropteryx macrotis (Wagner, 1843)	C	SA			
Rynchonycteris naso (Wied-Neuwied, 1820)	C	SA	Pr		
Saccopteryx bilineata (Temminck, 1838)	C	SA			
Saccopteryx leptura (Schreber, 1774)	C	SA	Pr		
FAMILY PHYLLOSTOMIDAE					
SUBFAMILY MACROTINAE					
Macrotus californicus Baird, 1858	C	NA			
Macrotus waterhousii Gray, 1843	IC	MA			
SUBFAMILY MICRONYCTERINAE					
Glyphonycteris sylvestris Thomas, 1896	C	SA			
Lampronycteris brachyotis (Dobson, 1879)	C	SA	T		
Micronycteris microtis Miller, 1898	IC	SA			
Micronycteris schmidtorum Sanborn, 1935	C	SA	T		
Trinycteris nicefori Sanborn, 1949	C	SA			
SUBFAMILY DESMODONTINAE					
Desmodus rotundus (È. Geoffroy St.-Hilaire, 1810)	C	SA			
Diaemus youngi (Jentink, 1893)	C	SA	Pr		
Diphylla ecaudata Spix, 1823	C	AM			
SUBFAMILY VAMPYRINAE					
Chrotopterus auritus (Peters, 1856)	C	SA	T		
Trachops cirrhosus (Spix, 1823)	C	SA	T		
Vampyrum spectrum (Linnaeus, 1758)	C	SA	E		NT
SUBFAMILY PHYLLOSTOMINAE					
TRIBE PHYLLOSTOMINI					
Lonchorhina aurita Tomes, 1863	C	SA	T		
Lophostoma brasiliense Peters, 1867	C	SA	T		
Lophostoma evotis (Davis & Carter, 1978)	C	MA	T		
Macrophyllum macrophyllum (Schinz, 1821)	C	SA	T		
Mimon cozumelae Goldman, 1914	C	SA	T		
Mimon crenulatum (È. Geoffroy St.-Hilaire, 1810)	C	SA	T		
Phylloderma stenops Peters, 1865	C	SA	T		
Phyllostomus discolor Wagner, 1843	C	SA			
Tonatia saurophila Koopman & Williams, 1951	IC	SA	T		
TRIBE GLOSSOPHAGINI					
Anoura geoffroyi Gray, 1838	C	SA			
Choeroniscus godmani (Thomas, 1903)	C	SA			
Choeronycteris mexicana Tschudi, 1844	C	NA	T		NT
Glossophaga commissarisi Gardner, 1962	C	SA			
Glossophaga leachii (Gray, 1844)	C	MA			
Glossophaga morenoi Martínez & Villa, 1938	C	MX			
Glossophaga soricina (Pallas, 1766)	C	SA			
Hylonycteris underwoodi Thomas, 1903	C	MA			
Leptonycteris nivalis (Saussure, 1860)	C	NA	T		EN
Leptonycteris yerbabuenae Martínez & Villa, 1940	IC	AM	T		VU
Lichonycteris obscura Thomas, 1895	C	SA			
Musonycteris harrisoni Schaldach & McLaughlin, 1960	C	MX	E		VU
TRIBE STENODERMATINI					
Artibeus hirsutus Andersen, 1906	C	MX			
Artibeus jamaicensis Leach, 1821	IC	SA			
Artibeus lituratus (Olfers, 1818)	IC	SA			
Carollia perspicillata (Linnaeus, 1758)	C	SA			

TAXA	DISTRIBUTION		CONSERVATION		
	Land/Marine	Continent	Mexico	CITES	IUCN
Carollia sowelli Baker et al., 2002	C	MA	*		
Carollia subrufa (Hahn, 1905)	C	MA			
Centurio senex Gray, 1842	C	SA			
Chiroderma salvini Dobson, 1878	C	SA			
Chiroderma villosum Peters, 1860	C	SA			
Dermanura azteca (Andersen, 1906)	C	MA			
Dermanura phaeotis (Miller, 1902)	IC	SA			
Dermanura tolteca (Saussure, 1860)	C	MA			
Dermanura watsoni (Thomas, 1901)	C	SA	Pr		
Enchisthenes hartii (Thomas, 1892)	C	SA	Pr		
Platyrrhinus helleri (Peters, 1866)	C	SA			
Sturnira hondurensis Anthony, 1924	C	MA			
Sturnira lilium (È. Geoffroy St.-Hilaire, 1810)	C	SA			
Uroderma bilobatum Peters, 1866	C	SA			
Uroderma magnirostrum Davis, 1968	C	SA			
Vampyressa thyone Thomas, 1909	C	SA			
Vampyrodes caraccioli (Thomas, 1889)	C	SA			
FAMILY MORMOOPIDAE					
Mormoops megalophylla (Peters, 1864)	IC	AM			
Pteronotus davyi Gray, 1838	IC	SA			
Pteronotus gymnonotus (Wagner, 1843)	C	SA	Pr		
Pteronotus parnellii (Gray, 1843)	IC	SA			
Pteronotus personatus (Wagner, 1843)	IC	SA			
FAMILY NOCTILIONIDAE					
Noctilio albiventris Desmarest, 1818	C	SA	Pr		
Noctilio leporinus (Linnaeus, 1758)	C	SA			
FAMILY THYROPTERIDAE					
Thyroptera tricolor Spix, 1823	C	SA	Pr		
FAMILY NATALIDAE					
Natalus lanatus Tejedor, 2005	C	SA			
Natalus mexicanus Miller, 1902	C	SA			
FAMILY MOLOSSIDAE					
SUBFAMILY MOLOSSINAE					
Cynomops mexicanus (Jones & Genoways, 1967)	C	MX	Pr		
Eumops auripendulus (Shaw, 1800)	C	SA			
Eumops ferox (Gundlach, 1862)	C	AM			
Eumops hansae Sanborn, 1932	C	SA			
Eumops nanus (Miller, 1900)	IC	SA	Pr		
Eumops perotis (Schinz, 1821)	C	AM			
Eumops underwoodi Goodwin, 1940	C	AM			
Molossus alvarezi González-Ruiz et al., 2011	C	MX			
Molossus aztecus Saussure, 1860	IC	MA			
Molossus coibensis J.A. Allen, 1904	C	SA			
Molossus molossus (Pallas, 1766)	C	SA			
Molossus rufus E. Geoffroy, 1805	C	SA			
Molossus sinaloae J.A. Allen, 1906	C	SA			
Nyctinomops aurispinosus (Peale, 1848)	C	SA			
Nyctinomops femorosaccus (Merriam, 1889)	C	NA			
Nyctinomops laticaudatus (È. Geoffroy St.-Hilaire, 1805)	C	SA			
Nyctinomops macrotis (Gray, 1840)	C	AM			
Promops centralis Thomas, 1915	C	SA			

TAXA	DISTRIBUTION		CONSERVATION	
	Land/Marine	Continent	Mexico	CITES IUCN
SUBFAMILY TADARINAE				
Tadarida brasiliensis (È. Geoffroy St.-Hilaire, 1824)	C	AM		
FAMILY VESPERTILIONIDAE				
SUBFAMILY MYOTINAE				
Myotis albescens (È. Geoffroy St.-Hilaire, 1806)	C	SA	Pr	
Myotis auriculus Baker & Stains, 1955	C	AM		
Myotis californicus (Audubon & Bachman, 1842)	C	AM		
Myotis carteri La Val, 1973	C	MX		
Myotis elegans Hall, 1962	C	MA		
Myotis evotis (H. Allen, 1864)	C	NA	*	
Myotis findleyi Bogan, 1978	I	MX		EN
Myotis fortidens Miller & Allen, 1928	C	MA		
Myotis keaysi J.A. Allen, 1914	C	SA		
Myotis melanorhinus (Merriam, 1890)	C	NA		
Myotis nigricans (Schinz, 1821)	C	SA	*	
Myotis occultus Hollister, 1909	C	NA		
Myotis peninsularis Miller, 1898	C	MX		EN
Myotis planiceps Baker, 1955	C	MX	E	EN
Myotis thysanodes Miller, 1897	C	NA		
Myotis velifer (J.A. Allen, 1890)	C	AM		
Myotis vivesi Menegaux, 1901	C	MX	E	VU
Myotis volans (H. Allen, 1866)	C	NA		
Myotis yumanensis (H. Allen, 1864)	C	NA		
SUBFAMILY VESPERTILIONINAE				
Corynorhinus mexicanus G.M. Allen, 1916	C	MX		NT
Corynorhinus townsendii (Cooper, 1837)	IC	NA		
Eptesicus brasiliensis (Desmarest, 1819)	C	SA		
Eptesicus furinalis (d'Orbigny & Gervais, 1847)	C	SA		
Eptesicus fuscus (Palisot de Beauvois, 1796)	C	AM		
Euderma maculatum (J.A. Allen, 1891)	C	NA	Pr	
Idionycteris phyllotis (G.M. Allen, 1916)	C	NA		
Lasionycteris noctivagans (Le Conte, 1831)	C	NA	Pr	
Lasiurus blossevillii (Lesson & Garnot, 1826)	IC	AM		
Lasiurus borealis (Müller, 1776)	C	NA		
Lasiurus cinereus (Palisot de Beauvois, 1796)	C	AM		
Lasiurus ega (Gervais, 1856)	C	AM		
Lasiurus intermedius H. Allen, 1862	C	NA		
Lasiurus xanthinus (Thomas, 1897)	C	NA		
Nycticeius humeralis (Rafinesque, 1818)	C	NA		
Parastrellus hesperus (H. Allen, 1864)	IC	NA		
Perimyotis subflavus (F. Cuvier, 1832)	C	NA		
Rhogeessa aeneus Goodwin, 1958	C	MX		
Rhogeessa alleni Thomas, 1892	C	MX		
Rhogeessa bickhami Baker et al., 2012	C	MX		
Rhogeessa genowaysi Baker, 1984	C	MX	Pr	EN
Rhogeessa gracilis Miller, 1897	C	MX		
Rhogeessa mira La Val, 1973	C	MX	Pr	VU
Rhogeessa parvula H. Allen, 1866	IC	MX		
Rhogeessa tumida H. Allen, 1866	C	SA		
FAMILY ANTROZOIDAE				
Antrozous pallidus (Le Conte, 1856)	IC	NA		
Bauerus dubiaquercus (Van Gelder, 1959)	IC	MA		NT

* Indicates that some subspecies are listed at least in one risk category of the Mexican endangered species norm (SEMARNAT, 2010).

Under the IUCN column are only those species listed in the following risk categories: vulnerable (VU), endangered (EN), critically endangered (CR), extinct in nature (EW), extinct (EX).

The following list explains the numbers found after the CITES IUCN designations for select species on pages 55 to 63.

1. Only the subspecies *Alouatta palliata mexicana*, the only one in Mexico.
2. Only the subspecies *Ateles geoffroyi yucatanensis*.
3. Only the subspecies *D. m. margaritae*.
4. Only the subspecies *P. y. cacomitli*.
5. Only the subspecies *L. p. albescens*.
6. Only the Mexican wolf *Canis lupus baileyi*.
7. Only the subspecies *U. a. nelsoni*.
8. Only the subspecies *E. b. senex*.
9. Only the subspecies *N. n. nelsoni*. This is now considered a species.
10. Only the subspecies *O. h. cerrosensis*.
11. The subspecies *A. a. penninsularis* is considered critically endangered, and *A. a. sonorensis* as vulnerable.
12. The subspecies *O. c. cremnobates* and *O. c. mexicana* are vulnerable, and *O. c. weemsi* is critically endangered.

Species Accounts

Order Didelphimorphia

Gerardo Ceballos

It has a bag between the breasts and belly
where it puts the cubs. It carries them inside
wherever it wants to go, and there they nurse.
— Fray Bernardino de Sahagún, 1570

Although most people think of marsupials as Australian animals, more than 90 species of marsupials are distributed on the American continent (Wilson and Reeder, 2005). Just a few years ago, all marsupials were classified in the order Marsupialia. Recent classifications based on genetic and morphological analyses, however, now divide the group into seven orders: four of these are exclusive to Australia, New Guinea, and nearby islands, while the other three are endemic to the Americas (Archer and Kirsch, 2006; Marshall et al., 1990; Nilsson et al., 2004). Most American marsupials belong to the order Didelphimorphia, which includes 1 family and 87 species that are commonly known as opossums. Most opossums inhabit South and Central America, but eight species are found in Mexico and one (*Didelphis virginiana*) in the United States.

Two of the most distinctive characteristics of marsupials are the peculiar morphology of their reproductive organs and the way they reproduce (Nowak, 1999b; Vaughan, 1978). In males, the scrotum is anterior to the penis, which lacks an os priapi or baculum. In many marsupial species, females have a pouch (or marsupium, hence the common name of the group), which encloses and protects the nursing young. The gestation period is much shorter than that of placental mammals: in effect, the young are born as embryos and continue their development inside the pouch, which they reach by climbing the mother's fur, unassisted, with their precociously well-developed forelegs.

Marsupials vary widely in size (from about 5 g to 70 kg) and morphology. The smallest American marsupials are about the size of a mouse, and the largest are about as big as a cat. The largest Australian marsupials, however, are human-sized gray kangaroos. Some marsupials are insect eaters, whereas others are carnivores, nectar eaters, omnivores, or herbivores. Marsupial metabolic rates are lower than those of most like-sized placental mammals, and their body temperatures are more variable. Most American marsupials are solitary, are nocturnal, and live in trees. Most Mexican species have a prehensile tail and an opposable large toe on the hind foot, adaptations that allow them to move nimbly in trees. The water opossum (*Chironectes minimus*), however, has webbed hind feet that are used for propulsion through water (Emmons and Feer, 1997).

Some species such as the Virginia opossum have proliferated due to human activities. In Mexico, however, the water opossum has disappeared from a vast part of its original distribution and is currently endangered. The brown four-eyed opossum (*Metachirus nudicaudatus*) has a marginal distribution in Mexico, restricted to the Lacandon Forest, which makes it vulnerable to being extirpated from the country.

(*Opposite*) Mexican mouse opposum (*Tlacuatzin canescens*). Tropical dry forest. Chamela-Cuixmala Biosphere Reserve. Photo: Gerardo Ceballos

Family Didelphidae

Subfamily Didelphinae

The subfamily Didelphinae includes 4 genera and 12 species of large opossums, most of which live in South America. In Mexico there are 7 species, representing 6 genera (*Chironectes*, *Didelphis*, *Marmosa*, *Metachirus*, *Philander*, and *Tlacuatzin*).

Chironectes minimus (Zimmermann, 1780)

Water opossum

Rodrigo A. Medellín

SUBSPECIES IN MEXICO
Chironectes minimus argyrodytes Dickey, 1928

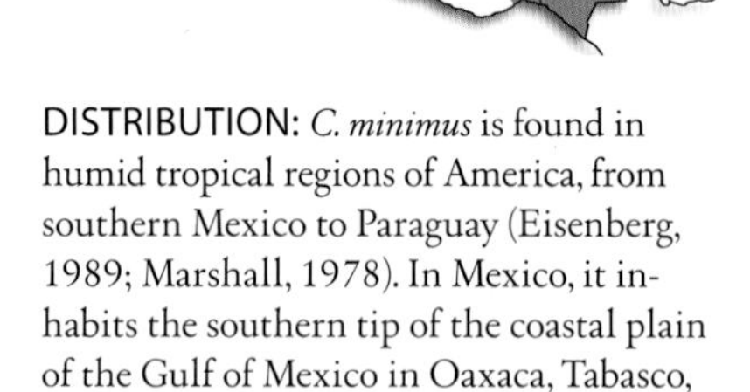

DISTRIBUTION: *C. minimus* is found in humid tropical regions of America, from southern Mexico to Paraguay (Eisenberg, 1989; Marshall, 1978). In Mexico, it inhabits the southern tip of the coastal plain of the Gulf of Mexico in Oaxaca, Tabasco, and Chiapas (Hall, 1981). It has been recorded in the states of CS, OX, and TB.

DESCRIPTION: *Chironectes minimus* is a medium-sized animal. The face is dark brown with a sallow stripe running from the top of the orbit to the base of the ear. The chin and the ventral part of the face are silvery white. The eyes are relatively large and the ears of moderate size. The hair is dense and thin. The back is dark brown, almost black, with three to four white bands running from the belly toward the back. The belly is pure white with silvery tones. The tail is covered with hair only on its base; only the last 10 cm are light-colored (Eisenberg, 1989; Marshall, 1978; Nowak, 1991). Its hind feet are webbed and are used for propulsion through water; the front legs have thin and globose fingers not linked by membranes and almost clawless. These features distinguish water opossums from other Neotropical marsupials. The female pouch opens from the back; it seals shut, with an oily substance and a strong sphincter, to prevent entrance of water (Marshall, 1978). The male has a pouch as well, but the sphincter is not sufficiently developed to close the bag. It provides the necessary protection for the testicles, which are withdrawn into the pouch when the animal is swimming (Marshall, 1978).

EXTERNAL MEASURES AND WEIGHT
TL = 645 to 745 mm; TV = 310 to 430 mm; HF = 55 to 73 mm; EAR = 22 to 31 mm.
Weight: 405 to 790 g.

DENTAL FORMULA: I 5/4, C 1/1, PM 3/3, F 4/4 = 50.

NATURAL HISTORY AND ECOLOGY: Water opossums are semi-aquatic animals, completely dependent on water bodies. They feed on crustaceans, mollusks, and fish caught overnight in the water (Eisenberg, 1989; Marshall, 1978; Nowak, 1991). Their dens are located on banks of streams under stones or tree roots. The entrance, found usually just above the water level, is a tunnel that leads to an underground chamber, where the animal builds a nest of leaves. Both sexes have pouches, but in the males they do not close completely. The female's pouch protects her offspring when she dives or swims. The offspring can survive several minutes at low oxygen levels, which allows the female to move freely in the water. Water opossums swim propelled by the hind legs, while the front legs are extended forward to touch and find their prey among rocks and other substrates at the bottom of streams and lakes. Home range varied in Brazil from 844 m to 3,388 m of river, with males having ranges up to 4 times larger than those of females (Galliez et al., 2009). The litter size varies from 3 to 5 offspring, with an average of 3.5. In Brazil births have been recorded in August, September, October (Galliez et al., 2009), December, and January (Collins, 1973). Juveniles were captured year round in the Mata Atlantica, however, suggesting that there is not a seasonal reproductive period (Galliez et al., 2009). Apparently, the offspring develop relatively early and rapidly. One individual lived 2 years and 11 months in captivity (Collins, 1973). Population densities recorded in the Mata Atlantica in Brazil are up to 1.34 individuals/km river (Galliez et al., 2009)

VEGETATIONAL ASSOCIATIONS AND ELEVATION RANGE: Water opossums inhabit tropical rainforests, montane rainforests, deciduous forests, and disturbed habitats, but are always associated with watercourses. They are found from sea level up to 1,860 m (Nowak, 1991).

CONSERVATION STATUS: *Chironectes minimus* is listed as endangered in Mexico (SEMARNAT, 2010). Almost two decades ago, Ceballos and Navarro (1991) included the water opossum in their list of endangered species because of its rarity and habitat destruction. In fact, little is known about this species. Apparently, it is rare throughout its distribution area and it appears to be susceptible to human disturbances. In Mexico the only recent records are from the Lacandona region in Chiapas, where they are protected by the Montes Azules Biosphere Reserve (Medellín, 1994).

Chironectes minimus. Tropical rainforest. Tuxtla Gutiérrez, Chiapas. Photo: Miguel Álvarez del Toro.

Didelphis marsupialis Linnaeus, 1758

Common opossum

Fernando Colchero, Georgina O'Farrill, and Rodrigo A. Medellín

SUBSPECIES IN MEXICO

Didelphis marsupialis caucae J.A. Allen, 1900

Didelphis marsupialis tabascensis J.A. Allen, 1901

DESCRIPTION: *Didelphis marsupialis* is approximately the size of a cat. The back is gray to black, with hair in two layers; one is dense, short, and pale yellow, and the other, long and black or gray (Emmons and Feer, 1997). The head is dark except on the bases of the whiskers, which are generally paler; the sides of the face are cream to orange. Black stripes come out of the nose, surround the eyes, and reach near the ears. The color of the belly is similar to that of the back but paler or even orange. The ears, the distal part of the legs, and the feet are black, as well as the base of the tail. The tail itself is totally naked. Females are generally smaller than males; they have a pouch with 13 nipples arranged in a circle (Nowak, 1991). The diploid chromosome number is 22 (Schneider, 1977). It can be confused with the Virginia opossum (*D. virginiana*), but the two species differ in the length of the dark area of the tail; in D. marsupialis it reaches more than half of the tail, while in *D. virginiana* it is shorter. The cheeks in *D. virginiana* are white, in stark contrast to the rest of the face, while in *D. marsupialis* the cheeks are cream in color (Gardner, 1973; McManus, 1974). The musk of *D. marsupialis* is brown, while that of *D. virginiana* is green (R. Medellín, pers. comm.). *D. marsupialis* does not simulate death (thanatosis) when at risk (Gardner, 1973).

Didelphis marsupialis. Photo: Gerardo Ceballos.

EXTERNAL MEASURES AND WEIGHT

TL = 324 to 425 mm; TV = 336 to 420 mm; HF = 51 to 70 mm; EAR = 46 to 58 mm. Weight: 565 to 1,610 g.

DENTAL FORMULA: I 5/4, C 1/1, PM 3/3, F 4/4 = 50.

NATURAL HISTORY AND ECOLOGY: Individuals from tropical forests are usually black (Emmons and Feer, 1997). Lighter-colored animals inhabit dry areas, although both colorations can be found in the same population. Common opossums are aggressive in stressful situations (Gardner, 1973). They are largely nocturnal, with a peak of activity between 19:00 hrs and 02:00 hrs (Sunquist et al., 1987), but they can also be active during the day (O'Connell, 1979). The young occupy tree substrates, while the adults, although being good climbers, are primarily terrestrial (Fonseca and Kierulff, 1988). They commonly nest in underground shelters, in hollow trees, under rocks, or under tree roots. Females use a nest much longer than males before moving to another refuge (Sunquist et al., 1987). Common opossums are solitary, antisocial, and aggressive (Nowak, 1999b). The average home range for males has been estimated as 11 ha during the dry season and 13 ha during the rainy season. Home range varies depending on the availability of resources, and overlaps extensively with that of other males and several females (Sunquist et al., 1987). The density is variable, but has been estimated as 0.09 to 1.32 individuals/ha. Higher densities are found in tropical rainforests (Fleming, 1972; Hunsaker, 1977; Malcolm, 1990; Medellín, 1992; O'Connell, 1979). Their abundance is higher during the wet season and the first part of the dry season, and is associated with richness and abundance of tree species. In the Lacandon Forest, common opossums represent 80% of the marsupials' biomass and 41% of all small mammals (Medellín, 1992). Males are polygenic (Sunquist and Eisenberg, 1993). Females are polyestric and have a life span of about 2.4 years (O'Connell, 1979). The mating season begins in January and extends until October (Fleming, 1973; Tyndale-Biscoe and Mackenzie, 1973). They usually have two litters a year, but there are reports of up to three reproductive peaks (Sunquist and Eisenberg, 1993). The average litter size is 7.1 to 8, and the number of offspring increases with latitude. Females in reproductive age range from 172 to 345 days. Litter size and proportion of sexes are correlated with food availability, favoring the nurturing of males under optimum conditions (Sunquist and Eisenberg, 1993). In the Lacandon Forest, the sex ratio is 1.71 males for each female (Medellín, 1992). Common opossums are generalists and feed mainly on birds, mammals, insects, fruit, grass, other plants, and garbage remains, although the highest proportion of their diet is small vertebrates (Fonseca and Kierulff, 1988; Medellín, 1991). They can prey on animals of similar size (Wilson, 1970). In the Lacandon Forest, the opossums prefer crustaceans and beetles, fruits of *Cecropia obtusifolia*, *Piper* sp., *Ficus* sp., hymenopterans, tadpoles, and mollusks (Clemente, 1994). During the dry season, the diet consists primarily of insects, whereas during the rainy season they are essentially vegetarians (Clemente, 1994). They are important seed dispersers, especially of *Cecropia* sp. (Medellín, 1994). They are intermediary hosts of *Besnoitia darlingi* and carriers of *Trypanosoma cruzi* and *Leishmania chagasi* (Carreira et al., 1996; Herrera and Urdaneta-Morales, 1992; Smith and Frenkel, 1984; Steindel et al., 1995; Travi et al., 1994). In Mexico they are the final host for *Gnathostoma* species, which caused gnathostomasis, an important emergent human disease (Diaz Camacho et al., 2010).

DISTRIBUTION: The common opossum is found throughout Central America, from northern and western Colombia, Ecuador, and east of northeastern Peru (Gardner, 1993). In Mexico, it is found in tropical regions from Tamaulipas to Chiapas and the Yucatan Peninsula. It has been recorded in the states of CA, CS, HG, OX, PU, SL, TA, TB, VE, and YU.

VEGETATIONAL ASSOCIATIONS AND ELEVATION RANGE: Common opossums inhabit tropical rainforests, dry habitats, gallery forests, secondary forests, and perturbed habitats. They are found below 1,350 masl, although they are most conspicuous around 700 m (Emmons and Feer, 1997; Gardner, 1973; Hunsaker, 1977; Medellín, 1992; Timm et al., 1989).

CONSERVATION STATUS: *D. marsupialis* is a common species, which is very adaptable to disturbed environments. It is hunted for its meat (Emmons and Feer, 1997), because it steals chickens and eggs from hen houses, and because some of its body parts are believed to have medicinal properties (Leopold, 1959).

Didelphis virginiana Kerr, 1792

Virginia opossum

Heliot Zarza and Rodrigo A. Medellín

SUBSPECIES IN MEXICO
Didelphis virginiana californica Bennett, 1833
Didelphis virginiana yucatanensis J.A. Allen, 1901

DISTRIBUTION: The Virginia opossum is distributed from the southeastern part of Canada and the eastern United States to northwestern Costa Rica (Gardner, 1973). In Mexico, is distributed throughout the country with the exception of the central part of the Central Plateau and the Baja California Peninsula (Ramírez-Pulido et al., 1986). In Baja California there is a single record from the city of Ensenada, Baja California Norte; it has not been not determined whether if it was an introduced individual or if it represents a local population (Mellink, 1998). *D. virginiana* is sympatric with *D. marsupialis* in southeastern Mexico and Central America (Hall, 1981). It has been recorded in the states of AG, CA, CH, CL, CO, CS, DU, GR, HG, JA, MI, MO, MX, NL, NY, OX, PU, QE, QR, SI, SL, SO, TA, TB, TL, VE, YU, and ZA.

DESCRIPTION: *Didelphis virginiana* is a relatively large-sized marsupial with a strong, robust body and a long, pointed face. *D. virginiana* can be confused with *D. marsupialis*, but differs from it because of the white cheeks, because of a shorter tail or one equal to the length of the body, and because the black portion of the tail is larger than the white portion (Emmons and Feer, 1997; McManus, 1974). The color of the musk in *D. virginiana* is green rather than brown ochre as in *D. marsupialis* (Medellín, 1992). The coloration of the dorsal portion of the body is gray or sallow (rarely dark), with pale guard hairs. The ventral part is white, cream, or yellow; the middle part of the basal tail, legs, and feet are black. The hair is long and rough. The ears are naked and black with a thin white line at the tip in the northern populations. The face is pale, with narrow black eye rings and a median pale facial line. The tail is prehensile, sharp, hairy in the base, and scaly in the rest (Ceballos and Galindo, 1984; Emmons and Feer, 1997; Reid, 1997). There is a strong geographic pattern of coloration between the subspecies of *D. virginiana.* The populations of the central and northern United States are characterized by a paler body color, while the southern populations (southwestern United States, Mexico, and Central America) have a darker coloration (Gardner, 1973).

EXTERNAL MEASURES AND WEIGHT

TL = 645 to 1017 mm; TV = 255 to 535 mm; HF = 48 to 80 mm; EAR = 45 to 60 m. Weight: 1,100 to 2,800 g.

DENTAL FORMULA: I 5/4, C 1/1, PM 3/3, F 4/4 = 50.

NATURAL HISTORY AND ECOLOGY: Virginia opossums have nocturnal, arboreal, and terrestrial habits. In the United States, the average population density is variable, with about 0.26 individual/ha, with an average home range of 20 ha (4.7 ha to 254 ha). In Mexico, an average population density of 1 individual/ha has been recorded (Ceballos and Galindo, 1984). *D. virginiana* is nomadic and stays on a site between six months and one year (Hunsaker and Shupe, 1977). They are not territorial animals, but can defend their space if warranted (McManus, 1974). Individuals reach sexual maturity between 6 and 8 months old. Females apparently only have

2 years of reproductive activity since very few opossums survive beyond the third year of life (Nowak, 1991). The reproductive season is long; there are two mating peaks a year. In North America, the first mating takes place in January and February and the second peak occurs in June and July. The gestation period is 12.5 to 13 days. The number of offspring per birth can be up to 21 (Ceballos and Galindo, 1984); however, the average number is 6.8 to 8.9 (McManus, 1974). The size of the litters is larger in the northern part of its range (8 to 9 offspring) than in the south (6 to 7 offspring). At birth, the offspring measure between 10 mm and 14 mm and weigh 0.13 g to 0.16 g. After birth, they migrate to the pouch, where they continue their development gripped to a nipple for about 50 days; thereafter they make brief incursions outside the pouch without leaving the mother. They are completely weaned and independent at the end of 3 to 5 months (Nowak, 1991). Opossums are omnivores and have a highly opportunistic feeding pattern. They consume insects, small vertebrates, carrion, and vegetable matter such as fruit and seasonal seeds (Emmons and Feer, 1997; Nowak, 1991). They can forage a distance of 1.6 km to 2.4 km in a night (Hunsaker, 1977). The period of increased activity is between 23:00 hrs and 02:00 hrs (Ceballos and Galindo, 1984). Northern populations accumulate fat during the autumn to withstand the severe winters, allowing the animal to remain idle in its den but not hibernating (Nowak 1991). Their shelters are usually found at ground level, between rocks, in tree holes, or in burrows used by other animals (Reid, 1997). A family or a female usually occupies one burrow. Opossums are not sociable, except during the breeding season (Gardner, 1973; McManus, 1974). They vocalize in combat or defense situations using whistles, shrieks, and grunts. Occasionally, Virginia opossums employ a passive defense tactic called thanatosis, remaining inert and pretending to be dead (McManus, 1974).

VEGETATIONAL ASSOCIATIONS AND ELEVATION RANGE: Virginia opossums inhabit a great variety of habitats, mainly lowlands and hillocks of deciduous forests near

Didelphis virginiana. Cloud forest. El Jabalí, Colima. Photo: Gerardo Ceballos.

creeks, estuaries, swamps, and marshes, as well as in regions of bush, farmland, and suburban areas (Ceballos and Miranda, 1986; Medellín, 1992). They are found from sea level up to 3,000 m (Reid, 1997).

CONSERVATION STATUS: *D. virginiana* is a common species and locally abundant in some regions (Reid, 1997).

Marmosa mexicana Merriam, 1897

Mexican mouse opossum

Rodrigo A. Medellín

SUBSPECIES IN MEXICO

Marmosa mexicana mayensis Osgood, 1913
Marmosa mexicana mexicana Merriam, 1897
Gutiérrez et al. (2010) suggested that *M. mexicana* from the lowlands of Mexico, Belize, and Guatemala represent a different species, probably *M. mayensis*. But they proposed to follow that change until more data are available.

DESCRIPTION: *Marmosa mexicana* is a small-sized marsupial similar to *Tlacuatzin canescens*. The sides of the face, the inside of the limbs, and the belly are whitish-yellow to light brown; a black ring of variable extension surrounds the eyes. The back is cinnamon or brownish-red, with tones ranging from intense to moderate. The tail is almost completely hairless and prehensile, and the coloration is lighter on the ventral side than on the dorsal side. The ears are large and thin. Females have 11 to 15 nipples and no functional pouch. This species can coexist in some regions of Mexico

Marmosa mexicana. Tropical rain forest. Photo: Gerardo Ceballos.

with *Tlacuatzin canescens*, and possibly near the border with Guatemala with *M. alstoni*. *M. mexicana* is different from *Tlacuatzin canescens* because its color is brownish-red and not gray. Furthermore, *Tlacuatzin canescens* has supraorbital, prominent ridges, fenestrae between the second upper molar, and normal palatal fenestrae. *M. mexicana* can be separated from *M. alstoni*, which is larger (total length > 400 mm).

DISTRIBUTION: *M. mexicana* inhabits humid tropical regions of Central and northern America, from Mexico to Panama. In Mexico, it is distributed in the coastal plain of the Gulf of Mexico and from Tamaulipas to Yucatan and at the southern tip of the Pacific slope from Oaxaca (Alonso-Mejía and Medellín, 1992; Cervantes et al., 2002; Hall, 1981; Vargas Contreras and Hernandez Huerta, 2001). It has been recorded in the states of CS, HG, OX, SL, TA, TB, VE, and YU.

EXTERNAL MEASURES AND WEIGHT
TL = 260 to 386 mm; TV = 140 to 205 mm; HF = 16 to 22 mm; EAR = 18 to 27 mm. Weight: 29 to 92 g.

DENTAL FORMULA: I 5/4, C 1/1, PM 3/3, F 4/4 = 50.

NATURAL HISTORY AND ECOLOGY: Mexican mouse opossums are nocturnal animals and live almost totally in trees; they rarely fall to the ground. They feed mainly on medium-sized and large insects, but they also eat fruit, eggs, and small vertebrates (Alonso-Mejía and Medellín, 1992; Hall and Dalquest, 1963). Their predators include owls and barn owls (*Tyto alba* and *Ciccaba virgata*), as well as rattlesnakes (Alonso-Mejía and Medellín, 1992). They use the interior of hollow trees, burrows in the ground, or birds' nests as refuges (Alonso-Mejía and Medellín, 1992; Hall and Dalquest, 1963). The nests are lined with dry leaves or feathers. The cubs are transported on their mother's back or grip onto their mother's abdomen. In captivity, a female raised a litter of 11 pups; litters range from 2 to 13 pups (Gewalt, 1968). It appears to reproduce during the first half of the year, between March and June (Alonso-Mejía and Medellín, 1992).

VEGETATIONAL ASSOCIATIONS AND ELEVATION RANGE: The Mexican mouse opossum inhabits tropical rainforests and deciduous forests, crops, and disturbed vegetation (Alonso-Mejía and Medellín, 1992; Espinosa-Medinilla et al., 2003, 2004; Vargas Contreras and Hernandez Huerta, 2001). It is found from sea level up to 1,800 m (Alonso-Mejía and Medellín, 1992). There is an isolated report, probably an inaccurate one, that falls outside this pattern, to 3,000 m in the Volcano Tacaná (Villa-R., 1948).

CONSERVATION STATUS: *M. mexicana* is an abundant species that can survive in perturbed habitats.

Metachirus nudicaudatus (È. Geoffroy St.-Hilaire, 1803)

Brown four-eyed opossum

Rodrigo A. Medellín

SUBSPECIES IN MEXICO
Metachirus nudicaudatus dentaneus Goldman, 1912

DESCRIPTION: Brown four-eyed opossums are medium-sized marsupials. They are brown with relatively short, wooly hair. The tail is long and thin. Although the base of the tail is dark and the tip whitish, there is no clear division between them. Unlike other species of the same family, they have no pouch. The legs and face are more slender than those of Philander opossum.

EXTERNAL MEASURES AND WEIGHT
TL = 530 to 600 mm; TV = 325 to 335 mm; HF = 39 to 52 mm; EAR = 30 to 33 mm. Weight: 300 to 450 g.

DENTAL FORMULA: I 5/4, C 1/1, PM 3/3, F 4/4 = 50.

NATURAL HISTORY AND ECOLOGY: Brown four-eyed opossums are omnivorous, and feed on insects, mollusks, fruits, eggs, and small vertebrates (Eisenberg, 1989; Hunsaker, 1977). They are generally terrestrial because their limbs are not well suited for climbing. Their nests are found in hollows between rocks or in burrows in the ground. Little is known about their reproductive cycle; they may have one to nine offspring. Apparently, it is a polyestric species. The two Mexican individuals were nursing females captured during the months of August and September (Medellín et al., 1992).

DISTRIBUTION: *M. nudicaudatus* occupies a large area from southern Mexico to Argentina (Eisenberg, 1989; Medellín et al., 1992; Redford and Eisenberg, 1992). It has been recorded in the state of CS.

VEGETATIONAL ASSOCIATIONS AND ELEVATION RANGE: The only two known individuals in Mexico were caught in a tropical rainforest. In other regions they have been found in scrublands and more open areas (Handley, 1976). It occupies tropical low-lying areas from sea level up to approximately 600 masl. The Mexican locality is located 120 masl.

CONSERVATION STATUS: *Metachirus nudicaudatus* is considered threatened in Mexico (SEMARNAT, 2010). The only known population in Mexican territory is located in an area currently facing a high rate of destruction.

Philander opossum (Linnaeus, 1758)

Four-eyed opossum

Iván Castro-Arellano and Rodrigo A. Medellín

SUBSPECIES IN MEXICO
Philander pallidus J.A. Allen opossum, 1901

DESCRIPTION: Philander opossum is a small-sized marsupial. The color of the body varies from pale gray to black; the face has darker hair in the form of a mask and a pair of white spots on each eye. All ventral portions, including the bottom of the cheeks, are whitish. The base of the tail (the first 50 mm to 80 mm) is covered by hair matching the color of that of the body; the remaining part is naked and bicolored, whitish in the distal region and dark in the proximal region. The pouch of females is stained with orange color if they have offspring (Emmons and Feer, 1997). The hair is relatively short. The limbs are quite robust, as well as the overall construction of the body. *Metachirus nudicaudatus* has a similar pattern of coloration, but the supraocular stains are cream color and the rest of the body is brown; females lack a pouch so it is easy to distinguish them from *P. opossum* (Eisenberg, 1989). In addition, *Metachirus nudicaudatus* is a more delicate animal, with a long tail (over 330 mm), slender limbs, and a long, sharp face (Eisenberg, 1989).

EXTERNAL MEASURES AND WEIGHT
TL = 489 to 610 mm; TV = 253 to 329 mm; HF = 35 to 50 mm; EAR = 33 to 41 mm.
Weight: 200 to 600 g.

DENTAL FORMULA: I 5/4, C 1/1, PM 3/3, F 4/4 = 50.

NATURAL HISTORY AND ECOLOGY: The four-eyed opossum is a nocturnal and terrestrial marsupial. It is the more agile member of the genus Philander; it has the ability to climb trees and to swim gracefully; it occupies the low and middle stratum of the forest (Emmons and Feer, 1997). They are usually solitary animals, although several animals have been seen separated by few tens of meters. In the south of the Lacandon Forest, Chiapas, a density of 0.48 individual/ha was found (Medellín, 1992). Other data of the density, reported 0.17 individual/ha in a primary forest in French Guyana (Julian-Laferrière, 1991), and 1.37 individuals/ha in a secondary forest in the same country (Atramentowicz, 1986). The maximum longevity observed in captive individuals is 2 years and 4 months (Collins, 1973 in Medellín, 1992). In a study of its population ecology in Mexico, the average movement between captures was 47 m; the largest movement recorded was an adult male who traveled 214 m (Medellín, 1992). This species is more common where water bodies occur and where dense vegetation covers the lower stratum, although it is not restricted to this type of habitat. The nests are usually located 8 to 10 m high, in cavities of trees or branches, and occasionally on the ground in hollow or fallen logs; some specimens have been found in nests built on palm roofs of abandoned houses (Hall and Dalquest, 1963). The four-eyed opossum is omnivorous, feeds on fruit and nectar, and is an active predator. As part of their diet they include fruits of Manilkara sapota (Hall and Dalquest, 1963), crabs, crustaceans, frogs, and carrion (Davis, 1944), and corn (Kuns and Tashian, 1954). In the Montes Azules Biosphere Reserve, Chiapas, the highest percentage of food components in stool samples, in descending order, were the family Scarabaeidae, crustaceans, seeds of *Cecropia obtusifolia*, *Piper* sp., cockroaches of the

Philander opossum. Tropical rainforest. Los Tuxtlas, Veracruz. Photo: Gerardo Ceballos.

DISTRIBUTION: Four-eyed opossums are found from Mexico to northern Argentina (Redford and Eisenberg, 1992). In Mexico, it is found from Tamaulipas throughout the Gulf coastal plain to northern and eastern Oaxaca, covering the Isthmus of Tehuantepec to Chiapas and the Yucatan Peninsula (Hall, 1981). It has been recorded in the states of CA, CS, OX, PU, QR, TA, TB, VE, and YU.

family Blattidae, hymenopterans, beetles of the families Passalidae and Gyrinidae, as well as seeds of *Ficus* sp. (Clemente, 1994). In French Guyana the diet is similar (Charles-Dominique et al., 1981). In the Lacandon Forest these animals disperse many seeds such as *Cecropia obtusifolia*. Along with *Didelphis marsupialis*, they provide higher-quality dispersion than that of other arboreal frugivores such as monkeys (*Ateles*), kinkajou (*Potos*), and ringtail cats (*Bassariscus*; Medellín, 1994). Reproduction probably takes place all year (Hall and Dalquest, 1963). There are reports of Mexican females with pups in March, April, June, and October. The litters vary from four to seven offspring (Davis, 1944; Jones et al., 1974b). The activity areas of adults overlap; the size depends on the availability of resources and there is no defense of the territory. The contact between adults is only to mate.

VEGETATIONAL ASSOCIATIONS AND ELEVATION RANGE: The four-eyed opossum mainly inhabits tropical rainforests and montane rainforests. It is commonly found from sea level up to 1,650 m (Hall and Dalquest, 1963).

CONSERVATION STATUS: *P. opossum* is a common species (Álvarez, 1963a; Davis, 1944; Kuns and Tashian, 1954; Villa-R., 1948), although Jones et al. (1974b) mentioned it as rare for the state of Yucatan. They are not threatened with extinction.

Tlacuatzin canescens (J.A. Allen, 1893)

Gray mouse opossum

Gerardo Ceballos

SUBSPECIES IN MEXICO

Tlacuatzin canescens canescens (J.A. Allen, 1893)
Tlacuatzin canescens insularis (Merriam, 1898)
Tlacuatzin canescens oaxacae (Merriam, 1897)
Tlacuatzin canescens sinaloae (J.A. Allen, 1898)
This species was considered part of the genus *Marmosa*; however, in a recent study based on morphological characteristics and genetic markers, Voss and Jansa (2003) propose to upgrade it to a generic level.

DESCRIPTION: The gray mouse opossum is a small-sized marsupial similar to a mouse, with big, dark eyes. The face is long. The opossum's conspicuous ears lack hair and are very thin. The tail, which is longer than the head and body, has only one color; it is prehensile and hairless. The thumb is opposable and has no claw. Unlike other marsupials, the female lacks a pouch and has up to 15 nipples (Ceballos and Miranda, 1986, 2000). The color of the back varies from brownish-yellow to reddish-brown. The belly is lighter colored, of yellow and white tones. It has a dark eye ring. The tail and feet are light brown and pink (Reid, 1997).

EXTERNAL MEASURES AND WEIGHT

TL = 260 to 310 mm; TV = 90 to 154 mm; HF = 16 to 22 mm; EAR = 18 to 30 mm. Weight: 20 to 60 g.

DENTAL FORMULA: I 5/4, C 1/1, PM 3/3, F 4/4 = 50.

Tlacuatzin canescens. Tropical dry forest. Chamela-Cuixmala Biosphere Reserve, Jalisco. Photo: Gerardo Ceballos.

NATURAL HISTORY AND ECOLOGY: *Tlacuatzin canescens* is a solitary, semi-arboreal, and primarily nocturnal species (Ceballos and Miranda, 1986). They build their dens in hollow trees, in abandoned bird nests such as that of the white-bellied wren (*Uropsila leucogastra*), and among the foliage of trees and shrubs such as iguanero (*Caesalpinia eriostachys*), orange flower (*Jaquinia pungens*), opuntia cactus (*Opuntia excelsa*), and algarroba (*Prosopis juliflora*; Ceballos, 1989, 1990). The nests are constructed with dry leaves; fine vegetable matter such as lichens and grasses; and feathers. In the tropical deciduous forest or lowland of the Chamela-Cuixmala Biosphere Reserve, the population density varies from 0.4 to 4.5 individuals/ha (Ceballos, 1990). They feed primarily on insects such as bugs (Hemiptera), cockroaches and praying mantises (Orthoptera), scarabiid, bruchid, and cerambycid beetles (Coleoptera). Occasionally, they prey on geckos, bird eggs, and chicks (Zarza et al., 2003). They have been observed feeding on nectar, and probably pollinating columnar cacti (Ibarra Cerdeña et al., 2007). They are preyed on by a wide variety of animals, including foxes, coatis, owls, snakes, and barn owls (Lopez-Forment and Urbano, 1977). They apparently reproduce throughout the year. In the Chamela-Cuixmala Biosphere Reserve, pregnant females have been found between July and September, and young individuals between February and July. The copula has been described as follows:

> *The male approached the female so that they were face to face. The male either passed food to the female or simply touched her snout; this could not be determined because of the vegetation between animals and observers. Both opossums started to make loud noises (resembling suction with saliva) for approximately 3 minutes. Then they went to a nearby branch, 4 cm in diameter, which was 1.8 m above the ground. They suspended themselves upside down by wrapping their tails around the branch. The tail was their only support, and there*

DISTRIBUTION: The gray mouse opossum is endemic to Mexico, where it is distributed along the tropical lowlands of the Pacific from Sinaloa to Chiapas, penetrating to central Mexico through the Balsas River Basin up to Puebla. There is a population in the Yucatan Peninsula, probably separated from the populations in the Pacific coast (Ceballos and Miranda, 1986; Hernández-Cardona et al., 2007; Vargas-Contreras et al., 2004; Voss and Jansa, 2003). It has been recorded in the states of CL, CS, DU, GR, JA, MI, MO, MX, NY, OX, PU, SI, SO, YU, and ZA.

was nothing between them and the ground. The male grabbed the female from the back and wrapped his forelimbs around her shoulders, secured her neck with a prolonged neck-bite that extended throughout copulation and used his legs to force the female to open her posterior limbs. The male introduced his penis three times, with an approximate duration of two minutes each time. (Valtierra Azotla and Garcia, 1998:146)

The litters can have up to 14 offspring that are born in an embryonic state; once they are weaned they move gripped to the back of their mother (Ceballos, 1989, 1990).

VEGETATIONAL ASSOCIATIONS AND ELEVATION RANGE: The gray mouse opossum lives in deciduous and montane rainforests, thorny forests, xeric shrubland, mangrove forests, crops, and disturbed vegetation (Ceballos and Miranda, 2000). It is found from sea level up to 2,300 m, although most of the records have been made in regions below 1,000 m (Zarza et al., 2003).

CONSERVATION STATUS: The gray mouse opossum is an abundant species, which survives even in severely disturbed environments such as crops. It is not at risk of extinction (Zarza et al., 2003).

Subfamily Caluromyinae

The subfamily Caluromyinae includes two genera and four species, one of which is distributed in Mexico. They inhabit tropical rainforests, montane rainforests, and secondary vegetation of the Gulf and Pacific coast of Oaxaca and Chiapas. The other species have a more equatorial distribution in South America.

Caluromys derbianus (Waterhouse, 1841)

Woolly opossum

Rodrigo A. Medellín

SUBSPECIES IN MEXICO

Caluromys derbianus aztecus (Thomas, 1913)
Caluromys derbianus fervidus (Thomas, 1913)

DESCRIPTION: *Caluromys derbianus* is a medium-sized marsupial, smaller than a domestic cat. The face is gray with a dark stripe between the eyes running from the crown to the nose. The ears are prominent and rounded, thin, and pink or purple. The hair is long, dense, and thin. The front of the back is reddish-brown or cinnamon with lead-colored areas between the shoulders and into the limbs. The lower parts are white. The tail is covered with grayish-brown hair until, at least, half its length, ending in white with brown, amorphous spots. Both the front and hind legs are prehensile, with thin fingers enlarged at the tip. It can be distinguished from other marsupials living in Mexico by the dense, thin hair, the pattern of coloration, the tail hair that covers the middle portion, and the brown stains from the end of the tail.

EXTERNAL MEASURES AND WEIGHT
TL = 587 to 760 mm; TV = 384 to 490 mm; HF = 30 to 48 mm; EAR = 35 to 40 mm. Weight: 245 to 370 g.

DENTAL FORMULA: I 5/4, C 1/1, PM 3/3, F 4/4 = 50.

NATURAL HISTORY AND ECOLOGY: Woolly opossums are almost entirely arboreal. They feed mainly on fruit, but also eat insects and small vertebrates (Bucher and Hoffman, 1980; Eisenberg, 1989). Apparently, they are completely nocturnal. The reproductive period covers several months of the dry season and the beginning of the rainy season. They can give birth from 1 to 6 offspring, with an average of 3.3 per birth, which suffer high mortality before being weaned (Bucher and Hoffman, 1980). They reach sexual maturity between 7 and 9 months old and, in captivity, can live more than 5 years (Bucher and Fritz, 1977).

DISTRIBUTION: The woolly opossum inhabits tropical humid regions from Mexico to Colombia and Ecuador (Eisenberg, 1989). In Mexico, it occupies the southern part of the coastal plain of the Gulf of Mexico from Veracruz and southern Oaxaca and the southern tip of the Yucatan Peninsula (Hall, 1981). It has been recorded in the states of CA, CS, OX, TB, and VE.

VEGETATIONAL ASSOCIATIONS AND ELEVATION RANGE: The woolly opossum inhabits forests in tropical evergreen and, at medium altitudes, deciduous forests, and habitats altered by human activities (Handley, 1976). It is found from sea level up to 2,500 m (Bucher and Hoffman, 1980; Emmons and Feer, 1997).

CONSERVATION STATUS: *C. derbianus* is considered threatened mainly because of the destruction of its habitat (SEMARNAT, 2010). Given the scarcity of Mexican specimens, their conservation status is not well determined. It is necessary to obtain information on the effects of habitat destruction on this species.

Caluromys derbianus. Tropical rainforest. Los Tuxtlas, Veracruz. Photo: Gerardo Ceballos.

American manatee (*Trichechus manatus*). Gulf of Mexico. Photo: Francois Gohier.

Order Sirenia

Gerardo Ceballos

> *Because of the frequent hunt for their meat, bone and skin, manatees have diminished significantly in recent years so that their numbers, in many localities [of the Yucatan Peninsula] where they previously abounded, were no longer found in that year (1917).*
>
> — G.M. Allen, 1942

The order Sirenia is a small, diverse order with two families and four species of aquatic habits (Wilson and Reeder, 2005). The family Dugongidae is represented only by the dugon (*Dugong dugon*), which lives in the tropical seas of Asia, Australia, and the Pacific (Nowak, 1999). Steller's sea cow (*Hydrodamalis gigas*), another species of this family and the largest in the order, lived in the cold seas of the Bering Sea, where it disappeared around 1768 (Allen, 1942). The family Trichechidae has an interesting distribution because it is found in West Africa and the east of the Americas, from Florida to Brazil.

The sirenians are aquatic mammals of voluminous bodies that weigh up to 1,000 kg. Their extremities are modified into fins; the hind limbs are fused into a single fin used as a locomotion system (Nowak, 1991). The body is covered with little hair. They have nictitating membranes, which protect their eyes when opened under water. The number of teeth varies, with few in dugones and many in manatees. They are herbivores and feed on submerged plants such as sea grass or plants suspended in the water surface such as the aquatic lily (Ripple and Perrine, 1999). They can be found solitary or forming small groups or couples. They are gentle, nonaggressive, and generally slow.

The three species of sirenians are considered endangered or threatened with extinction due to hunting, destruction of their habitat, and problems with outboard engines. For example, in Florida, the largest source of mortality of manatees is collisions with boats with outboard engines. The manatees and the dugong are hunted throughout their range of distribution.

Family Trichechidae

The family Trichechidae is represented in Mexico by the manatee (*Trichechus manatus*), which inhabits rivers, coastal lagoons, and seas of the coasts of the Gulf of Mexico and the Caribbean. It is a gentle and very interesting species that is considered at risk of extinction by the destruction of its habitat, pollution, and poaching.

Trichechus manatus Linnaeus, 1758

West Indian manatee

Benjamín Morales Vela and Luz del Carmen Colmenero Rolón

SUBSPECIES IN MEXICO
Trichechus manatus manatus (Linnaeus, 1758)

DESCRIPTION: Manatees have a fusiform body that is robust and circular in a transverse cut. They have a dorsal-ventrally flattened tail that is circular in shape. They lack hind limbs and the anterior limbs are flippers. The nipples are located in the armpits. Unique characteristics of the species are the six cervical vertebrae, pachyostosis, and a constant replacement of the molars. The head is short with a bloated face with wide, fleshy, and highly movable lips. On the tip of the nose it has two semi-circular nostrils that are closed when the animal is under water. In general, the body lacks hair but the lips have small and rigid whiskers. Their skin is thick, rough, and covered by barnacles, algae, and other epibionts. The coloration is a gray brown tonality, more obscure in the offspring. There is no apparent sexual dimorphism, but females reach a greater weight when they reach sexual maturity (Quiring and Harlan, 1953; Reeves et al., 1992).

EXTERNAL MEASURES AND WEIGHT
TL= 3.7 to 4.6 m. Weight: 800 to 1,500 kg.

DENTAL FORMULA: I 1/1, M 5–7/5–7 = 24–32.

NATURAL HISTORY AND ECOLOGY: The basic habitat of *Trichechus manatus* requires a temperature greater than 15°C, a depth of 1 m to 1.5 m, and abundant edible vegetation (Hartman, 1979). It undertakes seasonal movements in Tabasco (Arriaga Weiss and Contreras Sánchez, 1993; Colmenero, 1984; Irvine, 1983). In freshwater environments they feed on a wide variety of submerged vascular plants and/or float-

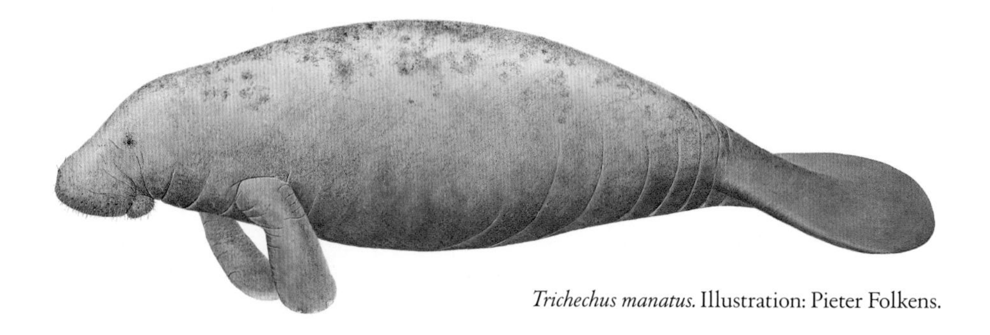

Trichechus manatus. Illustration: Pieter Folkens.

ing plants such as the camalote (*Paspalum*), grasses (*Chloris*), wild celery (*Vallisneria americana, Cabomba palaeofirmis, Ceratophyllum demersum*), and aquatic lily (*Eichornia crassipes*). In brackish environments they consume sea grass such as *Thalassia testudinum, Halodule wrightii*, and *Siringodium filiforme* as well as green and brown algae associated with the epiphytic fauna of the plants they consume (Colmenero, 1984; Colmenero and Hoz, 1986). An adult animal requires the equivalent of 8% to 15% of its body weight in daily food (Best, 1979). They do not show a very defined seasonality in their reproduction. In Tabasco, the breeding season coincides with the rainy season (Colmenero, 1984). Gestation lasts 12 to 14 months. Calves are born measuring close to 1 m and weighing around 30 kg. Nursing lasts 12 to 18 months and pups are weaned when they measure more than 2 m. Females reproduce every two to five years, with an average of three. Manatees reach sexual maturity at five years of age, but seem to reproduce after seven years of age in the case of females and after nine years in the case of males (Caldwell and Caldwell, 1985; Dekker, 1974; Reeves et al., 1992). Crocodiles and sharks can prey on them. The main health issues include skin infections, parasites, and injuries (Beck and Forrester, 1988; Buergelt et al., 1984; Caldwell and Caldwell, 1985; Irvine et al., 1980). A study on their genetic variation showed that there is a strong branch population, which is divided into two clades separated by the Caribbean Sea. The latter is inconsistent with the current division, which only considers the populations of Florida as distinct (*T. manatus latirostris*; García-Rodríguez et al., 1998). In Mexico, gene flow between manatees from Florida, the West Indies, and Central America has occurred. In addition, the genetic variation shows evidence of a drastic recent reduction of the population and fragmentation. It is estimated that the original population was made up of several hundreds of thousands of individuals (Morales Vela and Medrano González, 1997).

CONSERVATION STATUS: The manatees were used as a source of food since pre-Columbian times (Bradley, 1983; McKillop, 1985). After the Conquest, they were intensively hunted, increasing the trade of the products obtained from these animals and making them usual victual of sailing ships of the region (Reeves et al., 1992). This exploitation continued until the nineteenth century, and only the decline of its populations slowed their capture. Since the nineteenth century, the greater threats they face are pollution and destruction of their habitat. In Mexico, the abundance of manatees has been reduced particularly in the past 30 years as a result of poaching and the destruction of their habitat (Arriaga and Contreras, 1993; Colmenero, 1991). The most alarming situation in Florida is the affinity of manatees to congregate in places of industrial thermal discharge during winter, as well as the damage caused by boats. In Florida, a slightly negative population growth rate was estimated, and if the mortality rate increases 10%, the species could become extinct in that region (Marmontel et al., 1997). Events of massive mortality, which can mean fluctuations in the mortality rate with high risk of sudden extinction, have been observed on several occasions possibly in association with immunological deficiencies (IUCN, 2010). In Mexico there is no precise data on abundance; however, in 1976 a total of approximately 5,000 individuals was estimated (Heinsohn et al., 1976). It has been estimated that several hundred exist in Tabasco (Arriaga and Contreras, 1993; Colmenero and Hoz, 1986; Valencia, 1988) and around 100 in Quintana Roo (Colmenero and Zárate, 1990). Morales-Vela and Olivera-Gómez (1995) have counted 264 manatees in censuses in the Chetumal Bay and along the coast of Belize. This abundance appears to increase annually but whether this corresponds to an absolute increment of the population is unclear. Throughout its geographical distribution the manatee is a species in danger of extinction. The International Union for Conservation of Nature (IUCN) catalogues it as vulnerable. It is found in appendix I of the Convention on International Trade in Endangered Species (CITES), and in Mexico it is considered at risk of extinction (SEMARNAT, 2010).

DISTRIBUTION: The manatee inhabits freshwater environments (rivers, streams, and lagoons), brackish water (lagoons and coastal cenotes), and marine water (coves and bays). Their main habitats are the major rivers and lagoons adjacent to the sea. This species is distributed along the East Coast of the United States of America and the coasts of the Gulf of Mexico, Central America, the Antilles and South America, to Brazil. Due to their low metabolism, they show affinity to warm waters; during the summer they can be seen as far north as New York and in Florida they tend to associate in waters used as waste disposals (Bertram and Bertram, 1973; Caldwell and Caldwell, 1985; Colmenero, 1984, 1991; Heinsohn, 1976; Reeves et al., 1992). In Tamaulipas they have been found in the Pánuco River and adjacent lagoons (Lazcano and Packard, 1989). In Veracruz, it inhabits lagoons and streams of the Papaloapan Basin River (Colmenero, 1984; Colmenero and Hoz, 1986). In Tabasco and Campeche, it has been recorded in the major rivers, streams, and continental lagoons (Álvarez et al., 1988; Arriaga and Contreras, 1993; Colmenero, 1984; Colmenero and Hoz, 1986; Valencia, 1988). In Yucatan, it is found in coastal environments close to Progreso, and in Quintana Roo it is found in the bays, coves, and cenotes (Colmenero and Hoz, 1986; Colmenero and Zárate, 1990; Gallo, 1983; Zárate, 1993). In Chiapas it has been reported in the lagoons of Catazajá (Colmenero, 1984). It has been registered in the states of CA, CS, QR, TA, TB, VE, and YU and in the areas of the GM and CS.

Order Cingulata

Gerardo Ceballos

Armadillos, as it is well known, are intensively hunted because of its tasty meat, and if they are still abundant it is because of its secretive and nocturnal habits.

— Miguel Álvarez del Toro, 1977

The order Cingulata is exclusive to the American continent, and the greatest diversity of species is found in South America (Nowak, 1999b). It was considered part of the order Edentata, whose use was discontinued years ago (Wilson and Reeder, 1993). More recently, it was considered with the order Xenarthra with the species that now belong to the order Pilosa (Wilson and Reeder, 2005). The order includes 20 species belonging to a single family, Dasypodidae. The order is distributed mainly in the tropics, but one species, the nine-banded armadillo (*Dasypus novemcinctus*), is now widely distributed throughout the continent, from the southern United States to Argentina (Reid, 1997).

Homodont teeth characterize the order; all the species lack incisors and canines. Armadillos are terrestrial and specialize in digging; they have strong claws on two or three fingers of the forefeet. Their body is covered with a shell that varies in form between species. They are active during the day or night, and their diet is omnivorous. They can live alone or in groups (Layne and Glover, 1977). There are only two species in Mexico: the common armadillo (*Dasypus novemcinctus*), which is abundant and widely distributed in the country, and the naked-tailed armadillo (*Cabassous centralis*), a species with a marginal distribution known at few localities in the Lacandon Forest and recently recorded in Mexico and northern Guatemala (Cuarón et al., 1989).

(*Opposite*) Nine-banded armadillo (*Dasypus novemcinctus*). Tropical dry forest. La Parota, Guerrero. Photo: Gerardo Ceballos.

Family Dasypodidae

The family Dasypodidae is represented by 9 genera and 20 species distributed from the southern United States to Argentina. In Mexico only two species can be found (*Cabassous centralis* and *Dasypus novemcinctus*). The other species are from South America. The greatest diversity of species is found in Argentina and Bolivia.

Subfamily Dasypodinae

The subfamily Dasypodinae contains one genus with eight species. In Mexico there is one species (*Dasypus novemcinctus*).

Dasypus novemcinctus Linnaeus, 1758

Nine-banded armadillo

Angeles Mendoza Durán and Gerardo Ceballos

SUBSPECIES IN MEXICO

Dasypus novemcinctus mexicanus Peters, 1864
Dasypus novemcinctus davisi Russell, 1953

DESCRIPTION: Nine-banded armadillos are medium-sized mammals. They are the only mammals on the American continent whose body is covered with ossified dermal plates on the back, sides, tail, and top of the head, forming a turtle-like carapace. The carapace can have from 7 to 11 moveable transversal bands in the dorsal half. Their limbs have big claws that enable them to dig; the forelimbs have four fingers and the vestige of a fifth. They have no incisors or canines (Hall, 1981; McBee and Baker, 1982).

DISTRIBUTION: This species is found from the southern United States to southern Argentina. In Mexico it is present in most of the territory except in the Baja California Peninsula and the Central Altiplane (Hall, 1981; Wilson and Reeder, 2005). It has been registered in the states of CH, CL, CO, CS, DU, GR, HG, JA, MI, MO, MX, NL, NY, OX, PU, QE, QR, SI, SL, SO, TA, TB, TL, YU, and ZA.

EXTERNAL MEASURES AND WEIGHT

TL = 615 to 800 mm; TV = 245 to 370 mm; HF = 75 to 100 mm; EAR = 37 to 51 mm. Weight: up to 10 kg.

DENTAL FORMULA: 7–9/7–9 = 28–36.

NATURAL HISTORY AND ECOLOGY: Nine-banded armadillos are nocturnal or crepuscular. They construct their burrows underground, with nests of leaves and grasses. The mating season takes place in autumn. Gestation lasts 120 days and the average number of offspring is 4; these are formed from a single zygote, so they are genetically identical. They may have 1 or 2 litters a year and live up to 15 years (Buchanan, 1957; Chapman and Feldhamer, 1982; Olin and Thompson, 1982). In Florida their home range varies from 1.1 ha to 13.8 ha, and their movements are generally short (0.5 km/day). They are insectivorous, as their diet consists of small invertebrates, but they can also consume amphibians, reptiles, eggs, and carrion. Armadillos are an important prey for many species of predators, including jaguars, pumas, and coyotes (Aranda et al., 1995; Ceballos et al., 2007; Emmons and Feer, 1997). They have the habit of digging in the ground to catch food (Ceballos and Galindo, 1984; Layne and Glover, 1977). They are common in areas with soils

Dasypus novemcinctus. Tropical dry forest. Infiernillo, Michoacán. Photo: Gerardo Ceballos.

of sand and clay, and their abundance in a region may vary depending on the type of soil. They are hunted for their meat or for their shells, which are used in handicraft projects. Indeed, they are one of the most widely hunted mammals in rural communities in Mexico (Ceballos & Galindo, 1981; González-Bocanegra et al., 2011). Armadillos are used in folk medicine (Enríquez Vázquez et al., 2006). They are also valuable for use in medical research due to the fact that their immune system response is similar to that of the human (McBee and Baker, 1982).

VEGETATIONAL ASSOCIATIONS AND ELEVATION RANGE: Nine-banded armadillos are common in several types of vegetation such as grasslands, xeric scrublands, thorny forests, coniferous and oak forests, tropical rainforests, montane rainforests, deciduous forests, and temperate cloud forests (Ceballos and Galindo, 1984; Olin and Thompson, 1982). They live from sea level up to 3,000 m (Nowak and Paradiso, 1983).

CONSERVATION STATUS: *Dasypus novemcinctus* is a common species that can live in disturbed environments. It is not at risk of extinction.

Subfamily Tolypeutinae

The subfamily Tolypeutinae contains three extant genera with eight species. In Mexico there is one species (*Cabassous centralis*).

Cabassous centralis (Miller, 1899)

Foxtail armadillo

Alfredo D. Cuarón

SUBSPECIES IN MEXICO
C. centralis is a monotypic species.

DESCRIPTION: The foxtail armadillo has a wide head and a proportionately wide and short snout. The tail is short and lacks epidermal plates. The carapace has 12 or 13 flexible bands. The front legs have five toes with large claws, especially the third one.

EXTERNAL MEASURES AND WEIGHT
TL = 305 to 403 mm; TV = 130 to 183 mm; HF = 60 to 74 mm; EAR = 31 to 37 mm. Weight: 2,000 to 3,500 g.

DENTAL FORMULA: 10-8/7-8 = 30-36.

DISTRIBUTION: The foxtail armadillo is distributed from southern Mexico to Venezuela and Ecuador (Cuarón et al., 1989; Wetzel, 1985a, b). In Mexico, it is known exclusively in the Lacandon Forest in Chiapas (Cuarón et al., 1989). It has been registered in the state of CS.

NATURAL HISTORY AND ECOLOGY: These armadillos are solitary animals with nocturnal, underground, and digging habits. They feed on ants and termites and other arthropods associated with colonies of these insects (Redford, 1985). Apparently, their burrows are often in nests of ants (Attini, Myrmecinae) and termites. Their body temperature varies with the air temperature (McNab, 1985). Litters typically consist of a single young, which is born with its eyes and ears closed, weighing approximately 100 g (Merritt, 1975). They escape their predators by running, bending and leaving exposed only the carapace, or digging a hole quickly and disappearing (A. Cuarón, pers. comm.). "Foxtail" refers to the naked tail, similar to that of opossums (Didelphidae), which are also known as foxes in southern Mexico.

VEGETATIONAL ASSOCIATIONS AND ELEVATION RANGE: *Cabassous centralis* has been found in areas associated with tropical rainforests, including areas converted into pastures and secondary vegetation (Cuarón et al., 1989). The genus is found in a wide range of habitats, including various types of forests, savannah, and thorny scrublands (Wetzel, 1982). It is found from sea level up probably to 1,000 m. The known localities in Mexico are located at less than 500 m.

CONSERVATION STATUS: *C. centralis* is listed as endangered in Mexico (SEMARNAT, 2010) because it is very rare and has a restricted distribution. It is protected in the Montes Azules Biosphere Reserve in Chiapas. It appears in appendix III of the Convention on International Trade in Endangered Species (CITES).

Cabassous centralis. Tropical rainforest. Lacandona, Chiapas. Photo: Carlos Guichard.

Order Pilosa

Gerardo Ceballos

Anteaters do not attack or bother anyone, nor give rise to any significant damage, hence they are harmless animals. Regardless of their significant function, they are caught the moment they are discovered not even thinking of the number of termites they destroy.

— Miguel Álvarez del Toro, 1977

The order Pilosa is exclusive to the American continent, and the greatest diversity of species is found in South America (Nowak, 1999b). It was considered part of the order Edentata (Wilson and Reeder, 1993) and more recently of the order Xenarthra (Wilson and Reeder, 2005). Pilosa includes 10 species grouped into 4 families: Bradipodidae (sloths; 3 species), Megalonychidae (sloths; 2 species), Myrmecophagidae (anteaters; 4 species), and Cyclopediade (silky anteater; 1 species). The order is distributed mainly in the tropics from southern Mexico to Argentina (Reid, 1997). Homodont teeth characterize the order; all the species lack incisors and canines; the family of anteaters has no teeth at all. All families are highly specialized. The sloths are adapted to life in trees; they have powerful legs and claws and cryptic fur. They feed on leaves and spend most of their life in trees. Anteaters are arboreal, semi-arboreal, or terrestrial, and specialize in eating ants and termites, so they lack teeth and their tongues are long and covered with sticky saliva to catch their prey (Montgomery, 1983). Mothers carry their offspring hanging from their backs (Leopold, 1965). There are two species in Mexico: the common anteater and the silky anteater, with the latter apparently very scarce.

(*Opposite*) Silky anteater (*Cyclopes didactylus*). Tropical semi-green forest.
Photo: Gerardo Ceballos

Family Cyclopedidae

The family Cyclopedidae is represented by one genus and one species (*Cyclopes didactylus*), which is distributed from tropical regions in Mexico to northern Argentina.

Cyclopes didactylus (Linnaeus, 1758)

Silky anteater

Alfredo D. Cuarón

SUBSPECIES IN MEXICO
Cyclopes didactylus mexicanus Hollister 1914

DESCRIPTION: *Cyclopes didactylus* is the smallest anteater, characterized by a dense, glossy, golden-yellow coat, with a dark, thin line in the middle of the back. The legs are modified to be prehensile. The forelimbs have five fingers, one of them significantly larger with a big, acute nail. The hind feet have four fingers. The tail is prehensile. They have no teeth, and the tongue is long and thin (Reid, 1997).

EXTERNAL MEASURES AND WEIGHT
TL = 366 to 500 mm; TV = 165 to 295 mm; HF = 30 to 50 mm; EAR = 10 to 18 mm. Weight: 175 to 357 g.

DENTAL FORMULA: It lacks teeth.

NATURAL HISTORY AND ECOLOGY: Silky anteaters are strictly nocturnal and arboreal. They tend to move and rest within large masses of vines and in trees (Montgomery, 1985a; Sunquist and Montgomery, 1973). This is probably a mechanism to protect themselves from predators. Their legs are adapted to grab branches as slender as a pencil and to move and feed themselves (Taylor, 1985). To climb they grab the branches in a transversal position to the leg, and they move each leg in the opposite direction; for balance they use their prehensile tails. They roll up their tails in the branches as a preventative measure in case they slip (Montgomery, 1983). If the silky anteater feels threatened, it stands in a defensive posture on its hind legs, holding its forefeet close to its face so it can strike with its sharp claws (Álvarez del Toro, 1977). They feed exclusively on arboreal ants, mostly of the genera *Crematogaster*, *Solenopsis*, *Pseudomyrmex*, *Camponotus*, and *Zacryptocerus* (Álvarez del Toro, 1977; Best and Harada, 1985; Montgomery, 1985a, b). Most of the ants they eat are less than 4 mm long. Depending on an individual's sex and age, the daily consumption varies from 700 to 8,000 ants. They pinpoint their prey by scent. They usually feed on nests or covered roads of ants in the branches rather than isolated individuals. With their sharp claws they open up nests and pull out the ants to eat them (Álvarez del Toro, 1977; Montgomery, 1983). They spend much of their activity periods in search and consumption of ants. They visit many colonies in each feeding period. They feed in short sessions and take only a small proportion of the available ants in each colony, probably due to the many defense mechanisms of ants. They rarely cause serious damage to the nests. They are solitary. The home range of adults of the same sex do not overlap, but the area of an adult male overlaps with that of about three adult females (Mont-

DISTRIBUTION: *C. didactylus* is found from Mexico to Brazil (Haysen et al., 2012; Wetzel, 1982, 1985a, b). In Mexico, it has been registered in southeastern Veracruz, Oaxaca, Chiapas, and Tabasco (Cervantes and Yepes Mulia, 1995; Cuarón, 2000). It may also be found in the Yucatan Peninsula, especially in southern Campeche and Quintana Roo. It has been registered in the states of CS, OX, TB, and VE.

gomery, 1985a). Considering their size and the quantity of food potentially available, their population density is relatively low and their home range relatively large. In Panama, a density of 0.77 individual/ha and a biomass of 0.13 kg/ha was considered. The home range of males is about 11 ha and of females about 2.8 ha (Montgomery, 1985a). They usually have a single young per litter and sometimes two (Álvarez del Toro, 1977; Montgomery, 1983). The gestation period is unknown. In Mexico there have been births in March and April, plus a lactating female with an infant in May and a female with two fetuses in March (Álvarez del Toro, 1977; Cuarón, 2000). During the day the female carries the young on her back; during the night, she leaves the area where they spent the day and forages for food (Álvarez del Toro, 1977; Montgomery, 1983). Before dawn, the mother returns and takes her young to a site where they spend the day together, inactive, in a resting place in the shade and protecting vines. The young sleep between the folds of the abdomen of the mother, who stays hunched. They change sites almost every night. When the young has grown to approximately one-third of the size of the mother it starts to eat ants, but continues to breastfeed until it measures about two-thirds (Haysen et al., 2012). Before dispersing, the offspring live exclusively in the area occupied by the mother; they leave the mother's home range abruptly, moving apparently in a straight line until they find an empty territory (Montgomery, 1983, 1985a). This species is locally known as golden miquito, serafín, banana monkey, silky anteater, or *kisin* (Mayan *lacandón*).

Cyclopes didactylus. Tropical rainforest. Photo: Gerardo Ceballos.

VEGETATIONAL ASSOCIATIONS AND ELEVATION RANGE: The silky anteater's main habitat is the tropical rainforest (Álvarez del Toro, 1977; Eisenberg, 1989); it is not found in savannas and areas where the forest canopy is not continuous (Eisenberg, 1989). They can live in areas of secondary vegetation (McCarthy, 1982a; Montgomery, 1985a) and in orchards or fields of perennial crops such as cacao plantations (Álvarez del Toro, 1977). It has been found from sea level up to probably 800 m, although most of the localities known in Mexico are below 500 m.

CONSERVATION STATUS: Due to high transformation rates of their habitat (Ceballos and Navarro, 1991; Cuarón, 2000), silky anteaters are considered endangered in Mexico (SEMARNAT, 2010). It is not included in the lists of the International Union for Conservation of Nature (IUCN), the Convention on International Trade in Endangered Species (CITES), and the United States.

Family Myrmecophagidae

The family Myrmecophagidae is composed of two genera and three species distributed from the tropical regions of Mexico to northern Argentina. In Mexico there is one genus (*Myrmecophaga*) represented by one species.

Tamandua mexicana (Saussure, 1860)

Mexican anteater

Alfredo D. Cuarón

SUBSPECIES IN MEXICO
Tamandua mexicana hesperia (Davis, 1955)
Tamandua mexicana mexicana (Saussure, 1860)

DESCRIPTION: The Mexican anteater is a medium-sized animal similar in size to a small dog. It has a peculiar shape; the head and snout are elongated in a tubular form. The opening of the mouth is small. It lacks teeth. It has a long, thin tongue covered with sticky saliva produced by large salivary glands. The middle finger of the front digits has a big nail, and the other fingers are reduced in size. The prehensile tail is naked on the distal part. The Mexican anteater's fur is thick and rough with a characteristic coloration pattern, similar to a black or dark brown vest with a white or cream background.

DISTRIBUTION: The Mexican anteater is distributed from Mexico to northern South America and west of the Andes. In Mexico, it is found in the tropics, from Colima and Michoacan in the Pacific slope, and the Huasteca Potosina in the Gulf side, up to Chiapas and the Yucatan peninsula (Burton and Ceballos, 2006; Leopold, 1959; Ramirez Bravo, 2012). It has been recently recorded in Michoacan and Colima, and the species is apparently moving north along the Pacific slope (Burton and Ceballos, 2006; Núñez et al., 2011). It has been registered in the states of CA, CL, CS, GR, HG, MI, OX, PU, QR, SL, TA, TB, VE, and YU. It is possibly found in MOR.

EXTERNAL MEASURES AND WEIGHT
TL = 1015 to 1200 mm; TV = 485 to 675 mm; HF = 97 mm; EAR = 28 to 41 mm. Weight: 3,200 to 7,000 g.

DENTAL FORMULA: The Mexican anteater lacks teeth.

NATURAL HISTORY AND ECOLOGY: Mexican anteaters are characterized by a large variability in their individual habits (Lubin, 1983; Montgomery, 1985a, b). They can be active during the day, at night, or both. They move, feed, and rest in trees and on the ground (Montgomery, 1985a, b). The prehensile tail aids locomotion in trees. They use as dens hollow trees or dens of other animals (Eisenberg, 1989; Lubin, 1983). Apart from the mother-young and reproduction interactions, anteaters are solitary (Lubin, 1983). Home ranges do not overlap, except in the limits. In the rainy forests of Panama their home range measures about 25 ha. Its population density is 0.05 individual/ha and its biomass is 0.2 kg/ha (Montgomery, 1985a). They feed on termites and ants. The percentage of ants and termites in the diet varies seasonally and individually, but typically two-thirds of the diet is termites and one-third ants (Montgomery, 1985a). They avoid some species probably because of their chemical and physical defenses (Lubin, 1983). Individuals that spend more time in trees tend to eat more ants, and those that spend more time on the ground consume more termites (Lubin, 1983). The following termite genera are important in their diet: *Armitermes*, *Calcaritermes*, *Coptotermes*, *Leucotermes*, *Microcerotermes*, and *Nasutitermes*. Predominant genera of ants in their diet are *Azteca*, *Camponotus*, *Crematogaster*, *Dolichoderus*, *Pheidole*, *Procryptocerus*, *Solenopsis*, and *Zacryptocerus* (Montgomery, 1985a, b).

Consumed ants measure less than 4 mm. In addition, occasionally the Mexican anteater consumes some fruit pulp (Cuarón, 1987). Anteaters invest considerable time and energy seeking and consuming social insects. They pinpoint their prey by their scent and dig them up with their strong front claws. Once their prey is captured, they lick them with their long tongues. They only eat the nests and roads covered with ants and termites, and not isolated individuals (Montgomery, 1985b). They usually search for food on the nests of social insects (Lubin, 1983; Lubin and Montgomery, 1981). They visit 50 to 80 ant and termite colonies daily, and feed for periods of less than one minute (Montgomery, 1985a, b). They rarely cause major damage to the nests (Lubin, 1983; Lubin et al., 1977; Montgomery, 1985b). They usually give birth to a single young. Apparently, reproduction is not seasonal. In Mexico there have been births, newborn infants, or lactating females in March, May, and December (Cuarón, 1987). We do not know their gestation period. The first vaginal bleeding in a female—evidence of possible sexual maturity—was observed at six months old (Cuarón, 1987). The young remain with the mother until they are approximately half her size (Lubin, 1983). During the first growth stage, the young is placed in a nest, in a safe place usually inside a hollow tree; the mother returns to breastfeed after feeding herself (Eisenberg, 1989). When the young is bigger, the female moves from one place to another carrying it on her back. During these movements the female stops frequently to feed and the young dismounts to eat with her. This may be one way the young learns the diet and home range (Lubin, 1983). While adults rarely vocalize, infants do so frequently (Cuarón, 1987). When Mexican anteaters feel threatened, they stand on their hind legs and lift their forefeet, agitating them in a defensive way. They can cause serious injuries with the powerful grip of their forefeet. They are locally known as anteater, strong arm, *chupamiel*, beekeeper, or *chabad* (Mayan *lacandón*).

VEGETATIONAL ASSOCIATIONS AND ELEVATION RANGE: *Tamandua mexicana* inhabits predominantly tropical rainforests, tropical montane rainforests, tropical deciduous forests, temperate cloud forests, and mangroves. It is capable of living in areas used for agricultural activities (e.g., pastures), especially if they are near sites with natural vegetation (Gomez Zamora et al., 2011). It is found from sea level up to 2,000 m. Most of the localities are situated within 1,000 m.

CONSERVATION STATUS: The Mexican anteater is listed in appendix III of CITES. It is not on the lists of endangered species of IUCN and the United States. Due to habitat destruction and exploitation, the species populations are likely to decline (Cuarón, 1991). Although the species can survive in human-dominated landscapes highly impacted by human activities, it is considered endangered in Mexico (SEMARNAT, 2010). It is likely that it is not threatened in Mexico.

Tamandua mexicana. Tropical rainforest. Tuxtla Gutiérrez, Chiapas. Photo: Patricio Robles Gil.

Black howler monkey (*Allouata pigra*). Tropical semi-green forest. Calakmul Biosphere Reserve, Campeche. Photo: Gerardo Ceballos.

Order Primates

Gerardo Ceballos

There are a lot of monkeys in this land. They are paunchy animals, with long and coiled tails. They have hands and feet like men.

— Fray Bernardino de Sahagún, 1570

PRIMATES INCLUDE 250 species of apes and monkeys, distributed in tropical and subtropical regions of America, Asia, and Africa (Wilson and Reeder, 2005). The primates are divided into two old lineages formed by the species of the New and Old Worlds, respectively. Human beings belong to the Old World primates, and together with the gorilla (*Gorilla*), orangutan (*Pongo*), and chimpanzees (*Pan*) form the family Hominidae. The rest of the world's primates are grouped in 12 families, 2 of which are exclusive to America, with around 100 species (Wilson and Reeder, 2005).

Among the distinctive features of the members of Primates are the opposable thumbs of the anterior and posterior limbs, which give them the ability to climb and manipulate objects. Most of the species have a tail; in some species of America, such as the spider monkeys (*Ateles*), the tail is prehensile. The cranial box and skull are relatively well developed in all species, especially in the great apes of the family Hominidae (Ankel-Simons, 1999; Feagle, 1998). There is a great variation in the size of the species, from less than 100 g in some lemurs (*Microcebus murinus*) of Madagascar and pygmy marmosets (*Cebuella pygmaea*) of South America, to 200 kg gorillas. In the American continent the largest species weighs about 11 kg (Nowak, 1991). Most of the species are diurnal; however, the prosimians of Asia and Africa, which include species like the galagos (*Galago*), and monkeys of the genus *Aotus* in the American continent, are exclusively nocturnal (Reid, 1997).

In general, primates are gregarious, although there is a great variation in the size and composition of the groups, from family groups to large troops of dozens of individual such as in the case of the baboons and other species. The groups have well-defined hierarchies, and social communication is maintained through gestures, positions, vocalizations, and pheromones (Eisenberg, 1981). Their diet mainly consists of leaves and fruit, but some species also consume insects such as termites, eggs, small vertebrates, nectar, and resins. The chimpanzees hunt other species of monkeys (Eisenberg, 1981; Nowak, 1991). In most of the species, the mating season occurs in a well-defined period of the year and the number of offspring is low, one or two per litter. The offspring are generally altricial and the parental care is complex and lengthy (Ankel-Simons, 1999; Feagle, 1998).

Primates are among the most endangered mammals. They are very susceptible to the destruction of their habitat and many species are hunted as food, for the pet market, or to supply the pharmaceutical industry (Cowlishaw and Dunbar, 2000). The three species present in Mexico are threatened with extinction (Mittermeier et al., 1999; SEMARNAT, 2010).

Family Atelidae

The family Atelidae groups together all the primates of the Americas. It is represented by 58 species (Wilson and Reeder, 2005). In Mexico, there are three species of two genera (*Alouatta* and *Ateles*). Originally, the spider monkey (*Ateles geoffroyi*) was the species with the widest distribution, followed by the mantled howler monkey (*Alouatta palliata*). The black howler monkey (*A. pigra*) is a species with a more restricted distribution since it is endemic to the Yucatan Peninsula, including the north of Belize and Guatemala.

Subfamily Mycetinae

The subfamily Mycetinae contains one genus and fifteen species. Two species are known from Mexico (*Alouatta palliata* and *A. pigra*).

Alouatta palliata (Gray, 1849)

Mantled howler monkey

Gilberto Silva López

SUBSPECIES IN MEXICO
Alouatta palliata equatorialis Festa, 1903
Alouatta palliata coibensis Thomas, 1902
Alouatta palliata mexicana Merriam, 1902
Alouatta palliata palliata (Gray, 1849)

DISTRIBUTION: Mantled howler monkeys are distributed in the remnants of forest of southeastern Mexico, in Central America, and in the coastal rainforests of the Pacific in South America to 30 degrees south latitude (Neville et al., 1988). In Mexico, this species is found in Veracruz (in the region of Los Tuxtlas, in the Sierra de Santa Marta, and near the municipality of Jesus Carranza and Minatitlan), Chiapas (in Montes Azules and surroundings of Palenque), and Oaxaca (it is possible to find them in Los Chimalapas and near Juchitan). It has been registered in the states of CA, CS, OX, TB, and VE

DESCRIPTION: Mantled howler monkeys are large and robust. The arms and legs are unequal in size. The tail is prehensile. The face is naked and pigmented. The pelage is light brown, almost golden. The thumb is divergent and opposable; the genitals are prominent in both sexes. They have two axillary nipples. The hyoid bone is broad and included in a sack in the form of an egg that works as a resonance chamber (Eisenberg, 1981); it is more prominent in males than in females, which allows them to have a high vocalization similar to roars (Neville et al., 1988).

EXTERNAL MEASURES AND WEIGHT
TL = 700 to 1,400 mm; TV = 520 to 670 mm. HF = 140 to 180 mm.
Weight: 3,600 to 7,400 g.

DENTAL FORMULA: I 2/2, C 1/1, PM 3/3, M 3/3 = 36.

NATURAL HISTORY AND ECOLOGY: These monkeys are strictly arboreal, preferring the middle and upper strata of the forests; occasionally they come down to cross from one area to another or to pick up a young (Carpenter, 1965). They are diurnal (Carpenter, 1934). The complex vocalizations are issued by males and females at the beginning of dawn and are a way to establish territoriality and spacing, although they also indicate climatic changes, social dominance, gun shootings, potential predators, and stress by visitors (Baldwin and Baldwin, 1974; Carpenter, 1934; Lundy, 1954; Sekulic, 1983). Their locomotion and posture are varied, but they prefer quadrupedalism as a way to move in the canopy of trees (Cant, 1986; Mendel,

1976; Schön-Ybarra, 1984), while "seating" and "reclining" are frequent positions (Cant, 1986; Schön-Ybarra, 1984). They have lethargic habits after eating, sleeping up to 74% of the day (Carpenter, 1934; Chivers, 1969; Mendel, 1976; Richard, 1970; Smith, 1977). They mainly feed on leaves of about 50 species of trees, but also include fruits and flowers, twigs, petioles, buds, and seeds in their diet. They occasionally eat insects and other arthropods found in the tropical fruits or leaves (Estrada, 1984; Jiménez-Huerta, 1992; Milton, 1980; Neville et al., 1988; Silva-López et al., 1993). Howler monkeys are folivores, although, in times of fruiting, fruits compose almost 95% of their diet (Altmann, 1959). They act as important seed dispersers in the forest (Estrada and Coates-Estrada, 1984a). By licking the leaves they obtain water (Carpenter, 1934); and they also drink from small ponds formed in hollow trees (Glander, 1975, 1978). They are socially cohesive and form groups of a few individuals or groups composed of 3 or 4 adult males, 7 to 10 adult females and a number of juveniles, which on average makes up a group of 15 to 19 individuals (Carpenter, 1934). This is a function of the quality of their habitat; in fragments of forest, for example, groups from 3 to 10 individuals have been observed, in addition to solitary adult males (Silva-López, 1987; Silva-López et al., 1987). Births occur throughout the year. In general, they do not adapt to captivity, and the offspring die in a few days (Neville et al., 1988).

VEGETATIONAL ASSOCIATIONS AND ELEVATION RANGE: *Alouatta palliata* mainly inhabits the high perennial rainforests and medium subperennial rainforests of southeastern Mexico (Cuarón, 1991; Estrada, 1984; Silva-López et al., 1987). In the province of Guanacaste, Costa Rica, this species inhabits dry tropical and riparian forests (Freese, 1976; Glander, 1975). In the island of Barro Colorado, Panamá, its habitat has been described as tropical rainforest (Carpenter, 1934). It has also been

Alouatta palliata. Tropical rainforest. Montes Azules Biosphere Reserve, Chiapas. Photo: Claudio Contreras Koob.

studied in the medium forests of the coastal provinces of Chiriqui, Panama (Baldwin and Baldwin, 1972). In Mexico, it inhabits the tropical rainforest from sea level to 900 masl, although it is more common below 600 m.

CONSERVATION STATUS: *A. palliata* is not considered vulnerable in the lists of the International Union for Conservation of Nature (IUCN, 2000). It is considered at risk of extinction in Mexico (SEMARNAT, 2010). It is included in the appendix I of the Convention on International Trade in Endangered Species (CITES). The high rates of capture and pouching, together with the environmental degradation, deforestation, or fragmentation of its habitat, are the factors that determine the risk of a species (Cuarón, 1987; Estrada and Coates-Estrada, 1989; Silva-López, 1987). It is protected in several areas such as Los Tuxtlas and Santa Marta in Veracruz, Laguna de Terminos in Campeche, and Palenque and Montes Azules in Chiapas.

Alouatta pigra Lawrence, 1933

Mexican black howler monkey

Gilberto Silva López

SUBSPECIES IN MEXICO
A. pigra is a monotypic species.

DESCRIPTION: *Alouatta pigra* is a large and robust monkey, slightly smaller in body size and complexion than *A. pigra mexicana*; its black color also sets it apart. The hyoid bone is larger than that of *A. palliata*. The scrotum in the young is pink with stains (Horwich, 1983).

EXTERNAL MEASURES AND WEIGHT
TL = 1,000 to 1,500 mm; TV = 498 to 630 mm.
Weight: 5,500 to 8,000 g.

DENTAL FORMULA: I 2/2, C 1/1, PM 3/3, M 3/3 = 36.

NATURAL HISTORY AND ECOLOGY: *A. pigra* inhabits primary forests. In Guatemala, they have been observed in areas of shrubs and trees with canopies of not more than 7 m or 8 m (G. Silva-López, pers. obs.) and in riparian forests in several seasonal stages (Horwich and Lyon, 1987). They cross open spaces to move between the fragments of the forests. These monogamous monkeys are found in groups made up of a couple of adults and a young (Bolin, 1981). They form small groups of five to six individuals in average (Coelho et al., 1976; Horwich, 1989). Bolin (1981) has described the complex conduct pattern of parental care as a social system of monogamy. They consume leaves and fruits of 7 of the 36 species of trees most common in Tikal, Guatemala, selecting the abundant *Brosimum alicastrum* (ramon) and *Achras zapota* (sapodilla; Coelho et al., 1976). The number of plants used in their diet surely is greater, as recent data, obtained in mesophytic areas of Guatemala, suggest (Silva-López et al., 1995). These monkeys drink water from hollows in trees (Coelho et al., 1976). They have also been observed drinking from small pools formed in the top of the pyramids of Tikal (Coelho et al., 1976). The births occur throughout the year (Neville et al., 1988).

DISTRIBUTION: *A. pigra* is distributed from the forests of southern Mexico (Estrada and Coates-Estrada, 1984b, 1988; Horwich and Johnson, 1986; Rico-Gray and Watts, 1989; Watts and Rico-Gray, 1987) to Belize, the south of Guatemala (Neville et al., 1988; Silva-López et al., 1995), and the border between Guatemala-Honduras (Curds, 1993). It has been registered in the states of CA, CS, QR, TB, and YU.

Alouatta pigra. Tropical semi-green forest. Calakmul Biosphere Reserve, Campeche. Photo: Gerardo Ceballos.

VEGETATIONAL ASSOCIATIONS AND ELEVATION RANGE: *A. pigra* lives in tropical forest, including rainy forests, subdeciduous forests, tropical dry forests, gallery forests, and cloud forests (Curds, 1993; Horwich and Lyon, 1987; Neville et al., 1988). The altitude is a limiting factor in their distribution; they have not been observed over 330 masl. In general, they inhabit areas with an average temperature of more than 25°C and precipitation exceeding the annual average of 1,000 mm (Horwich, 1989).

CONSERVATION STATUS: According to IUCN, the situation of *A. pigra* is undetermined. In Mexico, its distribution is restricted (Estrada and Coates-Estrada, 1984b; Horwich and Johnson, 1986; Rico-Gray and Watts, 1989), and it is believed to be endangered (SEMARNAT, 2010). It is included in appendix II of CITES. The main problem that it faces is the destruction and fragmentation of its habitats. Relatively large, viable populations are found in the Yucatan Peninsula reserves such as Sian Ka'an, Yum Balam, and Calakmul.

Subfamily Atelinae

The subfamily Atelinae contains four genera of spider and woolly monkeys. One species is known from Mexico (*Ateles geoffroyi*).

Ateles geoffroyi Kuhl, 1820

Spider monkey

Gilberto Silva López

DISTRIBUTION: This species is found in the coastal forests of southeastern Mexico. It has also been reported in the south of Tamaulipas (Villa-R., 1958 in Hall, 1981), in the region of La Huasteca, within the limits of San Luis Potosí and Veracruz (Estrada and Coates-Estrada, 1989), and near Cihuatlán, Jalisco (Villa-R., 1958), although the latter locality is in doubt. Its range extends to the south of Honduras and El Salvador (Hall, 1981; Roosmalen and Klein, 1988). It has been recorded in the states of CA, CS, JA, OX, QR, SL, TA, TB, VE, and YU.

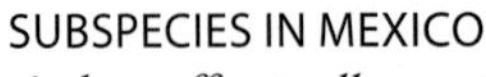

SUBSPECIES IN MEXICO

Ateles geoffroyi vellerosus Gray, 1866

DESCRIPTION: The body of *Ateles geoffroyi* is elongated, with the extremities particularly long and thin; it has a prominent abdomen. The head is small, with the muzzle well marked. The hair is variable in color, although the dark coloration dominates. The thumb is vestigial or absent, which prevents the spider monkey from handling food and objects like other primates. It favors the "hook" grip (Roosmalen and Klein, 1988). The spider monkey has pectoral mammary glands under the armpit. In females the clitoris is elongated and pendant, which seen at distance distinguishes them from the males (Napier and Napier, 1967).

EXTERNAL MEASURES AND WEIGHT

TL = 950 to 1,580 mm; TV = 630 to 823; HF = 159 to 194 mm.
Weight: 5,740 to 6,880 g.

DENTAL FORMULA: I 2/2, C 1/1, PM 3/3, M 3/3 = 36.

NATURAL HISTORY AND ECOLOGY: Spider monkeys have been observed in the middle and upper strata of the forests, although they cross secondary zones of vegetation 5 m in height (Silva-López et al., 1987). They group in bands of a few to 50 individuals (Carpenter, 1935; Izawa, 1976; Silva-López, 1987). Females with offspring are very aggressive; several groups can be temporarily separated or combined (Fedigan and Baxter, 1984). The interactions of affinity and aggressiveness indicate that spider monkeys have a sexually segregated society; males form cohesive and affiliative groups, while adult females are separated, accompanied by the offspring and young who depend on them. Several types of locomotion and positions have been identified; they are, however, mainly bimanual and quadrupedal (Mittermeier, 1978; Napier, 1976; Toledo-Cárdenas, 1993). The locomotion and positions vary depending on the availability and size of the substrates. Spider monkeys feed mainly on fruits of more than 40 species of tropical plants. They also include in their diet a proportion of primary and secondary meristems, flowers, trunks in state of decomposition, some types of crust, and, incidentally or directly, insects (Carpenter, 1935; Hladik and Hladik, 1969; Jiménez-Huerta, 1992; Roosmalen, 1980). The field studies do not indicate a defined reproductive season. Births occur throughout the year (Carpenter, 1935; Hall and Dalquest, 1963; Klein, 1971), although Milton (1981), in his birth studies, shows seasonality on the island of Barro Colorado, Panama. Spider monkeys play an important role as seed dispersers (Pacheco and Simonetti, 2000). Unlike other species, they consume the seeds of the fruit that are defecated later (Roosmalen, 1980).

VEGETATIONAL ASSOCIATIONS AND ELEVATION RANGE: In Mexico, spider monkeys are distributed in the tropical forests, including the high perennial forests and median subperennial forests of Veracruz (Estrada and Coates-Estrada, 1984b; Silva-López, 1987). It lives in mangrove forests in Chiapas (Álvarez del Toro, 1977; Eisenberg and Kuehn, 1966). In the Yucatan Peninsula, it has been observed in low forest with marshes, low subperennial forest, and vegetation of hammocks (Watts and Rico-Gray, 1987). In Mexico, it is found from sea level to 1,500 masl, although most locations are at 700 masl.

CONSERVATION STATUS: IUCN considers this species vulnerable due to environmental degradation and transformation of its natural habitat (Hilton-Taylor, 2000). In Mexico, it has been considered an endangered species (SEMARNAT, 2010). It is listed in appendix I of CITES. In Mexico its threats are poaching, the pet trade, as well as the use of its fat for traditional medicine (Cuarón, 1991; Silva-López et al., 1987, 1995). Many of the plants consumed by *Ateles* are also used by humans for food and other purposes (e.g., medicine, construction material, firewood) (Silva-López et al., 1993). The use of these natural resources can be adduced as an argument in favor of the conservation of the spider monkeys in rural areas close to their natural habitat. They are protected in several large reserves where they have viable populations such as Sian Kaán in Quintana Roo, Calakmul in Campeche, and Montes Azules in Chiapas.

Ateles geoffroyi. Tropical semi-green forest. Calakmul Biosphere Reserve, Campeche. Photo: Gerardo Ceballos.

Tehuanpetec jackrabbit (*Lepus flavigularis*). Grassland. Santa Maria del Mar, Oaxaca. Photo: Gerardo Ceballos.

Volcano rabbit (*Romerolagus diazi*). Grassland. Popo-Izta National Park, Distrito Federal. Photo: Gerardo Ceballos.

ORDER LAGOMORPHA

Gerardo Ceballos and Andrew T. Smith

Taking into account their rarity, restricted geographical distribution, and the lack of effective protection, it is unlikely that the volcano rabbit will survive for many years without proper legal measures of the Mexican Government.

— James Fisher, Noel Simon, and Jack Vincent, 1969

THE ORDER LAGOMORPHA comprises 91 species in 2 families, the pikas (Ochotonidae) and the rabbits and hares (Leporidae). Recently, a new species of rabbit was described in Laos and Vietnam (*Nesolagus timminsi*). This order is distributed throughout the entire world, with the exception of Madagascar, Australia, New Zealand, Antarctica, southern South America, and some islands (Hoffmann and Smith, 2005). All lagomorphs in Mexico are of the family Leporidae. One species (*Oryctolagus cuniculus*) has been introduced into Australia, where it has caused severe environmental impacts. Lagomorphs have a robust body, large ears, and long hind legs, and the tail is very small or absent. The upper lip is in the form of a Y, and all lagomorphs have two pairs of incisors in the upper jaw (Nowak, 1999b). The frontal incisors are larger than those behind, and most species' incisors have a longitudinal furrow. The skull has heavily fenestrated bones.

The majority of the species are terrestrial, but some are semi-aquatic. They are skillful runners and jumpers, and many species specialize in running away from their predators (Eisenberg, 1981). Depending on the species, activity might be diurnal, crepuscular, or nocturnal. Lagomorphs can be solitary as are the hares and rock-dwelling pikas, or form social groups as do some rabbits and burrowing pikas. They dig their own burrows or use the cover under rocks or vegetation. Most species are territorial, marking their territory with urine and other smelly compounds (Ceballos and Galindo, 1984; Chapman and Feldhamer, 1982).

Lagomorphs are strictly herbivorous, consuming a wide variety of leaves, stems, grasses, shrubs, and trees. They possess a peculiar digestive system and produce two types of feces: hard, dry pellets and soft, mucousy stools. The soft (or night) feces are re-ingested to achieve more efficient digestion; this process is known as coprophagy (Ceballos and Miranda, 2000; Chapman and Feldhamer, 1982). Many lagomorph species have great reproductive potential, with short periods of gestation, several offspring per litter, and several litters per year (Eisenberg, 1981). Other species have very small litter sizes. The majority of the species are subject to high rates of predation by birds and mammals, as well as by humans. Some island species of rabbits and hares (*Sylvilagus graysoni*), or with restricted distribution (*Lepus flavigularis, Ochotona iliensis*) are in danger of extinction.

FAMILY LEPORIDAE

The family Leporidae, which includes 11 genera and 61 species of hares and rabbits, is distributed throughout most of the world, with the exception of Australia and southern South America, regions in which several species have been introduced (Hoffmann and Smith, 2005). In Mexico, 3 genera and 15 species are found. The Tehuantepec jackrabbit (*Lepus flavigularis*) and the black jackrabbit (*L. insularis*), the Tres Marias cottontail (*Sylvilagus graysoni*), the Omilteme cottontail (*S. insonus*), the Mexican cottontail (*S. cunicularius*), the San Jose brush rabbit (*S. mansuetus*), and the zacatuche (also called the teporingo, or volcano rabbit; *Romerolagus diazi*) are endemic to Mexico. Most of these species are considered at risk of extinction; in fact, the International Union for Conservation of Nature (IUCN) (2010) lists five Mexican lagomorphs as endangered and three as near threatened.

Lepus alleni Mearns, 1890

Antelope jackrabbit

Julieta Vargas Cuenca and Fernando A. Cervantes

SUBSPECIES IN MEXICO
Lepus alleni alleni Mearns, 1890
Lepus alleni palitans Bangs, 1900
Lepus alleni tiburonensis Townsend, 1912

DESCRIPTION: The antelope jackrabbit is one of the largest hares in North America. The coloration from the neck through the hindquarters is yellowish-brown mixed with black; the fur on the sides of the body, including the sides of the anterior limbs, hips, and hindquarters, is white with black tips, which makes the coat appear light gray. The ventral part and sides of the legs are white; the inside and the back of the forefeet and sides of the neck are lighter in contrast with the yellowish chest. The hairless ears, with the exception of the tips, are white with a black stripe at the tips and edges, where the hairs are short and yellowish white. The back of the tail is black mixed with steel, extending to the legs. The sole of the foot is brown, while the instep is white. This hare is distinguished by its white sides, its yellow throat, and its enormous ears (Mearns, 1890).

DISTRIBUTION: In the United States *L. alleni* only inhabits south-central Arizona. In Mexico, it is found in Sonora, including Isla Tiburon, Sinaloa, southeast Chihuahua, and areas north of Nayarit (Hall, 1981). It has been registered in the states of CH, NY, SI, and SO.

EXTERNAL MEASURES AND WEIGHT
TL = 553 to 670 mm; TV = 48 to 76 mm; HF = 127 to 150 mm; EAR = 138 to 173 mm. Weight: 3 to 6 kg.

DENTAL FORMULA: I 2/1, C 0/0, PM 3/2, M 3/3 = 28.

NATURAL HISTORY AND ECOLOGY: *Lepus alleni* is nocturnal, although it is often seen at noon sitting under the shadow of cacti or mesquite (Flux and Angermann, 1990; Ingles, 1959). It feeds in the evening or very early in the day. In Arizona, its diet during the rainy season consists of herbs, while during times of drought it feeds on mesquite and cacti. Offspring can be found throughout the year, with the exception

Lepus alleni. Scrubland. Ensenada, Baja California. Photo: Eric Mellink.

of November; however, the percentage of pregnant females is greater in the rainy season. The same is true with the size of the litter, which varies from 1.5 to 3.1, with an average of 2.1. The age of sexual maturity is unknown, but females probably become reproductively active in their first year of life (Flux and Angermann, 1990).

VEGETATIONAL ASSOCIATIONS AND ELEVATION RANGE: In Arizona, the antelope jackrabbit is found in habitats where associations of cacti, dominated by mesquite, creosote bush (*Larrea*), cat's claw (*Acacia greggii*), and grasses, are common (Dunn et al., 1982). In Mexico, it inhabits thorny forest and areas of mesquite and pasture (Leopold, 1965); it is found from 123 masl to 500 masl (Hall, 1981).

CONSERVATION STATUS: The antelope jackrabbit is considered common, but research to determine its status has been recommended (Flux and Angermann, 1990). The subspecies *L. a. tiburonensis* is of Special Protection (SEMARNAT, 2010).

Lepus californicus Gray, 1837

Black-tailed jackrabbit

Fernando A. Cervantes and Norma A. Hernández

SUBSPECIES IN MEXICO

Lepus californicus altamirae Nelson, 1904
Lepus californicus asellus Miller, 1899
Lepus californicus bennettii Gray, 1843
Lepus californicus curti Hall, 1951
Lepus californicus deserticola Mearns, 1896
Lepus californicus eremicus J.A. Allen, 1894
Lepus californicus festinus Nelson, 1904
Lepus californicus magdalenae Nelson, 1907
Lepus californicus martirensis Stowell, 1895
Lepus californicus merriami Mearns, 1896
Lepus californicus sheldoni Burt, 1933
Lepus californicus texianus Waterhouse, 1848
Lepus californicus xanti Thomas, 1898

DESCRIPTION: The black-tailed jackrabbit is a rather large hare, and its coloration ranges from brown to gray, with a lighter belly. The ears are longer than the hind legs. Its common name derives from the presence of a remarkable black stripe along the back of the tail. It also can be distinguished by the black spot on the tip of each ear (Hall, 1981; Woloszyn and Woloszyn, 1982).

EXTERNAL MEASURES AND WEIGHT

TL = 523 to 606 mm; TV = 77 to 101 mm; HF = 113 to 135 mm; EAR = 99 to 129 mm. Weight: 1.5 kg.

DENTAL FORMULA: I 2/1, C 0/0, PM 3/2, M 3/3 = 28.

NATURAL HISTORY AND ECOLOGY: The black-tailed jackrabbit does not dig burrows, but uses depressions under the trees, or "beds," to hide from predators. They have crepuscular habits since they are active during the early hours of the day and evening. Their diet varies seasonally; during times of drought their diet is based on herbs, and in the rainy season they primarily eat grass and the bark of trees or shrubs. They are solitary animals that meet only to mate. Leverets are born with hair, open eyes, and ready to walk (Ceballos and Galindo, 1984). The females produce 10 to 15 offspring per year (Flux and Angermann, 1990). Larvae of a parasitic fly infest this species during some times of the year (Leopold, 1959). These hares are common prey of coyotes and bobcats (Woloszyn and Woloszyn, 1982). They can be very abundant, and thus they are considered a plague on pastures (Leopold, 1965); they prefer to feed in overexploited areas. The economic importance of the black-tailed jackrabbit lies in its use as meat and as a hunting trophy (Dunn et al., 1982).

VEGETATIONAL ASSOCIATIONS AND ELEVATION RANGE: *Lepus californicus* lives mainly in regions of xeric scrublands composed primarily of grasses, combined with dominant species such as *Agave*, *Hectia*, and *Yucca* (Rzedowski, 1978). It is also abundant in pastures, where the dominant plant genus is *Bouteloua* (Ceballos and Galindo, 1984; Rzedowski, 1978). The range of altitudes where this species inhabits varies from sea level up to 3,800 m (Flux and Angermann, 1990).

DISTRIBUTION: This hare is found in the arid and semi-arid areas of the peninsula and north of Baja California, the northwest and center of Mexico, extending to the northern part of the Valley of Mexico (Ceballos and Galindo, 1984; Ramírez-Pulido et al., 1983). Its distribution includes the deserts of the Southwest of the United States extending to the states of Washington and South Dakota (Flux and Angermann, 1990). It has been registered in the states of BC, BS, CH, CO, DU, GJ, HG, JA, MX, NL, QE, SL, SO, TA, TL, and ZA.

Lepus californicus. Grassland. Janos Biosphere Reserve, Chihuahua. Photo: Gerardo Ceballos.

CONSERVATION STATUS: The black-tailed jackrabbit is a very abundant species that is not threatened. Poaching has decreased its populations in the center of Mexico (Flux and Angermann, 1990). In some regions, such as the Valley of Mexico, the destruction of its habitat has caused the disappearance of this species (Ceballos and Galindo, 1984). *L. c. magdalenae* and *L. c. sheldoni* are of Special Protection (SEMARNAT, 2010).

Lepus callotis Wagler, 1830

White-sided jackrabbit

Fernando A. Cervantes, M. Carmen Resendiz, and Ana L. Colmenares

SUBSPECIES IN MEXICO

Lepus callotis callotis Wagler, 1830
Lepus callotis gaillardi Mearns, 1896

DESCRIPTION: *Lepus callotis* is a hare of relatively large size. The dorsal coloration of this hare is dark gray; the sides, the belly, and the limbs are white; and the tail is bicolored (the bottom is white and the upper part is black; Anderson and Gaunt, 1962). The back of the ears are yellowish, and the hair of the tip and the posterior edge are white (Mearns, 1895). Its diploid chromosome number (2n) is 48, but the fundamental number is 90 (Anderson, 1972; González, 1992).

Lepus callotis. Grassland. Charcas, San Luis Potosí. Photo: Juan Cruzado.

EXTERNAL MEASURES AND WEIGHT

TL = 542 to 550 mm; TV = 70 to 72 mm; HF = 129 to 131 mm; EAR = 120 to 126 mm.

Weight: 2 to 3 kg.

DENTAL FORMULA: I 2/1, C 0/0, PM 3/2, M 3/3 = 28.

NATURAL HISTORY AND ECOLOGY: The white-sided jackrabbit is commonly found in semi-arid areas, and only in western Sonora does it live in the desert where it frequents streams surrounded with vegetation rather than the naked desert plains (Leopold, 1959). It avoids mountainous areas, preferring areas with few shrubs and with an annual rainfall of approximately 383 mm (Bernardz and Cook, 1984). In New Mexico, it prefers habitats composed of 65% or more of pasture, 25% or less of herbaceous plants, and less than 1% of shrubs, which is very similar to the habitat of the white-sided jackrabbit in northwestern Chihuahua. It constructs shelters that are 37 cm long, 18 cm wide, and 6 cm deep on the surface of the ground and associated with grasses of the species *Hilaria mutica* (Findley, 1987). Each animal uses several shelters depending on the time of the year. They can also take shelter under the ground, although this behavior is rare. They are totally nocturnal, with their activity peak occurring between 22:00 hrs and 5:00 hrs (Bernardz and Cook, 1984). Running away is a typical behavior of this hare. When startled, these hares may flash their white sides by pushing the dorsal skin upward and lifting it to one side, while lifting the white skin in the opposite direction, close to the midline of the back. This has been observed when they are standing or moving at an average speed. This expansion of the white area occurs on the side facing the persecutor and alternates from side to side while the animal runs away in a zigzag pattern (Nelson, 1909). Another escape pattern this hare uses is to leap very high, extending the hind feet in the air and displaying its white sides. This behavior is evident when the hare is frightened or alarmed by a predator. A conspicuous trait of *L. callotis* is its tendency to occur in pairs (Anderson and Gaunt, 1962; Bernardz and Cook, 1984; Bogan and Jones, 1975; Conway, 1976). The pair usually consists of a male and a female that exhibit a strong affinity to stay together, which is more evident during the breeding season. With the exception of the pairing, its reproductive behavior is similar to that of the black-tailed jackrabbit (Lechleitner, 1958; Pontrelli, 1968). Once the pair has been established, the male becomes aggressive toward any male intruder. The members of the pair are usually within 5 m of each other, dominating a visual area of 5 m to 25 m, and they run together up to 500 m when approached by intruders (Bogan and Jones, 1975). In Zacatecas, it is almost always observed in pairs during summer (Matson and Baker, 1986). In Chihuahua, pairs have been observed from May until October (Anderson, 1972). The couple does not separate if the female is pregnant (Cook, 1986). *L. callotis* has three types of vocalizations. The alarm or fear reaction consists of a high-pitched scream. Another sound, emitted by males in a pair when approached by an outside intruding male, is a series of harsh grunts until the intruder leaves or is chased away. A third vocalization, consisting of a trilling grunt, is heard during the sexual chase of *L. callotis*; however, it is not known which member of the pair makes this sound. Additional information on their reproduction is limited. In New Mexico, the breeding season of the white-sided jackrabbit is a minimum of 18 weeks, occurring from mid-April to mid-August (Bernardz and Cook, 1984). Females can have several litters in a year since the gestation period lasts only six weeks, and they may deliver two to four offspring per litter. The number of leverets per litter and per year is probably correlated with the amount of food. Leverets are born in a "hollowed nest" outside the ground under a bush or a large pile of grass; they are born with open eyes and covered with hair. The mother covers the leverets with hair plucked from her belly (Leopold, 1959). Their diet consists of soft and green parts of growing plants; it might sometimes consume the bark of the branches of trees and shrubs. In arid regions, where crops are irrigated, the hares can live exclusively on alfalfa and other green crops. Sometimes they use their anterior claws to extract the bulbs of plants like *Cyperus rotundus* (Bernardz and Cook, 1984). The interactions between *L. callotis* and *L. californicus* are limited and probably only occur in areas of marginal habitat (Dunn et al., 1982). These hares usually occupy large tracts of grassland, in which *L. californicus* is not common (Conway, 1976; Findley, 1987). Bacteria that can infect *L. callotis* are *Pneurococcus*, *Pseudomonas pseudomallei*, *Klebsiella ozanae*, *Moraxella*, *Staphylococcus aureus*, *Streptococcus*, *Bacillus*, *Escherichia coli*, and *Yersinia pseudotuberculosis*. Their known ectoparasites are fleas (*Pulex simulans*) and mites (Bernardz and Cook, 1984). Their predators are birds and mammals such as the coyote, the fox, the golden eagle (*Aquila chrysaetos*), the hen harrier (*Cyrcus cyaneus*), the red-tailed hawk (*Buteo jamaicensis*), the Swainson's hawk (*Buteo swainsoni*), and the great horned owl (*Bubo virginianus*). In Chihuahua, they are used as food, and in Zacatecas, local residents consider this hare highly edible (Bogan and Jones, 1975; Matson and Baker, 1986).

DISTRIBUTION: This species is virtually endemic to Mexico, as the only population in the United States is restricted to an area of 120 km^2, located in southern Hidalgo County, New Mexico (Bernardz and Cook, 1984). In Mexico, it is found from Chihuahua to the center of Oaxaca, passing through the mountainous spurs of the eastern Sierra Madre Occidental and part of the Eje Volcanico Transversal (Anderson and Gaunt, 1962; González-Christien et al., 2002). It has been registered in the states of CH, DU, GJ, GR, HG, JA, MI, MO, OX, PU, SL, SO, VE, and ZA.

VEGETATIONAL ASSOCIATIONS AND ELEVATION RANGE: *L. callotis* dwells in open areas surrounded by pine and pine-oak forests. It is common in areas of mesquite, pastures, and thorny forest of semi-arid areas of the country (Leopold, 1959). In the United States, it is found in areas where *Hilaria mutica* is abundant (Findley, 1987) and in areas with short grasses (*Boutelova gracilis*, *B. eriopoda*). It is found at mid-elevations from 750 masl in Morelos to an elevation of 2,550 masl north of Puebla (Nelson, 1909).

CONSERVATION STATUS: The white-sided jackrabbit seems to be a rare species and although little is known about its abundance in Mexico, it is likely that modifications in its original habitat and overgrazing decrease its presence and encourage its displacement by the black-tailed jackrabbit (*L. californicus*). Increased biological knowledge could promote their management and promote the survival of their populations, as well as prevent the deterioration of their habitat (Dunn et al., 1982). Information on its status in Mexico is urgently needed (Flux and Angermann, 1990).

Lepus flavigularis Wagner, 1844

Tehuantepec jackrabbit

Fernando A. Cervantes, Andrea Cerecero Reyes,
Francisco J. Romero, and Ana L. Colmenares

SUBSPECIES IN MEXICO

L. flavigularis is a monotypic species.

DESCRIPTION: *Lepus flavigularis* is a medium-sized hare with long legs and ears. The back, the head, and the ears are yellowish-brown. The sides and the belly are white; the tail is black on its upper part and white underneath (Anderson and Gaunt, 1962; Hall, 1981). It is characterized by the yellowish color of its throat and the presence of two black stripes that extend from the base of each ear to the base of the nape of the neck (Hall, 1981). The bulla is smaller than that of other hares (Anderson and Gaunt, 1962). The coloration of its sides has been interpreted as evidence of kinship with *L. callotis* and *L. alleni*.

Lepus flavigularis. Grassland. Santa Maria del Mar, Oaxaca. Photo: Gerardo Ceballos.

EXTERNAL MEASURES AND WEIGHT
TL = 595 mm; TV = 133 mm; HF = 77 mm; EAR = 112 mm.
Weight: 3.5 kg.

DENTAL FORMULA: I 2/1, C 0/0, PM 3/2, M 3/3 = 28.

NATURAL HISTORY AND ECOLOGY: The Tehuantepec jackrabbit lives close to riparian vegetation, in sand dunes and thorny scrub along the shores of the nearby lagoons. It is nocturnal and coexists with *Sylvilagus floridanus*. Two females collected in February were pregnant; one of them contained two fetuses that measured 175 mm and 178 mm. There is no additional information on its natural history (Flux and Angermann, 1990).

VEGETATIONAL ASSOCIATIONS AND ELEVATION RANGE: *L. flavigularis* dwells in thorny bush with abundant grass and sand dunes. It is found from sea level up to 50 m (Cervantes, 1993b; Nelson, 1909).

CONSERVATION STATUS: This species is the most endangered species of hare and requires urgent attention (Ceballos and Navarro, 1991; Flux and Angermann, 1990). This is mainly due to the loss of its habitat by the increase of agriculture and poaching. In 1984, the staff of the U.S. Fish and Wildlife Service sighted six individuals in the same night, and emphasized the alarming reduction of its habitat (Ceballos and Navarro, 1991).

DISTRIBUTION: This species is endemic to Mexico; it is only found in Oaxaca, and is limited to the south of the Isthmus of Tehuantepec, in the land bordering the Laguna Superior, Laguna Inferior, and the Mar Muerto (Anderson and Gaunt, 1962). It has been registered in the state of OX.

Lepus insularis W. Bryant, 1891

Black hare

Fernando A. Cervantes, José H. Avila, and Francisco J. Romero

SUBSPECIES IN MEXICO
L. insularis is a monotypic species.

DESCRIPTION: *Lepus insularis* is a medium-sized hare that is distinguished by its black coloration. The back, head, and tail are glossy. The ventral part is pale cinnamon, and the sides are darker, merging with the black back. The ears are gray with a black spot in the tip and a white line in the lower margin. The black hare is closely related to *L. californicus* (Nelson, 1909). The skull is smaller and lighter than that of some subspecies of *L. californicus*. Its jugal bone, however, is thicker than in any of the subspecies of *L. californicus* that inhabit the low area of the Baja California Peninsula. In general, the skull of *L. insularis* is very much like the skull of *L. californicus xanthi*, but with a larger tympanic bullae (Nelson, 1909). The most conspicuous difference between *L. insularis* and *L. californicus*, however, is the coloration of the coat since *L. insularis* is darker; they also differ in their chromosomal number (Cervantes et al., 1999).

EXTERNAL MEASURES AND WEIGHT
TL = 574 mm; TV = 96 mm; HF = 121 mm; EAR = 105 mm.
Weight: 2.5 kg.

DENTAL FORMULA: I 2/1, C 0/0, PM 3/2, M 3/3 = 28.

DISTRIBUTION: This hare dwells exclusively on Espiritu Santo Island, Baja California Sur, in front of the bay of La Paz, in the Gulf of Baja California (Flux and Angermann, 1990). Therefore, the potential available habitat is no more than 99 km^2 (Moctezuma and Serrato, 1988). It has been registered in the state of BS.

NATURAL HISTORY AND ECOLOGY: Most black hares have been observed in valleys and low slopes on hillsides. Because of their coloration, they contrast with the substrate and are easily discernible, even when they are not moving (Nelson, 1909). The isolation of the hares on Espiritu Santo Island, combined with the virtual absence of natural predators, has eliminated the need for a protective coloration, and therefore this hare sports its distinctive coat (Nelson, 1909). The local fishermen hunt this hare for local consumption (Thomas and Best, 1994). The characteristics of its behavior, diet, and reproduction are unknown (Chapman et al., 1990).

VEGETATIONAL ASSOCIATIONS AND ELEVATION RANGE: Due to its volcanic origin, Espiritu Santo Island is mostly composed of alternating layers of black lava, volcanic ash, and other eruptive rocks (Lindsay, 1962). Most of the vegetation on the island corresponds to xeric scrub (Rzedowski, 1978). The vegetation of the coastal areas is typical of dunes with some low-lying species of pasture vegetation (Lindsay, 1962). In large part, however, the island is mountainous and has narrow valleys that retain moisture, enabling the development of tropical deciduous forest mixed with bushes (Rzedowsky, 1978; Thomas and Best, 1994). The black hare has been found from sea level up to an altitude of about 300 m (Nelson, 1909; Cervantes et al., 1996). Espiritu Santo Island has a maximum altitude of 595 masl (Thomas and Best, 1994).

CONSERVATION STATUS: The conservation status of the black hare is unknown (Angermann et al., 1990). It is believed that the hare is not abundant; hence, it should be considered rare (Chapman et al., 1990). On the other hand, during an expedition to the islands of the Gulf of California by a team of the Natural History Museum of San Diego and the Belvedere Science Foundation of San Francisco in 1962, two specimens of *L. insularis* were collected (Lindsay, 1962), but additional information was not provided. Unfortunately, there is no further information about its conservation status. It is unlikely that humans have occupied the island, thus endangering the populations of black hare (Flux and Angermann, 1990). Yet, it is urgent to define the status of this hare considering its distribution, population, habitat requirements, and the negative impact of exotic species in the island (Chapman et al., 1990; Flux and Angermann, 1990; Moctezuma and Serrato, 1988). It is considered as of Special Protection (SEMARNAT, 2010).

Lepus insularis. Scrubland. Isla Espíritu Santo, Baja California Sur. Photo: Gerardo Ceballos.

Zacatuche, Teporingo, Volcano rabbit

Francisco J. Romero and Fernando A. Cervantes

SUBSPECIES IN MEXICO

R. diazi is a monotypic species.

DESCRIPTION: The zacatuche, or teporingo, has a short and dense coat--ochre-colored mixed with black on the back, sides, and tail. The underfur and tips of guard-hairs are black, while the middle part is yellow; the top of the legs is buffy colored, and the ventral surface is pale brown (Rojas, 1951). *Romerolagus diazi*, like other species of *Sylvilagus*, has a triangle of yellow hair on the neck between the bases of the ears (Granados et al., 1980). The pectoral region is covered with long and soft hair that contrasts with the ventral coat, as is in other leporids (Corbet, 1982). The thighs and feet are short, the ears are small and rounded, and the tail is so small that it is not visible (Merriam, 1896b, 1897b).

EXTERNAL MEASURES AND WEIGHT

TL = 234 to 321 mm; TV = 18 to 31 mm; HF = 42 to 55 mm; EAR = 40 to 45 mm.
Weight: 386 to 602 g.

DENTAL FORMULA: I 2/1, C 0/0, PM 3/2, M 3/3 = 28.

Romerolagus diazi. Pine forest and grassland. Popo-Izta National Park, state of Mexico. Photo: Gerardo Ceballos.

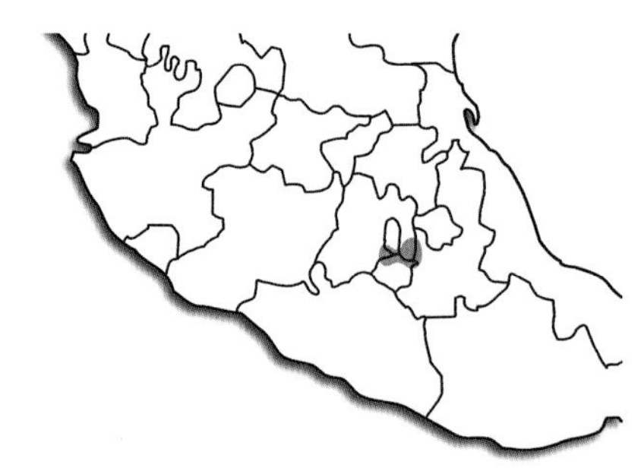

DISTRIBUTION: Compared with some other mammals of Mexico, this rabbit has a restricted distribution (Hall, 1981; Ingles, 1958; Nelson, 1909). It is an endemic species of Mexico that inhabits the central part of the Transverse Neovolcanic Belt; populations can be found on the slopes of the mountains of south and southeastern parts of the Valley of Mexico, and in the Nevado de Toluca, in a disjunct distribution pattern (Ceballos et al., 1998; Hoth et al., 1987; López-Forment and Cervantes-Reza, 1981; Romero et al., 1991; Velazquez, 1993; Velazquez et al., 1991). It is believed that the total area of distribution of *R. diazi* is 280 km^2, delimited by the Sierra Nevada and the Sierra Chichinautzin, particularly in the volcanoes Iztaccihuatl, Popocatepetl, Tlaloc, and Pelado (Bell et al., 1985; Fa and Bell, 1990; Hoth et al., 1987); however, recent studies show that the area of distribution comprises 386.5 km^2 due to the existence of 12 small areas surrounding these volcanoes (Romero et al., 1991; Velázquez et al., 1991). Since a specimen was collected approximately 40 km west of Coatepec, Mexico, in the Cerro Gordo of the Nevado de Toluca (Cervantes-Reza, 1980), other localities with similar altitude, climate, and vegetation were explored between 1893 and 1981, registering this species. These expeditions were conducted in the eastern part of the Transverse Neovolcanic Belt (Volcano La Malinche, Volcano Pico de Orizaba, and Volcano Cofre of Perote), and the western part (Volcano Nevado de Toluca, volcanoes of Colima, and mountains of Michoacán); however, there are no new records outside the Sierra Nevada and the Sierra Chichinautzin (Bell et al., 1985; Hoth et al., 1987). This suggests that *R. diazi* has disappeared from the areas of the center of the Transverse Neovolcanic Belt, and the eastern slope of the Volcano Iztaccihuatl and the Nevado de Toluca, mentioned respectively by Rojas (1951) and Granados (1981). Recently, however, it has been located in the Nevado de Toluca (Ceballos et al., 1998). Its distribution has been seriously affected by the fragmentation of its habitat, constant fires, urban development, and the destruction of forests for crop areas (Hoth et al., 1987; Romero et al., 1991; Velazquez, 1988; Velazquez, 1993; Portales et al., 1997). It has been registered in the states of DF, MO, MX, and PU.

NATURAL HISTORY AND ECOLOGY: The habitat of the zacatuche is warm and rainy during summer, and cold and dry during winter; the average annual temperature is 9.6°C and the average annual precipitation is 1,334 mm; the warmest month is May and the coldest is January, while August is the wettest month and February the driest (Cervantes-Reza, 1980). The zacatuche occupies rugged areas, where the substrate consists of abundant basaltic rocks, deep soil, permeable and dark ground, mainly of basalt (Cervantes-Reza, 1980; Cervantes et al., 1990). The zacatuche is gregarious and lives in groups of two to five individuals (Gaumer, 1913). It can be active both day and night, although its activities are mainly diurnal (Cervántes-Reza, 1980; Durrell and Mallison, 1968). Between 10:00 hrs and 14:00 hrs it is possible to find a large number of rabbits outside their shelters, but not very active; it is more common to see them feeding, playing, exploring, and interacting with each other at dawn and shortly before nightfall (Cervantes et al., 1990; López-Forment and Cervantes-Reza, 1981). A similar pattern has been observed in captivity (De Poorter and Van der Loo, 1981). They frequently use acute and strong vocalizations; among the lagomorphs few other leporids vocalize, except for the pikas of the family Ochotonidae, which have a wide variety of vocalizations (Cervantes-Reza, 1980; Conner, 1985; Kawamichi, 1981; Leopold, 1959). The vocalizations may become more frequent after rain and, apparently, they include a wide repertoire associated with the development of different behavioral patterns (Gaumer, 1913; A. Velazquez, pers. comm.). As many as five different calls have been described in captivity (De Poorter and Van der Loo, 1981). They produce a squeal or noise similar to scrubbing a plastic ball with the thumb, and a brief, sharp bark (Durrel and Mallison, 1968). When startled, they produce an acute vocalization and run away to a burrow or refuge without stopping (Leopold, 1959). They dig tunnels under the zacatonal bunch grass, which are used as burrows and that have different exits (Cervantes-Reza, 1980; Gaumer, 1913; Rojas, 1951). The entries are oblique and, on average, measure 10.8 cm high and 9.3 cm wide and have no particular orientation, while the tunnels, which measure on average 10.9 cm high, 11.1 cm wide, 500 cm in length, and 40 cm deep, have junctions and changes of direction due to the presence of rocks and roots (Cervantes-Reza, 1980; Cervantes et al., 1990; Gaumer, 1913; Rojas, 1951). These rabbits can also use abandoned burrows of pocket gophers (*Pappogeomys merriami*), rock squirrels (*Spermophilus variegatus*), armadillos, and badgers, or use temporarily any type of refuge such as holes or slots between rocks and stony mounds (Aranda et al., 1979; Cervantes-Reza, 1979, 1980; Cervantes-Reza et al., 1990). The nests are hidden under a grass clump, for example, under *Mulhembergia macroura* (Cervantes et al., 1990; J. Hoth, pers. comm.). The nests are shallow with an average size of 15 cm in diameter and 11 cm deep, and lined with fur and plant fragments from *Pinus* and *Alnus*, and herbs such as *Penstemon*, *Eryngium*, and *Gnaphalium*, as well as by fine fragments of zacatonal grass and a large amount of hair from the female, which takes up most of the volume of the nest. Nests are constructed only from April through September (Cervantes et al., 1990; Cervantes-Reza, 1982). The reproductive period covers the entire year, being more intense during the warm and moist summer and decreasing during the cold and dry winter. On average, the gestation period lasts 39 days (Cervantes-Reza, 1982), but a pregnancy of 38 to 40 days has been reported at the Jersey zoo in England and 40 days at the Antwerp zoo in Belgium (De Poorter and Van der Loo, 1981; Durrell and Mallison, 1968). Based on a study of 20 deliveries in 12 females, 35% of the pregnancies last 39 days, 50% 40 days, and 15% 41 days (Matsuzaki et al., 1982, 1985). The gestation period is longer than that of other species of the genera *Sylvilagus* and *Ochotona* (30 days), but shorter than in *Lepus*. The average size of the litter of the zacatuche is 2.1 bunnies, resembling hares of the genus *Lepus* that have small litters (Cervantes-Reza, 1982; Layne, 1967). Bunnies are born with hair, closed eyes, facial whiskers, and well-developed nails (Cervantes-Reza and López-Forment, 1981). In the newborn, the tail is clearly visible; they lack hair in the groin region, and the navel is prominent (Cervantes et al., 1990). The bunnies open their eyes between 4 and 8 days after birth, and remain in the nest until 14 days old (Cervantes et al., 1990). Field observations showed that young rabbits with an average total length of 154 mm and weight of 99.4 g enter and leave the nest, even though they are still nursing (Cervantes-Reza, 1980). In captivity, two-week-old rabbits start to react to observers by hiding under the hay; around the third week of age they start eating pieces of apple, jumping, and grooming, and in general their vitality is increased and they gradually become independent (De Poorter and Van der Loo, 1981; Durrel and Mallison, 1968; Matsuzaki et al., 1982). They feed primarily on grass clumps, in particular, *Festuca amplissima*, *F. rosei*, *Mulhembergia macroura*, *Symphyandra ichu*, and *Epicampes* sp. (Cervantes-Reza, 1980; Gaumer, 1913; Rojas, 1951). In an area of 1,225 m^2, where only 1,321 clumps of *M. macroura* and 734 clumps of *S. ichu* were present, the volcano rabbit's diet was composed of clumps of *M. macroura* (42.1%) and clumps of *S. ichu* (45.4%) (Cervantes, 1980). *R. diazi* prefers young soft and green leaves, and they usually consume pieces close to the base of the leaves and the lower parts of the zacatonal, thus helping the natural process in which the leaves fold and form a ceiling of dense coverage among the zacatonals (Cervantes et al., 1990). In addition to the pasture, zacatuches consume dicotyledons such as *Alchemilla procumbens* and *Donnell-smithia juncea* (Rojas, 1951). They also feed on young leaves of thorny grass (*Cirsium jorullense*, and especially Eryngium spp.), seeds of the annual vine Sicyos parviflorus, and the juicy bark of young birch trees (*Alnus firmifolia*; Cervantes-Reza, 1980). A fecal analysis showed that, in order of importance, the food items for one year were: *M. macroura*, *A. firmifolia*, *S. ichu*, *Buddleia parviflora*, *Geranium* spp., *F. amplissima*, and *Eryngium columnare* (Martínez-Vazquez, 1987). Villa-R. (1953) suggested that the perennial herb *Cunila lythrifolia* (probably the aromatic herb called "hierba buena," *Hedeoma piperitum*) was important in the diet of the zacatuches; however, this observation has not been recorded again. They also feed in agricultural fields next to their natural habitat without causing considerable damage (Cervantes-Reza, 1979; Gaumer, 1913). During the wet season, they disperse toward the crop fields; once established there, they commonly feed on foliage of young oats (*Avena sativa*) and maize (*Zea mays*; Cervantes-Reza, 1979, 1980). The feces of the volcano rabbit are clearly distinguishable from those of the other two rabbits with which it shares its habitat (*Sylvilagus floridanus* and *S. cunicularius*); pellets are discoid with the central part swollen, with an average diameter of 5 mm to 9 mm, ochre-colored, shiny, and with a smooth texture, becoming yellowish when dried (Aranda et al., 1979; Cervantes-Reza, 1980). The droppings are regularly found in groups of more than 25 pellets (latrines), under the cover of zacatonal bunch grass, and near the nests or entries of the burrows. In accordance with field observations, the place where they feed is the same place where they deposit their droppings, hence denoting their presence (Cervantes-Reza, 1980). Thus droppings have been used as indirect indicators of their presence and abundance (Bell et al., 1985; Hoth et al., 1987; Romero, 1993; Romero and López-Paniagua, 1991; Velazquez, 1993). To explore the relationship of local coexistence with the two species of rabbits (S. floridanus and S. cunicularius), droppings were used as indirect evidence of their abundance and patterns of habitat usage, and a negative correlation between the two genera was found (Fa et al., 1992). *R. diazi* is an important part of the food chain. Among the mammals that exist in its area of distribution, the weasel is one of its most common predators (Cervantes-Reza, 1981; Gaumer, 1913; Rojas, 1951). Likewise, the lynx is another important predator; the presence of hair and bone fragments has been found in its droppings (Cervantes-Reza, 1982). A fecal analysis of lynx, collected over the course of a year at Volcan Pelado, showed that the zacatuche was the most important food item for this feline (more than 80%) (Romero-R., 1987). Coyotes are also important predators; for example, of 12 droppings of coyote, collected proportionately during drought and the rainy season, 11 contained remains of zacatuches (Cervantes-Reza, 1980). In addition to mammals, in the region near Volcan Chichinautzin a dusky rattlesnake (*Crotalus triseriatus*) was collected, and two small zacatuches of 125 mm and 127 mm total length were recovered from its stomach (Cervantes-Reza, 1981). It has also been suggested that the red-tailed hawk (*Buteo jamaicensis*) and the horned owl (*Bubo virginianus*), as well as other mammals such as the gray fox, the cacomistle, and the badger, could be natural predators of the zacatuche (Cervantes-Reza, 1980). Unfortunately, domestic dogs and cats, as well as poaching, represent threats to the zacatuche, contributing significantly to the decline of its populations (Cervantes-Reza, 1979; Leopold, 1959; Velazquez, 1988).

VEGETATIONAL ASSOCIATIONS AND ELEVATION RANGE: According to Rzedowski and Rzedowski (1981), the vegetation types that occur in the distribution areas in which *R. diazi* is distributed are forests of *Abies* (2,700 masl to 3,500 masl), forests of *Pinus* (2,350 masl to 4,000 masl), forests of *Quercus* (2,350 masl to 3,100 masl), prairies of *Potentilla candicans* (2,900 masl to 3,500 masl), and subalpine and alpine zacatonals (3,000 masl to 4,300 masl), in which it is common to find different species of cluster grass that measure 60 cm to 120 cm tall (also called zacatonals or macollos). The zacatuches show marked patterns of habitat use, however, and can usually be found inhabiting the forests of *Pinus* where a high underbrush is present, or, in the zacatonals where the tree canopy is absent (Cervantes-Reza, 1979, 1980; Davis and Russell, 1953; Gaumer, 1913; Granados, 1980; Romero and López-Paniagua, 1991; Romero et al., 1987, Velazquez, 1993). In general, the canopy consists of forests of *Pinus montezumae* and *P. hartwegi*, with an average of 25 m in height, and where *P. rudis*, *P. teocote*, *P. patula*, and *P. pseudostrobus* can be present (Cervantes et al., 1990; Fa and Bell, 1990). The lower stratum of these forests is generally covered by a dense layer of zacatonals of up to 1.5 m high, mainly of *Mulhembergia macroura*, *M. quadridentata*, *Calamagrostis tolocensis*, *Festuca amplissima*, *F. tolucensis*, *F. rosei* and *Stipa ichu*, with other species of the herbaceous stratum, such as *Lupinus montanus*, *Penstemon campanulatus*, *Geranium potentillaefolium*, *Gnaphalium oxiphyllum*, *Alchemilla procumbens*, *Cirsium ehrenbergii*, *Eupatorium pazcuarense*, *Eryngium carlinae*, *E. columnare*, *E. proteiflorum*, *Commelina alpestris*, *Salvia elegans*, *Dahlia mercki*, and *Potentilla candicans*, among others, and with a bushy stratum with species such as *Solanum cervantesii*, *Senecio cinerarioides*, *S. salignus*, *Symphoricarpus microphyllus*, *Ribes ciliatum*, and *Fuchsia microphylla* (Cervantes et al., 1990; Fa and Bell, 1990; Romero and López-Paniagua, 1991; Romero et al., 1987; Velazquez, 1993). Research about the habitat shows that a high environmental heterogeneity is present in the distribution areas of the zacatuche, which reflects a mosaic of communities of plants (Velazquez, 1993); this has hampered the analysis of the vegetation to characterize the habitat of this species. It has been observed, however, that pastures (*Festuca tolucensis* or *Calamagrostis tolucensis-Festuca tolucensis*), open pine forests (*Pinus hartwegii-Festuca tolucensis*), mixed forests of pine-alder (*Pinus* spp.-*Alnus firmifolia-Mulhembergia macroura*), and forests of pine (*Pinus* spp.-*Mulhembergia quadridentata*) have the most favorable conditions for the presence and development of the populations of this rabbit (Romero and López-Paniagua, 1991; Romero et al., 1987; Velazquez, 1993). Other communities of vegetation, in which populations of this species are less abundant, are secondary forests of alder and colonies formed by individuals of *Furcraea bendinghausii* (Cervantes et al., 1990; Velazquez, 1993). Some small populations may be present in forests of *Abies religiosa* and prairies of *Potentilla* (*Symphyandra ichu-Potentilla candicans* or *Mulhembergia quadridentata-Potentilla candicans*), although these habitats are less favorable for the development of *R. diazi* (Romero, 1993; Velazquez, 1993). Zacatuches can colonize, temporarily, crops of oats (*Avena sativa*) when the plants form a dense coverage in July, remaining there until the harvest season in October (Cervantes-Reza, 1979; Cervantes et al., 1990). They generally inhabit areas ranging from 2,800 masl to 4,250 masl (Cervantes et al., 1990).

CONSERVATION STATUS: The zacatuche is a rare and endangered species (Smith, 2008). It has been listed as being at risk of extinction since 1972 in the Red Data Book of the IUCN (see Smith, 2008). In addition, it is listed in appendix I of the Convention on the International Treaty of Endangered Species (CITES) and in the Endangered Species Act in the United States. In Mexico, it is considered in danger of extinction (SEMARNAT, 2010).

Sylvilagus audubonii (Baird, 1858)

Audubon's cottontail

Julieta Vargas Cuenca and Fernando A. Cervantes

SUBSPECIES IN MEXICO

Sylvilagus audubonii arizonae (J.A. Allen, 1877)
Sylvilagus audubonii confinis (J.A. Allen, 1898)
Sylvilagus audubonii goldmani (Nelson, 1904)
Sylvilagus audubonii minor (Mearns, 1896)
Sylvilagus audubonii parvulus (J.A. Allen, 1904)
Sylvilagus audubonii sanctidiegi (Miller, 1899)

DESCRIPTION: The front legs of Audubon's cottontail are long, while the hind legs are thin and do not have the dense coat seen in other species of the same genus. It has long ears with little hair on the outside (Chapman and Wilmer, 1978). The back and

Sylvilagus audubonii. Grassland. Janos Biosphere Reserve, Chihuahua. Photo: Gerardo Ceballos.

tail are gray, and the belly is white (Chapman and Ceballos, 1990; Chapman et al., 1982). Unlike in other wild rabbits of Mexico, the bulla is highly developed and the supraorbital process is prominent (Hall, 1981).

DISTRIBUTION: Audubon's cottontail is distributed in the United States from Montana and the Pacific coast of California to Texas and the center of Mexico (Chapman and Ceballos, 1990). In Mexico, it is distributed in the north, northwest, and central parts of the country, from the Baja California Peninsula, Sonora, and Tamaulipas, extending to the northern part of the Valley of Mexico (Ceballos and Galindo, 1984; Cervantes, 1993). It has been registered in the states of AG, BC, BS, CH, CO, DU, GJ, HG, JA, MX, NL, PU, QE, SI, SL, SO, TA, TL, VE, and ZA.

EXTERNAL MEASURES AND WEIGHT

TL = 372 to 400 mm; TV = 39 to 60 mm; HF = 81 to 94 mm; EAR = 70 to 75 mm. Weight: 755 to 1250 g.

DENTAL FORMULA: I 2/1, C 0/0, PM 3/2, M 3/3 = 28.

NATURAL HISTORY AND ECOLOGY: In California, Audubon's cottontail is found in forests of pine and juniper; scrublands constitute their natural havens (Chapman and Wilmer, 1978). Its main food items are grasses, shrubs, leaf stems, and bark; occasionally, it consumes cultivated plants. The reproduction period does not seem to be restricted to any time of the year, although females with offspring have been reported in the summer. The litters have two to five offspring. When the climatic conditions are very severe, the number of offspring and the breeding season change (Ceballos and Galindo, 1984). *Sylvilagus audubonii* is active at dawn and in the early hours of the night, although it is possible to observe them at any time of the day, mainly in areas where shrubs and herbs are abundant (Ceballos and Galindo, 1984). Among its main predators are coyotes, foxes of the desert, badgers, and raptors (*Buteo* sp.).

VEGETATIONAL ASSOCIATIONS AND ELEVATION RANGE: This species dwells in bushes, forests, and pastures of arid and semi-arid areas. Occasionally, they can be found in agricultural fields such as crops of maguey (Ceballos and Galindo, 1984). In the United States, this species is distributed from sea level up to 1,829 masl in mountainous regions (Chapman and Ceballos, 1990). In Mexico, it has been found at elevations of 2,240 masl (Diersing and Wilson, 1980).

CONSERVATION STATUS: *S. audubonii* is a common species in Mexico and the United States, and none of its populations is threatened (Angermann et al., 1990; Chapman and Ceballos, 1990).

Sylvilagus bachmani (Waterhouse, 1839)

Brush rabbit

Fernando A. Cervantes, Consuelo Vázquez, and Ana L. Colmenares

SUBSPECIES IN MEXICO

Sylvilagus bachmani cerrosensis (J.A. Allen, 1898)
Sylvilagus bachmani cinerascens (J.A. Allen, 1890)
Sylvilagus bachmani exiguus Nelson, 1907
Sylvilagus bachmani howellii Huey, 1927
Sylvilagus bachmani peninsularis (J.A. Allen, 1898)
Sylvilagus bachmani rosaphagus Huey, 1940

DESCRIPTION: Among the species of the genus *Sylvilagus*, *S. bachmani* is a small-sized rabbit. Its coloration is dark brown or grayish-brown, except for the belly, the anal region, and the ventral part of the tail, which are white. The internal part of the ears is white; their edges are dark gray. The tarsus is dark brown, and the thighs are light brown. The hairs that cover the fingers are white. The coat is long, smooth, and woolly at its base. The claws are long and sharp (Waterhouse, 1838).

DISTRIBUTION: The distribution of *S. bachmani* in the United States includes all three states of the West Coast: California, Oregon, and Washington. In Mexico, it only inhabits the states of Baja California and Baja California Sur, through the Baja California Peninsula, including Cabo San Lucas and Isla Cedros, in the Pacific (Hall, 1981; Nelson, 1909). It has been registered in the states of BC and BS.

EXTERNAL MEASURES AND WEIGHT

TL = 300 to 375 mm; TV = 20 to 43 mm; HF = 64 to 81 mm; EAR = 50 to 64 mm. Weight: 517 to 843 g.

DENTAL FORMULA: I 2/1, C 0/0, PM 3/2, M 3/3 = 28.

NATURAL HISTORY AND ECOLOGY: *S. bachmani* inhabits thickets, hence its common name, brush rabbit. It adapts well and inhabits abandoned burrows of other species. Its nest is a cavity of approximately 75 mm to 150 mm lined with dry grass (Chapman et al., 1982). The home range is relatively small (Orr, 1940), being larger for males. Young males have the largest home range of any individual of the same age and sex (Chapman, 1971). In California, it mainly feeds on grasses, and in the summer on wild roses (*Rosa californiana*) and Mexican tea (*Chenopodium ambrosioides*). In the autumn it consumes flowers and leaves of blackberry (*Rubus* spp.) and crawling pasture of the genus *Eragrotis*. In the winter it eats tender grass and green clovers (Chapman, 1974). The gestation period lasts approximately 27 days, and a female can give birth to up to 15 pups in 6 litters, which are born without hair (altricial), but after a week they are covered in gray hair (Chapman et al., 1982). The bunnies are fed at night, and leave the nest when they are 13 to 14 days old. Females give birth outside the nest, but shortly after they move the litter to the nest (Orr, 1942). Among its predators are lynx, coyotes, gray foxes, weasels, hawks (*Buteo* sp.), owls (*Bubo virginianus*), eagles, and some snakes (*Crotalus* sp.). Their ectoparasites include fleas, worms of the genus *Taenia*, and nematodes (Chapman et al., 1982). The diploid chromosomal complement (2n) is 48, and the fundamental number (FN) is 80 (Worthington, 1970).

VEGETATIONAL ASSOCIATIONS AND ELEVATION RANGE: The brush rabbit inhabits dense xeric thickets (Huey, 1940) consisting of floristic associations of the genera *Fouquieria*, *Pachycormus*, *Agave*, *Pachycereus*, *Ambrosia*, and *Opuntia*, among others (Rzedowski, 1983). The distribution of *S. bachmani rosaphagus*, in the plateaus of North Baja California is related to the wild rose (*Rosa minutifolia*). In other regions of the peninsula, plants such as *Adenostoma*, *Arctostaphylos* spp., *Artemisia*, and cacti of the genus *Opuntia* are common. The cactus is the most abundant plant along the coasts, while *Adenostoma sparsifolium* is more abundant in the high areas of the mountains. In arid regions the vegetation is scattered, however, islets of *Franseria* and *Agave* spp. provide protection to the fauna (Huey, 1940). *S. bachmani* also inhabits coniferous forests and those of *Quercus* that are not common in the peninsula but occur in some mountainous areas of Sierra de Juarez and San Pedro Martir, as well as on Island Cedros (Huey, 1940). This rabbit is found at 1,830 masl in the mountains surrounding the Lagoon Hansen, and 2,135 masl in the mountains of San Pedro Martir, in northern Baja California. In the south of the peninsula it can be found up to 915 masl (Nelson, 1909).

CONSERVATION STATUS: There is no information on the status of *S. bachmani* for the Baja California Peninsula, Mexico, but it is considered to be abundant and without risk (Chapman and Ceballos, 1990). *S. b. cerrosensis* is considered as of Special Protection (SEMARNAT, 2010).

Mexican cottontail

Fernando A. Cervantes, Patricia Delgado, and Ana L. Colmenares

SUBSPECIES IN MEXICO

Sylvilagus cunicularius cunicularius (Waterhouse, 1848)
Sylvilagus cunicularius insolitus (J.A. Allen, 1890)
Sylvilagus cunicularius pacificus (Nelson, 1904)

DESCRIPTION: *Sylvilagus cunicularius* is the largest rabbit in Mexico. Its coat is rough, abundant, and grayish-brown. Its tail is short and gray with little dorsal pigmentation; its ventral part is white (Ceballos and Galindo, 1984; Chapman et al., 1982). In the west of its range, it has long ears and a short tail, and the coat is short and reddish in the back, while *S. floridanus* has a medium-sized tail and ears and a long coat of red to dark reddish coloration (Diersing and Wilson, 1980). Like *S. floridanus* and *S. graysoni*, the diploid chromosome number is 42 (Lorenzo et al., 1993).

EXTERNAL MEASURES AND WEIGHT

TL = 485 to 515 mm; TV = 54 to 68 mm; HF = 108 to 111 mm; EAR = 60 to 63 mm. Weight: 1.8 to 2.3 kg.

DENTAL FORMULA: I 2/1, C 0/0, PM 3/2, M 3/3 = 28.

NATURAL HISTORY AND ECOLOGY: The Mexican cottontail shares its habitat with other rabbits and wild hares. Along the south coast of Sinaloa, it coexists with *S.*

Sylvilagus cunicularius. Pine forest. Perote National Park, Puebla. Photo: Gerardo Ceballos.

floridanus, and in Guerrero with the hare *Lepus callotis*. In the central highlands of Mexico it shares habitat with the rabbits *Romerolagus diazi*, *S. floridanus*, and *S. audubonii*. In flat areas it co-occurs with the hares *L. callotis* and *L. californicus*. Unlike *S. floridanus*, it prefers the highest portions in the hills around the Valley of Mexico (Ceballos and Galindo, 1984; Cervantes et al., 1992; Davis, 1944; Villa-R., 1952). They are solitary animals, and their peak of activity occurs at twilight, at either dawn or dusk, but they can be active in the day and night. They are herbivores and feed on grasses (*Muhlenbergia macroura*, *Symphyandra ichu*, and *Festuca amplissima*), as well as sprouts and cultivated plants, such as oats, corn, and barley (Ceballos and Galindo, 1984; Ceballos and Miranda, 1986, 2000; Cervantes et al., 1992; Davis, 1944). At the end of June, young individuals have been found in the middle of their growth period (Davis, 1944). A pregnant female with five large embryos was caught at the same time of year, suggesting that this subspecies can have two reproductive periods (Davis, 1944). The reproduction period of this species probably occurs throughout the year. The average gestation period lasts 30 days, with 6 offspring per litter. They are preyed on, when young, by owls (*Bubo*), hawks (*Buteo*), and snakes (*Crotalus*), and as adults by lynx, coyotes, and foxes (Ceballos and Galindo, 1984; Gaona and López, 1991).

VEGETATIONAL ASSOCIATIONS AND ELEVATION RANGE: In the center of Mexico, *S. cunicularius* dwells in forests of pine and oak covered by zacatonals and abounds in pastures, valleys, and mountains. In the western part of the country it inhabits deciduous and semi-deciduous forests at low density (Ceballos and Miranda, 2000; Cervantes et al., 1992; Chapman and Ceballos, 1990). This species is found from sea level up to 4,300 m (Ceballos and Galindo, 1984; Chapman and Ceballos, 1990).

CONSERVATION STATUS: This rabbit is abundant throughout its range and does not require special measures for its conservation. Due to habitat destruction, overgrazing, uncontrolled logging, and poaching, however, some of its populations are declining (Chapman and Ceballos, 1990). For example, in the National Park Zoquiapan and its environs, it is the mammal most intensively hunted and its populations are scarce (Blanco et al., 1981).

DISTRIBUTION: This rabbit is endemic to Mexico and is distributed from Sinaloa to Oaxaca, along the Pacific coast, and toward the east through the highlands of the Transverse Volcanic Belt from Michoacán to Veracruz. In the center of the country it is common in the mountains of the Basin of Mexico at the Cerro Pelado, Popocatepetl, Iztaccíhuatl and Nevado de Toluca (Ceballos and Galindo, 1984; Chapman and Ceballos, 1990). It has been registered in the states of CL, DF, GR, HG, JA, MI, MO, NY, OX, PU, SI, TL, and VE .

Sylvilagus floridanus (J.A. Allen, 1890)

Eastern cottontail

Consuelo Lorenzo and Fernando A. Cervantes

SUBSPECIES IN MEXICO

Sylvilagus floridanus aztecus (J.A. Allen, 1890)
Sylvilagus floridanus chapmani (J.A. Allen, 1899)
Sylvilagus floridanus chiapensis (Nelson, 1904)
Sylvilagus floridanus connectens (Nelson, 1904)
Sylvilagus floridanus holzneri (Mearns, 1896)
Sylvilagus floridanus macrocorpus Diersing and Wilson, 1980
Sylvilagus floridanus orizabae (Merriam, 1893)
Sylvilagus floridanus robustus (Bailey, 1905)
Sylvilagus floridanus russatus (J.A. Allen, 1904)
Sylvilagus floridanus yucatanicus (Miller, 1899)

DESCRIPTION: Among its genus, *Sylvilagus floridanus* is a large-sized species. The coat is long and dense, brown to gray dorsally, and white on the belly, including the tail (Ceballos and Galindo, 1984; Chapman et al., 1982). Because of its wide distribution, the diagnostic characteristics vary according to the location; however, in local areas, in general, it is easily distinguished from other rabbits that inhabit the same place (Chapman et al., 1980).

EXTERNAL MEASURES AND WEIGHT
TL = 335 to 485 mm; TV = 21 to 73 mm; HF = 77 to 102 mm; EAR = 50 to 69 mm.
Weight: 0.9 to 1.8 kg.

DENTAL FORMULA: I 2/1, C 0/0, PM 3/2, M 3/3 = 28.

NATURAL HISTORY AND ECOLOGY: The eastern cottontail is distributed in various habitats, and its preferences vary from season to season and between latitudes and regions (Chapman et al., 1982). It is common to observe them in clearings and on agricultural land. It takes refuge in the weeds and in herbaceous and shrubby vegetation. It feeds on a wide variety of pasture, herbs, seedlings, vegetables, fruits, and grains. It seems to prefer sprouts, so at certain times of the year, it is common to find

Sylvilagus floridanus. Pine forest and grassland. Zoquiapan and Anexas National Park, state of Mexico. Photo: Patricio Robles Gil.

them in the crop fields. It feeds on herbs during summer and woody plants during winter. Reproduction varies among populations, depending on the climate, latitude, and altitude (Chapman and Ceballos, 1990); however, it is carried out throughout the year, happening briefly in higher altitudes and latitudes. The average gestation period is 28 days, producing offspring that vary in size from 3.06 cm to 5.06 cm. The female has an average of three or four litters per year with births of three to five offspring (Ceballos and Galindo, 1984; Chapman et al., 1982). The burrows are inclined holes in the earth, covered with grass or stems of herbs and hairs. The altricial offspring open the eyes in the fourth to fifth day after parturition, and leave the nest between 14 and 16 days (Chapman et al., 1982). Eastern cottontails show two phases of social behavior: (1) basic positions, movements, and vocalizations; in this category the male's position is alert, submissive, rapprochement, launch, courtship, mount, marking, scratch, and fights, while the behavior of the females is threat, attack, racket, and presentation; and (2) social interactions, which include reproductive interactions and dominant-subordinate interactions. The social organization is manifested by male dominance hierarchies that control the social structure of populations and are aggressive even toward other leporidae (Chapman and Ceballos, 1990; Chapman et al., 1982). The eastern cottontail is an important link in the food chain because it is preyed on by hawks (*Buteo*), ravens (*Corvus*), owls (*Bubo*), barn owls (*Tyto alba*), weasels, raccoons, coyotes, foxes, cacomistles, lynx, rattlesnakes (*Crotalus* spp.), and other species (Ceballos and Galindo, 1984; Chapman et al., 1982). They have a variety of parasites. The common ectoparasites are mites (*Ixodidae* and *Trombiculidae*), fleas (*Pulicidae* and *Leptopsyllidae*), and maggots (*Cuterebridae*). Their endoparasites are cestodes (*Taenia* sp. and *Multiceps* sp.), nematodes (*Dermatoxys* sp. and *Ascaris* sp.), and flukes (*Hasstilesia* sp.), among others (Chapman et al., 1980, 1982). They are reservoirs of tularaemia (*Francisella tularensis*). This species is hunted intensively in Mexico as a sport, as a food resource, and for its fur. They can sometimes cause damage to crops, mainly during the early growth of the seedlings (Ceballos and Galindo, 1984; Chapman and Ceballos, 1990).

DISTRIBUTION: The geographical distribution of this species is the most extensive of any other member of the genus *Sylvilagus*. It can be found from southern Canada to central and northwestern South America, including some islands north of Venezuela (Chapman et al., 1980). It has been introduced in North America and Europe. It dwells throughout Mexico with the exception of the Baja California Peninsula, the northern High Plateau, and the eastern portion of the Yucatan Peninsula (Ceballos and Galindo, 1984; Dowler and Engstrom, 1988; Ramírez-Pulido et al., 1986). It has been registered in the states of CA, CH, CO, CS, DF, DU, GJ, HG, JA, MI, MO, MX, NL, NY, OX, QE, SI, SL, SO, TA, TB, TL, VE, YU, and ZA.

VEGETATIONAL ASSOCIATIONS AND ELEVATION RANGE: *S. floridanus* dwells in valleys, plains, and mountains with coniferous and oak forests, tropical forests, grasslands, and xeric scrublands (Chapman and Ceballos, 1990; Nelson, 1907, 1909). It lives from sea level to approximately 3,200 masl (Chapman et al., 1980; Davis, 1944).

CONSERVATION STATUS: *S. floridanus* is a very common species, which benefits from anthropogenic disturbances (Angermann et al., 1990).

Sylvilagus gabbi truei (J.A. Allen, 1890)

Tropical cottontail

Fernando A. Cervantes, Verónica I. Zavala, and Ana L. Colmenares

SUBSPECIES IN MEXICO

Sylvilagus gabbi truei (J.A. Allen, 1890)

DESCRIPTION: Within the genus *Sylvilagus*, *Sylvilagus gabbi* is characterized by its small size. Its back is dark brown, and some individuals are completely gray or black. The belly is lighter, but the throat is also dark. The dorsal part of the tail is dark and,

DISTRIBUTION: This rabbit inhabits the southeastern states of Mexico, reaching Central America. In Mexico, it is found from the south of Tamaulipas to Chiapas and Campeche (Chapman and Ceballos, 1990; Hall, 1981). In Chiapas, it inhabits the north and some places of the center (Álvarez del Toro, 1977). It has been registered in the states of CA, CS, HG, OX, PU, QR, SL, TA, TB, VE, and YU.

unlike in other species of the genus *Sylvilagus*, with the exception of *S. insonus*, the ventral part of the tail has the same color as the back (Allen, 1890; Coates-Estrada and Estrada, 1986; Diersing, 1981; Hall, 1981). Its diploid chromosome number is the smallest (36) of the genus, and it is the only species whose chromosomes have 2 arms (Güereña et al., 1982).

EXTERNAL MEASURES AND WEIGHT
TL = 250 to 420 mm; TV = 8 to 21 mm; HF = 70 to 80 mm; EAR = 39 to 50 mm. Weight: 500 to 950 g.

DENTAL FORMULA: I 2/1, C 0/0, PM 3/2, M 3/3 = 28.

NATURAL HISTORY AND ECOLOGY: In some regions of southeast Mexico, *S. gabbi* shares its habitat with *S. floridanus*, although their local distributions appear to be mutually exclusive in northwest South America (Chapman and Ceballos, 1990; Diersing, 1981; Hall, 1981). In undisturbed areas the rabbits prefer marshes and the margins of rivers (Emmons, 1990). When disturbed, they bounce away and hide among the dense vegetation. This rabbit is solitary and nocturnal, although it is also active at dawn and dusk (Eisenberg, 1989; Emmons and Feer, 1997). Its diet consists of leaves, tender stems, and small seeds. They also, like all lagomorphs, engage in coprophagy, an activity that allows them to obtain essential vitamins that are not assimilated during the first digestion. Females are very prolific and can give birth to up to 25 offspring per year in 5 to 7 litters. They reach sexual maturity at two and a half months and can reproduce at any time of the year, depending on the rainy season (Eisenberg, 1989). In Chiapas, this species reproduces at the end of the year, with a gestation period of 28 days and litters of 3 to 8 individuals. Lactating females have been found in February and, in Tuxtlas, Veracruz, young rabbits are produced in May (Chapman and Ceballos, 1990; Coates-Estrada and Estrada, 1986). The male initiates mating by chasing the female and dancing on one leg around her. When more than one male competes for a female, fights occur between them (Mora and Moreira, 1984).

VEGETATIONAL ASSOCIATIONS AND ELEVATION RANGE: This rabbit inhabits dense tropical forests and hillsides of the Sierra Madre Oriental; it can also be found in rainforests, in deciduous and secondary-growth forests, in pastures near wooded habitats, and at the edge of the jungle (Chapman and Ceballos, 1990; Diersing, 1981). It is common in tropical perennial forests where the trees have heights of more than 50 m such as *Brosimum alicastrum* and *Nectandra ambigens*, in addition to precious woods such as the mahogany (*Swietenia macrophylla*) and the red cedar (*Cedrela mexicana*; Chapman and Ceballos, 1990; Rzedowski, 1978). In the region of Tuxtlas, Veracruz, and in Chiapas it is found in areas of secondary vegetation, crops, and along the edges of the jungle (Álvarez del Toro, 1977; Coates-Estrada and Estrada, 1986). It is found from sea level up to 1,600 m (Diersing, 1981; Güereña et al., 1982).

CONSERVATION STATUS: The tapeti is one of the species of Mexican rabbits whose hunting is permitted since it has been regarded as one of the species that show a great distribution across the country (Cervantes, 1993). Their populations are declining rapidly, however, due to the rapid disappearance of tropical forests. Unfortunately, knowledge about the impact of deforestation on the survival of this species is poor. In Mexico, its abundance apparently declines when extensive areas are deforested. *S. gabbi* survives well in some areas of secondary vegetation and pastures. The real status of the populations is unknown (Chapman and Ceballos, 1990; Cervantes, 1993).

Tres Marias cottontail

Fernando A. Cervantes, Gloria L. Portales, and Francisco J. Romero

SUBSPECIES IN MEXICO

Sylvilagus graysoni badistes Diersing and Wilson 1980
Sylvilagus graysoni graysoni (J.A. Allen, 1877)

DESCRIPTION: *Sylvilagus graysoni* is similar in appearance and size to *S. floridanus*. It is medium in size, and its dorsal coloration is brown to reddish, with reddish pale on the sides. The belly is whitish except for a brown portion in the throat (Diersing and Wilson, 1980; Wilson, 1991). Most of the later extension of the supraorbital process is linked to the cranial vault, as in *S. palustris*. Morphologically, it is closely related to *S. cunicularius*, but its ears are shorter and more reddish on the upper side. In addition, the nasal region of the Tres Marias cottontail is thinner (Diersing and Wilson, 1980; Wilson, 1991). On the other hand, both species have the same chromosome number (42; Chapman and Ceballos, 1990).

EXTERNAL MEASURES AND WEIGHT

TL = 480 mm; TV = 51 mm; HF = 99 mm; EAR = 57 mm.
Weight: unknown.

DENTAL FORMULA: I 2/1, C 0/0, PM 3/2, M 3/3 = 28.

Sylvilagus graysoni. Tropical dry forest. Maria Madre Island, Nayarit. Photo: Horacio Barcenas.

DISTRIBUTION: This rabbit dwells exclusively in the Islands Marias, Nayarit, in the Mexican Pacific between 21 degrees and 22 degrees north latitude and 106 degrees and 107 degrees west longitude. The subspecies *S. graysoni graysoni* is found on the islands María Madre, María Magdalena, and María Cleofas, while *S. graysoni badistes* is distributed only on the island of San Juanito (Chapman and Ceballos, 1990). It has been registered in the state of NY.

NATURAL HISTORY AND ECOLOGY: This rabbit is surprisingly gentle and easy to trap, and has little fear of humans. The etymology of the name of one of the subspecies, *badistes*, comes from the Greek *badio*, which means "step by step" and refers to its not running away when approached. This behavior is apparently due to its few predators, although the red-tailed hawk (*Buteo jamaicensis*) and the raccoon (*Procyon insularis*) feed occasionally on this rabbit (Wilson, 1991). A study in 1976 showed that this species is relatively more abundant on uninhabited islands (Chapman and Ceballos, 1990).

VEGETATIONAL ASSOCIATIONS AND ELEVATION RANGE: This species inhabits deciduous tropical forests, where dominant species such as stick of arco (*Lysiloma divaricata*), copal (*Cyrtocarpa procera*), and papelillo (*Bursera* spp.) occur. It lives from sea level up to about 305 m (Hall, 1981; Rzedowski, 1978).

CONSERVATION STATUS: The islands this species inhabits are being disrupted by intensive human activities. Large tracts of natural vegetation of the island María Madre and María Cleofas have been recently decimated. In Island Maria Magdalena the deliberate introduction of pigs, cattle, goats, and white-tailed deer (*Odocoileus virginianus*) and the accidental introduction of the rat (*Rattus rattus*) have caused major changes in the habitat of *S. graysoni* and threaten its survival (Ceballos and Navarro, 1991; Chapman and Ceballos, 1990). For all these reasons, the rabbit of the Marías Islands is currently considered in danger of extinction (SEMARNAT, 2010). The subspecies *S. graysoni graysoni* seems to be the most affected; a team of the University of Oxford could not find it during an expedition carried out in 1987 to the Island San Juanito (Dooley, 1988). As a result, it has been recommended to totally prohibit hunting and to transform some portions of their habitat into an ecological reserve, as has already been done in the Island Maria Magdalena (Dooley, 1988). It is listed as endangered by the IUCN (2010).

Sylvilagus insonus (Nelson, 1904)

Omiltemi cottontail

Fernando A. Cervantes, Norma A. Figueroa, Francisco J. Romero, and Ana L. Colmenares

SUBSPECIES IN MEXICO

S. insonus is a monotypic species.

DESCRIPTION: *Sylvilagus insonus* is a medium-sized rabbit. Its hair is rough and grayish-brown on the back and dark brown on the belly. Its tail is reddish-brown and, unlike in most wild rabbits, the belly is dark buffy. It is smaller than *S. cunicularius* (Diersing, 1981), but greater in size than *S. gabbi*, its closest relative (Hershkovitz, 1950).

EXTERNAL MEASURES AND WEIGHT

TL = 435 mm; TV = 42.5 mm; HF = 92.5; EAR= 60.9 mm.
Weight: probably around 500 to 950 g.

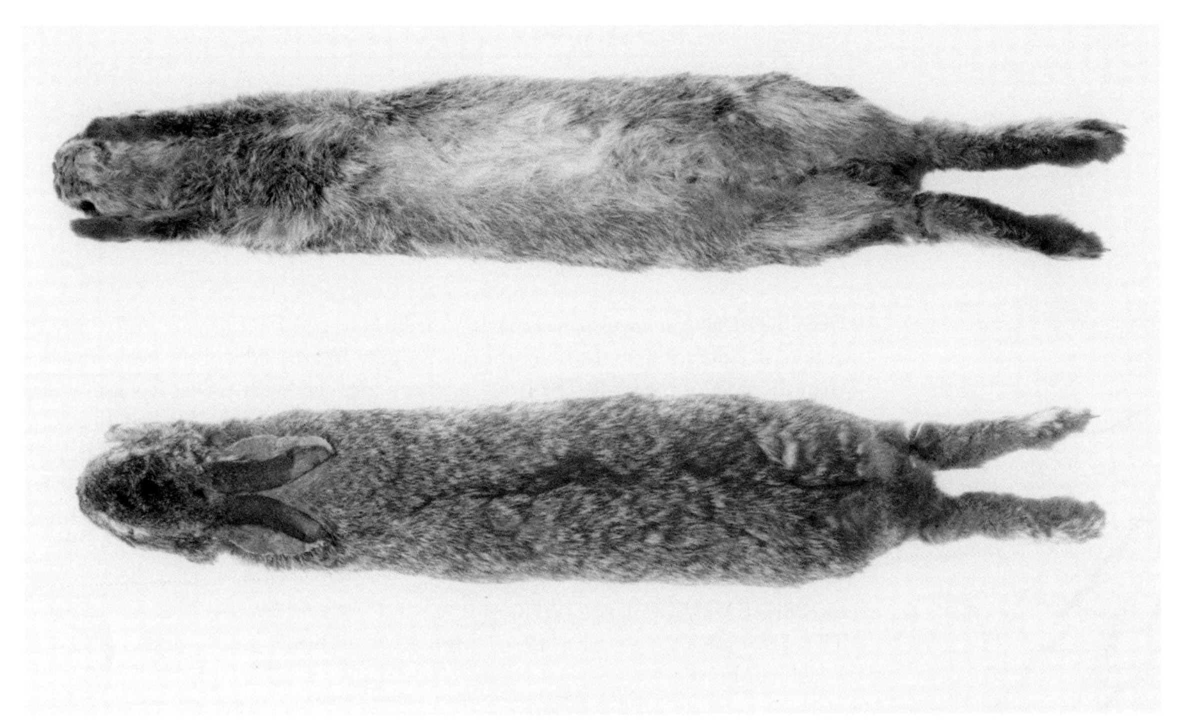

Syvilagus insonus. Typical specimen. Omiltemi, Guerrero. Photo: Don Wilson.

DENTAL FORMULA: I 2/1, C 0/0, PM 3/2, M 3/3 = 28.

NATURAL HISTORY AND ECOLOGY: This species lives in burrows in the forest and has nocturnal habits (Chapman and Ceballos, 1990). It shares its habitat with the Mexican cottontail (*S. cunicularius*; Cervantes, 1993), although it is unlikely that it coexists with Eastern cottontail (*S. floridanus*; Chapman and Ceballos, 1990). There is no additional information on its natural history.

VEGETATIONAL ASSOCIATIONS AND ELEVATION RANGE: The Omiltemi cottontail is mainly associated with coniferous forests containing *Pinus*, *Quercus*, and *Alnus*. It has also been found inhabiting associations of dense cloud forest (Chapman and Ceballos, 1990). This rabbit dwells in areas ranging in elevation from 2,133 m to 3,048 m (Nelson, 1909).

CONSERVATION STATUS: *Sylvilagus insonus* is a very rare species know by fewer than five specimens (Angermann et al., 1990; Cervantes et al., 2004). A recent record was made when a fresh skin of an individual killed by a local hunter was recovered after a century without sightings (Cervantes et al., 2004). It is listed as endangered by the IUCN (2010) and the Mexican Official Norm NOM-059 (SEMARNAT, 2010) because of its rarity and extremely restricted geographic range (less than 500 km^2) (Ceballos and Navarro, 1991). Although the Omiltemi region has been declared a nature reserve by the government of the state of Guerrero, the biggest threat to this rabbit is poaching and habitat destruction. Unfortunately, coniferous forests in this area have been modified and fragmented by intense deforestation (Chapman and Ceballos, 1990).

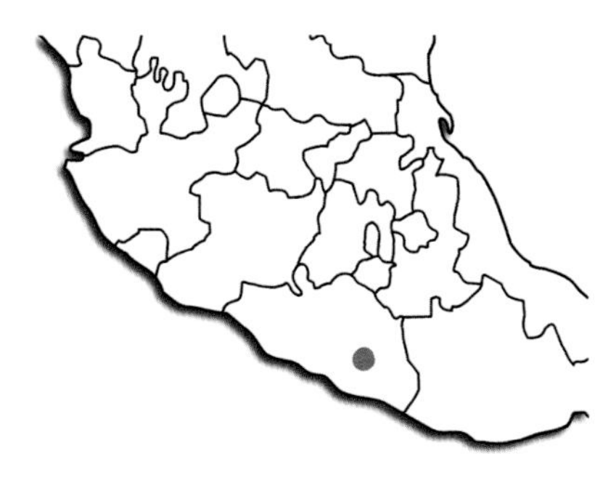

DISTRIBUTION: This rabbit dwells exclusively in the surrounding area of Omiltemi, in temperate and tropical climates, in the mountains of Guerrero, in the Sierra Madre del Sur (Diersing, 1981). It has been registered in the state of GR.

San Jose brush rabbit

Fernando A. Cervantes, Alejandro Rojas Viloria, and Francisco J. Romero

SUBSPECIES IN MEXICO
S. mansuetus is a monotypic species

DESCRIPTION: *Sylvilagus mansuetus* is a medium-sized rabbit of pale buffy-gray or yellowish coloration on the head; the ears are gray and the sides of the neck are paler than the sides of the body. The front legs are ochre clay with darker feet and white tonalities; the hind legs are brown to black and the front of the legs is white. The ventral part of the body is white (Nelson, 1909). This species is distinguished from *S. bachmani*, its closest relative, by its lighter coloration and longer ears (Nelson, 1909). It has a longer and narrower skull, with very elongated nasals, a wide and lifted supraorbital process, and a long jugal (Nelson, 1909).

DISTRIBUTION: This species lives exclusively on the Island San Jose in the southern Gulf of California, Baja California Sur, Mexico (Nelson, 1909). Due to the extent of the island (28 km long and 7.5 km wide); the potential habitat for S. mansuetus does not exceed 194 km^2. It has been registered in the state of BS.

EXTERNAL MEASURES AND WEIGHT
TL = 339 mm; TV = 44 mm; HF = 73 mm; EAR = 63 mm.
Weight: unknown.

DENTAL FORMULA: I 2/1, C 0/0, PM 3/2, M 3/3 = 28.

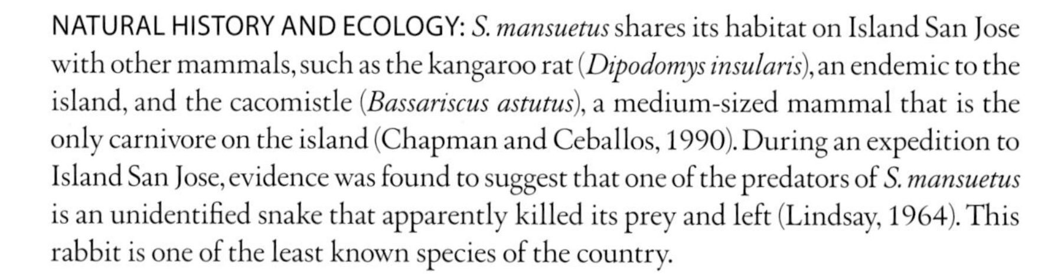

NATURAL HISTORY AND ECOLOGY: *S. mansuetus* shares its habitat on Island San Jose with other mammals, such as the kangaroo rat (*Dipodomys insularis*), an endemic to the island, and the cacomistle (*Bassariscus astutus*), a medium-sized mammal that is the only carnivore on the island (Chapman and Ceballos, 1990). During an expedition to Island San Jose, evidence was found to suggest that one of the predators of *S. mansuetus* is an unidentified snake that apparently killed its prey and left (Lindsay, 1964). This rabbit is one of the least known species of the country.

VEGETATIONAL ASSOCIATIONS AND ELEVATION RANGE: This rabbit inhabits all associations of plants on Island San Jose (Nelson, 1909). The coastal areas are characterized by the presence of mangrove, while the beach vegetation is typical of coastal dunes (Chapman and Ceballos, 1990). In the rocky areas, a species of thorny scrub grows, mixed with species of scrub without thorns and small leaves. In the brooks, patches of deciduous lowland jungle are found. The main plants that live in the island are mesquite (*Prosopis*), cholla (*Opuntia*), governor (*Larrea*), torote (*Bursera*), cudweed (*Ambrosia*), and romerito (*Suaeda*) (Rzedowski, 1978). This mammal lives from sea level up to the low mountains at an altitude not determined (Nelson, 1909). The maximum elevation of the island is 633 masl (Chapman and Ceballos, 1990).

CONSERVATION STATUS: It is believed that this species is common in its habitat (Angermann et al., 1990); however, an expedition in 1962 carefully examined several points of Island San Jose and found no individuals (Lindsay, 1962). Currently, the status of this rabbit is unknown, as several mammals of the islands of Baja California have been displaced by the introduction of exotic mammals such as cats and rats (Chapman et al., 1990). Studying this species is considered a high priority (Chapman and Ceballos, 1990). The IUCN lists this species as near threatened (2010). In Mexico it is endangered (SEMARNAT, 2010).

Robust cottontail

Fernando A. Cervantes and Irelia López-Hernández

SUBSPECIES IN MEXICO

S. robustus is a monotypic species.

It was considered a subspecies of *S. floridanus*, but an assessment of its morphology and genetics showed that they are different species (Ruedas, 1998).

DESCRIPTION: Unlike other members of its genus, *Sylvilagus robustus* is relatively large, even larger than *S. floridanus holzneri* (Nelson, 1909). Its dorsal coloration is buffy gray, slightly pale, and a little grayer than *S. f. holzneri*, but similar to *S. cognatus*, and has a patch of steel-gray in the dorsal region of the tail. *S. robustus* has light rust-colored or cinnamon rusty limbs, and large whitish legs with a thick layer of hair. The ears are large and gray. The skull is big, long, and proportionally narrow (Nelson, 1909). The face is long and thin. The bullae are relatively large. A dental feature distinguishes *S. robustus* from *S. floridanus*: in *robustus* the wall above the internal re-

Sylvilagus robustus. Pine forest. Maderas del Carmen Protected Area, Coahuila. Photo: Jonas A. Delgadillo V.

DISTRIBUTION: The geographical distribution of this species comprises west Texas and adjacent areas in New Mexico and several locations in Mexico in the vicinity of the village of Ocampo and the Sierra del Carmen, Coahuila, near the northern border (Ruedas, 1998). Therefore, its distribution is very restricted, especially in Mexico. It has only been registered in the state of CO.

entrant angle of enamel of the third premolar contains a thin layer of enamel, while in *floridanus* the layer is thick (Ruedas, 1998).

EXTERNAL MEASURES AND WEIGHT
TL = 375 to 463 mm; TV = 39 to 65 mm; HF = 103 mm; EAR = 87 to 104 mm. Weight: 745 to 1,200 g.

DENTAL FORMULA: I 2/1, C 0/0, PM 3/2, M 3/3 = 28.

NATURAL HISTORY AND ECOLOGY: Little is known about the biology of this species. In Texas, it inhabits forested regions of the Chisos Mountains, Davis Mountains, and Fort Davis (Nelson, 1909; Ruedas, 1998). In the Sierra of la Madera, in Coahuila, it inhabits plant associations of pine-oak, with several species of oaks (*Quercus hypolecucoides*, *Q. gravesii*, *Q. laceyi*, and *Q. rough*) and pines (*Pinus edulis*, *P. strobiformis*, and *P. ponderosa*). In the undergrowth of its habitat, herbaceous plants and dominant grasses include *Bouteloa gracilis*, *B. curtipendula*, *Schizachyrium neomexicanum*, and *Piptochaetum pringei*.

VEGETATIONAL ASSOCIATIONS AND ELEVATION RANGE: *S. robustus* dwells in forests of oaks dominated by evergreen oaks, especially *Quercus grisea*, and junipers, particularly *Juniperus depeana* (Ruedas, 1998). In the Sierra del Carmen it is found in the wettest canyons (Delgadillo Villalobos et al., 2005). It has been found between 1,500 masl and 2,000 masl, being more common above 1,800 masl (Ruedas, 1998).

CONSERVATION STATUS: *S. robustus* is an endemic species found in isolated mountains and is considered rare (Ruedas, 1998). It is assumed that it dwells in the mountains of Guadalupe in Texas, but currently it has been eradicated from that area. Ruedas (1998) noted that only one specimen has been collected in the Davis Mountains in the last 20 years. These circumstances, coupled with the human disturbance of their habitat, led to the recommendation that it be considered a protected species by the state of Texas and endangered by the IUCN (Ruedas, 1998; IUCN, 2010). On the Mexican side of its distribution, its conservation status is unknown (Ruedas, 1998).

Southern flying squirrel (*Glaucomys volans*). Oak forest. Chapa de Mota, state of Mexico. Photo: Gerardo Ceballos.

Order Rodentia

Gerardo Ceballos

The apparently fragile and harmless kitty was abandoned one afternoon on Estanque Island, an island near Angel de la Guarda Island . . . In the dead of night, alone and starving, a sudden movement caught his attention; he instinctively jumped and caught an endemic mouse (Peromyscus guardia guardia) *with his claws! The following months the cat managed to survive by eating these mice, until finally in the summer of 1997 the last specimen of this subspecies died.*

— G. Ceballos, 1999

With more than 2,200 living species, rodents are by far the largest order of mammals; 42% of the total species of mammals belong to the order Rodentia (Wilson and Reeder, 2005). They are found around the world, except on some very isolated islands and in the extreme polar regions. They have a wide variety of forms. There are, however, two distinctive features of the order: the continuous growth of the incisors and the lack of canines. The continuous growth of the incisors implies a constant wearing down to prevent excessive growth (Nowak, 1999b). In Mexico, this order is represented by 8 families and 243 species, which include squirrels (*Sciuridae*), beavers (*Castoridae*), gophers (*Geomyidae*), kangaroo rats (*Heteromyidae*), rats and mice (*Muridae*), acuchis (*Dasyproctidae*), agoutis (*Agoutidae*), and porcupines (*Erethizontidae*; List et al., 1999).

Some groups, such as the agoutis, are diurnal, but most, such as the flying squirrels, are nocturnal. The majority of species are terrestrial, but some groups specialize in other types of habitats, such as squirrels that live in trees, gophers that live underground, and beavers that are semi-aquatic. Some species, such as prairie dogs, form aggregations of thousands of individuals, and their colonies can occupy thousands of hectares (Ceballos et al., 1993). Flying squirrels (*Glaucomys volans*) are gregarious during winter in order to withstand cold temperatures (Nowak, 1999b).

Reproduction may be restricted to a defined period of the year or may occur throughout the year. The number of offspring is low in the larger or more specialized species and very high in generalist species. The offspring in most of the rats and mice are generally precocious, but those of some species are altricial (Eisenberg, 1981). Most species are herbivores, and their diet is based on seeds, bark, fruits, leaves, bulbs, and foliage. They usually supplement their diet with insects, other invertebrates, and even small vertebrates. With these eating habits they play an important role in all ecosystems, acting as dispersers and predators of seeds and seedlings, hence affecting the structure and function of plant communities (Ceballos and Galindo, 1994; Eisenberg, 1981). In general, most species do not face conservation problems because they have survived through changing conditions and have high birth rates. Specialized species or those with restricted distributions, however, are very susceptible to extinction because of anthropogenic activities.

Family Sciuridae

Gerardo Ceballos and Don E. Wilson

The family Sciuridae includes prairie dogs, marmots, and both terrestrial and flying squirrels. It has a wide distribution throughout the world, with the exception of Australia, Madagascar, part of South America, and remote islands. It consists of 60 genera and 278 species (Wilson and Reeder, 2005). In Mexico, it is represented by 11 genera and 36 species, of which 13 are endemic. The genera *Ammospermophilus*, *Callospermophilus*, *Cynomys*, *Glaucomys*, *Ictidomys*, *Otospermophilus*, *Tamias*, *Tamiasciurus*, and *Xerospermophilus* are limited to arid and temperate areas. Prairie dogs (*Cynomys mexicanus* and *C. ludovicianus*), flying squirrels (*Glaucomys volans*), and some squirrels such as *Tamiasciurus mearnsi* are at risk of extinction from destruction and fragmentation of their habitat.

Subfamily Sciurinae

The subfamily Sciurinae contains three genera of tree squirrels, chickarees, and flying squirrels. Thirty-six species are known from Mexico.

Ammospermophilus harrisii (Audubon and Bachman, 1854)

Harris's antelope squirrel

Reyna A. Castillo

SUBSPECIES IN MEXICO

Ammospermophilus harrisii harrisii (Audubon and Bachman, 1854)
Ammospermophilus harrisii saxicola (Mearns, 1896)

DESCRIPTION: *Ammospermophilus harrisii* is a medium-sized squirrel. There is geographic variation in body size, with larger individuals in populations of the northeast (Best et al., 1990a). The tail is medium to short (25% to 33% of the total length), and the ears are short and wide. The eyes are moderate in size and are surrounded by a white circle. The body is thin; the head is small and delicate; the neck and limbs are long. On the forefeet, the thumb is reduced to a small tubercle with a dull claw, and the second toe is the longest. On the hind feet, the middle digit is the longest. The paws are very hairy. The claws are slightly compressed and somewhat curved. In winter, the dorsal coat is pale brown to brownish-black, but in summer these colors are paler; the ventral part of the tail is a mixture of black and white (Best et al., 1990a). It can be distinguished from other squirrels by the two white lines on both sides of the back.

DISTRIBUTION: This species is found in the Sonoran desert from Arizona and New Mexico in the United States to Sonora, Mexico. It has been registered only in the state of SO.

EXTERNAL MEASURES AND WEIGHT

TL = 229 to 246 mm; TV = 70 to 85 mm; HF = 37 to 42 mm; EAR = 12 to 14 mm. Weight: 122 g.

DENTAL FORMULA: I 1/1, C 0/0, PM 2/1, M 3/3 = 22.

NATURAL HISTORY AND ECOLOGY: *A. harrisii* inhabits open sites of plains and valleys, and rocky sides of canyons and rugged areas (Best et al., 1990a). They dig

Ammospermophilus harrisii. Scrubland. Puerto Libertad, Sonora. Photo: Patricio Robles Gil.

burrows under trees and shrubs. The diet consists mainly of fruits and seeds of chollas (*Opuntia* spp.), biznagas (*Ferocactus* spp.), legumes, and pitahaya (*Stenocereus thurberi*), whose juice frequently stains them (Best et al., 1990a). They climb chollas and pitahaya and are notorious for this behavior. Their cheek pouches have considerable storage capacity. Reproduction takes place from January to June, producing a single litter per year of 6.5 individuals on average after a gestation period of approximately 30 days (Yensen and Valdés Alarcón, 1999). They are active even in the hottest hours of the day so it is very easy to find them in the field. Their densities vary from 0.80 to 0.36 individuals per hectare (Best et al., 1990a). In some sites, they are sympatric with *Spermophilus tereticaudus* (Yensen and Valdés Alarcón, 1999).

VEGETATIONAL ASSOCIATIONS AND ELEVATION RANGE: This species is limited to xeric vegetation, from sea level up to 1,350 m (Hoffmeister, 1956).

CONSERVATION STATUS: The current situation is unknown; these squirrels are widely distributed in regions with little anthropogenic impact. Apparently, this species is not threatened (G. Ceballos, pers. comm.)

Ammospermophilus insularis Nelson and Goldman, 1909

Espiritu Santo Island antelope squirrel

Aimeé de la Cerda and Eric Mellink

SUBSPECIES IN MEXICO

A. insularis is a monotypic species.
The taxonomic relationships between this species and the rest of the genus are still unclear (Best et al., 1990c; Mascarello and Boyes, 1980).

DESCRIPTION: *Ammospermophilus insularis* is a terrestrial squirrel of medium size; it has a short tail and small ears. The dorsal coloration is brownish-cinnamon with a white line on the side; the belly is yellowish-white. During summer, its coat is shorter, rougher, and paler than in winter (Best et al., 1990c; Hall, 1981).

DISTRIBUTION: *A. insularis* is an endemic species of Espiritu Santo and Partida islands off Baja California Sur (Yensen and Valdés-Alarcón, 1999). Despite the fact that it is generally considered exclusive to Espiritu Santo Island, there is a specimen from nearby Partida Island. Both islands are connected during low tide (Yensen and Valdés-Alarcón, 1999).

EXTERNAL MEASURES AND WEIGHT

TL = 210 to 240 mm; TV = 71 to 83 mm; HF = 36 to 42 mm; EAR = 8 to 11 mm.
Weight: 70 to 100 g.

DENTAL FORMULA: I 1/1, C 0/0, PM 2/1, M 3/3 = 22.

NATURAL HISTORY AND ECOLOGY: *A. insularis* is an exclusively diurnal squirrel. It is terrestrial, although it is commonly seen climbing in the vegetation. It is found in sandy areas with the typical vegetation of the islands it inhabits. The island of Espiritu Santo is 6 km from the coast of the Baja Peninsula, almost across from La Paz. The diet includes all kinds of seeds, as well as green parts of plants and fruit of cacti, shrubs, and herbaceous plants such as tomatillo (*Lycium breviceps*) (López-Forment et al., 1996). The alarm cry is a low-frequency sound that seems adapted to its rocky habitat (Bolles, 1988). On Espiritu Santo Island, they are sympatric with an endemic hare (*Lepus insularis*). The potential predators of this species include birds of prey and snakes.

VEGETATIONAL ASSOCIATIONS AND ELEVATION RANGE: This antelope squirrel inhabits desert scrublands characterized by herbaceous plants and shrubs such as palo blanco (*Lysilosoma candida*) and cacti such as cardones (*Pachycereus* sp.) and garambullos (*Lophocereus* sp.). It is distributed from sea level to the highest parts of the hills at 595 masl.

CONSERVATION STATUS: This species is considered threatened with extinction (Ceballos and Navarro, 1991; SEMARNAT, 2010) because of its extremely restricted distribution and because of the presence of domestic cats (Ceballos and Rodriguez, 1993; López-Forment et al., 1996; Yensen and Valdés-Alarcón, 1999). There have been no appropriate studies, however, to determine its current situation.

Ammospermophilus insularis. Scrubland. Isla Espíritu Santo, Baja California Sur. Photo: Eric Mellink.

Ammospermophilus interpres (Merriam, 1890)

Texas antelope squirrel

Guadalupe Téllez-Girón and Gerardo Ceballos

SUBSPECIES IN MEXICO

Ammospermophilus interpres is a monotypic species.

By the morphology of the skull and skeleton, *A. interpres* seems more like *A. nelsoni* and *A. leucurus* than its genetically closest species *A. harrisii* (Hafner, 1984).

DESCRIPTION: *Ammospermophilus interpres* is a terrestrial squirrel, medium in size. It can be distinguished by having tail and feet longer than those in all other species of the genus. It is slightly smaller and darker than *A. leucurus*. It has two black bands on the ventral part of the tail, perhaps the most distinguishing feature of the species. The back is grayish; the shoulders, hips, and backs of the hind feet and forefeet are ochre colored. It has two white bands that extend from the back to the hip. The belly is cream-colored (Baker, 1956; Best et al., 1990a). The skull is wide, short, slightly shallow, and flattened.

DISTRIBUTION: This species is found from west Texas and the center of New Mexico, in the United States, through Chihuahua, Durango, Coahuila, and Zacatecas, in northern Mexico (Best et al., 1990a). It has been registered in the states of CH, CO, DU, and ZA.

EXTERNAL MEASURES AND WEIGHT

TL = 220 to 235 mm; TV = 68 to 84 mm; HF = 36 to 40 mm; EAR = 8 to 11 mm.
Weight: 98 to 122 g.

DENTAL FORMULA: I 1/1, C 0/0, PM 2/1, M 3/3 = 22.

NATURAL HISTORY AND ECOLOGY: The habits of this antelope squirrel are diurnal, with two peaks of activity in summer, between 07:00 hrs and 19:00 hours, avoiding the hottest periods of the day (Baker et al., 1980; Blair and Miller, 1949; Hermann, 1950). They are cautious, shy, and very nervous (Baker, 1956; Borrel and Bryant, 1942). They often pose on the upper parts of trees or large shrubs. They run rapidly between cacti, shrubs, and rocks in search of food or shelter. They emit an alarm call much like that of the other species of the genus (Bailey, 1905; Bolles, 1988). Such calls have been interpreted in two ways. A traditional assumption is that it serves to alert other squirrels to the presence of a predator (Eisenberg, 1981). A more recent hypothesis indicates that, in reality, it is a way to draw the attention of the predator, indicating that it has already been detected (Hasson et al., 1989). In the Chihuahuan desert, *A. interpres* is always associated with communities of vegetation with dominant species such as gobernadora, tar, lechuguilla, and sotol. It also has been found sporadically in grasslands and forests of oak-cedar (Findley and Caire, 1977; Findley et al., 1975; Packard, 1977). It lives in canyons and on cliffs (Bailey, 1905; Baker, 1977); occupying abandoned holes of other rodents or holes between the rocks, they build their nests with various materials such as hair, feathers, and grasses (Davis, 1978). Their diet mainly consists of seeds, berries, and the soft parts and fruit of cactus. Some seeds are stored in their shelters, including species such as sotol (*Dasylirion*), mesquite (*Prosopis*), creosote (*Larrea*), yucca (*Yucca*), juniper (*Juniperus*), and some ripe fruits of cacti (Bailey, 1932; Davis, 1978).

VEGETATIONAL ASSOCIATIONS AND ELEVATION RANGE: *A. interpres* is found in highlands from 540 masl to 1,800 masl, but it is very common between 1,050 masl and 1,650 masl (Schmidly, 1977b). It inhabits the xeric thickets of the Chihuahuan desert, where species such as creosote (*Larrea*), lechuguilla (*Agave*), sotol (*Dasylirion*), and tar (*Flourensia* sp.) (Packard, 1977) are common. They are also found in pastures and associations of oak and junipers (Findley and Caire, 1977).

CONSERVATION STATUS: *A. interpres* is a relatively common species that presently does not face conservation problems.

Ammospermophilus leucurus (Merriam, 1889)

White-tailed antelope squirrel

Gerardo Ceballos and Guadalupe Téllez-Girón

SUBSPECIES IN MEXICO
Ammospermophilus leucurus canfieldae Huey, 1929
Ammospermophilus leucurus extimus Nelson and Goldman , 1929
Ammospermophilus leucurus leucurus (Merriam, 1889)
Ammospermophilus leucurus peninsulae (J.A. Allen, 1893)

DESCRIPTION: *Ammospermophilus leucurus* is a terrestrial squirrel that is medium in size. It has short ears and relatively long feet. Males are slightly larger than females. The hair color varies from gray to brown and cinnamon. It has a lateral cream or white band from the flanks to the hips. It has white rings around the eyes. The coat during winter is darker. The belly is creamy white. The tail is bicolored, dark dorsally and white ventrally (Hall, 1981; Hoffmeister, 1986; Howell, 1938).

DISTRIBUTION: *A. leucurus* is found from southeast Oregon and Idaho in the United States through the Peninsula of Baja California in Mexico (Hall, 1981; O'Farrell and Clark, 1984). The Colorado River represents a barrier in the distribution of the species; hence it is not found in Sonora. It has been registered in the states of BC and BS.

EXTERNAL MEASURES AND WEIGHT
TL = 194 to 239 mm; TV = 54 to 77 mm; HF = 37 to 41 mm; EAR = 7 to 10 mm.
Weight: 103 to 117 g.

DENTAL FORMULA: I 1/1, C 0/0, PM 2/1, M 3/3 = 22.

NATURAL HISTORY AND ECOLOGY: White-tailed antelope squirrels are diurnal, and they are most active in summer and early winter. In summer they show two peaks of activity, avoiding the hottest part of the day. In winter the activity is unimodal (Yensen and Valdés Alarcón, 1999). Alternating periods of activity of 20 to 30 minutes, interspersed with periods of rest, help them to avoid high temperatures. Although they do not hibernate, they do increase body mass during colder parts of the year (Kenagy and Bartholomew, 1979). Their home ranges cover between 3 ha and 8 ha (Belk and Smith, 1991). They are often seen in open areas of rocky surfaces with some shrubs. Their burrows are located in the ground and the entry is generally located at the base of a bush. They use abandoned burrows of kangaroo rats (*Dipodomys*) and other mammals (Bradley, 1967; Grinnell and Dixon, 1918). Their diet is mainly composed of seeds, fruit, leaves, insects, and occasional small vertebrates. Their diet depends on the availability of resources and water (Karasov, 1983, 1985). The breeding season takes place from February to June. Births occur in March and April, after a gestation period of 30 to 35 days (Belk and Smith, 1991). The average litter size is nine (Grinnell and Dixon, 1918). Their main predators are coyotes, foxes, badgers, lynx, weasels, and snakes.

VEGETATIONAL ASSOCIATIONS AND ELEVATION RANGE: *A. leucurus* is restricted to scrublands with succulents, mixed annual scrublands, and coniferous forests of cedars and pines (Hoffmeister, 1986; Honeycutt et al., 1981). They are found from sea level up to 1,000 m.

CONSERVATION STATUS: These squirrels do not face major conservation problems because they are found over a large area with few significant impacts of anthropogenic activities. Actually, they thrive in disturbed areas such as roads and camping sites in Baja California (Yensen and Valdés Alarcón, 1999). They are protected in El Vizcaino Biosphere Reserve and the San Pedro Martir National Park in Baja California.

Callospermophilus madrensis Merriam, 1901

Sierra Madre ground squirrel

Manuel Valdéz Alarcón

DISTRIBUTION: *C. madrensis* is an endemic to Mexico with a distribution restricted to the north of the Sierra Madre Occidental in the municipality of Temósachic, in the west-central Chihuahua and neighboring areas in Sonora, to southern Durango (Baker and Greer, 1962; Caire, 1997; Hall, 1981; List et al., 1998; López-Wilchis and López Jardines, 1998; Servin et al., 1996). It has been registered in the states of CH, DU, and SO.

SUBSPECIES IN MEXICO

C. madrensis is a monotypic species.

DESCRIPTION: *Callospermophilus madrensis* is a small arboreal squirrel. It resembles *C. lateralis* but it is smaller and its tail is shorter. Its coloration is reddish-brown mixed with gray tones. The hair is opaque. It has black bands on the sides that are short and diffuse, and white bands from the base of the tail. The skull is more arched and narrower than in *Spermophilus lateralis* (Hall, 1981; Howell, 1938).

EXTERNAL MEASURES AND WEIGHT

TL = 215 to 253 mm; TV = 40 to 82 mm; HF = 31 to 42 mm; EAR = 18 to 22 mm. Weight: 109 to 158 g.

DENTAL FORMULA: I 1/1, C 0/0, PM 2/1, M 3/3 = 22.

NATURAL HISTORY AND ECOLOGY: There is very little information on the biology of this squirrel. They are diurnal and can be observed easily on the edge of glades and

Callospermophilus madrensis. Conifer forest. Sierra Madre Occidental, Chihuahua. Photo: Rurik List.

roads. Pregnant females have been found in May and June, and lactating females in July (Anderson, 1972; Best and Thomas, 1991b; Webb and Baker, 1984).

VEGETATIONAL ASSOCIATIONS AND ELEVATION RANGE: *C. madrensis* is mainly found in mixed forests of coniferous trees with dominant species such as fir (*Pseudotsuga*), pine (*Pinus*), juniper (*Juniperus*), and poplars (*Populus*; Best and Thomas, 1991b). The elevational range varies from 2,700 masl to 3,750 masl.

CONSERVATION STATUS: There is very little information on this species, which is endemic to a relatively restricted area in the Sierra Madre Occidental. In some regions, it is relatively abundant even in disturbed areas. It is not likely to present serious problems of conservation. It is considered as of Special Protection (SEMARNAT, 2010).

Cynomys ludovicianus (Ord, 1815)

Black-tailed prairie dog

Jesús Pacheco and Gerardo Ceballos

SUBSPECIES IN MEXICO

Cynomys ludovicianus ludovicianus (Ord), 1815

DESCRIPTION: The black-tailed prairie dog is a very large rodent with a chubby body, short legs, and a short tail. The ears are small and round (Olin and Thompson, 1982). Females have 8 to 12 mammae (Hall, 1981; Nowak, 1999b). Males are on average 10% to 15% heavier than females (Hoogland, 1995). The forefeet are distinguished by having five digits with strong and elongated claws; the claws of the hind feet are smaller in size (Hall, 1981; Olin and Thompson, 1982). The dorsal coloration varies from yellowish-brown to reddish-brown with some black hairs; the belly is paler. The tip of the tail is black.

EXTERNAL MEASURES AND WEIGHT

TL = 305 to 430 mm; TV = 30 to 115 mm; HF = 52 to 67 mm; EAR = 10 to 14 mm. Weight: 700 to 1,400 g.

DENTAL FORMULA: I 1/1, C 0/0, PM 2/1, M 3/3 = 22.

NATURAL HISTORY AND ECOLOGY: Black-tailed prairie dogs are herbivores with semi-fossorial habits. They inhabit prairies and extensive plains. They are diurnal and gregarious, living in large colonies, or towns, of up to millions of individuals (Ceballos et al., 1993; Hoogland, 1995; King, 1955). They dig their own burrows, which are formed by elaborate tunnels of 100 mm to 130 mm in diameter, 4 m to 34 m long, and 1 m to 3 m deep. A sign of presence of prairie dogs consists of conical-shaped earth mounds that surround the entries of the burrows, looking like small volcanoes (Olin and Thompson, 1982). In Mexico, the average number of burrows per hectare varies from less than 10 to 52 (Ceballos et al., 1999; Ceballos et al., 2010b). They reproduce from February to March (Whitaker, 1980). Four to five pups are born in May, and by mid-summer they are capable of their own food (Olin and Thompson, 1982). Among their main predators are the coyote, lynx, desert fox, badger, black-legged ferret, eagles, hawks, and rattle snakes (Hoogland, 1995). Around 98% of their

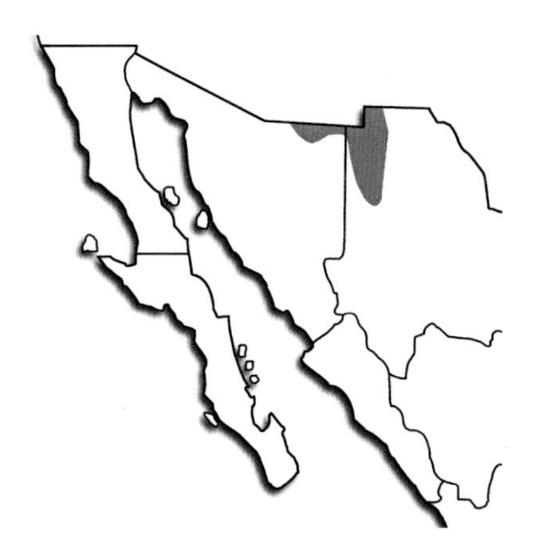

DISTRIBUTION: *Cynomys ludovicianus* occurs from the south of Canada, through the Great Plains of the center of the United States, to northern Mexico. In Mexico it is exclusively found in the pastures of northwestern Chihuahua and northeastern Sonora (Anderson, 1972; Ceballos et al., 1993). In Sonora, it is found only in a small area to the north of Cananea, near the border with Arizona. In the state of Chihuahua, it is found 50 km from the border with the United States, and is limited to the north and west by the Sierra Madre Occidental, and to the south and east by the desert of Chihuahua (Castillo-Gamez et al., 2005; Ceballos et al., 1999). It has been registered in the states of CH and SO.

Cynomys ludovicianus. Grassland. Janos-Casas Grandes, Chihuahua. Photo: Gerardo Ceballos.

diet consists of different grasses and herbs; occasionally, they consume grasshoppers and other insects (Foster and Hygnstrom, 1990). A wide variety of plants and animals inhabit the colonies because prairie dogs change the structure and composition of the vegetation with their feeding activities and conduct. Because of this they are considered key to the creation of environmental conditions that are essential for maintaining regional biodiversity (Ceballos et al., 1999; Ceballos et al. 2000; List and MacDonald, 1998; Manzano et al., 1999; Miller et al., 2000, Sharp and Uresk, 1990).

VEGETATIONAL ASSOCIATIONS AND ELEVATION RANGE: This species is closely associated with meadows and pastures, characterized by grasses and ground plants. The colonies are usually surrounded by high grasses or scrub with abundant shrubs (Ceballos et al., 1993, 1999). In Mexico it is found from 1,400 masl to 1,600 masl.

CONSERVATION STATUS: In Mexico, prairie dogs are considered threatened due to the significant reduction of their area of geographical distribution in recent decades (Ceballos et al., 1993; Miller et al., 1994). Currently, the species is restricted to a small region of Sonora and Chihuahua. The colonies in Chihuahua are, however, the most extensive in North America, as they encompass more than 35,000 ha in the region of Janos and Casas Grandes (Ceballos et al., 1993, 1999; Miller et al., 1994). The intense pressure on the Mexican populations, especially from the rapid deterioration of this species' habitat due to the encroachment of the agricultural frontier, has led to the disappearance of prairie dogs in extensive prairies formerly occupied by them (Ceballos et al., 1993). It is therefore imperative to establish efficient measures for their conservation. Recently, the Janos Biosphere Reserve, with more than half a million hectares, was decreed to protect the last colonies of this species in Mexico. The protection of that region offers, likewise, a unique opportunity to preserve and maintain the ecological and evolutionary processes of the grasslands of North America.

Mexican prairie dog

Jesús Pacheco and Gerardo Ceballos

SUBSPECIES IN MEXICO
C. mexicanus is a monotypic species (Hollister, 1916; Pizzimenti, 1975)

DESCRIPTION: *Cynomys mexicanus* is one of the largest species of the genus *Cynomys*, only slightly smaller than *C. ludovicianus*. In general, the appearance of both species is similar. They both have a robust body, short legs, and a short tail. The ears are small. The dorsal coloration of *C. mexicanus* is less red, with more interspersed black hair, which gives it a darker appearance. The color of the belly is paler and the tip of the tail is black. Females have eight mammae (four pectoral and four inguinal). The skull is wide and angular with expansive malar bones (Ceballos and Wilson, 1985). Pizzimenti (1975) found sexual dimorphism in 13 of 18 morphological characteristics.

EXTERNAL MEASURES AND WEIGHT
TL = 385 to 440 mm; TV = 88.7 mm; HF = 60.4 mm; EAR = 10 to 14 mm.
Weight: 700 to 1,400 g.

DENTAL FORMULA: I 1/1, C 0/0, PM 1/1, M 3/3 = 20.

NATURAL HISTORY AND ECOLOGY: Mexican prairie dogs are terrestrial rodents with semi-fossorial habits and are exclusively diurnal. They remain active throughout the year, and have one or two peaks of daily activity (Treviño-Villarreal, 1990). They build complicated tunnels; the surface of the land inhabited by prairie dogs is formed of mounds, which gives the landscape a peculiar appearance. Each mound serves as an observation and vigilance post against predators (Ceballos and Wilson, 1985). They are social animals, forming colonies that occupy large tracts of land. The colonies are well organized and are composed of several family groups with complex social

Cynomys mexicanus. Grassland. Tokio, Nuevo León. Photo: Gerardo Ceballos.

interactions between individuals (Ceballos and Wilson, 1985). Mating takes place in mid-January and early February, extending until April (Mellink and Madrigal, 1993). The gestation period lasts one month, as in other species of the genus. The reproductive cycle is annual, with only one litter per year and two to eight offspring per litter (Koford, 1958). Births occur in mid-February and early March; lactation lasts from February to April, and juveniles emerge in April (Mellink and Madrigal, 1993). Weaning occurs at 41 to 50 days after birth (Ceballos and Wilson, 1985). Their predators include the coyote, badger, desert fox, golden eagle (*Aquila chrysaetos*), red-tailed hawk (*Buteo jamaicensis*), and rattlesnakes (Ceballos and Wilson, 1985; Mellink and Madrigal, 1993). Although it occasionally feeds on cacti and yuccas, its main food consists of different species of grasses and herbs. The seasonal changes in its diet are attributed to changes in the phenology of plants (Medina and De la Cruz, 1976; Mellink and Madrigal, 1993). It is a key species since its activities affect significantly the structure, composition, and role of the ecosystem (Miller et al., 2000).

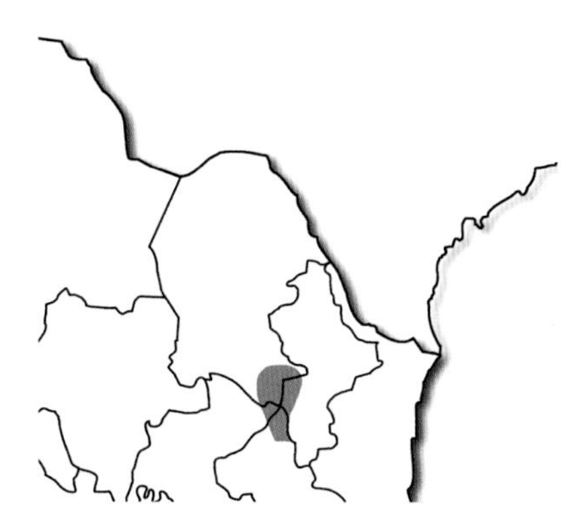

DISTRIBUTION: *C. mexicanus* is an endemic species of Mexico that probably originated as a population of *C. ludovicianus* that was isolated during the Pleistocene (Hoffman and Jones, 1970). The nearest population of that species is separated by 530 km (Baker, 1956). Its distribution is restricted to a small region of approximately 500 km^2, located within Coahuila, Nuevo Leon, Zacatecas, and San Luis Potosí (Ceballos and Wilson, 1985; Ceballos et al., 1993; Treviño-Villarreal and Grant, 1998; Treviño-Villarreal et al., 1998). It is limited to the north and east by the Sierra Madre Oriental, and to the south and west by hills and plains with more arid vegetation (Ceballos et al., 1993). The populations of Zacatecas have disappeared. It has the most southern distribution among the species of the genus. It has been registered in the states of CO, NL, SL, and ZA.

VEGETATIONAL ASSOCIATIONS AND ELEVATION RANGE: The Mexican prairie dog is closely associated with valleys and pastures in the mountains characterized by grasses and ground plants of less than 5 cm in height and well-drained soil (Ceballos and Mellink, 1990). Plant species common in the colonies include the liendrillo (*Muhlembergia repens*), *Psora decipens*, *Haliomolobus* sp., *Haplopappus spinosus*, *Physalis* sp., *Guara coccinea*, and *Calylophus hartwegii* (Ceballos et al., 1993). Frequently, the vegetation of the areas surrounding prairie dog colonies is more arid, which forms an impenetrable barrier for their dispersal (Baker, 1956). This species is found at elevations between 1,600 masl and 2,200 masl (Ceballos et al., 1993).

CONSERVATION STATUS: The Mexican prairie dog is a Mexican endemic, which is now classified as endangered by the International Union for Conservation of Nature (IUCN), the U.S. Fish and Wildlife Service, and SEMARNAT (Ceballos and Navarro, 1991; Hilton-Taylor, 2000; SEMARNAT, 2002). Its future is uncertain because there is intense human pressure throughout its range, particularly from agricultural activities. The decline and fragmentation of the habitat has caused the loss of 62% of its historic geographic area. The populations in Zacatecas are extinct and the colonies of San Luis Potosí currently occupy only 2% (10 km^2) of the geographical area they occupied 50 years ago (Ceballos and Mellink, 1990; Treviño-Villarreal and Grant, 1998).

Glaucomys volans (Linnaeus, 1758)

Southern flying squirrel

Gerardo Ceballos and Patricia Manzano

SUBSPECIES IN MEXICO

Glaucomys volans chontali Goodwin, 1961
Glaucomys volans goldmani (Nelson, 1904)
Glaucomys volans herreranus Goldman, 1936
Glaucomys volans guerreroensis Diersing, 1980
Glaucomys volans madrensis Goldman, 1936
Glaucomys volans oaxacensis Goodwin, *1961*

DESCRIPTION: *Glaucomys volans* is the smallest arboreal squirrel of Mexico. The dorsal coat varies from pale cinnamon brown to dark brown; the belly is cream-colored or white. A unique feature of these squirrels is the flap of loose skin called the patagium that connects the forelimbs and hind limbs and serves as a gliding membrane. The tail is very fluffy and the hair is flattened in the dorsoventral aspect. The eyes are large. The skeleton is light and the bullae are highly developed (Wells-Gosling, 1985).

DISTRIBUTION: The southern flying squirrel is distributed widely in North America, from Canada to Honduras. Its distribution in Mexico is wide but very fragmented, as it is limited to around 40 locations in some mountains of the Sierra Madre Oriental, Transverse Volcanic Axis, and the Sierra Madre del Sur (Ceballos and Galindo, 1983; Ceballos and Miranda, 1986; Ceballos et al., 2010; Gaona et al., 2000; Manzano, 1993). Only one record exists for the Sierra Madre Occidental of Chihuahua from the end of the nineteenth century, and it is dubious because of the lack of other records and because the original label of the specimen is missing (Anderson, 1972). There are records from the states of CS, GR, HG, MI, MX, OX, QE, SL, TA, and VE.

EXTERNAL MEASURES AND WEIGHT

TL = 198 to 255 mm; TV = 81 to 120 mm; HF = 21 to 33 mm; EAR = 13 to 23 mm. Weight: 45 to 85 g.

DENTAL FORMULA: I 1/1, C 0/0, PM 2/1, M 3/3 = 22.

NATURAL HISTORY AND ECOLOGY: This is the only nocturnal squirrel in Mexico. They are arboreal and move by gliding between trees, the feature that gives them their common name (Dolan and Carter, 1977). Generally, they glide distances from 6 m to 9 m, but glides of up to 90 m have been reported (Ceballos and Miranda, 1986; Dolan and Carter, 1977). In Mexico, they are found on the humid slopes of dense temperate forests, in which oaks are the dominant trees (Ceballos and Galindo, 1984; Ceballos et al., 2010; Manzano, 1993). In Pinal de Amoles, Queretaro, the forests are composed of about 10 species of trees, the most abundant of which is a type of oak (*Quercus laurina*) called "escobillo" (Manzano, 1993). Flying squirrels make their nests in the trees, in natural hollows or in abandoned nests of woodpeckers. The nests have been found at more than 1.5 m above the ground, mainly in oaks and pines, and less often in fir trees (*Abies religiosa*). The interior of the nest is covered by a semicircular structure constructed of mosses and lichens (Ceballos and Miranda, 1986; Manzano, 1993). The availability of nesting sites is apparently a factor limiting their distribution (Dolan and Carter, 1977; Manzano, 1993; Muul, 1974). During winter they are gregarious, and up to 20 individuals may share a nest in order to maintain

Glaucomys volans. Pine-oak forest. Chapa de Mota, Mexico. Photo: Gerardo Ceballos.

their body temperature. These aggregations have been observed in Pinal de Amoles, Queretaro, and Chapa de Mota in the state of Mexico (Ceballos and Miranda, 1986). In late spring, when mating begins, they become territorial, and couples defend a well-defined territory. Reproduction occurs in spring and summer. It is possible that females have two litters per year (Wells-Gosling, 1985). The gestation period lasts 40 days, and the average number of offspring per litter is 3 to 4 pups, which are born altricial, blind, and hairless, and weigh about 3 g (Ceballos and Miranda, 1985; Dolan and Carter, 1977). Despite being mainly herbivorous, *G. volans* is the most carnivorous squirrel in North America. They eat fruits and seeds of species such as oaks and pines, lichens, small birds, eggs, and insects. When acorns are abundant, these are virtually their only food item. Their major predators are owls, especially barn owls (Dolan and Carter, 1977).

VEGETATIONAL ASSOCIATIONS AND ELEVATION RANGE: Southern flying squirrels have been documented in temperate forests, mainly of oak, but also in mixed forests of pine, fir, and oak, and pine-oak woodlands. In Mexico, it occurs from 840 masl to 3,040 masl (Ceballos et al., 2010).

CONSERVATION STATUS: The species is listed as threatened with extinction, due to its fragmented distribution and the destruction of the forests it inhabits (Ceballos and Navarro, 1991; SEMARNAT, 2010). An assessment of its current situation in four regions of central Mexico showed that the species has disappeared from several places in the last decade (Manzano, 1993). For long-term conservation it is imperative to protect several populations throughout its distribution, including regions such as Pinal de Amoles (Querétaro) and Chapa de Mota (state of Mexico), which are the only sites where its biology has been studied (Ceballos and Galindo, 1984; Ceballos and Miranda, 1986; Ceballos et al., 2010; Manzano, 1993).

Ictidomys mexicanus (Erxleben, 1777)

Mexican ground squirrel

Manuel Valdés Alarcón and Gerardo Ceballos

SUBSPECIES IN MEXICO

This is a monotypic species. Until recently, it was considered to have two subspecies. Helgen et al. (2009) found out that *I. parvidens*, usually recognized as a subspecies of *I. mexicanus*, deserves species-level recognition within *Ictidomys*.

DESCRIPTION: *Ictidomys mexicanus* is medium in size with a relatively long, thin body. The ears are short and rounded. The extremities are also short and the length of the tail is less than half the total length of the body (Howell, 1938). The tail has a flattened aspect and the hair color is brown at the base but the tip has alternating bands of white-black-white, which gives it a dark shade. The hair color of the body is very characteristic. The back is sepia with nine rows of whitish quadrangular spots and a number of variable spots in each row. The sides are brown and the belly is grayish to pink. The head has the same coloring as the back, but it is slightly darker. Toward the nose, it tends to be yellowish. It has a very conspicuous ring around the eyes (Davis, 1944; Mearns, 1896; Schmidly, 1977b). It is morphologically very similar to *I. parvidens* but is markedly larger (Helgen et al., 2009).

EXTERNAL MEASURES AND WEIGHT
TL = 290 to 330 mm; TV = 100 to 125 mm; HF = 37 to 40 mm; EAR = 9 to 10 mm. Weight: 215 to 334 g.

DENTAL FORMULA: I 1/1, C 0/0, PM 2/1, M 3/3 = 22.

NATURAL HISTORY AND ECOLOGY: The Mexican ground squirrel prefers deep soils in intermountain grasslands of temperate areas and open scrublands. It is often seen in crop fields. They are neither arboreal nor social and do not form complex social groups. In some places, they might live in colonies, but their members live in individual burrows. The sociability is restricted to the mating season and males are very territorial during this time of the year (Edwards, 1946). They build their burrows in the ground in open areas; entrances are flat and sometimes located at the bases of the grass clumps or shrubs. The number of exits is variable (one to three). The burrows are long and winding with several chambers or nests (Valdez and Ceballos, 1991). The number of exits depends on the frequency of use. Females build spherical nests with grass, leaves, and twigs that measure approximately 180 mm to 200 mm in diameter. The nests are located in chambers in the deepest part of the tunnel (Bailey, 1932; Blair, 1952; Davis, 1944). They are basically herbivorous and consume seeds, grain crops, berries, some fruit, and annual herbs as well as insects and their larvae (Bailey, 1932; Davis, 1974; Edwards, 1946). The annual cycle of activity varies with latitude and altitude. They are active for 8 to 9 months and hibernate during the coldest months of the year (Valdez and Ceballos, 1991). Reproduction takes place during June and July, with a gestation period of 28 to 30 days, and results in an average of 5 offspring per litter (Davis, 1974; Edwards, 1946). This species is sympatric with pocket gophers (*Crateogeomys* spp.), voles (*Microtus mexicanus*), and other ground squirrels such as *Xerospermophilus spilosoma* and *Otospermophilus variegatus* (Aragon and Baudoin, 1990; Ceballos and Galindo, 1984). Its main predators are weasels, badgers, coyotes, dogs, cats, red-tailed hawks (*Buteo jamaicensis*), and rattlesnakes (Valdez and Ceballos, 1991).

DISTRIBUTION: *I. mexicanus* is endemic to the Central Plateau of Mexico, from Jalisco to Veracruz (Ceballos and Galindo, 1984; Helgen et al., 2009). The distribution of *I. mexicanus* is separated from that of *I. parvidens*, which is found in northeastern Mexico and Texas (Baccus, 1979; Baker et al., 1981; González-Christien et al., 2002; Young and Jones, 1982). It has been registered in the states of AG, DF, GJ, HG, JA, MX, PU, QE, TA, VE, and ZA.

Ictidomys mexicanus. Grassland. Zoquiapan and Anexas National Park, state of Mexico. Photo: Gerardo Ceballos.

VEGETATIONAL ASSOCIATIONS AND ELEVATION RANGE: *I. mexicanus* is mostly found in grassland and arid scrublands. In the Sierra Nevada, near Rio Frio, in the state of Mexico, it is found in intermountain meadows with plants such as *Potentilla candicans* and *Eryngium carlinae* (Valdez and Ceballos, 1991). It is found from 2,100 masl to 3,200 masl (Baker, 1956; Davis, 1944).

CONSERVATION STATUS: This squirrel is not found within any risk category (SEMARNAT, 2010). Its populations are relatively abundant. In some places it is considered a pest because it causes damage to crops of corn, alfalfa, and wheat (Ceballos and Galindo, 1984; Valdez and Ceballos, 1996). In some places, pups are caught and sold as pets.

Ictidomys parvidens (Mearns, 1896)

Spotted ground squirrel

Gerardo Ceballos

SUBSPECIES IN MEXICO

This is a monotypic species. Until recently, it was considered to be a subspecies of *I. mexicanus*. Helgen et al. (2009) found out it deserves species-level recognition within *Ictidomys*.

DESCRIPTION: *Ictidomys parvidens* is small to medium in size, very similar but smaller than *I. mexicanus*. It has a relatively long, thin body. The ears are short and

Ictidomys parvidens. Scrubland. Acuña, Nueva Léon. Photo: Juan Cruzado.

rounded. The extremities are also short and the length of the tail is less than half the total length of the body (Howell, 1938). The tail is flattened and the color is brown at the base but the tip has alternating bands of white-black-white, which gives it a dark shade. It is characterized by the striking coloration of the body. The back is sepia with nine rows of whitish quadrangular spots and a number of variable spots in each row. The sides are brown and the belly is grayish to whitish. The head has the same coloring as the back, but it is slightly darker. It has a conspicuous ring around the eyes (Davis, 1944; Mearns, 1896; Schmidly, 1977b). It is morphologically very similar to *I. mexicanus* but markedly smaller (Helgen et al., 2009).

EXTERNAL MEASURES AND WEIGHT
TL = 290 to 300 mm; TV = 100 to 110 mm; HF = 37 to 40 mm; EAR = 9 to 10 mm. Weight: 215 to 280 g.

DENTAL FORMULA: I 1/1, C 0/0, PM 2/1, M 3/3 = 22.

NATURAL HISTORY AND ECOLOGY: The ecology of *I. parvidens* is very similar to that of *I. mexicanus*. They prefer deep soils in grasslands and open scrublands and are often seen in crop fields. They are strictly terrestrial and solitary. In some places, they live in loose colonies, but their members live in individual burrows. The sociability is restricted to the mating season and males are very territorial during this time of the year (Edwards, 1946). They build their burrows in the ground in open areas; entrances are flat and sometimes located at the bases of the grass clumps or shrubs. The burrows are long and winding with several chambers or nests. The number of exits depends on the frequency of use. Females build spherical nests with grass, leaves, and twigs. The nests are located in chambers in the deepest part of the tunnel (Bailey, 1932; Blair, 1952; Davis, 1944). They are basically herbivorous and feed on seeds, grains, berries, small fruits, and annual herbs as well as insects and their larvae (Bailey, 1932; Davis, 1974; Edwards, 1946). The annual cycle of activity varies with latitude and altitude. *Ictinomys* squirrels are active for 8 to 9 months and hibernate during the coldest months of the year (Valdez and Ceballos, 1991). *Ictinomys parvidens* in southeastern New Mexico are active only five months during the year (Schwanz, 2006). Reproduction takes place in the summer, with a gestation period of 28 to 30 days, and results in an average of five offspring per litter (Davis, 1974; Edwards, 1946). The populations in Texas have a seasonal pattern with two peaks of reproduction, one in April and another in June (Matocha, 1968). Litter emergence coincides with the peak precipitation and plant productivity during the summer. Mating season lasts from one to two months (Schwanz, 2006). These squirrels are sympatric with gophers (*Crateogeomys* spp.) and other ground squirrels such as *Ictidomys tridecemlineatus*, *Xerospermophilus spilosoma*, and *Otospermophilus variegatus* (Aragon and Baudoin, 1990; Ceballos and Galindo, 1984). Their main predators are weasels, coyotes, badgers, dogs, cats, hawks, eagles, and rattlesnakes (Davis, 1974; Schmidly, 1977b; Valdez and Ceballos, 1991).

VEGETATIONAL ASSOCIATIONS AND ELEVATION RANGE: *I. parvidens* is mainly found in arid grassland and scrublands with species such as creosote (*Larrea tridentata*), mesquite (*Prosopis* sp.), and cacti (*Opuntia* sp.; Young and Jones, 1982). It is found from almost sea level to around 2,100 masl (Baker, 1956; Schmidly, 1977b).

CONSERVATION STATUS: This species is not considered at risk of extinction (SEMARNAT, 2010). It is abundant in many arid regions. In some places it is considered a pest because it causes damage to crops of corn, alfalfa, and wheat (Baker, Valdez, and Ceballos, 1996).

DISTRIBUTION: *I. parvidens* is distributed in northeastern Mexico and the southwestern United States (Helgen et al., 2009). There is an isolated population in Mapimi, Durango (Aragon et al., 1993), which likely represents a different species (G. Ceballos and M. Valdes, pers. obs.). The distribution of *I. parvidens* is separated in north central Mexico from *I. mexicanus*, which is found in the Central Plateau of Mexico (Baccus, 1979; Baker et al., 1981; González-Christien et al., 2002; Young and Jones, 1982). It has been registered in the states of DU, CO, NL, and TL.

Buller's chipmunk

Heliot Zarza and Guadalupe Tellez-Girón

SUBSPECIES IN MEXICO
N. bulleri is a monotypic species.

DESCRIPTION: *Neotamias bulleri* is a small squirrel, similar to *N. durangae*, but slightly larger and paler. The white lines on the head are wider and the grayish patches behind the ears are larger. The dorsal coloration is grayish-brown interspersed with reddish; it has five dark lines nuanced with brown on the back, and four whitish lines alongside some cinnamon lines on the sides. The hair color on the sides and the dorsal surface of the tail is pale olive gray mixed with black and reddish; there is marked sexual dimorphism (Bartig et al., 1993; Callahan, 1980; Levenson, 1990; Levenson et al., 1985).

DISTRIBUTION: This species is endemic to Mexico, distributed from the semi-arid eastern slope of the Sierra Madre Occidental, from the southern Durango and western Zacatecas (Sierra of Valparaiso) to northern Jalisco (Hall, 1981; Matson and Baker, 1986). It has been registered in the states of DU, JA, and ZA.

EXTERNAL MEASURES AND WEIGHT
TL = 119 to 147 mm; TV = 77 to 112 mm; HF = 30 to 38 mm; EAR = 18 to 22 mm. Weight: 70 g.

DENTAL FORMULA: I 1/1, C 0/0, PM 2/1, M 3/3 = 22.

NATURAL HISTORY AND ECOLOGY: Buller's chipmunks are diurnal and semi-arboreal, although most of their activity is carried out on the ground. During the day they are frequently seen among the trees and shrubs. They prefer rocky substrates, although it is common to find them along the shores of streams (Matson and Baker, 1986). It has been seen foraging for seeds, fruits, and flowers of oak, which are stored for winter, but they also feed on plants, fungi, insects, bulbs, and bird eggs (Nowak, 1999b). The mating season begins at the end of February and ends in early July; lactating females have been found in June and July. The genus *Neotamias* has a gestation period of 31 days and a lactation period of 37 to 41 days (Bartig et al., 1993). They reproduce only once a year, but if the litter fails, the female comes into estrus again and can have a second litter (Nowak, 1999b). It has an audible vocalization, absent in other species of the genus (Bartig et al., 1993). It has been found associated with *Otospermophilus variegatus*, *Sciurus nayaritensis*, *Peromyscus boylii*, *P. truei*, *P. melanotis*, *Microtus mexicanus*, and *Neotoma mexicana* (López-Vidal and Álvarez, 1993).

VEGETATIONAL ASSOCIATIONS AND ELEVATION RANGE: *N. bulleri* mainly inhabits forests of oak and pine-oak, from 2,240 masl to 2,610 masl (Baker and Greer, 1962; Bartig et al., 1993).

CONSERVATION STATUS: There is not enough information to assess the current status of this species. Its distribution is relatively wide and its tolerance to some degree of disturbance from anthropogenic activities can be used as indicators to consider it out of risk.

Cliff chipmunk

Reyna A. Castillo

SUBSPECIES IN MEXICO
Neotamias dorsalis carminis (Goldman, 1938)
Neotamias dorsalis dorsalis Baird, 1855
Neotamias dorsalis nidoensis (Lidicker, 1960)
Neotamias dorsalis sonoriensis (Callahan and Davis, 1977)

DESCRIPTION: *Neotamias dorsalis* is medium in size and the upper parts vary from smoky gray to neutral gray (Hart, 1992). It has a couple of dark brown dorsal lines, barely evident (Hoffmeister, 1986). Sexual dimorphism exists in the size of the body, as females are larger than males (Levenson et al., 1985). It differs from other sympatric squirrels by the black and white facial lines on the sides of the eyes and by the reddish fur on the sides.

EXTERNAL MEASURES AND WEIGHT
TL = 217 to 249 mm; TV = 85 to 115 mm; HF = 34 to 37 mm; EAR = 18 to 23 mm.
Weight: 64 to 74 g.

DENTAL FORMULA: I 1/1, C 0/0, PM 2/1, M 3/3 = 22.

NATURAL HISTORY AND ECOLOGY: *N. dorsalis* is found in rocky sites and the availability of this type of substrate is important in their distribution because they build their burrows and shelters there (Hart, 1992). They feed on roots, buds,

Neotamias dorsalis. Pine-fir forest. Janos Biosphere Reserve, Chihuahua. Photo: Gerardo Ceballos.

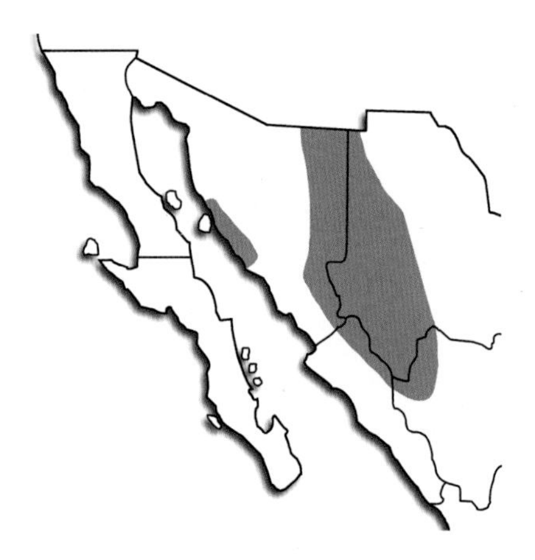

DISTRIBUTION: *N. dorsalis* is found from the southwest of the United States to the north of Mexico, where it occurs in the Sierra Madre Occidental in Chihuahua, Sonora, and Durango. There are two isolated subspecies, *N. dorsalis carminis* in the Sierra del Carmen in Coahuila and *N. dorsalis sonoriensis* in the Pacific coastline between Guaymas and Tastiota in Sonora (Callahan and Davis 1977; Hart, 1992; Jiménez-Guzman et al., 1999). It has been registered in the states of CH, CO, DU, and SO.

buttercups, flowers, fruits, seeds, and fungi (Callahan and Davis, 1976); it is not known with certainty if they consume food of animal origin (Hart, 1992). They often store seeds in underground reservoirs or in their burrow and use their pouches for transport (Hart, 1992). The populations that live in arid sites get water from succulent plants and the dew that is deposited on the plants (Callahan and Davis, 1976). The populations in wooded sites drink free water. The beginning of reproduction seems to depend on rainfall and minimum temperatures during winter, but it generally occurs from February to July, with a reproductive peak in May (Hart, 1992). They have a single litter per year, consisting of four to six offspring. It is common to hear an alarm call (acute and strident) and then observe them climbing quickly and nimbly up the rocky slope. When the squirrel is alert, it swishes its tail like a cat (R. Castillo, pers. obs.)

VEGETATIONAL ASSOCIATIONS AND ELEVATION RANGE: *N. dorsalis* is typically found in temperate forests of conifers and oaks, but *N. dorsalis sonoriensis* inhabits xeric scrublands (Callahan and Davis, 1977). *N. dorsalis sonoriensis* is found at sea level but the other subspecies are distributed in the range of 1,500 masl to 3,700 masl (Callahan and Davis, 1977; Findley, 1969; Hall, 1981).

CONSERVATION STATUS: The current status of this species is unknown, but *N. dorsalis sonoriensis* and *N. dorsalis carminis* have a very restricted distribution and might be at risk of extinction. Therefore, it is necessary to determine their distribution with more accuracy, as well as to assess the populations and the habitat in which they live.

Neotamias durangae (J.A. Allen, 1903)

Durango chipmunk

Arturo Hernández Huerta

SUBSPECIES IN MEXICO

Neotamias durangae durangae (J.A. Allen, 1903)
Neotamias durangae solivagus (A. Howell, 1922)
N. durangae was regarded as subspecies of *T. bulleri*. It is currently considered within the *amoenus* group of the genus *Neotamias* (Levenson et al., 1985).

DESCRIPTION: *Neotamias durangae* is a small chipmunk with sexual dimorphism—males are generally smaller than females (Best et al., 1993; Callahan, 1980). The back has a distinctive pattern of longitudinal stripes: five dark bands alternating with four pale bands. It has a very pale buffy coloration on the hind legs, the bottom of the tail, and the flanks; in addition, the light stripes on the back are pale cinnamon (Allen, 1903). These characteristics, the genital morphology, and its relatively greater size distinguish it from *N. bulleri*, the species with which it is sympatric in the south of Durango (Callahan, 1980).

EXTERNAL MEASURES AND WEIGHT

TL = 222 to 248 mm; TV = 87 to 110 mm; HF = 31 to 38 mm; EAR = 16 to 22 mm.
Weight: 75 g.

DENTAL FORMULA: 1/1, C 0/0, PM 2/1, M 3/3 = 22.

NATURAL HISTORY AND ECOLOGY: These squirrels are diurnal. They live and forage on the ground, although they have the ability to climb. They build their burrows in the earth between rocks or logs (Baker and Greer, 1962). They mainly feed on seeds, fruits, and flowers, although insects are an important part of their diet. They are generalists, consuming whatever food is available throughout the year, exploiting those resources that are more plentiful. Because they store very little body fat, during the winter they depend on their stockrooms to survive (Findley, 1987). From mid-summer through autumn, they store seeds, using their pouches to fetch acorns and grains. In the colder periods of winter they hibernate, but emerge sporadically to get food from their caches. Little is known about reproduction, but for most of the species of this genus the gestation period lasts 30 to 33 days. The litters have four to six offspring, which are born blind and without hair (Findley, 1987). The reproductive period for some northern species occurs from the end of February to early July (Jameson and Peeters, 1988). In Durango, females in reproductive status have been trapped at the end of June and mid-July (Baker and Greer, 1962), and in Coahuila, in late July and early August (Baker, 1956). It is unknown if they build nests in the trees, as other species of the same genus do. The offspring remain in the nest for nearly five weeks. Lactation lasts 39 to 45 days and juveniles become independent at 6 or 7 weeks of age. They have a single reproductive period during the year, but if a female loses her litter, she can enter estrus again and have a second litter. In general, after juveniles are weaned, females molt and do not become sexually active until the following year (Nowak, 1999b). Young animals mature sexually in a year. They molt twice a year, once in spring and once in autumn; the coat has a paler coloration in winter than in summer. The dark coat of summer makes the differences between this species and *N. bulleri* more evident, facilitating identification (Howell, 1929). Like other species from the same genus, they are likely to be territorial. They are more abundant in humid environments than in dry environments (Baker and Greer, 1962). In Coahuila, the highest abundances have been observed at the end of July and early August (Baker, 1956). There are no data on longevity, but some species live up to eight years.

DISTRIBUTION: *N. durangae* is an endemic species of Mexico. *N. durangae durangae* has a distribution restricted to the wet part of the Sierra Madre Occidental from the southwestern Chihuahua to the south of Durango, and *N. durangae solivagus* is isolated in the Sierra Madre Oriental, in the Sierra del Carmen in southeast of Coahuila. It has been registered in the states of CH, CO, and DU.

Neotamias durangae. Pine forest, Mapimi Biosphere Reserve, Durango. Photo: Patricio Robles Gil.

VEGETATIONAL ASSOCIATIONS AND ELEVATION RANGE: This chipmunk dwells in forests of pine-oak. In the Sierra Madre Occidental it is associated with communities whose dominant species are the sad pine (*Pinus lumholtzii*), the oak (*Quercus resinosa*), and the Sonoran oak (*Q. viminea*), along with junipers (*Juniperus*), madrones (*Arbutus*), and manzanita (*Arctostaphylos*; Baker and Greer, 1962). In the mountains of Coahuila, it is found in forests of Arizona pine (*Pinus arizonica*), fir (*Abies durangensis*), and poplars (*Populus*; Goldman, 1951). It is found from 1,980 masl to 2,590 masl in Durango and Chihuahua, and from 2,590 masl to 2,900 masl in Coahuila.

CONSERVATION STATUS: *N. durangae* is an endemic species of Mexico; it is not included in any list of endangered species because apparently it is locally abundant. The Sierra del Carmen is in excellent condition and has been decreed a natural protected area (SEMARNAT, 2010).

Neotamias merriami (J.A. Allen, 1889)

Merriam's chipmunk

Eric Mellink and Jaime Luévano

SUBSPECIES IN MEXICO

Neotamias merriami merriami J.A. Allen, 1889

DESCRIPTION: *Neotamias merriami* is a large chipmunk. The back is gray-brown and the belly is white. The pale and dark stripes of the sides are not very contrasting and have equal width. The dark stripes are gray or brown, and the light ones are grayish. They have black spots in front and in back of the eyes, and the line below the ear is brown. The tail is long, with white or cream edges.

DISTRIBUTION: This species has only been registered in the Valley of Nachogüero, in the north of the state of Baja California (Hall, 1981; Huey, 1964). This is the southern limit of its distribution; except for this locality, it is exclusively found in the state of California in the United States. It has been registered in the state of BC.

EXTERNAL MEASURES AND WEIGHT

TL = 233 to 277 mm; TV = 84 to 140 mm; HF = 28 to 40 mm; EAR = 16 to 18 mm. Weight: 60 to 82 g.

DENTAL FORMULA: I 1/1, C 0/0, PM 2/1, M 3/3 = 22.

NATURAL HISTORY AND ECOLOGY: Merriam's chipmunks tend to be active throughout the year, although they can hibernate. They are mainly terrestrial, but they may climb. They feed on seeds of conifers, manzanita (*Arctostaphylos* spp.), juniper (*Juniperus californica*), and other species of chaparral and different herbs, in addition to eating some insects. They can save food in warehouses for later consumption. They build their nests in fallen logs, stumps, standing dead trees, and burrows excavated in the earth. Apparently, they have a single litter per year, of between three and eight offspring, with an average of four, born at the end of April. The offspring reach sexual maturity before one year of age (Harvey and Polite, 1990; Jameson and Peeters, 1988).

VEGETATIONAL ASSOCIATIONS AND ELEVATION RANGE: This chipmunk is found in chaparral, dense shrubs, places with oaks (*Quercus* spp.), rocky areas with logs and stumps, open forests of conifers, forests of pine (*Pinus* spp.), and juniper thickets

Neotamias merriami. Pine forest. Photo: Gerardo Ceballos.

(*Juniperus californica*) (Harvey and Polite, 1990; Ingles, 1965). In Baja California, it has been registered only at 1,030 masl. In California, it is found at altitudes up to 2,700 masl (Jameson and Peeters, 1988).

CONSERVATION STATUS: In Baja California, *N. merriami* is restricted to one location, which constitutes the southern boundary of its distribution. It is common in California. Ceballos and Navarro (1991) considered it fragile. It was officially considered "rare" in the NOM-059-ECOL-1994 (SEMARNAT, 2010). Because of its restricted range, it must be regarded as endangered in Mexico (Ceballos et al., 1998).

Neotamias obscurus (J.A. Allen, 1890)

California chipmunk

Eric Mellink and Jaime Luévano

SUBSPECIES IN MEXICO

Neotamias obscurus meridionalis (Nelson and Goldman, 1909)
Neotamias obscurus obscurus J.A. Allen, 1890

Neotamias obscurus. Pine forest, San Pedro Martir National Park, Baja California. Photo: Patricio Robles Gil.

DISTRIBUTION: The distribution of this species is restricted almost entirely to the Baja California Peninsula; outside it only occurs in a small area of southern California. *N. obscurus obscurus* is found in Laguna Hanson and the nearby hills of the Sierra de Juarez, and in Rosarito in the Sierra de San Pedro Martir. *N. obscurus meridionalis* is distributed in San Pablo, Aguaje de San Esteban, and Rancho Las Calabazas, in the Sierra de San Francisco, and in Baja California Sur. There is a sighting, not supported by specimens, of its presence in the Sierra de San Borja, Baja California (Callahan, 1976, 1977). The distribution of *N. obscurus meridionalis* is relictual. It has been registered in the states of BC and BS.

DESCRIPTION: *Neotamias obscurus* is a big chipmunk of opaque coloration. The back is gray-brown and the belly is white, coloration very similar to that of *N. merriami*. It has gray stripes on the sides and behind the ears. Its separation from *N. merriami* is based on characteristics of the bacula of males (Callahan, 1977).

EXTERNAL MEASURES AND WEIGHT

TL = 208 to 240 mm; TV = 71 to 120 mm; HF = 30 to 37 mm; EAR = 16 to 19 mm. Weight: 100 g.

DENTAL FORMULA: I 1/1, C 0/0, PM 2/1, M 3/3 = 22.

NATURAL HISTORY AND ECOLOGY: This species is active throughout the year. Their reproduction begins in February and may be extended until the end of June (Callahan, 1976; Harris, 1990). It feeds on pine seeds (*Pinus* spp.), acorns (*Quercus* spp.), juniper berries (*Juniperus californica*), and manzanita (*Arctostaphylos* spp.), but includes many other food items (Harris, 1990). The subspecies *obscurus* does not require free water, and can survive on food only. They take refuge under rocks, fallen logs, shrubs, and trees. They reproduce between January and June and the average litter has four offspring. The desert subspecies gets water from permanent water bodies, fruits and flowers of cacti, other succulent food, and dew deposited on the plants (Callahan and Davis, 1976). Individuals of this subspecies build their nests in cavities excavated by woodpeckers in cacti (*Pachycereus pringlei*) several meters from the ground. Reproduction apparently occurs in the first half of the year.

VEGETATIONAL ASSOCIATIONS AND ELEVATION RANGE: *N. obscurus obscurus* dwells in chaparral and coniferous forest (Callahan, 1977); *N. obscurus meridionalis* inhabits xeric scrublands (Callahan and Davis, 1976), an atypical habitat for the genus *Neotamias*. The two subspecies are found at different altitudes. *N. obscurus meridionalis* occurs from 305 masl to 1,370 masl, while *N. obscurus obscurus* is found from 760 masl to 2,590 masl (Callahan, 1977).

CONSERVATION STATUS: The current status of this species in Mexico is unknown, but the subspecies *N. obscurus meridionalis* is restricted to an area of less than 40 km in diameter, so it probably is threatened with extinction.

Notocitellus adocetus (Merriam, 1903)

Tropical ground squirrel

Manuel Valdéz Alarcón and Gerardo Ceballos

SUBSPECIES IN MEXICO

Notocitellus adocetus adocetus (Merriam, 1903)
Notocitellus adocetus infernatus Álvarez and Ramirez-Pulido, 1968

DESCRIPTION: *Notocitellus adocetus* is medium in size, similar to *N. annulatus* but smaller and paler in color and the tail has no rings. The hair color is relatively uniform cinnamon brown intermingled with black dorsally. The tail has a greater proportion of black hair, and the head is darker than the body. The abdomen is pale yellowish extending inside the legs, which can be yellowish-red. Two white lines can be distinguished above and below the eye (Álvarez and Ramirez-Pulido, 1968; Howell, 1938; Merriam, 1903).

EXTERNAL MEASURES AND WEIGHT

TL = 286 to 366 mm; TV = 145 to 163 mm; HF = 41 to 50 mm; EAR = 14 to 20 mm. Weight: 163 to 250 g.

DENTAL FORMULA: I 1/1, C 0/0, PM 2/1, M 3/3 = 22.

Notocitellus adocetus. Crops in secondary tropical dry forest. Bejucos, state of Mexico. Photo: Gerardo Ceballos.

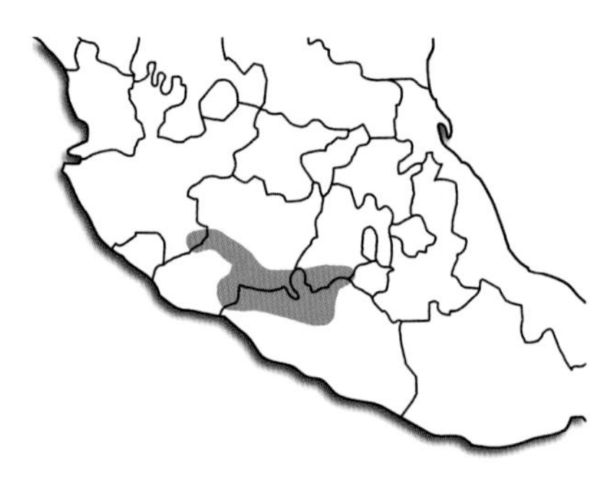

DISTRIBUTION: *N. adocetus* is endemic to Mexico; it is distributed in the tropical region of the Balsas Basin from Guerrero to the state of Mexico. The records from Mexico City, Hidalgo, and Tlaxcala (Hall, 1981; Villa-R. et al., 1981) are probably released pets because there is no evidence of wild populations in these states. It has been registered in the states of DF, GR, HG, JA, MI, MX, and TL.

NATURAL HISTORY AND ECOLOGY: *N. adocetus* is a diurnal ground squirrel, which occasionally climbs up several meters in the trees and shrubs in search of food. It frequently lives in gaps, under stones or shrubs along the canyons, and in fruit or corn crops adjacent to areas with low deciduous jungle (Ceballos and Miranda, 1986, 2000; Genoways and Jones, 1973). Its peak of activity is between 09:00 hrs and 11:00 hrs. It is active throughout the year, although Villa-R. (1943) speculated that it has a small period of aestivation in the hottest part of the year. It feeds on fruits, seeds, and sprouts of plants such as calabash (*Crescentia alata*), white thorn (*Acacia cochlyacanta*), huizache (*Acacia farnesiana*), mesquite (*Prosopis jugiflora* sp.), and guamuchil (*Pithecellobium dulce*). It commonly feeds on seeds of corn, beans, and sorghum. It can cause damage to the crops; hence, it is considered a plague in some regions (Villa-R. et al., 1968). There are no data on the reproduction period; however, it is possible that in farmlands, they reproduce throughout the year. In seasonal jungles reproduction surely occurs at the end of dry season (May–June). The tropical ground squirrel is sympatric with other species typical of dry forest such as *Marmosa canescens*, *Liomys pictus*, *Orthogeomys grandis*, *Osgoodomys banderanus*, and *Hodomys alleni* (Genoways and Jones, 1973).

VEGETATIONAL ASSOCIATIONS AND ELEVATION RANGE: *N. adocetus* dwells in low deciduous jungles, xeric scrublands, and farmlands (Howell, 1938). It is found from 200 masl to 1,200 masl.

CONSERVATION STATUS: *N. adocetus* is an abundant species in most of its range, which presents no problems of conservation. It is a species easy to maintain in captivity and is often marketed as a pet.

Notocitellus annulatus (Audubon and Bachman, 1842)

Ring-tailed ground squirrel

Manuel Valdéz Alarcón and Gerardo Ceballos

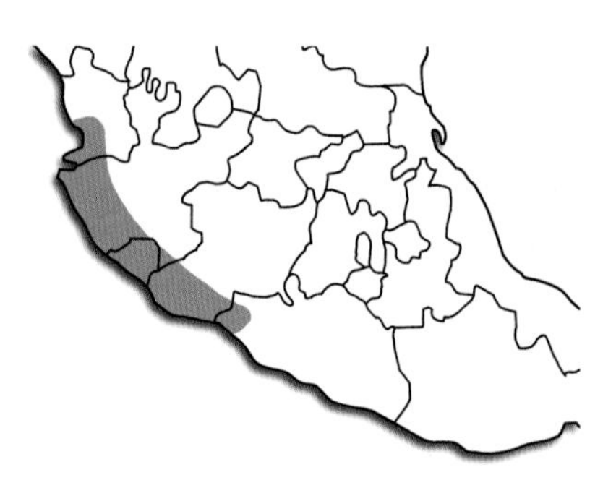

DISTRIBUTION: This species of squirrel is endemic to Mexico. It is distributed in the Pacific Rim, from the south of Nayarit to the northwest of Guerrero (Hall, 1981). It has been registered in the states of CL, GR, JA, MI, and NY.

SUBSPECIES IN MEXICO

Notocitellus annulatus annulatus (Audubon and Bachman, 1842)
Notocitellus annulatus goldmani Merriam, 1902

DESCRIPTION: *Notocitellus annulatus* is medium in size. The color of the back is uniform brown cinnamon interspersed with black. The head is darker and in some individuals it is predominantly black. The sides of the face, nose, and throat are dark ochre. The belly is paler. During the breeding season the chest and belly are dark orange, which gives this squirrel an unmistakable look. The tail is almost the same length as the head and body, and has around 15 bands of alternating pale and dark hair (Elliot, 1905b; Hall, 1981). The subspecies *Notocitellus annulatus goldmani* has shorter feet, the whitish ring around the eyes is more evident, and the body is darker (Howell, 1938).

EXTERNAL MEASURES AND WEIGHT

TL = 383 to 470 mm; TV = 187 to 228 mm; HF = 54 to 64 mm; EAR = 14 to 16 mm. Weight: 315 to 570 g.

Notocitellus annulatus. Tropical dry forest-cloud forest. El Jabalí Wildlife Refuge, Colima. Photo: Gerardo Ceballos.

DENTAL FORMULA: I 1/1, C 0/0, PM 2/1, M 3/3 = 22.

NATURAL HISTORY AND ECOLOGY: *N. annulatus* is diurnal and semi-arboreal. It spends part of its time in trees and shrubs, sometimes in the treetops, in search of food. It is common in natural rocky areas and in fences of rock. Their nests are among the rocks, in hollow trees, in branches or fallen palms, and in thorny vegetation of very dense acacias (Howell, 1938). It has been frequently observed in cultivated areas of agave, banana, lemon, mango, guava, and corn (Ceballos and Miranda, 1986, 2000; Téllez-Girón et al., 1997). It feeds on fruit and nuts of palm trees, seeds of mesquite, cacti, wild figs, and sometimes insects (Best, 1995b; Howell, 1938). They reproduce in the dry season, from December to June (Best, 1995b). It is sympatric with other rodents such as *Hodomys alleni, Xenomys nelsoni, Spermophilus variegatus, Sciurus aureogaster, Sciurus colliaei, Pappogeomys bulleri, Liomys pictus, Oryzomys melanotis, Osgodomys banderanus*, and *Peromyscus boylii* (Best, 1995b; Genoways and Jones, 1973; Téllez-Girón et al., 1997).

VEGETATIONAL ASSOCIATIONS AND ELEVATION RANGE: The ring-tailed ground squirrel is common in low deciduous forest, thorny scrub, and crop areas (Ceballos and Miranda, 1986, 2000; Howell, 1938). It is found from sea level up to 1,200 m (Best, 1995b).

CONSERVATION STATUS: *N. annulatus* is a relatively abundant species in most of its range. It has a limited distribution, however, and it is necessary to assess its current situation.

Baja California rock squirrel

Iván Castro Arellano and Gerardo Ceballos

DISTRIBUTION: *O. atricapillus* is an endemic to Mexico; its distribution is restricted to mountains of volcanic origin on the Baja California Peninsula south of the 28 degree parallel to San Pedro de la Presa, 430 km away (Hall, 1981; Yensen and Valdés-Alarcón, 1999). All the records come from oases in the Sierra de la Giganta, Sierra de San Pedro, and Sierra de San Francisco, on the east side of the Peninsula (Yensen and Valdés-Alarcón, 1999). It has been registered in the states of BC and BS.

SUBSPECIES IN MEXICO

O. atricapillus is a monotypic species.

DESCRIPTION: *Otospermophilus atricapillus* is a relatively large ground squirrel, similar to *O. beecheyi*. Females are smaller than males. The head is black with a white ring around the eyes. The anterior of the back is also black, but the hairs on the sides of the neck and the forefeet are pale cream at the tip. The posterior of the back is black at the base but switches to a tan and pale cream in the tip. The legs and belly are pale cream. The tail has a mixture of black and pale cream coloration (Bryant, 1889; Howell, 1938). There is considerable individual variation in specimens from the type locality (Bryant, 1889). This squirrel can be distinguished from *O. beecheyi* by its darker color in the anterior part, its longer tail, and its allopatric distribution because the two species are separated by approximately 64 km (Álvarez Castañeda et al., 1996; Hall, 1981; Howell, 1938).

EXTERNAL MEASURES AND WEIGHT

TL = 387 to 486 mm; TV = 156 to 217 mm; HF = 50 to 60 mm; EAR = 22 to 29 mm. Weight: 350 to 706 g.

DENTAL FORMULA: I 1/1, C 0/0, PM 2/1, M 3/3 = 22.

NATURAL HISTORY AND ECOLOGY: This ground squirrel is diurnal and remains active throughout the year (Álvarez Castañeda et al., 1996). It molts between September and October; a specimen captured at the end of September had almost completed

Otospermophilus atricapillus. Scrubland. San Francisco de la Sierra, Baja California Sur. Photo: Eric Mellink.

molting, and a female captured in mid-October already had her new coat (Howell, 1938). They build their burrows in cavities below logs or rocks as well as in cracks in cliffs (Nowak, 1999b). They mainly feed on fruits and seeds such as dates and corn. In agricultural areas they cause damage to crops so they are considered a pest. There are no data on their reproduction, but it probably resembles that of *O. beecheyi* and occurs at the end of spring.

VEGETATIONAL ASSOCIATIONS AND ELEVATION RANGE: This species mainly inhabits oases within xeric scrub containing abundant shrubs and palm trees near springs (M. Valdez, pers. obs.). Its altitudinal distribution varies from sea level to 600 masl.

CONSERVATION STATUS: *O. atricapillus* is a species with a very restricted distribution, abundant in some sites but scarce in most of the localities where it is known (Yensen and Valdés-Alarcón, 1999). In agricultural areas, it is considered a pest and has been eliminated. Because of these characteristics, the species should be considered threatened.

Otospermophilus beecheyi (Richardson, 1829)

California ground squirrel

César A. Loza Salas and Gerardo Ceballos

SUBSPECIES IN MEXICO

Otospermophilus beecheyi nudipes (Huey, 1931)
Otospermophilus beecheyi rupinarum (Huey, 1931)

DESCRIPTION: *Otospermophilus beecheyi* is a large squirrel with variable coloration. The back is gray, with white diffuse specks, and the belly is paler. In winter the coat is browner and denser. The head is brown with a white collar, the ears are rounded, and the tail is dense (Burt, 1976; Hall, 1981; Jameson and Peeters, 1988; Tomich, 1982).

EXTERNAL MEASURES AND WEIGHT

TL = 357 to 500 mm; TV = 145 to 200 mm; HF = 50 to 64 mm; EAR = 17 to 20 mm. Weight: 300 to 650 g.

DENTAL FORMULA: I 1/1, C 0/0, PM 2/1 M 3/3 = 22.

NATURAL HISTORY AND ECOLOGY: This squirrel lives in colonies and is active during the day. It inhabits underground burrows dug under rocks and fallen trees. The galleries have several exits and are composed of extensive tunnels. They build their nests with dried vegetation (Tomich, 1982) and reproduce in spring. Gestation lasts 25 to 30 days, and results in 3 to 10 offspring per litter; they may have more than one litter per year. The offspring are born in spring and summer (Jameson and Peeters, 1988). Birth varies according to the locality. In the wild they live up to 5 years (Burt, 1976; Hall, 1981; Jameson and Peeters, 1988). They feed on fruits, seeds, berries, grass, tubers, fungi, insects, eggs, and often carrion (Burt, 1976; Hall, 1981; Jameson and Peeters, 1988; Meier, 1983; Tomich, 1982). Hibernation occurs from late autumn until the beginning of spring. Juveniles remain active until autumn. The

DISTRIBUTION: *O. beecheyi* is found from the southwestern Washington, in the United States, to the northern part of Baja California, which is the southern limit of its distribution (Hall, 1981; Ramírez Pulido et al., 1983, 1986; Tomich, 1982). It has been registered in the state of BC.

Otospermophilus beecheyi. Grassland. Ensenada, Baja California. Photo: Gerardo Ceballos.

odoriferous glands play an important role in the social behavior because they delimit their territory with urine. They communicate with short, single note sounds, which sometimes serve as alarm signals (Tomich, 1982). The home range is 0.5 ha. Most activity is performed near the entrance of their burrows so they cover distances of no more than 100 m in their daily movements. They make very short forays in search of food, which is usually located in nearby places as they build their burrows where fresh grass and sprouts are found (Burt, 1976; Jameson and Peeters, 1988; Linsdale, 1946).

VEGETATIONAL ASSOCIATIONS AND ELEVATION RANGE: The California ground squirrel inhabits chaparral, bushes, pastures, coastal dunes, and hillsides. When they are found in pastures, usually these are not very high as this allows them to observe the surrounding area (Burt, 1976; Jameson and Peeters, 1988; Tomich, 1982). It is found from sea level up to 2,200 m (Jameson and Peeters, 1988).

CONSERVATION STATUS: *O. beecheyi* is a species with a limited distribution in Mexico. It is abundant and survives in disturbed areas. It is not considered at risk of extinction. It has economic importance because it is considered a pest on crops and pastures causing considerable losses to farmers and ranchers (Burt, 1976).

Rock squirrel

Manuel Valdéz Alarcón and Gerardo Ceballos

SUBSPECIES IN MEXICO

Otospermophilus variegatus couchii Baird, 1855
Otospermophilus variegatus grammurus (Say, 1823)
Otospermophilus variegatus rupestris (J.A. Allen, 1903)
Otospermophilus variegatus variegatus Erxleben, 1777

DISTRIBUTION: This squirrel has a wide distribution from the central United States to the highlands of Mexico (Hall, 1981). It has been registered in the states of AG, CH, CL, CO, DF, DU, GJ, GR, HG, JA, MI, MO, MX, NL, NY, OX, PU, QE, SI, SL, SO, TA, TL, VE, and ZA.

DESCRIPTION: *Otospermophilus variegatus* is a large squirrel with a long, hairy tail that is 44% of the length of the body. The dorsal coloration is gray interspersed with black, mottled with white. In general, the hind legs are black; the abdomen varies from grayish-white to cinnamon. The coloration in different populations, however, is highly variable. It has very distinctive white eye-rings; the eyes are very large, and the ears are longer than wide. In northern Mexico, completely black individuals are common (Bryant, 1945; Hall, 1981; Nowak, 1999b).

EXTERNAL MEASURES AND WEIGHT

TL = 30 to 540 mm; TV = 172 to 263 mm; HF = 53 to 65 mm; EAR = 15 to 29 mm. Weight: 681 to 817 g.

DENTAL FORMULA: I 1/1, C 0/0, PM 2/1, M 3/3 = 22.

NATURAL HISTORY AND ECOLOGY: This ground squirrel lives in semi-arid areas; it builds its burrows in rocky sites, but is not restricted to them since it often uses fissures between stonewalls, as well as soft soils under the bases of maguey plants

Otospermophilus variegatus. Scrubland-oak forest. Cuernavaca, Morelos. Photo: Gerardo Ceballos.

in farmlands. It feeds basically on plants and insects; however, it tends to be very opportunistic, consuming a wide variety of fruits and seeds, small invertebrates, and carrion of fresh or dried meat (Bailey, 1932; Burt, 1934).

VEGETATIONAL ASSOCIATIONS AND ELEVATION RANGE: The rock squirrel is very tolerant of different environmental conditions and is found in pine forests, oak forests, xeric scrublands, lowland jungle, riparian vegetation, disrupted areas, and crops, from sea level up to 3,600 m (Ceballos and Galindo, 1984; Hall, 1981; Howell, 1938).

CONSERVATION STATUS: *O. variegatus* is very common in natural and disturbed areas and can become a pest on crops. It has no conservation issues.

Sciurus aberti Woodhouse, 1853

Abert's squirrel

Manuel Valdéz Alarcón and Guadalupe Téllez-Girón

SUBSPECIES IN MEXICO
Sciurus aberti barberi J.A. Allen, 1904
Sciurus aberti durangi Thomas, 1893
Sciurus aberti phaeurus J. Allen, 1904

DESCRIPTION: *Sciurus aberti* is an arboreal squirrel of large size. The body is long and the length of the tail is approximately 42% of the total length of the body. The ears are long (up to 25 mm) and wide with an obvious tuft on the tip that may be absent in spring (Keith, 1965). The dorsal coloration and the upper part of the tail are grayish, and the belly is white. It has a distinctive black edge between the stomach and the back. Apparently, there is no sexual dimorphism in size (Allen, 1895; Armstrong, 1972; Hall and Kelson, 1959).

DISTRIBUTION: This squirrel has a discontinuous distribution, with populations in Colorado, New Mexico, and Arizona in the United States, and in the Sierra Madre Occidental in Durango and Chihuahua in the north of Mexico (Hall, 1981). It has been registered in the states of CH, DU, and SO.

EXTERNAL MEASURES AND WEIGHT
TL = 463 to 530 mm; TV = 210 to 255 mm; HF = 64 to 80 mm; EAR = 39 to 42.
Weight: 500 g.

DENTAL FORMULA: I 1/1, C 0/0, PM 2/2, M 3/3 = 20.

NATURAL HISTORY AND ECOLOGY: This squirrel is strictly diurnal. It begins and ends its activity during the hours of sunlight, and immediately takes refuge in its nest at dusk. They live in pine forests, where they search for food and shelter. They build their nests in the stronger branches of the pines using small branches and needles, usually at more than 4 m above the ground. At times, they have been seen taking refuge in other conifers (Keith, 1965). The nests generally are 300 mm to 1,000 mm in external diameter with two or three entries and with an internal diameter of 102 mm to 254 mm (Keith, 1965). The pines provide food throughout the year, as these squirrels consume the inner bark, seeds, buds, stamens, and flowers. They also feed on fleshy fungi and carrion (Keith, 1965). They are solitary and only get together during the breeding season. They reproduce from February to March,

Sciurus aberti. Pine forest. Sierra Madre Occidental, Chihuahua. Photo: Juan Cruzado.

with a gestation period of 40 to 46 days and a litter size of 2 to 4 offspring per year (Nash and Seaman, 1977).

VEGETATIONAL ASSOCIATIONS AND ELEVATION RANGE: Abert's squirrels are limited to forests of ponderosa pine (*Pinus ponderosa*), which they depend on for food and shelter (Findley et al., 1975; Nash and Seaman, 1977). They are found from 700 masl to 2,100 masl (Nash and Seaman, 1977).

CONSERVATION STATUS: These squirrels are vulnerable because of their limited distribution and close association with forests of *P. ponderosa*. Their populations are sensitive to the destruction of the habitat by logging and in some places they are relatively scarce. They are regarded as of Special Protection (SEMARNAT, 2010). The status of their populations in Mexico is unknown.

Sciurus alleni Nelson, 1898

Allen's squirrel

César A. Loza Salas

SUBSPECIES IN MEXICO

S. alleni is a monotypic species.

DESCRIPTION: Allen's squirrel is an arboreal species of relatively small size. The body coloration is yellowish-brown. This pattern is darker on the back than on the sides. The top of the tail is mottled with white and sometimes with a paler tone. The bottom of the tail is brown or yellowish-gray, lined with black and white.

Sciurus alleni. Pine-oak forest. Cumbres de Monterrey National Park, Nuevo León. Photo: Juan Cruzado.

EXTERNAL MEASURES AND WEIGHT

TL = 415 to 493 mm; TV = 175 to 235 mm; HF = 55 to 63 mm; EAR = 30 to 34 mm. Weight: 290 to 510 g.

DENTAL FORMULA: I 1/1, C 0/0, PM 1/1, M 3/3 = 20.

NATURAL HISTORY AND ECOLOGY: These squirrels have arboreal habits and they are only found in mountains with temperate forests. They are common in both pine and hardwood forests, where there is good canopy cover. They make their burrows in abandoned holes of woodpeckers or build them themselves at heights of more than 3 m. The diameter of the holes is 5 cm to 8 cm (Burton and Pearson, 1987). They mainly reproduce between March and April, although there is evidence of pregnancy and lactation throughout the year. They have two to four offspring per litter. Their diet consists of pine cones, oak sprouts, seeds, nuts, corn, oats, apples, tomatoes, peaches, grapes, plums, larvae, insects, and anurans. Their eating habits sometimes cause considerable damage to crops (Best, 1995b).

DISTRIBUTION: This species is endemic to the northeast of Mexico, where its distribution is restricted to temperate forests of the Sierra Madre Oriental, southeast of Coahuila, Nuevo Leon, west of Tamaulipas, and north of San Luis Potosí (Best, 1995b; Burton and Pearson, 1987; Dalquest, 1953b; Hall, 1981). It has been registered in the states of CO, NL, SL, and TA.

VEGETATIONAL ASSOCIATIONS AND ELEVATION RANGE: *Sciurus alleni* dwells exclusively in vegetation dominated by trees. It is very common in deciduous forests of oak and arbutus, as well as conifers; they are found from 600 masl to 2,550 masl (Best, 1995b; Hall, 1981).

CONSERVATION STATUS: Allen's squirrels are not threatened and are not included in any category of the IUCN (Hilton-Taylor, 2000) or of the Mexican Official Standard of Species at Risk (SEMARNAT, 2010). They are common in Cumbres de Monterrey National Park. Their populations could be declining due to deforestation of their habitats for agriculture and animal husbandry (Burton and Pearson, 1987).

Arizona gray squirrel

Reyna A. Castillo

SUBSPECIES IN MEXICO
Sciurus arizonensis huachuca J.A. Allen 1894

DESCRIPTION: *Sciurus arizonensis* is a large arboreal squirrel with a very hairy tail. Its coloration is silver gray with fulvous tufts, but varies greatly between seasons, which has wrongly been reported as geographic variation (Hall, 1981). In addition, the nuts that make up its diet stain the legs and abdomen, which has also produced erroneous descriptions of its coloring (Mearns, 1907). It has white rings around the eyes, and most of the tail has a white border. The legs are dotted with gray, and there is an ochre cast between the ears that becomes much more noticeable on the back (Brown, 1984). Unlike in other squirrels, black specimens are unknown.

DISTRIBUTION: The distribution of *S. arizonensis* is very restricted in Arizona, New Mexico, and northern Mexico, where it is exclusively known in north-central Sonora. It is known from the Sierra de los Ajos, Sierra Azul, Sierra de la Madera, Sierra Patagonia, and Sierra de los Piñitos (Caire, 1978). It has only been registered in SO.

EXTERNAL MEASURES AND WEIGHT
TL = 506 to 568 mm; TV = 240 to 310 mm; HF = 66 to 77 mm; EAR = 30 to 37 mm.
Weight: 605 to 706 g.

DENTAL FORMULA: I 1/1, C 0/0, PM 2/2, M 3/3 = 20.

NATURAL HISTORY AND ECOLOGY: *S. arizonensis* prefers riparian habitats at moderate altitudes (Caire, 1978). It can also be found in the streams of desert-like grasslands and chaparral, as long as riparian communities have some diversity of species and continuity in the coverage and height of the vegetation (Brown, 1984). The Arizona walnut (*Juglans major*) has been used as an indicator of its presence (Brown, 1984). These squirrels are difficult to find at the beginning of summer, when females are pregnant and nursing, and in winter, when the trees have no leaves. They are conspicuous when they are feeding in tree branches or chasing each other to mate. They "freeze" in the face of danger and may remain so for more than 45 minutes (Brown, 1984). Because of this behavior and the lack of habituation to human beings, a need for special protection was suggested many years ago (Cahalane, 1939). They are quiet compared to other squirrels and vocalize only when they are in the security of a tree or in the presence of an intruder (Hobbs, 1980). They vocalize more in summer than in winter, a behavior attributed to the characteristics of the habitat and to parental selection. They feed in trees of 18 m in height or more. They do not store food because their habitat is relatively free of snow. Sixty-seven percent of their diet is composed of fungi, nuts, acorns, pine nuts, juniper berries (*Juniperus* spp.), and Netleaf hackberry fruits (*Celtis reticulata*). Flowering parts are important seasonally and coincide with the reproductive season; it is suspected that these contain vitamin A and other vitamins that stimulate reproductive activity after a period of rest (Brown, 1984), which would be consistent with what has been reported in other species. At the end of the summer months, insects and other animals complement their diet. They build two types of nests: a flat, platform-type structure used to rest in summer, and a covered and more solid nest in winter, which is also used for breeding (Brown, 1984). Depending on the availability of suitable trees, each squirrel can have several nests or none at all. The nests differ from the nests of hawks by the position in the tree (built in the fork of two large branches or in the joint of a large branch and the trunk) as well as by the form of the dome. The nests are at 12 m to 18 m high in sycamores (*Platanus*), walnuts (*Juglans*), alders (*Alnus*), ash (*Fraxinus*), poplars (*Populus*), and pines (*Pinus*), but are more common in oaks (*Quercus*). The nest may be up to 6 m high in the tree and measures 30 cm in diameter; the entrance is 5 to 10 cm wide. They use leaves, twigs, and bark to make their nests. The beginning of the mating season coincides with flowering and the presence of flower parts in the diet (Theobald, 1983). Reproduction extends from January to June, but males are sexually active from November to July with a peak in April. The females are in estrus in April–May; not all females reproduce each year. Each litter has 3.1 pups on average, with a range of 2 to 4; there is no evidence of females producing more than one litter per year. The time of care of the offspring seems to be long and variable (Theobald, 1983). It is not an abundant species (Cahalane, 1939; Coues, 1867). There are no estimates of population densities, but it is believed that they fluctuate with the availability of food and that they vary from year to year (Monson, 1972). Its known predators are the red-tailed hawk (*Buteo jamaicensis*) and the wild cat (Brown, 1984).

VEGETATIONAL ASSOCIATIONS AND ELEVATION RANGE: *S. arizonensis* inhabits deciduous riparian vegetation, composed mainly of walnuts (*Juglans*), sycamores (*Platanus wrightii*), poplars (*Populus angustifolia* and *P. fremonti*), maple (*Acer grandidentatum*), and boxelder (*A. negundo*), as well as in canyons with forests of pine (*Pinus*) and oak (*Quercus*). They are found from 1,200 masl to 2,145 masl.

CONSERVATION STATUS: *S. arizonensis* is not abundant (Caire, 1978) and its riparian habitat is subject to strong anthropogenic pressure; hence it is considered threatened in Mexico (SEMARNAT, 2010). Populations exist in the biosphere reserves of El Pinacate and Great Desert of Altar, which has recently been established in Sonora.

Sciurus aureogaster F. Cuvier, 1829

Red-bellied squirrel

Manuel Valdéz Alarcón and Guadalupe Téllez-Girón

SUBSPECIES IN MEXICO

Sciurus aureogaster aureogaster Cuvier, 1829
Sciurus aureogaster nigrescens Bennett, 1833

DESCRIPTION: *Sciurus aureogaster* is one of the largest arboreal squirrels. Its coloration pattern shows great variation, and melanistic individuals are common. The back is gray splashed with white; the top of the nape of the neck, neck, hindquarters, and sides are interrupted with ochre brown. The belly varies from white to pale brown. The tail is long and fluffy, and varies in color from white to gray (Emmons and Feer, 1997; Hall, 1981; Musser, 1968).

EXTERNAL MEASURES AND WEIGHT

TL = 470 to 573 mm; TV = 235 to 276 mm; HF = 63 to 70 mm; EAR = 23 to 36 mm. Weight: 432 to 690 g.

DENTAL FORMULA: I 1/1, C 0/0, PM 2/2, M 3/3 = 20.

NATURAL HISTORY AND ECOLOGY: Red-bellied squirrels have diurnal and arboreal habits. Two peaks of activity during the day are present: the first from 07:00 hrs

Sciurus aureogaster. Tropical dry forest. Acapulco, Guerrero. Photo: Patricio Robles Gil.

to 09:00 hrs and the second from 15:00 hrs to 17:00 hrs. They are solitary and only get together during the mating season, forming groups of no more than four individuals (Coates-Estrada and Estrada, 1986). They build their nests with green leaves and needles of the trees where they live. Some nests are located in tree holes (Álvarez, 1963a; Hall and Dalquest, 1963). Their food includes a wide variety of cones, sprouts, buds, seeds, acorns, and fruit such as wild plums (*Spondias mombin*), green figs, tamarinds, and chicle tree. Its diet also includes fungi, insects, eggs, and chicks of birds. On the Pacific coast it feeds on coconuts (Álvarez and Aviña, 1963; Hall and Dalquest, 1963; Ramírez-Pulido and López-Forment, 1976). Reproduction probably occurs in spring and summer. The gestation period lasts approximately 44 days and females give birth up to as many as 4 offspring per litter (Ceballos and Galindo, 1984).

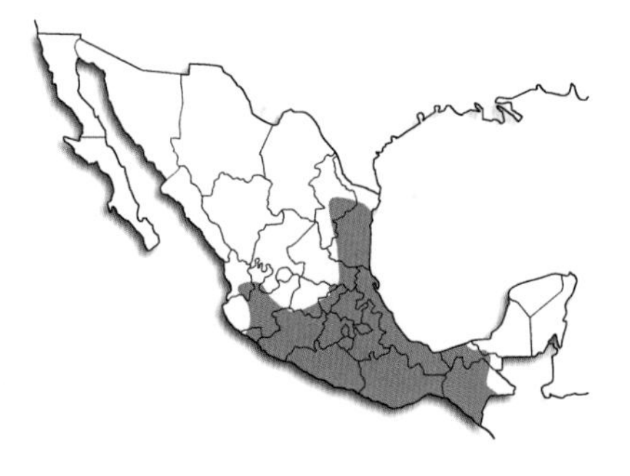

DISTRIBUTION: Red-bellied squirrels are found from Nuevo Leon and northern Tamaulipas to Tabasco, in the Gulf of Mexico, and from Nayarit to Chiapas and Guatemala in the Pacific. The species is found in the center of the country and is not present in the Yucatan Peninsula (Hall, 1981). It has been registered in the states of BC, CO, CS, DF, GR, HG, JA, MI, MO, MX, NL, NY, OX, PU, QE, SL, TA, TB, TL, VE, and ZA.

VEGETATIONAL ASSOCIATIONS AND ELEVATION RANGE: *S. aureogaster* is found in temperate forests of pine-oak, cedar, and fir, but also inhabits seasonal tropical forests and thorny bush. They are common in transformed areas such as coconut plantations and city parks (Musser, 1968; Ramírez-Pulido and López-Forment, 1976). They live from sea level up to 3,300 m (Goodwin, 1969; Hooper, 1961).

CONSERVATION STATUS: *S. aureogaster* is an abundant species with a wide distribution. In some regions of Mexico, their populations cause damage to crops and they are considered pests. The harm caused to coconut crops, however, is less than 5% (Ramírez-Pulido and López-Forment, 1976). They are used as a source of animal protein for human consumption and are acquired as pets (Ramírez-Pulido et al., 1977). They have no conservation issues.

Collie's squirrel

Vinicio J. Sosa and Joaquín Bello

SUBSPECIES IN MEXICO
Sciurus colliaei colliaei Richardson, 1839
Sciurus colliaei nuchalis Nelson, 1899
Sciurus colliaei sinaloensis Nelson, 1899
Sciurus colliaei truei Nelson, 1899

DESCRIPTION: *Sciurus colliaei* is a large-sized squirrel with a thin and flexible body. The eyes are big and the tail is longer than the body (Jones, 1963). The dorsal coloration is gray and the ventral coloration is white or gray. The dorsal part has a band of black hair that runs from the head to the base of the tail (Musser, 1968). In some regions *S. colliaei* is sympatric with *S. nayaritensis* and *S. aureogaster*. In general, it can be distinguished from these species because of its larger size and the combination of buffy and black on the back. *S. aureogaster* is smaller and gray from head to tail, and *S. nayaritensis* is characterized by its red or yellowish sides (Musser, 1968). The skull is compact with well-developed postorbital processes.

EXTERNAL MEASURES AND WEIGHT
TL = 440 to 578 mm; TV = 203 to 300 mm; HF = 57 to 71 mm; EAR = 30 to 32 mm. Weight: 380 to 490 g.

DENTAL FORMULA: I 1/1, C 0/0, PM 1/1, M 3/3 = 20.

NATURAL HISTORY AND ECOLOGY: This species generally inhabits tropical and subtropical vegetation, and rarely temperate vegetation, where other species of the genus are more common. On some coconut plantations they can become very abun-

Sciurus colliaei. Tropical dry forest. Chamela-Cuixmala Biosphere Reserve, Jalisco. Photo: Gerardo Ceballos.

dant (Anderson, 1962). They are diurnal with peaks of activity in the early hours of the morning and evening. They spend most of their time feeding or building spherical nests with leaves and branches (Leopold, 1965). Sometimes they crawl down to the ground in search of food or water, or to move to another tree. They are very agile climbers and can drop from a height of 5 m to 10 m without injury (Ceballos and Miranda, 1986, 2000). They use their tails as an aid to move from one place to another (Leopold, 1965). They are solitary, but can form groups to eat or during the mating season. They feed on a large amount of fruits and seeds. Their diet varies according to the type of vegetation (Leopold, 1965). In areas of pine-oak they consume acorns (*Quercus*) and pine nuts (*Pinus*), but in tropical forests they consume figs (*Ficus*), and plums (*Spondias*). In the Chamela-Cuixmala Biosphere Reserve, on the coast of Jalisco, they are more abundant in the medium jungle than in the low jungle; their densities vary from 0.9 to 4.3 individuals per km^2 (Mandujano, 1997). There are no precise data on their reproductive cycle, but reproduction probably occurs in the summer, at the beginning of the rains. In addition, the gestation period lasts 44 days with litters of 2 to 6 offspring (usually 4 to 5), which are nursed for 8 to 10 weeks (Ceballos and Miranda, 1986, 2000).

VEGETATIONAL ASSOCIATIONS AND ELEVATION RANGE: Throughout its distribution in the coastal plain of the Pacific, *S. colliaei* is located in several types of vegetation such as tropical deciduous and subdeciduous forests and palm groves (Musser, 1968). In the northern portion of its distribution, it dwells in the coastal plain, in mountainous areas with pine-oak forests, in the riparian vegetation that grows on some slopes of the Sierra Madre Occidental, and, in smaller proportions, in fir forests. It is found from sea level up to 2,134 m. The majority of the localities are below 1,000 masl.

CONSERVATION STATUS: This species is not in danger of extinction. It is not in any category of protection (SEMARNAT, 2010), but it is endemic to the country. Its wide range of distribution and the variety of its diet suggest that as long as wooded habitats, whether natural or cultured, are available this species will not be threatened with extinction.

DISTRIBUTION: *S. colliaei* is an endemic species of Mexico. Its distribution covers the northwestern portion of the coastal plain of the Pacific, from Sonora to Colima. It is found at high altitudes in some portions of the Sierra Madre Occidental in Chihuahua, Durango, Sinaloa, and part of the Mexican plateau, mainly in the state of Jalisco (Anderson, 1962; Hall, 1981). It has been registered in the states of CH, CL, DU, JA, NY, SI, and SO.

Sciurus deppei Peters, 1863

Deppe's squirrel

Livia León Paniagua and Hugo A. Ruiz Piña

SUBSPECIES IN MEXICO

Sciurus deppei deppei Peters, 1863
Sciurus deppei negligens Nelson, 1898
Sciurus deppei vivax Nelson, 1901

DESCRIPTION: *Sciurus deppei* is the smallest arboreal species in Mexico. Its dorsal coloration is variable, from brown-yellow to very dark brown. The front legs have a grayish coloration that, in some populations, extends to the shoulders. The belly is usually white, although in populations from Chiapas and Guatemala it is yellowish-red. They have spots on the bases of the ears; however, these spots are smaller than those in larger species. There is variation in the tonality of the dorsal coloration,

Sciurus deppei. Tropical evergreen forest. Calakmul Biosphere Reserve, Quintana Roo. Photo: Gerardo Ceballos.

and, in lesser degree, the ventral coloration. Dalquest (1953b) found large individual variation in populations from San Luis Potosí. Populations from the Yucatan Peninsula, including the Peten in Guatemala and Belize, have a pale tonality and some grayish spots that extend from the shoulders up to the back. There are no records of melanism in this species (Best, 1995b; Hall, 1981; Nelson, 1898, 1901).

EXTERNAL MEASURES AND WEIGHT

TL = 343 to 387 mm; TV = 155 to 197 mm; HF = 46 to 55 mm; EAR = 21 to 30 mm. Weight: 200 to 300 g.

DENTAL FORMULA: I 1/1, C 0/0, PM 2/2, M 3/3 = 20.

NATURAL HISTORY AND ECOLOGY: *S. deppei* mainly inhabits areas of dense vegetation and humid lowlands; it shuns open areas and disturbed woodlands. It is diurnal and for most of the day it can be sighted on the ground in search of food, especially in forests and rainforests with shorter trees. It is granivorous and folivorous. It feeds on palm nuts, wild fruit trees, and shrubs. In Veracruz, its diet is 95% fruits and 5% leaves (Estrada and Coates-Estrada, 1985). It can reproduce at any time of year but only gives birth once a year. The size of the litter varies from four to eight pups (Best, 1995b). Pregnant females and males with scrotal testes have been found throughout the year (Baker, 1951a; Dalquest, 1953b; Hall and Dalquest, 1963; Jones et al., 1983; León-Paniagua et al., 1990). It builds nests 2 m or 3 m high in the trees. In some regions of the country it is hunted because it causes damage to the cornfields.

VEGETATIONAL ASSOCIATIONS AND ELEVATION RANGE: *S. deppei* generally lives in forests and dense forests of pine, pine-oak, cloud forest, subperennial medium jungle, and perennial high jungle (Álvarez, 1963a; Hall and Dalquest, 1963; Hooper, 1953; Jones et al., 1983; León-Paniagua et al., 1990). It has been found most often at altitudes less than 1,800 masl, although it is sporadically found above 2,800 masl (Wilson, 1983). Most of the records are below 300 masl (Best, 1995b).

DISTRIBUTION: This species has a tropical distribution, which extends from Tamaulipas to the Yucatan Peninsula, Chiapas, and Costa Rica (Hall, 1981; Jones et al., 1983; León-Paniagua et al., 1990). It has been registered in the states of CA, CS, GJ, HG, NL, OX, PU, QE, QR, SL, TA, TB, VE, and YU.

CONSERVATION STATUS: This species is very susceptible to anthropogenic alterations to its habitat. Logging is one of the main factors that could affect the density of its populations. In spite of the absence of adequate information in this respect, it is not considered in danger of extinction because it survives in sunflower fields and disturbed jungles. Furthermore, there are some populations in several biosphere reserves such as Calakmul in Campeche, Los Tuxtlas in Veracruz, Sian Ka'an in Quintana Roo, and Montes Azules in Chiapas.

Sciurus griseus Ord, 1818

Western gray squirrel

Eric Mellink and Jaime Luévano

SUBSPECIES IN MEXICO
Sciurus griseus anthonyi Mearns, 1897

DESCRIPTION: *Sciurus griseus* is a large arboreal squirrel. The dorsal coloration is mottled dark gray to pale gray, with the abdomen whitish to creamy. The tail is gray and very large (Hall, 1981).

EXTERNAL MEASURES AND WEIGHT
TL = 510 to 570 mm; TV = 265 to 290 mm; HF = 74 to 80 mm; EAR = 28 to 36 mm.
Weight: 567 to 794 g.

DENTAL FORMULA: I 1/1, C 0/0, PM 2/2, M 3/3 = 20.

NATURAL HISTORY AND ECOLOGY: There is no information available on the biology of this species in Mexico. In other areas, its biology is known in some detail (Flyger and Gates, 1982a). This species is active throughout the year. Its activity is mainly in the morning and, although it is predominantly arboreal, it can be seen on the ground. During the summer it uses nests of bark and branches, located at least at 6 m above the ground. Its diet mainly consists of acorns (*Quercus* spp.), but it also feeds on pine nuts and other nuts, some fungi, blackberries, and insects. The reproductive season goes from January to May, during which up to two litters can be produced. Each litter consists of two to six offspring. The gestation period lasts 44 days. It is found in densities of up to five individuals per ha.

VEGETATIONAL ASSOCIATIONS AND ELEVATION RANGE: In Baja California, *S. griseus* has been found only in open forests of pine, but in other areas, it is found in relatively open forests of oak and pine-oak, from 1,600 masl to 1,700 masl (Flyger and Gates, 1982a).

CONSERVATION STATUS: Its current status is threatened (SEMARNAT, 2010), and *S. griseus* appears not to be abundant. Ceballos and Navarro (1991) considered it threatened. Part of the area of its distribution is located inside the National Park Constitución de 1857; the rest is on communal lands. In these areas logging was very intense in the 1950s. Now it is limited to the extraction of dead wood, as the dominant species of pine (*Pinus jeffreyi* and *P. adulterinum*) are subject to special protection (SEMARNAT, 2010).

DISTRIBUTION: *S. griseus* is distributed from California to Baja California. In Mexico, it has only been registered in the central part of the Sierra de Juarez, and there does not seem to be a current connection between these and the populations in southern California (Huey, 1964; Mellink and Contreras, 1993). It has been registered in the state of BC.

Mexican fox squirrel

Elizabeth E. Aragón

SUBSPECIES IN MEXICO
Sciurus nayaritensis apache J.A. Allen, 1893
Sciurus nayaritensis nayaritensis J.A. Allen, 1890

DESCRIPTION: *Sciurus nayaritensis* is a large arboreal squirrel. The back is gray with brown or yellow spots and the belly is ochre colored; the southern populations have white spots (Álvarez, 1961a; Álvarez and Aviña, 1963; Álvarez and Polaco, 1984). During the winter, the body is ash-gray, but in the spring the back is grayish-brown. Females have the bases of the hair dark red with white tips, which gives them a gray-haired look; in males, red is more conspicuous than white, giving them a darker look (Álvarez and Polaco, 1984).

EXTERNAL MEASURES AND WEIGHT
TL = 517 to 590 mm; TV = 235 to 292 mm; HF = 70 to 81 mm; EAR = 27 to 38 mm. Weight: 569 to 983 g.

DENTAL FORMULA: I 1/1, C 0/0, PM 2/2, M 3/3 = 20.

NATURAL HISTORY AND ECOLOGY: Mexican fox squirrels are diurnal animals that build their nests in the highest parts of pines (Álvarez and Aviña, 1963). In spring the periods of increased activity are during the morning and evening. Subsequently, in July and August, the squirrels remain close to their nests; their activity occurs in the branches of trees and they are not observed on the ground (Álvarez and Polaco,

Sciurus nayaritensis. Riparian forest. Janos Biosphere Reserve, Chihuahua. Photo: Gerardo Ceballos.

1984). Reproduction occurs in winter and summer. Pregnant females have been found in February and July, lactating animals in July, non-reproductive females in April, and males with scrotal testes in May. The average litter size is 2.3 (Álvarez and Polaco, 1984; Baker and Greer, 1962; Matson and Baker, 1986). They probably feed on pine nuts, as pine cones have been found under pine trees in which they live (Baker and Greer, 1962; Matson and Baker, 1986). Their diet varies seasonally. In February, they mainly feed on seeds (99%) and some fungi. In April, seeds (87.4 %), fungi, and pollen from oaks (12.6%) predominate. The daily intake per individual is about 9 g, which corresponds to 1% of its weight (Álvarez and Polaco, 1984).

DISTRIBUTION: The distribution of this species is restricted to the northwest of Mexico, in the Sierra Madre Occidental extending barely into the southwestern United States (Hall, 1981; Ramírez-Pulido, 1983). It has been registered in the states of AG, CH, CL, DU, JA, NY, SI, SO, and ZA.

VEGETATIONAL ASSOCIATIONS AND ELEVATION RANGE: These are arboreal squirrels that inhabit forests of pine-oak and occasionally in associations of pine with juniper (Álvarez and Aviña, 1963; Álvarez and Polaco, 1984; Matson and Baker, 1986). It is found from 900 masl to 2,700 masl.

CONSERVATION STATUS: *S. nayaritensis* is not considered endangered (SEMARNAT, 2010). Despite its limited distribution, apparently it is not at risk. It is hunted for sport and for local consumption.

Sciurus niger Linnaeus, 1758

Eastern fox squirrel

Gerardo Ceballos

SUBSPECIES IN MEXICO
Sciurus niger limitis Baird, 1855

DESCRIPTION: *Sciurus niger* is a large arboreal squirrel. The hair coloration of the back varies from gray to yellowish-brown. The belly is yellow or orange. The tail is bicolored, with a contrast similar to that of the body. The eyes are large and the ears are relatively small. A distinctive feature of this squirrel is that it has only a single upper premolar on each side. The face is short and wide.

EXTERNAL MEASURES AND WEIGHT
TL = 454 to 698 mm; TV = 200 to 330 mm; HF = 51 to 82 mm; EAR = 30 to 34 mm. Weight: 507 to 1,361 g.

DENTAL FORMULA: I 1/1, C 0/0, PM 2/2, M 3/3 = 20.

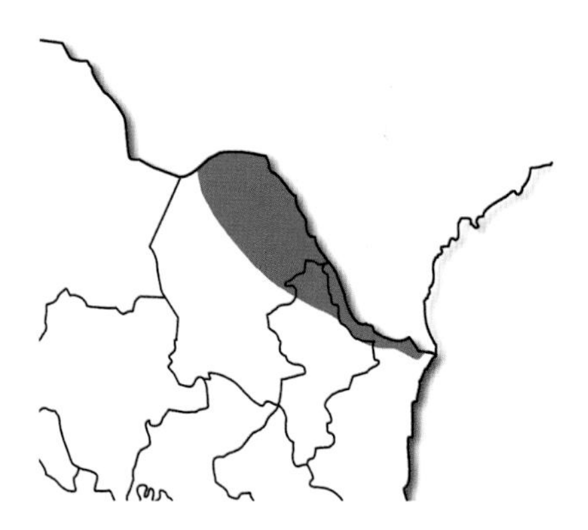

DISTRIBUTION: This squirrel is widely distributed in the eastern and central United States. Its distribution in Canada and Mexico is marginal. In Mexico, it is distributed throughout the Rio Bravo drainage in the northeast of Coahuila and Nuevo León (Koprowski, 1994). It has been registered in the states of CO and NL.

NATURAL HISTORY AND ECOLOGY: This squirrel has diurnal habits. In spring and summer it has several peaks of activity during the day. In winter its activity is reduced and is monotypic. They spend a large part of their time on the ground, in search of food (Wilson and Ruff, 1999). They shelter in abandoned holes of woodpeckers and other tree holes. They can also build nests between the high branches of the trees. As coating material they use fragments of leaves, twigs, and other small materials. The nests have a dimension of 34 cm in diameter with an entry of 7 to 9 cm in diameter. It has been observed that during the summer they build other types of nests, more fragile and platform-like, and during the year they may occupy more than nine nests (Baumgarther, 1939; Christen, 1985; Nixon and Hansen, 1987). Its area of activity is

Sciurus niger. Riparian forest. Río Bravo, Coahuila. Photo: Gerardo Ceballos.

very extensive, sometimes covering more than 20 ha. Immigration is constant, and adult females are an important factor in the structure and density of the population; subadults and young individuals of both sexes enter the population, but adult females that come from other areas are also incorporated (Hansen et al., 1986). Almost all year they mainly feed on acorns, but in winter and early spring they eat fruit, eggs, pine nuts, and flowers. Occasionally, they can eat insects such as beetles, moths, crickets, butterflies, and ants (Baumgarther, 1939; Korschgen, 1981; Nixon et al., 1968; Packard, 1956). The reproductive cycle is bimodal and occurs mainly from November to February and April to June. Females become reproductively active at 1.25 years of age, and remain so for at least 12 years on average (Koprowski et al., 1988). Females can have two to three pups, but only 2% of the females reproduce twice a year (Harnishfeger, 1978). Reproduction is closely related to the availability of food (Nixon and McClain, 1969). Males reach sexual maturity between 10 and 11 months of age and reproductive activity is also bimodal from November to February and May to July (Kirkpatrick, 1955). The main predators of fox squirrels are rattlesnakes, snakes, hawks, owls, opossums, weasels, raccoons, foxes, lynx, coyotes, dogs, and domestic cats (Flyger and Gates, 1982; Packard, 1956; Weigl et al., 1989).

VEGETATIONAL ASSOCIATIONS AND ELEVATION RANGE: An abundant squirrel population can be found in open forests of oak (*Quercus* sp.), pine (*Pinus* sp.), walnuts (*Juglans* sp.), and hickory nuts (*Carya* sp.). In the northern and central United States it is found in mixed and deciduous forests (Lowery and Davis, 1942). In Mexico, it is more common in gallery forests along the Rio Bravo and its tributaries.

CONSERVATION STATUS: The subspecies of Mexico could be considered fragile, but it is not considered within any risk category (SEMARNAT, 2010). Its distribution is restricted and marginal in Mexico, however; in addition, logging and excessive use of areas for agriculture in the forests of Coahuila and Nuevo Leon (Leopold, 1965) make the species vulnerable.

Peters's squirrel

Manuel Valdéz Alarcón and Guadalupe Téllez-Girón

SUBSPECIES IN MEXICO

Sciurus oculatus oculatus Peters, 1863
Sciurus oculatus shawi Dalquest, 1950
Sciurus oculatus tolucae Nelson, 1898

DESCRIPTION: *Sciurus oculatus* is a large-sized squirrel. The hair coloration varies slightly between the subspecies, but in general, the dorsal coloration is gray and toward the base becomes brown to black. The ventral part varies from white to pale ochre yellow. The dorsal coloration of the tail is black, while its ventral part is dark with an edge of hair with white tips. Its eye-ring is well differentiated, of whitish or cream color. The most important characteristic is the lack of the third upper premolar (Dalquest, 1953a; Hall, 1981).

EXTERNAL MEASURES AND WEIGHT

TL = 508 to 560 mm; TV = 256 to 269 mm; HF = 68 to 73 mm; EAR: 29 to 35 mm. Weight: 550 to 750 g.

DENTAL FORMULA: I 1/1, C 0/0, PM 1/1, M 3/3 = 20.

NATURAL HISTORY AND ECOLOGY: *S. oculatus* is an arboreal squirrel of diurnal habits. They are solitary, except in the mating season, when up to 20 individuals can be seen in one tree (G. Ceballos, pers. obs.). They move skillfully among the trees (Dalquest, 1953a). They are sighted with ease during summer, but during the colder

Sciurus oculatus. Pine-oak forest. Sierra Gorda Biosphere Reserve, Queretaro. Photo: Juan Cruzado.

DISTRIBUTION: *S. oculatus* is an endemic species of Mexico that is distributed along the transvolcanic belt, in central Mexico (Dalquest, 1953a; Hall, 1981). It has been registered in the states of DF, GJ, HG, MI, MO, MX, PU, QR, SL, and VE.

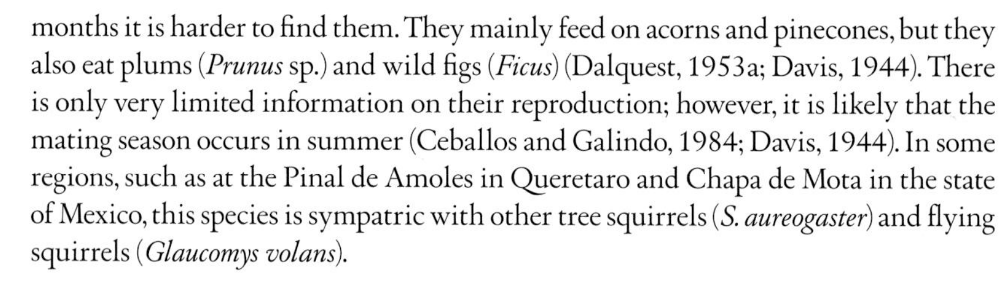

months it is harder to find them. They mainly feed on acorns and pinecones, but they also eat plums (*Prunus* sp.) and wild figs (*Ficus*) (Dalquest, 1953a; Davis, 1944). There is only very limited information on their reproduction; however, it is likely that the mating season occurs in summer (Ceballos and Galindo, 1984; Davis, 1944). In some regions, such as at the Pinal de Amoles in Queretaro and Chapa de Mota in the state of Mexico, this species is sympatric with other tree squirrels (*S. aureogaster*) and flying squirrels (*Glaucomys volans*).

VEGETATIONAL ASSOCIATIONS AND ELEVATION RANGE: *S. oculatus* inhabits more or less well-preserved pine, oak, and fir forests (Ceballos and Galindo, 1984; Nelson, 1899c). It is distributed in areas between 1,500 masl and 3,600 masl (Goldman, 1951; Hall, 1981).

CONSERVATION STATUS: Peters's squirrel is considered rare or fragile (Ceballos and Navarro, 1991; SEMARNAT, 2010). Immoderate logging of forests of the transvolcanic belt is the most severe factor that threatens its long-term conservation (Ceballos and Galindo, 1984).

Sciurus variegatoides Ogilby, 1839

Variegated squirrel

Gerardo Ceballos and Manuel Valdéz Alarcón

SUBSPECIES IN MEXICO

Sciurus variegatoides goldmani Nelson, 1898

DESCRIPTION: The variegated squirrel is an arboreal squirrel of medium size. Its coloration is variable. The dorsal coloration varies from dark brown to grayish-yellow, the central part of the neck is darker than the rest of the body, and it has a paler patch behind the ears. The tail is long with dense hair, black or frosted with white tips, which looks like a white edge around and over the tail. In Nicaragua and Costa Rica individuals exist with the tail and belly very pale colored. The belly varies from reddish-cinnamon to light cinnamon. There is great variation between individuals in the same population (Eisenberg, 1989; Emmons and Feer, 1997; Hall, 1981; McPherson, 1985).

EXTERNAL MEASURES AND WEIGHT

TL = 510 to 560 mm; TV = 240 to 305 mm; HF = 60 to 70 mm; EAR = 20 to 35 mm. Weight: 450 to 520 g.

DENTAL FORMULA: I 1/1, C 0/0, PM 2/2, M 3/3 = 20.

DISTRIBUTION: *S. variegatoides* is an endemic species of Mesoamerica, found from southern Chiapas, Mexico, to northwestern Panama (Emmons and Feer, 1997; McPherson, 1985). It has been registered in the state of CS.

NATURAL HISTORY AND ECOLOGY: *Sciurus variegatoides* is diurnal and generally solitary. It builds its nests in holes of the trees or with leaves between the high branches. It is a generalist (Glanz, 1984), feeding on soft and juicy fruits, nuts, seeds, sprouts, or buds of flowers of species such as *Schelea* sp., guacimo (*Guazuma ulmifolia*), *Mangifera indica*, plum (*Spondias mombin*), guava, and mango. They consume the fruit without affecting the seed. While eating they make sounds or snaps that may be alarm signals (Reid, 1997).

Sciurus variegatoides. Tropical dry forest. Photo: Gerardo Ceballos.

VEGETATIONAL ASSOCIATIONS AND ELEVATION RANGE: The variegated squirrel is often found in low and seasonal jungles, oak forests, and fruit orchards (Emmons and Feer, 1997). It is rarely seen in high perennial jungles (Reid, 1997). It has also been found in disturbed habitat, sometimes even in sunflowers with few trees. It occurs from sea level up to 1,800 m (Timm et al., 1989).

CONSERVATION STATUS: The current status of *S. variegatoides* in Mexico is unknown, but it is considered as of Special Protection (SEMARNAT, 2010). It probably does not face serious risks, however, because it can survive in disturbed areas, including parks and gardens (G. Ceballos, pers. obs.). It is frequently seen on fences around crops (Reid, 1997).

Sciurus yucatanensis J.A. Allen, 1877

Yucatan squirrel

Livia León Paniagua and Hugo A. Ruiz Piña

SUBSPECIES IN MEXICO

Sciurus yucatanensis baliolus Nelson, 1901
Sciurus yucatanensis myotis Goodwin, 1932
Sciurus yucatanensis yucatanensis J.A. Allen, 1877

Sciurus yucatanensis. Tropical semi-green forests. Calakmul Biosphere Reserve, Campeche. Photo: Gerardo Ceballos.

Three subspecies of *S. yucatanensis* are recognized in Mexico (Hall, 1981). Jones et al. (1974), however, considered there was no consistent differentiation between *S. baliolus* and *S. yucatanensis*, and suggested eliminating the name of baliolus. Recent analyses of geographic variation, using cranial characteristics and patterns of coloration, indicate that *S. yucatanensis* could be divided into two species, *S. yucatanensis* and *S. baliolus* (including *S. yucatanensis phaeopus*; Best et al., 1995; Ruiz-Piña, 1994).

DESCRIPTION: *Sciurus yucatanensis* and *S. deppei* are the smallest squirrels of the subgenus *Sciurus*. Dorsally, it has a grayish-yellow coloration. Northern populations are pale gray dorsally with a yellowish central stripe from the neck to the base of the tail. The ventral coloration is white, although on some occasions it has a yellowish tinge that gives the appearance of a dirty white. This coloration intensifies from north to south, where the populations have a darker and more intense color pattern. In addition, they have some small black spots on the ears and in some regions of the face, characteristics that led to the description of two subspecies. The presence of tufts of white hair on the tip of the ears is a feature that distinguishes *S. yucatanesis* from other species of Mexico and Central America (Goodwin, 1934; Hall, 1981; Nelson, 1899c; Ruiz-Piña, 1994).

EXTERNAL MEASURES AND WEIGHT
TL = 345 to 525 mm; TV = 220 to 280 mm; HF = 52 to 63 mm; EAR = 25 to 30 mm.
Weight: 320 to 540 g.

DENTAL FORMULA: I 1/1, C 0/0, PM 2/2, M 3/3 = 20.

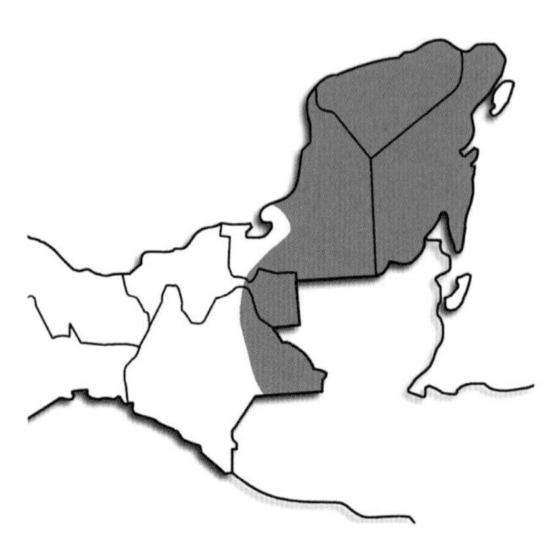

DISTRIBUTION: *S. yucatanensis* is a typically tropical species, endemic to the Yucatan Peninsula. Its range extends from the north of the Yucatan Peninsula, east of Tabasco and northeast of Chiapas, to Belize, El Peten, and the center of Guatemala (Hall, 1981; Musser, 1968; Nelson, 1899c; Ruiz-Piña, 1994). It has been registered in the states of CA, CS, QR, TB, and YU.

NATURAL HISTORY AND ECOLOGY: The Yucatan squirrel is shy and begins its activity in the early morning hours (Goodwin, 1934). It feeds on almost any wild fruit; however, it shows preference for fleshy fruits such as the nance, chicle tree, and mango. In some regions of the Yucatan Peninsula this squirrel is considered a plague because it invades crops, mainly of corn. Their reproductive season begins at the end of the rainy season and lasts about two to three months. Apparently, they only have a single litter per year, whose number varies from three to four offspring (Best et al., 1995; Ruiz-Piña, 1994). In some regions, it hybridizes with *S. deppei* (Gaumer, 1917; Leopold, 1959). Occasionally, this squirrel is part of the diet of rural communities or is traded for its skin (which is mainly decorative). It has been observed as a pet in some sites in northern Yucatan and south of Quintana Roo (Ruiz and Leon, pers. obs.). Its parasites include bed bugs (*Eutrombicula alfreddugesi* and *Hoffmannina suriana*) and lice (*Enderneinellus hondurensis*) (Kim, 1966; Loomis, 1969).

VEGETATIONAL ASSOCIATIONS AND ELEVATION RANGE: *S. yucatanensis* is found in the low deciduous jungle and shrubs in semi-arid regions of northern Yucatan, and in forests of pine-oak and high perennial jungles in Guatemala (Musser, 1968; Ruiz-Piña, 1994). Generally, this squirrel occurs from sea level up to 450 masl. In most of its distribution the localities are below 200 masl (Best et al., 1995).

CONSERVATION STATUS: This squirrel is not considered in any category of conservation. Its distribution is restricted to the Yucatan Peninsula. It seems to tolerate some degree of habitat disruption and apparently has not suffered a significant decline (A. Ruiz and L. Leon, pers. obs.).

Mearns's squirrel

Eric Mellink and Jaime Luévano

DISTRIBUTION: *T. mearnsi* is endemic to Mexico. It has been registered in Vallecitos, La Grulla, and San Pedro Martir, in the Sierra de San Pedro Martir, in Baja California (Lindsay, 1981). It has been registered in the state of BC.

SUBSPECIES IN MEXICO
T. mearnsi is a monotypic species.

DESCRIPTION: *Tamiasciurus mearnsi* is medium in size; the dorsal coloration is ochre, brown, or olive, and the abdomen is white. There are seasonal variations in the coloration. In the summer they have a black stripe on the sides, separating the colors of the back and belly. They have a small tuft of hair on the tip of the ears. There is no sexual dimorphism (Lindsay, 1981).

EXTERNAL MEASURES AND WEIGHT
TL = 201 mm; TV = 100 mm; HF = 51 mm; EAR = 35 mm.
Weight: 200 g.

DENTAL FORMULA: I 1/1, C 0/0, PM 2/1, M 3/3 = 22.

NATURAL HISTORY AND ECOLOGY: Very little is known of this squirrel's biology, which is probably similar to that of *T. douglasii*, its closest taxonomic relative. It is active throughout the year, both in trees and on the ground. It builds nests with branches, lichens, mosses, and bark, or occupies holes in the trees. It feeds on seeds of conifers, including pine and fir trees. It also eats blackberries, fungi, buds, leaves, and eggs of birds. It probably has two litters per year, with four to seven offspring. The first litter is born in June. The offspring remain with the parents most of the first year of age (Flyger and Gates, 1982b).

VEGETATIONAL ASSOCIATIONS AND ELEVATION RANGE: *T. mearnsi* inhabits coniferous forest at 2,100 masl to 2,400 masl.

CONSERVATION STATUS: The information on this matter is rather confusing. In 1921 *T. mearnsi* was considered widely distributed in the transition area. Subsequently, it has been reported as exclusively in forests of high altitude. The preferred habitat is protected within the National Park Sierra de San Pedro Martir. Although its populations have not been assessed, Ceballos and Navarro (1991) considered it threatened due to its restricted distribution. Officially, it is considered threatened (SEMARNAT, 2010).

Tamiasciurus mearnsi. Pine forest. San Pedro Martir National Park, Baja California. Photo: B.S. Pash.

Perote ground squirrel

Manuel Valdéz Alarcón and Gerardo Ceballos

SUBSPECIES IN MEXICO

X. perotensis is a monotypic species.

DESCRIPTION: *Xerospermophilus perotensis* is one of the largest squirrels in the genus. It has a robust body, the feet are short, and the tail is one-third of the total length of the body. The dorsal coloration is fulvous cinnamon, with small white or slightly pink spots. The spotting is irregular and not always evident. The head is darker than the rest of the body and there is a whitish ring around the eye. From the middle part of the tail to the distal end, there is a greater proportion of black interspersed with the base color. The ventral region, including the inner part of the legs, varies from pale pink to whitish gray. It has five pairs of mammae and there is no sexual dimorphism (Hall, 1981; Howell, 1938).

EXTERNAL MEASURES AND WEIGHT

TL = 250 to 270 mm; TV = 57 to 78 mm; HF = 36 to 40 mm; EAR = 8 to 11 mm.
Weight: 175 to 270 g.

DENTAL FORMULA: I 1/1, C 0/0, PM 2/1, M 3/3 = 22.

NATURAL HISTORY AND ECOLOGY: *X. perotensis* dwells on rocky hillsides and in flat areas of deep soil with halophilous grassland or associations of cacti, chollas (*Opuntia* sp.) and agave (*Agave oscura*). At the limit of its distribution, it can be found near forests of pine. It shelters in burrows built in the ground, but may also occupy abandoned tunnels of gophers (*Pappogeomys merriami*). The entry is flat, and sometimes

Xerospermophilus perotensis. Scrubland-grassland. Perote, Veracruz. Photo: Gerardo Ceballos.

located at the bases of prickly pears or chollas (Valdez and Ceballos, 1997). Its annual activity lasts approximately nine months, and in the coldest months of the year, usually from December to February, they hibernate. The breeding season starts in May. The gestation period goes from June to August, lasting between 28 and 30 days. They have an average of six embryos and four offspring per litter (Davis, 1944; Valdez and Ceballos, 1997). Juveniles are observed from July to October. Their diet is basically herbivorous, and they eat a wide variety of annual plants; however, they also consume insects. They live in colonies, but use individual burrows. Couples are only observed during the breeding period and during the first weeks after emerging from the burrow (Valdez and Ceballos, 1997). They are sympatric with other small mammals such as gophers (*Pappogeomys merriami*) and kangaroo rats (*Dipodomys phillipsi*). Its main predators are the weasel, birds of prey (*Buteo jamaicensis*), and dogs (Valdez and Ceballos, 1997).

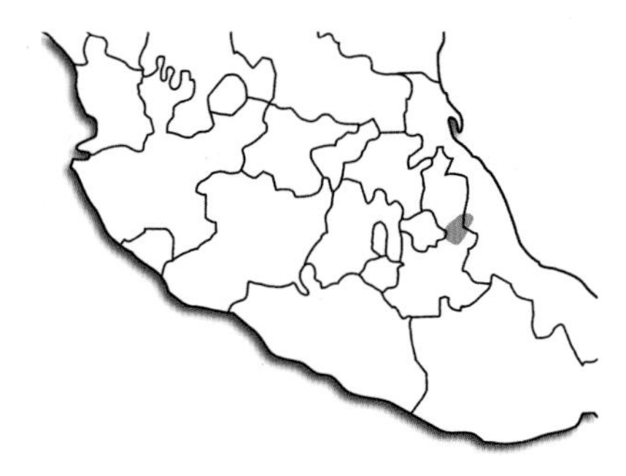

DISTRIBUTION: This species is only found in arid areas of the Valle del Oriental, in Puebla and Veracruz, east of the transvolcanic belt (Best and Ceballos, 1995a; Hall, 1981). It has been registered in the states of PU and VE.

VEGETATIONAL ASSOCIATIONS AND ELEVATION RANGE: Perote ground squirrels are found in halophytic pastures and xeric scrublands with grass, and on the edges of crops of oats and corn (Howell, 1938; Valdez and Ceballos, 1997). It is only found between 2,340 masl and 2,370 masl (Best and Ceballos, 1995).

CONSERVATION STATUS: *X. perotensis* is an endemic species in Mexico with a very restricted distribution. Its habitat is much deteriorated due to agriculture, livestock, and urban development. A recent study confirmed it as an endangered species (Valdez and Ceballos, 1997). It is considered threatened (SEMARNAT, 2010).

Xerospermophilus spilosoma (Bennett, 1833)

Spotted ground squirrel

Elizabeth E. Aragón

SUBSPECIES IN MEXICO

Xerospermophilus spilosoma altiplanensis Anderson, 1972
Xerospermophilus spilosoma ammophilus Hoffmeister, 1959
Xerospermophilus spilosoma bavicorensis Anderson, 1972
Xerospermophilus spilosoma cabrerai (Dalquest, 1951)
Xerospermophilus spilosoma canescens Merriam, 1890
Xerospermophilus spilosoma marginatus Bailey, 1902
Xerospermophilus spilosoma oricolus Álvarez, 1962
Xerospermophilus spilosoma pallescens (Howell, 1928)
Xerospermophilus spilosoma spilosoma Bennett, 1833

DESCRIPTION: The spotted ground squirrel is small and has a wide range in coloration, which varies mainly with the color of the soil (Anderson, 1972; Hall, 1981; Treviño, 1981). The back is brown with cream-colored spots, non-aligned along the spine. The belly is yellowish-white. The color of the tail is similar to that of the back but with black spots mixed with cream in the posterior dorsal region; the abdomen is cream colored, except for the rear, which is black and yellowish-white (Anderson, 1972; Baker, 1956; Hall, 1981; Jiménez-Guzman, 1966; Scott-M., 1984; Treviño, 1981).

Xerospermophilus spilosoma. Grassland. Janos Biosphere Reserve, Chihuahua. Photo: Rurik List.

EXTERNAL MEASURES AND WEIGHT
TL = 135 to 265 mm; TV = 40 to 100 mm; HF = 22 to 42 mm; EAR = 5 to 14 mm. Weight: 140 g.

DENTAL FORMULA: I 1/1, C 0/0, PM 2/1, M 3/3 = 22.

NATURAL HISTORY AND ECOLOGY: *Xerospermophilus spilosoma* is diurnal and terrestrial. Its burrows are tunnels of variable branching and depth. The tunnels are used for nesting, shelter, and escape routes (Streubel and Fitzgerald, 1978; Treviño, 1981). They also use burrows of other rodents. They molt twice a year; adults molt before and after hibernation, and subadults before hibernation (E. Aragon, pers. obs.). They are omnivores, and feed mainly on plant material such as seeds, flowers, pinnules, fruits, grasses (*Mulenbergia villiflora, Bouteloua chasei*), and insects such as coleopterans and larvae (Scott-M., 1984; Streubel and Fitzgerald, 1978). It commonly consumes ants, the fruits of cacti, and other plants such as *Agave asperrima, Lycium berlanderi, Larrea tridentata, Opuntia leptocaulis*, and *O. rastrera* (Serrano, 1982; Sosa and Serrano, 1987). Its annual cycle includes an active period and an inactive one (hibernation). The active period lasts nine months (March to November) and includes emergence, reproduction, recruitment of juveniles, and pre-hibernation. Then hibernation occurs (December to February), when it reduces its metabolism to a minimum and remains dormant (Aragon et al., 1993). It is an optional hibernator since young individuals have been caught during winter and in captivity they do not hibernate at all (Baker, 1956; Baudoin et al., 1991). Male and adult females enter hibernation first, and subadults and juveniles do so later. The period of lethargy ends first in subadults and adults (Aragon et al., 1993; Michener, 1984). Reproduction occurs from mid-

March to the end of July, but mating takes place mainly in May. They invest almost all their energy reproducing, males mainly during the search of a female to copulate with and females during pregnancy and care of the breeding. After that period they accumulate fat to hibernate (Aragon and Baudoin, 1990). Mating occurs within the burrow and the gestation period lasts 27 to 28 days (Streubel and Fitzgerald, 1978). The litter size varies depending on the altitude and quality of food. The average is 5.4, with a range of 4 to 9 offspring (Aragon and Baudoin, 1990; Baker, 1956; Baker and Greer, 1962; Matson and Baker, 1986; Streubel and Fitzgerald, 1978; Treviño, 1981). In the Chihuahuan desert, it is monoestrous. Late reproduction occurs in subadult females that reach sexual maturity that year (Aragon and Baudoin, 1990). Juveniles weigh between 28 g and 50 g when leaving the burrows; they gain weight rapidly and remain in the maternal area for one to two months (Aragon and Baudoin, 1990). *X. spilosoma* is physiologically adapted to arid areas; it has a low water loss from sweating and a high level of conductance (Hudson and Deavers, 1973). In the Mapimi Biosphere Reserve in Durango, density varies from 2.9 individuals/ha to 7.7 individuals/ha (Aragon et al., 1993; Grenot and Serrano, 1981; Serrano, 1987), according to the type of vegetation and environmental factors such as rain (Aragon et al., 1993). The sex ratio of males and females is 1:0.5 to 1:3.1, and the female reproductive potential determines the distribution of the males (Aragon and Baudoin, 1990; Aragon et al., 1993). These squirrels are solitary and territorial except in times of reproduction (Aragon and Baudoin, 1990). The area of activity is 0.08 ha to 3.5 ha (Aragon et al., 1993; Baudoin and Aragon, 1991; Grenot and Serrano, 1982). Variations in the size of the areas, according to sex and to the different periods of the annual cycle, have been documented. Males show greater movements and areas of activity during the breeding season (Baudoin and Aragon, 1991; Livorell et al., 1993). *X. spilosoma* is sympatric with other squirrels (*Ictidomys tridecemlineatus* and *I. mexicanus*), with which it has minimum interspecies competition due to temporal and spatial differences (Aragon et al., 1993; Grenot, 1983; Grenot and Serrano, 1982; Streubel, 1975). The daily activity goes from 09:00 hrs to 12:00 hrs and is related to day length, ambient temperature, and soil temperature in the shade. The peak of activity is when the temperature reaches 31°C to 35°C and decreases to 15°C. Hibernation occurs when the average temperature is under 20°C (Aragon et al., 1993). Locomotion and alert positions are hallmarks of the species (Aragon et al., 1993). In addition, maintenance (reproduction) and locomotion (recruitments) predominate. Agonistic behavior of females toward males is predominant during pregnancy and lactation (Aragon and Baudoin, 1990; Aragon et al., 1993). Their predators include rattlesnakes (*Crotalus scutulatus, C. atrox, C. molossus, C. lepidus*), Swainson's hawk (*Buteo swainsoni*), red-tailed hawk (*Buteo jamaicensis*), kestrel (*Falco sparverius*), coyotes, foxes, skunks, and badgers (E. Aragon, pers. obs.; Treviño, 1981).

DISTRIBUTION: *X. spilosoma* occurs from the center of the United States to Aguascalientes in Mexico, through the Central Highlands (Hall, 1981; Ramírez Pulido et al., 1983). It has been registered in the states of AG, CH, CO, DU, GJ, JA, NL, SL, SO, TA, and ZA.

VEGETATIONAL ASSOCIATIONS AND ELEVATION RANGE: *X. spilosoma* is a typical species of arid areas, which lives mainly in xeric scrublands and pasture, in association with thickets of species such as creosote (*Larrea tridentata*), tarbush (*Flourensia cernua*), mesquite (*Prosopis glandulosa*), and prickly pear (*Opuntia* sp.). It is also found in pastures, on dunes, at the margins of streams, in disturbed areas, and in crops of economic importance. It inhabits regions from 472 masl to 2,408 masl (Grenot and Serrano, 1981, 1982; Jiménez, 1966; Scott, 1984; Streubel and Fitzgerald, 1978; Treviño, 1981).

CONSERVATION STATUS: *X. spilosoma* is an important species in arid areas; it controls insects and undesired weeds and acts as a seed and fungal spore disperser, which has a profound impact on the maintenance of the structure of plant communities. It is not included in any risk category.

Round-tailed ground squirrel

Reyna A. Castillo

SUBSPECIES IN MEXICO

Xerospermophilus tereticaudus apricus (Huey, 1927)
Xerospermophilus tereticaudus neglectus Merriam, 1889
Xerospermophilus tereticaudus tereticaudus Baird, 1858

DESCRIPTION: *Xerospermophilus tereticaudus* is small. The head is small and round with big black eyes. The hind feet are long, wide, and covered with long rigid hair, with the exception of the soles. The coat is dark brown and does not have spots or stripes. It differs from other sympatric squirrels in that its tail is not shaggy, as well as by its small and round ears (Ernest and Mares, 1987).

EXTERNAL MEASURES AND WEIGHT
TL = 204 to 278 mm; TV = 60 to 112 mm; HF = 32 to 40 mm; EAR = 5 to 9 mm.
Weight: 110 to 170 g.

DENTAL FORMULA: I 1/1, C 0/0, PM 2/1, M 3/3 = 22.

NATURAL HISTORY AND ECOLOGY: The habitat of the round-tailed ground squirrel is characterized by extreme temperatures (Ernest and Mares, 1987). It is found in the plains of deserts and dunes. They excavate their burrows under shrubs or where dirt or sand accumulates, and they can also use burrows of kangaroo rats (*Dipodomys*). It is omnivorous and the proportion of vegetation (leaves, flowers, buds, and peels), seeds,

Xerospermophilus tereticaudus. Grassland. Nogales, Sonora. Photo: Gerardo Ceballos.

and insects varies with the season of the year. They cannot survive, however, on a diet of dry seeds without water (Schmidt-Nielsen and Schmidt-Nielsen, 1952). There is a direct relationship between the amount of rain in winter and the beginning of reproduction (Neal, 1965). In Arizona and California, the reproductive period is from March to June (Neal, 1965b, c); there are no studies for Mexico, but it is possible that the reproductive pattern is similar. The gestation period was estimated to be between 25 and 35 days (Neal, 1965c). The average size of the litter is 6.5 pups, with a range of 1 to 12 (Reynolds and Turkowsky, 1972). An increase in litter size has been attributed to the extension of the rains between October and February, where an increase of 2.5 cm increases the size of the litter by 1 (Reynolds and Turkowsky, 1972). They enter hibernation during autumn and winter (Neal, 1965a, b) and emergence from the burrow signals that the cold days are over (R. Castillo, pers. obs.). Their predators include coyotes, badgers, ravens, sparrowhawk, hawks (*Falco mexicanus*), and snakes such as *Pituophis melanoleucus*, *Masticophis flagellum*, and *Crotalus* spp. (Drabek, 1970; Dunford, 1977b; Jaeger, 1961). It is common to see them on their hind legs in a position of alert and running rapidly to the closest burrow at any sign of disturbance. It is also common to observe them "sitting" and manipulating food with their hands. They are active even in the hottest hours of the day.

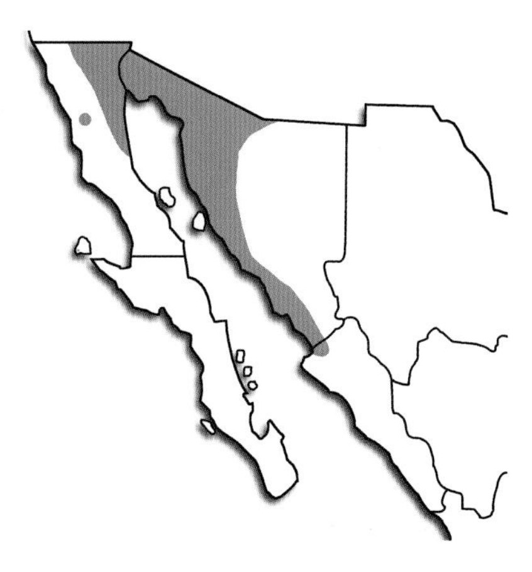

DISTRIBUTION: *X. tereticaudus* is found in desert areas of the southwestern United States and northwestern Mexico. Its distribution includes western Arizona, southeastern California, and southern Nevada, as well as northeastern Baja California and central northwestern Sonora. It has been registered in the states of BC, SI, and SO.

VEGETATIONAL ASSOCIATIONS AND ELEVATION RANGE: *X. tereticaudus* is found in xeric scrublands and thorn forests. They are also found in disturbed areas such as golf courses and gardens. They occur from 70 masl to 1,190 masl (Grinnell and Dixon, 1918).

CONSERVATION STATUS: The current situation of *X. tereticaudus* is unknown but apparently it does not present conservation problems in Mexico, where it is very abundant in some areas, despite having a restricted distribution.

Family Castoridae

Beavers make up a very small family of only two species with a Holarctic distribution; they are found in North America and the Palaearctic ecozone (Wilson and Reeder, 2005). They are conspicuous animals with a series of morphological specializations such as large webbed hind feet and a moderately long but highly flattened tail. One species (*Castor canadensis*) inhabits Mexico; it is considered in danger of extinction and distributed in the Rivers San Pedro, Bravo, Bavispe, Colorado, Conchos, and some of their tributaries.

Castor canadensis Kuhl, 1820

American beaver

Gerardo Ceballos and Karla Pelz-Serrano

SUBSPECIES IN MEXICO

Castor canadensis frondator Mearns, 1897
Castor canadensis mexicanus Bailey, 1913
Hall (1981) mentions that the beavers of the Colorado River, between Sonora and Baja California, are the subspecies *Castor canadensis repentinus* Goldman, 1932. Hoffmeister (1986), however, found they belong to the subspecies *frondator*. Therefore, only two subspecies are found in Mexico.

DISTRIBUTION: Beavers are widely distributed in Canada and the United States. Its distribution in Mexico is marginal, mainly along the border with the United States. Three separate populations exist (Ceballos, 1985; Hall, 1981; Leopold, 1959). A population is located along the Rio Bravo, the Rio Conchos, and other tributaries, from Chihuahua to Tamaulipas. The second population lives in the Bavispe River, with records in Mesa de Tres Ríos in Chihuahua and Cajon Bonito in Sonora (Ceballos, 1985; R. List, pers. obs.; Mearns, 1907). The third population lives in the Colorado River (Mellink and Luévano, 1998). It has been registered in the states of BC, CH, CO, NL, SO, and TA.

DESCRIPTION: Beavers are the largest rodents in North America. Their head and body are robust. The tail is flattened and covered with thick scales, features that distinguish it from other species of mammals. The eyes are protected by nictitating membranes. A membrane webs the hind fingers. The dorsal coloration varies from brown to light black, but is generally reddish-brown; the abdomen is lighter than the back (Jenkins and Busher, 1979).

EXTERNAL MEASURES AND WEIGHT

TL = 1,000 to 1,200 mm; TV = 258 to 325 mm; HF = 156 to 205 mm; EAR = 23 to 29 mm.
Weight: 11 to 39 kg.

DENTAL FORMULA: I 1/1, C 0/0, PM 1/1, M 3/3 = 20.

NATURAL HISTORY AND ECOLOGY: Beavers are active both day and night. They live in lodges dug in sandbanks of rivers or ponds near rivers. They build dams in streams and rivers of slow flow, thereby increasing the number of available habitats for other aquatic and semi aquatic species (Jenkins and Busher, 1979). In Mexico, this behavior has been registered at different locations such as Nuevo Leon and Sonora (Bernal, 1978; Gallo-Reynoso, 2002; Pelz-Serrano et al., 2005). The entry to the lodge is generally under water, and the access tunnel penetrates from 1.6 m to 6 m, finishing in a chamber of several meters in diameter (Leopold, 1965). They mainly feed on bark and cambium of trees and shrubs such as willow (*Salix*), poplars (*Populus*), ash trees (*Fraxinus*), oaks (*Quercus*), and mesquites (*Prosopis*); they also consume roots and tubers of herbs and tule (Hoffmeister, 1986; Leopold, 1965). They knock down trees, even very big ones, which are cut and carried into the water to feed on or to use in building their dams (Jenkins and Busher, 1979). Beavers reproduce once a year, in winter, between January and February. In every birth, mean litter size is three to four offspring, but when a large amount of food is available, beavers can have up to six offspring (Müller-Schwarze and Sun, 2003). In the Colorado River, in the border between Arizona and Sonora, litters have up to four kits (Hoffmeister, 1986). The offspring are born in May or June, after a gestation period of 107 days (Schmidly, 1977b). They are preyed on mainly by wolves and coyotes, though their predators in Mexico are unknown.

VEGETATIONAL ASSOCIATIONS AND ELEVATION RANGE: In Mexico, beavers live exclusively in riparian vegetation, gallery forests, and arid scrubland near permanent rivers. They are found from sea level up to 1,600 masl.

CONSERVATION STATUS: The American beaver is considered a species of least concern throughout most of its geographic range (IUCN, 2010). Beavers are scarce in Mexico. The two Mexican subspecies are considered in danger of extinction (Ceballos and Navarro, 1991; IUCN, 2010; SEMARNAT, 2010). A study on their populations conducted in 1978 revealed nearly 256 beavers in Nuevo Leon (Bernal, 1978). Its populations in Sonora were considered extirpated; however, recent records (1998, 2000, and 2004) were made in arroyo Cajon Bonito in the Sierra de San Luis (G. Ceballos and R. List, pers. comm.; Pelz-Serrano et al., 2005), in the Bavispe River, Sonora (Gallo-Reynoso, 2002), in the San Pedro River in Cananea (A.C. Naturalia, 2008, pers. comm.), and in the Colorado River in Baja California (Mellink and Luévano, 1998). The main causes of its rarity are indiscriminate hunting and the limited availability of environments for their survival (Leopold, 1965). If the species is protected, it could recover within a few years, as it happened in various regions of Canada and the United States (Jenkins and Busher, 1979).

Castor canadensis. Sonora-Arizona Desert Museum, Arizona. Photo: Gerardo Ceballos.

Family Heteromyidae

Gerardo Ceballos and Troy L. Best

The New World family Heteromyidae is widely distributed from southern Canada to northern South America. Heteromyidae and Geomyidae (pocket gophers) are the only two families of mammals with external, fur-lined cheek pouches. The cylindrical body, muscular jaws, head, neck, and chest, short tail, and long claws on the forefeet easily distinguish geomyids from the more slender, long-tailed, and more agile heteromyids. Heteromyidae includes 60 species and 6 genera of kangaroo mice (*Microdipodops*), kangaroo rats (*Dipodomys*), spiny pocket mice (*Liomys* and *Heteromys*), and pocket mice (*Chaetodipus* and *Perognathus*; Wilson and Reeder, 2005) that generally inhabit arid deserts, but a few species occur in coniferous forests, deciduous forests, and a variety of tropical habitats. In Mexico, the family is represented by 40 species in 5 genera: *Chaetodipus*, *Dipodomys*, *Heteromys*, *Liomys*, and *Perognathus*. Of these, 13 species are endemic to the country. A few species have broad geographic distributions (e.g., *Dipodomys merriami* and *D. ordii*), most occur over relatively small areas (e.g., *Chaetodipus nelsoni* and *Perognathus merriami*), and some have small ranges (e.g., *Dipodomys insularis* and *Chaetodipus goldmani*). One species, *Dipodomys gravipes*, which occupied a small geographic range in Baja California, probably is extinct because of habitat destruction and introduction of exotic species.

Subfamily Dipodomyinae

The subfamily Dipodomyinae contains two genera (*Dipodomys* and *Microdipodomys*). Ten species of the genus *Dipodomys* are known from Mexico.

Dipodomys compactus True, 1889

Gulf Coast kangaroo rat

Vinicio J. Sosa and Claudia Álvarez A.

SUBSPECIES IN MEXICO
Dipodomys compactus compactus True, 1889

DESCRIPTION: *Dipodomys compactus* is medium in size; its tail and hind feet are shorter than those of other species of the genus. The hind feet have five toes and are much larger than the forefeet. The tail is long and ends in a brush. Because the auditory bullae are narrower, the skull is narrower than that of other species of the genus. *D. compactus* has the greatest number of chromosomes in the genus (diploid number 2n = 74; Stock, 1974). Unlike most members of the genus, its pelage is paler on the sides of the head and hind legs, apparently as an adaptation to aridity of the region where it lives (True, 1888) or to the color of the soil. The Mexican subspecies displays two variants in dorsal coloration; orange on Island Padre, Texas, and gray on barrier islands of Laguna Madre, Tamaulipas. Lower parts of the body, forelimbs, and the hind limbs are white. Sides and most of the distal part of the tail are white; upper parts are yellowish-brown. Soles of the hind feet and fronts of the ears are brownish. Ears are covered with short white or gray hairs. Whiskers are white and brownish, and nails are white (Baumgardner, 1991; Hall, 1951b; Selander et al., 1962).

EXTERNAL MEASURES AND WEIGHT

TL = 101 to 109 mm; TV = 109 to 113 mm; HF = 34 to 37 mm; EAR = 12 to 16 mm. Weight: 44 to 60 g.

DENTAL FORMULA: I 1/1, C 0/0, PM 1/1, M 3/3, total = 20.

NATURAL HISTORY AND ECOLOGY: Little is known about the biology of this species. It is a nocturnal animal adapted to a diet comprised mostly of seeds. On barrier islands of the Laguna Madre, it coexists with land crabs that apparently outnumber them, and with other mammals such as the spotted ground squirrel (*Spermophilus spilosoma*), silky pocket mouse (*Perognathus flavus*), hispid cotton rat (*Sigmodon hispidus*), southern plains woodrat (*Neotoma micropus*), and Texas pocket gopher (*Geomys personatus*; Hall, 1951b; Selander et al., 1962). This species can be infested by nematodes that can penetrate the skull (Hall, 1951b). Little is known about its reproductive biology, but it probably reproduces in summer, as reproductively active males and females were reported in July and August (Baumgardner, 1991; Selander et al., 1962).

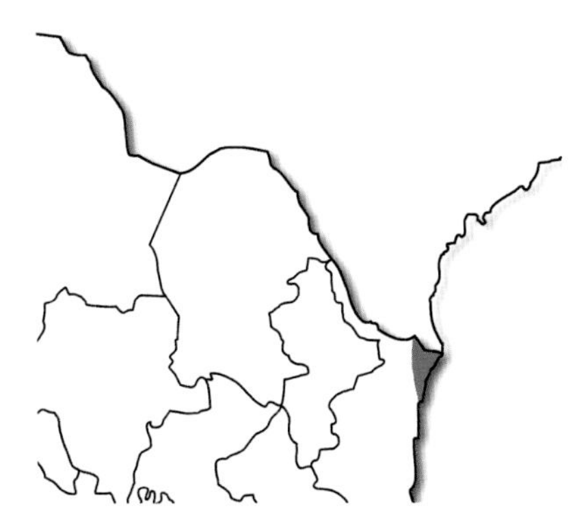

DISTRIBUTION: Most of the range is in southeastern Texas. *D. compactus* has a marginal distribution in Mexico, where it occurs only in northeastern Tamaulipas and on barrier islands of the Laguna Madre (Baumgardner and Schmidly, 1981). It has been recorded in the state of TA.

VEGETATIONAL ASSOCIATIONS AND ELEVATION RANGE: This species inhabits coastal-dune vegetation. On barrier islands of Laguna Madre, *D. compactus mexicana* generally is associated with flat dunes with *Croton punctatus* and *Fimbristylis castanea.* On the western edge of the dunes, a mixture of morning glories (*Ipomoea pescaprae*), *Croton*, *Lycium carolinianum*, and velvet mesquite (*Prosopis juliflora*) is common. On stabilized dunes on the western side of the islands, common plants include cacti (*Opuntia lindheimeri*), *Gaillardia pulchella*, *Iva*, *Flaveria oppositifolia*, *Enstoma exaltatum*, and *Croton capitatus* (Selander et al., 1962). Elevational range is sea level to a few meters.

CONSERVATION STATUS: *D. compactus* is not formally listed under any category of conservation concern, although it is considered vulnerable because of its restricted distribution (Ceballos et al., 1998). Barrier islands of Laguna Madre are under increasing pressure from developments for tourism and grazing by livestock. It is likely that the status of this locally common kangaroo rat is threatened. In the future, it would be desirable to establish a protected area to ensure its conservation. A protected area in its range also would protect other potentially endangered species with marginal distributions in Mexico (Ceballos, 1999; Ceballos et al., 1998) such as eastern moles (*Scalopus aquaticus*) and Texas pocket gophers (*Geomys personatus*).

Dipodomys deserti Stephens, 1887

Desert kangaroo rat

Eric Mellink and Jaime Luévano

SUBSPECIES IN MEXICO

Dipodomys deserti deserti Stephens, 1887

Dipodomys deserti sonoriensis Goldman, 1923

DESCRIPTION: *Dipodomys deserti* is large. It is creamy white without marks in the face. Usually, the tail does not have the dark ventral stripe that characterizes other kangaroo rats and the long tail ends in a white tip. Hind feet have four toes (Hoffmeister, 1986).

EXTERNAL MEASURES AND WEIGHT
TL = 305 to 377 mm; TV = 180 to 215 mm; HF = 50 to 58 mm; EAR = 15 to 18 mm. Weight: 83 to 138 g.

DENTAL FORMULA: I 1/1, C 0/0, PM 1/1, M 3/3 = 20.

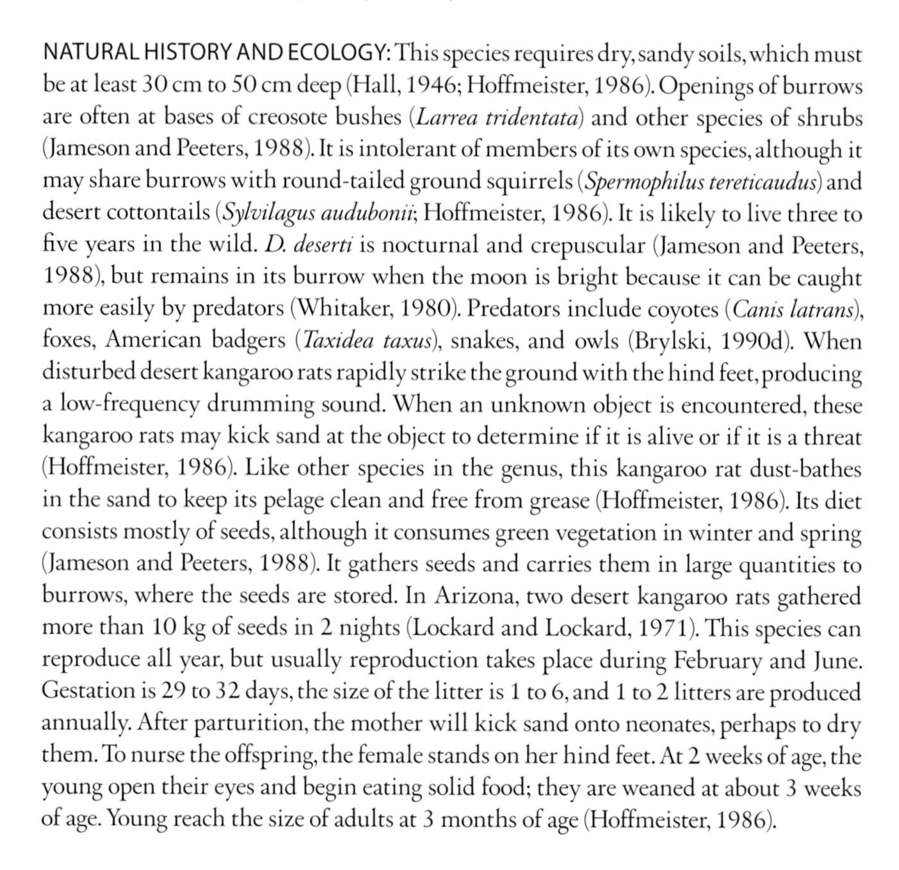

NATURAL HISTORY AND ECOLOGY: This species requires dry, sandy soils, which must be at least 30 cm to 50 cm deep (Hall, 1946; Hoffmeister, 1986). Openings of burrows are often at bases of creosote bushes (*Larrea tridentata*) and other species of shrubs (Jameson and Peeters, 1988). It is intolerant of members of its own species, although it may share burrows with round-tailed ground squirrels (*Spermophilus tereticaudus*) and desert cottontails (*Sylvilagus audubonii*; Hoffmeister, 1986). It is likely to live three to five years in the wild. *D. deserti* is nocturnal and crepuscular (Jameson and Peeters, 1988), but remains in its burrow when the moon is bright because it can be caught more easily by predators (Whitaker, 1980). Predators include coyotes (*Canis latrans*), foxes, American badgers (*Taxidea taxus*), snakes, and owls (Brylski, 1990d). When disturbed desert kangaroo rats rapidly strike the ground with the hind feet, producing a low-frequency drumming sound. When an unknown object is encountered, these kangaroo rats may kick sand at the object to determine if it is alive or if it is a threat (Hoffmeister, 1986). Like other species in the genus, this kangaroo rat dust-bathes in the sand to keep its pelage clean and free from grease (Hoffmeister, 1986). Its diet consists mostly of seeds, although it consumes green vegetation in winter and spring (Jameson and Peeters, 1988). It gathers seeds and carries them in large quantities to burrows, where the seeds are stored. In Arizona, two desert kangaroo rats gathered more than 10 kg of seeds in 2 nights (Lockard and Lockard, 1971). This species can reproduce all year, but usually reproduction takes place during February and June. Gestation is 29 to 32 days, the size of the litter is 1 to 6, and 1 to 2 litters are produced annually. After parturition, the mother will kick sand onto neonates, perhaps to dry them. To nurse the offspring, the female stands on her hind feet. At 2 weeks of age, the young open their eyes and begin eating solid food; they are weaned at about 3 weeks of age. Young reach the size of adults at 3 months of age (Hoffmeister, 1986).

DISTRIBUTION: The geographic range of *D. deserti* is the southwestern United States and northwestern Mexico. It inhabits areas of the Sonoran desert that are characterized by deep sandy soils from south of Hermosillo, Sonora, to near San Felipe, Baja California. It has been recorded in BC and SO.

VEGETATIONAL ASSOCIATIONS AND ELEVATION RANGE: The desert kangaroo rat occurs in microphyllous scrublands, with short and scattered shrubs. In Mexico, it occurs from sea level to 300 m. In the United States, it has been recorded at elevations to 1,710 masl (Hall, 1981).

CONSERVATION STATUS: Current status of populations is unknown. The size of populations probably has not changed substantially since the arrival of Europeans because the deep sandy soils where this species occurs are not desirable as agricultural croplands.

Dipodomys gravipes Huey, 1925

San Quintín kangaroo rat

Eric Mellink and Jaime Luévano

SUBSPECIES IN MEXICO
D. gravipes is a monotypic species.

DESCRIPTION: The San Quintín kangaroo rat is relatively large and has a stocky body and small ears. Males are larger than females. Individuals near El Rosario, in the southern part of the range, are slightly larger than those to the north. The underside of body, forefeet and paws, upper surface of the hind feet, and sides of hind legs are white. The rest of the body is cream-reddish in color. *Dipodomys gravipes* has white spots above the eyes and around the ears. The tail is thick and moderately long, with a black tip. The hind feet are large with five toes (Best and Lackey, 1985). Based on characteristics of the baculum (the bone in the penis) there is a close relationship with *D. stephensi* (Lidicker, 1960) and both species could be similar to the ancestral form of the *heermanni* group of kangaroo rats; the diploid number is 2n = 70 (Best and Lackey, 1985; Stock, 1974).

DISTRIBUTION: *D. gravipes* is endemic to Mexico and is known from about 1,000 km^2 along the coastal plain between San Telmo and El Rosario in the San Quintín Valley of western Baja California (Best and Lackey, 1985; Hall, 1981). It has been recorded in BC.

EXTERNAL MEASURES AND WEIGHT

TL = 312 mm; TV = 168 to 180 mm; HF = 43 to 44 mm; EAR = 11 to 16 mm.
Weight: 69 g.

DENTAL FORMULA: I 1/1, C 0/0, PM 1/1, M 3/3 = 20.

NATURAL HISTORY AND ECOLOGY: Little is known about the biology of this kangaroo rat (Best and Lackey, 1985). *D. gravipes* inhabits flat areas, with lowland vegetation in sandy soils. It constructs burrows that are up to 50 cm deep, with an average of 4.7 openings, which include a main tunnel and several secondary passages. On average, there are 3.3 nests in each burrow, with up to 10 food chambers. No openings to burrows are under shrubs. Although its diet has not been studied, it is likely that it is primarily made up of seeds, but green plants are consumed when available. It may be parasitized by the protozoan *Eimeria scholtysecki* (Stout and Duszynski, 1983).

VEGETATIONAL ASSOCIATIONS AND ELEVATION RANGE: The San Quintín kangaroo rat occurs in communities of coastal xeric shrubs in alluvial soils, from sea level to about 30 m (Best and Lackey, 1985).

CONSERVATION STATUS: In 1925, when this species was described, large colonies were present. Since then, the whole area has been transformed into agricultural fields, which resulted in virtual extinction of this species (Best, 1983; E. Mellink, pers.

Dipodomys gravipes. Scrubland. San Quintin Valley, Baja California. Photo: Troy L. Best.

comm.). Ceballos and Navarro (1991) considered *D. gravipes* to be in danger of extinction and noted that the population probably did not exceed 30 individuals. The last two known colonies have disappeared (Mellink, 1992a), and intensive searches have not been successful (E. Mellink, pers. obs.). It is likely that the species is extinct. Officially, it is considered to be extinct (SEMARNAT, 2010).

DISTRIBUTION: *D. insularis* is endemic to Mexico. It occurs only on San José Island in the Gulf of California, which is north of La Paz, Baja California Sur (Hall, 1981). It has been recorded in BCS.

Dipodomys insularis Merriam, 1907

San José Island kangaroo rat

Eric Mellink and Jesús Ramírez Ruíz

SUBSPECIES IN MEXICO

D. insularis is a monotypic species.

The taxonomic status of *D. insularis* is not clear, as some authors consider it a subspecies of *D. merriami* (Best and Janecek, 1992; Williams et al., 1993a; Wilson and Reeder, 2005), while others consider it a distinct species (Best, 1993; Best and Thomas, 1991a; Merriam, 1907).

DESCRIPTION: *Dipodomys insularis* is among the smallest of kangaroo rats. *D. insularis* is distinguished from *D. merriami* by its larger ears, more robust body, and the dorsal line on the tail that is wide and dark. Dorsal color is yellow with black hairs interspersed, which gives *D. insularis* a more grayish look than *D. merriami*. Markings around the eyes are paler than the rest of the body and the ears are also pale. The skull is more triangular, the parietal plate is narrower, and the mastoid bullae are smaller than in *D. merriami* (Best and Thomas, 1991a; Hall, 1981; Merriam, 1907).

EXTERNAL MEASURES AND WEIGHT

TL = 243 to 258 mm; TV = 146 to 150 mm; HF = 38 to 40 mm; EAR = 13 to 13.5 mm. Weight: 36 to 47 g.

DENTAL FORMULA: I 1/1, C 0/0, PM 1/1, M 3/3 = 20.

NATURAL HISTORY AND ECOLOGY: This species occurs along bottoms of canyons and near beaches in open habitats. Associated species of mammals are the mule deer (*Odocoileus hemionus*), San José cottontail (*Sylvilagus mansuetus*), spiny pocket mouse (*Chaetodipus spinatus*), desert woodrat (*Neotoma lepida*), and ringtail (*Bassariscus astutus*; Best and Thomas, 1991a; Nelson, 1992). There are no data on diet or reproductive biology, but a juvenile was caught in May indicating that could have been born in February or March (Best and Thomas, 1991a).

VEGETATIONAL ASSOCIATIONS AND ELEVATION RANGE: San José Island is of volcanic origin and has vegetation similar to that on the mainland, which is crassicaule scrubland, with most vegetation in arroyos between hills (Best and Thomas, 1991a). *D. insularis* occurs from sea level to a few meters in elevation (Best and Thomas, 1991a).

CONSERVATION STATUS: The San José Island kangaroo rat is an endangered species (SEMARNAT, 2010) and is vulnerable to introduction of exotic species such as feral cats.

Merriam's kangaroo rat

Reyna A. Castillo

SUBSPECIES IN MEXICO
Dipodomys merriami ambiguus Merriam, 1890
Dipodomys merriami annulus Huey, 1951
Dipodomys merriami arenivagus Elliot, 1904
Dipodomys merriami atronasus Merriam, 1894
Dipodomys merriami brunensis Huey, 1951
Dipodomys merriami llanoensis Huey, 1951
Dipodomys merriami mayensis Goldman, 1928
Dipodomys merriami margaritae Merriam, 1907
Dipodomys merriami melanurus Merriam, 1893
Dipodomys merriami merriami Mearns, 1890
Dipodomys merriami mitchelli Mearns, 1897
Dipodomys merriami olivaceus Swarth, 1929
Dipodomys merriami platycephalus Merriam, 1907
Dipodomys merriami quintinensis Huey, 1951
Dipodomys merriami semipallidus Huey, 1927
Dipodomys merriami trinidadensis Huey, 1951

DESCRIPTION: *Dipodomys merriami* is among the smallest of kangaroo rats. It has four toes on the hind feet. The tail is thin and moderately long, and ends in a tuft of hair. Dorsal color varies greatly according to subspecies, but the underside, back of the legs, supraorbital region, behind the eyes, and bands on the sides of the tail and legs are white (Lidicker, 1960). Sexual dimorphism is present and males are larger than females in several morphological characteristics (Best, 1993; Lidicker, 1960). In addition, there is geographic variation in sexual dimorphism (Best, 1993).

Dipodomys merriami. Grassland. Janos Biosphere Reserve, Chihuahua. Photo: Gerardo Ceballos.

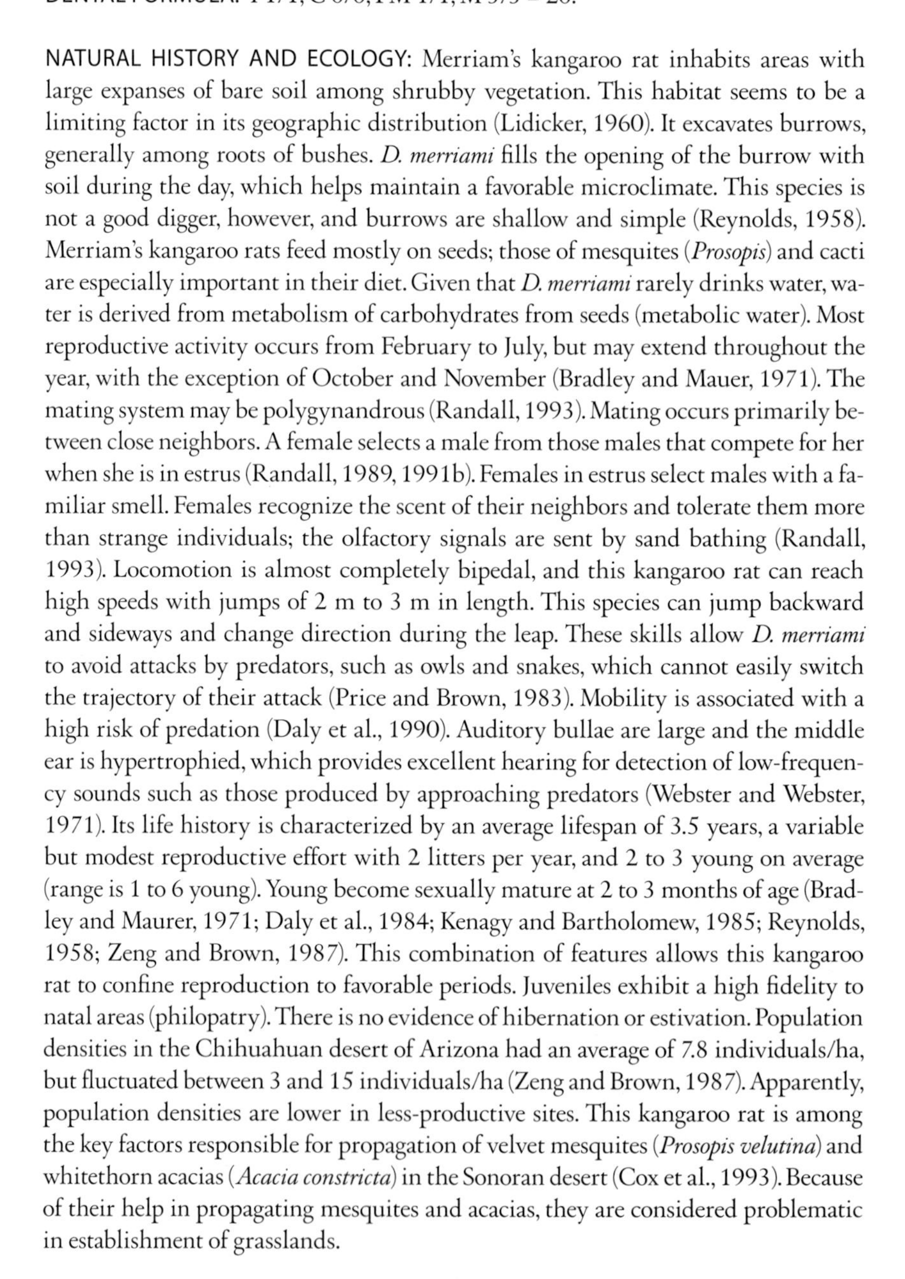

EXTERNAL MEASURES AND WEIGHT
TL = 234 to 259 mm; TV = 135 to 161 mm; HF = 36 to 41 mm; EAR = 12 to 16 mm. Weight: 38 to 47 g.

DENTAL FORMULA: I 1/1, C 0/0, PM 1/1, M 3/3 = 20.

NATURAL HISTORY AND ECOLOGY: Merriam's kangaroo rat inhabits areas with large expanses of bare soil among shrubby vegetation. This habitat seems to be a limiting factor in its geographic distribution (Lidicker, 1960). It excavates burrows, generally among roots of bushes. *D. merriami* fills the opening of the burrow with soil during the day, which helps maintain a favorable microclimate. This species is not a good digger, however, and burrows are shallow and simple (Reynolds, 1958). Merriam's kangaroo rats feed mostly on seeds; those of mesquites (*Prosopis*) and cacti are especially important in their diet. Given that *D. merriami* rarely drinks water, water is derived from metabolism of carbohydrates from seeds (metabolic water). Most reproductive activity occurs from February to July, but may extend throughout the year, with the exception of October and November (Bradley and Mauer, 1971). The mating system may be polygynandrous (Randall, 1993). Mating occurs primarily between close neighbors. A female selects a male from those males that compete for her when she is in estrus (Randall, 1989, 1991b). Females in estrus select males with a familiar smell. Females recognize the scent of their neighbors and tolerate them more than strange individuals; the olfactory signals are sent by sand bathing (Randall, 1993). Locomotion is almost completely bipedal, and this kangaroo rat can reach high speeds with jumps of 2 m to 3 m in length. This species can jump backward and sideways and change direction during the leap. These skills allow *D. merriami* to avoid attacks by predators, such as owls and snakes, which cannot easily switch the trajectory of their attack (Price and Brown, 1983). Mobility is associated with a high risk of predation (Daly et al., 1990). Auditory bullae are large and the middle ear is hypertrophied, which provides excellent hearing for detection of low-frequency sounds such as those produced by approaching predators (Webster and Webster, 1971). Its life history is characterized by an average lifespan of 3.5 years, a variable but modest reproductive effort with 2 litters per year, and 2 to 3 young on average (range is 1 to 6 young). Young become sexually mature at 2 to 3 months of age (Bradley and Maurer, 1971; Daly et al., 1984; Kenagy and Bartholomew, 1985; Reynolds, 1958; Zeng and Brown, 1987). This combination of features allows this kangaroo rat to confine reproduction to favorable periods. Juveniles exhibit a high fidelity to natal areas (philopatry). There is no evidence of hibernation or estivation. Population densities in the Chihuahuan desert of Arizona had an average of 7.8 individuals/ha, but fluctuated between 3 and 15 individuals/ha (Zeng and Brown, 1987). Apparently, population densities are lower in less-productive sites. This kangaroo rat is among the key factors responsible for propagation of velvet mesquites (*Prosopis velutina*) and whitethorn acacias (*Acacia constricta*) in the Sonoran desert (Cox et al., 1993). Because of their help in propagating mesquites and acacias, they are considered problematic in establishment of grasslands.

VEGETATIONAL ASSOCIATIONS AND ELEVATION RANGE: Merriam's kangaroo rat occurs in xeric scrublands, open pasturelands, and thorny forests from sea level to 2,200 m.

CONSERVATION STATUS: Current status is unknown, but *D. merriami* is locally abundant and widely distributed, which decreases threats to populations. The subspecies *D. merriami mitchelli*, which is endemic to Tiburón Island in the Gulf of California, is listed as threatened (SEMARNAT, 2010).

DISTRIBUTION: *D. merriami* is among the most widely distributed species in the genus. It occurs from northern Nevada southward through most of Baja California, including Margarita and Tiburón islands. On the mainland, it occurs from Sonora to the foothills of the Sierra Madre Oriental in Nuevo León and southward into Aguascalientes and San Luis Potosí. It has been recorded in AG, BC, BS, CH, CO, DU, NL, QE, SI, SO, SL, TA, and ZA.

Nelson's kangaroo rat

Elizabeth E. Aragón

SUBSPECIES IN MEXICO

D. nelsoni is a monotypic species.

Morphologically, *D. nelsoni* is similar to *D. spectabilis cratodon* (Matson, 1980), but most morphometric studies indicate that *D. nelsoni* is a separate species. Studies using genetic variables also support its specific status (Best, 1988a; Williams et al., 1993a, b).

DISTRIBUTION: *D. nelsoni* is a species endemic to Mexico; its distribution is restricted to the northern part of the Chihuahuan desert from Chihuahua into Nuevo León (Anderson, 1972). *D. nelsoni* has been recorded in CH, CO, DU, NL, SL, and ZA.

DESCRIPTION: *Dipodomys nelsoni* is similar and closely related to *D. spectabilis*. Both species are large, but *D. nelsoni* is smaller and the white tip of the tail is shorter than in *D. spectabilis*. Sexual dimorphism is present; females are larger than males in some morphological characteristics (Anderson, 1972; Baker, 1956; Best, 1993; Best et al., 1988; Hall, 1981; Petersen, 1978). This kangaroo rat has a large head, eyes, and hind feet, a short neck, and a tail that is longer than the head and body (Anderson, 1972). The mastoids are large, the maxillary arch is moderate with a well-developed and strong angle, the auditory bullae are relatively large, the supraoccipital and interparietal are small, and the external opening of the auditory meatus is oval (Nader, 1978). Coloration is dark brownish-buffy, more intense on the sides, with reddish shades. The upper surface of the distal one-half of the tail is black and there are white lateral stripes for almost two-thirds of the length of the tail (Merriam, 1907).

EXTERNAL MEASURES AND WEIGHT

TL = 258 to 333 mm; TV = 122 to 199 mm; HF = 44 to 50 mm; EAR = 12 to 17 mm. Weight: 55 to 102 g.

DENTAL FORMULA: I 1/1, C 0/0, PM 1/1, M 3/3 = 20.

NATURAL HISTORY AND ECOLOGY: *D. nelsoni* inhabits open, sandy soils and rocky areas. It excavates burrows among shrubs such as creosotebushes (*Larrea*) and mesquites (*Prosopis*), and it forages along wind-blown hummocks or roadways where seeds dispersed by wind accumulate (Best, 1988a; Best et al., 1988). Burrows contain caches of seeds and leaves (Grenot and Serrano, 1982; Serrano, 1987). Burrows are within mounds that average 2.7 m in diameter and 0.6 m high; they have 1 to 6 openings often interconnected with tunnels. In bare soil, these mounds cover a radius of 3 m to 7 m. Only one to two individuals inhabit each mound (Baker, 1956; Baker and Greer, 1962; Matson and Baker, 1986). Nelson's kangaroo rats are granivores, but also feed on plant material and insects. They can be opportunistic but often select beans of mesquites (*Prosopis*); 75% of stored food within burrows is pods of this legume (Grenot and Serrano, 1980). Reproduction occurs throughout the year, although there is a peak during April–September. The average size of a litter is two, with a range two to three offspring (E. Aragon, pers. obs.; Best, 1988a). In Mapimi, Durango, population density was 8.5 to 28 individuals/ha in 1978 and 1979 (Serrano, 1987). Abundance varies according to habitat, with up to 6 individuals/ha in areas dominated by magueys, cacti, and shrubs, and up to 3 individuals/ha in grasslands (Rogovin et al., 1991). Annual movements cover 40 m to 180 m and activity areas are 0.52 ha to 1.56 ha annually (Grenot and Serrano, 1982).

VEGETATIONAL ASSOCIATIONS AND ELEVATION RANGE: This kangaroo rat occurs in xeric scrublands, thorny forests, and open pasturelands. It lives in areas with thorny vegetation, such as *Larrea tridentata*, *Agave asperrima*, *Opuntia rastrera*, *Cordia greggi*, *Acacia*, *Prosopis*, *Fouqueria splendens*, *Yucca*, *Myrtillocactus geometrizans*, *Suaeda*, and *Hilaria mutica*, as well as in areas denuded of vegetation. *D. nelsoni* occurs 549 masl to 2,100 masl.

CONSERVATION STATUS: In Mexico, the species has not been included on any list needing protection (SEMARNAT, 2002). Because *D. nelsoni* is an endemic species with a restricted distribution, however, its current status should be evaluated to determine if it is at risk.

Dipodomys ordii Woodhouse, 1853

Ord's kangaroo rat

Gisselle Oliva Valdés

SUBSPECIES IN MEXICO
Dipodomys ordii durranti Setzer, 1949
Dipodomys ordii extractus Setzer, 1949
Dipodomys ordii obscurus (J.A. Allen, 1903)
Dipodomys ordii ordii Woodhouse, 1853
Dipodomys ordii palmeri (J.A. Allen, 1881)
Dipodomys ordii pullus Anderson, 1972

DESCRIPTION: *Dipodomys ordii* is a medium-sized kangaroo rat. The hind feet have five toes, which distinguish *D. ordii* from *D. merriami* and *D. phillipsii*. The tail is short, hairy, dark-colored dorsally, and white ventrally, ending in a brush. Hair is long, silky, and brown, reddish or blackish depending on the subspecies, but with the abdomen and dorsal surfaces of the hind feet always white. Auditory bullae are 14.6 mm to 16.7 mm across (Garrison and Best, 1990; Hall, 1981; Jones, 1985; Whitaker, 1980).

Dipodomys ordii. Grassland. Janos Biosphere Reserve, Chihuahua. Photo: Gerardo Ceballos.

EXTERNAL MEASURES AND WEIGHT
TL = 208 to 281 mm; TV = 100 to 163 mm; HF = 35 to 40 mm; EAR = 12 to 16 mm. Weight: 50 to 96 g.

DENTAL FORMULA: I 1/1, C 0/0, PM 1/1, M 3/3 = 20.

NATURAL HISTORY AND ECOLOGY: Ord's kangaroo rats are nocturnal, active throughout the year, solitary, and territorial (Ceballos and Galindo, 1984; Eisenberg, 1963; Garrison and Best, 1990). Males are more active than females. They build burrows in sandy soils. They are mainly granivores, but they also may include plant materials and small insects in the diet. They consume seeds of many species of grasses, weeds, and shrubs. Reproduction occurs in spring and summer (Flake, 1974; Hall, 1946; Johnston, 1956). The size of the litter is 1 to 6 young, which are born after a gestation of 28 to 32 days (Day et al., 1956; Duke, 1944; Hall, 1946)

VEGETATIONAL ASSOCIATIONS AND ELEVATION RANGE: This kangaroo rat is associated with grasslands and a variety of xeric scrublands; they are occasionally found in oak forests (Bailey, 1930; Blair, 1943; Hall, 1946; Hallett, 1982). *D. ordii* occurs from 660 masl to 2,025 masl.

CONSERVATION STATUS: The current status of this species in Mexico is unknown. Due to the broad distribution of *D. ordii* and because it often inhabits regions with little disturbance by humans, it is considered not to be at risk.

DISTRIBUTION: *D. ordii* is distributed from the high plains of southern Canada into Hidalgo in central Mexico (Baumgardner and Schmidly, 1981; Garrison and Best, 1990; Hall, 1981; Setzer, 1949). *D. ordii* has been recorded in AG, CH, CO, DU, GJ, HG, JA, NL, QE, SL, SO, TA, and ZA.

Dipodomys phillipsii Gray, 1841

Phillips' kangaroo rat

Gisselle Oliva Valdés

SUBSPECIES IN MEXICO
Dipodomys phillipsii oaxacae Hooper, 1947
Dipodomys phillipsii ornatus Merriam, 1894
Dipodomys phillipsii perotensis Merriam, 1894
Dipodomys phillipsii phillipsii Gray, 1841

DESCRIPTION: *Dipodomys phillipsii* is medium in size. It has four toes on the hind feet; the tail is relatively long, dorsally dark, and white on the bottom. The tip of the tail is usually white, but may be black. Auditory bullae are relatively small, the skull is flattened, the jaw is wide, the post-rostral region is square, the maxillary plate is protected at the second and third molars, and the face is relatively narrow. Dorsal coloration varies from brownish-orange to dark cinnamon with interspersed black hairs (Genoways and Brown, 1993; Genoways and Jones, 1971; Hall, 1981; Jones and Genoways, 1975).

EXTERNAL MEASURES AND WEIGHT
TL = 230 to 304 mm; TV = 149 to 192 mm; HF = 36 to 45 mm; EAR = 12 to 16 mm. Weight: 53 g.

DENTAL FORMULA: I 1/1, C 0/0, PM 1/1, M 3/3 = 20.

Dipodomys phillipsii. Scrubland. Valle del Oriental, Totalco, Veracruz. Photo: Gerardo Ceballos

DISTRIBUTION: *D. phillipsii* is endemic to Mexico. It occurs in highlands from central Durango into northern Oaxaca (Jones and Genoways, 1975; Ramírez-Pulido et al., 1983). It has been recorded in DF, DU, GJ, HG, JA, MX, OX, PU, QE, SL, TL, VE, and ZA.

NATURAL HISTORY AND ECOLOGY: Burrows of this species are built in open areas or near bases of shrubs. Each burrow has several openings with a slight slope. *D. phillipsii* is nocturnal, and its diet is seeds, leaves, and seedlings. In places where there is no available water for drinking, they derive metabolic water from plant material that they consume. Data on reproductive biology are scarce; young may be present through most of the year (except April, August, and November), which indicates a long period of reproduction (Genoways and Jones, 1971; Jones and Genoways, 1975). Each litter contains one to six young (Ceballos and Galindo, 1984).

VEGETATIONAL ASSOCIATIONS AND ELEVATION RANGE: Phillips' kangaroo rat inhabits arid areas with sandy soil where dominant vegetation may be grasslands or xeric shrubs (Davis, 1944; Hall and Dalquest, 1963; Jones and Genoways, 1975). Elevational distribution varies from 950 m in Oaxaca to 2,850 masl in Veracruz (Jones and Genoways, 1975).

CONSERVATION STATUS: Little is known about the current status of populations. Probably they are not in danger of extinction. Some populations, however, such as those in the Valley of Mexico and in Oaxaca, are threatened by destruction of their habitat (Ceballos and Galindo, 1984). It is considered as of Special Protection (SEMARNAT, 2010).

Dipodomys simulans (Merriam, 1904)

Dulzura kangaroo rat

Gisselle Oliva Valdés and Edmundo Huerta Patricio

SUBSPECIES IN MEXICO

Dipodomys simulans peninsularis (Merriam, 1907)
Dipodomys simulans simulans (Merriam, 1904)
This species was considered a subspecies of *D. agilis*. Morphometric and genetic stud-

ies, however, have shown they are different species (Best, 1983; Sullivan and Best, 1997). All subspecies of *D. agilis* in Mexico and *D. paralius* are synonyms of *D. simulans simulans* (Best, 1983; Patton and Álvarez-Castañeda, 1999; Sullivan and Best, 1997).

DISTRIBUTION: *D. simulans* occurs from southern California in the United States and northwestern Mexico, across the Sierra de San Pedro Mártir, the Vizcaino desert, and on to the Magdalena Plain in Baja California Sur (Álvarez-Castañeda and Patton, 1999; Best, 1983; Hall, 1981; Sullivan and Best, 1997). It has been recorded in BC and BS.

DESCRIPTION: *Dipodomys simulans* is medium in size; its dorsal coloration is dark brown, and the ventral surface is white. The tail is hairy with a dark line in the lower surface that extends to the tip. The ears are small (> 16 mm) and they have 5 toes with claws on each foot (Hall, 1981; Whitaker, 1980; Williams et al., 1993). They usually move by bipedal locomotion, as the hind feet are much longer than the forefeet and aid in hopping locomotion (Álvarez, 1960). The maxillary processes of *D. similans* are broad and thick, and those of *D. gravipes* are narrower and thinner (Álvarez, 1960). The width of the skull between maxillary arches is less than 55% of the total length of the skull (Williams et al., 1993a), the size of the baculum (bone in the penis) is less than 9.40 mm (Best, 1981), and the diploid number of chromosomes is 2n = 60 (Best et al., 1986). Unlike *D. ordii*, the forefeet are less than 43 mm and length of the skull is up to 39.6 mm (Álvarez, 1960).

EXTERNAL MEASURES AND WEIGHT

TL = 265 to 319 mm; TV = 155 to 203 mm; HF = 43 to 46 mm; EAR = 12 to 16 mm.
Weight: 45 to 77 g.

DENTAL FORMULA: I 1/1, C 0/0, PM 1/1, M 3/3 = 20.

NATURAL HISTORY AND ECOLOGY: The Dulzura kangaroo rat occurs in arid or semi-arid plains with clay and sandy soils, and in extreme deserts where water is scarce (Álvarez, 1960; Best, 1981; Williams et al., 1993a). It is nocturnal and solitary. Adults generally define territories and defend their burrows. Females maintain and defend their territories and those of other females during the reproductive season (Jones, 1993). It probably is similar to other species of the genus with gestation lasting 27 to 30 days and the size of the litter being 2 to 3 offspring (Jones, 1993). It feeds on seeds collected and carried in its external cheek pouches (Álvarez, 1960).

VEGETATIONAL ASSOCIATIONS AND ELEVATION RANGE: This species mainly inhabits xeric scrublands, grasslands, and coastal chaparral. Occasionally, it occupies forests of pine-oak and fir (Lackey, 1967; Williams et al., 1993a). It occurs from 810 masl to 2,250 masl (Hall, 1981).

CONSERVATION STATUS: This species is not included in any list of conservation concern (Ceballos et al., 2002). Although its range is increasingly impacted by humans, it appears abundant and has a relatively broad geographic distribution.

Dipodomys spectabilis Merriam, 1890

Banner-tailed kangaroo rat

Gisselle Oliva Valdés

SUBESPECIES IN MEXICO

Dipodomys spectabilis cratodon Merriam, 1907
Dipodomys spectabilis intermedius Nader, 1965
Dipodomys spectabilis perblandus Goldman, 1933

Dipodomys spectabilis spectabilis Merriam, 1890
Dipodomys spectabilis zygomaticus Goldman, 1923
Nader (1978) considered *D. nelsoni* a subspecies of *D. spectabilis*, but most authors have not accepted such an arrangement (Best, 1988b; Williams et al., 1993a).

DISTRIBUTION: The geographic range of *D. spectabilis* extends from the southwestern United States into northern Mexico, where it has two separate populations, one in Sonora and Chihuahua, and another in Aguascalientes and San Luis Potosí (Best, 1988b; Goldman and Moore, 1946). It has been recorded in AG, CH, SO, SL, and ZA.

DESCRIPTION: *Dipodomys spectabilis* is among the largest species in the genus. The tail is long and covered with short hairs, and in its distal part the hair is long and forms a brush. Dorsal coloration is pale orange-brown mixed with some black hairs that are paler in the middle. Dorsal and lateral sides of the hind feet, lateral lines, tip of the tail, and supraorbital and postauricular spots are white. Proximally, the tail is dark gray and progressively becomes black or almost black just before the long, white tip. The white lateral lines that stem from the suborbital spots narrow gradually to the middle of the tail and disappear in the terminal part. Juveniles are grayish, almost nut-brown colored and slightly paler on the hind feet (Dalquest, 1953; Nader, 1978). Auditory bullae are large and the auditory system is highly specialized (Best, 1988b).

EXTERNAL MEASURES AND WEIGHT
TL = 310 to 349 mm; TV = 180 to 208 mm; HF = 47 to 51 mm; EAR = 9 to 18 mm.
Weight: 97 to 170 g.

DENTAL FORMULA: I 1/1, C 0/0, PM 1/1, M 3/3 = 20.

NATURAL HISTORY AND ECOLOGY: These are solitary animals (Randall, 1986). Mating occurs February to April. Males frequently visit neighboring burrows of adult females in spring and summer. Gestation is 22 to 27 days (Bailey, 1931; Jones, 1984). They are mainly granivores and store large quantities of seeds in their underground burrows, which may be more than 50 cm deep (Reichman et al., 1985; Vorhies and Taylor, 1922). In San Luis Potosí, it occupies open areas where the soil is deep and sandy (Dalquest, 1953a). It has numerous attributes that allow it to survive in arid regions; for example, the urine is concentrated and alkaline; they consume seeds, stems, and leaves of succulent plants; and their burrows provide shelter from high temperatures on the surface and a subterranean environment that is cool and humid to prevent loss of water (Best, 1988b; Vorhies and Taylor, 1922). Its main predators are the American badger (*Taxidea taxus*), red fox (*Vulpes vulpes*), bobcat (*Lynx rufus*), coyote (*Canis latrans*), and owls (*Bubo virginianus*, *Tyto alba*). They are eaten by humans in some regions of San Luis Potosí (Mellink et al., 1986)

Dipodomys spectabilis. Grassland. Janos Biosphere Reserve, Chihuahua. Photo: Gerardo Ceballos.

VEGETATIONAL ASSOCIATIONS AND ELEVATION RANGE: Banner-tailed kangaroo rats live in xeric vegetation and grasslands with shrubs. The vegetation in northern Sonora is dominated by mesquites (*Prosopis*) and pasturelands dominated by grasses of the genera *Aristida* and *Bouteloua* (Dice and Blossom, 1937; Monson, 1943; Vorhies and Taylor, 1922). It occurs from 570 m in Sonora to 2,100 masl in Zacatecas.

CONSERVATION STATUS: This species faces no significant threat because of its use of disturbed areas and its wide geographic distribution.

Subfamily Heteromyinae

The subfamily Heteromyinae contains two genera (*Heteromys* and *Liomys*) with seven species known from Mexico.

Heteromys desmarestianus Gray, 1868

Desmarest's spiny pocket mouse

Iván Castro Arellano and Mery Santos G.

SUBSPECIES IN MEXICO

Heteromys desmarestianus desmarestianus Gray, 1868
Heteromys desmarestianus goldmani Merriam, 1902
Heteromys desmarestianus temporalis Goldman, 1911
Although Rogers and Schmidly (1982) considered *H. goldmani* different from *H. demarestianus*, Rogers (1990) detected no difference between the two species, hence considering *goldmani* a subspecies of *demarestianus*.

DESCRIPTION: *Heteromys desmarestianus* is a large spiny pocket mouse. Dorsal coloration is gray, with guard hairs that are orangish-brown. The underside is white and a lateral line occasionally is present but never prominent. The tail is longer than the head and body, almost naked, dark dorsally, and white ventrally. The skull is large and the bullae are small (Goldman, 1911; Hall, 1981; Williams et al., 1993a). Based on genetic, chromosomal, and morphological features, populations of *H. desmarestianus* in Mexico can be separated into two groups (Rogers, 1990).

EXTERNAL MEASURES AND WEIGHT

TL = 255 to 347 mm; TV = 130 to 199 mm; HF = 31 to 42 mm; EAR = 15 to 23 mm.
Weight: 61 to 83 g.

DENTAL FORMULA: I 1/1, C 0/0, PM 1/1, M 3/3 = 20.

NATURAL HISTORY AND ECOLOGY: Desmarest's spiny pocket mouse builds burrows, which may contain food-storage chambers (Fleming, 1974). Population density varies considerably in space and time. For example, within one year at Tuxtlas, Veracruz, there were 2 to 50 individuals/ha, with an average of 21 individuals/ha

(Sánchez-Cordero, 1993). In Costa Rica, density was 10 to 18 individuals/ha and activity areas were 1,060 m^2 for males and 883 m^2 for females (Fleming, 1974). Gestation lasts 28 days. Females are polyestric and may have up to 5 litters per year. Sexual maturity is reached at about 8 months of age in males and 8.6 months of age in females (Fleming, 1974). Breeding season at Tuxtlas, Veracruz, is 8 months with 2 litters per year and 3 offspring per litter (Sánchez-Cordero, 1993). In Costa Rica, the breeding season covers 9 months (July–March), with 3 litters and an average of 3 offspring per litter (Fleming, 1974). Maximum lifespan is about 2 years (Sánchez-Cordero, 1993). It is mainly granivorous, but it also consumes seeds and fruits of trees and shrubs such as palm chocho (*Astrocaryum mexicanum*), *Bactris tricophylla*, ramón (*Brosimum alicastrum*), guazumo (*Cecropia obtusifolia*), palm camedor (*Chamaedora tepejilote*), *Cymbopetalum baillonii*, amate (*Ficus yoponensis*), *Guarea glabra*, *Nectandra ambigens*, *Pentaclethra macroloba*, *Pithecellobium*, *Pleuranthodendron mexicana*, *Poulsenia armata*, *Pseudolmedia oxyphylaria*, *Psychotria*, *Pterocarpus*, *Rauwolfia tetraphylla*, *Sapindus*, *Socratea durissima*, *Spondias*, *Vatairea*, *Virola sebifera*, *Turpina occidentalis*, and *Welfia georgii* (Fleming, 1974; Martínez-Gallardo and Sánchez-Cordero, 1993; Sánchez-Cordero and Fleming, 1993). Of these species, *P. armata*, *C. baillonii*, *B. alicastrum*, *N. ambigens*, and *A. mexicanum* have a high dietary preference by captive individuals (Martínez-Gallardo and Sánchez-Cordero, 1993). Villa-R. (1948) captured two animals whose cheek pouches were filled with leaves of herbaceous plants. They have numerous parasites such as mites, fleas, protozoans, and phoretic moths (Whitaker et al., 1993). Predators include the big-eared woolly bat *Chrotopterus auritus* (Medellín, 1988), coyotes (*Canis latrans*), tayras (*Eira barbara*), skunks, white-nosed coatis (*Nasua narica*), long-tailed weasels (*Mustela frenata*), opossums (*Didelphis*), ocelots (*Leopardus pardalis*), owls (*Ciccaba virgata*, *Tyto alba*), hawks (*Leucopternis semiplumbea*), boas (*Boa constrictor*), and other snakes, including *Spilotes pullatus* and *Bothrops atrox* (Fleming, 1974; Sánchez-Cordero and Fleming, 1993).

DISTRIBUTION: *H. desmarestianus* occurs from southern Veracruz and Tabasco, Oaxaca, Chiapas, and south of the Yucatan Peninsula, to northern Colombia (Hernández-Huerta et al., 2000, Jones et al., 1974a; Patton, 1993a; Williams et al., 1993a). It has been recorded in CA, CS, OX, QR, TB, VE, and YU.

VEGETATIONAL ASSOCIATIONS AND ELEVATION RANGE: This species is associated with tropical rainforests, coffee plantations, and agricultural crops at 45 masl to 1,860 masl (Hall and Dalquest, 1963; Jones et al., 1974a; Rogers and Schmidly, 1982; Villa-R., 1948; Williams et al., 1993a).

Heteromys desmarestianus. Cloud forest. Photo: Gerardo Ceballos.

CONSERVATION STATUS: This species is abundant in natural and disturbed environments; hence it is not at risk.

Heteromys gaumeri J.A. Allen and Chapman, 1897

Gaumer's spiny pocket mouse

Ma. de Lourdes Romero and Cornelio Sánchez H.

SUBSPECIES IN MEXICO

H. gaumeri is a monotypic species.

Apparently, this species could represent a branch of early lineage that leads to *Heteromys*. It is so different that it could be removed from the *desmarestianus* group and placed into a separate subgenus (Allen and Chapman, 1897; Engstrom et al., 1987; Genoways, 1973).

DESCRIPTION: *Heteromys gaumeri* is medium in size for the genus. Dorsal coloration varies from dark brownish to gray, mixed in large proportion with orange hair, which gives it an orange-brown appearance. It has a wide lateral line, bright and ochre-colored, which extends from the cheek pouches to the base of tail. The underside and legs are white; the tail is bicolored and covered with brownish-gray hair on top and whitish on the bottom. The hair on the tip of the tail ends in a brush. Ears are dark, with the edges slightly white. The back of the body is covered with two types of hair, one soft and another that is stiff and spiny. The feet have six plantar tubers and there is hair on their underside (Genoways, 1973; Goldman, 1911). The baculum (bone in the penis) has a wide base that is about one-third of its total length (Burt, 1960). Unlike other species of the genus, it has hair on soles of the hind feet, which are naked in the other species (Genoways, 1973; Goldman, 1911).

Heteromys gaumeri. Tropical semi-green forest. Caret, Quintana Roo. Photo: Jesús Pacheco.

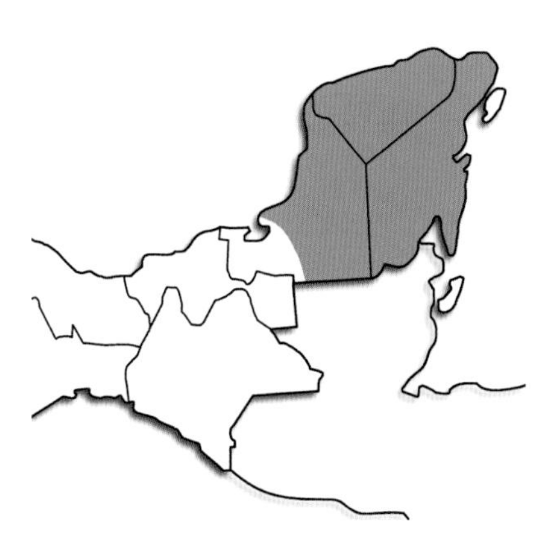

DISTRIBUTION: *H. gaumeri* is restricted to the Yucatan Peninsula in southeastern Mexico, to El Peten in Guatemala, and to northern Belize (Engstrom et al., 1987). In Mexico, it is sympatric with *H. desmarestianus* on the southern part of the Yucatan Peninsula (Dowler and Engstrom, 1988; Engstrom et al., 1987; Jones et al., 1974a). It has been recorded in CA, QR, TB, and YU.

EXTERNAL MEASURES AND WEIGHT

TL = 250 to 302 mm; TV = 129 to 193 mm; HF = 30 to 34 mm; EAR = 14 to 16 mm. Weight: 43 to 70 g.

DENTAL FORMULA: I 1/1, C 0/0, PM 1/1, M 3/3 = 20.

NATURAL HISTORY AND ECOLOGY: Gaumer's spiny pocket mouse feeds on seeds. *Ipomoea* has been found in the pouches (Hatt and Villa-R., 1950). Reproductive activity begins almost at the end of the dry season, continuing in the rainy season, and probably extending to the end of January. Apparently, there is asynchrony in reproduction (Schmidt et al., 1989). The size of litters is two to four offspring (Birney et al., 1974; Genoways, 1973). Four pregnant females were observed at the end of April and early May (Birney et al., 1974), another female had two embryos in October (Hatt, 1938), and another was pregnant in December (Genoways, 1973). None of seven females collected in July in Tabasco were pregnant, but some males in the same population had scrotal testes (Dowler and Engstrom, 1988).

VEGETATIONAL ASSOCIATIONS AND ELEVATION RANGE: Gaumer's spiny pocket mice occur in tropical subperennial and subdeciduous habitats in the southeastern Yucatan Peninsula, and in tropical deciduous forests and thorny shrubs in the north and northwestern sections of the peninsula. It is common in agricultural areas and in the herbaceous vegetation along roads and sugarcane fields (Engstrom et al., 1987; Schmidt et al., 1989). It inhabits elevations from sea level to 100 m.

CONSERVATION STATUS: There is no precise information on current status. Because of its wide distribution and its resistance to disturbance, however, this species is not at risk of extinction.

Heteromys nelsoni Merriam, 1902

Nelson's spiny pocket mouse

Ma. de Lourdes Romero and Cornelio Sánchez H.

SUBSPECIES IN MEXICO

H. nelsoni is a monotypic species.

In his assessment of genetic and enzymatic data, Rogers (1989, 1990) suggested that the species is more similar to some species of the *desmarestianus*-group than to *H. oresterus*, and that its position within the subgenus *Xylomys* is doubtful.

DESCRIPTION: *Heteromys nelsoni* is the largest species of the genus. The back of the body is dark gray to blackish, darker along the middle part; the belly, lips, and cheek pouches are white, the external parts of the hind feet and forefeet are dark gray; they have no lateral line; the tail is bicolored except at the tip, which is blackish on top and whitish on the bottom (Goldman, 1911; Merriam, 1902; Williams et al., 1993a). Compared to other species of the genus, the hair is long, thin, and soft (Homan and Genoways, 1978). The ears are long, dark, almost without hair and do not have a whitish edge. The hind feet have six plantar tubers. The tail is longer than the length of head and body and is almost naked (Goldman, 1911; Merriam, 1902). The skull is long, thin, high, rather smooth, and rounded (Merriam, 1902; Rogers and Rogers,

1992; Williams et al., 1993a). The diploid number of chromosomes is 2n = 42, with a fundamental number of FN = 72 (Rogers, 1989). It is distinguished from other species of the genus in Mexico because the coat in adults has soft bristles, while in other species the coat is spiny (Williams et al., 1993a).

DISTRIBUTION: *H. nelsoni* has a limited distribution, living in the southeastern part of Chiapas and eastern Guatemala (Goldman, 1911; Rogers and Rogers, 1992; Williams et al., 1993b). It has been recorded in CS.

EXTERNAL MEASURES AND WEIGHT
TL = 328 to 356 mm; TV = 174 to 211 mm; HF = 39 to 44 mm; EAR =19 to 23 mm. Weight: 60 to 110 g.

DENTAL FORMULA: I 1/1, C 0/0, PM 1/1, M 3/3 = 20.

NATURAL HISTORY AND ECOLOGY: Nelson's spiny pocket mouse occurs in humid environments, with abundant vegetation, mosses, and ferns. It probably reproduces in late winter and early spring. Of five males observed in mid-December, four had scrotal testes 21 mm to 26 mm in length, and the fifth was a subadult; also in December, a subadult female was present. In February, a post-lactating adult female was recorded (Rogers and Rogers, 1992). *H. nelsoni* is sympatric with *Peromyscus boylii*, *P. guatemalensis*, *Reithrodontomys mexicanus*, and *R. tenuirostris* (Rogers and Rogers, 1992).

VEGETATIONAL ASSOCIATIONS AND ELEVATION RANGE: *H. nelsoni* occupies only cloud forests that are 2,500 masl to 2,850 masl.

CONSERVATION STATUS: Although *H. nelsoni* is considered as of Special Protection (SEMARNAT, 2010), the lack of records possibly is due to the scarcity of studies conducted in cloud forests within its geographic range.

Liomys irroratus (Gray, 1868)

Mexican spiny pocket mouse

Leticia A. Espinosa and Catalina Chávez Tapia

SUBSPECIES IN MEXICO
Liomys irroratus alleni (Coues, 1881)
Liomys irroratus bulleri (Thomas, 1893)
Liomys irroratus guerrerensis Goldman, 1911
Liomys irroratus irroratus (Gray, 1868)
Liomys irroratus jaliscensis (J.A. Allen, 1906)
Liomys irroratus texensis Merriam, 1902
Liomys irroratus torridus Merriam, 1902

DESCRIPTION: *Liomys irroratus* is a medium-sized mouse, and as in other heteromyids, it has a pair of external cheek pouches. Its coat is grayish-brown in the back with a tenuous and pale lateral stripe of pink to buffy yellow coloration. The abdomen is white. Hind feet have only five pads (Dowler and Genoways, 1978; Hall, 1981).

EXTERNAL MEASURES AND WEIGHT
TL = 194 to 300 mm, TV = 95 to 169 mm, HF = 22 to 36 mm; EAR = 12 to 15 mm. Weight: 34 to 50 g.

DENTAL FORMULA: I 1/1, C 0/0, PM 1/1, M 3/3 = 20.

DISTRIBUTION: *L. irroratus* has a wide distribution from southern Texas in the United States to central Mexico, where it occurs in the eastern Sierra Madre Occidental from Chihuahua into Michoacán, from central Mexico into Oaxaca, and from Tamaulipas into Veracruz (Genoways, 1973; Ramírez and Castro, 1994; Ramírez et al., 1982; Wilson and Reeder, 2005). It has been recorded in AG, CH, DF, DU, GJ, GR, HG, JA, MI, MO, MX, NL, NY, OX, PU, QE, SL, TA, TL, VE, and ZA.

NATURAL HISTORY AND ECOLOGY: The Mexican spiny pocket mouse lives in rocky areas, where it builds burrows under logs, rocks, and shrubs that are used for refuge or to store food. Diet consists mainly of seeds that are transported in the cheek pouches; the species also may consume plants and invertebrates. It has many physiological and behavioral adaptations that allow survival in arid places. *L. irroratus* is nocturnal and solitary with little social tolerance. This species coexists with *L. pictus* in central and southern Mexico. Associated species include *Chaetodipus hispidus*, *Perognathus flavus*, *Baiomys taylori*, *B. musculus*, *Neotoma mexicana*, *N. micropus*, *Onychomys leucogaster*, *Oryzomys melanotis*, *O. couesi*, *Peromyscus boylii*, *P. difficilis*, *P. leucopus*, *P. pectoralis*, *P. truei*, *Reithrodontomys fulvescens*, *R. sumichrasti*, *Sigmodon hispidus*, *Sorex saussurei*, *Cryptotis goldmani*, and *C. parva* (Ceballos and Galindo, 1984; Chávez and Espinosa, 1993; Dowler and Genoways, 1978; Genoways and Brown, 1993). Reproduction occurs throughout the year, with greatest activity during August to November. Litters contain two to seven offspring, with an average of four (Dowler and Genoways, 1978). Chávez and Espinosa (1993) noted differences in reproductive activity in two locations in Hidalgo. They reported seasonal reproductive activity in autumn and winter in an area with secondary shrubs, and in crassicaule scrublands, reproductive activity occurred throughout the year with a peak during August to February. Population density was one to six individuals/ha, with peaks during spring and autumn. Lifespan for a male and female after they were first captured was 379 and 480 days, respectively.

VEGETATIONAL ASSOCIATIONS AND ELEVATION RANGE: This species mainly inhabits xeric scrublands and thorny forests, where annual precipitation is more than 500 mm; it also occurs in grasslands, coniferous forests, oak forests, agricultural crops, and open pasturelands (Dowler and Genoways, 1978) from sea level in coastal Tamaulipas and Veracruz to 3,050 masl in the Cerro San Felipe and Mount Zempoaltepec, Oaxaca.

CONSERVATION STATUS: Currently, *L. irroratus* is not in peril because it is widely distributed. It may become abundant enough to damage agricultural crops (Ceballos and Galindo, 1984; González-Romero, 1980).

Liomys irroratus. Scrubland. Xochitepec, Distrito Federal. Photo: Daniel Navarro.

Painted spiny pocket mouse

Yolanda Domínguez and Gerardo Ceballos

SUBSPECIES IN MEXICO

Liomys pictus annectens (Merriam, 1902)
Liomys pictus hispidus (J.A. Allen, 1897)
Liomys pictus pictus (Thomas, 1893)
Liomys pictus plantinarensis Merriam, 1902

DESCRIPTION: *Liomys pictus* is a medium-sized spiny pocket mouse with body and tail about the same length. The coat is coarsely haired. Dorsal coloration is brownish-ochre and white or cream in the abdomen; an ochre or yellow line on the side divides the underside from the upper parts (Ceballos and Miranda, 1986, 2000; Hall, 1981). The tail is bicolored, dark on top, pale on the bottom, and covered with hair (Ceballos and Miranda, 1986, 2000). It has a pair of external pouches, which are folds of skin on the cheeks (McGhee and Genoways, 1978; Reichman and Price, 1993). There is sexual dimorphism; males are larger than females (Hall, 1981). The skull is narrow and the auditory bullae are well developed (Ceballos and Miranda, 1986, 2000; McGhee and Genoways, 1978). *L. pictus* differs from *L. irroratus* by having six plantar tubers (McGhee and Genoways, 1978) and from *L. salvini* and *L. adspersus* by the lateral ochre line, by the dorsal hair that is not incurved, and by no obvious coarse spines in the pelage. Finally, *L. spectabilis* has shorter hind feet and a shorter skull (Hall, 1981).

EXTERNAL MEASURES AND WEIGHT

TL = 218 to 264 mm; TV = 105 to 116 mm; HF = 26 to 31 mm, EAR = 13 to 15 mm. Weight: 30 to 80 g.

DENTAL FORMULA: I 1/1, C 0/0, PM 1/1, M 3/3 = 20.

Liomys pictus. Tropical dry forest. Chamela-Cuixmala Biosphere Reserve, Jalisco. Photo: Gerardo Ceballos.

DISTRIBUTION: *L. pictus* occurs in areas with marked seasonality from southern Texas in the United States into Central America (Hall, 1981). In Mexico, it occurs in arid and semi-arid areas from Sonora to Chiapas, and also in northwestern Veracruz (Hall, 1981; McGhee and Genoways, 1978; Wilson and Reeder, 2005). It has been recorded in CH, CO, CS, DU, GR, JA, MI, MX, NY, OX, SI, SO, and VE.

NATURAL HISTORY AND ECOLOGY: *L. pictus* is nocturnal, territorial, and aggressive (Eisenberg, 1963). The species is solitary, but forms pairs during the mating season until the birth of offspring (Jones, 1985). *L. pictus* can reproduce all year, but has a peak that coincides with abundance of food. Gestation lasts 26 days and the size of the litter is 2 to 6 (Ceballos, 1989; Ceballos and Miranda, 1986, 2000; Eisenberg, 1963; Pinkham, 1973). Newborns weigh about 2 g and they are weaned at 23 days of age (Ceballos, 1989; Romero, 1993). This species tends to be bipedal and hops. It can live without drinking water, although it needs foods that provide large amounts of water to maintain its metabolic processes and to regulate temperature (Pinkham, 1973). In the Chamela-Cuixmala Biosphere Reserve, *L. pictus* is abundant and occurs at densities of 2 to 71 individuals/ha in lowland forests and 2 to 49 individuals/ha in open forests (Ceballos, 1990; Mendoza, 1997). The species is granivorous and the diet consists mainly of seeds of trees and vines; they help in maintaining the structure and dispersal of vegetation where they occur (Fleming and Brown, 1975; Janzen, 1982c; Matson and Christian, 1977; Mendoza D., 1997; Sanchez-Cordero and Fleming, 1993). In Chamela, more than 150 species of plants have been recorded in the diet, including *Nissolia fructicosa*, *Ipomoea*, and *Panicum*; occasionally, in the rainy season, they consume insects and mollusks (Ceballos, 1989; Mendoza, 1997; Pérez, 1978). Food is carried in the cheek pouches until it is deposited in burrows, which are underground and are composed of tunnels, chambers, and nests. Chambers are used as latrines and to store food. Nests contain a large amount of leaves and are used as shelters or to raise the young (Meadows, 1991a; Reichman and Smith, 1990). Burrows can be as simple as one tunnel or a complex formed by many chambers, nests, and tunnels. There may be 1 to 23 chambers and 1 to 3 nests, and depth of burrows may be 18 cm to 75 cm (Domínguez, 2000; Hernández, 2000).

VEGETATIONAL ASSOCIATIONS AND ELEVATION RANGE: *L. pictus* inhabits xeric shrublands, thorn forests, pine-oak forests, disturbed vegetation, agricultural crops, and orchards (Ceballos, 1989; Ceballos and Miranda, 1986, 2000; Matson and Christian, 1977). It occurs from sea level to 2,045 m (Hall, 1981).

CONSERVATION STATUS: The painted spiny pocket mouse is abundant in undisturbed and disturbed areas; it is not in danger of extinction.

Liomys salvini (Thomas, 1893)

Salvin's spiny pocket mouse

Luis Arturo Peña

SUBSPECIES IN MEXICO
Liomys salvini crispus Merriam, 1902

DESCRIPTION: In both cranial and external measurements, *Liomys salvini* is a small species within the genus. The pelage is rough, with rigid hair interspersed with soft hair. Dorsal coloration varies from brownish-gray to chocolate brown, the abdomen is white, and it has no lateral stripes. *L. salvini* has a pair of cheek pouches and six plantar tubers. Sexual dimorphism is present, as males are one-third heavier than females (Fleming, 1983).

EXTERNAL MEASURES AND WEIGHT
TL = 196 to 235 mm; TV = 88 to 110 mm; HF = 25 to 30 mm; EAR = 13 to 17 mm. Weight: 45 g.

DENTAL FORMULA: I 1/1, C 0/0, PM 1/1, M 3/3 = 20.

NATURAL HISTORY AND ECOLOGY: These mice are nocturnal and active throughout the year. They feed on seeds and insects (Fleming, 1974). In Costa Rica, an important food item is seeds of *Cochlospermum vitafolium*. Cheek pouches serve to carry seeds, construction material, and offspring. Materials are carried into the burrow, which is used for storage, nests, and refuge (Fleming, 1983). The reproductive period is January to mid-June, gestation lasts 28 days, and females give birth to 1 to 2 litters per year. The size of the litter is 2 to 6 offspring, with an average of 3.8 (Fleming, 1974). Few individuals of *L. salvini* live to 18 months of age. Research on the genus and species suggests that there is little sociability among individuals; in general, they are solitary except in the mating season (Eisenberg, 1963; Fleming, 1974). Size of home range in lowland forests is 0.20 ha. Predators include mammalian carnivores, birds of prey, and snakes.

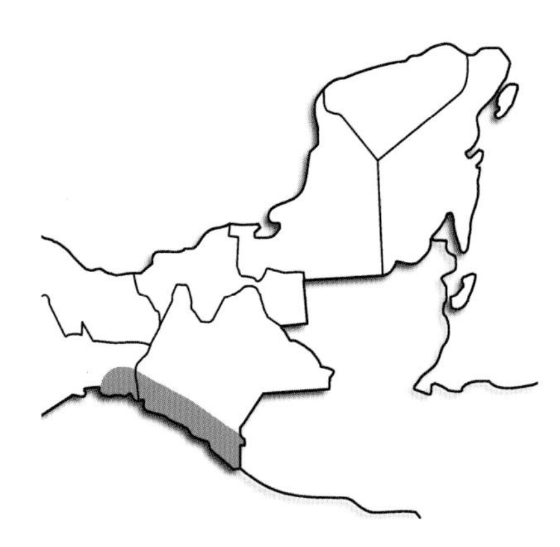

DISTRIBUTION: *L. salvini* occurs from the vicinity of Reforma in Oaxaca to Costa Rica (Williams et al., 1993b). It has been recorded in CS and OX.

VEGETATIONAL ASSOCIATIONS AND ELEVATION RANGE: Salvin's spiny pocket mouse occurs in lowland deciduous forests from sea level to 1,500 m (Schmidly et al., 1993).

CONSERVATION STATUS: Apparently, this species tolerates disturbed habitats, as it has been reported in areas with secondary vegetation. It does not appear to be in danger of extinction.

Liomys spectabilis Genoways, 1971

Jaliscan spiny pocket mouse

Ma. de Lourdes Romero and Cornelio Sánchez H.

SUBSPECIES IN MEXICO
L. spectabilis is a monotypic species.

DESCRIPTION: Within its genus, *Liomys spectabilis* is large in both external and cranial measurements. Dorsal coloration is brown with a lateral line that is ochre-colored and bright; the belly is white. The hair is hispid. The feet have six plantar tubers. The skull is similar to that of *L. pictus* but larger (Genoways, 1971). The baculum (bone in the penis) is long with a small rounded base; the head of sperm is longer and wider than that of *L. pictus* (Genoways, 1973). It can be distinguished from *L. irroratus* by the six plantar tubers; from *L. salvini* and *L. adspersus* by the lateral line, by dorsal hairs that are not curved toward the top, and by spines that are not so obvious in the pelage; and from *L. pictus* by the longer hind feet and skull (Hall, 1981).

EXTERNAL MEASURES AND WEIGHT
TL = 242 to 280 mm; TV = 122 to 142 mm; HF = 30 to 32 mm; EAR = 16 to 18 mm. Weight: 65 g.

DENTAL FORMULA: I 1/1, C 0/0, PM 1/1, M 3/3 = 20.

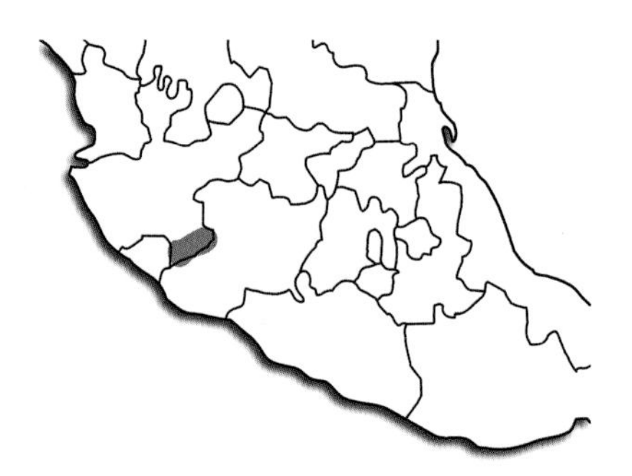

DISTRIBUTION: *L. spectabilis* is endemic to Mexico, where it is restricted to southeastern Jalisco (Genoways, 1971). It has been recorded in JA.

NATURAL HISTORY AND ECOLOGY: The natural history of this species is unknown, but it is probably similar to that of *L. pictus*. Only 1 of 6 adult females observed in September was pregnant; the 5 embryos were 4 mm in length. Average size of testes of 4 adult males in September was 21.5 mm (range, 21 mm to 22 mm; Genoways, 1971). It is sympatric with the closely related *L. pictus* (Genoways, 1971, 1973).

VEGETATIONAL ASSOCIATIONS AND ELEVATION RANGE: *L. spectabilis* inhabits tropical deciduous forests and disturbed environments. It has been captured along roads and fences, with weeds, shrubs, and trees (Genoways, 1971). It occurs from 880 masl to 1,500 masl (Genoways, 1971).

CONSERVATION STATUS: Current status is not well known because it is sympatric with other species of the same genus that can be reliably differentiated only by genetic traits. The geographical range is small, so it could be at risk. *L. spectabilis* tolerates anthropogenic disturbance, however. A detailed study to determine its conservation status is needed.

Subfamily Perognathinae

The subfamily Perognathinae contains two genera with eighteen species in the genus *Chaetodipus* and five species in the genus *Perognathus* known from Mexico.

Chaetodipus anthonyi (Osgood, 1900)

Anthony's pocket mouse

Eric Mellink and Jaime Luévano

SUBSPECIES IN MEXICO

C. anthonyi is a monotypic species.

The species has been considered a subspecies of *C. fallax*, but this arrangement is tenuous (Williams et al., 1993a).

DESCRIPTION: *Chaetodipus anthonyi* is average in size for the genus. It has a pair of cheek pouches and relatively small pinnae. The species closely resembles *C. fallax*, which is more reddish and is smaller. *C. anthonyi* has a less-arched skull, with a more robust face, a smaller mastoid, a smaller and shorter interparietal, and wider zygomatic bones (Anthony, 1925; Osgood, 1900).

EXTERNAL MEASURES AND WEIGHT

TL = 168 mm; TV = 92 mm; HF = 23.5 mm; EAR = 7 to 11 mm.

Weight: 16.4 g.

DENTAL FORMULA: I 1/1, C 0/0, PM 1/1, M 3/3 = 20.

Chaetodipus anthonyi. Sand dunes. Isla Cedros, Baja California. Photo: Eric Mellink.

NATURAL HISTORY AND ECOLOGY: Little is known about the biology of this species. It seems to be restricted to places with sandy soils. In sand dunes, it is abundant, but it does not live in adjacent rocky areas. Reproduction occurs in spring and summer (Mellink, 1992a).

VEGETATIONAL ASSOCIATIONS AND ELEVATION RANGE: Anthony's pocket mouse mainly inhabits shrubs in sand dunes, but uses other communities with sandy substrate. It occurs from sea level to 700 m.

CONSERVATION STATUS: Ceballos and Navarro (1991) considered *C. anthonyi* fragile. Due to its restricted range, however, it should probably be considered to be in danger of extinction. Risk factors include domestic carnivores (cats and dogs) and extraction of sand for building materials. Officially, it is considered to be threatened (SEMARNAT, 2010).

DISTRIBUTION: Anthony's pocket mouse is endemic to Cedros Island, which is located off the western coast of Baja California. The island has an area of 367 km^2. Although it has been reported occupying a large part of the island (Anthony, 1925), this seems incorrect. In recent sampling, it was only detected in sand dunes to the south and in a small orchard of olive trees and other fruit trees in the Sierra de Vargas (Mellink, 1993a). It is likely that *C. anthonyi* occurs at other sites with sandy soil, but there has been only limited fieldwork. It has been recorded in BC.

Chaetodipus arenarius (Merriam, 1894)

Little desert pocket mouse

Eric Mellink and Jaime Luévano

SUBSPECIES IN MEXICO

Chaetodipus arenarius arenarius (Merriam, 1894)
Chaetodipus arenarius albescens (Huey, 1926)
Chaetodipus arenarius ambiguus (Nelson and Goldman, 1929)
Chaetodipus arenarius mexicalis (Huey, 1939)
Chaetodipus arenarius albulus (Nelson and Goldman, 1923)
Chaetodipus arenarius helleri (Elliot, 1903)
Chaetodipus arenarius ramirezpulidoi Álvarez-Castañeda & Cortés-Calva, 2004
Chaetodipus arenarius siccus (Osgood, 1907)
Recently, *Ch. a. siccus* has been proposed to have specific status (Álvarez-Castañeda and Rios, 2012), but this has not been widely accepted.

DESCRIPTION: *Chaetodipus arenarius* is a medium-sized mouse. The pelage is moderately silky and the color of the back varies from pale gray or pale yellowish to dark brown. In darker individuals, the hair may end in a black tip, which gives them a speckled appearance. The sides may be paler than the back. The ventral surface is white. The tail is bicolored and has a small terminal fleck. In some populations, there is a pale lateral line. Ears are brown and, in some subspecies, there is a small band of white hairs at the base. In one subspecies, there are two phases of coloration, gray and yellow (Hall, 1981; Huey, 1926, 1964; Lackey, 1991).

DISTRIBUTION: *C. arenarius* is endemic to the Baja California Peninsula, Mexico (Hall, 1981). It has been recorded in BC and BS.

EXTERNAL MEASURES AND WEIGHT
TL = 136 to 182 mm; TV = 70 to 103 mm; HF = 20 to 23 mm; EAR = 8 mm.
Weight: 15 g.

DENTAL FORMULA: I 1/1, C 0/0, PM 1/1, M 3/3 = 20.

NATURAL HISTORY AND ECOLOGY: This species occurs in arid environments, primarily in sandy soils (Lackey, 1991). On Cerralvo Island, it was in soil that was finer and not as sandy (Banks, 1964a). Reproduction occurs mainly during summer. In a field study conducted in northeastern Baja California, the largest number of juveniles and greatest numbers of males with scrotal testes were found in July (E. Mellink and J. Plaguing, pers. obs.). In Baja California Sur, the reproductive season was April to late August (Cortés and Álvarez-Castañeda, 1997). On Cerralvo Island, a pregnant female with two embryos was present at the end of May (Banks, 1964a).

VEGETATIONAL ASSOCIATIONS AND ELEVATION RANGE: *C. arenarius* is found in arid scrubs and sandy soils, from sea level to 600 masl (Álvarez, 1958; Lackey, 1991).

CONSERVATION STATUS: Three subspecies (*C. arenarius albulus*, *C. arenarius ammophilus*, and *C. arenarius siccus*) are believed to be threatened (SEMARNAT, 2010) because they live on islands, but this assumption does not apply to *C. arenarius albulus* (Álvarez, 1958). *C. arenarius sublucidus* is restricted to a small area isolated from the rest of the species south of Bahía de la Paz. This subspecies may require protection to ensure its survival.

Chaetodipus artus (Osgood, 1900)

Narrow-skulled pocket mouse

Eric Mellink and Sergio Méndez Moreno

SUBSPECIES IN MEXICO
C. artus is a monotypic species.

DESCRIPTION: The upper parts of *Chaetodipus artus* are brownish, obscured by black hair along the midline, especially on the rump. There is also a reddish-buffy and diffuse line that separates the colors of the sides from the white abdomen. This mouse has a well-defined lateral buffy stripe. The pelage is thick, with little or no bristles; the legs have thick and short bristles. Ears are large and rounded (Williams et al., 1993a). *C. artus* is distinguished from *C. goldmani* by its smaller size, a less hairy tail, and a wider dorsal stripe on the tail (Hall, 1981).

EXTERNAL MEASURES AND WEIGHT
TL = 170 to 200 mm; TV = 80 to 106 mm; HF = 24 mm; EAR = 10 to 12 mm.
Weight: 13 to 27 g.

DENTAL FORMULA: I 1/1, C 0/0, PM 1/1, M 3/3 = 20.

NATURAL HISTORY AND ECOLOGY: There are few data on the biology of *C. artus*. It is granivorous and its diet consists almost exclusively of small seeds (Best, 1992). It probably reproduces all year, with peaks of births at the end of the dry season and beginning of the rainy season (May–July). The size of litters is two to five young (Álvarez-Castañeda and Patton, 2000). It dwells in the foothills of the Sierra Madre Occidental in thorny scrublands and low-elevation forests. In general, it inhabits sites where riparian vegetation occurs, generally following riverbeds, streams, ravines, and edges of agricultural fields (Best, 1992). In Sonora, it is sympatric with *C. goldmani* and occurs in sandy areas with mesquites (*Prosopis*) along the Río Mayo. In northern Sinaloa and southern Sonora, this species occurs at higher elevations that are characterized by higher humidity and more tropical flora than along coastal areas (Anderson, 1964).

DISTRIBUTION: *C. artus* is endemic to northwestern Mexico, where it is on the coastal plain of southern Sonora into Nayarit (Anderson, 1964; Hall, 1981). It has been recorded in CH, DU, NY, SI, and SO.

VEGETATIONAL ASSOCIATIONS AND ELEVATION RANGE: This species inhabits xeric shrublands, thorny forests, riparian vegetation, and tropical deciduous forest (Best, 1992). It occurs from sea level to 1,900 m.

CONSERVATION STATUS: *C. artus* is endemic and rare in Mexico. Current status of populations is unknown. Because it is endemic to Mexico, rare, and with a restricted distribution affected by agricultural development, it may be threatened with extinction.

Chaetodipus baileyi (Merriam, 1894)

Bailey's pocket mouse

Sergio Méndez Moreno and Gerardo Ceballos

SUBSPECIES IN MEXICO
Chaetodipus baileyi baileyi (Merriam, 1894)
Chaetodipus baileyi insularis (Townsend, 1912)
Taxonomic relationships among subspecies of *C. baileyi* have been evaluated. Despite evidence that two groups existed within this species based on differences in karyotypes and allozymes (Patton and Rogers, 1993; Patton et al., 1981), the two groups were assigned to the same species for many years. Populations west of the Colorado River, however, have been separated from *C. baileyi* and are now recognized under *C. rudinoris*, which includes the subspecies *extimus*, *fornicatus*, *hueyi*, and *mesidios* (Riddle et al., 2000a).

DESCRIPTION: *Chaetodipus baileyi* is among the largest species of the genus. The tail is heavily crested and is larger than the length of head and body. Ears are relatively large. The pelage is smooth and without spines. Dorsal coloration and sides are washy-gray, mixed with ochre buffy or pale buffy, and slightly striped with black (Hall, 1981; Paulson, 1988). *C. baileyi* can be distinguished from other species of the genus because it lacks spines, and by its larger size and dorsal coloration (Hall, 1981). Although its morphology is similar to that of *C. rudinoris*, it is distinguished by genetic differences and allopatric distribution. *C. baileyi* is exceeded in size of head and body only by *C. hispidus*, which can be distinguished from *C. baileyi*. *C. baileyi* has a more grayish coloration, its tail is longer and crested, and the tail of *C. hispidus* is not crested and is shorter than the length of head and body (Williams et al., 1993a).

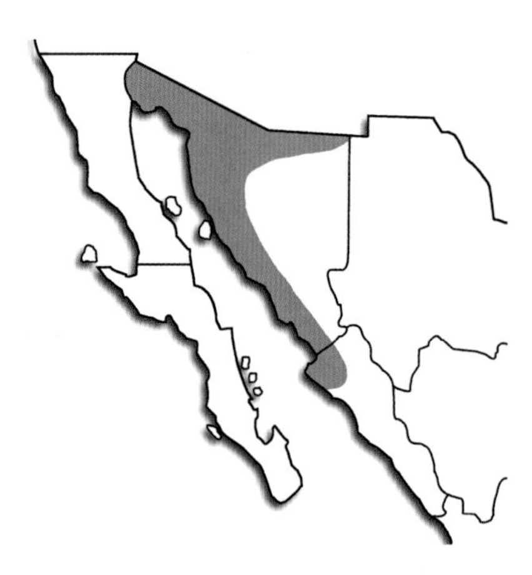

DISTRIBUTION: *C. baileyi* has a limited distribution in southern Arizona and northwestern Mexico in Sonora and Sinaloa (Álvarez-Castañeda and Patton, 1999; Riddle et al., 2000a). It has been recorded in SI and SO.

EXTERNAL MEASURES AND WEIGHT
TL = 188 to 223 mm; TV = 103 to 124 mm; HF = 24 to 27 mm; EAR = 7 to 11 mm. Weight: 24 to 28 g.

DENTAL FORMULA: I 1/1, C 0/0, PM 1/1, M 3/3 = 20.

NATURAL HISTORY AND ECOLOGY: Bailey's pocket mouse is nocturnal and terrestrial. Burrows are under rocks or shrubs, where temperature is lower and relative humidity is higher. This species is more common in the ecotone between the slopes of hills and desert plains (Álvarez-Castañeda and Patton, 1999). They reach densities of up to 86 individuals/ha. Its diet is primarily seeds, but it also consumes insects and green vegetation (Paulson, 1988). They are generalists and consume seeds of many species such as jojoba (Reichman, 1973). They can survive long periods without drinking water. In general, they mate twice a year during June to October. The average size of a litter is three to four offspring (Reichman, 1973). They are preyed on by foxes, coyotes (*Canis latrans*), hawks, owls, and snakes. To avoid predators, activities are carried out under vegetation. Its color resembles that of the soil, and it uses sudden-prolonged immobility to avoid predators (Reynolds, 1949).

VEGETATIONAL ASSOCIATIONS AND ELEVATION RANGE: *C. baileyi* primarily occurs in transition areas of lowlands of the Sonoran desert and desert grasslands, often in plains and rocky slopes. It occurs in xeric shrubs with such dominant elements as cresotebush (*Larrea divaricata*), palo verde (*Cercidium praecox*), sahuaros (*Carnegiea gigantea*), and ocotillo (*Fouquieria*), from sea level to 1,100 m (Hall, 1981). Its distribution coincides in large part with that of jojoba (*Simmondsia chinensis*; Sherbrooke, 1976).

CONSERVATION STATUS: This species has a relatively wide geographic distribution in large, undisturbed areas, and is not considered at risk. *C. baileyi insularis* is in danger of extinction because it is endemic to Tiburón Island (SEMARNAT, 2010).

Chaetodipus californicus (Merriam, 1889)

California pocket mouse

Jaime Luévano and Eric Mellink

SUBSPECIES IN MEXICO
Chaetodipus californicus femoralis (J.A. Allen, 1891)
Chaetodipus californicus mesopolius (Elliot, 1903)

DESCRIPTION: *Chaetodipus californicus* is a large mouse. Coloration is a mixture of yellow and black hair on the back and a yellowish-white abdomen. The legs have rigid hairs that look like spines. In general, the pelage is conspicuously rough. The tail has a distinctive tuft and is bicolored. The soles of the hind feet are naked. The skull is strongly arched and auditory bullae are separated anteriorly (Brylski, 1990b; Jameson and Peeters, 1988).

EXTERNAL MEASURES AND WEIGHT
TL = 190 to 235 mm; TV = 103 to 143 mm; HF = 24 to 29 mm; EAR = 9 to 14 mm. Weight: 16 to 21 g.

DENTAL FORMULA: I 1/1, C 0/0, PM 1/1, M 3/3 = 20.

NATURAL HISTORY AND ECOLOGY: *C. californicus* is mainly nocturnal, solitary, and active throughout the year, but it ceases activity in cold weather. Diet is annual grasses, weeds, and shrubs, such as *Salvia*, but at certain times of the year, California pocket mice consume insects. Water is obtained mainly from seeds and leaves. Foraging activity usually takes place on the ground, but they may climb onto bushes (Brylski, 1990b; Jameson and Peeters, 1988). They are most common in areas with shrubs, but also occupy pasturelands, and largest populations occur in ecotones between pasturelands and chaparral. Populational densities are high. Grazing by domestic cattle eliminates vegetational cover that allows this mouse to hide from predators (Brylski, 1990b). *C. californicus* typically produces one litter per year (occasionally two) during April–July. The size of a litter is two to seven and averages four young (Brylski, 1990b; Jameson and Peeters, 1988).

VEGETATIONAL ASSOCIATIONS AND ELEVATION RANGE: *C. californicus* occupies various types of chaparral, although localities where it occurs in Baja California are coniferous forests. It occurs from sea level to 1,800 m (Huey, 1964).

CONSERVATION STATUS: It is possible that grazing by livestock has affected some populations. There is no study to clarify this effect, but given the size and density of shrubs in the habitat, overall effect of grazing probably is minimal. The species does not appear to be at risk.

DISTRIBUTION: The California pocket mouse occurs in southern California and northern Baja California, where it occupies Mediterranean plant communities from the coast between Tijuana and Ensenada to highlands of the Sierra de Juarez and the Sierra de San Pedro Mártir (Hall, 1981; Huey, 1964). It has been recorded in BC.

Chaetodipus dalquesti (Roth, 1976)

Dalquest's pocket mouse

Lourdes Martinez-Estevez and Gerardo Ceballos

SUBSPECIES IN MEXICO
Chaetodipus dalquesti ammophilus (Osgood, 1907)
Chaetodipus dalquesti dalquesti (Roth, 1976)
Chaetodipus dalquesti sublucidus (Nelson and Goldman, 1929)
It was originally described as a species (Roth, 1976), but then some authors considered it as a subspecies of *Ch. arenarius*. Presently it is considered a full species (Álvarez-Castañeda and Rios, 2010).

DESCRIPTION: *Chaetodipus dalquesti* is a medium-sized mouse. The pelage is grizzled; ears are large and with black margins; the bullae is moderately inflated; the posterior palatal pits are smaller; and it has long nasal bones (Álvarez-Castañeda and Rios, 2010). Males are larger than females. Its pelage is moderately silky and the color of the back varies from pale gray or pale yellowish to dark brown. The hair may end in a black tip, which gives it a speckled appearance. The venter is white. The tail is bicolored and has a small terminal fleck. Ears are brown. It is very similar to *Chaetodipus arenarius*, but larger, with a longer and darker tail and the rump covered with weak spine bristles (Álvarez-Castañeda and Rios, 2010).

DISTRIBUTION: The type locality of *Ch. dalquesti* is four miles southeast of Migriño, Baja California Sur, Mexico (23°10′N, 110°07′W). It is found along the Pacific coastal area from Lopez Mateos to the Cape Region of the Baja California Sur and Margarita Island (Álvarez-Castañeda and Rios, 2010).

EXTERNAL MEASURES AND WEIGHT
TL = 172 mm; TV = 96; HF = 24; EAR = 8.9.
Weight: 14 g.

DENTAL FORMULA: I 1/1, C 0/0, PM 1/1, M 3/3 = 20.

NATURAL HISTORY AND ECOLOGY: This species occurs in moister areas of the coastal plain of the southern part of the Baja California Peninsula (Álvarez-Castañeda and Rios, 2010). The species is poorly known and there is little information on its ecology.

VEGETATIONAL ASSOCIATIONS AND ELEVATION RANGE: *Ch. dalquesti* is found in desert scrubland near sea level which, because of the coastal fogs, also contains epiphytes such as *Gongylocarpus* sp., *Opuntia pycnacantha*, *Ferocactus santa-maria*, *Stenocereus eruca*, and *Harfordia macroptera* (Álvarez-Castañeda and Rios, 2010; Roth, 1976).

CONSERVATION STATUS: *Ch. dalquesti* is an endemic species considered as of Special Protection (SEMARNAT, 2010) and is listed as vulnerable by IUCN (2013). Its habitat is under threat because of tourism development and the populations are fragmented along the range.

Chaetodipus eremicus (Mearns, 1898)

Chihuahuan pocket mouse

Yolanda Domínguez and Gerardo Ceballos

SUBSPECIES IN MEXICO
C. eremicus is a monotypic species.
Lee et al. (1996) raised the subspecies *C. penicillatus eremicus* and *C. penicillatus atrodorsalis* to species level, under the name of *C. eremicus*, based on differences in mitochondrial DNA, karyotypes, and nuclear genes (Patton, 1970; Patton et al., 1981)

DESCRIPTION: Compared to *Chaetodipus penicillatus*, this pocket mouse is medium sized. The pelage is coarser than in other species of the genus. It lacks guard hairs on the rump, the tail ends in a brush, dorsal coloration varies from yellowish-brown to cinnamon-brown peppered with black and dark brown, the venter is white, and there is a well-defined lateral line (Hoffmeister and Lee, 1967). The tail is bicolored, white ventrally and brown dorsally. Hind feet are small in comparison with those of other species of the genus. The skull is medium-sized, but nasal passages are long and the interorbital is wide (Anderson, 1972; Hoffmeister and Lee, 1967). The diploid number of chromosomes is 2n = 46 (Patton, 1967).

Chaetodipus eremicus. Scrubland. Charcas, San Luis Potosí. Photo: Juan Cruzado.

EXTERNAL MEASURES AND WEIGHT
TL = 147 to 179 mm; TV = 77 to 98 mm; HF = 21 to 25 mm; EAR = 7 to 8 mm.
Weight: 13 to 19 g.

DENTAL FORMULA: I 1/1, C 0/0, PM 1/1, M 3/3 = 20.

NATURAL HISTORY AND ECOLOGY: In Chihuahua, *Chaetodipus eremicus* occurs in sandy or alluvial soils, and in thickets dominated by mesquites (*Prosopis*) where it is sympatric with *C. nelsoni.* In Bolson de Mapimi, Durango, it is sympatric with *Dipodomys nelsoni, D. merriami,* and *Neotoma albigula* (Grenot and Serrano, 1981). It is granivorous, as are all species of the genus (Grenot and Serrano, 1981; Hoffmeister and Lee, 1967). The Chihuahuan pocket mouse builds underground burrows with many chambers at the base of cacti such as *Opuntia rastrera.* Burrows are used as nests and to store food; in addition, they provide favorable microclimatic conditions (Grenot and Serrano, 1981). There is little information on reproductive biology of this species; in Coahuila, pregnant females have been observed in February and June; nonreproductive females have been reported during February–August and in December (Baker, 1956). Of 12 females examined in Durango in June and July, 9 had no sign of reproductive activity and 3 had embryos (Baker and Greer, 1962).

VEGETATIONAL ASSOCIATIONS AND ELEVATION RANGE: *C. eremicus* occurs in pasturelands, areas with *Opuntia rastrera,* dunes with *Larrea divaricata* and *Prosopis,* and up to 1,700 masl (Anderson, 1972; Grenot and Serrano, 1981).

CONSERVATION STATUS: Current status of populations is unknown. The Chihuahuan pocket mouse is not listed as endangered and its wide distribution, which covers large expanses with little disturbance, indicates it faces no risk.

DISTRIBUTION: *C. eremicus* is distributed from southern New Mexico and western Texas in the United States across desert regions of Chihuahua into Durango and western San Luis Potosí (Anderson, 1972; Hall, 1981; Hoffmeister and Lee, 1967; Williams et al., 1993a). It has been recorded in CH, CO, DU, SL, TA, and ZA.

San Diego pocket mouse

Iván Castro Arellano and Gerardo Ceballos

DISTRIBUTION: *C. fallax* occurs from southern California in the United States to Bahía Tortuga and San Bartolomé in northwestern Baja California (Hall, 1981; Huey, 1964). It has been recorded in BC and BS.

SUBSPECIES IN MEXICO

Chaetodipus fallax fallax (Merriam, 1889)
Chaetodipus fallax inopinus (Nelson and Goldman, 1929)
Chaetodipus fallax majusculus (Huey, 1960)
Chaetodipus fallax pallidus (Mearns, 1901)
Chaetodipus fallax xerotrophicus (Huey, 1960)
Some authors consider *C. anthonyi* a subspecies of *C. fallax* (Wilson and Reeder, 2005), primarily because their karyotypes are similar. This taxonomic designation is not widely accepted in Mexico, however (Arita and Ceballos, 1997).

DESCRIPTION: Compared to other members of the genus, *Chaetodipus fallax* is medium in size. Dorsal coloration is brown with the rump darker. The distal part of guard hairs is pale and the belly is creamy colored. Pinnae of ears are dark dorsally and grayish on the underside. The tail is bicolored (Osgood, 1900). There is variation from dark tones to pale tones, however, probably as a result of color of substrates that are occupied (Huey, 1960). The diploid number of chromosomes is 2n = 44 (Patton and Rogers, 1993). *C. fallax* is distinguished from *C. californicus* by its smaller size and more rounded ears.

EXTERNAL MEASURES AND WEIGHT

TL = 176 to 200 mm; TV = 88 to 118 mm; HF = 23 mm; EAR = 9 mm.
Weight: 20 g.

DENTAL FORMULA: I 1/1, C 0/0, PM 1/1, M 3/3 = 20.

Chaetodipus fallax. Scrubland. Valle de San Quintin, Baja California. Photo: Eric Mellink.

NATURAL HISTORY AND ECOLOGY: The San Diego pocket mouse is nocturnal and inhabits rocky substrates in scrublands. These mice are active throughout the year, but are inactive on cold nights. They may remain in burrows for several weeks (Álvarez-Castañeda and Patton, 1999). They feed mainly on seeds, but also consume green vegetation and insects. They can survive for extensive periods on a diet of dried seeds (Macmillen, 1964; Morton, 1979). Its home range is small, about 0.3 ha. Population density is variable, but may reach more than 50 individuals/ha in years with abundant food. They reproduce in spring and summer; there is an average of 4 offspring per litter and gestation is 25 days (Jones, 1993). Predators include barn owls (*Tyto alba*; Banks, 1965).

VEGETATIONAL ASSOCIATIONS AND ELEVATION RANGE: Vegetation within the geographic range of this species includes a variety of xeric scrublands from sea level to 1,350 m (Hall, 1981; Huey, 1964).

CONSERVATION STATUS: This species has a relatively wide distribution, including large regions with little anthropogenic disturbance. It is not considered to be at risk.

Chaetodipus formosus (Merriam, 1889)

Long-tailed pocket mouse

Sergio Méndez Moreno

SUBSPECIES IN MEXICO
Chaetodipus formosus cinerascens (Nelson and Goldman, 1929)
Chaetodipus formosus infolatus (Huey, 1954)
Chaetodipus formosus mesembrinus (Elliot, 1904)

DESCRIPTION: *Chaetodipus formosus* is medium in size. Its pelage is smooth, sepia in color with gray hair on the back, and the underside and legs are white. The tail is long; it is buffy on the ventral surface, it has a crest of hair, and it ends with a darker tuft than the rest of the pelage. Ears are blackish with a tuft of black and white hairs on the bases. The subauricular spot is small. *C. formosus* is similar to other species of the genus that do not have bristles or spines (Williams et al., 1993a).

DISTRIBUTION: *C. formosus* occurs from California and Utah to northwestern Baja California from the Sierra de San Pedro Mártir to near Santa Rosalia (Hall, 1981; Williams et al., 1993b). It has been recorded in BC and BS.

EXTERNAL MEASURES AND WEIGHT
TL = 104 mm; TV = 118 mm; HF = 25 mm; EAR = 5 to 8 mm.
Weight: 18 g.

DENTAL FORMULA: I 1/1, C 0/0, PM 1/1, M 3/3 = 20.

NATURAL HISTORY AND ECOLOGY: The long-tailed pocket mouse is nocturnal. During the day, they shelter in underground burrows, whose openings are plugged with soil or soft sand to prevent entry of predators such as snakes and to maintain high relative humidity. Openings of burrows are 2.5 cm to 4 cm in diameter (Eisenberg, 1963). Diet consists mainly of seeds from xeric plants. They are capable of using moisture that is contained within seeds. They mate once a year, in spring or summer, and give birth after four weeks of gestation to a litter of two to six offspring that weigh 1 g on average. Young are weaned about three weeks later and reach sexual maturity at one year of age. This species hibernates for about three months beginning in November. Males emerge from hibernation two weeks before females (Kenagy and Bartholomew, 1985). In general, they only reproduce once a year, but in favorable conditions, they may produce two litters (Duke, 1957).

VEGETATIONAL ASSOCIATIONS AND ELEVATION RANGE: *C. formosus* inhabits xeric scrublands in communities dominated by shrubs and forbs such as *Atriplex corfertifolia*, *A. canescens*, *Sarcobatus*, and sagebrush (*Artemisia*). It is present in flat and rocky lands with abundant plants such as *Opuntia* (Eisenberg, 1963). They are not common on flat and sandy areas (Kenagy and Bartholomew, 1985). It inhabits elevations from sea level to 2,100 m.

CONSERVATION STATUS: Neither the current size nor current status of populations is known. It is probable that *C. formosus* does not face serious conservation issues.

Chaetodipus goldmani (Osgood, 1900)

Goldman's pocket mouse

Lorena Morales Pérez and Iván Castro Arellano

SUBSPECIES IN MEXICO

C. goldmani is a monotypic species.

DISTRIBUTION: *C. goldmani* is endemic to northwestern and southern Sonora, southwestern Chihuahua, and northern Sinaloa, Mexico (Patton, 1993). It has been recorded in CH, SI, and SO.

DESCRIPTION: *Chaetodipus goldmani* is a medium-sized mouse. The back is brown and the rump is darker. The underside is creamy colored; the tail is markedly bicolored, dark dorsally and white ventrally (Osgood, 1900). One population on a lava field near Moctezuma, Sonora, is especially dark colored (Findley, 1967). Ears are blackish with interspersed white hairs, distal edges are whitish, and there is a subauricular spot (Lackey and Best, 1992). Compared with *C. pernix*, Goldman's pocket mouse is larger. *C. baileyi* has darker hair in its back, and *C. goldmani* has a greater occipitonasal length, a wider zygomatic bone, and larger interparietals (Anderson, 1972; Lackey and Best, 1992). The diploid number of chromosomes is 2n = 52 (Patton, 1967).

EXTERNAL MEASURES AND WEIGHT

TL = 194 mm; TV = 106 mm; HF = 24 mm; EAR = 11 mm.

Weight: 19 to 23 g.

DENTAL FORMULA: I 1/1, C 0/0, PM 1/1, M 3/3 = 20.

NATURAL HISTORY AND ECOLOGY: This mouse is abundant in alluvial soils of southern Sonora and northern Sinaloa; in contrast, in northern and central Sonora, it occurs in shrubby vegetation on a lava field (Burt and Hooper, 1941). In its northern range, it lives on rocky soils, allegedly displacing *C. penicillatus* to areas with sandy soils (Patton, 1969). Reproductive biology is unknown, but in *C. pernix*, a sympatric species of the same genus, young are born October–April and the size of a litter usually is seven (Best and Lackey, 1992). Diet also is unknown, but they do consume grass seeds (Burt, 1938).

VEGETATIONAL ASSOCIATIONS AND ELEVATION RANGE: *C. goldmani* inhabits xeric scrublands with shrubs and mesquites (*Prosopis*); it also occurs near agricultural fields (Burt, 1938; Burt and Hooper, 1941). This species has been reported from sea level to 120 m (Hall, 1981).

CONSERVATION STATUS: This species is not included in any list as being in need of conservation (Arita and Ceballos, 1997). Because it is endemic to Mexico and its distribution is relatively restricted, it is necessary to determine if it is at risk of extinction.

Chaetodipus hispidus (Baird, 1858)

Hispid pocket mouse

Ella Vázquez

SUBSPECIES IN MEXICO
Chaetodipus hispidus hispidus (Baird, 1858)
Chaetodipus hispidus paradoxus (Merriam, 1889)
Chaetodipus hispidus zacatecae (Osgood, 1900)

DESCRIPTION: *Chaetodipus hispidus* is a medium-sized to large mouse. Hair is short and dense with thorny awns on the rump (Hall, 1981). The tail lacks a distal ridge, there is long hair on the tip, the top is brown, the bottom is white, and the tail is shorter than the length of body and head (Hoffmeister, 1986). The soles of the forefeet are naked. Dorsally, the body is brown to gray and the abdomen is whitish. The hind feet are larger and longer than the forefeet, which allows them to hop on two feet

Chaetodipus hispidus. Grassland. Janos Biosphere Reserve, Chihuahua. Photo: Gerardo Ceballos.

DISTRIBUTION: The hispid pocket mouse occurs from the plains of south-central North Dakota in the United States to Mexico, where it occurs from Chihuahua and Tamaulipas into Guanajuato, Zacatecas, and Jalisco (Ramírez-Pulido et al., 1983; Williams et al., 1993b). It has been recorded in AG, CH, CO, DU, GJ, HG, JA, NL, SL, SO, TA, and ZA.

(Anderson, 1972; Williams et al., 1993b). This species represents a monophyletic group with a diploid number of chromosomes of 2n = 34 (Patton, 1967).

EXTERNAL MEASURES AND WEIGHT
TL = 152 to 230 mm; TV = 72 to 113 mm; HF = 22 to 29 mm; EAR = 10 to 13 mm. Weight: 15 to 60 g.

DENTAL FORMULA: I 1/1, C 0/0, PM 1/1, M 3/3 = 20.

NATURAL HISTORY AND ECOLOGY: Hispid pocket mice are solitary, terrestrial, and nocturnal (Jones, 1993). They occur in a variety of substrates, from rocky to sandy soils (Álvarez-Castañeda and Patton, 1999). They are granivorous, but they also consume vegetation and insects. Burrows are underground with nests and chambers to store seeds (Brown and Harney, 1993; Hoffmeister, 1986). They reproduce in spring and summer. Females can produce two litters per breeding season. Litters consist of five young on average, with a range of two to nine. Maximum lifespan is about 2 years and they reach sexual maturity at 60 days old (Jones, 1993).

VEGETATIONAL ASSOCIATIONS AND ELEVATION RANGE: This species usually is encountered in arid pasturelands that contain moderately dense stands of grasses with *Cassava* and *Agave* (Hoffmeister, 1986; Petersen, 1980). It also occurs in scrublands dominated by *Prosopis* (Álvarez and Álvarez-Castañeda, 1991; Petersen, 1980). In Tamaulipas, it lives in lowland-deciduous forests and thorny scrublands. It occurs from 600 masl to 2,700 masl (Hall, 1981).

CONSERVATION STATUS: *C. hispidus* is not included in any category of risk (Ceballos and Navarro, 1991). Existing studies do not report populations or habitats in need of conservation efforts (Genoways and Brown, 1993).

Chaetodipus intermedius (Merriam, 1889)

Rock pocket mouse

Iván Castro Arellano and Gerardo Ceballos

SUBSPECIES IN MEXICO
Chaetodipus intermedius intermedius (Merriam, 1889)
Chaetodipus intermedius lithophilus (Huey, 1937)
Chaetodipus intermedius minimus (Burt, 1932)
Chaetodipus intermedius pinacate (Blossom, 1933)

DESCRIPTION: The rock pocket mouse is a medium-sized *Chaetodipus*. Dorsal coloration is highly variable, from pale gray to black brownish; the sides are paler than the back and the belly varies from creamy white to darker shades. The tail is long, and is darker in its dorsal and distal parts. Hairs are coarse with weak spines on the rump (Hall, 1981; Williams et al., 1993a). *Chaetodipus intermedius* is geographically sympatric with *C. penicillatus*, with which it often is confused by similarity in size and proportions. *C. intermedius* differs from *C. penicillatus* by the presence of rump spines and smaller hind feet (< 22 mm), and by its thinner face (Hoffmeister, 1986; Williams et al., 1993a). It can be distinguished from other sympatric species such as *C. baileyi*, *C.*

formosus, and *C. hispidus* because these are larger and have no spines (Williams et al., 1993a). The diploid number of chromosomes is 2n = 46 (Patton and Rogers, 1993).

DISTRIBUTION: *C. intermedius* occurs from Utah and Texas in the United States into northern Sonora and north-central Chihuahua in Mexico (Hall, 1981; Patton, 1993; Williams et al., 1993b). It also occurs on Tiburón Island, Sonora (Hoffmeister, 1974). It has been recorded in CH and SO.

EXTERNAL MEASURES AND WEIGHT
TL = 152 to 180 mm; TV = 83 to 103 mm; HF = 19 to 24 mm; EAR = 7 to 12 mm. Weight: 10 to 19 g.

DENTAL FORMULA: I 1/1, C 0/0, PM 1/1, M 3/3 = 20.

NATURAL HISTORY AND ECOLOGY: This species is strongly associated with rocky environments and xeric scrublands; it is so linked with this substrate that its color is nearly the same. For this reason, several subspecies of darker color have been described; these are restricted to lava flows (Hoffmeister, 1986; Williams et al., 1993b). In the desert of northeastern Sonora, it is present only in rocky areas and canyons (May, 1976). They reproduce in spring and summer. In Arizona, pregnant females occur April–July, with a peak in the reproductive activity in June. Number of offspring is one to seven per litter (Hoffmeister, 1986). It is a granivorous species. In Arizona, 82% of diet consists of seeds and 16% of insects, with virtually no presence of vegetation in the diet (Hoffmeister, 1986; Morton, 1979).

VEGETATIONAL ASSOCIATIONS AND ELEVATION RANGE: In Mexico, vegetation within the geographic range of this species is xeric scrublands and pasturelands. It occurs from sea level to 1,900 m (Hall, 1981).

CONSERVATION STATUS: The species has a restricted distribution but encompasses vast regions with little impact by humans. Protected populations inhabit the biosphere reserves of Pinacate and Alto Golfo in Sonora. The subspecies *C. intermedius minimus*, which is restricted to Turner Island in Sonora, is regarded as threatened (SEMARNAT, 2010).

Chaetodipus intermedius. Grassland. Janos Biosphere Reserve, Chihuahua. Photo: Gerardo Ceballos.

Chaetodipus lineatus (Dalquest, 1951)

Lined pocket mouse

Jorge Ortega R. and Héctor T. Arita

SUBSPECIES IN MEXICO
C. lineatus is a monotypic species.

DISTRIBUTION: *C. lineatus* is endemic to the Mexican Plateau. Only five records exist for the state of San Luis Potosí and one for the state of Zacatecas (Hall, 1981; Matson and Baker, 1986). It has been recorded in SL and ZA.

DESCRIPTION: *Chaetodipus lineatus* is small. The pelage is smooth, which is a feature that distinguishes it from *C. nelsoni*. Its dorsal coloration is dark gray and the underside is white; it has a buffy line in the mid-part of the head (Dalquest, 1951). From the 15 species of the genus, *C. lineatus* is morphologically closest to *C. spinatus* (Claire, 1976).

EXTERNAL MEASURES AND WEIGHT
TL = 174 mm; TV = 95 to 98 mm; HF = 23 mm; EAR = 8 to 10 mm.
Weight: 17 g.

DENTAL FORMULA: I 1/1, C 0/0, PM 1/1, M 3/3 = 20.

NATURAL HISTORY AND ECOLOGY: As in other members of the genus, the lined pocket mouse has a pair of external cheek pouches in which to carry seeds. Its distribution and habitat suggest that its diet is based on seeds and insects. Reproductive and gestation periods are unknown, but *C. penicillatus*, a similar species, gives birth twice a year with an average of 2 to 8 offspring per litter and a gestation period of 23 days (Nowak, 1999b).

VEGETATIONAL ASSOCIATIONS AND ELEVATION RANGE: The species has been observed exclusively in xeric scrublands 1,600 masl to 2,400 masl.

CONSERVATION STATUS: Current status of this species is unknown; however, it is fragile or vulnerable as records show a low population density and a restricted distribution. As it is endemic to Mexico, it should be a priority to study its natural history and current status.

Chaetodipus nelsoni (Merriam, 1894)

Nelson's pocket mouse

Elizabeth E. Aragón

SUBSPECIES IN MEXICO
Chaetodipus nelsoni canescens (Merriam, 1894)
Chaetodipus nelsoni nelsoni (Merriam, 1894)

DESCRIPTION: *Chaetodipus nelsoni* is a small species. Dorsal coloration is brownish-black with the underside white or buffy, and there is a conspicuous lateral stripe. The tail has a thick black ridge on top and is white on the bottom. Coloration varies according to color of soils that it inhabits (Baker and Greer, 1962). Its pelage is dense and spiny. This species is similar to *C. intermedius*, differing in the larger size, more conspicuous spines on the back, and thicker pelage.

EXTERNAL MEASURES AND WEIGHT
TL = 152 to 204 mm; TV = 72 to 117 mm; HF = 18 to 28 mm; EAR = 7 to 10 mm. Weight: 12 to 20 g.

DENTAL FORMULA: I 1/1, C 0/0, PM 1/1, M 3/3 = 20.

NATURAL HISTORY AND ECOLOGY: *C. nelsoni* is a granivorous species that also consumes insects and vegetation (Grenot and Serrano, 1981; Petersen and Petersen, 1979). It dwells on bare soils, often occurring in rocky areas on hills, hillsides, and slopes, and in pasturelands with grasses and shrubs. Dominant plants in its habitat are honey mesquites (*Prosopis glandulosa*), cresotebushes (*Larrea tridentata*), sotols (*Dasylirion wheeleri*), magueys (*Agave asperrima*), prickly pears (*Opuntia*), leatherstems (*Jatropha dioica*), and various cacti. It is common in habitats with coarse soils and less common in slopes where the soil has mixed texture and many microenvironments (Álvarez, 1963a; Baker, 1956; Baker and Greer, 1962; Grenot, 1983; Grenot and Serrano, 1981; Matson and Baker, 1986; Serrano, 1987). Associated rodents include kangaroo rats (*Dipodomys*), deermice (*Peromyscus*), southern grasshopper mice (*Onychomys torridus*), and white-throated woodrats (*Neotoma albigula*; Baker and Greer, 1962; Serrano, 1987). Occasionally, it is sympatric with other pocket mice such as *C. penicillatus* and *Perognathus flavus* (Grenot and Serrano, 1981). Reproduction occurs at the end of summer. Pregnant females have been observed in late March and early August, and nonreproductive females have been noted during January–April, June, July, November, and December. Average number of young per litter is 2.8, with a range of 1 to 4 offspring (Baker, 1956; Baker and Greer, 1962; Dahlquest, 1953b; Matson and Baker, 1986). Populational densities vary seasonally, and up to 93 individuals/ha have been reported, with greatest densities in scrublands of cresotebush (*Larrea*; Grenot, 1983). This variation in densities occurs relative to availability of resources during the year. In wet years and during summer, forbs and grasses increase, which is reflected in

DISTRIBUTION: The geographic range of *C. nelsoni* includes plains and arid areas of the Chihuahuan desert from the southern United States into northern Mexico, where it is present on the Mexican Plateau into Aguascalientes (Hall, 1981). It has been recorded in AG, CH, CO, DU, JA, NL, SL, TA, and ZA.

Chaetodipus nelsoni. Scrubland. Santo Domingo, San Luis Potosí. Photo: Juan Cruzado.

an increase in number of juveniles in populations. The population decreases significantly in winter (Whitford, 1976). In early summer, when there is an absence of green sprouts, grasses, and insects, survival and recruitment are related to caloric content of foliage (Serrano, 1987). In Mapimi, Durango, length of movements and size of home ranges are 28 m to 118 m and 0.04 ha to 0.86 ha, respectively; while in autumn–winter these are 28 m to 130 m and 0.01 ha to 0.45 ha (Grenot and Serrano, 1981, 1982).

VEGETATIONAL ASSOCIATIONS AND ELEVATION RANGE: *C. nelsoni* occurs in xeric scrublands, thorny forests, and pasturelands at 372 masl to 2,450 masl.

CONSERVATION STATUS: Nelson's pocket mouse is not at risk in Mexico (SEMARNAT, 2002). On the Mexican Plateau, it is one of the most common species.

Chaetodipus penicillatus (Woodhouse, 1852)

Desert pocket mouse

Iván Castro Arellano and Jorge Iván Uribe J.

SUBSPECIES IN MEXICO
Chaetodipus penicillatus angustirostris (Osgood, 1900)
Chaetodipus penicillatus penicillatus (Woodhouse, 1852)
Chaetodipus penicillatus pricei (J.A. Allen, 1894)
Chaetodipus penicillatus seri (Nelson, 1912)
Chaetodipus penicillatus sobrinus (Goldman, 1939)
Chaetodipus penicillatus stephensi (Merriam, 1894)
The subspecies *eremicus* and *atrodorsalis* were elevated to species level, under the name of *C. eremicus*, because of genetic differences in nuclear genes and karyotypes (Lee et al., 1996).

DISTRIBUTION: The desert pocket mouse occurs from southeastern California, Nevada, Arizona, New Mexico, and Texas in the United States into northeastern Baja California and most of Sonora (Hall, 1981; Patton, 1993). The subspecies *C. penicillatus seri* occupies most of Tiburón Island, Sonora (Burt, 1938). It has been recorded in BC and SO.

DESCRIPTION: Within the genus, *Chaetodipus pencillatus* is rather large. Dorsally it possesses a variable coloration of yellowish-brown to yellowish-gray hair; in both cases it is dotted with black hair. Sides, face, and cheeks have the same coloration, but under the ears the color is darker; the underside is white. Usually, it lacks guard hair in the rump. The pelage is smooth. The tail is long and bicolored, white ventrally and brown dorsally. The soles of the hind feet are naked to the heel (Hoffmeister and Lee, 1967; Osgood, 1900). In general, males are larger than females and there is geographic variation in size (Hoffmeister and Lee, 1967). *C. penicillatus* can be confused with *C. intermedius*, with which it is geographically sympatric; however, it occurs on sandy substrates and it can be distinguished by lack of guard hairs on the rump, longer hind feet (usually greater than 22 mm), more-roughened and less-exposed mastoid bone in dorsal view, interparietal with a distinguishable anteriomedian angle, and a wider face (Hoffmeister and Lee, 1967). The diploid number of chromosomes is 2n = 46 (Patton, 1967).

EXTERNAL MEASURES AND WEIGHT
TL = 162 to 216 mm; TV = 83 to 129 mm; HF = 22 to 27 mm; EAR = 7 to 10 mm. Weight: 16.5 g.

DENTAL FORMULA: I 1/1, C 0/0, PM 1/1, M 3/3 = 20.

Chaetodipus penicillatus. Scrubland. Puerto Libertad, Sonora. Photo: Horacio Barcenas.

NATURAL HISTORY AND ECOLOGY: The desert pocket mouse occupies sandy or alluvial soils with mesquites (*Prosopis*) instead of rocky soil and open areas dominated by shrubs such as *Larrea* and *Atriplex*. It is a granivore (Hoffmeister and Lee, 1967; Morton, 1979). Little is known about its reproductive biology in Mexico. Apparently, reproduction occurs in late spring and summer, coinciding with the rainy season. Gestation is about 26 days and the size of a litter is 2 to 8, with an average of 4 offspring (Eisenberg, 1993; Jones, 1993). These mice reach densities of up to 50 individuals/ha. Predators of this species include owls (*Tyto alba* and *Bubo virginianus*; Baker, 1953).

VEGETATIONAL ASSOCIATIONS AND ELEVATION RANGE: *C. pencilliatus* occurs in rosetophilous scrublands, mesquitals, microphyllous scrublands, meadows (Álvarez and Álvarez-Castañeda., 1991a), and desert scrublands dominated by *Larrea tridentata* and *Atriplex* (Baker, 1856; Hoffmeister and Lee, 1967). It lives 70 masl to 1,800 masl (Baker, 1956; Baker and Greer, 1962; Hall, 1981).

CONSERVATION STATUS: This species has a broad distribution and has been reported to be common, but the subspecies *C. penicillatus seri* is restricted to Tiburón Island in the Gulf of California and is listed as threatened (SEMARNAT, 2010).

Chaetodipus pernix (J.A. Allen, 1898)

Sinaloan pocket mouse

Iván Castro Arellano

SUBSPECIES IN MEXICO

Chaetodipus pernix pernix (J.A. Allen, 1898)
Chaetodipus pernix rostratus (Osgood, 1900)

DISTRIBUTION: *C. pernix* is endemic to Mexico and is distributed in a narrow band that extends from south-central Sonora into northern Nayarit (Hall, 1981). It has been recorded in NY, SI, and SO.

DESCRIPTION: *Chaetodipus pernix* is a small-sized mouse. Dorsal coloration is yellowish-brown with dark hair. Sides are paler and the belly is white. The tail is brown, but white ventrally. Ears are dark with a white spot on the lower margins (Allen, 1898; Osgood, 1900). Males are larger than females, and there is variation in size among subspecies (*C. pernix rostratus* is the largest; Best and Lackey, 1992b). The diploid number of chromosomes is 2n = 38 (Patton, 1967).

EXTERNAL MEASURES AND WEIGHT
TL = 157 mm; TV = 81 mm; HF = 21 mm; EAR = 10 mm.
Weight: 17 g.

DENTAL FORMULA: I 1/1, C 0/0, PM 1/1, M 3/3 = 20.

NATURAL HISTORY AND ECOLOGY: The Sinaloan pocket mouse lives in areas with alluvial soil, almost without rocky material, and with a dense shrubby coverage (Patton and Jones, 1972). In Sonora, births have been reported in October and November (Burt, 1938), and in Sinaloa, pregnant animals have been observed in early April (Hooper, 1955). Based on number of placental scars, the usual number of offspring is seven (Patton and Soulé, 1967). This species is a granivore. Remains of seeds of *Opuntia* and grasses have been noted in contents of cheek pouches (Burt, 1938; Morton, 1979).

VEGETATIONAL ASSOCIATIONS AND ELEVATION RANGE: *C. pernix* inhabits the coastal plain of Sonora and Sinaloa. It inhabits shrubs and thorny forests characterized by legumes, arboreal cacti, and prickly pears. Wilson (1985) defined this type of vegetation as tropical semi-arid forest. Some species of plants common to this habitat are *Acacia cymbispina*, *Ipomea arborescens*, and columnar cacti (*Pachycerus pecten-arboriginum*). It also occurs along edges of agricultural fields, in prickly pear plantations (*Opuntia*), and near trees of the genus *Ficus* (Best and Lackey, 1992; Burt, 1938; Patton and Jones, 1972). It occurs from sea level to 90 m (Best and Lackey, 1992).

CONSERVATION STATUS: This species is common (Burt, 1938) and does not appear to be at risk. Its current status is unknown. Because it is endemic to Mexico and because it has a limited distribution, it must be considered a priority in assessing its conservation status.

Chaetodipus rudinoris (Elliot, 1903)

Baja California pocket mouse

Gerardo Ceballos

SUBSPECIES IN MEXICO
Chaetodipus rudinoris extimus (Nelson and Goldman, 1930)
Chaetodipus rudinoris fornicatus (Burt, 1932)
Chaetodipus rudinoris hueyi (Nelson and Goldman, 1929)
Chaetodipus rudinoris mesidios (Huey, 1964)
All subspecies were considered as subspecies of *C. baileyi*. Due to differences in karyotypes, allozymes (Patton and Rogers, 1993; Patton et al., 1981), and other features, however, populations west of the Colorado River have been assigned to *C. rudinoris* (Riddle et al., 2000).

DESCRIPTION: *Chaetodipus rudinoris* is one of the largest species in the genus; it is similar to *C. baileyi*, which can be distinguished by its allopatric distribution and genetic characteristics. The tail is longer than the length of head and body, and the tail is conspicuously crested. The ears are large. The pelage is smooth and it has bristles. Dorsal coloration varies from yellowish-gray to buffy ochre or pale buffy (Hall, 1981; Paulson, 1988). It can be distinguished from other species by its soft pelage that lacks bristles, large size, and dorsal coloration (Hall, 1981). Within the genus, *C. rudinoris* and *C. baileyi* are only surpassed in size by *C. hispidus* (Williams et al., 1993a).

DISTRIBUTION: *C. rudinoris* occurs from California in the United States to the southern tip of the Baja California Peninsula in Mexico (Riddle et al., 2000). The Colorado River separates it from *C. baileyi*. The subspecies *C. rudinoris fornicatus* dwells on Montserrat Island. It has been recorded in BC and BS.

EXTERNAL MEASURES AND WEIGHT

TL = 188 to 223 mm; TV = 103 to 124 mm; HF = 24 to 27 mm; EAR = 7 to 11 mm. Weight: 24 to 28 g.

DENTAL FORMULA: I 1/1, C 0/0, PM 1/1, M 3/3 = 20.

NATURAL HISTORY AND ECOLOGY: Baja California pocket mice are nocturnal and terrestrial. Burrows are under rocks or shrubs. Within burrows, temperature is lower and more stable and humidity is higher than on the surface, which assists *C. rudinoris* in retaining water and energy. Apparently, they are common on slopes in the ecotone between slopes of foothills and desert plains (Álvarez-Castañeda and Patton, 1999). Densities up to 86 individuals/ha have been reported. They feed mainly on seeds, but also consume insects and green vegetation (Paulson, 1988). They are generalists and consume seeds of many species such as jojoba (Reichman, 1973). If there is no water available, they can survive long periods generating metabolic water. They usually mate twice a year during June–October. Average size of a litter is three to four offspring (Reichman, 1973). They are preyed on by a variety of vertebrates, including foxes, coyotes (*Canis latrans*), hawks, owls, and snakes. To avoid predators, they are active under dense vegetation. Coloration usually is similar to that of the substrate and sudden-prolonged immobility helps them avoid predators (Reynolds, 1949).

VEGETATIONAL ASSOCIATIONS AND ELEVATION RANGE: The Baja California pocket mouse mainly inhabits shrublands in arid areas on the plains of the Baja California Peninsula, but on hillsides of the Sierra de San Pedro Mártir it occupies areas that may be rocky. Where it is present on Montserrat Island in the Gulf of California, dominant vegetation is xeric scrubland. It occurs from sea level to 250 m.

CONSERVATION STATUS: This species is not at risk, as it has a wide distribution, including large areas with little disturbance by humans.

Chaetodipus spinatus (Merriam, 1889)

Spiny pocket mouse

Miguel A. Briones and Julia P. López

SUBSPECIES IN MEXICO

Chaetodipus spinatus broccus (Huey, 1960)
Chaetodipus spinatus bryanti (Merriam, 1894)
Chaetodipus spinatus corderoi (Benson, 1930)
Chaetodipus spinatus evermanni (Nelson and Goldman, 1929)

Chaetodipus spinatus guardiae (Burt, 1932)
Chaetodipus spinatus latijugularis (Burt, 1932)
Chaetodipus spinatus lorenzi (Banks, 1967)
Chaetodipus spinatus magdalenae (Osgood, 1907)
Chaetodipus spinatus macrosensis (Burt, 1932)
Chaetodipus spinatus margaritae (Merriam, 1804)
Chaetodipus spinatus occultus (Nelson, 1912)
Chaetodipus spinatus oribates (Huey, 1960)
Chaetodipus spinatus peninsulae (Merriam, 1894)
Chaetodipus spinatus prietae (Huey, 1930)
Chaetodipus spinatus pullus (Burt, 1932)
Chaetodipus spinatus serosus (Burt, 1932)
Chaetodipus spinatus spinatus (Merriam, 1889)

DISTRIBUTION: *C. spinatus* occurs from the southwestern United States in Arizona and California to nearly all of the Baja California Peninsula in Mexico (Eric, 1981; Loomis, 1971; Patton et al., 1981). It has been recorded in BC and BS.

DESCRIPTION: *Chaetodipus spinatus* is a relatively large species. The muzzle is long; the ears are short and rounded. On both sides of the mouth it has external cheek pouches that are used to transport food to its burrow. The tail is long with short hairs and ends in a small brush. The pelage is rough and brownish-yellow; the underside is white or cream colored. The tail is bicolored, brownish above and white below. The legs and base of tail have long hair with coarse spines. Forefeet are shorter than hind feet, allowing bipedal locomotion (Woloszyn and Woloszyn, 1982).

EXTERNAL MEASURES AND WEIGHT
TL = 85 to 89 mm; TV = 107 to 124 mm; HF = 25 to 26 mm; EAR = 9 to 10 mm.
Weight: 20 to 22 g.

DENTAL FORMULA: I 1/1, C 0/0, PM 1/1, M 3/3 = 20.

NATURAL HISTORY AND ECOLOGY: Spiny pocket mice inhabit arid areas with xeric scrublands. They dig burrows under shrubs or between rocks. The species is strictly

Chaetodipus spinatus. Scrubland. Isla Margarita, Baja California. Photo: Eric Mellink.

nocturnal and feeds on seeds that are stored in burrows. It drinks no water; in captivity it eats green parts of plants to provide its water requirements. Reproduction occurs during June–September. Females have two to six young per litter. Natural enemies include owls, foxes, skunks, and rattlesnakes (Woloszyn and Woloszyn, 1982).

VEGETATIONAL ASSOCIATIONS AND ELEVATION RANGE: In Sierra de la Laguna, *C. spinatus* mainly inhabits thorny scrublands, plateaus, and slopes of the mountain covered by lowland deciduous forests, and it reaches upper-elevational limits in oak forests. It also may occupy pine-oak forests. It occurs from sea level to 2,000 m.

CONSERVATION STATUS: The species is not considered at risk of extinction. All subspecies on islands, which include *C. spinatus bryanti*, *C. spinatus corderoi*, *C. spinatus evermanni*, *C. spinatus guardiae*, *C. spinatus latijugularis*, *C. spinatus lorenzi*, *C. spinatus marcosensis*, *C. spinatus margaritae*, *C. spinatus occultus*, *C. spinatus pullus*, and *C. spinatus seorsus*, however, are listed as endangered, threatened, and extinct (SEMARNAT, 2010).

Perognathus amplus Osgood, 1900

Arizona pocket mouse

Adrián Quijada Mascareñas and Jorge Ortega R.

SUBSPECIES IN MEXICO

Perognathus amplus amplus Osgood, 1900
Perognathus amplus taylori Goldman, 1932,
The taxonomic status of Mexican subspecies is unclear (Hoffmeister, 1986).

DESCRIPTION: *Perognathus amplus* is a small-sized mouse. Coloration varies among subspecies but, in general, it is salmon-ochre on the back and yellowish-white on the abdomen; the orbital region is paler than the back (Hall, 1981). *P. amplus* is closely related to *P. longimembris*. These species can be separated by the larger skull of *P. amplus*, and by its well-developed mastoid bones and small bullae (Hoffmeister, 1986). There are also external features that distinguish the two species (Williams et al., 1993a).

DISTRIBUTION: The Arizona pocket mouse occurs from western and central Arizona in the United States into northwestern Sonora in Mexico. It has been recorded in SO.

EXTERNAL MEASURES AND WEIGHT

TL = 123 to 170 mm; TV = 79 to 95 mm; HF = 17 to 22 mm; EAR = 7 to 12 mm.
Weight: 17 to 21 g.

DENTAL FORMULA: I 1/1, C 0/0, PM 1/1, M 3/3 = 20.

NATURAL HISTORY AND ECOLOGY: This species occurs in areas with loose and soft soil. It is nocturnal and diet is mainly seeds. In a study conducted in Arizona, diet included 94.3% seeds, 3.7% insects, and 2% green vegetation (Reichman, 1975). They forage indiscriminately, and seeds collected are transported to the burrow (Reichman and Oberstein, 1977). Reproduction occurs from late winter into early summer. In each litter, three to five offspring are born (Reichman and Van de Graaff, 1973). When temperature decreases in winter, Arizona pocket mice stay in burrows and lower their body temperature.

VEGETATIONAL ASSOCIATIONS AND ELEVATION RANGE: Ecological distribution is limited to xeric scrublands, associated with mesquites (*Prosopis*), cresotebushes (*Larrea tridentata*), and palo verdes (*Cercidium*). This species occurs from sea level to 500 m.

CONSERVATION STATUS: Current status of populations is unknown in Mexico. Due to its restricted distribution, it probably is a fragile species. It is protected in the El Pinacate Biosphere Reserve in Sonora (Caire, 1985).

Perognathus flavescens Merriam, 1889

Plains pocket mouse

Jesús Pacheco R.

SUBSPECIES IN MEXICO
Perognathus flavescens melanotis Osgood, 1900

DESCRIPTION: *Perognathus flavescens* is a small-sized mouse. It has a pair of external cheek pouches situated on both sides of the muzzle. The pelage is smooth. Dorsal coloration of the body is ochre intermingled with black, particularly in the middle region; the underside is white (Osgood, 1900). Post-auricular spots are inconspicuous and yellow (Hoffmeister, 1986). The tail is shorter than the length of body (less than 92 %), dark on top, and white on the bottom (Hoffmeister, 1986; Osgood, 1900; Williams et al., 1993a). Across its range, it shows wide variation in coloration, depending on that of the local substrate and indirectly on amount of annual precipitation. The skull is short and wide, and the nasal passages are long (Hall, 1981). It is slightly larger than *P. flavus* and *P. merriami*, the tail and feet are relatively larger, the pelage is paler, and it has more interspersed black hairs in the back (Findley, 1987); in addition, the ears are relatively smaller (Anderson, 1972). It differs from other species of *Perognathus* in cranial features such as greater length of the skull and nasals (Anderson, 1972; Hoffmeister, 1986).

DISTRIBUTION: *P. flavescens* is a widely distributed species in the Great Plains and prairies of the United States from North Dakota into New Mexico. In Mexico, it has a marginal distribution in Chihuahua, where it is restricted to two small areas in sand dunes and scrublands of Samalayuca and pasturelands and shrublands in Janos-Casas Grandes (Pacheco et al., 1999; Patton, 1993a; Williams, 1978a; Williams et al., 1993b). It has been recorded in CH.

EXTERNAL MEASURES AND WEIGHT
TL = 113 to 154 mm; TV = 56 to 73 mm; HF = 15 to 21 mm; EAR = 6.5 mm.
Weight: 8 to 11 g.

DENTAL FORMULA: I 1/1, C 0/0, PM 1/1, M 3/3 = 20.

NATURAL HISTORY AND ECOLOGY: Plains pocket mice are nocturnal. The diet primarily is seeds of a variety of species, which are transported in the cheek pouches for storage and later consumption. Insects and green plants also are part of the diet (Wilson and Ruff, 1999). They rarely drink water, as they obtain water from the food they eat. Burrows are small and built under shrubs. During the day, they cover the main entrance and leave less-conspicuous openings unplugged. Presumably, this is done to maintain a lower temperature and higher humidity in the burrow compared to on the surface of the ground. The size of the home range is 0.04 ha. Reproduction takes place in April–July, with four to five young per litter. They probably mate twice a year (Burt and Grossenheider, 1976; Findley, 1987). When it occurs sympatrically with *P. flavus*, *P. flavescens* usually occupies sandy soils (Findley, 1987). This species lives in dry environments with sandy soils or dunes. Apparently, the nature of the substrate is more important than vegetation, which consists of grasslands, ephedrines (*Ephedra*), shrubs (*Chrysothamnus*, *Gutierrezia*), and yuccas (*Yucca*; Hoffmeister, 1986). Sometimes the ground where it occurs is completely bare (Hoffmeister, 1986).

VEGETATIONAL ASSOCIATIONS AND ELEVATION RANGE: The plains pocket mouse mainly inhabits arid grasslands and xeric scrublands; that is, arid, semi-arid, and open areas with scattered vegetation and sandy soils (Anderson, 1972; Findley, 1987; Williams et al., 1993b). Occasionally, it may occur in pine forests (Hoffmeister, 1986). In Chihuahua, it inhabits sand dunes and scrublands in Samalayuca, as well as grasslands and scrublands in Janos-Casas Grandes at 1,000 masl to 1,350 masl (Anderson, 1975; Pacheco et al., 1999).

CONSERVATION STATUS: Status of this species is unknown. In Mexico, it probably is fragile or vulnerable because it has a marginal distribution that is restricted to a small region in northwestern Chihuahua (Ceballos et al., 1998; Pacheco et al., 2000). An assessment of its status is needed to determine current conservation needs.

Perognathus flavus Baird, 1855

Silky pocket mouse

Gisselle A. Oliva Valdés

SUBSPECIES IN MEXICO

Perognathus flavus flavus Baird, 1855
Perognathus flavus fuscus Anderson, 1972
Perognathus flavus medius Baker, 1954
Perognathus flavus mexicanus Merriam, 1894
Perognathus flavus pallescens R.H. Baker, 1954
Perognathus flavus parviceps R.H. Baker, 1954
Perognathus flavus sonoriensis Nelson and Goldman, 1934

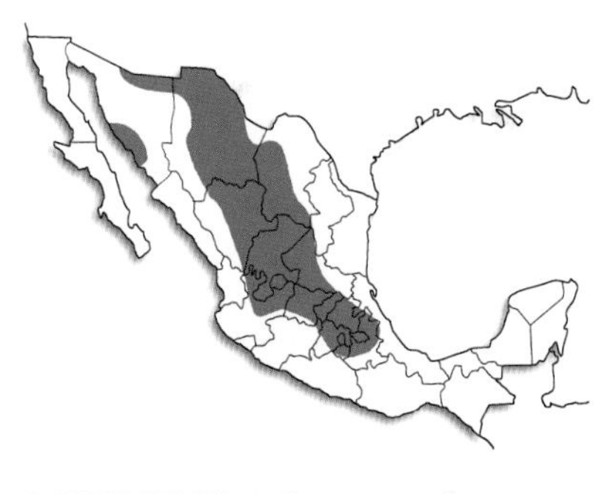

DISTRIBUTION: *P. flavus* occurs from Wyoming in the United States to Mexico, where it occurs across the Mexican Plateau to Jalisco, Morelos, Puebla, and a small area in east-central Veracruz (Anderson, 1972; Hall, 1981; Matson and Baker, 1983; Ramírez Pulido et al., 1983; Sanchez-Hernandez et al., 1999; Wilson, 1973c). It has been recorded in AG, CH, CO, DF, DU, GJ, HG, JA, MI, MO, MX, NL, PU, QE, SL, SO, TA, TL, VE, and ZA.

DESCRIPTION: *Perognathus flavus* is the smallest species of the family Heteromyidae in Mexico. It has a pair of external, fur-lined cheek pouches. The back has two fine black lines that are distinct from the ochre coloration; some subspecies may also have yellow and pink hairs. The lateral stripe is buffy and fades in the post-auricular area, where the spots become paler. The underside is completely white. The pelage is smooth and the soles of the hind feet are somewhat furry. The tail is sparsely haired, white at the tip and gray at the base. Auditory bullae are medium sized. The interparietal bone is small, pentagonal, and symmetrical (length equal to width), the face is thin, and the interorbital space is compressed (Ceballos and Galindo, 1984; Findley, 1987; Hall, 1981; Hoffmeister, 1986; Osgood, 1900).

EXTERNAL MEASURES AND WEIGHT

TL = 100 to 122 mm; TV = 44 to 60 mm; HF = 15 to 18 mm; EAR = 5 to 6 mm.
Weight: 6 to 9 g.

DENTAL FORMULA: I 1/1, C 0/0, PM 1/1, M 3/3 = 20.

Perognathus flavus. Grassland. Janos Biosphere Reserve, Chihuahua. Photo: Gerardo Ceballos.

NATURAL HISTORY AND ECOLOGY: These animals are nocturnal. Activities are conducted among patches of vegetation and near rocks. They carry seeds and green vegetation in their cheek pouches. They build burrows at the foot of shrubs and trees or between cracks in rocks. They mainly feed on seeds, but also include vegetation and some insects. Predators are ringtails (*Bassariscus astutus*), long-tailed weasels (*Mustela frenata*), and owls (*Bubo virginianus, Tyto alba*). They reproduce during March–August. Each female has two or more litters per year. Gestation period is unknown, but probably is 28 to 32 days. Each litter consists of two to six offspring (Baker, 1954, 1956; Ceballos and Galindo, 1984).

VEGETATIONAL ASSOCIATIONS AND ELEVATION RANGE: The silky pocket mouse inhabits areas with xeric vegetation, including scrublands and grasslands. It also occurs in agricultural crops and in areas devoid of vegetation; they rarely occur in rocky areas or where vegetation is dense. They tolerate a wide range of conditions, selecting open spaces (Baker, 1954, 1956; Ceballos and Galindo, 1984). Elevations occupied range from 975 m in Coahuila to 2,400 m in Puebla.

CONSERVATION STATUS: Silky pocket mice are abundant, with a wide distribution and a great tolerance to anthropogenic disturbances, so they are not considered at risk of extinction.

Perognathus longimembris (Coues, 1875)

Little pocket mouse

Jaime Luévano and Eric Mellink

SUBSPECIES IN MEXICO

Perognathus longimembris aestivus Huey, 1928
Perognathus longimembris bombycinus Osgood, 1907

Perognathus longimembris internationalis Huey, 1939
Perognathus longimembris kinoensis Huey, 1935
Perognathus longimembris venustus Huey, 1930

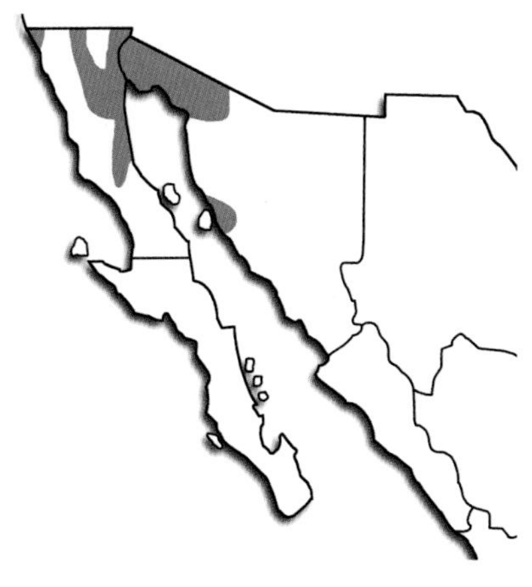

DISTRIBUTION: *P. longimembris* occurs in the southwestern United States and northwestern Mexico. In northwestern and central Baja California, it is known only from small isolated localities in the valleys of Jacumé, San Rafael (now Ojos Negros), La Trinidad, and San Augustín. It also occurs on the plains of the Colorado River, south of San Felipe, and in part of the Great Desert (Sonora) into El Pinacate. The subspecies *P. longimembris kinoensis* occupies coastal Sonora from Puerto Lobos to Estero Tastiota near Guaymas (Caire, in press; Hall, 1981; Huey, 1964). It has been recorded in BC and SO.

DESCRIPTION: As its common name indicates, *Perognathus longimembris* is a small mouse, with a pink-creamy coloration. The tail is long, greater than the length of body and head, bicolored or pale, hairier on the distal one-third, and with a tuft of hair at the tip. The posterior one-third of the soles of the hind feet is hairy. The baculum (bone in the penis) has a small hook on its distal part. It has 56 pairs of chromosomes (Hall, 1981; Hoffmeister, 1986; Jameson and Peeters, 1988).

EXTERNAL MEASURES AND WEIGHT
TL = 110 to 115 mm; TV = 53 to 83 mm; HF = 15 to 20 mm; EAR = 5 to 7 mm.
Weight: 7 to 10 g.

DENTAL FORMULA: I 1/1, C 0/0, PM 1/1, M 3/3 = 20.

NATURAL HISTORY AND ECOLOGY: The little pocket mouse inhabits places with sandy or stony soil, where it builds burrows. It shows greater activity during spring, two to five hours after sunset, with less activity immediately before dawn; it is affected by moonlight. During the day, it plugs entrances to the burrow with soil. When food is scarce or when the temperature is low, they become lethargic. During winter, they lower activity and body temperature, and they consume seeds stored during summer and autumn (Brylski, 1990g; Hoffmeister, 1986; Jameson and Peeters, 1988). The diet consists mainly of seeds of desert plants, and contents of cheek pouches have included seeds of plantains (*Plantago*), goosefoots (*Chenopodium*), desert trumpets (*Eriogonum inflatum*), and fescues (*Festuca*). They also consume green vegetation and insects. They search for seeds under shrubs. Temperature, amount of food, and plant phenology influence reproduction, which occurs January–August with peaks during March and May. Gestation is 21 to 31 days, with a litter of 2 to 8 offspring per year. Young are weaned 30 days after birth. Females reach sexual maturity at 50 days old and males at 150 days (Brylski, 1990g; Hoffmeister, 1986; Jameson and Peeters, 1988).

VEGETATIONAL ASSOCIATIONS AND ELEVATION RANGE: This species inhabits plant communities, including desert scrublands in Sonora and central and northwestern Baja California, and grasslands in the intermountain valleys of northern Baja California. In Mexico, it occurs from sea level to 750 m.

CONSERVATION STATUS: Despite the fact that much of its range in northwestern and central Baja California are now agricultural areas, the rest of its geographic range is little changed. Therefore, it is not considered at risk of extinction.

Perognathus merriami J.A. Allen, 1892

Merriam's pocket mouse

Astrid Frisch Jordán and Héctor T. Arita

SUBSPECIES IN MEXICO
Perognathus merriami merriami J.A. Allen, 1892
Perognathus merriami gilvus Osgood, 1900

P. merriami is similar to *P. flavus* (Mayr, 1978) and phylogenetically they are closely related (Steyskal, 1972). In fact, *P. merriami* was considered a subspecies of *P. flavus* until Lee and Engstrom (1991) conducted a genetic study, which revealed two distinct species.

DESCRIPTION: *Perognathus merriami* is a small-sized rodent. The pelage of Merriam's pocket mouse is soft and brownish-yellow or yellowish-pink, the belly is whitish, and the tail is yellow dorsally and white beneath. It has conspicuous yellowish post-auricular spots and white subauricular spots. Juveniles are gray (Anderson, 1972; Hall, 1981; MacMahon, 1985). It differs from *P. flavus* by its slightly paler coloration and less contrast between dorsal and lateral coloration. The tendency to develop a dark dorsal stripe is not seen in *P. merriami*. Compared to *P. merriami*, *P. flavus* has auditory bullae that are smaller, the skull is narrower, the interparietal bone is wider, the interorbital space is larger (4.80 mm versus 4.25 mm), and probably the width of the mastoid is larger (Anderson, 1972). The hind feet are longer than the forefeet, which have long nails for digging.

DISTRIBUTION: *P. merriami* occurs from New Mexico and Texas in the United States into northern Mexico, from Chihuahua into Tamaulipas (Hall, 1981; Wilson and Ruff, 1999). It has been recorded in CH, CO, NL, and TA.

EXTERNAL MEASURES AND WEIGHT
TL = 100 to 122 mm; TV = 44 to 60 mm; HF = 15 to 18 mm; EAR = 5 to 6 mm.
Weight: 7 to 10 g.

DENTAL FORMULA: I 1/1, C 0/0, PM 1/1, M 3/3 = 20.

NATURAL HISTORY AND ECOLOGY: *P. merriami* is a digging and nocturnal rodent that only leaves its burrow to mate and to collect seeds that are carried in its cheek pouches. Although it is granivorous, it also consumes insects and plant material. It rarely drinks water, as it relies on metabolic water and that obtained from its food (Nowak, 1999b). These mice primarily consume seeds of *Salsola*, *Chenopodium*, *Festuca*, *Cryptantha*, *Amaranthus*, *Opuntia*, *Oryzopsis*, and *Sphaeralcea* (Forbes, 1962). Burrows consist of nest chambers, tunnels for storage of food, and places to deposit feces. In addition, they build several escape burrows scattered throughout the home range. These escape burrows are simple tunnels without an exit that are used to escape from predators (MacMahon, 1985; Nowak, 1999b). Various openings to burrows, which generally are hidden under shrubs, are plugged with soil to maintain low temperature and high humidity inside the burrow. Burrows usually are at bases of vegetation, probably because the root system gives stability to the opening and so that the scattered soil does not make the opening conspicuous to predators (Chapman and Packard, 1974; Nowak, 1999b). During winter, they enter torpor for up to 48 hours but remain active and consume food at other times. In Texas, reproduction occurs in April–November and each female produces more than two litters (Chapman and Packard, 1974). In general, members of this genus produce one or more litters per year, each one with two to seven offspring and four on average. Population density can vary widely, with an average of up to 10/ha (Britt, 1972; Findley, 1975).

VEGETATIONAL ASSOCIATIONS AND ELEVATION RANGE: This species inhabits grasslands, arid plains, and deserts (Findley, 1975; Lee and Engstrom, 1993; MacMahon, 1985). It occurs from 54 masl to 1,181 masl (Álvarez, 1963a).

CONSERVATION STATUS: Current status in Mexico is unknown. It has a wide distribution, which includes extensive regions with little disturbance by humans, so it probably is not at risk.

Family Geomyidae

Gerardo Ceballos and Mark S. Hafner

Pocket gophers are native to North America and reach their highest diversity in Mexico, where they are known from every state in the Republic. Representative species of all 6 extant genera and 19 of the 38 extant species of geomyids (Patton, 2005) occur in Mexico, and 1 of the genera (*Zygogeomys*) and 12 of the 19 Mexican species are endemic to Mexico. Most lines of evidence suggest that pocket gophers originated in Mexico and underwent a rapid phyletic radiation approximately 5 million years ago (Hafner et al., 2005; Spradling et al., 2004). As the only fossorial herbivore in North America, pocket gophers are agricultural pests over much of their geographic range. Their distribution, however, is typically patchy and local populations are usually small (< 20 individuals), even in areas of seemingly homogeneous habitat.

All pocket gophers share a large number of morphological, ecological, and behavioral characteristics, so it is not necessary to repeat these characteristics in each of the following species accounts. Morphologically, all pocket gophers have a fusiform body, large head with no discernible neck, robust claws on the forefeet, and loose skin. Their eyes and pinnae (external ears) are small, and all species have a pair of fur-lined cheek pouches that open adjacent to, but independent of, the mouth. These cheek pouches are used to transport food, not dirt. The skulls of pocket gophers are heavily built, angular, broad, and somewhat flattened. The lips can be closed behind the large and powerful incisors to prevent entry of soil when the incisors are used for digging. The tail is short and stout, often nearly hairless.

Most species of pocket gopher can be active at any time of the day or night, but most show peaks of activity in the early morning, especially in hot climates. Pocket gophers dig extensive systems of tunnels and galleries often exceeding 100 m in total length. The loose earth that accumulates as they dig these tunnels occasionally is pushed to the surface, forming the scattered network of earthen mounds characteristic of pocket gophers. Tunnel entrances always are plugged when not in use, and most species are quick to plug collapsed burrows or other openings into their burrow systems. These systems usually have one or two main tunnels with many lateral branches extending several meters in multiple directions. Many of the lateral branches are only 1 cm to 5 cm beneath the ground and are used for harvesting plant roots, tubers, bulbs, and rhizomes ("feeding tunnels"). The main tunnels are deeper (sometimes as deep as 2 m, if soil depth permits) and are used for food storage and nesting (Baker, 1956; Davidow-Henry et al., 1989; Hickman, 1977). Most species of pocket gophers occasionally exit their burrow systems to eat surface vegetation in close proximity to the burrow entrance. Others, such as the Michoacán pocket gopher (*Zygogeomys trichopus*), are thought to rarely, if ever, leave their burrow systems. All species are believed to disperse above ground, although evidence of above-ground dispersal in most species is anecdotal. All pocket gophers require non-flooding soils that are deep enough (usually 1 m or deeper) to provide protection from digging predators (Baker, 1956; Best, 1973; Hafner et al., 2005; Russell, 1968). Considering that more than 60 species of vertebrates and invertebrates are known to be facultative or obligate occupants of pocket gopher burrow systems (Cervantes et al., 1993; Hafner et al., 2005; Hubbell and Goff, 1940), pocket gophers play an important role in ecosystem diversity (Turner et al., 1973).

Yellow-faced pocket gopher

Iván Castro Arellano

SUBSPECIES IN MEXICO

Cratogeomys castanops castanops (Baird, 1852)
Cratogeomys castanops consitus Nelson and Goldman, 1934
This species was recently revised by Hafner et al. (2008), who reduced the number of subspecies in *C. castanops* from a total of 26 to the 2 subspecies listed above.

DISTRIBUTION: This species is patchily distributed from southeastern Colorado and southwestern Kansas in the United States south into the high-elevation desert grasslands of the Chihuahuan desert to the Sierra Guadalupe, Sierra Parras, and Mayrán Basin of southern Coahuila, and across the Río Nazas and west of the Río Aguanaval in Durango to the northeastern slopes of the Sierra de Yerbanís, Durango (Hafner et al., 2008). *C. castanops* is known from the states of CH, CO, DU, NL, and TA.

DESCRIPTION: *Cratogeomys castanops* is medium-sized for the genus, and sexual dimorphism in body size is pronounced (150 g to 300 g, adult females; 200 g to 450 g, adult males). As in other pocket gophers, body size is heavily influenced by local climate, with smaller animals in drier regions with sparse vegetation and larger animals in wetter regions with denser vegetation (Hafner et al., 2008). Dorsal coloration varies from pale yellowish-cream to dark reddish-brown with a mixture of black guard hairs. The abdomen varies from white to a mixture of shiny ochre-cream. The eyes are relatively large compared with those of *Geomys* and *Thomomys* (Davidow-Henry et al., 1989; Russell, 1968). *C. castanops* is larger than its sister species, *C. goldmani*, in all cranial dimensions, with relatively longer palatines and nasals and relatively narrower auditory bullae (Hafner et al., 2008). As in all members of the genus *Cratogeomys*, the anterior surface of each upper incisor contains a single, longitudinal groove. The diploid number of *C. castanops* is 2n = 46 and the fundamental number is FN = 86 (Lee and Baker, 1987).

EXTERNAL MEASURES AND WEIGHT

TL = 144 to 305 mm; TV = 50 to 105 mm; HF = 27 to 42 mm; EAR = 5 to 7 mm.
Weight = 150 to 450 g.

DENTAL FORMULA: I 1/1, C 0/0, PM 1/1, M 3/3 = 20.

NATURAL HISTORY AND ECOLOGY: The natural history of *C. castanops* is typical of most pocket gophers (see chapter introduction). Two periods of reproduction have been documented for *C. castanops* in Coahuila, one in summer and another in winter (Baker, 1956). In other parts of their range, individuals of *C. castanops* reproduce throughout the year (Hedgal et al., 1965; Smolen et al., 1980). The number of offspring varies from one to five, and the young are born blind and hairless (Davidow-Henry et al., 1989). Hawks, owls, and carnivorous mammals are known predators of *C. castanops* (Davidow-Henry et al., 1989).

VEGETATIONAL ASSOCIATIONS AND ELEVATION RANGE: This species of pocket gopher inhabits a wide range of habitats, including grasslands, deserts, and agricultural fields. Typical vegetation associated with these environments includes *Prosopis juliflora*, *Florensia* sp., *Larrea tridentata*, and *Microrhamnus ericoides* (Davidow-Henry et al., 1989). The elevational range of *C. castanops* extends from approximately 10 m (near the mouth of the Río Bravo) to 2,150 m (Hafner et al., 2008).

CONSERVATION STATUS: *C. castanops* is abundant over much of its geographic range and, as a result, is not included in the list of threatened or endangered Mexican mammals (SEMARNAT, 2010). These pocket gophers are considered local pests in parts of their range because they often live in agricultural fields, where they can cause considerable damage to the underground parts of many crops (Davidow-Henry et al., 1989).

Oriental Basin pocket gopher

Mark S. Hafner

SUBSPECIES IN MEXICO

Cratogeomys fulvescens is a monotypic species.

The systematic revision of *C. merriami* by Hafner et al. (2005) divided *C. merriami* into three species: *C. merriami*, *C. fulvescens*, and *C. perotensis*.

DESCRIPTION: *Cratogeomys fulvescens* is medium-sized for the genus, and sexual dimorphism in body size is pronounced (250 g to 350 g, adult females; 250 g to 550 g, adult males). Pelage coloration distinguishes *C. fulvescens* from all nearby geomyid species. Dorsally, *C. fulvescens* has a grizzled yellowish-brown appearance with a strong mixture of black-tipped hairs imparting a "salt-and-pepper" appearance (Russell, 1968). The underparts are similar in color, but much paler. The skull is relatively small for the genus (cranial width usually < 26 mm in adults), and the anterior edge of the jugal is broad (usually > 2 mm) compared to those of *C. merriami* and *C. perotensis* (Hafner et al., 2005). The anterior surface of each upper incisor contains a single, longitudinal groove. The diploid number of *C. fulvescens* is 2n = 40, and the fundamental number is FN = 72 (Hafner et al., 2005).

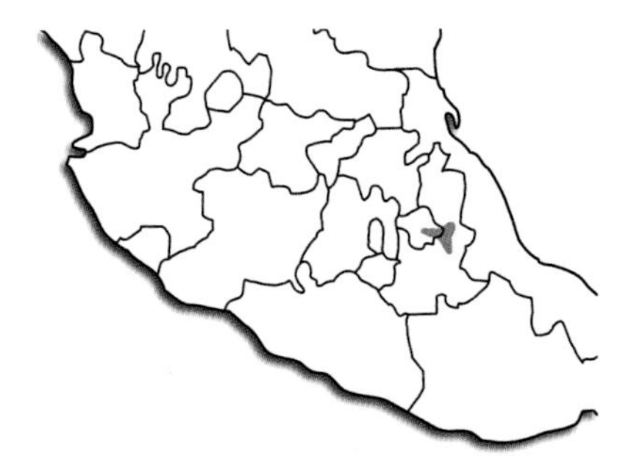

DISTRIBUTION: *C. fulvescens* is endemic to Mexico. This species is patchily distributed around the Oriental Basin of eastern Puebla from Esperanza, Puebla, north to Perote, Veracruz, and west to the base of Volcán La Malinche in Tlaxcala. *C. fulvescens* is known from the states of PU, TL, and VE.

EXTERNAL MEASURES AND WEIGHT

TL = 290 to 350 mm; TV = 94 to 112 mm; HF = 36 to 39 mm; EAR = 6 to 8 mm.
Weight = 250 to 550 g.

DENTAL FORMULA: I 1/1, C 0/0, PM 1/1, M 3/3 = 20.

NATURAL HISTORY AND ECOLOGY: *C. fulvescens* inhabits dry, sandy soils in the Oriental Basin (Cuenca Oriental), where it is surrounded on the north and east by *C. perotensis* and on the west by *C. merriami*. *C. fulvescens* and *C. merriami* have not been found in contact, although their ranges approach each other north of the city of Puebla and west of Volcán La Malinche. Where *C. fulvescens* and *C. perotensis* are in contact near the town of Perote, Veracruz, these similar-sized species show no obvious habitat specialization. This locality, however, is the lowest known elevation for *C. perotensis* and among the highest known capture sites for *C. fulvescens*, so the two species may show elevation-related specializations. Specimens of *C. fulvescens* have been collected in agricultural fields near Ciudad Serdán (M.S. Hafner, pers. obs.), and it is possible that they are regarded as agricultural pests in parts of their range. Little is known of the breeding biology of *C. fulvescens*, except that pregnant females have been captured in December through February (M.S. Hafner, pers. obs.).

VEGETATIONAL ASSOCIATIONS AND ELEVATION RANGE: *C. fulvescens* occurs in the dry, sparsely vegetated desert habitats surrounding the Oriental Basin. These areas are dominated by alkaline-tolerant plants, including *Distichlis spicata*, *Actinella chrysanthemoides*, *Atriplex pueblensis*, and *Bouteloua breviseta*. Cacti (*Mammillaria* sp.), prickly pear (*Opuntia* sp.), and agave (*Agave obscura*) also are common in this region. *C. fulvescens* also can be found in agricultural fields surrounding the Oriental Basin, where they feed on plant roots, bulbs, and rhizomes. The elevational range of *C. fulvescens* extends from approximately 2,300 m to 2,700 m.

CONSERVATION STATUS: *C. fulvescens* (previously recognized as *C. merriami fulvescens*) is not included in the list of threatened or endangered Mexican mammals (SEMARNAT, 2010). Because the geographic range of this species is small relative to those of other pocket gophers of Mexico, the conservation status of *C. fulvescens* should be investigated in the near future.

Cratogeomys fumosus (Merriam, 1892)

Smoky pocket gopher

Mark S. Hafner and Gerardo Ceballos

SUBSPECIES IN MEXICO

Cratogeomys fumosus angustirostris (Merriam, 1903)
Cratogeomys fumosus fumosus (Merriam, 1892)
Cratogeomys fumosus imparilis (Goldman, 1939)
Cratogeomys fumosus tylorhinus (Merriam, 1903)

Hafner et al. (2004) revised pocket gophers of the *C. gymnurus* species group (now known as the *C. fumosus*-group) and synonymized *C. gymnurus*, *C. neglectus*, *C. tylorhinus*, and *C. zinseri* under *C. fumosus*. This new taxonomy was suggested in earlier research by Honeycutt and Williams (1982), DeWalt et al. (1993), Monterrubio et al. (2000), and Demastes et al. (2003).

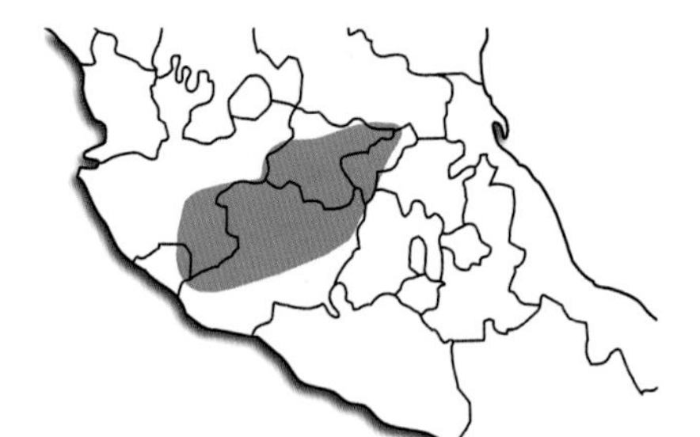

DISTRIBUTION: *C. fumosus* is endemic to Mexico. This species is patchily distributed over widespread regions of the Trans-Mexican Volcanic Belt and Mesa Central. The large geographic range of *C. fumosus* extends from northeastern Querétero southward to the southern edges of the Valle de México, and from eastern regions of the state of Mexico westward to the eastern slopes of the Sierra Madre del Sur in Jalisco and Colima. *C. fumosus* is known from the states of CO, DF, GJ, HG, JA, MI, MX, and QE.

DESCRIPTION: *Cratogeomys fumosus* now includes four taxa previously recognized as separate species: *C. gymnurus*, *C. neglectus*, *C. tylorhinus*, and *C. zinseri* (Hafner et al., 2004; Honeycutt and Williams, 1982). *C. fumosus* is extremely variable in body size, but most individuals are large for the genus (250 g to 350 g, adult females; 300 g to 1,030 g, adult males). Sexual dimorphism in body size is pronounced, and a 1,030 g male captured 8 km east of Opopeo, Michoacán, is the largest pocket gopher ever recorded from Mexico (M.S. Hafner, pers. obs.). Pelage coloration in *C. fumosus* is equally variable, ranging from light brown to almost black dorsally and lighter ventrally. The English common name, smoky pocket gopher, applies only to certain individuals of the subspecies *C. fumosus fumosus* near the city of Colima, which have dorsal pelage that is unusually dark (smoky gray) for the species. The occipital region of the skull of *C. fumosus* is flattened, and the paraoccipital processes are usually flat and broad when viewed from the ventral perspective (Hafner et al., 2004). The anterior surface of each upper incisor contains a single, longitudinal groove. The diploid number of *C. fumosus* is 2n = 40, and the fundamental number is FN = 76 (Berry and Baker, 1972; León Paniagua et al., 2001). Lee and Baker (1987) reported a diploid number of 2n = 38 for a specimen of *C. gymnurus* (now *C. fumosus*), which appears to be an error (Hafner et al., 2004).

EXTERNAL MEASURES AND WEIGHT

TL = 212 to 375 mm; TV = 75 to 123 mm; HF = 35 to 59 mm; EAR = 5 to 10 mm. Weight = 250 to 1,030 g.

DENTAL FORMULA: I 1/1, C 0/0, PM 1/1, M 3/3 = 20.

NATURAL HISTORY AND ECOLOGY: Near Pátzcuaro, Michoacán, *C. fumosus* occurs in habitats also occupied by *Thomomys umbrinus* and, occasionally, *Zygogeomys trichopus*.

Cratogeomys fumosus. Sugar cane crop. Coquimatlán, Colima. Photo: Gerardo Ceballos.

In these areas of contact, specimens of *C. fumosus* reach their largest size (> 1,000 g), whereas *Thomomys* individuals are small for the genus (< 70 g) and *Zygogeomys* individuals are intermediate in size (ca. 500 g). Although average burrow depths have not been quantified rigorously, preliminary observations suggest that the three genera tunnel at different depths in the soil, with *Thomomys* tunneling near the surface (ca. 10 cm deep), *Zygogeomys* at intermediate depths (20 cm to 30 cm deep), and *Cratogeomys* still deeper (> 30 cm deep). There is no evidence that the tunnels of these species connect. Pregnant females of *C. fumosus* have been captured in February, April, and July (M.S. Hafner, pers. obs.), and it is likely that this species breeds opportunistically throughout the year. Litter size varies from one to five young (Davidow-Henry et al., 1989). Natural predators of *C. fumosus* include foxes, badgers, coyotes, large raptors, and snakes (*Pituophis deppei* and *Crotalus aquilus*; López-Forment, 1968; Monterrubio, 1995; Sosa, 1981).

VEGETATIONAL ASSOCIATIONS AND ELEVATION RANGE: *C. fumosus* occurs in a wide variety of habitats from low-elevation deciduous forests to high-elevation pine-oak woodlands. Most populations at higher elevations are found in small clearings and other open areas of the forest with deep soils of volcanic origin. Densely forested areas and areas with shallow or rocky soils are avoided. Lowland populations are found in arid habitats with sandy soils supporting a thin cover of grasses and shrubs. *C. fumosus* is relatively common in agricultural fields, including alfalfa, beans, sorghum, corn, and sugar cane, where they feed on plant roots, bulbs, and rhizomes. Populations of *C. fumosus* near Pinal de Amoles, Querétaro, occur in apple orchards. The elevational range of *C. fumosus* extends from approximately 300 m near the city of Colima to more than 3,000 m in the Trans-Mexican Volcanic Belt.

CONSERVATION STATUS: Populations of *C. fumosus* near the city of Colima (now restricted to the subspecies *C. fumosus fumosus*; Hafner et al., 2004) are listed as "threatened" in the Mexican Endangered Species Act (SEMARNAT, 2010). This subspecies is considered at risk because its already limited distribution continues to decline due to expanding urbanization in the region (Arita and Ceballos, 1997). A recent assessment revealed that a large portion of *C. fumosus fumosus* habitat near the city of Colima has been destroyed by human settlements, livestock grazing, and introduction of agricultural crops, including sugar cane, sorghum, and corn (G. Ceballos, pers. obs.; Téllez-Girón et al., 1997). Similarly, the isolated population of *C. fumosus* near Pinal de Amoles, Querétaro (formerly recognized as *C. neglectus*; Demastes et al., 2003; Hafner et al., 2004), is also listed as "threatened" in the Mexican Endangered Species Act (SEMARNAT, 2010). Continued expansion of agriculture in this region may lead to the extirpation of *C. fumosus* in this portion of Querétaro (Arita and Ceballos, 1997; León Paniagua et al., 2001; Monterrubio, 1995).

Cratogeomys goldmani Merriam, 1895

Goldman's yellow-faced pocket gopher

Joaquín Arroyo-Cabrales and Robert J. Baker

SUBSPECIES IN MEXICO

Cratogeomys goldmani goldmani Merriam, 1895 (in part)
Cratogeomys goldmani subnubilus Nelson and Goldman, 1934 (in part)
This species was recently revised by Hafner et al. (2008), who provided a formal description of *C. goldmani* and recognized the above two subspecies.

DISTRIBUTION: *C. goldmani* is endemic to Mexico. This species is patchily distributed throughout the arid, high-elevation Mesa del Norte south of the Mayrán Basin, Sierra Parras, and Sierra Guadalupe of southern Coahuila from the drainage of the Río Aguanaval (Coahuila, Durango, and Zacatecas) to the western flanks of the Sierra Madre Oriental, and south to the Río Verde in San Luis Potosí (Hafner et al., 2008). *C. goldmani* is known from the states of CO, DU, NL, SL, TA, and ZA.

DESCRIPTION: *Cratogeomys goldmani* shows extreme variation in body size (150 g to 300 g, adult females; 150 g to 400 g, adult males), and the body size of most specimens is medium to small for the genus. As in other species of *Cratogeomys*, body size in *C. goldmani* is influenced heavily by local climate, with smaller animals in drier regions and larger animals in wetter regions (Hafner et al., 2008). Dorsal coloration is yellowish-brown or dark reddish-brown, generally with more ochraceous coloration than in the sister species, *C. castanops*. Belly coloration varies from white to cream. As noted by Dalquest (1953) and Russell (1968), populations of *C. goldmani* from eastern Zacatecas and western San Luis Potosí show a high incidence (89%; Hafner et al., 2008) of white spots on the belly, sides, and rump. Compared to *C. castanops*, *C. goldmani* is smaller in all dimensions of the cranium and has relatively shorter palatines and nasals and relatively broader auditory bulla. The anterior surface of each upper incisor contains a single, longitudinal groove. The diploid number of *C. goldmani* is 2n = 42, and the fundamental number is FN = 78 (Lee and Baker, 1987).

EXTERNAL MEASURES AND WEIGHT

TL = 178 to 278 mm; TV = 55 to 90 mm; HF = 29 to 38 mm; EAR = 5 to 6 mm.
Weight = 150 to 400 g.

DENTAL FORMULA: I 1/1, C 0/0, PM 1/1, M 3/3 = 20.

NATURAL HISTORY AND ECOLOGY: The natural history and ecology of *C. goldmani* are very similar to those of *C. castanops*. As with all other species of pocket gophers, *C. goldmani* requires relatively deep (1 m to 2 m) soils that do not flood (Baker, 1956; Best, 1973). *C. goldmani* and *C. castanops* likely occur in sympatry (current or intermittent) along the lower reaches of the Río Aguanaval between La Unión, Durango, and La Flor de Jimulco, Coahuila, where they have been collected within a few kilometers of each other with no intervening barrier (Hafner et al., 2008). Pregnant females have been collected in April (M.S. Hafner, pers. obs.) and August (Matson and Baker, 1986).

Cratogeomys goldmani. Scrubland. Santo Domingo, San Luis Potosí. Photo: Juan Cruzado.

VEGETATIONAL ASSOCIATIONS AND ELEVATION RANGE: *C. goldmani* inhabits arid and semi-arid areas covered with sandy soil or sandy loam with desert vegetation consisting of xeric-adapted shrubs and grasses (Russell, 1968). The elevational range of *C. goldmani* extends from approximately 750 m to 2,650 m (Hafner et al., 2008).

CONSERVATION STATUS: *C. goldmani* is abundant over much of its geographic range and, as a result, is not included in the list of threatened or endangered Mexican mammals (SEMARNAT, 2010).

Cratogeomys merriami (Thomas, 1893)

Merriam's pocket gopher

Gisselle Oliva Valdés and Gerardo Ceballos

SUBSPECIES IN MEXICO

C. merriami is a monotypic species.
The systematic revision of *C. merriami* by Hafner et al. (2005) divided *C. merriami* into three species, *C. merriami*, *C. fulvescens*, and *C. perotensis*.

DESCRIPTION: *Cratogeomys merriami* is relatively large for the genus (450 g to 700 g, adults), and sexual dimorphism in body size is less pronounced in *C. merriami* than in most other pocket gophers, with males being only slightly larger than females (Ceballos and Galindo, 1986; Hall, 1981). Cranial width is usually > 26 mm in adults. Fur color is highly variable, from dull chestnut brown to slate black dorsally. Ventral coloration is similar to dorsal coloration, but paler. The anterior surface of each up-

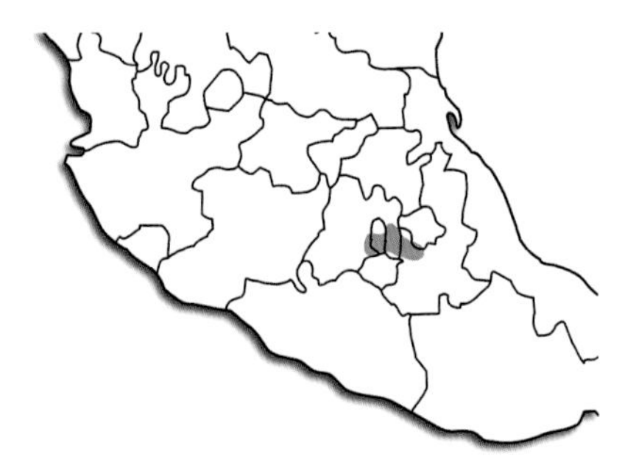

DISTRIBUTION: *C. merriami* is endemic to Mexico. This species is patchily distributed in the southern part of the Valley of Mexico, Sierra de Las Cruces, Sierra de Ajusco, and Mt. Popocatépetl, Mt. Iztaccíhutl, and from Lerma at the eastern end of the Valley of Toluca eastward into western Puebla (Mt. Popocatépetl; Hafner et al, 2005). *C. merriami* is known from the states of DF, MO, MX, and PU.

per incisor contains a single, longitudinal groove. The diploid number of *C. merriami* is 2n = 36, and the fundamental number is FN = 68 (Hafner et al., 2005).

EXTERNAL MEASURES AND WEIGHT
TL = 320 to 360 mm; TV = 100 to 120 mm; HF = 42 to 53 mm; EAR = 6 to 10 mm. Weight = 450 to 700 g.

DENTAL FORMULA: I 1/1, C 0/0, PM 1/1, M 3/3 = 20.

NATURAL HISTORY AND ECOLOGY: The natural history of *C. merriami* is much like that of other large-bodied pocket gophers. The soil mounds they produce are large (approaching 1 m in diameter) because of the large diameter (80 mm to 120 mm) of the tunnels they dig and the expansiveness of their burrow systems (often encompassing 400 m^2 of surface area). *C. merriami* appears to reproduce throughout the year, and each female can have several litters with one to three offspring each. Peak reproduction occurs in late fall and early spring. Pocket gophers of this species are eaten by weasels, foxes, badgers, and larger birds of prey. It has been reported that *C. merriami* is eaten by local people in certain parts of its range (Ceballos and Galindo, 1984; Villa-R., 1986).

VEGETATIONAL ASSOCIATIONS AND ELEVATION RANGE: *C. merriami* inhabits grasslands at lower elevations and clearings in oak, pine, and fir woodlands at higher elevations. This species often is abundant in agricultural fields, where it is considered a pest. Elevational range is approximately 2,000 m to 4,000 m (Ceballos and Galindo, 1984; Villa-R., 1986).

CONSERVATION STATUS: *C. merriami* is abundant over much of its geographic range and, as a result, is not included in the list of threatened or endangered Mexican mammals (SEMARNAT, 2010).

Cratogeomys merriami. Croplands, Mexico D.F. Photo: Vinicio Sosa.

Cofre de Perote pocket gopher

Mark S. Hafner

SUBSPECIES IN MEXICO

C. perotensis is a monotypic species.

The systematic revision of *C. merriami* by Hafner et al. (2005) divided *C. merriami* into three species, *C. merriami*, *C. fulvescens*, and *C. perotensis*.

DESCRIPTION: *Cratogeomys perotensis* is medium size for the genus (400 g to 650 g, adults), and sexual dimorphism in body size is less pronounced in *C. perotensis* than in most other pocket gophers, with males being only slightly larger than females. Fur coloration ranges from light to dark brown dorsally, and the underparts are similar, but paler. Most, perhaps all, specimens of *C. perotensis* have one or more small patches of white fur near the base of the tail. Patch size varies from approximately 1 cm^2 to more than 2 cm^2. In study skins, the skin at the base of the tail often is folded under the skin in the rump region, thereby obscuring the white patch at the base of the tail. The skull of *C. perotensis* is relatively small for the genus (cranial width usually < 26 mm in adults), and the anterior edge of the jugal is narrow (usually < 2 mm) compared to those of *C. merriami* and *C. fulvescens* (Hafner et al., 2005). The anterior surface of each upper incisor contains a single, longitudinal groove. The diploid number of *C. perotensis* is 2n = 38, and the fundamental number is FN = 72 (Hafner et al., 2005).

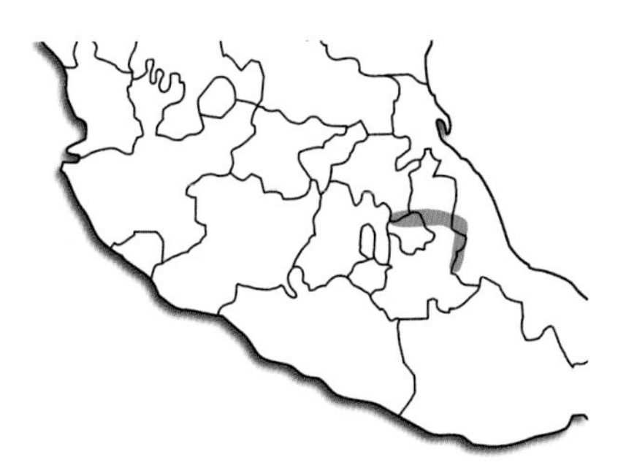

DISTRIBUTION: *C. perotensis* is endemic to Mexico. This species is patchily distributed in the Plain of Apan in southern Hidalgo eastward through the mountains of northern Puebla and the pine forest zone of west-central Veracruz, southward to Cofre de Perote and Mt. Orizaba. *C. perotensis* is known from the states of HG, PU, TL, and VE.

EXTERNAL MEASURES AND WEIGHT

TL = 300 to 360 mm; TV = 82 to 110 mm; HF = 35 to 44 mm; EAR = 6 to 9 mm.
Weight = 400 to 650 g.

DENTAL FORMULA: I 1/1, C 0/0, PM 1/1, M 3/3 = 20.

NATURAL HISTORY AND ECOLOGY: Little is known about the natural history of *C. perotensis*. This species is found in contact with *C. fulvescens* near the town of Perote, Veracruz, where the two species can be trapped only meters apart (see *C. fulvescens* account for additional information about this contact zone). Little is known of the breeding biology of *C. perotensis*, except that pregnant females have been captured in December (M.S. Hafner, pers. obs.).

VEGETATIONAL ASSOCIATIONS AND ELEVATION RANGE: *C. perotensis* occurs in a wide range of arid habitats. In the vicinity of Ciudad Sahagún, Hidalgo, and Perote, Veracruz, this species appears to be near the lower end of its elevational range and can be found in dry, sparsely vegetated, desert habitat dominated by alkaline-tolerant plants. At higher elevations, *C. perotensis* occurs in open areas surrounded by pine, oak, spruce, and fir forests. The open areas often contain bacharris (*Bacharris* sp.), ragworts (*Senecio* sp.), lupin (*Lupinus* sp.), and bunch grasses ("zacatón"; *Muhlenbergia* sp., *Festuca* sp.). The elevational range of *C. perotensis* extends from approximately 2,400 m to 4,000 m.

CONSERVATION STATUS: *C. perotensis*, which now includes pocket gophers previously recognized as *C. merriami estor*, *C. merriami irolonis*, *C. merriami peraltus*, and *C. merriami perotensis* (Hafner et al., 2005), is not included in the list of threatened or endangered Mexican mammals (SEMARNAT, 2010).

Cratogeomys planiceps (Merriam, 1895)

Volcán de Toluca pocket gopher

Mark S. Hafner

SUBSPECIES IN MEXICO

C. planiceps is a monotypic species.

C. planiceps is a member of the *C. fumosus*-group, which was revised by Hafner et al. (2004). In that revision, Hafner et al. resurrected the species name *C. planiceps*, which was originally described as *Platygeomys planiceps* by Merriam (1895) and subsequently recognized as *C. tylorhinus planiceps* by Hooper (1947).

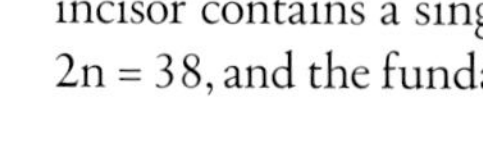

DISTRIBUTION: *C. planiceps* is endemic to Mexico. This species is patchily distributed on the northern slopes of Volcán de Toluca, the southeastern slopes of Valle de Bravo, and the forested hills north of Valle de Bravo (Hafner et al., 2004). *C. planiceps* is known only from the state of MX.

DESCRIPTION: *Cratogeomys planiceps* is medium size for the genus (400 g to 600 g, adults), and sexual dimorphism in body size is not evident based on specimens examined by Hafner et al. (2004). Dorsal pelage usually is dark (approaching black in some individuals) with a light-brown wash laterally. The skull of *C. planiceps* is flattened in the occipital region, but less so than in *C. fumosus*. The anterior edge of the jugal is narrow, and the depth of the skull at the occipital plate is usually greater than 90% of the least width of the occipital plate. The anterior surface of each upper incisor contains a single, longitudinal groove. The diploid number of *C. planiceps* is 2n = 38, and the fundamental number is FN = 72 (Hafner et al., 2004).

EXTERNAL MEASURES AND WEIGHT

TL = 290 to 350 mm; TV = 100 to 110 mm; HF = 40 to 44 mm; EAR = 7 to 9 mm. Weight = 400 to 600 g.

DENTAL FORMULA: I 1/1, C 0/0, PM 1/1, M 3/3 = 20.

NATURAL HISTORY AND ECOLOGY: *C. planiceps* occurs in open spaces within pine forests and in meadows and croplands surrounded by pine forests. Little is known of the breeding biology of *C. planiceps*. Specimens captured in January through March were not breeding (M.S. Hafner, pers. obs.).

VEGETATIONAL ASSOCIATIONS AND ELEVATION RANGE: *C. planiceps* is restricted to open areas surrounded by pine forests from approximately 2,500 m elevation to near timberline.

CONSERVATION STATUS: *C. planiceps* is not included in the list of threatened or endangered Mexican mammals (SEMARNAT, 2010).

Geomys arenarius Merriam, 1895

Desert pocket gopher

Gerardo Ceballos

SUBSPECIES IN MEXICO

Geomys arenarius arenarius Merriam, 1895.

This species was considered conspecific with *G. bursarius* by Hafner and Geluso (1983), but the phylogenetic separation of *G. arenarius* and *G. bursarius* was established by Jolley et al. (2000). Phylogenetic relationships among species of the genus *Geomys* have been investigated recently based on molecular characters (Sudman et al., 2006). *G. arenarius* belongs to the *G. bursarius*-group, which includes *G. bursarius*, *G. knoxjonesi*, *G. lutescens*, and *G. texensis*.

DESCRIPTION: *Geomys arenarius* is medium size for the genus, and males are noticeably larger than females (198 g to 254 g, adult males; 165 g to 207 g, adult females; Davis, 1974). The anterior surface of each upper incisor contains two longitudinal grooves. As in all pocket gophers, the neck in *G. arenarius* is poorly differentiated from the head, the pinnae are reduced, and the eyes small. Dorsal coloration is uniformly brown with scattered black hairs, and the underside is lighter. *G. arenarius* is morphologically very similar to *G. tropicalis* and *G. bursarius*, but the three species can be distinguished by genetic and cranial characteristics (Álvarez, 1963a; Williams and Baker, 1974). *G. arenarius* has a diploid number of 2n = 70 and a fundamental number of FN = 102 (Davis et al., 1971).

DISTRIBUTION: *G. arenarius* is patchily distributed from south-central New Mexico and extreme western Texas in the United States to northern Mexico. In Mexico, this species is restricted to five disjunct localities in northeastern Chihuahua (Anderson, 1972; Hall, 1981). *G. arenarius* is known only from the state of CH.

EXTERNAL MEASURES AND WEIGHT
TL = 225 to 280 mm; TV = 63 to 95 mm; HF = 29 to 35 mm; EAR = 5 mm.
Weight = 165 to 254 g.

DENTAL FORMULA: I 1/1, C 0/0, PM 1/1, M 3/3 = 20.

NATURAL HISTORY AND ECOLOGY: *G. arenarius* inhabits deep and sandy soils near the edge of rivers, ponds, and other bodies of water (Williams and Baker, 1974). In Mexico, this species is known only from the vicinity of the Río Bravo and the dunes of Samalayuca (Anderson, 1972). Patches of habitat suitable for *G. arenarius* generally are isolated by areas of dry, inhospitable habitats, which constitute a major barrier to pocket gopher dispersal. It is thought that *G. arenarius* reproduces in the summer, with two to six offspring per litter (William and Baker, 1974). Other species of *Geomys* are known to reproduce throughout the year, and each female can produce two litters annually with up to four offspring per litter (Williams, 1982). Predators of *G. arenarius* include rattlesnakes, birds of prey, and mammalian carnivores such as weasels, coyotes, and badgers.

VEGETATIONAL ASSOCIATIONS AND ELEVATION RANGE: *G. arenarius* occurs only in scrublands and grasslands adjacent to bodies of water. Elevational range is approximately 1,200 m to 1,600 m.

CONSERVATION STATUS: *G. arenarius* is not included in the list of threatened or endangered Mexican mammals (SEMARNAT, 2010). Ceballos and Navarro (1991), however, consider this species in danger of extirpation in Mexico because of its limited geographic range, which has been reduced further by overgrazing of livestock, industrial development, and urban growth, especially in the vicinity of Ciudad Juárez, Chihuahua. A designated conservation area in the dunes of Samalayuca would protect this and other species of mammals and reptiles in this region that are threatened with extinction.

Texas pocket gopher

Gerardo Ceballos

SUBSPECIES IN MEXICO

Geomys personatus megapotamus Davis, 1940

Phylogenetic relationships among species of the genus *Geomys* have been investigated recently based on molecular characters (Sudman et al., 2006). *G. personatus megapotamus* belongs to the *G. personatus*-group, which includes *G. attwateri*, *G. personatus*, *G. streckeri*, and the Mexican endemic, *G. tropicalis* (see following account).

DESCRIPTION: *Geomys personatus* is medium size for the genus (180 g to 380 g, adults). Mexican populations of *G. personatus* are uniformly light brown dorsally and paler ventrally. As in all species of the genus *Geomys*, the anterior surface of each upper incisor contains two longitudinal grooves. The diploid number of *G. personatus* ranges from 2n = 70 to 72, and the fundamental number ranges from FN = 76 to 102 (Williams, 1982).

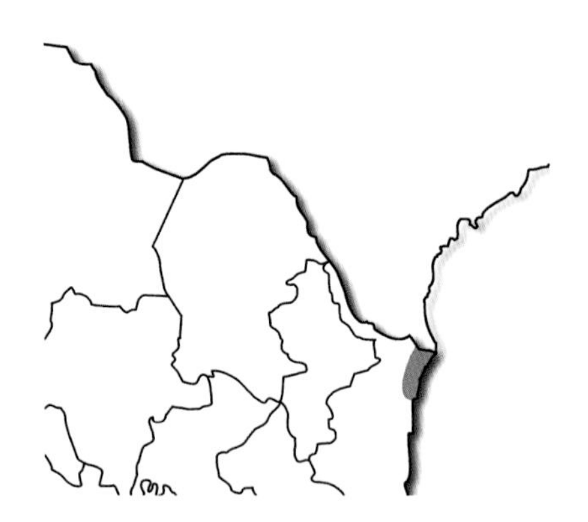

DISTRIBUTION: This species is patchily distributed throughout southern Texas, but has a marginal distribution in Mexico, where it is restricted to a small coastal region in northeastern Tamaulipas. In Mexico, *G. personatus* is known only from the state of TA.

EXTERNAL MEASURES AND WEIGHT

TL = 269 to 310 mm; TV = 69 to 103 mm; HF = 31 to 37 mm; EAR = 5 mm.
Weight = 180 to 380 g.

DENTAL FORMULA: I 1/1, C 0/0, PM 1/1, M 3/3 = 20.

NATURAL HISTORY AND ECOLOGY: *G. personatus* occurs in deep, sandy soils. It is entirely absent from the silt loams of the flood plains of the Río Bravo and also from gravelly, stony, or clayey soils scattered throughout its range. The Río Bravo has been a major barrier to the southward dispersal of this species, except in the region of Matamoros, Tamaulipas, where this species is thought to have colonized Mexico during the Pleistocene (Selander et al., 1962). In Mexico, *G. personatus* is found exclusively in the cambisol soils that separate Laguna Madre from the sea (Álvarez, 1963a). Known dietary items include the roots of several species of perennial grasses of the genera *Paspalum*, *Cynodon*, and *Cenchrus*, and the roots, stems, and leaves of the composite *Helianthus* (Schmidly and Davis, 2004). Breeding occurs between October and February, with females producing two litters of three to four young each year (Williams, 1982).

VEGETATIONAL ASSOCIATIONS AND ELEVATION RANGE: *G. personatus* inhabits thorny scrubland and vegetation around sand dunes. Habitats occupied by *G. personatus* often are dominated by mesquite (*Prosopis juliflora*). In Mexico, *G. personatus* occurs from near sea level to 30 m elevation.

CONSERVATION STATUS: Because of its extremely limited range in Mexico, *G. personatus* is listed by the Mexican government as threatened (SEMARNAT, 2010). The range of this species has been further reduced by human activities in the coastal region of northeastern Tamaulipas (Ceballos and Navarro, 1991). The current geographic distribution and conservation status of *G. personatus* populations in Mexico should be investigated in the near future.

Tropical pocket gopher

Roberto Márquez Huitzil and Gerardo Ceballos

SUBSPECIES IN MEXICO

G. tropicalis is a monotypic species.

Phylogenetic relationships among species of the genus *Geomys* have recently been investigated using molecular characters (Sudman et al., 2006). *G. tropicalis* is a member of the *G. personatus*-group, which includes *G. attwateri, G. personatus, G. streckeri*, and *G. tropicalis.*

DESCRIPTION: *Geomys tropicalis* is small for the genus (200 g to 300 g, adults); sexual dimorphism in body size is pronounced, with males larger than females. As in all species of the genus *Geomys*, the anterior surface of each upper incisor contains two longitudinal grooves. Dorsal coloration is uniformly brown; the abdomen is lighter (Álvarez, 1963a). *G. tropicalis* is very similar to *G. personatus* morphologically, but the two species can be distinguished by cranial features and karyotypes. The diploid number of *G. tropicalis* is 2n = 38, and the fundamental number is FN = 72 (Baker and Williams, 1974; Williams and Genoways, 1973).

EXTERNAL MEASURES AND WEIGHT

TL = 212 to 270 mm; TV = 61 to 93 mm; HF = 28 to 35 mm; EAR = 5 mm.
Weight = 200 to 300 g.

DENTAL FORMULA: I 1/1, C 0/0, PM 1/1, M 3/3 = 20.

Geomys tropicalis. Sand dunes. Altamira, Tamaulipas. Photo: Gerardo Ceballos.

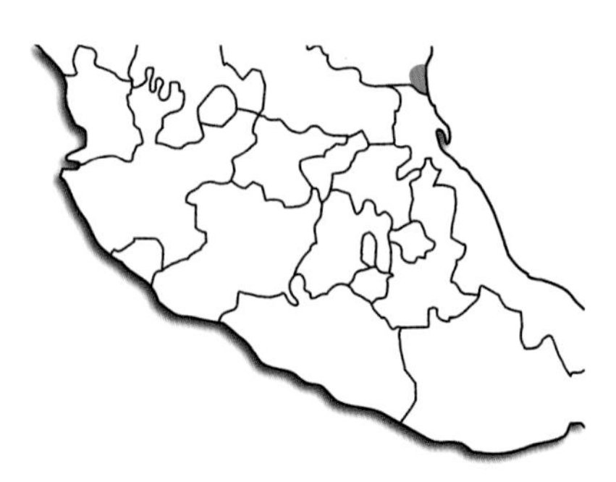

DISTRIBUTION: *G. tropicalis* is endemic to Mexico. This species is patchily distributed within a small (< 150 km^2), coastal area of southeastern Tamaulipas (Ceballos and Oliva, 2005; Hall, 1981; Márquez-Huitzil, 1994). *G. tropicalis* is known only from the state of TA.

NATURAL HISTORY AND ECOLOGY: *G. tropicalis* is the southernmost distributed species of the genus *Geomys*. It is restricted to soils with a high proportion (70% to 83%) of sand, a low proportion (11% to 21%) of clay, and a depth of at least 40 cm (Márquez-Huitzil, 1994). These soils have good drainage and appropriate compaction and consistency for burrowing by pocket gophers (Wilkins and Swearing, 1990). As in other species of pocket gophers, individuals of *G. tropicalis* maintain exclusive territories, but no other information exists on their social structure. Similarly, little is known about reproduction in *G. tropicalis*. A lactating female was captured in December, a scrotal male was captured in May, and a captive female was in estrus from May to August, suggesting that *G. tropicalis* may breed year-round. Although populations of the much larger pocket gopher, *Orthogeomys hispidus*, occur just 3 km south of the *G. tropicalis* range, these two species are separated by the Río Pánuco, which appears to be an effective barrier to dispersal for both species (Márquez-Huitzil, 1994).

VEGETATIONAL ASSOCIATIONS AND ELEVATION RANGE: *G. tropicalis* is found in sandy soils that support a heterogeneous mixture of trees, including oak (*Quercus oleoides*), mesquite (*Prosopis* sp.), acacias (*Acacia cornifera*), and guacimo (*Guazuma ulmifolia*). This species also is known from low-elevation deciduous forest, thorny scrublands, natural grasslands, livestock pastures, and agricultural fields containing watermelons, pineapples, and orange trees (Baker and Williams, 1977; Márquez-Huitzil, 1994). The elevational range of *G. tropicalis* extends from near sea level to approximately 100 m.

CONSERVATION STATUS: Because of its extremely limited geographic range, *G. tropicalis* was considered endangered by Ceballos and Navarro (1991) and currently is listed by the Mexican government as threatened (SEMARNAT, 2010). The distribution of *G. tropicalis* was reduced by approximately 60% during the twentieth century, mainly due to growth and expansion of the cities of Tampico, Ciudad Madero, and Altamira in coastal Tamaulipas (Márquez-Huitzil, 1994).

Orthogeomys cuniculus Elliot, 1905

Oaxacan pocket gopher

Beatriz Villa Cornejo and Eduardo Espinoza Medinilla

SUBSPECIES IN MEXICO

O. cuniculus is a monotypic species.

Studies are needed to determine the taxonomic status of *O. cuniculus*. When Elliot (1905b) described this species, he lacked sufficient samples of the nearby species, *O. grandis*, to compare with his two specimens of *O. cuniculus* (one a juvenile). Now that additional specimens of *O. grandis* are available for comparison, *O. cuniculus* appears to fit well within the range of skin and skull measurements and fur color and density observed within *O. grandis* (M.S. Hafner, pers. obs.).

DESCRIPTION: *Orthogeomys cuniculus* is very similar in appearance to the giant pocket gopher, *O. grandis*, but is somewhat smaller in overall size. *O. cuniculus* is medium-sized for the genus (ca. 400 g, adults), and its dorsal pelage is coarse, sparse, and predominantly brown in color. Fur on the underparts is especially sparse, such that the surface of the skin of the belly and legs is easily visible through the fur. The tail is relatively long and naked, and the nasal pad is small or absent (Goodwin, 1969). As in all species of the genus *Orthogeomys*, the anterior surface of each upper incisor contains a single longitudinal groove. The karyotype of *O. cuniculus* is unknown.

DISTRIBUTION: *O. cuniculus* is endemic to Mexico and is known only from the vicinity of Zanatepec, Ingenio Santo Domingo, and Unión Hidalgo in coastal southeastern Oaxaca (Elliot, 1905b; Goodwin, 1969). *O. cuniculus* is known only from the state of OX.

EXTERNAL MEASURES AND WEIGHT
TL = 320 to 330 mm; TV = 95 to 100 mm; HF = 44 to 48 mm; EAR = 8 mm.
Weight = 400 g.

DENTAL FORMULA: I 1/1, C 0/0, PM 1/1, M 3/3 = 20.

NATURAL HISTORY AND ECOLOGY: There is no published information on the natural history of *O. cuniculus*, other than the fact that it lives in the arid tropical lowlands of southeastern Oaxaca (Goodwin, 1969).

VEGETATIONAL ASSOCIATIONS AND ELEVATION RANGE: Vegetation in the region occupied by *O. cuniculus* is dominated by dry, thornscrub forests. The elevational range of *O. cuniculus* extends from near sea level to 40 m.

CONSERVATION STATUS: *O. cuniculus* is listed by the Mexican government (SEMARNAT, 2010) as threatened because of its extremely limited geographic range in southeastern Oaxaca.

Orthogeomys grandis (Thomas, 1893)

Giant pocket gopher

Beatriz Villa Cornejo, Ivonne Barrios A., and Eduardo Espinosa Medinilla

SUBSPECIES IN MEXICO
Orthogeomys grandis alleni Nelson and Goldman, 1930
Orthogeomys grandis alvarezi Schaldach, 1966
Orthogeomys grandis annexus Nelson and Goldman, 1933
Orthogeomys grandis carbo Goodwin, 1956
Orthogeomys grandis felipensis Nelson and Goldman, 1930
Orthogeomys grandis guerrerensis Nelson and Goldman, 1930
Orthogeomys grandis huixtlae Villa, 1944
Orthogeomys grandis nelsoni Merriam, 1895
Orthogeomys grandis scalops (Thomas, 1894)
Orthogeomys grandis soconuscensis Villa, 1949

DESCRIPTION: *Orthogeomys grandis* is among the largest pocket gophers in Mexico (only a few specimens of *Cratogeomys fumosus* are larger). Adult *O. grandis* of both sexes can approach 1,000 g body mass. At higher elevations (ca. 1,000 m to 2,700 m), the dorsal pelage is moderately dense and dark brown to almost black in color. The fur of the belly is somewhat sparser and similarly colored. At lower elevations and in drier habitats, the fur is extremely sparse over the entire body. The tail is naked, and the feet generally have less hair than the rest of the body and occasionally have a whitish dorsal coloration. Hairs around the muzzle are sometimes whitish-brown, and some specimens have white, irregularly shaped spots on the back of the head. The nasal

Orthogeomys grandis. Maize crop. Zanacatepec, Oaxaca. Photo: David J. Hafner.

pad on certain individuals, especially those at lower elevations, is sparsely haired or naked. As in all species of the genus *Orthogeomys*, the anterior surface of each upper incisor contains a single longitudinal groove. The diploid number of *O. grandis* is 2n = 58 and the fundamental number is FN = 108 (Hafner, 1979).

DISTRIBUTION: The geographic range of *O. grandis* extends southward along the Pacific coast of Mexico from just south of the city of Colima, Colima, and Jilotlán de Dolores, Jalisco, into Guatemala. The range of this species also extends eastward from the Pacific coast of Guerrero along the valleys of the Balsas, Mezcala, and Amacuzac rivers to the Nexapa river valley in southwestern Puebla, then southeasterly to the mountains east of Oaxaca city (Hall, 1981). *O. grandis* is known from the states of CH, CL, GR, JA, MI, OX, and PU.

EXTERNAL MEASURES AND WEIGHT

TL = 335 to 405 mm; TV = 95 to 140 mm; HF = 44 to 55 mm; EAR = 9 to 13 mm. Weight = 480 to 985 g.

DENTAL FORMULA: I 1/1, C 0/0, PM 1/1, M 3/3 = 20.

NATURAL HISTORY AND ECOLOGY: Like other geomyids, pocket gophers of the species *O. grandis* spend most of their lives inside their underground burrows, occasionally leaving their holes to feed on nearby surface vegetation. Both nocturnal and diurnal activity has been reported for *O. grandis*, which feeds primarily on the roots and other underground parts of plants such as guapinol (*Hymenaca courbaril*). This species often is considered a pest in plantations of cocoa (*Theobroma cacao*), bananas (*Musa* sp.), rubber (*Hevea brasiliensis*), corn (*Zea mays*), rambutan (*Nephelium lappaceum*), and rice (*Oryza sativa*). Reproductively active individuals have been captured in January through March (M.S. Hafner, pers. obs.), and the typical litter size is two.

VEGETATIONAL ASSOCIATIONS AND ELEVATION RANGE: *O. grandis* inhabits perennial tropical forests, tropical deciduous and subdeciduous forests, thorny xeric scrublands, and medium- to high-elevation cloud forests (Rzedowski, 1981). The elevational range of *O. grandis* extends from near sea level to 2,700 m.

CONSERVATION STATUS: *O. grandis* is abundant over much of its geographic range and, as a result, is not included in the list of threatened or endangered Mexican mammals (SEMARNAT, 2010).

Hispid pocket gopher

Beatriz Villa Cornejo

SUBSPECIES IN MEXICO

Orthogeomys hispidus chiapensis (Nelson and Goldman, 1929)
Orthogeomys hispidus concavus (Nelson and Goldman, 1929)
Orthogeomys hispidus hispidus (Le Conte, 1852)
Orthogeomys hispidus isthmicus (Nelson and Goldman, 1929)
Orthogeomys hispidus latirostris (Hall and Álvarez, 1961)
Orthogeomys hispidus negatus (Goodwin, 1953)
Orthogeomys hispidus teapensis (Goldman, 1939)
Orthogeomys hispidus tehuantepecus (Goldman, 1939)
Orthogeomys hispidus torridus (Merriam, 1895)
Orthogeomys hispidus yucatanensis (Nelson and Goldman, 1929)

DESCRIPTION: Although it is the smallest species of the genus *Orthogeomys* in Mexico, *Orthogeomys hispidus* is a relatively large pocket gopher (ca. 600 g, adult males; adult females somewhat smaller). The dorsal pelage of *O. hispidus* usually is sparse and bristled, but specimens taken at higher elevations, such as the type locality of *O. hispidus concavus* (2,400 m), have moderately dense pelage (M.S. Hafner, pers. obs.). Hair on the ventrum is usually sparser than dorsal hair. Pelage coloration varies from dark brown to yellow ochre dorsally, and is somewhat lighter ventrally. The frequency of individuals with partial or complete lumbar belts (a 5 mm to 50 mm wide belt of white fur in the lumbar region) reaches 100% in certain populations of *O. hispidus*, but is less frequent in most (Hafner and Hafner, 1987). As in all species of the genus *Orthogeomys*, the anterior surface of each upper incisor contains a single longitudinal groove. The diploid number of *O. hispidus* is 2n = 52 and the fundamental number is FN = 88 (Hafner and Hafner, 1987).

Orthogeomys hispidus. Cacao crop. Reforma, Chiapas. Photo: Gerardo Ceballos.

DISTRIBUTION: *O. hispidus* ranges from southern Tamaulipas through the mountains and coastal lowlands of Veracruz, eastern Oaxaca, central Chiapas, and most of the Yucatan Peninsula southward into Guatemala. *O. hispidus* is known from the states of CA, CS, OX, PU, QE, QR, SL, TA, TB, VE, and YU.

EXTERNAL MEASURES AND WEIGHT
TL = 300 to 350 mm; TV = 75 to 85 mm; HF = 40 to 45; EAR = 8 to 10 mm.
Weight = 450 to 600 g.

DENTAL FORMULA: I 1/1, C 0/0, PM 1/1, M 3/3 = 20.

NATURAL HISTORY AND ECOLOGY: The extensive underground galleries of *O. hispidus* can reach 60 m in length and > 1 m in depth. Near the ruins of Chichen Itza in Yucatan, it is common to find *O. hispidus* burrows in very shallow soils. Reproductive activity seems to occur year-round, but increased reproductive activity occurs between October and June. The average number of offspring per litter is two individuals. *O. hispidus* is widely considered to be an agricultural pest. In Yucatan, this species lives on sisal (*Agave sisaliana*) plantations, where it causes extensive damage to young plants.

VEGETATIONAL ASSOCIATIONS AND ELEVATION RANGE: *O. hispidus* inhabits perennial tropical forests, arid thornscrub habitats, and agricultural fields. Its elevational range extends from near sea level to approximately 2,500 m.

CONSERVATION STATUS: *O. hispidus* is abundant over much of its geographic range and, as a result, is not included in the list of threatened or endangered Mexican mammals (SEMARNAT, 2010).

Orthogeomys lanius (Elliot, 1905)

Big pocket gopher

Beatriz Villa Cornejo

SUBSPECIES IN MEXICO
O. lanius is a monotypic species.
Studies are needed to determine the taxonomic status of *O. lanius*. When Elliot (1905a) described this species, he lacked sufficient samples of the nearby species, *O. hispidus*, to compare with two specimens of *O. lanius*. Now that additional specimens of *O. hispidus* are available for comparison, it appears that the specimen described as *O. lanius* may simply be a somewhat large and more densely furred specimen of *O. hispidus*. Body size in pocket gophers is known to be a remarkably plastic trait (Patton and Brylski, 1987; Smith and Patton, 1988; Wilkins and Swearingen, 1990) and is heavily dependent on local food availability (Hafner et al., 2008). As noted in the *O. hispidus* species account, specimens of *O. hispidus* taken from higher elevations tend to have moderately dense pelage, which may explain the soft and woolly pelage of the *O. lanius* specimen, which was collected at 2,500 m.

DESCRIPTION: As its common name suggests, *Orthogeomys lanius* is a large pocket gopher. Total length of the holotype is 361 mm, which is slightly larger than the largest known specimens of *O. hispidus* (ca. 350 mm). The two known specimens of *O. lanius* are dark brown, almost black, dorsally and dark brown ventrally. The belly has many white hairs, especially near the base of the tail. The pelage is soft and woolly, unlike most specimens of *O. hispidus*. As in all species of the genus *Orthogeomys*, the anterior surface of each upper incisor contains a single longitudinal groove. The diploid and fundamental numbers of *O. lanius* are unknown.

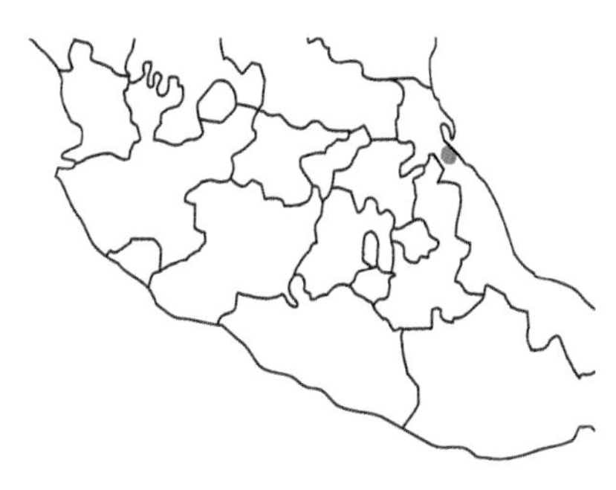

DISTRIBUTION: *O. lanius* is endemic to Mexico. This species is known only from the type locality of Xuchil, Veracruz. The exact location of the type locality is unknown, but it may be near the present-day town of El Xuchitl, Veracruz (18.89 degrees north, 97.24 degrees west). *O. lanius* is known only from the state of VE.

EXTERNAL MEASURES AND WEIGHT

TL = 361 mm; TV = 90 mm; HF = 54 mm; EAR = 10 mm.
Weight = probably 450 to 600 g.

DENTAL FORMULA: I 1/1, C 0/0, PM 1/1, M 3/3 = 20.

NATURAL HISTORY AND ECOLOGY: Nothing is known about the natural history of *O. lanius*, but this species is probably very similar to *O. hispidus* ecologically (see *O. hispidus* account).

VEGETATIONAL ASSOCIATIONS AND ELEVATION RANGE: *O. lanius* is known only from the type locality of Xuchil, Veracruz, the exact location of which is unknown. The type locality may be near the present-day town of El Xuchitl, Veracruz, which is surrounded by pine-oak woodlands at 2,500 m elevation.

CONSERVATION STATUS: *O. lanius* is listed by the Mexican government (SEMARNAT, 2010) as threatened because of its extremely limited geographic range in northwestern Veracruz.

Pappogeomys bulleri (Thomas, 1892)

Buller's pocket gopher

Jorge Ortega R.

SUBSPECIES IN MEXICO

Pappogeomys bulleri albinasus Merriam, 1895
Pappogeomys bulleri alcorni Russell, 1957
Pappogeomys bulleri bulleri (Thomas, 1892)
Pappogeomys bulleri burti Goldman, 1939
Pappogeomys bulleri nayaritensis Goldman, 1939
Demastes et al. (2003) recommended that *P. alcorni* be recognized as a subspecies of *P. bulleri* based on examination of mitochondrial DNA and morphometric data. A recent systematic revision of *Pappogeomys* by Hafner et al. (2009) concurred with Demastes et al. (2003) and reduced the number of subspecies of *P. bulleri* from nine to five.

DESCRIPTION: *Pappogeomys bulleri* is a medium to small pocket gopher (90 g to 150 g, adult females; 130 g to 270 g, adult males). As in other pocket gophers, the legs of *P. bulleri* are short and fitted with long, powerful claws used for digging (Ceballos and Miranda, 1986, 2000). Dorsal pelage is of average length for pocket gophers (ca. 10 mm), but can be quite sparse in specimens living in arid or low-elevation habitats (M.S. Hafner, pers. obs.). Coloration varies from smoky black to reddish-brown dorsally, with ventral pelage similar in color or occasionally somewhat lighter. The anterior surface of each upper incisor contains a single longitudinal groove in all specimens of *P. bulleri*. The diploid number of *P. bulleri* is 2n = 56 and the fundamental number is FN = 106 (Hafner et al., 2009).

Pappogeomys bulleri. Cloud forest. El Jabalí Wildlife Refuge, Colima. Photo: Gerardo Ceballos.

EXTERNAL MEASURES AND WEIGHT
TL = 195 to 275 mm; TV = 60 to 98 mm; HF = 26 to 36 mm; EAR = 4 to 6 mm.
Weight = 90 to 270 g.

DISTRIBUTION: *P. bulleri* is endemic to Mexico. This species is patchily distributed from the mountains northwest of Tepic, Nayarit, to the vicinity of Guadalajara, Jalisco, then southward through much of the Sierra Madre del Sur to the coast near Armería, Colima. An isolated population (formerly recognized as *P. alcorni*) exists in the mountains south of Lago de Chapala near the town of Mazamitla, Jalisco. *P. bulleri* is known from the states of CL, JA, and NY.

DENTAL FORMULA: I 1/1, C 0/0, PM 1/1, M 3/3 = 20.

NATURAL HISTORY AND ECOLOGY: Most specimens of *P. bulleri* dig their burrows in sandy soil with little clay (Russell, 1968), but this species also is known from fairly shallow, rocky soils near the summit of Volcán Tequila (M.S. Hafner, pers. obs.). Reproduction is thought to occur throughout the year, usually with two offspring (but sometimes as many as eight) per litter (Ceballos and Miranda, 1986, 2000). Where they occur sympatrically with *Cratogeomys fumosus*, *P. bulleri* specimens are usually in forested habitats surrounding open fields occupied by *C. fumosus*.

VEGETATIONAL ASSOCIATIONS AND ELEVATION RANGE: *P. bulleri* is known from a wide variety of habitats ranging from hot, humid coastal lowlands, through tropical deciduous and subdeciduous forests, to oak and pine forests at higher elevations (Soler-Frost et al., 2003). Where they occur in forested habitats, the above-ground mounds produced by *P. bulleri* often are hidden from view by a dense cover of pine needles and other leaf litter. Unlike other species of pocket gophers, *P. bulleri* seems reluctant to enter large agricultural fields in certain parts of its range. For example, *P. bulleri* could not be found in the large alfalfa fields near the town of Mascota, Jalisco, but was collected in the surrounding hillsides covered with native deciduous thornscrub vegetation (M.S. Hafner, pers. obs.). The elevational range of *P. bulleri* is from near sea level to approximately 3,000 m.

CONSERVATION STATUS: One subspecies of *P. bulleri* (*P. bulleri alcorni*) is subject to special protection by the Mexican government (SEMARNAT, 2010) because of its limited geographic range in the vicinity of Mazamitla, Jalisco. Fieldwork near Mazamitla in January 2009 showed this subspecies to be abundant and widespread in the forested hillsides south and west of Mazamitla (M.S. Hafner, pers. obs.). Elsewhere in its range, *P. bulleri* is usually abundant and appears to face no conservation problems.

Botta's pocket gopher

Iván Castro Arellano and Lorena Morales Pérez

SUBSPECIES IN MEXICO

Thomomys bottae abbotti Huey, 1928
Thomomys bottae albatus Grinnell, 1912
Thomomys bottae analogus Goldman, 1938
Thomomys bottae angustidens R.H. Baker, 1953
Thomomys bottae anitae J.A. Allen, 1898
Thomomys bottae aphrastus Elliot, 1903
Thomomys bottae basilicae Benson and Tillotson, 1940
Thomomys bottae borjasensis Huey, 1945
Thomomys bottae brazierhowelli Huey, 1960
Thomomys bottae cactophilus Huey, 1929
Thomomys bottae camoae Burt, 1937
Thomomys bottae catavinensis Huey, 1931
Thomomys bottae convergens Nelson and Goldman, 1934
Thomomys bottae cunicularius Huey, 1945
Thomomys bottae divergens Nelson and Goldman, 1934
Thomomys bottae estanciae Benson and Tillotson, 1939
Thomomys bottae homorus Huey, 1949
Thomomys bottae humilis R.H. Baker, 1953
Thomomys bottae jojobae Huey, 1945
Thomomys bottae juarezensis Huey, 1945
Thomomys bottae lucidus Hall, 1932
Thomomys bottae martirensis J.A. Allen, 1898
Thomomys bottae modicus Goldman, 1931
Thomomys bottae nigricans Rhoads, 1895
Thomomys bottae perditus Merriam, 1901
Thomomys bottae proximarius Huey, 1945
Thomomys bottae retractus R.H. Baker, 1953
Thomomys bottae rhizophagus Huey, 1949
Thomomys bottae ruricola Huey, 1949
Thomomys bottae russeolus Nelson and Goldman, 1909
Thomomys bottae siccovallis Huey, 1945
Thomomys bottae simulus Nelson and Goldman, 1934
Thomomys bottae sinaloae Merriam, 1901
Thomomys bottae sturgisi Goldman, 1938
Thomomys bottae toltecus J.A. Allen, 1893 (part)
Thomomys bottae vanrosseni Huey, 1934
Thomomys bottae varus Hall and Long, 1960
Thomomys bottae villai R.H. Baker, 1953
Thomomys bottae winthropi Nelson and Goldman, 1934
Thomomys bottae xerophilus Huey, 1945

Hoffmeister (1986) and Patton and Smith (1990) reduced the number of Mexican subspecies of *T. bottae* from 47 (Hall, 1981) to 45. Rios and Álvarez-Castañada (2007) and Trujano-Álvarez and Álvarez-Castañada (2007) reviewed the systematic status of populations of *T. bottae* in Baja California Sur and further reduced the number of *T. bottae* subspecies in Mexico from 45 (Patton, 2005) to the current 40. Future studies using molecular data likely will reduce the number of subspecies still further.

DISTRIBUTION: *T. bottae* is patchily distributed throughout the western United States from southern Oregon to southern Colorado and western Texas into northern Mexico. In Mexico, *T. bottae* is found on the Baja California Peninsula and on the mainland from Sonora east to Nuevo Leon (Burt, 1938; Patton, 1993b). *T. bottae* is known from the states of BC, BN, CH, CO, NL, SI, and SO.

Thomomys bottae. Grassland. Janos Biosphere Reserve, Chihuahua. Photo: Gerardo Ceballos.

DESCRIPTION: Mexican subspecies of *Thomomys bottae* are generally small in body size (130 g to 250 g, adult males; 70 g to 150 g, adult females). Dorsal coloration varies widely from near black to light reddish-brown. The ventrum usually is lighter, often with dark underfur and white tips. The tail varies from yellowish-brown to gray and is sparsely haired. Unlike all other genera in the Geomyidae, the genus *Thomomys* lacks longitudinal grooves on the anterior surfaces of the upper incisors. Mexican subspecies of *T. bottae* can be distinguished from the much more widespread species, *T. umbrinus*, by the presence of two pairs of pectoral mammae (rather than one) in females. A few specimens of *T. bottae* have been reported to have two pectoral mammae on one side and one on the other (Dunnigan, 1967). The baculum in *T. bottae* usually is longer than 11 mm, whereas that of *T. umbrinus* is shorter. The diploid number in Mexican subspecies of *T. bottae* is 2n = 76, although one individual from near Choix, Sinaloa, was reported to have 2n = 74 (Patton and Dingman, 1970). The fundamental number in *T. bottae* varies from FN = 118 to 148 (Hafner et al., 1987; Hoffmeister, 1986; Nelson and Goldman, 1934; Patton, 1968; Whitaker, 1980).

EXTERNAL MEASURES AND WEIGHT
TL = 150 to 300 mm; TV = 42 to 105 mm; HF = 24 to 35 mm; EAR = 4 to 8 mm. Weight = 70 to 250 g.

DENTAL FORMULA: I 1/1, C 0/0, PM 1/1, M 3/3 = 20.

NATURAL HISTORY AND ECOLOGY: The natural history of *T. bottae* is much like that of all other species of pocket gophers (see chapter introduction). Where they live near coastal mangrove swamps, individuals of this species have been known to cover the entry of their burrows with shells, pebbles, or small branches (Haskins, 1912). Where food is abundant, *T. bottae* can reproduce year-round (Jones and Baxter, 2004), but the main breeding season appears to be in early spring. Pregnant females were captured in the Sierra Madre Oriental in late June and mid-August (M.S. Hafner, pers. obs.). The gestation period is approximately 18 days (Schramm, 1961), and the number of offspring per litter varies from 1 to 12 (more commonly 3 to 6). Reproduction appears to be synchronized with the growth of grasses and herbs (Hoffmeister, 1986; Patton, 1973). *T. bottae* feeds on a wide variety of herbaceous plants, grasses, and the bulbs, tubers, and roots of shrubs (Hoffmeister, 1986).

VEGETATIONAL ASSOCIATIONS AND ELEVATION RANGE: In northwestern Mexico, *T. bottae* is abundant in plains with deep and sandy soils, desert scrublands, and sites close to rivers and streams. In north-central Mexico, this species extends from lower-elevation desert scrublands into forest clearings at, or above, 2,500 m in the Sierra Madre Oriental. *T. bottae* often invades agricultural fields, where it can damage crops and cause leaks in irrigation canals (Nowak, 1999b). The elevational range of *T. bottae* extends from near sea level to approximately 3,000 m (Allen, 1899; Hall, 1981).

CONSERVATION STATUS: *T. bottae* is abundant over much of its geographic range and, as a result, is not included in the list of threatened or endangered Mexican mammals (Arita and Ceballos, 1997; SEMARNAT, 2010).

Thomomys umbrinus (Richardson, 1829)

Southern pocket gopher

Lorena Morales Pérez and Iván Castro Arellano

SUBSPECIES IN MEXICO

Thomomys umbrinus arriagensis Dalquest, 1951
Thomomys umbrinus atrodorsalis Nelson and Goldman, 1934
Thomomys umbrinus atrovarius J.A. Allen, 1898
Thomomys umbrinus camargensis Anderson, 1972
Thomomys umbrinus chihuahuae Nelson and Goldman, 1934
Thomomys umbrinus crassidens Nelson and Goldman, 1934
Thomomys umbrinus durangi Nelson and Goldman, 1934
Thomomys umbrinus enixus Nelson and Goldman, 1934
Thomomys umbrinus eximius Nelson and Goldman, 1934
Thomomys umbrinus goldmani Merriam, 1901
Thomomys umbrinus juntae Anderson, 1972
Thomomys umbrinus madrensis Nelson and Goldman, 1934
Thomomys umbrinus musculus Nelson and Goldman, 1934
Thomomys umbrinus nelsoni Merriam, 1901
Thomomys umbrinus newmani Dalquest, 1951
Thomomys umbrinus parviceps Nelson and Goldman, 1934
Thomomys umbrinus potosinus Nelson and Goldman, 1934
Thomomys umbrinus pullus Hall and Villa, 1948
Thomomys umbrinus sonoriensis Nelson and Goldman, 1934
Thomomys umbrinus supernus Nelson and Goldman, 1934
Thomomys umbrinus toltecus J.A. Allen, 1893
Thomomys umbrinus umbrinus (Richardson, 1829)
Thomomys umbrinus zacatecae Nelson and Goldman, 1934

Castro-Campillo and Ramírez-Pulido (2000) reviewed the systematic status of *T. umbrinus* populations in the Trans-Mexican Volcanic Belt and reduced the number of subspecies of *T. umbrinus* in Mexico from 30 to the current 23. Future studies using molecular data likely will reduce the number of subspecies further. Ongoing molecular and chromosomal studies of *T. umbrinus* in Mexico by Mark S. Hafner and colleagues show strong evidence for recognition of additional species of *Thomomys* in Mexico.

DISTRIBUTION: In the United States, *T. umbrinus* occurs in extreme southern Arizona and New Mexico. In Mexico, this species can be found from the Pacific coastal plain of Sinaloa and Nayarit eastward to the western edge of the Altiplano. It also occurs in the Trans-Mexican Volcanic Belt from near Pátzcuaro, Michoacán, eastward to the western slopes of Volcán Pico de Orizaba. *T. umbrinus* has been recorded from the states of AG, CH, CO, DF, DU, GJ, HG, JA, MI, MX, NY, PU, SI, SL, TL, and ZA.

DESCRIPTION: *Thomomys umbrinus* is a small species (70 g to 150 g, adults). Dorsal coloration varies widely from light brown to near black in melanistic individuals, and the ventrum is usually lighter. Most specimens of *T. umbrinus* have a faint dorsal stripe extending from the nose to the base of the tail. This stripe, which is difficult to see in darker specimens, is ca. 1 cm to 2 cm wide and darker than surrounding pelage. The nearly naked tail varies in color from brown to yellowish-gray (Hoffmeister, 1986). As in all species of *Thomomys*, *T. umbrinus* lacks longitudinal grooves on the anterior surfaces of the upper incisors. Characters used to distinguish between *T. umbrinus* and *T. bottae* in Mexico are listed in the *T. bottae* account. Populations of *T. umbrinus* in the Sierra Madre Occidental and along the Pacific coast of Sinaloa and Nayarit have a diploid number of 2n = 76. All other populations of *T. umbrinus* in Mexico have 2n = 78 (Hafner et al., 1987). Fundamental numbers in *T. umbrinus* range from FN = 96 to FN = 152.

EXTERNAL MEASURES AND WEIGHT

TL = 130 to 270 mm; TV = 45 to 80 mm; HF = 22 to 31 mm; EAR = 5 to 8 mm.
Weight = 70 to 150 g.

DENTAL FORMULA: I 1/1, C 0/0, PM 1/1, M 3/3 = 20.

NATURAL HISTORY AND ECOLOGY: *T. umbrinus* is usually abundant wherever it occurs. As in all pocket gophers, it prefers deep soils in areas resistant to flooding. It is often captured near irrigation canals and in nearby agricultural fields (Nowak, 1999b). *T. umbrinus* feeds on a wide variety of herbaceous plants, grasses, and tubers, as well as the roots of trees and shrubs (Hoffmeister, 1986). Pregnant females have

Thomomys umbrinus. Pine forest-bunch grassland. Nevado de Toluca National Park, State of Mexico. Photo: Gerardo Ceballos.

been captured in the months of December through July, and it is possible that *T. umbrinus* may reproduce year-round where food resources are abundant. The average litter size is 4 to 5 young (Hoffmeister, 1986; Patton, 1973).

VEGETATIONAL ASSOCIATIONS AND ELEVATION RANGE: *T. umbrinus* inhabits a wide variety of habitats in Mexico, including grasslands, scrublands, dry thornscrub forests, and pine-oak woodlands. The elevational range of *T. umbrinus* extends from approximately 50 m in coastal Sinaloa and Nayarit to approximately 3,800 m in the Trans-Mexican Volcanic Belt (Hafner et al., 1987; Hall, 1981). Preliminary molecular and chromosomal evidence suggests that the low-elevation populations of *T. umbrinus* along the Pacific coast may be specifically distinct from *T. umbrinus* populations elsewhere in Mexico (M.S. Hafner, pers. obs.).

CONSERVATION STATUS: *T. umbrinus* is abundant over much of its geographic range and, as a result, is not included in the list of threatened or endangered Mexican mammals (Arita and Ceballos, 1997; SEMARNAT, 2010).

Zygogeomys trichopus Merriam, 1895

Michoacán pocket gopher

Osiris Gaona and Rodrigo A. Medellín

SUBSPECIES IN MEXICO

Zygogeomys trichopus trichopus (Merriam, 1895)
Zygogeomys trichopus tarascensis (Goldman, 1938)

DESCRIPTION: *Zygogeomys trichopus* is a medium-sized pocket gopher (up to 580 g, adult males; up to 400 g, adult females). The fur of *Z. trichopus* is shorter and finer than that of other pocket gopher species, giving it a shiny appearance, much like that of moles (Talpidae). Dorsally, color varies from smoky gray to lustrous brown, and the ventrum usually is slightly lighter in color. The belly often has irregular white spots (2 cm to 4 cm in diameter), especially in the throat region. The tail is completely naked, and the skin of the tail is white and almost translucent. Above the nose is a conspicuous nose pad. The pad itself is naked, but it is surrounded dorsally and laterally by a narrow band of off-white fur. The anterior surface of each upper incisor has two longitudinal grooves, reminiscent of the condition seen in *Geomys*. The diploid number of *Z. trichopus* is 2n = 40 and the fundamental number is FN = 74 (Hafner and Barkley, 1984).

EXTERNAL MEASURES AND WEIGHT

TL = 270 to 370 mm; TV = 95 to 115 mm; HF = 38 to 48 mm; EAR = 5 to 9 mm.
Weight = 190 to 580 g.

DENTAL FORMULA: I 1/1, C 0/0, PM 1/1, M 3/3 = 20.

NATURAL HISTORY AND ECOLOGY: It is possible that the foraging behavior of *Z. trichopus* is unlike that of any other geomyid species. The surface mounds produced by *Z. trichopus* are significantly taller (ca. 23 cm) and more conical than those of other

Zygogeomys trichopus. Oak forest. Patzcuaro, Michoacan. Photo: Mark S. Hafner.

pocket gophers (Hafner and Hafner, 1982). The top of these conical mounds lacks the characteristic opening seen in the mounds of other pocket gopher species, and it appears that the animals rarely, if ever, exit the mounds to forage on the surface. Partial excavation of a *Z. trichopus* burrow system on Cerro Patambán found the (probable) nest chamber more than 2 m beneath the ground (Hafner and Barkley, 1984). Whereas most other field-captured pocket gophers are extremely pugnacious, specimens of *Z. trichopus* were reported to be more docile and usually made no attempt to bite their captor (Hafner and Hafner, 1982). Very little is known about reproduction in *Z. trichopus*, although testes were small (< 5 mm) in males captured in March and August and much larger (> 14 mm) in specimens captured in December. A pregnant female with a single embryo was captured in mid-December (Hafner and Barkley, 1984; Hafner and Hafner, 1982).

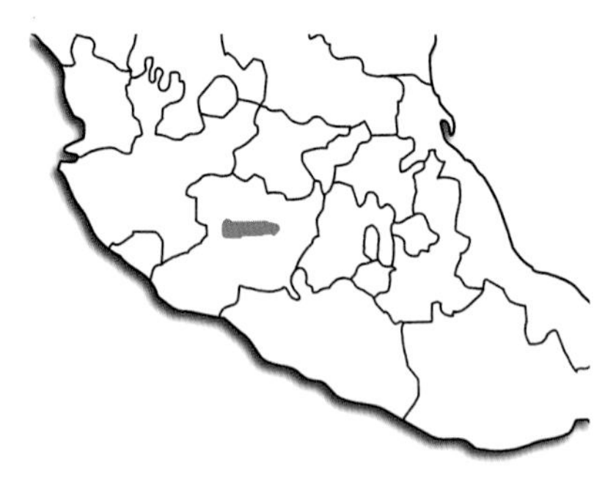

DISTRIBUTION: *Z. trichopus* is known from only four localities in the Sierra Madre of Michoacán. These include the type locality, Nahuatzén (where it is now extirpated), Cerro Patambán, Cerro Tancítaro, and the vicinity of Pátzcuaro (Hafner and Barkley, 1984). *Z. trichopus* may be extirpated in the immediate vicinity of Pátzcuaro, Michoacán, where *T. umbrinus* and *C. fumosus* are now common. Surface mounds characteristic of *Z. trichopus* have been observed in the mountains southeast of Pátzcuaro, near Cerro Burro (M.S. Hafner, pers. obs.). *Z. trichopus* is known only from the state of MI.

VEGETATIONAL ASSOCIATIONS AND ELEVATION RANGE: *Z. trichopus* inhabits small clearings within forests of pine, fir, spruce, and alder. This species also has been taken in agricultural clearings in forested regions and in avocado (*Persea americana*) orchards (M.S. Hafner, pers. obs.). The known elevational range of *Z. trichopus* is 2,000 m to 3,000 m.

CONSERVATION STATUS: *Z. trichopus* is listed as in danger of extinction by the Mexican government (SEMARNAT, 2010). This species was extirpated many decades ago at its type locality (Nahuatzén, Michoacán), where it has been replaced by the larger and more aggressive species, *Cratogeomys fumosus*. Expansion of agriculture into previously forested areas has destroyed much of the native habitat of *Z. trichopus* and favored range expansion of *Thomomys umbrinus* and *C. fumosus* (Hafner and Barkley, 1984). *Z. trichopus* and *C. fumosus* have now been captured less than 7 km apart with no apparent ecological or environmental barriers between them (M.S. Hafner, pers. obs.). Fieldwork by Mark S. Hafner in 1993 found *Z. trichopus* abundant in the avocado (*Persea americana*) orchards that now cover large portions of the western slopes of Cerro Tancítaro.

Family Cricetidae

Gerardo Ceballos, Robert D. Bradley, and Livia León Paniagua

The family Cricetidae has a broad distribution, being absent only from some islands. It consists of 130 genera and 681 species of rats and mice (Musser and Carleton, 2005). Until recently, most mice occurring in the Americas were classified in the family Muridae; however, the current arrangement that is widely accepted entails recognition of the native New World rats and mice within the family Cricetidae. A specific morphological character in all the members of the family is a series of two tubercules in the upper molars. In Mexico, the family is represented by 23 genera and 142 species. The species richness of the genera *Peromyscus*, *Neotoma*, and *Reithrodontomys* is extremely high; in fact, *Peromyscus*, with 51 species, is the most diverse rodent genus in Mexico and one of the most diverse throughout the world. Cricetids can be distinguished by their restricted distributions, in many cases of less than 10,000 km^2. Cricetids are the group that has suffered the greatest number of extinctions in Mexico. Species such as *Peromyscus guardia* and *Neotoma insularis* have disappeared mainly by the introduction of exotic species, and species such as *Xenomys nelsoni* and *Microtus pennsylvanicus* are threatened by the habitat destruction (List et al., 2010).

Subfamily Arvicolinae

The subfamily Arvicolinae contains twenty-eight genera of voles, lemmings, and muskrats. Two genera (*Microtus* and *Ondatra*) are known from Mexico.

Microtus californicus (Peale, 1848)

California vole

Eric Mellink and Jaime Luévano

SUBSPECIES IN MEXICO

Microtus californicus aequivocatus Osgood, 1928
Microtus californicus grinnelli Huey, 1931
Microtus californicus huperuthrus Elliot, 1903
Microtus californicus sanctidiegi R. Kellogg, 1918

DESCRIPTION: *Microtus californicus* is medium sized with a compact body, small ears that are rounded and nearly concealed by the pelage, and a tail of shorter length than the head and body. Hair is long and dark brown. The tail is moderately bicolored and the feet are lighter than the body. Females have eight mammae (Jameson and Peeters, 1988).

EXTERNAL MEASURES AND WEIGHT
TL = 157 to 211 mm; TV = 39 to 68 mm; HF = 20 to 25 mm; EAR= 11 to 16 mm.
Weight: 35 to 72 g.

DENTAL FORMULA: I 1/1, C 0/0, PM 0/0, M 3/3 = 16.

DISTRIBUTION: This species has a wide distribution in the United States. In Mexico, however, it is exclusively found in northwestern Baja California (Huey, 1964). It has been recorded in wet areas of the River Tijuana, in the Valley of San Rafael, at the foot of Sierra de Juarez, in the Valley of La Grulla, Vallecitos, Concepcion, and San Ramon, in the Sierra de San Pedro Martir, and in the western coastal plain of Sierra de San Pedro Martir (San Jose, San Antonio, San Telmo, El Rosario). It has been recorded in the state of BC.

NATURAL HISTORY AND ECOLOGY: *M. californicus* is a nocturnal mouse that is active throughout the year. This species is not strongly territorial and uses areas of up to 1 ha. It mainly feeds on the foliage of grasses and other herbaceous plants. While in captivity they drink water; in the wild water is obtained from their food. They take refuge under dense zacates, under piles of branches, under trunks, and in underground burrows in soft soils. They reproduce throughout the year, producing two to five litters per year. Reproduction increases when cover and food are more plentiful. The gestation period is 21 days. Litters of 1 to 9 young can be produced, but the average is 4 young. The pups are quickly weaned at 21 days old, and females reach sexual maturity at 29 days (Brylski, 1990a; Jameson and Peeters, 1988).

VEGETATIONAL ASSOCIATIONS AND ELEVATION RANGE: These mice inhabit swamps, both freshwater and saltwater, and grasslands of mountains with high levels of moisture. In California, they found in fields of alfalfa (Jameson and Peeters, 1988). They are found from sea level to 2,400 m.

CONSERVATION STATUS: The status of this species is unknown, but populations have diminished due to alterations of their habitat (Ceballos and Navarro, 1991). The swamps of the Sierras, with high grasslands, have been subjected to intense grazing. The River Tijuana has been modified in large part of its extension. The Valley of San Rafael is now transformed into agricultural area, as well as the coastal plain of San Quintin. Ceballos and Navarro (1991) consider it endangered. It is officially considered in danger of extinction (SEMARNAT, 2010).

Microtus guatemalensis Merriam, 1898

Guatemalan vole

Jesús Martínez and Fernando A. Cervantes

DISTRIBUTION: *M. guatemalensis* is restricted to the northwestern region of Guatemala and the central part of Chiapas, Mexico. In Mexican territory it has only been recorded in Tzontehuitz, 11 km north of San Cristobal de las Casas (Smith and Jones, 1967). It has been recorded in the state of CS.

SUBSPECIES OF MEXICO

M. guatemalensis is a monotypic species.

DESCRIPTION: *Microtus guatemalensis* is a medium-sized mouse within the genus. It has a robust body, a short tail, and small ears. The dorsal coloration is dark brown. It has five plantar tubercles, and the skull is elongated and somewhat arched. It differs from other species by possessing five closed triangles in the occlusal surface of the third upper molar (Merriam, 1898b; Smith and Jones, 1967).

EXTERNAL MEASURES AND WEIGHT

TL = 143 to 158 mm; TV = 30 to 36 mm; HF = 20 to 22 mm; EAR = 12 to 16 mm. Weight: 34 to 44 g.

DENTAL FORMULA: I 1/1, C 0/0, PM 0/0, M 3/3 = 16.

NATURAL HISTORY AND ECOLOGY: Information on the natural history of *M. guatemalensis* is scarce (Smith and Jones, 1967). They live in the soil and leaflitter and burrow between roots in rocky soil. Other species of small mammals living in the same habitat are *Oryzomys alfaroi*, *Reithrodontomys sumichrasti*, *R. microdon*, *Peromyscus beatae*, and *P. zarhynchus* (Smith and Jones, 1967).

VEGETATIONAL ASSOCIATIONS AND ELEVATION RANGE: The Guatemalan vole inhabits cloud forests with pines (Smith and Jones, 1967), from 2,500 masl to 3,200 masl (Smith and Jones, 1967). The Mexican locality is at 2,910 masl.

CONSERVATION STATUS: There is no knowledge about the status of the populations of this rodent. The populations are fragmented and the species' distribution area is marginal. It is threatened in Mexico (SEMARNAT, 2010).

Microtus mexicanus (Sassure, 1861)

Mexican vole

Rosa Ma. González and Fernando A. Cervantes

SUBSPECIES IN MEXICO

Microtus mexicanus fulviventer Merriam, 1898
Microtus mexicanus fundatus Hall, 1948
Microtus mexicanus madrensis Goldman, 1938
Microtus mexicanus mexicanus (Saussure, 1861)
Microtus mexicanus neveriae Hooper, 1955
Microtus mexicanus ocotensis Álvarez and Hernandez-Chavez, 1993
Microtus mexicanus phaeus (Merriam, 1892)
Microtus mexicanus salvus Hall, 1948
Microtus mexicanus subsimus Goldman, 1938

DESCRIPTION: In contrast with other mice of temperate zones, *Microtus mexicanus* is medium in size, its hair is long and smooth, the dorsal coloration is dark brown, the sides are lighter, and the belly is grayish. The tail is slightly bicolored. It is recognized by its flat snout, small tail, and short and round ears, almost obscured by the coat. It has six plantar tubercles (Bailey, 1900; Camacho, 1940; Ceballos and Galindo, 1984; Goldman, 1938; Hall, 1981). The diploid number of chromosomes is 2n = 48 and the fundamental number is FN = 58 (Uribe Álvarez et al., 1977).

EXTERNAL MEASURES AND WEIGHT

TL = 121 to 152 mm; TV = 24 to 35 mm; HF = 17 to 21 mm; EAR= 12 to 15 mm.
Weight: 26 to 43 g.

DENTAL FORMULA: I 1/1, C 0/0, PM 0/0, M 3/3 = 16.

NATURAL HISTORY AND ECOLOGY: These mice are active both day and night; they use paths and underground tunnels to move. They are very social and live in family groups formed by two to eight individuals (Camacho, 1940). Their burrows are 10 m to 20 m deep; their tunnels are in a horizontal direction, destroying roots in their paths (Camacho, 1940; Hooper, 1955). They mainly feed on roots, stems, and leaves of herbaceous plants. They reproduce throughout the year but mainly from May to August. Each litter has one to four offspring with an average of three (Gardener and Cervantes, 1989). They coexist with other small mammals such as shrews (*Cryptotis* spp., *Sorex* spp.) and mice (*Reithrodontomys megalotis*, *Neotomodon alstoni*, *Peromyscus maniculatus*, *P. melanotis*, *P. levipes*, *P. beatae*, *P. hylocetes*, and *P. boylii*). They are preyed on mainly by coyotes, lynx, owls, barn owls, and rattlesnakes (*Crotalus* spp.; Ceballos

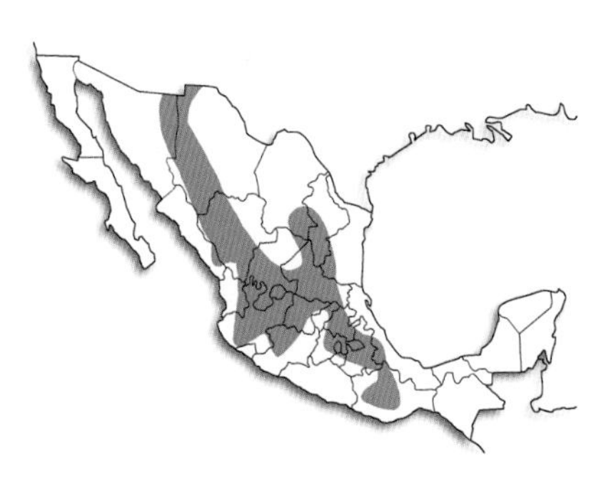

DISTRIBUTION: *M. mexicanus* ranges from the Southwest of the United States to the center and south of Mexico, occupying the Sierras Madre Occidental and Oriental, the Eje Volcanico Transversal, and Sierras North and Ixtlan of Oaxaca (Ceballos and Galindo, 1984; Choate and Jones, 1979; Hall, 1981). It has been recorded in the states of CH, CO, CS, DF, DU, HG, JA, MI, MO, MX, OX, PU, QE, TA, TL, and VE.

Microtus mexicanus. Pine forest. Ajusco National Park, Distrito Federal. Photo: Gerardo Ceballos.

and Galindo, 1984). The more common ectoparasites are mites (*Laelops*, *Hystrionyssus*), coleopterans (*Amblyopinus*), some dipterans (*Hypoderma*), and siphnapterans of great ecological and economic importance in Mexico (*Ctenophthalmus*; Machado-Allison and Barrera, 1964). They are considered herbivores and have a high reproductive potential, which contributes to their reputation as a plague species (Gardener and Cervantes, 1989).

VEGETATIONAL ASSOCIATIONS AND ELEVATION RANGE: This species occurs mainly in grasslands, forests of pine-oak, forests of fir, and forests of oak. It has also been found in disturbed environments such as golf courses, ditches, and crops of alfalfa and potato (Ceballos and Galindo, 1984). It has been collected at elevations from 2,220 masl to 4,115 masl (Ceballos and Galindo, 1984).

CONSERVATION STATUS: This species faces no conservation concerns, as it has a wide distribution and great ability to invade agriculture fields.

Tarabundí vole

Yolanda Hortelano and Fernando A. Cervantes

SUBSPECIES IN MEXICO

M. oaxacensis is a monotypic species (Goodwin, 1969).
It has been placed in the subgenus *Pytimys* with *M. guatemalensis* and *M. quasiater* (Martin, 1987).

DESCRIPTION: Within its genus, *Microtus oaxacensis* is of average size. It has small eyes, a robust body, short feet, and small and round ears covered with hair. The coat is long, smooth, and dense. The back is blackish-brown with completely black hair mixed with hair with ochraceous tips. The belly is slightly paler, and the legs are completely black, contrasting with the color of the nails, which are whitish-yellow. The tail is short, longer than the hind feet and shorter than the length of the head and body; it is slightly darker on the back and has little hair (Goodwin, 1966; Johnson and Johnson, 1982; Jones and Genoways, 1969). Unlike *M. mexicanus*, it has two pairs of pectoral mammary glands and five plantar tubercles (Jones and Genoways, 1967). In addition, *M. oaxacensis* has an elongated skull that is less angled than that of most members of the genus; its molars are long, wide, and with acute angles. The upper molar has five closed triangles unlike *M. umbrosus*, which has only two (Goodwin, 1966).

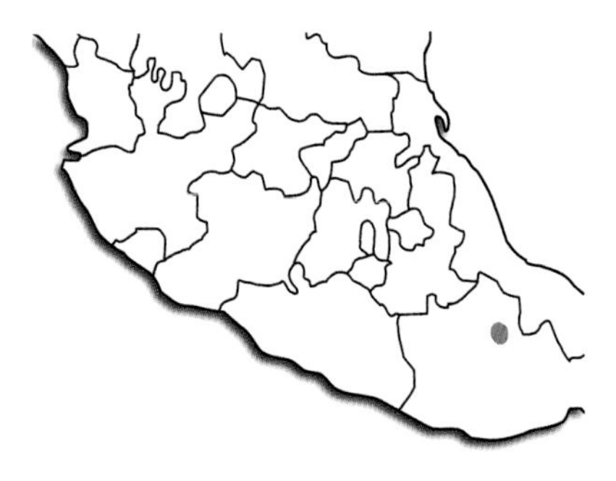

DISTRIBUTION: *M. oaxacensis* is an endemic species of Mexico, which is only distributed in the state of Oaxaca, on the outskirts of Vista Hermosa, in the Sierra de Juarez. This town is located 135 km north of the city of Oaxaca (Goodwin, 1966; Jones and Genoways, 1967). It has been recorded in the state of OX.

EXTERNAL MEASURES AND WEIGHT

TL = 159 to 163 mm; TV = 31 to 38 mm; HF = 21 to 23 mm; EAR= 12 to 15 mm.
Weight: 37 to 43 g.

DENTAL FORMULA: I 1/1, C 0/0, PM 0/0, M 3/3 = 16.

NATURAL HISTORY AND ECOLOGY: These mice live in humid forests with mesophile vegetation mixed with pines (Goodwin, 1966; Hoffmann and Koeppl, 1985; Jones and Genoways, 1967). Their populations are restricted to the high mountains (Johnson and Johnson, 1982), where they inhabit cold and wet environments (Getz, 1985). Apparently, type of vegetation, humidity, and vegetation cover are the most important factors in their local distribution (Getz, 1985). The herbaceous vegetation provides cover and food (Getz, 1985). The genus *Microtus* is characterized by high reproductive rates; however, there is no particular information on this species (Johnson and Johnson, 1982). It shares its habitat with other small mammals such as shrews (*Sorex veraepacis*, *Cryptotis magna*, and *C. mexicana*) and mice (*Oryzomys alfaroi*, *Reithrodontomys microdon*, *Peromyscus melanocarpus*, and *Megadonthomys cryophilus*; Jones and Genoways, 1967).

VEGETATIONAL ASSOCIATIONS AND ELEVATION RANGE: *M. oaxacensis* inhabits cloud forests with little grasses, from 1,500 masl to 2,500 masl (Getz, 1985; Goodwin, 1966; Jones and Genoways 1967).

CONSERVATION STATUS: *M. oaxacensis* is considered threatened, primarily because of its very restricted distribution of less than 10,000 km^2 (SEMARNAT, 2010).

Microtus pennsylvanicus (Ord, 1815)

Meadow vole

Rosa Ma. González and Fernando A. Cervantes

SUBSPECIES IN MEXICO

Microtus pennsylvanicus chihuahuensis Bradley and Cockrum, 1968

DESCRIPTION: *Microtus pennsylvanicus* is distinguished by its flat snout, small tail, and short and rounded ears that are almost covered by the coat. It is medium in size. Its hair is long and soft, brown, with lighter sides and grayish belly. It differs from other species of the genus by having six plantar tubercles as well as five closed triangles in the occlusal surface of the first inferior molar (Reich, 1981). It is darker than specimens of *M. pennsylvanicus modestus* and *M. pennsylvanicus aztecus* from nearby towns in New Mexico and southern Colorado (Anderson and Hubbard, 1971). With some exceptions, in these taxa there is an increment in the size of the skull that identifies *M. pennsylvanicus chihuahensis* as the largest subspecies (Anderson and Hubbard, 1971).

DISTRIBUTION: The distribution of *M. pennsylvanicus* extends from Canada and the northern and eastern regions of the United States to the northern part of Mexico. In Mexico it is only known from a locality at 4.8 km southeast of Galeana, northwest of the city of Chihuahua (Anderson, 1972; Hall, 1981), in an area of no more than 50 ha (Ceballos and Navarro, 1991). This distribution is a relic of wet periods of the Pleistocene (G. Ceballos, pers. comm.). Probably in the late Wisconsinian, the distribution was continuous with other neighboring populations in the United States (Anderson and Hubbard, 1971). Currently, *M. pennsylvanicus chihuahensis* is the most southern population of the species and is distributed about 430 km from the nearest population in the United States (Anderson and Hubbard, 1971). It has been recorded in the state of CH.

EXTERNAL MEASURES AND WEIGHT

TL = 152 to 199 mm; TV = 38 to 57 mm; HF = 21 to 25 mm; EAR= 11 to 16 mm. Weight: 71 g.

DENTAL FORMULA: I 1/1, C 0/0, PM 0/0, M 3/3 = 16.

NATURAL HISTORY AND ECOLOGY: The meadow vole mainly inhabits grasslands and areas with abundant coverage of herbs, as well as forests of oak (*Quercus*) and places with nearby water (Bailey, 1900). In Ojo de Galeana, Chihuahua, common plant species include sedge, grasses, and submerged herbs (G. Ceballos, pers. comm.). This spring is surrounded by more dry grasslands and by arid areas. The natural history of this species in Mexico is unknown. In other parts of North America, it mainly feeds on grasses (*Poa compresa*, *Panicun capillare*, and *Muhlenbergia sobolifera*) and occasionally insects such as beetles (Coleoptera) and larvae of butterflies (Lepidoptera; Zimmerman, 1965). The gestation period lasts 21 days, and the size of litter varies from 4 to 11 individuals with an average of 5 (Innes, 1978). It coexists with other small mammals such as *Reithrodontomys megalotis*, *R. montanus*, *Sigmodon fulviventer*, and *Peromyscus maniculatus* (Bradley and Cockrum, 1968). It is preyed on by hawks, owls, carnivores, and in some cases by snakes (Madison, 1978).

VEGETATIONAL ASSOCIATIONS AND ELEVATION RANGE: This species is commonly found in grasslands and coniferous forests and forests of oaks (*Quercus*; Reich, 1981; Zimmerman, 1965). In Chihuahua, it is known in the vegetation surrounding Ojo de Galeana (G. Ceballos, pers. comm.), which is composed of grasslands and sedges, at an elevation of about 1,400 masl (Bradley and Cockrum, 1968).

CONSERVATION STATUS: *M. pennsylvanicus chihuahuensis* is in danger of extinction due to its reduced distribution (Ceballos and Navarro, 1991, SEMARNAT, 2010). Its habitat is threatened by overgrazing and human settlements (G. Ceballos, pers. comm.). The species has not been recorded in recent years, despite the fact that systematic samplings have been carried out; it may currently be extinct in Mexico (G. Ceballos, pers. comm.).

Jalapan vole

Alondra Castro Campillo, Jesús Martínez,
Fernando A. Cervantes, and José Ramírez-Pulido

SUBSPECIES IN MEXICO
M. quasiater is a monotypic species.

DESCRIPTION: *Microtus quasiater* is a medium-sized species with short tail and ears; its dorsal coloration is brownish-olive (Coues, 1874; Hall, 1981). Unlike in *M. mexicanus*, the third molar has two closed triangles instead of three, its coat is much darker and uniform, and the interorbital is greater than 3.5 mm; the interparietal tends to be more pentagonal. Females have four mammary glands (Hall, 1981; Ramírez-Pulido et al., 1991).

EXTERNAL MEASURES AND WEIGHT
TL = 112 to 137 mm; TV = 17 to 25 mm; HF = 17 to 19 mm; EAR= 11 to 18 mm. Weight: 26 to 30 g.

DENTAL FORMULA: I 1/1, C 0/0, PM 0/0, M 3/3 = 16.

NATURAL HISTORY AND ECOLOGY: Jalapan voles live in burrows and delineated paths beneath the stubble in the wetlands near water bodies. They feed mainly on roots and the juicy parts of herbs of the underbrush. Little is known about their reproductive biology, but scrotal males have been collected in all seasons of the year (Ramírez-Pulido et al., 1991). Females give birth to up to four offspring, but the average is two (Hall and Dalquest, 1963; Keller, 1985; Ramírez-Pulido et al., 1991). It is sympatric with other species of mammals such as *Marmosa mexicana*, *Cryptotis* spp., *Sorex vagrans*, *Oligoryzomys fulvescens*, and *Microtus mexicanus* (Baker and Villa, 1953; Davis, 1944; Hall and Dalquest, 1963; Hooper, 1957; Ramírez-Pulido et al., 1991).

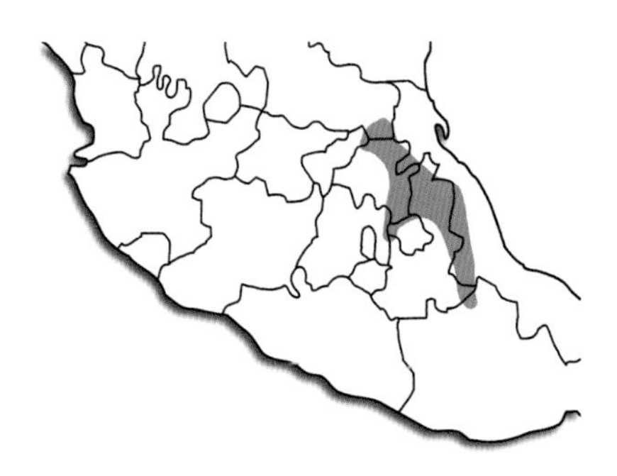

DISTRIBUTION: *M. quasiater* is an endemic to Mexico distributed along a narrow portion of the eastern slope of the Sierra Madre Oriental from Xilitla, San Luis Potosí, to Huautla de Jimenez, Oaxaca, in temperate zones (Baker and Villa, 1953; Goodwin, 1969; Hall, 1981; Hall and Cockrum, 1953; Hall and Dalquest, 1963; Ramírez-Pulido et al., 1991). It has been recorded in the states of HG, OX, PU, QE, SL, and VE.

Microtus quasiater. Cloud forest. Naolinco, Veracruz. Photo: Gerardo Ceballos.

VEGETATIONAL ASSOCIATIONS AND ELEVATION RANGE: Jalapan voles live in oak forests (León et al., 1990), mainly in cloud forests composed of liquidambar and in the ecotone of pine-oak forests, both with a dense undergrowth of perennial grasses, shrubs, and small trees (Hall and Cockrum, 1953; Hall and Dalquest, 1963; Hoffmann and Koeppl, 1985). It has been collected from rocky hillsides, chaparral, and stone walls (Hall and Dalquest, 1963), as well as in the edges of maize fields, grasslands, summer grassland, and orchards surrounding the forests mentioned above (Baker and Villa, 1953; Dalquest, 1950; Davis, 1944; Goodwin, 1969; Hall and Dalquest, 1963; Ramírez-Pulido et al., 1991). It is found from 700 masl to 2,150 masl (Ramírez-Pulido et al., 1991).

CONSERVATION STATUS: Because of the high rate of urbanization, populations in Xalapa seem to be extinct: however, in other parts of this species' distribution, populations can be considered abundant (Ramírez-Pulido et al., 1991). Because of its restricted distribution it is considered as of Special Protection (SEMARNAT, 2010).

Microtus umbrosus Merriam, 1898

Zempoaltépec vole

Yolanda Hortelano Moncada and Fernando A. Cervantes

SUBSPECIES IN MEXICO

M. umbrosus is a monotypic species.

It is the only member of the subgenus *Orthriomys* (Bailey, 1900; Musser and Carleton, 1991).

DESCRIPTION: *Microtus umbrosus* is a medium-sized mouse. Its body is robust; the feet are short and the ears small and rounded and almost covered by the coat. The hair is long and smooth. The dorsal coloration is dark and the belly is lighter. Among the characteristics that distinguish it from other species of the genus are the relatively long tail (35% of the total length), which is slightly bicolored and with little hair, and the presence of five plantar tubercles with remnants of the sixth and four pectoral mammary glands. The skull is long and narrow with small bullae. The teeth are typical of the genus *Microtus*; molars do not have roots and have external re-entrant and internal angles that are approximately equal. Furthermore, the surface of the third upper molar has two closed triangles and one open triangle (Hall, 1981; Johnson and Johnson, 1982; Merriam, 1898). The diploid number of chromosomes is 2n = 56 and the fundamental number is FN = 60 (Cervantes et al., 1994)

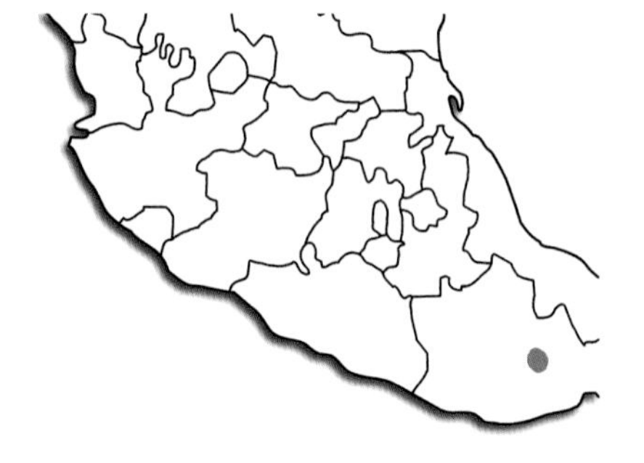

DISTRIBUTION: *M. umbrosus* is an endemic to Mexico, whose distribution is restricted to the slopes of Volcan Zempoaltépetl, near Totontepec in Oaxaca (Goldman, 1951; Goodwin, 1969). It is one of the species of the genus *Microtus* with the southernmost distribution (Hoffmann and Koeppl, 1985; Johnson and Johnson, 1982). It has been recorded in the state of OX.

EXTERNAL MEASURES AND WEIGHT

TL = 177 mm; TV = 61 mm; HF = 23 mm; EAR = 18 mm.

Weight: 42 g.

DENTAL FORMULA: I 1/1, C 0/0, PM 0/0, M 3/3 = 16.

NATURAL HISTORY AND ECOLOGY: This species' natural history is not well known. It occurs in wetlands of oak forests and open areas dominated by grasses (Bailey, 1900; Goodwin, 1969; Hoffmann and Koeppl, 1985). They live in burrows; small pathways are constructed among grasses to move from burrow to burrow or to search for food. Its diet consists basically of green vegetation, roots, and bark of trees. They are preyed on by snakes, owls, hawks, weasels, coyote, badgers, skunks, and wild cats (Bailey, 1900; Johnson and Johnson, 1982). It has been collected with species such as *Peromyscus melanocarpus* and *Habromys lepturus*.

VEGETATIONAL ASSOCIATIONS AND ELEVATION RANGE: *M. umbrosus* inhabits humid areas in forests of oak (*Quercus*) with gaps of grassland (Goldman, 1951; Goodwin, 1969). It is found between 1,700 masl and 2,400 masl.

CONSERVATION STATUS: *M. umbrosus* is considered as of Special Protection (Ceballos and Navarro, 1991; SEMARNAT, 2010) because of its restricted distribution and the alteration of its habitat. It is considered Special Protection (SEMARNAT, 2010).

Ondatra zibethicus (Linnaeus, 1766)

Common muskrat

Eric Mellink and Jaime Luévano

SUBSPECIES IN MEXICO
Ondatra zibethicus bernardi Goldman, 1932
Ondatra zibethicus ripensis (Bailey, 1902)

DESCRIPTION: The muskrat, or water rat, is large in size and semi-aquatic with a dense, brilliant coat consisting of two layers: soft underfur and dense outer fur formed by protective, stronger, and longer hairs. The first layer provides waterproofing and is dark brown on the back, lighter below, and almost white in the throat. The external coat varies from reddish-brown to very dark (almost black). Due to genetic variations, white (albino) and black individuals can be found. Females have three or four pairs of mammae. The tail is long, scaly, and vertically flattened, and ends in a tip. The

Ondatra zibethicus. Yellowstone, Wyoming. Photo: Gerardo Ceballos.

DISTRIBUTION: Muskrats are widely distributed in North America, from Canada to Mexico. Its distribution in Mexico is marginal, however. They are found in two isolated populations, the first along the Colorado River, between Baja California and Sonora, in the Valley of Mexicali and in the Delta of Colorado River (Leopold, 1959; Mearns, 1907; Mellink, 1995). The second population is found along Rio Bravo, in the state of Chihuahua, and the western half of the border of Coahuila with the United States (Anderson, 1972; Baker, 1956). It was introduced to Lake Texcoco (Villa-R., 1953), where it has currently disappeared (G. Ceballos, pers. comm.). It has been recorded in the states of BC, CH, and SO.

hind feet are larger than the forefeet and have a partial membrane, which assists in swimming. Muskrats have 54 chromosomes (Hoffmeister, 1986; Perry, 1982; Willner et al., 1980).

EXTERNAL MEASURES AND WEIGHT

TL = 409 to 422 mm; TV = 180 to 295 mm; HF = 64 to 88 mm; EAR =19 mm. Weight: 1,000 to 1,500 g.

DENTAL FORMULA: I 1/1, C 0/0, PM 0/0, M 3/3 = 16.

NATURAL HISTORY AND ECOLOGY: Muskrats are mainly active at night. They are excellent swimmers; they propel themselves with the hind feet and use the tail as a rudder. Their physiological adaptations allow them to spend up to 20 minutes underwater (Errington, 1963; Willner, 1980). In the north of Mexico, they mainly inhabit riversides. The entrance to the burrow is underwater and extends a few meters, slightly upward. Only a few burrows have nests; others are only for refuge. Introduced muskrats in Lake Texcoco build conical burrows, which are typical of areas with sedge (Leopold, 1959). The area of activity varies according to the social characteristics of the colony, habitat, time of year, and environmental conditions. Inside the average diameter was 61 m. Muskrats living in rivers may move more than 180 m along the shoreline, and through its entire width (Errington, 1963; Perry, 1982). They feed mainly on plant material, although sometimes they consume animal material (invertebrates, fish, and turtles). The main food items of their diet are roots and stems of aquatic plants. The mouth closes behind the incisors, allowing muskrats to feed underwater. Most of the feeding takes place less than 5 m to 10 m from the burrow (Perry, 1982; Willner et al., 1980). Bullrush (*Typhas* spp.) constitutes the main food item north of Mexico (Leopold, 1959). Muskrats can establish lax reproductive links with a single partner or be promiscuous. Females enter estrus several times during the year. Intercourse takes place in the water. Each female can produce between 1 and 5 litters per year of 1 to 11 offspring per litter, with an average of 4 to 7. They mate when females are still lactating. The gestation period lasts between three and four weeks. In Mexico and the southern United States they reproduce throughout the year, with a peak in winter. Most juveniles become sexually active the first spring of their life (Grinnell et al., 1937; Leopold, 1959; Perry 1982; Willner et al., 1980). In the United States this species provides the greatest economic income of all species used in fur trade. In Mexico and Canada more than 8 million muskrats were caught in 1975-1976. The meat is edible (Grinnell et al., 1937; Perry, 1982). Probably its commercial value promoted its introduction to Lake Texcoco in the 1940s. They are sometimes considered pests as they cause damage to the dikes of irrigation systems, (Grinnell et al., 1937; Willner et al. 1980). Muskrats have many predators such as coyotes, fox, wildcat, weasels, snakes, and birds of prey (Leopold, 1959); and harbor several diseases and parasites (Perry, 1982).

VEGETATIONAL ASSOCIATIONS AND ELEVATION RANGE: *Ondatra zibethicus* is found in swamps, fresh as well as brackish and salty, riversides, and lagoons. In Mexico it is found from sea level to 800 masl.

CONSERVATION STATUS: Muskrats are widely distributed in the Valley of Mexicali and the delta of the Colorado River. In this area they are found in irrigation channels and flood areas in the Rivers Hardy and Colorado and in the Swamp of Santa Clara, in very varied water conditions (Mellink, 1995). They are not at risk in this area. In other regions of the country the status of their populations is unknown. Ceballos and Navarro (1991) consider the species threatened in part of its distribution. It is officially considered threatened (SEMARNAT, 2010).

Subfamily Neotominae

Twenty genera of the subfamily Neotominae (New World rats and mice) are known from Mexico.

Baiomys musculus (Merriam, 1892)

Southern pygmy mouse

Catalina B. Chávez and Leticia A. Espinosa

SUBSPECIES IN MEXICO

Baiomys musculus brunneus (J.A. Allen and Chapman, 1897)
Baiomys musculus infernatis (Hooper, 1952)
Baiomys musculus musculus (Merriam, 1892)
Baiomys musculus nigrescens (Osgood, 1904)
Baiomys musculus pallidus Russell, 1952

DESCRIPTION: *Baiomys musculus* is the smallest mouse in Mexico. It is distinguished from *B. taylori* by the size of the foot (16 mm or more), as well as by the size of the baculum, which is greater in this species (3 mm to 3.9 mm) (Eshelman and Cameron, 1987; Packard and Montgomery, 1978). The dorsal coloration varies from brown red to dark brown; the underside is pale or white. Young individuals are uniformly gray (Hall, 1981).

EXTERNAL MEASURES AND WEIGHT

TL = 100 to 135 mm; TV = 35 to 56 mm; HF = 14.1 to 17 mm; EAR = 9 to 12 mm.
Weight: 6 to 10 g.

Baiomys musculus. Tropical dry forest. Chamela-Cuixmala Biosphere Reserve, Jalisco. Photo: Gerardo Ceballos.

DENTAL FORMULA: I 1/1, C 0/0, PM 0/0, M 3/3 = 16.

NATURAL HISTORY AND ECOLOGY: *B. musculus* inhabits tropical areas and semi-tropical areas, with abundant grasses and shrubs. It also is found in abandoned fields of maize and various cereals and in edges close to plantations of sugar cane and coconut palms (Ceballos and Miranda, 1986, 2000; Cervantes and Gardener, 1991). They construct their nests with remnants of grasses in underground burrows, located between the rocks or grasses. They feed mainly on herbs and fresh grass, in addition to seeds and small insects. In the regions where it coexists with *B. taylori*, it shows a preference for tropical lowlands, whereas *B. taylori* predominates in temperate highlands. Hooper (1952) recorded both species in common habitats in three locations in Jalisco. Other sympatric species are *Liomys irroratus*, *L. pictus*, *L. salvini*, *Neotoma mexicana*, *Orizomys couesi*, *Olygorizomys fulvescens*, *Peromyscus beatae*, *P. levipes*, *P. hylocetes*, *P. gymnotis*, *P. melanophrys*, *P. mexicanus*, *P. gratus*, *Reithrodontomys fulvescens*, *R. gracilis*, *R. sumichrasti*, *Sigmodon toltecus*, *Tylomys nudicaudus*, and *Mus musculus* (Packard and Montgomery, 1978). They have diurnal and crepuscular habits. They reproduce throughout the year, with less activity during winter and spring (Packard, 1960). Gestation lasts 20 to 25 days and lactation an average of 20 days. The size of the litter varies from one to five offspring, with three as an average. Its area of activity comprises 30 m in diameter, and population densities are between 15 to 20 individuals/ha (Ceballos and Miranda, 1986, 2000). Amman and Bradley (2004) depicted a genetic subdivision in *B. musculus* that corresponds to the coastal regions of Michoacan, suggested that two species may be represented, and suggested further studies.

DISTRIBUTION: This species is found from southern Nayarit and central Veracruz to the south with the exception of the Yucatan Peninsula, to northwestern Nicaragua. It has been recorded in the states of CL, CS, GR, JA, MI, MO, NY, OX, PU, and VE.

VEGETATIONAL ASSOCIATIONS AND ELEVATION RANGE: *B. musculus* is found in tropical deciduous forests, thorn forests, xeric scrublands, coastal dunes, grasslands, and croplands. It is distributed from sea level to 2,000 m.

CONSERVATION STATUS: These mice are abundant in croplands and grasslands; hence they are not considered endangered (Ceballos and Miranda, 2000).

Baiomys taylori (Thomas, 1887)

Northern pygmy mouse

Catalina B. Chávez and Leticia A. Espinosa

SUBSPECIES IN MEXICO

Baiomys taylori allex (Osgood, 1904)
Baiomys taylori analogous (Osgood, 1909)
Baiomys taylori paulus (J.A. Allen, 1903)

DESCRIPTION: *Baiomys taylori* is one of the smallest species of mice in North America. The dorsal coloration varies from reddish-brown to dark gray or black. The belly is gray with white or cream. The color of the tail differs between subspecies and may be gray or bicolor. This species differs from *B. musculus* by the size of the hind feet (less than 16 mm), as well as in the average length of the baculum, which is significantly shorter and thinner (Eshelman and Cameron, 1987; Hall, 1981; Packard, 1960).

EXTERNAL MEASURES AND WEIGHT
TL = 87 to 123 mm; TV = 34 to 53 mm; HF = 12 to 15 mm; EAR = 9 to 12 mm.
Weight: 6 to 9 g.

DENTAL FORMULA: I 1/1, C 0/0, PM 0/0, M 3/3 = 16.

NATURAL HISTORY AND ECOLOGY: These mice have nocturnal and crepuscular habits. They are found mainly in dense grassland and rocky areas that confer greater protection. They feed on seeds of grasses, leaves, and roots of grass. In semi-arid habitats they consume tunas (*Opuntia lindheimeri*) and legume seeds (*Prosopis juliflora*; Villa-R., 1953). In captivity, they have been fed insects and small snakes (*Leptotyphlops dulcis* and *Tropidoclonion lineatum*). They build runways similar to those of *Microtus* but smaller. Their nests of vegetable fiber are located generally under trunks, cacti, or herbs. They reproduce throughout the year with a maximum peak in late autumn and early spring. The data obtained from animals in captivity show it is polyestric, with an estrus period of approximately 7.5 days. The gestation period is 20 to 23 days. The size of the litter varies from one to five offspring, with an average of two. Both parents participate in the care of the offspring; an approximate period of 27.6 days occurs between litters. Lactation lasts 17 to 24 days. Males reach sexual maturity between 70 and 80 days after birth and females between 60 and 90 days, with pregnancies at an early age of 28 and 44 days. They molt the post-juvenile and adult fur annually at the beginning of the rainy season (Packard, 1960). In Texas, densities of 1 to 84 mice per ha have been recorded in areas of xeric scrub or oak with abundant vegetation; lower densities occur during summer and the highest during autumn and winter. The densities in abandoned fields were two mice per ha (Eshelman and Cameron, 1987). In the Basin of Mexico densities of 15 to 20 mice per ha have been recorded (Ceballos and Galindo, 1984). In a study conducted in Hidalgo, in an area of xeric scrub, the first 11 months no mice were recorded, but the density started increasing during the second year with a maximum of 28 mice per hectare in summer and autumn, and a gradual decline until its disappearance during the third

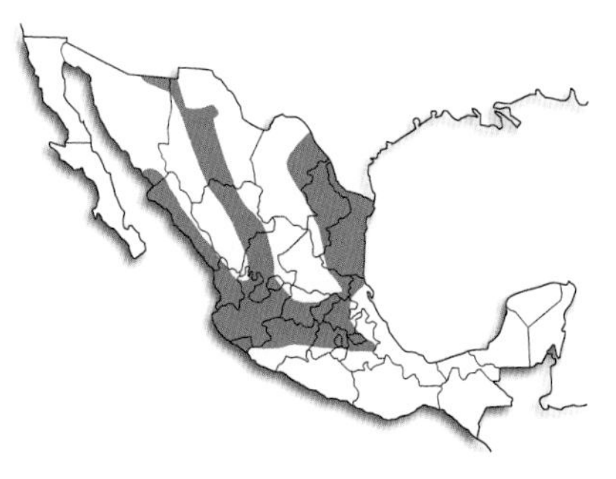

DISTRIBUTION: *B. taylori* is distributed from southeastern Arizona, New Mexico, and Texas in the United States, to Mexico, where it is found from Sonora and Chihuahua along the Pacific coast to Michoacán, and on the Gulf from the north of Coahuila, Nuevo Leon, and Tamaulipas to Hidalgo and Veracruz (Anderson, 1972; Phillips, 1965; Ramírez-Pulido and Castro, 1994; Ramírez-Pulido et al., 1986). It has been recorded in the states of AG, CH, CL, CO, DF, DU, GJ, HG, JA, MI, MO, MX, NL, NY, QE, SI, SL, SO, TA, VE, and ZA.

Baiomys taylori. Grassland. Apan, Hidalgo. Photo: Gerardo Ceballos.

year. Males constituted more than 80% of the population, with a sexual proportion of 1:1; the largest longevity recorded was 572 days in males and 344 days in females (Chávez and Espinosa, 1993). Natural populations in Texas show an average survival of 5 months and the highest survival was of 59 weeks. The roads and highways act as a barrier to their dispersal; its home range varies between 45 m^2 and 729 m^2 (Eshelman and Cameron; 1987). It coexists with *Liomys irroratus*, *Perognathus flavus*, *P. hispidus*, *Peromyscus boylii*, *P. difficilis*, *P. leucopus*, *P. melanotis*, *P. pectoralis*, *P. gratus*, *Reithrodontomys sumichrasti*, *Olygorizomys fulvescens*, *Orizomys texensis*, *Sigmodon toltecus*, *Mus musculus*, *Onychomys leucogaster*, *O. torridus*, *Neotoma micropus*, and *Cryptotis parva* (Chávez and Espinosa 1993; Eshelman and Cameron, 1987; Hall and Dalquest, 1963). Its main predators are snakes, owls, and mammals such as coyotes, raccoons, and skunks (Ceballos and Galindo, 1984; Eshelman and Cameron, 1987; Hall and Dalquest, 1963).

VEGETATIONAL ASSOCIATIONS AND ELEVATION RANGE: *B. taylori* is mainly found in grasslands, xeric scrublands, thorn forests, forests of oak and conifers, and croplands, from sea level to 2,450 m.

CONSERVATION STATUS: *B. taylori* is a common species, with a very wide distribution that faces no conservation concerns.

Habromys chinanteco (Robertson and Musser, 1976)

Chinanteco crested tail mouse

Livia León and Esther Romo

SUBSPECIES IN MEXICO

H. chinanteco is a monotypic species.

A recent revision of the genus concludes that this species is very close to, possibly conspecific with, *H. simulatus* (Carleton et al., 2002). León-Paniagua et al. (2007) based on molecular evidence found a closer phylogenetic relationship between *H. lepturus* and *H. chinanteco* than between *H. simulatus* and *H. chinanteco*.

DESCRIPTION: *Habromys chinanteco* is a small mouse in the genus, a little longer than *H. simulatus* but smaller than *H. lophurus*. The dorsal fur is grayish-brown, similar to that of *Peromyscus* subadults. The abdomen is grayish-white and the sidelines are well marked; the coat is smooth and thin both dorsally and ventrally. The toes are white, while the region of the metatarsal and metacarpal is gray. It is distinguished by a dark ring around the eye (Robertson and Musser, 1976). The tail is longer than the length of the body, is solid in color, and is covered with long brownish-gray hair (hairs on the tip of the tail are approximately 7 mm) and gives the appearance of a brush. The skull of *H. chinanteco* is medium sized and elongated with a wide and short nose. It is similar to that of *H. simulatus* in size, but the length of the palatal bone in *H. chinanteco* is always greater than 3.75 while that of *H. simulatus* is less than 3.57. The first molar is larger than that of *H. simulatus*.

EXTERNAL MEASURES AND WEIGHT

TL = 192 to 212 mm; TV = 103 to 121 mm; HF = 23 to 24 mm; EAR = 16 to 18 mm.
Weight: 17 to 22 g.

DENTAL FORMULA: I 1/1, C 0/0, PM 0/0, M 3/3 =16.

NATURAL HISTORY AND ECOLOGY: Like other species of the genus, *H. chinanteco* inhabits cloud forests of pine-oak cloud forests of high to moderate elevation. Robertson and Musser (1976) described the trees where this species lives covered with lichens, mosses, orchids, and bromeliads. The specimens used in the description were collected at night, around fallen trees and between tree roots. Robertson (1975) found a high degree of arboreability in this species, obtained through the radius of the tail and body length. It apparently feeds on seeds and plant matter. Reproductive activity has been reported in February and July (Robertson and Musser, 1976). In the town of Vista Hermosa *H. chinanteco* occurs sympatrically with *H. ixtlani, Peromyscus mexicanus, Megadontomys nelsoni, Microtus mexicanus,* and *Oryzomys alfaroi* (Robertson and Musser, 1976).

VEGETATIONAL ASSOCIATIONS AND ELEVATION RANGE: *H. chinanteco* inhabits pine-oak forests with orchids, ferns, and bromeliads, with dense underbush and high humidity throughout the year (Rickart, 1977). It is found from 2,080 m to 2,650 m.

CONSERVATION STATUS: *H. chinanteco* is an endemic species to Mexico that is only known to the type locality and its surroundings so that the population status is unknown. Arita et al. (1997) mentioned that some of the species in Mexico with a very restricted distribution like that of *H. chinanteco* should be included in the Mexican endangered species list.

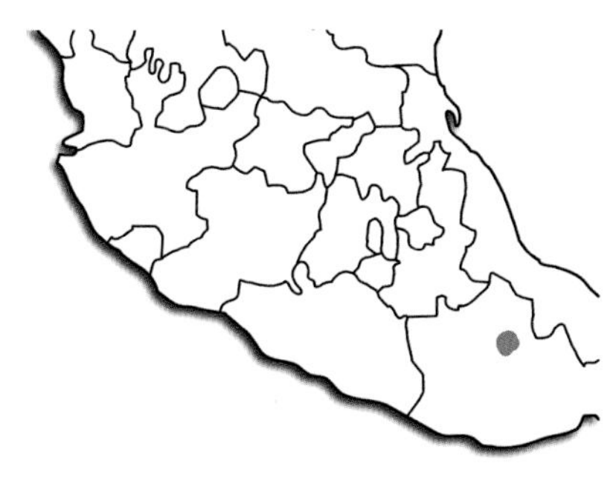

DISTRIBUTION: *H. chinanteco* is an endemic to Mexico, which is only known from the slopes of mountains along the Gulf of Mexico, in Cerro Pelón in Sierra de Juarez, District of Ixtlán, Oaxaca between 2,080 masl and 2,650 masl. It has been recorded in the state of OX.

Habromys delicatulus Carleton, Sánchez and Urbano-Vidales, 2002

Delicate crested tail mouse

Óscar Sánchez

SUBSPECIES IN MEXICO

H. delicatulus is a monotypic species.

DESCRIPTION: *Habromys delicatulus* is a very small-sized mouse. The coat is soft, delicate, and dense; the dorsal coloration varies from dark brown to brownish without an abrupt darker area on the middle of the back. The sides are ochraceous-orange, sometimes forming a more notorious stripe toward the hindquarters. The sides of the face have a dark area that extends to the eye, in the way of a mask with very diffuse edges. The forefeet are white to the carpal bones. The hind feet have a darker coloration on the back, extending to the metatarsal, as a spot. The belly is light grayish with the tips of the hair of lustrous appearance. The hair of the chin and throat are white to the base. The tail is almost as long as the head and body together, bicolored, and relatively hairy (which partially obscures the scales), and ends in a brush. The ears are relatively long, wide, and prominent, with dense fur and coloration similar to that of the back. All of these traits clearly distinguish it from other species of the genus. The skull is tiny and resembles a reduced version of *H. simulatus*; it is gracile and the face is shorter and thinner. The zygomatic arches are very delicate and narrow. The interparietal bone is relatively narrow and its edges are widely separated from the scaly bone. The bullae are comparatively expanded. The previous description is based on the original description provided by Carleton et al. (2002).

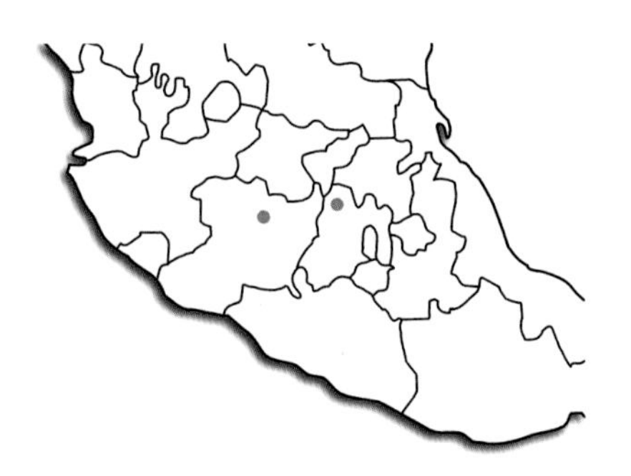

DISTRIBUTION: *H. delicatulus* is an endemic species to Mexico. It was exclusively known from the type locality, in a ravine (La Ermita) in the northwestern part of the state of Mexico. Recently, it was, however, recorded in Michoacan (Rogers et al., 2007). It has been recorded in the states of MI and MX.

EXTERNAL MEASURES AND WEIGHT
TL = 148 to 165 mm; TV = 74 to 81; HF = 18 to 20 mm; EAR = 18 to 20 mm.
Weight: 10 to 19 g.

DENTAL FORMULA: I 1/1, C 0/0, PM 0/0, M 3/3 = 16.

NATURAL HISTORY AND ECOLOGY: The information on the natural history and ecology of *H. delicatulus* is still scarce. The species seems to be linked in some degree with the arboreal environments (Carleton et al., 2002; Sanchez, 1996). Other species of rodents have been collected at the type locality of this species. *Nelsonia goldmani*, which represents the most eastern record of this species, was discovered in virtually the same site that *H. delicatulus* (Engstrom et al., 1992). Other species are *Peromyscus levipes*, *P. difficilis*, *Reithrodontomys fulvescens*, and *Neotoma mexicana*.

VEGETATIONAL ASSOCIATIONS AND ELEVATION RANGE: The area this species inhabits is part of the northern slope of the Eje Volcanico Transversal at an elevation of 2,570 m. The general area is covered by oak (*Quercus* spp.) with small patches of fir (*Abies* sp.) In some areas, there are relict tree elements of the cloud forest such as *Ilex*, *Garrya*, and *Cornus* (Carleton et al., 2002). Other detailed features of the vegetation are currently under study (O. Sanchez, pers. obs.).

CONSERVATION STATUS: *H. delicatulus* is one of the species of mammals of Mexico more recently described; hence its biology and conservation status are little known. Its micro-endemic situation makes it a species susceptible to extinction by environmental changes in its distribution area. The current state of its habitat, in terms of its needs, could seem moderately favorable, but the degree of inherent vulnerability of the species and the balance of its interactions with humans are still unknown. Currently, a first evaluation is being developed to learn about its conservation status. This is part of a proposal for the protection of the only area where it is known, which is also shared with *Nelsonia goldmani*. Meanwhile, it can be considered threatened.

Habromys ixtlani (Goodwin, 1964)

Ixtlán crested tail mouse

Gerardo Ceballos and Livia León

SUBSPECIES IN MEXICO
H. ixtlani is a monotypic species.
This species originally was described as *P. ixtlani* (Goodwin, 1964). Subsequently, *Habromys* was elevated to generic level. Musser (1969) considered that the differences between *H. ixtlani* and *H. lepturus* were not consistent and gave a subspecies level to *ixtlani*. Recently, however, Carleton et al. (2002) found sufficient evidence to separate the two taxa and assign *ixtlani* to species level. Molecular evidence confirms it to species level (León-Paniagua et al., 2007; Rogers et al., 2007).

DESCRIPTION: *Habromys ixtlani* is a medium-sized mouse, externally similar to *H. lepturus* but larger, with a longer bicolor tail and without a dark spot on the back. The hair on the back is smooth, dense, and conspicuously long, 5 mm to 6 mm in length.

The back is blackish-brown sprinkled with cinnamon. The belly is cream with gray at the base of the hairs. The ears are dark, thin, and covered with brown hairs. The legs are white. The tail is long, nearly the length of the head and body, slightly hairy, bicolored, with the top darker. The skull and molars are longer than in *H. lepturus* (Carleton et al., 2002; Goodwin, 1964; León-Paniagua et al., 2007).

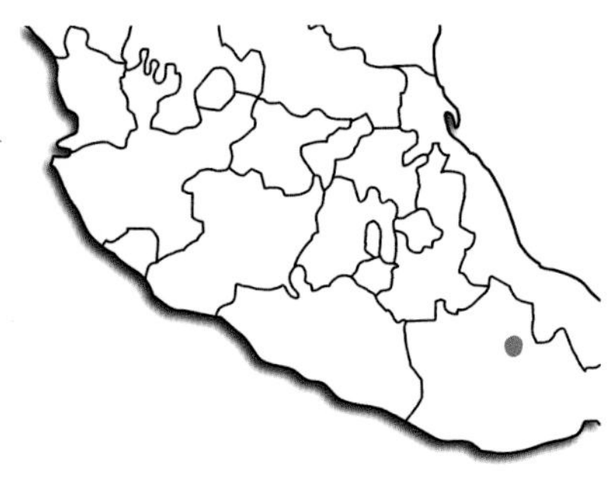

DISTRIBUTION: This species is endemic to the state of Oaxaca, and is known exclusively from Cerro Machín, in the Sierra de Juarez. This hill is located in the northern part of the Sierra of Zempoaltepec, which is geographically isolated (Carelton et al., 2002). *H. ixtlani* and *H. chinanteco* are in the same slope of Sierra de Juárez. It has been recorded in the state of OX.

EXTERNAL MEASURES AND WEIGHT
TL = 210 to 280 mm; TV = 100 to 147 mm; HF = 27 to 31 mm; EAR = 20 to 24 mm. Weight: 26 to 55 g.

DENTAL FORMULA: I 1/1, C 0/0, PM 0/0, M 3/3 =16.

NATURAL HISTORY AND ECOLOGY: The biology of *H. ixtlani* is not well known; apparently, it has semi-arboreal habits and is commonly found on the ground. It inhabits humid forests with dense undergrowth. Apparently, it is scarce, as it is only known from a few localities and a few individuals (Carleton et al., 2002). It has been collected with *Megadontomys chyophylus*, *Peromyscus beatae*, *P. mexicanus*, *Oryzomys rostratus*, and *O. alfaroi* (G. Ceballos, pers. obs.).

VEGETATIONAL ASSOCIATIONS AND ELEVATION RANGE: *H. ixtlani* inhabits cloud forests and other humid forests of oak and pine, with dense undergrowth with ferns, mosses, bromeliads, lichens, and other epiphytes (Robertson, 1975). It has been collected from 2,500 masl to 3,000 masl (Carleton et al., 2002).

CONSERVATION STATUS: The current status of this species is unknown. Its distribution is restricted to forests that have suffered a significant impact by agriculture and livestock; it must be considered endangered.

Habromys lepturus (Merriam, 1898)

Zempoaltepec crested tail mouse

Livia León and Esther Romo

SUBSPECIES IN MEXICO
H. lepturus is a monotypic species.
Recently, Carleton et al. (2002) found morphological and morphometric differences between *H. lepturus lepturus* and *H. lepturus ixtlani*, and assigned it a species level. Therefore, *H. lepturus* is a monotypic species. Molecular evidence confirms its species level (León-Paniagua et al., 2007; Rogers et al., 2007).

DESCRIPTION: *Habromys lepturus* is a medium-sized mouse, second in size among the species of the genus *Habromys*, of dark coloration, dotted with tan and a more pronounced line in the sides starting on the shoulders; the belly is white with black underfur and beige tips. The ears are dark, thin, and covered with brown short hair. The forefeet are white and the hind feet dark. The tail is long, almost the length of the head and body, hairy, and bicolored, although it may be uniformly dark (Goodwin, 1969; Osgood and Merriam, 1909; Robertson, 1975). The skull is large and robust, similar in measurements to those of *H. ixtlani*. The length of the palate is larger in *H. ixtlani* than in *H. lepturus* (León-Paniagua et al., 2007).

Habromys lepturus. Cloud forest. La Soledad, Oaxaca. Photo: Gerardo Ceballos.

EXTERNAL MEASURES AND WEIGHT
TL = 207 to 262 mm; TV = 103 to 146 mm; HF = 24 to 28 mm; EAR = 20 to 23 mm. Weight: 32 g.

DENTAL FORMULA: I 1/1, C 0/0, PM 0/0, M 3/3 =16.

NATURAL HISTORY AND ECOLOGY: *H. lepturus* has semi-arboreal habits and inhabits humid environments where epiphytes predominate. Apparently, they use bromeliads as refuge (A. Peña-Hurtado, pers. obs.). Although their eating habits are unknown, they are omnivores, eating fruits, seeds, and insects (Robertson, 1975). Reproduction is seasonal, and takes place at the end of the dry season and beginning of the rainy season (May and June). The average size of the litter is 1.9. It has been collected with *Peromyscus beatae*, *P. aztecus*, and *P. melanocarpus* (Musser, 1969; Robertson, 1975), and *Megadontomys chyophylus*, *Peromyscus mexicanus*, *Oryzomys rostratus* and *O. alfaroi* (L. León-Paniagua, pers. obs.).

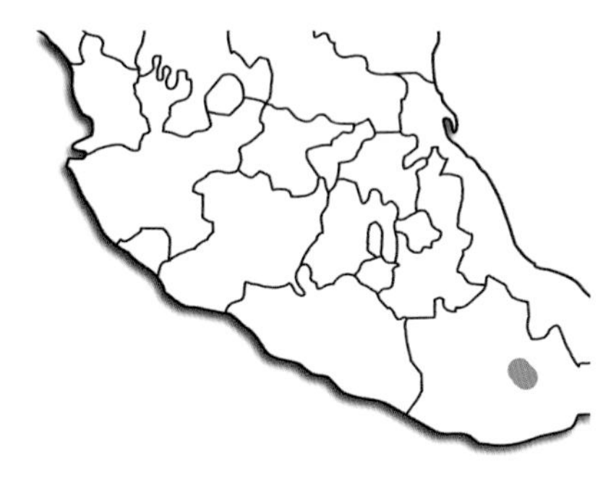

DISTRIBUTION: This species is endemic to the state of Oaxaca, distributed only in the highest parts of the Volcan Zempoaltepec (Carleton et al., 2002; Musser, 1969). It has been recorded in the state of OX.

VEGETATIONAL ASSOCIATIONS AND ELEVATION RANGE: *H. lepturus* is found in humid forests of oak and pine, and cloud forests, where trees are covered with bromeliads, mosses, lichens, and other epiphytes (Robertson, 1975). It has been collected from 2,500 masl to 3,000 masl (Carleton et al., 2002).

CONSERVATION STATUS: Apparently, this species is common in its restricted range. The accelerated deterioration of cloud forests throughout the country, and the strong requirement for this type of habitat, make this species vulnerable, however. It should be considered threatened with extinction (G. Ceballos, per. obs.).

Crested-tailed mouse

Livia León and Esther Romo

SUBSPECIES IN MEXICO
H. lophurus is a monotypic species.

DESCRIPTION: *Habromys lophurus* is a very similar rodent to *H. lepturus*, but smaller and of lighter coloration; the females are bigger than the males. The tail is long and hairy and ends in a brush. Its coat is silky and lustrous, brown-colored, tan, and fulvous in the upper parts, with a small dark area on the back; the ventral region is white. The tail is brown sepia. The forefeet are white and the hind feet are dark brown (Osgood, 1904, 1909). The skull is similar to that of *H. lepturus* (Osgood, 1904) but smaller. This species is intermediate in size in the genus *Habromys*, although generally larger than *H. simulatus* and *H. chinanteco*. The interparietal is large, the interorbital width is less than that of *H. simulatus* and *H. chinanteco*, and the face is short. The jaw of *H. lophurus* is as high as that of *H. lepturus* and *H. ixtlani*; however, the length of the teeth is much smaller. Compared with *Peromyscus beatae*, the skull of *H. lophurus* is shorter, with a shorter nose and a wider infraorbital region (León-Paniagua et al., 2007; Osgood, 1904).

EXTERNAL MEASURES AND WEIGHT
TL = 187 to 230 mm; TV = 92 to 115 mm; HF = 23 to 25 mm; EAR = 17 to 19 mm. Weight: 24 to 41 g.

DENTAL FORMULA: I 1/1, C 0/0, PM 0/0, M 3/3 =16.

NATURAL HISTORY AND ECOLOGY: Very little is known about the biology of this species; however, the habits of other species of the genus suggest that it is arboreal or semi-arboreal. It has been collected with *Peromyscus mayensis*, *P. guatemalensis*, and *P. oaxacensis* in Santa Eulalia in Guatemala (Carleton and Huckaby, 1975). Some individuals collected in May showed evidence of reproductive activity (Carleton and Huckaby, 1975).

VEGETATIONAL ASSOCIATIONS AND ELEVATION RANGE: *H. lophurus* inhabits humid and cold forests of oaks and conifers. In El Triunfo, it is found in the cloud forest (Hooper, 1947; Villa-R., 1948). In El Salvador, it has been found on humid slopes with large oaks and covered with epiphytes, ferns, and mosses (Burt and Stirton, 1961). The elevation interval where it occurs ranges from 1,950 masl in El Triunfo to 3,100 masl in Calel, Guatemala (Carleton and Huckaby, 1975; Villa-R., 1948).

CONSERVATION STATUS: *H. lophurus* is the only species of the genus that is not endemic to Mexico. Its conservation status is unknown. Due to its restricted distribution, it is likely to be fragile or threatened.

DISTRIBUTION: This species has been recorded from the high mountains of Chiapas, Guatemala, and El Salvador. León-Paniagua et al. (2007) maintain that the populations of Sierra de las Minas, Guatemala differ from *H. lophurus*. In Mexico it is distributed in the upper parts of the Sierra Madre in Chiapas in El Triunfo, Pinabete, Tacaná, and San Cristobal (León-Paniagua et al., 2007; Musser, 1969). It has been recorded in the state of CS.

Schmidly's crested tail mouse

Livia León

SUBSPECIES IN MEXICO
H. schmidlyi is a monotypic species.

DESCRIPTION: *Habromys schmidlyi* is one of the smallest *Habromys*, only slightly larger than *H. delicatulus*, and similar in body proportions to *H. simulatus* and *H. chinanteco*. The dorsal fur is dense, silky, and reddish brown with a darker narrow line in the middle of the back; the lateral line is orange. Like other species of *Habromys*, *H. schmidlyi* has dark rings around the eyes. The ventral fur is shorter than the dorsal fur and white. Most specimens have a yellowish stain at the base of the neck. The ears are dark and medium size, smaller than the size of the legs. The forelegs are relatively short and white; the hind legs are long. The tail is long. The sides of the head and body are covered with long hair, dark on the dorsum and clear in the ventral part. The tail is clear and well developed with a brush at the tip. The skull and teeth are very small; the braincase is low and wide at the height of the mastoid region; the zygomatic plate is as wide as the region postpalatal; the mesopterigoidea fossa is short and narrow. The most distinctive feature is that the first molars are wide, in comparison with the small size of the skull (Romo et al., 2005).

EXTERNAL MEASURES AND WEIGHT
TL = 144 to 167 mm; TV = 72 to 89 mm; HF = 19 to 21 mm; EAR = 16 to 21 mm.
Weight: 10 to 15 gr.

DENTAL FORMULA: I 1/1, C 0/0, PM 0/0, M 3/3 = 16.

Habromys schmidlyi. Oak forest. Taxco, Guerrero. Photo: Ricardo Paredes León.

NATURAL HISTORY AND ECOLOGY: There is little information about this species' natural history and ecology. *H. schmidlyi* is arboreal and restricted to forests with abundant epiphytes. It is not an abundant species (12 individuals collected in approximately 2,500 traps, 9 of them in traps placed on trees; Romo et al., 2005), following the method suggested by Sánchez (1996). Bromeliads are used as nests and fungi are part of their diet (Romo et al., 2005). Other rodents that have been collected with *H. schmidlyi* are *Peromyscus aztecus*, *P. levipes*, *Reithrodontomys megalotis*, and *R. sumichrasti* (Romo et al., 2005).

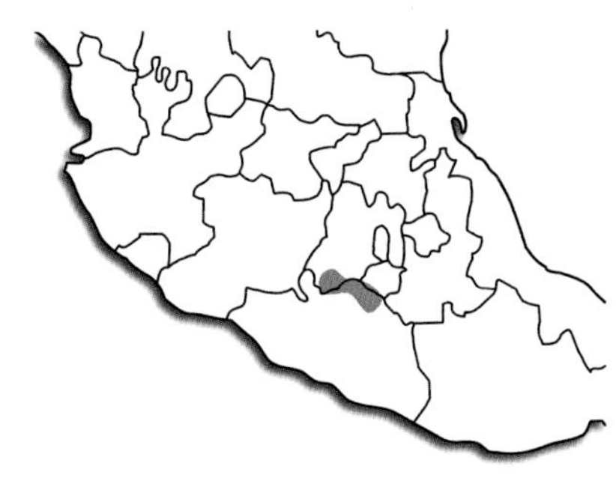

DISTRIBUTION: *H. schmidlyi* is endemic to Mexico. It was known from the type locality and two more nearest localities in the northern part of Sierra de Taxco, Guerrero. It has been recorded in the states of GR and MX.

VEGETATIONAL ASSOCIATIONS AND ELEVATION RANGE: The region where Schmily crested tail mice are found is part of the northern slope of the Transvolcani belt in central Mexico. The dominant plant community is an oak (*Quercus* spp.) forest with small patches of fir (*Abies* sp.) forest. There are relict tree elements of the cloud forest such as *Ilex*, *Garrya*, and *Cornus* (Carleton et al., 2002). Other detailed features of the vegetation are currently under study (O. Sanchez, pers. obs.).

CONSERVATION STATUS: *H. schmidlyi* is the species of the genus most recently described (June 2005). Therefore, little is known of its biology and conservation status. Because of its extremely small geographic range and arboreal habits it is likely to be at some risk of extinction.

Habromys simulatus (Osgood, 1904)

Jico crested tail mouse

Livia León and Esther Romo

SUBSPECIES IN MEXICO

H. simulatus is a monotypic species.

DISTRIBUTION: This species is known from only two type specimens collected in the towns of Xico and Zacualapan, Veracruz, on the hillside east of the Sierra Madre Oriental (Robertson and Musser, 1976; Hall, 1981). Carleton et al. (2002) and Rogers et al. (2007) considered that it was widely distributed along the slopes (1,830 m to 2,200 m) of the Sierra Madre Oriental, from southern and central Hidalgo, Veracruz, to Oaxaca. Recent studies, however, suggest that the populations in Oaxaca represent a new undescribed species (León-Paniagua et al., 2007). It has been recorded in the states of HG and VE.

DESCRIPTION: *Habromys simulatus* is a very small mouse with dark brown dorsal coloration and a white belly. It is distinguished from other mice of this genus by its small size, a thin white line in the bottom of the tail, and dark brown feet (Musser, 1969). The feet are small and have well-marked spots in the metatarsal region. The ears are medium and dark. The braincase is inflated and is smaller than that of medium-sized species of the genus (*H. lophurus* and *H. chinanteco*). The zygomatic plate and the face are wide, the zygomatic arches are not as delicate as those of *H. delicatulus* and *H. schmidlyi*, the interorbital constriction is wide, and the teeth, especially the third molar, are small (Hall, 1981; Osgood, 1904, 1909).

EXTERNAL MEASURES AND WEIGHT

TL = 168 to 203 mm; TV = 78 to 111 mm; HF = 21 to 24 mm; EAR = 16 to 19 mm. Weight: 17 to 19 g.

DENTAL FORMULA: I 1/1, C 0/0, PM 0/0, M 3/3 =16.

NATURAL HISTORY AND ECOLOGY: These arboreal mice are restricted to cloud forests. They feed on mosses and bromeliads, which are also used as shelters. It has been collected in trees up to 10 m and 12 m high (León-Paniagua et al., 2007). It has

Habromys simulatus. Cloud forest. Puerto de la Soledad, Oaxaca. Photo: Susette Castañeda Rico.

been collected with *Peromyscus levipes*, *Reithrodontomys mexicanus*, and *Megadontomys nelsoni*. Data on their reproduction are unavailable.

VEGETATIONAL ASSOCIATIONS AND ELEVATION RANGE: This species inhabits cold and humid cloud forests with dense vegetation, composed mainly of oaks. It has only been found at 2,000 masl.

CONSERVATION STATUS: *Habromys simulatus* is an endemic species of Mexico and considered a relict taxon (Carleton, 1990). The status of its populations is unknown. The original vegetation where this species was found has been replaced by coffee plantations (*Cofea arabiga*); it could be considered in danger of extinction.

Hodomys alleni (Merriam, 1892)

Allen's woodrat

Ticul Álvarez and Juan Carlos López-Vidal

SUBSPECIES IN MEXICO

Hodomys alleni alleni (Merriam, 1892)
Hodomys alleni elattura Osgood, 1938
Hodomys alleni guerrerensis Goldman, 1938
Hodomys alleni vetula Osgood, 1894

Hodomys alleni was described by Merriam (1892) and subsequently assigned to *Hodomys* by the same author (1894). The range of genus or subgenus of *Hodomys* has been widely discussed; however, most authors consider it a valid genus (Carleton, 1980; Edwards and Bradley, 2002; Ramírez-Pulido et al., 1983; Schaldach, 1960).

DESCRIPTION: Allen's woodrats are relatively large rats whose size and color vary markedly within the subspecies. In general, the dorsal coloration varies from reddish-cinnamon to fawn, with greater or lesser mixture of dark hair. The belly is opaque white. The legs are whitish, although the hind feet have a mixture of dark brown hair in the dorsal surface. The tail is solid or clearly bicolored (Genoways and Jones, 1973; Goldman, 1938; Merriam, 1892). The skull is long and narrow, the face is elongated and thin, the bullae are small, and the molars are S-shaped. The number of chromosomes is 2n = 48, FN = 52–53. This species is related to the genus *Neotoma*, but is distinguished mainly by the S shape of the third inferior molar (Merriam, 1894).

DISTRIBUTION: *H. alleni* is an endemic species of Mexico with a distribution restricted to the Pacific Rim from El Rosario, Sinaloa (Genoways and Birney, 1974), to Acapulco, Guerrero (Osgood, 1938). It has been recorded in the states of CL, GR, JA, MI, MO, NY, OX, PU, and SI.

EXTERNAL MEASURES AND WEIGHT

TL = 300 to 446 mm; TV = 25 to 33 mm; HF =140 to 224 mm; EAR= 28 to 33 mm. Weight: 119 to 452 g.

DENTAL FORMULA: I 1/1, C 0/0, PM 0/0, M 3/3 =16.

NATURAL HISTORY AND ECOLOGY: These rats prefer rocky areas, mainly in ravines, marking their paths and leaving accumulations of excrement or debris to mark their presence. They do not construct large piles of debris like other species of rats; however, Schaldach (1960) asserted that in Colima they form large nests with branches. Their feeding habits are unknown. Births occur in August, September, and October. A female with embryos was captured in September (Uribe-Peña et al., 1981), and lactating females were reported in August (Genoways and Jones, 1973) and September (Birney and Jones, 1972). In addition, juveniles have been caught in September and December (Allen, 1897). At the beginning of the year, all encountered specimens are adults. They have been generally captured in the same area with *Liomys pictus* and *Osgoodomys banderanus*, although in some localities they have also been caught with *Xenomys nelsoni*, *Neotoma mexicana*, *Megasorex gigas*, *Spermophilus adocetus*, *Peromyscus hylocetes*, and *P. maniculatus* (Ceballos and Miranda, 1986). Remains of this species have been found in owl pellets (*Tyto alba*; Ramírez-Pulido and Sánchez-Hernández, 1972). Apparently, this rat emits sounds similar to those emitted by squirrels (Schaldach, 1960) or by *Ochotona princeps* (Birney and Jones, 1972).

Hodomys alleni. Tropical dry forest. Tierra Colorada, Guerrero. Photo: Gerardo Ceballos.

VEGETATIONAL ASSOCIATIONS AND ELEVATION RANGE: *Hodomys alleni* inhabits tropical deciduous forests, thorn scrublands and chaparral, secondary vegetation, and crops such as coconut, mango, cornfields, or melon fields. It is found from sea level to 1,800 m (Winkelmann, 1962a).

CONSERVATION STATUS: *H. alleni* is not recorded in any risk category of the Convention on International Trade in Exotic Species (CITES), the International Union for Conservation of Nature (IUCN) (2010), or the Mexican Official Standard (SEMARNAT, 2010). It is relatively rare, judging by the small number of captures; however, because of its wide distribution in Mexico it is not considered threatened.

Megadontomys cryophilus (Musser, 1964)

Oaxacan big-toothed deermouse

Luis Arturo Peña and Yolanda Domínguez

SUBSPECIES IN MEXICO
M. cryophilus is a monotypic species.

DESCRIPTION: Compared to mice of the genus *Peromyscus*, *Megadontomys cryophilus* is rather large. Its face is wide, the ears are small, and the hind feet are large. The hair is long, thick, and silky; the hair of the ears is very short, making them look nude. The dorsal coloration is dark brown; the belly is creamy white or light gray. The nose, the bases of the whiskers, and the orbital area are black. The feet are white. The nasals are particularly long, the zygomatic region is wide, and the length of upper molar toothrow is more than 6 mm (Werbitsky and Kilpatrick, 1987).

EXTERNAL MEASURES AND WEIGHT
TL = 300 to 350 mm; TV = 155 to 188 mm; HF = 31 to 34 mm; EAR= 20 mm.
Weight: 57 g.

Megadontomys cryophilus. Cloud forest. La Soledad, Oaxaca. Photo: Gerardo Ceballos.

DENTAL FORMULA: I 1/1, C 0/0, PM 0/0, M 3/3 = 16.

NATURAL HISTORY AND ECOLOGY: Little is known about this species' natural history. It is a nocturnal species, which feeds on seeds. When startled they seek shelter in trees. Males with evidence of reproductive activity were caught in April, June, July, and November. Lactating females have been caught in March, May, July, and November, indicating that they are polyestric. The litter generally consists of three offspring. It has been captured together with *Peromyscus mexicanus*, *P. aztecus*, *Tylomys nudicaudus*, and *Oryzomys alfaroi*.

VEGETATIONAL ASSOCIATIONS AND ELEVATION RANGE: *M. cryophilus* has been collected in cloud forests, pine forests, and forests of pine-oak, from 2,400 masl to 3,500 masl.

CONSERVATION STATUS: *M. cryophilus* is considered as threatened, as it is endemic to the Sierras of Oaxaca and its distribution is restricted (SEMARNAT, 2010).

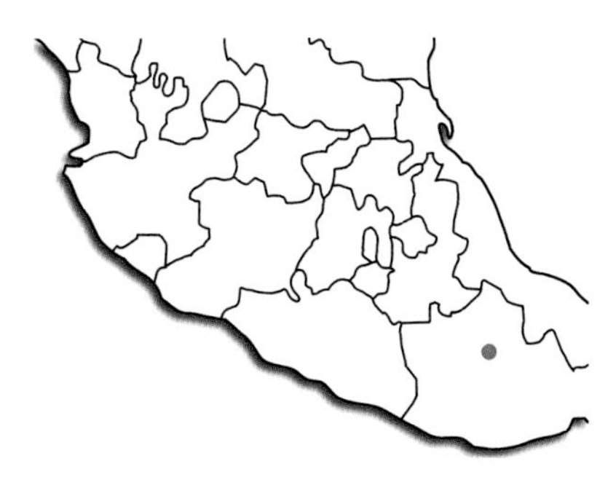

DISTRIBUTION: *M. cryophilus* is an endemic species of Mexico, distributed in the high-elevation regions in the Sierra de Juarez. It has been recorded in the state of OX.

Megadontomys nelsoni (Merriam, 1898)

Nelson's big-toothed deermouse

Luis Arturo Peña and Beatriz Hernández

SUBSPECIES IN MEXICO

M. nelsoni is a monotypic species.

DESCRIPTION: Morphologically, *Megadontomys nelsoni* has intermediate characters between *M. cryophilus* and *M. thomasi*, but it is considered closer to the latter (Musser, 1964). The differences between species in the genus are at the genetic level. It is a large mouse with a slightly elongated face. The ears are small, the tail is longer than the rest of the body, and the hind feet are long. The dorsal coloration is dark brown; the back is slightly darker than the sides, and the abdomen is creamy white or light gray. The legs are white. The nose, the bases of the whiskers, and the orbital area are black.

EXTERNAL MEASURES AND WEIGHT

TL = 302 to 318 mm; TV = 170 to 172 mm; HF = 32 to 35 mm; EAR = 20 mm.
Weight: 57 g.

DENTAL FORMULA: I 1/1, C 0/0, PM 0/0, M 3/3 = 16.

NATURAL HISTORY AND ECOLOGY: This species has nocturnal habits. Its diet is based on seeds. The reproductive period extends from March to November, suggesting it is polyestric. The litter comprises two to three offspring. It is an important species because it participates in the dispersal of seeds and fungal spores and in the removal of forest soil. It has been captured together with *Habromys simulatus*, *Microtus quasiater*, *Peromyscus levipes*, *P. furvus*, and *Oligoryzomys fulvescens*.

VEGETATIONAL ASSOCIATIONS AND ELEVATION RANGE: *M. nelsoni* inhabits cloud forests and forests of pine-oak, from 2,000 masl to 3,500 masl.

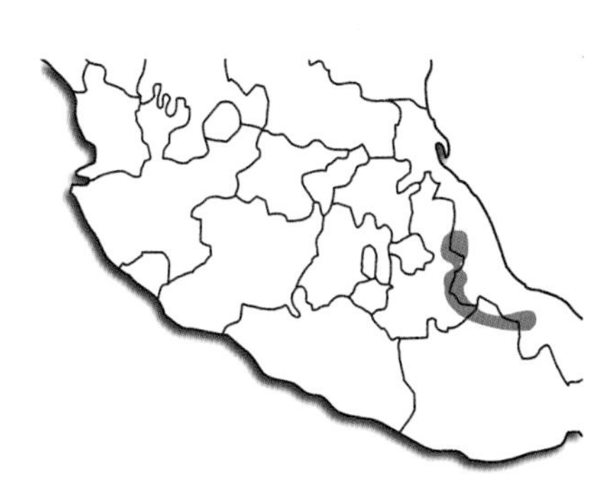

DISTRIBUTION: *M. nelsoni* is distributed in the high-elevation regions of the Sierra Madre Oriental. It has been recorded in the states of OX, PU, and VE.

CONSERVATION STATUS: This species is considered threatened, as it is endemic to Mexico, has a very restricted distribution, and may be susceptible to habitat destruction (SEMARNAT, 2010).

Megadontomys thomasi (Merriam, 1898)

Thomas's big-toothed deermouse

Luis Arturo Peña and Yolanda Domínguez

SUBSPECIES IN MEXICO

M. thomasi is a monotypic species.

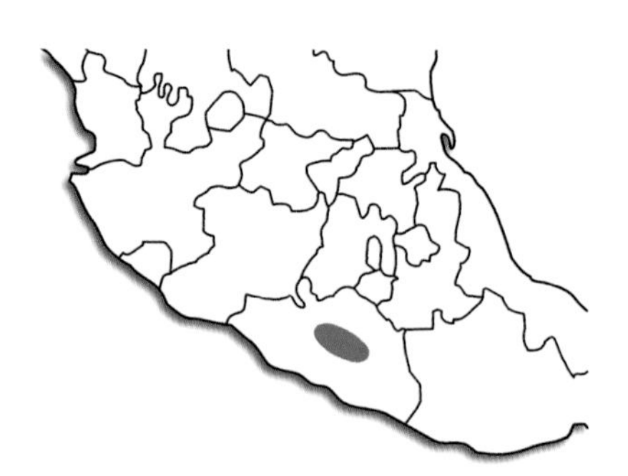

DISTRIBUTION: *M. thomasi* is endemic to Mexico, and is distributed in the high-elevation regions of the Sierra Madre del Sur in Guerrero. It has been recorded in the state of GR.

DESCRIPTION: *Megadontomys thomasi* is the largest species in the genus. It is distinguished by the very long tail, which is longer than the body; the ears are very small. The coat is long, thick, and smooth; the ears have tiny hair and appear naked. The dorsal coloration is dark brown or fulvous, mixed with black in the upper part of the body; the belly is creamy white or light gray. The nose, bases of the whiskers, and orbital area are black. The feet are white.

EXTERNAL MEASURES AND WEIGHT

TL = 300 to 350 mm, TV = 155 to 188 mm; HF = 31 to 34 mm; EAR = 21 to 25 mm. Weight= 58 g.

DENTAL FORMULA: I 1/1, C 0/0, PM 0/0, M 3/3 = 16.

NATURAL HISTORY AND ECOLOGY: *M. thomasi* is nocturnal and is active throughout the year. It feeds on seeds. When threatened it climbs trees. Reproductively active males have been captured during April, June, July, and November. Lactating females were found in March, May, July, and November, which indicate it is polyestric, giving birth to two offspring per litter. It has been captured together with *Peromyscus beatae*, *P. aztecus*, *P. megalops*, *Neotoma mexicana*, and *Oryzomys alfaroi*. Ectoparasites include species of *Siphonaptera*.

VEGETATIONAL ASSOCIATIONS AND ELEVATION RANGE: *M. thomasi* has mainly been found in cloud forests, oak forests, pine forests, and pine-oak forests, from 3,000 masl to 3,500 masl.

CONSERVATION STATUS: *M. thomasi* is rare and endemic to the mountains of central Guerrero. It is considered Special Protection (SEMARNAT, 2010).

Goldman's diminutive woodrat

Mark. D. Engstrom

SUBSPECIES IN MEXICO

Nelsonia goldmani cliftoni Genoways and Jones, 1968
Nelsonia goldmani goldmani Merriam, 1903
Initially, the genus *Nelsonia* included a single species, *N. neotomodon*; however, Engstrom et al., (1992) found sufficient evidence to separate *N. goldmani* from *N. neotomodon.*

DESCRIPTION: *Nelsonia goldmani* is a small-sized rat. The tail is thick and hairy, the ears are rounded, and the whiskers are large. It is similar to the largest species of *Peromyscus*, but can be distinguished by its thick and hairy tail, the long and rough facial hair, and the larger hind feet. Both subspecies are distinguished by their coloration; *N. goldmani goldmani* is dark gray and the lateral line is poorly defined. In contrast, *N. goldmani cliftoni* is more yellowish and the lateral line is wide and buffy. Both subspecies are distinguished from *N. netomodon* by the darker back, larger and darker feet, and the coloration of the tail that in *N. netomodon* is clearly bicolored with a white tip.

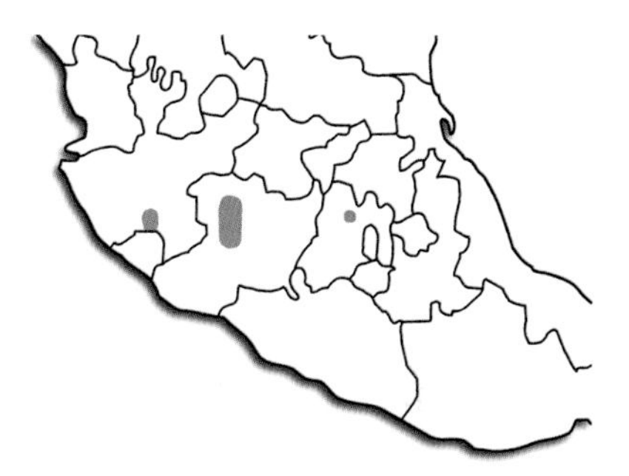

DISTRIBUTION: This rat is endemic to Mexico. It inhabits several isolated localities in the high-elevation regions of the Eje Volcanico Transversal, from the eastern state of Mexico to Colima. It has been recorded in the states of CL, JA, MI, and MX.

EXTERNAL MEASURES AND WEIGHT

TL = 225 to 259 mm; TV = 108 to 126 mm; HF = 25 to 30 mm; EAR = 22 to 26 mm. Weight: 43 to 57 g.

DENTAL FORMULA: I 1/1, C 0/0, PM 0/0, M 3/3 =16.

NATURAL HISTORY AND ECOLOGY: This rat has been found in temperate forests of pine-oak and fir (*Abies*) of the west of the Eje Volcanico Transversal and in cloud forests in the state of Mexico. It is generally found in wet environments associated with rocky slopes, ravines, and gullies. Most known specimens have been caught on the ground, frequently under rocks with mosses. Once, a tail was found in a trap placed 2.5 m from the ground, in a tree, which indicates it is partially arboreal (Engstrom et al., 1992). Details of its diet are unknown; however, stomach content had a green mass, which had the appearance of pine needles (Hooper, 1964). Very little is known about its reproduction. Non-pregnant females have been captured in October and July (Genoways and Jones, 1968a). Juveniles have been found in Michoacán in February (Merriam, 1903). It has been captured together with *Peromyscus hylocetes*, *Microtus mexicanus*, and *Reithrodontomys sumichrasti.*

VEGETATIONAL ASSOCIATIONS AND ELEVATION RANGE: *N. goldmani* mainly inhabits forests of conifers and oak (*Quercus*) and cloud forests from 2,000 masl to 3,100 masl.

CONSERVATION STATUS: This rat is endemic to Mexico, it is quite rare and apparently, it is a specialist. It could be capable of withstanding certain habitat disturbances and deforestation, but within the restricted area and habitat requirements it could become vulnerable to extinction. In the Official Mexican Standard it is regarded as of Special Protection (SEMARNAT, 2010). The preferred habitats of this rat are similar to those used by monarch butterflies in winter (Glendinning, 1992). The efforts to preserve the butterfly's sanctuaries in Michoacán could benefit indirectly this species of rat.

Nelsonia neotomodon Merriam, 1897

Western diminutive woodrat

Oscar Sánchez

SUBSPECIES IN MEXICO

N. neotomodon is a monotypic species.

Two subspecies were thought to exist. Recently, the taxonomic status was assessed and two species (*goldmani* and *neotomodon*) were recognized (Engstrom et al., 1992).

DISTRIBUTION: This species is endemic to Mexico. The southern limit of its distribution seems to coincide with the southern ridges of the Sierra Madre Occidental, from Durango to Jalisco, where apparently, there is a discontinuity of relict mountain environment that reappear in the Eje Volcanico Transversal, where *N. goldmani*, sister species of *N. neotomodon*, substitutes for it (Engstrom et al., 1992; Muñiz-Martinez and Arroyo-Cabrales, 1996). It has been recorded in the states of AG, DU, JA, and ZA.

DESCRIPTION: *Nelsonia neotomodon* is a medium-sized rodent. The dorsal coloration is light brown with cinnamon shades, darker toward the midline of the back. The back of the hind feet is usually white, the back of the tail is dark and is clearly delimited by the light coloration of the underside; usually, the tip of the tail is white. The molars have a prismatic appearance and the prisms alternate instead of being opposed (Baker and Greer, 1962; Engstrom et al., 1992; Hall, 1981).

EXTERNAL MEASURES AND WEIGHT

TL = 227 to 238 mm; TV = 109 to 117 mm; HF = 25 to 30 mm; EAR = 23 to 25 mm.
Weight: 43 to 55 g.

DENTAL FORMULA: I 1/1, C 0/0, PM 0/0, M 3/3 = 16.

NATURAL HISTORY AND ECOLOGY: The close relation of this species and its mountain environments proves its relict nature, and thereby, its dependence on local vegetation climax. According to current evidence, they are locally scarce (Baker and Greer, 1962; Matson and Baker, 1986), and share their habitat with other rodents such as *Peromyscus difficilis*, *P. boylii*, and *Neotoma mexicana* (Baker and Greer, 1962; Matson and Baker, 1986). Females caught in Durango in July show no evidence of reproductive activity (Baker and Greer, 1962).

VEGETATIONAL ASSOCIATIONS AND ELEVATION RANGE: *N. neotomodon* has been recorded in pine-oak forests, fir forests, and sometimes forests with juniper and poplar, on wet and freezing slopes facing north, in areas where mosses and herbs prevail on rocky areas (Baker and Greer, 1962; Engstrom et al., 1992). It has been found from 2,225 masl to 2,985 masl (Baker and Greer, 1962; Engstrom et al., 1992; Hall, 1981).

CONSERVATION STATUS: *N. neotomodon* is protected, and is included in the category of Special Protection and endemic species (SEMARNAT, 2010). Although its distribution area is relatively large (ca. 25,000 km^2), there is a real need to protect the wet ravines with climax vegetation, especially in those localities where the species is known.

White-throated woodrat

Eric Mellink and Jaime Luévano

DISTRIBUTION: This species is found from the western United States to northwestern Mexico. In Mexico, its distribution includes northeastern Baja California, Sonora, and northern Sinaloa and Chihuahua, west of the Rio Conchos. It has been recorded in the states of BC, CH, DU, SI, and SO.

SUBSPECIES IN MEXICO

Neotoma albigula albigula Hartley, 1894
Neotoma albigula melanura Merriam, 1894
Neotoma albigula seri Townsend, 1912
Neotoma albigula sheldoni Goldman, 1915
Neotoma albigula venusta True, 1894

The subspecies that distribute on the eastern side of the River Conchos in Chihuahua have been recently raised to a species level under the name *N. leucodon*, based on genetic differences (Edwards et al., 2001).

DESCRIPTION: *Neotoma albigula* is a rather large rat. The dorsal coloration is grayish-brown, with light brown and orange shades. The coloration turns ochraceous with age. In some habitats, such as lava fields, its coloration is very dark. Unlike in *N. lepida*, the hair of the throat is white at the base. The tail is bicolored with scarce hair. The feet are covered with hair, except for the soles. The eyes are big and black (Hall, 1981; Hoffmeister, 1986). Edwards et al. (2001) revised *N. albigula* by removing eastern populations and placing them in *N. leucodon.*

EXTERNAL MEASURES AND WEIGHT

TL = 282 to 400 mm; TV = 76 to 185 mm; HF = 30 to 39 mm; EAR = 28 to 30 mm. Weight: 145 to 200 g.

DENTAL FORMULA: I 1/1, C 0/0, PM 0/0, M 3/3 = 16.

NATURAL HISTORY AND ECOLOGY: *N. albigula* is nocturnal but is occasionally seen during the day. It is terrestrial but can be found climbing bushes and columnar cacti.

Neotoma albigula. Scrubland. Janos Biosphere Reserve, Chihuahua. Photo: Gerardo Ceballos.

It builds its burrows in the base of trees, shrubs, rocks, and cacti. For construction they use thorny plants (chollas, prickly pear, mesquite), leaves, stones, feathers, bones, feces of other animals, and even garbage. In general, the burrows are conical and have several exits; they measure up to 118 cm high and 300 cm in diameter. In the entrance, stones of several colors and shiny objects such as traps used for their capture have been found (Brylski, 1990b; Hoffmeister, 1986; Jameson and Peeters, 1988). Their density is related to the habitat. The diet varies according to the habitat and season, but is based on succulents such as prickly pear (*Opuntia* sp.) and other cacti; they also eat herbs, leaves of izote (*Yucca* sp.), fruits, and inflorescences. Most animals are prone to severe physiological problems when their diet consists only of cacti, as these plants are rich in oxalic acids. This species can consume large amounts of prickly pear, however, thanks to its unique intestinal flora. In the densest environments these rats are larger and fatter; hence in this habitat, and in those with less quality, they suffer seasonal variation in their nutritional condition (Rangel and Mellink, 1993). They do not require liquid water as most of it is obtained from succulents (Brylski, 1990b; Hoffmeister, 1986; Jameson and Peeters, 1988). They reproduce between January and July, and reproduction apparently depends on the rain. They can have more than a litter per year with one to three offspring. The offspring are born in spring, after a gestation period of 38 days, and are weaned within 27 to 40 days. Sexual maturity is reached at 80 and 100 days of age in females and males, respectively. Locals consume the meat of these rats along its distribution area, and the O'odham people value it highly. Woodrats defecate and urinate in a certain spot, where different elements of the environment (twigs, seeds, leaves, and bones) are also deposited. When the burrows are constructed in crevices or caves, and when the nests are protected from atmospheric conditions, these dunghills can be fossilized; some have even been collected, producing valuable information since they are several thousand years old. From these analyses, the local vegetation of the areas has been reconstructed for the past 20,000 to 30,000 years.

VEGETATIONAL ASSOCIATIONS AND ELEVATION RANGE: *N. albigula* is found in a great variety of environments, including variations of bushes in arid areas and thorn scrublands, and in lesser densities in forests of conifers and grasslands with some bushes. It is found from sea level to 2,475 masl.

CONSERVATION STATUS: *N. albigula* and its different subspecies are very tolerant of environmental variations and anthropogenic activities. They have a wide distribution, which probably ensures the survival of all subspecies. This is true even for the island subspecies *N. albigula seri* of Tiburon Island, although it is officially classified as threatened (SEMARNAT, 2010).

Neotoma angustapalata Baker, 1951

Tamaulipan woodrat

Emilio Daniel Tobón García

SUBSPECIES IN MEXICO

N. angustapalata is a monotypic species.

DESCRIPTION: *Neotoma angustapalata* is a large-sized rat. The upper parts of the body are brownish. The cheeks are grayish, and the underside is white. The gular region and groin are completely white. The tail is slightly black in the upper part and white underneath. The tail is long, almost the same length as the head and body; it is robust and covered with thin hair. The ears and eyes are large (Álvarez, 1963a; Baker 1951a; Birney, 1973; Hall, 1981; Hooper, 1953; Hutterer, 1993).

DISTRIBUTION: This rat is endemic to Mexico. It is known from a few localities in the Sierra Madre Oriental, in southern Tamaulipas, and in San Luis Potosí (Álvarez, 1963a; Birney, 1973, Hutterer, 1993). It has been recorded in the states of SL and TA.

EXTERNAL MEASURES AND WEIGHT
TL = 380 to 420 mm; TV = 195 to 200 mm; HF = 39 to 44 mm; EAR = 30 to 32 mm. Weight: 180 to 240 g.

DENTAL FORMULA: I 1/1, C 0/0, PM 0/0, M 3/3 = 16.

NATURAL HISTORY AND ECOLOGY: This species has been collected in crevices in a small hill with a dense cover of humus. Inside a cave, a female with two young were captured. It builds its burrows inside caves (Álvarez, 1963a; Birney, 1973; Reddell, 1981). Usually, they are solitary, nocturnal, and active throughout the year. They mainly feed on seeds and other parts of plants and occasionally on invertebrates. Apparently, they reproduce throughout the year. The size of the litter is two to three offspring in spring, and one to two in autumn. The gestation period lasts 30 to 40 days; the offspring become independent within 4 weeks and reach sexual maturity near 8 months of age (Novak, 1991).

VEGETATIONAL ASSOCIATIONS AND ELEVATION RANGE: *N. angustapalata* is distributed in pine-oak forests (Álvarez, 1963a; Birney, 1973). There is a record at an elevation of 1,200 masl in El Encino, Tamaulipas, and another at 1,150 masl (Álvarez, 1963a; Birney, 1973).

CONSERVATION STATUS: This species is not considered endangered. It is endemic and scarce so it should be regarded as threatened. Nonetheless, accurate assessments are required to determine its current status.

Neotoma bryanti Merriam, 1887

Bryant's woodrat

Eric Mellink, Sergio Ticul Álvarez, and Jaime Luévano

SUBSPECIES IN MEXICO
Neotoma bryanti anthonyi J. A. Allen, 1898
Neotoma bryanti bryanti Merriam, 1887
Neotoma bryanti intermedia Rhoads, 1894
Neotoma bryanti macrosensis Burt, 1932
Neotoma bryanti martinensis Goldman, 1905
Patton et al. (2007) reviewed the systematics of the genus. They concluded that *N. bryanti* is widespread in the Baja California Peninsula, that *N. bryanti* has several subspecies instead of being a monotypic species endemic to an island, and that *N. anthonyi* and *N. martinensis* are part of its subspecies instead of different species. They also concluded that *N. bunkeri* is not a different species but a synonym of *N. bryanti*.

DESCRIPTION: *Neotoma bryanti* is a very large rat with large ears and a moderately long and bicolored tail. The dorsal coloration is creamy white with the underside

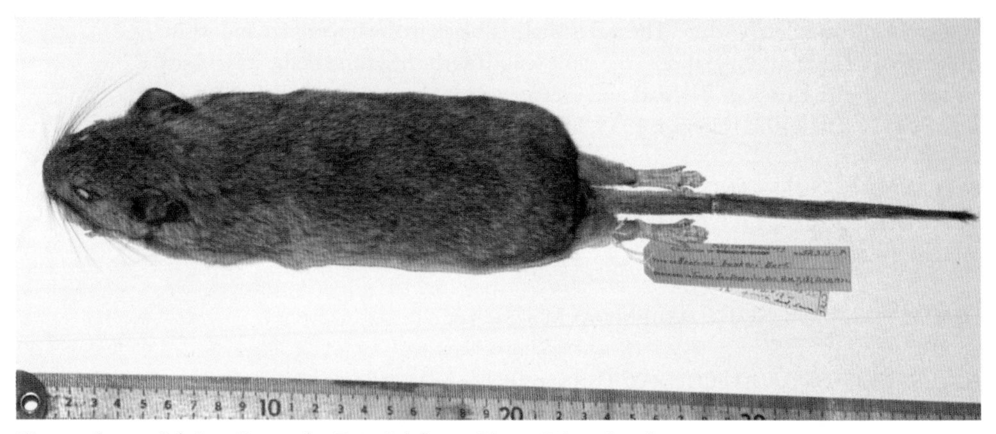

Neotoma bryanti. Isla Los Coronados, Baja California. Photo: Felissa Smith.

whitish. The outer parts of the feet have conspicuous black marks resembling those of *N. bryanti anthonyi* and *N. bryanti martinensis*. The skull is similar to that of *N. lepida*, but is larger and angular; the frontals widen abruptly in the anterior part, between the lachrymals; the interpterigoidal fossa is narrower and the teeth are more robust (Goldman, 1910; Merriam, 1887). Its taxonomic status has been resolved with genetic techniques (Patton et al., 2007). Apparently, it derived from *N. lepida* (Mascarello, 1978) when the island drifted from the mainland approximately 9,000 years ago (Smith, 1992).

EXTERNAL MEASURES AND WEIGHT

TL = 330 to 369 mm; TV = 155 to 176 mm; HF = 36 to 40 mm; EAR= 30 to 31 mm. Weight: 100 to 190 g.

DENTAL FORMULA: I 1/1, C 0/0, PM 0/0, M 3/3 = 16.

NATURAL HISTORY AND ECOLOGY: Like other species of the genus, *N. bryanti* builds its burrows in crevices and under bushes. The burrows resemble those of *N. lepida* and are built with twigs and other materials. The mating season is unknown, but it includes spring–summer. During this time the animals that have been caught were healthy and without ectoparasites (Mellink, 1993a).

VEGETATIONAL ASSOCIATIONS AND ELEVATION RANGE: *N. bryanti* inhabits desert bush, chaparrals, forests of pine, and small orchids (Mellink, 1993a). It is found from sea level to 1,000 m.

CONSERVATION STATUS: Island populations, considered as species until recently, are threatened by predation by feral cats (Ceballos and Navarro, 1991; SEMARNAT 2010). It is necessary to eradicate the cats to ensure the species' long-term conservation. Populations from Todos Santos Island (*anthonyi*), San Martin Island (*martinensis*), and Coronados islands (*bunkeri*) are considered extinct (Álvarez-Castañeda and Cortes-Calva, 1999; Ceballos and Navarro, 1991).

DISTRIBUTION: Based on the revision of Patton et al. (2007) the distribution of this species now includes California and all of the Baja California Peninsula. The species has been recorded in BC and BS.

Arizona woodrat

Yolanda Domínguez

SUBSPECIES IN MEXICO

N. devia is a monotypic species.

Based on chromosomal banding (C and G) of the genus *Neotoma*, Koop et al. (1985) noted that chromosome 6 of *Neotoma devia* and *N. lepida* has a pericentric inversion, which used to bond both species; however, the G-band karyotype in *N. lepida* is different from the other species, including *N. devia*; hence they were separated. Patton et al. (2007) used morphologic and genetic data to revise this and other species of the *N. lepida*-group.

DESCRIPTION: *Neotoma devia* is a small-sized rat similar to *N. lepida*. The dorsal coloration varies from gray to reddish-brown; the underside is usually white, as well as the feet (Nowak, 1999b). The ears are relatively large and furry; as in other members of the genus the tail is hairy (Nowak, 1999b). See the revision by Patton et al. (2007) for more details.

EXTERNAL MEASURES AND WEIGHT

TL = 150 to 230 mm; TV = 75 to 240 mm; HF = 28 to 41 mm; EAR = 23 to 25 mm. Weight: 100 to 199 g.

DENTAL FORMULA: I 1/1, C 0/0, PM 0/0, M 3/3 = 16.

NATURAL HISTORY AND ECOLOGY: This rat is nocturnal and occasionally crepuscular, and is active throughout the year (Miller and Stebbins, 1964; Stones and Hayward, 1968). They are solitary and aggressive (Macmillen, 1964). Females can be very prolific, having up to five litters per year (Burt and Grossenheider, 1980). The gestation period lasts 30 to 36 days, and the size of the litter varies from 1 to 5 offspring (Egoscue, 1957). Their diet includes vegetable matter such as thorny cacti, pods, berries, pine nuts, seeds, and leaves (Meserve, 1974). They also eat succulents as a metabolic water resource. The burrows are built with rocks, roots of cacti, and chollas, and are used to avoid extreme temperatures, hide from predators, and store food (Nowak, 1999b).

DISTRIBUTION: *N. devia* is found in the western part of Arizona, and northern Sonora (Goldman, 1932; Patton et al., 2007; Riddle et al., 2000; Wilson and Reeder, 2005). It has been recorded in the state of SO.

VEGETATIONAL ASSOCIATIONS AND ELEVATION RANGE: *N. devia* inhabits desert areas with xeric bushes, from sea level to 1,600 m (Miller and Kellogg, 1955; Wilson and Reeder, 2005).

CONSERVATION STATUS: This species has been poorly studied. The current status of its populations is unknown, but it may be fragile. Some populations inhabit El Pinacate Biosphere Reserve.

Goldman's woodrat

Emilio Daniel Tobón García

SUBSPECIES IN MEXICO
N. goldmani is a monotypic species.

DISTRIBUTION: *N. goldmani* is an endemic species of Mexico. It is found in the northern part of the Altiplane, from Coahuila, southeast Chihuahua, to Nuevo Leon, San Luis Potosí and Zacatecas (Baker and Greer, 1962; Matson and Baker, 1986; Musser and Carleton, 1993). It has been recorded in the states of CH, CO, DU, NL, SL, and ZA.

DESCRIPTION: *Neotoma goldmani* is a medium-sized rat. The upper part of the body is light yellow; it is paler on the head and darker on the rear; the underside is white (Hall, 1981).

EXTERNAL MEASURES AND WEIGHT
TL = 279 mm; TV = 128 mm; HF = 30 mm; EAR = 28 mm.
Weight: 91 g.

DENTAL FORMULA: I 1/1, C 0/0, PM 0/0, M 3/3 = 16.

NATURAL HISTORY AND ECOLOGY: This species inhabits arid areas and rocky sites (Baker and Greer, 1962; Matson and Baker, 1986). In general, it is nocturnal, solitary, and active throughout the year. It mainly feeds on roots, stems, leaves, and seeds; occasionally, it eats invertebrates. The size of the litter ranges from 2 to 4 offspring, which reach sexual maturity at 6 to 8 months of age; the gestation period lasts 30 to 40 days (Nowak, 1999b). In general, the populations do not alter crops; hence they can be considered of little economic importance.

VEGETATIONAL ASSOCIATIONS AND ELEVATION RANGES: The vegetation types *N. goldmani* occupies are not completely known, but in general it inhabits forests of conifers and oaks. The elevation of the type locality is 1,525 masl; the altitudinal interval goes from 1,150 masl to 1,830 masl (Anderson, 1972; Hall, 1981; Musser and Carleton, 1993).

CONSERVATION STATUS: *N. goldmani* is not included in any list of endangered species, but the alteration and modification of its habitat can determine its conservation status, taking into account its endemism in Mexico.

Neotoma goldmani. Scrubland. Santo Domingo, San Luis Potosí. Photo: Juan Cruzado.

Ángel de la Guarda woodrat

Matthew R. Mauldin and Robert D. Bradley

SUBSPECIES IN MEXICO

N. insularis is a monotypic species.

A distinctly supported clade derived from cytochrome b mitochondrial data indicates substantial genetic divergence from *N. lepida* (Patton et al., 2007) and led to the re-elevation to species status.

DESCRIPTION: *Neotoma insularis* is a medium-sized woodrat closely resembling *N. lepida*. The dorsal pelage is pinkish-buff to buffy gray; the venter is a lighter gray. The bases of hairs are gray, while the tips are lighter. The tail is short (approximately 71% of head and body length) and bicolored with dark gray on top and paler gray underneath. The glans penis is short and broad. Spines cover most of the glans, with the only exception being the distal tip. The rostrum is visibly shorter and stouter than in *N. lepida*. The zygomatic arches are squared (broad both anteriorly and posteriorly). The septum of the incisive foramen consists of a short vomerine portion and an elongated vacuity. The contact area between the frontal and lacrimal bones is much greater than the contact area between the maxillary and lacrimal bones. The auditory bullae are long and narrow. Dentition is heavier than in *N. lepida*, and the anteromedial flexus of anteroloph of M1 shallow except in young individuals. Previously considered a member of the *N. lepida*-complex, recent mtDNA data and cranial morphology (Patton et al., 2007) has provided adequate evidence to elevate *N. insularis* to the species level.

DISTRIBUTION: *N. insularis* is an endemic species to Mexico, known exclusively from the Isla Ángel de la Guarda located in the Sea of Cortez. The few localities of specimens are from the northern and southern tips of the island. This may be due to difficulty accessing the center of the island.

EXTERNAL MEASURES AND WEIGHT

TL = 290 mm; TV = 120 mm; HF = 35 mm; EAR = N/A.

Weight = N/A.

DENTAL FORMULA: I 1/1, C 0/0, PM 0/0, M 3/3= 16.

NATURAL HISTORY AND ECOLOGY: Little is known about this species in general. It has been recorded only from the Isla Ángel de la Guarda (the second largest island in the Gulf of California). It is located 53 km northeast of Bahía de los Ángeles, and is separated from Punta Remedios of the peninsula by 13 km by the Canal de Ballenas. The island is very arid (receiving 10 cm rain annually), and vegetation includes cacti, chollas, palo verde, and elephant trees (Case et al., 2002).

VEGETATIONAL ASSOCIATIONS AND ELEVATION RANGE: Little is known about the habitat preferences of *N. insularis*; however, the elevation on the island ranges from 0 m to 1,300 m. Vegetation on the island consists of arid-adapted plants such as cacti, grasses, shrubs, succulents, and boojums.

CONSERVATION STATUS: *N. insularis* is considered to be threatened (SEMARNAT, 2010) as recent attempts have failed to procure specimens, leading some experts to believe this species may already be extinct (Álvarez-Castañeda and Cortes Calva, 1999).

Isthmian woodrat

Cody W. Thompson and Robert D. Bradley

DISTRIBUTION: *N. isthmica* occurs in Mexico and Guatemala. It is known from southeastern Guerrero, central and coastal Oaxaca (Edwards and Bradley, 2002; Goldman, 1904; Goodwin, 1969), the southern portions of the Isthmus of Tehuantepec (Edwards and Bradley, 2002) eastward to the Chiapas River valley in the state of Chiapas (Edwards and Bradley, 2002; Goldman, 1904, 1910). In these areas, *N. isthmica* is confined to the montane and arid coastal regions near the Pacific Ocean (Goldman, 1910). It is not known whether the distribution extends south into neighboring countries (Edwards and Bradley, 2002).

SUBSPECIES IN MEXICO

N. isthmica is a monotypic species.

Originally described by Goldman in 1904, this species was later relegated to a subspecies of *N. ferruginea* (Goldman, 1910). Hall (1955) subsequently moved *N. ferruginea* containing *N. f. isthmica* into *N. mexicana*. Recently, Edwards and Bradley (2002) re-elevated *N. isthmica* to species status based on genetic data. A revision is needed to determine whether other subspecies currently included in *N. mexicana* should be assigned to this taxon.

DESCRIPTION: *Neotoma isthmica* is a large rat with orange-rufous to ferruginous color on its upper parts. The color fades to grayish fulvous on the forearms and hind legs. The face, top of the head, and back have scattered blackish hairs. Ventral areas, including the upper lip and lower sides of the face, are white. The forefeet are white, whereas the hind feet are dusky to the toes. The toes are white. The tail is long and thick but sparsely haired to scaly and indistinctly bicolored (brownish above, paler below). Ears are medium-sized. *N. isthmica* is similar in color to *N. picta* but coarser and generally larger in size. Their tails are both indistinctly bicolored, but *N. picta* has a thinner and more haired tail. The skull is larger, longer, and narrower than in *N. picta*. The zygomatic arch is more arched at its anterior. The frontal is flatter posteriorly. The braincase is less rounded (Goldman, 1904).

EXTERNAL MEASURES AND WEIGHT

TL = 355 to 390 mm; TV = 166 to 198 mm; HF = 35 to 39 mm; EAR: 24 to 27 mm. Weight = N/A.

DENTAL FORMULA: I 1/1, C 0/0, PM 0/0, M 3/3 = 16.

NATURAL HISTORY AND ECOLOGY: In general, *N. isthmica* is a montane species associated with open woodland or shrub vegetation types (Cornely and Baker, 1986). In addition, it has been recorded in rocky outcroppings of montane and cloud forest areas (R.D. Bradley, pers. comm.) Little is known of the natural history and ecology of this species.

VEGETATIONAL ASSOCIATIONS AND ELEVATION RANGE: The type locality is Huilotepec, 13 km south of Tehuantepec, Oaxaca, Mexico. The elevation is approximately 30 m (Goldman, 1904). Vegetation is typical of the arid tropical and lower Sonoran zones.

CONSERVATION STATUS: There is no known conservation risk for this species.

Desert woodrat

Jaime Luévano and Eric Mellink

SUBSPECIES IN MEXICO
Neotoma lepida lepida Thomas, 1893

DESCRIPTION: *Neotoma lepida* is a relatively small rat. The dorsal coloration varies according to the subspecies, but in general is gray. The hairs in the gular region, unlike in *N. albigula*, are dark in the base. The tail is bicolored, gray above and light gray below, and although it looks very hairy it is not. The upper part of the feet is white. The ears are large and hairy. The baculum (bone of the penis) is long and thin. The skull is robust and the bullae are inflated (Hall, 1981; Hoffmeister, 1986; Jameson and Peeters, 1988; Patton et al., 2007).

DISTRIBUTION: *N. lepida* is found from the southwestern United States and in small parts of northeastern British Columbia and northwestern Sonora. In Mexico, it inhabits rocky areas in sandy desert in a small part of northeastern Sonora. It has been recorded in the states of BC and SO.

EXTERNAL MEASURES AND WEIGHT
TL = 225 to 383 mm; TV = 95 to 185 mm; HF = 28 to 41 mm; EAR = 23 to 25 mm.
Weight: 100 to 190 g.

DENTAL FORMULA: I 1/1, C 0/0, PM 0/0, M 3/3 = 16.

NATURAL HISTORY AND ECOLOGY: This rat is mainly nocturnal; but sometimes crepuscular. It builds its burrows in crevices, but also in the base of shrubs and cacti or under tree branches; for construction they use twigs, pieces of wood, parts of cacti, and rocks, depending on the availability. In some areas the entrances are constructed with parts of chollas (*Opuntia* spp.). The burrows are used to rest, as nests, to store food, and to escape from predators. It can have a close relationship with *N. bryanti*, *N. albigula*, and *N. mexicana* (Brylski, 1990e; Hoffmeister, 1986; Jameson and Peeters, 1988; Patton et al., 2007). The diet is similar to those of other members of the genus, and includes pods and leaves of mesquite, herbs, leaves and seeds, chollas, cacti fruits and canutillo (*Ephedra* sp.), and locoweed (*Astragalus* sp.). In large part it depends on prickly pear fruits and succulent plants as water resources, which is why it is associated with this vegetation. Reproduction takes place in October and May, depending on the habitat. Each female can produce from one to five litters per year, but generally two. The gestation period lasts 30 to 36 days. On average, the litters have 2.7 offspring (ranging from 1 to 5). The weight of newborns and the age at which they are weaned depends on the size of the litter. If the litter has only 2 offspring, these are weaned within 21 days, but if it has 4, they are weaned within 34 days. Females become sexually active between two and three months old (Brylski, 1990e; Hoffmeister, 1986; Jameson and Peeters, 1988).

VEGETATIONAL ASSOCIATIONS AND ELEVATION RANGE: *Neotoma lepida* is widely distributed in scrubland including sagebrush and creosote bush, desert grasslands, and conifer and oak forests from sea level to 2,700 masl.

CONSERVATION STATUS: This species is widely distributed and has no conservation problems. The subspecies *abbreviata*, *insularis*, *latirostra*, *marcosensis*, *nudicauda*, *perpallida*, and *vicina* are considered as threatened (SEMARNAT, 2010).

White-toothed woodrat

Gerardo Ceballos and Eric Mellink

SUBSPECIES IN MEXICO

Neotoma leucodon durangae J.A. Allen, 1903
Neotoma leucodon latifrons Merriam, 1894
Neotoma leucodon leucodon Merriam, 1894
Neotoma leucodon subsolana Álvarez, 1962
This species was considered part of *N. albigula* but genetic studies proved they are separated taxa; they are isolated by the Rivers Bravo and Conchos (Edwards et al., 2001)

DESCRIPTION: *Neotoma leucodon* is a rather large rat, the biggest of the genus in its distribution area. Morphologically, it resembles *N. albigula*, but can be distinguished by the allopatric distribution and by genetic features. The dorsal coloration is grayish-brown, with light brown and orange shades. In some environments, such as lava fields, they can be very dark. The hairs of the throat are white on the base. The tail is bicolored with scarce hair. The feet are white and are covered with hair, except for the soles, which are naked. The eyes are very conspicuous (Hall, 1981; Hoffmeister, 1986). This species was elevated from subspecies status in *N. albigula* by Edwards et al. 2001 based on genetic data.

EXTERNAL MEASURES AND WEIGHT

TL = 283 to 400 mm; TV = 76 to 185 mm; HF = 30 to 40 mm; EAR = 25 to 30 mm. Weight: 145 to 200 g.

DENTAL FORMULA: I 1/1, C 0/0, PM 0/0, M 3/3 = 16.

Neotoma leucodon. Scrubland. Maderas del Carmen Wildlife Refuge, Coahuila. Photo: Gerardo Ceballos.

NATURAL HISTORY AND ECOLOGY: White-toothed woodrats are mainly nocturnal and terrestrial, but can climb bushes and cacti. Its burrows are very conspicuous and built on the bases of bushes and trees, between rocks and crevices. They are very large and measure up to 3 m or 4 m in diameter, and more than 1 m high. These woodrats use remnants of chollas, prickly pear, mesquite, twigs, leaves, rocks, feathers, bones, and feces of other animals in their construction (Brylski, 1990e; Hoffmeister, 1986; Jameson and Peeters, 1988). In some regions its distribution is positively related to the plant density of chollas and prickly pear, which serves as refuge and food. The larger densities in Mexico, of 72 active burrows per ha, were recorded in areas with prickly pear (*Opuntia*), mesquite (*Prosopis*), and izote (*Yucca* sp.). In the environments where densities are higher, individuals are also bigger (Rangel and Mellink, 1993). They do not drink water as they obtain it metabolically (Brylski, 1990e; Hoffmeister, 1986; Jameson and Peeters, 1988). The mating season takes place from the end of winter until summer. In years with good food availability, they can have more than one litter per year. Each litter comprises up to 3 offspring, which are born in spring after a gestation period of 38 days and which are weaned within 27 to 40 days. Sexual maturity is reached at 80 and 100 days, in females and males, respectively. These rats are used for human consumption in San Luis Potosí and Zacatecas, where the meat is highly valuable (Mellink et al., 1986). Apparently, this activity does not present serious health problems.

DISTRIBUTION: This species is distributed in the southeastern United States and northern Mexico. In Mexico, it is found in the southern part of Rio Bravo and the eastern part of Rio Conchos, from Chihuahua to the center of the Altiplane in Aguascalientes. It has been recorded in the states of AG, CO, DU, GJ, HG; JA, MI, NL, SL, and ZA.

VEGETATIONAL ASSOCIATIONS AND ELEVATION RANGE: Though *N. leucodon* is widely distributed, it is only found in scrubland and desert grasslands, forests of juniper (*Juniperus*), and sometimes in forests of oak and some perennial crops such as prickly pear. It has been found from 1,000 masl to 2,700 masl.

CONSERVATION STATUS: This species is widely distributed and faces no conservation problems.

Neotoma macrotis Thomas, 1893

Big-eared woodrat

Jaime Luévano and Eric Mellink

SUBSPECIES IN MEXICO

N. macrotis is a monotypic species.

DESCRIPTION: *Neotoma macrotis* is a medium-sized rat. The dorsal coloration varies from gray to brown and the belly is white. The feet are brown at the base and white in the mid-part. The tail is slightly bicolored and almost naked. The ears are wide and have a moderate amount of hair (Hall, 1981; Jameson and Peeters, 1988). This species was separated from *N. fuscipes* by Matocq (2002).

EXTERNAL MEASURES AND WEIGHT

TL = 335 to 468 mm; TV = 158 to 241 mm; HF = 32 to 47 mm; EAR = 31 to 34 mm. Weight: 184 to 358 g.

Dental formula: I 1/1, C 0/0, PM 0/0, M 3/3 = 16.

DISTRIBUTION: *N. macrotis* is distributed in the United States and Mexico, where it is found in the Mediterranean region of the northeastern part of Baja California. It has been recorded in the state of BC.

NATURAL HISTORY AND ECOLOGY: This rat is mainly nocturnal, but reduces its activities on bright nights and rainy nights. Its average activity area covers 0.2 ha. In chaparral a density of 18.8 individuals/ha has been reported, but its density depends on the type of habitat. They prefer areas with an average vegetal cover and with enough material to construct their burrows, which are built with twigs, leaves, and other material, at the base of trees and bushes or in the base of hills. Their burrows can measure up to 2.4 m high with the same diameter (Brylski, 1990f; Jameson and Peeters, 1988). Their diet includes a variety of plants. They eat leaves, flowers, berries, underground fungi, grasses, and acorns. They drink water, but can survive on tender vegetation and fungi (Brylski, 1990f; Jameson and Peeters, 1988). Reproduction extends from December to September, but with a peak in mid-spring. They usually have a litter per year, though they can have more, and rarely five. Two to three offspring are born within each litter, ranging from one to four. Females are probably promiscuous (Brylski, 1990f; Jameson and Peeters, 1988).

VEGETATIONAL ASSOCIATIONS AND ELEVATION RANGE: *N. macrotis* inhabits Mediterranean plant communities, including coastal bush, chaparral, and forests of conifers, from sea level to 2,550 m.

CONSERVATION STATUS: This species is distributed along the area of Mediterranean climate in Baja California. The plant communities, except for an area of coastal bush, have not been strongly altered; hence *N. macrotis* is considered not at risk.

Neotoma mexicana Baird, 1855

Mexican woodrat

Heliot Zarza and Gerardo Ceballos

SUBSPECIES IN MEXICO
Neotoma mexicana chamula Goldman, 1909
Neotoma mexicana distincta Bang, 1903
Neotoma mexicana erimita Hall, 1955
Neotoma mexicana griseoventer Dalquest, 1951
Neotoma mexicana inopinota Goldman, 1933
Neotoma mexicana mexicana Baird, 1855
Neotoma mexicana navus Merriam, 1903
Neotoma mexicana ochracea Goldman, 1905
Neotoma mexicana parvidens Goldman, 1904
Neotoma mexicana sinaloae Allen, 1898
Neotoma mexicana tenuicauda Merriam, 1892
Neotoma mexicana torquata Ward, 1891
Neotoma mexicana tropicalis Goldman, 1904
Neotoma mexicana vulcani Sanborn, 1935
Recently, Edwards and Bradley (2002) proposed that the subspecies *isthmica* and *picta* should be regarded as species — see their accounts.

DESCRIPTION: *Neotoma mexicana* has an average size compared to other members of the genus. The dorsal coloration varies from light gray to reddish-brown, depending on the subspecies; usually the back is dark due to the presence of intermingled black hair. On the sides it has a brownish-orange coloration. The belly is white or yellowish, intermingled with grayish hair, except for some subspecies with the groin and gular areas completely white. The hair is thick and sometimes rough. The ears are elongated, almost naked, and brown. The eyes are big and are ringed by a dark line. It has thick and long whiskers (Cornely and Baker, 1986; Hall, 1981; Reid, 1997). It is distinguished by the angle of the antero-internal entrance of M1, beyond half of the crown (Hall, 1981). The skull is light and softened, the face is thin, the interorbital region is narrow, and the upper incisors are small (Cornely and Baker, 1986). The Mexican woodrat can be distinguished from the house rat (*Rattus* sp.) by the hairy tail (Reid, 1997).

DISTRIBUTION: The Mexican woodrat has a wide distribution that ranges from northern Colorado to western Honduras (Hall, 1981). In Mexico it is found in the northeastern, central, and southern states of the country, with the exception of the Baja California Peninsula and eminently tropical environments (Ceballos and Galindo, 1984). It has been recorded in the states of AG, CH, CL, CO, CS, DF, DU, GJ, GR, HG, JA, MI, MO, MX, NY, PU, SI, SL, SO, TL, VE, and ZA.

EXTERNAL MEASURES AND WEIGHT

TL = 285 to 421 mm; TV = 133 to 216 mm; HF = 33 to 41 mm; EAR = 25 to 30 mm. Weight: 151 to 253 g.

DENTAL FORMULA: I 1/1, C 0/0, PM 0/0, M 3/3 = 16.

NATURAL HISTORY AND ECOLOGY: This rat is nocturnal, with peaks of activity during dusk and midnight. They tend to be more active during the dry season than in the rainy season (López-Vidal and Álvarez, 1993). They prefer rocky environments, slopes, and cliffs, with open forests and bushes (Cornely and Baker, 1986; López-Vidal and Álvarez, 1993). The average population density in the La Michilía Biosphere Reserve, Durango, was 13.3 individuals/ha with an average activity area of 350 m2, and an interval of 250 m2 to 524 m2. Males lingered less time in the area (López-Vidal and Álvarez, 1993). The population density may depend on the availability of food and the rainy season (López-Vidal and Álvarez, 1993). Two annual peaks of activity have been recorded, one from May to June, and the second from September to November (López-Vidal and Álvarez, 1993). Males with testicles in the scrotum have been found from June to August and November (Cornely and Baker, 1986). The gestation period lasts 33 days with an interval of 31 to 34 days (Olsen, 1968). The litter size varies from 1 to 3 offspring (Cornely and Baker, 1986). In La Michilía, 2.3 young per female have been reported (López-Vidal and Álvarez, 1993). Unlike other species, females do not build complex burrows, preferring cavities in trees, crevices, abandoned constructions, or abandoned nests of other vertebrates (Cornely and Baker,

Neotoma mexicana. Fir forest. Temascaltepec, state of Mexico. Photo: Gerardo Ceballos.

1986). The interior of the nests is covered with vegetable matter such as sawdust of bark and pine spines (Finely, 1958). They molt three times a year. They feed on green vegetables but also consume fruits, seeds, acorns, fungi, and cacti (Álvarez and Polaco, 1985; Cornely and Baker, 1986). In smaller amounts, they consume insects such as ants and may use undigested plant matter of other animals (Álvarez and Polaco, 1985). The number of species in its diet can be very high (Polaco et al., 1982). Small mammals associated with this rat include *Sorex oreopulus*, *Reithrodontomys megalotis*, *Peromycus maniculatus*, *P. boylii*, *P. difficilis*, *P. truei*, and *Sigmodon leucotis* (Ceballos and Galindo, 1984; López-Vidal and Álvarez, 1993).

VEGETATIONAL ASSOCIATIONS AND ELEVATION RANGE: The Mexican woodrat has a considerable ecological plasticity. It inhabits pine forests, oak forests, manzanitas, bushy oaks, thorn forests, and grasslands. It is found from sea level up to 4,045 m (Ceballos and Galindo, 1984; López-Vidal and Álvarez, 1993).

CONSERVATION STATUS: *N. mexicana* is a common species that faces no conservation concerns.

Neotoma micropus Baird, 1855

Southern plains woodrat

Eric Mellink

SUBSPECIES IN MEXICO

Neotoma micropus canescens J.A. Allen, 1891
Neotoma micropus litoralis Goldman, 1905
Neotoma micropus micropus Baird, 1855
Neotoma micropus planiceps Goldman, 1905
N. micropus litoralis was disqualified by Birney (1973), though this study was not revised by Hall (1981); *N. micropus planiceps* requires taxonomical revision.

DESCRIPTION: *Neotoma micropus* is a large-sized rat; males are larger than females. The dorsal coloration is gray, darker than in other species, but it can be brown. The belly is light gray; the chest and throat are white. The hair of the chest, throat, and feet is white in the base. The ears are large, and the tail is relatively short, blackish above and gray below (Davis, 1974; Hall, 1981).

EXTERNAL MEASURES AND WEIGHT

TL = 300 to 476 mm; TV = 120 to 185 mm; HF = 34 to 41 mm; EAR = 23 to 25 mm. Weight: 200 to 317 g.

DENTAL FORMULA: I 1/1, C 0/0, PM 0/0, M 3/3 = 16.

NATURAL HISTORY AND ECOLOGY: *N. micropus* inhabits valleys, plains, alluvial fans, smooth slopes, and, rarely, very rocky areas. The absence of the species in rocky sites is more evident in places where it occurs sympatrically with *N. albigula*. In Tamaulipas, it can be found near the beach (Álvarez, 1963a; Bailey, 1932; Baker, 1956; Davis, 1974; Findley, 1987; Findley et al., 1975; Finley, 1958; Schmidly, 1977). The nests are vegetal matter accumulations up to 1.2 m high and in diameter, and include abundant cladodes of cacti and many other materials. Sometimes these woodrats complement the burrow with underground tunnels. The preferred areas for the burrows are cacti accumulations or thorn forests. Occasionally, they use rocks or crevices (Bailey, 1932; Birney, 1973; Cockrum, 1952; Davis, 1974; Finley, 1958; Schmidly, 1977). The diet is almost completely vegetarian, with cacti and prickly pear fruits as dominant food items. Other food items include the base of leaves of sotol, acorns, pods, and seeds of mesquite. Water is obtained from the diet (Bailey, 1932; Davis, 1974; Finley, 1958; Schmidly, 1977). Its reproductive features are poorly known. They probably have only one litter per year, and the reproductive season may be restricted to the beginning of spring (Davis, 1974; Schmidly, 1977). Although the reproductive season may be restricted to one peak only, pregnant females have been found in December, May, June, July, and August, and juveniles in March, April, June, July, September, and November (Álvarez, 1963a; Bailey, 1932; Baker, 1956; Dice, 1937). Each litter comprises 2 to 4 offspring, which are born after a gestation period of approximately 33 days. Newborns weigh 10 g, but they grow quickly and are weaned within 30 days. At three months old, they have already reached their complete development, and weigh 85% of their adult weight. The following year they are sexually mature (Álvarez, 1963a; Davis, 1974; Schmidly, 1977). Its predators include birds, lynx, cacomistles, foxes, coyotes, sparrow hawks, owls, and barn owls (Bailey, 1932; Baker, 1956). Like other woodrats, its meat was highly valuable in Coahuila (Baker, 1956).

DISTRIBUTION: This species is distributed in the southwestern United States and northern Mexico, where it is found in northern Chihuahua to Tamaulipas and San Luis Potosí. It has been recorded in the states of CH, CO, NL, SL, and TA.

VEGETATIONAL ASSOCIATIONS AND ELEVATION RANGE: *N. micropus* inhabits a variety of arid and semi-arid environments, including grasslands, oaks, bushes, and mesquite (Álvarez, 1963a; Baker, 1956; Birney, 1973; Davis, 1974; Findley, 1987). It is found from sea level to 1,700 m.

CONSERVATION STATUS: No conservation problems have been reported for this species, and it is not included in any list of endangered species (SEMARNAT, 2010). The taxonomic status of *N. micropus planiceps* needs to be revised, however (see comments in Dalquest, 1953a, and Birney, 1973). If its subspecies status is proved, its population status should be revised as it is only known in the type locality, Rio Verde, San Luis Potosí, and only the type specimen exists. Dalquest (1953a) failed to collect it in his study of mammals of San Luis Potosí.

Neotoma nelsoni Goldman, 1905

Nelson's woodrat

Alberto Rojas Martínez

SUBSPECIES IN MEXICO

N. nelsoni is a monotypic species.

DESCRIPTION: *Neotoma nelsoni* is a large-sized rat of dark coloration, with a thick tail covered with short hair. The dorsal coloration is pale cinnamon, intensely mixed with dark brown; this coloration predominates along the back and extends to the head and nose. The cheeks are grayish-brown. The underside is bicolored, with gray underfur and white tips. It has a white pectoral stain. The tail is slightly darker on the upper part and is uniformly colored (Goldman, 1910). This rat is very similar to *N. albigula* but differs by the darker coloration (Goldman, 1905).

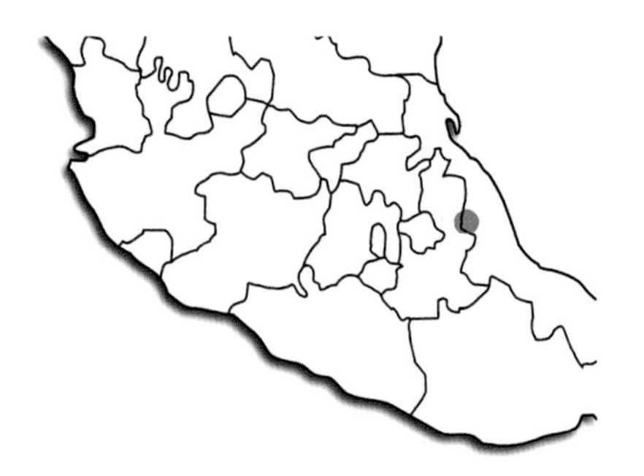

DISTRIBUTION: *N. nelsoni* is an endemic species of Mexico, whose known distribution was only the type locality in Perote, Veracruz. It was recently found 40 km from southeastern Perote (González-Christen et al., 2002). It has been recorded in the state of VE.

EXTERNAL MEASURES AND WEIGHT
TL = 349 mm; TV = 154 mm; HF = 38 mm; EAR = not reported.
Weight: 198 g.

DENTAL FORMULA: I 1/1, C 0/0, PM 0/0, M 3/3 = 16.

NATURAL HISTORY AND ECOLOGY: The natural history of this species is completely unknown. Hall and Dalquest (1963) indicate that this rat inhabits desert plains dominated by cacti in Valle del Oriental in Perote, Veracruz. This valley is isolated from other arid regions and maintains other endemic mammals such as *Spermophilus perotensis* and *Peromyscus bullatus* (Valdéz and Ceballos, 1997). The original vegetation has been severely modified by overgrazing, agriculture, and invasion of exotic species.

VEGETATIONAL ASSOCIATIONS AND ELEVATION RANGE: This rat has been found in thorn forests, an ecotone of xeric scrubland and pine-oak forest, and high perennial tropical forests. It has been captured from 950 masl to 2,300 masl (González-Christen et al., 2002).

CONSERVATION STATUS: *N. nelsoni* is a threatened species. After the fieldwork of 1906, it was recently captured again (González Christen et al., 2002). Hall and Dalquest (1963) failed to capture it, and indicated that the desert region of Perote had been modified to fields of maize and maguey. The transformation of the natural vegetation in Valle del Oriental has been severe. In fact, *Spermophilus perotensis* is also considered at risk of extinction because of similar factors (Valdéz and Ceballos, 1997). The recent record was made in a medium tropical forest 40 km from southeastern Perote, which indicates a wider distribution.

Neotoma palatina Goldman, 1905

Bolaños woodrat

Ma. Lourdes Romero A. and Cornelio Sánchez H.

SUBSPECIES IN MEXICO
N. palatina is a monotypic species.

DESCRIPTION: *Neotoma palatina* is a large-sized rat. The coat is short and rough. The mid-back is dark due to the presence of black hair. The sides and cheeks are pale cinnamon; the belly is whitish. The chest, groin, and snout are grayish-brown, and the feet are white. The ears are small; the tail has little hair and is bicolor, blackish in the upper part and whitish-gray in the bottom. It differs from *N. micropus*, *N. albigula*, and *N. goldmani* by the sphenopalatine vacuities; from *N. nelsoni* by the uniformly colored tail; and from *N. albigula* and *N. lepida* by their contrasting bicolor tails. *N. mexicana* and *N. angustapalata* have a re-entering antero-internal angle in M1 that extends beyond half the tooth, while in *N. palatina* it extends less than half (Hall, 1981; Hall and Genoways, 1970).

EXTERNAL MEASURES AND WEIGHT
TL = 326 to 404 mm; TV = 144 to 180 mm; HF = 34 to 39 mm; EAR = 28 to 32 mm.
Weight: 198 g.

DENTAL FORMULA: I 1/1, C 0/0, PM 0/0, M 3/3 = 16.

NATURAL HISTORY AND ECOLOGY: *N. palatina* is found in riparian vegetation, frequently associated with pronounced slopes along sandbanks and rocky environments, including rock walls. A female with an embryo was captured in August (Matson and Baker, 1986).

VEGETATIONAL ASSOCIATIONS AND ELEVATION RANGE: *N. palatina* inhabits xeric scrubland and has been recorded from 938 masl to 1,904 masl (Matson and Baker, 1986).

CONSERVATION STATUS: Because of its restricted distribution and scarce sightings, this rat is considered rare. It is necessary to evaluate its current status to determine if it is endangered.

DISTRIBUTION: *N. palatina* is an endemic species of Mexico that is only known in a small area of the west-central Mexican Altiplane. It has been recorded in the states of JA and ZA.

Neotoma phenax (Merriam, 1903)

Sonoran woodrat

Reyna A. Castillo

SUBSPECIES IN MEXICO

N. phenax is a monotypic species.

This is the only species of the subgenus *Teanopus*; formerly it was regarded as a genus (Merriam, 1903).

DESCRIPTION: *Neotoma phenax* is a large-sized rat. It differs from other rats with which it shares a habitat by the very large ears, hairy tail, and inflated bullae. Sexual dimorphism is evident as males are slightly larger than females. The dorsal coloration is brownish to gray, depending on the age and season, but the head is always gray (Jones and Genoways, 1978).

EXTERNAL MEASURES AND WEIGHT

TL = 330 to 431 mm; TV = 149 to 220 mm; HF = 22 to 40 mm; EAR= 28 to 35 mm.
Weight: 188 to 279 g.

DENTAL FORMULA: I 1/1, C 0/0, PM 0/0, M 3/3 = 16.

NATURAL HISTORY AND ECOLOGY: In Sonora, *N. phenax* builds its burrows in small shrubs and short trees along rivers (Burt, 1938). In Sinaloa it nests on the ground and in shrubs, as well as in columnar cacti (*Pachicereus pecten-arborigeum*) and thorn trees 1.6 m to 6.6 m high (Birney and Jones, 1972). Females have two litters per year with an average of two offspring per litter (Jones and Genoways, 1978). Apparently, fruits of persimmon (*Dyospirus* spp.) are important in its diet (Burt, 1938).

VEGETATIONAL ASSOCIATIONS AND ELEVATION RANGE: *N. phenax* has only been found in desert scrub and pine-oak forest, from sea level to 150 m.

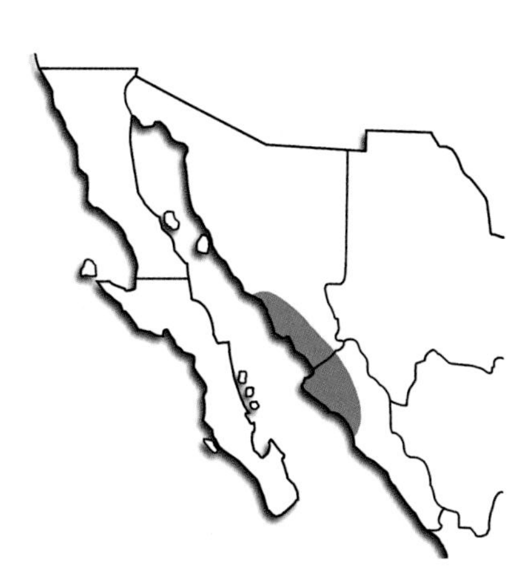

DISTRIBUTION: *N. phenax* is an endemic species of Mexico. It is distributed along the Pacific coast from Guaymas, Sonora, to El Dorado in the central region of Sinaloa. It has been recorded in the states of SI and SO.

CONSERVATION STATUS: As it is endemic to the country and its distribution is restricted, *N. phenax* is considered as of Special Protection (SEMARNAT, 2010). It is important to establish its current status, as deforestation is probably threatening it (G. Ceballos, pers. comm.).

Neotoma picta Goldman, 1904

Painted woodrat

Cody W. Thompson and Robert D. Bradley

SUBSPECIES IN MEXICO

N. picta is a monotypic species.

Goldman originally described the species in 1904. Goldman (1910) later arranged the species as a subspecies of *N. ferruginea*. Hall (1955) subsequently moved *N. ferruginea* containing the subspecies *N. ferruginea picta* into *N. mexicana*. In 2002, Edwards and Bradley re-elevated *N. picta* to species status based on genetic data. A revision is needed to determine whether other species currently included in *N. mexicana* should be assigned to this taxon.

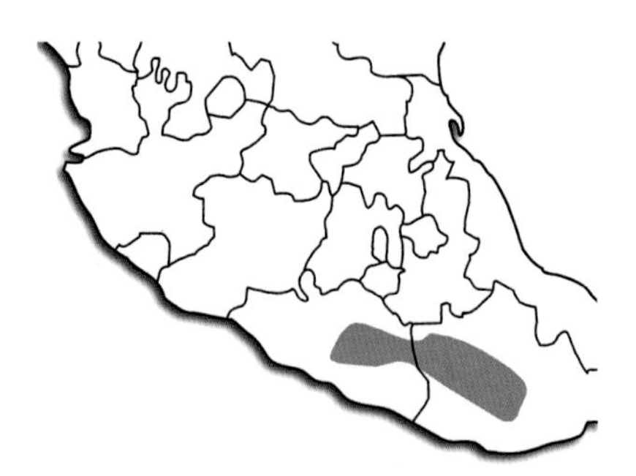

DISTRIBUTION: *N. picta* is endemic to Mexico. The species is known from the type locality in the mountains near Chilpancingo, Guerrero, Mexico (Goldman, 1904) to west-central Oaxaca (Goldman, 1910; Goodwin, 1969). It has been recorded in GR and OX.

DESCRIPTION: *Neotoma picta* is a medium-sized rat similar in color to *N. isthmica* with orange-rufous to ferruginous coloration along the cheeks, shoulders, and sides (Goldman, 1904). It is smaller than *N. isthmica* (Goldman, 1910). The face, top of the head, and back are darkened by black-tipped hairs. The ventral parts are nearly white to salmon with a plumbeous basal color. The tail is long and slender and covered with short hairs indistinctly bicolored (dusky above, paler below). The ears are relatively small with short, dusky hairs. The forefeet are yellowish-white, while the hind feet are dusky or pale fulvous. The toes are white (Goldman, 1904). The skull is less robust and less arched than that of *N. isthmica*. The zygomatic arch is less arched anteriorly (Goldman, 1910).

EXTERNAL MEASURES AND WEIGHT

TL = 338 to 355 mm; TV = 166 to 182 mm; HF = 33 to 35.5 mm; EAR= N/A.
Weight = N/A.

DENTAL FORMULA: I 1/1, C 0/0, PM 0/0, M 3/3 = 16.

NATURAL HISTORY AND ECOLOGY: In general, *N. picta* is a montane species associated with woodlands and shrub vegetation types (Cornely and Baker, 1986). Typically, it is found along rocky outcroppings in cloud forest habitats (R.D. Bradley, pers. comm.). Little else is known of the natural history and ecology of this species.

VEGETATIONAL ASSOCIATIONS AND ELEVATION RANGE: Type locality is found in the mountains near Chilpancingo, Guerrero, Mexico. Elevation was 3,048 m (Goldman, 1904). Vegetation in this area is typical of the transition and Canadian zones (Goldman, 1910).

CONSERVATION STATUS: There is no known conservation risk for this species.

Volcano deermouse

Catalina B. Chávez T.

SUBSPECIES IN MEXICO
N. alstoni is a monotypic species.

DESCRIPTION: *Neotomodon alstoni* is a medium-sized mouse of robust appearance and docile behavior. The dorsal coloration is dark gray; on the sides it is yellowish-ochraceous and white on the underside. Most of the specimens have a small yellowish pectoral spot. Juveniles are uniformly gray. The species is distinguished by the medium size when compared to the genus *Peromyscus*; the tail is relatively short (shorter than the length of the head and body) and slightly bicolor. The dorsum is darker, The diploid chromosome number is 2n = 48 and the fundamental number is FN = 66 (Williams et al., 1985).

EXTERNAL MEASURES AND WEIGHT
TL = 174 to 233 mm; TV = 78 to 105 mm; HF = 23 to 27 mm; EAR = 19 to 23 mm. Weight: 40 to 60 g.

DENTAL FORMULA: I 1/1, C 0/0, PM 0/0, M 3/3 = 16.

NATURAL HISTORY AND ECOLOGY: *N. alstoni* is a nocturnal mouse associated with grasses. It builds its burrows between clusters of roots, showing preference for a microhabitat of zacatonal and forest-zacatonal (Rojas, 1984). In natural conditions its diet includes stems, sprouts, and leaves of herbs (Prieto, 1988). These mice are very docile and will breed in captivity. In the wild as in captivity they reproduce throughout the year. Newborns are naked and pink, and have whiskers and small hairs around the mouth. They measure 52.8 mm to 57 mm and weigh around 3.8 g. A peak of births has been recorded from April to September. The gestation period lasts 25 to 35 days and the offspring are weaned within 20 to 30 days (Martin and Álvarez, 1982). The vaginal aperture occurs at 49 days (+/-10) and estrus lasts 4.5 days. Fe-

Neotomodon alstoni. Bunch grassland. Nevado de Toluca National Park, state of Mexico. Photo: Gerardo Ceballos.

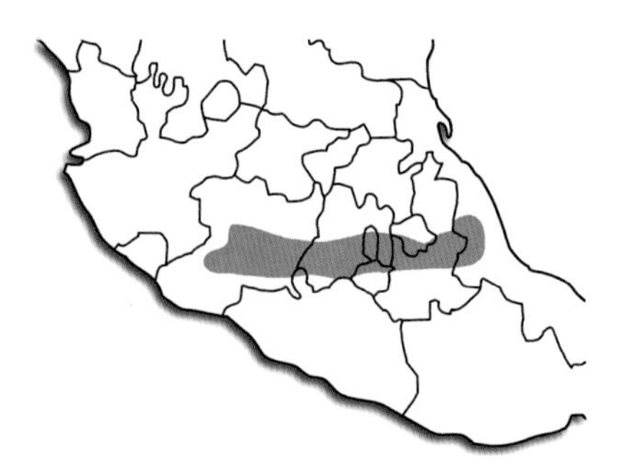

DISTRIBUTION: *N. alstoni* is an endemic species of Mexico with a distribution restricted to the Eje Volcanico Transversal from central Michoacán to central Puebla and Veracruz. It has been recorded in the states of DF, MI, MO, MX, PU, TL, and VE.

males copulate with pregnancy success at 65 days and have litters with an average of 3.1 offspring (Oliver et al., 1986). *N. alstoni* has been widely studied in captivity, including its growth (Granados and Ramírez, 1986) and reproduction (Granados and Hoth, 1987; Granados and Luis, 1987; Luis and Granados, 1990, Luis et al., 1993, 2000). Chávez (1998) conducted a population study in an area of 1.2 ha during a period of 4 years and 7 months in a forest of conifers integrated by *Pinus motezumae*, *P. hartwegii*, *P. patula*, and *P. radiata*, zacatonals of clusters of grass such as *Muhlenbergia macroura* and *Stipa ichu*, and herbs and bushes such as *Baccharis glutinosa*, *Sisyrinchom angustifolium*, and *Gnaphalium americana*. He found an annual behavior characterized by an average density of 29 individuals/ha, with a continuous polyestric annual pattern. The reproductive peak occurred from spring to summer, with postpartum estrus, and 70% of inactive adults during winter. The density shows significant fluctuations. From the first summer it increased continuously with significant differences between the minimum observed (11 individuals/ha) and the maximum (54 individuals/ha). The analysis of time series of rain in summer during 23 years (1961-1984) displayed a seasonal cyclic behavior with peaks every 12 years, matching the peak of 1981 with the phase change of a year, with the maximum peak observed during autumn of 1982. The maximum density was not seasonal or periodical (Chávez and Gallardo, 1993). The life expectancy for adults was 142 days for females and 106 days for males. Juveniles showed a shorter life expectancy than subadults of 155 and 160 days, respectively. More than 50% of the population does not exceed 1 year of age. Approximately 16% of the juveniles die before reaching 50 days old, and 27% of subadults do not reach 75 days of life. The maximum longevity was a juvenile female captured in autumn, with 960 days and 869 days for a subadult male captured in spring. The net rate of reproduction (Ro) fluctuated from 1.55 to 2.45. The generation interval (T) fluctuated from 203 to 276 days. They coexist with small mammals as *Cryptotis aticola*, *Sorex sassurei*, *Sigmodon leucotis*, *Peromyscus levipes*, *P. melanotis*, *Reithrodontomys megalotis*, and *Microtus mexicanus* (Ceballos and Galindo, 1984; Chávez, 1988). Aguirre and Ulloa (1982) studied mosses in the feces of this rodent.

VEGETATIONAL ASSOCIATIONS AND ELEVATION RANGE: *N. alstoni* mainly inhabits grasslands of clusters of grass (*Muhlenbergia macroura*), but is also found in forests of conifers and oak. It is found from 2,400 masl to 4,960 masl.

CONSERVATION STATUS: Although it is an endemic species, *N. alstoni* does not face conservation concerns. Its distribution is relatively wide. It is abundant in diverse localities and has populations in protected areas such as the national park Popo-Izta and Nevado de Toluca.

Nyctomys sumichrasti (Sassure, 1860)

Sumichrast's vesper rat

Luis Arturo Peña, Yolanda Domínguez, and Beatriz Hernández

SUBSPECIES IN MEXICO

Nyctomys sumichrasti colimensis Laurie, 1953
Nyctomys sumichrasti pallidullus Goldman, 1937
Nyctomys sumichrasti salvini Thomes, 1862
Nyctomys sumichrasti sumichrasti Sassure, 1860

Nyctomys sumichrasti. Tropical dry forest. Chamela-Cuixmala Biosphere Reserve, Jalisco. Photo: Gerardo Ceballos.

DESCRIPTION: *Nyctomys sumichrasti* is a small-sized rat of cinnamon brown or orange-brown dorsal coloration. The sides are lighter, and the belly is white or creamy white. The eyes are big and ringed by dark hair. The feet are generally white, and the digits and plantar tubercles are well developed. The tail is hairy brown and ends in a tuft; the length is almost the same as that of the fur on the body; the ears are small and almost naked. Females have four mammae (Ceballos and Miranda, 1986, 2000).

EXTERNAL MEASURES AND WEIGHT
TL = 208 to 286 mm; TV = 85 to 156 mm; HF = 17 to 26 mm; EAR = 16 to 19 mm. Weight: 40 to 55 g.

DENTAL FORMULA: I 1/1, C 0/0, PM 0/0, M 3/3 = 16.

NATURAL HISTORY AND ECOLOGY: *N. sumichrasti* is a strictly nocturnal rat. It is arboreal and the hind feet are modified for its habits. It builds external burrows in hollow branches, similar to those of squirrels of the genus *Sciurus* (Hall, 1981). Its diet is based on fruits and seeds; in Jalisco it mainly feeds on fruits of *Jacquinia pungens*. Animals in captivity survived with seeds and fruits of *Spondias purpurea*, *Cresentia alata*, and *Terminalia catappa* (Ceballos, 1990). They reproduce throughout the year; each female can have several litters per year (Genoways and Jones, 1972). Gestation lasts 30 to 38 days, and the litter varies from 1 to 4 individuals. The offspring are weaned within the first five weeks of life (Ceballos, 1990). They live up to five years in captivity. They coexist with other rodents such as *Lyomys pictus*, *Osgoodomys banderanus*, *Peromyscus perfulvus*, and *Xenomys nelsoni*.

DISTRIBUTION: *N. sumichrasti* is distributed in tropical areas in the Eje Volcanico Transversal, the Pacific Rim to Jalisco, and the Gulf side to Veracruz. It has been recorded in the states of CO, CS, GR, JA, MI, OX, and VE.

VEGETATIONAL ASSOCIATIONS AND ELEVATION RANGE: *N. sumichrasti* has been recorded in dense tropical forests, moderately dense tropical forests, and less dense in low tropical forests, from sea level to 1,500 m.

CONSERVATION STATUS: This species has no special protection, and its distribution is wide. Its habitat is affected by human activities, however, which could represent a risk for their survival.

Fulvous colilargo

Iván Castro Arellano and Mery Santos G.

SUBSPECIES IN MEXICO

Oligoryzomys fulvescens engraciae Osgood, 1945
Oligoryzomys fulvescens fulvescens Saussure, 1860
Oligoryzomys fulvescens lenis Goldman, 1915
Oligoryzomys fulvescens mayensis Goldman, 1918
Oligoryzomys fulvescens pacificus Hooper, 1952

DESCRIPTION: *Oligoryzomys fulvescens* is a small-sized mouse of delicate features, with large ears and a tail larger than the head and body with scarce hairs (Goldman, 1918). The dorsal coloration varies from light creamy ochraceous to darker, especially in the rear. The face, head, and back are slightly darker due to the presence of black guard hair. The underside varies from white to very light creamy-ochraceous. The feet are white, and the tail is dark brown above and lighter below (Goldman, 1918). It is very similar to members of the genus *Reithrodontomys*, from which it can be distinguished by the smooth surface of the upper incisors (Álvarez, 1986; Álvarez et al., 1987; Hall and Dalquest, 1963). The diploid number of chromosomes has been described as 2n = 60 and 2n = 54 (Gardner and Patton, 1976; Haiduk et al., 1979) in specimens from Costa Rica and Veracruz, Mexico, respectively. It is possible that two species are represented or that geographical variation exists (Haiduk et al., 1979).

EXTERNAL MEASURES AND WEIGHT

TL = 168 to 253 mm; TV = 96 to 130 mm; HF = 20 to 25 mm; EAR = 10 to 15 mm.
Weight: 9 to 15 g.

Oligoryzomys fulvescens. Grassland. Chamela-Cuixmala Biosphere Reserve, Jalisco. Photo: Gerardo Ceballos.

DENTAL FORMULA: I 1/1, C 0/0, PM 0/0, M 3/3 = 16.

NATURAL HISTORY AND ECOLOGY: Little is known about the biology of this species. It is common in dry tropical forests of western Mexico and it is associated with riparian habitat (Álvarez, 1986; Álvarez et al., 1987; Genoways and Jones, 1973). In Veracruz, it inhabits areas with tall grasses, bushes, crops of sugar cane and maize, and patches of succulents and arboreal ferns (Hall and Dalquest, 1963). Apparently, they reproduce throughout the year. Females with embryos have been collected in November, December, and January in Veracruz, in July in Tabasco, and in December in Campeche. The number of embryos varied from three to four (Dowler and Engstrom, 1988; Hall and Dalquest, 1963; Jones et al., 1974). Hall and Dalquest (1963) found out that juveniles are more numerous than adults, which suggests they live less than a year, like other species of the genus (Nowak, 1999b).

DISTRIBUTION: This species is widely distributed in the Neotropical region from the warm regions of eastern and western Mexico, through Mesoamerica, to Ecuador and the northern part of Brazil and Guyanas in South America (Hall, 1981; Musser and Carleton, 1993). It has been recorded in the states of CA, CL, CS, GR, HG, JA, MI, NL, NY, OX, PU, QR, TA, TB, VE, and YU.

VEGETATIONAL ASSOCIATIONS AND ELEVATION RANGE: *O. fulvescens* mainly inhabits tropical environments such as low tropical forests, medium tropical forests, and thorn forests, but can also be found in forests of oak and pine-oak. It is very abundant in crops and induced grasslands (Álvarez, 1986; Álvarez et al., 1987; Dowler and Engstrom, 1988; Jones et al., 1974). It is found from sea level to 1,550 m (Hall, 1981).

CONSERVATION STATUS: These mice face no conservation concerns as they are abundant and their distribution are wide; they even inhabit disturbed areas.

Onychomys arenicola Mearns, 1896

Chihuahuan grasshopper mouse

Héctor Godínez Álvarez

SUBSPECIES IN MEXICO

O. arenicola is a monotypic species.

Formerly, it was regarded as part of *Onychomys torridus torridus*. Hinesley (1979), however, evaluated the chromosomes, finding a new variety with a different karyotype. Based on these findings and a morphological examination, this author proposed to recognize it at species level. This arrangement is widely accepted. For a long time, it was regarded as a sister taxon of *O. torridus*; however, Sullivan et al. (1986), based on genetic and morphological analysis, proposed that the phylogenetic relationship between *O. arenicola* and *O. leucogaster* is closer than between *O. arenicola* and *O. torridus*.

DISTRIBUTION: *O. arenicola* is found in the desert of Chihuahua, from southeastern Arizona, south-central New Mexico, and western Texas, to Zacatecas and Aguascalientes, and in the central part of Mexico (Baker, 1962; Hinesley, 1979; Riddle and Honeycutt, 1990). It has been recorded in the states of AG, CH, DU, NL, SL, and ZA.

DESCRIPTION: *Onychomys arenicola* is a small-sized mouse, similar to *O. torridus*, but of smaller size, with the posterior edge of the palate concave or truncated. The dorsal coloration is opaque brown with darker hair, which makes it look grayish; the belly is white. The small but wide tail varies from gray to brown, except for the tip, which is white. The ears are brown with white hair on the edges (Hinesley, 1979). The diploid number of chromosomes is 2n = 48 and the karyotype has 5 pairs of subtelocentric chromosomes, 8 pairs of submetacentric and metacentric chromosomes, and 10 pairs of acrocentric chromosomes. The chromosome X is submetacentric and Y is acrocentric (Hinesley, 1979).

Onychomys arenicola. Grassland. Janos Biosphere Reserve, Chihuahua. Photo: Gerardo Ceballos.

EXTERNAL MEASURES AND WEIGHT

TL = 131 to 159 mm; TV = 46 to 55 mm; HF = 18 to 22 mm; EAR = 16 to 22 mm. Weight: 21 to 29 g.

DENTAL FORMULA: I 1/1, C 0/0, PM 0/0, M 3/3 = 16.

NATURAL HISTORY AND ECOLOGY: *O. arenicola* has been captured mainly in slopes or in foothills, where the substrate is sandy or rocky and the dominant vegetation includes bushes of cresote bush (*Larrea tridentata*) and other bushes (Hinesley, 1979). It is a rare species with densities of approximately 2.5 individuals/ha (Findley, 1987). It feeds on insects, beetles, and other arthropods (Findley, 1987; Nowak, 1999b). It has nocturnal habits and builds its burrows at ground level, where one male and two females with offspring live. Males are territorial. Sexual maturity in females is reached at eight weeks old (Findley, 1987). Data on its reproduction biology are scarce; however, they apparently reproduce from March to September (Anderson, 1972). Males participate in the parental care of the offspring, which are weaned within 20 to 23 days after birth (Findley, 1987). Other common rodents collected along with this species are *Dipodomys merriami*, *Perognathus intermedius*, *Peromyscus eremicus*, and *P. maniculatus*.

VEGETATIONAL ASSOCIATIONS AND ELEVATION RANGE: *O. arenicola* is only found in grassy areas and xeric scrubland (Findley, 1987), from 1,300 masl to 1,580 masl (Hinesley, 1979).

CONSERVATION STATUS: *O. arenicola* is not endangered as it has a wide distribution that includes disturbed areas.

Northern grasshopper mouse

Emilio Daniel Tobón García

SUBSPECIES IN MEXICO
Onychomys leucogaster albescens Merriam, 1904
Onychomys leucogaster longipes Merriam, 1889
Onychomys leucogaster ruidosae Stone and Rehn, 1903

DESCRIPTION: *Onychomys leucogaster* is a small-sized mouse that is distinguished by a short and bicolored tail, of shorter length than the head and body. The coat is smooth. The head and upper part of the body vary from brown to pink cinnamon, with the back having a darker color. The underside and tail are completely white. It differs from members of the genus *Peromyscus* by the shorter tail and longer feet. The length of the tail is less than half of the total length (Baker, 1956; Hall, 1981; McCarty, 1978; Nowak, 1999b). It differs from other species of the genus in Mexico by its allopatric distribution.

EXTERNAL MEASURES AND WEIGHT
TL = 119 to 190 mm; TV = 29 to 62 mm; HF = 17 to 25 mm; EAR = 12 to 17 mm. Weight: 30 to 60 g.

DENTAL FORMULA: I 1/1, C 0/0, PM 0/0, M 3/3 = 16.

NATURAL HISTORY AND ECOLOGY: *O. leucogaster* inhabits prairies with short grasses and desert scrubland in high plateaus. They are basically carnivores but occasionally eat plants. This genus is the only one in North America whose feeding behavior is based on other animals. It naturally regulates insect populations that can become plagues (McCarty, 1978). In natural conditions it feeds on insects and small mice such as *Peromyscus*, *Perognathus*, and *Microtus*. On average, the densities are low (2.3 individuals/ha), with these mice probably living isolated or in couples with large home ranges (McCarty, 1978). They are basically nocturnal and are active throughout the year, without hibernating. Courtship and mating behaviors are among the most complex observed in rodents (McCarty, 1978). The gestation period has been observed in captivity, with an interval of 26 to 37 days for non-lactating females and 32 to 47 days for lactating females. Estrus takes place a few hours up to two days after giving birth. The size of the litter varies from three to four offspring. These mice build burrows for upbringing, shelter, food storage, defecation, and marking their territories. They are preyed on by barn owls (*Tyto alba*), owls (*Bubo virginianus*), hawks (*Buteo jamaicensis*), foxes, and coyotes. It interacts with 34 different species of mammals, but due to its low densities and to its feeding behavior, this species can minimize potential competition for food, space, and other limiting resources (McCarty, 1978; Nowak, 1999b).

DISTRIBUTION: *O. leucogaster* is found in the northern part of the continent, from Canada to the north of Mexico, Coahuila, Nuevo León, and Tamaulipas (Álvarez, 1963a; Anderson, 1972; Baker, 1956; Hall, 1981; McCarty, 1978; Musser and Carleton, 1993). It has been recorded in the states of CO, NL, and TA.

VEGETATIONAL ASSOCIATIONS AND ELEVATION RANGE: *O. leucogaster* is mainly found in xeric scrubland and in different types of grasslands, prairies with short grasses, areas with mesquite, and high grasslands that are not plowed constantly, from 152 masl to 1,400 masl (Álvarez, 1963a; Anderson, 1972; Baker, 1956; McCarty, 1978).

CONSERVATION STATUS: This species is not endangered since it has a wide distribution and is tolerant of human activities.

Southern grasshopper mouse

Gerardo Ceballos

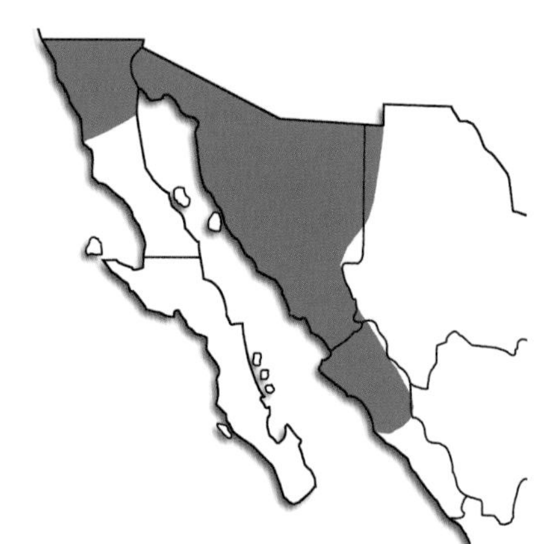

DISTRIBUTION: *O. torridus* is distributed in the West and Southwest of the United States, from California to Arizona, and the northeast of Mexico, in the Baja California Peninsula and the Pacific Rim in Sonora and Sinaloa. It has been recorded in BC, SI, and SO.

SUBSPECIES IN MEXICO

Onychomys torridus ramona Rhoads, 1893

Onychomys torridus yakesis Merriam, 1904

The populations in the Chihuahuan desert in the United States and Mexico that were regarded as part of *O. torridus torridus* were elevated to species level based on genetic evidence such as karyotypes (Hinesley, 1979). These populations are currently known as *O. arenicola*.

DESCRIPTION: *Onychomys torridus* is a relatively small mouse that is distinguished by its short tail (shorter than the length of the head and body). Morphologically it is very similar to *O. arenicola*. The coat is soft and dense. The ears are medium in size, and the eyes are large. The underside of the tail, the feet, and the belly are white. The rest of the body and the tail are light brown to grayish-brown. It is distinguished from other species of the genus in Mexico by its genetic characteristics and by its allopatric geographic distribution. It is distinguished from the species of the genus *Peromyscus* by its short tail (Hall, 1981; Riddle, 1995).

EXTERNAL MEASURES AND WEIGHT

TL = 131 to 159 mm; TV = 46 to 55 mm; HF = 18 to 22 mm; EAR = 16 to 22 mm.
Weight: 21 to 29 g.

DENTAL FORMULA: I 1/1, C 0/0, PM 0/0, M 3/3 = 16.

NATURAL HISTORY AND ECOLOGY: These mice are efficient predators of invertebrates and small vertebrates that inhabit arid environments. Even though they are carnivores, they can occasionally eat vegetable matter. They are able to feed

Onychomys torridus. Grassland. Janos Biosphere Reserve, Chihuahua. Photo: Gerardo Ceballos.

on dangerous prey such as scorpions and spitter beetles. An interesting behavior when eating beetles is that after they capture them, they keep them buried in the sand with their tails, which avoids the sprinkle of aromatic substances that the beetles use for self-defense. Their densities are low, generally less than 2 individuals per hectare (McCarty, 1978). They are basically nocturnal and are active throughout the year, including the cold months of winter. The gestation period lasts 26 to 47 days. The size of the litter varies from three to four offspring. They build complex burrows with several chambers. They are preyed on by a great variety of vertebrates such as foxes, lynx, skunks, barn owls (*Tyto alba*), and owls (*Bubo virginianus*).

VEGETATIONAL ASSOCIATIONS AND ELEVATION RANGE: *O. torridus* is found in scrublands and arid grasslands, chaparrals, and semi-tropical thorn forests, from sea level to 1,400 m.

CONSERVATION STATUS: This species does not face conservation concerns as it has a wide geographical distribution in areas with little human disturbance. It has protected populations in the Alto Golfo Biosphere Reserve and Pinacate.

Oryzomys albiventer Merriam, 1901

White venter rice rat

Joaquin Arroyo-Cabrales

SUBSPECIES IN MEXICO
O. albiventer is a monotypic species.

DESCRIPTION: *Oryzomys albiventer* is a large and long-tailed *Oryzomys*. The upper parts are brightly ochraceous, becoming grayer toward the front. The hairs on the underparts are pale gray near the bases and white in the outer half, so that the underparts appear pale grayish. The tail is dark above and light below. The skull and molars are relatively robust. Compared to its lowland relative *O. mexicanus*, *O. albiventer* is larger and more brightly colored and has larger molars.

EXTERNAL MEASURES AND WEIGHT
TL = 245 to 314 mm; TV = 129 to 173 mm, HF = 33 to 40 mm; EAR = 13 to 18 mm.
Weight: 4.5 to 7.1 g.

DENTAL FORMULA: I 1/1, C 0/0, PM 0/0, M 3/3 = 16.

NATURAL HISTORY AND ECOLOGY: Very little is known about the ecology and natural history of this species.

VEGETATIONAL ASSOCIATIONS AND ELEVATION RANGE: *O. albiventer* occupies deciduous tropical woodlands, from 1,200 masl to 1,800 masl.

CONSERVATION STATUS: *O. albiventer* is an inhabitant of highly populated regions in central Mexico, where most of its habitat has been converted into crops. It may be at risk of extinction.

DISTRIBUTION: *O. albiventer* is a tropical species that is distributed from central and eastern Jalisco, northern Michoacán, and southern Guanajuato. It has been registered in the states of GA, JA, and MI.

Alfaro's oryzomys

Esther Romo V.

DISTRIBUTION: *O. alfaroi* is distributed in tropical areas from southern Tamaulipas and Guerrero to western Colombia (Musser and Carleton, 1993). It has been recorded in the states of CS, GR, OX, TA, and VE.

SUBSPECIES IN MEXICO

Oryzomys alfaroi agrestis Goodwin, 1959
Oryzomys alfaroi gloriaensis Goodwin, 1956
Oryzomys alfaroi palatinus Merriam, 1901
The subspecies *O. alfaroi chapmani*, *rhabdops*, and *saturatior* were elevated to a species level by Musser and Carleton (1993).

DESCRIPTION: *Oryzomys alfaroi* is a small-sized mouse with short and fine hair. The dorsal coloration varies from ochraceous-brown to brown and orange. The belly is grayish or whitish. The tail is long, with little hair, and dark. Some individuals have a bicolored tail. The face is short. The ears are small and rounded, lined by black or orange hair (Reid, 1997). The digits have fur that protrudes beyond the claws (Hall, 1981). Its external form resembles that of *Peromyscus*; however, unlike that species, its scaly tail is ringed with scattered hair. The rear legs are proportionally larger in *Oryzomys*. They may also be confused with members of the genus *Sigmodon*, but differ from them by the thickness of the coat and the strength of the tail.

EXTERNAL MEASURES AND WEIGHT

TL = 174 to 265 mm; TV = 94 to 114 mm; HF = 25 to 29 mm; EAR = 15 to 19 mm. Weight: 20 to 44 g.

DENTAL FORMULA: I 1/1, C 0/0, PM 0/0, M 3/3 = 16.

NATURAL HISTORY AND ECOLOGY: These mice are nocturnal and feed on seeds and tender sprouts. It has been collected with *Peromyscus beatae*, *P. levipes*, *P. zarhynchus*,

Oryzomys alfaroi. Tropical semi-green forest. Yucatan. Photo: Brian and Annika Keeley.

and *P. melanocarpus* in rocky slopes, wet forests, crops, and zacatonals (Baker et al., 1971). They reproduce throughout the year. The litter size varies from two to four offspring. They build their burrows under trunks, under rocks, or between roots. The nests are spherical, covered with grass and peels, and are 13 cm wide and 8 cm long (Hall and Dalquest, 1963).

VEGETATIONAL ASSOCIATIONS AND ELEVATION RANGE: *O. alfaroi* mainly inhabits wet mountainous slopes in cloud forests, oak forests, and pine-oak forests, from 860 masl to 2,350 masl.

CONSERVATION STATUS: There are no accurate data on the status of these mice in Mexico, but some in localities they can be abundant, and so they are not regarded at risk.

Oryzomys chapmani Thomas, 1898

Chapman's oryzomys

Esther Romo-V.

SUBSPECIES IN MEXICO
Oryzomys chapmani caudatus Merriam, 1901
Oryzomys chapmani chapmani Thomas, 1898
Oryzomys chapmani dilutior Merriam, 1901
Oryzomys chapmani guerrerensis Goldman, 1915
Oryzomys chapmani huastecae Dalquest, 1951
The species was regarded as subspecies of *O. alfaroi* by Goldman (1918); however, Musser and Carleton (1993) regarded it as a polytypic species. Goodwin (1969) recognized *O. caudatus* as different from *O. alfaroi* in the north of Oaxaca, but by the first arrangement of Merriam (1901) *O. chapmani* has priority over this Mexican species.

DESCRIPTION: *Oryzomys chapmani* is a medium-sized rat within the genus *Oryzomys*. The coat is short and dark, ranging from brown to brown mixed with black. The tail has the same length as the head and body; it is uniformly colored and has little hair. The digits have fur that protrudes beyond the claws. The face is short and wide, and the ears are relatively small.

EXTERNAL MEASURES AND WEIGHT
TL = 174 to 265 mm; TV = 94 to 114 mm; HF = 25 to 29 mm; EAR = 15 to 19 mm. Weight: 20 to 44 g.

DENTAL FORMULA: I 1/1, C 0/0, PM 0/0, M 3/3 = 16.

NATURAL HISTORY AND ECOLOGY: This species is terrestrial and nocturnal; it feeds on grains and tender sprouts. It has been collected together with *P. beatae*, *P. aztecus*, and *Megadonthomys nelsoni* in a variety of environments ranging from rocky areas to grasslands and crops.

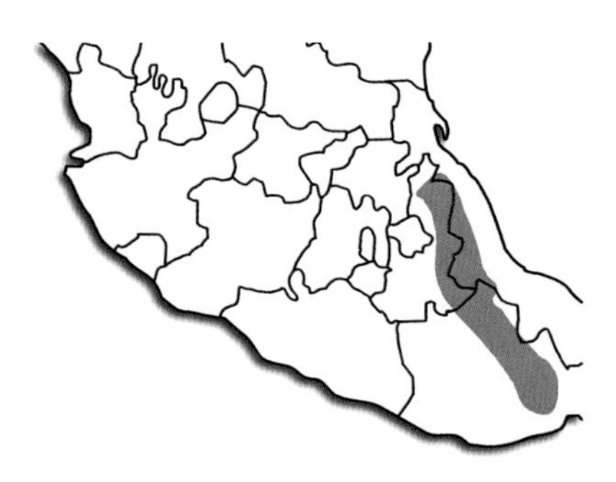

DISTRIBUTION: *O. chapmani* is distributed from the Sierra Madre Oriental and Sierra Madre del Sur, from Hidalgo and Guerrero to Central America. It has been recorded in the states of GR, HG, OX, PU, SL, and VE.

VEGETATIONAL ASSOCIATIONS AND ELEVATION RANGE: *O. chapmani* mainly inhabits wet areas of cloud forests and grassy areas inside forests of pine-oak. The elevational interval where it is found in Mexico ranges from 1,550 masl to 2,500 masl.

CONSERVATION STATUS: Apparently, this species faces no conservation concerns since it has a relatively wide distribution and tolerates human perturbations. However, it is considered as of Special Protection (SEMARNAT, 2010).

Oryzomys couesi (Alston, 1877)

Coues' oryzomys

Xavier López Medellín and Rodrigo A. Medellín

SUBSPECIES IN MEXICO

Oryzomys couesi albiventer Merriam, 1901
Oryzomys couesi aquaticus J.A. Allen, 1891
Oryzomys couesi couesi Alston, 1877
Oryzomys couesi cozumelae Merriam, 1901
Oryzomys couesi fulgens Thomas, 1893
Oryzomys couesi peninsulae Thomas, 1893
Oryzomys couesi peragrus Merriam, 1901

O. couesi was considered a subspecies of *O. texensis*, but genetic studies demonstrated they are different species. It has 26 pairs of acrocentric autosomes (Benson and Gehlbach, 1979) and has a reduced number of chromosomes unlike species that are distributed in South America and are considered primitive (Sánchez-Cordero and Valadez-Azúa, 1989).

DISTRIBUTION: *O. couesi* is distributed from southern Texas along the east coast of Mexico to southern Nicaragua. Only *O. couesi* occurs in tropical dry environments corresponding to tropical deciduous tropical forests (Ceballos and Miranda, 1986, 2000; Ramírez-Pulido et al., 1986; Ramírez-Pulido and Castro-Campillo, 1993; Sánchez-Cordero and Valadez-Azúa, 1989; Wilson and Reeder, 2005). It has been recorded in the states of BS, CA, CO, CS, DF, GR, JA, MI, MO, NL, NY, OX, PU, QR, SI, SL, SO, TA, TB, TL, VE, and YU.

DESCRIPTION: *Oryzomys couesi* is a medium- to large-sized rat. The ears are small and covered with hair, and the tail has the same length as the head and body. The dorsal coloration is grayish-brown with a small amount of black, the sides are paler, and the belly is white or light yellow. The tail is bicolored, darker on the top and lighter underneath, and naked and scaly (Ceballos and Miranda, 1986, 2000; Goodwin, 1969; Hall, 1981). Formerly, the distribution of *O. couesi* included the coastal regions of western Mexico, but recently Hanson et al. (2010) assigned the western population to *O. mexicanus*.

EXTERNAL MEASURES AND WEIGHT

TL = 242 to 294 mm; TV = 120 to 145 mm; HF = 28 to 40 mm; EAR= 13 to 18 mm. Weight: 40 to 80 g.

DENTAL FORMULA: I 1/1, C 0/0, PM 0/0, M 3/3 = 16.

NATURAL HISTORY AND ECOLOGY: This is a nocturnal and terrestrial rat with great ability to swim; it is commonly found in flooded environments such as mangrove swamps. In general, they build their nests in hollow trees and shrubs. In the Chamela-Cuixmala Biosphere Reserve, Jalisco, they are rare to common in tropical forests but abundant in wetlands and mangroves. In this region, their densities vary considerably; sometimes these rats disappear for several years before repopulating the area (Ceballos and Miranda, 2000). They are omnivores, feeding on seeds, fruits, herbs, small fish, crustaceans, and other invertebrates (Juárez, 1992). In some places they are considered a plague, as their populations can increase very rapidly. It is an opportunistic species (Sánchez-Cordero and Valadez-Azúa, 1989). They reproduce throughout the year and have up to eight offspring per litter. The gestation period

Oryzomys couesi. Wetlands. Xochimilco, Distrito Federal. Photo: Alberto Gonzalez.

oscillates from 21 to 28 days. They reach sexual maturity within seven weeks and have a short lifecycle (Sánchez-Cordero and Valadez-Azúa, 1989; Svihla, 1931).

VEGETATIONAL ASSOCIATIONS AND ELEVATION RANGE: *O. couesi* inhabits low deciduous tropical forests, median subperennial tropical forests, pine-oak forests, temperate deciduous forests, thorn forests, tropical deciduous and perennial tropical forests, mangroves, and riparian vegetation (Ceballos and Galindo, 1984; Ceballos and Miranda, 1986, 2000; Sánchez-Cordero and Valadez-Azúa, 1989). It is a common species in crops, grasslands, orchids, and coconut fields. It is found from sea level to 2,300 m (Hall, 1981).

CONSERVATION STATUS: This species faces no conservation problems and in many places it is considered a plague. It has a wide distribution in a variety of environments.

Oryzomys melanotis Thomas, 1893

Black-eared oryzomys

Guillermo Téllez and Rodrigo A. Medellín

SUBSPECIES IN MEXICO

Oryzomys melanotis colimensis Goldman, 1918
Oryzomys melanotis melanotis Thomas, 1893

O. melanotis is similar to *O. rostratus*. Almost since the beginning of this century, the existence of the two species was acknowledged (Goldman, 1918). Hooper (1953), however, only considered the existence of *O. melanotis*, and regarded *rostratus* as a subspecies within the group *melanotis*. Engstrom (1984) returned it to the category of species, arguing the chromosomal, genetic, and morphological differences between the two species. Chromosomally, *O. melanotis* has a predominantly telocentric karyotype and a diploid number of chromosomes 2n = 60; in contrast, *O. rostratus* has an acrocentric karyotype and a diploid number of chromosomes from 2n = 62 to 2n = 64. The somatic and cranial measures are smaller for *O. melanotis*, particularly for those populations located in the southern part of its distribution (Engstrom, 1984).

DISTRIBUTION: This species is endemic to Mexico; on the side of the Pacific Ocean it is distributed from the southern part of Sinaloa, to the southwest of Oaxaca (Medellín, 1992; Musser and Carleton, 1993). It has been recorded in the states of CL, GR, JA, MI, NY, OX, and SI.

DESCRIPTION: *Oryzomys melanotis* is a relatively small-sized rat. The dorsal coloration is yellowish-brown with dark underfur, but sometimes old individuals are reddish (Hall, 1981). The belly is white with diffuse gray and gray underfur. They have a light spot on the back and less than 40 supraorbital whiskers. The tips of the ears are light yellow. The tail has the same bicolored pattern as the body (Hall, 1981). The skull is delicate, with the infraorbital opening vertically elongated, and the palatine is fenestrated (Ceballos and Miranda, 1986, 2000; Engstrom, 1984). In *O. rostratus* the belly is white with light yellow, the back is yellow with diffuse brown, which turns reddish-yellow in some individuals, and the hair on the ears is yellowish-red. A white spot on the back is common in *O. melanotis* and is very rare or absent in *O. rostratus* (Engstrom, 1984).

EXTERNAL MEASURES AND WEIGHT
TL = 186 to 253 mm; TV = 99 to 135 mm; HF = 24 to 28 mm; EAR= 16 to 19 mm. Weight: 25 g.

DENTAL FORMULA: I 1/1, C 0/0, PM 0/0, M 3/3 = 16.

NATURAL HISTORY AND ECOLOGY: These rats are nocturnal and live in underground burrows. They are terrestrial but excellent swimmers. They feed on seeds, but can consume fruits, succulent leaves, and occasionally insects (Ceballos and Miranda, 1986, 2000). It has been collected with *Liomys pictus*, *O. alfaroi*, *O. couesi*, *Olygorizomys fulvescens*, *Baiomys musculus*, *Nyctomys sumichrasti*, *Peromyscus spicilegus*, *P.*

Oryzomys melanotis. Mangrove. Chamela-Cuixmala Biosphere Reserve, Jalisco. Photo: Gerardo Ceballos.

hylocetes, *P. levipes*, *P. megalops*, *Reithrodontomys fulvescens*, *Sigmodon mascotensis*, and *S. toltecus* (Juárez, 1992; Medellín, 1992; Nuñez et al., 1980). The average size of the litter is four; pregnant females have been found in March and July, and scrotal males in February and July (Juárez, 1992).

VEGETATIONAL ASSOCIATIONS AND ELEVATION RANGE: *O. melanotis* inhabits tropical deciduous forest, thorn forests, and tropical perennial forests, from sea level to 2,000 m (Engstrom, 1984; Medellín, 1992; Sánchez-Cordero and Valadez, 1996).

CONSERVATION STATUS: *O. melanotis* is classified as a non-endangered species (Ceballos and Navarro, 1991; Medellín, 1992), as it faces no conservation concerns and inhabits natural and disturbed environments.

Oryzomys mexicanus Allen, 1897

Mexican oryzomys

J. Delton Hanson

SUBSPECIES IN MEXICO

Oryzomys mexicanus albiventer (Merriam, 1901)
Oryzomys mexicanus aztecus (Merriam, 1901)
Oryzomys mexicanus bulleri (Allen,1897)
Oryzomys mexicanus crinitus (Merriam, 1901)
Oryzomys mexicanus lambi (Burt, 1934)
Oryzomys mexicanus mexicanus (Allen, 1897)
Oryzomys mexicanus molestus (Elliot, 1903)
Oryzomys mexicanus regillus (Goldman, 1915)
Oryzomys mexicanus zygomaticus (Merriam, 1901)

Recent work investigating relationships within the genus suggested three subclades in *mexicanus* (Hanson et al., 2010). In addition, morphological work by Carleton and Arroyo-Cabrales (2009) suggested that *O. mexicanus albiventer* should be considered a unique species (including *O. mexicanus molestus*). Furthermore, *O. mexicanus aztecus* was named as a subspecies of *O. mexicanus crinitus* prior to *crinitus* being subsumed into *mexicanus*. Considering this level of confusion, an extensive molecular examination of this species (including individuals from the Mexican highlands and the coasts of Sonora and Sinola) is necessary to establish the validity of the named subspecies.

DESCRIPTION: *Oryzomys mexicanus* is a medium- to large-sized mouse with a nearly naked, long, scaly tail, which is longer than the head and body. The ears are short and hairy, often lined with orange hairs. The upper parts are ochraceous or tawny, usually grizzled with black guard hairs. The dorsal pelage is moderately long (7 mm to 9 mm at the shoulders). The feet are white. The under parts are white to buffy. The karyotype is 2n = 56, FN = 56-60; however, this may be conserved throughout the genus. Considered a subspecies of *couesi* until recently (Hanson et al., 2010), the two species are very difficult to distinguish, although in general *mexicanus* is more robust than *couesi*. Other species that may be confused with *mexicanus* are *Rattus*, which has larger ears, a longer, more naked tail, as well as more obvious guard hairs and a less woolly coat, and *Sigmodon*, which has a much shorter tail, darker feet (not white), and an overall more brown and less ochraceous hue.

DISTRIBUTION: *O. mexicanus* is a widespread species found along the western and southern coast of Mexico, occurring from southern Sonora to the Guatemalan border. In the southern part of its distribution it does not occur very far inland; however, in the north its range extends to the Mexican Highlands. It has been recorded in the states of CH, CO, GR, JA, MI, MO, MX, NY, OX, PU, SI, and SO.

EXTERNAL MEASURES AND WEIGHT
TL = 230 to 295 mm; TV = 120 to 156 mm; HF = 25 to 31 mm; EAR = 14 to 18 mm.
Weight: 33 to 70 g.

DENTAL FORMULA: I 1/1, C 0/0, PM 0/0, M 3/3 = 16.

NATURAL HISTORY AND ECOLOGY: Not much is known regarding the natural history of this mouse except what has been published for *O. couesi*. Although these mice do not appear to have overt swimming adaptations, they often occupy a semi-aquatic niche. They are generalists and are usually associated with marshy areas, where they build nests in grasses and standing vegetation. They are good swimmers, using primarily their back legs and tail for propulsion, and have been observed to dive repeatedly (Cook et al., 2001). This ability to utilize a more aquatic habitat may provide niche separation between this species and cotton rats that are often caught in the same trap line as *mexicanus*. *O. mexicanus* has been collected with *Baiomys musculus*, *Liomys pictus*, *L. salvini*, *Sigmodon toltecus*, *S. mascotensis*, *Oligoryzomys fulvescens*, *Reithrodontomys gracilis*, *R. fulvescens*, *Pappogeomys*, *Marmosa*, and *Cryptotis goldmani*. Ecological investigations examining the reproductive habits and natural history of this mouse are needed to better characterize differences and similarities between it and *O. couesi*.

VEGETATIONAL ASSOCIATIONS AND ELEVATION RANGE: *O. mexicanus* is found primarily in grassy or shrubby habitats near water. It has been collected in grasslands near small creeks, as well as in thorn scrubland habitat next to the coast. Elevation ranges are from sea level to 2,500 masl.

CONSERVATION STATUS: This mouse is found throughout western Mexico. It is readily captured in both native and disturbed habitats and as such should be considered common.

Oryzomys nelsoni Merriam, 1898

Nelson's oryzomys

Iván Castro Arellano and Jorge Iván Uribe

SUBSPECIES IN MEXICO
O. nelsoni is a monotypic species.

DESCRIPTION: *Oryzomys nelsoni* is a large-sized rat of the genus *Oryzomys*; the tail is longer than the head and body. The back is creamy-ochraceous with a darker shade in the rear; this rat is slightly darker in the face, head, and anterior area, due to the presence of black hair. The belly is white; the tail is almost naked and pale brown except for the ventral-proximal region, which is a pale yellowish; the feet have little white hair. The skull is elongated and thin, but robust (Goldman, 1918; Merriam, 1898; Wilson, 1991). This species is apparently related to *O. couesi* as illustrated by Hershkovitz (1971), who considered it a subspecies of *O. couesi*. It is not possible to resolve its taxonomic position because only four specimens exist (Wilson, 1991).

EXTERNAL MEASURES AND WEIGHT
TL = 320 to 324 mm; TV = 185 to 191 mm; HF = 37 to 39 mm; EAR = 17 mm.
Weight: 40 to 80 g.

DENTAL FORMULA: I 1/1, C 0/0, PM 0/0, M 3/3 = 16.

NATURAL HISTORY AND ECOLOGY: This species is only known from four specimens found in a limited area located near the top of the highest elevation of Maria Madre Island, in the Marías islands. At the site, small springs and abundant herbaceous vegetation exist; this area is known as a zacatonal (Nelson, 1899; Wilson, 1991). Due to its rarity, the complete natural history of *O. nelsoni* is unknown, but its biology is probably similar to that of other species of the genus *Oryzomys*; hence its diet probably consists of succulent parts of grassland, seeds, and insects. It probably reproduces throughout the year. Its gestation period is probably between 25 to 30 days and the litter size is 2 to 5 offspring.

VEGETATIONAL ASSOCIATIONS AND ELEVATION RANGE: The vegetation of the type locality has been described as a zacatonal, with numerous herbaceous plants and grasses (Nelson, 1899; Wilson, 1991). The dominant vegetation on the island is low and medium deciduous tropical forests. The elevation of the type locality is 540 masl (Nelson, 1899).

CONSERVATION STATUS: It is almost certain that this species is extinct (Ceballos, 1999; Ceballos and Navarro, 1991; Wilson, 1991). Wilson mentioned that in intense fieldwork instead of collecting this species in the type locality only introduced rats (*Rattus*) were captured. A recent assessment (2001) carried out by specialists from the group Ecology and Conservation of Islands only recorded introduced rats in the island; as a result, *O. nelsoni* is probably extinct (Larsen et al., 2002).

DISTRIBUTION: This species is endemic to Mexico; its distribution is restricted to the type locality in Maria Madre Island, in the archipelago of the Marías Islands, located opposite of the coast of Nayarit (Goldman, 1918; Nelson, 1899). It has been recorded in the state of NY.

Oryzomys rhabdops Merriam, 1901

Highland oryzomys

Esther Romo V.

SUBSPECIES IN MEXICO
Oryzomys rhabdops angusticeps Merriam, 1901
O. rhabdops was regarded as a subspecies of *Oryzomys alfaroi* (Goldman, 1918); however, Musser and Carleton (1993) assigned it to species status.

DESCRIPTION: *Oryzomys rhabdops* is a medium-sized mouse, with thin ears, a long and thin face, and a black line running from the nose to the eye. The coat is up to 11 mm long. Its coloration is brownish-ochraceous on the sides and cheeks; the back and shoulders are darker due to black hairs that are less conspicuous than in other species. The dorsal coloration is white mixed with black; the tail is dark brown on top and yellowish below (Merriam, 1901). It differs from other members of the group *chapmani* by unique marks on the face and by the long and woolly hair.

DISTRIBUTION: The distribution of this species is restricted to the central and southern highlands of Chiapas and central Guatemala. It has been recorded in the state of CS.

EXTERNAL MEASURES AND WEIGHT
TL = 255 mm; TV = 141 mm; HF = 25 to 29 mm; EAR = 15 to 19 mm.
Weight: 22 to 44 g.

DENTAL FORMULA: I 1/1, C 0/0, PM 0/0, M 3/3 = 16.

NATURAL HISTORY AND ECOLOGY: *O. rhabdops* is a nocturnal species that inhabits coffee plantations and feeds on succulent plants of low growth along the limits of rivers or fields covered with herbs and grasslands.

VEGETATIONAL ASSOCIATIONS AND ELEVATION RANGE: This species dwells in the upper parts of the mountains of central and south Chiapas, in cloud forests and mixed forests of pine-oak (Villa-R., 1948). In Mexico it is found from 1,300 masl to 2,000 masl.

CONSERVATION STATUS: There are no data on the current status of this species of restricted distribution. It is not considered in any category of conservation, and is able to survive in disturbed environments.

Oryzomys rostratus Merriam, 1901

Long-nosed oryzomys

Mark D. Engstrom

SUBSPECIES IN MEXICO
Oryzomys rostratus carrorum Lawrence, 1947
Oryzomys rostratus rostratus Merriam, 1901
Oryzomys rostratus salvadorensis Felten, 1958
Oryzomys rostratus yucatanensis Merriam, 1901
O. rostratus has been regarded as a subspecies of *O. melanotis* (Hall, 1981), but after a thorough review Engstrom (1984) considered it a different species.

DESCRIPTION: *Oryzomys rostratus* is a medium-sized mouse. Its dorsal coloration is reddish-brown; the belly is white contrasting with the back. The tail is bicolored and longer than the body. The ears are blackish, protruding from the coat, and are covered on the inside by thin reddish hair. It is very similar to *O. couesi* and *O. alfaroi*, with which it is sympatric. In *O. rostratus* and *O. alfaroi* the hair on the digits of the hind feet spread beyond the tip of the nails, whereas in *O. couesi* it is short. The ears in *O. couesi* are furry and are slightly hidden in the coat, while the ears in the other two species are almost naked and protrude from the coat. The hair in *O. rostratus* and *O. alfaroi* is short and thin and in *O. couesi* is long and rough. Adults of *O. rostratus* are larger than those of *O. alfaroi*, and are reddish with reddish hair inside the ear instead of gray. There is individual variation in both species, however, and sometimes it is difficult to distinguish them from each other without genetic methods.

EXTERNAL MEASURES AND WEIGHT
TL = 220 to 290 mm; TV = 120 to 150 mm; HF = 27 to 33 mm; EAR = 17 to 22 mm.
Weight: 30 to 55 g.

DENTAL FORMULA: I 1/1, C 0/0, PM 0/0, M 3/3 = 16.

NATURAL HISTORY AND ECOLOGY: *O. rostratus* is a rare species; its natural history is not well known. They are nocturnal, terrestrial, and live in a variety of areas associated with forests. They have been collected in sugar cane fields, corn crops, coffee plantations, acahuales, cenotes, secondary forests, ecotones in forests, and mature forests. In arid localities, they generally live in riparian environments or in low and wet areas (Engstrom, 1984). It feeds on plants and animals, but details of its diet are unknown. Dalquest (1953b) reported remnants of plants in the stomach contents of an individual from San Luis Potosí. There are no reports on their reproduction. Pregnant females have been collected throughout the year, although reproduction is probably seasonal at a local level. Females do not mature synchronically, so not all females are sexually active in the mating season. They reach sexual maturity at juvenile or subadult stage, after leaving the nest. The litter size has four offspring with an average of two to six (Disney, 1968, M. Engstrom, pers. obs.). Its behavior is little known.

VEGETATIONAL ASSOCIATIONS AND ELEVATION RANGE: *O. rostratus* mainly inhabits tropical perennial forests, subdeciduous forests, and deciduous forests from sea level to 1,500 m. It has also been found in mixed forests.

CONSERVATION STATUS: *O. rostratus* is not included in any category of conservation because it has a wide distribution and can survive in disturbed environments.

DISTRIBUTION: *O. rostratus* inhabits forested areas in low-lying and intermediate areas from Tamaulipas to western Nicaragua. In Mexico, *O. rostratus carrorum* dwells in central Tamaulipas, and *O. rostratus rostratus* is distributed over most of the Gulf side of Mexico, to Tabasco and Chiapas. *O. rostratus salvadorensis* inhabits the Pacific rim of Chiapas, whereas *O. rostratus yucatanensis* is endemic to the Yucatan Peninsula (Engstrom, 1984). It has been recorded in the states of CA, CS, HG, OX, PU, QR, SL, TA, TB, VE, and YU.

Oryzomys rostratus. Cloud forest. La Soledad, Oaxaca. Photo: Gerardo Ceballos.

Cloud forest oryzomys

Esther Romo V. and J. Delton Hanson

SUBSPECIES IN MEXICO

Oryzomys saturatior hylocetes Merriam, 1901
Oryzomys saturatior saturatior Merriam 1901
Until recently, this species was considered a subspecies of *O. alfaroi* (Goldman, 1918). Musser and Carleton (1992) elevated it to species level.

DESCRIPTION: *Oryzomys saturatior* is a small-sized mouse with a short face and small, rounded ears. The coat is smooth and very short, ochraceous to dark brown mixed with black; the belly is gray. The tail is uniformly colored and naked. The ears are small and uniformly lined with black hairs. The claws on the hind feet have hair that extends past the end of the claws. This species is phylogenetically more closely related to *O. rhabdops* and *O. chapmani*, which are two other highland species, than it is to other members of the *O. alfaroi*-group. Other species with which it may be confused include *O. mexicanus* and *O. couesi*, which lack hairs extending past the claws on the hind feet. It may also be confused with *O. rostratus*, which is lighter, has orange hairs in the ears, and is generally found at lower elevations.

DISTRIBUTION: *O. saturatior* is distributed from the central highlands of Chiapas to northern Nicaragua. In Mexico, it has only been found in four localities in the central part of Chiapas. It has been recorded in the state of CS.

EXTERNAL MEASURES AND WEIGHT

TL = 220 to 290 mm; TV = 100 to 140 mm; HF = 15 to 19 mm; EAR =25 to 29 mm.
Weight: 22 to 44 g.

DENTAL FORMULA: I 1/1, C 0/0, PM 0/0, M 3/3 = 16.

NATURAL HISTORY AND ECOLOGY: *O. saturatior* is a nocturnal rodent that feeds on seeds and sprouts of herbs. It inhabits wetlands of the cloud forest in areas of steep and rocky slopes. It also inhabits grasslands.

VEGETATIONAL ASSOCIATIONS AND ELEVATION RANGE: *O. saturatior* primarily occurs in moist sylvan habitats such as cloud forests. They are found in association with mature evergreen and semi-deciduous trees and tree ferns and are readily trapped near fallen logs and along trail and stream edges. They seem to prefer primary forests and are not usually captured in disturbed habitats or recent secondary forests; however, they will occasionally occupy wet fields. In Mexico, it has been found from 1,800 masl to 2,500 masl.

CONSERVATION STATUS: There are no current data on populations of this mouse in Mexico. Focused inventories are necessary to determine its conservation status. Due to habitat destruction and only a few known localities in the country, this species in Mexico is probably at risk.

Rice rat

Mark D. Engstrom

SUBSPECIES IN MEXICO

This rat has been regarded as *O. couesi* (Hall, 1981). Schmidt and Engstrom (1994), however, clarified that the two species are valid and coexist in southern Texas and the northeast of Tamaulipas. Hanson et al. (2010) presented evidence that two species might be present within *O. texensis*, so populations in the southwestern United States and northeastern Mexico are recognized as *O. texensis*.

DESCRIPTION: *Oryzomys texensis* is a relatively large-sized rat. The back is grayish-brown, and the belly is white. The tail is as long as the body or a little shorter and has little hair. The ears are furry and partially covered by hair. In the northeast of Tamaulipas it is sympatric with *O. couesi*, a related species. *O. texensis* is distinguished by its smaller size, proportionately smaller tail, and lighter belly. From house rats (*Rattus rattus*) it differs by the naked tail and almost naked ears that protrude clearly from the coat, from the cotton rat (genus *Sigmodon*) by the dark gray instep, its longer coat that forms spines, and the shorter tail. Hanson et al. (2010) presented evidence that two species might be present within *O. texensis*, so populations in the southwestern United States and northeastern Mexico are recognized as *O. texensis*.

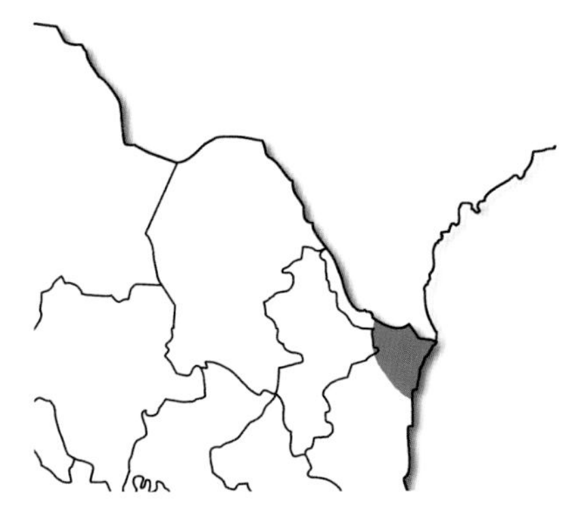

DISTRIBUTION: *O. texensis* inhabits lowlands from Texas in the United States to the northeastern tip of Mexico, usually associated with wetlands. In Mexico, it is only known in the region of Matamoros, Tamaulipas (Schmidt and Engstrom, 1994). It has been recorded in the state of TA.

EXTERNAL MEASURES AND WEIGHT

TL = 220 to 280 mm; TV = 100 to 135 mm; HF = 26 to 31 mm; EAR = 12 to 16 mm. Weight: 50 to 90 g.

DENTAL FORMULA: I 1/1, C 0/0, PM 0/0, M 3/3 = 16.

NATURAL HISTORY AND ECOLOGY: This rat is semi-aquatic and is generally associated with wetlands, mostly coastal marshes. It inhabits very wet or flooded sites such as shores of lakes, ponds, streams, ponds next to roads, or flooded crop fields. Occasionally, they are found in forests, but always in association with bodies of water. It is an opportunistic species whose diet varies throughout the year. In some seasons they mainly feed on invertebrates such as insects, crabs, and snails. In other periods their diet is based on vegetable matter and fungi. They occasionally feed on eggs of birds, turtles, and small fish and carrion. Reproduction may occur at any time of the year, but the pattern depends on the location and the population density. They show a reproductive peak in winter (November to March), which coincides with the stronger rains, and an inactive period in the hot months of summer. The size of the litter varies with the density, but the average fluctuates from four to six, with an interval of two to nine. The gestation period lasts 21 to 28 days. The population density may reach 50 individuals/ha, although the density varies annually and seasonally. The activity area varies from 0.2 ha to 0.4 ha. If the wet habitat is temporary, individuals can move great distances in search of favorable sites. They are nocturnal and very good swimmers, capable of moving underwater up to 10 m. They can be captured with traps placed on aquatic vegetation in sites where the water depth reaches 2 m. They build rounded nests, usually within or under fallen or floating logs or under bushes. In flooded plains, they hang their nests 1 m from the water in the stems of aquatic plants. Among its known predators are owls, hawks, snakes (especially *Agkistrodon piscivorous*), raccoons, foxes, weasels, and skunks. Although they have been considered pests of crops, their carnivorous habits would suggest they play a beneficial role in agriculture.

VEGETATIONAL ASSOCIATIONS AND ELEVATION RANGE: *O. texensis* is associated with aquatic and subaquatic vegetation in riparian areas and wetlands. It is found from sea level up to 500 m. Most localities are at sea level.

CONSERVATION STATUS: In Mexico, this species is only found in a small area near the coast of the Gulf of Mexico in Tamaulipas, which constitutes the southern limit of its distribution. It practically inhabits any patch of available swampy habitat, and it is possible that the species persists in the country unless there is a massive destruction of wetlands in the region.

Osgoodomys banderanus (J.A. Allen, 1897)

Osgood's deermouse

Joaquín Arroyo-Cabrales

SUBSPECIES IN MEXICO

Osgoodomys banderanus banderanus (J.A. Allen, 1897)

Osgoodomys banderanus vicinior (Osgood, 1904)

After its original description, this mouse was allocated to the genus *Peromyscus* until Carleton (1980) noted that several of the characters that differentiate the subgenus *Osgoodomys*, described by Hooper and Musser (1964), are synapomorphic, which justified its acknowledge at the generic level. The proposal was supported by Corbet and Hill (1991) and Musser and Carleton (1993). Other researchers, based on different characters from those proposed by Carleton (1980), consider that the differences between *Osgoodomys* and *Peromyscus* only divide at the subgeneric level (Bradley et al., 2007; Fuller et al., 1984; Rogers et al., 1984).

Osgoodomys banderanus. Tropical dry forest. Chamela-Cuixmala Biosphere Reserve, Jalisco. Photo: Gerardo Ceballos.

DESCRIPTION: *Osgoodomys banderanus* is medium in size; the dorsal coloration is cinnamon brown to dark brown and the ventral coloration is creamy white. The tail is ventrally naked, scaly, and of greater length than the body; the soles are partially or completely naked. The skull differs from those in other related genera by the developed supraorbitals, forming an edge on the orbits (Álvarez and Hernández-Chávez, 1990; Álvarez et al., 1987).

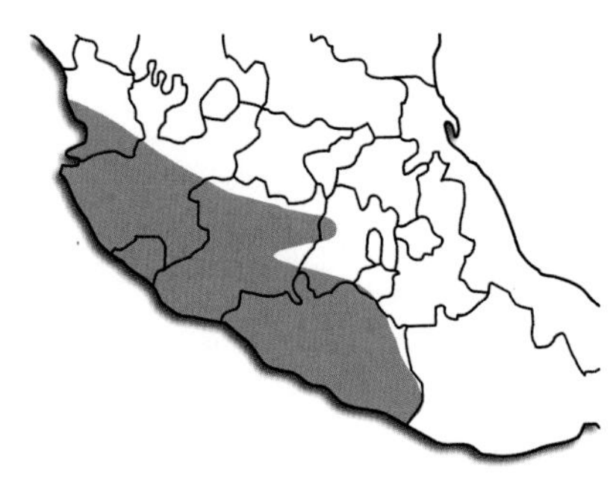

DISTRIBUTION: This species is endemic to western Mexico; it is distributed from the tip of southern Nayarit to the south of Guerrero, along the coastal slope and in the internal arid valleys and basins (Álvarez and Hernández-Chávez, 1990; Carleton, 1989). It has been recorded in the states of CL, GR, JA, MI, and NY.

EXTERNAL MEASURES AND WEIGHT
TL = 200 to 274 mm; TV = 96 to 125 mm; HF = 22 to 25 mm; EAR = 16 to 23 mm. Weight: 39 to 58 g.

DENTAL FORMULA: I 1/1, C 0/0, PM 0/0, M 3/3 = 16.

NATURAL HISTORY AND ECOLOGY: *O. banderanus* has nocturnal habits and is semi-arboreal. It is frequently found in rocky habitats with stubble and brush (Ceballos, 1990). The activity pattern was greater than that of other species observed in the Chamela-Cuixmala Biosphere Reserve, Jalisco; the distance traveled by two individuals between captures was 47.8 m; the mean of the greater distance between successive captures was 68.3 m (Collett et al., 1975). It feeds on seeds, insects, and fruits. They are solitary most of their life. Reproduction occurs from February to October; females are polyestric (Álvarez et al., 1987; Ceballos and Miranda, 1986, 2000; Núñez and Pastrana, 1990; Núñez et al., 1981). They are preyed on by owls (*Tyto alba*), mammals, and snakes (López-Forment and Urban, 1977; Ramírez-Pulido and Sánchez-Hernández, 1972). *O. banderanus* has been collected with *Liomys pictus*, *Peromyscus perfulvus*, *Xenomys nelsoni*, and *Tlacuatzin canescens* (Ceballos, 1989).

VEGETATIONAL ASSOCIATIONS AND ELEVATION RANGE: Osgood's deermouse has been collected in tropical deciduous forests, xeric scrublands, grasslands, forests of pine-oak (*Pinus-Quercus*), and riparian vegetation from sea level to 1,400 m (Ceballos and Miranda, 1986, 2000; Lechuga and Núñez, 1992).

CONSERVATION STATUS: *O. banderanus* is common in its restricted area in the west of Mexico. It survives in disturbed areas, which diminishes its vulnerability. Logging tropical forests may threaten this species' conservation in the long term. Detailed ecological studies must be conducted to identify the effect of overharvesting forests on the populations of these mice. Abundant populations survive in the Chamela-Cuixmala Biosphere Reserve in the coast of Jalisco (Ceballos and Miranda, 1986, 2000).

Otonyctomys hatti Anthony, 1932

Yucatan Vesper Rat

Julio R. Juárez-G.

SUBSPECIES IN MEXICO
O. hatti is a monotypic species.

DESCRIPTION: *Otonyctomys hatti* is a small to medium-sized mouse. The face is short, with two dark, very conspicuous eye rings. The ears are short with the apex rounded. The dorsal coloration varies from cinnamon to light brown, almost uniform, with the

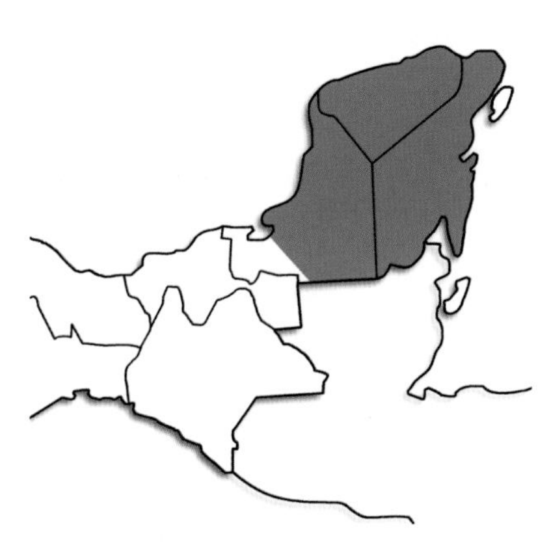

DISTRIBUTION: This species is restricted to the Yucatan Peninsula, in Mexico, Belize, and northern Guatemala. It has been recorded in the states of CA, QR, and YU.

rear darker; the sides are tawny to ochraceous tawny; the underside is white bathed with cream. The tail is medium in size and thick, dark brown, and with a tuft of hair on the tip. Females have two pairs of inguinal mammae (Hall, 1981; Reid, 1997). The rostrum is relatively short; the auditory bullae are disproportionately large and occupy most of the basicranial region (McSwiney et al., 2009). The diploid number is 2n = 50 and fundamental number is FN = 62 (Genoways et al., 2005).

EXTERNAL MEASURES AND WEIGHT
TL = 160 to 231 mm; TV = 85 to 127 mm; HF = 21 to 22 mm; EAR = 13 to 15 mm. Weight: 34 to 84 g.

DENTAL FORMULA: I 1/1, C 0/0, PM 0/0, M 3/3 =16.

NATURAL HISTORY AND ECOLOGY: *O. hatti*, a nocturnal species, seems to have strongly arboreal habits. It has been trapped on dead logs (Genoways et al., 2005), in trees (Hernández-Huerta et al., 2000; Jones et al., 1974), and from the top of a coconut palm (Peterson, 1966), usually 1 m to 2 m from the ground (McSwiney et al., 2009). It builds its burrows in hollows between branches of trees. Its diet may be restricted to sprouts of leaves and fruit (Genoways et al., 2005). Data on its reproduction have not been published; a female with three pups strongly adhered to the inguinal nipples was captured in Quintana Roo in June, and an adult male with undescended testicles was captured in November. A lactating female was present in February in Quintana Roo (Aranda et al., 1997). *O. hatti* has been caught in the same trap lines with other species of rodents, including *Heteromys gaumeri*, *Oryzomys melanotis*, *Ototylomys phyllotis*, and *Peromyscus yucatanicus* (Genoways et al., 2005; Hatt, 1938b; McSwiney et al., 2009).

VEGETATIONAL ASSOCIATIONS AND ELEVATION RANGE: *O. hatti* has been found in medium subperennial tropical forests and high perennial tropical forests. In Mexico, it has been collected between 25 masl and 50 masl.

CONSERVATION STATUS: *O. hatti* is considered a rare species throughout all of its geographic range (Genoways et al., 2005) and a threatened species by the NOM-ECOL059 (SEMARNAT, 2010)

Ototylomys phyllotis Merriam, 1901

Big-eared climbing rat

Hector T. Arita

SUBSPECIES IN MEXICO
Ototylomys phyllotis connectens Sanborn, 1935
Ototylomys phyllotis phyllotis Merriam, 1901

DESCRIPTION: *Ototylomys phyllotis* is a large-sized rat. There is marked geographic variation in size, with individuals of the Yucatan Peninsula (subspecies *phyllotis*) being smaller than those of Tabasco and Chiapas (subspecies *connectens*). The dorsal coloration varies from brown in Yucatan to grayish or blackish in Chiapas and Central America. The eyes and ears are large. The tail is long, scaly, and virtually devoid of hair. It differs from rats of the genus *Tylomys* by being smaller, by the proportionately larger ears, and by a slightly shorter tail that is uniformly colored, without the whitish tip characteristic of *Tylomys* (Kuns and Tashian, 1954; Lawlor, 1982a; Nowak, 1999b).

DISTRIBUTION: *O. phyllotis* is an endemic species of Mesoamerica. It is found in the Yucatan Peninsula, Tabasco, Chiapas, and in Central America to Costa Rica (Hall, 1981; Lawlor, 1982a). There is also a record of remnants of skulls of *O. phyllotis* in owl pellets in Guerrero (Ramírez-Pulido and Sánchez-Hernández, 1972). This record is probably false, due to its remoteness to other locations and its indirect provenance. It has been recorded in the states of CA, CS, QR, TB, and YU.

EXTERNAL MEASURES AND WEIGHT

TL = 95 to 190 mm; TV = 100 to 190 mm; HF = 26 to 29 mm; EAR = 24 to 25 mm. Weight: 80 to 120 g.

DENTAL FORMULA: I 1/1, C 0/0, PM 0/0, M 3/3 = 16.

NATURAL HISTORY AND ECOLOGY: This species prefers wooded and rocky sites (Lawlor, 1982a). In the Lacandon Forest it apparently prefers open sites (Medellín, 1992). In Yucatan it has been found in sinkholes, caves, and other rocky places (Jones et al., 1974). They are nocturnal and good climbers, which seek their food, consisting of fruits and leaves, between the branches of trees (Lawlor, 1982a). Its nesting sites are unknown (Lawlor, 1982a), although in Yucatan, females with young have been found, during the day, in cracks inside caves (Arita, 1992). Apparently, they reproduce throughout the year, although in Belize a decrease in the reproductive activity has been documented during the dry season (Disney, 1968). The gestation varies from 49 to 69 days (Helm, 1975) and the size of litter varies from 1 to 4 offspring, with averages ranging between 2 and 3 (Lawlor, 1982a). It does not frequent farmlands, so it is not considered a plague. It has been demonstrated that this rat is the main reservoir of leishmaniasis (*Leishmania mexicana*) in Belize (Disney, 1968). *O. phyllotis* has been caught in the same trap lines with other species of rodents, including *Heteromys gaumeri*, *Oryzomys melanotis*, and *Peromyscus yucatanicus* (Genoways et al., 2005; Hatt, 1938; McSwiney et al., 2009). In Yaxchilán, Chiapas has been caught with *Heteromys desmarestianus* and *Oryzomys melanotis* (Escobedo-Morales et al., 2005).

Ototylomys phyllotis. Tropical semi-green forest. Yucatan. Photo: Brian and Annika Keeley.

VEGETATIONAL ASSOCIATIONS AND ELEVATION RANGE: *O. phyllotis* inhabits tropical perennial forests, subperennial forests, and deciduous forests (Lawlor, 1982a). It is found from sea level to 2,000 m (Nowak, 1999b). The majority of the localities in Mexico are found in lowlands, below 1,000 m.

CONSERVATION STATUS: *O. phyllotis* is not included in any official list of endangered species. The fact that it shows preference for wooded sites, and that its population density is low, for example, 0.59 individual/ha in the Lacandon Forest, Chiapas (Medellín, 1992) indicates that this rat would be particularly vulnerable to deforestation in tropical areas of Mexico. Its wide distribution and presence in protected areas of Chiapas and the Yucatan Peninsula, however, indicate it would not be a priority species for conservation programs.

Peromyscus aztecus (Saussure, 1860)

Aztec deermouse

José Ramírez-Pulido and Claudia Aguilar

SUBSPECIES IN MEXICO

Peromyscus aztecus aztecus (Saussure, 1860)
Peromyscus aztecus evides Osgood, 1904
Peromyscus aztecus oaxacensis Merriam, 1898
The systematics of this species varied in the past two decades (see Tiemann-Boege et al., 2000 for a revision). The subspecies *P. aztecus hylocetes* was first considered a species (Merriam, 1897), then a subspecies of *P. aztecus* (Carleton, 1977); currently it is considered a species (Tiemann-Boege et al., 2000).

DISTRIBUTION: *P. aztecus* has a distribution restricted mainly to the mountainous area from central and southeast Mexico, to the northern part of Honduras and southern El Salvador. The populations of Mexico are distributed as follows: *P. aztecus aztecus* dwells in a small portion of the Sierra Madre Oriental, in the states of Puebla and Veracruz. *P. aztecus evides* is located in the mountains of Guerrero and Oaxaca (Carleton, 1989). It has been recorded in the states of CS, GR, OX, PU, SL, and VE.

DESCRIPTION: *Peromyscus aztecus* is a large-sized mouse within its genus. The dorsal coloration varies from cinnamon brown to pale ochraceous, mixed with black to reddish, depending on the subspecies. The belly is creamy and the feet are white. The tail is bicolored and has the same length as the head and body together (Hall, 1981; Hooper, 1947; Merriam, 1898; Osgood, 1909; Ramírez-Pulido et al., 1977; Saussure, 1860). By its morphology, it can be confused with some other species of the genus. It is distinguished from *P. beatae*, because its color is more intense, it is larger, the feet are smaller, and the soles of the hind feet are hairy; however, the main characteristic that distinguishes these two species is the greater length of teeth in *P. aztecus*, particularly the length of the second molar. It differs from *P. maniculatus* and *P. melanotis* by its greater size and hind feet with a length of 25 mm; the ears of *P. truei* are smaller than the feet and the ears of *P. difficilis* measure less than 20 mm. Finally, it is easily distinguishable from *P. melanophrys* because its external measures are smaller (Álvarez, 1961b; Bradley and Schmidly, 1987; Ceballos and Galindo, 1984; Goodwin, 1969; Hall, 1981; Hall and Dalquest, 1963; Hooper and Musser, 1964).

EXTERNAL MEASURES AND WEIGHT

TL = 203 to 238 mm; TV = 102 to 121 mm; HF = 24 to 27 mm; EAR = 15 to 18 mm.
Weight: 32 to 45 g.

DENTAL FORMULA: I 1/1, C 0/0, PM 0/0, M 3/3 = 16.

NATURAL HISTORY AND ECOLOGY: This species inhabits temperate forests in the mountainous regions of central and southeast Mexico. At times they have been found in inhospitable places with little vegetation cover and little food such as lava fields (Baker and Phillips, 1965; Ceballos and Galindo, 1984). They have been collected with *Megasorex gigas*, *Heteromys desmarestianus*, *Liomys pictus*, *Osgoodomys banderanus*, *P. beatae*, *P. furvus*, *Reithrodontomys sumichrasti*, *Sigmodon alleni*, and *S. mascotensis* (Goodwin, 1969; Winkelmann, 1962). It is mainly herbivorous, although it supplements its diet with insects (Carleton, 1973). Reproduction seems to be carried out in September and October. It is preyed on by coyotes, lynx, weasels, and nocturnal raptors (Baker et al., 1971; Ceballos and Galindo, 1984; Ramírez-Pulido et al., 1977).

VEGETATIONAL ASSOCIATIONS AND ELEVATION RANGE: *P. aztecus* inhabits tropical subdeciduous forests, cloud forests, pine-oak forests, oak-pine-fir forests, forests of liquidambar, and wet forests of pine-oak in soil and rocky soil with plenty of humus and dense undergrowth. It has also been found in disturbed environments, coffee plantations, sugar cane fields, and mango. It has been collected between 500 m and 3,200 m (Baker and Phillips, 1965; Baker et al., 1971; Bradley et al., 1990; Carleton, 1979; Goodwin, 1969; Hooper, 1957; Hooper and Musser, 1964; Musser, 1964, Ponce Ulloa and Llorente Boursquets, 1993; Schaldach, 1966; Webb and Baker, 1969; Winkelmann, 1962).

CONSERVATION STATUS: *P. aztecus* has a broad distribution that apparently presents no conservation concerns.

Peromyscus beatae Thomas, 1903

Orizaba deermouse

Alberto Enrique Rojas Martínez

SUBSPECIES IN MEXICO

Peromyscus beatae beatae Thomas, 1903

Peromyscus beatae sacarensis Dickey, 1928

P. beatae was regarded as a subspecies of *P. boylii*, but Rennert and Kilpatrick (1987) and Schmidly et al. (1988) found biochemical and chromosomal evidence indicating a specific range. The distribution of the chromosome variants is complex, however, and does not coincide with the geographical limits recognized for the subspecies of the group *boylii* (Schmidly and Schoeter, 1974). Houseal et al. (1987) considered that the specimens of Jalapa, Veracruz, do not belong to this species. On the other hand, Schmidly et al. (1988) suggested that the specimens of Oaxaca and Guerrero, considered by Houseal et al. (1987), do not belong to *P. beatae*. Bradley et al. (2000) proposed the division of the subspecies *beatae* and *sacarensis*.

DESCRIPTION: *Peromyscus beatae* is a medium-sized mouse; the length of the tail is similar to that of the head and body. The dorsal coloration is dark brown, with a blackish line along the mid-back. The sides are shiny tobacco-brown with a tenuous lateral line of yellowish-orange color that extends to the cheeks and hindquarters. The belly is bicolored, white on the tip and black-slate in the base; the tail is grayish-black in the back and white ventrally, with a tuft of hair on the tip. The hind feet are white but obscure above the ankles. The ears are blackish-brown (Schmidly et

DISTRIBUTION: *P. beatae* is found from the center of Veracruz and center of Guerrero to Oaxaca and Chiapas in Mexico, southward through the highlands of Guatemala, El Salvador, and Honduras. It has been recorded in the states of CS, GR, HG, OX, PU, and VE.

al., 1988). Osgood (1909) considered it indistinguishable from *P. levipes*. Schmidly et al. (1988) suggested that *P. boylii rowleyi*, *P. levipes*, and *P. beatae* are cryptic, as they cannot be distinguished based on their external morphological characteristics. In general, however, *P. beatae* is larger and darker and has a larger tail (Álvarez, 1961b; Schmidly et al., 1988).

EXTERNAL MEASURES AND WEIGHT
TL = 178 to 250 mm; TV = 90 to 132 mm; HF = 17 to 26 mm; EAR = 14 to 21 mm. Weight: 26 g.

DENTAL FORMULA: I 1/1, C 0/0, PM 0/0, M 3/3 = 16.

NATURAL HISTORY AND ECOLOGY: *P. beatae* commonly inhabits mesic forests on the east side of the Sierra Madre Oriental. It is common in the humid forests (Schmidly et al., 1988). Rocky soil, volcanic spills, and thickets are also appropriate for this species (Hall and Dalquest, 1963). It is nocturnal and can coexist with *P. melanotis* and *P. levipes* (Hall and Dalquest, 1963; Houseal et al., 1977). With regard to its reproduction, the only known information is from reproductively active individuals caught from October to November (Hall and Dalquest, 1963).

VEGETATIONAL ASSOCIATIONS AND ELEVATION RANGE: This species inhabits subhumid and humid environments where coniferous forests, mixed forests of pine-oak, and forests of oak grow (Schmidly et al., 1988), as well as tropical evergreen forests (Álvarez, 1961b). It is found from 1,371 masl to 3,810 masl, but most localities are located above 1,800 masl.

CONSERVATION STATUS: This species was originally thought to be found exclusively in Veracruz, but since its distribution in Mexico has been proved to be wider, it can be said that it faces no conservation concerns.

Peromyscus boylii (Baird, 1855)

Brush deermouse

Leticia A. Espinosa A. and Catalina B. Chávez T.

SUBSPECIES IN MEXICO
Peromyscus boylii glasselli Burt, 1932
Peromyscus boylii rowleyi (J.A. Allen, 1893)

DESCRIPTION: Compared to other species of the genus, *Peromyscus boylii* is a small-sized mouse. The tail is longer than the length of the head and body; the feet are small and the ears are medium-sized (Anderson, 1972; Osgood, 1909). The dorsal coloration is reddish-olive and the sides are light brown with a tenuous lateral line of ochraceous-orange color. The underside and feet are white. The tail is bicolored, dorsally brown and ventrally white. The ears are dark gray. The skull is medium-sized (the greater length of the skull is 25.8 mm to 27 mm). At a chromosomal level all subspecies of *P. boylii* have a diploid number of 2n = 48, with some differences in the essential number, since *P. b. rowleyi* has FN = 52 and *P. b. glasselli* FN = 52-54. *P. boylii* differs from *P. levipes* because the latter is larger in size and is also regarded as a

Peromyscus boylii. Grassland. Janos Biosphere Reserve, Chihuahua. Photo: Gerardo Ceballos.

polymorphic species with a FN = 58-60 (Houseal et al., 1987; Schmidly et al., 1988; Tiemann-Boege et al., 2000).

EXTERNAL MEASURES AND WEIGHT
TL = 151 to 220 mm; TV = 68 to 115 mm; HF = 19 to 23 mm; EAR = 16 to 24 mm. Weight: 19 to 30 g.

DENTAL FORMULA: I 1/1, C 0/0, PM 0/0, M 3/3 = 16.

NATURAL HISTORY AND ECOLOGY: *P. boylii* is found both in semi-arid regions of the country, where grasslands and bush vegetation with cacti and mesquite abound, and in the temperate-subhumid regions of the mountainous areas of the Sierra Madre Occidental, which are covered by forests of pine-oak. They show preference for rocky areas (Anderson, 1972; Baker and Greer, 1962; Schmidly et al., 1988). In Durango a density of 4.37 individuals/ha was found in an activity area of males and females, of 0.06 ha and 0.04 ha, respectively, during the months of June and July. Lactating and pregnant females with an average of three embryos, with a range of one to four, were also found at that time (Baker and Greer, 1962). Several studies on gene, chromosomal, and morphometric variation have been conducted to establish the taxonomic differences between *P. boylii* and groups with which they can be sympatric or allopatric, and the results are consistent with regard to the distribution and characteristics of habitats for this species (Houseal et al., 1987; Rennert and Kilpatrick, 1986; Schmidly et al., 1988). Bradley et al. (2004) assigned populations from the pine-oak forests of western Durango to *P. schmidlyi*; populations in the more arid regions of Durango remained in *P. boylii*. Other rodents recorded in the same localities are *P. difficilis*, *P. maniculatus*, *P. gratus*, *Neotoma mexicana*, *Nelsonia neotomodon*, *Reithrodontomys megalotis*, *Microtus mexicanus*, *Sigmodon leucotis*, *Eutamias bulleri*, and *Sorex oreopolus* (López-Vidal and Álvarez, 1993).

VEGETATIONAL ASSOCIATIONS AND ELEVATION RANGE: *P. boylii* inhabits grasslands, xeric scrublands, and forests of conifers and oaks, from 914 masl to 2,590 masl.

CONSERVATION STATUS: This species is very common and widely distributed, and has no conservation problems. The subspecies *P. b. glasselli* is threatened (SEMARNAT, 2010).

DISTRIBUTION: *P. boylii* has a wide distribution from Southern California and western Oklahoma to Mexico, from Baja California Norte and southern Sonora through the Mexican Altiplane to Querétaro and the western part of Hidalgo (Burt, 1932; Hall, 1981; Houseal et al., 1987; Osgood, 1909; Ramírez-Pulido and Castro-Campillo, 1994; Rennert and Kilpatrick, 1986; Wilson and Reeder, 2005). It has been recorded in the states of AG, BC, CH, CO, DU, HG, QE, SL, SO, and ZA.

Peromyscus bullatus Osgood, 1904

Perote deermouse

Alberto Enrique Rojas Martínez

SUBSPECIES IN MEXICO

P. bullatus is a monotypic species.

DESCRIPTION: *Peromyscus bullatus* is a medium-sized mouse with very large ears. The length of the tail is smaller than that of the body and head together (Osgood, 1904). Its coat is dense, dorsally dark yellowish, with a blackish stripe along the back reaching the nasal region. The coat is grayish between the ears and eyes. The eyes are ringed with black hair. The underside is creamy white. The feet are white, but obscured above the ankles. The tail is bicolor (Osgood, 1904). It is similar to *P. truei* and *P. difficilis*, but is distinguished by the size of the ears and bullae, which are the largest within the genus (Osgood, 1904).

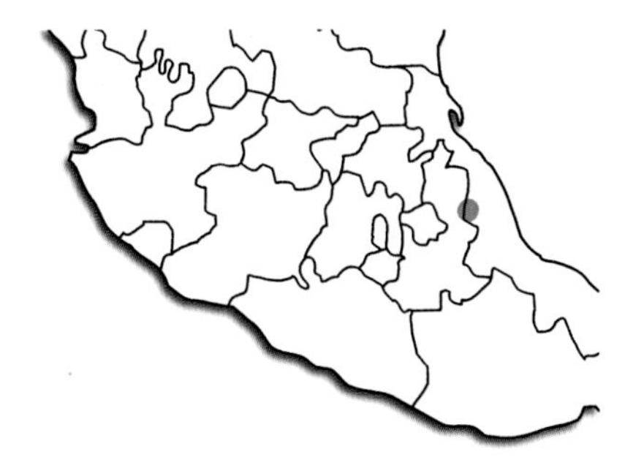

DISTRIBUTION: This species is endemic to Valle Oriental between Puebla and Veracruz. It is only known in Perote and El Limon, in Veracruz. It has been recorded in the state of VE.

EXTERNAL MEASURES AND WEIGHT

TL = 175 to 223 mm; TV = 87 to 119 mm; HF = 19 to 25 mm; EAR = 23 to 27 mm. Weight: 20 to 36 g.

DENTAL FORMULA: I 1/1, C 0/0, PM 0/0, M 3/3 = 16.

NATURAL HISTORY AND ECOLOGY: *P. bullatus* is a mouse that is rare in regards to occurrence. It inhabits plains and dunes of fine sand, covered by weeds and scarce xeric grasses. Reproduction apparently takes place during spring and early summer, so at the end of September it has already ended. It coexists with *P. maniculatus*, which is very abundant in its habitat (Hall and Dalquest, 1963).

VEGETATIONAL ASSOCIATIONS AND ELEVATION RANGE: This species inhabits xeric scrublands in sandy soils (Hall and Dalquest, 1963). It has only been caught at 2,377 masl.

CONSERVATION STATUS: The current status of this species is unknown; however, it is regarded as fragile and of Special Protection (Ceballos and Navarro, 1991; SEMARNAT, 2010). It is likely to be threatened because it is an endemic species with a highly restricted area of distribution and because of the replacement of native thorny vegetation by maize and maguey crops in the surrounding area of Perote, Veracruz (Hall and Dalquest, 1963; Valdéz and Ceballos, 1997).

California deermouse

Jaime Luévano and Eric Mellink

SUBSPECIES IN MEXICO

Peromyscus californicus insignis Rhoads, 1895

DESCRIPTION: *Peromyscus californicus* is a large-sized mouse. The dorsal coloration varies geographically, but in general it is dark brown on the back and ochraceous on the sides. Its coat is long and loose, and the tail is hairy with scaly rings. The baculum has the shape of a vase; its base is wide, narrowing near the middle, and has depressions in its distal and proximal sections. This mouse has 48 chromosomes (Hall, 1981; Jameson and Peeters, 1988; Merritt, 1978).

EXTERNAL MEASURES AND WEIGHT

TL = 220 to 285 mm; TV = 117 to 156 mm; HF = 24 to 31 mm; EAR = 20 to 26 mm. Weight: 33 to 55 g.

DENTAL FORMULA: I 1/1, C 0/0, PM 0/0, M 3/3 = 16.

NATURAL HISTORY AND ECOLOGY: *P. californicus* is mainly nocturnal and has a peak of activity just before dawn. It is a common resident of all Mediterranean communities, from coastal scrubland to coniferous forest, passing through different types of chaparral and communities of pine-oak. The higher density is located in chaparral, where very often it is associated with *Neotoma macrotis* since it uses its abandoned burrows. In laboratory conditions, it is tolerant of other species, but totally incompatible with the kangaroo rat *Dipodomys agilis* (Brylski, 1990d; Jameson and Peeters, 1988; Merritt, 1978). Its diet consists mainly of fruits, flowers, and seeds of a wide variety of plants, such as *Rhus integrifolia*, *Lotus scoparius*, and *Salvia apiana*, which are common components of coastal scrubland. Other food items are fungi and arthropods. In wooded environments they can eat acorns (*Quercus* sp.). They are good climbers, and probably forage on trees and shrubs. Water is probably obtained from the vegetation and dew (Brylski, 1990d; Jameson and Peeters, 1988; Merritt, 1978). Females are polyestric and, unlike other species of the same genus, they do not have estrus postpartum. The estrus cycle lasts about seven days. Mating can take place throughout the year, but occurs mainly between March and September and the majority of pups born between April and October. The litter comprises 2 to 4 offspring that are born after a gestation period of 21 to 25 days. Commonly, females have three to four litters per year, under natural conditions, and six or more in captivity. If the number of litters is high, the number of offspring reduces to two. Males devote the same amount of time to parental care. They live in couples even during pregnancy and lactation. Females reach sexual maturity at 11 weeks old and males a little later (Brylski, 1990d; Jameson and Peeters, 1988; Merritt, 1978).

VEGETATIONAL ASSOCIATIONS AND ELEVATION RANGE: This species inhabits Mediterranean habitats, where it occupies chaparral, coastal scrublands, and forests. It is found from sea level to 1,800 m in the Sierra of San Pedro Martir.

CONSERVATION STATUS: This species has no conservation concerns since it tolerates different environmental conditions and because it is distributed in many environments that are not altered.

DISTRIBUTION: This species is found in the southwestern United States and Baja California, where numerous records have been reported in the Sierras de Juárez and San Pedro Mártir (Hall, 1981; Huey, 1964). It has been recorded in the state of BC.

Monserrat Island deermouse

Eric Mellink and Jaime Luévano

SUBSPECIES IN MEXICO

P. caniceps is a monotypic species.

Recently, Hafner et al. (2001) considered *P. caniceps* to be a subspecies of *P. fraterculus*. Their study is only based on molecular evidence, however. We have decided to recognize it as species until morphological evidence and karyotypes demonstrate the opposite.

DISTRIBUTION: This species is endemic to Monserrat Island (25 degrees 38 minutes north, 111 degrees 2 minutes west), on the coasts of the Gulf of California. It has been recorded in the state of BC.

DESCRIPTION: *Peromyscus caniceps* is a medium-sized species. The back and flanks are ochraceous-yellowish; the head is gray and contrasting. The underside is white with some yellowish coloration. The lateral line is not very conspicuous. The tail is slightly bicolored, and the ears are brownish. Young are gray (Burt, 1932; Sánchez-Hernández et al., 1997).

EXTERNAL MEASURES AND WEIGHT

TL = 180 to 220 mm; TV = 100 to 124 mm; HF = 20 to 22 mm; EAR = 16 to 20 mm. Weight: 13 to 25 g.

DENTAL FORMULA: I 1/1, C 0/0, PM 0/0, M 3/3 = 16.

NATURAL HISTORY AND ECOLOGY: Trapping this species in Monserrat Island has not been very successful (Álvarez-Castañeda and Gómez-Machorro, 1998; Lindsay, 1964) and its biology is unknown. Like other rodents of the genus, in other islands in the Gulf of California, this species probably prefers rocky areas. A lactating female was captured in October (Álvarez-Castañeda and Gómez-Machorro, 1998), and it probably reproduces mostly in summer and autumn. Vegetation (leaves, flowers, fruit, and seeds) and insects compose its diet. On the island, four species of snakes eat rodents.

VEGETATIONAL ASSOCIATIONS AND ELEVATION RANGE: *P. caniceps* is exclusively found in xeric scrubland, from sea level to 220 m, in the highest parts of the island.

CONSERVATION STATUS: This species is apparently rare. It is endemic to an island and for that reason is classified as of Special Protection in Mexico (SEMARNAT, 2010). The island has important human presence, especially fishermen, and there are apparently introduced rodents.

Canyon deermouse

Heliot Zarza

SUBSPECIES IN MEXICO
Peromyscus crinitus delgadilli Benson, 1940
Peromyscus crinitus disparilis Goldman, 1932
Peromyscus crinitus pallidissimus Huey, 1931
Peromyscus crinitus stephensi Mearns, 1897

DESCRIPTION: Within its genus, *Peromyscus crinitus* is a small-sized mouse. It shows clinal variation in the size, as northern individuals are larger than those to the south; however, along its distributions, females are larger than males. The snout is pointed, the eyes are prominent, and the ears are as long as the feet. The coat is long and soft. The colorations vary according to the geographic distribution, and sometimes resemble the substrate, generating microgeographic races. The dorsal coloration varies from dark brown to light yellowish-red. The underside is white, except in *P. crinitus delgadilli*. The dorsal hair is long and loose, with three bands with the following pattern: dark-light-dark. The forefeet and hind feet are white, except in *P. crinitus delgadilli*, which are black. The tail is bicolored and is covered with soft and long hairs that sometimes seem discontinuous, and is much longer than the head and body. The skull of *P. crinitus* is long and wide; the width of the zygomatic arch is smaller than the width of the cranial box; M1 and M2 lack wrinkles (Hall, 1981; Johnson and Armstrong, 1987).

EXTERNAL MEASURES AND WEIGHT
TL = 62 to 192 mm; TV = 82 to 118 mm; HF = 17.5 to 23 mm; EAR = 15.3 to 21.5 mm. Weight: 17 to 24 g.

DENTAL FORMULA: I 1/1, C 0/0, PM 0/0, M 3/3 = 16.

NATURAL HISTORY AND ECOLOGY: This species inhabits rocky environments such as cliffs, slopes, and desert with a layer of gravel. It has been captured in associations of pine (*Pinus ponderosa*, *P. monophylla*, *Juniperus osteosperma*, *Larrea* spp., among others; Johnson and Armstrong, 1987). This species is sympatric with *P. eremicus*, *Perognathus fallax*, and *Neotoma lepida* (Eidemiller, 1980). The species is seasonally polyestric; it has an average of 2.5 litters per year with a size of the litter of 1 to 5 offspring (Egoscue, 1964). It molts twice a year (post-juvenile and post-adult stages); the post-adult fur is spotted and irregular, appearing first on the sides, hindquarters, and mid-back (Johnson and Armstrong, 1987). *P. crinitus* is omnivororous and feeds on seeds, fruits, insects, and small animals (Morton, 1979).

VEGETATIONAL ASSOCIATIONS AND ELEVATION RANGE: *P. crinitus* mainly inhabits rocky environments; the vegetation has little effect on its local distribution (Egoscue, 1964; Johnson and Armstrong, 1987). It is found from sea level to 3,230 m (Hall, 1981; Johnson and Armstrong, 1987).

CONSERVATION STATUS: At the species level, *P. crinitus* is not included in any risk category; however, the subspecies *P. crinitus pallidissimus* is considered threatened by the Mexican government (SEMARNAT, 2010).

DISTRIBUTION: *P. crinitus* has a highly discontinuous distribution due to its habitat preference (Hall, 1981). It is found in the northeastern part of Baja California Norte and the northeastern part of Sonora. The subspecies *P. crinitus delgadilli* and *P. crinitus pallidissimus* are endemic to the Sierra del Pinacate, Sonora, and to San Luis Gonzaga Island, Gulf of California, respectively (Hall, 1981; López-Forment et al., 1996). It has been recorded in the states of BC and SO.

Dickey's deermouse

Eric Mellink and Jaime Luévano

SUBSPECIES IN MEXICO

P. dickeyi is a monotypic species.

Recently Hafner et al. (2001) considered *P. dickeyi* to be a subspecies of *P. merriami*. Their work is based solely on molecular evidence, however, so we decided to recognize the species until there is further evidence.

DISTRIBUTION: This species is endemic to Tortuga Island (27 degrees 21 minutes north, 111 degrees 54 minutes west) in the Gulf of California. It has been recorded in the state of BS.

DESCRIPTION: *Peromyscus dickeyi* is a large-sized mouse, with the length of the tail smaller than the length of the head and body. The back and flanks are dark brown with pinkish-cinnamon; the ears are dark. There is a lateral line, the tail is bicolored, and the soles of the hind feet are naked. The belly is white, although in some individuals a tenuous pectoral spot of pinkish-cinnamon coloration can exist. Young and adult individuals that have recently molted are pinkish-cinnamon throughout the ventral surface. Individuals that recently molted in June are darker and grayer than individuals with worn overcoat. The offspring are gray (Burt, 1932a).

EXTERNAL MEASURES AND WEIGHT

TL = 191 mm; TV = 90.5 mm; HF = 22 mm; EAR = 20 mm.

Weight = 15 to 20 g.

DENTAL FORMULA: I 1/1, C 0/0, PM 0/0, M 3/3 = 16.

NATURAL HISTORY AND ECOLOGY: It has been reported that this species prefers rocky slopes (Álvarez-Castañeda and Cortes Calva, 1999) and seems to be more abundant at the southern end of the island (Bourillón et al., 1988). We were successful capturing it in the upper parts of the island (E. Mellink and M. Gonzalez-Jaramillo, pers. obs.). Their reproduction takes place, at least, during summer and early fall. At the end of October 1999 five females, including a pregnant one, were caught, two of which were lactating; one was inactive and the other was a juvenile. Two males were also caught, one with abdominal testicles and one a juvenile. In Tortuga Island the rattlesnake (*Crotalus tortugensis*), which is very abundant, and the false coralillo (*Lampropeltis getula*) may be their predators.

VEGETATIONAL ASSOCIATIONS AND ELEVATION RANGE: *P. dickeyi* exclusively inhabits xeric scrublands, from sea level to about 300 masl, which is the maximum height of the island.

CONSERVATION STATUS: *P. dickeyi* is of Special Protection (SEMARNAT, 2010). The island has little human presence, as it is far from the coast and lacks anchorages. In addition, the high population of rattlesnakes in this island can help prevent the establishment of exotic species. In fact, introduced rodents or domestic cats have not been reported in this island (Mellink et al., 2002). Thus, the risks facing this mouse seem to be much lower than those facing rodents in other islands.

Southern rock deermouse

J. Cuauhtémoc Chavez T. and Gerardo Ceballos

SUBSPECIES IN MEXICO

Peromyscus difficilis amplus Osgood, 1904
Peromyscus difficilis difficilis (J.A. Allen, 1891)
Peromyscus difficilis felipensis Merriam, 1898
Peromyscus difficilis petricola Hoffmeister and de la Torre , 1959
Peromyscus difficilis saxicola Hoffmeister and de la Torre, 1959

The taxonomy was revised by Hoffmeister and Tower (1961), who included *P. nasutus* as subspecies. Biochemical and karyotype evidence suggested that in reality they are sister species (Carleton, 1989; Warn et al., 1979; Zimerman et al., 1975, 1978).

DESCRIPTION: *Peromyscus difficilis* is a medium-sized mouse. The dorsal coloration can include the following variants: grayish-brown, yellowish-brown, and brownish-ochraceous on a gray background. The sides are yellowish, and the belly is whitish with gray underfur; occasionally, it has an orange pectoral mark. The tail is bicolored and longer than the head and body. The ears are large. The back of the hind feet is whitish-colored in the metatarsal region (Ceballos and Galindo, 1984; Hall, 1981; Hernández, 1990). The edge of the frontals is rounded, without elevations in the interorbital region; the bullae are inflated; the face is long; and the cranial box is flattened. It can be differentiated from *P. melanophrys*, which is bigger and has a larger and thicker tail.

EXTERNAL MEASURES AND WEIGHT

TL = 180 to 260 mm; TV = 91 to 145 mm; HF = 22 to 28 mm; EAR= 17 to 28 mm. Weight: 24 to 32 g.

DENTAL FORMULA: I 1/1, C 0/0, PM 0/0, M 3/3 = 16.

DISTRIBUTION: This species is endemic to Mexico; it is distributed from western Chihuahua and southeastern Coahuila, toward south-central Oaxaca. It inhabits elevations of more than 2,000 masl. *P. difficilis amplus* is distributed in central Mexico; its southern limits are the mountainous areas of Oaxaca and western Veracruz, reaching the Eje Neovolcanico in the state of Mexico. *P. difficilis difficilis* is distributed in western Chihuahua and the Sierra Madre Occidental to the Eje Volcanico Transversal. *P. difficilis felipensis* has a fragmented distribution, with isolated populations in the Eje Volcanico Transversal in the state of Mexico, Mexico City, and Morelos and in the mountains of Oaxaca. *P. difficilis petricola* is located in the central region of the Sierra Madre Oriental, between Coahuila, Tamaulipas, and Zacatecas. Finally, *P. difficilis saxicola* is found in Queretaro, Hidalgo, and Veracruz. It has been recorded in the states of AG, CH, CO, DF, DU, GJ, MI, MO, MX, NL, OX, PU, QE, SL, TA, TL, VE, and ZA.

Peromyscus difficilis. Scrubland. Totalco, Veracruz. Photo: Gerardo Ceballos.

NATURAL HISTORY AND ECOLOGY: These mice are herbivores and most of their diet consists of seeds, but they may also feed on vegetable matter such as stems and roots. They build their nests with plants and other materials between crevices or in tree stumps, trunks, and holes of trees; they usually have more than one burrow (Ceballos and Galindo, 1984). They have the ability to climb trees (Horner, 1954). Their large pinnae allow them to perceive ultrasounds, such as those made by shrews and bats, and perhaps emit sound as well. Reproduction occurs from June to November. The number of offspring varies from two to three per litter. They have a certain specificity of habitat, generally preferring rocky sites (Fa et al., 1990). Its density varies from 27 to 58.4 individuals per ha (Romero, 1994). Its top predators are coyotes, foxes, cacomistles, skunks, birds of prey such as eagles and owls, and some reptiles (Villa-R., 1953). In the drier regions of Querétaro it has been collected with *Peromyscus gratus*, *P. pectoralis*, and *Neotoma leucodon*, while in the humid areas it coexists with *P. boylii*, *Reithrodontomys megalotis*, and *Microtus mexicanus* (L. León, pers. comm.)

VEGETATIONAL ASSOCIATIONS AND ELEVATION RANGE: *P. difficilis* generally inhabits rocky environments and forests of pine (*Pinus*) and oaks (*Quercus*). It has also been collected in desert scrubland and grassland. The range of elevations that it inhabits varies from 1,200 masl to 3,700 masl.

CONSERVATION STATUS: *P. difficilis* is rare. Due to its broad distribution and the variety of environments it inhabits, it is not at risk of extinction (G. Ceballos, pers. obs.).

Peromyscus eremicus (Baird, 1858)

Cactus deermouse

Jaime Luévano and Eric Mellink

SUBSPECIES IN MEXICO

Peromyscus eremicus alcorni Anderson, 1972
Peromyscus eremicus anthonyi (Merriam, 1887)
Peromyscus eremicus collatus Burt, 1932
Peromyscus eremicus eremicus (Baird, 1858)
Peromyscus eremicus papagensis Goldman, 1917
Peromyscus eremicus phaeurus Osgood, 1904
Peromyscus eremicus pullus Blossom, 1933
Peromyscus eremicus sinaloensis Anderson, 1972
Peromyscus eremicus tiburonensis Mearns, 1897

Recently, Riddle et al. (2000) separated the populations of *P. eremicus* into two species, naming *P. fraterculus* to those on the west of the Colorado River and *P. eremicus* to those on the east side of the river. The allocation of the subspecies in both species is attempted in this chapter, as Riddle and collaborators did not attempt to delineate subspecies boundaries.

DISTRIBUTION: This species is distributed in the United States and Mexico. It occupies a large area in Mexico, which includes Sonora and Sinaloa and much of the North Altiplane. It has been recorded in the states of CH, CO, DU, NL, SI, SL, SO, and ZA.

DESCRIPTION: *Peromyscus eremicus* is a medium-sized mouse, with the tail longer than the body and small ears. It occurs sympatrically with other species of *Peromyscus* and although there are subtle external differences, the best feature to distinguish them is the baculum (bone of the penis), which is relatively short, wide, and dorsally curved, with a square and very small base. Females are bigger and heavier than males.

Peromyscus eremicus. Scrubland. Charcas, San Luis Potosí. Photo: Juan Cruzado.

The coat is soft and silky, with a tuft in the tip of the tail. In general, the coloration is ochraceous to cinnamon, mixed with lighter hair in the top. Along the outer surface of the forefeet there is an ochraceous stripe. The tail is slightly bicolored, dark on top and whitish on the bottom, finely ringed, and covered with short hairs. The soles of the hind feet are completely naked. The head and sides are slightly gray. The coloration varies among subspecies and even among different populations. Females appear to be a little paler than males. They have 48 pairs of chromosomes (Hall, 1981; Hoffmeister, 1986; Jameson and Peeters, 1988; Veal and Caire, 1979).

EXTERNAL MEASURES AND WEIGHT

TL = 169 to 218 mm; TV = 92 to 118 mm; HF = 18 to 22 mm; EAR = 13 to 18 mm. Weight: 13 to 18 g.

DENTAL FORMULA: I 1/1, C 0/0, PM 0/0, M 3/3 = 16.

NATURAL HISTORY AND ECOLOGY: This mouse has nocturnal habits. Much of its activity occurs in trees or shrubs, and it has been seen foraging on mesquite beans, berries (*Celtis* sp.), and chollas fruits (*Opuntia* spp.). Its activity is affected in bright nights. They build their burrows in soft soils, but also live in abandoned burrows of other mammals, including those of woodrats (*Neotoma* sp.). It is often found associated with the kangaroo rat *Dipodomys merriami*. They can enter into torpor for 12 hours when the temperature exceeds 30°C, and resume their activity when the temperature drops to 15°C or less. In summer, with high temperature, little water, and moisture, they can enter into aestivation and it is difficult finding them (Brylski, 1990a; Hoffmeister, 1986; Jameson and Peeters, 1988; Veal and Caire, 1979). Their diet consists mainly of fruits and flowers of shrubs, but they also consume seeds, insects, and green vegetation in different amounts, depending on their abundance. Insects are an important part of their food. They are adapted to arid conditions and do not need to drink water, surviving on the water obtained from their food (Brylski, 1990a; Hoffmeister, 1986; Jameson and Peeters, 1988; Veal and Caire, 1979). In some places, they reproduce throughout the year, but reproduction is affected by the hot and dry periods of summer. The gestation period lasts 21 days, with an average of 5 to 8 offspring per litter. The average number of litters per year is 1 to 3. Under laboratory conditions they can have up to 12 litters per year. Most reproductive activity occurs between April and October (Brylski, 1990a; Hoffmeister, 1986; Jameson and Peeters, 1988; Veal and Caire, 1979).

VEGETATIONAL ASSOCIATIONS AND ELEVATION RANGE: *P. eremicus* is found in a wide range of habitats from desert scrubland, in its many variants, to coniferous forests. It is found from sea level to 2,330 m.

CONSERVATION STATUS: This species is widely distributed and resilient; it is found in a wide range of environments, and in most cases poses no conservation concerns, although, all island subspecies are classified as threatened (SEMARNAT, 2010).

Peromyscus eva Thomas, 1898

Southern Baja deermouse

Eric Mellink and Jaime Luévano

SUBSPECIES IN MEXICO

Peromyscus eva eva Thomas, 1898
Peromyscus eva carmeni Townsend, 1912

DISTRIBUTION: This species is endemic in Mexico. It is distributed in the southeastern part of the state of Baja California, to about 29 degrees north, and in almost all of Baja California Sur, including Carmen Island. The population of Carmen Island constitutes an endemic subspecies. It has been recorded in the states of BC and BS.

DESCRIPTION: *Peromyscus eva* is very similar to *P. fraterculus*, but is distinguished externally by the proportionally longer tail. In addition, it has cranial differences, reflected in a more robust head, and differences in the coat and the morphology of the baculum (Burt, 1960; Lawlor, 1971a). The coat is short with mixtures of russet, yellowish, and brown, and without dark brown or black hair. The dorsal coat varies from russet-sandy in the south to ochraceous-brown in the north. *P. eva carmeni*, of Island Carmen, has a coat similar to that of *P. fraterculus* (Hall, 1981; Lawlor 1971a, b). The offspring are born without hair and with the eyes closed. Juveniles are gray (Woloszyn and Woloszyn, 1982).

EXTERNAL MEASURES AND WEIGHT

TL = 185 to 218 mm; TV = 100 to 128 mm; HF = 20 to 21 mm; EAR = 15 to 17 mm. Weight = 13 to 20 g.

DENTAL FORMULA: I 1/1, C 0/0, PM 0/0, M 3/3 = 16.

NATURAL HISTORY AND ECOLOGY: *P. eva* is nocturnal, and excavates burrows in the soil or between rocks and trees. Its diet is almost completely vegetarian and includes green vegetation, seeds, and fruits. It reproduces throughout the year, although with more intensity during the rainy season, summer–autumn. Pregnant animals that have been caught had one to four embryos (Woloszyn and Woloszyn, 1982). The size of the individuals decreases significantly from north to south. This variation in size reflects a displacement of characters. The largest individuals of *P. eva* occur in the north; they are sympatric with *P. fraterculus* and, in turn, these individuals are smaller than in nearby sites where *P. eva* is absent (Lawlor 1971a, b).

VEGETATIONAL ASSOCIATIONS AND ELEVATION RANGE: *P. eva* inhabits desert scrublands and palm groves and, more rarely, forests of pine-oak and oak, and the edges between wet grasslands and forests. It has been suggested that its absence in pine forests is due to the competition with *P. truei* (Banks, 1967; Woloszyn and Woloszyn, 1982). It is found from sea level to 1,650 m, in the upper parts of the Sierra of La Laguna in Baja California Sur (Banks, 1967).

CONSERVATION STATUS: *P. eva* faces no conservation concerns, but the subspecies *P. eva carmeni* of Isla del Carmen is listed as threatened (SEMARNAT, 2010). On this island there are exotic rodents, cats, dogs, donkeys, and goats, and, most recently, bighorn sheep have been introduced (Mellink, 2002).

Peromyscus fraterculus (Miller, 1892)

Northern Baja deermouse

Gerardo Ceballos

SUBSPECIES IN MEXICO

Peromyscus fraterculus avius Osgood, 1909
Peromyscus fraterculus cedrosensis J.A. Allen, 1898
Peromyscus fraterculus cinereus Hall, 1931
Peromyscus fraterculus fraterculus (Miller, 1892)
Peromyscus fraterculus insulicola Osgood, 1909
Peromyscus fraterculus polypolius Osgood, 1909
Riddle et al. (2000) recently separated the populations of *P. eremicus* in two species, *P. fraterculus* the one on the west of the Colorado River and *P. eremicus* the one on the east of the river. The allocation of the subspecies in both species is tentative since Riddle et al. (2000) did not mention anything about the situation of the subspecies.

DISTRIBUTION: This species is distributed in the southwestern United States and the Baja California Peninsula in Mexico. It has been recorded in the states of BC and BS.

DESCRIPTION: *Peromyscus fraterculus* is a mouse with a tail longer than the body and ending in a brush. The ears are relatively large. It is very similar to *P. eremicus*, although the two have an allopatric distribution. The hair is short, soft, and silky. Females are bigger and heavier than males. The dorsal coloration varies from ochraceous to cinnamon mixed with lighter hair on the top. The outer surface of the forefeet has an ochraceous band. The tail is slightly bicolored, dark on the top and lighter on the bottom. The soles of the hind feet are completely naked. The head and sides are slightly gray. There is variation in the coloration in accordance to the substrate. Females appear to be paler than males (Hall, 1981).

EXTERNAL MEASURES AND WEIGHT

TL = 169 to 218 mm; TV = 92 to 118 mm; HF = 18 to 22 mm; EAR = 13 to 18 mm. Weight: 13 to 18 g.

DENTAL FORMULA: I 1/1, C 0/0, PM 0/0, M 3/3 = 16.

NATURAL HISTORY AND ECOLOGY: *P. fraterculus* is semi-arboreal and nocturnal. It commonly climbs trees or shrubs. It feeds on flowers, fruits, seeds, buds, insects, and occasionally small lizards. Burrows are constructed in soft soils, under rocks or roots; occasionally, they occupy abandoned burrows of other mammals such as woodrats (*Neotoma* spp.). In summer, when the temperature is very high and there is little water and humidity they can enter into aestivation. They do not drink water. They reproduce throughout the year, if abundant food is available. The gestation period is approximately 21 days. In each litter between four and eight offspring are born. The average number of litters per year is one to three. Most reproductive activity occurs between April and October (Brylski, 1990a; Hoffmeister, 1986; Jameson and Peeters, 1988; Veal and Caire, 1979).

VEGETATIONAL ASSOCIATIONS AND ELEVATION RANGE: *P. fraterculus* is a generalist species found in a wide range of environments from desert scrublands to coniferous forests and coastal scrublands (Mediterranean). It is found from sea level to 2,300 m.

CONSERVATION STATUS: *P. fraterculus* is an abundant species with a broad distribution that presents no concerns for conservation. None of the terrestrial subspecies face conservation risks and even the island subspecies are probably not at risk, with the exception of *P. fraterculus cedrosensis* (Mellink, 1993a), for which sufficient fieldwork does not exist. All island subspecies are legally classified as endangered (SEMARNAT, 2010).

Peromyscus furvus J.A. Allen and Chapman, 1897

Blackish deermouse

Alondra Castro-Campillo, Elsa González Cruz,
Hugo Martínez Paz, and José Ramirez-Pulido

SUBSPECIES IN MEXICO

P. furvus is a monotypic species.

DESCRIPTION: *Peromyscus furvus* is one of the largest species of the genus; the dorsal coloration ranges from dark brown to blackish; completely melanistic individuals have been found. The sides are slightly reddish, the belly is grayish, and the feet are white, but the tarsal articulation has blackish spots. The tail varies from completely black to irregularly bicolored and is spotted in the ventral portion. The skull is narrow, especially between the orbits; the supraorbital edge is not rounded and is a distinctive feature of the species; the anterior part of the nasal is heavily wider (Dalquest, 1950; Hall, 1981; Huckaby, 1980). It differs from *Megadontomys nelsoni* by being smaller and darker (Allen and Chapman, 1897; Hall, 1981). The tail of *H. simulatus* is almost bicolored and relatively hairy and ends in a brush; the skull and maxillary teeth are small; finally, *P. difficilis* is smaller, with the tail larger than the body, the bullae are very inflated, the nasals do not spread anteriorly, and the teeth show a tendency to divide in the anterocone (Huckaby, 1980).

EXTERNAL MEASURES AND WEIGHT

TL = 229 to 300 mm; TV = 114 to 162 mm; HF = 26 to 33 mm; EAR = 20 to 23 mm. Weight: 40 to 60 g.

DENTAL FORMULA: I 1/1, C 0/0, PM 0/0, M 3/3 = 16.

NATURAL HISTORY AND ECOLOGY: *P. furvus* inhabits rocky walls of canyons, hollow logs, and roots. It feeds on fruits such as blackberries and pokeweed (*Phytolacca*). It coexists with *Marmosa mexicana, Habromys simulatus, Peromyscus leucopus, P. aztecus, Oryzomys alfaroi, O. chapmani, Microtus quasiater, Megadontomys nelsoni*, and *P. mexicanus* in the lower parts of the east and with *P. difficilis* in dry habitats of the west (Davis, 1944; Hall and Dalquest, 1963; Heaney and Birney, 1977; Hooper, 1957; Huckaby, 1980). It is nocturnal and apparently reproduces throughout the year. Male juveniles have been in the scrotal condition, and pregnant females (with one embryo per female) or lactating in the last part of October (Hall and Dalquest, 1963); during July subadults have been captured (Davis, 1944). Juveniles have been caught in March April, July, and September (Dalquest, 1953).

VEGETATIONAL ASSOCIATIONS AND ELEVATION RANGE: *P. furvus* is found in cloud forests, pine forests, and pine-oak forests (Hooper, 1957). It inhabits crop areas and scrublands with dense herbaceous vegetation (Hall, 1968). It has been found from 650 masl to 2,950 masl (Dalquest, 1950; Hall, 1968).

CONSERVATION STATUS: This species has a restricted distribution. Its current status is not entirely known. It is necessary to assess their susceptibility to the fragmentation of their natural habitat and their conservation status.

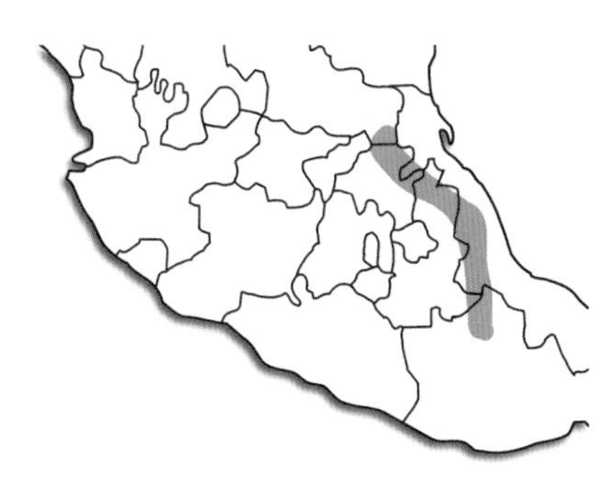

DISTRIBUTION: This species is endemic to Mexico; it is distributed in temperate and wet climates of a reduced area in the Sierra Madre Oriental, from southwestern San Luis Potosí to the mountains of northern Oaxaca (Dalquest, 1950; Hall, 1981; Huckaby, 1980; Ramírez-Pulido et al., in press). It has been recorded in the states of HG, OX, PU, QE, SL, and VE.

Peromyscus furvus. Cloud forest. Naolinco, Veracruz. Photo: Gerardo Ceballos.

Saxicoline deermouse

J. Cuauhtémoc Chávez T.

SUBSPECIES IN MEXICO

Peromyscus gratus erasmus Finley, 1952
Peromyscus gratus gentilis Osgood, 1904
Peromyscus gratus gratus Merriam, 1898
Peromyscus gratus zapotecae Hooper, 1957

Differences in their genetics and karyotypes have separated the species into two groups: *P. truei* for populations in the United States from northern New Mexico to Baja California, and *P. gratus* for populations of southern New Mexico and the rest of Mexico (Durish et al., 2005; Janecek, 1990; Mody and Lee, 1984).

DESCRIPTION: *Peromyscus gratus* is a medium-sized mouse that can possess three types of coloration: light gray slightly mixed with yellowish on a gray background, ochraceous slightly mixed with brown on a gray background, or brown slightly mixed with ochraceous on a gray background (Ceballos and Galindo, 1984; Hernández, 1990). It has a lateral line of buffy coloration. The legs are white. The tail is bicolored, brown above and whitish below. The ears are very large, a feature that distinguishes it from *P. maniculatus*, *P. melanotis*, and *P. aztecus* (Anderson, 1972; Ceballos and Galindo, 1984; Chávez, 1993; Goodwin, 1969; Hernández, 1990).

EXTERNAL MEASURES AND WEIGHT

TL = 171 to 231 mm; TV = 76 to 123 mm; HF = 20 to 27 mm; EAR = 18 to 25 mm.
Weight: 20 to 33 g.

DENTAL FORMULA: I 1/1, C 0/0, PM 0/0, M 3/3 = 16.

NATURAL HISTORY AND ECOLOGY: These mice build their burrows in cracks of rocks. They are semi-arboreal and their climbing skills are greater than that of other mice. Their large ears allow them to efficiently detect predators in the habitat where the vegetation cover is sparse. Their reproduction in Mexico takes place from May to December, but more often in the rainy season (Anderson, 1972; Chávez, 1993; Chávez and Ceballos, 1994). The gestation period lasts 25 to 27 days, with an average 3 offspring per birth. Young are weaned between 21 and 28 days old, and after 50 days they are sexually mature. The offspring emit a series of sounds that play an important role in adult-infant relations. Between the fifth and fourteenth week of their development they change color from gray to ochraceous-brown (Ceballos and Galindo, 1984). In a study conducted in scrubland of pittocaulon (*Senecio praecox*) in El Pedregal de San Angel, Mexico City, an average density of 33.7 individuals/ha was found, with a maximum density in December (59 individuals/ha) and a minimum in August with 20 individuals/ha. The survival rate, or the percentage of individuals that survive, was three months for 50% of the individuals caught; only 3.1% survived one year (Ceballos and Chávez, 1994). Its diet consists primarily of green plant material, seeds, and fungi, although sometimes they can eat insects and other invertebrates. In Durango a high correlation between the availability of fallen fruits of cedar (*Juniperus deppeana*) and the density of rodents was found, with an asynchronous response by the mice to the availability of the resource (Servin et al., 1994).

DISTRIBUTION: *P. gratus* is found from New Mexico in the United States to central and southeastern Mexico, including the Altiplane and Neovolcanic Axis. It has been recorded in the states of AG, CH, CO, DU, DF, GJ, HG, JA, MI, MX, OX, PU, QE, SL, and ZA.

Peromyscus gratus. Scrubland. El Pedregal Ecological Reserve, Distrito Federal. Photo: Gerardo Ceballos.

VEGETATIONAL ASSOCIATIONS AND ELEVATION RANGE: *P. gratus* is found in brushy habitats of arid regions, rainforests, forests of pine-oak, lava fields, open valleys, and crop fields, from 1,710 masl to 2,700 masl.

CONSERVATION STATUS: *P. gratus* is abundant in all its range, especially in rocky environments, and faces no conservation problems.

Peromyscus guardia Townsend, 1912

La Guarda deermouse

Eric Mellink and Jaime Luévano

SUBSPECIES IN MEXICO

Peromyscus guardia guardia Townsend, 1912
Peromyscus guardia harbisoni Banks, 1967
Peromyscus guardia mejiae Burt, 1932

DESCRIPTION: These mice are very similar to *Peromyscus eremicus* (Townsend, 1912). They are gray in the back and white in the ventral surface. The tail is slightly bicolored and with few hairs. The ears have almost no hair. The skull is very arched and the face is long. The anterior palatine foramen does not reach the molars. The interpterygoid fossa is wide and the bullae are big.

Peromyscus guardia. Scrubland. Isla Estanque, Baja California. Photo: Gerardo Ceballos.

DISTRIBUTION: *P. guardia* is endemic to the Angel de la Guarda Island (*P. guardia guardia*) and its islets: Mejía (*P. guardia mejia*), Granito (*P. guardia harbisoni*), and Estanque. In Angel de la Guarda it has only been recorded in the northern tip, in the surrounding area of Puerto Refugio, and in the northeast in Los Cantiles. It has been recorded in the state of BC.

EXTERNAL MEASURES AND WEIGHT
TL =189 to 223 mm; TV = 93 to 123 mm; HF = 23 to 25 mm; EAR = 20 to 23 mm. Weight: 40 g.

DENTAL FORMULA: I 1/1, C 0/0, PM 0/0, M 3/3 = 16.

NATURAL HISTORY AND ECOLOGY: The natural history of *Peromyscus guardia* is not well known. It is found in different microhabitats on islands, in rocky areas with vegetation, on open plains with chollas (*Opuntia* spp.), and in dense bushes near the beach. A burrow lined with fine plant material was found under a rock. They reproduce in spring (Banks, 1967). They are preyed on by owls, snakes, and domestic cats (Mellink et al., 2002).

VEGETATIONAL ASSOCIATIONS AND ELEVATION RANGE: *P. guardia* is found in xeric scrublands, including areas with hylophytes a few meters above sea level. The maximum elevation of the islands is 1,315 masl.

CONSERVATION STATUS: This mouse is considered endangered (SEMARNAT, 2010). Recently, it has been considered as critically endangered or extinct since no recent records have been made in the four islands where it is located (Mellink et al., 2002). The endemic subspecies of Granito seems to be extinct and the one in Mejia is in danger of extinction or has also disappeared (Mellink, 1992). The only recent records in Angel de la Guarda relate to two specimens collected in 1993 (A. Zavala, pers. comm.) and a sighting made in 1992 (L. Grismer, pers. comm.). Therefore, *P. guardia guardia* may also be considered in danger of extinction. These comments contrast with the abundance of the species recorded by T. Lawlor (pers. comm., field notes) and R. C. Banks (pers. comm., field notes). The most plausible reason for these reductions is the introduction (voluntary and involuntary) of domestic animals by fishermen. In Granito there is a high population of black rats (*Rattus rattus*) and possibly cats in the past. In Mejía Islands, there are domestic cats and mice (Mus musculus). And in Angel de la Guarda there are domestic cats (Mellink et al., 2002). A single cat eliminated the population of Estanque (Vázquez-Domínguez, et al., 2004).

Guatemalan deermouse

Anna Horváth

SUBSPECIES IN MEXICO

Peromyscus guatemalensis guatemalensis Merriam, 1898

DESCRIPTION: *Peromyscus guatemalensis* is the second largest mouse of the genus *Peromyscus* in Mexico. The coat is soft, the back is gray, sometimes with dark grayish-brown, but gray is always the dominant color. The underside is whitish, and pale, sometimes buffy toward the sides. The feet are white with the proximal parts dark. The tail is long, generally longer than the body and head, slightly hairy, and bicolored. The eye rings are well marked (Hall, 1981; Reid, 1997).

DISTRIBUTION: This species is distributed from the Sierra Madre in southern Chiapas, to central and southeastern Guatemala, in temperate mountains (Huckaby, 1980; Reid, 1997; Wilson and Reeder, 2005).

EXTERNAL MEASURES AND WEIGHT

TL = 150 to 290 mm; TV = 125 to 154 mm; HF = 29 to 32 mm; EAR = 22 to 25 mm. Weight: 40 to 68 g.

DENTAL FORMULA: I 1/1, C 0/0, PM 0/0, M 3/3 = 16.

NATURAL HISTORY AND ECOLOGY: The natural history and ecology of this species are not well known. They are terrestrial and nocturnal. They have been found in primary forests with a large number of epiphytes and structural elements of habitat such as fallen logs (Reid, 1997) and with a deep layer of stubble (A. Horváth, pers. obs.). In El Triunfo, Chiapas, most pregnant females were caught in beginning of the rainy season (A. Horváth, pers. obs.).

VEGETATIONAL ASSOCIATIONS AND ELEVATION RANGE: This species is found in humid forests of the mountains, in cloud forests and forests of oak, at elevations of 1,300 masl to 3,000 masl (Reid, 1997).

Peromyscus guatemalensis. Cloud forest. Lagunas de Montebello National Park, Chiapas. Photo: Anna Horváth.

CONSERVATION STATUS: The species has no official status. Locally, they can be relatively abundant (Reid, 1997); however, since they are only found in well-preserved forests and cannot tolerate modification of their habitat, it is important to promote the conservation of cloud forests in the Sierra Madre of Chiapas and Guatemala. In Chiapas, an important part of its distribution area coincides with the core zones of the El Triunfo Biosphere Reserve, which seems to ensure the long-term survival of these populations.

Peromyscus gymnotis Thomas, 1894

Naked-eared deermouse

J. Cuauhtémoc Chávez T.

SUBSPECIES IN MEXICO
Peromyscus gymnotis gymnotis Thomas, 1894
Peromyscus gymnotis allophylus Osgood, 1904
P. gymnotis was classified as a subspecies of *P. mexicanus* by Osgood (1904), but Musser (1971) raised *gymnotis* to species level and included the subspecies *allophylus*. Huckaby (1980) and Jones and Yates (1983) give the elements to separate morphologically *P. gymnotis* from *P. mexicanus*.

DISTRIBUTION: This species is distributed along the coastal regions of the Pacific side of Chiapas, Mexico, reaching the west of Lake Managua in Nicaragua. It has been recorded in the state of CS.

DESCRIPTION: *Peromyscus gymnotis* is a small-sized mouse. The dorsal coloration is dark buffy and the ventral region is whitish; the tail is dark buffy and slightly shorter than the head and body together. The ears have scattered hair and the soles of the feet have little hair (Hall, 1981; Musser, 1971). This species is smaller, in most of the external measures, than *P. guatemalensis* and *P. mexicanus*; *P. guatemalensis* has the buffy coat shorter than the gray-based coat. It does not have pectoral mammae like *P. beatae*, *P. oaxacensis*, and *Habromys lophurus* (Huckaby, 1980; Jones and Yates, 1983; Musser, 1971).

EXTERNAL MEASURES AND WEIGHT
TL = 202 to 243 mm; TV = 90 to 110 mm; HF = 24 to 27 mm; EAR = 17 to 20 mm.
Weight: 31 to 50 g.

DENTAL FORMULA: I 1/1, C 0/0, PM 0/0, M 3/3 = 16.

NATURAL HISTORY AND ECOLOGY: In Mexico, *P. gymnotis* is a rare species; however, in Nicaragua it is abundant, especially in forests of secondary growth and in the edges of coffee plantations. Pregnant females have been collected in March, April–June, and August, pups have been collected in April and August, and juveniles have been caught in March, June, July, and August (Jones and Yachts, 1984). The maturation of juveniles is similar to what has been described for other species of the genus (Jones and Yachts, 1984).

VEGETATIONAL ASSOCIATIONS AND ELEVATION RANGE: *P. gymnotis* inhabits tropical deciduous forests, tropical subperennial forests, grasslands, and coffee plantations. It is found at 50 masl in southwestern Mapastepec and 1,775 masl in Cerro Ovando near Escuintla in Chiapas.

CONSERVATION STATUS: In Mexico, *P. gymnotis* is a rare species that has only been caught at 11 locations in southern Chiapas. It is necessary to assess the real status of their populations since it is a species with a marginal distribution in Mexico.

Peromyscus hooperi Lee and Schmidly, 1977

Hooper's deermouse

Iván Castro Arellano

SUBSPECIES IN MEXICO

P. hooperi is a monotypic species.

Schmidly et al. (1985) classified it within the subgenus *Peromyscus*, but Fuller et al. (1984) suggested that it represents a subgeneric form between the subgenera *Peromyscus* and *Haplomylomys*. Bradley et al. (2007) suggest that it represents its own species group with *Peromyscus*.

DESCRIPTION: *Peromyscus hooperi* is a medium-sized mouse. The back, face, and head are grayish with brownish colors that range from scarce to moderate. Its coloration is similar to the grayish coat of juveniles of some species of *Peromyscus*. The underside is creamy colored, although the underfur is gray. The tail is generally bicolored, gray on top and white on the bottom; the feet and inner part of the legs are white (Lee and Schmidly, 1977). It has two pairs of inguinal mammae, and the diploid number of chromosomes is 2n = 48 (Lee and Schmidly, 1977; Schmidly et al., 1985).

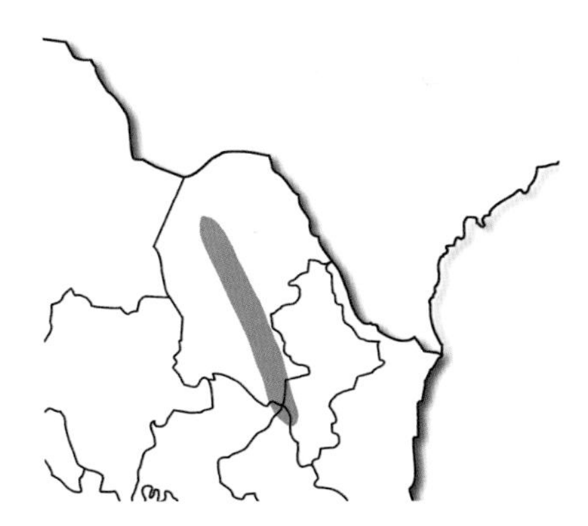

DISTRIBUTION: *P. hooperi* is an endemic mouse of Mexico, which is distributed from central Coahuila to northeastern Zacatecas and San Luis Potosí (Álvarez and Álvarez-Castañeda, 1991; Musser and Carleton, 1993). It has been recorded in the states of CO, SL, and ZA.

EXTERNAL MEASURES AND WEIGHT

TL = 201 mm; TV = 110 mm; HF = 21 mm; EAR = 19 mm.

Weight: 19 to 30 g.

DENTAL FORMULA: I 1/1, C 0/0, PM 0/0, M 3/3 = 16.

NATURAL HISTORY AND ECOLOGY: The species is restricted to a type of habitat characterized by a transition grassland with dominant plants such as *Dasyliron* sp., *Nolina* sp., *Yucca* spp., *Agave* spp., and *Bouteloua* sp. Under laboratory conditions, births occur from June to December with an average of 2.9 offspring per litter; however, in wild populations pregnant females have been collected from March to May with an average of 3.5 embryos per female (Schmidly et al., 1985). The offspring open their eyes 13 days after they are born. The gestation period lasts 33.5 days, and the average age of the first estrus is 69 days for laboratory individuals (Schmidly et al., 1985). It is nocturnal and feeds on seeds, nuts, insects, and other small invertebrates.

VEGETATIONAL ASSOCIATIONS AND ELEVATION RANGE: In Coahuila, *P. hooperi* has been collected in grasslands with yucca (*Yucca* sp.), *Nolina* sp., and *Dasylirion* sp. (Lee and Schmidly, 1977). In San Luis Potosí, it is associated with the rosetophilous bush, where the dominant species are *Agave lechugilla*, *A. striata*, *Hechtia glomerta*, *Opuntia* sp., *Senecio* sp., *Acacia* sp., *Yucca* sp., and *Dasylirion* sp. (Álvarez and Álvarez-Castañeda, 1991). It is found from 1,155 masl to 2,100 masl (Álvarez and Álvarez-Castañeda, 1991; Lee and Schmidly, 1977), although in most localities it is found between 1,400 masl and 1,700 masl (Schmidly et al., 1985).

CONSERVATION STATUS: *P. hooperi* is common in the central part of Coahuila in the type of habitat to which apparently it is restricted (Schmidly et al., 1985). Matson and Baker (1986) were not able to collect it in the limits of Coahuila and Zacatecas in a similar type of habitat. It is not included in any conservation list, but since it has a restricted distribution it is susceptible to changes in their environment, so it must be regarded as fragile or threatened.

Peromyscus hylocetes Merriam, 1898

Transvolcanic deermouse

Juan Cruzado and Gerardo Ceballos

SUBSPECIES IN MEXICO
P. hylocetes is a monotypic species.
Originally, Merriam (1898) described it as *P. hylocetes*, but Osgood (1909) considered it a subspecies of *P. aztecus*. Recent studies made by Sullivan and Kilpatrick (1991) and Sullivan et al. (1997) suggest assigning a species level, which has been widely accepted (e.g., Tiemann-Boege et al., 2002).

DESCRIPTION: Within the genus *Peromyscus hylocetes* is medium in size. The dorsal coloration is ochraceous-pale, mixed with black hair; the coat is reddish-brown. The belly is white to slightly creamy. It possesses a black eye ring. The feet are white, with a dark line that stretches from the tarsus to the metatarsal. The tail is bicolored, of the same length as the head and body (Ceballos and Galindo, 1984; Hall, 1981; Merriam, 1989; Osgood, 1909; Vázquez et al., 2001). It differs from *P. gratus* and *P. difficilis* by the smaller size of the ears; from *P. melanotis* and *P. maniculatus* by its greater size and greater length of the tail; and from *P. boylii* and *P. levipes* by the greater length of the feet (Ceballos and Galindo, 1984; Hall, 1981).

Peromyscus hylocetes. Fir forest. Miguel Hidalgo National Park, Distrito Federal. Photo: Alberto González Romero.

EXTERNAL MEASURES AND WEIGHT
TL = 203 to 238 mm; TV = 102 to 117 mm; HF = 25 to 27 mm; EAR = 15.5 to 18.5 mm. Weight: 22 to 36 g.

DENTAL FORMULA: I 1/1, C 0/0, PM 0/0, M 3/3 = 16.

NATURAL HISTORY AND ECOLOGY: *P. hylocetes* is terrestrial and mainly nocturnal. These mice feed on grasses and seeds, although they supplement their diet with insects (Ceballos and Galindo, 1984). In a study conducted in western Mexico, their diet in a cloud forest was dominated by fruits (51.4%), leaves (21%), and seeds (21%). The average density was 9 individuals per ha, but clear fluctuations of lower density were recorded in the dry season (June; 3 individuals/ha) and the highest peaks at the end of the rainy season (October; 12 individuals/ha) and the beginning of the dry-cold season (January and February; 15 individuals/ha). Reproduction takes place in the rainy season, from August to October, and in the dry-hot season (May) (Vázquez et al., 2000). It is sympatric with *P. boylii*, *P. difficilis*, *P. gratus*, *P. levipes*, *P. maniculatus*, *P. melanotis*, *Reithrodontomys chrysopsis*, *R. sumichrasti*, *Microtus mexicanus*, and *Sigmodon leucotis*.

DISTRIBUTION: *P. hylocetes* is an endemic species of Mexico, which is distributed along the Eje Volcanico Transversal, from Colima to Mexico City. It has been recorded in the states of CO, DF, JA, MI, MO, and MX.

VEGETATIONAL ASSOCIATIONS AND ELEVATION RANGE: *P. hylocetes* inhabits middle and upper elevations of the Eje Volcanico Transversal, mainly in forests of oak, pine-oak, pine, and fir, and cloud forests. Occasionally, it has been found in places with little vegetation cover such as lava fields. It has been collected in heights ranging from 2,300 masl to 2,700 masl (Ceballos and Galindo, 1984; Vázquez et al., 2000, 2001).

CONSERVATION STATUS: Despite being an endemic species of the Eje Volcanico Transversal, it is relatively abundant in some environments and presents no problems of conservation.

Peromyscus interparietalis Burt, 1932

San Lorenzo deermouse

Jesús Ramírez Ruíz and Rafael Ávila Flores

SUBSPECIES IN MEXICO
Peromyscus interparietalis interparietalis Burt, 1932
Peromyscus interparietalis lorenzi Banks, 1967
Peromyscus interparietalis ryckmani Banks, 1967
Recently, Hafner et al. (2001) have considered *P. interparietalis* as a subspecies of *P. eremicus*. Their work is based solely on molecular evidence, however, so we have decided to recognize the species until there is morphological evidence and additional genetic confirmation to identify with greater certainty its taxonomic position.

DESCRIPTION: *Peromyscus interparietalis* is a medium-sized mouse. The dorsal coat has yellowish-brown underfur and blackish-brown tips, resulting in a speckled grayish appearance; the coloration of the belly and legs varies from white to yellowish-white. The tail is longer than the length of the head and body, and is usually bicolored, reddish-brown on the back and whitish on its ventral part (Brand and Ryckman,

DISTRIBUTION: *P. interparietalis* is an endemic species of Mexico, whose distribution is restricted to the Salsipuedes, Las Animas, and San Lorenzo islands, in the north of the Gulf of California (Hall, 1981). It has been recorded in the state of BC.

1969; Hall, 1981; López-Forment et al., 1996). It is similar in size and color to *P. guardia*, but unlike this species the hind feet and tail are shorter; the main differences can be found in the skull. In *P. interparietalis* the interparietal bone extends to the suture between the parietal and squama temporalis, the edge of the nasal is square (not rounded), the zygomatic widens toward the squama temporalis, the interpterygoid fossa is narrower, the palatal bone is shorter, and the incisive foramina reaches the first molar (Banks, 1967; Burt, 1932a; Hall, 1981). As *P. interparietalis* presents some morphological (coloration, cranial features, reproductive tract of the male, structure of the penis, etc.), karyotype, and molecular similarities with *P. eremicus* (coupled with the fact that these two species are capable of producing fertile individuals), some authors consider that the latter species is the most similar to *P. interparietalis*; thus, it has been considered that *P. eremicus* led through evolutionary lines to *P. interparietalis* and *P. guardia* (Banks, 1967; Brand and Ryckman, 1969; Lawlor, 1971a, b, c; Linzey and Layne, 1969). Protein-banding patterns in *P. interparietalis* and *P. guardia*, however, suggest a much closer kinship between these two species (Warn et al., 1974a, b).

EXTERNAL MEASURES AND WEIGHT
TL = 173 to 215 mm; TV = 87 to 117 mm; HF = 22 mm; EAR =18 mm.
Weight: 20 to 22 g.

DENTAL FORMULA: I 1/1, C 0/0, PM 0/0, M 3/3 = 16.

NATURAL HISTORY AND ECOLOGY: These mice are nocturnal animals that live in rocky beaches with sparse vegetation, valleys, and low plains, but can also be found in the other microhabitats of the islands (Banks, 1967; Bourillón et al., 1988; Brand and Ryckman, 1968; López-Forment et al., 1996). Their burrows are found in hollow spaces between large rocks. Their nests are constructed with sticks, dry leaves, and other materials. They feed on seeds, sprouts, flowers, and fruits, as well as various insects (López-Forment et al., 1996). Their predators include snakes and feral cats (Banks, 1967; J. Ramirez, pers. obs.). The peak of the reproductive activity occurs in early spring, during March, although it may be extended to summer (Banks, 1967). The gestation period lasts on average 32 days and females give birth to 4 offspring, which are born altricial (the sex ratio is almost 1:1). The offspring open their eyes during the second week of age, and are completely weaned 25 days later; they molt 35 to 55 days after the birth. The females enter into oestrus after the delivery (Brand and Ryckman, 1968).

VEGETATIONAL ASSOCIATIONS AND ELEVATION RANGE: The islands occupied by this mouse are very dry, and the dominant vegetation is xeric scrubland, with chollas and cardones as dominant elements (Banks, 1967; Bourillón et al., 1988). It is found mainly in the lower parts of the islands, but can reach 485 masl.

CONSERVATION STATUS: This species is considered threatened in Mexico (SEMARNAT, 2010). Although it seems to be abundant in the islands where it is distributed, the survival rate of *P. interparietalis ryckmany* has been seriously threatened due to the presence of feral cats in the Salsipuedes Island (López-Forment et al., 1996; J. Ramirez, pers. obs.).

White-footed deermouse

Catalina B. Chávez and Leticia A. Espinosa

SUBSPECIES IN MEXICO

Peromyscus leucopus affinis (J. A. Allen, 1891)
Peromyscus leucopus arizonae (J. A. Allen, 1894)
Peromyscus leucopus castaneus Osgood, 1904
Peromyscus leucopus cozumelae Merriam, 1901
Peromyscus leucopus incensus Goldman, 1942
Peromyscus leucopus lachiguiriensis Goodwin, 1956
Peromyscus leucopus mesomelas Osgood, 1904
Peromyscus leucopus noveboracensis (Fischer, 1829)

DESCRIPTION: *Peromyscus leucopus* is a small-sized rodent. The dorsal coloration varies from brown to cinnamon or gray. Some animals have a darker dorsal line and a yellowish speck on the chest. The belly is whitish with dark gray underfur. The legs are white and the ears are dark. In places where it occurs sympatrically with *P. manicuatus*, it differs by its larger tail, which is markedly bicolored (Goodwin, 1969; Hall, 1981).

EXTERNAL MEASURES AND WEIGHT

TL = 130 to 205 mm; TV = 45 to 100 mm; HF = 17 to 25 mm; EAR = 13 to 16 mm. Weight: 20 to 23 g.

DENTAL FORMULA: I 1/1, C 0/0, PM 0/0, M 3/3 = 16.

NATURAL HISTORY AND ECOLOGY: *P. leucopus* is mainly nocturnal but may occasionally be active during the day. It is omnivorous; seeds constitute 43% of its diet, insects 30%, and vegetation 25%. When temperatures drop below 3°C individuals enter into torpor. It is considered semi-arboreal because of its ability to climb trees. In wooded environments they can use narrow branches and stay for a long time at

Peromyscus leucopus. Pastureland. Janos-Casas Grandes, Chihuahua. Photo: Gerardo Ceballos.

DISTRIBUTION: *P. leucopus* ranges from southern Canada and the western United States to Mexico, where it is found from the northeastern tip of Sonora toward the central region of Chihuahua, through Coahuila and northeastern Durango, and along the east of the country to the Isthmus of Tehuantepec and northwestern portion of the Yucatan Peninsula (Álvarez, 1963a; Anderson, 1972; Baker and Greer 1962; Baker and Phillips, 1965; Dalquest, 1953b; Ramírez-Pulido et al., 1986; Ramírez-Pulido and Castro-C, 1994; Wilson and Reeder, 2005). It has been recorded in the states of CA, CH, CO, DU, HG, NL, OX, PU, QR, SL, SO, TA, TB, VE, YU, and ZA.

considerable heights. Nests have been found above ground level, on the ground, and under piles of rocks, tree stumps, or fallen or hollow trees; their nests are constructed with herbs, leaves, hair, seeds, peel, and mosses (Lackey et al., 1985). Its populations are abundant in the forests, whereas in brushy areas, grasslands, fields of cereal grains, and sugar cane crops densities are low. Density and activity are positively related to the vegetation cover (Birney et al., 1974). In Hidalgo densities of up to 4.8 individuals per ha have been reported in xeric scrubland dominated by *Myrtillocactus geometrizans*, *Stenocereus dumortieri*, and *Opuntia cantabrigens* (Chávez and Espinosa, 1993). In tropical areas it is common in sugar cane fields (Hall and Dalquest, 1963). In tropical regions it can reproduce throughout the year. In northern regions reproduction is seasonal with peaks in spring and the end of summer. Females reach sexual maturity at 38 days old, and have spontaneous ovulation and postpartum estrus. The estrus cycle has an average length of 6 days. The gestation period lasts 23 days and in lactating females it extends for 14 days. The size of the litter varies from 3 in northern Tamaulipas to 5 in Canada, with a variation of 3.4 to 5.4 for southern Mexico. Both sexes participate in the care of the offspring, which complete their growth in approximately six weeks. Young individuals explore the surrounding area near the nest at 16 to 25 days old, and reach adult body length at 25 weeks of age. Females abandon the litter after 20 to 40 days postpartum (Álvarez, 1963a; Birney et al., 1974; Lackey et al., 1985). The size of the home range varies seasonally, with greater areas during the reproductive season. Schug et al. (1991) reported areas for individuals with long and short life expectancy of 2,562 m^2 and 1,155 m^2 in males and 591 and 459 m^2 for females, as well as lifespan in forests of oak of 669 and 779 days for females and males, respectively.

VEGETATIONAL ASSOCIATIONS AND ELEVATION RANGE: This species has a high ecological plasticity; it is found in arid, temperate, and tropical environments, which include thorn scrublands, grasslands, pine forests, oak forests, tropical deciduous forests, and tropical perennial forests. It is found from sea level to 3,000 m.

CONSERVATION STATUS: *P. leucopus* is a common species, with a wide distribution and a great tolerance to different environmental conditions; hence, it faces no problems of conservation. The subspecies *P. l. cozumelae* is threatened (SEMARNAT, 2010).

Peromyscus levipes Merriam, 1898

Nimble-footed deermouse

J. Cuauhtémoc Chávez Tovar

SUBSPECIES IN MEXICO

Peromyscus levipes ambiguus Álvarez, 1961
Peromyscus levipes levipes Merriam, 1898
P. levipes was considered a subspecies of *P. boylii*, but Houseal et al. (1987), Rennert and Kilpatrick (1987), Schmidly et al. (1988), Smith (1990), and Castro-Campillo et al. (2000) found biochemical and chromosomal evidence that indicates that it is a different species.

DESCRIPTION: *Peromyscus levipes* is a mouse of average size within the genus. The tail has the same length as the head and body. The foot and ear are medium in size in comparison to those of other species of the genus. The dorsal coloration is slightly dark brown-auburn. It has an ochraceous to brown-orange line on the sides, which contrast sharply with the coloration of the underside, which is white to gray. The tail is hairy on the tip and bicolored, dorsally brown and whitish ventrally. The ankles are dark; the feet are white and the ears sepia. The skull is elongated, twice as long as wide, with the zygomatic arches complete and the bullae medium in size. The karyotype possesses 12, 13, or 14 bi-armed chromosomes; 32, 33, or 34 acrocentric chromosomes; chromosome X is long and submetacentric and chromosome Y is small and submetacentric.

DISTRIBUTION: The distribution of this endemic species includes the western Sierra Madre Oriental in Nuevo Leon to the Mesa Central in the states of Morelos and Mexico. The subspecies *P. levipes ambiguus* is distributed in the Sierra Madre Oriental, in the states of Nuevo Leon, Coahuila, Tamaulipas, and San Luis Potosí. *P. levipes levipes* is located in the western Sierra Madre Oriental in the states of San Luis Potosi, Queretaro, and Hidalgo, continuing to the south within the Mesa Central of the south of Queretaro, Hidalgo, Mexico, Tlaxcala, Puebla, and east to the vicinity of Jalapa and Jilotepec, Veracruz. It has been recorded in the states of CO, DF, HG, MO, MX, PU, QE, SL, TL, and VE.

EXTERNAL MEASURES AND WEIGHT

TL = 180 to 220 mm; TV = 91 to 115 mm; HF = 22 to 24 mm; EAR = 19 to 22 mm. Weight: 17 to 24 g.

DENTAL FORMULA: I 1/1, C 0/0, PM 0/0, M 3/3 = 16.

NATURAL HISTORY AND ECOLOGY: These mice are strictly nocturnal herbivores, and feed mainly on seeds, fruits, stems, and sprouts; they also eat worms, crustaceans, mollusks, and small vertebrates (Ceballos and Galindo, 1984). Insects constitute about 60% of their diet during summer; the rest of the year, they consume vegetative material such as seeds of pine (*Pinus* sp.), acorns of oak (*Quercus* sp.), and green parts of plants (Bradford, 1974). They build their nests with plants and other materials between cracks in the rocks; in tree stumps, trunks, and holes of trees; or under piles of branches (Ceballos and Galindo, 1984; Hernández, 1990). Mating season takes place from May to November; however, if food is plentiful it may extend to February (Ceballos and Galindo, 1984; Hernández, 1990; Sánchez-Hernández and Romero, 1997). The number of offspring varies from one to six with an average of four (Romo, 1993). At 5 weeks old they start molting and reach sexual maturity at 50 days old. The offspring open their eyes around the third week of age. In an altitudinal distribution study conducted in Ajusco, Mexico City, these mice were found in low densities and only at 2,700 masl, in forests of oaks (Baca, 1982). In the northeastern Queretaro they were abundant, as they represented 30% of all captures in forest of oak and pine at elevations of 1,400 masl to 2,500 masl (Romo, 1993). Its main predators are coyotes, lynx, foxes, birds of prey, and some reptiles like snakes (*Pituophis melanoleucus* and *Masticophis taeniatus*; Ceballos and Galindo, 1984). They have been collected with *Liomys irroratus*, *Peromyscus difficilis*, *P. melanotis*, *P. gratus*, *P. hylocetes*, *P. melanophrys*, *P. perfulvus*, *Osgoodomys banderanus*, *Neotomodon alstoni*, *Reithrodontomys sumichrasti*, *R. megalotis*, *R. fulvescens*, *Baiomys taylori*, *B. musculus*, *Neotoma mexicana*, *Microtus mexicanus*, and *Sigmodon leucotis* (Baca, 1982; Ceballos and Galindo, 1984; Hernández, 1990).

VEGETATIONAL ASSOCIATIONS AND ELEVATION RANGE: In Mexico, *P. levipes* is usually found in rocky habitats in pine forests, pine-oak forests, gallery forests, scrublands of oaks (*Quercus* sp.), and tropical deciduous forests. It has been collected from 690 masl in Cola de Caballo, Nuevo Leon, to 3,100 masl in Cerro del Ajusco, D.F.

CONSERVATION STATUS: The current status of this species is undetermined, but it is abundant in some regions. The limits of its geographic range and that of the other chromosomal forms of *P. boylii*-species group require attention (Musser and Carleton, 1993).

Tres Marias deermouse

Don E. Wilson

SUBSPECIES IN MEXICO

P. madrensis is a monotypic species.

DISTRIBUTION: *P. madrensis* is an endemic species of Mexico. It is only found in the Marias Islands (Maria Madre, Maria Cleofas, and San Juanito), near the coast of the state of Nayarit. Apparently, the populations of the Maria Magdalena Island have disappeared (Wilson, 1991). It has been recorded in the state of NY.

DESCRIPTION: *Peromyscus madrensis* is medium in size. It differs from other species of the genus in Mexico by the large skull but relatively small teeth and bullae, and by the coloration of the tail, which is uniform and not lighter on the bottom, as in other species (Carleton et al., 1982). In addition, its karyotype (2n = 48 and FN = 54) distinguishes it from most other species.

EXTERNAL MEASURES AND WEIGHT

TL = 203 to 250 mm; TV = 99 to 130 mm; HF = 23 to 28 mm; EAR = 17 to 20 mm. Weight: probably 25 to 40 g.

DENTAL FORMULA: I 1/1, C 0/0, PM 0/0, M 3/3 = 16.

NATURAL HISTORY AND ECOLOGY: *P. madrensis* takes refuge mainly in hollows of fallen logs, tree roots, and rocks. They inhabit tropical dry forests with a deep layer of fallen leaves, They are also found in the shores of streams with rocky substrate. Its diet is unknown, but it is likely to be herbivore-granivore, like other species of the genus. Its pattern of reproduction has not been studied, but in March juveniles, scrotal males, and pregnant and lactating females have been found (Wilson, 1991).

VEGETATIONAL ASSOCIATIONS AND ELEVATION RANGE: *P. madrensis* is mainly found in tropical deciduous forests, but also inhabits thorn forests and xeric scrublands, from sea level to 600 m.

CONSERVATION STATUS: This species is considered vulnerable because it has been restricted to three islands where invasive species such as the Norway rat (*Rattus nor-*

Peromyscus madrensis. Tropical dry forest. Isla Maria Madre, Nayarit. Photo: Juan Cruzado.

vegicus) have been introduced. It has been reported that the populations have probably declined because of the introduction of cats and rats. For example, these mice were abundant in Maria Magdalena Island in 1897 (Nelson, 1899a, b), but in 1976 no specimen was found (Wilson, 1991). Another expedition did not find it, and *Rattus rattus* was very abundant, suggesting that *P. madrensis* was critically endangered or extinct (Larsen et al., 2000). Another species native to these islands, the rice rat (*Oryzomys nelsoni*), was probably extinguished by the same causes (Ceballos and Navarro, 1991; Wilson, 1991). Fieldwork in 2010, however, showed that the species is still alive and probably relatively abundant (J. Cruzado, pers. comm.). It is considered threatened (SEMARNAT, 2010).

Peromyscus maniculatus (Wagner, 1845)

North American deermouse

José Ramírez-Pulido, Adalinda Sánchez, Ulises Aguilera, and Alondra Castro-Campillo

SUBSPECIES IN MEXICO

Peromyscus maniculatus assimilis Nelson and Goldman, 1931
Peromyscus maniculatus blandus Osgood, 1904
Peromyscus maniculatus cineritius J.A. Allen, 1898
Peromyscus maniculatus coolidgei Thomas, 1898
Peromyscus maniculatus dorsalis Nelson and Goldman, 1931
Peromyscus maniculatus dubius J.A. Allen, 1898
Peromyscus maniculatus exiguus J.A. Allen, 1898
Peromyscus maniculatus fulvus Osgood, 1904
Peromyscus maniculatus gambel (Baird, 1858)
Peromyscus maniculatus geronimensis J.A. Allen, 1898
Peromyscus maniculatus hueyi Nelson and Goldman, 1932
Peromyscus maniculatus labecula Elliot, 1903
Peromyscus maniculatus magdalenae Osgood, 1909
Peromyscus maniculatus margaritae Osgood, 1909
Peromyscus maniculatus rufinus (Merriam, 1890)
Peromyscus maniculatus sonoriensis (Le Conte, 1853)

DESCRIPTION: *Peromyscus maniculatus* is a small-sized mouse. The dorsal coloration varies from grayish-buffy to reddish-brown; the belly and legs are white. The tail, which is clearly bicolored, is dorsally dark and ventrally light, is shorter than the head and body combined, and is covered with short, thin hair. The length of the nose is less than 11 mm. The skull is inflated and delicate; the face is thin, short, and conical (Hall, 1981). In the upper parts of their geographical distribution, *P. maniculatus* may be confused with *P. melanotis* by its size and color; however, the coat is shorter, more hirsute, and less brilliant than in the latter. The distinguishing feature between the two species is the absence of a dark spot at the base of the ears that is the diagnostic trait for *P. melanotis*. Another species with which *P. maniculatus* can be confused is *P. leucopus*, especially in areas where they cohabit; however, the second species is larger, its hair is sparse, short, and hirsute, the tail is not so clearly bicolored, and the length of the foot is of greater dimensions.

Peromyscus maniculatus. Pine forest. El Ajusco National Park, Distrito Federal. Photo: Gerardo Ceballos.

DISTRIBUTION: *P. maniculatus* has a wide geographical distribution in North America, which extends from the limits of Canada with Alaska to the southwest of Mexico. In Mexico it is found from the Baja California Peninsula, the Central Plateau, and the Eje Volcanico Transversal to the central portion of Oaxaca (Carleton, 1989; Hall, 1981). There are 16 subspecies, of which 9 inhabit the islands surrounding the Baja California Peninsula (*P. maniculatus assimilis, P. maniculatus cineritius, P. maniculatus coolidgei, P. maniculatus dorsalis, P. maniculatus dubius, P. maniculatus exiguus, P. maniculatus geronimensis, P. maniculatus hueyi*, and *P. maniculatus margaritae*) and the others are continental. It has been recorded in the states of BC, BS, CH, CL, CO, DF, DU, GJ, HG, JA, MI, MO, MX, NL, NY, OX, PU, SI, SL, SO, TB, TL, VE, and ZA.

EXTERNAL MEASURES AND WEIGHT

TL = 121 to 222 mm; TV = 46 to 123 mm; HF = 17 to 25 mm; EAR = 12 to 20 mm. Weight: 17 to 28 g.

DENTAL FORMULA: I 1/1, C 0/0, PM 0/0, M 3/3 = 16.

NATURAL HISTORY AND ECOLOGY: *P. maniculatus* is nocturnal and begins its activities shortly after sunset and reaches its peak of activity half an hour after sunset, continuing until midnight, when its activities decline. Generally, they live in burrows built in soft soils or sandy soils. They can be found in the gaps of fallen logs, but they are also good at climbing and they have been caught in stone walls and plantations of maguey. Within their burrows, they build circular nests (100 mm in diameter), usually with grass, but abandon them relatively often, so they construct several nests through the year. They are territorial, especially during the breeding season. They are highly opportunistic and their diet has seasonal and regional variations. They feed on seeds, tender vegetable matter, insects, mollusks (snails and terrestrial slugs), and worms. *P. maniculatus* is considered a serious threat in areas of plant regeneration since they feed on seeds, especially of conifers (Banfield, 1974). Moreover, they are also considered dispensers of microhizae fungi (Ceballos and Galindo, 1984). They reproduce throughout the year, although most often from June to August. In temperate and cold environments they can have up to 9 offspring per litter; under laboratory conditions up to 11 young have been born. The gestation period in non-lactating females varies from 22.4 to 25.5 days, whereas in lactating females it is 24.1 to 30.6. The first estrus occurs at 48.7 days old in average. In captivity, pups open their eyes at 13.2 days old (Banfield, 1974; Drickamer and Berstein, 1972; Drickamer and Vestal, 1973; Millar, 1989). The average longevity in natural conditions is two years, but this time can be extended to eight years and four months under laboratory conditions (Banfield, 1974). In this regard, *P. maniculatus* is used as an excellent laboratory model for genetic, physiology, and biochemistry studies because they adapt easily to laboratory conditions, feed well with food concentrates, reproduce with success, and grow rapidly (Banfield, 1974). It has been caught with *Notiosorex crawfordi, Chaetodipus hispidus, Perognathus flavus, Dipodomys ordii, D. phillipsii, D. spectabilis, D. merriami, Hodomys alleni, Liomys irroratus, Baiomys taylori, Microtus mexicanus, Neotoma albigula,*

N. martinensis, *Onychomys torridus*, *P. bullatus*, *P. difficilis*, *P. melanotis*, *P. melanophrys*, *P. pectoralis*, *Reithrodontomys fulvescens*, and *R. megalotis* (Ceballos and Galindo, 1984; Genoways and Jones, 1973; Hall and Dalquest, 1963; Petersen, 1973; Schulz et al., 1970). Remnants of these mice have been found in owl pellets (Anderson and Long, 1961; Anderson and Nelson, 1960; Banks, 1965).

VEGETATIONAL ASSOCIATIONS AND ELEVATION RANGE: *P. maniculatus* inhabits a wide range of habitats such as mixed forests, pine forests, grasslands, xeric scrublands, deserts, arid areas, and in the vicinity of or in crops (Ceballos and Galindo, 1984). It has been captured from 60 masl to 3,800 masl, but most records are below 2,500 masl (Carleton et al., 1982; Hall, 1981; Hooper, 1947; Koestner, 1944).

CONSERVATION STATUS: In general, continental populations of *P. maniculatus* do not face conservation problems. Even when it has a wide geographical distribution and although it has the ability to live in various types of vegetation, it is clear that constant deforestation and urbanization in their natural habitat can cause pressure on its populations, especially in the center of the country. On the other hand, island populations have a reduced geographical distribution and their environment is fragile; hence, those populations can be categorized as threatened or extinct (SEMARNAT, 2010).

Peromyscus megalops Merriam, 1898

Broad-faced deermouse

Heliot Zarza and Gerardo Ceballos

SUBSPECIES IN MEXICO

Peromyscus megalops auritus Merriam, 1898
Peromyscus megalops megalops Merriam, 1898
Hall (1981) included *P. megalops azulensis* and *P. megalops melanurus* as subspecies of *megalops*; however, Huckaby (1980) considered *P. megalops azulensis* a subspecies of *P. mexicanus* and raised *P. megalops melanurus* to a species level, which is widely accepted (Ramírez-Pulido et al., 1996; Wilson and Reeder, 2005).

DISTRIBUTION: This species is endemic to Mexico; it is distributed in the Sierra Madre del Sur of Guerrero and Oaxaca (Hall, 1981) and in a hill adjacent to the south of the Nevado de Toluca (G. Ceballos, pers. obs.; Hernández-Chavez, 1990). Two collections outside Mexico have specimens identified as *P. megalops* from Jalisco and Michoacán (López-Wilchis and Lopez, 1998); however, such specimens are, probably, incorrectly identified. It has been recorded in the states of GR, MX, and OX.

DESCRIPTION: Among the genus, *Peromyscus megalops* is rather large. The dorsal coloration varies from reddish-brown to brown ochraceous, and is darker between the ears and along the back. The sides are lighter. The belly varies from cream to pale yellow with dark underfur. The pectoral and axillary regions are reddish-brown. The legs are white, which contrasts with the dark parts of the shank and ankle. The tail is bicolored, dark above and spotted below, and is much longer than the head and body. The coat is long and thick. The skull is long and wide, with well-developed supraorbital crests; the frontal region is ribbed and the tear duct is thickened (Hall, 1981; Merriam, 1898). The alveolar length of the upper molars ranges from 5.0 mm to 5.5 mm, and there is a long diastem that ranges from 9.1 mm to 9.8 mm (Hernández-Chavez, 1990).

EXTERNAL MEASURES AND WEIGHT

TL = 277 mm; TV = 145 mm; HF = 30 mm; EAR = 23 mm.
Weight: 30 to 50 g.

DENTAL FORMULA: I 1/1, C 0/0, PM 0/0, M 3/3 = 16.

NATURAL HISTORY AND ECOLOGY: This species is not well known; however, it is typical of temperate forests of conifers and oaks and cloud forests. It is found in humid microhabitats with underbrush of humus layer, stubble, mosses, and trunks, with rocky outcrops coated with a dense layer of moss. They are sympatric with *Peromyscus aztecus*, *P. beatae*, *P. maniculatus*, *Reithrodontomys sumichrasti*, *R. megalotis*, *Megadontomys thomasi*, *Microtus mexicanus*, and *Neotoma mexicana* (Hernandez-Chavez, 1990; Musser, 1964). Pregnant and lactating females have been caught in June (G. Ceballos, pers. obs.) and August in the state of Mexico (Hernández, 1990).

VEGETATIONAL ASSOCIATIONS AND ELEVATION RANGE: This species is typical of temperate environments, which is mainly found in highlands dominated by pine forests, pine-oak forests, and gallery forests (Musser, 1964). It has been collected in riparian environments, on the shores of rivers and streams (Hernández-Chavez, 1990), from 1,500 masl to 3,000 masl (Hall, 1981; Musser, 1964).

CONSERVATION STATUS: This species is not included in any category of conservation. Despite its limited distribution, it might not face immediate risk of extinction as it tolerates human modifications to its habitat.

Peromyscus mekisturus Merriam, 1898

Puebla deermouse

Alondra Castro-Campillo, Hugo Martínez-Paz, and José Ramírez-Pulido

SUBSPECIES IN MEXICO

P. mekisturus is a monotypic species.

DESCRIPTION: The type specimen of *Peromyscus mekisturus* is an adult female, with a small head and body, but an extraordinarily long tail, moderately covered with hair (Merriam, 1898; Osgood, 1909). The ears are large and the feet are average in size. The hair of the body is long, smooth, and gray fawn; gray in the dorsal part and fawn toward the tailbone; it lacks dark dorsal and longitudinal lines. The nose is gray and the eyes are ringed with black. The forefeet are white, and two-thirds of the metatarsals are dark, but the rest, to the fingers, are white. The tail is dorsally dark and lighter on the bottom. The skull is small, and the face is short and narrow. The frontals are constrained above and lack supraorbital indentation. The skull is wide and round, the bullae are small, and the upper incisors are large. The dorsal hair from the head to the tailbone is brown and shiny, and the underside is lighter. The forefeet are white and the hind feet are dark and white. The tail is covered with short, uniformly colored hair. In accordance with Carleton (1989), of the three species that make up the group *melanophrys* (*P. melanophrys*, *P. mekisturus*, and *P. perfulvus*), *P. mekisturus* is the most divergent species and differs in the orthodontics of the incisors, the lack of supraorbital indentation, short and wide mesopterigoid fossa, and the development of an incisive capsule in the dentary.

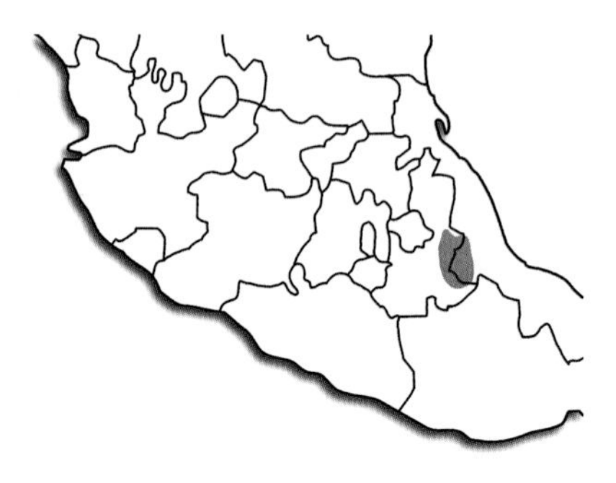

DISTRIBUTION: This species is endemic to the state of Puebla (Carleton, 1989; Hall, 1981; Merriam, 1898; Musser and Carleton, 1993; Ramírez-Pulido et al., 1983), of which only two specimens have been collected in the southeast of Puebla (Chalchicomula is now City Serdan and Tehuacán). It has been recorded in the state of PU.

EXTERNAL MEASURES AND WEIGHT

TL = 94 to 222 mm; TV = 135 to 249 mm; HF = 22 to 24 mm; EAR = 17 to 19 mm. Weight: 25 to 58 g.

DENTAL FORMULA: I 1/1, C 0/0, PM 0/0, M 3/3 = 16.

NATURAL HISTORY AND ECOLOGY: The biology of this species is very poorly known. They are arboreal (Merriam, 1898), which is supported by the exceptional size of the tail (equivalent to 160% of the length of the head and body), as well as the length of the fifth finger of the hind feet. It has been collected with *P. melanophrys*.

VEGETATIONAL ASSOCIATIONS AND ELEVATION RANGE: *P. mekisturus* inhabits forests of oak and pine in San Andres Chalchicomula (Goldman, 1951). It has been collected at elevations of 1,700 masl in Tehuacán (Hooper, 1947) and 2,545 masl in the vicinity of Chalchicomula, today City Serdan (Merriam, 1898).

CONSERVATION STATUS: *P. mekisturus* is a rare species with a very restricted geographical distribution. It is threatened (SEMARNAT, 2010) by its rarity, distribution, and habitat destruction. It is essential to assess the species' current status using appropriate methods to sample in trees, which apparently is its natural habitat (Carleton, 1989).

Peromyscus melanocarpus Osgood, 1904

Black-wristed deermouse

Iván Castro Arellano

SUBSPECIES IN MEXICO

P. melanocarpus is a monotypic species.

DESCRIPTION: Among the subgenus *Peromyscus*, *Peromyscus melanocarpus* is a large species. The coat is long and smooth, dorsally dark brown and ventrally grayish mixed with white. The tail is covered with short and grayish hair, and is a bit lighter near the base; the forefeet and hind feet are brownish to the bases of the thumbs

Peromyscus melanocarpus. Cloud forest. La Soledad, Oaxaca. Photo: Gerardo Ceballos.

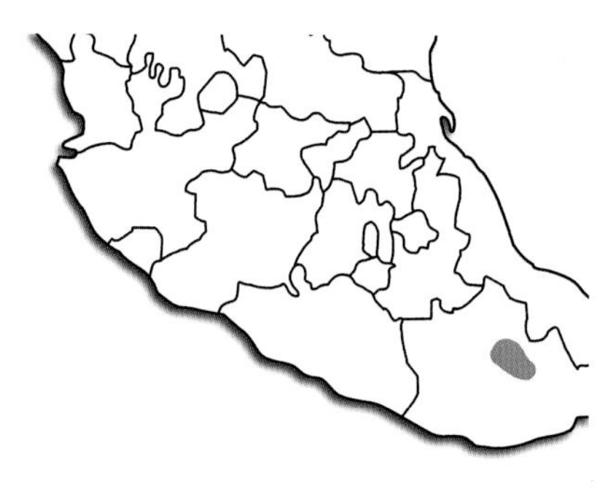

DISTRIBUTION: *P. melanocarpus* is endemic to Mexico and is restricted to cloud forests in north-central Oaxaca. Its distribution is discontinuous and covers two segments, one in Cerro Zempoaltepec and another in the northern slope of the Sierra de Juarez (Rickart and Robertson, 1985). It has been recorded in the state of OX.

(Goodwin, 1969; Osgood, 1909; Rickart and Robertson, 1985). There is no apparent sexual dimorphism in size (Rickart, 1977). The large size, absence of pectoral mammae, and supraorbital seams with little projections serve to distinguish it from *P. boylii*, *Habromys lepturus*, and *H. chinateco*, with which it shares its habitat. *P. mexicanus* can be distinguished because the latter occupies low-elevation areas and its forefeet and hind feet are white (Huckaby, 1980). The belly of *P. melanocarpus* is very similar to that of *P. melanurus*, with a glandular area slightly bent (Carleton, 1973).

EXTERNAL MEASURES AND WEIGHT
TL = 100 to 130 mm; TV = 100 to 133 mm; HF = 28 to 30 mm; EAR = 18 to 22 mm. Weight: 59 g.

DENTAL FORMULA: I 1/1, C 0/0, PM 0/0, M 3/3 = 16.

NATURAL HISTORY AND ECOLOGY: The forests that *P. melanocarpus* inhabits are characterized by marked slopes, lateritic soils, high rainfall, and a cover of moss in the forest floor. *P. melanocarpus* commonly uses exposed tree roots as refuge (Rickart and Robertson, 1985). Even though a seasonal increase in reproduction has been found, they probably reproduce throughout the year. The average size of the litter is 2.3 and the gestation period has been estimated at 30 days (Rickart and Robertson, 1985). Mating behavior has been studied under laboratory conditions (Deswbury, 1979). The development of pups is much slower than in other species of the genus; at 10 weeks old the size is still less than 90% of the adult size (Rickart, 1977). In captivity, it has been observed that the male participates in the care of the offspring and the nest, which indicates a possible monogamous structure (Rickart, 1977). Acorns and other seeds are perhaps an important part of its diet (Rickart and Robertson, 1985). They are sympatric with *Habromys lepturus*, *Peromyscus beatae*, *Megadontomys nelson*, and *Microtus umbrosus*.

VEGETATIONAL ASSOCIATIONS AND ELEVATION RANGE: This species is found in cloud forests, characterized by *Engelhardia mexicana*, *Liquidambar styraciflua*, *Cyathea mexicanus*, and a variety of ferns, orchids, and bryophytes. It is also found in cloud forests composed of *L. styraciflua*, *Bambusa* sp., *Pinus strobes*, and several species of *Quercus* (Rickart and Robertson, 1985), from 900 masl to 2,800 masl (Rickart and Robertson, 1985).

CONSERVATION STATUS: The current status of *P. melanocarpus* is unknown, but because of its restricted distribution, it is vulnerable to anthropogenic activity.

Peromyscus melanophrys (Coues, 1874)

Plateau deermouse

Elizabeth E. Aragón

SUBSPECIES IN MEXICO
Peromyscus melanophrys coahuilensis R.H. Baker, 1952
Peromyscus melanophrys consobrinus Osgood, 1904
Peromyscus melanophrys melanophrys (Coues, 1874)
Peromyscus melanophrys micropus R.H. Baker, 1952

Peromyscus melanophrys xenurus Osgood, 1904
Peromyscus melanophrys zamorae Osgood, 1904

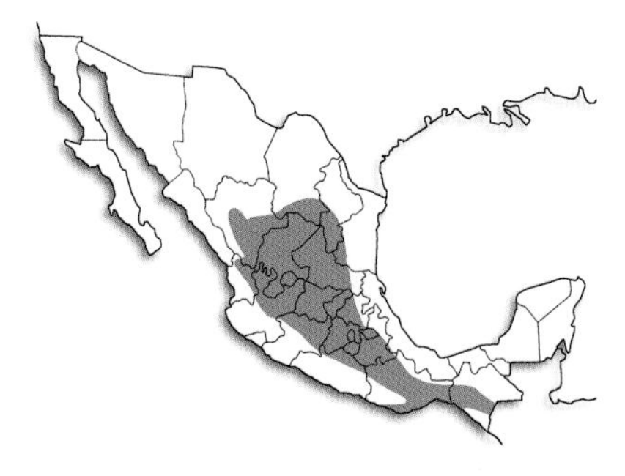

DISTRIBUTION: This species is endemic to Mexico. It is distributed from the north-central portions to the southern regions of the country through the Plateau, from Durango to Oaxaca and Chiapas (Hall, 1981). It has been recorded in the states of AG, CO, CS, DF, DU, GJ, GR, HG, JA, MI, MO, MX, OX, PU, QE, SL, TB, and ZA.

DESCRIPTION: *Peromyscus melanophrys* is one of the largest species of the genus *Peromyscus*. The dorsal coloration varies from ochraceous, gray, brown, yellowish, to cinnamon. The cheeks are light colored and the belly is beige with lead-gray spots on the base of the hair (Hall, 1981). The variation of the coloration is related to the substrate on which they live, being darker in volcanic areas (Baker, 1960). The supraorbital angle is well defined and has a smooth dorsal arch, the temporal region is large, the nasals are parallel, and the bullae are smaller than in the group *truei* but larger than in *mexicanus* (Hall, 1981). It differs from other species of *Peromyscus* in Mexico by the longer tail and thin brain lobes (Hall, 1981).

EXTERNAL MEASURES AND WEIGHT
TL = 185 to 292 mm; TV = 112 to 172 mm; HF = 24 to 30 mm; EAR = 11 to 32 mm.
Weight: 26 to 58 g.

DENTAL FORMULA: I 1/1, C 0/0, PM 0/0, M 3/3 = 16.

NATURAL HISTORY AND ECOLOGY: These mice prefer arid regions associated with yucca, cholla, ocotillo, prickly pear, mesquite, and some cacti, building their nests in some of these plants (Baker, 1956; Dalquest, 1953b; Jones and Webster, 1977; Matson and Baker, 1986). They are abundant in rocky sites. Reproduction occurs from February to March and from June to October. The average size of the litter is two to five offspring (Baker, 1956; Matson and Baker, 1986). They are granivores and can consume 1,646 g of seeds per year under controlled conditions (Petersen and Petersen, 1979). They are preyed on by owls (*Tyto alba*) and barn owls (*Bubo virginianus*; Petersen and Petersen, 1979; Sylpheed, 1978). Mites of genus *Mesostigmata* have been found parasitizing this species (Eric, 1981; Hoffman et al., 1972). Other rodents recorded in the same localities are *P. difficilis*, *P. maniculatus*, *P. gratus*, *Neotoma leucodon*, and *Liomys irroratus*.

Peromyscus melanophrys. Scrubland. Zacatecas. Photo: Gerardo Ceballos.

VEGETATIONAL ASSOCIATIONS AND ELEVATION RANGE: These mice are mainly found in xeric scrublands, thorn forests, and grasslands, in associations of desert scrublands and chaparral (oaks), typical of the contact area between arid and temperate climates. They have also been found in tropical valleys and coniferous forests (Álvarez, 1963a; Baker, 1960; Baker and Greer, 1962; Baker et al., 1981; Hall, 1981; Ramírez-Pulido et al., 1983; Winkelmann, 1962), from 50 masl to 2,700 masl.

CONSERVATION STATUS: In Mexico, *P. melanophrys* is not included in any conservation list (SEMARNAT, 2010), as it has a wide distribution, which includes large areas with little disturbance.

Peromyscus melanotis J.A. Allen and Chapman, 1897

Black-eared deermouse

Alondra Castro-Campillo, Matías Martínez-Coronel,
Ulises Aguilera, and José Ramírez-Pulido

SUBSPECIES IN MEXICO

P. melanotis is a monotypic species.

DESCRIPTION: *Peromyscus melanotis* is a small-sized mouse. It owes its name to the distinctive tuft of dark hair at the base of the ear. The dorsal coloration is ochraceous-brown interspersed with dark gray, especially in the mid-dorsal region, with a very conspicuous lateral line (Hall, 1981). The ears are dark with white edges. The belly and extremities are white (Hall, 1981). The tail is clearly bicolored (dorsally dark; Hall, 1981). Melanic animals exist (Martínez-Colonel et al., 1991; Osgood, 1909). It can be easily confused with *P. maniculatus*, but is distinguished by the well-defined

Peromyscus melanotis. Pine forest and grassland. Nevado de Toluca National Park, state of Mexico. Photo: Gerardo Ceballos.

dorsal line, and the dark bases of the ears (Martínez-Coronel et al., 1991). The face and nasals are thin and elongated. The nasals are compact anteriorly, the skull is round and inflated, and the bullae are relatively small (Hall, 1981).

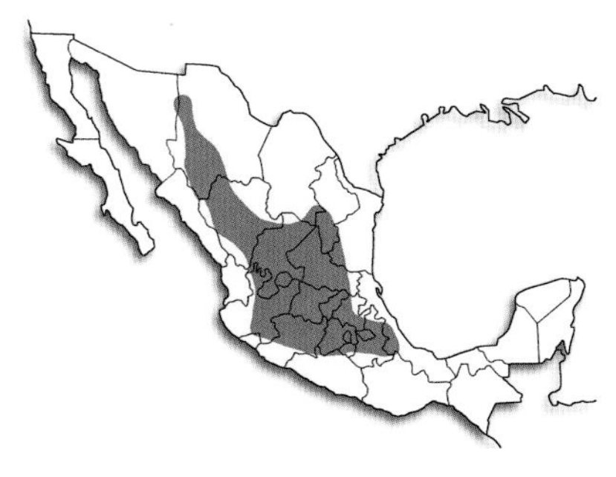

DISTRIBUTION: This species is endemic to Mexico with a very wide distribution, but discontinuous, covering a strip of the Sierra Madre Occidental in the states of Chihuahua and Durango, the middle portion of the Mexican Plateau, the Sierra Madre Oriental, and the Eje Volcanico Transverso (Allen and Chapman, 1897; Anderson, 1972; Baker and Greer, 1962; Baker and Phillips, 1965; Hall and Dalquest, 1963; Martínez-Coronel et al., 1991; Osgood, 1904; Thomas, 1903). It has been recorded in the states of AG, CH, CL, DF, DU, HG, JA, MI, MO, MX, NL, PU, SL, TB, TL, VE, and ZA.

EXTERNAL MEASURES AND WEIGHT

TL = 132 to 175 mm; TV = 58 to 81 mm; HF = 17 to 22 mm; EAR = 16 to 18 mm. Weight: 17 to 28 g.

DENTAL FORMULA: I 1/1, C 0/0, PM 0/0, M 3/3 = 16.

NATURAL HISTORY AND ECOLOGY: *P. melanotis* inhabits zacatonals and hollow logs of coniferous forests (Martínez-Coronel et al., 1991). It is crepuscular and feeds on seeds of seasonal herbs and grasses. Juveniles have a greater morphological variation than subadults, which is related to their differential growth rate. Females are larger than males (Martínez-Coronel et al., 1991). Young individuals are homogeneously dark until shortly before they reach sexual maturity (Martínez-Coronel et al., 1991). Individuals captured during winter had a lighter coloration than individuals caught in summer (Hall, 1981; Martínez-Coronel et al., 1991; Osgood, 1909). The populations of the Eje Volcanico Transverso have two to four offspring per litter. *P. melanotis* lives associated with *Sorex milleri*, *Neotoma mexicana*, *Microtus mexicanus*, *Neotomodon alstoni*, and *P. maniculatus* (Baker, 1956; Hall and Dalquest, 1963).

VEGETATIONAL ASSOCIATIONS AND ELEVATION RANGE: This species lives in temperate and semi-cold areas in cloud forests and coniferous forests with zacatonals and prairies of high mountains. It may also be found in the edges of native vegetation with crops close to the forests, and has even have been captured in lakes and rocky areas (Baker, 1956; Martínez-Coronel et al., 1991). It is distributed between 1,097 masl and 4,300 masl (Allen and Chapman, 1897; Davis, 1944; Martínez-Coronel et al., 1991). Most records occur above 2,000 m.

CONSERVATION STATUS: In general, *P. melanotis* presents no conservation concerns. Constant deforestation and urbanization in its distribution area can cause pressures, especially in the center of the country.

Peromyscus melanurus Osgood, 1909

Black-tailed deermouse

Iván Castro Arellano

SUBSPECIES IN MEXICO

P. melanurus is a monotypic species.

In the original description (Osgood, 1909) it was considered a subspecies of *P. megalops*; however, Huckaby (1980) verified its species level.

DESCRIPTION: *Peromyscus melanurus* is a large-sized *Peromyscus*. The dorsal coloration varies from brown to ochraceous. The feet and belly are white (Osgood, 1909). The tail is considerably longer than the body and is uniformly black. It is similar to *P. megalops*, but smaller, with a shorter and lighter coat, in addition to a smaller skull (Goodwin, 1969). The supraorbital seams have many projections, which together

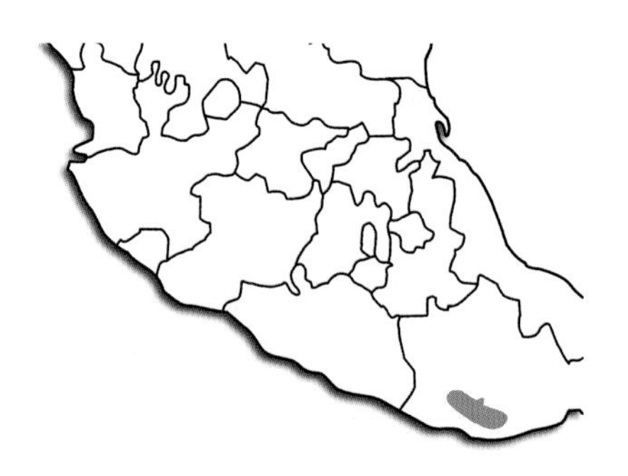

DISTRIBUTION: *P. melanurus* is an endemic species of Mexico, known only from the slopes of the Sierra Madre del Sur toward the Pacific in Oaxaca (Musser and Carleton, 1993). It is likely to be found in the mountains west of Chilpancingo, Guerrero (Huckaby, 1980). It has been recorded in the state of OX.

with the absence of a pectoral stain and pectoral mammae, serve to separate this species from others with which it is sympatric such as *P. beatae*, *P. aztecus*, and *P. melanophrys*. It can be distinguished from *Osgoodomys banderanus* and *P. megalops* by the narrower and oval skull of the first species, as well as by the greater size and presence of a pectoral stain cream-colored in the second species (Huckaby, 1980). A comparative study of the gastric morphology showed profound differences between *P. melanurus* and *P. megalops* (Carleton, 1973). The different karyotypes between the species of the group *mexicanus* can be distinguished by the X chromosome that has a submetacentric shape with a heterochromatic variation of the short arm (Smith et al., 1986).

EXTERNAL MEASURES AND WEIGHT
TL = 238 to 278 mm; TV = 127 to 145 mm; HF = 26 to 28.5 mm; EAR = 15 to 18 mm. Weight: 40 to 60 g.

DENTAL FORMULA: I 1/1, C 0/0, PM 0/0, M 3/3 = 16.

NATURAL HISTORY AND ECOLOGY: Little is known about the biology of this species since it has mostly been studied for its taxonomic features, recognition of its distribution area, and karyotypes (Baker and Womochel, 1966; Goodwin, 1969; Huckaby, 1980; Osgood, 1909). Like other species of genus, it is probably nocturnal and its diet may be based on seeds, acorns, fruit, insects, and other small invertebrates (Nowak, 1999b). It probably reproduces throughout the year. The gestation period must be within the range of 21 to 40 days. The average number of offspring per litter reported for the genus is 3.4, but in a close species, *P. melanocarpus*, it is 2.3, which is likely more accurate (Rickart and Robertson, 1985). Other species recorded in Oaxacan localities are *P. beatae*, *Nictomys sumichrasti*, *Reithrodontomys sumichrasti*, and *Tlacuatzin canescens* (L. León, pers. obs.).

VEGETATIONAL ASSOCIATIONS AND ELEVATION RANGE: The type of vegetation in which *P. melanurus* has been found is pine-oak forests in the highest parts of its distribution and tropical subdeciduous forests in lower elevations (Baker and Womochel, 1966). It is found from 700 masl to 1,900 masl (Huckaby, 1980).

CONSERVATION STATUS: The current status of this species is unknown, but as it is endemic and its distribution is very restricted, it is likely to be vulnerable (Ceballos and Rodriguez, 1993).

Peromyscus merriami Mearns, 1896

Merriam's deermouse

Adrián Quijada Mascareñas and Jorge Ortega R.

SUBSPECIES IN MEXICO
Peromyscus merriami goldmani Osgood, 1904
Peromyscus merriami merriami Mearns, 1896

DESCRIPTION: *Peromyscus merriami* is small in size. It frequently has a pectoral dark brown stain and the belly is whitish. The dorsal color is variable. The species has unique cranial features such as a robust skull with a wide zygomatic arch and large interorbital channel (Caire, 1985). *P. merriami merriami* has a smaller body and its coloration is lighter than in *P. merriami goldmani*. The dentary, the mastoid, and the cranial length are of the same magnitude in both subspecies (Hall, 1981; Hoffmeister, 1986; Hoffmeister and Diersing, 1973; Hoffmeister and Lee, 1963; Lawlor, 1971a, b, c).

EXTERNAL MEASURES AND WEIGHT
TL = 183 to 223 mm; TV = 94 to 126 mm; HF = 20 to 24 mm; EAR = 17 to 23 mm. Weight: 15 to 20 g.

DENTAL FORMULA: I 1/1, C 0/0, PM 0/0, M 3/3 = 16.

NATURAL HISTORY AND ECOLOGY: There is relatively little information about this species. It is closely associated with mesquite (Hoffmeister, 1986; Hoffmeister and Lee, 1963). Pregnant and lactating females have been captured throughout the year, suggesting reproduction is not seasonal. The litters comprise two to four offspring on average (Hoffmeister, 1986). It is nocturnal and its diet is mainly based on seeds.

VEGETATIONAL ASSOCIATIONS AND ELEVATION RANGE: *P. merriami* is associated with xeric scrublands, thorn forests, and grasslands, but especially with sites with dense mesquite (*Prosopis* spp.) (Caire, 1985; Hoffmeister, 1986; Hoffmeister and Lee, 1963). It is distributed from sea level to 800 masl.

CONSERVATION STATUS: *P. merriami* is not in any list of species threatened with extinction. Despite the fact that there is little information on the matter, the close association with forests of mesquite suggests a vulnerability to local extinction where mesquite is being overexploited (Hoffmeister, 1986).

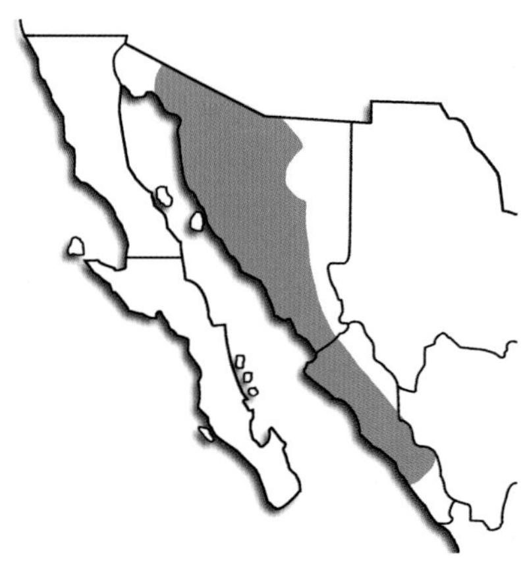

DISTRIBUTION: This species is distributed in the southwestern United States and northwestern Mexico. In Mexico, it is found in Sonora and northern Sinaloa (Hoffmeister and Diersing, 1973; Hoffmeister and Lee, 1963; Lawlor, 1983). It has been recorded in the states of SI and SO.

Peromyscus mexicanus (Saussure, 1860)

Mexican deermouse

Anna Horváth

SUBSPECIES IN MEXICO
Peromyscus mexicanus angelensis Osgood, 1904
Peromyscus mexicanus azulensis Goodwin, 1956
Peromyscus mexicanus mexicanus (Saussure, 1860)
Peromyscus mexicanus putlaensis Goodwin, 1964
Peromyscus mexicanus saxatilis Merriam, 1898
Peromyscus mexicanus teapensis Osgood, 1904
Peromyscus mexicanus totontepecus Merriam, 1898

DESCRIPTION: *Peromyscus mexicanus* is a medium to moderately large mouse. The size of the body, length of the coat, and coloration vary with the elevation, humidity, and season. It may be ochraceous-fulvous to light ochraceous on the sides in warm and humid areas, or opaque orange-brown to ochraceous-buffy in warm and dry areas: The belly is white. The ears are large and visibly naked. The eye ring is moderate to markedly dark. The tail is usually bicolored and spotted below, but in some individuals is dark and almost naked. The hind feet are dark at the top one-third or less than the length of the foot; the fingers are white (Hall, 1981; Reid, 1997).

Peromyscus mexicanus. Cloud forest. Soconusco, Chiapas. Photo: Gerardo Ceballos.

EXTERNAL MEASURES AND WEIGHT

TL = 213 to 77 mm; TV = 105 to 140 mm; HF = 25 to 28 mm; EAR = 19 to 24 mm. Weight: 29 to 50 g.

DENTAL FORMULA: I 1/1, C 0/0, PM 0/0, M 3/3 = 16.

NATURAL HISTORY AND ECOLOGY: *P. mexicanus* is mainly terrestrial, and a bit of a climber. It is shy and rarely sighted; however, it is one of the most easily captured species. Its diet is mainly composed of arthropods such as spiders, ants, and beetles, seeds, and green plant material. They live in burrows under roots or trunks (Reid, 1997). Reproduction can take place at any time of year, but varies depending on weather conditions; there is usually a peak in the beginning of the rainy season (A. Horváth, pers. obs.), as well as a pause at the beginning of the dry season, especially in areas with marked dry and wet seasons. The size of the litter may be two to three offspring (Reid, 1997).

VEGETATIONAL ASSOCIATIONS AND ELEVATION RANGE: *P. mexicanus* is widely distributed and found between 600 masl and 2,000 masl, in a variety of vegetation types, especially in tropical forests, coniferous forests, cloud forests, semi-deciduous forests, secondary forests, and riparian vegetation (Reid, 1997). It is also frequently found in ecotones, acahuals, cacao plantations, and edges of crop areas (Horváth et al., 2001).

CONSERVATION STATUS: *P. mexicanus* is a common and abundant species throughout its range, with great ecological plasticity.

DISTRIBUTION: This species' geographical range extends from San Luis Potosí, Guerrero, and Oaxaca, to Costa Rica and Panama, excluding the lowlands of the Yucatan Peninsula and the Caribbean (Huckaby, 1980; Reid, 1997; Wilson and Reeder, 2005).

Northern rock deermouse

J. Cuauhtémoc Chávez T.

SUBSPECIES IN MEXICO

Peromyscus nasutus nasutus (J.A. Allen, 1891)

Peromyscus nasutus penicillatus Mearns, 1896

P. nasutus was considered a distinct species (Osgood, 1909), but Hoffmeister and Torre (1961) included it as a subspecies of *P. difficilis*. Subsequently, Zimmerman et al. (1975, 1978) recognized it as a sister species separated from *P. difficilis*.

DISTRIBUTION: *P. nasutus* is distributed from northern Colorado and Oklahoma to northern Mexico. In Mexico, it is commonly found in the mountainous areas (pine-oak) in the Sierra Madre Occidental in Chihuahua and the Sierra Madre Oriental in Coahuila. It has been recorded in the states of CH and CO.

DESCRIPTION: *Peromyscus nasutus* is medium in size. The dorsal coloration is grayish-brown with gray underfur. The sides are ochraceous-yellow and the belly is whitish with gray underfur; occasionally, it has an orange pectoral stain. The tail is bicolored and longer than the head and body; the ears are large; the back of the hind feet has a whitish coloration in the metatarsal region (Baker, 1956; Hall, 1981). It differs from *P. difficilis* by a more tenuous coloration and smaller size (Hoffmeister and Torre, 1961).

EXTERNAL MEASURES AND WEIGHT

TL = 220 to 241 mm; TV = 121 to 134 mm; HF = 22 to 26 mm; EAR = 21 to 26 mm.
Weight: 25 to 35 g.

DENTAL FORMULA: I 1/1, C 0/0, PM 0/0, M 3/3 = 16.

NATURAL HISTORY AND ECOLOGY: These mice are herbivores, and mainly feed on acorns, stems, and roots. They build their burrows in cracks of rocks, tree stumps, trunks, and holes of trees (Baker, 1956). They are very skillful climbers (Horner, 1954). The mating season takes place from July to December. The number of offspring varies from three to four, and lactating females have been collected in December (Baker, 1956).

VEGETATIONAL ASSOCIATIONS AND ELEVATION RANGE: *P. nasutus* inhabits rocky environments in forests of pine (*Pinus* sp.) and oaks (*Quercus* sp.), desert scrublands, and grasslands, at 2,000 masl in the Sierra del Carmen to 3,117 masl in San Antonio de las Alazanas.

CONSERVATION STATUS: The status of this species is undetermined, but apparently it is scarce where it has been collected. It is necessary to assess the status of its populations in Mexico.

Peromyscus oaxacensis Merriam, 1898

Oaxacan deermouse

Ryan M. Duplechin and Robert D. Bradley

SUBSPECIES IN MEXICO

P. oaxacensis is a monotypic species.

DISTRIBUTION: The distribution of this species presumably includes the montane regions south of the Isthmus of Tehuantepec including much of Chiapas, Guatemala, El Salvador, and Honduras (Hall, 1981; Sullivan, 1997).

DESCRIPTION: *Peromyscus oaxacensis* is a medium-size mouse with rich tawny, dull fulvous, or cinnamon brown upper parts darkest along the mid-dorsal line with an admixture of black hairs (Hall, 1981; Merriam, 1898). The orbital rings are narrow and dusky to black; the underparts and forefeet are white to creamy white; the tail is sharply bicolored, dusky to black above and white to whitish below (Merriam, 1898). The skull resembles that of *P. beatae*, but is larger and more robust, and contains cheek teeth and palatine foramina that are larger on average (Hall, 1981). The postpalatal notch is much broader than that of *P. mexicanus* (Merriam, 1898). Initially described as a species by Merriam (1898), this taxon was placed as a subspecies of *P. aztecus* by Carleton (1979); however, Sullivan et al. (1997) used genetic data to infer that populations south of the Isthmus of Tehuantepec were substantially different from *P. aztecus*. It is unclear, however, whether populations in Oaxaca are referrable to *P. aztecus*, *P. aztecus evides*, or *P. oaxacensis*.

EXTERNAL MEASURES AND WEIGHT
TL = 242 mm; TV = 122; HF = 27; EAR = 15.8 to 17.5 mm.
Weight= N/A.

DENTAL FORMULA: I 1/1, C 0/0, PM 0/0, M 3/3 = 16.

NATURAL HISTORY AND ECOLOGY: Little is known about the biology and history of this species. It occurs in areas dominated by moist pine-oak forests and cloud forest habitats. It is frequently collected with *P. beatae* and other high-elevational species. Presumably its habits and ecology resemble those of *P. aztecus* and *P. hylocetes*, with which it is most closely related.

VEGETATIONAL ASSOCIATIONS AND ELEVATION RANGE: *P. oaxacensis* is associated with pine-oak and cloud forests at moderate to high elevations throughout the Sierra Madre Oriental and Sierra Madre del Sur (Sullivan, 1997). Its distribution also extends southward into the Sierra Madre de Chiapas and southward throughout the Central American Highlands to Honduras.

CONSERVATION STATUS: The current status of *P. oaxacensis* is unknown although the disappearance of montane pine-oak forests is reason for potential concern.

Peromyscus ochraventer Baker, 1951

El Carrizo deermouse

Arturo Hernández Huerta

SUBSPECIES IN MEXICO
P. ochraventer is a monotypic species.
P. ochraventer was previously included within the group *mexicanus* of the subgenus *Peromyscus* (Hooper and Musser, 1964; Huckaby, 1980), currently, it is provisionally considered within the group *furvus*, together with *P. mayensis* and *P. furvus* (Carleton, 1989) since their karyotypes possess a chromosomal arrangement different from the basic pattern shown in the species of the group mexicanus (Robbins and Baker, 1981). Its cranial and dental features, gastric anatomy, and number of mammary glands coincide closely to those of *P. furvus* (Carleton, 1973, 1989; Huckaby, 1980). Musser

and Carleton (2005) considered it to be an indeterminate species group and Bradley et al. (2007) placed it sister to a clade containing *nasutus*, *difficilis*, and *attwateri*.

DISTRIBUTION: This species is endemic to Mexico; its distribution is restricted to the wet areas of the east of the Sierra Madre Oriental, in the portions corresponding in southern Tamaulipas and northeastern San Luis Potos. It has been recorded in the states of SL and TA.

DESCRIPTION: *Peromyscus ochraventer* is a medium-sized field mouse. Its fur is brown; the back is ochraceous-fawn, with lighter shades toward the sides. The cheeks, neck, and shoulders are ochraceous-orange, and the belly is cinnamon buffy. It has blackish eye ring and the ears are dark; the tail has a scaly look and is slightly bicolored, dorsally dark and lighter underneath (Baker, 1951). Young individuals, like in other species of the genus, have a gray coloration. Adults have different shades of color and females have three pairs of mammae. The morphology of the interorbital area, the complexity of the teeth, and the color of the belly allow differentiation from other species (Huckaby, 1980). In the field, it is possible to distinguish *P. ochraventer* from other sympatric species such as *P. levipes*, *P. pectoralis*, and *P. leucopus* by its relatively larger size and by the typical brown coloration of the belly. It differs from *P. mexicanus*, which has a similar size and inhabits geographic areas adjacent to San Luis Potosí, by possessing a pair of pectoral mammae.

EXTERNAL MEASURES AND WEIGHT

TL = 96 to 122 mm; TV = 103 to 129 mm; HF = 22 to 25 mm; EAR = 16 to 21 mm. Weight: 24 to 40 g.

DENTAL FORMULA: I 1/1, C 0/0, PM 0/0, M 3/3 = 16.

NATURAL HISTORY AND ECOLOGY: These mice live mainly on the ground but are also skillful climbers, surpassing *P. levipes*, a species with which it is sympatric in the region of Gomez Farías, Tamaulipas, in agility. They are mainly nocturnal. They make their burrows between rocks and near fallen logs; some individuals of Carrizo, Tamaulipas, were captured near excavated land (Baker, 1951), but in Gomez Farías no evidence of such ability was found, possibly because the substrate is predominantly karst. These animals are docile and easy to handle. They are frequent found in rocky sites with little vegetation, and are not common in milpas or houses (A. Hernandez, pers. obs.). In Tamaulipas, it inhabits forests of oak in Gomez Farías (1,500 masl to

Peromyscus ochraventer. Cloud forest. El Cielo Biosphere Reserve, Tamaulipas. Photo: Gerardo Ceballos.

2,200 masl), which is characterized by dominant species as *Quercus glabrescens*, *Q. affinis*, *Pinus patula*, and *P. pseudostrobus*. It also is abundant in cloud forests (800 masl to 1,400 masl) with dominant species such as ocotillo (*Liquidambar styraciflua*), white oak (*Q. germana*), dark oak (*Q. sartorii*), and magnolia (*Magnolia schiedeana*), as well as in tropical forests (300 masl to 800 masl) with dominant species such as mulato (*Bursera simaruba*), ramon (*Brosimum alicastrum*), and amates (*Ficus*; Martin, 1958; Puig et al., 1983; Sosa et al., in press). In the state of San Luis Potosí, it dwells in forests of oak (Baker and Phillips, 1965). They are generalists and feed on seeds, fleshy fruits such as *Hoffmania*, insects, and fungi; during autumn and winter their diet consists basically of acorns. Its main predators are rattlesnakes (*Crotalus molossus* and *C. lepidus*), weasels, gray fox, margay, tropical owl (*Ciccaba virgata*), and probably the crowned owl (*Bubo virginanus*; A. Hernandez, pers. obs.). The mating season extends from the end of September to mid-February. Litters consist of two to three young that are born naked and with their eyes closed. In the cloud forests their reproductive period coincides with the production of acorns. They are very sensitive to low temperatures. In El Carrizo and the cloud forests of Gomez Farías, it coexists with *Neotoma angustapalata*, which is endemic to the state of Tamaulipas; other species with which it cohabits in the cloud forest are the rice rat (*Orizomys alfaroi*), mice (*Reithrodontomys* sp.), the flying squirrel (*Glaucomys volans*), and shrews of the genus *Cryptotis* (Hooper, 1953). In San Luis Potosí, it has been captured with *Peromyscus leucopus* and *P. pectoralis* (Baker and Phillips, 1965).

VEGETATIONAL ASSOCIATIONS AND ELEVATION RANGE: *P. ochraventer* inhabits forests of oak and pine-oak, cloud forests, and tropical subdeciduous forests, from 200 masl to 2,300 masl.

CONSERVATION STATUS: *P. ochraventer* is endemic to Mexico. It is not included in any list of endangered species. It is locally abundant in undisturbed cloud forests; however, its special affinity for this type of vegetation, its restricted geographical distribution, and its preference for undisturbed habitats make this species fragile. The cloud forest occupies a very reduced area and is being strongly transformed by human activities. Protected populations exist in the El Cielo Biosphere Reserve in southern Tamaulipas.

Peromyscus pectoralis Osgood, 1904

White-ankled deermouse

Gisselle Oliva

SUBSPECIES IN MEXICO

Peromyscus pectoralis collinus Hooper, 1952
Peromyscus pectoralis laceianus Bailey, 1906
Peromyscus pectoralis pectoralis Osgood, 1904

DESCRIPTION: *Peromyscus pectoralis* is a medium-sized mouse with small ears and elongated body. The coat is scarce and moderately long and straight. The back is brown, and the feet and belly are white. Young individuals are gray. The tail is longer

Peromyscus pectoralis. Scrubland. Santo Domingo, San Luis Potosí. Photo: Juan Cruzado.

than the total length of the body and has little hair. The tail is bicolored, dark at the top and light in the bottom (Schmidly, 1974). *P. pectoralis* can be distinguished from most species of *Peromyscus* by the presence of white hairs from the heel upward to the ankle; hence the common name white-ankled mouse. This species is similar to *P. boylii* but differs by being smaller in size.

EXTERNAL MEASURES AND WEIGHT
TL = 187 mm; TV = 95 mm, HF = 22 mm; EAR =16 mm.
Weight: 24 to 39 g.

DENTAL FORMULA: I 1/1, C 0/0, PM 0/0, M 3/3 = 16.

NATURAL HISTORY AND ECOLOGY: This species is nocturnal and feeds on seeds, nuts, cacti, and sometimes insects. Their diet changes according to their range and to the season. In the Sierra Madre Occidental in Tamaulipas it mainly feeds on prickly pears (Hall, 1981). It builds its burrows in rocky areas. Its breeding season is not well known; females have 3 to 7 offspring, the gestation period lasts 23 days. In Queretaro *P. pectoralis* cohabits with *P. difficilis*, *P. gratus*, and *Neotoma leucodon* (L. León, pers.com.)

VEGETATIONAL ASSOCIATIONS AND ELEVATION RANGE: *P. pectoralis* inhabits semi-arid regions, with xeric shrubs, grasslands, forests of pine-oak, and forests of fir. In San Luis Potosí it has been reported in desert mountains and in neighboring regions to tropical areas of the Sierra Madre (Schmidly, 1974). It is found from 100 masl to 2,500 masl.

CONSERVATION STATUS: This species is widely distributed in the Mexican Plateau and has no conservation concerns.

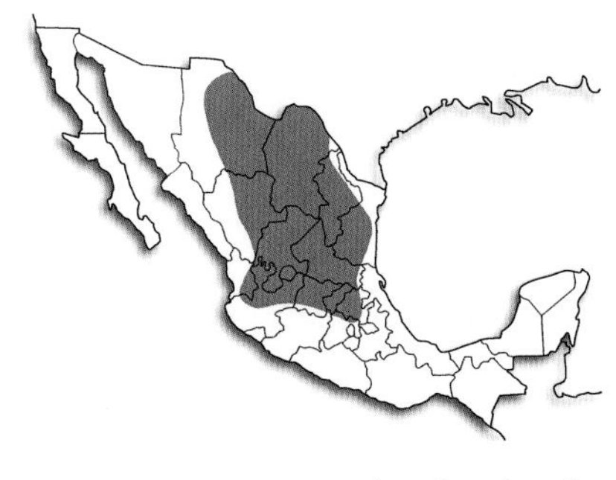

DISTRIBUTION: *P. pectoralis* is distributed from south-central Oklahoma, Texas, and New Mexico to central Mexico (Schmidly, 1974). In Mexico, it can be found in the Mexican Plateau from central Chihuahua to southern Queretaro in the Eje Neovolcanico Transversal. It has been recorded in the states of AG, CH, CO, DU, GJ, HG, JA, NL, NY, QE, SL, TA, and ZA.

Pemberton's deermouse

Aimeé de la Cerda and Eric Mellink

DISTRIBUTION: This species is endemic to the San Pedro Nolasco Island (latitude 27 degrees 58 minutes north, longitude 111 degrees 24 minutes west), in the Gulf of California, coast of Sonora, Mexico. It has only been found in SO.

SUBSPECIES IN MEXICO
P. pembertoni is a monotypic species.

DESCRIPTION: *Peromyscus pembertoni* has a large body and short tail. The dorsal coloration is light cinnamon mixed with black hair. The head is paler than the back. The tail is bicolored, dorsally brown and ventrally white. The belly is white (Burt, 1932). Evolutionarily, it seems it derived from *P. merriami* (Lawlor 1971a, b, c).

EXTERNAL MEASURES AND WEIGHT
TL = 210 mm; TV = 100 mm; HF = 24; EAR =18.
Weight: 15 to 20 g.

DENTAL FORMULA: I 1/1, C 0/0, PM 0/0, M 3/3 = 16.

NATURAL HISTORY AND ECOLOGY: The only time *P. pembertoni* was captured, in 1932, it was found in a steep slope of the west of the island, covered with grass (Burt, 1938); however, it may have been distributed throughout the island. Its biology is virtually unknown, although it might feed on seeds and green parts of plants. It is sympatric with *P. eremicus.*

VEGETATIONAL ASSOCIATIONS AND ELEVATION RANGE: These mice inhabit dry and xeric scrublands, from sea level to 326 m.

CONSERVATION STATUS: Although the Mexican Official Regulation (SEMARNAT, 2010) considers it extinct, as all attempts to capture it for more than 30 years have been unsuccessful (Álvarez-Castañeda and Cortes-Calva, 1999; Ceballos and Navarro, 1991; Ceballos and Rodriguez, 1993; O. Ward, pers. comm.) The habitat of the island is in good condition and shows no great alteration (Álvarez-Castañeda and Cortes-Calva, 1999). It is very possible that the cause of their extinction was the introduction of exotic species (Mellink, 2002).

Peromyscus perfulvus Osgood, 1945

Tawny deermouse

Gerardo Ceballos and Iván Castro-Arellano

SUBSPECIES IN MEXICO
Peromyscus perfulvus perfulvus Osgood, 1945
Peromyscus perfulvus chrysopus Hooper, 1955

DESCRIPTION: *Peromyscus perfulvus* is a medium-sized mouse. The body is tan with darker hair thinly scattered, more abundant in the back. The head is grayish and contrasts with the back. The belly is whitish and the feet are white. The eyes are surrounded by a slim dark brown line (Osgood, 1945). The tail is long, covered with little hair and uniformly sepia. The skull is similar to that of *P. levipes*. *P. perfulvus* has a long face and small bullae. It can be distinguished from *Osgoodomys banderanus* because it lacks supraorbital edges. The plantar tubercles are very developed, which probably favors their movement in the branches (Ceballos and Miranda, 1986, 2000; Hall, 1981; Helm et al., 1974). The gastric morphology is similar to that of most species of the genus since it has a bilocular and discoglandular stomach (Carleton, 1973). The morphology of the baculum is unique within the genus *Peromyscus* (Burt, 1960). The diploid number of chromosomes is 2n = 48 (Lee and Elder, 1977) and its karyotype is very similar to that of *P. melanophrys*, with which it is believed to be closely related (Stangl and Baker, 1984a, b).

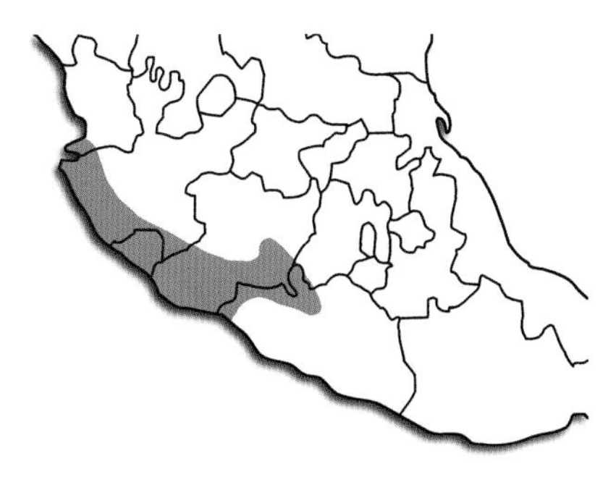

DISTRIBUTION: *P. perfulvus* is an endemic species of Mexico, which is distributed in the coastal area of Jalisco to Guerrero. It inhabits the country along the Rio Balsas to south of the state of Mexico (Álvarez and Hernández-Chavez, 1990; Musser and Carleton, 1993). It has been recorded in the states of CO, GR, JA, MI, and MX.

EXTERNAL MEASURES AND WEIGHT

TL = 208 to 254 mm; TV = 110 to 138 mm; HF = 23 to 26 mm; EAR = 18 to 26 mm. Weight: 25 to 40 g.

DENTAL FORMULA: I 1/1, C 0/0, PM 0/0, M 3/3 = 16.

NATURAL HISTORY AND ECOLOGY: This mouse carries out most of its activity in trees (Ceballos, 1989, 1990). It has been collected in traps placed in trunks of trees between 2 m and 5 m above the ground but is also active at ground level (Helm et al., 1974; Stephen et al., 1975). On the coast of Jalisco it is sympatric with *Nyctomys sumichastri* and *Osgoodomys banderanus* (Collet et al., 1975; Helm et al., 1974). Sixty-four percent of the annual captures were made in trees, so that *N. sumichrasti* is more arboreal and *O. banderanus* more terrestrial (Ceballos, 1990). Its burrows are located in hollow trees, under rocks, or underground. The nests are spherical and built with plant fibers (Ceballos and Miranda, 1986, 2000). Its diet is based on fruits, seeds, and insects. Its population density varies from 2 individuals/ha to 14 individuals/has (Ceballos and Miranda, 1986, 2000). It moves extensively throughout riparian forests (at a distance of 21 m between successive captures, with the longest being 1,400 m; Ceballos, 1989). In Chamela-Cuixmala the proportion of sexes was biased toward males in an annual survey. Reproduction is in summer and early autumn, coinciding

Peromyscus perfulvus. Tropical dry forest. Chamela-Cuixmala Biosphere Reserve, Jalisco. Photo: Gerardo Ceballos.

with the rainy season, although active individuals have been collected in January and May. Under laboratory conditions the average number of offspring per birth was 2.6. The offspring are born underdeveloped, with eyes closed, without hair, and with a weight of 2 g to 3 g (Ceballos, 1989). They open their eyes at 15 days old, are weaned at 25 days old, and reach sexual maturity at 6 weeks old (Ceballos, 1990; Helm et al., 1974; Stephen et al., 1975).

VEGETATIONAL ASSOCIATIONS AND ELEVATION RANGE: This species exclusively inhabits humid places in the dry tropics such as the tropical subperennial forests and riparian forests. It has also been found in coconut plantations, fruit trees, and fields of sugar cane. It is found from sea level to 1,300 masl (Ceballos and Miranda, 2000; Hall and Villa, 1949b; Helm et al., 1974; Hooper, 1955; Nuñez et al., 1980; Osgood, 1945).

CONSERVATION STATUS: *P. perfulvus* is very abundant in disturbed areas such as plantations of mango in the coast of Jalisco, and is not considered at risk. There are populations in the Chamela-Cuixmala Biosphere Reserve in Jalisco (Ceballos and Miranda, 1986, 2000).

Peromyscus polius Osgood, 1904

Chihuahuan deermouse

Iván Castro Arellano

SUBSPECIES IN MEXICO

P. polius is a monotypic species.

DESCRIPTION: *Peromyscus polius* is a large mouse with brown dorsal coloration; the head is grayish with marked cheeks; a line of the same color surrounds the eyes.

Peromyscus polius. Pine-oak forests. Mesa de San Luis, Janos Biosphere Reserve, Chihuahua. Photo: Gerardo Ceballos.

The abdomen, wrists, and ankles are white. The tail is bicolored, dorsally brown and ventrally white (Anderson, 1972; Hall, 1981; Osgood, 1904). The diploid number of chromosomes is 2n = 48 (Kilpatrick and Zimmerman, 1975).

EXTERNAL MEASURES AND WEIGHT
TL = 210 to 234 mm; TV = 111 to 120 mm; HF = 25 to 26 mm; EAR = 17 to 19 mm.
Weight: 22 to 36 g.

DENTAL FORMULA: I 1/1, C 0/0, PM 0/0, M 3/3 = 16.

NATURAL HISTORY AND ECOLOGY: Details of the biology of this species are scarce, but like other species of the genus it should be nocturnal and granivorous (Nowak, 1999b). It could be a prey of barn owls (*Tyto alba*; Anderson and Long, 1961).

VEGETATIONAL ASSOCIATIONS AND ELEVATION RANGE: *P. polius* mainly inhabits pine-oak vegetation, grasslands, and xeric scrublands, from 2,110 masl to 2,720 masl (Anderson, 1972).

CONSERVATION STATUS: *P. polius* is not included in any list of endangered species, but its current status is unknown; however, because of its restricted distribution it could be vulnerable.

DISTRIBUTION: This rodent is endemic to Mexico and its distribution is restricted to the western and central parts of Chihuahua (Musser and Carleton, 1993). It is likely to be found in Sonora. It has been recorded in the state of CH.

Peromyscus pseudocrinitus Burt, 1932

Coronados deermouse

Jesús Ramírez Ruiz and Rafael Ávila Flores

SUBSPECIES IN MEXICO
P. pseudocrinitus is a monotypic species.

DESCRIPTION: *Peromyscus pseudocrinitus* is the darkest species of the genus *Peromyscus* that inhabits the islands of the Gulf of California. The dorsal coloration is gray-black with cinnamon shades; the underside is white. The tail is long, has few hairs, and is bicolored in the first two-thirds, dark on top and light below. The general form of the skull is similar to that of *P. crinitus*, but longer (average length: 25.6 mm) and with less bulging bullae. The nasals are wide and have longitudinal and parallel edges, which are rounded in the posterior edge; the premaxilla extends slightly behind the nasal; the projection of the palatal bone is shorter than the length of the jaw; in addition, the accessory cusps of the first two upper molars (located between the primary cusps) are more prominent than those of *P. crinitus* (Burt, 1932; Hall, 1981). This species has been part of the subgenus *Peromyscus* based on their teeth characteristics (Hall, 1981); however, the baculum, glans, and other external characters are very similar to those of *P. eremicus*, which belongs to the subgenus *Haplomylomys* (Burt, 1932, 1960; Lawlor, 1971a, b, c).

EXTERNAL MEASURES AND WEIGHT
TL = 194 mm; TV = 110 mm; HF = 21 mm; EAR = 16 mm.
Weight: probably 40 g.

DISTRIBUTION: *P. pseudocrinitus* exclusively inhabits the Coronados Island, in the Gulf of California, close to port Loreto, in Baja California Sur. It has been recorded in the state of BS.

DENTAL FORMULA: I 1/1, C 0/0, PM 0/0, M 3/3 = 16.

NATURAL HISTORY AND ECOLOGY: This species has nocturnal habits. It inhabits the rocky areas of the island. Like other members of the genus, it feeds on seeds, fleshy parts of plants, and insects. Its predators are introduced cats and snakes such as *Masticophis flagellum* (Arnaud and Troyo, 1995; Bourillón et al., 1988; López-Forment et al., 1996).

VEGETATIONAL ASSOCIATIONS AND ELEVATION RANGE: The vegetation of the island is xeric scrublands with abundant cacti, such as chollas and cardones, which in the slopes of the volcanic cone are interspersed with elements of tropical deciduous forests. In addition, there is a small area of coastal dunes (Bourillón et al., 1988). *P. pseudocrinitus* is found from sea level to 283 m.

CONSERVATION STATUS: This species is considered threatened with extinction (SEMARNAT, 2010). The available data show that the population of *P. pseudocrinitus* is very small, despite its wide distribution in the island (Arnaud and Troyo, 1995). In addition, this species is in danger of disappearing due to the presence of feral cats in the island (Arnaud and Troyo, 1995; López-Forment et al., 1996).

Peromyscus sagax Elliot, 1903

Michoacan deermouse

Juan Cruzado

SUBSPECIES IN MEXICO

P. sagax is a monotypic species.

In 1903 Elliott described this species, but in a revision of the genus *Peromyscus*, Osgood (1909) considered it as a subspecies of *P. truei gratus*. Subsequently, Hoffmeister (1946) examined the type specimen and placed the species within the group *boylii* as a synonym of *P. boylii levipes*. Recent morphological and morphometric studies, as well as genetic studies, recognize it as a species (Bradley et al., 1997).

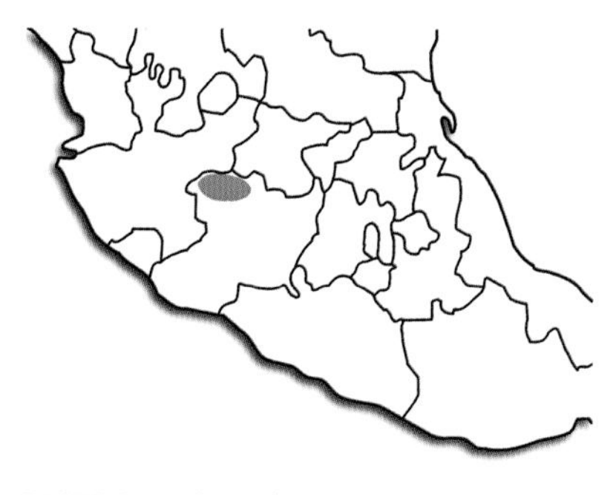

DISTRIBUTION: This species is endemic to Mexico; it is restricted to the northern and central part of Michoacán, in the region of Los Reyes (Bradley et al., 1997).

DESCRIPTION: The dorsal coloration of *Peromyscus sagax* is grayish-brown; the belly and legs are white. The tail is slightly larger than or equal to the length of the head and body. Morphologically, this species resembles *P. levipes* and *P. gratus*.

EXTERNAL MEASURES AND WEIGHT

TL = 190 to 210 mm; TV = 98 to 109 mm; HF = 22.5 to 23.5 mm; EAR = 19 to 21 mm. Weight: 22 to 36 g.

DENTAL FORMULA: I 1/1, C 0/0, PM 0/0, M 3/3 = 16.

NATURAL HISTORY AND ECOLOGY: Very little is known about the natural history and ecology of this species but it is very similar to other species of the genus. They are nocturnal and feed mainly on seeds, fruits, sprouts of plants, and sometimes invertebrates.

VEGETATIONAL ASSOCIATIONS AND ELEVATION RANGE: *P. sagax* inhabits pine-oak forests, as well as crops and scrublands, in mid-elevations in the central part of Michoacán, from 1,300 masl to 1,700 masl (Bradley et al., 1997; Hall, 1981).

CONSERVATION STATUS: The current status of this species is totally unknown, but because only a few individuals have been caught and because its distribution is very restricted it can be considered threatened (Bradley et al., 1997).

Peromyscus schmidlyi Bradley et al., 2004

Schmidly's deermouse

Nicté Ordóñez-Garza and Robert D. Bradley

SUBSPECIES IN MEXICO

P. schmidlyi is a monotypic species.

This species was previously referred to as *P. boylii rowleyi* or *P. boylii levipes* from the Sierra Madre Occidentals in northwestern Mexico (Bradley et al., 2004; Cabrera et al., 2007).

DESCRIPTION: *Peromyscus schmidlyi*, a medium-sized mouse, resembles *P. boylii levipes* in size and coloration. *P. schmidlyi* is larger and darker compared to *P. boylii rowleyi* (Bradley et al., 2004). The orbital region is hourglass and not angular as compared to that of *P. spicilegus*. There are not sufficient diagnostic morphological differences to separate *P. schmidlyi* from other species of the *P. boylii*-species group occurring in western Mexico, however (Cabrera et al., 2007). In terms of size, average cranial measurements from specimens of Sonora are smaller than those from the type locality in Durango, with the exception of the lengths of the rostrum and molar tooth row (Cabrera et al., 2007). The dorsal pelage is mummy brown at the tips and plumbeous-black at the bases; the sides are cinnamon-rufus; the ventral pelage is white at the tips and plumbeous-black at the bases; the feet have an iron gray stripe extending slightly past the ankle; the toes are white; the tail is bicolored, blackish-brown above and white below, scantily haired at the base, and tuffed at the tip; the ears are dark neutral gray, and the vibrissae are black. Chromosomally, Bradley et al. (2004) found that karyotypes ranged from FN = 54 to FN = 56 in fundamental number, differing from other members of the *P. boylii*-species group and having a unique karyotype; *P. schmidlyi* has a biarmed condition for chromosome 9 (Bradley et al., 2004). At the genetic level (mitochondrial cytochrome b gene sequences), *P. schmidlyi* differs from other members of the *P. boylii*-species group (*P. beatae*, *P. boylii*, *P. levipes*, *P. medrensis*, *P. stimulus*, *P. stephani*). To date *P. schmidlyi* can only be recognized using karyotypic or molecular analyses (Cabrera et al., 2007).

DISTRIBUTION: *P. schmidlyi* is an endemic to Mexico, known only from Durango, Sinaloa (Bradley et al., 2004), and Sonora (Cabrera et al., 2007). The Sonoran record (Cabrera et al., 2007) indicates that the geographical distribution may extend throughout the Sierra Madre Occidental, and probably continues to the western part of Chihuahua, northeastern Nayarit, northwestern Zacatecas, and northern Jalisco. It has been recorded in the states of DU, SI, and SO.

EXTERNAL MEASURES AND WEIGHT

TL = 186 mm; TV = 84 mm; HF = 20 mm; EAR = 20 mm.

Weight: NA.

DENTAL FORMULA: I 1/1, C 0/0, PM 0/0, M 3/3 = 16.

NATURAL HISTORY AND ECOLOGY: The biology of this species is little known. *P. schmidlyi* occupies the central highlands of the Sierra Madre Occidental, bounded by the Mesa Central to the east and the tropical deciduous forest to the west (Bradley et al., 2004). It may have a range similar to that of the other endemic species of this region, *Spermophilus madrensis*, *Sciurus nayaritensis*, and *Nelsonia neotomodon* (Cabrera et al., 2007).

VEGETATIONAL ASSOCIATIONS AND ELEVATION RANGE: Specimens have been collected in oak-juniper forest, including one seeded-juniper (*Juniperus monosperma*), Apache pine (*Pinus engelmani*), Ponderosa pine (*Pinus ponderosa*), Yécora pine (*Pinus yecorensis*), Arizona white oak (*Quercus arizonica*), Sípuri (*Quercus durifolia*), silverleaf oak (*Quercus hypoleucoides*), and kittle lemonhead (*Coreocarpus arizonicus*) (Cabrera et al., 2007), at elevations greater than 2,000 m (Bradley et al., 2004).

CONSERVATION STATUS: The current status of this species is unknown but it is likely that it is not under threat.

Peromyscus sejugis Burt, 1932

Santa Cruz deermouse

Jesús Ramírez R.

SUBSPECIES IN MEXICO

P. sejugis is a monotypic species.

DESCRIPTION: *Peromyscus sejugis* is an insular mouse, medium in size. It differs from other mice of the genus *Peromyscus* by the grayish and opaque coloration of the back and by the brownish ears. The belly is white; the tail has scales of approximately 1.5 mm wide and is bicolored, blackish in the top and whitish underneath (Burt, 1932; Hall, 1981).

EXTERNAL MEASURES AND WEIGHT

TL = 187 mm; TV = 89 mm; HF = 23 mm; EAR = 16 mm.
Weight: 22 to 36 g.

DENTAL FORMULA: I 1/1, C 0/0, PM 0/0, M 3/3 = 16.

DISTRIBUTION: This species inhabits only the Santa Cruz and San Diego Islands, located in the Gulf of California at 20 km and 25 km from Punta Botella, respectively, in Baja California Sur, Mexico. It has been recorded in the state of BS.

NATURAL HISTORY AND ECOLOGY: This species lives in ravines and small valleys of the islands, near the coastal area. When these mice colonized the islands, the availability of food and the absence of competitors and carnivores led to the increase in the body size of *P. sejugis* (Lawlor, 1982, 1983). That is why the mice of this species are larger than *P. maniculatus* of the Baja California Peninsula, with which they are closely related (Lawlor, 1982.). In 1986, there was a high density on Santa Cruz island, perhaps because of an abundance of food and the lack of competitors. Of 10 adult females captured during August 1986 in Santa Cruz Island, 4 were sexually active (J. Ramirez, pers. obs.). Currently, no more information about its biology is known.

VEGETATIONAL ASSOCIATIONS AND ELEVATION RANGE: *P. sejugis* is found in xeric scrublands, which is the dominant vegetation in the islands of the Gulf of California. It is found from sea level to about 100 m.

CONSERVATION STATUS: In Mexico, *P. sejugis* does not have a particular status. It can be considered, like many species that inhabit the islands, as a fragile species, as it inhabits very restricted areas that are susceptible to introduced species such as cats and rats.

Peromyscus simulus Osgood, 1904

Sinaloan deermouse

Clemente R. Melgar

SUBSPECIES IN MEXICO

P. simulus is a monotypic species.

For a long time this species was considered a subspecies of *P. boylii* (Osgood, 1909). In 1977, based mainly on characteristics of the penis, skull, and karyotype, it was elevated to species (Carleton, 1977; Schmidly and Shroeter, 1974).

DESCRIPTION: *Peromyscus simulus* is the smallest species of the group *boylii* (Schmidly and Bradley, 1995). The dorsal coloration is golden brown and sometimes ochraceous. The back has hair with dark tips, uniformly scattered. Sometimes the black hair forms a poorly defined dark stripe on the spine. The sides have a black and wide line that is not very contrasting. The ventral coat is whitish, generally with grayish shades as longitudinal stripes, and often has a large, dark pectoral spot. The eyes are ringed by black hair, extending slightly to the top and forming a gray area between the eye and the base of the ear (Osgood, 1909). Furthermore, it also has tufts of dark hair partly hidden in the anterior base of the ears, which are dark with or without light yellow edges. The feet are white, but have black hair near the metatarsal area; the tail is dorsally blackish, lighter ventrally, and sometimes bicolored (Osgood, 1909). Descriptions of the karyotype allow identification from other forms of the group *P. boylii* (Schmidly and Schroeter, 1974). This mouse also is distinguished by the smaller external measures and by the limited development of the supraorbital region, the shape of the zygomatic arch, and the small row of molars that distinguish it from *P. spicilegus* and *P. boylii rowleyi* (Baker, 1950a; Carleton, 1977; Osgood, 1904). It also has unique features of the penis, mainly in the glans (Bradley and Schmidly, 1987; Carleton, 1977; Schmidly and Bradley, 1995).

DISTRIBUTION: This species is endemic to Mexico; it has a limited distribution, ranging from San Ignacio, north of Sinaloa, to Las Varas, in Nayarit, near the limits with Jalisco (Bradley, and Schmidly, 1995). It has been recorded in the states of NY and SI.

EXTERNAL MEASURES AND WEIGHT

TL = 208 mm; TV = 90 mm; HF = 23 mm; EAR = 15 mm.
Weight: 22 to 36 g.

DENTAL FORMULA: I 1/1, C 0/0, PM 0/0, M 3/3 = 16.

NATURAL HISTORY AND ECOLOGY: Little is known about the natural history of this species. It has been noted that individuals of Sinaloa are smaller in size than those of Nayarit, and due to differences in the coloration, they are suspected to be isolated populations (Hooper, 1955). The color of *P. simulus* changes over a gradient due to differences in humidity. The specimens of the north are yellowish-brown, whereas in the south individuals have a similar coloration but with ochraceous shades (Carleton, 1977). Pregnant animals have been collected in April, with a variation of one and three embryos per female (Hooper, 1955). The species has been captured in thorn forests and tropical deciduous forests; it has also been found near areas with disturbed vegetation and crops of bananas (English, 1958; Hooper, 1955). Apparently, it is more common in humid environments and the populations have wide fluctuations in density (Schmidly and Bradley, 1995).

VEGETATIONAL ASSOCIATIONS AND ELEVATION RANGE: *P. simulus* inhabits tropical forests and forests of the coastal plain such as the thorn forests and tropical deciduous forests (Carleton, 1977), but is more abundant in tropical forests and humid forests, palm groves, and mangroves, from sea level to 244 m (Bradley and Schmidly, 1987; Carleton, 1977; Schmidly and Bradley, 1995).

CONSERVATION STATUS: This species is endemic to Mexico. It should be regarded as a vulnerable species due to its restricted range and because its habitat is quite disturbed (Ceballos and Navarro, 1991; Ceballos and Rodriguez, 1993; Schmidly and Bradley, 1995).

Peromyscus slevini Mailliard, 1924

Catalina deermouse

Jesús Ramírez R.

SUBSPECIES IN MEXICO

P. slevini is a monotypic species.

DESCRIPTION: *Peromyscus slevini* is a medium-sized mouse, which is distinguished by the pale tan coloration of the upper parts with intermingled brown hair. The ventral region is white and the pectoral area is pale brown. The hind feet are creamy white and the forefeet are pale cinnamon. The tail is bicolored, dark above and white beneath. The coat is short and the ears small (Hall, 1981; Mailliard, 1924).

DISTRIBUTION: This species is endemic to Mexico. It lives exclusively on Santa Catalina Island (43 km^2), which is located 23 km east of Punta San Marcial in the Gulf of California, Baja California Sur. It has been recorded in the state of BS.

EXTERNAL MEASURES AND WEIGHT

TL = 225 mm; TV = 120 mm; HF = 27 mm; EAR =16.5 mm.
Weight: NA

DENTAL FORMULA: I 1/1, C 0/0, PM 0/0, M 3/3 = 16.

NATURAL HISTORY AND ECOLOGY: *P. slevini* is the only native species of the island. It inhabits rocky ravines and ecotones in the mountains with temporary streams, where the vegetation offers shelter and food. Its body is larger than that of *P. maniculatus* from the Baja California Peninsula, with which it is related. *P. slevini* is larger in size, up to 22 mm, than other species related to *P. maniculatus*, which live in islands such as Santa Cruz and San Diego in the Gulf of California. The body size is related to the absence of competitors and the availability of food in Isla Santa Catalina (Lawlor, 1982, 1983). Human impact is reflected in the recent introduction of feral cats by fishermen. In stool samples, remnants of reptiles, birds, and rodents have been found, which proves the impact that introduced fauna has on the native and endemic species threatened with extinction (Bourillón et al., 1988; Luke, 1986). They are part of the food chain of the rattlesnake *Crotalus catalinensis*, the only rattlesnake that lacks a rattle.

VEGETATIONAL ASSOCIATIONS AND ELEVATION RANGE: *P. slevini* inhabits xeric scrublands, from sea level to 100 m.

CONSERVATION STATUS: In Mexico, this species does not have a special status. During four random samplings conducted in 1986, however, only two adult specimens were collected (J. Ramirez, pers. obs.). It is necessary, therefore, to determine whether *P. slevini* still inhabits other sites of the island. Stool samples of feral cats indicate that it should be considered in danger of extinction, along with other species of mice and rats of the islands in the Gulf of California (Ceballos and Navarro, 1991; Mellink et al., 2002). It is considered threatened (SEMARNAT, 2010).

Peromyscus spicilegus J.A. Allen, 1897

Gleaning deermouse

Erika Marcé-Santa and Iván Castro Arellano

SUBSPECIES IN MEXICO

Peromyscus spicilegus spicilegus J.A. Allen, 1897

DESCRIPTION: *Peromyscus spicilegus* is medium in size. The dorsal coloration is yellowish-brown mixed finely with dark hair, more conspicuously in the mid-back, forming a slightly darker area. The lateral line is wide and ochraceous colored, which contrasts with the white abdomen. The forefeet are completely white, while the hind feet are dark brown to the tarsus. The ears are almost naked and dark. The tail is bicolored, brown on the top and white on the bottom (Allen, 1897; Osgood, 1909; Sánchez-Cordero and Villa-R., 1988). The skull is similar to that of *P. boylii*, but in *P. spicilegus* the skull is wider and forms a supraorbital platform and is more angled (Carleton, 1979; Osgood, 1909).

EXTERNAL MEASURES AND WEIGHT

TL = 189 to 210 mm; TV = 95 to 108 mm; HF = 23 to 25 mm; EAR = 15 to 17 mm. Weight: 22 to 36 g.

DENTAL FORMULA: I 1/1, C 0/0, PM 0/0, M 3/3 = 16.

NATURAL HISTORY AND ECOLOGY: The burrows of this species are built under bushes near rocks and hillsides (Nuñez et al., 1981). Details of its reproduction are unknown; however, females with embryos have been caught in August and October, as well as lactating females in November. Nonetheless it is unknown if they reproduce throughout the year. The number of embryos reported in two females is three (Nuñez et al., 1981). Its diet is unknown but it is possible that, like other members of the genus *Peromyscus*, it consumes seeds, insects, and vegetable matter (Nowak, 1999b).

VEGETATIONAL ASSOCIATIONS AND ELEVATION RANGE: *P. spicilegus* is mainly associated with forests of pine-oak and medium tropical forests (Hooper, 1955; Nuñez et al., 1981). It has been collected from 260 masl to 2,160 masl (Hall, 1981; Nuñez et al., 1981). It ranges through much of the pine-oak forest of the Sierra Madre Occidental and at lower elevations in the tropical deciduous forests (Bradley et al., 1996).

DISTRIBUTION: This species is endemic to Mexico and is restricted to intermediate elevations of the Sierra Madre Occidental from southern Sonora and Chihuahua to Michoacán (Bradley et al., 1996; Hall, 1981; Musser and Carleton, 1993; Osgood, 1909). It has been recorded in the states of CH, CO, DU, JA, MI, NY, SI, and ZA.

CONSERVATION STATUS: This species is not included in any list of conservation (Ceballos et al., 2002). Because it is an endemic species, it is important to assess its conservation status.

Peromyscus stephani Townsend, 1912

San Esteban deermouse

Aimeé de la Cerda and Eric Mellink

SUBSPECIES IN MEXICO

P. stephani is a monotypic species.

Based on anatomical, serological, and cytological characters, this species seems to be in the group *boylii*, of *Peromyscus* (Álvarez-Castañeda and Cortes-Calva 1999, Hall, 1981; Lawlor, 1971b; López-Forment et al., 1996). In fact, recently, Hafner et al. (2001) based on genetic information suggested that *P. stephanie* is actually a subspecies of *P. boylii*. Morphological and morphometric information is needed to support their results, however. Bradley et al. (2007) used genetic data to show that *P. stephani* is a valid species.

DISTRIBUTION: This species is endemic to the San Esteban Island (latitude 28 degrees 43 minutes north, longitude 112 degrees 19 minutes west), Gulf of California, Sonora, Mexico. It has only been recorded in that state.

DESCRIPTION: *Peromyscus stephani* is characterized by the short tail and very large hind feet. The dorsal coloration is light yellowish-brown, with dark gray hair, which gives them a gray appearance. The tail has the same color as the back; the belly and feet are almost white (Álvarez-Castañeda and Cortes-Calva, 1999; Hall, 1981; López-Forment et al., 1996; Sánchez-Hernández et al., 1997; Townsend, 1911).

EXTERNAL MEASURES AND WEIGHT

TL = 195 to 207; TV = 100 to 124; HF = 18-22; EAR =16 to 20.

Weight: 16 to 25 g.

DENTAL FORMULA: I 1/1, C 0/0, PM 0/0, M 3/3 = 16.

NATURAL HISTORY AND ECOLOGY: The preferred habitats of this mouse are rocky areas or slopes. Its diet includes seeds and green parts of plants, as well as flowers, fruits, and large quantities of insects.

VEGETATIONAL ASSOCIATIONS AND ELEVATION RANGE: *P. stephani* occurs only on San Esteban Island in the xeric scrublands from sea level to 540 m.

CONSERVATION STATUS: This species is considered threatened in Mexico (SEMARNAT, 2010). San Esteban Island is of considerable size, and although the population of these mice is large, they are threatened by the presence of domestic cats (*Felis catus*) and black rats (*Rattus rattus*). Since the 1930s rats have been reported on the island; hence it can be assumed that somehow *P. stephani* has adapted to the presence of those invasive species (Álvarez-Castañeda and Cortes-Calva, 1999; Burt, 1938, López-Forment et al., 1996; Mellink, 2002).

Piñon mouse

J. Cuauhtémoc Chávez T.

SUBSPECIES IN MEXICO

Peromyscus truei lagunae Osgood, 1909
Peromyscus truei martirensis (J.A. Allen, 1893)
Based on evidence derived from cytogenetic studies, the presence of two cytotypes in *P. truei* was confirmed, separating this species in two taxa: *P. truei* for populations in northern New Mexico and Baja California, and *P. gratus* for populations in southern New Mexico and Mexico (Janecek, 1990; Mody and Lee, 1984).

DESCRIPTION: *Peromyscus truei* is a medium-sized mouse. The tail is approximately the same length as the head and body. The length of the ears is slightly larger than the feet. Its dorsal coloration is grayish-brown, bathed with yellow; the ventral hair is white with dark underfur (Woloszyn and Woloszyn, 1982). It has an ochraceous line on the sides, which contrasts sharply with the belly and back. The tail is hairy and bicolored, dorsally brown and ventrally whitish.

EXTERNAL MEASURES AND WEIGHT

TL = 205 to 221 mm; TV = 101 to 127 mm; HF = 23 to 24 mm; EAR = 21 to 24 mm. Weight: 15 to 35 g.

DENTAL FORMULA: I 1/1, C 0/0, PM 0/0, M 3/3 = 16.

NATURAL HISTORY AND ECOLOGY: *P. truei* is semi-arboreal and has nocturnal habits. It feeds on invertebrates and plants. Insects comprise up to 60% of their diet during summer, and the rest of the year they consume vegetative material such as seeds of pine (*Pinus* sp.), acorns from oak (*Quercus* sp.), and green parts of plants (Bradford, 1974). They make nests with plants and other materials in crevices, tree stumps, trunks, and holes of trees; they usually have more than one burrow (Douglas, 1969; Woloszyn and Woloszyn, 1982). Mating season extends from April to September. The number of offspring varies from three to six, with an average of four (Douglas, 1969). The offspring begin to open their eyes around the third week. Juveniles can be seen in summer and early fall, and reach adult size after 50 days of birth. At the end of spring and summer, the population density is increased, reaching its peak in October (62 individuals/ha), then declines at the end of the winter and early spring (28 to 29 individuals/ha) (Woloszyn and Woloszyn, 1982). Their metabolism is specialized to absorb water from the plants they eat, enabling them to inhabit places where water is scarce (Bradford, 1974; Douglas, 1969). Their predators include coyotes, wildcats, foxes, birds of prey such as eagles and owls, and some snakes such as *Pituophis melanoleucus* and *Masticophis taeniatus* (Douglas, 1969).

DISTRIBUTION: *P. truei* ranges from the southwestern United States to the Baja California Peninsula in Mexico. The subspecies *P. truei martinensis* is found in the southern United States and northern Mexico, in the state of Baja California Norte, mainly in the Sierra de San Pedro Martir. *P. truei lagunae* is distributed only in the Sierra de La Laguna, in the state of Baja California Sur. It has been recorded in the states of BC and BS.

VEGETATIONAL ASSOCIATIONS AND ELEVATION RANGE: In Mexico, *P. truei* is usually found in rocky habitats of pine forests (*Pinus* sp.) and sometimes in association with junipers (*Juniperus* sp.) and scrub oaks (*Quercus* sp.), from 900 masl to 2,250 masl (Woloszyn and Woloszyn, 1982).

CONSERVATION STATUS: The status of this species is undetermined, but because of its wide geographical distribution and its abundance it is considered not at risk.

Coalcomán deermouse

Clemente R. Melgar

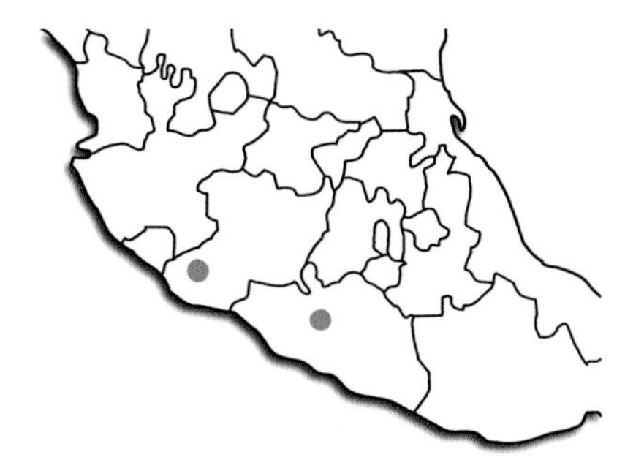

DISTRIBUTION: It is possible that the distribution of this species extends at least from the Cordillera Transvolcanica in southwestern Michoacán to the Sierra Madre del Sur in Guerrero (Smith et al., 1989; Sullivan and Kilpatrick, 1991). It has been recorded in the states of GR and MI.

SUBSPECIES IN MEXICO

P. winkelmanni is a monotypic species.

DESCRIPTION: *Peromyscus winkelmanni* is a medium-sized mouse. The dorsal coloration is yellowish-brown mixed with black, and on the midline the coat is significantly darker. The cheeks, sides, and legs are shiny and fulvous, with a cinnamon shade. The underside is dirty white, with an ochraceous spot in the pectoral region. The tail is bicolored, dorsally grayish and gradually changes to white (Álvarez et al., 1987; Carleton, 1977). The species belongs to the *aztecus* species group due to its size, the shape of the interorbital area, and the length of the glans (Bradley and Schmidly, 1987; Bradley et al., 2007; Carleton, 1977). The skull is similar, but slightly larger than that of *P. hylocetes* and *P. oaxacensis* (Carleton, 1977). The interorbital bump is not so pronounced as in *Osgoodomys banderanus* and *P. oaxacensis* (Carleton, 1977), and the stomach is discoglandular, which is unusual for the genus.

EXTERNAL MEASURES AND WEIGHT

TL = 230 to 265 mm; TV = 117 to 140 mm; HF = 25 to 29 mm; EAR = 20 to 23 mm. Weight: 44 to 55 g.

DENTAL FORMULA: I 1/1, C 0/0, PM 0/0, M 3/3 = 16.

NATURAL HISTORY AND ECOLOGY: Very little is known about the natural history of this species. It inhabits open and humid forests, pine forests, and oak and pine forests with dense undergrowth dominated by *Bacharis* sp.; in these sites lichens, orchids, and bromeliads are also abundant (Carleton, 1977). It has nocturnal habits and has been captured with *P. levipes* and *Reithrodontomys fulvescens* (Álvarez et al., 1987).

VEGETATIONAL ASSOCIATIONS AND ELEVATION RANGE: *P. winkelmanni* lives in temperate environments such as pine forests and oak and pine forests, from 2,040 masl to 2,438 masl (Álvarez et al., 1987; Bradley and Schmidly, 1987; Carleton, 1977).

CONSERVATION STATUS: This species is considered of Special Protection, due to its limited range and because the habitat it occupies shows signs of disturbance mainly by logging activities (Álvarez et al., 1987; Ceballos and Navarro, 1991; Ceballos and Rodriguez, 1993; SEMARNAT, 2010).

Peromyscus yucatanicus J.A. Allen and Chapman, 1897

Yucatan deermouse

Jorge Ortega R. and Héctor T. Arita

SUBSPECIES IN MEXICO

Peromyscus yucatanicus badius Osgood, 1904
Peromyscus yucatanicus yucatanicus Allen and Chapman, 1897

DESCRIPTION: *Peromyscus yucatanicus* is medium in size. The dorsal coloration depends on the subspecies. *P. yucatanicus yucatanicus* is ochraceous, whereas *P. yucatanicus badius* is dark brown; in both subspecies the belly is yellowish-white (Young and Jones, 1983). The forefeet are completely white from the wrist, and the hind feet from the base of the tarsus. The eyes are ringed by a narrow line, and the ears are large, brown, and without hair. The tail is long, gray on the top and yellow on the bottom (Young and Jones, 1983).

EXTERNAL MEASURES AND WEIGHT
TL = 208 to 232 mm; TV = 85 to 117 mm; HF = 20 to 24; EAR = 8 to 12 mm.
Weight: 15 to 18 g.

DENTAL FORMULA: I 1/1, C 0/0, PM 0/0, M 3/3 = 16.

NATURAL HISTORY AND ECOLOGY: Information on the biology of this species is scarce. Pregnant or lactating females have been captured throughout the year (Lawlor, 1965). This rodent is nocturnal, and its diet is based on fruits and seeds. They are caught with *Ototylomys phyllotis* and *P. leucopus*.

VEGETATIONAL ASSOCIATIONS AND ELEVATION RANGE: *P. yucatanicus* is distributed in tropical perennial forests, tropical subdeciduous forests, and deciduous forests. It has been collected in dense and wet shrubs, although it also has been found in disturbed areas such as crops of maize (Jones et al., 1974 a, b). It is found from sea level to 500 m.

CONSERVATION STATUS: There is little information about the status of this species. As it is endemic to Mexico, it is important to assess its current situation. It has been collected in farmlands, which suggests it is less vulnerable to extinction.

DISTRIBUTION: This rodent is endemic to the Yucatan Peninsula, with a single record from Guatemala (Zarza et al., 2002). It has been recorded in the states of CA, QR, and YU.

Peromyscus yucatanicus. Tropical semi-green forest. Sian Kaan Biosphere Reserve, Quintana Roo. Photo: Jesús Pacheco.

Chiapan deermouse

Esther Romo V. and Anna Horváth

SUBSPECIES IN MEXICO

P. zarhynchus is a monotypic species.

DESCRIPTION: *Peromyscus zarhynchus* is one of the larger species within the genus *Peromyscus*. The dorsal coloration is dark brown and grayish, darker in the middle. The sides are yellowish-brown; the belly is paler, whitish or buffy. The feet are white with brown proximal areas. The tail is long, thick, slightly hairy, and clearly bicolored, but can have spots on the base (Hall, 1981; Reid, 1997). The eyes are large, dark, and ringed. The muzzle is large, narrow, and elongated; the ears are big in relation to the size of the body. The whiskers around the nose and the eyelashes are dense and very long, the coat is very soft, which, in addition to the size, clearly distinguishes it from the other species of the genus (A. Horváth, pers. obs.). The lengths of head-body and tail are very similar. The long but not robust skull, with very elongated nasals and face, can distinguish it; the dental pattern is complex, with the anterocone undivided.

EXTERNAL MEASURES AND WEIGHT

TL = 303 to 327 mm; TV = 157 to 178 mm; HF = 33 to 38 mm; EAR = 21 to 24 mm. Weight: 45 to 90 g.

DENTAL FORMULA: I 1/1, C 0/0, PM 0/0, M 3/3 = 16.

NATURAL HISTORY AND ECOLOGY: These mice have terrestrial and crepuscular habits. They are common in humid ravines with trees, where the cracks of the rocks and roots provide sites for their burrows (A. Horváth, pers. obs.). They live in steep slopes, near the top of mountains covered with fog; they find refuge between rocks with mosses and other characteristic vegetation of the rainforest. Their feeding habits

DISTRIBUTION: *P. zarhynchus* is distributed in separate populations in the state of Chiapas, in the northern mountains in central at the center in Los Altos, and in southeastern Lagos de Montebello (Huckaby, 1980; Reid, 1997). To date, it has only been recorded in Mexico; however, it is possible that it inhabits some parts of the Cuchumatanes in Guatemala, an adjacent area to the region of Lagos de Montebello, Chiapas (Horváth and Navarrete Gutierrez, 1997). It has been recorded in the state of CS.

Peromyscus zarhynchus. Cloud forest. Montebello National Park, Chiapas. Photo: Anna Horváth.

are little known. It is likely to feed on fruits, seeds, mycorrhizae, and some arthropods (A. Horváth, pers. obs.). As for the population dynamics, in the area of Lagos de Montebello few changes in its abundances throughout the year were observed, with higher peaks of capture during the rainy season and the highest reproductive activity (Horváth, et al., 2001; Kings-Martinez, 2001; Sarmiento-Aguilar, 1999). This species has a relatively long lifespan compared with other species of rodents, with some individuals living in the field for at least four years (A. Horváth, pers. obs.).

VEGETATIONAL ASSOCIATIONS AND ELEVATION RANGE: *P. zarhynchus* is only found in cloud forests and forest of pine-oak-liquidambar, from 1,400 masl to 2,900 masl. It is strongly associated with primary forests with high structural complexity (Horváth et al., 2001).

CONSERVATION STATUS: This species is endemic to Mexico, but is not classified as endangered (Arita and Ceballos, 1997). Their populations seem to be abundant, and even dominant in local communities of rodents (Horváth et al., 2001; Kings-Martinez, 2001; Sarmiento-Aguilar, 1999). Because of their restricted geographical distribution, the high degree of isolation of the populations, and their dependence on cloud forests and pine-oak liquidambar forest in good condition, the future of *P. zarhynchus* depends on the preservation of these forests in Chiapas.

Reithrodontomys bakeri Bradley, Mendez-Harclerode, Hamilton and Ceballos, 2004

Baker's harvest mouse

Gerardo Ceballos and Robert D. Bradley

SUBSPECIES IN MEXICO

R. bakeri is a monotypic species.

This species, which was recently elevated to species level, was originally described as *R. microdon wagneri*. However, an assessment of cytochrome b and other morphological and genetic characteristics showed it was a new species, endemic to Mexico.

DESCRIPTION: *Reithrodontomys bakeri* is medium-sized, very similar in size to *R. microdon wagneri*. Its fur is cinnamon-brown with gray underfur. The sides are beige. The belly is whitish with gray underfur. The hind feet have a dark band from the ankle to the fingers. The tail is solid, dark gray, with little hair at the base and dense hair in the tip. The whiskers are black (Bradley et al., 2003). It is distinguished from *R. microdon wagneri* because the face is longer and wider, the jawbone is longer, the zygomatic arch is wider, and the mesopterigoid fossa is also wider. Genetically, it differs from *R. microdon* (Bradley et al., 2003).

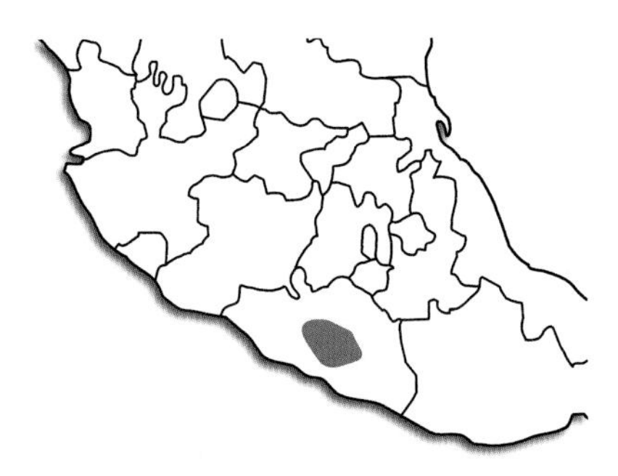

DISTRIBUTION: This species is endemic to Mexico. It is only known from Filo de Caballo and Omiltemi in the Sierra Madre del Sur. It has been recorded in the state of GR.

EXTERNAL MEASURES AND WEIGHT

TL = 185 mm; TV = 107 mm; HF = 19 mm; EAR = 18 mm.
Weight: 20 g.

DENTAL FORMULA: I 1/1, C 0/0, PM 0/0, M 3/3 = 16.

NATURAL HISTORY AND ECOLOGY: Very little is known about the biology of this species. It is found in temperate, cold, and wet forested regions. Despite the fact that they have been collected in traps on the ground, it is possible that they have semi-arboreal habits. Other species of the genus feed on seeds, fruits, and leaves. The Sierra Madre del Sur in Guerrero is a long chain of mountains relatively close to each other. It is one of the more isolated sierras in Mexico (Carleton et al., 2002). Because of this, it has a great number of endemic species of flora and fauna (Ceballos and Navarro, 1991; Luna Vega and Llorente, 1993). Some of the endemic species of the region include salamanders (*Pseudoerycea ahuitzotl, P. mixcoatl, Thorius grandis, T. omiltemi*), frogs (*Eleutherodactylus omiltemanus, E. saltator, Rana omiltemana*), lizards (*Anolis omiltemanus*), snakes (*Geophis omiltemanus*), 20 species and subspecies of birds (which it shares with Oaxaca), and several mammals such as the flying squirrel (*Glaucomys volans guerreroensis*) and the rabbit of Omiltemi (*Sylvilagus insonus*) (Adler, 1996; Ceballos and Navarro, 1991; Hanken et al. 1999; Luna Vega and Llorente, 1993).

VEGETATIONAL ASSOCIATIONS AND ELEVATION RANGE: This species inhabits the mountainous regions of central Guerrero, in forests of pine-oak associated with cloud forests.

CONSERVATION STATUS: Because of its restricted distribution and the destruction of the forests it inhabits *R. bakeri* should be regarded as fragile and vulnerable. There is not enough information to resolve the status of this species, however.

Reithrodontomys burti Benson, 1939

Sonoran harvest mouse

Oscar Sánchez

DISTRIBUTION: This species is endemic to Mexico; its distribution is restricted and ranges from west-central Sonora to central Sinaloa (Wilson and Reeder, 2005. It has been reported in two localities in each of these states. In Sonora, it is found in the type locality, Rancho San Francisco de Costa Rica, located in the coastal plain of Sonora, near the Sonora River, to 18 km west of Hermosillo, Sonora (Benson, 1939; Hooper, 1952a, d). In Sinaloa, it has been collected 6 km from Terrero, 27 km south-southeast of Guamúchil, and 1.6 km south of Pericos (Russell and Alcorn, 1957). It has been recorded in the states of SI and SO.

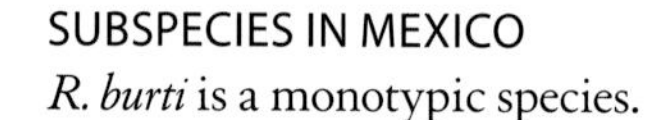

SUBSPECIES IN MEXICO

R. burti is a monotypic species.

DESCRIPTION: *Reithrodontomys burti* is a small-sized mouse. The tail is short and reaches 70% to 97% of the length of the head and body (Hall, 1981). The hind feet are relatively short, and the ears are long. The coat is pale-colored with yellowish shades in the cheeks and a light-colored area around the ear. In the anterior base of the ear it has a pale tuft. The tail is practically devoid of pigment (Benson, 1939). The top is ochraceous-creamy; the hair has black tips and is more intense on the sides and rear. The lower parts are whitish with a gray shade at the base of the hair. The feet are white and on the back they may have a fine dark line. The skull is similar to that of *R. montanus*, but differs by the different width of the interorbital area; the zygomatic process of the maxilla is wider, the interorbital foramina is longer, and the nasals are relatively larger (Benson, 1939; Hall, 1981). A distinctive character of the species is the baculum, which is curved, wide, and flat, and distinguishes it from other species with which it is geographically linked such as *R. montanus, R. megalotis, R. humulis*, and *R. fulvescens* (Benson, 1939).

EXTERNAL MEASURES AND WEIGHT

TL = 129 mm; TV = 59 mm; HF = 16 mm; EAR = 15 mm.
Weight: 101 g.

DENTAL FORMULA: I 1/1, C 0/0, PM 0/0, M 3/3 = 16.

NATURAL HISTORY AND ECOLOGY: Very little is known about the natural history of this species. Its reproductive habits, diet, and overall biology are unknown. Hooper (1952a, d) considers *R. burti* a relict species related to *R. montanus*, although it may be geographically and environmentally closer to *R. megalotis* and *R. humulis* (Benson, 1939). Although it has been collected associated with *R. fulvescens* there might not be hybridization between these species, even when they appear to be closely related. The only reported predators are owls (*Tyto alba*), but it is possible that it is preyed on by several carnivores (Bradshaw and Hayward, 1960).

VEGETATIONAL ASSOCIATIONS AND ELEVATION RANGE: In Sonora this species is found in scrublands of mesquite, palo fierro, palo verde and grasses, and cultivated fields with abundant stubble of wheat (Benson, 1939). In Sinaloa, it has been collected in thorn scrub forests with trees of up to 5 m high, in wildland with shrubs, and in crops of agave with mesquite and grass (Russell and Alcorn, 1957). It can be found from 60 masl to 182 masl (Csuti, 1980; Miller and Kellogg, 1955).

CONSERVATION STATUS: Although this species is not included in the official list of endangered species (SEMARNAT, 2010), it can be considered fragile since it is endemic and has a restricted distribution.

Reithrodontomys chrysopsis Merriam, 1900

Volcano harvest mouse

Irma E. Lira and Salvador Gaona

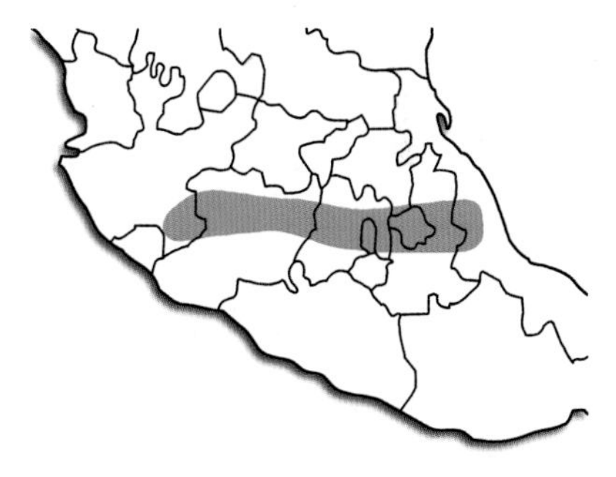

DISTRIBUTION: This species is endemic to Mexico and is found in the high mountains of central Mexico from Veracruz to Jalisco. Populations exist in the east-central parts of the country from Veracruz to the mountains surrounding the west of the Valley of Mexico, which are separate from the Nevado of Colima and southeastern Jalisco. It has been recorded in the states of DF, JA, MI, MO, MX, PU, and VE.

SUBSPECIES IN MEXICO

Reithrodontomys chrysopsis chrysopsis Merriam, 1901

Reithrodontomys chrysopsis perotensis Merriam, 1901

DESCRIPTION: *Reithrodontomys chrysopsis* is a medium-sized mouse. The dorsal coloration is black interspersed with yellowish-orange to reddish-orange; the black color is dominant in the back, head, and snout. Guard hairs are black underneath and silver in the tips. The ventral region is cinnamon-pink. The ears are big and black. The tail is bicolored, black on the top and light below.

EXTERNAL MEASURES AND WEIGHT

TL = 165 to 201 mm; TV = 90 to 116 mm; HF = 17 to 21 mm; EAR = 17 to 21 mm. Weight: 14 g.

DENTAL FORMULA: I 1/1, C 0/0, PM 0/0, M 3/3 = 16.

NATURAL HISTORY AND ECOLOGY: These mice live in high, wooded, cold, and wet regions, showing a very restricted tolerance. It is arboreal and nocturnal. It molts annually and feeds on seeds, grains, and green parts of grasses and other plants. They build spherical nests of grass (Ceballos and Galindo, 1984).

VEGETATIONAL ASSOCIATIONS AND ELEVATION RANGE: This mouse can be found in hillsides and canyons with forests of pine-fir, pine-oak, zacatonals, and areas covered by ferns, mosses, and shrubs, from 1,830 masl in Monte Tacítaro in Michoacán to about 4,100 masl in Popocatépetl, Mexico.

CONSERVATION STATUS: This mouse is endemic to Mexico and has a restricted distribution. Their populations are considered naturally small. Humans sparsely populate the areas they inhabit, but they harvest timber, and as the distribution of the species is quite restricted and is subjected to strong pressures, its populations are declining rapidly. It is not included in any list of endangered species.

Reithrodontomys fulvescens J.A. Allen, 1894

Fulvous harvest mouse

Oscar Sánchez and Gisselle A. Oliva

SUBSPECIES IN MEXICO

Reithrodontomys fulvescens amoenus (Elliot, 1905)
Reithrodontomys fulvescens canus Benson, 1939
Reithrodontomys fulvescens chiapensis H. Howell, 1914
Reithrodontomys fulvescens difficilis Merriam, 1901
Reithrodontomys fulvescens fulvescens J.A. Allen, 1894
Reithrodontomys fulvescens griseoflavus Merriam, 1901
Reithrodontomys fulvescens helvolus Merriam, 1901
Reithrodontomys fulvescens infernatis Hooper, 1950
Reithrodontomys fulvescens intermedius J.A. Allen, 1895
Reithrodontomys fulvescens mustelinus H. Howell, 1914
Reithrodontomys fulvescens nelsoni H. Howell, 1914
Reithrodontomys fulvescens tenuis J.A. Allen, 1899
Reithrodontomys fulvescens toltecus Merriam, 1901
Reithrodontomys fulvescens tropicalis Davis, 1944

DISTRIBUTION: This species has a wide distribution, ranging from Missouri, Mississippi, and Texas in the United States to Costa Rica. In Mexico, it is distributed throughout the country, with the exception of Sonora and Baja California Peninsula in the north, and the coastal plain of the Gulf from Veracruz and Yucatan Peninsula in the southeast (Hall, 1981; Sánchez, 1993). It has been recorded in the states of AG, CH, CL, CO, CS, DF, DU, GJ, GR, HG, JA, MI, MO, MX, NL, NY, OX, PU, QE, SI, SL, SO, TA, TL, VE, and ZA.

DESCRIPTION: *Reithrodontomys fulvescens* is a small-sized mouse, with remarkable genetic and geographic diversification. As in all *Reithrodontomys*, the upper incisors have a prominent longitudinal furrow. Its fur is rather rough, reddish-brown and blackish, since the guard hair is black and the underfur is lighter and fitted with recognizable bands (Hooper, 1952a, d). The tail is long and can exceed between 10% and 50% the length of the head and body, according to the subspecies (Hall, 1981). The tail tends to be paler on the bottom, often with a clear demarcation with the dorsal coloration. The face is robust, especially in the tip of the nasal passages, and the zygomatic bone is wider than the mesopterigoid fossa.

EXTERNAL MEASURES AND WEIGHT

TL = 134 to 200 mm; TV = 72 to 116 mm; HF = 16 to 22 mm; EAR = 11 to 17 mm.
Weight: 8 to 12 g.

DENTAL FORMULA: I 1/1, C 0/0, PM 0/0, M 3/3 = 16.

Reithrodontomys fulvescens. Tropical dry forest. Chamela-Cuixmala Biosphere Reserve, Jalisco. Photo: Gerardo Ceballos.

NATURAL HISTORY AND ECOLOGY: These mice occupy ecotones with grassland, or rocky outcrops with patches of scrub or with other traits that provide protection (Hooper, 1952a, d; Petersen, 1978). It seems that these mice are strictly nocturnal (Cameron et al., 1979b). They are semi-arboreal. They are omnivores, and during the rainy season their diet can include almost 90% invertebrates, whereas at other times of the year seeds comprise the most significant part of their diet (Kincaid, 1975); the diet may vary latitudinally. In central Mexico the populations have an average density of 0.6 individual/ha (Chavez, 1993). The size of the litter varies from two to four offspring, which can occur in at least two different times of the year, without being able to establish the reproductive peaks (Spencer and Cameron, 1982). These rodents are born in spherical nests made of grass and placed several mm in the soil (Svihla, 1930) between shrubby vegetation. They reproduce throughout the year. In Guerrero pregnant females have been found at the end of May (Ramírez-Pulido et al., 1977). After a gestation period of about 22 days, 2 to 8 offspring are born, with an average of 3 (Ceballos and Miranda, 2000). The offspring are born altricial and are weaned at 15 days old (Packard, 1968).

VEGETATIONAL ASSOCIATIONS AND ELEVATION RANGE: This species expresses a clear preference for forests of pine-oak and is less frequently found in tropical forests, deciduous forests, thorn scrublands, and xeric shrublands (Ceballos and Miranda, 2000; Sánchez, 1993). It has been found from sea level to 2,600 m; in this interval, it appears to be frequently associated with elevations between 500 masl and 1,800 masl (Sánchez, 1993).

CONSERVATION STATUS: This species has a very wide distribution, and poses no conservation concerns. The current situation of the micro-endemic subspecies such as *amoenus*, *infernatis*, *difficilis*, and *nelsoni*, in that order of priority, should be evaluated.

Slender harvest mouse

Guadalupe Téllez-Girón and Oscar Sánchez

SUBSPECIES IN MEXICO
Reithrodontomys gracilis gracilis J.A. Allen and Chapman, 1897
Reithrodontomys gracilis insularis Jones, 1964
Reithrodontomys gracilis pacificus Goodwin, 1932
This species belongs to the subgenus *Aporodon* and the group *mexicanus*. Its subspecies are allopatric and their populations are geographically isolated (Sánchez, 1993).

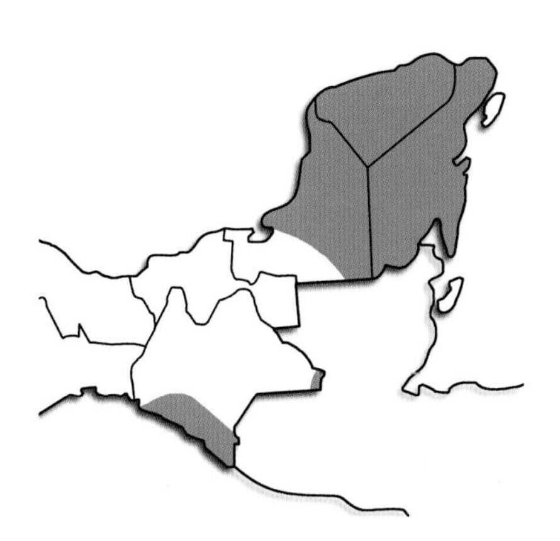

DISTRIBUTION: This species is distributed from southern Mexico to northeastern Costa Rica. In Mexico its distribution extends from the Yucatan Peninsula to southern Chiapas. The subspecies *R. gracilis insularis* is found only in the del Carmen Islands on the coast of Campeche (Hooper, 1952a; Jones, 1964; Young and Jones, 1984). It has been recorded in the states of CA, CS, QR, and YU.

DESCRIPTION: *Reithrodontomys gracilis* is a small mouse with a thin tail, bicolored and longer than the head and body (97% to 155% larger). It has a light line on the sides of the body. The ears are darker than the body. One of the distinguishing features is the presence of scarce hairs in the ears and the absence of the dark ring around the eye. The sole and back of the feet are generally white, but can be brownish. This coloration varies seasonally and geographically (Hooper, 1952a). The dorsal coloration varies from cinnamon-pink to shinny ochraceous-brown. The belly varies from white to bright orange-cinnamon (Hall, 1981; Hooper, 1952a; Young and Jones, 1984).

EXTERNAL MEASURES AND WEIGHT
TL = 163 to 185 mm; TV = 87 to 107 mm; HF = 17 to 20 mm; EAR =13 to 16 mm.
Weight: 10 to 11 g.

DENTAL FORMULA: I 1/1, C 0/0, PM 0/0, M 3/3 = 16.

NATURAL HISTORY AND ECOLOGY: This mouse is found in a variety of environments, including tropical deciduous forests, perennial forests, and secondary vegetation, and it has even been found on the branches of trees and shrubs. The data on its reproduction are scarce; however, pregnant females have been found in Yucatan and in del Carmen Islands in July and August, and males with scrotal testicles in April, June, and July. In Central America females with embryos have been found in March, June, and July. The breeding season seems to extend from the beginning of the winter to mid-autumn (Young and Jones, 1984). In the Yucatan Peninsula this mouse is associated with other species of rodents such as *Heteromys gaumeri*, *Oryzomys couesi*, *Ototylomys phyllotis*, *Peromyscus leucopus*, *P. yucatanicus*, and *Sigmodon hispidus* (Jones et al., 1974b). At times it has also been found with *R. fulvescens* in xeric scrublands (Hooper, 1952a).

VEGETATIONAL ASSOCIATIONS AND ELEVATION RANGE: *R. gracilis* is mainly found in tropical deciduous forests, subdeciduous forests, tropical forests, perennial forests, and, in lesser degree, in thorn forests and xeric scrublands. It also has been collected in the edges of areas with pine forests and in plantations of coconut and coffee (Goodwin, 1932; Hooper, 1952a; Sánchez, 1993). The elevation for this species is from sea level to 1,370 masl (Anderson and Jones, 1960; Burt and Striton, 1961; Goodwin, 1946; Hooper, 1952a; Jones and Genoways, 1970; Sánchez, 1993).

CONSERVATION STATUS: This mouse is common; however, the subspecies *R. gracilis insularis* is considered threatened because of its restricted distribution (SEMARNAT, 2010).

Hairy harvest mouse

Salvador Gaona and Irma E. Lira

SUBSPECIES IN MEXICO
R. hirsutus is a monotypic species.

DESCRIPTION: *Reithrodontomys hirsutus* is a moderately large mouse for the genus. The dorsal coloration is cinnamon with few black blotches distributed evenly on the back and sides. It has a lateral line. The ventral region varies from pinkish-cinnamon to light pinkish-cinnamon. The throat is pale pinkish-yellow. The forefeet are pinkish-yellow or whitish. The hind feet are dark gray with pinkish-yellow, and the fingers are pinkish-yellow. The tail possesses scales, and is monochromatic, dorsally dark gray, and slightly less dark beneath.

EXTERNAL MEASURES AND WEIGHT
TL = 175 to 202 mm; TV = 100 to 115 mm; HF = 20-22; EAR =16 to 17 mm.
Weight: 20 g.

DENTAL FORMULA: I 1/1, C 0/0, PM 0/0, M 3/3 = 16.

NATURAL HISTORY AND ECOLOGY: The natural history and ecology of this species are not well known. It has been found in rocky areas with little or no vegetation cover. *R. fulvescens* is the only species of the genus that occurs in the same habitat as *R. hirsutus*. They are nocturnal and probably semi-arboreal. Their diet consists of insects and seeds.

VEGETATIONAL ASSOCIATIONS AND ELEVATION RANGE: *R. hirsutus* inhabits rocky hillsides in xeric scrublands with little coverage and ravines with tropical deciduous forests. It has been found from 915 masl near San Jose del Conde to 1,680 masl in Ameca.

CONSERVATION STATUS: This species is endemic to Mexico, and has a very restricted distribution; hence it should be regarded as vulnerable.

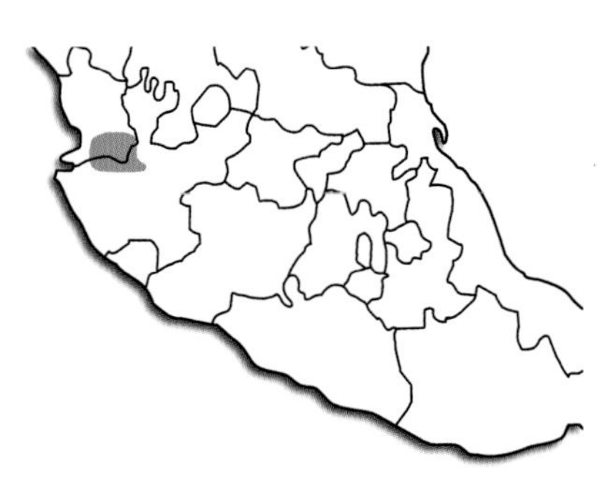

DISTRIBUTION: This mouse is endemic to Mexico and occurs in a small segment of the west-central part of the country, in the mountains of western Jalisco and southern Nayarit. It has been recorded in the states of JA and NY.

Reithrodontomys megalotis (Baird, 1858)

Western harvest mouse

Oscar Sánchez and Gisselle A. Oliva

SUBSPECIES IN MEXICO
Reithrodontomys megalotis alticolus Merriam, 1901
Reithrodontomys megalotis amoles H. Howell, 1914
Reithrodontomys megalotis hooperi Goodwin, 1954
Reithrodontomys megalotis longicaudus (Baird, 1858)
Reithrodontomys megalotis megalotis (Baird, 1858)

Reithrodontomys megalotis. Grassland. Janos Biosphere Reserve, Chihuahua. Photo: Gerardo Ceballos.

Reithrodontomys megalotis peninsulae (Elliot, 1903)
Reithrodontomys megalotis saturatus J.A. Allen and Chapman, 1897

DESCRIPTION: *Reithrodontomys megalotis* is a small-sized mouse. Dorsally, the coat tends to be light brown; the guard hair, relatively dense, is dark grayish or black. The positive identification of *R. megalotis* requires detailed comparison with specimens of close species, as several characters often overlap, but in general, the tail is slightly longer than the body and the width of the skull is less than 10.7 mm.

EXTERNAL MEASURES AND WEIGHT
TL = 135 to 154 mm; TV = 64; HF = 15 to 18 mm; EAR = 13 to 15 mm.
Weight: 8 to 13 g.

DENTAL FORMULA: I 1/1, C 0/0, PM 0/0, M 3/3 = 16.

NATURAL HISTORY AND ECOLOGY: These mice are nocturnal and mainly terrestrial. Their activity areas are large, around 3,500 m^2 (Meserve, 1977). The densities of their populations fluctuate from 4 to 11 individuals/ha, but in exceptionally good years they can increase to 60 individuals/ha after rains (Gray, 1943; Whitford, 1976). In central Mexico densities of between 0.7 individual/ha and 15.8 individuals/ha have been reported in open sites and fields (Fa et al., 1990). Males tend to be polygamous and maintain certain dominance (Fisler, 1965). The offspring are born in spherical nests, approximately 12 cm in diameter, constructed with grass and other vegetative material. The litters tend to be numerous, but with a wide range of variation, from one to nine offspring (Ceballos and Galindo, 1984; Long, 1962). The length of the gestation extends for about 25 days; the offspring are born altricial, and after about 25 days of lactation they are weaned (Webster and Jones, 1982). There is evidence that the substitution of individuals in a population can be almost total in the span of a year (Fisler, 1971). They are mainly granivores, but can consume leaves and stems, as well as some invertebrates (Meserve, 1976b).

DISTRIBUTION: These mice have one of the wider distributions in North America; they can be found from southern Canada to central Mexico. In Mexico, it is absent in the southern part of the Baja California Peninsula, coastal plains of the Pacific and Gulf of Mexico, and the Yucatan Peninsula. De la Riva Hernandez (1989) cited the presence of *R. megalotis* in Aguascalientes, but the elevation and coordinates of their records do not seem to correspond to the geographical area occupied by *R. megalotis*. It has been recorded in the states of AG, BC, CH, CO, DF, DU, GJ, HG, JA, MI, MO, MX, NL, OX, PU, QE, SL, SO, TA, TL, VE, and ZA.

VEGETATIONAL ASSOCIATIONS AND ELEVATION RANGE: *R. megalotis* is basically associated with open forests of pine-oak and with natural grasslands (Sánchez, 1993). Even in forests of pine-oak, they select gaps covered by grass and shrubs (Hooper, 1952a, b; Webster and Jones, 1982). This is probably the most widespread species in the genus *Reithrodontomys* relative to different elevations (Sánchez, 1993), as it has been found from sea level to almost 4,000 masl (Hall, 1981; Hooper, 1952a, d).

CONSERVATION STATUS: This species has a wide distribution; it is not in danger of extinction. Because it is endemic to a very restricted area, it is necessary to assess the conservation status of the three Mexican subspecies: *amoles*, *hooperi*, and *peninsulae*.

Reithrodontomys mexicanus (Saussure, 1860)

Mexican harvest mouse

Oscar Sánchez and Gisselle A. Oliva

SUBSPECIES IN MEXICO

Reithrodontomys mexicanus howelli Goodwin, 1932
Reithrodontomys mexicanus mexicanus (Saussure, 1860)
Reithrodontomys mexicanus riparius Hooper, 1955
Reithrodontomys mexicanus scansor Hooper, 1950

DESCRIPTION: *Reithrodontomys mexicanus* is a relatively large mouse, very similar to other species of the genus. The upper incisors have a prominent longitudinal furrow (Goodwin, 1969; Hooper, 1952a, d). The dorsal coloration varies from yellow-brown to orange-brown, but with great variation between the subspecies. The belly varies from creamy white to light cinnamon or cinnamon orange. The tail is longer than the body, monochromatic, and uniformly dark (Hall, 1981).

Reithrodontomys mexicanus. Cloud forest. Naolinco, Puebla. Photo: Gerardo Ceballos.

DISTRIBUTION: *R. mexicanus* is distributed from Mexico to Panama. In Mexico is it found from southern Tamaulipas and Michoacán to Oaxaca and Chiapas (Hall, 1981). It has been recorded in the states of CS, GR, HG, MI, OX, PU, SL, TA, and VE.

EXTERNAL MEASURES AND WEIGHT
TL = 150 to 195 mm; TV = 90 to 115 mm; HF = 17 to 21 mm; EAR = 12 to 18 mm. Weight: 19 g.

DENTAL FORMULA: I 1/1, C 0/0, PM 0/0, M 3/3 = 16.

NATURAL HISTORY AND ECOLOGY: The local distribution of *R. mexicanus* may be limited by the availability of shrubby cover, as well as grasses and other elements of the herbaceous stratum (Hooper, 1952a, d). They are nocturnal and arboreal. In Villa Flores, Chiapas, they have been caught on flowering trees of colorin (*Erythrina* sp.). In arid regions, they are found throughout the rivers and other water bodies where figs and other trees can be found (Hooper, 1955). The size of the litter varies from three to five offspring, and the mating season takes place from June to August (Jones and Genoways, 1970).

VEGETATIONAL ASSOCIATIONS AND ELEVATION RANGE: *R. mexicanus* inhabits, in descending order of importance, forests of pine-oak, cloud forests, and tropical deciduous forests (Sánchez, 1993). In the deciduous forest, the more stable conditions of the riparian environments make its existence possible (Hooper, 1955). In Mexico, it has been recorded between 90 masl in Gomez Farías, Tamaulipas, and 1,800 masl in Totontepec, Oaxaca, and Altotonga, Veracruz (Goodwin, 1969; Hall, 1981; Hooper, 1952a, d; Sánchez, 1993).

CONSERVATION STATUS: This species is not specially protected, nor has evidence of any imminent threat in Mexico been documented.

Reithrodontomys microdon Merriam, 1901

Small-toothed harvest mouse

Carolina Müdespacher and Salvador Gaona

DISTRIBUTION: The distribution of *R. microdon* is found in disjunct populations from central Mexico to the north of Guatemala. It is found in the high mountains and volcanoes of the Valley of Mexico and Michoacán, Oaxaca, and Chiapas. It has been recorded in the states of CS, DF, MI, and OX.

SUBSPECIES IN MEXICO
Reithrodontomys microdon albilabris Merriam, 1901
Reithrodontomys microdon microdon Merriam, 1901
Reithrodontomys microdon wagneri Hooper, 1950

DESCRIPTION: *Reithrodontomys microdon* is a medium-sized mouse. The coat is thick cinnamon-orange on the dorsum. The ventral region varies from whitish to pinkish-cinnamon. It possesses black eye rings. The hind feet are blackish dotted with white, with white phalanges. The tail is solid black (Hooper, 1950, 1952a).

EXTERNAL MEASURES AND WEIGHT
TL = 169 to 187 mm; TV = 101 to 117 mm; HF = 19 to 21 mm; EAR = 16 to 19 mm. Weight: 20 g.

DENTAL FORMULA: I 1/1, C 0/0, PM 0/0, M 3/3 = 16.

NATURAL HISTORY AND ECOLOGY: *R. microdon* occurs in wooded regions, at high elevations and in cold and wet habitats. It has little tolerance to alterations of its habitat. It is semi-arboreal and is the most specialized species of the genus. It molts annually. It feeds on seeds, grains and green grasses, and other plants (Hooper, 1950, 1952a).

VEGETATIONAL ASSOCIATIONS AND ELEVATION RANGE: These mice dwell on hillsides and canyons with forests of pine-fir, forests of pine-oak, zacatonals, and areas covered by ferns, mosses, and shrubs. It is found from 2,225 masl in the surrounding area of San Cristobal de las Casas in Chiapas to near 3,050 masl in Canyon Contreras in Mexico City and Cerro San Felipe in Oaxaca (Sánchez, 1993).

CONSERVATION STATUS: Forests that comprise the habitat of this species have been drastically reduced as a result of human activity; hence, it is believed that the number of individuals is not very high. Their populations are increasingly isolated. In addition, the reduction of its population as well as its arboreal habits makes *R. microdon* a vulnerable species. It is threatened in the list of species at risk (SEMARNAT, 2010).

Reithrodontomys montanus (Baird, 1855)

Plains harvest mouse

Guadalupe Téllez Girón and Oscar Sánchez

SUBSPECIES IN MEXICO
Reithrodontomys montanus montanus (Baird, 1855)

DESCRIPTION: *Reithrodontomys montanus* is a small rodent whose upper incisors show a very visible longitudinal furrow. The dorsal coloration tends to be light grayish-brown, with a prominent dark middle-dorsal area. In sites where it is sympatric with *R. megalotis*, it can be recognized by the proportions of the tail, which is equal to or shorter than the body; the tail has a longitudinal, narrow, bicolored, and dark stripe, which is clearly distinct from the sides (Anderson, 1972; Wilkins, 1986).

DISTRIBUTION: This mouse is found from the south of the United States to northwestern Mexico, from Sonora to Durango. It has been recorded in the states of CH, DU, and SO.

EXTERNAL MEASURES AND WEIGHT
TL = 109 to 130 mm; TV = 54 to 65 mm; HF = 15 to 17 mm; EAR = 13 to 17 mm.
Weight: 7 to 9 g.

DENTAL FORMULA: I 1/1, C 0/0, PM 0/0, M 3/3 = 16.

NATURAL HISTORY AND ECOLOGY: These mice are nocturnal and terrestrial. Their densities tend to be higher in wild meadows, with abundant blue gramma (*Bouteloua gracilis*) and other species of perennial grasses (Maxwell and Brown, 1968; Moulton et al., 1981). Their activity areas may comprise up to approximately 2,300 m^2, and their densities of around 1.5 individuals per ha are considerably low (Wilkins, 1986). Their nests are oval, 110 mm − 80 mm, built with grass, lined with soft leaves, and generally located between shrubs, under fallen logs, and between rocks (Brown, 1946; Davis, 1974). On average they give birth to four offspring per litter, with a range of one to nine. The young are born altricial after 21 days of gestation, and are weaned within 2 weeks (Wilkins, 1986). They reach the adult size in five weeks and at two months they are sexually mature (Davis, 1974; Schmidly, 1983). They can be considered omnivores, consuming grasshoppers, grass, seeds, and fruits of cacti, among other things (Brown, 1946). Under natural conditions they live up to one year (Waggoner, 1975).

VEGETATIONAL ASSOCIATIONS AND ELEVATION RANGE: *R. montanus* is associated with grasslands, especially in the edges with pine and oak forests. It has been found in riparian vegetation and abandoned crop fields (Hooper, 1952a, d; Wilkins, 1986). In Mexico, it is found from 1,200 masl in Agua Prieta, Sonora, up to 1,920 masl in Cannula, Durango (Hall, 1981; Hooper, 1952a, d). Most of the records are around the lower limit of its altitudinal interval (Sánchez, 1993).

CONSERVATION STATUS: The current status of this species in Mexico has not been properly assessed. Even so, taking into account that about half of its distribution area is found in Mexico, and that this rodent depends on native grassland, intense overgrazing, which is common in the north of Mexico, is the most important threat to their survival.

Reithrodontomys spectabilis Jones and Lawlor, 1965

Cozumel harvest mouse

Guadalupe Téllez-Girón and Oscar Sánchez

SUBSPECIES IN MEXICO

R. spectabilis is a monotypic species.

This species belongs to the subgenus *Aporodon*, located within the group *mexicanus* (Jones, 1982).

DISTRIBUTION: This species is endemic to Mexico. It is only known in Cozumel Island, about 16 km east of the Yucatan Peninsula. This island is about 45 km long and almost 14 km wide. It has been recorded in the state of QR.

DESCRIPTION: *Reithrodontomys spectabilis* is a medium-sized mouse. The tail is bicolored, dark on the top and pale below, with hair quite visible between the scales. The dorsal coloration is dark ochraceous; the belly varies from white to gray, with some hairs with white tips. The hind feet have a dark stripe in the tarsus (Hopper, 1952a; Jones and Lawlor, 1965). It is similar to *R. gracilis* and *R. mexicanus*; however, it is much bigger, is darker, has a more robust skull, and the length of the tail exceeds by almost 34% the length of the body.

EXTERNAL MEASURES AND WEIGHT

TL = 205 to 221 mm; TV = 121 to 132 mm; HF = 20 to 22 mm; EAR = 16 to 19 mm. Weight: 18 to 21 g.

DENTAL FORMULA: I 1/1, C 0/0, PM 0/0, M 3/3 = 16.

NATURAL HISTORY AND ECOLOGY: Data on this species are scarce. It has been caught on the branches and at the top of trees, which possibly indicates arboreal habits. Juveniles, subadults, and lactating females have been caught in August; at the same time adult scrotal males with testicles of 9.1 mm, 13 mm, and 14 mm in length were collected. It has been found with *Oryzomys couesi* and *Peromyscus leucopus* in secondary vegetation and mature deciduous forests (Engstrom et al., 1989; Jones and Lawlor, 1965; Jones et al., 1974a, b).

VEGETATIONAL ASSOCIATIONS AND ELEVATION RANGE: *R. spectabilis* is mainly found in tropical deciduous forests, in bushes and secondary acahuales. The maximum height of Cozumel Island is 10 masl (Engstrom et al., 1989; Jones and Lawlor, 1965).

CONSERVATION STATUS: Because *R. spectabilis* is an endemic species with a restricted distribution it has been regarded as threatened (SEMARNAT, 2010). Its current status is critical. In an assessment conducted in 1997, no specimens were captured (G. Ceballos and J. Pacheco, pers. obs.). Recently, only 7 specimens were caught in more than 30,000 trap-nights, which indicate its rarity (A. Cuarón, pers. obs.). There are no known factors affecting the species, but it is assumed that introduced rats and boas may be important risk factors.

Reithrodontomys sumichrasti (Saussure, 1861)

Sumichrast's harvest mouse

José Ramírez-Pulido, Ramón Quijano Pérez,
Ulises Aguilera, and Alondra Castro-Campillo

SUBSPECIES IN MEXICO

Reithrodontomys sumichrasti dorsalis Merriam, 1901
Reithrodontomys sumichrasti luteolus Howell, 1914
Reithrodontomys sumichrasti nerterus Merriam, 1901
Reithrodontomys sumichrasti sumichrasti (Saussure, 1861)

DISTRIBUTION: This species is found from the central part of Mexico to Panama. In Mexico it is found in the Eje Volcanico Transverso, in a small portion in the Sierra Madre Oriental between Queretaro and Puebla, and in the Sierra Madre del Sur between Guerrero and Chiapas (Davis and Russell, 1953, 1954; Goodwin, 1969; Hall, 1981; Hall and Dalquest, 1963; Hooper, 1952a, d; Mass et al., 1981; Ponce-Ulloa and Llorente, 1993; Ramírez-Pulido, 1969a; Smith and Jones, 1967; Weber and Roguin, 1983). It has been recorded in the states of CL, CS, DF, GR, HG, JA, MI, MO, MX, OX, PU, QE, and VE.

DESCRIPTION: *Reithrodontomys sumichrasti* is characterized by the groove on the incisors, by its long ears, by its prominent eyes, and, above all, by its long tail and smaller size compared to other genera of field mice. The dorsal coloration varies with the subspecies. The color of the back varies from dark cinnamon to gray and orange. The cheeks and sides are darker. The belly is generally lighter and varies from cinnamon to brown or buffy pink to cinnamon-pink. The tail is brown or dark brown in the back and lighter underneath. In the extremities, the tarsal and carpal are dark, but the back of the hind feet varies from white, cinnamon, and pink to cinnamon-pink. A dark line in the dorsal region characterizes the subspecies *R. sumichrasti dorsalis* (Goodwin, 1969; Hall, 1981; Hall and Dalquest, 1963; Hooper, 1952a; Smith and Jones, 1967).

EXTERNAL MEASURES AND WEIGHT

TL = 75 to 88 mm; TV = 81 to 113 mm; HF = 18 to 21 mm; EAR = 12 to 18 mm.
Weight: 12 to 16 g.

DENTAL FORMULA: I 1/1, C 0/0, PM 0/0, M 3/3 = 16.

NATURAL HISTORY AND ECOLOGY: This species lives in a wide variety of habitats; it can be found in forests of subtropical or tempered climate, and dry, rocky, sandy, or very wet substrate. It can be sympatric with *R. chrysopsis*, *R. megalotis*, *R. fulvescens*, *R. mexicanus*, and *R. microdon*. Other species with which it has also been caught are *Oryzomys alfaroi*, *O. couesi*, *Microtus quasiater*, *Peromyscus difficilis*, *P. beatae*, *P. boylii*, *P. melanotis*, and *P. zarhynchus*. In its vertical distribution it is above *R. megalotis* and below *R. chrysopsis* (Baker and Phillips, 1965; Hall and Dalquest, 1963; Hooper, 1952a; Smith and Jones, 1967). A pregnant female was found in April (Heaney and Birney, 1977).

VEGETATIONAL ASSOCIATIONS AND ELEVATION RANGE: *R. sumichrasti* lives in temperate regions, grasslands, shrubby vegetation, forests of pine and oak, cloud forests, subtropical forests with abundance of ferns and mosses, maize fields, and sugar cane and mango plantations. The species is more abundant in cold and wet forests, as well as in the vicinity of streams. It is found between 800 masl and 3,200 masl (Baker and Phillips, 1965; Baker and Womochel, 1966; Burt, 1961; Goodwin, 1969; Heaney and Birney, 1977; Hooper, 1952a, d, 1961; Ponce-Ulloa and Llorente, 1993; Smith and Jones, 1967).

CONSERVATION STATUS: Despite the fact that considerable portions of its distribution area have been impacted, *R. sumichrasti* is not considered at risk of extinction due to its tolerance to anthropogenic disturbances.

Reithrodontomys tenuirostris Merriam, 1901

Narrow-nosed harvest mouse

Oscar Sánchez

SUBSPECIES IN MEXICO
R. tenuirostris is a monotypic species.

DESCRIPTION: *Reithrodontomys tenuirostris* is part of the subgenus *Aporodon*, which includes larger mice than those species of the nominal subgenus *Reithrodontomys*. In fact, in the scope of its geographical distribution, *R. tenuirostris* has the largest size. The dorsal coat tends to be long (10 mm) and dense, brown, with rich orange hues; it is more reddish toward the rear of the body. The overall general appearance is dark due to the presence of a large number of black guard hairs. The belly is pinkish-cinnamon and does not have a different coloration from the back. They have narrow black eye rings, and the ears also have a blackish shade (Arellano and Rogers, 1994; Hooper, 1952a). The skull is globular and proportionally large, while the nasal bones are long and narrow compared with those of other species of *Reithrodontomys* (Hooper, 1952a). As the rest of the mice of this genus, the upper incisors have an obvious longitudinal furrow. Its karyotype is identical to that of other species in the subgenus *Aporodon* (Arellano and Rogers, 1994; Rogers et al., 1983).

DISTRIBUTION: *R. tenuirostris* is distributed from southern Mexico to the border between Guatemala and Honduras. In Mexico it has a marginal distribution, and has only been recorded in Cerro Mozotol and Cerro Tzontehuitz (Rogers et al., 1983), and Volcán Tacaná (L. León, pers. obs.). It has been recorded in the state of CS.

EXTERNAL MEASURES AND WEIGHT
TL = 200 to 231 mm; TV = 120 to 129 mm; HF = 20 to 25 mm; EAR = 15 to 17 mm. Weight: 18 to 23 g.

DENTAL FORMULA: I 1/1, C 0/0, PM 0/0, M 3/3 = 16.

NATURAL HISTORY AND ECOLOGY: Since its discovery in Mexico (Rogers et al., 1983) little information on the natural history of *R. tenuirostris* has been accumulated; it has been found sympatrically with *Heteromys nelsoni*, *Peromyscus guatemalensis*, *Peromyscus beatae,* and *Reithrodontomys mexicanus* (Rogers, 1992). In Guatemala, there is evidence of their potential semi-arboreal habits or perhaps arboreal (Hooper, 1952a). In the Tacaná volcano, Chiapas, they have been captured up to 10 m high in trees (H. Olguín, pers. obs.). Other aspects of its natural history remain unknown.

VEGETATIONAL ASSOCIATIONS AND ELEVATION RANGE: These mice are found in temperate and wet climates, and inhabit cloud forests, from 2,800 masl to 2,910 masl (Arellano and Rogers, 1994).

CONSERVATION STATUS: *R. tenuirostris* is not currently protected, although its distribution area in Mexico has been reduced in recent years.

Reithrodontomys zacatecae Merriam, 1901

Zacatecan harvest mouse

Oscar Sánchez

SUBSPECIES IN MEXICO

R. zacatecae is a monotypic species.

This species has been considered a subspecies of *R. megalotis*. Hooper (1952a, d) proposed that *zacatecae* could be a distinct species, but not until 1984, when Hood et al. (1984) found evidence in karyotypes was it placed at a species level. This arrangement has been widely followed (Arita and Ceballos, 1997; Matson and Baker, 1986; Musser and Carleton, 1993). Bell et al. (2001) used genetic data to support its recognition as a species.

DESCRIPTION: The dorsal and ventral coloration of *Reithrodontomys zacatecae* is reddish-brown, notoriously more intense than that of *R. megalotis*. At least in the north of its distribution, *zacatecae* can be distinguished from *megalotis* by the longer tail (on average 75.4 mm) and a minor alveolar length (average, 3.26 mm; range, 3 mm to 3.5 mm; Anderson, 1972) and darker dorsal pelage. In the specimens from Jalisco and Michoacán the zygomatic arch is comparatively narrower and the mesopterigoid fossa is wider than that of *megalotis* (Hooper, 1952a, d). In accordance with Hood et al. (1984), the karyotype of *zacatecae* is different because the diploid number is 2n = 50 and the fundamental number is FN = 96 (unlike *megalotis*, which displays a diploid number of 2n = 42 and a fundamental number of FN = 80). Bell et al. (2001) used genetic data to confirm its distinction from *R. megalotis*.

DISTRIBUTION: *R. zacatecae* is an endemic species to Mexico that lives in the Sierra Madre Occidental and mountain systems adjacent to the southern portions of the Sierras (Anderson, 1972; Hall, 1981). It has been recorded in the states of AG, CH, DU, JA, MI, and ZA.

EXTERNAL MEASURES AND WEIGHT

TL = 128 to 157 mm; TV = 67 to 87 mm; HF = 17 to 20 mm; EAR = 13 to 18 mm. Weight: 9 to 13 g.

DENTAL FORMULA: I 1/1, C 0/0, PM 0/0, M 3/3 = 16.

NATURAL HISTORY AND ECOLOGY: *R. zacatecae* occupies sites of higher elevation, parapatrically or allopatrically with *R. megalotis*. In San Luis, Durango, Hooper found individuals of *zacatecae* using well-established roads between the macollos of grass (Hooper, 1955). In Durango, it has been found sympatrically with *Neotoma mexicana* (López-Vidal and Álvarez, 1993). A specimen of Patzcuaro, Michoacán, is apparently the only known case of *Reithrodondomys* with supernumerary teeth (Hooper, 1952a, d). Given the prior taxonomic controversy with regard to *R. zacatecae* and *R. megalotis*, it is difficult to assign correctly some data on reproduction that appear in literature. Other aspects of the natural history are currently unknown.

VEGETATIONAL ASSOCIATIONS AND ELEVATION RANGE: *R. zacatecae* usually inhabits open grasslands in pine-oak forests, with an average annual rainfall between 600 mm and 800 mm (Anderson, 1972; Hooper, 1952a; Matson and Baker, 1986). It has been found from 1,828 masl in Los Conejos, Michoacán, to 2,590 masl in Nahuatzin, Michoacán (Anderson, 1972; Baker and Greer, 1962; Hooper, 1952a; Matson and Baker, 1986).

CONSERVATION STATUS: *R. zacatecae* is not legally protected, nor does it appear to be in imminent threat; however, future changes in the land of the sites where it inhabits, particularly logging, could threaten it.

Rheomys mexicanus Goodwin, 1959

Mexican water mouse

Gerardo Ceballos

DISTRIBUTION: *R. mexicanus* is a species endemic to Mexico, known in Guelatao, San Jose Lachiguirí, Union Hidalgo, and Totontepec in the Pacific rim of Oaxaca, Oaxaca (Goodwin, 1969; Santos Moreno et al., 2003). It has been recorded in the state of OX.

SUBSPECIES IN MEXICO

R. mexicanus is a monotypic species.

DESCRIPTION: *Rheomys mexicanus* is relatively small. The body is robust and the ears are very small and covered with hair. The color of the body is dark brown with the belly grayish. The tail is longer than the body, bicolored, dark on the top and white on the bottom (Goodwin, 1969; Voss, 1988). The feet are large and wide, with a fringe of long hair. It can be distinguished from *R. thomasi* because the latter has visible ears and a bicolored tail, but less contrasting (Voss, 1988).

EXTERNAL MEASURES AND WEIGHT

TL = 270 to 300 mm; TV = 140 to 171 mm; HF = 32 to 42 mm; EAR = 6 to 8 mm. Weight: 40 g.

DENTAL FORMULA: I 1/1, C 0/0, PM 1/1, M 3/3 = 20.

NATURAL HISTORY AND ECOLOGY: *R. mexicanus* is a rare species; it has underground habits and is highly specialized. Its habitat requirements are very strict, and it is found exclusively in the margins of permanent rivers and streams, with continuous flow, clear, and well-oxygenated water (Goodwin, 1969; Voss, 1988). It is carnivorous and nocturnal; its diet mainly consists of aquatic insects, and, in smaller proportions, of crustaceans, small amphibians, and fish (Hooper, 1968; Voss, 1988). Apparently, they hunt in the water. There are no data on its reproduction; however, other similar species have up to two offspring per litter.

VEGETATIONAL ASSOCIATIONS AND ELEVATION RANGE: *R. mexicanus* mainly inhabits cloud forests, coniferous forests, and oak forests, but also tropical deciduous forests (Goodwin, 1969; Voss, 1988), near sea level to approximately 2,200 masl (Santos Moreno et al., 2003).

CONSERVATION STATUS: *R. mexicanus* is rare and has a limited distribution in highly specialized habits. Because of this is and the destruction and pollution of rivers and streams that constitute its habitat, this species has been regarded as endangered (Ceballos and Navarro, 1991). It is necessary to conduct further research to determine its current distribution and the conservation status of its populations.

Rheomys thomasi Dickey, 1928

Thomas' water mouse

Gerardo Ceballos

SUBSPECIES IN MEXICO

Rheomys thomasi stirtoni Dickey, 1928

Hooper (1947) described the subspecies *R. thomasi chiapensis* based on a young specimen of the Sierra Madre de Chiapas; this was accepted by other authors (Hall, 1981). Voss (1988), however, found that this taxon is indistinguishable from *R. t. stirtoni*, which has priority.

DESCRIPTION: *Rheomys thomasi* is very similar to *R. mexicanus* but slightly smaller. The body is robust and the ears are small, visible, and covered with hair. The dorsal coloration is dark brown, with the belly grayish. The tail is longer than the body, bicolored, dark on the top and grayish below (Hall, 1981; Voss, 1988). The feet are long and wide, with a fringe of long hair. It can be distinguished from *R. mexicanus* because the latter is bigger, with the ears more visible and the tail more contrasting (Voss, 1988).

DISTRIBUTION: This species is endemic to Mesoamerica. It can be found from Mexico to Costa Rica. In Mexico it is only known in two localities in Chiapas.

EXTERNAL MEASURES AND WEIGHT

TL = 208 to 270 mm; TV = 110 to 140 mm; HF = 30 to 35 mm; EAR = 8 to 12 mm. Weight: 40 g.

DENTAL FORMULA: I 1/1, C 0/0, PM 1/1, M 3/3 = 20.

NATURAL HISTORY AND ECOLOGY: *R. thomasi* is a rare species of subaquatic habits, and with morphological adaptations to move in aquatic environments (Stirton, 1944). Its habitat requirements are very strict, as it is exclusively found in rivers and streams of mountains, with continuous flow and clear and well-oxygenated water (Voss, 1988). Its activity is nocturnal. They are carnivores, feeding mainly on aquatic insects, and, in smaller proportions, on snails, crabs, frogs, and fish (Álvarez del Toro, 1991; Hooper, 1968; Voss, 1988). There are no data on its reproduction; however, other similar species have up to two offspring per litter (Voss, 1988).

VEGETATIONAL ASSOCIATIONS AND ELEVATION RANGE: *R. thomasi* is found exclusively in cloud forests (Hooper, 1947; Stirton, 1944; Voss, 1988), in areas from 360 masl to 1,100 masl in Mexico and 2,200 masl in El Salvador (Stirton, 1944; Voss, 1988).

CONSERVATION STATUS: *R. thomasi* has a limited distribution and a small population size. It is regarded as rare because of the fragility of its habitat (Ceballos and Navarro, 1991). It is likely to be threatened. It is necessary to carry out detailed studies on its ecology and vulnerability to extinction. It is protected in the El Triunfo Biosphere Reserve in Chiapas (Espinoza et al., 1998).

Scotinomys teguina (Alston, 1877)

Short-tailed singing mouse

Mark D. Engstrom

SUBSPECIES IN MEXICO
Scotinomys teguina teguina (Alston, 1877)

DISTRIBUTION: *S. teguina* is distributed at intermediate elevations of temperate or subtropical areas from southeastern Oaxaca to western Panama. In Mexico, it is only known in Altos de Chiapas in northern Chiapas and southwestern Oaxaca (Müllerried, 1957). It has been recorded in the states of CS and OX.

DESCRIPTION: *Scotinomys teguina* is a small-sized mouse. The dorsal coloration is chocolate brown and virtually no difference can be found between the back and the underside; within its range, this is the only small mouse with this coloration. Other mice potentially sympatric that are the same size, such as *Baiomys* and *Mus*, can be distinguished by their darker coloration (especially in the belly) and by their relatively longer tails.

EXTERNAL MEASURES AND WEIGHT
TL = 115 to 138 mm; TV = 46 to 58 mm; HF = 15 to 19 mm; EAR = 12 to 15 mm.
Weight: 10 to 13 g.

DENTAL FORMULA: I 1/1, C 0/0, PM 1/1, M 3/3 = 20.

NATURAL HISTORY AND ECOLOGY: This mouse is mainly terrestrial and is associated with temperate and wet forests. They are insectivorous. Adult insects, mainly beetles, constitute about 80% of their diet. It is one of the few diurnal mice found in its range. Their peak of activity occurs in the morning, depending mainly on the light to locate their prey. They emit particular vocalizations that are easily audible in the field and demonstrate their presence. These vocalizations last 7 to 10 seconds and include both ultrasonic and audible sounds; the latter are very high-pitched sounds, similar to those made by insects. This species reproduces throughout the year, and the gestation period lasts 31 days on average. The litter usually comprises two to three young but varies from one to five according to the locality, elevation, and

Scotinomys teguina. Cloud forest. Photo: Bernal Rodriguez.

the age of the mother. Both males and females participate in the construction of the nests, which are complex, and corridors. Apparently, males also contribute in caring for the offspring. Because of their insectivorous habits, it is likely that these mice contribute in the control of insect populations and therefore are important components of the ecosystem.

VEGETATIONAL ASSOCIATIONS AND ELEVATION RANGE: *S. teguina* is found mainly in wet temperate forests, such as those of oak (*Quercus*) and conifers, and in cloud forests. It has been found in primary forests, in open areas covered with grass and shrubs, in the edges of forests, under logs in grasslands, in coffee plantations, and in milpas close to human settlements. It has been recorded from 1,000 masl to 2,940 masl, but most records are below 2,400 masl.

CONSERVATION STATUS: It has a very restricted distribution and the species is considered as of Special Propection (SEMARNAT, 2010). It seems to tolerate moderate disturbances of its habitat, so its populations could disappear only in the event of a large deforestation and disappearance of wet and temperate microenvironments.

Sigmodon alleni Bailey, 1902

Allen's cotton rat

Jesús Ramírez and Cuauhtémoc Chávez Tovar

SUBSPECIES IN MEXICO
Sigmodon alleni alleni Bailey, 1902
Sigmodon alleni planifrons Nelson and Goldman, 1933
Sigmodon alleni vulcani J.A. Allen, 1906

DESCRIPTION: *Sigmodon alleni* is an average-sized rat with no evidence of sexual dimorphism. Its dorsal coloration is brown; the belly is white with dark gray underfur. The feet are yellowish-brown and the tail is bicolored, black on top and brown below (Baker, 1969). It differs from other species of the group *fulviventer* (*S. fulviventer*, *S. ochrognathus*, and *S. leucotis*) by the strongly curved incisors (opisthodont; Shump and Baker, 1978). The distinctive features of the skull are the inflamed appearance when seen laterally, the capsular projection slightly bulging for the upper incisors, and the slightly bent paraoccipital process (Baker, 1969). *S. alleni* differs from the group *hispidus* (*S. hispidus*, *S. arizonae*, and *S. mascotensis*) by having smaller, inconspicuous scales on the tail (0.50 mm to 0.75 mm wide), a very hairy tail, strongly curved incisors (opisthodont), and a moderately deep palatal keel (Baker, 1969; Goodwin, 1969; Zimmerman, 1970).

DISTRIBUTION: *S. alleni* is an endemic species of Mexico, the most tropical species of the *fulviventer* group. It is found in the plains and mountains of the Pacific, from Sinaloa to Oaxaca. It has been recorded in the states of CL, GR, JA, MI, NY, OX, and SI.

EXTERNAL MEASURES AND WEIGHT
TL = 224 to 266 mm; TV = 88 to 127 mm; HF = 27 to 34 mm; EAR = 20 to 22 mm. Weight: 178 to 183 g.

DENTAL FORMULA: I 1/1, C 0/0, PM 0/0, M 3/3 = 16.

NATURAL HISTORY AND ECOLOGY: Cotton rats are active day and night. Their runways are evident in areas covered by grasslands, but indistinguishable in areas with a significant shrubby coverage and at ground level. They build their burrows with grass in a variety of places, including cracks of rocks, tree stumps, trunks, and holes of trees or beneath the ground (Shump and Baker, 1978). They are omnivores, feeding on stems, leaves, seeds, insects, lizards, and eggs of birds (Baker, 1969). Among its top predators are weasels and raptors. Its distribution seems to overlap in the limits of tropical environments and tropical boreal environments (pine-oak) in the slopes of the west of Sierra Madre Occidental, Durango. It is possible that *S. alleni* and *S. fulviventer* are sympatric near Patzcuaro, Michoacán, where *S. alleni* occupies scrublands, and *S. fulviventer* and *S. hispidus* grassland (Hall and Villa-R., 1949; Nelson and Goldman, 1890). The population densities of this species are low, on average 0.023 individual/ha and 0.52 individual/ha in cloud forests and dismantled areas for agricultural purposes in the Biosphere Reserve Manantlán Sierra, Jalisco (Vázquez, 1997). It has been caught with *Peromyscus hylocetes*, *P. aztecus*, *Reithrodontomys fulvescens*, *R. sumichrasti*, *Hodomys alleni*, *Oryzomys couesi*, and *Liomys pictus* (Ramírez-Pulido et al., 1977; Vázquez, 1997). They mate throughout the year (Ceballos and Miranda, 2000). Females are fertile at 75 days old. The gestation period lasts 27 days. The offspring are extremely precocious and are weaned within 7 days (Baker, 1969; Shump and Baker, 1978).

VEGETATIONAL ASSOCIATIONS AND ELEVATION RANGE: This cotton rat inhabits the tropical deciduous forests and pine-oak forests, on humid slopes covered by vines and shrubs, in the lowlands of the Pacific coast, and in the temperate-tropical ecotone of the slopes of the Pacific of the Sierra Madre Occidental and Sierra Madre del Sur. In Jalisco, it inhabits mixed areas of grasses and bushes, from sea level to 3,050 masl (Baker, 1969; Shump and Baker, 1978).

CONSERVATION STATUS: The status of this species is undetermined. They are vulnerable, however, because livestock, which destroys their burrows and changes the plant structure, displaces them. It is necessary to assess the current situation of their populations, especially in areas used extensively for livestock.

Sigmodon arizonae Mearns, 1890

Arizona cotton rat

Jesús Ramírez and Cuauhtémoc Chávez Tovar

SUBSPECIES IN MEXICO

Sigmodon arizonae cinegae B. Howell, 1919
Sigmodon arizonae major Bailey, 1901
Analyses of features such as chromosomal morphology, coat, cranial morphology, and continuity of its distribution led to *S. arizonae* being removed as a subspecies of *S. hispidus* and being reinstated as a species by Zimmerman (1970). Studies carried out by Fuller et al. (1984), Elder and Lee (1985), and Peppers and Bradley (2000) concluded, through immunological and cytogenetic data and DNA sequences, that *S. arizonae* and *S. mascotensis* are closely related and probably derived recently from an ancestor similar to *hispidus.*

DESCRIPTION: *Sigmodon arizonae* is a large-sized rat. Sexual variation can only be found in the interparietal width, where males have a greater size (Ramirez, 1984). The dorsal coloration is pale gray; the nose is slightly yellowish; the belly is whitish.

The feet are gray. The tail is bicolored, gray underneath and blackish dorsally. In general, it is very similar to *S. hispidus*, but the distance between the occipital crest and temporal bone is bigger, the oval foramen is large, and the nasal edges are straight (Hall, 1981; Zimmerman, 1970).

DISTRIBUTION: This species inhabits the southwestern United States and western Mexico, from Sonora to Nayarit. It has been recorded in the states of DU, NY, SI, and ZA.

EXTERNAL MEASURES AND WEIGHT
TL = 231 to 345 mm; TV = 95 to 155 mm; HF = 34 to 42 mm; EAR = 17 to 24 mm. Weight: 140 to 200 g.

DENTAL FORMULA: I 1/1, C 0/0, PM 0/0, M 3/3 = 16.

NATURAL HISTORY AND ECOLOGY: These rats are active both day and night. They live in desert areas, usually characterized by mesquites and tumbleweeds with a small amount of grasses, or in less arid areas along banks and small rivers or streams with shrubs and woody plants. They are omnivores, and most of their diet consists of seeds, but they may also consume stems and roots, insects, and other animals (Hoffmeister, 1986). They move through trails more or less defined in the grasslands. Their burrows are located in dense vegetation, cracks of rocks, tree stumps, trunks, and holes of trees, or beneath the ground. Mating season occurs throughout the year; pregnant females have been collected from February to April and from August to October. The gestation period is 27 days and the number of offspring per litter varies from 1 to 12, with 5 on average. The offspring are extremely precocious, and are weaned within 7 days. This species is usually not very abundant (Hoffmeister, 1986). Some of its predators are domestic cats; other carnivores include lynx and raptors.

VEGETATIONAL ASSOCIATIONS AND ELEVATION RANGE: In Mexico this rodent has been mainly collected in desert environments, although it can be found in tropical deciduous forests and disturbed areas. They are more common in introduced grasslands and in annual crops such as maize. It inhabits areas from sea level in San Blas, Nayarit, to 2,000 masl in St. Lucia, Sinaloa.

CONSERVATION STATUS: The situation of this species is undetermined, but it is relatively common, even in areas modified by humans.

Sigmodon fulviventer J.A. Allen, 1889

Tawny-bellied cotton rat

Jesús Ramírez and Cuauhtémoc Chávez Tovar

SUBSPECIES IN MEXICO
Sigmodon fulviventer fulviventer J.A. Allen, 1889
Sigmodon fulviventer minimus Mearns, 1894
Sigmodon fulviventer melanotis Bailey, 1902

DESCRIPTION: *Sigmodon fulviventer* is a medium-sized rat. The coat is hirsute. The back is mottled with hair of different colors, including brown, black, black with brown tips, black with brown underfur, an ivory stripe, and black tips (McWhirter et al., 1974). The belly is very conspicuous brown-orange, contrasting with the back. The tail is dark and has few hairs. The tail, feet, and ears are moderately long

(Anderson, 1972). The ears are covered by hair. The brown color of the belly and the mottled color of the back, together with its size, distinguish it from other members of the group *fulviventer*. Distinctive cranial features include the narrow anterior tip of the mesopterigian fossa, the long oval foramen (higher or equal to the width of M3), the deeply marked palatal fossa, and the well-developed median keel of the palatine (Baker and Shump, 1978). Unlike other species of the group *hispidus*, *S. fulviventer* has smaller (0.50 mm) scales in the tail and very scattered hair (Baker, 1969; Zimmerman, 1970).

DISTRIBUTION: *S. fulviventer* is distributed from central New Mexico and southeastern Arizona to central Mexico, through the east side of the Sierra Madre Occidental, and through the Eje Volcanico Transversal in Jalisco and Michoacán. It has been recorded in the states of CH, DU, GJ, JA, MI, SO, and ZA.

EXTERNAL MEASURES AND WEIGHT

TL = 216 to 278 mm; TV = 101 to 118 mm; HF = 26 to 31 mm; EAR = 19 to 23 mm. Weight: 82 to 136 g.

DENTAL FORMULA: I 1/1, C 0/0, PM 0/0, M 3/3 = 16.

NATURAL HISTORY AND ECOLOGY: These rats are active both day and night. They use well-defined paths, which can be partially observed in dense grasses. Their burrows are built between the densest vegetation, crevices, tree stumps, trunks, and holes in trees, or beneath the ground. They also build external nests in grassy areas (Baker, 1969; Hoffmeister, 1986; Whitaker, 1980). They are sympatric with *S. hispidus*, *S. leucoti*, and *S. ochrognathus* (Baker, 1969). In Durango, the removal of *S. fulviventer* caused an increase in population levels, survival rates, and size of home range of S. hispidus, suggesting that *S. fulviventer* is a better competitor (Harris and Petersen, 1979; Petersen, 1979; Petersen and Helland, 1978). Their densities can reach up to 28 individuals/ha (Petersen 1973). Along the east side of the Sierra Madre Occidental, where grasslands have deep alluvial soils, presumably *S. hispidus* also excludes or dominates other species of cotton rats (Baker and Greer, 1962). When *S. leucotis* or *S. ochrognathus* occur sympatrically with *S. fulviventer*, the first two species are found in open areas, in the periphery, and where the vegetation is scattered. When *S. fulviventer* is absent, they can occupy all habitats. It has been estimated that *S. fulviventer* can potentially consume approximately 0.267 kg of *Bouteloua gracilis*/ha/day, which is

Sigmodon fulviventer. Grassland. Ojo de Galeana, Chihuahua. Photo: Gerardo Ceballos.

equivalent to 0.04 cow/ha/day (Mitchell, 1964). These data suggest that competition can exist between livestock and this species (Petersen, 1993). In captivity they reach sexual maturity within 75 days, with a gestation period of 25 days (Baker, 1969). In Atotonilco of Campa, Durango, litters comprise on average 4.5 pups (Petersen, 1978). An age range suggests that until 75 days old a rat is within the range of juveniles, young adults between 76 and 200 days, adults between 201 and 300 days, and old adults more than 300 days (Baker, 1969). Their predators include owls (*Bubo* and *Tyto*), other raptors, mammals, and reptiles (Petersen, 1979b).

VEGETATIONAL ASSOCIATIONS AND ELEVATION RANGE: In Mexico, *S. fulviventer* has been collected in thickets such as mesquite-grassland and grasslands (Baker and Greer, 1962; Hooper, 1955). Some species of dominant grasses are *Muhlenbergia* sp., *Aristida* sp., *Cynodon dactylon*, and *Sporoblus wrightii* (Baker and Shump, 1978). In Mexico, it has been collected at elevations ranging from 1,290 masl to 2,475 masl.

CONSERVATION STATUS: This species faces no conservation issues, as it tolerates anthropogenic disturbances.

Sigmodon hirsutus (Burmeister, 1854)

Southern cotton rat

Ryan M. Duplechin and Robert D. Bradley

SUBSPECIES IN MEXICO

S. hirsutus is a monotypic species.

Peppers et al. (2002), Carroll et al. (2005), Bradley et al. (2008), and Henson and Bradley (2009) all have corresponding genetic evidence elevating *S. hirsutus* to the species level with *S. alleni* as the sister species.

DISTRIBUTION: This species occurs from central Chiapas (including the Sierra Madre de Chiapas) and southern Oaxaca through Central America to Venezuela (Bradley et al., 2008). Near Ocozocoautla in central Chiapas, *S. hirsutus* is reported as being sympatric with *S. toltecus* (Carroll et al., 2005). These two species are sympatric in north-central Chiapas, where the lowlands of the Depression Central de Chiapas rise in elevation to become the Sierra Madre de Chiapas (Bradley et al., 2008; Carroll et al., 2005).

DESCRIPTION: Blackish or dark brown hairs constitute the dorsal pelage of *Sigmodon hirsutus* with grayish hairs interspersed; the sides are slightly paler than the dorsal pelage; the underparts are usually pale to grayish, sometimes faintly buffy; the tail is dark and sparsely haired with annulations. The skull is long and narrow with a long and broad basioccipital bone and shallow palatal pits (Baker, 1969). The chromosome diploid number and the fundamental number are both 52 (Swier et al., 2009). In the past, *S. hirsutus* was not recognized as a valid species. Recently, molecular studies have revealed a high degree of genetic divergence between *S. hispidus* and *S. hirsutus* (Bradley et al., 2008; Carroll et al., 2005; Peppers and Bradley, 2000). *S. hirsutus* has since been elevated to species status and resides in the *hispidus* species group (Carroll and Bradley, 2005).

EXTERNAL MEASURES AND WEIGHT

TL = 224 to 365 mm; TV = 81 to 166 mm; HF = 28 to 41 mm; EAR = 16 to 24 mm. Weight: 110 to 225 g (male), 100 to 200 g (female).

DENTAL FORMULA: I 1/1, C 0/0, PM 0/0, M3/3 = 16.

NATURAL HISTORY AND ECOLOGY: *S. hirsutus* is found in Mexico mostly at moderate to high elevations across the Sierra Madre de Chiapas. It occurs primarily in the montane grasslands of the Central American Highlands, although *S. hirsutus* has been collected in lowland grasslands adjacent to the Sierra Madre de Chiapas (Carroll et al., 2005). Feeding habits and natural history characteristics are probably similar to those of *S. hispidus* in that grasses and vegetation are the primary food source.

VEGETATIONAL ASSOCIATIONS AND ELEVATION RANGE: *S. hirsutus* occupies mixed grass and brush, including perennial grasses and forbs, at medium to high elevations. In addition to its primary occurrence at higher elevations, *S. hirsutus* has been collected in grasslands or coastal grasslands closer to sea level (Bradley et al., 2008; Carroll et al., 2005).

CONSERVATION STATUS: This species is of least concern according to the IUCN (2010).

Sigmodon hispidus Say and Ord, 1825

Hispid cotton rat

Jesús Ramírez, Cuauhtémoc Chávez Tovar, and Gisselle A. Oliva Valdés

SUBSPECIES IN MEXICO
Sigmodon hispidus berlandieri Baird, 1855
Sigmodon hispidus eremicus Mearns, 1897
Sigmodon hispidus furvus Bangs, 1903
Sigmodon hispidus microdon Bailey, 1902
Sigmodon hispidus saturatus Bailey, 1902
Sigmodon hispidus obvelatus Russell, 1952
Sigmodon hispidus solus Hall, 1951
Sigmodon hispidus tonalensis Bailey, 1902
Sigmodon hispidus villae Goodwin, 1958

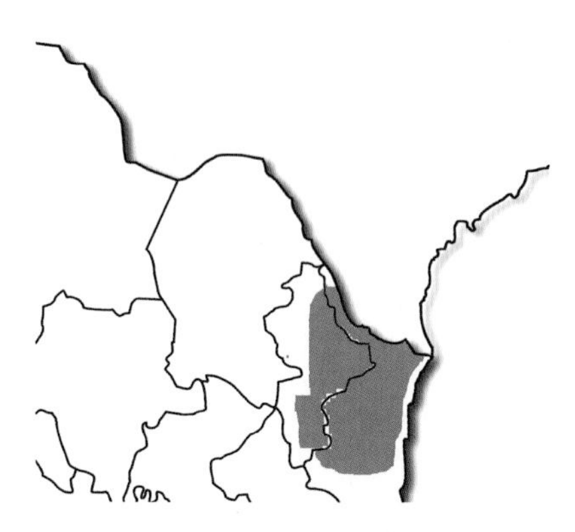

DISTRIBUTION: *S. hispidus* has a wide distribution, from the western United States to the northern portions of Tamaulipas. It has been recorded in TA.

DESCRIPTION: Compared to the other species of the genus, *Sigmodon hispidus* is medium in size. Sexual dimorphism is evident since males are larger than females (Chipman, 1965; Jiménez, 1971; Ramírez, 1984). A greater morphological variation has been recorded in populations of the north (Kansas) than in Mexico (McClenaghan, 1977). Its dorsal coloration is usually gray or dark brown mixed with gray hair, and darker in accordance with the subspecies. The sides are lighter; the belly is dark gray or gray and sometimes brown. The tail is usually ringed with hairs that do not hide the scales. The skull is relatively long and thin, with the mastoid generally wide, but less than 46% of the basal length (Cameron and Spencer, 1981; Hall, 1981). Its tail differs from that of the group *fulviventer* by the large proportion of scales (0.75 mm), scattered hairs, and a very hairy tip. *S. hispidus* differs from other members of the group *hispidus* by having shorter feet (less than 34 mm, on average 32 mm), a relatively short distance between the temporal crest and occipital crest (less than 3.6 mm, on average 3.2 mm), and diameter of the oval foramen smaller than the width of M3. It has a well-developed crest in the back of palatine (Zimmerman, 1970). Peppers and Bradley (2000) split *S. hispidus* into three species (*hirsutus*, *hispidus*, and *toltecus*) with *hispidus* being restricted to the United States and perhaps the extreme northern parts of Tamaulipas (see Bradley et al., 2008 for a discussion of the distribution). Also, Carleton et al. (1999) reassigned some populations of *S. hispidus* to other taxa.

EXTERNAL MEASURES AND WEIGHT

TL = 224 to 300 mm; TV = 81 to 166 mm; HF = 25 to 33 mm; EAR = 16 to 24 mm. Weight: 110 to 225 g.

DENTAL FORMULA: I 1/1, C 0/0, PM 0/0, M 3/3 = 16.

NATURAL HISTORY AND ECOLOGY: *S. hispidus* is active both day and night, and its patterns of activity are influenced by biotic and abiotic factors (Calhoun, 1945). In Texas, they are active at twilight, with peaks of activity at 19:00 hrs and 09:00 hrs, and a minimum activity at 23:00 hrs to 05:00 hrs (Cameron et al., 1979b). They are good swimmers; hence water is not a barrier for their dispersal (Esher et al., 1978). Their runways are evident in areas covered by grasslands, but indistinguishable in areas with a dense shrubby coverage. They build their burrows with grass in a variety of places, including crevices, tree stumps, trunks, and holes of the trees, or beneath the ground (Baar et al., 1974; Dawson and Lang, 1973; Halloran, 1942; Shump, 1978). They are omnivores, and their diet include stems, leaves, seeds, insects, lizards, and birds' eggs, but they mainly consume grasses (Fleharty and Olson, 1969). Its main predators are weasels and raptors; on some occasions they are the main food item of species such as owls (*Tyto alba*), red-tailed hawks (*Buteo jamicensis*), and nauyacas (*Bothrops asper*), among others (Korschgen and Stuart, 1972; Martínez-Gallardo and Sánchez-Cordero, 1997; Ramírez-Pulido and Sanchez-Hernandez, 1972; Raun, 1960; Schnell, 1968; Weigert, 1972). It has also been suggested that predation by birds and mammals can control their social behavior (Roberts and Wolfe, 1974). The gestation period is approximately 27 days and the size of the litter ranges from 1 to 15 offspring (Meyer and Meyer, 1944; Randolph et al., 1977). The average life expectancy is more than 3 months, with longer periods of up to 10 months (Cameron, 1977). The average activity area is 0.35 ha for adults, 0.22 ha for subadults, and 0.35 ha for juveniles. Males have larger activity areas than females. Females, however, have exclusive home ranges (Cameron et al., 1979b; Erickson, 1949; Fleharty and Seas, 1973; Goertz, 1964; Layne, 1974). The temporary differences in their movements seem to

Sigmodon hispidus. Secondary vegetation. Montes Azules, Chiapas. Photo: Heliot Zarza.

be influenced by the reproductive stages and/or availability of coverage (Bigler and Jenkins, 1975; Briese and Smith, 1974; Smith and Vrieze, 1979). Daily movements of 13 m have been reported (combining genus and age); males move an average of 17 m, and females an average of 6.6 m (Cameron et al., 1979b). A positive correlation between dispersion and density, independent of the sex ratio and age structure of the dispersers, has been found (Joule and Cameron, 1975). It has also been observed that interspecific interactions affect the density and movement patterns (Bigler et al., 1977; Ramsey and Briese, 1971). *S. hispidus* can exclude *Microtus mexicanus* (Terman, 1974). It is sympatric with *Oryzomys couesi*, *Onychomys torridus*, *Perognathus flavus*, *P. hispidus*, *P. penicillatus*, *Dipodomys ordii*, *D. merriami*, *Peromyscus maniculatus*, *P. eremicus*, *P. leucopus*, *Neotoma albigula*, *Reithrodontomys fulvescens*, and *R. megalotis* (Birney et al., 1974; Mellink, 1986).

VEGETATIONAL ASSOCIATIONS AND ELEVATION RANGE: In Mexico, this species is more prevalent in habitats dominated by grasses, but is also found in mixed areas of grassland-grass-shrubs, dry bushes, tropical thorn forests, medium subperennial tropical forests, and crops, especially of alfalfa and sugar cane. It has been found only in northern Tamaulipas.

CONSERVATION STATUS: This species is widely distributed and its populations are abundant; it can even become a pest in crops.

Sigmodon leucotis Bailey, 1902

White-eared cotton rat

Jesús Ramírez, Cuauhtémoc Chávez Tovar, and Gisselle A. Oliva Valdés

SUBSPECIES IN MEXICO

Sigmodon leucotis leucotis Bailey, 1902
Sigmodon leucotis alticola Bailey, 1902

DESCRIPTION: *Sigmodon leucotis* is a medium-sized rat. Its dorsal coloration is grayish-brown; the belly is white with dark gray underfur. The feet are grayish-brown. The scales of the tail are smaller than in *S. hispidus*. The tail is shorter than the head and body, scaly and practically hairless, dark on the back and lighter on the bottom (Shump and Baker, 1978). It differs from other members of the group *fulviventer* (*S. fulviventer*, *S. alleni*, and *S. ochrognathus*) by having a prominent premaxillary depression on each side of the face and an absent or very reduced lingual root of the first molars. The distinctive cranial characters are the small interparietal bone with the length of the median line less than 2 mm; the anterior portion of the mesopterigian fossa and parallel angular process of the jaw are slightly hooked, as well as rounded (Baker, 1969). It is distinguished from the *hispidus* group (*S. hispidus*, *S. arizonae*, and *S. mascotensis*) by the scales of the tail less than 0.50 mm wide, by the shallow palatal fossa, and by the slightly developed median keel in the palatal (Baker, 1969; Zimmerman, 1970).

EXTERNAL MEASURES AND WEIGHT

TL = 225 to 252 mm; TV = 84 to 105 mm; HF = 27 to 30 mm; EAR = 16 to 21 mm. Weight: 86 to 140 g.

DENTAL FORMULA: I 1/1, C 0/0, PM 0/0, M 3/3 = 16.

NATURAL HISTORY AND ECOLOGY: These rats are diurnal and have two activity periods, one in the morning, between 08:00 hrs and 11:00 hrs, and another in the afternoon between 17:00 hrs and 21:00 hrs (Álvarez and Polaco, 1984). They use well-defined paths in the grassland, which are indistinguishable in shrubby vegetation (Ceballos and Galindo, 1984). The burrows are shallow and almost horizontal between the densest vegetation, crevices, tree stumps, trunks, and holes of trees or beneath the ground (Álvarez and Polaco, 1984). They are omnivores and feed on stems, leaves, seeds, insects, lizards, and eggs of birds (Ceballos and Galindo, 1984). In a sample of 24 stomach contents, in February 70% was starch, 20% remnants of green material, and 10% insects. In April 70% was plant tissue, such as stems and leaves, 9.6% seeds, and 20.4% remnants of insects. In May 94.6% was green vegetable matter, 4.4% seeds, and 1% insects. In August 33% was green material, 25.5% seeds, 11.9% yeasts, and 30.6% insects. Finally, in November 88.6% was green plant material, 8.7% seeds, and 2.7% remnants of insects (Álvarez and Polaco, 1984). Among its top predators are weasels and raptors (Ceballos and Galindo, 1984). They are relatively abundant in the Michilía Biosphere Reserve, Durango. They inhabit the edges of farmlands, rocky areas, and grasslands, and flat areas of pine forests, especially in those places where fallen logs abound (Álvarez and Polaco, 1984). In the cone of Volcano Pelado, in Mexico City, in alpine grassland at 3,600 masl, densities with an average of 1.1 individuals/ha have been found (Gomez, 1990). They reproduce throughout the year. After a gestation period of 35 days, 5 to 12 offspring are born (Ceballos and Galindo, 1984). Apparently, there is a reproductive peak in summer (Matson and Baker, 1986). It has been caught with *Peromyscus melanotis*, *P. difficilis*, *P. levipes*, *P. gratus*, *Microtus mexicanus*, *Reithrodontomys megalotis*, *Neotomodon alstoni*, *Neotoma Mexicana*, and *Thomomys umbrinus* (Ceballos and Galindo, 1984).

DISTRIBUTION: This species is endemic to Mexico and is distributed from 25 degrees north latitude in the Sierra Madre Occidental and in the Sierra Madre Oriental southward to the Neovolcanic Axis and the Sierra Madre del Sur in Oaxaca. It has been recorded in the states of AG, CH, DF, DU, GJ, MO, MX, NL, OX, PU, QE, and ZA.

VEGETATIONAL ASSOCIATIONS AND ELEVATION RANGE: *S. leucotis* inhabits grasslands, forests of pine-oak, pine, meadows, and rocky areas restricted to the mountains (Ceballos and Galindo, 1984). Some dominant plants species are *Ceanothus fendleri*,

Sigmodon leucotis. Pine forest and grassland. Zoquiapan and Anexas National Park, state of Mexico. Photo: Gerardo Ceballos.

Senecio actinella, *S. pinnatisectum*, *Symphyandra*, and *Muhlenbergia* (Baker, 1969; Webb and Baker, 1962). In Mexico, it has been collected at elevations ranging from 1,800 masl to 2,623 masl.

CONSERVATION STATUS: The status of this species is undetermined, but they are displaced by livestock in places with dense coverage, which destroy their burrows and change the vegetative vegetable structure (Ceballos and Galindo, 1984). It is necessary to assess the situation of their populations, especially in areas used for livestock.

Sigmodon mascotensis J.A. Allen, 1897

West Mexican cotton rat

Jesús Ramírez and Cuauhtémoc Chávez Tovar

SUBSPECIES IN MEXICO

Sigmodon mascotensis mascotensis J.A. Allen 1897
Sigmodon mascotensis inexoratus Elliot 1903
Sigmodon mascotensis ischyrus Goodwin 1956

Based on its morphology, chromosome characteristics, meiotic conduct, coat, and cranial features, it was removed as subspecies of *S. hispidus* and reinstated as a species by Zimmerman (1970). Studies carried out by Severinghaus and Hoffmeister (1978), Fuller et al. (1984), Elder and Lee (1985), and Peppers and Bradley (2000) concluded, through immunological and cytogenetic data, that *S. arizonae* and *S. mascotensis* are closely related and probably derived recently from an ancestor similar to *hispidus*.

DISTRIBUTION: *S. mascotensis* is endemic to Mexico; it is distributed on the Pacific Rim from Nayarit and Zacatecas to Oaxaca (Chavez and Ceballos, 1998; Hall, 1981). It has been recorded in the states of CL, GR, JA, MI, MX, NY, OX, and ZA.

DESCRIPTION: *Sigmodon mascotensis* is a large-sized rat. The coat is hirsute and has a very distinctive coloration, a mixture of black, long, and thin hair and brown, thick, and short hair with shades that vary from yellowish to reddish; the fur is paler on the sides and cheeks. The belly is grayish-white, the bases of the hair are dark gray, and the tips are white. The tail is shorter than the head and body, scaly, and scantily haired, brown on top and light beneath. The feet are yellowish-white. The ears are small and covered with hair. *S. mascotensis* is very similar to the rats of the genus *Oryzomys*, from which it is distinguished by having three types of hair on the back of the feet. It differs from *S. arizonae* and *S. hispidus* by having larger feet and a greater distance between the occipital and temporal crests, by the length of the skull, and because they lack a crest in the palatine (Hall, 1981; Zimmerman, 1970). The taxonomy of this group of species has been revised recently (Bradley et al., 2008; Carleton et al., 1999).

EXTERNAL MEASURES AND WEIGHT

TL = 240 to 345 mm; TV = 136 to 166 mm; HF = 30.7 to 40 mm; EAR = 17 to 23 mm. Weight: 49 to 135 g.

DENTAL FORMULA: I 1/1, C 0/0, PM 0/0, M 3/3 = 16.

NATURAL HISTORY AND ECOLOGY: *S. mascotensis* is omnivorous; a large portion of its diet is based on seeds, although it can eat vegetation such as stems and roots, insects, and other animals (Ceballos and Miranda, 1986, 2000). They are active both day and night. They use well-defined runways to move through dense vegetation. Their burrows are built in dense vegetation, crevices, tree stumps, trunks, and holes of the trees, or beneath the ground (Ceballos and Miranda, 1986, 2000). They reproduce throughout the year. Pregnant females have been found in March, May, July, and December, and juveniles in February and July (Álvarez et al., 1987; Juárez, 1992; Matson and Baker, 1986). The gestation period lasts 27 days. The litter has 1 to 12 offspring, with 5 on average. The offspring are extremely precocious and are weaned within 7 days. In a study conducted in Michoacán, at a total of 31 localities, with different types of vegetation such as pine-oak forests, tropical deciduous forests, and tropical subdeciduous forests, it was captured exclusively in tropical deciduous forests. It is sympatric with *Oryzomys couesi, O. alfaroi, O. melanotis, Osgoodomys banderanus, Reithrodontomys fulvescens, R. sumichrasti, Peromyscus perfulvus, P. spicilegus, P. levipes, P. megalops, Baiomys musculus, Liomys pictus*, and *Marmosa canescens* (Álvarez et al., 1987; Ceballos and Miranda, 2000; Ramírez-Pulido et al., 1977). *S. mascotensis* is scarce in tropical deciduous forests, maize fields close to this environment, and grasslands close to pine-oak forests (Collet et al., 1975; Juárez, 1992). Apparently, it is not abundant in fields of coconuts, corn, or bananas, in Costa Grande de Guerrero (Ramírez-Pulido et al., 1977); however, it has been reported that this species is abundant in the coconut fields of Costa Chica (Barrera, 1952). In Chamela, Jalisco, the average distance traveled by two males in one night was 18.1 m in tropical deciduous forest (Collet et al., 1975).

VEGETATIONAL ASSOCIATIONS AND ELEVATION RANGE: These mice can be found in tropical deciduous forests, pine forests, and disturbed areas. They are more common in induced grasslands and annual crops (such as corn). It is found at elevations ranging from sea level in Tecpan, Guerrero to 2,550 masl in Mazamitla, Jalisco.

CONSERVATION STATUS: *S. mascotensis* is not at risk of extinction since it survives in crops and other disturbed environments.

Sigmodon ochrognathus Bailey, 1902

Yellow-nosed cotton rat

Jesús Ramírez and Cuauhtémoc Chávez Tovar

SUBSPECIES IN MEXICO

S. ochrognathus is a monotypic species.

DESCRIPTION: *Sigmodon ochrognathus* is a large rat without sexual dimorphism. The coat is hirsute and has a very distinctive coloration; the base is dark gray and the tips are brown, the tips of some hairs are dotted with yellow, which gives them a grayish appearance. The belly is white with dark gray underfur. The tail is bicolored, dark on top and gray below. Each side of the nose has a yellow spot, and the eyes are ringed (Anderson, 1972; Baker, 1969). These external features distinguish it from the other species of *Sigmodon*. It is also distinguished by the following cranial features: a moderately pronounced keel in the basioccipital; a small and elongated bulla; and an obvious bulging of the capsular projections of the upper incisors. It is distinguished from the *hispidus* group by the smaller scales (0.50 mm) and very hairy tail (Baker, 1969; Zimmerman, 1970).

Sigmodon ochrognathus. Scrubland. Ojo de Galeana, Chihuahua. Photo: Juan Cruzado.

EXTERNAL MEASURES AND WEIGHT

TL = 192 to 243 mm; TV = 85 to 110 mm; HF = 26 to 30 mm; EAR = 16 to 21 mm. Weight: 41 to 133 g.

DENTAL FORMULA: I 1/1, C 0/0, PM 0/0, M 3/3 = 16.

NATURAL HISTORY AND ECOLOGY: *S. ochrognathus* is active both day and night. These rats use well-defined runways to move between the dense vegetation or indiscriminately in rocky places with scattered vegetation (Baker and Greer, 1962; Baker and Shump, 1978b). These runways are within the radius of the nest or burrow, avoiding obstacles and allowing the transit of two or more rats (Hoffmeister, 1986). They can occupy abandoned burrows of pocket gophers (*Thomomys umbrinus*), in which they store grasses (Baker and Greer, 1962). They build complicated and spherical nests in grassy areas (Hoffmeister, 1963, 1986). They are mainly found in the periphery of the habitats they inhabit, as other cotton rats displace them (Baker, 1969). Although the distribution of *S. ochrognathus* and *S. leucotis* overlap in the latitude of Canatlán-Tepehuanes, west-center of Durango, the two species are not micro-sympatric. This species appears to be confined to low and dry areas with slopes (below 1,950 masl). Apparently, it reproduces throughout the year, but pregnant females have been found in April and October, with litters of 2 to 9 offspring; the gestation period lasts from 33 to 36 days (Baker, 1956, 1986; Baker and Shump, 1974). The activity area of an adult male was 2,830 m^2 in the Chiricahua Mountains, Arizona (Hoffmeister, 1986).

DISTRIBUTION: *S. ochrognathus* is distributed in the east side of the Sierra Madre Occidental from Arizona and New Mexico toward the south, to central Durango. It has been recorded in the states of CH, CO, DU, and SO.

VEGETATIONAL ASSOCIATIONS AND ELEVATION RANGE: *S. ochrognathus* is the most xeric species of the genus; it inhabits dry parts and rocky slopes of pine-juniper environments, along the east side of the Sierra Madre Occidental. In Mexico, it has been collected in thickets such as mesquite-grassland and rangeland in rocky environments. It is associated with *Bouteloua gracilis*, *Muhlenbergia*, *Agave*, and *Quercus*. In Mexico, it has been collected at elevations of 990 masl in San Geronimo, Coahuila, to 2,375 masl in Sierra del Pino, Coahuila (Baker, 1956, 1959; Denyes, 1956; Findley and Jones, 1960; Hoffmeister, 1963).

CONSERVATION STATUS: The status of this species is undetermined, but apparently it is scarce. Because *S. ochrognathus* shows preference for grasslands, its populations are only found in grasslands (Whitaker, 1980).

Sigmodon planifrons Nelson and Goldman, 1933

Miahuatlán cotton rat

Megan S. Corley and Robert D. Bradley

SUBSPECIES IN MEXICO
Sigmodon planifrons minor Goodwin (1955)
Sigmodon planifrons setzeri Goodwin (1959)

DESCRIPTION: *Sigmondon planifrons* is one of the smallest members of the genus *Sigmodon* and closely resembles *S. alleni* in color, but cranial characters are more slender with a more depressed rostrum and less recurved incisors. Coloration of the dorsum is between an ochraceous-buff and ochraceous-tawny, and mixed with black. The dark hairs are most numerous on the top of the head and back, thinning out along the sides. The rump is suffused with light tawny, and the underparts are overlaid with dull white; the under color is dark plumbeous. The ears are covered with finely mixed black and buffy banded hairs like the body. The feet are buffy grayish, and the tail is bicolored with dark brown above and lighter brown below. The diploid number of chromosomes is 2n = 52 (Carleton et al., 1999). Nelson and Goldman (1933) originally described *S. planifrons*; however, Baker (1969) placed *S. planifrons* as a subspecies of *S. alleni*. The form is distinct from *S. alleni*, however, as suggested by the diminutive size of the type series (Nelson and Goldman, 1933) and the material and sympatry reported by Goodwin (1955, 1969). Carleton et al. (1999) formally resurrected *S. planifrons* as a species and suggested that it may be most closely related to *S. alleni*.

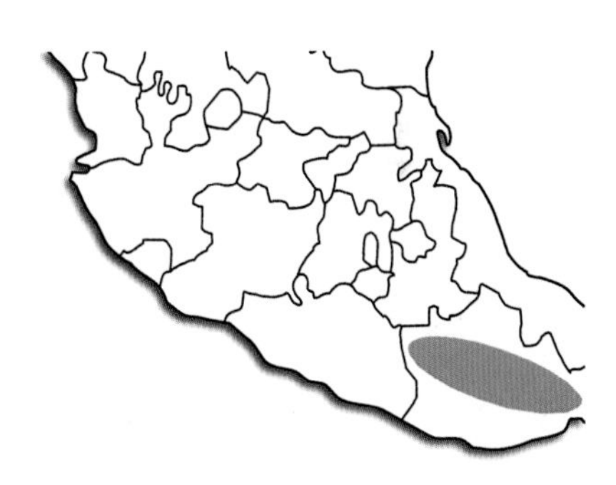

DISTRIBUTION: *S. planifrons* is known only from the type locality in the mountains of southwestern Oaxaca (Nelson and Goldman, 1933).

EXTERNAL MEASURES AND WEIGHT
TL = 207 to 228 mm, TV = 88 to 128 mm, HF = 27 to 34 mm, EAR = NA
Weight: NA.

DENTAL FORMULA: I 1/1, C 0/0, PM 0/0, M 3/3 = 16.

NATURAL HISTORY AND ECOLOGY: This species prefers moist slopes covered with vines and shrubs and constructs grass nests in a variety of places, including rotten pine logs. Runways are generally evident in grassy areas, but indistinct in brushy areas with little ground cover (Baker, 1969).

VEGETATIONAL ASSOCIATIONS AND ELEVATION RANGE: The type locality is Oaxaca state, Mexico (1,524 m). This species prefers a mosaic of vegetation, including cloud forests and pine forests.

CONSERVATION STATUS: This species is threatened by habitat fragmentation and habitat transformation for tourism development and small landowner agriculture. *S. planifrons* is not found in any protected areas (IUCN, 2010).

Toltec cotton rat

Sheri Westerman and Robert D. Bradley

SUBSPECIES IN MEXICO

S. toltecus is a monotypic species.

Peppers et al. (2002), Carroll et al. (2005), and Bradley et al. (2008) suggested that *S. toltecus* be elevated to species level based on genetic data. Bradley et al. (2008) found genetic support for two clades within *S. toltecus*: a southern geographical group corresponding to individuals from coastal lowlands of southern Mexico and a northern group corresponding to individuals from Tamaulipas and Veracruz, Mexico; however, no subspecies have been proposed.

DESCRIPTION: The dorsal pelage of *Sigmodon toltecus* is gray with blackish or dark brownish hairs, interspersed with buffy hairs. It is similar in appearance to *S. hispidus* except for reddish-gray ventral coloration. The sides are slightly paler than the underside; the tail is darkly colored and sparsely haired, and has large scales. The skull is long and narrow, in contrast with the short, broad skulls of the *S. fulviventer*-group (Baker 1969). The chromosome diploid number is 2n = 52 and the fundamental number is FN = 52 (Swier et al. 2009). Previously, *S. toltecus* was ranked as a subspecies of *S. hispidus*; however, recent molecular studies have revealed a high degree of genetic divergence (Bradley et al., 2008; Carroll et al., 2005; Peppers and Bradley, 2000) and *S. toltecus* subsequently has been elevated to species rank.

DISTRIBUTION: This species occurs in the lowlands and foothills of eastern and southeastern Mexico into northern Honduras. Its distribution ranges from the Rio Grande east of the Sierra Madre Oriental, south across the Isthmus of Tehuantepec into the Yucatan Peninsula, Chiapas, Belize, and Honduras. Two mountain ranges (Montañas del Norte de Chiapas and Sierra Madre de Chiapas) may restrict the habitat of *S. toltecus* to the northeast along the coastal regions of the Yucatan Peninsula, and the Cordillera Nombre de Dios may restrict the distribution to the east. It has been recorded in sympatry with *S. hirsutus* in north-central Chiapas, where the lowland grasslands of the Depression Central de Chiapas transition into the moderate-elevation levels of the Sierra Madre de Chiapas (Bradley et al., 2008; Carroll et al., 2005).

EXTERNAL MEASURES AND WEIGHT

TL = 224 to 365 mm; TV = 81 to 166 mm; HF = 28 to 41 mm. EAR = 16 to 24 mm. Weight: 110 to 225 g (male), 100 to 200 g (female).

DENTAL FORMULA: I 1/1, C 0/0, PM 0/0, M3/3 = 16.

NATURAL HISTORY AND ECOLOGY: *S. toltecus* is found in coastal and moist lowland grassland and foothill habitats, including the subhumid and humid subtropical zones east of the Sierra Madre Oriental. Its diet consists primarily of grasses (Fleharty and Olson, 1969). In Mexico, a maximum density of 51 individuals/ha occurred in the autumn and a minimum of 25 individuals/ha occurred in the summer (Petersen, 1973). *S. toltecus* is most commonly caught in grass-dominated habitat. The gestation period is approximately 27 days (Randolph et al., 1977), and the offspring are well developed at birth (Meyer and Meyer, 1944).

VEGETATIONAL ASSOCIATIONS AND ELEVATION RANGE: *S. toltecus* inhabits humid coastal grasslands and foothills between lowland grasslands and montane grasslands, particularly perennial grasslands near edges of ponds. This species ranges from sea level in Tamaulipas, Veracruz, and the Yucatan Peninsula, to a few hundred meters in Chiapas and northern Honduras.

CONSERVATION STATUS: Due to its large population and wide distribution throughout Mexico, *S. toltecus* is listed as least concern by the IUCN (2010).

Zanjón cotton rat

Megan S. Corley and Robert D. Bradley

SUBSPECIES IN MEXICO

Sigmodon zanjonensis villae Goodwin (1958)

The status and relationship of these highland Central American taxa remain uncertain (Carleton et al., 1999). The geographic and altitudinal ranges are improbably broad and likely include lowland populations of *S. zanjonensis* referable to *S. toltecus* and possibly *S. hirsutus*; however, limits require critical analysis and vouchered documentation (Musser and Carleton, 2005).

DESCRIPTION: Goodwin (1932) originally described *Sigmondon zanjonensis* from specimens collected in Zanjón, Guatemala. Coloration of the dorsum, including the outer surfaces of the forearms and hind legs, is dull yellowish-brown, darkened by long black guard-hairs, some of which are tipped with yellow on the rump and sides. Cheeks, legs, and shoulders are slightly less darkened by the black guard-hairs. The eyes have indistinct rings of yellow, and the ears are sparsely covered with fine white-tipped hairs. The tail is distinctly bicolored with blackish-brown on the dorsal surface and white on the ventral surface. Underparts are white with the plumbeous at the base. The diploid number of chromosomes is 2n = 52 (Cameron and Spencer, 1981).

DISTRIBUTION: *S. zanjonensis* is known from the highlands of Chiapas, Mexico, and western Guatemala.

EXTERNAL MEASURES AND WEIGHT

TL = 280 mm; TV = 124 mm; HF = 30 mm; EAR = 16 mm.

Weight: NA.

DENTAL FORMULA: I 1/1, C 0/0, PM 0/0, M 3/3 = 16.

NATURAL HISTORY AND ECOLOGY: As with most species of *Sigmodon*, *S. zanjonensis* is found predominantly in grassy habitats in which grass height and density are important components. They also may occur in perennial grasses and forbs near edges of ponds. Cotton rats build and utilize nests made of woven grasses that can range from cup-shaped to hollow ball-shaped with one entrance. Litter sizes range from 1 to 15, with smaller litters occurring in moister habitats and larger litters in drier habitats; areas with marked seasonality in temperature or rainfall yield females with larger litters. The gestation period lasts approximately 27 days. Neonates are well developed at birth. Young cotton rats grow rapidly and can be weaned at 10 to 15 days. Males reach sexual maturation between 2 weeks and 1 month of age, and females generally begin to show signs of estrus at 50 days of age, with cycles lasting 8 to 9 days. Female body weights are on average heavier than those of males during breeding seasons but not non-breeding seasons. Cotton rats eat primarily grasses and seeds, but show seasonal utilization of insects. Average home range is 0.35 ha for adults, 0.22 ha for subadults, and 0.35 ha for juveniles. Males have larger home ranges than females, but females have exclusive home ranges. Cotton rats are active at all hours of the day and night; they exhibit a labile activity pattern influenced by biotic and/or abiotic factors. Seasonal differences in movements are influenced by reproductive stage or the availability of cover (Cameron and Spencer, 1981).

VEGETATIONAL ASSOCIATIONS AND ELEVATION RANGE: *S. zanhonensis* is found in montane grasslands and cloud forests. It is found up to 2,700 m.

CONSERVATION STATUS: The conservation status of this species is unknown.

Chiapan climbing rat

Eduardo E. Espinoza Medinilla

DISTRIBUTION: *T. bullaris* is endemic to Mexico. It is only known in the central valley of Tuxtla Gutierrez. It has been recorded in the state of CS.

SUBSPECIES IN MEXICO
T. bullaris is a monotypic species.

DESCRIPTION: *Tylomys bullaris* is a large-sized rat similar to *T. nudicaudus*. The type specimen is an immature individual. The dorsal coloration is gray; the upper lip and a patch on the nose are white. The belly is white, the feet are dark, and the nails are white. The tail is bicolored, dark at the base and white on the tip. The zygomatic arch is very wide, and the bulla is rounded at the top (Hall, 1981).

EXTERNAL MEASURES AND WEIGHT
TL = 320 to 495 mm; TV = 189 to 230 mm; HF = 33 to 40 mm; EAR = 26 to 30 mm. Weight: 182 to 370 g.

DENTAL FORMULA: I 1/1, C 0/0, PM 0/0, M 3/3 = 16.

NATURAL HISTORY AND ECOLOGY: Very little is known about the natural history of *T. bullaris*. In general, the species of the genus *Tylomys* are nocturnal, solitary, arboreal, or terrestrial (Emmons and Feer, 1997). They have been found in hollow trees, lianas, and rocks in well-preserved forest. They feed on fruit, sprouts, and bark (Álvarez del Toro, 1977).

VEGETATIONAL ASSOCIATIONS AND ELEVATION RANGE: These rats are only found in undisturbed tropical deciduous forests, approximately 600 masl.

CONSERVATION STATUS: *T. bullaris* is threatened (SEMARNAT, 2010). Its habitat has been severely impacted in the central valley of Chiapas and its populations are scarce. A small population inhabits the forest of the zoo in Tuxtla Gutierrez, and it is possible that it inhabits the Cañon del Sumidero.

Tylomys bullaris. Tropical semi-green forest. Ocozocuautla, Chiapas. Photo: Miguel Alvarez del Toro.

Peter's climbing rat

Eduardo E. Espinoza Medinilla

SUBSPECIES IN MEXICO

Tylomys nudicaudus gymnurus Villa, 1941
Tylomys nudicaudus microdon Goodwin, 1955
Tylomys nudicaudus nudicaudus (Peters, 1866)
Tylomys nudicaudus villai Schaldach, 1966

DISTRIBUTION: *T. nudicaudus* is distributed from the southeastern Mexico, throughout Central America to Costa Rica; it occupies a large part of Belize and Guatemala (Hall, 1981). In Mexico, it occurs on the Gulf side, from Puebla to Veracruz, and on the Pacific Rim from Guerrero to Oaxaca and Chiapas. It has been recorded in the states of CS, GR, OX, PU, and VE.

DESCRIPTION: *Tylomys nudicaudus* is one of the largest rats in Mexico. The coat is relatively long and smooth. The dorsal coloration is cinnamon-red, giving the rat a reddish look; the belly is yellowish-white. The tail is almost as long as the body, with two-thirds dark and the remaining yellowish-white. The ears are large, dark, and rounded. The feet are dark to the metacarpal, and the phalanges are white. The feet are covered with white hairs that protrude beyond the nails. Immature individuals are dark gray on the back and white underneath. The tail is not bicolored. Lactating females have a prominent udder with four teats.

EXTERNAL MEASURES AND WEIGHT

TL = 300 to 500 mm; TV = 160 to 400; HF = 38.6 to 45 mm; EAR = 20 to 23.4 mm. Weight: 251 to 370 g.

DENTAL FORMULA: I 1/1, C 0/0, PM 0/0, M 3/3 = 16.

NATURAL HISTORY AND ECOLOGY: *T. nudicaudus* is nocturnal. During the day it takes refuge in hollow trees such as ceibas, rocks, and caves. They have been captured near rocky areas on trees, in walls of rocks, and in the interior of caves, in both humid and dry regions, although the presence of water is a major requirement for its activities (Álvarez del Toro, 1977; Ramírez-Pulido and Sánchez-Hernandez, 1971; Webb and Baker, 1969). It feeds on seeds and fruits. They have two to three pups in two seasons, one that extends from March to May and another in November and December; in only five months pups acquire the adult coloration (Álvarez del Toro, 1977). They have been caught with other species of rodents such as *Reithrodontomys fulvescens*, *Oryzomys couesi*, *Peromyscus mexicanus*, and *Sigmodon hirsutus*. Its predators include barn owls (*Tyto alba*) and other birds of prey (Baker and Petersen, 1965; Ramírez-Pulido and Sánchez-Hernandez, 1971). In some regions of northern Chiapas, it is known as the squirrel mouse. In these regions they are hunted and consumed by locals (Álvarez del Toro, 1977).

VEGETATIONAL ASSOCIATIONS AND ELEVATION RANGE: *T. nudicaudus* inhabits tropical deciduous forests, tropical perennial forests, and cloud forests, from sea level to 1,600 m (Baker et al., 1971; Espinoza et al., 1999; Ramírez-Pulido and Sánchez-Hernandez, 1971; Webb and Baker, 1969).

CONSERVATION STATUS: The current status of the populations of this species is unknown. It requires habitat in good condition and is relatively scarce; hence, it should be considered threatened.

Tumbalá climbing rat

Eduardo E. Espinoza Medinilla

DISTRIBUTION: *T. tumbalensis* is endemic to Mexico; it is only known in the type locality in Tumbalá, Central Highlands of Chiapas (Hall, 1981). It has been recorded in the state of CS.

SUBSPECIES IN MEXICO
T. tumbalensis is a monotypic species.

DESCRIPTION: The only known specimen of *Tylomys tumbalensis* is a young subadult. It is very large, similar to *T. nudicaudus*. The body is robust; the tail is naked, scaly, and white on the tip. The dorsal coloration varies from dark-gray to blackish. The cheek, gular part, and groin are white, while the rest of the belly is gray. The feet are gray with obscure nails. The skull is slimmer and elongated; the bulla is less rounded and slightly longer (Hall, 1981).

EXTERNAL MEASURES AND WEIGHT
TL = 448 mm; TV = 234 mm; HF = 46 mm; EAR = 22 mm.
Weight: 250 g.

DENTAL FORMULA: I 1/1, C 0/0, PM 0/0, M 3/3 = 16.

NATURAL HISTORY AND ECOLOGY: There are no accurate data on the natural history of *T. tumbalensis*. The species of this genus are nocturnal, arboreal, and terrestrial. They are solitary, and are frequently found between rocks, in fallen logs, and in caves. They depend on undisturbed and dense forests. Its diet possibly consists of green plants (Emmons and Feer, 1997).

VEGETATIONAL ASSOCIATIONS AND ELEVATION RANGE: The habitat type is pine-oak forests, at approximately 1,000 masl.

CONSERVATION STATUS: *T. tumbalensis* is regarded as rare (SEMARNAT, 2010). Due to the steady increase in the reduction of its habitat, its restricted distribution, and the scant knowledge of its populations, it should be regarded as threatened.

Xenomys nelsoni Merriam, 1892

Magdalena woodrat

Gerardo Ceballos

SUBSPECIES IN MEXICO
X. nelsoni is a monotypic species.
Its systematic relationships are little known; however, the genus is closely related to *Nelsonia*, *Hodomys*, and *Neotoma* (Carleton, 1980; Carleton and Musser, 1993; Edwards and Bradley, 2002).

DESCRIPTION: *Xenomys nelsoni* is a large-sized rat. The body is robust and the tail is long and relatively hairy (Ceballos and Miranda, 1986, 2000). The ears are large, with few hairs. The feet have well-developed plantar tubercles. The dorsal coloration is yellowish-brown, and the belly is white. The hair of the back is gray at the base and yellowish in the terminal portion. It has two conspicuous white spots above the eyes and behind the ears. The skull is robust, with large bullae, molars without apparent lateral cusps, and with the last molar in the shape of an S (Ceballos and Miranda, 2000; Ceballos et al., 2002; Schaldach, 1960).

EXTERNAL MEASURES AND WEIGHT
TL = 249 to 345 mm; TV = 143 to 170 mm; HF = 27 to 32 mm; EAR = 21 to 23 mm. Weight: 70 to 160 g.

DENTAL FORMULA: I 1/1, C 0/0, PM 0/0, M 3/3 = 16.

NATURAL HISTORY AND ECOLOGY: This rat is strictly nocturnal and arboreal. In the region of Chamela-Cuixmala, on the coast of Jalisco, 65% (22) of the captures were in trees and the rest at the base of trees (Ceballos, 1989, 1990). In Colima, they have been collected between the branches of trees in low tropical forests and thorn forests (Schaldach, 1960). They feed on vegetative material such as leaves of mojo (*Trichilia* sp.; Schaldach, 1960). The stomach contents of several specimens of Chamela had leaves and, probably, fruit, nuts, and finely shredded seeds; at the same site, some individuals survived in captivity in excellent condition, with a diet of seeds and wild fruits (Ceballos, 1989). They live in areas with dense vegetation, where the canopy is closed. The only study on its population was carried out in the Chamela-Cuixmala Biosphere Reserve, Jalisco (Ceballos, 1989, 1990; Mendoza, 1997). Their densities are low, and vary from 0.3 to 1 individual/ha in low tropical forests and adjacent medium tropical forests, respectively. The area of activity of males is greater than that of females. The habitat in which the largest population density has been recorded is characterized by a large amount of vines, which help them move between the trees. In general, they coexist with species such as the gray mouse opossum (*Tlacuatzin canescens*) and several species of rodents (*Nyctomys sumichrasti*, *Peromyscus perfulvus*, *Osgoodomys banderanus*) (Ceballos 1989; 1990; Helm et al. 1974). They build their burrows in holes of trees such as the iguanero (*Caesalpina eriostachys*), barcino (*Codia eleagnoides*), *Coupeia polyandra*, and probably mojo (*Trichilia* sp.). The nests are spherical and built with twigs, leaves, and fibers of fruits such as *Ceiba* (Ceballos, 1990; López-Forment et al., 1971; Schaldach, 1960). There is little information on their reproductive biology, but apparently they mate at the end of the dry season and in

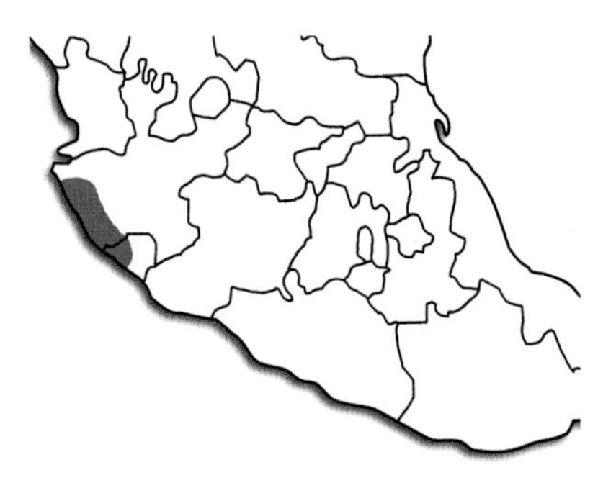

DISTRIBUTION: *X. nelsoni* is endemic to Mexico. It is only known in few localities located in the coastal plain of the Pacific, Colima, and Jalisco. It has been recorded in the states of CL and JA.

Xenomys nelsoni. Tropical dry forest. Chamela, Cuixmala Biosphere Reserve, Jalisco. Photo: Gerardo Ceballos.

the rainy season (May–November). Pregnant females have been captured from May to September, and scrotal males between February and October. Each litter has up to two offspring that weigh 5 g and are naked; their eyes are closed; they remain glued to the mammae for several hours, even when the mother moves (Ceballos, 1990).

VEGETATIONAL ASSOCIATIONS AND ELEVATION RANGE: *X. nelsoni* has been recorded exclusively in low deciduous tropical forests, low subdeciduous tropical forests, and thorn forests (Ceballos and Miranda, 1986, 2000; Schaldach, 1960). In Colima, it was found in an orchard of mamey (*Calocarpum mammosum*) adjacent to a thorn forests (Schaldach, 1960). It is found from sea level to 600 masl.

CONSERVATION STATUS: *X. nelsoni* belongs to one of the 11 genera of endemic mammals of Mexico. It is one of the most threatened rodents in the world, and is regarded as endangered by its restricted distribution and the destruction of the dry tropical forests (Ceballos et al., 1998; Ceballos and Miranda, 1986, 2000). It is threatened (SEMARNAT, 2010).

Family Erethizontidae

The family Erethizontidae is endemic to the American continent, where it is represented by 4 genera and 12 species (Wilson and Reeder, 1993). In Mexico, it is represented by two species. Porcupines are very large rodents that live in trees or are semi-arboreal and nocturnal. They are found in forests, jungles, and xeric scrublands from Canada to Argentina. The northern porcupine (*Erethizon dorsatum*) is at risk of extinction in Mexico, although it is very common in Canada and the United States.

Erethizon dorsatum (Linnaeus, 1758)

North American porcupine

Eduardo Ponce and Gerardo Ceballos

SUBSPECIES IN MEXICO

Erethizon dorsatum couesi Mearns, 1897

DESCRIPTION: *Erethizon dorsatum* is one of the largest rodents in Mexico. It is a robust animal that can reach 11 kg in weight and up to 90 cm long, with the head and feet short. A feature that separates it from virtually all other mammals of Mexico, with the exception of *Sphiggurus mexicanus*, is the body covered by around 30,000 quills or spines, which are modified hairs mainly distributed on the back and sides of body; the largest are located on lower part of the back and tail and can reach up 75 mm in length and 2 mm wide. The underside does not have quills and is mostly uncovered by hair. In many regions there is the popular idea about a quill shooting; however, this is not true. The quills are controlled by muscles that release them only when they make contact with a surface. The color of the body is brown-yellowish, dotted with darker shades (Woods, 1973). It differs from tropical porcupines (*C. mexicanus*) in many ways, including its larger size and non-prehensile tail.

EXTERNAL MEASURES AND WEIGHT

TL = 645 to 1,030 mm; TV = 145 to 300 mm; HF = 75 to 91 mm; EAR = 25 to 42 mm. Weight: 3 to 12 kg.

DENTAL FORMULA: I 1/1, C 0/0, PM 1/1, M 3/3 = 20.

NATURAL HISTORY AND ECOLOGY: North American porcupines mostly live in conifers and oak forests. In northwestern Mexico and the southwestern United States, however, they are found in riparian habitats along rivers that cross the extensive arid plains and in arid grasslands (Benson, 1953; Jones and Genoways, 1968; Taylor, 1935). In these regions it is also found in shrublands and grasslands with an abundance of cresote bush (*Larrea tridentata*), mesquites (*Prosopis* spp.), chollas (*Opuntia* spp.), desert sumac (*Rhus microphyla*), and other shrubs (List et al., 1999, Ponce et al., in press). When available, they prefer rocky slopes with dense vegetation habitats (Schmidly, 1977a; Taylor, 1935). In the Janos-Casas Grandes region, Chihuahua, as well in arid and temperate regions of southern Arizona, its density varies from 1.2 to 5.8 animals per km^2 (Ponce et al., in press; Taylor, 1935). In Mexico, their densities are very low, despite having the appropriate conditions for their proliferation (Jones and Genoways, 1968; Leopold, 1965). They are solitary most of the year, with the exception of the mating season, which occurs from autumn to early winter. They are relatively sedentary and their area of activity comprises between 5 ha and 14 ha, with daily movements between 300 m and 500 m. Activity areas and movements are larger in the summer probably because of low food availability (Ponce et al., in press; Woods, 1973). Their diet consists almost exclusively of plant material, which is degraded by bacteria in their intestines. In winter, they feed on buds, sprouts, and bark of trees. The inner bark of trees such as oak (*Quercus* sp.), pine (*Pinus cembroides*), willow (*Salix* sp.), juniper (*Juniperus* sp.), and netleaf hackberry (*Celtis reticulata*) is obtained removing the external bark with their incisors, which have an intense orange color due to their iron coating that help them chew the outer bark. In other seasons their diet includes roots, stems, fruits, seeds, and flowers (Schmidly, 1977b; Taylor, 1935; Woods, 1973). They find refuge in underground burrows, trees, shrubs, and cracks on rocky slopes. In Janos-Casas Grandes they use poplar (*Platanus wrightii*), walnut (*Juglans major*), neatleaf hackberry (*Celtis reticulata*), velvet ash (*Fraxinus* spp.), narrowleaf cottonwood (*Populus angustifolia*), and mesquites (*Prosopis glandulosa*) as refuges (List et al., 1999; Ponce et al., in press). They have few natural predators as their body is covered with quills (spines) that act as a formidable protection. In many regions there is the popular idea about quill shooting; however, this is not true. When another animal approaches they only can do a fast tail movement to hit and nail a few quills in the intruder's body. When young they can be attacked by coyotes, mountain lions, bobcats, and great horned owls. They are also attacked by domestic dogs. Gestation lasts approximately seven months, after which a female gives birth to one or two offspring. They are born between February and June and stay with the mother for at least a year (Ponce et al., in press; Schmidly, 1977b; Woods, 1973). From birth the offspring are covered by soft quills that harden in minutes; they weigh around 500 kg, increasing 500 g each month to reach the adult size and sexual maturity at a year and a half (Woods, 1973). They have high survival rates, and can live up to 10 years on captivity. In the wild tree cutting and the low food availability in winter are the main natural causes of mortality among porcupines.

DISTRIBUTION: The North American porcupine is widely distributed in North America, and is known from Canada to the north of Mexico. Its distribution in Mexico is, however, very limited since it is known in less than 20 locations, scattered in the north of the country (Benson, 1953; Genoways and Jones, 1967; Hall, 1981; Jiménez G. and Zúñiga-R, 1992; List et al., 1999). It has been registered in the states of CH, CO, DU, NL, SI, and SO.

VEGETATIONAL ASSOCIATIONS AND ELEVATION RANGE: The North American porcupine inhabits a vast region from the tundra in Alaska to desert habitats on northern Mexico. Most records in Mexico are from semi-arid regions with riparian

Erethizon dorsatum. Riparian forest. Janos Biosphere Reserve, Chihuahua. Photo: Rurik List.

vegetation and scrubland. There are few records in temperate forests. It is found from 1,000 masl to 2,800 masl (List et al., 1999).

CONSERVATION STATUS: In Mexico *E. dorsatum* is considered endangered due to its low densities (Ceballos and Navarro, 1991; SEMARNAT, 2010). Paul Martin (pers. comm.) considers that the restricted distribution of porcupines in Mexico is due to the reduction of their populations by poaching. Also, their restricted populations can be the result of loss of suitable habitat for the species such as riparian corridors that have been transformed by livestock management and water overexploitation on most parts of Mexico (List et al., 1999; Ponce et al., in press). In order to determine the most appropriate measures for their protection, it is necessary to evaluate their current distribution and the status of their populations. The protection of this species in the Southwest of the United States has allowed the restoration of their populations, and a notable increment in their distribution and abundance (Schmidly, 1977b). In Mexico, porcupines are protected in several reserves such as Maderas del Carmen in Coahuila, and Mapimi in Durango. The largest population is located in the recently created (2009) Janos Biosphere Reserve (List et al., 1999).

Sphiggurus mexicanus (Kerr, 1792)

Mexican porcupine

Julio R. Juárez-G.

SUBSPECIES IN MEXICO

Sphiggurus mexicanus mexicanus (Kerr, 1792)
Sphiggurus mexicanus yucataniae (Thomas, 1902)

DESCRIPTION: *Sphiggurus mexicanus* is a large-sized rodent with a robust body and approximately the size of a rabbit. The face is short and wide. The ears are very small. The paws have four fingers (the fifth is reminiscent) fitted with long and curved nails. The body is covered by white-yellowish spines with dark tips, mixed with long hair that is dorsally yellow to dark brown; the underside has shorter spines, which are flexible and scarce, mixed with light gray fur. The tail is long and has few spines and black hair in the first one-third; the tip is naked and prehensile.

DISTRIBUTION: *S. mexicanus* is exclusively found from Mexico to Costa Rica and Panama (Reid, 1997). In Mexico, it is found in two coastal plains, from San Luis Potosí to the Yucatan Peninsula in the Gulf, and Guerrero to Chiapas in the Pacific (Hall, 1981). It has been registered in the states of CA, CS, GR, QE, QR, OX, SL, VE, and YU.

EXTERNAL MEASURES AND WEIGHT
TL = 625 to 900 mm; TV = 295 to 440 mm; HF = 70 to 95 mm; EAR = 22 to 24 mm. Weight: 1,500 to 3,000 g.

DENTAL FORMULA: I 1/1, C 0/0, PM 1/1, M 3/3 = 20.

NATURAL HISTORY AND ECOLOGY: Porcupines are nocturnal and apparently solitary. Their shelters are usually found in hollow trees and between the branches of dense foliage. Its diet consists of leaves, bark, and fruits. They give birth to one young per litter and reproduce at the end of winter and spring. In Mexico, a female with an embryo was captured in January. Pregnant females and births of several species of the genus occur in January, February–March, May, July, August, and September. With regard to the males, there is no published data on the testicular measures.

VEGETATIONAL ASSOCIATIONS AND ELEVATION RANGE: Porcupines live in different types of vegetation that include high perennial jungle, medium subperennial jungle, tropical deciduous forest, and cloudy forest, from sea level up to 2,350 m.

CONSERVATION STATUS: Because they depend on forests and jungles, these porcupines are considered as threatened (SEMARNAT, 2010). They are more tolerant of anthropogenic disturbances than other species, making them vulnerable. There are protected populations in reserves such as Calakmul, Sian Káan, Tuxtlas, Burial, Huatulco, and Montes Azules.

Sphiggurus mexicanus. Tropical rainforest. Coatzacoalcos, Veracruz. Photo: Gerardo Ceballos.

Family Dasyproctidae

The family Dasyproctidae is exclusive to the tropical regions of America. It contains 13 species in 2 genera (Wilson and Reeder, 1993). The acuchis and agoutis are large terrestrial and nocturnal rodents. In Mexico, two species can be found: *Dasyprocta punctata*, which has a wide distribution, and *D. mexicana*, which is endemic to the country and is vulnerable because of the disappearance of the tropical forests that it inhabits.

Dasyprocta mexicana Saussure, 1860

Mexican agouti

Héctor T. Arita

SUBSPECIES IN MEXICO

D. mexicana is a monotypic species.

The black guaqueque could only be a geographical variety (subspecies) of *D. punctata*. It is likely that many of the 10 or 11 species recognized for *Dasyprocta* are only subspecies (Woods, 1993). Emmons and Feer (1997) reported a possible hybrid between *D. mexicana* and *D. punctata* in Tabasco. Similarly, in northern Chiapas, individuals with an intermediate coloration between these two species have been observed (M. Aranda, pers. comm.)

DESCRIPTION: *Dasyprocta mexicana* is a medium-sized hystricognath. The body is elongated, although the typical position of the animal makes it look rounded. The ears are short, the tail is very small and barely visible, and the hind feet are comparatively long and have three digits. The dorsal coat is black or brown, very dark but

Dasyprocta mexicana. Tropical rainforest. La Venta, Tabasco. Photo: Gerardo Ceballos.

with white tips. The belly is lighter, especially in the neck, where it is almost white. It differs from *D. punctata* solely by its darker color.

EXTERNAL MEASURES AND WEIGHT
TL = 515 to 560 mm; TV = 20 to 30 mm; HF = 116 to 127 mm; EAR = 35 to 47.
Weight: 2 to 5 kg.

DENTAL FORMULA: I 1/1, C 0/0, PM 1/1, M 3/3 = 20.

NATURAL HISTORY AND ECOLOGY: Very little is known about the biology of this species. In some cases, it is difficult to discern if the information available corresponds to this species or to *D. punctata*. It is terrestrial and diurnal, although it can be seen at night, especially in areas with a lot of human activity (Hall and Dalquest, 1963). They are solitary animals, although they sometimes form couples. They build their burrows inside or below fallen logs or between the buttresses of big trees of the tropical forests. The typical refuge consists of a simple tunnel between 3 m and 5 m long and between 10 cm and 25 cm wide, usually at a depth of between 30 cm and 1 m (Hall and Dalquest, 1963). Between these shelters and their feeding sites there is narrow path used regularly. When disturbed, it remains motionless for a moment and then runs to take refuge in any hole or tunnel. When it flees, the back hairs bristle and emit a very audible sound (Hall and Dalquest, 1963). Its diet is mainly based on fruits and seeds. In a study conducted in Los Tuxtlas, Veracruz, they were observed eating fruit from trees such as *Ficus* spp., *Spondias mombin*, *Brosimum alicastrum*, and *Pseudolmedia oxyphyllaria* (Navarro, 1981). There are no systematic data on its pattern of reproduction, although in Los Tuxtlas young have been found from January to May and in the vicinity of the river Coatzacoalcos breast-feeding females or with embryos were found at the end of April (Coates-Estrada and Estrada, 1986; Hall and Dalquest, 1963). Its predators include medium-sized carnivores such as ocelots (Hall and Dalquest, 1963). Hunters and farmers who argue that it damages the maize crops kill it.

DISTRIBUTION: This species is endemic to Mexico. It is found in tropical lowlands of southern Veracruz and the northern parts of Oaxaca and Chiapas. It has been introduced to Cuba (Hall, 1981). It has been registered in the states of CS, OX, and VE.

VEGETATIONAL ASSOCIATIONS AND ELEVATION RANGE: The guaqueque is found exclusively in tropical perennial forest, from sea level up to about 600 m.

CONSERVATION STATUS: Despite being an endemic species to Mexico and its restricted distribution, *D. mexicana* is not included in any official list of endangered species. Because of its association with the tropical perennial forest it could be threatened by deforestation of the tropical forests of southeast Mexico.

Dasyprocta punctata Gray, 1842

Central American agouti

Karina Santos del Prado and Héctor T. Arita

SUBSPECIES IN MEXICO
Dasyprocta punctata chiapensis Goldman, 1913
Dasyprocta punctata yucatanica Goldman, 1913

DESCRIPTION: *Dasyprocta punctata* is a medium-sized hystricognath and is one of the largest rodents in Mexico. The body is thin and elongated, the ears are short, the tail is small and barely visible, and the hind feet are relatively longer than the forefeet and have three digits (Emmons and Feer, 1997; Hall, 1981; Nowak, 1999b). Usually, the coat is rough and glossy, reddish-brown, yellowish-brown, or yellowish-gray more or less uniform on the back and sides and with fine blackish lines (Leopold, 1965). The hair of the nape and legs does not differ from that on the back but on the hindquarters it is generally longer than on any other part of the body. The chest is gray and the chin and groin are orange or white in animals with a yellowish-gray coloration. Females have four pairs of ventral mammae. The penis is coated by small spines, and has a pair of longitudinal bones (Navarro et al., 1991).

EXTERNAL MEASURES AND WEIGHT

TL = 490 to 620 mm; TV = 10 to 35 mm; HF = 110 to 133 mm; EAR = 36 to 47 mm. Weight: 1.3 to 4.0 kg.

DENTAL FORMULA: I 1/1, C 0/0, PM 1/1, M 3/3 = 20.

NATURAL HISTORY AND ECOLOGY: The Central American agouti is a diurnal animal, although in areas with great human activity it can also be seen at night (Leopold, 1965). It is adapted to a cursorial life and can jump vertically almost 2 m. In Barro Colorado Island, Panama, a population density of 1.0 individual per hectare has been reported. At Tikal, Guatemala, the reported density 0.1 individual per hectare (Nowak, 1999b). This rodent is strongly associated with water bodies. It builds burrows to eat or rest in various sites, usually located between limestone rocks, in cavities or below fallen logs, among the roots of trees, or under vegetation. In addition, it builds nests and paths (Smythe, 1978). Its diet is mainly based on fruits, soft seeds, tender stems from plants found in the soil of the tropical forests, roots, peel, and grains (Leopold, 1965). They can also eat flowers, leaves, fungi, and insects. The fruit is collected and taken to feeding sites. When food is abundant, they carefully bury seeds in holes 2 cm to 8 cm deep, to use as food when fruit is scarce or not in season. This behavior is important in the dispersal of the seeds of many species such as almond (*Dipteryx panamensis*), avocado (*Persea* spp.), and corozo (*Scheelea* spp.); they can transport the seeds up to 150 me from the trees, saving them from the attack of preda-

Dasyprocta punctata. Tropical semi-green forest. El Sumidero National Park, Chiapas. Photo: Gerardo Ceballos.

tors and promoting their germination (Smythe, 1986). When fruit is scarce, they eat the stored seeds; there is a report of guaqueques who ate a male adult mouse *Liomys pictus* (Smythe, 1978, 1986). Agoutis feed by sitting on their hind legs and holding their food in their forepaws. Individuals often follow bands of monkeys and pick up fruit dropped from trees. The basic social unit of the agouti is made up of a pair that mates with one another for life. Each pair occupies a territory of approximately 1 ha to 2 ha, which is defended aggressively, especially when fruit is scarce. Males are more active in the defense of the territory and are more aggressive with other males than with females. It is likely that neighboring territories are occupied by their offspring; thus a territorial dominance hierarchy exists. The simple fact that a guaqueque walks toward another constitutes an act of aggression: if the situation is highly aggressive, the aggressor tosses its head toward the other, erects the long hairs of its rump, shows its anal glands, and sits in front of its opponent to begin the combat. Agoutis devote considerable time to grooming to remove parasites, ticks, and mites. The forefeet are used to rake the hair and draw it within reach of the incisors, which are then used as a comb. Courtship begins with nasal contact between the female and male; the male then sprays the female with urine, which causes her to go into a "frenzy dance" and then the male mounts. Females have an estrus of 34 days and a gestation period of 104 to 120 days (Weir, 1974a). Usually, a litter normally contains one to two offspring (Álvarez del Toro, 1991), sometimes three, and there is only one report of four (Smythe, 1978). The birth occurs in any of the burrows that the mother uses to sleep in. The newborn are fully furred, their eyes are open, and they are able to run in their first hour of life. The second day they are ready to eat solids. The female stimulates the offspring, urinating and defecating on them, thereby reinforcing, probably, the mother-offspring link. During the first three weeks, the female does not allow any other guaqueque in the burrow, but the third week, the male comes closer when the female is foraging and juveniles can leave the burrow to interact with him. When the young are five to seven weeks of age they follow their mother through the whole foraging area, staying up to four or five months. The mother usually nurses for 20 weeks. Offspring become completely separated from the mother upon the arrival of a new litter because of parental aggression or due to lack of food. Sexual maturity is reached around six months of age (Smythe, 1978). The reproduction can be seasonal and continuous (Asdell, 1964; Weir, 1974a). In Barro Colorado Island, a peak of juveniles can be found from March to July, when fruit is abundant (Smythe, 1978). Visual communication plays an important role in the aggressive behavior; odors play an important role in agouti communication. Both males and females have anal scent glands that secrete a substance with a powerful and lasting smell that is used to mark their territory and their paths. They emit acute alarm sounds only in circumstances of danger, while running, but never when they are seated. They growl during combat and shriek when some predator catches them. It is one of the main prey of ocelot and jaguarondi, and occasionally of jaguar, puma, tayra, and raccoon.

DISTRIBUTION: This species is found in Central America and South America from the northwest of Venezuela and Colombia to the north of Argentina and has been introduced in the Cayman Islands. In Mexico, it is distributed in the tropical lowlands of the southeast, from the eastern half of Tabasco and southern Chiapas to the Yucatan Peninsula (Emmons and Feer, 1997; Hall, 1981; Nowak, 1999b). It has been registered in the states of CA, CS, QR, TB, and YU.

VEGETATIONAL ASSOCIATIONS AND ELEVATION RANGE: *D. punctata* inhabits the tropical perennial forest, tropical subdeciduous forest, and tropical deciduous forest (Eisenberg, 1989; Emmons and Feer, 1997; Smythe, 1986). It can be occasionally found in chaparral and secondary vegetation (Álvarez del Toro, 1991). It is found from sea level to approximately 1,500 m (Emmons and Feer, 1997).

CONSERVATION STATUS: *D. punctata* is not included in any official list of endangered species. Arita et al. (1990) consider it locally abundant and with a wide distribution. In some areas, however, the population number has declined because of the habitat destruction; they are intensively hunted for their meat, but even so, some remaining populations still survive in many areas of their distribution (Emmons and Feer, 1997; Nowak, 1999b; Vickers, 1991).

Family Cuniculidae

The family Cuniculidae is endemic to the Neotropical region, where it is represented by a single genus and species (Wilson and Reeder, 2005). The pacas are very large, terrestrial, and nocturnal rodents. They are found in rainforests from Mexico to northern Argentina. They are animals prized for their meat, and thus have disappeared from large tracts. In some regions they are at risk of extinction, but the species in general is not at risk.

Cuniculus paca (Linnaeus, 1766)

Lowland paca, Tepezcuintle

Jorge Ortega-R. and Héctor T. Arita

SUBSPECIES IN MEXICO

Cuniculus paca nelsoni Goldman, 1913

Woods (1993) indicated that the name *Cuniculus* was invalid. Recently, however, it has been recommended that it be kept since it is widely used (Ceballos et al., 2002).

DISTRIBUTION: *C. paca* is distributed from Mexico and Central America to southern Brazil and northern Argentina (Redford and Eisenberg, 1992). In Mexico it is found in the lowlands of the Gulf of Mexico, from the southeast of Tamaulipas to Chiapas, including the Yucatan Peninsula. It has been registered in the states of CA, CS, OX, QR, SL, TA, TB, VE, and YU.

DESCRIPTION: *Cuniculus paca* is the largest rodent of the Mexican tropics. The coat is light brown, with four longitudinal rows of white spots on the sides. The body is robust and the ears and tail are short; it has four digits on the forefeet and five on the hind feet. The cheeks are prominent due to the expanded zygomatic arch, which is used as a resonating chamber. This is a unique feature among mammals (fenestration), and is more developed in males (sexual dimorphism) and apparently amplifies emitted sounds (Nowak, 1999b).

EXTERNAL MEASURES AND WEIGHT

TL = 622 to 705 mm; TV = 24 to 27 mm; HF = 110 to 115 mm; EAR = 43 to 56 mm.
Weight: 6 to 12 kg.

DENTAL FORMULA: I 1/1, C 0/0, PM 1/1 M 3/3 = 20.

NATURAL HISTORY AND ECOLOGY: The paca is nocturnal and spends the day in its burrow, which is formed by simple tunnels 2 m to 6 m deep; the entrance is covered by dead leaves (Hall and Dalquest, 1963). In the Yucatan Peninsula, they often use caves and cenotes as burrows (Gaumer, 1917). They are active during the early hours of the night. When food is found, they transport it to the entrance of the burrow for later consumption (Leopold, 1959). They are solitary, but social interaction in captivity has been observed, even among individuals of the same sex (Aguirre and Fey, 1981; Smythe, 1991). They are territorial and mark their paths with secretions from the odoriferous glands located in the anal region (Smythe, 1983). They have been observed in population densities of 27.5 individuals/km^2, although under certain circumstances higher densities can be reached (Robinson and Redford, 1986). They are frugivores and granivores; their diet is based on stems, leaves, seeds, and fruits. In the Lacandon Forest they feed on fruits of mamey (*Pouteria sapota*), ramon (*Brosimum alicastrum*), pomo (*Eugenia uliginosa*), guapaque (*Dialium guianense*), and guatapil

Cuniculus paca. Tropical semi-green forest. El Sumidero National Park, Chiapas. Photo: Gerardo Ceballos.

(*Chamaedora tepejilote*), varying their diet in accordance with the availability of food (Hen, 1981). In Venezuela, the reproductive period occurs from March to May; single young are usual and twins are rare. Gestation has been reported to last 118 days (Kleiman et al., 1979). In Mexico, a female with a terminal embryo was captured in April in Veracruz (Hall and Dalquest, 1963). Its typical habitat is related to rivers. They are excellent swimmers and if they are threatened they often use water as escape route (Nowak, 1999b; Redford and Eisenberg, 1992). Pacas are one of the favorite prey of hunters, as they are relatively easy to catch and their meat is highly valued (Emmons and Feer, 1997). These animals are one of the five species hunted by natives and settlers (Ayres et al., 1991; Redford and Robinson, 1987; Vickers, 1991). Breeding them in captivity is difficult as they are highly territorial and show little reproductive potential (Emmons, 1987b; Smythe, 1978, 1991).

VEGETATIONAL ASSOCIATIONS AND ELEVATION RANGE: In Mexico pacas mainly live in tropical subdeciduous forest and tropical perennial forest. In the Yucatan Peninsula, they have been observed in a variety of habitats, from tropical subdeciduous forest to tropical deciduous forest, although apparently they do not inhabit extremely arid sites (Jones et al., 1974b). The elevation interval of this species in Mexico goes from sea level up to 1,800 masl (Hall and Dalquest, 1963).

CONSERVATION STATUS: This species has a broad distribution and high density (Arita et al., 1990). For Honduras, it is included within appendix III of CITES (Woods, 1993). In Mexico, it is possible that the species is overexploited and disappearing.

Order Soricomorpha

Gerardo Ceballos and Leslie N. Carraway

They are extremely voracious tiny animals with a fast metabolism, therefore they need to consume huge quantities of food.
Miguel Álvarez del Toro, 1977

Soricomorphans are a diverse order of mammals that includes 4 families, 45 genera, and 428 species (Hutterer, 2005). The order used to be called Insectivora. They are distributed virtually all around the world except for Australia, the southern Polar Regions, and most of South America (Nowak, 1999b). Of the four families, only two, Soricidae and Talpidae, are distributed in the North American continent. Two more families are represented in the Caribbean islands.

They are small, similar in size to a rat or mouse; in fact, the smallest mammal, which weighs about 2 g, is a shrew (*Suncus etruscus*). Generally, they have an elongated snout and each paw has five digits with claws. The hair is thick and glossy. In hedgehogs, some hairs are modified to keels or spines, similar to those that cover the body of porcupines. The eyes are small and in some species of subterranean habits there is no external opening. The tail is long, and in many instances serves as a specialized tactile organ, however, there are species that lack a tail (Vaughan, 1978). The number of nipples varies between 1 and 12 pairs.

They feed on animal matter, mainly insects, earthworms, snails, and small shellfish. Some species can trap bigger prey such as mice (Nowak, 1999b). Their metabolism is so high that most species have to consume food almost continuously. This is especially true for shrews which, due to their small size, have a larger surface area in relation to their mass, so they lose a lot of energy as heat (Eisenberg, 1981).

Most species have nocturnal habits. Shrews however, have both diurnal and nocturnal habits because of their high demand for food. Usually, they are terrestrial, although species such as the desmans (*Desmana*, *Galemys*) from Europe and some shrews (*Neomys fodiens*) are semi-aquatic and many of their activities are conducted in the water. Litter size is variable, with 2 to 10 young born per litter (Nowak, 1999b).

(*Opposite top*) Giant shrew (*Megasorex gigas*). Cloud forest. El Jabalí, Colima. Photo: Gerardo Ceballos.
(*Bottom*) Desert shrew (*Notiosorex crawfordi*). Pastureland. Janos, Chihuahua. Photo: Rurik List.

Family Soricidae

The family Soricidae is represented by 26 genera and 376 species; it is the most diverse family of soricomorphans. It is distributed mainly in the temperate regions of North America, northern South America, Eurasia, and Africa (Hutterer, 2005). In Mexico there are 38 species of 6 genera (*Cryptotis, Notiosorex, Megasorex, Sorex, Scalopus,* and *Scapanus*).

Cryptotis alticola (Merriam, 1895)

Popocatépetl shrew

Gerardo Ceballos and Gerardo Carréon Arroyo

SUBSPECIES IN MEXICO

C. alticola is a monotypic species.

It was considered a subspecies of *C. goldmani*, but was elevated to species level (Woodman and Timm, 1999). It belongs to the complex of species that make up the *C. goldmani*-subgroup of *C. mexicana.*

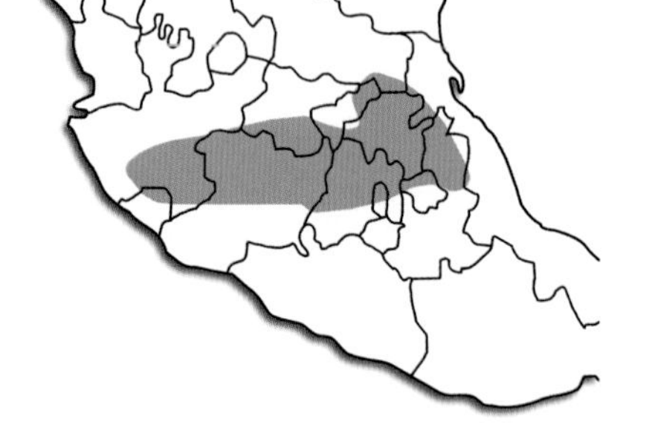

DISTRIBUTION: *C. alticola* is an endemic species of Mexico restricted to the Eje Volcanico Transversal from Colima to Puebla (Woodman and Timm, 1999). It has been recorded in the states of CL, DF, JA, HG, MI, MO, MX, and PU.

DESCRIPTION: *Cryptotis alticola* is a species very similar to *C. goldmani.* Among the species of the genus, it has a medium size. Its shape is similar to that of a small mouse, with an elongated face and small eyes. The pinnae are concealed. The tail is long, about 33% of the length of the head and body (Woodman and Timm, 2000). Its coloration varies from light gray to dark gray; the belly is paler than the back (Ceballos and Galindo, 1984). The hairs on the back are tricolor, silver gray at the base, with a brown stripe and a gray tip (Woodman and Timm, 2000). The front legs are exceptionally long and wide, with long, well-reinforced claws. The posterior part of the legs is of lighter color than the anterior part (Reid, 1997). The face is relatively long and the cranial box is angular (Choate, 1970). It differs from other species of the "Mexican"-group mainly by features of the skull (Woodman and Timm, 1999). It has pigmented teeth.

EXTERNAL MEASURES AND WEIGHT

TL = 90 to 117 mm; TV = 23 to 30 mm; HF = 11 to 16 mm.

Weight: 8 to 16 g.

DENTAL FORMULA: I 3/1, C 1/1, PM 2/1, M 3/3 = 30.

NATURAL HISTORY AND ECOLOGY: The Popocatépetl shrew typically inhabits temperate forests of pines in the center of the country, where winter temperatures are low and where there are frequent frosts and occasional snowfall. They use galleries built below the layer of moss and litter as dens. It has been found sharing habitat with other shrew species such as *Sorex oreopolus* and *S. saussurei* and the mice *Peromyscus aztecus, Reithrodontomys sumichrasti, Neotomodon alstoni,* and *Microtus mexicanus.* They live in temperate rainforests with dense coverage of plants, grass, and moss (Woodman and Timm, 1999). Reproduction occurs throughout the year. Three to six offspring are born in each litter. Their diet is based on insects and other invertebrates such as earthworms (Choate, 1970).

VEGETATIONAL ASSOCIATIONS AND ELEVATION RANGE: *C. alticola* has been found in temperate forests of pine and oak mixed with fir, often where bunch grasses dominate the underbrush (Ceballos and Galindo, 1984; Choate, 1970). It is known in environments from 2,460 m to 4,400 m (Fa, 1989; Woodman and Timm, 1999).

CONSERVATION STATUS: *C. alticola* is considered to be in the category of special protection, which means that it is likely threatened but there is no accurate information about its status (SEMARNAT, 2010). Because of the species' wide tolerance of different types of forests, probably it is not threatened; however, there is very little information about the species to define it accurately.

Cryptotis goldmani (Merriam, 1895a)

Goldman's small-eared shrew

Leslie N. Carraway

SUBSPECIES IN MEXICO

Cryptotis goldmani goldmani (Merriam, 1895a)
Cryptotis goldmani machetes (Merriam, 1895a)

DESCRIPTION: The dorsal pelage of *Cryptotis goldmani* is dark brown, the venter has light-tipped hairs, and the tail is slightly bicolored (Carraway, 2007). In *C. goldmani goldmani*, the venter hairs are white-tipped; in *C. goldmani machetes*, the venter hairs are blond-tipped. Goldman's small-eared shrew is considered a member of the *C. mexicana*-group of long-clawed shrews (Choate, 1970; Woodman, 2005; Woodman and Timm, 1993). The claw of the middle digit of the manus is 3.2 mm to 3.8 mm (2.9% to 3.6% of the total length) for *C. goldmani goldmani*; it is 2.7 mm to 3.5 mm (2.5% to 3.8% of the total length) for *C. goldmani machetes*. All the teeth are pigmented (dark red). U1–U3 have a second cusp and U4 is partially obscured or not visible at lateral view (Carraway, 2007). The i1s have deep interdenticular spaces and two denticles; in lateral view, it extends posteriorly beneath the paraconid of m1 (Carraway, 1995). The area between the condylar processes is shallowly emarginate. The upper condylar facet is tipped dorsally at the labial edge relative to the lower condylar facet (Carraway, 2005). The zygomatic processes are elliptic in shape and "extend posteriorly and ventrolaterally to below the occlusal surface of the teeth" (Carraway, 2007:19).

DISTRIBUTION: An endemic to Mexico, Goldman's small-eared shrew is distributed in Guerrero (*C. goldmani goldmani*) and Oaxaca (*C. goldmani machetes*; Fa and Morales, 1993). The identity of one record of *C. goldmani machetes* from La Sepultura Biosphere Reserve, Chiapas (Espinoza Medinilla et al., 2002), could not be confirmed. It has been recorded in the states of GR and OX.

EXTERNAL MEASURES AND WEIGHT

TL = 92 to 113 mm; Head and body l. = 63 to 84 mm; TV = 26 to 36 mm; HF = 12.5 to 15 mm; Claw l. = 2.7 to 3.8.
Weight: 5 to 14 g.

DENTAL FORMULA: I 3/1, C 1/1, PM 2/1, M 3/3 = 30.

NATURAL HISTORY AND ECOLOGY: In Guerrero, Goldman's small-eared shrews occur in temperate cloud forests (Choate, 1970) up into fir forests with a deep layer of leaves overlaying a deep humus layer on the forest floor (Davis and Lukens, 1958). It has also been found in dense pine-oak forests under rotting logs. In Oaxaca, it occurs along streams in woodlands composed of deciduous trees (including open canopy oaks) with a dense understory of shrubs, ferns, and herbaceous flora (Musser, 1964). In particular, they occur in areas that include damp leaf litter and rotting logs covered with moss. No reproductive information is available. The only mammalian associates reported for *C. goldmani goldmani* are *Peromyscus aztecus*, *P. megalops*, and *Reithrodontomys* sp. (P.L. Clifton, 1964, University of Kansas field notes and catalogue). Reported mammalian associates of *C. goldmani machetes* include *C. peregrina*, *C. phillipsi*, *Microtus mexicanus*, *Oryzomys chapmani*, *Peromyscus megalops*, *Reithrodontomys mexicanus*, *Rattus* sp., *Sorex veraecrucis oaxacae*, and *S. veraepacis mutabilis* (P.B. Robertson, 1970, University of Kansas field notes and catalogue; Schaldach, 1966).

VEGETATIONAL ASSOCIATIONS AND ELEVATION RANGE: Habitats in which Goldman's small-eared shrews have been found are woodlands and forests that include oaks and other deciduous trees, pines, shrubs, ferns, and herbaceous vegetation (Choate, 1970; Davis and Lukens, 1958; Musser, 1964). Also, their typical environment includes deep and damp leaf litter and rotting logs. Known localities from which they have been collected occur at elevations ranging from about 1,550 m to 3,000 m for *C. goldmani goldmani* and 2,250 m to 3,200 m for *C. goldmani machetes*.

CONSERVATION STATUS: *C. goldmani* has not been designated any level of conservation (SEMARNAT, 2010). Its limited distribution, however, does make local populations susceptible to extirpation from habitat changes related to human activities, for example, agriculture and logging.

Cryptotis goodwini Jackson, 1933

Goodwin's small-eared shrew

Heliot Zarza and Xavier López

DISTRIBUTION: *C. goodwini* is distributed from the south of the Sierra Madre of Chiapas, along the Sierra Madre of Guatemala to the west of El Salvador and Honduras (Choate, 1970; Choate and Fleharty, 1974; Woodman and Timm, 1999). In Mexico it is known only in two locations in the state of CS (Horvarth et al., 2008).

SUBSPECIES IN MEXICO

Cryptotis goodwini goodwini Jackson, 1933

This species was considered monotypic until recently; a new subspecies in Honduras recently was described (Woodman and Timm, 1999). It belongs to the complex of species that make up the *C. goldmani*-subgroup of *C. mexicana*.

DESCRIPTION: *Cryptotis goodwini* is a rather large shrew. The forefeet with long claws characterize it. The tail is short, approximately 35% of the total length. The face is relatively thin and long. This species has the greatest reduction of the teeth of all the species in the genus *Cryptotis* (Choate, 1969, 1970). The coat changes color with the seasons. In winter it is black, with numerous and exuberant vermiform ornamentations, whereas in summer it is lighter, with ornamentations that are present only when the coat is new. The ventral part is pale because the tips of the hairs are light colored (Choate and Fleharty, 1974; Hall, 1981). It differs from *C. goldmani* by being larger and black colored; the third molar talonid is reduced, short, and comparable only with the hypoconid limited anterio-posteriorly. The hypoconid frequently is vestigial (Choate, 1970).

EXTERNAL MEASURES AND WEIGHT

TL = 103 to 128 mm; TV = 36 to 44 mm; HF = 14 to 17 mm.
Weight: 7 g.

DENTAL FORMULA: I 3/1, C 1/1, PM 2/1, M 3/3 = 30.

NATURAL HISTORY AND ECOLOGY: *C. goodwini* has large forefeet, as well as claws, indicating digging habits (Choate and Fleharty, 1974). It has been captured together with *Marmosa mexicana*, *C. parva pueblensis*, *Nyctomys sumichrasti*, *Sorex veraecrucis cristobalensis*, *S. veraepacis chiapensis*, *Heteromys goldmani*, *Oryzomys couesi*, *O. alfaroi*, *Reithrodontomys megalotis*, *R. mexicanus*, *Peromyscus mexicanus*, *P. aztecus*, *P. boylii*, and *P. guatemalensis* (Carraway, 2007; Hutterer, 1980). It feeds on invertebrates, lizards, and toads. Probably all species of this genus are active at any time of the day and throughout the year (Nowak, 1999b).

VEGETATIONAL ASSOCIATIONS AND ELEVATION RANGE: Goodwin's small-eared shrews mainly inhabit temperate cloud forests, but are also found in forests of pine or oak, associated with cypress, fir, and sometimes mixed with rootbrom (Choate and Fleharty, 1974). It has been found from 915 m to 3,350 m (Choate and Fleharty, 1974).

CONSERVATION STATUS: The lack of information does not allow determination of the conservation status of *C. goodwini*. Because it has a very restricted distribution in Mexico, it should be considered rare or vulnerable.

Cryptotis griseoventris Jackson, 1933

Dark Mexican shrew

Gerardo Ceballos

SUBSPECIES IN MEXICO

C. griseoventris is a monotypic species. Until recently, it was considered a synonym of *C. goldmani* (Choate, 1970). It belongs to the complex of species that make up the *C. goldmani*-subgroup of *C. mexicana*; its recognition at species level is recent (Woodman and Timm, 1999).

DESCRIPTION: Among the species of the genus *Cryptotis*, the dark Mexican shrew is of average size, with the tail about 38% of the length of the head and body. Its face is long with small eyes. The ears are low key. The coloration on the back is brown with hair from 6 mm to 8 mm long, with 3 bands of color in every alternate hair; the belly is paler. It has seasonal variation in coloration, being darker in winter (Choate, 1970). The front legs are exceptionally long and wide, with long, well-reinforced claws. The posterior part of the legs is lighter in color than the anterior part (Reid, 1997). The face is relatively long and the cranial box is angular (Choate, 1970). It differs from other species of the "Mexican"-group by cranial characteristics (Woodman and Timm, 1999). The teeth are pigmented.

EXTERNAL MEASURES AND WEIGHT

TL = 90 to 117 mm; TV = 24 to 36 mm; HF = 11 to 16 mm.
Weight: 5 to 14 g.

DENTAL FORMULA: I 3/1, C 1/1, PM 2/1, M 3/3 = 30.

DISTRIBUTION: *C. griseoventris* is found in the highlands of central Chiapas and adjacent regions of Guatemala. It has been recorded in the state of CS.

NATURAL HISTORY AND ECOLOGY: Little is known about the biology of *Cryptotis griseoventris*. It inhabits temperate forests in the region of San Cristobal in Chiapas, which are heterogeneous. Depending on the elevation, slope, humidity, and other environmental factors, the predominant vegetation is pine forest (*Pinus* sp.), oaks (*Quercus* sp.), or mixed temperate cloud forests. The average annual temperature and precipitation in San Cristobal is 13°C to 14°C and 1,160 mm, respectively (Camacho Cruz et al., 2000). These forests are characterized by dense undergrowth, with large quantities of herbaceous plants and litter. In wetter areas there is a good coverage of moss. There are also high densities of epiphytic plants such as orchids and bromeliads. Among the conspicuous tree species, typical of the temperate cloud forests are *Cornus*, *Magnolia*, *Liquidambar*, *Persea*, and *Ternstroemia*. The temperature in winter can be low and frosts are frequent (Woodman and Timm, 1999). There are no known details on feeding or reproduction. It is possible that, like other shrews of the same genus, it feeds on small invertebrates and reproduces all year.

VEGETATIONAL ASSOCIATIONS AND ELEVATION RANGE: *C. griseoventris* inhabits pine, oak, mixed, or temperate cloud forests at altitudes higher than 2,000 m (Woodman and Timm, 1999).

CONSERVATION STATUS: *C. griseoventris* is not included in any list of conservation, probably because of its recent recognition. It is likely threatened with extinction because in the region of San Cristobal de las Casas in Chiapas, oak forests and temperate cloud forests have been reduced by more than 75% of their original distribution (Ochoa-Gaona and González-Espino, 2000). Modification and destruction of forests have been caused by the growth of agriculture, animal husbandry, and selective logging of pines and oaks for firewood and charcoal (Ochoa-Gaona and González-Espino, 2000). To assess the state of conservation of this species intensive study of its current distribution and habitat requirements needs to be conducted. In addition, it is necessary to protect some of the forest remnants, which have populations of other endangered species of mammals such as the flying squirrel (*Glaucomys volans*). On the outskirts of San Cristobal de las Casas there is a small reserve called Huitepec that protects about 100 ha of mixed forests in a good state of conservation.

Cryptotis magna (Merriam, 1895)

Big small-eared or big Mexican shrew

Xavier López and Heliot Zarza

SUBSPECIES IN MEXICO

C. magna is a monotypic species.

DESCRIPTION: *Cryptotis magna* can be distinguished from other members of the genus by its large size. Its body is robust; the tail is visible; and it has a pointed and elongated face. The eyes are small and the ears are hardly visible. Its teeth show moderate cusps (Choate and Rickart, 1974; Nowak, 1999b; Robertson and Rickart, 1975). The winter coat is dark brown on the back and a little lighter on the venter; during summer the back is black and the belly is pale (Hall, 1981). The upper unicuspid teeth have very small lingual tubercles; the long upper premolar is small and wide, with the internal-anterior angle circular and wide (Robertson and Rickart, 1975).

EXTERNAL MEASURES AND WEIGHT
TL = 123 to 141 mm; TV = 37 to 53 mm; HF = 16.0 to 17.5 mm.
Weight: 7 g.

DENTAL FORMULA: I 3/1, C 1/1, PM 2/1, M 3/3 = 30.

NATURAL HISTORY AND ECOLOGY: *C. magna* commonly inhabits the upper shrub layer of pine-oak forests and temperate cloud forests. It is sympatric with *C. mexicana*, *Sorex veraepacis mutabilis*, *S. veraecrucis oaxacae*, *Peromyscus boylii*, *P. furvus*, *P. lepturus*, *P. melanocarpus*, *P. mexicanus*, *Oryzomys alfaroi*, *O. caudatus*, *Reithrodontomys mexicanus*, *R. microdon*, *Habromys chinenteco*, *H. lepturus*, *Megadontomys cryophilus*, *Microtus mexicanus*, and *M. oaxacensis*. Also, it inhabits forests where there are tree rats such as *Tylomys nudicaudatus* and *Nyctomys sumichrasti* (Robertson and Rickart, 1975). The southern species of the genus *Cryptotis* breed all year (Nowak, 1999b). We have captured pregnant females with three embryos and males with prominent testicles in May and October. There might be a reproductive cycle during the wet season and another during the dry season (Choate, 1973). They make their burrows under logs or rocks or use mouse burrows. The nests they build are spherical, 50 mm to 150 mm in diameter, and are composed of herbs and dried leaves. It feeds on invertebrates, small reptiles, amphibians, and carrion (Nowak, 1999b).

DISTRIBUTION: *C. magna* is endemic to Mexico, with a distribution restricted to north-central Oaxaca (Robertson and Rickart, 1975). It has been recorded in the state of OX.

VEGETATIONAL ASSOCIATIONS AND ELEVATION RANGE: *C. magna* mainly inhabits temperate cloud forests, but is also found in pine-oak forests (Robertson and Rickart, 1975). It has been found from 1,300 m to 3,000 m (Hall, 1981).

CONSERVATION STATUS: In Mexico *Cryptotis magna* is considered a rare species (SEMARNAT, 2010). It is imperative to know the current status of its populations in order to assess its actual situation.

Cryptotis mayensis (Merriam, 1901)

Mayan small-eared shrew

Gerardo Carreón Arroyo and Gerardo Ceballos

SUBSPECIES IN MEXICO
C. mayensis is a monotypic species.

DESCRIPTION: *Cryptotis mayensis* is a small shrew, with the tail about 33% of the length of the head and body (Reid, 1997). The legs and claws are small. The face is stretched, the eyes unnoticeable, and the ears small. The back is light gray and the belly is lighter. Back hairs measure 3 mm to 4 mm. The skull is relatively short and the cranial box is not particularly angular. The jaw is large, with a deep, horizontal branch (Choate, 1970; Hall, 1981; Woodman and Timm, 1993).

EXTERNAL MEASURES AND WEIGHT
TL = 85 to 120 mm; TV = 24 to 33 mm; HF = 11 to 13 mm.
Weight: 4-6 g.

DENTAL FORMULA: I 3/1, C 1/1, PM 2/1, M 3/3 = 30.

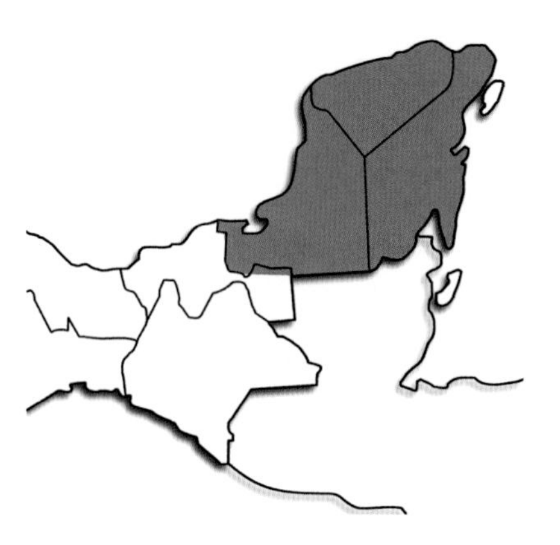

DISTRIBUTION: *C. mayensis* is an endemic species of the Yucatan Peninsula in Mexico, Guatemala, and Belize (Choate, 1970; Reid, 1997). There is an isolated report in the Cañon del Zopilote in Guerrero, which is probably erroneous. It has been recorded in the states of CA, QR, and YU.

NATURAL HISTORY AND ECOLOGY: Very little is known about the biology of *C. mayensis*. They have mining habits, as they move under the litter of the underbrush. They feed on snails, insects, and earthworms. Their extremely high metabolism and their voraciousness force them to consume the equivalent of their mass in one day. They are known to occur with *Grison canaster*, *Heteromys gaumeri*, *Marmosa mexicanus mayensis*, *Mus musculus*, *Oryzomys melanotis*, *Ototylomys hatti*, *O. phyllotis*, and *Peromyscus yucatanicus*. They have a rapid heart rate, which reaches about 1,200 beats per minute. They are solitary, getting together only in times of estrous. The gestation period is unknown, but in other shrews is about 21 to 22 days. Females give birth to 3 to 6 young, which are born without fur, with the eyes closed, and with an average mass of 0.3 g. Hawks, owls, and snakes prey on them. Their only defense are their anal glands, which expel a fetid-smelling liquid (Carraway, 2005).

VEGETATIONAL ASSOCIATIONS AND ELEVATION RANGE: *C. mayensis* mainly inhabits tropical deciduous forests, semi-deciduous forests, tropical rainforests, riparian vegetation, scrublands, and savannahs (Choate, 1970; Goldman, 1951). In some of the deciduous forests and tropical rainforests, sabal and other palms are the dominant species. Its elevational range is up to 100 m (Choate, 1970; Woodman and Timm, 1993).

CONSERVATION STATUS: *C. mayensis* is included in the list of Endangered Species of Mexico in the category of rare (SEMARNAT, 2010). Throughout its wide distribution, however, there are still huge extensions of forest. It has been recorded in the Calakmul Biosphere Reserve and Sian Ka'an.

Cryptotis merriami Choate, 1970

Merriam's small-eared shrew

Gerardo Carreón Arroyo and Gerardo Ceballos

SUBSPECIES IN MEXICO

C. merriami is a monotypic species.

DESCRIPTION: *Cryptotis merriami* is a small shrew, with an elongated face, small ears, and tiny eyes. The color of the back is olive brown and the belly is slightly lighter. The hair on the back measures about 4 mm long and has two color bands. The face is large and moderately long (Choate, 1970; Woodman and Timm, 1993). The cusps of the teeth range from more in the Yucatan Peninsula to fewer cusps in Costa Rica (Choate, 1970; Redford and Eisenberg, 1989).

EXTERNAL MEASURES AND WEIGHT

TL = 90 to 120 mm; TV = 24 to 33 mm; HF = 10 to 14 mm.
Weight: 4 to 6 g.

DENTAL FORMULA: I 3/1, C 1/1, PM 2/1, M 3/3 = 30.

NATURAL HISTORY AND ECOLOGY: This shrew is little known. It lives in the underbrush of temperate forests, in sites with good vegetation and litter cover. It feeds on invertebrates. They have an extremely high metabolism, and their voraciousness forces them to consume the equivalent of their mass every day.

DISTRIBUTION: Merriam's small-eared shrew is located in the highlands of southeastern Mexico and Guatemala, Honduras, El Salvador, and northern Costa Nicaragua. In Costa Rica there is an isolated population in the highlands of Tilarán (Choate, 1970). It has been recorded in CS.

VEGETATIONAL ASSOCIATIONS AND ELEVATION RANGE: *C. merriami* inhabits pine, oak, and mixed forests as well as in temperate cloud forests from 1,000 m to 1,650 m (Choate, 1970; Goldman, 1951; Woodman and Timm, 1993).

CONSERVATION STATUS: *C. merriami* is classified as rare (SEMARNAT, 2010). Apparently, it is unusual in its range (Reid, 1997).

Cryptotis mexicana (Coues, 1877)

Mexican small-eared shrew

Iván Castro-Arellano and Gerardo Ceballos

SUBSPECIES IN MEXICO

C. mexicana is a monotypic species
Recently, *C. mexicana nelsoni*, *C. mexicana obscura*, and *C. mexicana peregrina* were re-elevated to species status (Woodman and Timm, 1999).

DESCRIPTION: Among the species of the genus *Cryptotis*, the Mexican small-eared shrew is medium-sized. Pinnae are not obvious. The back varies from sepia to dark brown; the belly is slightly paler, with the tip of each hair pale cream (Choate, 1973). There is no sexual dimorphism. There is seasonal variation in color and texture of the fur (Choate, 1970). The tail is shorter than the body length (32% to 42%). The skull is relatively robust and the face is short. The external and cranial measurements show geographical variation, being greater in the northeastern part of its distribution with a decline toward the southeast (Choate, 1973).

EXTERNAL MEASURES AND WEIGHT

TL = 83 to 112 mm; TV = 22 to 33 mm; HF = 11 mm.
Weight: 8.2 g.

Cryptotis mexicana. Cloud forest. Sierra Gorda Biosphere, Querétaro. Photo: Juan Cruzado.

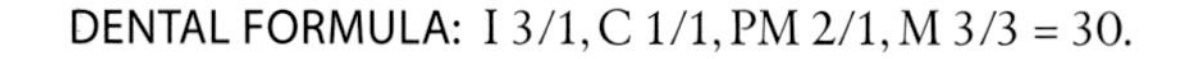

DENTAL FORMULA: I 3/1, C 1/1, PM 2/1, M 3/3 = 30.

DISTRIBUTION: *C. mexicana* is endemic to Mexico, where it is distributed from southwestern San Luis Potosí across the Sierra Madre Oriental into Veracruz southward to the Sierra Madre del Sur in Oaxaca (Carraway, 2007; Choate, 1973). Fossil specimens have been recorded in San Josecito Cave, near Aramberri, Nuevo Leon (Findley, 1953). It has been recorded in the states of CS, HG, OX, PU, SL, and VE.

NATURAL HISTORY AND ECOLOGY: *Cryptotis mexicana* mainly inhabits temperate cloud forests but is not restricted to undisturbed forests, as it has been found in other habitats such as riparian areas and edges of crops (Choate, 1973; Hall and Dalquest, 1963; Musser, 1964). In Hidalgo, it has been collected near fallen tree trunks, which probably are used as shelters (Jones et al., 1983). If all populations from different elevations and latitudes are considered, they are likely to reproduce throughout the year (Choate, 1973). It is known to occur with a great variety of small shrews and mice (Carraway, 2007). Their activity is both diurnal and nocturnal. They build spherical nests (50 mm to 120 mm in diameter) with leaves and dry grasses under logs, rocks, and other objects that serve as a refuge. It feeds on invertebrates and small vertebrates such as lizards (Nowak, 1999b). In Veracruz, stomach contents included the remains of species of insects and worms (Hall and Dalquest, 1963). Koopman and Martin (1959) reported an owl (*Strix virgata*) as its predator in Tamaulipas.

VEGETATIONAL ASSOCIATIONS AND ELEVATION RANGE: *C. mexicana* inhabits forests of pine-oak and is abundant in temperate cloud forests with mosses, lichens, orchids, and bromeliads (Choate, 1970; Goodwin, 1954a; Hall and Dalquest, 1963). It has been found from 520 m to 3,200 m (Choate, 1970).

CONSERVATION STATUS: Apparently, *C. mexicana* is a common species throughout its distribution (Hall and Dalquest, 1963; Heaney and Birney, 1977).

Cryptotis nelsoni (Merriam, 1895)

Nelson's shrew

Gerardo Ceballos and Joaquín Arroyo-Cabrales

SUBSPECIES IN MEXICO

C. nelsoni is a monotypic species.

It belongs to the complex of species that make up the *Cryptotis mexicana*-group; its recognition at species level is recent (Woodman and Timm, 2000).

DESCRIPTION: *Cryptotis nelsoni* is similar to *C. mexicana*, but the postpalate notch is especially short and wide. It resembles a small mouse with an elongated face and small eyes. The ears are very inconspicuous. The tail is long, about 30% of the length of the head and body (Woodman and Timm, 2000). Its coloration varies from light gray to dark gray.

EXTERNAL MEASURES AND WEIGHT

TL = 106 mm; TV = 29 mm, HF = 13.3 mm.

Weight: 3 to 6 g.

DENTAL FORMULA: I 3/1, C 1/1, PM 2/1, M 3/3 = 30.

NATURAL HISTORY AND ECOLOGY: *C. nelsoni* inhabits a region characterized by lush vegetation, including a large number of epiphytic plants such as orchids and bromeliads, with trees averaging 35 m in height and some up to 50 m high

(Cervantes and Guevara, 2010). The lower stratum of the forest has a large amount of litter and seedlings, and is dominated by thorny palms such as the *chocho* (*Astrocarium mexicanum*). The middle and upper strata of the forest are dominated by trees such as the ficus (*Ficus tecolutensis*), laurel (*Nectandra ambigens*), elm (*Ulmus mexicana*), and cork (*Omphalea oleifera*), among many others (Cervantes and Guevara, 2010). The main components of the soil are sand and volcanic ash (Choate, 1970). The Tuxtlas region has a high endemism of vertebrates, suggesting isolation during a substantial portion of the Pleistocene. It is characterized by tropical humid climate, with annual rainfall of about 5,000 mm and monthly average temperature of 23°C (Cervantes and Guevara, 2010). This species is sympatric with other small mammals like mice (*Peromyscus mexicanus* and *Heteromys desmarestianus*) and shrews (*Cryptotis parva*). It is probably the prey of a wide variety of mammals, birds, and reptiles.

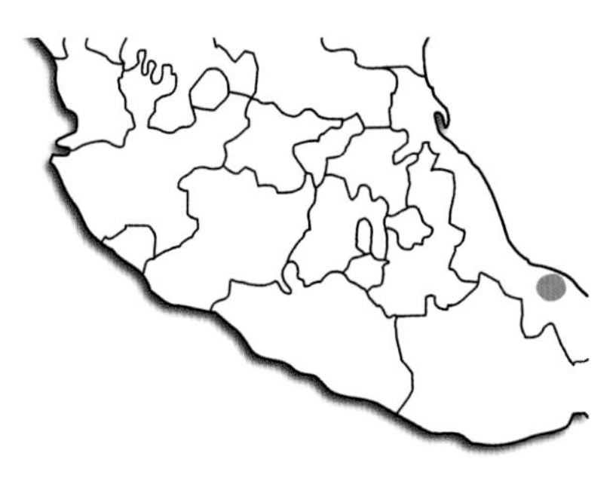

DISTRIBUTION: *C. nelsoni* is an endemic species of the region of Los Tuxtlas, in the state of Veracruz (Woodman and Timm, 2000). It has been recorded in the state of VE.

VEGETATIONAL ASSOCIATIONS AND ELEVATION RANGE: Nelson's shrew is only known in tropical rainforests, between 1,000 m and 2,550 m.

CONSERVATION STATUS: Known from only 12 specimens collected in 1894, *C. nelsoni* recently was rediscovered in the area of the type locality on the south face of Volcán San Martín, Veracruz (Cervantes and Guevara, 2009). As this species has a very restricted distribution, it should be considered fragile. It inhabits a region that has suffered alarming rates of deforestation; the Volcano of St. Martín and surrounding areas have lost about 90% of the original forest cover (Dirzo and Garcia, 1992). Since 1970, around 700 ha of the region have been partially protected by the Los Tuxtlas Biological Station of the National Autonomous University of Mexico (Cervantes and Guevara, 2010). Recently, it was decreed Los Tuxtlas Biosphere Reserve, with an area of nearly 120,000 ha. The reserve seeks to preserve the last 40,000 ha of forest remnants of "Los Tuxtlas" and "Santa Marta" Sierras, as well as to protect thousands of plant and animal species. As a subspecies of *C. mexicana*, *Cryptotis nelsoni* is considered rare within the Mexican Official Regulation (SEMARNAT, 2010).

Cryptotis obscura (Merriam, 1895)

Grizzled shrew

Joaquín Arroyo-Cabrales and Gerardo Ceballos

SUBSPECIES IN MEXICO

C. obscura is a monotypic species.

It belongs to the complex of species that make up the *Cryptotis mexicana*-group; until recently it was considered a subspecies (Woodman and Timm, 2000).

DESCRIPTION: These are smaller and paler shrews of the *mexicana*-group (Woodman and Timm, 2000). Like other shrews of the same genus, this species has a long face and small eyes. The ears are very inconspicuous. The tail is long, but shorter than the length of the head and body. The coloration of the back is brown (Merriam, 1897a; Woodman and Timm, 2000).

EXTERNAL MEASURES AND WEIGHT

TL = 106 mm; TV = 26 mm, HF = 12 mm.

Weight: 3 to 5 g.

DISTRIBUTION: *C. obscura* is an endemic species of Mexico, distributed in the Sierra Madre Oriental from southwestern Tamaulipas to northeastern Queretaro and northern Hidalgo. It has been recorded in the states of HG, PU, QE, SL, TA, and VE.

DENTAL FORMULA: I 3/1, C 1/1, PM 2/1, M 3/3 = 30.

NATURAL HISTORY AND ECOLOGY: *Cryptotis obscura* inhabits temperate and subtropical cloud forests of pine and oak of the Sierra Madre Oriental that are characterized by dense underbrush, with abundant herbs, grasses, and mosses. They feed mainly on insects and other small invertebrates. Non-pregnant females have been reported from mid-July and females with embryos from early June, July, and August; testicles of three males caught in mid-August measured 4 mm, 6 mm, and 6 mm; (Heaney and Birney, 1977; Jones et al., 1983). There is evidence of molting (cross line along the back) from late July to mid-August (Jones et al., 1983). This species is sympatric with many small rodents, including *Peromyscus furvus*, *Microtus quasiater*, and *M. mexicanus*.

VEGETATIONAL ASSOCIATIONS AND ELEVATION RANGE: The grizzled shrew is found in pine-oak forests, oak forests, and temperate cloud forests, from 1,060 m to 2,550 m.

CONSERVATION STATUS: As a subspecies of *C. mexicana, C. obscura* is considered rare within the Mexican Official Regulation (SEMARNAT, 2010).

Cryptotis parva (Say, 1823)

Least shrew

Iván Castro-Arellano

SUBSPECIES IN MEXICO

Cryptotis parva berlandieri (Baird, 1858)
Cryptotis parva pueblensis Jackson, 1933
Cryptotis parva soricina (Merriam, 1895)
Recently, *C. parva tropicalis* was re-elevated to species status (Woodman and Croft, 2005).

DESCRIPTION: This small shrew has a brown back with black tones; the belly is lighter, almost white in adults. The tail is short, the ears are unnoticeable, the eyes are tiny and black, and it has an elongated and sharp snout. The fur is short, dense, and smooth. It can be distinguished from other species by its relatively short tail, which is less than 45% of the length of the head and body, by its small forefeet, and by the long condylobasal of the skull of 15.3 mm to 18.4 mm (Whitaker, 1974).

EXTERNAL MEASURES AND WEIGHT

TL = 67 to 103 mm; TV = 12 to 22 mm; HF = 9 to 13 mm.
Weight: 5 g.

DENTAL FORMULA: I 3/1, C 1/1, PM 2/1, M 3/3 = 30.

NATURAL HISTORY AND ECOLOGY: The soil type is not a determining factor of the presence of this species, but a dense herbaceous cover, especially grasses, is (Whitaker, 1974). To move, this shrew uses "corridors" made by other small mammals such

Cryptotis parva. Grassland. Distrito Federal. Photo: Alberto González.

as *Microtus*, *Sigmodon*, and *Oryzomys*. It occurs with a great variety of small shrews and mice (Carraway, 2007). It is highly social and frequently found in groups within nests built with leaves and dry grasses, beneath rocks or fallen logs. It also constructs nests in burrows, up to 1.5 m long, which are dug in different soil types such as clay, sandy soils, and sandbars (Davis, 1944; Goodwin, 1954a; Whitaker, 1974). It probably reproduces throughout the year, but has a reproductive peak between June and September (Choate, 1970). Litter size varies from 2 to 7 offspring, with an average of 4.9 (Choate, 1970; Conaway, 1958; Whitaker, 1974). In captivity, a life span of 21 months has been recorded (Pfeiffer and Gass, 1963). It feeds primarily on invertebrates such as butterfly larvae, earthworms, spiders, crickets, grasshoppers, larvae of beetles, centipedes, and mollusks; it also can consume frogs and lizards (Hamilton, 1944; Hatt, 1938a; Whitaker, 1974; Whitaker and Mumford, 1972b). Many ectoparasites have been reported for this species (Bassols, 1981; Whitaker, 1974; Whitaker and Mumford, 1972b). Their predators include snakes, owls, barn owls, skunks, foxes, and weasels (López-Forment, 1997; Whitaker, 1974).

VEGETATIONAL ASSOCIATIONS AND ELEVATION RANGE: Given its wide distribution *Cryptotis parva* inhabits a variety of environments. It is found mostly in grasslands (Whitaker, 1974), but it also has been collected in plains with mesquite, yucca, agave, and pastures, and is occasionally found in oak forests, pine-oak forests, temperate cloud forests, and tropical rainforests (Choate, 1970). It occurs from sea level in the coastal plains to 2,950 m, in the Sierra of Coalcomán, Michoacán (Choate, 1970).

CONSERVATION STATUS: *C. parva* has a wide distribution and is considered relatively common. One subspecies — *C. parva soricina*--in the center of the country has a very restricted distribution and is classified as rare (SEMARNAT, 2010).

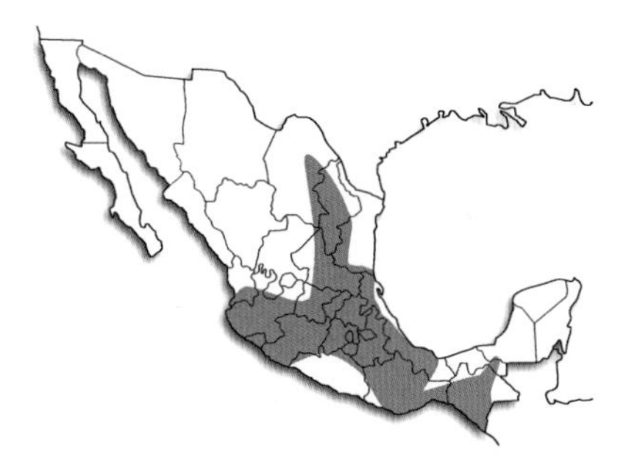

DISTRIBUTION: This shrew has a broad distribution that includes the far southeastern part of Canada southward to Panama (Hall, 1981; Hutterer, 1993). In Mexico it is found from the coast of Nayarit, Jalisco, and Veracruz to the Isthmus of Tehuantepec and Chiapas. The subspecies *C. Parva soricina* is found in the state of Mexico and Mexico City, isolated from the rest of the species distribution. It has been recorded in the states of AG, CH, CO, CS, GJ, HG, JA, MI, MX, NL, NY, OX, PU, QE, SL, TA, TB, TL, and VE.

Cryptotis peregrina (Merriam, 1895)

Mexican mountain shrew

Gerardo Ceballos and Joaquín Arroyo-Cabrales

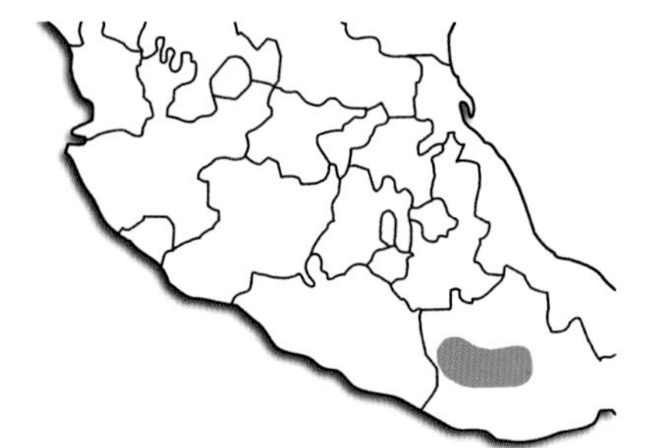

DISTRIBUTION: *C. peregrina* is an endemic species of Mexico known only from 25 specimens at 4 localities of the Sierra de Cuatro Venados and in the Sierra Yucuy-agua in Oaxaca. It has been recorded in the state of OX.

SUBSPECIES IN MEXICO

C. peregrina is a monotypic species.

It belongs to the complex of species that make up the *C. Mexicana*-group; its recognition at species level is recent (Woodman and Timm, 2000).

DESCRIPTION: This shrew can be distinguished from other shrews of the *mexicana*-group by the narrow zygomatic plate. The front feet are robust with long and wide claws (Woodman and Timm, 2000).

EXTERNAL MEASURES AND WEIGHT

TL = 106 mm; TV = 30 mm; HF = 12 mm.

Weight: 3 to 5 g.

DENTAL FORMULA: I 3/1, C 1/1, PM 2/1, M 3/3 = 30.

NATURAL HISTORY AND ECOLOGY: Very little is known about the biology of *Cryptotis peregrina*. It inhabits mixed forests of pine and oak, with a dense layer of litter and underbrush, with a high density of epiphytes such as orchids and bromeliads. It is found in the most humid parts of the region, generally with a dense moss cover. It feeds on insects and other invertebrates. The mating season is not known.

VEGETATIONAL ASSOCIATIONS AND ELEVATION RANGE: The Mexican mountain shrew inhabits pine-oak forests and pine and oak forests, from 2,900 m to 3,200 m.

CONSERVATION STATUS: *C. peregrina* has a very restricted distribution, its habitat is negatively impacted by deforestation and farming. As a subspecies of *C. mexicana* it is considered rare within the Mexican Official Regulation (SEMARNAT, 2010). It is probably at risk of extinction, but additional information is required to determine its precise current situation.

Cryptotis phillipsii (Schaldach, 1966)

Phillip's shrew

Gerardo Ceballos and Joaquín Arroyo-Cabrales

SUBSPECIES IN MEXICO

C. phillipsii is a monotypic species.

It belongs to the complex of species that make up the *C. Mexicana*-group; its recognition at species level is recent (Woodman and Timm, 2000)

DESCRIPTION: Among the species of the genus *Cryptotis*, Phillip's shrew is medium-sized, similar to *C. mexicana*. It has no visible pinnae. The back is dark brown and the belly is paler. The tail is shorter than the body (32% to 42%). It has short claws and small pads in the forefeet. A considerable number of specimens have a reduction of the fourth unicuspid, and in some instances, they lack it (Woodman and Timm, 2000).

EXTERNAL MEASURES AND WEIGHT
TL = 83 to 112 mm; TV = 22 to 33 mm; HF = 11 mm.
Weight: 8.2 g.

DENTAL FORMULA: I 3/1, C 1/1, PM 2/1, M 3/3 = 30.

NATURAL HISTORY: Very little is known of *Cryptotis phillipsii*, which lives in temperate forests and low to upper areas of cloud forest with dense underbrush (Schaldach, 1966). They feed on insects and other invertebrates. This species is known to occur with *C. goldmani machetes*, *Sorex veraecrucis oaxacae*, and *S. veraepacis mutabilis* (Carraway, 2007). We do not know the mating season.

VEGETATIONAL ASSOCIATIONS AND ELEVATION RANGE: *C. phillipsii* inhabits pine-oak forests and temperate cloud forests. It is found from 1,000 m to 2,550 m.

CONSERVATION STATUS: *C. phillipsii* has a very restricted distribution; its habitat is impacted by high rates of deforestation and overgrazing. Possibly it is threatened with extinction, but we do not have enough information to determine its current situation.

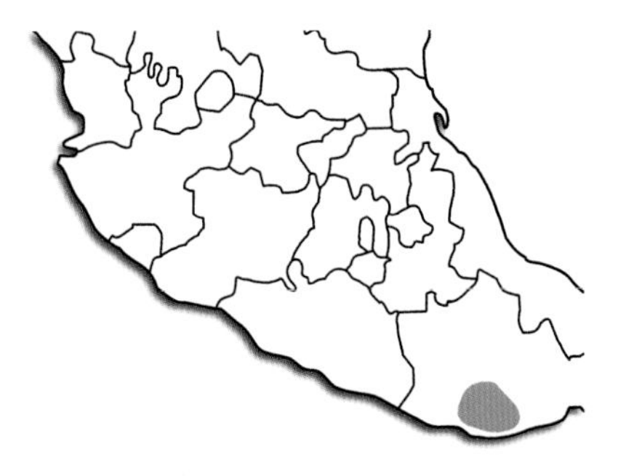

DISTRIBUTION: *C. phillipsii* is an endemic species of Mexico, known only from high-elevation forests in the Sierra de Miahuatlán and the southern part of Oaxaca. It has been recorded in the state of OX.

Cryptotis tropicalis (Merriam, 1895a)

Tropical shrew

Leslie N. Carraway

SUBSPECIES IN MEXICO
C. tropicalis is a monotypic species.
Recently re-elevated to species status (Carraway, 2007), it was considered a subspecies of *C. parva* (Woodman and Croft, 2005).

DESCRIPTION: *Cryptotis tropicalis* is a medium-sized shrew with dorsal pelage tipped dark reddish-brown, venter with blond-tipped hairs, and tail slightly bicolored (Carraway, 2007). All the teeth are pigmented (dark red). U1-U3 have a second cusp and U4 is partially obscured or not visible at lateral view (Carraway, 2007). The i1s have deep interdenticular spaces and two denticles; in lateral view, it extends posteriorly beneath the paraconid of m1 (Carraway, 1995). The area between the condylar processes is shallowly emarginate. The upper condylar facet is tipped dorsally at the labial edge relative to the lower condylar facet (Carraway, 2005). The zygomatic processes are sharply pointed and "extend posteriorly and ventrolaterally to below the occlusal surface of the teeth" (Carraway, 2007:12).

DISTRIBUTION: The tropical shrew is distributed from eastern Chiapas southward into Belize, Guatemala, and Honduras. It has been recorded in the state of CH.

EXTERNAL MEASURES AND WEIGHT
TL = 97 to 99 mm; Head to body L = 70 to 71 mm; TV = 26 to 29 mm; HF = 12 to 13 mm.
Weight: probably around 6 to 7 g.

DENTAL FORMULA: I 3/1, C 1/1, PM 2/1, M 3/3 = 30.

NATURAL HISTORY AND ECOLOGY: Very little is known of *C. tropicalis*, which lives in tropical rainforests (Álvarez del Toro, 1977; Choate, 1970; Hooper, 1947). They feed on insects and other invertebrates. No information concerning mammalian associates or reproduction is known.

VEGETATIONAL ASSOCIATIONS AND ELEVATION RANGE: *C. tropicalis* inhabits tropical rainforests of the high plains and inland slopes of the Sierra Madre (Álvarez del Toro, 1977). It is found from about 1,110 m to 1,400 m.

CONSERVATION STATUS: Because of its limited distribution, the tropical shrew is listed as protected under *C. parva tropicalis* (SEMARNAT, 2010).

Megasorex gigas (Merriam, 1897)

Mexican giant or Merriam's desert shrew

Gerardo Ceballos and Ángeles Mendoza Durán

DISTRIBUTION: *M. gigas* is an endemic species of Mexico. It is distributed along the Pacific coast, from Nayarit to Oaxaca, and in the Balsas River Basin in the states of Michoacán and Mexico (Armstrong and Jones, 1972; Goodwin, 1969; Hall, 1981; López-Forment and Urbano, 1977; Tellez-G. et al., 1997). It has been recorded in the states of CL, GR, JA, MI, MX, NY, and OX.

SUBSPECIES IN MEXICO

M. gigas is a monotypic species.

Although some authors such as Hall (1981) consider this species within the genus *Notiosorex* the majority, like Armstrong and Jones (1972) and Hutterer (2005), consider it a distinct genus, finding support in enzymatic and morphological studies.

DESCRIPTION: These shrews are similar in size to a mouse. They are similar to but larger than *Notiosorex crawfordi*. The ears are relatively conspicuous. The tail is hairy and shorter than the head and body. The fur coloration is leaden gray, and barely paler in the ventral region (Ceballos and Miranda, 2000). There are reports of specimens with a white spot on the back, either at the height of the shoulders or on the rear (Davis, 1957). They have a relatively short tail; the eyes are small and not very conspicuous. It differs from *Notiosorex crawfordi* and other shrews by the absence of pigment in the teeth and the presence of subsequent straight edges in p4, m1, and m2 (Hall, 1981). The molars are bulky and tight with each other, unexcavated posteriorly, with three unicuspid teeth forming a uniform set, with the third tooth half again longer than the second (Hall, 1981; Jones, 1966a; Orr, 1963).

EXTERNAL MEASURES AND WEIGHT

TL = 101 to 133 mm; TV = 32 to 49 mm; HF = 12 to 15 mm; EAR = 9 to 15 mm.
Weight: 9 to 12 g.

DENTAL FORMULA: I 3/2, C 1/0, PM 1/1, M 3/3 = 28.

NATURAL HISTORY AND ECOLOGY: Very little is known about the biology of *Megasorex gigas*. Because of its high metabolism, it is likely to be active during the day and night, but most specimens have been collected at night. In seasonal forests and scrublands it is only found in the wettest areas. For example, in the Chamela-Cuixmala Biosphere Reserve it has been collected in the montane rainforests and vegetation along seasonal streams. Lowland records are scarce. In the reserve, trees

Megasorex gigas. Cloud forest. El Jabalí, Colima. Photo: Gerardo Ceballos.

of the montane rainforest have an average height of 25 m; underbrush is relatively open with an abundance of vines and significant accumulation of litter. The most abundant trees include the ceiba (*Ceiba aescucifolia*), primavera (*Tubebuia donell-smithi*), rosa morada (*Tabebuia rosea*), iguanero (*Caesalpinia eriostachys*), and Culebra (*Astronium graveolens*). It has been collected in the dense temperate cloud forests of Colima, where abundant epiphytes and closed underbrush grow, with dominant tree species such as *Magnolia* and *Juglans* (Téllez-G. et al., 1997). In the seasonal dry forest of Jalisco, this shrew is sympatric with *Osgoodomys banderanus*, *Peromyscus perfulvus*, *Xenomys nelsoni*, *Nyctomys sumichrasti*, and *Oryzomys melanotis* (Ceballos, 1990). In Colima, they share habitat with *Spermophilus annulatus*, *Peromyscus* sp., and *Hodomys alleni* (Téllez-G. et al., 1997). In other regions, it has been collected in areas with abundant underbrush, rocks, and litter near margins of water bodies and under piles of wood (Fisher and Bogan, 1977; Orr, 1963). Some of its predators include mammals, owls, snakes, and barn owls (Lopez-Forment and Urbano, 1977).

VEGETATIONAL ASSOCIATIONS AND ELEVATION RANGE: *M. gigas* inhabits deciduous and montane rainforests, scrublands, pine and oak forests, and temperate cloud forests; it has also been found in coffee crops (Ceballos and Miranda, 1986, 2000; Tellez-G. et al., 1997). It has been captured in the Pacific coastal plain in the vicinity of rocks and rocky terrain. It has been found from sea level up to 1,700 m (Armstrong and Jones, 1972; Davis, 1957; Davis and Lukens, 1958; Jones, 1966a; Tellez-G. et al., 1997; Winkelman, 1962a).

CONSERVATION STATUS: *M. gigas* is considered an endangered species (SEMARNAT, 2010) because of its rarity and the destruction of its habitat (Ceballos and Miranda, 2000).

Cockrum's gray shrew

Leslie N. Carraway

SUBSPECIES IN MEXICO
N. cockrumi is monotypic.

DESCRIPTION: Hairs of the dorsal pelage of *Notiosorex cockrumi* are banded, proximally to distally, dark grayish-brown, pinkish-white, to very dark grayish-brown. Hairs of the venter are gray to white. The tail is colored uniformly very dark grayish-brown as for the dorsum. As for all *Notiosorex*, the smooth shield of the curly overhairs has a single series of slightly indented notches (Ducommun et al., 1994). Cockrum's gray shrews are intermediate in size between Crawford's and large-eared gray shrews. Within *Notiosorex*, Cockrum's gray shrews have the shortest (actual and relative) claw on the middle digit of the manus: 1.0 mm to 1.2 mm, 1.28% to 1.52% of total length. A prominent lateral ridge on the cranium is formed by an extension of the roof of the glenoid fossa (Carraway, 2007). The zygomatic plate is intermediate in width: 1.7 mm to 2.2 mm (10% to 14% of condylobasal length). As for all *Notiosorex*, the tail is 33% of length of head and body; only a light-orangish pigment occurs on I1, U1-U3, and occasionally on P4, i1, c1, and p1; and, the alveolus of i1 extends beneath the paraconid of m1 (Carraway, 1995). Unlike shrews of the genera *Cryptotis* and *Sorex*, a locking mechanism is present in the upper glenoid furrow to hold the condylar processes of the mandible in place (Carraway, 2005). For all members of the genus *Notiosorex*, a portion of the side of the ovale groove is composed of the labial edge of the alisphenoid.

DISTRIBUTION: Cockrum's gray shrew occurs in Arizona southward into central Sonora (Baker et al., 2003). It has been recorded in the state of SO.

EXTERNAL MEASURES AND WEIGHT
TL = 78 to 86 mm; Head and body l. = 52 to 61 mm; TV = 19 to 27 mm; HF = 9 to 11 mm; EAR = 7 to 9 mm; Claw l. = 1.0 to 1.2 mm.
Weight: 3 to 6.3 g.

DENTAL FORMULA: I 3/2, C 1/0, PM 1/1, M 3/3 = 28.

NATURAL HISTORY AND ECOLOGY: The little that is known of the ecological needs of Cockrum's gray shrews was recorded in field notes of collectors. They occur in riparian zones with deciduous tree overstories and mesquite grasslands containing giant sacaton. No reproductive information is available. In Arizona, Cockrum's gray shrews are known to occur in the same areas as Crawford's gray shrews (Baker et al., 2003).

VEGETATIONAL ASSOCIATIONS AND ELEVATION RANGE: Habitats in which Cockrum's gray shrews have been found include those with overstories of Arizona walnut and ash in riparian zones and grasslands with mesquite and giant sacaton (Baker et al., 2003). Known localities from which they have been collected occur at elevations ranging from 677 m to 1,195 m.

CONSERVATION STATUS: *N. cockrumi* had not been identified as a distinct species when the last status list was published (SEMARNAT, 2010). Its very limited distribution makes it extremely susceptible to extirpation from Mexico due to habitat changes related to human activities, for example, agriculture and logging.

Crawford's gray or desert shrew

Leslie N. Carraway

SUBSPECIES IN MEXICO
N. crawfordi is monotypic.

DESCRIPTION: In summer, hairs of the dorsal pelage of *Notiosorex crawfordi* are banded, proximally to distally, dark gray to very dark grayish-brown; hairs of the venter are banded very dark gray to pinkish-white. In winter, hairs of the dorsum are banded very dark grayish-brown, pinkish-white, to very dark grayish-brown; hairs of the venter are banded gray to white (Carraway and Timm, 2000). The tail is colored uniformly dark grayish-brown like the dorsum. Crawford's gray shrews are the smallest within the genus *Notiosorex*. The claw on the middle digit of the manus is average in length (1.3 mm to 1.6 mm; 1.57% to 1.82% of total length) for members of the genus *Notiosorex*. Small paroccipital processes lie against the exoccipitals (Carraway, 2007). The coronoid processes are slender relative to their height (Carraway and Timm, 2000). The zygomatic plate is intermediate in width: 1.7 mm to 2.0 mm (10% to 13% of condylobasal length).

DISTRIBUTION: *N. crawfordi* is known from the southwestern United States southward to northern Sinaloa, eastern Jalisco, and Hidalgo (Álvarez and González-Ruíz, 2001; Carraway and Timm, 2000; Rodríguez Vela, 1999). One record from San Martin Island is known. It has been recorded in the states of BC, BS, CH, CO, DU, HG, JA, NL, SI, SL, SO, and ZA.

EXTERNAL MEASURES AND WEIGHT
TL = 75 to 100 mm; Head and body l. = 53 to 68 mm; TV = 25 to 35 mm; HF = 9 to 12 mm; EAR = 7 to 9 mm; Claw l. = 1.3 to 1.6 mm.
Weight: 3 to 6.3 g.

DENTAL FORMULA: I 3/2, C 1/0, PM 1/1, M 3/3 = 28.

NATURAL HISTORY AND ECOLOGY: Crawford's gray shrews inhabit a variety of ecosystems, from desert, semi-desert, mesquite mud flats, grasslands intermixed with yuccas and junipers to pine-oak forests (Álvarez, 1963a; Armstrong and Jones, 1971; Baker and Greer, 1962; Dalby and Baker, 1967; Delgadillo Villalobos et al., 2005; Fisher, 1941; Petersen and Petersen, 1979; Yensen and Clark, 1986). A wide

Notiosorex crawfordi. Scrubland.
Photo: David J. Schmidly.

range of habitats occupied by Crawford's gray shrews is possible because of their unique physiological adaptations for very dry environments (Lindstedt, 1980) and the smooth structure of the shield of their curly overhairs (Carraway, 2007:44; Ducommun et al., 1994). In Baja California Sur, a lactating female was found in October (Woloszyn and Woloszyn, 1982). Crawford's gray shrews are known to be associated with an extraordinary array of mammals (> 55 species), which can be attributed to the broad variety of habitats in which they occur (Carraway, 2007). Barn owls (*Tyto alba*) and great-horned owls (*Bubo virginianus*) commonly prey on Crawford's gray shrews.

VEGETATIONAL ASSOCIATIONS AND ELEVATION RANGE: Crawford's gray shrews inhabit dry ecosystems containing *Larrea tridentata, Ambrosia chenopodifolia, A. dumosa, Opuntia echinocarpa, Idira columnaris*, and *Pachycereus pringlei* (Yensen and Clark, 1986). Also, they inhabit deserts dominated by lechuguilla, algerita, catclaw, prickly pear, and barrel cactus (Dalby and Baker, 1967); pine-oak forests (Álvarez, 1963a); and mixed desert-shrub and mesquite-grasslands (Armstrong and Jones, 1971; Baker and Greer, 1962; Petersen and Petersen, 1979). In the Maderas del Carmen, Coahuila, at 1,415 m, they occupy grasslands interspersed with *Dasylirion leiophyllum, Juniperus* sp., and *Yucca rostrata* (Delgadillo Villalobos et al., 2005). Known localities occur at elevations from near sea level on Peninsula de Baja California to 2,317 m in the Sierra Moroni of Zacatecas.

CONSERVATION STATUS: *N. crawfordi* is considered a threatened species (SEMARNAT, 2010). The variety of habitats occupied by Crawford's shrews is so broad, however, that the probability of extirpation from any one state within Mexico is unlikely.

Notiosorex evotis (Coues, 1877)

Long-eared gray shrew

Leslie N. Carraway

SUBSPECIES IN MEXICO

N. evotis is a monotypic species.

Until recently it was considered a subspecies of *N. crawfordi* (Carraway and Timm, 2000).

DESCRIPTION: In summer, hairs of the dorsum of *Notiosorex evotis* are banded, proximally to distally, dark gray, pinkish-white, to very dark grayish-brown; hairs of the venter are banded very dark gray to pinkish-white (Carraway and Timm, 2000). In winter, hairs of the dorsum are very dark gray to grayish-brown; hairs of the venter are dark gray to light yellowish-brown. The tail is colored uniformly very dark grayish-brown like the dorsum. Although large-eared gray shrews are the largest members of the genus *Notiosorex*, the claw on the middle digit of the manus is of average size for the genus (1.2 mm to 1.8 mm, 1.35% to 1.67% of total length). The paroccipitals are low-set and extend at an oblique angle from the cranium (Carraway, 2007). The coronoid processes are broad relative to their height. A prominent lateral ridge on the cranium is formed by an extension of the roof of the glenoid fossa (Carraway, 2007). The zygomatic plate is wider, actually (2.1 mm to 2.4 mm), and relatively longer (13% to 15% of condylobasal length) than for other members of the genus.

EXTERNAL MEASURES AND WEIGHT
TL = 84 to 98 mm; Head and body l. = 54 to 73 mm; TV = 23 to 32 mm; HF = 11 to 13 mm; EAR = 7 to 9 mm; Claw l. = 1.2 to 1.8 mm.
Weight: 5.0 to 6.3 g.

DENTAL FORMULA: I 3/2, C 1/0, PM 1/1, M 3/3 = 28.

NATURAL HISTORY AND ECOLOGY: Large-eared gray shrews are known to inhabit fallow fields, areas with thorny shrubs with scattered trees and annuals, and damp places under rocky ledges (Armstrong and Jones, 1971; Escalánte et al., 2003; Fisher and Bogan, 1977; Jones et al., 1962; Schlitter, 1973). One record from July 1962 of a female with five embryos and prominent mammae is known (Armstrong and Jones, 1971). Large-eared gray shrews are known to be associated with *Baiomys taylori*, *Chaetodipus pernix*, *Cryptotis parva berlandieri*, *Liomys irroratus*, *L. pictus*, *Mus musculus*, *Neotoma mexicana*, *Oryzomys palustris*, *Peromyscus maniculatus*, *P. melanophrys*, *Reithrodontomys fulvescens*, *R. megalotis*, *Sigmodon hispidus*, and *S. mascotensis* (Hernández-Chávez, 1997; Jones et al., 1962).

DISTRIBUTION: An endemic to Mexico, the long-eared gray shrew occurs along the western coast from northern Sinaloa to southern Jalisco (Carraway and Timm, 2000). Most of the records are located in the state of Sinaloa (Carraway and Timm, 2000). It has been recorded in the states of JA, MI, NY, and SI.

VEGETATIONAL ASSOCIATIONS AND ELEVATION RANGE: Habitats inhabited by large-eared gray shrews include fairly dry areas scattered with cacti, thornbush, and mesquite (Armstrong and Jones, 1971; Escalánte et al., 2003; Fisher and Bogan, 1977; Jones et al., 1962; Schlitter, 1973). Known localities occur at elevations ranging from 3 m along the Pacific coast to 550 m in the Sierra Madre Occidental of Sinaloa.

CONSERVATION STATUS: The current status of the populations of *N. evotis* is unknown. It is classified as endangered as a subspecies of *N. crawfordi* (SEMARNAT, 2010).

Notiosorex villai Carraway and Timm, 2000

Villa's gray shrew

Leslie N. Carraway

SUBSPECIES IN MEXICO
N. villai is monotypic.

DESCRIPTION: In summer, hairs of the dorsum of *Notiosorex villai* are banded, proximally to distally, dark gray to silvery-gray; hairs of the venter are dark gray to silvery-gray (Carraway and Timm, 2000). In winter, hairs of the dorsum are very dark gray, pinkish-white, to very dark grayish-brown; hairs of the venter are very dark gray to pinkish-white. The tail is colored uniformly very dark grayish-brown. Villa's gray shrews are intermediate in size between large-eared and Cockrum's gray shrews. The claw on the middle digit of the manus is average in length (1.2 mm to 1.3 mm; 1.33% to 1.44% of total length) for members of the genus *Notiosorex*. There is no prominent ridge on the lateral side of the cranium (Carraway, 2007). As for Crawford's gray shrews, the paroccipital processes lie against the exoccipitals and the coronoid processes are slender relative to their height (Carraway and Timm, 2000). The zygomatic plate is narrower, actually (1.7 mm) and relatively (8% of condylobasal length), than for *N. evotis* or most *N. crawfordi*.

DISTRIBUTION: An endemic to Mexico, Villa's gray shrew is known from only three localities in southwestern Tamaulipas (Carraway and Timm, 2000). It has been recorded in the state of TA.

EXTERNAL MEASURES AND WEIGHT
TL = 90 mm; Head and body l. = 59 to 62 mm; TV = 28 to 31 mm; HF = 11 to 11.5 mm; EAR = 7 to 9 mm; Claw l. = 1.2 to 1.3 mm.
Weight: 5 to 6.3 g.

DENTAL FORMULA: I 3/2, C 1/0, PM 1/1, M 3/3 = 28.

NATURAL HISTORY AND ECOLOGY: The three localities from which Villa's gray shrews have been collected are within isolated mountain valleys in an ecotone of "low meseta-like folded mountains characterized by closely spaced ridges" (Ferrusquía-Villafranca, 1993:33; see also Álvarez, 1963; Baker, 1971). The broad range of ecotypes inhabited by Villa's gray shrew suggests that its distribution is greater than presently known. Two lactating females were collected in July 1953. Villa's gray shrews are known to be associated with *Antrozous pallidus*, *Baiomys taylori*, *Bassariscus astutus*, *Cryptotis parva berlandieri*, *Desmodus rotundus*, *Lasiurus borealis*, *L. cinereus*, *Liomys irroratus*, *Mephitis macroura*, *M. mephitis*, *Mormoops megalophylla*, *Mus musculus*, *Mustela frenata*, *Myotis californicus*, *Onychomys arenicola*, *Oryzomys couesi*, *Peromyscus boylii*, *P. leucopus*, *P. pectoralis*, *Reithrodontomys fulvescens*, *Sciurus alleni*, *S. aureogaster*, *Sigmodon hispidus*, and *Tadarida brasiliensis* (Schmidly and Hendricks, 1984).

VEGETATIONAL ASSOCIATIONS AND ELEVATION RANGE: Vegetation of inhabited areas includes pine-oak forests, tropical forests, and riparian zones (Álvarez, 1963a; Baker, 1971; Ferrusquía-Villafranca, 1993:33). Known localities occur at elevations ranging from about 579 m to 1,342 m.

CONSERVATION STATUS: *N. villai* was not identified as a distinct species when the last status list was published (SEMARNAT, 2010). Because it is known only from two isolated mountain valleys it should be afforded some level of protection at least until more is known of its distribution.

Sorex arizonae Diersing and Hoffmeister, 1977

Arizona shrew

Leslie N. Carraway

SUBSPECIES IN MEXICO
S. arizonae is monotypic.

DESCRIPTION: Hairs of the dorsal pelage of *Sorex arizonae* are banded, proximally to distally, blond to dark brown, giving an overall medium brown appearance. Hairs of the venter are white tipped. The tail is bicolored medium brown dorsally and white ventrally. The hips and rump have brown, clear-tipped guard hairs extending 1.1 mm beyond the dorsal pelage. As for all *Sorex*, the curly overhairs have deep grooves with a central ridge and a superimposition of two rather elongated notches (Ducommun et al., 1994:635). Arizona shrews are of intermediate size with a condylobasal length of 16.25 mm and cranial breadth of 7.99 mm. Median tines are located within the pigment on the body of the I1s. First lower incisors have two denticles, pigment in one section, and a long strip of pigment at the anteromedial edge, and are angled 10 degrees from the horizontal ramus of the dentary (Carraway, 2007). As for all

Sorex, all teeth are pigmented, U1-U3 have no second cusp, the alveolus of i1 does not extend beneath m1 (Carraway, 1995), the upper glenoid furrow has no locking mechanism and is rotated 10 degrees from the horizontal of the skull and the area between the condylar processes is not emarginate (Carraway, 2005). Also, the tail is 40% of length of head and body. As a member of the subgenus *Sorex*, *S. arizonae* can be distinguished from congeners by U3 ≥ U4, i1s with deep interdenticular spaces and condylobasal l. < 17 mm.

DISTRIBUTION: The Arizona shrew occurs in Arizona and New Mexico southward into Chihuahua (Caire et al., 1978; Conway and Schmitt, 1978; Diersing and Hoffmeister, 1977). It has been recorded in the state of CH.

EXTERNAL MEASURES AND WEIGHT

TL = 110 mm; Head and body l. = 66 mm; TV = 44 mm; HF = 11 mm; EAR = 7 mm; Claw l. = 1.3 mm.
Weight: 1.9 to 5.2 g.

DENTAL FORMULA: I 3/1, C 1/1, PM 3/1, M 3/3 = 32.

NATURAL HISTORY AND ECOLOGY: *S. arizonae* is known from the Sonoran Desert among rocky outcrops on north-facing slopes (Caire et al., 1978; Escalánte et al., 2003). No reproductive information is available. Arizona shrews are known to be associated with *Neotoma mexicana*, *Peromyscus boylii*, *P. truei*, and *Thomomys umbrinus* (Caire et al., 1978; Conway and Schmitt, 1978).

VEGETATIONAL ASSOCIATIONS AND ELEVATION RANGE: The flora at the one known locality included *Arbutus* sp., *Juniperus* sp., *Pinus* sp., and *Quercus* sp. (Caire et al., 1978; Escalánte et al., 2003). The only known locality for the Arizona shrew is at 2,592 m elevation (Caire et al., 1978).

CONSERVATION STATUS: *S. arizonae* is considered an endangered species by SEMARNAT (2002). It is considered vulnerable by the International Union for Conservation of Nature (IUCN) (Ceballos et al., 2002).

Sorex emarginatus Jackson, 1925

Zacatecas or Sierra Madre long-tailed shrew

Rafael Ávila Flores

SUBSPECIES IN MEXICO

S. emarginatus is a monotypic species.

DESCRIPTION: *Sorex emarginatus* is a small shrew of the subgenus *Sorex*. It can be differentiated from its most closely related species, *S. saussurei* and *S. arizonae*, by its size, by having the first molar of the upper jaw narrow and indented, and by having the U4 remarkably smaller than the U3 (Jackson, 1925). The I1s have a median tine located within the pigment. The i1s have two denticles, shallow interdenticular spaces, pigment in one section, and a long strip of pigment at the anteromedial edge (Carraway, 2007). It is very similar to *S. ventralis*, even in size, so the only concrete difference between them appears to be its geographical distribution, as they are allopatric species (Diersing and Hoffmeister, 1977; Junge and Hoffmann, 1981). The pelage is grayish-brown with paler shades on the abdomen (Álvarez and Polaco, 1984). Hips and rump have medium brown, clear-tipped guard hairs extending 1.2 mm to

1.4 mm beyond the dorsal pelage (Carraway, 2007). Like other shrews of the subgenus *Sorex*, it has postmandibular foramen and canals; it lacks pigmentation of the surface of the higher occlusal unicusps (Junge and Hoffmann, 1981). The great similarity of the morphology (external and cranial) between *S. emarginatus*, *S. ventralis*, and *S. arizonae* indicates that, with the capture and analysis of more specimens, it could be proved that these three taxa actually are a single species (Junge and Hoffmann, 1981).

DISTRIBUTION: *S. emarginatus* is an endemic species of Mexico. Its distribution is restricted to some areas of the Sierra Madre Occidental, northward from Durango and Zacatecas, and from there, to northern Jalisco (Hall, 1981; Matson and Baker, 1986). Specimens originally reported to be *S. emarginatus* (Koestner, 1941) in Cerro Potosi, Nuevo Leon, actually are *S. milleri* (Carraway, 2007).

EXTERNAL MEASURES AND WEIGHT

TL = 88 to 107 mm; TV = 38 to 44 mm; HF = 11 to 13 mm; EAR = 5 to 7 mm.
Weight: 5.0 to 6.5 g.

DENTAL FORMULA: I 3/1, C 1/1, PM 3/1, M 3/3 = 32.

NATURAL HISTORY AND ECOLOGY: These shrews have been collected in humid areas covered with litter, grass, and moss, located within forests, in canyons, and along streams (Álvarez and Polaco, 1984; Baker and Greer, 1962; Koestner, 1944; Matson and Baker, 1986). Apparently, they are active both day and night and probably, like other species of the genus, they do not have periods of inactivity during the year (Álvarez and Polaco, 1984; Matson and Baker, 1986; Nowak, 1999b). Components of their diet include a wide variety of insects, worms, and other small invertebrates, and do not rule out other types of food such as vertebrate carrion and plant matter (Álvarez and Polaco, 1984; Nowak, 1999b). They are known to occur with *Peromyscus melanotis* and *P. boylii* (Baker and Greer, 1962). There is one documented instance of predation by a water snake (*Thamnophis elegans*; Baker and Webb, 1976). As with other shrews, it is likely they also are preyed on by other snakes, owls, and some carnivores (Ceballos and Galindo, 1984). The mating season occurs between July and August (Baker and Greer, 1962; Koestner, 1941; Matson and Baker, 1986).

VEGETATIONAL ASSOCIATIONS AND ELEVATION RANGE: *S. emarginatus* mainly inhabits cold and temperate forests in the highest and humid parts of the mountains, including fir forests, pine-oak forests, and pine forests. It also has been found in subtropical vegetation near forests of pine-oak (Baker and Greer, 1962), a mixed forest of oak-manzanita-juniper-arbutus (Matson and Baker, 1986), and a type of vegetation between the transitional forest pine-oak and tropical deciduous forest (Webb and Baker, 1962). It inhabits areas between 1,830 m and 3,660 m, but most records are between 2,300 m and 2,900 m.

CONSERVATION STATUS: The current status of the populations of *S. emarginatus* is unknown. It is provided no protection under the most recent list of species at risk (SEMARNAT, 2010).

Sorex ixtlanensis Carraway, 2007

Ixtlan shrew

Leslie N. Carraway

SUBSPECIES IN MEXICO

S. ixtlanensis is monotypic.

DESCRIPTION: Hairs of the dorsal pelage of *Sorex ixtlanensis* are banded, proximally to distally, reddish blond to dark brown; hairs of the venter have light reddish-brown tips (Carraway, 2007). The tail is colored uniformly dark brown like the dorsum. Hips and rump have pale reddish-brown and dark brown guard hairs extending 2.2 mm to 2.8 mm beyond the dorsal pelage (Carraway, 2007). First lower incisors have three denticles, deep interdenticular spaces, and pigment in one section, and are angled < 8 degrees from the horizontal ramus of the dentary (Carraway, 2007). *S. ixtlanensis* is a member of the subgenus *Otisorex*. Ixtlan shrews differ from Verapaz shrews by a long strip of pigment at the anteromedial edge and generally a larger-sized of mandible (length of mandible usually 8.2 mm to 9.4 mm, length of c1-m3 usually 5.4 mm to 5.9 mm and height of articular condyle 2.8 mm to 3.3 mm; Carraway, 2007). Ixtlan shrews are medium-sized *Sorex* and can be distinguished from congeners by pelage characteristics, median tines located above the pigment on the body of the I1s, U3 < U4, condylobasal l. 18.24 mm to 19.71 mm, and cranial breadth usually 9.2 mm to 10.38 mm (Carraway, 2007). Specimens now considered *S. ixtlanensis* were included within *S. veraepacis mutabilis*. It was determined that Ixtlan shrews have Ils with median tines located above the pigmented area on the body of the I1s as opposed to that of *S. veraepacis mutabilis*, which has median tines located well within the pigment on the body of the I1s (Carraway, 2007:48).

DISTRIBUTION: An endemic to Mexico, the Ixtlan shrew occurs from central Guerrero to northeastern Oaxaca (Carraway, 2007). It has been recorded in the states of GR and OX.

EXTERNAL MEASURES AND WEIGHT

TL = 117 to 135 mm; Head and body l. = 66 to 77 mm; TV = 45 to 67 mm; HF = 14 to 16 mm; Claw l. = 1.1 to 1.9 mm.
Weight: probably around 6 g.

DENTAL FORMULA: I 3/1, C 1/1, PM 3/1, M 3/3 = 32.

NATURAL HISTORY AND ECOLOGY: Ixtlan shrews inhabit high-elevation forests in the Canadian zone (Goldman, 1951). The ground in the shady areas of the dense forests is covered with a deep layer of leaves overlaying a deep layer of humus (Davis and Lukens, 1958; Musser, 1964). No reproductive information is available. No mammal associates have been documented.

VEGETATIONAL ASSOCIATIONS AND ELEVATION RANGE: Occupied habitats contain pine-fir or pine-oak forests (Davis and Lukens, 1958; Musser, 1964). Known localities occur at elevations ranging from 1,920 m to 3,000 m.

CONSERVATION STATUS: *S. ixtlanensis* was not identified as a distinct species when the last status list was published (SEMARNAT, 2010). *S. veraepacis mutabilis*, from which it was derived, was considered a protected subspecies. Thus, *S. ixtlanensis* should be afforded the same level of protection.

Sorex macrodon Merriam, 1895

Large-toothed shrew

Rafael Avila Flores

SUBSPECIES IN MEXICO

S. macrodon is a monotypic species.

DESCRIPTION: *Sorex macrodon* is a large shrew of the subgenus *Otisorex*. Hairs on the back are very dark brown, the venter hairs are only slightly paler, and the tail is uniformly colored like the back. Hips and rump have dark brown guard hairs extending 1.5 mm to 1.8 mm beyond the dorsal pelage (Carraway, 2007). It is characterized by well-developed median tines located on the first upper incisors, not very pigmented and quite high on the tooth surface (Junge and Hoffmann, 1981). It has a large and massive skull, the face and nostrils are relatively wide in its anterior portion, and the edges of the premaxilla are relatively thick (Hall, 1981; Hall and Dalquest, 1963). The third unicuspid tooth is smaller than the fourth and, as in other members of the subgenus *Otisorex*, lacks postmandibular foramen and canals, and has a pigmented edge on the surface of the occlusal unicuspids (Hall, 1981; Junge and Hoffmann, 1981). The i1s have three denticles, deep interdenticular spaces, and pigment in one section, and lack a long strip of pigment at the anteromedial edge (Carraway, 2007). *S. macrodon* is morphologically very similar to *S. veraepacis*, differing only in terms of the position of tine on I1s, i1s with pigment in one section, and uniform coloration of the tail (Carraway, 2007). With the collection and analysis of more specimens we could determine whether these taxa are conspecific (Junge and Hoffmann, 1981).

DISTRIBUTION: An endemic species of Mexico, *S. macrodon* is distributed exclusively in mountainous parts of central-west Veracruz, northern Oaxaca, and northeast Puebla. It has been recorded in the states of OX, PU, and VE.

EXTERNAL MEASURES AND WEIGHT

TL = 128 to 136 mm; TV = 49 to 52 mm; HF = 14.0 to 15.5 mm; EAR = 5 to 7 mm. Weight: 9.8 to 11.3 g.

DENTAL FORMULA: I 3/1, C 1/1, PM 3/1, M 3/3 = 32.

NATURAL HISTORY AND ECOLOGY: These shrews tend to move over the most humid parts of the forests, mainly in banks of moss and grass, under bushes, and on river edges. They are basically solitary and are active during both day and night (Nowak, 1999b). Although they have been found using mouse trails (*Microtus mexicanus*; Hall and Dalquest, 1963), it is likely that they also build their own paths (Nowak, 1999b). Also, they are known to occur with *Cryptotis mexicana*, *Liomys irroratus*, *Megadontomys thomasi*, *Oryzomys alfaroi*, *Peromyscus furvus*, *Plecotus mexicanus*, and *Sorex veraecrucis* (Hall and Dalquest, 1963; Heaney and Birney, 1977; Prieto Bosch and Sánchez-Cordero, 1993). There is almost no information about their diet. Probably their usual diet includes insects and other invertebrates, as well as carrion vertebrate and miscellaneous plant material. Among their predators must be owls, reptiles, and carnivores, as with other shrews of the genus (Ceballos and Galindo, 1984). As for the reproduction period, there is knowledge of only one lactating female captured in April (Heaney and Birney, 1977).

VEGETATIONAL ASSOCIATIONS AND ELEVATION RANGE: *S. macrodon* mainly inhabits wet forests and scrublands of pine and oak, some of which have mesophile elements (Hall and Dalquest, 1963; Heaney and Birney, 1977). It is distributed between 1,280 m and 2,600 m.

CONSERVATION STATUS: *S. macrodon* is known from 10 specimens from 5 locations very close together; there are no reports of it since the late 1950s. For these reasons, even though it is not included in published lists of species at risk, *S. macrodon* could be regarded as an endangered species. It is necessary to assess their current situation to determine appropriate measures to preserve them.

Jalisco shrew

Leslie N. Carraway

SUBSPECIES IN MEXICO
S. mediopua is monotypic.

DESCRIPTION: Jalisco shrews are medium-sized *Sorex*. Hairs of the dorsal pelage are banded, proximally to distally, medium silvery gray, white or blond, to dark brown, giving an overall medium brown appearance (Carraway, 2007). Hairs of the venter are medium silvery gray with blond tips. Hips and rump have medium or dark brown guard hairs extending 1.0 mm to 1.5 mm beyond the dorsal pelage (Carraway, 2007). First lower incisors have two denticles, shallow interdenticular spaces, pigment in one section, and a long strip of pigment at the anteromedial edge, and usually are angled < 6 degrees from the horizontal ramus of the dentary (Carraway, 2007). *Sorex mediopua* is a member of the subgenus *Sorex*. It can be distinguished from sympatric congeners by a tail colored uniformly dark brown like the dorsum, medium to large median tines located at the interface of the pigmented/unpigmented areas on the body of the I1s, U3 < U4 and overall larger mandible (length of mandible 7.7 mm to 9.2 mm, length of c1-m3 5.2 mm to 5.6 mm; Carraway, 2007). Specimens now considered *Sorex mediopua* were once included within *S. saussurei saussurei*. It was determined that Jalisco shrews have Ils with well-pigmented median tines at the interface of the pigmented/unpigmented areas on the body of the I1s unlike those of *S. saussurei saussurei* that have large well-pigmented median tines located well within the pigment on the body of the I1s (Carraway, 2007:48-49).

DISTRIBUTION: An endemic to Mexico, the Jalisco shrew occurs within the Transvolcanic Belt from western Jalisco eastward into Mexico (Carraway, 2007). It has been recorded in the states of GR, JA, MI, MO, and MX.

EXTERNAL MEASURES AND WEIGHT
TL = 102 to 120 mm; Head and body l. = 61 to 75.5 mm; TV = 38 to 49 mm; HF = 12 to 15 mm; CLAW = 1.3 to 1.9 mm.
Weight: probably around 6 g.

DENTAL FORMULA: I 3/1, C 1/1, PM 3/1, M 3/3 = 32.

NATURAL HISTORY AND ECOLOGY: Jalisco shrews inhabit a variety of high-elevation woodlands and wooded moist canyons with a deep layer of leaves or needles overlaying a deep layer of humus (Baker, 1956; Baker and Greer, 1962; Davis and Lukens, 1958). No reproductive information is available. Jalisco shrews are known to be associated with *Microtus mexicanus*, *Neotamias dorsalis*, *Reithrodontomys sumichrasti*, and *Sorex milleri* (Baker, 1956; Davis and Lukens, 1958).

VEGETATIONAL ASSOCIATIONS AND ELEVATION RANGE: The flora of occupied canyons includes pine-oak or Douglas fir-juniper woodlands with understories of *Bacheria* and *Senecio*. Upland habitats contain yellow pine-alder or pine-oak-juniper woodlands (Baker, 1956; Baker and Greer, 1962; Davis and Lukens, 1958). Known localities occur at elevations ranging from 1,875 m to 3,048 m.

CONSERVATION STATUS: *S. mediopua* was not identified as a distinct species when the last list status list was published (SEMARNAT, 2010). In that listing *S. saussurei saussurei*, from which it was derived, was not afforded any level of protection. Given the limited state of knowledge of the ecology of *S. mediopua*, it is unknown what level of protection it should be afforded.

Mount Carmen shrew

Iván Castro-Arellano and Gerardo Ceballos

SUBSPECIES IN MEXICO

S. milleri is a monotypic species.

DESCRIPTION: *Sorex milleri* is a small shrew that belongs to the subgenus *Otisorex*. The color of the back is between brown and brownish-gray, with the abdomen pale smoke gray; the tail is bicolored, pale brown in the dorsal part and cream at the bottom (Jackson, 1947). Hips and rump have dark brown, clear-tipped guard hairs extending 0.33 mm beyond the dorsal pelage (Carraway, 2007). The skull has an elongated face and an inflated cranial box. The third unicuspid tooth is larger than the fourth (Hall, 1981; Junge and Hoffman, 1981). The I1s have a median tine located within the pigment. The i1s have three denticles, deep interdenticular spaces, pigment in three sections, and a long strip of pigment at the anteromedial edge (Carraway, 2007). This species is closely related to *S. cinereus*, and likely is a relictual population of this species that was isolated in the late Pleistocene (Findley, 1955b). While most authors (Findley, 1955b; Hall, 1981; Junge and Hoffman, 1981) consider it as a valid species, Van Zyll de Jong and Kirkland (1989) suggested that is a subspecies of *S. cinereus*, basing their theory on a morphometric analysis of the skull.

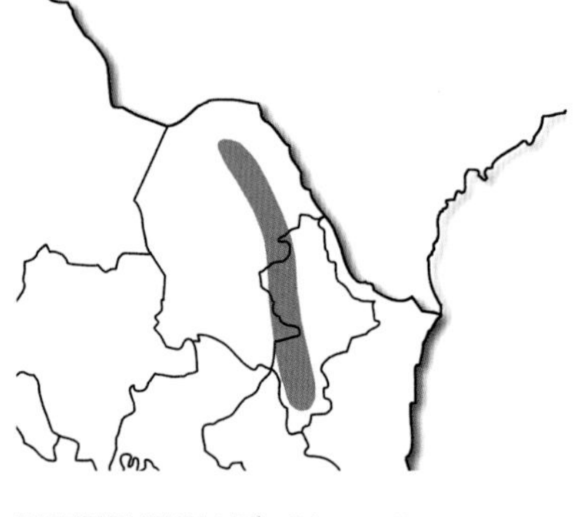

DISTRIBUTION: The Mount Carmen shrew is endemic to Mexico and is restricted to the Sierra Madre Oriental from the northwest of Coahuila to the south of Nuevo Leon (Hall, 1981; Hutterer, 1993). It has been recorded in the states of CO and NL.

EXTERNAL MEASURES AND WEIGHT

TL = 93 to 95 mm; TV = 37 to 44 mm; HF = 11 to 12 mm; EAR = 5 to 7 mm.
Weight: 2.9 to 4.8 g.

DENTAL FORMULA: I 3/1, C 1/1, PM 3/1, M 3/3 = 32.

NATURAL HISTORY AND ECOLOGY: Apparently, the Mount Carmen shrew is restricted to pine forests of the highest parts of the Sierra Madre Oriental. In Coahuila, it has been collected in traps placed around fallen tree trunks, in damp and shady wooded hillsides. In this zone, it occurs sympatrically with *Peromyscus boylii*, *P. melanotis*, *Microtus mexicanus*, *Reithrodontomys megalotis*, *Sigmodon ochrognathus ochrognathus*, and *Thomomys bottae* (Baker, 1956; Delgadillo Villalobos et al., 2005; Jiménez-G. et al. 1999). We are unaware of the specifics of their biology, but like other species of the genus, their diet probably consists of worms, insects, and other small invertebrates. Females have their young after a gestation period of nearly a month. The young are separated from the mother after a month and a half of life and live up to two years (Nowak, 1999b). In Coahuila, pregnant females have been collected in July and October (Baker, 1956).

VEGETATIONAL ASSOCIATIONS AND ELEVATION RANGE: *S. milleri* inhabits mountain forests of pine, oak, and fir where the dominant species are pine (*Pinus* sp.), juniper (*Cupressus* sp.), oaks (*Quercus* sp.), poplar (*Populus* sp.). and fir (*Abies* sp.; Baker, 1956). It has been recorded between 2,400 m and 2,805 m (Hall, 1981).

CONSERVATION STATUS: This shrew has a restricted distribution and there are few records of it; however, *S. milleri* is not provided any level of protection in published lists of species at risk (SEMARNAT, 2010).

Dusky or montane shrew

Emilio Daniel Garcia Tobón

SUBSPECIES IN MEXICO

S. monticolus is a monotypic species.

The original name for this species was given by Merriam (1890), but subsequently it has also been treated as *Sorex vagrans monticola* (Baker and Greer, 1962; Hall, 1981), *S. obscurus* (Hennings and Hoffman, 1977), or *S. melanogenys* (Hall, 1981). An analysis of differences in dental morphology and cranial proportions between *S. vagrans* and *S. monticolus*, previously considered within the subspecies complex of *S. vagrans* (Findley, 1955b), it was found that both species are biologically and ecologically different (Hennings and Hoffman, 1977).

DESCRIPTION: *Sorex monticolus* is a medium-sized shrew consistently larger than *S. vagrans* when they are sympatric (Junge and Hoffman, 1981). It has a color pattern composed of one to three colors ranging from red to gray hair in the summer and black to pale gray in winter. Hips and rump have dark brown, clear-tipped guard hairs extending 0.33 mm beyond the dorsal pelage (Carraway, 2007). The first upper incisors have large median tines covered with red pigment; U3 < U4; and the i1s have three denticles, deep interdenticular spaces, pigment in one section, and a long strip of pigment at the anteromedial edge (Carraway, 2007). *S. monticolus* belongs to the subgenus *Otisorex*.

DISTRIBUTION: *S. monticolus* is distributed from Alaska and Canada, throughout the western part of the United States and Mexico; it is located in the states of Chihuahua and Durango (Anderson, 1972; Baker and Greer, 1962; Gardner, 1965; Hutterer, 1993; Junge and Hoffman, 1981; Nowak, 1999b). It has been recorded in the states of CH and DU.

EXTERNAL MEASURES AND WEIGHT

TL = 90 to 153 mm; TV = 31 to 67 mm; HF = 13 mm; EAR = 5 to 7 mm.
Weight: 7.3 g.

DENTAL FORMULA: I 3/1, C 1/1, PM 3/1, M 3/3 = 32.

NATURAL HISTORY AND ECOLOGY: There is no specific information about this shrew's natural history, although some features are known at the generic level. They live in moist prairies with high grasses (Hoffmeister, 1986). They may be active both day and night and are active throughout the year. They feed mainly on insects, worms, and other small invertebrates and carrion; occasionally, they eat plant material (Nowak, 1999b). They are basically solitary and often quite aggressive with one another. The gestation period is not well known, but generally lasts 18 to 28 days. The litter size is an average of 4 to 7 young, which are independent at 4 or 5 weeks old (Nowak, 1999b). When the population density is low, some females may mate the first year, but in general they wait until the second. They live approximately two years (Nowak, 1999b). There may be two reproductive periods per year, which does not imply that the same female necessarily produces two litters per year (Hoffmeister, 1986). An important feature of these shrews is that they can lower their body temperature almost to ambient temperature, a phenomenon known as heterothermia (Fleming, 1979). Small mammals use hypothermia as an adaptation to reduce energy costs in times of food scarcity (Fleming, 1979).

VEGETATIONAL ASSOCIATIONS AND ELEVATION RANGE: *S. monticolus* is primarily located in forests of oaks and conifers in mountain areas (Anderson, 1972; Baker and Greer, 1962; Gardner, 1965; Junge and Hoffman, 1981). They are known to occur from about 2,000 m to 2,600 m in Durango (Anderson, 1972; Baker and Greer, 1962; Gardner, 1965; Hall, 1981).

CONSERVATION STATUS: In the Mexican Official Regulation *S. monticolus* is listed as rare (SEMARNAT, 2010) under the name of *S. vagrans monticola*. The disruption of their habitat by human activity can seriously affect their populations.

Sorex oreopolus Merriam, 1892

Volcano or Mexican long-tailed shrew

Iván Arellano-Castro

SUBSPECIES IN MEXICO

S. oreopolus is a monotypic species.

DESCRIPTION: *Sorex oreopolus* is a small shrew that belongs to the subgenus *Otisorex*. The back is sepia-brown and the whole belly is gray brown. The tail is bicolored following the pattern of the body, except for the most distal lower part, which is a little more obscure (Merriam, 1892). Hips and rump have dark brown, clear-tipped guard hairs extending 1.75 mm to 2.5 mm beyond the dorsal pelage. The I1s have a median tine located above the pigment; U3 < U4; and the i1s have two denticles, deep interdenticular spaces, pigment in one section, and a long strip of pigment at the anteromedial edge (Carraway, 2007). The subspecies *S. orizabae orizabae*, originally described as *S. orizabae* by Merriam (1895), was included as a subspecies of *S. vagrans* by Hennings and Hoffmann (1977) and as a subspecies of *S. oreopolus* by Junge and Hoffmann (1981). In 2007, it again was recognized as the distinct species *S. orizabae* (Carraway, 2007).

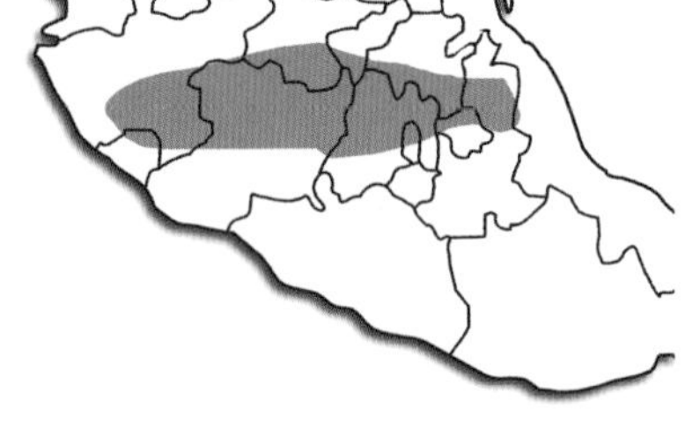

DISTRIBUTION: *S. oreopolus* is endemic to Mexico and is distributed from the far southwest of Jalisco to the east of Puebla and western Veracruz, always in highland areas (Hutterer, 1993; Junge and Hoffman, 1981). It has been recorded in the states of DF, JA, MO, MX, PU, and TX.

EXTERNAL MEASURES AND WEIGHT

TL = 88 to 112 mm; TV = 35 to 46 mm; HF = 12.5 to 14 mm EAR = 5 to 7 mm. Weight: 5 g.

DENTAL FORMULA: I 3/1, C 1/1, PM 3/1, M 3/3 = 32.

NATURAL HISTORY AND ECOLOGY: *S. oreopolus* was caught in Morelos between broomroots that were part of the lower stratum of a pine forest (Davis and Russell, 1954a). I am unaware of the specifics of their biology, but like other species of the genus their diet probably consists of worms, insects, and other small invertebrates. Jackson (1928) reported hymenopterans, beetles, dipterans (larvae and adults), locusts, crickets, spiders, mice, mosses, seeds, and other plants as part of the stomach contents of shrews of the genus *Sorex*. It is known to occur with *Neotoma Mexicana*, *Neotomodon alstoni*, *Peromyscus difficilis*, *P. maniculatus*, *P. melanotis*, *P. spicilegus*, *P. truei*, *Reithrodontomys chrysopsis*, and *R. megalotis* (Fa et al., 1990). The gestation period likely lasts between 18 and 28 days. Neonates become independent in 4 or 5 weeks after birth, and longevity is 1 to 2 years (Nowak, 1999b). A known predator of this species is the red-tailed hawk (*Buteo jamaicensis*; Davis, 1944).

VEGETATIONAL ASSOCIATIONS AND ELEVATION RANGE: *S. oreopolus* inhabits mountain forests dominated by oaks and conifers (*Pinus*, *Quercus*, and *Abies*) as well as broomroots (Hooper, 1955; Ramírez-Pulido, 1969a). It has been found from 2,000 m to 4,550 m (Davis, 1944; Hall, 1981).

CONSERVATION STATUS: The current status of *S. oreopolus* is unknown, but it is not shown in any list of species with conservation problems.

Sorex orizabae Merriam 1895b

Orizaba shrew

Leslie N. Carraway

SUBSPECIES IN MEXICO

S. orizabae is a monotypic species.

Recently re-elevated to species status, it was considered a subspecies of *S. oreopolus* (Carraway, 2007).

DESCRIPTION: Orizaba shrews are medium-sized *Sorex* with hairs of the dorsal pelage banded white medially with long, dark brown tips giving an overall grizzled appearance (Carraway, 2007). Hairs of the venter are medium brown with white tips. The tail is always bicolored. Hips and rump have medium or dark brown guard hairs extending 1.75 mm to 2.5 mm beyond the dorsal pelage (Carraway, 2007). First lower incisors have three denticles, deep interdenticular spaces, pigment in two sections, and a long strip of pigment at the anteromedial edge, and, usually are angled 5 degrees to 10 degrees from the horizontal ramus of the dentary (Carraway, 2007). *Sorex orizabae* is a member of the subgenus *Otisorex*. It can be distinguished from sympatric congeners by a tail bicolored like the dorsum, median tines located above pigmented areas on the body of the I1s, U3 < U4, cranial breadth < 8.5 mm, and least interorbital breadth < 4 mm (Carraway, 2007).

DISTRIBUTION: Endemic to Mexico, the Orizaba shrew is distributed across the Transvolcanic Belt. It has been recorded in the states of DF, ME, MI, MO, PU, TL, and VE.

EXTERNAL MEASURES AND WEIGHT

TL = 92 to 108 mm; Head and body l. = 54 to 71 mm; TV = 33 to 43 mm; HF = 12 to 14 mm.

Weight: Probably around 6 g.

DENTAL FORMULA: I 3/1, C 1/1, PM 3/1, M 3/3 = 32.

NATURAL HISTORY AND ECOLOGY: *S. orizabae* is known to occur with *Mephitis macroura macroura*, *Microtus mexicanus mexicanus*, *Neotomodon alstoni alstoni*, *Peromyscus maniculatus fulvus*, *P. maniculatus labecula*, *P. melanotis*, *Reithrodontomys megalotis saturatus*, *Sorex saussurei*, *Tadarida Mexicana*, and *Thomomys umbrinus peregrinus* (Davis and Russell, 1953, 1954a). No reproductive information is available. Predators include red-tailed hawks (*Buteo jamaicensis calurus*; Davis, 1944).

VEGETATIONAL ASSOCIATIONS AND ELEVATION RANGE: Throughout its distribution, the Orizaba shrew can be found in high-elevation montane valleys associated with boreal pine forests with muhly bunchgrass understory and above the timberline in talus slopes. Known localities from which they have been collected occur at elevations ranging from about 2,200 m to 4,215 m.

CONSERVATION STATUS: *S. orizabae* has not been designated any level of conservation (SEMARNAT, 2010).

Ornate or Tule shrew

Emilio Daniel Garcia Tobón

SUBSPECIES IN MEXICO

Sorex ornatus lagunae Nelson and Goldman, 1909
Sorex ornatus ornatus Merriam, 1895
Sorex ornatus juncencis Nelson and Goldman, 1909.
The subspecies *juncencis* has been regarded as a species (Hall, 1981), or subspecies (Junge and Hoffmann, 1981).

DISTRIBUTION: *S. ornatus* is found from the western part of the state of California to the south side of the Baja California Peninsula, Mexico (Hutterer, 1993; Owen and Hoffmann, 1983). There are records on the southern tip of Baja California Sur and Catalina Island and in the Sierra de San Pedro Martir and San Quentin (Owen and Hoffman, 1983). The type locality of *Sorex ornatus lagunae*, is in the Sierra de la Laguna, Baja California Sur (Hall, 1981). This subspecies, if it still exists, could be one of the rarest of this region (Woloszyn and Woloszyn, 1982). It has been recorded in the states of BC and BS.

DESCRIPTION: *Sorex ornatus* is a small shrew that belongs to the subgenus *Otisorex*. The upper parts of body are grayish brown, with pale lower parts. The tail is indistinctly bicolored and is short relative to the length of the head and body (Hall, 1981; Owen and Hoffmann, 1983). Hips and rump have dark brown guard hairs extending 2 mm beyond the dorsal pelage (Carraway, 2007). The I1s have a median tine located within the pigment; U3 < U4; and the i1s have three denticles, deep interdenticular spaces, pigment in one section, and a long strip of pigment at the anteromedial edge (Carraway, 2007).

EXTERNAL MEASURES AND WEIGHT

TL = 80 to 110 mm; TV = 28 to 46 mm; HF = 9 to 14 mm; EAR = 5 to 7 mm.
Weight: 5.12 g.

DENTAL FORMULA: I 3/1, C 1/1, PM 3/1, M 3/3 = 32.

NATURAL HISTORY AND ECOLOGY: The reproductive period occurs during late February until early October, although there may be variation among the different subspecies (Owen and Hoffman, 1983). A litter consists of four to six offspring. It is estimated that the gestation period lasts approximately 21 days (Owen and Hoffmann, 1983). These shrews most commonly occur in dense and low vegetation, where they can find a number of invertebrates (which are part of their main food), shelter, and a place to build their dens (Owen and Hoffmann, 1983). They also inhabit wetlands along streams and on the slopes of hills (Woloszyn and Woloszyn, 1982). Apparently, their populations are not constrained by food availability, although it seems they can be affected by the density and diversity of invertebrates (Owen and Hoffmann, 1983). This species has a life span of between 12 and 16 months, so that populations are replaced annually (Owen and Hoffmann, 1983). Individuals remain active during day and night, although nocturnal activity predominates, especially during reproduction (Owen and Hoffmann, 1983). Bone remnants of this species have been found in an owl nest (Owen and Hoffmann, 1983).

VEGETATIONAL ASSOCIATIONS AND ELEVATION RANGE: *S. ornatus* mainly inhabits forests of conifers, tropical deciduous forests, xeric scrublands, and salt lakes near the sea from sea level up to 2,000 m (Owen and Hoffmann, 1983).

CONSERVATION STATUS: One of the subspecies (*S. ornatus ornatus*) was categorized as rare in the Mexican Official Regulation (SEMARNAT, 2010) mainly because of its restricted distribution and degradation of its habitat.

Saussure's shrew

Leslie N. Carraway

SUBSPECIES IN MEXICO
S. saussurei is monotypic.

DESCRIPTION: Saussure's shrews are medium-sized *Sorex* with hairs of the dorsal pelage banded, proximally to distally, medium silvery gray, white, or blond, to dark brown, giving an overall medium brown appearance. Hairs of the venter are medium silvery gray with white or blond tips. The tail is colored uniformly dark brown like the dorsum or is slightly bicolored. Hips and rump have dark brown guard hairs extending 1.0 mm to 1.5 mm beyond the dorsal pelage (Carraway, 2007). First lower incisors have two denticles, pigment in one section, and a long strip of pigment at the anteromedial edge and usually are angled 8 degrees to 10 degrees from the horizontal ramus of the dentary (Carraway, 2007). *Sorex saussurei* is a member of the subgenus *Sorex*. It can be distinguished from sympatric congeners by pelage characteristics; large, well-pigmented median tines located well within the pigmented area on the body of the I1s; U3 < U4; and i1s with shallow interdenticular spaces (Carraway, 2007). *S. saussurei* was considered to include the nominal subspecies plus *cristobalensis*, *oaxacae*, and *veraecrucis* (Carraway, 2007:48-49). Within *S. saussurei saussurei*, however, a portion of the specimens possessed I1s with well-pigmented median tines at the interface of the pigmented/unpigmented areas on the body of the I1s. On this basis, the latter specimens were placed in the new species *S. mediopua* as distinct from *S. saussurei* (Carraway, 2007:48-49). A second portion within *S. saussurei saussurei* possessed I1s with small, unpigmented or slightly pigmented median tines located high above the pigment like *S. veraecrucis* and were placed in the new subspecies *S. veraecrucis altoensis*. Additionally, specimens of the subspecies *cristobalensis*, *oaxacae*, and

Sorex saussurei. Bunch grassland. Zoquiapan and Anexas Nacional Park, state of Mexico. Photo: Carlos Galindo.

veraecrucis also possess I1s with small, unpigmented or slightly pigmented median tines located high above the pigment, thus were designated as subspecies of *Sorex veraecrucis* (Carraway, 2007:48-49).

DISTRIBUTION: An endemic to Mexico, Sorex saussurei occurs within the Trans-volcanic Belt from western Jalisco eastward into Puebla (Carraway, 2007). It has been recorded in the states of CL, DF, JA, MI, MO, MX, and PU.

EXTERNAL MEASURES AND WEIGHT
TL = 101 to 123 mm; Head and body l. = 59 to 76 mm; TV = 39 to 51 mm; HF = 12 to 16 mm; EAR = 5 to 7 mm; CLAW = 1.4 to 2.1 mm.
Weight: 6.7 g.

DENTAL FORMULA: I 3/1, C 1/1, PM 3/1, M 3/3 = 32.

NATURAL HISTORY AND ECOLOGY: Most commonly, Saussure's shrews inhabit higher-elevation woodlands and wooded moist canyons with a deep layer of leaves or needles overlaying a deep layer of humus. They also are known to inhabit high-elevation agricultural fields with nearby tall grasses, rocky outcrops, and water (Álvarez and Sánchez-Casas, 1997a; Cervantes et al., 1995; Davis, 1944; Davis and Russell, 1954a; Hall and Villa-R., 1949a; Orduña Trejo et al., 1999-2000). No reproductive information is available. Saussure's shrews are known to be associated with *Cryptotis parva soricina*, *Liomys pictus*, *Microtus mexicanus*, *Neotamias dorsalis*, *Neotomodon alstoni*, *Peromyscus difficilis*, *P. levipes*, *P. melanotis*, *P. spiculegus*, *Reithrodontomys megalotis*, *R. sumichrasti*, *Sciurus aureogaster*, and *Sorex orizabae* (Cervantes et al., 1995; Davis and Russell, 1954a; Orduña Trejo et al., 1999-2000).

VEGETATIONAL ASSOCIATIONS AND ELEVATION RANGE: The flora of occupied canyons includes pine-oak or Douglas fir-juniper woodlands with understories of *Bacheria* and *Senecio*. Upland habitats contain sacaton grasses near agricultural fields and yellow pine-alder or pine-oak-juniper woodlands (Álvarez and Sánchez-Casas, 1997a; Cervantes *et al.*, 1995; Davis, 1944; Davis and Russell, 1954a; Hall and Villa-R., 1949a; Orduña Trejo et al., 1999-2000). Known localities occur at elevations ranging from about 2,100 m to 3,650 m.

CONSERVATION STATUS: *S. saussurei* is not provided any level of protection (SEMARNAT, 2010). Following the July 2003 eruption of Cerro Nevado de Colima (Volcán de Colima), it is unknown if Saussure's shrews still inhabit the area of the type locality.

Sorex sclateri Merriam, 1897

Sclater's shrew

Iván Arellano-Castro

SUBSPECIES IN MEXICO
S. sclateri is a monotypic species.

DESCRIPTION: Sclater's shrew is a large shrew. The back is brown with finely blended sepia-brown, and darker on the posterior part, the abdomen is paler brown. The tail is paler in its ventral part (Merriam, 1897a). Hips and rump have black guard hairs extending 1.7 mm to 1.9 mm beyond the dorsal pelage (Carraway, 2007). *Sorex sclateri* a member of the subgenus *Sorex*. The I1s have a median tine located above the pigment; U3 is as large or larger than the U4; and the i1s have three denticles, shallow interdenticular spaces, pigment in two sections, and a long strip of pigment at the anteromedial edge (Carraway, 2007).

DISTRIBUTION: Sclater's shrew is endemic to Mexico and is known only from the type locality (Tumbala) and San Antonio Buenavista, in the state of Chiapas (Álvarez et al., 1984; Hall, 1981; Hutterer, 1993).

EXTERNAL MEASURES AND WEIGHT
TL = 125 mm; TV = 53 mm, HF = 16 mm; EAR = 5 to 7 mm.
Weight: 7 g.

DENTAL FORMULA: I 3/1, C 1/1, PM 3/1, M 3/3 = 32.

NATURAL HISTORY AND ECOLOGY: We are unaware of the details of the natural history of this species. It is possible that as in other shrews of the genus *Sorex*, their diet consists of insects (locusts, crickets, beetles, larvae, and adult diptera), worms, and other small invertebrates (arachnids) as well as carrion and occasionally plant material consisting of mosses and seeds (Jackson, 1928; Nowak, 1999b). The gestation probably lasts from 2.5 to 4 weeks (Nowak, 1999b).

VEGETATIONAL ASSOCIATIONS AND ELEVATION RANGE: The vegetation of the type locality corresponds to cloud forest and pine forest. The height of the type locality is 1,500 m (Hutterer, 1993).

CONSERVATION STATUS: *S. sclateri* is considered rare in the ecological regulation (SEMARNAT, 2010). No data exist on the current status of its populations, but its restricted distribution makes it highly susceptible to changes in their environment. The advance of agriculture and animal husbandry has destroyed most of the vegetation in the type locality.

Sorex stizodon Merriam, 1895

San Cristóbal or pale-toothed shrew

Iván Arellano-Castro and Gerardo Ceballos

SUBSPECIES IN MEXICO
S. stizodon is a monotypic species.

DESCRIPTION: *Sorex stizodon* is a small shrew. The color of the back is dark brown with a tendency to blackish tones; only the ventral part is paler. Hips and rump have black guard hairs extending 1 mm beyond the dorsal pelage (Carraway, 2007). The skull is wide and flattened, the face is relatively short and wide and the teeth have very little pigmentation (Hall, 1981; Merriam, 1895). The I1s have a median tine located above the pigment; U3 < U4; and the i1s have three denticles, shallow interdenticular spaces, and pigment in one section and lacks a long strip of pigment at the anteromedial edge (Carraway, 2007). *S. stizodon* is a member of the subgenus *Sorex*. Junge and Hoffmann (1981) consider that *S. stizodon* could be conspecific with *S. ventralis*.

EXTERNAL MEASURES AND WEIGHT
TL = 107 mm; TV = 41 mm, HF = 13.5 mm; EAR = 6 to 8 mm.
Weight: 7 g.

DISTRIBUTION: *S. stizodon* is endemic to Mexico. It is only known in San Cristóbal de las Casas. It has been recorded in the state of CS.

DENTAL FORMULA: I 3/1, C 1/1, PM 3/1, M 3/3 = 32.

NATURAL HISTORY AND ECOLOGY: The biology of *S. stizodon* is almost totally unknown. It feeds on small invertebrates and occasionally plant matter (Nowak, 1999b). It is possible that their gestation period lasts 18 to 28 days, the offspring become independent in 4 or 5 weeks, and the longevity is between 1 and 2 years (Nowak, 1999b).

VEGETATIONAL ASSOCIATIONS AND ELEVATION RANGE: The vegetation of the type locality corresponds to forests of pine and oak. The elevation of the type locality is 2,743 m (Merriam, 1895).

CONSERVATION STATUS: The current status of this species is not known, but its range is so restricted that it makes it highly vulnerable to impacts of anthropogenic activities such as deforestation. It is classified as protected within the ecological regulation (SEMARNAT, 2010).

Sorex ventralis Merriam, 1895

Chestnut-bellied shrew

Iván Castro-Arellano and Gerardo Ceballos

SUBSPECIES IN MEXICO

S. ventralis is a monotypic species.

DESCRIPTION: *Sorex ventralis* is a small shrew. The back is brown to dark sepia, the abdomen is only a little lighter. Hips and rump have dark brown guard hairs extending 0.75 mm beyond the dorsal pelage (Carraway, 2007). Its third upper unicuspid tooth is smaller than the fourth (Merriam, 1895). The I1s have a median tine located within the pigment. The i1s have two denticles, shallow interdenticular spaces, pigment in one section, and a long strip of pigment at the anteromedial edge (Carraway, 2007). In areas where it occurs sympatrically with *S. saussurei*, it can be distinguished by the skull size, as it has condylobasal length less than 17.4 mm (Junge and Hoffmann, 1981).

DISTRIBUTION: This shrew is endemic to Mexico. Its distribution ranges from Puebla to Oaxaca (Hutterer, 1993). It has been recorded in the states of DF, MX, OX, PU, and TL.

EXTERNAL MEASURES AND WEIGHT

TL = 97 to 112 mm; TV = 36 to 43 mm; HF = 13 mm; EAR = 5 to 7 mm.
Weight: 5 g.

DENTAL FORMULA: I 3/1, C 1/1, PM 3/1, M 3/3 = 32.

NATURAL HISTORY AND ECOLOGY: We are unaware of the details of the biology of the chestnut-bellied shrew. The only references of habitat type are those where it has been captured in a conifer forest (Hooper, 1961). Their diet probably includes invertebrates, some plant material (seeds, mosses), and occasionally carrion (Jackson, 1928). It is known to occur with *Habromys lepturus*, *Peromyscus azrecus oaxacensis*, and *P. boylii* (Hooper, 1961). The gestation period probably lasts slightly less than a month and its longevity is one to two years (Nowak, 1999b).

VEGETATIONAL ASSOCIATIONS AND ELEVATION RANGE: This species inhabits mature forests of oaks, pines, and firs (Hooper, 1961). It has been found from 2,760 m to 3,000 m (Goodwin, 1969; Hall, 1981).

CONSERVATION STATUS: The chestnut-bellied shrew is not listed in any endangered species list, but the current status of its populations is unknown. In some localities is apparently abundant.

Sorex veraecrucis Jackson, 1925

Veracruz shrew

Leslie N. Carraway

SUBSPECIES IN MEXICO
Sorex veraecrucis veraecrucis Jackson, 1925
Sorex veraecrucis altoensis Carraway, 2007
Sorex veraecrucis cristobalensis Jackson, 1925
Sorex veraecrucis oaxacae Jackson, 1925

DESCRIPTION: As a species, Veracruz shrews have hairs on the dorsum with an overall medium brown appearance and i1s with two denticles and a long strip of pigment at the anteromedial edge (Carraway, 2007). It can be distinguished from congeners by a tail uniformly colored like the dorsum; I1s with small, unpigmented or slightly pigmented median tines located high above the pigment; U3 < U4; condylobasal length < 19 mm; and cranial breadth < 9 mm (Carraway, 2007). *S. veraecrucis altoensis* differs from other subspecies of *S. veraecrucis* by hairs of the venter with blond or white tips; hips and rump with dark brown guard hairs extending 1.0 mm to 1.5 mm beyond the dorsal pelage; i1s with shallow interdenticular spaces, pigment in one section, and angled 5 degrees to 13 degrees from the horizontal ramus of the dentary; and length of tail 38 mm to 51 mm (Carraway, 2007). *S. veraecrucis cristobalensis* has hairs of the venter with light reddish-brown tips; hips and rump with light brown clear-tipped guard hairs extending 1.7 mm beyond the dorsal pelage; i1s with deep interdenticular spaces, pigment in two sections, and angled 10 degrees from the horizontal ramus of the dentary; condylobasal length 18.29 mm; and length of tail ≥ 46.5 mm (Carraway, 2007). *S. veracrucis oaxacae* has hairs of the venter with white tips; hips and rump with dark brown clear-tipped guard hairs extending 1.9 mm beyond the dorsal pelage; i1s with deep interdenticular spaces, pigment in one section, and angled 10 degrees to 14 degrees from the horizontal ramus of the dentary; and condylobasal length 17.86 mm to 18.15 mm (Carraway, 2007). *S. veraecrucis veraecrucis* has hairs of the venter with blond or white tips; hips and rump with dark brown guard hairs extending 1.0 mm to 1.5 mm beyond the dorsal pelage; i1s with shallow interdenticular spaces, pigment in one section, and angled 5 degrees to 13 degrees from the horizontal ramus of the dentary; and length of tail usually 50 mm to 62 mm (Carraway, 2007). *S. veraecrucis* is a member of the subgenus *Sorex*. Originally, the subspecies *cristobalensis*, *oaxacae*, and *veraecrucis* were described as subspecies of *S. saussurei*. All three subspecies have I1s with small, unpigmented or slightly pigmented median tines located high above the pigment, whereas Saussure's shrews have I1s with large, well-pigmented median tines located well within the pigmented area on the body (Carraway, 2007).

EXTERNAL MEASURES AND WEIGHT

Sorex veraecrucis veraecrucis

TL = 112 to 133 mm; Head and body l. = 46 to 62 mm; TV = 62 to 75 mm; HF = 13 to 16 mm; CLAW = 1.3 to 2.1 mm.

Weight: probably around 6 g.

Sorex veraecrucis altoensis

TL = 102 to 119 mm; Head and body l. = 60 to 76 mm; TV = 39 to 54 mm; HF = 12 to 14.5 mm; CLAW = 1.4 to 2.1 mm.

Weight: probably around 6 g.

Sorex veraecrucis cristobalensis

TL = 109 to 116 mm; Head and body l. = 60 to 69.5 mm; TV = 46.5 to 49 mm; HF = 13 to 14 mm; CLAW = 1.7 to 1.8 mm.

Weight: probably around 6 g.

Sorex veraecrucis oaxacae

TL = 110 to 119 mm; Head and body l. = 65 to 72 mm; TV = 45 to 49 mm; HF = 12 to 15 mm; CLAW = 1.3 to 1.5 mm.

Weight: probably around 6 g.

DENTAL FORMULA: I 3/1, C 1/1, PM 3/1, M 3/3 = 32.

DISTRIBUTION: The most widely distributed soricid in Mexico, the Veracruz shrew occurs from Coahuila southward into Chiapas and from Pacific to Atlantic coasts (Hall, 1981). *S. veraecrucis cristobalensis* is distributed east of the Isthmus de Tehuantepec. It has been recorded in the states of CL, CO, CS, DF, DU, GJ, GR, HG, JA, MI, MO, MX, NL, OX, PU, QE, TA, and VE.

NATURAL HISTORY AND ECOLOGY: *S. veraecrucis altoensis* inhabits the highest elevations known for the species. Habitats for Veracruz shrews include pine-oak-juniper and yellow-pine-alder woodlands, as well as pine-oak and Douglas fir-juniper woodlands in moist montane canyons with shrubby understories and a deep layer of leaves or needles overlaying a deep layer of humus. Additionally, they can be found near high-elevation agricultural fields with tall grasses, rocky outcrops, and water nearby (Álvarez and Sánchez-Casas, 1997a; Baker and Greer, 1962; Cervantes et al., 1995; Davis, 1944:376; Davis and Lukens, 1958; Davis and Russell 1954a; Hall and Villa-R., 1949a; Orduña Trejo et al., 1999b, 2000). Reproductive information is available only for *S. veraecrucis altoensis*: 2 embryos, with 13 mm crown-rump length, were collected in August in Michoacán (Álvarez and Sánchez-Casas, 1997a). It is known to be associated with *Cryptotis parva soricina*, *Microtus mexicanus*, *Peromyscus difficilis*, *P. levipes ambiguus*, *P. melanotis*, *Reithrodontomys megalotis*, *R. sumichrasti*, *Sorex milleri*, *S. orizabae*, and *Tamias bulleri* (Baker, 1956; Cervantes et al., 1995; Davis and Lukens, 1958; Davis and Russell, 1954a; Jiménez-G. et al., 1999). *S. veraecrucis cristobalensis* inhabits the Transition Zone between the Upper Austral and Canadian zones in the central highlands of Chiapas (Álvarez del Toro, 1977; Goldman, 1951). It is known to be associated with *Cryptotis goodwini* and *S. veraepacis chiapensis* (Espinoza Medinilla et al., 1998). *S. veracrucis oaxacae* inhabits the edges of moss-covered rocky outcrops surrounded by an overstory of shrubs with moist and dense low-growing vegetation. It is known to be associated with *Cryptotis goldmani machetes*, *C. magna*, *C. mexicana mexicana*, *C. phillipsii*, *Habromys chinanteco*, *H. lepturus*, *Heteromys desmarestianus lepturus*, *Megadontomys cryophilus*, *Microtus mexicanus*, *M. oaxacensis*, *Oryzomys alfaroi*, *O. caudatus*, *Peromyscus boylii*, *P. megalops*, *P. melanocarpus*, *P. mexicanus*, *Reithrodontomys mexicanus*, *R. microdon*, and *S. veraepacis mutabilis* (Jones and Genoways, 1967b; Musser, 1964; Rickart, 1977; Robertson and Rickart, 1975; Schaldach, 1966). *S. veraecrucis veraecrucis* primarily inhabits higher areas along the Atlantic drainage from central Veracruz to northeastern Oaxaca (Jackson, 1928). These habitats include moist areas with low vegetation, dry woodlands, rainforests, and hedges near agricultural fields (Carraway, 2007; Hall and Dalquest, 1963). It is known to be associated with *Cryptotis mexicana*, *Microtus mexicanus*, *Neotoma mexicana torquata*, *Peromyscus boylii*, *P. melanotis*, and *Sorex macrodon* (Hall and Dalquest, 1963).

VEGETATIONAL ASSOCIATIONS AND ELEVATION RANGE: Habitats occupied by *S. veraecrucis altoensis* include pine-oak-juniper, yellow pine-alder, pine-oak, or Douglas fir–juniper woodlands, with *Bacheria*, *Senecio*, and sacaton grasses (Jones and Genoways, 1967; Musser, 1964; Rickart, 1977; Robertson and Rickart, 1975; Schaldach, 1966). Known localities occur at elevations ranging from 2,100 m to 3,650 m. Habitats occupied by *S. veraecrucis cristobalensis* include a blending of many floral elements from both the Upper Austral and Canadian zones (Goldman, 1951). Known localities occur at elevations ranging from 1,900 m to 2,560 m. Habitats occupied by *S. veraecrucis oaxacae* include dense herbaceous material and ferns among rocky outcrops (Musser, 1964; Webb and Baker, 1969). Known localities occur at elevations ranging from 1,600 m to 3,000 m. Habitats occupied by *S. veraecrucis veraecrucis* include short grasses, low ferns, bracken moss, liverworts, pine forests, and pine-oak woodlands (Carraway, 2007; Hall and Dalquest, 1963). Damp mosses are the most common floral element of almost all occupied habitats. Known localities occur at elevations ranging from 1,800 m to 2,860 m.

CONSERVATION STATUS: *S. veraecrucis veraecrucis* is considered a protected subspecies under *S. saussurei veraecrucis*. *S. veraecrucis cristobalensis* is considered a protected subspecies under *S. saussurei cristobalensis*. *S. veraecrucis oaxacae* is considered a protected subspecies under *S. saussurei oaxacae* (SEMARNAT, 2010). *S. veraecrucis altoensis* was not recognized as a distinct subspecies when the last status list was published (SEMARNAT, 2010); furthermore, as one of the most widely distributed taxa of shrews within Mexico, it is doubtful that it requires any level of protection.

Sorex veraepacis Alston, 1877

Verapaz shrew

Leslie N. Carraway

SUBSPECIES IN MEXICO
Sorex veraepacis chiapensis Jackson, 1925
Sorex veraepacis mutabilis Merriam 1895b

DESCRIPTION: *Sorex veraepacis* is a medium-sized *Sorex* and a member of the subgenus *Otisorex*. *S. veraepacis chiapensis* has hairs of the dorsum banded, proximally to distally, medium silvery gray, blond, to dark brown, giving an overall dark brown appearance (Carraway, 2007). Hairs of the venter are medium silvery gray to medium dark brown, thus the venter is only slightly paler than the dorsum in overall appearance. *S. veraepacis mutabilis* has hairs of the dorsum banded, reddish blond, to dark brown; hairs of the venter are light reddish-brown (Carraway, 2007). First lower incisors have three denticles and deep interdenticular spaces (Carraway, 2007). *S. veraepacis chiapensis* and *S. veraepacis mutabilis* can be distinguished by tail slightly bicolored–tail uniform in color of dorsum; hips and rump with dark brown–pale reddish-brown/dark brown guard hairs extending 1.0 mm to 1.3 mm–2.2 mm to 2.8 mm beyond the dorsal pelage; median tine located above–well within pigment on body of I1s; pigment on i1s in one–three sections; i1s angled 5 degrees to 10 degrees–3 degrees to 5 degrees from the horizontal ramus of the dentary; and breadth across M2-M2 4.8 mm to 5.1 mm–4.5 mm to 4.8 mm, respectively (Carraway, 2007). Verapaz shrews differ from Ixtlan shrews by no long strip of pigment at the anteromedial edge, and generally smaller size of mandible (length of mandible usually 7.1 mm to 8.3 mm, length of c1-m3 4.9 mm to 5.6 mm and height of articular condyle 2.3 mm to 2.9 mm; Carraway, 2007). They can be distinguished from other congeners by pelage and i1 characteristics, U3 < U4, and generally smaller or larger size (Carraway, 2007:78). Originally, *S. veraepacis mutabilis* included two morphotypes: one matching the holotype for the subspecies (I1 with median tine well-within pigment) and the other that subsequently was elevated to species level (*S. ixtlanensis*; I1 with median tine above pigment; Carraway, 2007:48).

DISTRIBUTION: The Verapaz shrew occurs from western Guerrero southward through Chiapas into Guatemala (Hall, 1981; Sanchez Cordero et al., 2005). *S. veraepacis mutabilis* is endemic to Oaxaca, Mexico. *S. veraepacis chiapensis* is distributed east of the Isthmus de Tehuantepec. It has been recorded in the states of CS, GR, OX, and VE.

EXTERNAL MEASURES AND WEIGHT

Sorex veraepacis chiapensis

TL = 105 to 126 mm; Head and body l. = 61 to 75 mm; TV = 44 to 51 mm; HF = 14 to 15 mm; CLAW = 1.4 mm.

Weight: probably around 6 g.

Sorex veraepacis mutabilis

TL = 116 to 130 mm; Head and body l. = 61 to 76 mm; TV. = 48 to 61 mm; HF = 13 to 16 mm.

Weight: probably around 6 g.

DENTAL FORMULA: I 3/1, C 1/1, PM 3/1, M 3/3 = 32.

NATURAL HISTORY AND ECOLOGY: Both *S. veraepacis chiapensis* and *S. veraepacis mutabilis* inhabit the Canadian Zone at high elevations (Goldman, 1951). *S. veraepacis chiapensis* shrews are found in forests dominated by firs and in cloud forests (Medellín, 1988). *S. veraepacis mutabilis* shrews are found where the ground in shady areas of dense forests is covered with a deep layer of leaves overlaying a deep layer of humus (Davis and Lukens, 1958; Musser, 1964). No reproductive information is available. *S. veraepacis chiapensis* shrews are known to be associated with *Cryptotis goodwini*, *Heteromys desmarestianus goldmani*, *Nyctomys sumichrasti*, *Peromyscus aztecus oaxacensis*, *P. guatemalensis*, *Reithrodontomys megalotis*, *R. mexicanus*, and *S. veraecrucis cristobalensis* (Espinoza Medinilla et al., 1998; Medellín, 1988). It is known to be preyed on by *Chrotopterus auritus* (false vampire bat; Medellín, 1988). *S. veraepacis mutabilis* shrews are known to be associated with *Cryptotis goldmani machetes*, *C. magna*, *C. mexicana mexicana*, *C. phillipsi*, *Heteromys lepturus*, *Microtus mexicanus*, *M. oaxacensis*, *Oryzomys alfaroi*, *O. caudatus*, *Peromyscus boylii*, *P. chinanteco*, *P. lepturus*, *P. melanocarpus*, *P. mexicanus*, *P. thomasi cryophilus*, *Reithrodontomys mexicanus*, *R. microdon albilabris*, and *S. veraecrucis oaxacae* (Jones and Genoways, 1967; Musser, 1964; Rickart, 1977; Robertson and Rickart, 1975; Schaldach, 1966).

VEGETATIONAL ASSOCIATIONS AND ELEVATION RANGE: Woodland habitats of *S. veraepacis chiapensis* include *Abies religiosa*, *Pinus ayacahuite*, *Pseudotsuga mucronata*, and *Salvia* sp. Cloud forest habitats include *Cedrela mexicana*, *Matudaea trinervia*, and *Quercus* sp. (including *Q. crispifolia*; Medellín, 1988). Woodland habitats occupied by *S. veraepacis mutabilis* include pine-fir or pine-oak communities (Davis and Lukens, 1958; Musser, 1964). Known localities occur at elevations ranging from 1,218 m to 2,900 m.

CONSERVATION STATUS: Both *S. veraepacis chiapensis* and *S. veraepacis mutabilis* are listed as protected subspecies (SEMARNAT, 2010). As of 2005, only 75.84% of suitable habitat for Verapaz shrews still existed (Sánchez Cordero et al., 2005). The remaining 24.16% had been converted for agricultural use or urban development.

Family Talpidae

The family Talpidae is represented by 17 genera and 39 species of moles, with a restricted distribution in North America and Eurasia (Hutterer, 2005). In Mexico, three species of two genera have been recorded (*Scalopus* and *Scapanus*). They have very restricted distributions in the country (Ceballos et al., 1998).

Scalopus aquaticus (Linnaeus, 1758)

Eastern mole

Iván Castro-Arellano and Gerardo Ceballos

SUBSPECIES IN MEXICO
Scalopus aquaticus inflatus Jackson, 1914
Scalopus aquaticus montanus R.H. Baker, 1951

DESCRIPTION: *Scalopus aquaticus* possesses a robust and flattened body. The hair is dense, smooth, and silky, with coloration varying from black to silver. The tail is almost naked. The nose is elongated in a distinctive snout with the nostrils located at its top. The eyes are small and there are no pinnae. The legs are large and fleshy; at the top they have little hair and are naked below. The thumbs are webbed, the palms are wider than long, and the anterior legs are modified for digging (Yates and Schmidly, 1978).

Scalopus aquaticus. Photo: Gerardo Ceballos.

EXTERNAL MEASURES AND WEIGHT
TL = 128 to 208 mm; TV = 18 to 38 mm; HF = 15 to 22 mm; EAR = 0.
Weight: 38 g.

Dental formula: I 3/2, C 1/0, PM 3/3, M 3/3 = 36.

NATURAL HISTORY AND ECOLOGY: The eastern mole is a digging mammal with broad areas of activity due to the need to move to find food. Males have larger areas of activity than rodents of larger size (Yates and Schmidly, 1977). Males require larger home ranges than do females (Harvey, 1976). Two types of tunnels are built: shallow ones to search for food and deep and permanent ones to use as burrows (Hisaw, 1923a). Eastern moles most commonly are found in wet sandy soil and not heavy clay, rocky soils or gravel lots. Rivers may not be barriers for dispersal, but because they are very hard to dig, clays associated with some riparian systems are (Yates and Schmidly, 1978). The mole feeds mainly on worms and insects, but also eats vegetable matter. Daily they consume the equivalent of 100% of their body mass (Hisaw, 1923b). In captivity they have been fed ground beef (Hisaw, 1923b; Yates and Schmidly, 1978). It reproduces once a year and has litters of 2 to 5 young. The gestation period is estimated to be of 30 to 45 days (Yates and Schmidly, 1978).

VEGETATIONAL ASSOCIATIONS AND ELEVATION RANGE: In Mexico, this species has been collected in a forest characterized by a mixture of *Quercus gravesii*, *Pinus cembroides*, and *Juniperus pachypholea* (Baker, 1956) and thorny scrub (Jackson, 1915). In the United States due to its extensive distribution it inhabits different types of vegetation, where soil type is a determinant of their presence (Yates and Schmidly, 1978). In Mexico, its altitudinal distribution reaches 1,633 m in Coahuila (Baker, 1951b) and from sea level to 200 m in Tamaulipas (Jackson, 1914).

CONSERVATION STATUS: In the eastern United States the eastern mole is widely distributed. In Mexico it is an endangered species, mainly because of its limited distribution (Ceballos et al., 1998). The population of Tamaulipas has disappeared because of habitat changes related to agriculture; the status of the population in Coahuila is unknown and eastern moles have not been caught since 1953 (Ceballos and Navarro, 1991).

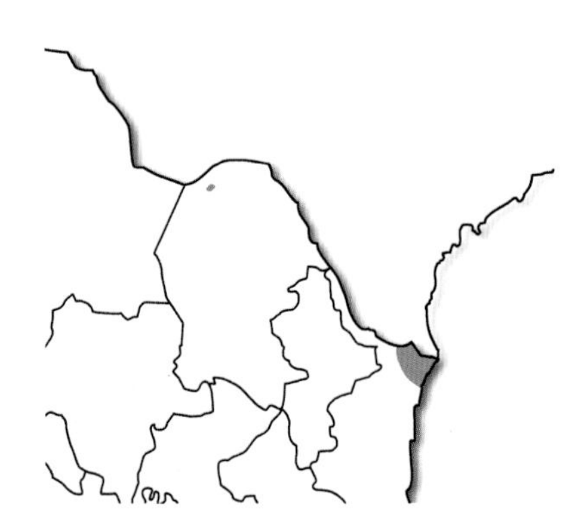

DISTRIBUTION: *S. aquaticus* has a wide distribution in the United States; in Mexico, its distribution is marginal and comprises two relictual populations representing the two subspecies in the country. A population of *S. aquaticus inflatus* was found in northeastern Tamaulipas and *S. aquaticus montanus* in northern Coahuila (Baker, 1956; Hall, 1981; Jackson, 1915; Yates and Schmidly, 1977). It has been recorded in the states of CO and TA.

Scapanus anthonyi J.A. Allen, 1893

Blind mole

Jorge Salazar-Bravo and Terry L. Yates

SUBSPECIES IN MEXICO
S. anthonyi is a monotypic species.
It was originally described as a species but relegated to a subspecies of *S. latimanus*. A recent study confirmed that it is actually a different species (Yates and Salazar, 2004).

DESCRIPTION: *Scapanus anthonyi* is the smallest mole of the genus. The body is robust; the tail is short (less than the length of the head and body) and covered with fine hair. The head is cone-shaped, with small eyes and the nostrils in a dorsal position. The webbed front legs are as wide as long. The anterior digits have an interdigital membrane, which is not present in the posterior digits. In all 12 specimens known of this species, the coloration of the body is fairly uniform. The dorsal region is cinnamon brown with streaks of silver and the ventral region is a lighter lead color. It has marked sexual dimorphism. Compared with *S. latimanus occultus*, of northern localities (Laguna Hanson specifically), *S. anthonyi* differs in its much smaller size (cranial and external measures) and qualitative characteristics. In particular, it differs from *S. latimanus occultus* by not possessing the extra metastyle in M1 and in the shape of the temporal fossa. In addition, there are two dental characters that support the taxonomic differentiation of *S. anthonyi* and *S. latimanus occultus*: the reduction of P2 (always missing in *S. anthonyi* and ever-present in *S. latimanus occultus*) and the lesser degree of hipodontya in *S. anthonyi* (Yates and Salazar, 2004).

DISTRIBUTION: *S. anthonyi* is an endemic species of Mexico, which is known from only 12 specimens and 4 locations in the Sierra of San Pedro Martir, Baja California. It has been recorded in the state of BC.

EXTERNAL MEASURES AND WEIGHT
TL = 133 to 135 mm; TV = 21 to 25 mm, HF = 18 to 20 mm; EAR = 0.
Weight: 30 to 35 g.

DENTAL FORMULA: I 3/3, C 1/1; PM3/3 or 2/2, M 3/3 = 36 or 40.

NATURAL HISTORY AND ECOLOGY: The natural history of the blind mole is little known. Its distribution is allopatric with that of *S. latimanus*. In one of the several

Scapanus anthonyi. Conifer forest. San Pedro Mártir, Baja California. Photo: Gerardo Ceballos.

attempts to capture animals of this species, one of us (TLY) collected a pair of animals at the foot of large boulders after several days of intermittent rain. Because they live in very arid regions, it is probable that they may use the wetter and more stable microclimates present below these cliffs. In one particular instance, an active tunnel was intersected and eventually converged with that of a pocket gopher, a behavior observed in other regions and with other species of pocket gophers and moles. For example, Baker (1951b) noted that tunnels of *S. aquaticus* converged with those of a pocket gopher (*Thomomys umbrinus*) in the Sierra del Carmen in Coahuila.

VEGETATIONAL ASSOCIATIONS AND ELEVATION RANGE: Regarded by many as an "island of forest" in a sea of "desert," the Sierra of San Pedro Martir rises from 1,000 m to 3,000 m. The dominant vegetation of the lower stratum is chaparral (1,150 m), pine (*Pinus jeffreyi*), and other conifers in the highlands. The holotype was collected at 2,120 m in a forest of Aliso (*Populus tremuloides*). *S. anthonyi* is known from locations of between 1,150 m and 2,600 m.

CONSERVATION STATUS: The current status of the species is unknown. Ceballos and Navarro (1991) suggested that it should be treated as endangered due to its restricted distribution. The Mexican Official Regulation listed it as endangered (SEMARNAT, 2010). Because of their limited range and the fact that logging concessions have been established in the Sierra of San Pedro Martir, it is possible that the habitat of the species and hence its long-term survival are in jeopardy. San Pedro Martir is a national park where the federal government is driving a strong conservation program, which could help the conservation of this and other endemic species of the region.

Scapanus latimanus (Bachman, 1842)

Broad-footed mole

Iván Castro-Arellano and Gerardo Ceballos

SUBSPECIES IN MEXICO

Scapanus latimanus occultus Grinnell and Swarth, 1912

Recently, the subspecies *S. latimanus anthonyi*, originally described as species and then relegated as subspecies, was elevated to a species category again, after a detailed analysis of external morphological traits and cranial characteristics (Yates and Salazar, 2004).

DESCRIPTION: *Scapanus latimanus* is a small mole (Palmer, 1937). The color of the back varies from brown to brownish; the abdomen is slightly lighter. There is seasonal and distribution variation of the coloration (Palmer, 1937). It has body features adapted to underground life: a robust and depressed body, a cone-shaped head, small eyes, and anterior limbs modified to dig (Hartman and Yates, 1985). It differs from *S. anthonyi* by its larger size (cranial and external measures) and qualitative characteristics such as lacking the extra metastyle in M1 and the shape of the temporal fossa. Additionally, the P2 is always absent in *S. anthonyi* and always present in this species. It has a higher degree of hipsodontya than *S. anthonyi* (Yates and Salazar, 2005).

EXTERNAL MEASURES AND WEIGHT
TL = 132 to 192 mm; TV = 21 to 45 mm; HF = 18 to 25 mm; EAR = 0.
Weight: 30 to 40 g.

DENTAL FORMULA: I 3/3, C 1/1, PM 4/4, F 3/3 = 44.

NATURAL HISTORY AND ECOLOGY: The distribution of the broad-footed mole is mainly determined by the existence of soil moisture (Palmer, 1937). They most commonly occur in slightly sandy soil where they can dig easily; they avoid clayey, very rocky, or gravel soils. The profound lack of invertebrates in dry and sandy soils perhaps is the reason for their rarity in this type of soils (Jackson, 1915). In the Sierra de La Laguna, *S. latimanus* occurs under boulders of arid chaparrals, where humidity is higher than the surrounding vegetation. In the course of digging, they form very conspicuous mounds on the surface; during their movements they also form lumps of earth or "mole hills," which are very similar to those made by the pocket gopher (Jackson, 1915). The gallery systems consist of a series of tunnels located 25 cm to 45 cm deep and other tunnels located just below the surface, which are used to search for food (insects and worms). During the dry season the tunnels have a greater depth (Hartman and Yates, 1985; Jackson, 1915). They build globular nests located 30 cm to 45 cm below the surface, usually below the roots of shrubs, and made from grasses, leaves, and small roots. The number of young in a litter varies from two to five and probably there is only one litter per year during the spring (Jackson, 1915).

VEGETATIONAL ASSOCIATIONS AND ELEVATION RANGE: *S. latimanus* inhabits pine-oak forests and thorny scrublands. In Mexico its elevational range goes from 1,000 m to 2,100 m (Hall, 1981).

CONSERVATION STATUS: The subspecies is classified as threatened, in the Mexican Official Regulation of species at risk (SEMARNAT, 2010). This is due to their extremely limited distribution and the impact of grazing in their habitat.

DISTRIBUTION: This species has a wide distribution in the United States stretching from the state of Oregon to California. In Mexico it is distributed marginally in the most northern part of the state of Baja California, in Laguna Hanson, in the Sierra de Juarez (Hall, 1981; Huey, 1963; Yates and Salazar, 2004). It has been recorded in the state of BC.

Grison (*Galictis vittata*). Tropical rainforest. Villahermosa, Tabasco. Photo: Gerardo Ceballos.

Black bear (*Ursus americanus*). Oak forest. Monterr
Nuevo Leon. Photo: Gerardo Ceballos.

ORDER CARNIVORA

Gerardo Ceballos and Rurik List

The Mexican wolf is a great tradition of the country, little spirit would be shown if it couldn't be found in at least one part of Mexico, where the wolf can survive safely from the incessant persecution of the ranchers.

— A.S. Leopold, 1959

THE ORDER CARNIVORA is represented by 271 species; it has a cosmopolitan distribution, with the exception of Australia, New Zealand, and some Pacific islands (Wilson and Reeder, 2005). The carnivores are characterized by specialized structures for a diet based on meat: their canines are highly developed, their premolars and molars are adapted for cutting and triturating, and their jaws are powerful. Carnivores' senses of hearing, sight, and smell are very developed, which makes them efficient predators of all kind of animals, from small insects to large mammals (Nowak, 1999b).

The Carnivora shows the greatest variation in body mass among mammals; the smallest species, the least weasel (*Mustela nivalis*), weighs around 30 g and the elephant seals (*Mirounga* spp.) weigh more than 3 tons, a difference of 5 orders of magnitude (Nowak, 1999b). Carnivores are species of solitary habits or live in social groups of up to 50 individuals such as the coati (*Nasua narica*) and some mongoose (Gittlemann, 1989; Valenzuela and Ceballos, 2000). Some species such as the lion (*Panthera leo*) form groups that hunt and defend their territory as a whole, with well-defined hierarchies (Gittleman, 1996). They are mainly nocturnal and crepuscular, but there are many diurnal species. Because most are predators, they have been able to establish themselves wherever there is available prey, so they occupy all terrestrial habitats, including marine and aquatic environments. The home range varies according to the body and group sizes, from few hundreds of square meters in weasels (*Mustela* spp.) to hundreds of square kilometers in species such as jaguars (*Panthera onca*) and wolves (*Canis lupus*; Nowak, 1999b).

Although most are meat eaters, the diet is highly variable, and many species are omnivores, like black bears (*Ursus americanus*) and red foxes (*Vulpes vulpes*); some, like the panda (*Ailuropoda melanoleuca*), are mainly herbivores (Nowak, 1999b). Most of the species reproduce once a year and the litter size varies from 1 to 15 offspring. Some families, like the Mustelidae, have delayed implantation, a phenomenon in which the fertilized egg remains dormant without developing the embryo for a prolonged period. The offspring are most often altricial, with the eyes closed and incapable of surviving alone. Parental care is complex and many behaviors, like hunting, are learned (Gittleman, 1996).

As predators, many carnivore species often come in conflict with humans; some are deleterious to animal production industries, like the ringtail (*Bassariscus astutus*) to poultry and sea lions (*Zalophus* spp.) to fisheries; some cause property damage, like black bears (*Ursus americanus*) to cars and houses; some even take human lives, like lions (*Panthera leo*). These situations result in hunting, trapping, and poisoning, with subsequent reduction in their numbers, which in conjunction with habitat loss, trade, and the increased human activities in their habitats, have caused severe decline and even extinction to some species like the Caribbean monk seal (*Monachus tropicalis*) and the Falkland Island fox (*Dusicyon australis*). This is particularly worrying for a group with many keystone species like the wolf (*Canis lupus*) and sea otter (*Enhydra lutris*), whose disappearance results in impoverished ecosystems (Estes, 1995; Ripple and Beschta, 2003).

Family Felidae

The family Felidae groups 36 species of felids, including the domestic cat (Wilson and Reeder, 2005). They are distributed in all continents with the exceptions of Oceania and Antarctica. In Mexico there are six species representing five genera (*Leopardus*, *Lynx*, *Panthera*, *Herpailurus*, and *Puma*). The puma (*Puma concolor*) is the species with greater distribution; in historical times it was found throughout the country. The jaguar (*Panthera onca*), the ocelot (*Leopardus pardalis*), and the margay (*L. wiedii*) are distributed primarily in tropical regions, and the bobcat (*Lynx rufus*) in temperate regions. The jaguar, ocelot, and margay are at risk of extinction.

Subfamily Felinae

The subfamily Felinae contains twelve extant genera of cats with thirty-nine species. Four of the genera and five species are known from Mexico.

Herpailurus yagouaroundi (Lacépède, 1809)

Jaguarundi

Marcelo Aranda and Arturo Caso

SUBSPECIES IN MEXICO
Herpailurus yagouaroundi cacomitli (Berlandier, 1859)
Herpailurus yagouaroundi fossata (Mearns, 1901)
Herpailurus yagouaroundi tolteca (Thomas, 1898)
Some authors have proposed that the generic name be *Puma* instead of *Herpailurus*, but the change has not been widely accepted.

DESCRIPTION: Among the felids, *Herpailurus yagouaroundi* is small in size; its coloration is uniform but there are two basic colors, gray and reddish-brown, with a variety of shades. The body is slim and elongated, with a small head and a long tail.

EXTERNAL MEASURES AND WEIGHT
TL = 888 to 1,372 mm; TV = 330 to 609 mm; HF = 120 to 152 mm; EAR = 25 to 40 mm.
Weight: 3.5 to 9 kg.

DENTAL FORMULA: I 3/3, C 1/1, PM 3/2, M 1/1 = 30.

DISTRIBUTION: *H. yagouaroundi* inhabits tropical and subtropical regions, humid and dry, from the north of Mexico to the north of Argentina (Tewes and Schmidly, 1987). In Mexico, it is distributed on the Pacific coast and the Gulf of Mexico, from Sonora and Tamaulipas to the south of Chiapas and the Yucatan Peninsula. It has been recorded in the states of CA, CL, CS, GR, HG, JA, MI, MX, NY, OX, PU, QR, SI, SL, SO, TA, TB, VE, and YU.

NATURAL HISTORY AND ECOLOGY: The jaguarundi inhabits the areas of ecotone between forests and open habitats, where it takes shelter in inner forests, small caves, or hollow tree trunks. It is a solitary hunter, diurnal, and skillful on land; it sometimes climbs trees, but is not as well suited to arboreal life as the margay (*Leopardus wiedii*). Its diet includes a variety of food items, including reptiles, birds, and small mammals. Some authors believe that birds are their main food source, especially forest fowl; however, using road-killed specimens in Venezuela, Mondolfi (1986) found that cotton rats (*Sigmodon hispidus*) were their main staple prey. Near human settlements, it can be an active predator of domestic birds (Álvarez del Toro, 1991; Konecny, 1989; Leopold, 1959; Tewes and Schmidly, 1987). Mating occurs throughout the year, although it is more common in January and March. In captivity, gestation lasts 72 to

Herpailurus yagouaroundi. Specimen in captivity. Arizona-Sonora Desert Museum, Arizona. Photo: Rurik List.

75 days (Hulley, 1976). The female usually gives birth to four offspring, which are born with different stages of coloring. The area of activity in Belize of 2 male adults was 88.3 km^2 and 99.9 km^2 and 20.1 km^2 for an adult female (Konecny, 1989). In Tamaulipas, Mexico, jaguarundis have home range values of 8.6 km^2 for males and 8.5 km^2 for females (Caso, 1994).

VEGETATIONAL ASSOCIATIONS AND ELEVATION RANGE: *H. yagouaroundi* is mainly found in tropical rainforests, subdeciduous forests, and deciduous forests. It is also found in thorn-scrubs, mangroves, cloud forests, xeric scrublands, and occasionally oak and conifer forests. It is found from sea level to 2,000 masl; however, most of the records are below 1,000 masl.

CONSERVATION STATUS: *H. yagouaroundi* has been included in appendix I of CITES (1982); the IUCN status is least concern with population trends decreasing (IUCN, 2010). In Mexico, it is classified as threatened and hunting is prohibited (SEMARNAT, 2010). Of the tropical felids in Mexico, the jaguarundi is the species that best withstands environmental impacts, and it can live in transformed areas with secondary vegetation. It seems that it is never abundant.

Leopardus pardalis (Linnaeus, 1758)

Ocelot

Ricardo Moreno and Marcelo Aranda

SUBSPECIES IN MEXICO
Leopardus pardalis albescens (Pucheran, 1855)
Leopardus pardalis nelsoni (Goldman, 1925)
Leopardus pardalis pardalis (Linnaeus, 1758)
Leopardus pardalis sonoriensis (Goldman, 1925)

DESCRIPTION: *Leopardus pardalis* is a medium-sized felid. The color of the body is light sandy brown to pale yellow and grayish-white in the interior parts of the limbs. All the body is covered with black spots, which on the flanks become elongated rosettes with brown centers; these spots are commonly oriented in oblique directions. The head is rounded and the tail is relatively short in comparison to the body length. Most ocelots weigh 8 kg to 13 kg, but some ocelots have been recorded to weigh 18.6 kg and 20 kg (Kays, 2006; Sunquist and Sunquist, 2002). It differs from the margay (*L. wiedii*) in its greater size, shorter tail, and spots pattern. It is distinguished from the jaguar (*Panthera onca*) by its smaller size and spots pattern.

EXTERNAL MEASURES AND WEIGHT
TL = 920 to 1,367 mm; TV = 270 to 400 mm; HF = 130 to 180 mm; EAR = 30 to 45 mm.
Weight: 6 to 18 kg.

DENTAL FORMULA: I 3/3, C 1/1, PM 3/2, M 1/1 = 30.

NATURAL HISTORY AND ECOLOGY: *L. pardalis* is associated with habitats of dense vegetation, but can use different habitat types throughout its range. It shelters in natural caves, hollow trunks, areas with shrubs, and occasionally dense tree branches (Emmons, 1988; Moreno et al., 2006a). It is a terrestrial hunter that feeds mainly on small and medium-sized mammals, but with a broad feeding spectrum that also includes invertebrates, reptiles, birds, and other large mammals such as monkeys, collared peccary, and red brocket deer (Aliaga-Rossel et al., 2006; Bustamante, 2008; Chinchilla, 1997; De Villa Meza et al., 2002; Emmons, 1987a; Miranda et al., 2005; Moreno et al., 2006b; Tewes and Schmidly, 1987). Mating occurs throughout the year, but in the northern areas of its range, mating is more frequent in September and November (Leopold, 1959; Tewes, 1986). The gestation period varies from 70 to 80 days and the litter commonly consists of 1 or 2 offspring, very rarely 3. It is a lone hunter with home ranges between 3.5 km^2 and 43 km^2 for males, and between 0.7 km^2 and 18.3 km^2 for females (Crawshaw, 1995; Crawshaw and Quigley, 1989; Di Bitetti et al., 2006; Dillon and Kelly, 2008; Emmons, 1988; Konecny, 1989; Ludlow and Sunquist, 1987; Maffei and Noss, 2008; Moreno et al., 2006a; Navarro, 1985; Tewes, 1986). Its predators include big cats, snakes, and crocodiles. It is mainly nocturnal,

Leopardus pardalis. Tropical semideciduous forest. Calakmul Biosphere Reserve, Campeche. Photo: Cuauhtémoc Chavez and Gerardo Ceballos.

but can be active throughout the day. It uses urine, scats, and scrapes in the ground to mark its territory (Bustamante, 2008; Emmons, 1988; Moreno and Bustamante, 2009; Moreno and Giacalone, 2006, 2008; Sunquist and Sunquist, 2002).

Ocelots, like other species of wildcats, make latrines (Bustamante, 2008; Moreno and Giacalone, 2006, 2008; Sunquist and Sunquist, 2002). These latrine sites can be used not only to delimit a cat's home range, but also to obtain intraspecific information about sex, age, reproductive status, and presumably the identity of individuals (Moreno and Giacalone, 2008). Radio-telemetry has played a very important role in the study of the ocelot's ecology, helping scientists understand and learn about various aspects of its population numbers. In the last decade, the use of camera traps has been vital for generating information about how the populations of this species are in different habitat types and different countries. Combining radio-telemetry and camera traps we now know that the population of ocelots within 100 km^2 can vary from 2.3 individuals in a pine forest in Belize to 80 in a rainforest in Peru (Crawshaw, 1995; Di Bitetti et al., 2006; Emmons, 1987a; Haines et al., 2006; Maffei et al., 2005; Moreno and Bustamante, 2007, 2009; Moreira et al., 2007; Trolle and Kery, 2003, 2005). These variations in the densities are based on the abundance of prey and the presence of larger predators such as pumas, jaguars, crocodiles, and humans.

VEGETATIONAL ASSOCIATIONS AND ELEVATION RANGE: The ocelot mainly inhabits tropical rainforests, subdeciduous and deciduous forests, and mangroves. It has also been found in cloud forests and occasionally in thorn forests and xeric scrublands. It is found from sea level up to 2,800 masl or even a little higher (Reid, 1997). Most of the records are located below 1,000 masl.

CONSERVATION STATUS: Ocelots are under strong pressure from poaching because their skin is considered valuable. It is listed in appendix II of CITES (1982), and it is considered as least concern by IUCN, with population trends decreasing (IUCN, 2010). In Mexico, it is classified at risk of extinction and its hunting is prohibited (SEMARNAT, 2010). Like the jaguar, its greatest threats are the destruction of its habitat and poaching. The international ban on the trade of the ocelot has helped the recovery of its populations in recent decades.

DISTRIBUTION: Ocelots are found in tropical and subtropical regions from the south of Texas in the United States to the north of Argentina (Sunquist and Sunquist, 2002; Tewes and Schmidly, 1987). In Mexico they are distributed along the Pacific coastal plains and the Gulf of Mexico, from Sinaloa and Tamaulipas, all the way to southern Mexico, including the Yucatan Peninsula (Chávez and Ceballos, 1998; Hall, 1981; Sánchez et al., 2002). It has been recorded in the states of CA, CL, CS, GR, HG, JA, MI, MX, NL, NY, OX, PU, QR, SI, SL, SO, TA, TB, VE, and YU.

Leopardus wiedii (Schinz, 1821)

Margay

Marcelo Aranda and Octavio Monroy

SUBSPECIES IN MEXICO

Leopardus wiedii cooperi Goldman, 1943
Leopardus wiedii glaucula Thomas, 1903
Leopardus wiedii oaxacensis Nelson and Goldman, 1931
Leopardus wiedii salvinia (Pocock)
Leopardus wiedii yucatanica Nelson and Goldman, 1931

DESCRIPTION: *Leopardus wiedii* is a medium-sized feline with brown-yellowish coloration, with white hair on the chest and on the internal part of the limbs. The body is covered with irregular black or dark brown spots, without rosettes. The head is small and rounded, the legs and tail are relatively elongated. It can be distinguished

DISTRIBUTION: The margay inhabits tropical and subtropical regions from the north of Mexico to the north of Argentina (Tewes and Schmidly, 1987). In Mexico, it is distributed in the coastal plains of the Pacific and Gulf of Mexico, from Sinaloa and Tamaulipas to Chiapas in the south, and in the Yucatan Peninsula (Chávez and Ceballos, 1998; Hall, 1981; Sánchez et al., 2002). It has been recorded in the states of CA, CL, CO, CS, CH, DU, GR, HG, JA, MI, MO, MX, NL, NY, OX, PU, QR, SI, SL, SO, TA, TB, VE, and YU.

from the ocelot (*L. pardalis*) by its smaller size, compact body, proportionately longer tail, and the marks on the neck. It has a long tail (52% to 53% of head-body length), large dark spots on the external hind leg area, and only one whirl of hair between the shoulders (Eisenberg, 1989; Emmons and Feer, 1997; Sánchez et al., 2002).

EXTERNAL MEASURES AND WEIGHT
TL = 805 to 1,300 mm; TV = 330 to 510 mm; HF = 89 to 132 mm; EAR = 40 to 55 mm. Weight: 3 to 5 kg.

DENTAL FORMULA: I 3/3, C 1/1, PM 3/2, M 1/1 = 30.

NATURAL HISTORY AND ECOLOGY: *L. wiedii* is the most arboreal felid of Mexico; its biology and ecology are little documented. It feeds on invertebrates, birds, reptiles, amphibians, and both arboreal and small mammals, mainly rodents (Álvarez, 1991; Dominguez-Castellanos and Ceballos, 2005; Konecny, 1989; Oliveira, 1998; Tewes and Schmidly, 1987). It is solitary and usually nocturnal; it hunts either in the trees or on the ground. When a prey is captured, the margay commonly climbs a tree to eat it. There is little information on its reproduction in the wild. In captivity margays live up to 20 years (Oliveira, 1998). The gestation period lasts around 70 days (Guggisberg, 1975), but in captivity it has been reported to last up to 81 days (Paintiff and Anderson, 1980). The litter consists of one or two kits. The home range goes from an average of 4 km^2 in El Cielo Biosphere Reserve, Mexico (Carbajal-Villareal, 2005) to 10.9 km^2 for an adult male in Belize (Konecny, 1989). Its valuable fur makes it one of the most hunted cats.

VEGETATIONAL ASSOCIATIONS AND ELEVATION RANGE: *L. wiedii* inhabits tropical rainforests, subdeciduous and deciduous forests, mangroves, and cloud forests. It is most commonly found from sea level to 1,000 m, but it can be found up to 3,000 m.

CONSERVATION STATUS: *L. wiedii* is included in appendix I of CITES, and it is considered as near threatened by the IUCN, with population trends decreasing (IUCN, 2010). In Mexico it is considered endangered and its hunting is prohibited (SEMARNAT, 2010). There are no data on the current status of its populations in Mexico, but it is common in tropical and subtropical and dry forests not altered by humans.

Leopardus weidii. Specimen in captivity. Arizona-Sonora Desert Museum, Arizona. Photo: Gerardo Ceballos.

Bobcat

Horacio V. Bárcenas and Federico Romero R.

SUBSPECIES IN MEXICO

Lynx rufus baileyi Merriam, 1980
Lynx rufus californicus Mearns, 1897
Lynx rufus escuinapae J.A. Allen, 1903
Lynx rufus oaxacensis Goodwin, 1963
Lynx rufus peninsularis Thomas, 1898
Lynx rufus texensis J.A. Allen, 1829

DESCRIPTION: *Lynx rufus* is medium in size, approximately twice the size of a domestic cat (*Felis catus*), and is, on average, slightly smaller than the Canada lynx (*L. lynx*). The bobcat, or red lynx, has long legs relative to its body length and a short tail (hence the name bobcat). The fur is dense, short, and soft. The color is light gray, yellowish-brown, buff, brown, and reddish with mottled brown, gray, or black. The middle of the back is frequently darker. The underparts and inside of the legs are generally white with black or dark brown spots or bars. Ears are tipped with a short tuft of black hair; the dorsal surface of the ears is black with a prominent white spot. A ruff of fur flares from the animal's checks and neck. The tail color is whitish below with a broad black band on the upper tip and several less distinct black bands toward the rump (Hall, 1981; Hansen, 2007; Leopold, 1959; Rolley, 1999; Sunquist and Sunquist, 2002).

EXTERNAL MEASURES AND WEIGHT

TL = 508 to 1,003 mm; TV = 90 to 198 mm; HF = 143 to 223 mm; EAR = 61 to 77 mm. Weight: 6.2 to 26.8 kg.

DENTAL FORMULA: I 3/3, C 1/1, PM 2/2, M 1/1 = 28.

NATURAL HISTORY AND ECOLOGY: Bobcats are native predators that have adjusted very well to land settlement (Leopold, 1959), but prefer areas with dense cover or uneven, broken terrain (Sunquist and Sunquist, 2002). Like many other felids, bobcats are solitary and are active day and night with peaks of activity during early morning and late afternoon (Hamilton, 1982; Rolley, 1999). The male and female interact almost exclusively during the mating season. Bobcats rarely vocalize, although they often yowl and hiss during the mating season (Hamilton, 1982). They often mark their territory with scent-marks, a combination of feces, urine, and excretion from anal glands. Uncovered feces are often repeatedly deposited in a specific site (Kight, 1962). These marking locations are usually near travel routes or den sites and demarcate home ranges (Bailey, 1974). The home range of a male can overlap with

Lynx rufus. Scrubland. Hermosillo, Sonora. Photo: Gerardo Ceballos.

that of several females and may also overlap one of another male (Nowak, 1999b). Home range varies widely, from 0.6 km^2 to 201 km^2 (Larivière and Walton, 1977; Rolley, 1999; Sunquist and Sunquist, 2002). The territory of a bobcat increases as food abundance decreases, and it is smaller for females than for males (Larivière and Walton, 1977). Density estimation of this felid varies across the distribution range and according to the environment. Reported bobcat densities in United States ranged from 0.04/km^2 to 1.53/km^2 (Heilbrun et al., 2003; Jones and Smith, 1979; Koehler and Hornocker, 1989; Lawhead, 1984; Lembeck, 1978; Rolley, 1985; Zezulak and Schwab, 1979). In Mexico, bobcat densities ranged from 0.05/km^2 to 0.53/km^2 (Medellín and Bárcenas, 2010). The higher population densities in the United States are from rocky areas and dense vegetation in California, while in Mexico the higher densities are located in riparian forest in Chihuahua and in the tropical deciduous forest in the coast of Sinaloa (H. Bárcenas and R. Medellín, unpublished data). Bobcats are strictly meat eaters. Across much of its geographic range the main diet is based on small mammals. Lagomorphs (*Sylvilagus* spp.) and rodents (Heteromyidae and Muridae) are the most common prey items (Aranda et al., 2002; Delibes and Hiraldo, 1987; Delibes et al., 1997; Jones and Smith, 1979; Litvaitis et al., 1986; Luna and López, 2006; Medellín and Bárcenas, 2010; Romero, 1993; Rolley, 1999; Sunquist and Sunquist, 2002). Birds and reptiles are of minor importance in their diet (Delibis and Hiraldo, 1987). In some sites in the northern United States, bobcats consume a larger prey, the white-tailed deer (*Odocoileus virginianus*; Kitchings and Story, 1979; Litvaitis et al., 1984, 1986; McLean et al., 2005). Dens are found in caves, cavities of rocks, and hollow trees, and even in grasslands or shrubs when these are tall and dense (Bailey, 1974; Leopold, 1959; McCord, 1974). Breeding appears to peak between February and April (Sunquist and Sunquist, 2002). There is seasonal variation in their reproductive period, however, apparently due to latitude, longitude, climate, and food availability (McCord and Cardoza, 1982; Sunquist and Sunquist, 2002). Gestation lasts approximately 63 days; average litter size ranges from 2.5 to 3.9 (Fritts and Sealander, 1978; McCord and Cardoza, 1982). The young are nursed for about three months; kittens begin to accompany the mother on hunts when they are four to six months old and are dependent on the mother until about seven months of age (Kitchings and Story, 1984; Rolley, 1983). Coyotes (*Canis latrans*) and pumas (*Puma concolor*) are known to kill bobcats occasionally (Fedriani et al., 2000; Koehler and Hornocker, 1989). Bobcats are considered efficient in controlling populations of rodents and other mammals considered pests to agriculture (Aranda, 2000; Leopold, 1959).

DISTRIBUTION: Bobcats are distributed from southern Canada to Oaxaca, Mexico. In Mexico, there are some records in tropical regions, both on the Pacific coast from Sonora to Oaxaca, as well as on the coast of the Gulf of Mexico and in the Yucatan Peninsula. It has been recorded in the states of AG, BC, BS, CH, CO, DF, DU, GJ, GR, HG, JA, MI, MO, MX, NL, NY, OX, PU, QE, SI, SL, SO, TA, TL, VE, and ZA.

VEGETATIONAL ASSOCIATIONS AND ELEVATION RANGE: Bobcats are found in a wide variety of habitats such as arid scrubland, pine forest, oak forest, mixed pine-oak forest, grassland, riparian vegetation, and tropical deciduous forest. It is found from sea level to 3,600 masl (Hall, 1981; Larivière and Walton, 1977; López-Wilchis and López-Jardines, 1998).

CONSERVATION STATUS: Bobcats are considered as least concern by IUCN, with population trends stable (IUCN, 2010). In some states of the United States, Canada, and Mexico the bobcat is classified as a game species or as a furbearer and is subject to regulation on harvest (Rolley, 1999; SEMARNAT, 2010). The bobcat's habitat has deteriorated due to intensive farming and human settlements (Ceballos and Galindo, 1984; McCord and Cardoza, 1982). Bobcats are listed in appendix II of CITES (Nowell and Jackson, 1996).

Puma

Cuauhtémoc Chávez Tovar and Gerardo Ceballos

SUBSPECIES IN MEXICO

Puma concolor couguar

Recent studies have shown a high level of genetic similarity among the North American cougar populations, suggesting that they all are fairly recent descendants of a small ancestral group. Based on genetic data (nuclear and mitochondrial DNA), Culvert et al. (2000) and Wilson and Reeder (2005) concluded that there are six subspecies (*azteca, browni, californica, mayensis, stanleyana, improcera*).

DESCRIPTION: *Puma concolor* is a large felid. The coloration of the back and head is yellowish-brown or sandy, varying to reddish-brown; ventrally it is whitish. The fur is short and dense. The tips of the ears and tail are black. It has clear facial markings, with a white spot around the muzzle and a black patch on the base of the whiskers. The legs are long; the forepaws are robust with five fingers; the posterior paws only have four. The claws are long, strong, and retractile. The offspring are spotted from birth to between 6 and 10 months. This species shows a wide range of shades, sizes, and weights, according to the subspecies; in general, the northern and southern subspecies are larger than those of Central America. In Mexico, the average weight is 60 kg in males and 40 kg in females (Álvarez del Toro, 1991); however, the individuals of the tropical forest are smaller. In the Calakmul region the weight of one female was 38 kg and the average weight was 42 kg for 8 males (C. Chavez, unpubl. data).

EXTERNAL MEASURES AND WEIGHT

TL=1,100 to 2,200 mm; TV=620 to 960 mm; HF=220 to 270 mm; EAR=55 to 85 mm. Weight: 38 to 110 kg.

Puma concolor. Specimen in captivity. Arizona-Sonora Desert Museum, Arizona. Photo: Rurik List.

DENTAL FORMULA: I 3/3, C 1/1, PM 3/2, M 1/1 = 30.

DISTRIBUTION: The puma is one of the mammals with the widest distribution in the world. It is found from the Canadian province of British Columbia and the northern United States to Argentina and Chile. Historically it has been found in every Mexican state.

NATURAL HISTORY AND ECOLOGY: Pumas are solitary except during the mating season when males and females pair temporarily, separating before the birth of cubs. They are terrestrial but are agile tree climbers and can jump from the ground up to a height of 5.5 m (Nowak, 1999b). They usually hunt on the ground, but sometimes in the trees (Aranda and March, 1987). Pumas are good swimmers but usually avoid water (Nowak, 1999b). Their dens are usually found in caves and other natural hollows in broken terrain (Ceballos and Galindo, 1984). Pumas can tolerate human presence better than jaguars, sometimes living in areas with important human activity if they have good shelters such as rocky areas or deep ravines; they become nocturnal in these situations. They hunt by lurking, and in temperate areas feed primarily on deer (De la Torre and De la Riva, 2009; Rosas-Rosas, 2006); however, in tropical areas, they frequently prey on large rodents, armadillos, deer, peccary, and sometimes rats and rabbits (Álvarez del Toro, 1991). Occasionally, they steal goats, lambs, or foals. In areas where jaguars have been extirpated, pumas shift to the largest species (Moreno et al., 2006). Distinctive features of its hunting behavior are the deep bites left in the neck and nape of its prey. When the prey is captured, the puma takes it to a safe place, generally hiding it in shrubs or rocks, where it is gutted and the viscera discarded; it often covers the carcass with leaves or soil (Ceballos and Galindo, 1984). When deer abound, pumas may kill an average of one per week, thus helping to control deer populations (Whitaker, 1980). During the exploratory movements during the breeding season, they emit acute screams and caterwauls (Álvarez del Toro, 1991). Except for humans, they have few enemies (Álvarez del Toro, 1991). The first reproduction can occur after the second year, but usually occurs after females have established a home range, possibly around the third year. Estrus lasts 9 days, and gestation between 82 to 98 days (Whitaker, 1980). Mating can take place any time of the year. They have a litter every two years, and most of the births occur shortly before the rainy season. The litter size varies from 1 to 6 kittens (Wolonszyn and Wolonszyn, 1982), with an average of 3 for temperate areas and an average of 1.5 for tropical areas. The offspring remains with the mother around 15 months (Aranda and March, 1987; Eisenberg, 1989). The track of the puma is slightly longer but smaller than that of the jaguar (Aranda, 2000; Whitaker, 1980). When they urinate or defecate, they usually cover the site with litter, as domestic cats do (Álvarez del Toro, 1991). It is common to find marks of their claws on tree trunks (Ceballos and Galindo, 1984). In central Mexico and the southwestern United States their activity is mainly crepuscular, but they can be active any time of day depending on many factors such as human influence, activity patterns of prey, and air temperature (C. Chávez pers. obs.; Monrroy et al., 2009; Sweanor et al., 2000). Most hunting takes place before dusk. They may travel long distances within 24 hours, between 5 km and 40 km, depending on the vegetation type. Their population densities are highly variable, being highest in pine and pine-oak forest, especially in northern areas (Ceballos and Galindo, 1984). Their densities are largely dependent on the availability and abundance of prey. The home range varies from 66 km^2 to 685 km^2 for females and 152 km^2 to 1,148 km^2 for males (Bailey, 1974; Berg, 1981; McCord and Cardoza, 1982; Zezulak and Schwab, 1981). The home range size depends on the density and dispersion of prey, seasonality of the site, and the sex of the individuals (Grigione et al., 2002). In the Chamela-Cuixmala Biosphere Reserve of the state of Jalisco, the home range of males is from 60 km^2 to 90 km^2 and from 25 km^2 to 62 km^2 for females (Nuñez et al., 1997). In the Calakmul Biosphere Reserve in the state of Campeche these areas are 135 km^2 and 108 km^2 for males and females, respectively (Ceballos et al., 2002; Chávez and Ceballos, 2006). The home ranges of males normally overlap with one or more females. In the tropics, pumas' home ranges overlap with that of the jaguars (*Panthera onca*), a usually dominant species, so the pumas adjust their movements to avoid contact (Ceballos et al., 2011; Chávez and Ceballos, 2006; Chávez et al., in review; Schaller and Crawshaw, 1980).

VEGETATIONAL ASSOCIATIONS AND ELEVATION RANGE: Pumas are found in all types of natural vegetation of the country. They are more abundant in conifer and oak forests in northern Mexico. They can be found in tropical deciduous, subdeciduous, and perennial forests, thorny forests, xeric scrublands, and cloud forests. Pumas are found from sea level to 3,500 masl, but they are more common between 1,500 masl and 2,500 masl.

CONSERVATION STATUS: Pumas are frequently persecuted as predators of livestock; however, there are no studies to support such reports. For example, of more than 50 records of livestock killed in the Yucatan Peninsula none was confirmed to have been caused by pumas (C. Chávez, unpubl. data). Similarly, in Iguacú, Brazil, 52 puma scats did not have livestock remains (Cascelli de Azevedo, 2008). It seems that the availability and abundance of wild prey is a determinant factor influencing livestock depredation. Pumas are legally hunted in many areas, although only one male per hunter is allowed per season. In Mexico it is classified as a species under special protection (SEMARNAT, 2010). Its current situation in many parts of the country is unknown. Within the Transversal Volcanic Belt its situation is critical and it has been extirpated from most areas. It is considered as least concern by IUCN, with population trends decreasing (IUCN, 2010), but certain subspecies are in danger of extinction.

Subfamily Pantherinae

The subfamily Pantherinae contains three genera (*Panthera, Uncia,* and *Neofelis*) containing seven extant species. Only one species is known from Mexico.

Panthera onca (Linnaeus, 1758)

Jaguar

Cuauhtémoc Chávez and Gerardo Ceballos

SUBSPECIES IN MEXICO

Panthera onca hernandesii (Gray, 1858)

Recent work with skull morphology suggests that all jaguar subspecies recognized were not valid (Larson, 1997). Posterior analysis with molecular genetics supported that conclusion (Eizirik et al., 2001, 2006). Mitochondrial DNA analysis weakly supported two phylogeographic groups of jaguars, north and south of the Amazon River in South America, with evidence of continued gene flow between the two groups (Eizirik et al., 2002; Ruiz-Garcia et al., 2006).

DESCRIPTION: Jaguars are the largest felids of the American continent. The color of the fur varies from pale yellow to reddish-brown and changes to white on the jowls, chest, and internal parts of the limbs. All the body is covered with black spots, which on the sides change to rosettes; within these, there may be one or more small spots. There are two color morphs, including black jaguars, or *panteras* as they are known in Spanish, which are melanic or blackish brown, with visible marks in oblique light

Panthera onca. Tropical semi-green forest. Calakmul Biosphere Reserve, Campeche. Photo: Gerardo Ceballos.

(Hoogesteijn et al., 1993). In some regions of South America, the melanistic jaguars have been observed regularly, but in Mexico there are no confirmed records. These coloration patterns can serve to identify individuals since two individuals do not have the same coloration pattern and rosettes.

EXTERNAL MEASURES AND WEIGHT
TL = 1,574 to 2,419 mm; TV = 432 to 675 mm; HF = 225 to 302 mm; EAR = 50 to 85 mm.
Weight: 36 to 158 kg.

DENTAL FORMULA: I 3/3, C 1/1, PM 3/2, M 1/1 = 30.

NATURAL HISTORY AND ECOLOGY: Jaguars shelter in caves and areas with dense vegetation. They are terrestrial hunters and skillful swimmers, and they climb trees with ease. *Panthera onca* has a broad spectrum of prey; thus it is considered an opportunistic carnivore. The diet basically depends on the density and availability of the prey (Seymour, 1989). More than 85 species have been reported in their diet, including invertebrates, fish, reptiles, birds, and mammals. Mammals weighting more than 1 kg and some reptiles and birds are the most common prey throughout most of its geographic range (Emmons, 1987a; Seymour, 1989; Tewes and Schmidly, 1987). In Calakmul, Campeche, the most important prey are white-collared peccary, coati, and armadillo (Aranda, 1993; Chávez et al., 2007). In Jalisco, the most consumed species are white-tailed deer, coati, collared peccary, and armadillo (Nuñez et al., 2000). The mating season varies geographically; kittens in South America have been found in June, August, November, and December (Seymour, 1989). In areas with marked seasonality kittens are born more frequently when the food is abundant. In Mexico, births occur between July and September (Leopold, 1959). The gestation period lasts 100 days on average and the litter size varies from 1 to 4 kittens, but 2 is the average. The kittens are born with spots and are altricial; they born with the eyes closed and weigh around 800 g. After 1.5 or 2 months, they begin to follow the mother, staying with her 15 to 24 months, reaching sexual maturity between 2 and 3 years (Seymour, 1989). Longevity in the wild is 10 to 12 years, while in captivity they can live up to 22 years. Jaguars are solitary, with the exception of the mating season. Females have smaller home ranges than males and usually a male includes one or more females in its territory. Home range is highly variable and it is related primarily to the abundance and availability of food. Males' home range varies from 28 km^{2}to 262 km^2 (Cavalcanti, 2008; Crawshaw and Quigley, 1991). Females' home range ranges from 10 km^2 to 97 km^2 (Cavalcanti, 2008; Schaller and Crawshaw, 1980). In Chamela, Jalisco, the home range of a female jaguar was estimated to be 25 km^2 during the dry season and 65 km^2 during the rainy season (Nuñez et al., 2000). In Latin America the reported densities range from 1 individual/9 km^2 to 1 individual/100 km^2 (Crawshaw and Quigley, 1991; Rabinowitz and Nottingham, 1986; Schaller and Crawshaw, 1980; Silver et al., 2004). In Mexico the density varies from 1 individual/13 km^2 to 1 individual/100 km^2. In the Yucatan Peninsula, density varies from 1 individual/13 km^2 to 1 individual/54 km^2 (Aranda, 1998; Ceballos et al., 2002; Chávez et al., 2007; Faller et al., 2007; Grigione et al., 2001). In the Chamela-Cuixmala Biosphere Reserve, Jalisco, density is 1 individual/58 km^2 (Nuñez et al., 2000). In northern Mexico the density is 1 individual/66 km^2 to 1 individual/100 km^2 (Grigione et al., 2001; Rosas-Rosas, 2006). Home ranges in Calakmul are up to 500 km^2 for males and up to 133 km^2 for females (Ceballos et al., 2002; Chávez et al., 2007). Jaguars have been declining in number and geographic range for the last 50 years. It has been estimated that there were around 25,000 jaguars in historic times, and that there are now around 4,000 in Mexico. The largest population is found in the Yucatan Peninsula (Ceballos et al., 2002, 2005, 2011; Chávez et al., 2007). Major causes of their decline are habitat destruction and poaching (Medellín et al., 2002; Sanderson et al., 2002; Zarza et al., 2007). The causes of natural mortality include fratricide, the mortality caused by the largest individual in the litter, which seizes the food provided by the mother before smaller kittens can. Although jaguars have few enemies, they can be preyed on when they are young by adult jaguars and crocodiles. There are few data on mortality caused by infectious diseases; however, they must be important factors in the wild. The mortality during the process of dispersion/migration can be considerable. Finally, the abundance of prey and water may be important factors of mortality, as well as the reproductive success of the species (Chávez et al., 2007; Smith and McDougal, 1991; Smith et al., 1987).

VEGETATIONAL ASSOCIATIONS AND ELEVATION RANGE: Jaguars live mainly in tropical rainforests, subdeciduous forests, deciduous forests, and mangroves. They are found less frequently in cloud, conifer, and oak forests, and xeric oak-scrublands. They live from sea level up to 2,000 masl. Most of the records come from localities of less than 1,000 masl.

CONSERVATION STATUS: Jaguars are included in appendix I of CITES (1982). In Mexico, this species is at risk of extinction and its hunting is prohibited (SEMARNAT, 2010). It is considered as near threatened by IUCN, with population trends decreasing (IUCN, 2010). Habitat destruction and poaching are the major threats to the species. In Mexico their populations have been reduced and fragmented (Ceballos et al., 2007). Apparently, there are viable populations in the reserves of Sian Ka'an (Quintana Roo), Calakmul (Campeche), and Montes Azules (Chiapas), and in the region of the Chimalapas (Oaxaca) (Ceballos et al., 2007, 2011). For example, in the Calakmul Biosphere Reserve, a population of at least 500 individuals has been estimated (Ceballos et al., 2002). The populations in other reserves are probably smaller because they have smaller areas. An appropriate strategy for the conservation of jaguars in southeastern Mexico is to maintain a linkage between Calakmul and Sian Ka'an to the east and Montes Azules to the west.

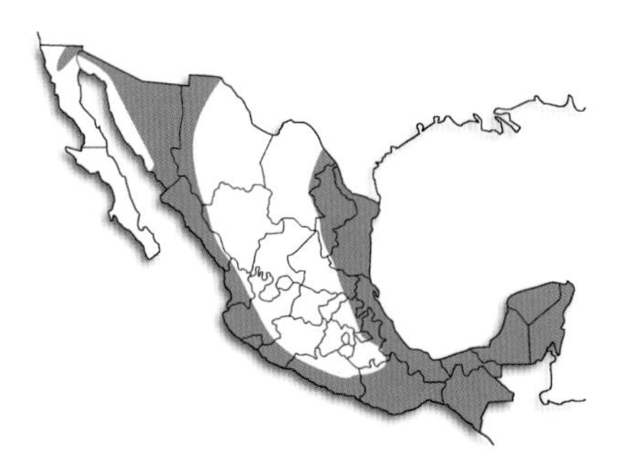

DISTRIBUTION: The historical distribution of jaguars was almost continuous from the southern United States to central Argentina (Hall, 1981; Seymour, 1989; Swank and Teer, 1989). Currently, the species is extinct in El Salvador, Uruguay, and Chile. There is not a resident population in the United States. Few individuals are likely resident; most are transient jaguars. In Mexico, the historical distribution followed the distribution of the tropical forest and subtropical shrublands. On the Pacific side jaguars are distributed from the U.S. border along the coastal plain and foothills of the Sierra Madre Occidental and Sierra Madre del Sur to the Isthmus of Tehuantepec. Along the Gulf coast its distribution ranges from central Tamaulipas to the Yucatan Peninsula, Oaxaca, and Chiapas (Chávez and Ceballos, 2005; Hall, 1981; Seymour, 1989; Swank and Teer, 1989). It has been recorded in the states of CA, CL, CS, GJ, GR, JA, MI, MX, NL, NY, OX, PU, QR, QE, SI, SL, SO, TA, TB, VE, and YU.

Family Canidae

The family Canidae is a family that groups together 35 wild species, and includes the domestic dog (Wilson and Reeder, 2005). They are widely distributed throughout the world, with the exception of Australia, Antarctica, and many islands. In Mexico there are three genera (*Canis*, *Urocyon*, *Vulpes*) with four species. The gray fox (*U. cinereoargenteus*) and coyote (*C. latrans*) are widely distributed. The kit fox (*V. macrotis*) is found in the arid areas of northern Mexico, and the wolf (*C. lupus*), a holartic species, has been extirpated.

Canis latrans Say, 1823

Coyote

Jorge Servín, Elias Chacón, and Rurik List

SUBSPECIES IN MEXICO

Canis latrans cagottis (Hamilton-Smith, 1839)
Canis latrans goldmani Merriam, 1904
Canis latrans jamesi Townsend, 1912
Canis latrans peninsulae Merriam, 1897
Canis latrans vigilis Merriam, 1897
Canis latrans clepticus Elliot, 1903
Canis latrans impavidus J.A. Allen, 1903
Canis latrans microdon Merriam, 1897
Canis latrans texensis V. Bailey, 1905
Canis latrans mearnsi Merriam, 1897

According to Hall (1981), there are nine subspecies in Mexico; however, due to its dispersion capacity and inbreeding, it could be impractical to use this classification (Bekoff and Wells, 1981).

DISTRIBUTION: *Canis latrans* has a wide distribution, ranging from Alaska and western Canada to Panama (Hall, 1981; Reid, 1997). In Mexico, it is found throughout the country and recently it has been recorded in Yucatan (Sosa-Escalante et al., 1997). It has been recorded in the states of AG, BC, BS, CH, CL, CO, CS, DF, DU, GJ, GR, HG, JA, MI, MO, MX, NL, NY, OX, PU, QE, SO, SI, SL, TA, TB, TL, VE, YU, and ZA.

DESCRIPTION: Coyotes are medium-sized canids, very similar to a thin German shepherd dog. The snout is elongated, and the eyes small and relatively close together. The pelage varies from gray to reddish, passing through brown tones; the tail has a black tip. The underfur is always lighter. The ears are large and pointed.

EXTERNAL MEASURES AND WEIGHT

TL = 1,075 to 1,150 mm; TV = 270 to 375 mm; HF = 177 to 220 mm; EAR = 107 to 110 mm.
Weight: 8 to 16 kg.

DENTAL FORMULA: I 3/3, C 1/1, PM 4/4, M 2/3 = 42.

NATURAL HISTORY AND ECOLOGY: Coyotes are social animals with crepuscular activity (Bekoff and Wells, 1980). The social organization varies from solitary nomadic animals to stable groups, whose average size varies from 1.4 individuals in summer to 3.9 in winter (Andelt, 1985; Bekoff and Wells, 1980; Bowen, 1981). They are monogamous with a mating season that goes from January to April; gestation lasts nine weeks, at the end of which six offspring are born that stay with their parents or become independent (Bekoff and Wells, 1981). In the nurturing of pups, other members of the group are also involved (Andelt, 1985). Their diet is generalist with seasonal variations, including lagomorphs, rodents, ungulates, fruits, insects, reptiles, and birds (Delibes et al., 1985; List et al., 2003; Servin, 1991). In some arid areas lagomorphs have been reported as being more important, especially during winter (Andelt, 1985; Delibes et al., 1985), but in grasslands, rodents and prairie dogs (*Cynomys ludovicianus*) and kangaroo rats (*Dipodomys*) predominate (List et al., 2003). In temperate forests the diet includes more rodents such as *Sigmodon*, *Neotoma*, and *Peromyscus* (Delibes et al., 1985; Servin, 1991). Close to human habitation garbage can be the main component of their diet (Hidalgo-Mihart et al., 2004). Among resident individuals the home range of adults is greater than that of juveniles (5.0 km^2 and 2.4 km^2), whereas nonresident individuals have larger home ranges but do not have a territory (Andelt, 1985). Only the groups or families are territorial and their territories do not overlap (Bekoff and Wells, 1980). In Mexico, home range of 4 males and 3 females in the grasslands of northern Chihuahua ranged from 32 km^2 to 111 km^2 (List, 1997; Moehrenschlager et al., 2004). The importance of coyotes in rural areas, especially in ranching areas, is subject to controversy because their presence is always associated, without clear foundations, with economic losses due to predation of domestic animals, while at the same time their value controlling populations of competitors of cattle like lagomorphs and kangaroo rats is neglected.

VEGETATIONAL ASSOCIATIONS AND ELEVATION RANGE: In Mexico, coyotes live in all plant communities, especially in plains with xeric scrub and pastures. They are found from sea level to over 3,000 m.

CONSERVATION STATUS: In the grasslands of northern Mexico, coyotes are abundant (List, 1997). In forest and tropical areas they are less abundant, but if anthropogenic food is available their numbers increase locally (Hidalgo-Mihart et al., 2004). The International Union for Conservation of Nature (IUCN) classifies the coyote as least concern with the population trend expanding (IUCN, 2010).

Canis latrans. Grassland. Janos Biosphere Reserve, Chihuahua. Photo: Gerardo Ceballos.

Mexican gray wolf

Rurik List

SUBSPECIES IN MEXICO

Canis lupus baileyi Goldman, 1944

Traditionally, two subspecies of gray wolf were identified: *C. lupus baileyi* and *C. lupus monstrabilis*. Recently *monstrabilis* was included within *baileyi* (Bogan and Melhop, 1983; Mech, 1974; Nowak, 1999b).

DESCRIPTION: *Canis lupus* is the largest wild canid in Mexico. Physically, it is similar to a German shepherd dog, but larger, with a big head and a short and thick snout with a wide nasal pad. The eyes are small and separated; the ears are short, rounded, and erect; the neck and forefeet are wide; there is a clear slope from the shoulders to the back legs (Leopold, 1959; McBride, 1980). The pelage on the back is long and varies from brown or gray and is lighter on the abdomen and legs; in general, the coloration ranges from white to black (Nowak, 1999b). The wolves from Chihuahua seem to have a grayish dorsal coloration and on the sides, while the wolves in south Durango are more yellowish and bright (McBride, 1980).

Canis lupus. Specimen in captivity. Pine forest. San Cayetano, state of Mexico. Photo: Gerard Ceballos.

EXTERNAL MEASURES AND WEIGHT

TL = 1,000 to 1,200 mm; TV = 390 to 410 mm; HF = 220 to 310 mm; EAR = 115 to 120 mm.

Weight: 25 to 41 kg.

DENTAL FORMULA: I 3/3, C 1/1, PM 4/4, M 2/3 = 42.

NATURAL HISTORY AND ECOLOGY: The data on the natural history of the Mexican wolf are scarce and often anecdotal because this subspecies was extirpated before scientific research was properly carried out (Brown, 1983). For this reason, the information presented here often corresponds to studies of Canada and the United States. The wolf is a social animal that normally forms packs, which are made up of family groups of a couple that mates for life and its descendants (Mech, 1970). Pack sizes vary with the size of the prey, normally including 5 to 8 individuals, but packs of up to 36 animals have been reported (Mech, 1974). In Spain, where the size of the prey is similar to those reported for the Mexican subspecies, the common size of the packs before the breeding season is 3 or 4 animals, sometimes 5, but rarely larger than this (Vila and Castroviejo, 1990). This coincides with observations of groups of three and five wolves in Mexico (McBride, 1980). The order in the pack is maintained by a hierarchical domination, of the adult male toward the adult female and the pups, the adult female toward the pups, and a lineal order among the pups. In larger packs, an arrangement of males and females is formed, but the leader is almost always a male known as the alpha male (Mech, 1974). The dynamics of the pack, the social ranking, movements, and certain aspects of seasonal habitat use are affected by the reproductive process (Carbyn, 1978). As the mating season approaches, the packs split; in January the couples are formed and mating occurs in February. After a gestation period of 60 to 63 days, the offspring are born between March and May (Leopold, 1959; McBride, 1980). The litters have 3 to 9 pups, with 5.6 on average. The pups are cared for by both parents; the Mexican subspecies does not appear to be as gregarious as the northern subspecies, where all the pack is involved in caring for the pups. When the pups are three months old, they leave the den and the family hunts together (Leopold, 1959). The pups are weaned in October, and in December they begin to travel away from their parents (McBride, 1980). Communication among wolves seems to have three main components: vocalizations, in which howling stands out; visual exhibitions that include body postures and face gestures; and scent marking (Mech, 1974). In Mexico, the home range of the packs of wolves is unknown, but among the reintroduced population in Arizona and New Mexico home range size ranged from 264 km^2 to 1,378 km^2 (Carroll et al., 2006); elsewhere in North America it varies from 64 km^2 on Vancouver Island for a pack of 10 animals (Scott and Shackleton, 1982) to 2,600 km^2 for a pack of 9 animals in northern Alaska (Stephenson and James, 1982). The home range varies seasonally; for example, in northeastern Alberta, Canada, it can be as small as 247 km^2 in summer and as large as 1,779 km^2 in winter (Fuller and Keith, 1980). The main natural food items of the Mexican wolf prior to its extirpation from the wild were white-tailed deer (*Odocoileus virginianus*) and mule deer (*Odocoileus hemionus*), and, less important, pronghorn (*Antilocapra americana*), peccary (*Pecari tajacu*), bighorn sheep (*Ovis canadensis*), rabbit (*Sylvilagus* spp.), and rodents, but cattle, as an abundant prey and relatively easy to catch, significantly replaced the deer (Leopold, 1959; McBride, 1980). The reintroduced Mexican wolf in Arizona and New Mexico feeds mainly on elk (*Cervus elaphus*) and to a lesser extent on deer (Carrera et al., 2008; Reed et al., 2006). The disappearance of the natural prey of the wolf, the reduction of its habitat, and the introduction of domestic cattle caused deep changes in the composition of its diet (Brown, 1983; Leopold, 1959; McBride, 1980). Wolves can be active any time of the day but are mainly nocturnal. In most areas where wolves still exist in North America, the main mortality factor is related to human activity, although natural factors such as diseases, poor diet, fights with other wolves, and predation also influence mortality (Carbyn, 1987; Carroll et al., 2006).

VEGETATIONAL ASSOCIATIONS AND ELEVATION RANGE: Wolves are found in all types of vegetation, except in rainforests and deserts, which lack the suitable prey (Carbyn, 1987). The elevational interval in Mexico begins at 1,500 masl and extends to higher areas in the mountains (Brown, 1983; McBride, 1980).

CONSERVATION STATUS: The Mexican wolf is considered reintroduced in the U.S. Southwest and highly endangered in Mexico by the IUCN (Mech and Boitani, 2004) and is included in appendix I of the Convention on International Trade in Endangered Species (CITES) (Thornback and Jenkins, 1982). Since 1976 the Mexican subspecies is included in the U.S. Endangered Species Act (Siminski, 1990), and in Mexico, it is considered endangered by the Mexican Red List (SEMARNAT, 2010). The reduction of the populations began in 1915, with the Division of Predator and Rodent Control of the United States (Ammes, 1982). In Mexico, control by farmers began in the 1930s. In the 1950s, under the pretext of controlling rabies, the U.S. Fish and Wildlife Service and the Pan American Sanitary Bureau trained ranchers and veterinarians in the use of the poison 1080, which caused the largest reduction of wolf populations (Ames, 1980; Baker and Villa-Ramírez, 1960; McBride, 1980; Villa-R., 1961). The few wolves that remained were eventually exterminated. The first effort to determine the population size of the wolves in Mexico took place between 1978 and 1980, with an estimation of no more than 50 couples in the region encompassing

DISTRIBUTION: The gray wolf has the widest natural distribution of any other terrestrial living mammal, except humans (Nowak, 1999b). It was originally distributed through the entire Northern Hemisphere above the 20 degree latitude N (Mech, 1974) and North America, except in Baja California and some islands (Carbyn, 1987). In Mexico, it was distributed from Sonora and Tamaulipas in the north to Oaxaca in the south (Hall, 1981; Leopold, 1959). It is currently extinct in the states of AG, CH, CO, DU, GR, GJ, MI, MX, NL, PU, QE, SO, TL, VE, and ZA.

Chihuahua, Durango, and possibly Sonora and Zacatecas (Ammes, 1982; McBride, 1980). Since then, the reports on wolves in Mexico have been scarce but constant; most have been footprints and howling near the hills of Sonora, Chihuahua, and Durango (Carrera, 1990; Hernández and Lafón, 1986; Servin, 1986). At least in the last decade, no report has been confirmed, so it can be considered that no viable population exists in Mexico, and probably the species has been extirpated from the national territory.

Since 1980 there has been a captive breeding program conducted by the U.S. Fish and Wildlife Service and the Mexican Ministry of the Environment; this program has about 300 individuals (descendants of 7 individuals) in 45 public and private facilities of both countries (Ammes, 1982; Siminski, 1990). In March 1998 the first reintroduction was conducted, with the release of 11 wolves in the Apache National Forest in eastern Arizona. Nine subsequent reintroductions of 92 wolves, captures, and translocations of individuals and of whole packs because of cattle predation or because they left the Recovery Zone resulted in a minimum of 52 free-ranging wolves and 2 breeding pairs in Arizona and New Mexico in 2008. In 1999, the Technical Advisory Subcommittee for the Recovery of the Mexican Wolf was created, which brings together representatives of the federal government, specialists, individuals, and organizations interested in the conservation of the wolf. The subcommittee oversees captive management and conducts environmental education, public relations, and research geared toward a future reintroduction of the species in Mexico. After the identification of suitable areas within the historic range of the species in Mexico, the first reintroduction of five individuals took place in October 2011 in northwestern Sonora (Araiza et al., 2012). Four individuals were found dead with traces of poison. An additional male was released in March 2012, whose radio signal was lost a week later. The surviving female was lost in the summer of 2012. Further reintroductions are planned (Comisión Nacional de Areas Naturales Protegidas, pers. comm.).

Urocyon cineroargenteus (Schreber, 1775)

Gray fox

Jorge Servín and Elías Chacón

SUBSPECIES IN MEXICO

Urocyon cineroargenteus fraterculus Elliot, 1896
Urocyon cineroargenteus madrensis Burt and Hooper, 1941
Urocyon cineroargenteus peninsularis Huey, 1928
Urocyon cineroargenteus colimensis Mearns, 1938
Urocyon cineroargenteus guatemalae Miller, 1899
Urocyon cineroargenteus nigrirostris (Lichtenstein, 1850)
Urocyon cineroargenteus scottii Mearns, 1891
Urocyon cineroargenteus orinosus Goldman, 1938

DISTRIBUTION: *U. cineroargenteus* is found from the United States to Central America (Hall, 1981; Leopold, 1965). In Mexico, it has been recorded in every state.

DESCRIPTION: *Urocyon cineroargenteus* is a medium-sized canid. The throat is white and the face gray; the sides of the neck, the abdomen, and the base of the tail are reddish. The back is grayish. The tail is also gray on the top, with a black distal end and a mid-dorsal line of the same color. The colors of the upper and lower sides are delimited by a brown opaque band that runs along each side of the body (Hall, 1981; Leopold, 1965).

Urocyon cineroargenteus. Tropical semi-green forest. Calakmul Biosphere Reserve, Campeche. Photo: Gerardo Ceballos.

EXTERNAL MEASURES AND WEIGHT
TL = 500 to 600 mm; TV = 300 to 400 mm; HF = 100 to 150 mm; EAR = 74 to 81 mm. Weight: 3 to 5 kg.

DENTAL FORMULA: I 3/3, C 1/1, PM 4/4, M 2/3 =42.

NATURAL HISTORY AND ECOLOGY: Gray foxes inhabit forested and brushy areas, especially those with disturbed vegetation (Leopold, 1965). The favorite sites for the construction of their shelters are hollow logs, roots of fallen trees, rocks or exposed soil, and occasionally the base of living trees (Nicholson et al., 1985). They have a high reproductive capacity, so their populations have a rapid recovery. The sex ratio is 1:1, and between 1.8 and 2.2 young individuals per adult exist. Mating occurs between late February and early March; the pups are born after approximately 45 days of gestation (Carey, 1982). The female is responsible for rearing the young, while the male probably does not participate in this activity (Nicholson et al., 1985). The offspring leave the refuge in autumn, when they are 10 to 13 weeks old; they become completely independent in the early winter (Nicholson et al., 1985), dispersing 18 km to 83 km from the natal den (Carey, 1982). Young females tend to stay in their place of origin. Females can reproduce in their first year, but the age at which males reach sexual maturity is unknown (Carey, 1982). They are monogamous, but survival of both members is difficult due to high mortality rates, caused especially by rabies (Nicholson et al., 1985). Their diet is opportunistic, in relation to abundance. Important food sources include rodents, lagomorphs, fruits, and insects (Carey, 1982). The home range varies from 1 km^2 to 8 km^2 depending on the season, the population density, and the quality of the habitat. Data obtained from radio-locations at La Michilia Biosphere Reserve, Durango, indicate a home range of 1.5 km^2 to 3 km^2. They frequently use oaks for resting places, as they are good climbers. They are important vectors of rabies (Carey, 1982).

VEGETATIONAL ASSOCIATIONS AND ELEVATION RANGE: *U. cineroargenteus* inhabits all vegetational associations. It has been found from sea level to 3,500 masl (Hall, 1981; Nowak, 1999b).

CONSERVATION STATUS: The gray fox is a very abundant carnivore that benefits from anthropogenic disturbances. It has no conservation problems. The IUCN status is least concern with stable populations (IUCN, 2010).

Kit fox

Rurik List and Arturo Jiménez Guzmán

SUBSPECIES IN MEXICO

Vulpes macrotis arsipus (Eliott, 1903)
Vulpes macrotis devia (Nelson and Goldman, 1909)
Vulpes macrotis macrotis (Merriam, 1888)
Vulpes macrotis neomexicana (Merriam, 1902)
Vulpes macrotis tenurostris (Nelson and Goldman, 1931)
Vulpes macrotis zinseri (Benson, 1938)

V. macrotis has been regarded as conspecific to *V. velox* by its morphological similarity and by the results of studies of protein electrophoresis (Clutton-Brock, et al. 1976; Dragoo et al., 1990; Ewer, 1973; Ginsberg and Macdonald, 1990; Hall and Kelson, 1980; May, 1981; Wilson and Reeder 1993). Or *V. macrotis* has been regarded as a different species by the results of multidimensional morphologic studies, and more recently, mitochondrial DNA (Egoscue, 1979; Maldonado et al., 1997; McGrew, 1979; Mercure et al., 1993; O'Farrell, 1987; Stromberg and Boyce, 1986). For Mexico, the recent studies have included samples of individuals only from Coahuila and Nuevo León, so their subspecifies status is still uncertain.

DISTRIBUTION: *V. macrotis* is distributed in the deserts and arid areas of North America, from southern Oregon and Indiana in the United States to central Mexico. It is found in the northern parts of the central Mexican plateau, in the northwest, and in the Baja California Peninsula (Hall and Kelson, 1980; Jiménez-G. and López-S., 1992; Leopold, 1959). It has been recorded in the states of BC, BS, CH, CO, DU, NL, SO, and ZA.

DESCRIPTION: Kit foxes are among the smallest American foxes. Their appearance is similar to that of the gray fox, but kit foxes are smaller and have remarkably large ears, which is the most conspicuous feature. The coat is short, with the head, back, and sides dirty yellow to grayish; the tip of tail is black; the shoulders and outer sides of the legs are brown-yellowish; the belly and inner side of the legs is white-yellowish. The hair between the feet pads is dense.

EXTERNAL MEASURES AND WEIGHT

TL = 600 to 840 mm; TV = 303 mm; HF = 111 to 132 mm; EAR = 69 to 95 mm.
Weight: 1.4 to 3 kg.

DENTAL FORMULA: I 3/3, C 1/1, PM 4/4, M 2/3 = 42.

NATURAL HISTORY AND ECOLOGY: Kit foxes are well adapted to life in the desert; they share many characteristics of other species of the genus *Vulpes* adapted to arid habitats (O'Farrell, 1987; Sheldon, 1992). They live in dens throughout the year, which are often excavated by other species, especially prairie dogs (*Cynomys* spp.) and kangaroo rats (*Dipodomys* spp.), so they are commonly associated with populations of both species (Cotera, 1996; Jiménez-G. and López-S., 1992; List and Macdonald, 2003). Most dens are in flat or alluvial ground, and have two to seven entries (Egoscue, 1962). Rodents are its main source of food; however, insects (Coleoptera, Hymenoptera, Orthoptera) and lagomorphs (*Lepus californicus*, *Sylvilagus audubonii*) are important elements of its diet (Cotera, 1996; Jiménez-G. and López-S., 1992; List et al., 2003). Their habits are predominantly nocturnal, with little crepuscular activity, although they are commonly seen resting outside the den during the day (Jiménez-G. and López-S., 1992; List et al., 2003). In Chihuahua, they breed from mid-December to January. Offspring are born in mid-February to mid-March and emerge from late March to late April. They become independent between July and August, and dispersal takes place toward the end of August (Moehrenschlager et al., 2004). The litters have 3 to 5 offspring, which are born after a gestation of 49 to 55 days (Egoscue, 1956). Some couples are monogamous and breed for life, while others change partners frequently (Egoscue, 1956, 1962; Morrell, 1972). The density of foxes reported for Chihuahua is 0.32 individual/km^2 to 0.8 individual/km^2 (List and Macdonald, 2003). The average area of activity in Nuevo León was 4.9 km^2 for males and 4.0 km^2 for females (Cotera, 1996), while in Chihuahua it was 11.4 km^2 for males and 10.2 km^2 for females (List and Macdonald, 2003). The main cause of death is predation by coyotes (Cypher and Spencer, 1998; Moehrenschlager et al., 2004; Ralls and White, 1995). A higher survival rate reported in the grasslands of Chihuahua versus a contemporary study in the prairies of Alberta, Canada suggested that the higher availability of prairie dog burrows provided more escape routes from coyote predation to the kit foxes (Moehrenschlager et al., 2004). In the state of Nuevo Leon collision with cars was the main cause of mortality (Cotera, 1996).

VEGETATIONAL ASSOCIATIONS AND ELEVATION RANGE: Kit foxes inhabit arid and semi-arid areas (O'Farrel, 1997; Sheldon, 1992). In Mexico, it has been reported in desert scrubs, halophytic vegetation, and natural grasslands and prairie dog towns (Cotera, 1996; Jiménez-G. and López-S., 1992; List and Macdonald, 1998, 2003). It has been recorded from 400 m to 1,900 m (List and Cypher, 2004).

CONSERVATION STATUS: In Mexico, this species is considered vulnerable (SEMARNAT, 2010). The IUCN status is least concern, with decreasing population trend (IUCN, 2010). The conservation status varies for some subspecies such as *V. macrotis macrotis*, which became extinct (McGrew, 1979); *V. macrotis mutica*, considered at risk of extinction (O'Farrell, 1987), shows its vulnerability to extinction. The transformation of the habitat is the main cause of the reduction of the populations of kit foxes in Mexico, especially conversion to agricultural land (List and Cypher, 2004). Locally, the eradication of the prairie dogs (*Cynomys mexicanus* and *C. ludovicianus*) has a negative effect on the kit fox.

Vulpes macrotis. Grassland. Janos Biosphere Reserve, Chihuahua. Photo: Rurik List.

FAMILY URSIDAE

The family Ursidae is represented by eight extant species of bears (Garshelis, 2009). Six of these species are included in the subfamily *Ursinae*, and are the brown bear, polar bear, Asiatic black bear, sloth bear, sun bear, and the American black bear. Panda bears belong in the subfamily Ailurpodinae, and Andean bears belong in the subfamily Tremarctinae. Mexico is represented by one genus, *Ursus*, and two species: the Mexican brown bear or grizzly (*Ursus arctos horribilis*) and the American black bear (*Ursus americanus*). The Mexican grizzly once occupied some parts of northern Mexico (complete range unknown) and was hunted until the late 1960s (Brown, 1985, Treviño and Jonkel, 1986). Tracks and other sign of grizzly bears were reported as late as the mid-1970s (Gallo-Reynoso et al., 2008; Treviño and Jonkel, 1986).

Ursus americanus Pallas, 1780

American black bear

Diana Doan-Crider and Oscar Moctezuma Orozco

SUBSPECIES IN MEXICO

Ursus americanus machetes Elliot, 1903
Ursus americanus eremicus Merriam, 1904
In Mexico, the two subspecies of *U. americanus eremicus* and *U. americanus machetes* have been reported. Current distribution maps (Baker and Greer, 1962; Hall, 1981; Leopold, 1959) are based on generalized reports, and genetic comparisons between the subspecies have not been conducted (Garshelis, 2009).

DESCRIPTION: The black bear is one of the largest carnivores in Mexico. The black bear has a straight facial profile, which distinguishes it from the grizzly bear's more dished appearance (Garshelis, 2009).The black bear does not have a distinguished hump on its back, and its claws are short (less than 50 mm in length). Body length varies from 120 cm to 190 cm, and the tail measures less than 12 mm. Color is generally black with a tan muzzle (sometimes called "Golondrino" in Mexico), but some color phases may include cinnamon or dark brown. In spring and early summer, pelage may be short, dry, and splotchy, indicating poor nutrition (Doan-Crider, 1995a) but will likely change to a more shiny, dark black after early autumn feeding begins.

EXTERNAL MEASURES AND WEIGHT

TL = 1,000 to 2,300 mm; TV = 100 to 130 mm; HF = 215 to 280 mm; EAR = 120 to 140 mm.
Weight: adult males: 60 to 225 kg; adult females: 40 to 180 kg. (Garshelis, 2009).

DENTAL FORMULA: I 3/3, C 1/1, PM 4/4, M 2/3 = 42.

NATURAL HISTORY AND ECOLOGY: Black bears are generalists and opportunists, and can readily adapt to a wide range of environments. Although bears are classified as carnivores, black bears are omnivorous in nature, and have a wide-ranging and adaptable diet throughout the year. Approximately 97% of their diet is composed of vegetation, but bears will also consume animal matter such as insects and carrion and prey on small animals. While foods may vary across regions, bears in Mexico normally consume large quantities of berries, nuts, and cacti (Baker, 1956; Herrero, 1985; Leopold, 1959). Bears have high energy requirements and must acquire as many calories as possible during spring, summer, and autumn (Hewitt and Doan-Crider, 2008). They hibernate during winter to avoid times of scarcity, so foods rich in lipids and carbohydrates (such as acorns) allow them to significantly increase their fat reserves, enabling them to survive during their winter sleep (Herrero, 1985; Hewitt and Doan-Crider, 2008). During spring, black bears will feed on grasses and on succulents such as yucca (*Yucca* spp.) and sotol (*Daylirion* spp.; Doan-Crider, 1995b; Hellgren, 1993). During drought in Coahuila, bears learned to prey on cattle, especially during the calving season (Hewitt and Doan-Crider, 2008). During summer, bears will consume berries and fruits such as chokecherry (*Prunus virginianus*), prickly pear fruit (*Opuntia* spp.), and wild grapes (*Vitis* spp.). In autumn, bears enter into what is known as "hyperphagia" and begin consuming large amounts of acorns (*Quercus* spp.), manzanita berries (*Arctostaphylos* spp.), Pinyon nuts (*Pinus* spp.), and madrone berries (*Arbutus* spp.). Bears have a highly sensitive sense of smell, and are readily attracted to non-natural food sources. They are easily "habituated" or conditioned to garbage, domestic animal feed, bird feeders, and other attractants, which, ultimately, may lead to increased mortality of bears (Masterson, 2006). In the Serranias del Burro, Coahuila, only pregnant females denned from December to April, and other bears remained semi-active throughout the winter (Doan-Crider, 1996, 2003). In subtropical regions of Florida, in the United States, they do not hibernate (McDaniel, 1979), which may be the case in the lower and warmer parts of the sierras of northern Mexico (Leopold, 1959). Bear dens in Mexico can be found in "nest-like" holes below trees, in shallow cavities in rocks, under ledges, and, in some cases, in deep caves (Doan-Crider, 1995, 2003). Bears acquire sexual maturity between 3 and 3.5 years of age, and even later, depending on the quality of habitat (Jonkel and Cowan, 1971). In Coahuila, 4.5 years

Ursus americanus. Pine-oak forest. Janos Biosphere Reserve, Chihuahua. Photo: Gerardo Ceballos.

DISTRIBUTION: The black bear inhabits the three countries of North America. Baker and Greer (1962), Hall (1981), and Leopold (1959) reported the historic range of the black bear in Mexico to include the states of Sonora, Chihuahua, Coahuila, Tamaulipas, Nuevo Leon, Durango, Sinaloa, and Nayarit. Most information pertaining to the black bear historic range is anecdotal, however, so true representation is unknown. Recent reports also include the states of Jalisco and Zacatecas (Hewitt and Doan-Crider, 2008). Prior to the 1970s, black bear populations were reported to be in serious decline in most of its range (Baker and Greer, 1962; Hall, 1981; Leopold, 1959). Recent governmental protection, landowner conservation programs, and changes in public attitude have resulted in a natural expansion and apparent recovery of the species in Coahuila and Nuevo Leon (Hewitt and Doan-Crider, 2008) and reports are increasing for other states as well. Doan-Crider and Hellgren (1996) and Doan-Crider (2003) reported very high densities of black bears in the Serranias del Burro, Coahuila, between 1991 and 2001.

was the average age of primiparity (Doan-Crider, 2003; Doan-Crider and Hellgren, 1996). The breeding period generally takes place between June and August in Mexico (Doan-Crider, 2003), but implantation does not take place until November–December. Females have delayed implantation, and while it was once hypothesized that implantation would not take place if fat reserves were insufficient, Hellgren (1998) found that cubs are born regardless, but then die and are consumed by the mother in the den. Birth and nursing of the cubs take place during the denning period. Cub production varies regionally from east to west in the United States and Canada (2.4 cubs/litter, 1.8 cubs/ litter, respectively; Beck, 1991). In Coahuila, Mexico, litter sizes ranged from 2.75 to 3.3 cubs/litter (Doan-Crider, 2003). No cubs survived to emerge from dens during a severe drought between 2000 and 2001. Cubs in Mexico are generally born in January (Leopold, 1959; Doan-Crider, 2003) blind and with very little hair, and weigh about 250 g. Cubs grow fast and upon den emergence in April may weigh more than 2 kg. Lactation may last up to 8 months in other areas of North America (Eiler et al., 1989). Lactation in Coahuila, Mexico, may last up to 10 months (D. Doan-Crider, unpubl. data). During the first 1.5 years, cubs remain with the mother, who cannot enter into estrus while nursing. During the second summer, cubs disperse from the mother, who enters into estrus again. Bears are solitary in nature, but come together to breed (Garshelis, 2009). Family groups have been observed in Coahuila, Mexico, but appear to be related bears (D. Doan-Crider, unpubl. data). Females generally establish their home ranges next or close to their mothers, but males will disperse away from natal areas. There is a clear hierarchy among black bears, which is established through direct interactions, and possibly through the marking of trees. Occupancy of areas by multiple bears likely shifts with availability of food resources (Rogers, 1977). In Minnesota, the activity area of a male overlapped with that of another male and enveloped between 7 and 15 territories of females (Rogers, 1987).

Black bears are generally diurnal, with activity peaking in early mornings and evenings (Garshelis, 2009). Bears may sometimes become nocturnal during autumn hyperphagia in order to increase food intake, or in human-populated areas in order to take advantage of non-natural foods such as garbage (Garshelis, 2009). Natural enemies for bears include mountain lions, wolves, or, most commonly, other bears. Intraspecific predation on black bears in Mexico is common, especially on cubs (Doan-Crider, 2003). Females will avoid adult male bears, and may select for habitats with sufficient cover, such as escape trees, to protect cubs from predation (Mollohan, 1987). Black bear densities vary from region to region, and are also dependent on temporal resource availability (Doan-Crider, 2003; Jonkel and Cowan, 1971). Beck (1991) published density estimates for 22 different black bear populations in North America, and ranged from 1.3 to 16.7 km^2/bear. Many estimates are obtained using different methodologies, are not always comparable, and contain many errors (Garshelis, 1992). In the Serranias del Burro, Mexico, densities were reported to be one of the highest in North America (Doan-Crider and Hellgren, 1996). Density estimation in desert ecosystems is precarious, however, due to the dynamic nature of black bears where food production is patchy, and temporal and spatial movement is highly variable (Hewitt and Doan-Crider, 2008). Current population estimates for Mexico are highly unreliable. Black bears occupy habitats ranging from low-lying deserts (McKinney and Pittman, 2000) to tropical rainforests (O. Rosas-Rosas, pers. comm.), to high montane elevations (Baker and Greer, 1962; Hall, 1981; Hewitt and Doan-Crider, 2008; Leopold, 1959). Food and water availability is likely a principal factor in habitat suitability (Hewitt and Doan-Crider, 2008), but very little is known about these driving mechanisms. Bear movements are correlated with seasonal food production, and can vary within elevations or even between mountain ranges. Prior to the 1970s, the black bear was heavily hunted. Law enforcement was sparse, and little management information was available due to lack of funding and trained personnel (Doan-Crider, 2008). Bears have commonly been seen as problematic in the agricultural industry, with both livestock and crops. Recent advances in conflict management, however, have resulted in increased coexistence between bears and agricultural producers (Hewitt and Doan-Crider, 2008), Currently, the black bear is protected in Mexico, but illegal hunting does persist in isolated areas (Ceballos and Navarro, 1991; SEMARNAT, 2010). Bears are coveted for their hide and meat. In addition, black market trade does exist for black bear gall bladders, whose bile is used in traditional eastern medicine. Since Asiatic black bear bile is the most valuable, pressure on American black bears does not appear to be significant (D. Garshelis, IUCN Bear Specialist Group, pers. comm.). In the United States and Canada, black bears are an attraction in many national parks, which contributes to the generation of significant economic revenues. Managing humans and bears in these environments is challenging because of bear behavior and their susceptibility to conflict with humans and conditioning to non-natural foods (Masterson, 2006), Costs associated with conflict management in national parks and for state agencies are significant. The most recommended approaches are preventative management and public education.

CONSERVATION STATUS: The level of direct mortality (uncontrolled hunting and poisoning) on bear populations was likely the primary factor affecting previous declines, and now, the long-term survival of the species. Because direct mortality levels have been significantly reduced in Mexico over the past 40 years, many bear populations appear to be recovering at rapid rates. Females are capable of producing between 20 and 30 cubs during a lifetime, so when direct mortality is reduced, numbers may rapidly increase (D. Garshelis, IUCN Bear Specialist Group, pers. comm.). In addition, despite the fact that many areas of Mexico are experiencing severe habitat fragmentation, bear populations are still able to expand and recolonize into new areas (Hewitt and Doan-Crider, 2008). Some subpopulations are still at risk, and require more focused management and conservation actions. Successful models throughout Mexico, however, indicate that if consistently applied, the species should recover throughout its range. Because Mexico is becoming more urban and less rural, bear populations appear to be benefiting from changing land-tenure patterns. Unlike the neighboring American state of Texas, where large ranches are rapidly being subdivided, large landholdings in northern Mexico continue to remain intact, thus allowing for large expanses of bear habitat. Conflicts are increasing in urban areas such as Monterrey, Nuevo Leon, and Saltillo, Coahuila, where human and bear habitats dramatically overlap and attractant management is lacking. Prevention and coexistence programs will help resolve conflicts in both urban and rural areas. In Mexico the black bear from Sierra El Burro, Coahuila is considered under special protection, while other populations are considered endangered (SEMARNAT, 2010). It is considered least concern by IUCN, with population trends increasing (IUCN, 2010).

Ursus arctos Linnaeus, 1758

Grizzly bear

Oscar Moctezuma Orozco and Gerardo Ceballos

SUBSPECIES IN MEXICO

Ursus arctos californicus Merrian, 1896

In the past, three subspecies for Mexico were recognized: *nelsoni* for Chihuahua; *magister* for Baja California; and *kennerleyi*, for Sonora (Merriam, 1918). Hall (1984), in a taxonomic revision of this species in North America, determined that based on the skulls of 2,478 specimens there are no differences among subspecies and that these obey geographic variations, therefore proposing nine subspecies for North America, leaving *nelsoni* and *kennerleyi* in *horribilis*, and *magister* in *californicus*.

DESCRIPTION: The grizzly bear was the largest carnivore in Mexico. Its head was large and bulky, with small ears and a high forehead, which gave it a concave look (Leopold, 1959). Adult animals had a prominent hump on the shoulders, which is absent on black bears (Leopold, 1959). The claws of the forefeet were also a distinctive feature, being very long (more than 80 mm) and slightly curved and of lighter coloration that those of the black bear, which are short, black, and hooked (Leopold, 1959). Their coat was brown, but could vary from pale gold to dark brown (Brown, 1985).

EXTERNAL MEASURES AND WEIGHT
TL = 1,400 to 2,200 mm; TV= 110 to 140 mm; HF = 230 to 300 mm; EAR= 130 to 150 mm.
Weight: 110 to 270 kg.

DENTAL FORMULA: I 3/3, C 1/1, PM 4/4, M 2/3 = 42.

NATURAL HISTORY AND ECOLOGY: The grizzly was commonly found in the ecotone between the grassland and oak (*Quercus* spp.) or juniper (*Juniperus* spp.) forest in hilly terrain at high altitudes. They used creeks and canyons at high altitudes, covered with pines and oaks (Brown, 1985). Their diet included acorns, nuts, pine nuts, berries, roots, tubers, various fruits like the manzanita (*Arctostaphylos*), or fruits of the prickly pear cactus (*Opuntia* spp.), insects, honey, small rodents, carrion, and occasionally cattle and other large animals (Brown, 1985; Leopold, 1959). To escape the stress of food scarcity during the winter, the grizzly bears hibernated in their dens. During the autumn, they ingested large quantities of food to increase their reserves of fat, used to survive during winter. In Mexico, they often dug their dens in the colder and most remote areas of the

Ursus arctos. One of the last Mexican grizzly bears. Sierra del Nido, Chihuahua, October 24, 1955. Photo courtesy of David E. Brown.

sierras, using them for several consecutive years, and sometimes using natural cavities (Brown, 1985; Craighead, 1979). In mid-November, most bears were already prepared to hibernate, emerging from their dens in late March or mid-April, depending on the winter (Leopold, 1959). Sexual maturity was reached at four or five years of age, but females reached maturity at an older age (Brown, 1985; Leopold, 1959). The reproductive period included a few weeks from the beginning of summer until mid-summer (Herrero and Hamer, 1977). Females had delayed implantation (Craighead, 1979): if the female had the adequate reserves of fat, the egg implanted in the wall of the uterus and gestation started. The cubs were born during the hibernation, in mid-January or February, and there were generally two cubs in the litter (Brown, 1985; Craighead et al., 1974; Leopold, 1959). For two years the mother cared for her cubs and entered in estrus while nursing her offspring (Brown, 1985). The cubs became independent at the age of 2 years but generally stayed together for 1 to 3 more years (when the litters had more than a cub) and completed their development at 8 or 10 years of life (Brown, 1985; Leopold, 1959). In the wild, bears living up to 25 years have been recorded, but more commonly less than 20 years (Brown, 1985; Craighead et al., 1974). The grizzly bear does not defend a territory, and usually the home ranges of several individuals overlap. The territories of the males are usually two to four times larger than those of females, and their size varies depending on the type of habitat and the characteristics of each bear (Brown, 1985). There are no data on the size of the home range of grizzly bears in Mexico, but an estimation from the northern Rocky Mountains of the United States, falls into an average over 46,620 ha for males and 25,000 ha for females, with a density of 1 bear for every 2,000 ha to 4,662 hectares (Brown, 1985; Craighead, 1979). Grizzlies are not a social animal; adults are solitary except during heat or when females are accompanied by cubs (Leopold, 1959; Murie, 1981); however, a complex hierarchy based on the strength and aggressiveness of each individual has been identified (Herrero, 1985). Grizzly bears avoid contact with humans, acquiring nocturnal habits at sites where this contact is inevitable and dangerous for the bear (Herrero, 1985); they establish resting sites where they hide during the day and well-defined trails to move to places with food and that serve as escape paths from potential dangers (Brown, 1985). They lack natural enemies. Only cubs are vulnerable to the attack by other animals; however, at this time the mothers protect them fiercely, thus virtually no predator attacks them. Their worst natural enemies are other bears because males frequently kill the offspring and sometimes the mothers as well (Herrero, 1985). When these large bears were common in northern Mexico, they were considered one of the most prized trophies because tracking and killing them implied great difficulty and danger; however, its habit of eating livestock carrion and occasionally killing calves was the factor that triggered an intense persecution against them by farmers, who used shotguns, traps, and poisons to kill them (Brown, 1985; Leopold, 1967). In other parts of the world, where these bears still live, they constitute a tourist attraction that generates significant amounts of economic resources, either through sport or photo hunting.

DISTRIBUTION: Originally, this species was distributed in temperate and cold areas of Europe, Asia, North America, and the mountain range of the Atlas, in the extreme northwest of Africa. In Mexico, its presence was documented in temperate zones of the northern region of the Sierra Madre Occidental and adjacent mountains such as the Sierra de San Luis and the Sierra de Juárez in northern Baja California (Leopold, 1959). It was recorded in the states of BC, CH, DU, and SO.

VEGETATIONAL ASSOCIATIONS AND ELEVATION RANGE: Grizzly bears inhabited conifer forests, pastures, and forests of oaks (Brown, 1985). They were found from 1,400 masl to 2,800 masl.

CONSERVATION STATUS: Grizzly bears are extirpated in Mexico (Brown, 1985; Ceballos and Navarro, 1991; Koford, 1969). The last confirmed grizzly was killed in Sierra del Nido, in the state of Chihuahua (Anderson, 1972; Brown, 1985; Leopold, 1959), but there is a much more recent record from a possible grizzly bear in Sonora in 1976 (Gallo-Reynoso et al., 2008). It is considered extinct in Mexico (SEMARNAT, 2010), and at a global level the status of the grizzly bear by the IUCN is least concern with population stable (IUCN, 2010), but extirpated from some locations, like northern Mexico and central and southern United States. It is listed in appendix I of CITES.

Family Otariidae

The family Otariidae is a little, diverse family, which has 14 species of wide distribution in the cool temperate seas around the globe (Wilson and Reeder, 2005). In Mexico it is represented by five species of three genera. The Guadalupe fur seal (*Arctocephalus townsendi*) almost became extinct a century ago, but its current population is reproducing in the Guadalupe Island and has several thousand individuals. The Galapagos sea lion (*Zalophus wollebaeki*) and the Galapagos fur seal (*A. galapagoensis*) have been recorded in both the Baja California and Chiapas coasts (Aureoles et al., 1993; Ceballos et al., 2010). The sea lion (*Zalophus californianus*) has never been at the edge of extinction; currently hundreds of thousands of individuals live in waters from the Gulf of California to the coast of Guerrero.

Arctocephalus galapagoensis Heller, 1904

Galapagos fur seal

Sandra Pompa Mansilla

SUBSPECIES IN MEXICO
A. galapagoensis is a monotypic species.

DESCRIPTION: The Galapagos fur seal is the smallest of otariids (eared Pinnipedia). They have a grayish-brown fur coat. The adult males of the species average 1.5 m in length and 64 kg in weight. The females average 1.2 m in length and 28 kg in weight. The Galapagos fur seal spends more time out of the water than almost any other seal. On average, 70% of their time is spent on land. Most seal species spend 50% of their time on land and 50% in the water. This small species exhibits less sexual dimorphism than other otariids, and is well adapted to equatorial climatic conditions (Reeves et al., 1992; Repenning et al., 1971; Trillmich 1984).

EXTERNAL MEASURES AND WEIGHT
LT = adult males to 1.6 m, adult females to 1.3 m.
Weight: males to 70 kg, females to 40 kg.

Arctocephalus galapagoensis.
Illustration: Pieter Folkens.

DENTAL FORMULA: I 3/2, C 1/1, PM 4/4, M 2/1 = 36.

DISTRIBUTION: The Galapagos fur seal is endemic to the Galapagos Islands, like most species found there, meaning they cannot be found anywhere else in the world (Reeves et al., 1992; Salazar, 2002). The Galapagos are a chain of islands found approximately 972 km west of Ecuador. Lactating females make trips of relatively short duration, suggesting they do not go far from their colonies. Foraging by males outside the breeding season is unknown. Most breeding colonies are located in the western and northern parts of the Archipelago, close to productive upwelling areas offshore. The Galapagos fur seal is now no longer only found on the Galapagos Islands; a colony has relocated to northern Peru, according to ORCA (Organisation for Research and Conservation of Aquatic Animals). Galapagos fur seals have also been recorded in the south of Colombia, with an occurrence pattern similar to that in Ecuador (Capella et al., 2002). Two other individuals of this species were recorded even farther north, in southwest Mexico, during the 1997-1998 El Niño (Aureoles-Gamboa et al., 2004).

NATURAL HISTORY AND ECOLOGY: These seals live on the rocky, usually western shores of the islands, leaving only to feed. Six to 10 Galapagos fur seals may occupy an area of about 100 m^2 (Nowak, 1999b). Grouping together in this manner may be largely due to the rarity of suitable rocky sites, but it also makes females less vulnerable to predation or harassment because they are in large groups (Bowen et al., 2002). Breeding males establish larger territories, around 200 m^2, which encompass a number of females (Nowak, 1999b). These seals do not migrate and remain near the islands their entire lives, which average about 20 years. The Galapagos fur seals live in large colonies on the rocky shores. These colonies are divided up into territories by the female seals during breeding season, which is Mid-August to mid-November. Every mother seal claims a territory for herself and raises her pups there. This seal has a longer nursing period than any other species of seal. Females have been known to nurse up to three-year-old pups, but usually it takes somewhere from one to two years before the pups leave and live on their own. The reason some pups take longer than others to develop is because when there is a shortage of food, which can be caused by El Niño, the mothers cannot properly nourish their young. El Niño has a devastating effect on the fur seals of the Galapagos. During El Niño years food supplies can drop to extremely low levels, causing many seals to die from starvation. This is because El Niño raises the temperatures of the waters around the Galapagos, causing the seals' food supply to migrate to cooler waters. During the nursing years, females will leave for up to four days at a time to forage for food. They then return to rest and feed the pup for just one day and then return back to the sea and repeat the process. While the mother is away, the pup must be careful because other female seals are extremely violent against pups that are not their own. Female seals will defend their territories to the death. They do not want to lose their area because they will not be able to breed then. If a pup wanders into another female's territory, the female sees this as a threat and will attack or may even kill the pup (Bowen et al., 2002; Nowak, 1999b). The Galapagos fur seal feeds primarily on fish and cephalopods. They feed relatively close to shore and near the surface, but have been seen at depths of 169 m (Arnould, 2002; Jefferson et al., 1993; Reijnders et al., 1993). They primarily feed at night because their prey is much easier to catch then. During normal years, food is relatively plentiful. During an El Niño year, however, there can be fierce competition for food and many young pups die during these years. The adult seals feed themselves before their young and during particularly rough El Niño years, most of the young seal populations will die. The Galapagos fur seal has virtually no constant predators to be wary of. Occasionally, sharks and orca whales have been seen feeding on the seals, but this is very rare. Sharks and orca whales are the main predators of most other seal species, but their migration path does not usually pass the Galapagos.

CONSERVATION STATUS: The Galapagos fur seals have had a declining population since the nineteenth century. Thousands of these seals were killed for their fur in the 1800s by poachers. Starting in 1959, Ecuador established strict laws to protect these animals. The government of Ecuador declared the Galapagos Islands a national park and since then no major poaching has occurred. Despite the laws, another tragic blow to their population occurred during the 1982–1983 El Niño weather event. Almost all of the seal pups died, and about 30% of the adult population was wiped out. The population is relatively stable now and is on the rise. Since 1983, no major calamity has occurred to decrease their population significantly. Fur seals were protected under Ecuadorian law in the 1930s, and since 1959 with the establishment of the Galapagos National Park, by the administration of the park. The waters around the islands are also protected by a 40 nautical mile no fishing zone. Tourism is regulated and most visitors are escorted by a trained park naturalist. *Arctocephalus galapagoensis* is listed in appendix II of CITES. It is considered endangered by IUCN (iucnredlist.org v. 3.1), and the Mexican Red List classifies it at risk of extinction (SEMARNAT, 2010).

Arctocephalus townsendi Merriam, 1897

Guadalupe fur seal

Alejandro Torres García

SUBSPECIES IN MEXICO

A. townsendi is a monotypic species.

DESCRIPTION: *Arctocephalus townsendi* is a medium-sized pinniped with remarkable sexual dimorphism. Six age categories can be differentiated: offspring, juveniles of phases 1 and 2, subadult males, breeding females, and adult males (Torres, 1991). The coat of males, which varies from dark brown to grayish-brown, has individual differences in both coloration and pattern. The body of the male is disproportionate; it has a corpulent thoracic region that becomes thinner toward the posterior region of the body. A distinctive feature of the species, and especially of males, is the presence of a very elongated muzzle with a pointed nose (Repenning et al., 1971). In females, the body has a more uniform shape with a more rounded head and less elongated snout. Females have a lighter coloration with more golden tones that vary according to the humidity.

EXTERNAL MEASURES AND WEIGHT

LT = adult males to 2.2 cm, adult females to 190 cm.
Weight: males to 170 kg, females to 55 kg.

DENTAL FORMULA: I 3/2, C 1/1, PM 4/4, M 2/1 = 36.

Artocephalus towsendi. Guadalupe Island off Baja California. Photo: Alejandro Torres.

NATURAL HISTORY AND ECOLOGY: Guadalupe fur seals inhabit the eastern coast of the Guadalupe Island, in rocky sites of lava formations at the bottom of cliffs. The island is volcanic, with an approximate area of 269 km^2 and a maximum elevation of 1,300 masl. It has a rugged topography with high cliffs and few sandy and rocky beaches. The reproductive cycle is annual. The breeding season takes place in summer when they congregate in large numbers along the east coast. These congregations are formed by territories guarded by a single male with a group of up to 11 females. The offspring are born between June and July, after a gestation period of 11 months. Females ovulate and copulate between 7 and 10 days after giving birth; delayed implantation occurs (King, 1983; Torres, 1990). Their diet is little known, but fish of the family Myctophidea and squid of the family Loliginidae have been identified. The Guadalupe fur seal is sympatric with other species of pinnipeds such as the common sea lion (*Zalophus californianus*) and the sea elephant (*Mirounga angustirostris*), but do not compete for space since the preference of substrate of these three species is very different.

DISTRIBUTION: The historical distribution of this species went from the Revillagigedo Islands, in Mexico, to California in the United States. Due to overexploitation, their distribution was restricted to a few individuals on the island of Guadalupe, Baja California (Fleischer, 1987; Hubbs, 1956). Almost all of the population is located on the Island of Guadalupe, but isolated animals were recorded in southern Baja California in 1981, 1985, 1986, and 1990 (Aguayo, 1990; Aurioles et al., 1993b), and sightings have been made in the coast of California (Stewart et al., 1987). It has been recorded in the state of BC and in the northern Pacific.

CONSERVATION STATUS: This species was intensively exploited in the last century for the commercial value of its skin (Scammon, 1874). According to an analysis of captures, it is estimated that the original population was of about 200,000 animals (Hubbs, 1956). In the Guadalupe Island, rocks polished by the movements of a large number of sea lions have been observed, estimating that up to 20,000 animals lived in those coasts. The latest captures were made in 1894, when the species was believed to be extinct. In 1926, 2 fishermen found a colony of 60 animals and caught them; again, the species was thought to have vanished. In 1954 a colony of 14 animals was found on the east coast of the Guadalupe Island, where the population continues to recover (Hubbs, 1956). A census conducted in 1988 recorded 3,259 individuals; in an analysis of the population growth in 1991 estimates exceeded 7,000 animals (Torres, 1991). A genetic study showed that Guadalupe fur seals have a high genetic variation but whether this variation is old or if it indicates equilibrium associated with bottlenecks is unknown (Bernardi et al., 1998). It is considered as near threatened by IUCN, with population trends increasing (IUCN, 2010). It is included in appendix I of CITES, and the Mexican Red List classifies it at risk of extinction (SEMARNAT, 2010).

Callorhinus ursinus (Linnaeus, 1758)

Northern fur seal

Sandra Pompa Mansilla

SUBSPECIES IN MEXICO

Callorhinus ursinus is a monotypic species.

DESCRIPTION: Northern fur seals have extreme sexual dimorphism, with males being 30% to 40% longer and more than 4.5 times heavier than adult females. Adult males are stocky in build, and have an enlarged neck that is thick and wide and a very short snout. A mane of coarse longer guard hairs extends from the top of the head to the shoulders and covers the nape, neck, chest, and upper back. While the skull of adult males is large and robust for their overall size, the head appears short because of the combination of a short muzzle and the back of the head behind the ear pinnae being obscured by the enlarged neck. Adult males have an abrupt forehead formed by the elevation of the crown from the sagittal crest and thicker fur on the top of the head. Adult females, subadults, and juveniles are moderate in build. It is difficult to distinguish the sexes until about age 5. They have 10 pairs of teeth in the upper jaw and 8 pairs in the lower jaw. The hind flippers are the largest of any otariid pinniped, with the tips of the digits particularly long and extending well beyond the small claws. Pups are blackish at birth, with variable oval areas of buff on the sides, in the axillary area, and on the chin and sides of the muzzle. After three to four months, pups molt to the color of adult females and subadults.

DISTRIBUTION: Northern fur seals are a widely distributed pelagic species in the waters of the North Pacific Ocean and the adjacent Bering Sea, Sea of Okhotsk, and Sea of Japan. They range from northern Baja California, Mexico, north and offshore across the North Pacific to northern Honshu, Japan. Their range is nearly the same as that of the Steller sea lion, another eared seal in the family *Otariidae*, which is larger than the northern fur seal, but similar in many respects. Its range also overlaps slightly that of its close relative, the Guadalupe fur seal, which is found along the Baja California coast, and especially on Guadalupe Island of Mexico. The southern limit of their distribution at sea is approximately 35 degrees north. Vagrants reach the Yellow Sea in the west and eastern Beaufort Sea in the Arctic. The vast majority of the population breeds on the Pribilof Islands, with substantial numbers on the Commander Islands as well. Still other sites are used, including San Miguel Island in California, Bogoslof Island in the Bering Sea, and Robben Island off Sakhalin Island in Russia. Many animals, especially juveniles, migrate from the Bering Sea south to southern California or the waters off Japan, to spend the winter feeding (IUCN, 2010). In November 2008, a northern fur seal was registered off the coast of Colima, Mexico (Ceballos et al., 2010).

EXTERNAL MEASURES AND WEIGHT

TL = adult males to 2.1 m, adult females to 1.5.
Weight: males to 270 kg, females to 60 kg.

DENTAL FORMULA: I 3/2, C 1/1, PM 4/4, M 2/1 = 36.

NATURAL HISTORY AND ECOLOGY: The distribution of the northern fur seal is limited to the North Pacific Ocean, the Bering Sea, and the Sea of Okhotsk; primary breeding colonies are in the Bering Sea. Adult males begin leaving colonies in September, and most are thought to remain in the Bering Sea and North Pacific in winter and early spring. Females return to sea by late October, migrating into the central North Pacific and south along the California coast (Aurioles-Gamboa et al., 1993; Nowak, 1999b) to feed until spring. Fur seals are opportunistic feeders, primarily feeding on pelagic fish and squid, depending on local availability. Identified fish prey include hake, anchovy, herring, sand lance, capelin, pollock, mackerel, and smelt. Their feeding behavior is primarily solitary. Northern fur seals are preyed on primarily by sharks and orcas. Occasionally, very young animals will be eaten by Steller sea lions. Sporadic predation on live pups by arctic foxes has also been observed. They form relatively dense aggregations during the breeding season; due to very high densities of pups on reproductive rookeries and the early age at which mothers begin their foraging trips, mortality can be relatively high. Consequently, pup carcasses are important in enriching the diet of many scavengers, in particular, gulls and arctic foxes. Breeding seals mostly occupy rocky beaches, although young males may move several hundreds of meters inland to rest and spar.

CONSERVATION STATUS: Recently there has been increased concern about the status of fur seal populations, particularly in the Pribilof Islands, where there has been a roughly 50% decrease in pup production since the 1970s and a continuing drop in pup production of about 6% to 7% per year. This has caused them to be listed as vulnerable under the U.S. Endangered Species Act and has led to an intensified research program into their behavioral and foraging ecology. The species is currently not included in CITES and is not included in the Mexican Official Regulation (SEMARNAT, 2010). It is protected in Mexico, as all marine mammals inhabiting its waters are, in frameworks such as the ones in the Federal Law on Fisheries (Diario Oficial de la Federación, 2002).

California sea lion

David Aurioles Gamboa and María del Carmen García Rivas

SUBSPECIES IN MEXICO

Zalophus californianus californianus (Lesson, 1828)

DESCRIPTION: The coloration of *Zalophus californianus* varies depending on the sex and age. The offspring and individuals under one year of age are steel gray; juveniles of both sexes and adult females are light brown, grayish-brown, and cream. The adult and subadult males can be dark brown, dark gray, or black. A distinctive feature of males is the presence of a sagittal crest that in the live animal is reflected as a bump of lighter coloration than the rest of the body (Peterson and Bartholomew, 1967). The forelegs are relatively large and its fur extends distally to the wrist through the dorsal part (Reeves et al., 1992).

EXTERNAL MEASURES AND WEIGHT

TL = adult males to 2.4 m, adult females to 2.0 m.

Weight: males to 390 kg, females to 110 kg.

DENTAL FORMULA: I 3/2, C 1/1, PM 4/4, M 2-1/1 = 35-36.

NATURAL HISTORY AND ECOLOGY: The diet of the California sea lion consists of diverse fish and cephalopods, while crustaceans are of minor importance. Because the California sea lion is a coastal predator and some of its prey have commercial interest, it is common to find fisheries-sea lion interactions along the entire geographic distribution. The feeding spectrum varies seasonally and spatially, but they show preference for some pelagic fish such as anchovies (*Engraulis mordax*, *Cetengraulis misticetus*), sardines (*Sardinops sagax*), herring (*Opisthonema*), mackerels (*Scomber japonicus*), and hake (*Merluccius productus*, *M. angustimanus*). Among the cephalopods, squid (*Loligo opalescens*, *Dosidicus gigas*) and octopus are common prey (Antonelis et al., 1984; Garcia and Aurioles, 2004; Lowry et al., 1991). In some regions, like the southern Gulf of California, the diet is based on deep-sea fish (Aurioles et al., 1984). The sea lions are highly gregarious, gathering in sites protected from terrestrial predators and forming dense herds or colonies organized on a polygynous breeding system (Peterson and Bartholomew, 1967). Both sexes reach sexual maturity around 5 years of age, but males do not reproduce until they reach 9 or 10 years old (Orr et al., 1970). Lifespan for females is up to 25 years whereas for males it is 19 years (Hernandez-Camacho et al., 2008). The breeding season takes place in June, July, and part of August; the reproductive period begins with the establishment of territories by males (Peterson and Bartholomew, 1967). The breeding season in the Gulf of California is longer (13 weeks) than in California (9.5 weeks; Garcia-Aguilar and Aurioles-Gamboa, 2004). Territorial defense in these animals is the longest recorded in any pinniped (63 days); adult males perform vocalizations and physical exhibitions (Garcia-Aguilar and Aurioles-Gamboa, 2004). Females give birth, and after three weeks they copulate again. In the Gulf of California 97 % of the mating takes the place in the water (Garcia-Aguilar and Aurioles-Gamboa, 2004). Females breed once year; the gestation period lasts 11 months and they can be impregnated during lactation (King, 1983). The adults of both sexes are more abundant on the islands during the summer. After breeding, adult males leave their rookeries to undertake long migrations; the ones living in the Pacific coast go north (Peterson and Bartholomew, 1967) and the ones from the Gulf of California go south (Aurioles et al., 1983). Females and their young normally stay

DISTRIBUTION: Three subspecies are recognized; these inhabit Japan and nearby islands (*Z. californianus japonicus*), the Galapagos Islands (*Z. californianus wollebaeki*), and the area of California and Baja California (*Z. californianus californianus*) (Rice, 1998). Recent skull morphological and genetic analyses, however, suggest that *Z. californianus* and *Z. wollebaeki* are valid separate species (Schramm et al., 2009; Wolf et al., 2008). The California sea lion is mainly found in islands and islets from the border with the United States in the Pacific coast to Bahia de Banderas in Jalisco, including the Gulf of California (Aurioles, 1988; Le Boeuf et al., 1983; Zavala, 1990). There are occasional records farther south of this region (Gallo and Ortega, 1986), which are not considered significant for their normal distribution. The sea lion is typically found along the coast, on the continental shelf; the Island of Guadalupe is the most oceanic region in Mexico it inhabits. In the Gulf of California, the distribution has a clear relationship with the abundance of food (Aurioles-Gamboa and Zavala-González, 1994; Porras-Peters et al., 2008). Several studies indicate that the Gulf of California sea lion population is isolated from its remaining distribution (Gonzalez-Suarez et al., 2009; Maldonado et al., 1995; Schramm et al., 2009). It has been recorded in the areas of GC, NP, and TP.

Zalophus californianus. Espíritu Santo Island off Baja California Sur. Photo: Gerardo Ceballos.

throughout the year in their natal areas along with some juveniles. Subadult males arrive at both coasts of the southern Baja California Peninsula during the winter (Aurioles, 1988; Aurioles et al., 1983). Among its predators are killer whales (*Orcinus orca*) and several species of sharks (King, 1983). Its most common predator in the Gulf of California is the bull shark (*Carcharinus leucas*). Seagulls, coyotes, and dogs can kill the offspring (Aurioles and Llinas, 1987; Sánchez, 1987).

CONSERVATION STATUS: The sea lion has been exploited in Mexico since prehistoric times by different cultures, most notoriously the Seri. This species was subject to commercial exploitation in Mexico from the nineteenth century to 1969 (Lluch, 1969), but there is no evidence that the population is at critical levels or about its genetic variability suggesting vulnerability. The California sea lion population on the west side of the Baja California Peninsula is estimated to be 75,000 to 87,000 (Lowry and Maravilla, 2005) whereas the population of the Gulf of California is close to 30,000, with some colonies in the central Gulf declining approximately 35% in the last 15 years (Szteren et al., 2006). In Mexico, this species has been protected since 1982 and the Mexican Red List considers them in the category of special protection (SEMARNAT, 2010). It is not included in the CITES, and it is considered least concern by IUCN, with population trends increasing (IUCN, 2010).

Zalophus wollebaeki (Sivertsen, 1953)

Galapagos sea lion

Sandra Pompa Mansilla

SUBSPECIES IN MEXICO

Z. wollebaeki is a monotypic species.

DESCRIPTION: The Galapagos sea lion is found in the Galapagos Archipelago where it is one of the most conspicuous and numerous marine mammals. Adult males tend to have a thicker, more robust neck, chest, and shoulders in comparison to their slender abdomen. Females are somewhat opposite males with a longer, more slender neck and thick torso. Once sexually mature, a male's sagittal crest enlarges, forming a small, characteristic bump-like projection on the forehead. Galapagos sea lions, compared to California sea lions, have a slightly smaller sagittal crest and a shorter muzzle. Adult females and juveniles lack this physical characteristic altogether, with a nearly flat head and little or no forehead. Both male and female sea lions have a pointy, whiskered nose and somewhat long, narrow muzzle. It swims using its strong and well-developed fore flippers. Adult males are much larger than females and are brown in color while females are a lighter tan. Adult males are also distinguished by their raised foreheads, and the hair on the crest may be a lighter color (Nowak, 1999b). Juveniles are chestnut brown in color and measure around 75 cm at birth.

DISTRIBUTION: Galapagos sea lions can be found on each of the different islands of the Galapagos Archipelago. They have also colonized just offshore of mainland Ecuador at Isla de la Plata, and can be spotted from the Ecuadorian coast north to Isla Gorgona in Colombia. Records have also been made of sightings on Isla del Coco, which is about 500 km southwest of Costa Rica (Macdonald, 2001). The species has been recorded twice in Mexico, first off Baja California (Aurioles Gamboa et al., 1993) and then registered off the coasts of Chiapas, Mexico (Ceballos et al., 2010).

EXTERNAL MEASURES AND WEIGHT
TL: males to 2.5 m, females to 1.5 m.
Weight: males to 300 kg, females to 100 kg.

DENTAL FORMULA: I 3/2, C 1/1, PM 4/4, M 1/1 = 34.

NATURAL HISTORY AND ECOLOGY: The Galapagos sea lion is essentially a coastal animal and is rarely found more than 16 km out to sea (Nowak, 1999b). Individuals are active during the day and hunt in relatively shallow waters (up to about 200 m deep) where they feed on fish, octopus, and crustaceans. Sea lions and seals are also capable of making extraordinarily deep dives of up to 200 m for 20 minutes or more, then rapidly surfacing with no ill effects. On land this sea lion prefers sandy or rocky flat beaches where there is vegetation for shade, tide pools to keep cool, and good access to calm waters. It also spends much of its time in the cool, fish-rich waters that surround the Galapagos Islands. They are extremely gregarious and pack together on the shore even when space is available (Nowak, 1999b). Each colony is dominated by one bull that aggressively defends his territory from invading bachelor males. This territorial activity occurs throughout the year and males hold their territories for only 27 days or so before being displaced by another male (Nowak, 1999b). The breeding season is not dependent on migration patterns, as seen in other sea lion species, since the Galapagos sea lion remains around the Galapagos Archipelago all year round. In fact the breeding season is thought to vary from year to year in its onset and duration, though it usually lasts 16 to 40 weeks between June and December (Nowak, 1999b). Like other sea lions this species relies on cooperation within the group. They are careful to keep the young pups out of deep water where they may be eaten by sharks.

CONSERVATION STATUS: The majority of the Galapagos sea lion population is protected, as the islands are a part of the Ecuadorian National Park surrounded by a marine resources reserve. Although the Galapagos Islands are a popular tourist destination, there are strict rules protecting all wildlife from disturbance. Fluctuating between 20,000 and 50,000 sea lions, the population does have a few threatening factors. During El Niño events, the population tends to decrease due to die-offs, cessation of reproduction, and collapses in marine life the seals are dependent on. Sharks are the main predator to the sea lion, and killer whales are presumed to be another predator as well. In Mexico, *Zalophus wollebaeki* is not included in the Mexican Official Regulation (SEMARNAT, 2010) and, as all marine mammals inhabiting Mexican waters, this species is protected in frameworks such as the article 420-I of the Federal Penal Code (Diario Oficial de la Federación, 2009). *Z. wollebaeki* is not included in CITES and considered endangered by IUCN (2010).

Family Phocidae

The family Phocidae comprises 19 species of seals and elephant seals, distributed in tropical and temperate seas around the world (Wilson and Reeder, 2005). In Mexico, only three species of three genera are represented. The elephant seal (*Mirounga angustirostris*), distributed in Baja California, was almost extinguished at the end of the nineteenth century, but currently its population exceeds thousands of individuals. The common seal (*Phoca vitulina*) has a marginal distribution in Mexico, and is only found in Baja California. Finally, the monk seal (*Monachus tropicalis*) inhabited the Caribbean sea and became extinct in the 1950s.

Mirounga angustirostris (Gill, 1866)

Northern elephant seal

María del Carmen García Rivas and Juan Pablo Gallo R.

SUBSPECIES IN MEXICO
M. angustirostris is a monotypic species in the Northern Hemisphere.

DESCRIPTION: *Mirounga angustirostris* is the largest phocid, with remarkable sexual dimorphism; males are up to 10 times larger than females. The adult males have a unique feature, the proboscis, and a very developed neck with thick, rough, wrinkled skin. The pelage coloration in adults varies from gray to brown; newborns are black and molt between the fourth and sixth week after birth to a silver (dorsal) and gray (ventral) pelage color. Molting occurs annually and involves the epidermis (Le Boeuf and Laws, 1994).

DISTRIBUTION: Elephant seals inhabit islands and coasts from Oregon to Baja California Sur. In Mexico, its distribution is on the islands of the Pacific off Baja California Peninsula and it is a visitor to some islands in the Gulf of California (Granito San Pedro Mártir, Islotes, and some other islands). During breeding season in Mexican islands, elephant seals are found at Islands Coronados, San Martín, Guadalupe, San Benito, Cedros, and Natividad (Aurioles et al., 1993b; Mesnick et al., 1998; Stewart et al., 1994). It has been recorded in the areas of GC and NP.

EXTERNAL MEASURES AND WEIGHT
TL= adult males to 4.3 m, adult females to 3.0 m.
Weight: males to 2,000 kg, females to 600 kg.

DENTAL FORMULA: I 2/1, C 1/1, PM 4/4, M 1/1 = 30.

NATURAL HISTORY AND ECOLOGY: Northern elephant seals have amphibian and gregarious habits. During reproduction (December–March) they prefer fine sand beaches where moving and thermoregulation are achieved easily (Odell, 1974). Elephant seals are polygamous, forming harems where a male can defend up to 100 females through the frequent exhibition of aggressive moves that involve jostling, onslaughts, and bites (Le Boeuf and Peterson, 1969). Females reach sexual maturity at 4 years of age and are active for 15 years on average; males become sexually mature at 6 to 7 years of age and are active for 9 to 10 years (Le Boeuf and Reiter, 1988). Females give birth to a single pup each year. The nursing of the pup has a duration of 27 days; adoption of foster pups is frequent (Le Boeuf et al., 1989). Elephant seals fast throughout the breeding season; in February they abandon the beaches to start a migration in search of food, usually to the Gulf of Alaska. Their diet consists mainly of squid and octopus, but fish, sharks, and rays are also consumed (Reeves et al., 1992). During foraging migrations, diving is a common activity performed continuously;

Mirounga angustirostris. Guadalupe Island off Baja California. Photo: Alejandro Torres.

the depth and time spent underwater vary depending on age and sex. Females dive at depths between 370 m and 480 m for 13 to 22 minutes. The deepest dive reported for a male was at a depth of 1,581 m (Le Boeuf, 1994). The main cause of death of the offspring is starvation as a result of separation from its mother. Reported predators include white sharks (*Carcharodon charcharias*) and killer whales (*Orcinus orca*; King, 1964).

CONSERVATION STATUS: Due to excessive hunting, this species was considered extinct in 1880. In 1892, however, a few specimens were found at Isla de Guadalupe. In 1911, the Mexican government prohibited their capture. The protection of this population led to exponential growth that in 1991 reached a population numbering 115,000 animals (Reeves et al., 1992). Today, the elephant seal is considered a threatened species according to the Mexican Red List (SEMARNAT, 2010). In the United States, the species is protected by the Marine Mammal Protection Act (Reeves et al., 1992). This species was erased off the lists of CITES in 1992, and it status according to the IUCN is least concern, with population trends increasing (IUCN, 2010).

Caribbean monk seal

María de Jesús Vázquez Cuevas and Luis Medrano González

SUBSPECIES IN MEXICO
M. tropicalis was a monotypic species.

DISTRIBUTION: The Caribbean monk seal was very abundant in the Gulf of Mexico and the Caribbean Sea. It was distributed from the Bahamas Islands to the Yucatan Peninsula and in the Caribbean coasts of Central America and the Antilles. In Mexico, monk seals were found in the coast of Campeche, Yucatan, and Quintana Roo (Kenyon, 1977; King, 1983; Reeves et al. 1992). They were recorded in the Gulf of Mexico and Caribbean Sea.

DESCRIPTION: Biological information about *Monachus tropicalis* is scarce and incomplete. The appearance of this species was similar to that of other monk seals, this is, with a relatively slender body. Males were slightly larger than females. The offspring were born with a size close to 1 m and covered with a soft black pelage. The dorsal coloration in adults was grayish-brown or black, dimming on the flanks to a beige color on the ventral region. These seals had light green tones due to algae growing on their skin. Their whiskers were white (King, 1953, 1983; Reeves et al., 1992; Ward, 1887; Ximénez, 1722).

EXTERNAL MEASURES AND WEIGHT
TL= 2.0 to 2.4 m.
Weight: not known though we have estimated around 200 kg based on an alometric analysis of monachids.

DENTAL FORMULA: I 2/2, C 1/1, PM 4/4, M 1/1 = 32.

NATURAL HISTORY AND ECOLOGY: The Caribbean monk seals formed large groups in sandy or rocky beaches (Allen, 1887; Díaz del Castillo, 1521; Ximénez, 1722). It is believed that these seals fed on fish (King, 1983). Births occurred in early December (King, 1953, 1983). These animals spent most of their time on the beaches sleeping and vocalizing (Allen, 1887; Díaz del Castillo, 1521; Vilchez, 1978; Ward, 1887); they did not fear the human presence but, with the excessive hunting after the Spanish conquest, their behavior became more elusive though ineffective (Ward, 1887).

Monachus tropicalis. Specimen in captivity. New York Zoo. Photo: courtesy of Wildlife Conservation Society, New York.

CONSERVATION STATUS: The monk seal was hunted since pre-Columbian times. Intense hunts started with the arrival of Columbus to the Americas and Spanish and English colonizers, who overexploited the seal population for four centuries to obtain oil, skin, and meat (Allen, 1887; Díaz del Castillo, 1521; King, 1983; Vílchez, 1978; Villa et al., 1986; Ximénez, 1722). At the end of the nineteenth century and first half of the twentieth century, fishermen eliminated most of these seals as they represented competition (Kenyon, 1977). The overexploitation that Caribbean monk seals suffered during the colonial times in combination with the destruction of their habitats during the twentieth century slowly drove this species to extinction. No scientific reports have documented the existence of these animals since 1952 (Kenyon, 1977; Villa et al., 1986). Its disappearance represents the latest recorded extinction of a marine mammal. The Caribbean monk seal is not included in the list of protected species (SEMARNAT, 2010); since 1975 it has been included in appendix I of CITES, and the status on the IUCN is extinct (IUCN, 2010).

Phoca vitulina Linnaeus, 1758

Common seal

Alejandro Torres García

SUBSPECIES IN MEXICO
Phoca vitulina richardsi (Gray, 1864)

DESCRIPTION: *Phoca vitulina* is a small-sized pinniped with little sexual dimorphism. It has variation in both the spots' pattern and hair color, which varies from grayish-white to dark brown. The body has numerous spots that rarely merge and that are surrounded by light rings (Bigg, 1981; King, 1983). It is a chubby animal that can be distinguished by its rounded head and short nasal region that results in a concave profile (Padilla, 1990).

EXTERNAL MEASURES AND WEIGHT
TL= adult males 150 to 190 cm, adult females up to 170 cm.
Weight: adult males 73 to 170 kg, adult females 59 to 130 kg.

DENTAL FORMULA: I 3/2, C 1/1, PM 4/4, M 1/1 = 34.

NATURAL HISTORY AND ECOLOGY: In Mexico, the reproductive season occurs from January to April. Common seals form numerous congregations in the rocky beaches and difficult sites of the coasts where they can share sites with the common sea lion (*Zalophus californianus*) and the elephant seal (*Mirounga angustirostris*; Gallo and Aurioles, 1984). This species is not migratory and is found throughout the year near the congregation sites. They reproduce annually; births occur at the beginning of February. Around six weeks after giving birth, females copulate. The gestation period lasts between 10 and 11 months, including the period of delayed implantation (Bonner, 1979). There is no information on the diet of this species in Mexico, but in studies conducted farther north, they consume a wide variety of fish and shellfish, some of commercial importance such as the sole, herring, and hake (Antonelis and Fiscus, 1980). On the coast of the United States, seals show a marked division of their

DISTRIBUTION: Common seals inhabit both the northern Pacific and northern Atlantic. The subspecies *P. vitulina richardsi* is distributed on the northeastern coast of the Pacific, from the Bering Sea, the Gulf of Alaska, and the coasts of Canada and the United States, having its most southern distribution in the Pacific of northern Mexico (Scheffer, 1958; Shaughnessy and Fay, 1977). In Mexico, some colonies have been found in sites of the coast like San Sebastián Vizcaíno, Negro Punta Banda, and Bay of San Quintín (Gallo and Aurioles, 1984; Padilla, 1990). The insular distribution includes the Coronado Islands, Todos los Santos, San Martín, San Jerónimo, San Benito, Cedros, and the southernmost location in the Navidad Island (Mate, 1977; Padilla, 1990). Occasional sightings have been made farther south at the Creciente Island or within the Gulf of California at the sites known as Los Frailes and Los Islotes (Aurioles et al., 1993b; Gallo and Aurioles, 1984). It has been recorded in the areas of GC and NP.

populations, in which isolation by distance seems to be a result of the dispersion of seals associated with the melting of glaciers (Lamont et al., 1996).

CONSERVATION STATUS: The common seal has been exploited for several centuries for its skin and fat. Its populations are recovering in the areas where they are protected, and in others they are used as food (Newby, 1973; Reeves et al., 1992). This species is vulnerable to urban and industrial pollution, so they can act as indicators of environmental disturbances (Padilla, 1990). Censuses conducted between 1982 and 1986 in Mexico estimated a population of 1,715 seals (Padilla, 1990). It is not listed as an endangered species in CITES or IUCN (Thornback and Jenkins, 1982); the Mexican Red List considers them under special protection (SEMARNAT, 2010). It is considered as least concern by IUCN, with population trends stable (IUCN, 2010). The marked population structure found in the United States (Lamont et al., 1996) suggests that the common seal in Mexico may constitute an independent management unit, and as this is a species with a very restricted distribution in Mexico, it is affected by tourism and fisheries. The common seal should be considered as vulnerable in Mexico (Torres, 1994).

Phoca vitulina. Pacific Ocean off Baja California. Photo: Gerardo Ceballos.

Family Mustelidae

The family Mustelidae groups 55 species of weasels, otters, tayras, and grisons (Wilson and Reeder, 2005). Until recently, this family included skunks (family Mephitidae). They are widely distributed in America, Asia, and Africa. Mexico is represented by eight species of the genera *Eira, Galictis, Mustela, Enhydra, Taxidea,* and *Lontra.* They can be divided into species of wide distribution in Mexico such as the weasel (*Mustela frenata*) and southern river otter (*Lontra annectens*), and restricted species such as the tayra (*Eira barbara*), grison (*Galictis vittata*), and Canadian otter (*Lontra canadensis*). The sea otter (*Enhydra lutris*), once extirpated from the shores of the Pacific of Baja California, is returning to Mexico.

Subfamily Lutrinae

The subfamily Lutrinae contains seven extant genera of otters, with thirteen species. Two of the genera and three species are known from Mexico.

Enhydra lutris (Linnaeus, 1758)

Sea otter

Juan Pablo Gallo R.

SUBSPECIES IN MEXICO
Enhydra lutris nereis Merriam, 1904

DESCRIPTION: *Enhydra lutris* is a large otter with a long and robust body. The width of the neck is smaller than the width of the skull. The head is rounded. The anterior limbs are short without interdigital membranes. The posterior limbs are long with large interdigital membranes. The tail is short and cylindrical in shape; the pinna of the ear is small. The muzzle is short but wide; the teeth are very different from those of other species of otters, with flat areas specialized for crushing shells and carapaces and lacking sharp cusps. The coloration of the fur is dark brown to reddish-brown. The head is creamy white extending to the middle of the throat and neck (Hall, 1981).

DISTRIBUTION: The sea otter is found in the North Pacific coast, from Alaska to Canada (*E. lutris*) and from Washington to Baja California (*E. lutris nereis*). It has been registered in the states of BC and BS.

EXTERNAL MEASURES AND WEIGHT
TL = 675 to 1,630 mm; TV = 273 to 360 mm; HF = 220 to 222 mm.
Weight: 19 to 34 kg.

DENTAL FORMULA: I 3/2, C 1/1, PM 3/3, M 1/2 = 32.

NATURAL HISTORY AND ECOLOGY: Sea otters are completely aquatic; they do not require dens for refuge or giving birth. They feed mainly on sea urchins, clams, abalone, fish, shellfish, octopus, and occasionally birds. They capture their prey at depths of up to 20 m. The California population of sea otters reproduces throughout the year; females give birth to a single young after a gestation period of four months. Females give birth every two years, although there is evidence that they can give birth each year. They can live up to 15 or 20 years.

Enhydra lutris. Monterey Bay, California. Photo: Gerardo Ceballos.

VEGETATIONAL ASSOCIATIONS AND ELEVATION RANGE: *E. lutris* is a marine species. It inhabits areas associated with giant kelp (*Macrocystis pyrifera*).

CONSERVATION STATUS: This species is considered at risk of extinction (SEMARNAT, 2010) it is protected by the General Law on Wildlife. It is included in appendix I of CITES as a species in danger of extinction. It is widely protected in the United States, where it is considered threatened. In Mexico, their populations disappeared due to overharvest but at present the species is recovering (Ceballos and Navarro, 1991). The last surveys indicate that the species is present in Mexico, perhaps due to the expansion of the population in California (Gallo-Reynoso and Rathbun, 1997); births have been reported in the coast of the Baja California Peninsula (Rodríguez Jaramillo and Gendron, 1996). The El Vizcaíno Biosphere Reserve, in the Baja California Peninsula, is the ideal place to promote its conservation (Gallo-Reynoso and Rathbun, 1997).

Lontra canadensis (Schreber, 1777)

Northern river otter

Juan Pablo Gallo R. and Ma. Antonieta Casariego

SUBSPECIES IN MEXICO

Lontra canadensis lataxina Cuvier, 1823

Lontra canadensis sonorae Rhoads, 1898

Some authors believe that the otters of the New World should be *Lutra* (Van Zyll de Jong, 1972; Wozencraft, 1993); however, such designation is no longer used because it minimizes the close relationship between the Old and New World otters (Davis, 1978).

DESCRIPTION: *Lontra canadensis* is a medium-sized otter. The body is slim with a cylindrical trunk; the width of the neck is equal to the width of the skull; the head is flat and rounded; the legs are short with interdigital membranes. The tail is wide at the base and cylindrical in transverse view; the pinna of the ear is small. The muzzle is short but wide. The fur is short and dense. The color of the back is beige to dark brown; the ventral area is grayish-brown but changes to cream color in the throat. The color is similar in both sexes and age classes. An important feature is that the soles of the feet have hair (Hall, 1981).

DISTRIBUTION: This species is found from Alaska, Canada, and the United States to northern Mexico. There are only four records of it in Mexico, all in the large tributaries of the Rio Grande (*L. canandensis lataxina*) to its delta in the Gulf of Mexico and in the delta of the Colorado River (*L. canandensis sonorae*) in the Gulf of California. It has been recorded in the states of BC, CO, NL, SO, and TA.

EXTERNAL MEASURES AND WEIGHT
TL = 1,000 to 1,530 mm; TV = 300 to 507 mm; HF = 100 to 146; EAR = 25 to 40 mm. Weight: 7 to 12 kg.

DENTAL FORMULA: I 3/3, C 1/1, PM 4/3, M 1/2 = 36.

NATURAL HISTORY AND ECOLOGY: This species is found in rivers, streams, lakes, dams, and irrigation canals. In the regions with severe winters, they move to inner valleys with a milder climate during this season (Melquist and Hornocker, 1983). Their dens are close to the water and have one access through the water and another by land, located among vegetation, fallen logs, and beaver dams. The population size varies according to the habitat, from 0.28 individual/km to 0.85 individual/km of habitat (Bowyer et al., 1995; Testa et al., 1994). They feed on fish, crustaceans, amphibians, reptiles, rodents, birds, and insects (Toweill and Tabor, 1982). Reproduction takes place during the spring, with delayed implantation in winter or early spring. Pups are born between January and May, depending on the latitude, and after a gestation period of 60 to 65 days. Litter size varies from two to four pups, which remain with the mother until reaching sexual maturity (two years of age in females and later in males). The few reports indicate that they live 10 to 15 years in the wild and up to 25 years in captivity (Melquist and Dronkert, 1987). The home range is about 155 km^2, but varies in relation to food and den availability, being larger in mountainous areas and smaller in flatter areas like coastal zones, where food and refuges are more

Lontra canadensis. Specimen in captivity, Arizona-Sonora Desert Museum, Arizona. Photo: Gerardo Ceballos.

evenly distributed. They frequently travel distances up to 96 km within their territory along rivers and lakes. Males can travel up to 16 km in one night. Their main potential enemies are carnivores, but in the coastal zones they are also preyed on by orcas (*Orcinus orca*) and bald eagles (*Haliaeetus leucocephalus*) (Hill, 1994; Melquist and Dronkert, 1987). *L. canadensis* is hunted for its fur, possibly to the extent that it affects its population. For example, in Louisiana the population has decreased dramatically in the last 30 years; during 1976 and 1977, 11,900 pelts were recorded (Knaus et al., 1983).

VEGETATIONAL ASSOCIATIONS AND ELEVATION RANGE: *L. canadensis* is found in riparian vegetation associated with thorn forest, tropical deciduous forest, and xeric scrubland. It is found from sea level up to 700 m. Most localities are below 1,000 m.

CONSERVATION STATUS: This otter is not listed in any category of risk in Mexico (SEMARNAT, 2002), where the distribution is marginal and their populations are seriously threatened and/or extinct (Gallo-Reynoso, 1997). Ceballos and Navarro (1991) considered it at risk of extinction.

Lontra longicaudis (Olfers, 1818)

Neotropical river otter

Juan Pablo Gallo R. and Ma. Antonieta Casariego

SUBSPECIES IN MEXICO

Lontra longicaudis annectens Major, 1897

DESCRIPTION: *Lontra longicaudis* is a medium-sized otter. The body is long and slim, with a cylindrical trunk; the width of neck is equal to the width of the skull; the head is flat and rounded; the legs are short with interdigital membranes. The tail is wide at the base, long, and oval from a transverse view; the pinna of the ear is small. The muzzle is short but wide. An important feature is the naked soles of the feet (Gallo, 1989). The fur is soft and fine, arranged in two layers; the first one consists of long and short hair, the second has soft and abundant short hair that protects the skin from the water. The fur is dark brown to pale reddish; ventrally the coat is grayish-brown, more yellowish on the throat, and creamy on the pectoral region. The nose is narrow without exceeding the height of the nostrils. Males are approximately 20% to 25% larger than females (Parera, 1996).

DISTRIBUTION: *L. longicaudis* is found in the slopes of the lowlands of both the Pacific and Gulf coasts of Mexico, from Sonora and Tamaulipas through the humid and tropical regions to Argentina. It has been recorded in the states of CA, CH, CL, CO, CS, DU, GR, HG, JA, MI, MO, MX, NY, OX, PU, QR, SI, SL, SO, TA, TB, VE, YU, and ZA.

EXTERNAL MEASURES AND WEIGHT

TL = 1,000 to 1,700 mm; TV = 375 to 600 mm; HF = 100 to 146 mm; EAR = 20 to 35 mm.

Weight: 15 kg.

DENTAL FORMULA: I 3/3, C 1/1, PM 4/3, M 1/2 = 36.

NATURAL HISTORY AND ECOLOGY: These otters are found in rivers, streams, lakes, dams, coastal lagoons, and irrigation canals, but always with dense vegetation cover, where they can find refuge among the roots of the trees. It has also been reported that in places like northern Argentina otters remain in the coast during winter and move to the marshes during the summer (Gori et al., 2003). They are solitary animals but maintain active communication through visual and olfactory signals with adjacent conspecifics by depositing scats in conspicuous places like rocks and tree trunks. They are mainly diurnal but can be active at night. The only time they are in groups is during the rearing period, when the pups stay with the mother. Their dens are located close to rivers, usually with an aquatic entry and one on land on higher ground; they can den in most available sites, from caves to abandoned refuges, without showing preference for any particular site (Pardini and Trajano, 1999). They feed on river crustaceans, fish, amphibians, reptiles, small mammals, birds, insects, and even fruits (they are considered good seed dispersers). Several studies in different regions have found similarities in the main genera consumed by otters, which include the crayfish of the genera *Macrobachium* spp. and *Atya* sp., the fish *Agonostomus monticola* and species of the genera *Awaous* and *Sycidium*, as well as the river crab *Pseudotelphusa*, among other genera specific to each region (Casariego Madorell et al., 2008; Gallo, 1989; Macías-Sánchez and Aranda, 1999; Quadros and Monteiro-Filho, 2000; Spínola and Vaughan, 1995). In most environments there is no defined reproductive season; males and females stay together for a couple of days during estrus. Although there are no detailed studies, it appears that the gestation period is about 60 days; the female gives birth to 1 to 4 offspring, 3 being most common. The average weight of pups at birth is 320 g. Most of the time the pups are together; they play with the mother until gradually venturing farther away once they ingest solid food and after an arduous hunting learning process; the pups become independent at around one year of age (Gallo, 1989; Parera, 1996). Population density is highly variable, from 0.25 individual/km for Quintana Roo (Orozco-Meyer and Morales-Vela, 1998) to 0.34/km in Sonora (Gallo, 1996), with large variations within the same region (< 0.01/km to 1/km of river in Oaxaca) depending on the conditions of the habitat (Casariego Madorell et al., 2008). Higher densities, 1.47/km to 2.14/km, have been reported for Argentina given that the habitat is not completely linear, but has a large number of ponds that increase the availability of refuges and food (Parera, 1996). From isolated observations of the river otter, its biology has been said to be comparable to that of the European otter (*L. lutra*), which has reported home ranges of 15 km in diameter; these otters travel up to 10 km in one day, and only one female

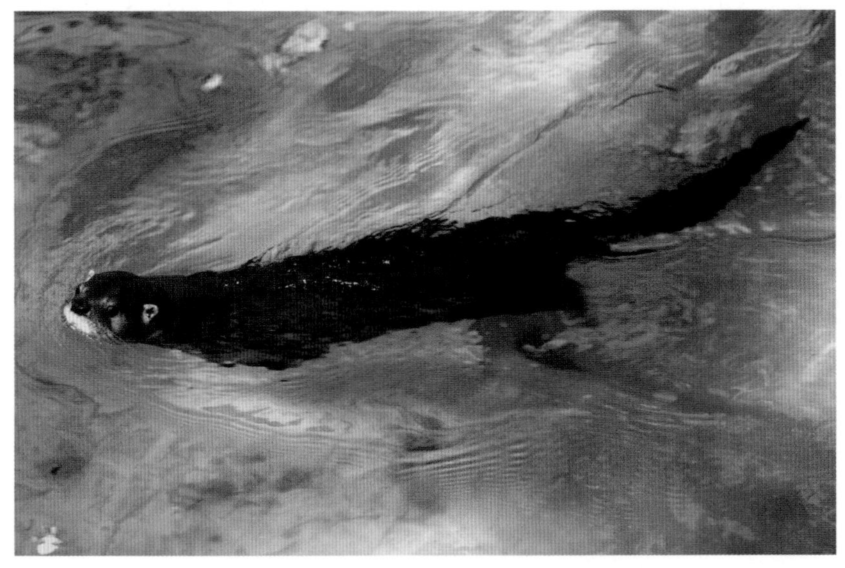

Lontra longicaudis. Wetlands. Villahermosa, Tabasco. Photo: Gerardo Ceballos.

with pups lives within that home range (Chebez, 1994). Predators of the river otter in countries like Brazil include the anaconda (*Eunectes* sp.) and jaguars (*Panthera onca*), although they can also be preyed on by cayman (*Caiman* sp.), dogs, and some birds of prey. Their populations have declined steeply because of habitat destruction, water pollution, and hunting, given the value of the pelts, and the pups are sold in the pet market. In 1990, the price of a pelt fluctuated from $25 to $90 U.S. (Aranda, 1991; Chebez, 1994; Parera, 1996).

VEGETATIONAL ASSOCIATIONS AND ELEVATION RANGE: *L. longicaudis* is found in riparian vegetation associated with cloud forests, tropical rainforests, deciduous rainforests, and subdeciduous rainforests (Gallo, 1989; Polechla et al., 1987). It is distributed from sea level to 1,800 m. Most of the locations are in the coastal watersheds below 1,700 m. In the northern mountain ranges of Oaxaca it has been recorded at 2,000 m and rarely has been reported higher than 3,000 m in Costa Rica and Argentina (Lariviére, 1999).

CONSERVATION STATUS: *L. longicaudis* is included in the list of endangered species in Mexico (SEMARNAT, 2010). It is listed in appendix I of CITES, and as data deficient by the IUCN (2010). The populations of the Neotropical otter in Mexico have a better conservation status, depending on the distance from towns and cities, industries, and intensive agriculture areas (Gallo-Reynoso, 1997).

Subfamily Mustelinae

The subfamily Mustelinae contains seven extant genera of otters, with thirteen species. Three of the genera and four species are known from Mexico.

Eira barbara (Linnaeus, 1758)

Tayra

Cuauhtémoc Chávez Tovar

SUBSPECIES IN MEXICO
Eira barbara senex (Thomas 1900)

DESCRIPTION: *Eira barbara* is a medium-sized mustelid with an elongated body. In general, the coloration of the body is black with the head and upper portions of the shoulders grayish-brown or yellowish, from which it gets its name, since the head seems to be gray-haired. On the chest it has a white spot of variable size (Álvarez del Toro, 1991). Reid (1997) indicated, however, that the pale-brown head form, with a diamond on the throat and a dark brown body, is found only from Honduras to Panama. Krumbiegel (1942) stated that the diamond-shape spot on the throat is not a reliable feature to separate the subspecies because its presence is part of intrapopulation variation. Melanistic and albino individuals have been reported in South America (Krumbiegel, 1942; Tortato and Althoff, 2007).

Eira barbara. Tropical semi-green forest. Calakmul Biosphere Reserve, Campeche. Photo: Cuauhtemoc Chavez and Gerardo Ceballos.

EXTERNAL MEASURES AND WEIGHT

TL = 999 to 1,125 mm; TV = 365 to 470 mm; HF = 90 to 123 mm; EAR = 30 to 42 mm. Weight: 2.7 to 7.0 kg.

DENTAL FORMULA: I 3/3, C 1/1, PM 3/3, M 1/2 = 34.

NATURAL HISTORY AND ECOLOGY: Tayras are generally solitary, but also form family groups of up to 5 individuals (Álvarez del Toro, 1991) and occasionally up to 15 to 20 individuals (Leopold, 1965). The adult males are 30% larger than females (Kaufmann and Kaufmann, 1965). They have semi-arboreal habits, traveling both on land and in trees (Álvarez del Toro, 1991). They are active day and night but have periods of increased activity in the early hours of the morning and evening (Aranda and March, 1987; Konecny, 1989; Pacheco et al., 2006; Sunquist et al., 1989). They spend the day in caves or in holes of old trees. They generally inhabit tropical and subtropical forests, secondary forests, gallery forests, gardens, plantations, and dry scrub forests, and are rarely seen (Emmons and Feer, 1997). In Costa Rica, tayras have been observed on highly disturbed coffee plantations and in fragments of riparian forests (Pacheco et al., 2006). Although one female tayra did not show habitat preference within her home range, the use of secondary forest and grassland was consistent with previous studies (Konecny, 1989). This species has also been observed in highly disturbed and fragmented forest landscapes of southern and eastern Amazonia (Cabrera and Yepes, 1960; Michalski et al., 2006). They can be considered omnivores because they feed on fruits, invertebrates (many of which are arboreal), honey, and vertebrates as big as a brocket deer (Álvarez del Toro, 1991; Cabrera and Yepes, 1960). Analyzing the contents of the digestive tracts of two tayras in the Isle of Barro Colorado, Panama, Enders (1935) found that they only fed on fruit pulp and insect pupas. They feed on fruit trees of Cecropia (*Cecropia mexicana*) and mamey (*Calocarpum mammosum*; Kaufmann and Kaufmann, 1965). In a study conducted in the plains of Venezuela during the dry season, the frequency of fruits and vertebrates consumed was similar. The speckled tree rat (*Echimys semivillosus*) was the most consumed vertebrate, whereas the seeds of *Genipa americana* represented a significant proportion of the plant component (Sunquist et al., 1989). A study conducted in Cockscomb Basin, Belize, found that fruits (67%) and arthropods (58%) were the categories with greater frequency in a sample of 31 feces (Konecny, 1989). Sightings indicate that it can feed on iguanas (*Iguana iguana*), monkeys, and agouties (Galef et al., 1976). The tayra can live near urban areas and consume foods available near human habitations, in gardens, cane, and corn fields (Hall and Dalquest, 1963; Heshkovitz, 1972). The reproductive season takes place from March to June. The gestation period lasts 63 to 70 days (Poglayen-Neuwall, 1975; Vaughan, 1974). Each litter has one to four offspring, most commonly two (Álvarez del Toro, 1991; Aranda and March, 1987). Five stages of development have been defined (Poglayen-Neuwall and Poglayen-Neuwall, 1976): (1) infant stage (birth to 2 months), in which the cubs take exclusively milk; (2) fledgling stage (2 to 2.5 months), in which the mother provides them solid food and milk; (3) weaning stage (2.5 to 3.5 months), in which the cubs make excursions near the den eating insects and fruits; (4) transitional stage (3.5 to 6.5 months), in which the shelter of birth is abandoned and the cubs hunt with the mother; (5) dispersion (6.5 to 10 months), in which the familiar group dissolves. They can be preyed on by jaguars and pumas (Ceballos et al., 2007, 2011; Chavez et al., in review). The home range, calculated with the minimum convex polygon (MPC) method, of two radio-collared females in Brazil had a 5.3 km²; another female in the plains of Venezuela, had a home range of 9 km (Michalski et al., 2006; Sunquist et al., 1989), whereas, in Belize the size of the home range of a female was 16 km² and 24 km² for a male (Konecny, 1989). They can travel daily distances of 2 km to 8 km (Emmons and Feer, 1997). Tayras that can cross large non-forested areas are able to subsist in agricultural mosaics retaining some forest cover, and can supplement their diet with easy pickings from household orchards, such as ripe papaya, and by preying on small livestock such as chickens (C. Chávez, unpubl. data; F. Michalski, pers. obs.).

DISTRIBUTION: *E. barbara* is found from southern Mexico to the south of Argentina. In Mexico it is located in the tropical areas of San Luis Potosí to Quintana Roo. In the Pacific Coast, separate populations are found in the states of Chiapas and Oaxaca and in the southern portion of Sinaloa. It has been recorded in the states of CA, CO, CS, PU, OX, QR, SL, TA, TB, and VE and possibly HG and CL.

VEGETATIONAL ASSOCIATIONS AND ELEVATION RANGE: In Mexico, the species has been recorded in tropical rainforests, semigreen forests, dry forests, cloud forests, and sometimes in pine forests. It is generally associated with low altitudes; the majority of the records are of localities below 1,000 masl.

CONSERVATION STATUS: In Mexico *E. barbara* is considered at risk of extinction due to the fragmentation and loss of habitat caused by intensive agriculture and conversion to pastures (Ceballos and Navarro, 1991; SEMARNAT, 2010; Schreiber et al., 1989). The subspecies *Eira barbara inserta* of Honduras is listed in appendix III of CITES (in danger of extinction). It is considered as least concern by IUCN, with population trends decreasing (IUCN, 2010).

Galictis vittata (Schreber, 1776)

Grison

Cuauhtémoc Chávez Tovar

SUBSPECIES IN MEXICO

Galictis vittata canaster Nelson, 1901

DESCRIPTION: *Galictis vittata* is a mustelid of similar size to a domestic cat but with a long body and very short legs. The neck is long; the head is flat with small and rounded ears. The tail is short. The color on the back is gray (white-haired), reaching the head; there is a thin white band on the forehead; the rest of the face, neck, belly, and limbs is black. The coat is short and dense (Álvarez del Toro, 1991).

DISTRIBUTION: *G. vittata* inhabits tropical and subtropical areas of America, from the south of Tamaulipas, and Oaxaca, Mexico to Misiones, northern Argentina. It has been recorded in the states of CA, CS, PU, OX, QE, QR, SL, VE, and YU.

EXTERNAL MEASURES AND WEIGHT

TL = 600 to 960 mm; TV = 135 to 195 mm; HF = 75 to 97 mm; EAR = 20 to 30 mm. Weight: 1.4 to 3.5 kg.

DENTAL FORMULA: I 3/3, C 1/1, PM 3/3, M 1/2= 34.

NATURAL HISTORY AND ECOLOGY: Grisons are very active during the early hours of the day, after evening, and at night. They use as dens holes under exposed roots, hollows in rocks, and caves abandoned by other animals (Eisenberg, 1989); however, they are capable of digging their own shelters (Álvarez del Toro, 1991). The gestation period is around 40 days (Aranda, 2000; Eisenberg and Redford, 1999; Hernández-Huerta, 1992). They can have one to four cubs that are born in October, March, August, or September (Leopold, 1959). Family groups of three or four individuals have been observed, most likely formed by the mother and her offspring (Aranda and March, 1987; Eisenberg, 1989), or possibly males following a female in heat (Álvarez del Toro, 1991). Grisons swim frequently; therefore, they are more commonly found in the vicinity of water, like rivers and streams (Álvarez del Toro, 1991), however, they have been seen in many environments, from palm savanna to closed forest, secondary forest, open land, and plantations (Yensen and Tarifa, 2003). In Costa Rica, grisons are apparently scarce in the region of San Vito (Pacheco et al., 2006). Although grisons can climb, they generally forage on the ground (Kaufmann and Kaufmann, 1965). They feed on wild fruit, insects, reptiles, small mammals, birds, and fish (Álvarez del Toro, 1991; Eisenberg, 1989). Grisons are cursorial mustelids and semi-fossorial; foraging habitats that were widely reported through interviews reported them as being able to traverse open areas and use even the most hostile matrix of actively managed grasslands and cultivated areas. This is entirely consistent with observations on habitat use by this species elsewhere (Eisenberg and Redford, 1999; Konecny, 1989). It is unlikely, however, that the deforestation process has increased the population density of grisons, as they rely primarily on forest habitat and have an eclectic diet of fruits and small forest vertebrates (Konecny, 1989). A female with a radio-collar spent 28% of her time in open habitat. An analysis of nine digestive tracts in Los Llanos, Venezuela, showed that they are mainly carnivorous, preying on small mammals such as mice (78%) and, in smaller proportions, reptiles (Sunquist et al., 1989). In the south of Mexico, reproduction takes place during the summer. Females give birth from 1 to 4 offspring after a gestation period of 39 days (Álvarez del Toro, 1991; Eisenberg, 1989). Females build their nests carrying branches, leaves, and litter to a deep cave (Álvarez del Toro, 1991). In Venezuela, the reported home range for a female was 4.2 km^2 and she had daily movements of 2 km to 3 km in a 24-hour period (Sunquist et al., 1989). Density estimates are 1 to 2.4 animals per km^2 (Eisenberg et al., 1979), but the radio-tracking data suggest much lower density, and that they are uncommon and rare (Timm et al., 1989).

VEGETATIONAL ASSOCIATIONS AND ELEVATION RANGE: The main vegetational associations in which this species is found include the tropical rainforests, subdeciduous forests, and deciduous forests; they are occasionally found in secondary growth and in cloud forests. Most of the records are below 500 masl, although there are records up to 1,200 masl.

CONSERVATION STATUS: In Mexico, this species is considered threatened with extinction (SEMARNAT, 2010). It is included in appendix III of CITES for Costa Rica and it is considered as least concern by IUCN, with population trends stable (IUCN, 2010). The conservation status of the grison is not clear; therefore, it requires protection.

Long-tailed weasel

Gerardo Ceballos and Gisselle Oliva Valdés

SUBSPECIES IN MEXICO
Mustela frenata frenata (Lichtenstein, 1831)
Mustela frenata goldmani (Merriam, 1896)
Mustela frenata latirostra (Hall, 1936)
Mustela frenata leucoparia (Merriam, 1896)
Mustela frenata macrophonius (Elliot, 1905)
Mustela frenata neomexicana (Barber and Cockerell,1898)
Mustela frenata perda (Merriam, 1902)
Mustela frenata perotae (Hall, 1936)
Mustela frenata tropicalis (Merriam, 1896)

DISTRIBUTION: Weasels have a very broad distribution, as they are found from southeastern Canada to Argentina. In Mexico, the species is found in virtually all the territory, with the exception of the Peninsula of Baja California and part of Sonora. It has been registered in the states of BN, CH, CO, CS, DF, DU, GR, JA, MI, MO, MX, NL, NY, OX, PU, QR, SL, SO, TB, TM, VE, YU, ZA.

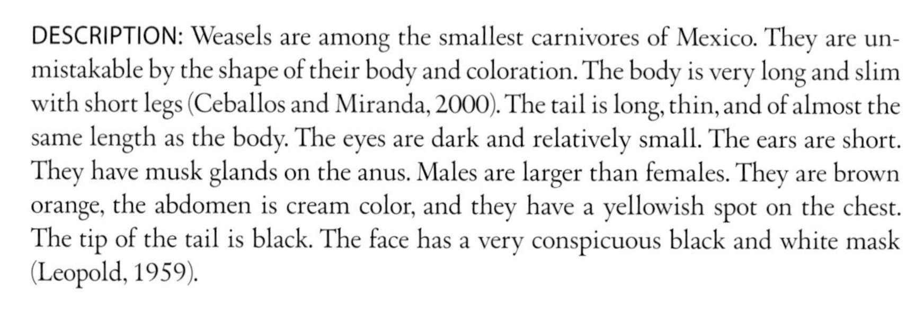

DESCRIPTION: Weasels are among the smallest carnivores of Mexico. They are unmistakable by the shape of their body and coloration. The body is very long and slim with short legs (Ceballos and Miranda, 2000). The tail is long, thin, and of almost the same length as the body. The eyes are dark and relatively small. The ears are short. They have musk glands on the anus. Males are larger than females. They are brown orange, the abdomen is cream color, and they have a yellowish spot on the chest. The tip of the tail is black. The face has a very conspicuous black and white mask (Leopold, 1959).

EXTERNAL MEASURES AND WEIGHT
TL = 215 to 350 mm; TV = 115 to 205; HF = 32 to 54 mm; EAR = 14 to 22 mm.
Weight: 85 to 340 g.

DENTAL FORMULA: I 3/3, C 1/1, PM 3/3, M 1/2 = 34.

NATURAL HISTORY AND ECOLOGY: Weasels tolerate various natural and disturbed ecological conditions, and they prefer open sites, with shrubby or herbaceous vegetation near sources of water (Nowak, 1999). Due to their small size they can be unnoticed even in sites with little vegetation. They are active during the day and night (Ceballos and Miranda, 2000). They build their dens in tunnels, in cavities between fallen rocks and logs, or in holes. They are territorial and generally solitary. Its area of activity, which overlaps with that of other individuals, varies from 4 to 120 hectares, depending on the region, time of year, and the individual's age. The density of the populations varies between one and ten individuals per square kilometer (Nowak, 1999; Svendsen, 1982). They are exclusively carnivorous. They feed on rodents like mice, pocket gophers, shrews, squirrels, rabbits, and birds. They are very clever and often attack much larger prey (Leopold, 1959). Due to their shape, they show a high relation between the surface and the volume of their body, so their metabolism is much higher than that of other mammals of similar size (Brown and Lasiewski, 1972). However, their shape allows them to pursue their prey through underground tunnels and galleries. They breed in the summer; but births occur in March and April, after a gestation period of 205 to 337 days, as they present delayed implantation. The number of average offspring per litter is six, with a range of three to nine. The offspring are born altricial: with the eyes closed, naked, and with a weight of around a gram. They develop quickly and are weaned around 30 days after birth (Nowak, 1999). Their predators include coyotes and birds of prey (Ceballos and Miranda, 2000).

Mustela frenata. Pine Forest, Nevado de Toluca, State of Mexico. Photo: Gerardo Ceballos.

VEGETATIONAL ASSOCIATIONS AND ELEVATION RANGE: Weasels are found in a variety of vegetations, including the perennial rainforest, deciduous rainforest, xeric scrubland, pasture, forests of oaks, and different types of forests of conifers and alpine tundra (Leopold, 1959; Wilson et al., 1999). They are also found in fields, orchards, and suburban areas. They are one of the carnivorous species with greater tolerance to anthropogenic impact. They have been registered from sea level to 4,200 masl. However, most of the records in Mexico are below 3,000 masl.

CONSERVATION STATUS: This species is not at any risk, since it is abundant, with a wide distribution and a great tolerance to anthropogenic impact.

Mustela nigripes (Audubon and Bachman, 1851)

Black-footed ferret

Jesús Pacheco, Rurik List, and Gerardo Ceballos

SUBSPECIES IN MEXICO

Mustela nigripes is a monotypic species.

DESCRIPTION: The black-footed ferret is a medium-sized mustelid, characterized by a very long body, with short legs and a tail of shorter length than that of the head and body (Hall, 1981; Miller et al., 1996). Females are smaller (10%) than males. The coloration of the dorsal fur is yellowish-brown; the face has a distinctive black "mask"; the legs, the tip of the tail, and the nose are black, and the abdomen is light cream color. The elongated body and the short and flexible joints of the legs are specialized features that enable the ferret to move easily through the tunnels of prairie dogs.

EXTERNAL MEASURES AND WEIGHT

TL = 500 to 567 mm; TV = 114 to 135 mm; HF = 60 to 73 mm; EAR = 16 mm.
Weight: up to 1,050 g.

DENTAL FORMULA: I 3/3, C 1/1, PM 3/3, M 1/2 = 34.

Mustela nigripes. Reintroduced specimen. Grassland, Janos Biosphere Reserve, Chihuahua. Photo: Gerardo Ceballos.

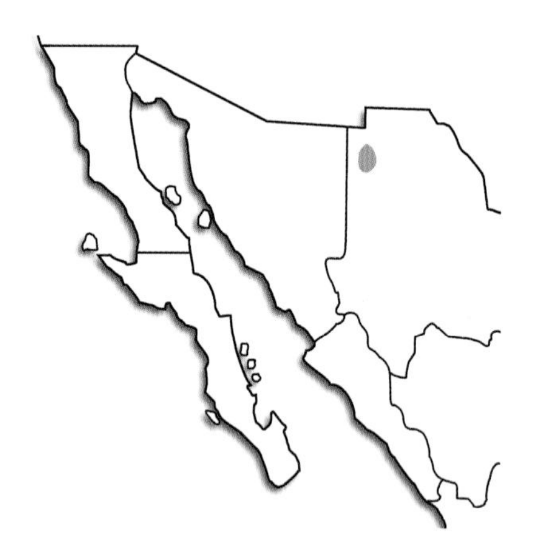

DISTRIBUTION: The historical distribution of *Mustela nigripes* included a large part of the Great Plains of North America, from southern Canada in the provinces of Alberta and Saskatchewan to northern Mexico in Chihuahua and Sonora. The distribution of the black-footed ferret coincides almost entirely with that of the prairie dogs (*Cynomys* spp.). Subfossil remnants have been found in Jimenez, Chihuahua (Harris, 1977; Messing, 1986). Currently there are 18 wild populations, all of them the result of reintroduction efforts in the United States, with the exception of a population in Janos-Casas Grandes in northwestern Chihuahua (Pacheco et al., 2000).

NATURAL HISTORY AND ECOLOGY: The black-footed ferret has nocturnal habits. They are solitary animals. The mating season occurs in March to April. The gestation period lasts from 42 to 45 days. Females can have an average of three offspring per year; young are weaned at around six weeks of age. They are active throughout the year, emerging from their dens every two to six days to look for another den occupied by prairie dogs, which are hunted underground by the ferrets. Although black-footed ferrets are very effective predators underground, on the surface its role changes, as they are prey of coyotes, foxes, badgers, and owls. They have a close association with the prairie dogs of the genus *Cynomys*, which constitute 90% of their diet. In addition, ferrets live in prairie dog burrow systems. They mark their territory with feces, urine, and aromatic substances. The territory of the males is larger than that of the females; thus, several females can occupy a male's territory.

VEGETATIONAL ASSOCIATIONS AND ELEVATION RANGE: The black-footed ferret is a highly specialized animal, which only inhabits grasslands occupied by prairie dogs in the Great Plains of North America, between 1,400 masl and 1,670 masl (Hall, 1981; Miller et al., 1994).

CONSERVATION STATUS: This is one of the most threatened mammal species in the world. It is listed as endangered by the U.S. Endangered Species Act, it is included in appendix I of CITES, and it is considered as endangered by IUCN (iucnredlist.org v. 3.1). In fact, in the early 1980s it was considered extinct (Clark, 1989). Several factors have influenced the reduction of black-footed ferret populations, including the poisoning of prairie dogs, the expansion of agriculture and livestock over the grasslands, epidemics of bubonic plague and distemper, and urbanization. Currently, the distribution of the prairie dogs, species with which it maintains a very close relationship, has been reduced by 98% (Miller et al., 1994). At the end of 1974, after a disastrous combination of diseases such as distemper and bubonic plague and the extensive poisoning of the colonies of prairie dogs, the last wild specimens were captured in an attempt to save them. The last of these ferrets succumbed to cancer in January 1979 (Ceballos and Pacheco, 2000). In 1980, this species was declared as possibly extinct. Fortunately, in September 1981, almost 3 years later, a population of more than 100 ferrets in the region of Meeteetse, Wyoming, was found (Miller et al., 1996). Once again, diseases decimated this population, and all survivors (18 individuals) were captured in the autumn of 1985. The recovery program in captivity has been a success, and by 2009, almost 7,000 kits had been born in the captive breeding program since 1987. The first specimens were released in Shirley Basin, Wyoming, in 1991, and then in Montana, South Dakota, Arizona, Colorado, and Utah. Today there are between 800 and 1,000 ferrets in the wild. Most of the reintroductions have faced problems due to the small size of the colonies of prairie dogs and to epidemics of bubonic plague, which in a few weeks can destroy an entire prairie dog town. In 1988, the discovery of the largest remaining colonies of prairie dogs in North America, in the region of Janos-Casas Grandes, Chihuahua, led to the proposal that this site could serve as a potential area for the reintroduction of black-footed ferrets (Ceballos et al., 1993). The Black-footed Ferret Recovery Team in the United States supported the establishment of a population in Mexico, and the first reintroduction occurred in September 2001. By 2008, a total of 299 ferrets had been released in 2 prairie dog towns of Janos, Chihuahua, and a minimum of 11 wild-born ferrets had been confirmed. The reintroduction of these ferrets marked a new stage in the conservation of wildlife in Mexico, and undoubtedly opens up a new opportunity for the recovery of this species.

Subfamily Taxidiinae

The subfamily Taxidiinae contains one extant genera and species.

Taxidea taxus (Schreber, 1777)

Badger

Arturo Jiménez Guzmán and Rurik List

DISTRIBUTION: *Taxidea taxus* is distributed from northern Alberta and the south of British Columbia in Canada, and through most of the western part of the United States to the center of Mexico. It has been recorded in the states of BC, BS, CH, CO, DU, HI, MX, NL, PU, SL, SI, SO, TA, VE, and ZA.

SUBSPECIES IN MEXICO

Taxidea taxus berlandieri Baird 1858

DESCRIPTION: The badger has a flattened body with a short neck, almost as wide as the head. The legs are strong and short, with long, curved claws in the forelegs and short, shovel-shaped claws in the front paws; the ears are short and rounded; and the eyes are small. The dorsal fur has three colors, yellow on the base, black on the subterminal region, and white on the tips, which in general gives it a grayish tone; the ventral region is yellowish. The neck, lower jaw, throat, cheeks, and sides are whitish. There is a black patch in the form of a triangle between the eye and the ear, and two parallel black bands from the nose to the base of the skull; a white dorsal band extends from the nose to the base of the tail. The tip of the tail and the legs are black (Leopold, 1959; Lindzey, 1982; Long, 1973).

EXTERNAL MEASURES AND WEIGHT

TL = 500 to 889 mm; TV = 98 to 174 mm; HF = 88 to 155 mm; EAR = 50 to 53 mm. Weight: 3,500 to 14,000 g.

DENTAL FORMULA: I 3/3, C 1/1, PM 3/3, M 1/2 = 34.

NATURAL HISTORY AND ECOLOGY: The badger is the only truly digging carnivore of North America. Its local distribution and activity depends on the presence of rodents with underground habits (Messick, 1987). They use dens throughout the year, either as a resting site during the day, for food storage, and as foraging and delivery sites (Messick, 1987). They excavate dens or modify those of other animals (Nowak, 1999b). The mating season occurs in summer and early autumn, but the implantation is delayed until January or February; females give birth to from one to five offspring between March and April (Messick, 1987; Messick and Hornocker, 1981; Nowak, 1999b). They have solitary habits and are mainly nocturnal, but may be active at any time of day (Lindzey, 1982). Badgers do not hibernate, but remarkably reduce their activity, remaining in their dens during the winter and in response to the availability of prey (Lindzey, 1982; Messick, 1987). They are adapted for the capture of fossorial prey; their diet is mainly composed of rodents such as squirrels (*Spermophilus* spp.), prairie dogs (*Cynomys* spp.), and pocket gophers, although they can also prey on other mammals, birds, reptiles, and insects and feed on carrion, so they are considered opportunistic (Leopold, 1959; Messick and Hornocker, 1981; Minta et al., 1992). The main causes of mortality appear to be associated with human activities such as secondary poisoning for the control of coyotes, poisoning of their prey, collisions with vehicles, and poaching, although the role of the natural causes of mortality such as predation and diseases can be underestimated.

VEGETATIONAL ASSOCIATIONS AND ELEVATION RANGE: In Mexico the badger inhabits open areas, semi-arid pastures, mesquite shrublands, grasslands, and pine-oak forests (Leopold, 1959; List and Macdonald, 1998). It is found from sea level up to 3,660 masl (Long, 1973).

CONSERVATION STATUS: The badger is considered threatened in Mexico (SEMARNAT, 2010) and as least concern by IUCN, with population trends decreasing (IUCN, 2010). The conversion to agricultural land, hunting, and the poisoning of its prey are probably the major causes of the reduction of the populations of this species (Ceballos and Navarro, 1991).

Taxidea taxus. Grassland. Janos Biosphere Reserve, Chihuahua. Photo: Gerardo Ceballos.

Family Mephitidae

The family Mephitidae was recently separated from the family Mustelidae (Dragoo and Honeycutt, 1997). It groups around 12 species of skunks and stink badgers (genus *Myadus*). They are distributed in America and Asia. In Mexico, seven species of three genera are represented. *Mephitis macroura, M. mephitis, Conepatus leuconotus*, and *Spilogale angustifrons* have a wide distribution, and *S. putorius, S. pigmaea*, and *C. semistriatus* have a more restricted distribution. The pygmy skunk (*S. pigmaea*) is the smallest of the species of skunks in Mexico, and is exclusively distributed in the dry forests of the Pacific.

Conepatus leuconotus (Lichtenstein, 1832)

Hog-nosed skunk

Angeles Mendoza Durán and Gerardo Ceballos

SUBSPECIES IN MEXICO

Conepatus leuconotus leuconotus (Lichteinstein, 1832)

Conepatus leuconotus telmalestes Bailey, 1905

Until recently, *C. mesoluecus* and *C. leuconotus* were considered different species (Hall, 1981; Wilson and Reeder, 2005). Dragoo et al. (2003), however, assessed the patterns of coloration, which was the feature used for recognizing both species, analyzed the morphometric variation and the mitochondrial DNA, and concluded that they were the same species. The name *C. leuconotus* has priority.

DISTRIBUTION: Hog-nosed skunks are found from Arizona, Colorado, and Texas to Nicaragua. In Mexico, they are found throughout the country, with the exception of the Baja California Peninsula, northwestern of Sonora, the Yucatan Peninsula, and southern Chiapas (Dragoo et al., 2003). It has been recorded in the states of AG, CH, CL, CO, CS, DF, DU, GR, HG, JA, MI, MO, MX, NL, OX, PU, SI, SL, SO, TA, TL, VE, and ZA.

DESCRIPTION: This skunk has a size similar to that of a large cat. The coloration is highly variable, and basically consists of a black body with a wide white dorsal band that goes from the top of the head to the tail. The white band can be very restricted in some individuals (Dragoo et al., 2003). The nose is long, naked, and flexible. The tail covers more than one-third of the total length and is black in its ventral and proximal parts (Hall, 1981; Olin and Thompson, 1982).

EXTERNAL MEASURES AND WEIGHT

TL = 410 to 633 mm; TV = 165 to 350 mm; HF = 59 to 79 mm; EAR = 19 to 33 mm. Weight: 900 to 4,500 g.

DENTAL FORMULA: I 3/3, C 1/1, PM 2/3, M 1/2 = 32.

NATURAL HISTORY AND ECOLOGY: Hog-nosed skunks are nocturnal animals that can dig their own dens or use hollow logs and holes between the rocks. The breeding season starts at the end of winter, and after two months of pregnancy four offspring are born; these can take care of themselves at three or four months of age (Ceballos and Galindo, 1984; Findley, 1987). They are solitary animals. In Mexico, densities of up to four individuals per km^2 have been reported. They feed on invertebrates such as insects (mainly beetles) and worms, fruits, small vertebrates, and, occasionally, carrion. They find their food through the sense of smell, and use their claws to dig it up (Reid, 1997). They have few predators because the secretion of their anal glands is an effective defense. In some regions, they are used in traditional medicine since their fat is considered to have healing properties (Ceballos and Galindo, 1984). They

Conepatus leuconotus. Specimen in captivity. Tropical rainforest. Tuxtla Gutierrez Zoo, Chiapas. Photo: Oscar Mendoza Mayorga, Chiapas Zoo (ZooMAT).

can readily climb trees to escape predators (Brashear et al., 2010). They are sympatric in some regions with *M. macroura*, *M. mephitis*, *Spilogale putorius*, and *S. angustifrons*.

VEGETATIONAL ASSOCIATIONS AND ELEVATION RANGE: Hog-nosed skunks inhabit a variety of temperate, arid, and tropical environments, including thorn forests, tropical rainforests, tropical dry forests, cloud forests, arid shrublands, grasslands, pine-oak forests, and croplands (Ceballos and Galindo, 1984; Hall and Dalquest, 1963; Leopold, 1965). They have been collected from sea level up to 2,500 masl.

CONSERVATION STATUS: These skunks are relatively abundant, both in undisturbed habitat and in agricultural areas. In Texas, they are considered seriously threatened by the destruction of their habitat and the excessive use of pesticides. Its IUCN status is least concern, with population trends decreasing (IUCN, 2010).

Conepatus semistriatus (Boddaert, 1785)

Striped hog-nosed skunk

Ángeles Mendoza Durán

SUBSPECIES IN MEXICO
Conepatus semistriatus conepatl Gmelin, 1788
Conepatus semistriatus yucatanicus Goldman, 1943

DESCRIPTION: These skunks are of medium size. Like in other species of the genus, females are smaller than males. Its general coloration is black, except for the tail, which is white. It has two white lines that start united on the neck and continue on the back, fading or disappearing toward the tail. The tail is less than a third of the total length (Hall, 1981). Differences in the shape and measures of the skull distinguish this skunk from other species of the genus.

EXTERNAL MEASURES AND WEIGHT

TL = 340 to 530 mm; TV = 160 to 310 mm; HF = 70 to 82 mm; EAR = 25 mm.
Weight: 1,400 to 3,500 g.

DENTAL FORMULA: I 3/3, C 1/1, PM 2/3, M 1/2 = 32.

NATURAL HISTORY AND ECOLOGY: Striped hog-nosed skunks den in rocky areas and hollow trees. Their diet is based on insects and other invertebrates, although they may consume fruits and small vertebrates. In general, little is known about its natural history although it probably does not differ much from that of the other species of the genus. The breeding season can vary, but more studies are necessary. They give birth to four offspring after two months of pregnancy. As in all skunks, the secretion of its anal glands constitutes its main defense (Álvarez del Toro, 1977; Aranda and March, 1987; Genoways and Jones, 1975; Reid, 1997).

VEGETATIONAL ASSOCIATIONS AND ELEVATION RANGE: This species mainly inhabits the tropical rainforest, although it is also found in the tropical dry forest, cloud forest, and thorn scrubs. It is found from sea level up to 300 masl.

CONSERVATION STATUS: It is classified as under special protection by the Mexican Red List (SEMANAT, 2002). Its IUCN status is least concern, with population trends unknown (IUCN, 2010). It can survive in disturbed areas.

DISTRIBUTION: *Conepatus semistriatus* is found from southeastern Mexico to northern Peru and eastern Brazil. In Mexico, it is distributed along the coast of the Gulf of Mexico, from southern Veracruz to the Yucatan Peninsula, including part of the Tehuantepec Isthmus. It has not been recorded in Oaxaca, but it may be found in that state (Goodwin, 1969; López-Wilchis et al., 1992). In Chiapas it has been collected in the Montes Azules Biosphere Reserve (Medellín, 1992). It has been recorded in the states of CA, CS, QR, TB, VE, and YU.

Mephitis macroura Lichtenstein, 1832

Hooded skunk

Jesús Pacheco R.

SUBSPECIES IN MEXICO

Mephitis macroura eximius Hall and Dalquest, 1950
Mephitis macroura macroura Lichtenstein, 1832
Mephitis macroura milleri Mearns, 1897

DESCRIPTION: *Mephitis macroura* is medium in size. Its body is elongated, with short legs and with a long tail covered with long hair (Godin, 1982). The head is small and elongated (Ceballos and Galindo, 1984; Hwang and Lariviere, 2001). Females are smaller than males (Hall, 1981). They are black with two white lines on the back. The tail is black mixed with white hairs. The coloration is highly variable, in particular, in the length and width of the stripes. It always has a white vertical line on the face (Leopold, 1965). It differs from *M. mephitis* by its coloration pattern, by the relatively longer tail, and by its larger size (Anderson, 1972; Hoffmeister, 1986). The skunks of the genus *Conepatus* are larger and do not have the white line on the face, the nose is longer, and the tail is shorter (Reid, 1997). They are larger than the skunks of the genus *Spilogale* (Hwang and Lariviere, 2001).

Mephitis macroura. Pine-oak forest. Janos Biosphere Reserve, Chihuahua. Photo: Gerardo Ceballos.

EXTERNAL MEASURES AND WEIGHT
TL = 558 to 790 mm; TV = 275 to 435 mm; HF = 58 to 73 mm; EAR = 28 to 32 mm. Weight: 1,700 to 2,000 g.

DENTAL FORMULA: I 3/3, C 1/1, PM 3/3, M 1/2 = 34.

NATURAL HISTORY AND ECOLOGY: Hooded skunks are solitary animals, mainly nocturnal; however, on cloudy or cold days they may be active during the day (Ceballos and Miranda, 1986). In Mexico, its density has been estimated to be 9 individuals per km^2 (Cervantes et al., 2002). Its home range is confined to 800 m^2 to 3,000 m^2. It dens underground, although it can also occupy crevices between rocks, hollow trees, and even abandoned houses (Ceballos and Galindo, 1984; Hoffmeister, 1986). They are omnivores and spend much time foraging and feeding, mainly on insects and their larvae, amphibians, snakes, lizards, birds and their eggs, small mammals, and vegetable matter (roots, seeds, and fruits). Around 80% to 90% of its diet is composed of animal matter, primarily insects obtained by digging (Godin, 1982). It has few natural predators, among which are the puma, badger, coyote, and some birds of prey (Leopold, 1965). Mating occurs at the end of winter (February and March), and the offspring are born in the spring (May and June). The gestation period lasts 8 weeks (55 to 70 days); females give birth to 3 to 8 offspring. After five months, the young become independent (Hoffmeister, 1986). Hooded skunks have a couple of glands on both sides of the anus, which expel a smelly liquid used as defense when disturbed; the spraying is frequently accompanied by growling and banging the ground with the forefeet.

DISTRIBUTION: *M. macroura* is distributed from the southwestern United States to northern Costa Rica (Reid, 1997). In Mexico it is widely distributed, with the exception of the Baja California Peninsula, northwestern Mexico, and the Yucatan Peninsula. It has been recorded in the states of CH, CL, CO, CS, DF, DU, GR, HG, JA, MI, MO, MX, NL, NY, OX, PU, SI, SL, SO, TA, TL, VE, and ZA.

VEGETATIONAL ASSOCIATIONS AND ELEVATION RANGE: This species inhabits a variety of temperate and tropical environments such as conifer forest, grassland, xeric scrubland, deciduous tropical forest, and riparian vegetation (Godin, 1982). They are common in agricultural and suburban areas (Ceballos and Galindo, 1984). It is found from the sea level to 3,000 m in the National Park "Lagunas de Zempoala" in Morelos (Ramírez-Pulido, 1969a).

CONSERVATION STATUS: This skunk has benefited from anthropogenic activities since it is now abundant in crops and even logged areas and garbage dumps (Ceballos and Miranda, 2000). IUCN classified this species as least concern and the population trend is increasing (IUCN, 2010).

Striped skunk

Jesús Pacheco R.

SUBSPECIES IN MEXICO
Mephitis mephitis estor Merriam, 1890
Mephitis mephitis holznery Mearns, 1897
Mephitis mephitis varians Gray, 1837

DESCRIPTION: *Mephitis mephitis* is the size of a domestic cat. It shows sexual dimorphism, with males being larger than the females (Godin, 1982). The shape of the head is triangular; the ears are small and rounded. The legs are short and robust; feet are plantigrade with five partially webbed toes. The front paws have long, curved claws, whereas the claws of the hind feet are shorter (Wade-Smith and Verts, 1982). Females have 10 to 14 mammae, usually 12. They have a couple of glands located on each side of the anus (Verts, 1967). The coloration of the body is black with a white band that starts on top of the head covering the dorsal part of the animal; on half of the back the white stripe forks, covering the sides of the animal and extending to the rear. It has a very thin white stripe from the forehead to the base of the nose. The tail is long and fluffy. Morphologically, it is very similar to *M. macroura*; however, the tympanic bulla is smaller in *M. mephitis*, the white lines of the back are not mixed with black hairs, and the tail is shorter (Anderson, 1972; Hall, 1981; Hoffmeister, 1986; Whitaker, 1980).

EXTERNAL MEASURES AND WEIGHT
TL = 575 to 800 mm; TV = 184 to 393 mm; HF = 60 to 90 mm. EAR = 27 to 32.
Weight: 2,700 to 6,300 kg.

DENTAL FORMULA: I 3/3, C 1/1, PM 3/3, M 1/2 = 34.

NATURAL HISTORY AND ECOLOGY: Striped skunks breed during February or March and the gestation period varies from 62 to 66 days (Godin, 1982). The births occur in May or at the end in June. The number of offspring varies from 2 to 10 (Bailey,

Mephitis mephitis. Grassland. Janos Biosphere Reserve, Chihuahua. Photo: Rurik List.

DISTRIBUTION: Striped skunks are widely distributed in southern Canada and almost all of the United States to northern Mexico. In Mexico, they are found in all the Mexico-U.S. border states and in Durango (Godin, 1982; Wozencraft, 1993). It has been recorded in the states of BC, CH, CO, DU, TA, NL, and SO.

1971). The reported longevity in wild individuals is of 2 to 4 years (Verts, 1967), while in captivity it is 10 years (Schwartz and Schwartz, 1959). Offspring feed themselves at two months old (Hoffmeister, 1986). They are omnivores and opportunistic animals that feed primarily on insects (grasshoppers, beetles, and larvae of butterflies and wasps), spiders, earthworms, and snails. Occasionally they feed on vegetable matter, as well as vertebrates such as small mammals, birds, salamanders, frogs, and toads (Hoffmeister, 1986). They den in underground burrows they dig themselves, although sometimes they use abandoned dens of armadillos, prairie dogs, or badgers (Hwang and Lariviere, 2001). They are solitary and nocturnal, with occasional crepuscular activity. Its home range has been estimated at 511 ha for males and 378 ha for females (Storm, 1972). The estimated population density varies from 1.8 individuals/km^2 to 4.8 individuals/km^2 (Bailey, 1971; Verts, 1967). Their predators include puma, coyote, badger, fox, lynx, owls, and eagles. When these skunks are disturbed they react defensively, arching the body, raising the tail, bristling the hair of the head and tail, and banging on the ground. If the aggression persists, the skunk arches its body in an attacking position, looking toward its aggressor, exposes the anus, and expels a volatile musky and pestilent liquid toward the intruder, at a distance of up to 6 m (Wade-Smith and Verts, 1976). The musk can be expelled as a continuous dew or as small discharges aimed at the aggressors, sometimes causing temporary blindness (Hofmeister, 1986; Verts, 1967). It is one of the wild species in which the highest incidence of rabies has been recorded (Hwang and Lariviere, 2001; Parker, 1975; Verts, 1967).

VEGETATIONAL ASSOCIATIONS AND ELEVATION RANGE: Striped skunks are found in a variety of vegetational associations such as grasslands, thorny forests, oak forests, and agricultural areas (Anderson, 1972). In the United States, it has been recorded from sea level up to 4,200 m (Nelson, 1930); however, it is usually found below 1,800 masl (Grinnell et al., 1937). In Mexico, the highest altitude at which it has been recorded is in Chihuahua at 2,250 m (Goldman, 1951).

CONSERVATION STATUS: In some places in the United States striped skunks are considered a plague (Wade-Smith and Verts, 1982) and IUCN classified this species as least concern with a stable population trend. In Mexico, they are common in the north of the country. These skunks tolerate well the environmental impact caused by humans.

Spilogale angustifrons Howell, 1902

Spotted skunk

Gerardo Ceballos

SUBSPECIES IN MEXICO

Spilogale angustifrons amphalus Dickey, 1929
Spilogale angustifrons leucoparia Merriam, 1890
Spilogale angustifrons lucasana Merriam, 1890
Spilogale angustifrons martinensis Elliot, 1903

The taxonomic position of *S. angustifrons* and *S. putorius* is complex because since the revision of Van Gelder (1959a) both taxa were considered as the same species. Each taxon is now considered valid, however, since they differ in morphology, number of chromosomes, and physiology (Dragoo et al., 1993; Hsu and Mead, 1969; Kinlaw, 1995). We follow Wozencraft (2005) in recognizing two subspecies in Mexico.

DISTRIBUTION: This species is found from central Mexico to Costa Rica. The distribution of this species and that of *S. gracilis* in central Mexico are not well known. It has been recorded in the states of DF, CA, CS, CO, GR, JA, MI, MO, MX, OX, PU, QR, TL, VE, YU, and QR.

DESCRIPTION: *Spilogale angustifrons* is a small skunk with three longitudinal black bands, three vertical white bands, and a white spot on the forehead; the tail is black in the base and white in its last third (Verts et al., 2001). They differ from other skunks by size and coloration. Morphologically, they are very similar to *S. angustifrons*, but are distinguished by the white spot between the eyes, which is smaller in size, and by having a higher number of chromosomes (64 vs. 60, respectively). They are larger than *S. pigmaea*, which in addition is white yellowish.

EXTERNAL MEASURES AND WEIGHT
TL = 345 to 500 mm; TV = 80 to 142 mm; HF = 37 to 59 mm; EAR = 28 mm.
Weight: 227 to 750 g.

DENTAL FORMULA: I 3/3, C1/1, PM 3/3, M 1/2 = 34.

NATURAL HISTORY AND ECOLOGY: Spotted skunks are strictly nocturnal. In the cold months of the year their activity is reduced (Dalquest, 1948). They build their dens under logs or rocks, in hollow trees, and between the roots of trees and shrubs; they also use abandoned dens of armadillos and other mammals (Ceballos and Miranda, 2000; Reid, 1997). Their home range covers up to 64 ha. In productive ecosystems their density is about eight individuals/km^2. In Mexico, densities of up to five individuals/km^2 have been reported (Cervantes et al., 2002). They mainly feed on insects and small mammals, but also include amphibians and wild fruits in their diet. Fifty percent of their diet consists of invertebrates, and the rest of vertebrates and plants (Baker and Baker, 1975; Crabb, 1941; Howard and Marsh, 1982). The mating season occurs from September to the first weeks of spring. They have delayed implantation (200 to 220 days) and a gestation period of about 200 days (Mead, 1968). Most of the births occur in May, and the litters vary from two to nine offspring (Verts et al., 2001). Their main predators are owls, foxes, badgers, lynx, coyote, and dogs. They are

Spilogale angustifrons. Tropical forest. Tuxtla Gutierrez, Chiapas. Photo: Oscar Mendoza Mayorga, ZooMAT.

skillful climbers (Ceballos and Galindo, 1984; Howard and Marsh, 1982). As defense they spray their urine, whose odor is more penetrating than that of other skunks. Sometimes they try to intimidate by banging with their forefeet. They are useful species for agriculture, since they control invertebrates like rodents that destroy crops (Leopold, 1965; Villa-R., 1953).

VEGETATIONAL ASSOCIATIONS AND ELEVATION RANGE: Spotted skunks inhabit a variety of vegetatal associations such as the tropical deciduous forest, thorn scrub, xeric shrubland, pine-oak forest, fir forest, and grasslands. They are common in disturbed areas and agricultural fields (Baker and Baker, 1975; Villa-R., 1953). They have been found from sea level to 2,744 masl in the Sierra de San Pedro Martir, Baja California (Verts et al., 2001).

CONSERVATION STATUS: This species has a wide distribution and a high tolerance to anthropogenic impact, so it is not at any risk of extinction, being considered as least concern by IUCN, with population trends stable (IUCN, 2010).

Spilogale gracilis Merriam, 1890a

Spotted skunk

Federico Romero R.

SUBSPECIES IN MEXICO
Spilogale gracilis amphialus Dickey, 1929
Spilogale gracilis leucoparia Merriam, 1890b
Spilogale gracilis lucasana Merriam, 1890b
Spilogale gracilis martirensis Elliot, 1903

DISTRIBUTION: Spotted skunks are widely distributed in southern Canada and almost all of the United States to north-central Mexico. In Mexico, the species is found from the Mexico-U.S. border states to Queretaro (Verts et al., 2001; Wozencraft, 1993). It has been recorded in the states of AG, BC, BS, CH, CO, DU, GJ, HG, JA, NL, PU, QE, SL, SO, TA, VE, and ZA.

DESCRIPTION: *Spilogale gracilis* is a small-sized skunk. The dorsal pelage is black, with four white bands, more or less parallel along the back and interrupted on the hip. It has another white band on each side that extends to the tail. On the forehead, it has a white spot (Hall, 1981; Van Gelder, 1959a; Wilson, 1993). The tail is short with black hair.

EXTERNAL MEASURES AND WEIGHT
TL = 310 to 610 mm; TV = 80 to 280 mm; HF = 30 to 59 mm; EAR= 28 mm.
Weight: 207 to 885 g.

DENTAL FORMULA: I 3/3, C1/1, PM 3/3, M 1/2 = 34.

NATURAL HISTORY AND ECOLOGY: This species was known as *S. putorius*, so much of the information on its biology was recorded for that species (Verts et al., 2001). They are solitary animals, mainly nocturnal. They feed on insects and small mammals, but also ingest carrion, eggs, poultry, and plant material (Kinlaw, 1995). Coyotes, dogs, and raptors prey on them. They reproduce in the spring and summer (Mead, 1968). Mating occurs between March and April. Apparently, there is no delayed implantation unlike in *S. angustifrons* and other species of skunks. The offspring are born between May and July, blind and with little hair, after a gestation period of 50 to 65 days (Kinlaw, 1995). The litters can have up to 5 pups, which are weaned after 60 days. The population density can exceed 20 individuals per ha. Its home range is variable, but it has been estimated at around 25 ha. In a year, however, they can move in areas of up to 4,000 ha (Verts et al., 2001). Their highest densities have been reported in environments with a dense vegetal cover and in agricultural fields. Like other skunks, they are very susceptible to rabies.

VEGETATIONAL ASSOCIATIONS AND ELEVATION RANGE: Spotted skunks live in arid, temperate, and tropical environments, including thorny forests, oak forests, and tropical rainforests. It also inhabits disturbed environments, crops, and suburban areas. It has been recorded from sea level to 3,000 masl.

CONSERVATION STATUS: This species is not at risk of extinction. Agricultural fields have benefited their populations. They may be affected by pesticides, however. It is considered as least concern by IUCN, with population trends decreasing (IUCN, 2010).

Spilogale pygmaea Thomas, 1898

Pygmy skunk

Gerardo Ceballos, Rafael Ávila Flores, and Rodrigo A. Medellín

SUBSPECIES IN MEXICO

Spilogale pygmaea australis Hall, 1938
Spilogale pygmaea intermedia López-F.and Urbano-V., 1979
Spilogale pygmaea pygmaea Thomas, 1898

DESCRIPTION: *Spilogale pygmaea* is the smallest skunk in Mexico. It is characterized by a black and white coloration pattern that gives the appearance of spots (Medellín et al., 1998). Observed with more care, these spots are in fact white and black stripes that alternate on the back and sides, along the longitudinal axis of the body with the exception of the legs and the rear of the flanks, where the stripes are vertical. The shades of both colors vary from white to light cream or yellowish white, and dark brown to black (Ceballos and Miranda, 1986; López-F. and Urban-V., 1979; Van Gelder, 1959a). It differs from the other two species of the genus by its smaller size, by the two continuous white dorsal lines to the hindquarters, by the nasal spot joined with the pre-auricular spots, and by the four legs totally white in their anterior half (Hall, 1981; Van Gelder, 1959a). It has sexual dimorphism, with females being smaller than males (López-F. and Urban-V., 1979).

DISTRIBUTION: *S. pygmaea* is an endemic species of Mexico distributed along a narrow tropical coastal strip of Pacific lowlands, from northern Sinaloa, to the Tehuantepec Isthmus in Oaxaca (Ceballos and Miranda, 1986; Hall, 1981). It has been recorded in the states of CL, GR, JA, MI, NY, SI, and OX.

EXTERNAL MEASURES AND WEIGHT

TL = 250 to 291 mm; TV = 58 to 84 mm; HF = 34 to 38 mm; EAR = 23 to 25 mm. Weight: 156.8 to 313 g.

DENTAL FORMULA: I 3/3, C 1/1, PM 3/3, M 1/2 = 34.

NATURAL HISTORY AND ECOLOGY: The pygmy skunk has crepuscular and nocturnal habits (Ceballos and Miranda, 1986; Rosatte, 1987), and apparently performs much of its activities hidden in the understory (Goodwin, 1956). They usually use trails and dry streams that lead to water bodies, and it is not uncommon to find them around the dens of small rodents in search of food (Genoways and Jones, 1968c; Jones et al., 1962). Among its main prey are, in addition to the rodents and other small vertebrates, a variety of arthropods such as beetles, grasshoppers, spiders, scorpions, prawns, small birds, eggs, and fruit and vegetable matter (Medellín et al., 1998). Although there is only one report of a barn owl eating a pygmy skunk (*Tyto alba*; López-F. and Urbano-V., 1976); other owls are likely predators, as well as some larger carnivores and snakes (Rosatte, 1987). Their reproductive period occurs from May to August; the offspring born partly covered by a fine white hair after an average gestation period of 48 days, forming litters that vary from 1 to 6 offspring (Teska et al., 1981).

VEGETATIONAL ASSOCIATIONS AND ELEVATION RANGE: *S. pygmaea* mainly inhabits areas of tropical deciduous forest, subperennial rainforest, and xeric scrubland; it has also been observed in thorn shrublands and sand dunes with herbaceous vegetation. It is found from sea level to 1,630 masl, but generally below 350 masl (Ceballos and Miranda, 2002; Genoways and Jones, 1968a; Greer and Greer, 1970; López-F. and Urban-V., 1979).

CONSERVATION STATUS: The pygmy skunk is considered threatened (SEMARNAT, 2010). Although it is locally abundant in certain areas of the country (Ceballos and Miranda, 2000), the accelerated destruction of its habitat threatens it (Zarza et al., 2000). It is considered vulnerable by IUCN, with population trends decreasing (IUCN, 2010). It is protected in the Chamela-Cuixmala Biosphere Reserve in Jalisco, the Veladero (Guerrero), and Huatulco National Park.

Spilogale pygmaea. Tropical dry forest. Chamela-Cuixmala Biosphere Reserve, Jalisco. Photo: Gerardo Ceballos.

Family Procyonidae

The family Procyonidae comprises 18 species of medium-sized carnivores, including kinkajous, ring-tailed cats, and olingos, distributed in tropical and temperate dry regions of the Americas (Wilson and Reeder, 2005). In Mexico seven species of four genera are represented. Two of them, a coati (*Nasua nelsoni*) and a racoon (*Procyon pygmaeus*), are endemic to Cozumel Island. The racoon in the Tres Marias Islands (*P. lotor insularis*) was considered a different species until recently.

Subfamily Potosinae

The subfamily Potosinae contains five species in two genera, with one species known from Mexico.

Potos flavus (Schreber, 1774)

Kinkajou

Fernanda Figueroa and Héctor T. Arita

SUBSPECIES IN MEXICO
Potos flavus chiriquensis J.A. Allen, 1904
Potos flavus prehensilis Kerr, 1792

DESCRIPTION: Kinkajous are large procyonids; they show sexual dimorphism since males are larger than females (Emmons and Feer, 1997; Kortlucke, 1973; Nowak, 1999b). They have characteristics similar to those of other mammals of frugivorous and arboreal habits such as the primates. It is considered the most specialized procyonid (Ford and Hoffmann, 1988; Kortlucke, 1973). They have elongated bodies with short legs, a rounded head with large eyes, a small muzzle, rounded ears placed very low and to the sides of the head; and a long, fully prehensile tail that slims toward the tip; the eyes are round, separated and directed toward the forehead as in the primates. The jaw resembles the jaw of a primate rather than that of a carnivore (Emmons and Feer, 1997; Ford and Hoffmann, 1988; Kortlucke, 1973). The coat is short, woolly, and soft. It has a reddish-brown or grayish-brown dorsal coloration; the ventral color is dark yellow to orange; and the muzzle is dark brown. Some individuals have a dark antero-posterior stripe that runs along the center of the back (Emmons and Feer, 1997; Ford and Hoffmann, 1988). Kinkajous have a geographic variation in their coloration and size. Individuals tend to be lighter and larger in the northern part of their distribution (Kortlucke, 1973). The legs are dark with five nimble fingers, with very sharp and curved claws. The posterior part of the sole has dense fur and the pads are naked. They have an interdigital membrane along the first third of the length of the digits. They have the ability to manipulate objects with ease because the five fingers of the hands converge at the same point in the center of the palm when closed (McClearn, 1992). Kinkajous have distinctive features that distinguish them from the rest of the procyonids, among which are the existence of a long and extensible tongue, as well as a couple of ventral cutaneous glands; they have a baculum with four terminal processes, two directed toward the sides and back and two located anteriorly (Ford and Hoffmann, 1988).

Potos flavus. Tropical rain forests. Montes Azules Biosphere Reserve, Chiapas. Photo: Fulvia Eccardi.

EXTERNAL MEASURES AND WEIGHT
TL = 820 to 1,330 mm; TV = 392 to 570 mm; HF = 70 to 140 mm; EAR = 30 to 55 mm. Weight: 1,400 to 4,600 g.

DENTAL FORMULA: I 3/2, C 1/1, PM 4/4, M 1/1 = 34.

NATURAL HISTORY AND ECOLOGY: Kinkajous are almost completely arboreal and nocturnal (Nowak, 1999b). During the day they reduce their metabolic rate sleeping in hollow trees or niches formed by branches and vines in the canopy (Ford and Hoffmann, 1988; Kortlucke, 1973). This species is very sensitive to extreme temperatures and always maintains a metabolic rate lower than that expected for its body size (Ford and Hoffmann, 1988). It is mainly frugivorous. The second most important component of its diet is flowers, but it also consumes honey, nectar, and insects (Eisenberg, 1989; Kortlucke, 1973; Nowak, 1999b). The kinkajou especially forages in the middle and high strata of the canopy, in a variety of species of trees (Ford and Hoffmann, 1988). Its activity begins after the sunset and ends before dawn (Julian-Laferriére, 1993). This nocturnal forage is carried out jumping from tree to tree with great agility. During their movements the backbone can turn 180 degrees from the hip to the neck (McClearn, 1992). They use the prehensile tail and the ability to turn the legs at

180 degrees to hang upside down and descend the same way; these features also give them the ability to forage in the tips of the branches and to feed upside down manipulating the food with the forefeet (Ford and Hoffmann, 1988; McClearn, 1992; Nowak, 1999b). The reproductive pattern of this species is not fully understood. It has been suggested that it is polyestric, without a particular reproductive period, as pregnant females have been found at different times of the year (Ford and Hoffmann, 1988; Nowak, 1999b). Males and females reach sexual maturity at 1.5 and 2.2 years, respectively (Nowak, 1999b). The gestation period lasts from 98 to 120 days; females generally give birth to a single young per litter, and rarely two (Eisenberg, 1989; Hall and Dalquest, 1963). The offspring are born in a hollow trunk and weigh from 150 g to 200 g. They open their eyes at seven to nine days after birth, and approximately seven weeks later they eat solid food and have the strength to hang from their tails (Ford and Hoffmann, 1988; Nowak, 1999b). They are solitary and territorial, although couples or small groups, feeding together, have been observed. Population densities of 12.5 to 74 individuals per km^2 have been estimated (Emmons, and Feer 1997; Julian-LaFerriére, 1993; Nowak, 1999b). In Surinam, the area of activity has been calculated between 15 ha and 17 ha for females and from 26 ha to 39 ha for males (Julian-LaFerriére, 1993; Kays and Gittleman, 2001). Their feeding habits make them important dispersers of seeds. They visit a large number of species of trees in one night. In Surinam, of 10 species visited, the seeds of at least 7 are effectively dispersed, passing intact through the digestive tract of the animal (Julian-La Ferriére, 1993). They are important dispersers of species such as *Ficus* (figs), *Virola*, and *Luga* (Eisenberg, 1989; Ford and Hofmann, 1988). They can also play an important role in cross-pollination (Eisenberg, 1989). Their predators are little known. There are reports in South America of a harpy eagle (*Harpia harpyja*) eating a kinkajou. They can also be preyed on by the jaguar and the puma (*Panthera onca* and *Puma concolor*; March and Aranda, 1992). Its main predator is the human. Ectoparasites include lice and ticks. Occasionally, they have skin infections produced by *Leishmania brasiliensis*. In natural conditions, the endoparasites that have been detected are ascarids, coccidia, hematozoans, bacteria, and a virus (Ford and Hoffmann, 1988). This species is hunted for consumption by tribes from Surinam (Robinson and Redford, 1991) and by the Lacandons in Chiapas (March, 1987). It is also sold as a pet; however, the main reason it is hunted is because of the value of its skin. There is an important traffic in its fur, especially in South America. In Peru, for example, more than 100 live animals and hundreds of skins are exported every year (Ford and Hoffmann, 1988). There is knowledge of skin traffic in Chiapas even though this state forbids their hunting (Robinson and Redford, 1991).

DISTRIBUTION: *P. flavus* is distributed along the Pacific and Gulf lowlands, from west-central Mexico, along the coast of Michoacan to the coast of Guerrero, and through the Gulf of Mexico from the south of Tamaulipas, to the area of Matto Grosso in the central part of Brazil (Ford and Hoffmann, 1988; Hall, 1981; Martínez-Meyer et al., 1998; Ramírez-Pulido et al., 1983). The presence of this species in Tamaulipas was questioned by Álvarez (1963a); however, it was recently recorded near Goméz Farías (Moreno Valdéz, 1996). It has been recorded in the states of CA, CS, GR, HG, MI, OX, QE, SL, TA, TB, VE, and YU.

VEGETATIONAL ASSOCIATIONS AND ELEVATION RANGE: Kinkajous are mainly found in perennial rainforests. They have been reported in medium forests, deciduous low forests, riparian forests, secondary and disturbed high forests, and occasionally orchards. The kinkajou prefers undisturbed sites with mature stages of tropical perennial forest (Emmons and Feer, 1997; Ford and Hoffmann, 1988). It is found from sea level to 1,750 masl, but it is rarely found above 500 masl (Emmons and Feer, 1997).

CONSERVATION STATUS: *Potos flavus* is listed in appendix III of CITES for Honduras (Emmons and Feer, 1997; Wozencraft, 1993). Ceballos and Navarro (1991) consider it a fragile species in Mexico. The Mexican Red List classifies it as under special protection (SEMARNAT, 2010), and it is considered least concern by IUCN, with population trends decreasing (IUCN, 2010). It is necessary to take into account that it prefers high perennial rainforests, habitat that is seriously threatened. We know very little about the populations of kinkajous in Mexico, so it is necessary to carry out population studies in areas where it is distributed. There are protected populations in reserves such as Calakmul in Quintana Roo and Montes Azules in Chiapas.

Subfamily Procyoninae

The subfamily Procyoninae contains nine species in four genera, with three of the genera and six species known from Mexico.

Bassariscus astutus (Lichtenstein, 1830)

Ring-tailed cat

Gerardo Ceballos and Virginia Nava Vargas

SUBSPECIES IN MEXICO

Bassariscus astutus astutus Lichtenstein, 1830
Bassariscus astutus bolei Goldman, 1945
Bassariscus astutus consitus Nelson and Goldman, 1932
Bassariscus astutus flavus Rhoads, 1893
Bassariscus astutus insulicola Nelson and Goldman, 1909
Bassariscus astutus macdougalli Goodwin, 1956
Bassariscus astutus palmarius Nelson and Goldman, 1909
Bassariscus astutus saxicola Merriam, 1897

DISTRIBUTION: *B. astutus* is distributed from the south of the United States into Central America. In Mexico, it is found throughout the north and center of the country; it is only absent in the slope of the Gulf of Mexico, the Yucatan Peninsula, Chiapas, and part of Oaxaca (Ceballos and Miranda, 1986; Hall, 1981; Leopold, 1959; Poglayen-Neuwall and Toweill, 1986; Ramírez and López, 1982). It has been registered in the states of BN, BS, CH, CO, DF, DU, GR, HI, JA, MI, MO, NL, OX, PU, SI, TM, VE.

DESCRIPTION: A medium-sized carnivore. The eyes are large and are surrounded by black or dark brown rings. The ears are narrow and rounded, white to pink with brown patches, the body is long and slender, the tail is of equal size to the body with fluffy hair and with seven to eight black rings interspersed with white rings (Nowak, 1999). The back legs are longer and more robust than the forefeet. The second, third, fourth, and fifth fingers of the legs and hands are densely covered by hair. The claws are short and semi-retractable (Hall, 1981; Leopold, 1959). The dorsal pelage is thick, generally gray, with brown-yellowish tones. The ventral part is smoother and whitish (Poglayen-Neuwall and Toweill, 1988). The skull is small and lacks sagittal crest. The carnassials are not well developed, the canines are rounded, and the molars present high and sharp cusps (Hall, 1981).

EXTERNAL MEASURES AND WEIGHT

TL = 616 to 811 mm; TV = 310 to 438 mm; HF = 57 to 78 mm; EAR = 44 to 50 mm.
Weight: 870 to 1,100 g.

DENTAL FORMULA: I 3/3, C1/1, PM 4/4, M 2/2 = 40.

NATURAL HISTORY AND ECOLOGY: Ring-tailed cats inhabit mountainous areas and slopes of rugged ground. They make their dens in hollow trees, among rocks and roots. They are very agile and excellent climbers; they are able to rotate the forearm 180° (Trapp, 1972). They are omnivores and feed mainly on small mammals, insects, fruits, birds, reptiles, and occasionally, nectar (González, 1982; Kuban and Schwartz, 1985; Mead and Van Devender, 1981; Taylor, 1954; Toweill and Teer, 1977; Trapp, 1978; Wood, 1954). They are solitary animals with nocturnal habits. Their area of activity is variable and depends directly on the habitat, season of the year, and sex. In central Mexico it is between 7 and 10 hectares (Castellanos and List, 2005). The breeding season is from February to May; the gestation period lasts about 8 weeks and births occur between April and June. The litter size varies from one to four offspring (González, 1982; Poglayen-Neuwall and Toweill, 1988).

Bassariscus astutus. Oak forest, Picacho, Nuevo León. Photo: Marcelo Sada.

VEGETATIONAL ASSOCIATIONS AND ELEVATION RANGE: The ring-tailed cat is found in xeric scrublands, pine, fir, oak, and juniper forests, in semi-arid tropics, scrublands, chaparrals, and even in city parks (Ceballos and Galindo, 1984; Trapp, 1972). It can be found from sea level to 2,880 masl (Kaufmann, 1982; Poglayen-Neuwall and Toweill, 1988).

CONSERVATION STATUS: Ring-tailed cats can survive even in suburban areas. The species is rather common, although two subspecies (*B. astutus insulicola* and *B. astutus saxicola*) are considered threatened (SEMARNAT, 2010).

Bassariscus sumichrasti (Saussure, 1860)

Central American cacomistle

Gerardo Ceballos and Virginia Nava Vargas

SUBSPECIES IN MEXICO

Bassariscus sumichrasti campechensis Nelson and Goldman, 1932
Bassariscus sumichrasti latrans Davis and Lukness, 1958
Bassariscus sumichrasti oaxacensis Goodwin, 1956
Bassariscus sumichrasti sumichrasti Saussure, 1860

DESCRIPTION: This species is similar but larger than the ring-tailed cat (*Bassariscus astutus*) (Leopold, 1959). It is distinguished from *B. astutus* by having muzzle and legs of darker coloration. The ears are longer and the hair is straight and soft. The tail is long and hairy with nine rings with black rings interspersed with gray rings; the tip is black. The second, third, fourth, and fifth fingers of the hands are completely naked and the pads are elongated and narrow. The claws are long, curved, and not retractable (Leopold, 1959; Poglayen-Neuwall and Toweill, 1988).

Bassariscus sumichrasti. Tropical rainforest, Tuxtla Gutiérrez, Chiapas. Photo: Oscar Mendoza Mayorga. ZooMAT.

EXTERNAL MEASURES AND WEIGHT
TL = 790 to 1,003 mm; TV = 390 to 530 mm; HF = 82 to 90 mm; EAR = 44 to 45 mm. Weight: 600 to 1,600 g.

DENTAL FORMULA: I 3/3, C1/1, PM 4/4, M 2/2 = 40.

NATURAL HISTORY AND ECOLOGY: Central American cacomistles are strictly nocturnal and arboreal. Their dens are built in the holes of trunks. They eat mainly fruits, but they also feed on insects and small vertebrates (Aranda, 1981; Emmons and Feer, 1990). Reproduction is concentrated from February through June, but estrus can occur at any time (Aranda and March, 1987; Emmons and Feer, 1997). Interestingly, females are receptive for only one day. Gestation lasts around two months, after which one young is born. Newborns are born naked and blind. They open their eyes at around 34 days and are weaned at three months. Captive animals can live up to 23 years (Poglayen-Neuwall, 1989).

DISTRIBUTION: *B. sumichrasti* is found from southeastern Mexico to Panama. In Mexico, it is sympatric with the ring-tailed cat (*B. astutus*) in the states of Veracruz, Guerrero, and Oaxaca (Poglayen-Neuwall and Toweill, 1988). It has been registered in the states of CA, CS, GR, OX, QR, VE, and YU.

VEGETATIONAL ASSOCIATIONS AND ELEVATION RANGE: They live in tropical forests, cloudy forest, and in the most humid and dense forests of pine-oak in tropical regions. Generally they are found in the canopy of medium- and high-mature tropical forests. They have not been found in disturbed areas (Aranda and March, 1987; Coates-Estrada and Estrada, 1986; Emmons and Feer, 1990; Nowak, 1999). Usually they are found at altitudes from sea level to 2,900 masl (Hall, 1981; Poglayen-Neuwall and Toweill, 1988).

CONSERVATION STATUS: The species is classified in appendix III of CITES and is considered of special protection in Mexico (SEMARNAT, 2010).

Nasua narica (Linnaeus, 1766)

White-nosed coati

David Valenzuela Galván

SUBSPECIES IN MEXICO
Nasua narica molaris Merriam, 1902

Nasua narica narica (Linnaeus, 1766)
Nasua narica nelsoni Merriam, 1901
Nasua narica yucatanica J.A. Allen, 1904
The taxonomic status of the Cozumel's coati, or dwarf coati, it is not definitive and it has been considered as a subspecies of the white-nosed coati (*N. narica nelsoni*) or as a separate species (*N. nelsoni*); in either case it is clear that it is threatened. Although data from genetic analysis based on mitochondrial DNA are not conclusive about the taxonomic status of the dwarf coati, it is clear that its population should be managed as distinct management units, whether as a subspecies of *N. narica* or a species by itself (McFadden et al., 2008).

DESCRIPTION: Coatis are similar in size to a medium-sized dog. The body is long and slender. The tail is long and often held erect. The muzzle is long and sharp and the tip is highly movable (Gompper, 1995; Hall, 1981; Kaufmann, 1987). They have well-developed and strong claws in each of the five fingers of the forefeet. Males are slightly larger than females. The dorsal coloration varies from dark-chestnut to reddish-brown or golden brown; the neck and shoulders tends to be golden chestnut. The populations of northeastern and northwestern Mexico have golden brown coloration, while the southerners are dark brown. The hair around the eyes, edge of the ears, throat, chin, and tip of the snout are yellowish and lighter than the rest of the body. They have a dark brown mask. The tail often has darker rings (Hall, 1981; Kaufmann, 1987; Leopold, 1959).

DISTRIBUTION: *N. narica* is distributed from the western United States (Arizona and New Mexico) to northern Colombia. In Mexico, it occupies all states, except the Baja California Peninsula and part of central Mexico, from southern Chihuahua northeastern Michoacán and the northwestern part of the state of Mexico (Hall, 1981). It has been recorded in states of CA, CH, CO, CL, CS, DF, DU, GR, HG, JA, MI, MO, MX, NL, NY, OX, PU, QE, QR, SI, SL, SO, TA, TB, TL, VE, and YU.

EXTERNAL MEASURES AND WEIGHT
TL = 850 to 1,340 mm; TV = 420 to 680 mm; HF = 95 to 122 mm; EAR = 38 to 44 mm.
Weight: 4 to 6 kg.

DENTAL FORMULA: I 3/3, C 1/1, PM 4/4, M 2/2 = 40.

NATURAL HISTORY AND ECOLOGY: White-nosed coatis have diurnal and terrestrial habits; sometimes they show nocturnal activities, particularly when or where hunting pressure is intense. Coatis are plantigrade with shorter forelegs than hind legs; when running they can reach speeds of 27 km per hour. It is also a good swimmer that only enters the water when it is forced to do so (Kaufmann, 1962). Their anatomical features allow them to climb trees easily. Nocturnal resting sites are caves, cracks, holes, or branches of trees. It is an omnivorous gatherer, which feeds mainly on fruit, invertebrates such as insects (coleopterans, orthopterans, lepidopterans, hymenopterans, and isopterans), millipedes, spiders, crustaceans, and worms (Delibes et al., 1989; Saénz, 1994; Valenzuela, 1998). It also occasionally hunts small terrestrial vertebrates such as rodents, amphibians, and reptiles. They consume fruits such as fig (*Ficus*), bonete (*Jacaratia*), plum (*Spondias*), ramon (*Brosimum*), and others (*Guapira, Astrocaryum, Cecropia, Dypteryx, Morisonia, Jacquinia*; Kaufmann, 1962; Russell, 1982; Smythe, 1970; Valenzuela, 1998). Coatis are important seed dispersers (Alves-Costa et al., 2004; Saénz, 1994); it is estimated that they can carry the seeds in their digestive tract for almost 72 hours (Janzen and Wilson, 1983). The mating season occurs between January and April; in southern latitudes it starts earlier. Gestation lasts 10 to 11 weeks, and the litter may have 2 to 7 offspring, which can forage independently at 5 weeks of age. They reach sexual maturity at two years old (Kaufmann, 1962; Russell, 1982). In natural conditions, coaties live up to 7 years; in captivity, some individuals have reached up to 17 years (Gompper, 1995). The populations of coati have a high degree of reproductive synchrony since mating only occurs during one or two weeks of the year (Gompper, 1995). This seems to be connected with cycles of abundance of arthropods and fruiting of several species of trees (Russell, 1982; Smythe, 1970). The coati is highly social; it forms groups of up to 20 individuals, but groups up to 38 individuals have been re-

Nasua narica. Tropical dry forest. Chamela-Cuixmala Biosphere Reserve, Jalisco. Photo: Gerardo Ceballos.

ported (Kaufmann et al., 1976). The groups consist almost exclusively of adult females and juveniles, organized in a matriarchal social system. Males remain in the groups until two years of age; after that they are expelled or leave the groups and live solitary lives, which probably prevents attacks of males on the offspring (Smythe, 1970) or is a strategy to avoid competition for limited resources (Gompper, 1996). Males are found with females only during the short reproductive periods that occur annually, in which they are tolerated by the adult females. The females in the groups cooperate in the vigilance against predators and in the rearing of young (Gompper, 1995; Russell, 1982). The groups spend most of the time in search of food between the brushes; solitary males exhibit similar behavior, but they travel greater distances and show nocturnal activity (Russell, 1982; Smythe, 1970; Valenzuela and Ceballos, 2000). Grouping behavior in coatis and a set of different behaviors like foraging with the juveniles in the center of the group, sharing vigilance, alarm calling, mobbing and attacking predators, and a highly synchronous birth season suggest that predation has played an important role in the social behavior of this species (Hass and Valenzuela, 2002). The population density is higher and the size of the home range of the groups is smaller in the southern part of its distribution than in the semi-arid areas of the north. On the island of Barro Colorado, Panama, densities between 24 and 42 individuals per km^2 and areas of activity of 50 ha per group have been reported. In Los Tuxtlas, Veracruz, the densities are 33 individuals per km^2. In Arizona, densities between 0.4 and 1.2 individuals per km^2 and average areas of activity between 600 ha and 1,357 ha have been reported (Hass, 2002). In the deciduous rainforest of the Chamela-Cuixmala Biosphere Reserve, Jalisco, Mexico, it is one of the carnivores with highest density, with reported average density of 42.9 ± 16.8 individuals per square km^2 and an average home range of 383.0 ± 32.8 ha (Valenzuela, 1998; Valenzuela and Ceballos, 2000). These values have inter- and intra-annual fluctuations due to diseases or variations in the availability of food and water (Ceballos and Miranda, 1986; Coates-Estrada and Estrada, 1986; Glanz, 1982, 1990; Gompper, 1995; Kaufmann, 1962; Kaufmann et al., 1976; Valenzuela, 1999). The offspring and young are more vulnerable to the attack of certain predators such as eagles (*Aquila chrysaetos*, *Harpia harpyja*), hawks (*Buteo jamaicensis*), boas (*Constrictor constrictor*), and occasionally monkeys (*Cebus capucinus*; Gompper, 1995; Perry and Rose, 1994). Adults are preyed on by black bears, pumas, and jaguars, and probably by tayras, lynx, ocelots, and jaguarundis (Fedigan, 1990; Gompper, 1995; Hass and Valenzuela, 2002; Janzen, 1970; Janzen and Wilson, 1983; Kaufmann, 1962; Newcomer and DeFarcy, 1985; Valenzuela, 1998). Predation rates have been reported to be significantly higher on solitary coatis than on coatis in groups, a result that is consistent with the hypothesis that group living provides significant anti-predator advantages for coatis (Hass and Valenzuela, 2002). Coatis are susceptible to rabies and distemper. Parasites found in them include cestoda, trematoda, larvae of flies, and ticks (Gompper, 1995; Kaufmann, 1962, 1983, 1987, 1987; Nowak, 1999b). On the coast of Jalisco in the 1990s an outbreak of scabies seriously affected the population density (Valenzuela et al., 2000). Coatis are valued as pets in rural areas, so the capture of young is common. Given the low economic value of its skin the coati has little hunting value (Kaufmann, 1987). Even so, in some places of Mexico, such as in Chiapas, they are often hunted for the skin (Aranda, 1991). Coatis are an important species used as food or hunted to prevent the damage they cause to certain crops such as corn.

VEGETATIONAL ASSOCIATIONS AND ELEVATION RANGE: Coatis are mainly found in deciduous and subdeciduous rainforests and perennial rainforests along the coasts of Mexico. They also are common in pine and pine-oak forests and xeric scrublands. It is found from sea level to 2,900 masl, but is distributed mainly in the lowlands of the Gulf of Mexico and the Pacific.

CONSERVATION STATUS: Coatis are not considered endangered in Mexico (SEMARNAT, 2010). Even when in many places it is very abundant, practically no information on the status of its populations exists. In northern Mexico hunting has caused significant reductions in their populations (Gompper, 1995). During the hunting season that took place from 1993 to 1994, hunting was allowed in 18 states with a maximum limit of 2 animals per season in almost all the states. The subspecies *Nausa narica nelsoni* was considered at risk of extinction by Ceballos and Navarro (1991), and threatened in the Mexican Red List (SEMARNAT, 2010), but at the species level is not listed in Mexico, and it is considered as least concern by IUCN, with population trends decreasing (IUCN, 2010). In New Mexico, the coati is under legal protection (Kaufmann, 1987) and in Honduras it is included in appendix III of CITES (Gompper, 1995).

Nasua nelsoni (Merriam, 1901)

Dwarf coati or Cozumel's coati

David Valenzuela-Galván, Alfredo Cuarón, María Eugenia Copa-Alvaro, and Ella Vázquez

SUBSPECIES IN MEXICO

N. nelsoni is a monotypic species.

DESCRIPTION: *Nasua nelsoni* is a procionid similar to a small dog in size. It resembles its congeneric species *N. narica* but is notoriously smaller and has smaller teeth (Hall, 1981). It has a slender body, a pointed snout, and a large tail, usually carried erect during foraging. It has strong claws on each of the five fingers of feet; the ears are short. Males are larger than females. Coat color is brown, varying from pale to yellowish or reddish. The fur on the muzzle, postauricular patches, and tips of ears is white. The tail has faint rings of darker color.

EXTERNAL MEASURES AND WEIGHT
TL = 785, 861; TV: 348, 328; HF = 84, 79.
Weight = 1,864 ± 170 g and 2,600 ± 135 g.
(Copa and Valenzuela, unpubl. data; males and females).

DENTAL FORMULA: I 3/3, C 1/1, PM 4/4, M 2/2 = 40.

DISTRIBUTION: *N. nelsoni* is a species endemic to Cozumel Island (Quintana Roo), located 17 km west of the Yucatan Peninsula. The island surface of nearly 477 km^2 is covered mainly by semi-evergreen forest, mangroves, and tropical dry forest (Romero-Najera, 2004). Cozumel is the largest Mexican island in the Caribbean Sea.

NATURAL HISTORY AND ECOLOGY: Little is known about the dwarf coati's natural history. It has been recorded at the central portion of the island and at the archaeological site of San Gervasio, in semi-evergreen forest, and at Punta Sur Park in tropical dry forest (Cuarón et al., 2004). Sightings confirm that it has diurnal activity. Like its congeneric species *N. narica*, it is plantigrade and capable of arboreal activity. There is no information about its diet but presumably it is an omnivorous collector that feeds mainly on fruit and arthropods. Not much information is available about its social behavior, but it has been recorded forming groups of individuals. During 1994-1995, several dwarf coati individuals were recorded in line transects at the semi-evergreen forest in the central portion of the island, and a density of 0.43 ± 0.27 dwarf coatis/km^2 was estimated, data that in turn, assuming that coatis were present throughout Cozumel tropical forests, produced an initial population estimation of 150 ± 95 individuals (Cuarón et al., 2004). In 1962 an adult female was captured 3.5 km north of San Miguel Town (Jones and Lawlor, 1965). Intensive trapping efforts (> 6,600 trap-days) during 2001-2003 at 19 sites throughout the island resulted in only 1 adult male dwarf coati captured (McFadden et al., 2010). In 2006, a carcass (skull and bones) of an adult male dwarf coati was collected at the San Gervasio archaeological site (D. Valenzuela, pers. obs.). Also, 11 adult females and 9 adult males of dwarf coati were captured during 2006 with a total trapping effort of 3,547 trap-days at semi evergreen forest in the central portion of the island (M. Copa and D. Valenzuela, unpubl. data).

Copa-Alvaro (2007) used intensive track census data to evaluate differences in the Activity Index (AI) before and after a major hurricane impact on Cozumel Island. The AI value after hurricane disturbance was nearly three times greater than before at the same study sites, indicating that hurricanes can alter its local density and/or change its activity patterns as well as its foraging behavior. Gestating and lactating females have been captured in May and July, and young and juvenile individuals in July and August (M. Copa-Alvaro and D. Valenzuela-Galván, unpubl. data); scant information that suggests its reproductive period occurs from the end of spring and during summer. At Cozumel there are no native predators that prey on *N. nelsoni*; however, there are feral dogs that have been reported to prey on pygmy raccoons *(Procyon pygmaeus)* (Bautista, 2006) and hence could easily be preying on dwarf coati as well. Also, introduced *Boa constrictor* (Martínez-Morales and Cuarón, 1999) could prey on dwarf coatis. Little is known about diseases and parasites affecting *N. nelsoni*, but it is likely that, as its congeneric species *N. narica*, it is susceptible to rabies, scabies, and canine distemper (Gompper, 1995; Valenzuela et al. 2000) and that a serious threat to this endemic carnivore could be the high risk of pathogen and disease spillover from feral and domestic animals (Mena, 2007). Navarro and Suárez (1989) considered that hunting pressure could affect endemic carnivores at Cozumel; however, we have not found evidence that currently could represent a problem (Cuarón et al., 2009).

VEGETATIONAL ASSOCIATIONS AND ELEVATION RANGE: Dwarf coatis have been recorded mostly in the semi-evergreen forest of Cozumel's interior, but there have also been scarce sightings in coastal and mangrove areas (Cuarón et al., 2009). Cozumel Island has an altitude range between 0 masl and 30 masl.

CONSERVATION STATUS: In Mexico *N. nelsoni* is included in the official Mexican list of threatened species (SEMARNAT, 2002) as endemic and threatened. It has been considered in danger of extinction by Ceballos and Navarro (1991). Its species status was questioned and proposed to be considered a subspecies as *N. narica nelsoni* (Decker, 1991). The available evidence clearly supports species-level recognition for *N. nelsoni* (McFadden et al., 2008); it is geographically and reproductively isolated, and genetically and morphologically distinct, from its mainland congener. *N. nelsoni* should be treated as a species and as a separate management unit (McFadden et al., 2008). We have proposed that it should be recognized as a species by the *IUCN Red List of Threatened Species* and as critically endangered (Cuarón et al., 2009).

Procyon lotor (Linnaeus, 1758)

Common Raccoon

David Valenzuela Galván

SUBSPECIES IN MEXICO

Procyon lotor fuscipes Mearns, 1914
Procyon lotor grinnelli Nelson and Goldman, 1930
Procyon lotor hernándezii Wagler, 1831
Procyon lotor insularis Merriam, 1898
Procyon lotor mexicanus Baird, 1858
Procyon lotor pallidus Merriam, 1900
Procyon lotor psora Gray, 1842
Procyon lotor shufeldti Nelson and Goldman, 1931
The subspecies of the Tres Marias Islands (*P. lotor insularis*) was regarded as a species but a recent study has shown that it is better considered a subspecies of *P. lotor* (Helgen and Wilson, 2005).

DESCRIPTION: *Procyon lotor* is a medium-sized procyonid with a robust body and short legs. The hind feet are larger than the forefeet, which have five long and widely separated thin fingers. The claws are short, curved, and not retractable (Kaufmann, 1987; Lotze and Anderson, 1979). The dorsal pelage is long, with a coloration ranging from grayish to blackish with yellowish tones or diffuse brown. Ventrally the color varies from brown-yellowish to grayish. The face has a black mask that covers the eyes and cheeks and extends from the nose to the forehead moving to the middle of the eyes. This mask is clearly delimited by white and grayish fur that covers the rest of the face and muzzle. The tail has four to seven dark rings alternating with gray rings. The tip of the tail is black as well as the sides of the legs (Kaufmann, 1987; Lotze and Anderson, 1979; Nowak, 1999b). Males are 10% to 15% larger than females (Kaufmann, 1987).

EXTERNAL MEASURES AND WEIGHT

TL = 603 to 950 mm; TV = 192 to 405 mm; HF = 83 to 138 mm; EAR = 59 to 62 mm. Weight: 3 to 9 kg.

DENTAL FORMULA: I 3/3, C 1/1, PM 4/4, M 2/2 = 40.

DISTRIBUTION: Raccoons are widely distributed from southern Canada to Panama, including the islands near the coast (Lotze and Anderson, 1979). They have been introduced in Germany, France, the Netherlands, and the territory of the former Soviet Union (Nowak, 1999b; Sanderson, 1987). In Mexico, they are found throughout the country in places with permanent rivers or bodies of water (Leopold, 1959; Navarro et al., 1991; Ramírez-Pulido and Castro-C., 1990; Ramírez-Pulido et al., 1983, 1986). They have been recorded in the states of BC, BS, CA, CH, CL, CS, DF, DU, GJ, GR, HG, JA, MI, MO, MX, NL, NY, OX, QE, QR, SI, SL, SO, TA, TB, VE, YU, and ZA.

Procyon lotor. Mangrove. Chamela-Cuixmala Biosphere Reserve, Jalisco. Photo: Gerardo Ceballos.

NATURAL HISTORY AND ECOLOGY: Raccoons have crepuscular and nocturnal habits (Ceballos and Miranda, 1986; Coates-Estrada and Estrada, 1986). Their feet are semi-plantigrade to plantigrade and, like other species of the family, raccoons can climb trees easily (McClearn, 1992). They are strong swimmers and often cross rivers or bodies of water of up to 300 m in width (Kaufmann, 1987). They den in hollow trees, cracks, or small caves in rocky walls and even in abandoned dens of other animals. Raccoons are omnivorous and consume large varieties of animals and plants. Plant matter constitutes from 48% to more than 70% of their diet (Sanderson, 1987). It tends to be selective when food is abundant (Lotze and Anderson, 1979). In the United States, it consumes fruits and seeds of various wild plants such as grapes, cherries, apples, and acorns (*Quercus* spp.). In Mexico, it also consumes acorns and in tropical regions, figs (*Ficus* spp.), mangos, and other fleshy fruits. In the vicinity of agricultural sites, it can consume large quantities of corn, wheat, sorghum, and oats (Kaufmann, 1987; Sanderson, 1987). It consumes more invertebrates than vertebrates such as insects and their larvae (orthopterans, coleopterans, hymenopterans, lepidopterans), terrestrial crabs, river shrimps (*Cambarus* and *Astacus*), and, in lesser and variable proportions throughout the year, annelids and mollusks. The vertebrates consumed in smaller proportions are fish, frogs, turtles, eggs of sea turtles and caimans, small birds and their eggs, mice, squirrels, rabbits, hares, as well as carrion from larger animals such as deer, cows, and horses (Erickson and Scudder, 1947; Kaufmann, 1987; Lotze and Anderson, 1979; Sanderson, 1983, 1987; Urbán, 1970). It is a popular belief that the raccoon "washes" its food; the truth is that more than washing, it checks it; its forefeet are well adapted to manipulate and grasp objects (Lotze and Anderson, 1979; McClearn, 1992) and have abundant innervations that give them a highly developed sense of touch (Kaufmann, 1987; Lotze and Anderson, 1979). In addition, because it lacks salivary glands it needs to soak its food. Mating occurs during winter, from

December to March, and gestation lasts about 63 days. The number of offspring per litter is one to seven, with an average of four. In Los Tuxtlas, Veracruz, they usually have three or four offspring (Coates-Estrada and Estrada, 1986). At 2 months old they are already capable of obtaining their own food and are weaned around 7 to 16 weeks of age. Juveniles disperse at 9 to 12 months of age (Fritzell, 1978; Kaufmann, 1987; Leopold, 1965; Lotze and Anderson, 1979; Nowak, 1999b). They can live up to 16 years, although the majority live less than 7 years (Kaufmann, 1987; Lotze and Anderson, 1979). The basic social unit is a female with her offspring, but in general, they are considered solitary even when small groups of short duration consisting of females, adult males, and juveniles can be found in common dens. It is also possible to find temporary groups feeding together (Kaufmann, 1987; Leopold, 1965; Lotze and Anderson, 1979). Home range size varies from few to more than 4,000 ha; typically, home range averages 65 ha for males and 39 ha for females (Nowak, 1999b). In the basin of Mexico their home range comprises 160 ha to 320 ha (Ceballos and Galindo, 1984). Raccoons are not territorial; however, unrelated animals tend to avoid each other. The typical population densities range from 2.32 animals/km^2 to 20 animals/km^2 (Lotze and Anderson, 1979). In Chamela, Jalisco, densities of 4 individuals/km^2 to 20 individuals/km^2 have been reported (Ceballos and Miranda, 1986). The leading causes of mortality of this species are related to humans (hunting, trapping, and cars). Malnutrition is due to periods of food scarcity, diseases, and parasites (Hasbrouck et al., 1992; Kaufmann, 1987; Sanderson, 1987). With some regularity, high mortality rates occur in the populations due to distemper, chronic pleurisy, rabies, and pneumonia (Kaufmann, 1987; Lotze and Anderson, 1979). Raccoons can be reservoirs of nearly 13 pathogens that cause disease in humans, which include leptospirosis, rabies, Chagas disease, tularemia, and tuberculosis (Kaufmann, 1987; Lotze and Anderson, 1979). It can be parasitized by helminths, nematodes, cestodes, trematodes, and mites (Kaufmann, 1987; Lotze and Anderson, 1979). Its predators include the puma, bobcat, wolf, coyote, fox, owls, and crocodiles (Hasbrouck et al., 1992; Kaufmann, 1987). Raccoons defecate in the sites where they feed, but there are several reports of sites in which large amount of feces can accumulate, which indicates that such area has been used for long periods (Kaufmann, 1987). The skin of the raccoon has commercial value, and it has been valued at $26 U.S. per raccoon pelt. In the United States, this is the wild species with greater importance in the fur trade (Nowak, 1999b; Sanderson, 1987). Given its economic importance, it has been introduced in European countries like France, Germany, and Holland (Nowak, 1999b). In Mexico, raccoons are not hunted much, but are used as food (Leopold, 1965). In some states, like Chiapas, their skins can be sold for up to U.S. 1.5 dollars (Aranda, 1991).

VEGETATIONAL ASSOCIATIONS AND ELEVATION RANGE: Raccoons have adapted to live in a variety of habitats with permanent water bodies. They are more abundant in tropical perennial forests, tropical deciduous and subdeciduous forests, mangroves, and areas of aquatic and underwater vegetation associated with marshes, swamps, and wetlands. In the Chamela-Cuixmala Biosphere Reserve, Jalisco, the highest densities are located in the mangroves and swamps contiguous to the coast (Ceballos and Miranda, 1986). They can also be found in areas of xeric scrub and in pine-oak forests (Ceballos and Galindo, 1984; Sanderson, 1987; Woloszyn and Woloszyn, 1982). They can be found from sea level to almost 3,000 masl (Hall, 1981). They are more abundant in the coastal plains along the Pacific and the Gulf of Mexico and are scarce in mountain areas.

CONSERVATION STATUS: In Mexico, the raccoon does not have a special status, although the destruction of the suitable habitats for this species may cause reductions in their populations (Ceballos and Galindo, 1984; Leopold, 1959). It is considered least concern by IUCN, with population trends increasing (IUCN, 2010).

Pygmy raccoon

David Valenzuela Galván and Alfredo D. Cuarón

SUBSPECIES IN MEXICO

P. pygmaeus is a monotypic species.

DESCRIPTION: *Procyon pygmaeus* is the smallest species of the genus. Compared to *P. lotor*, the pygmy raccoon is 15% to 18% smaller, nearly 50% lighter, and with a tail 28% to 37% shorter (García-Vasco, 2005; McFadden, 2004). It has sexual dimorphism with males being larger than females. Its external and skull measures are markedly shorter; the teeth are smaller with a lighter coloration than the ones of its continental relative *P. lotor* (Cuarón et al., 2004; Genoways and Jones, 1951; Hall, 1981). Its pelage resembles that of *P. lotor*, but with some distinctive details. In the upper parts, it has a gray to lighter coloration, the middle dorsal part is gray mixed with buffy hair, the external lower parts have light buffy tones and toward the interior, light brown coloration (Hall, 1981; Navarro et al., 1991). The facial mask is browner than in *P. lotor* and is surrounded by white lines. A dark brownish to black patch crosses the throat; the tail is golden yellowish or ochraceous buffy with six or seven dark rings that are fainter on the underside, and there is a pronounced orange coloration on the neck scruff region and dorsal part of the tail that is more noticeable in males (Cuarón et al., 2004). Their feet are semi-plantigrade or plantigrade.

EXTERNAL MEASURES AND WEIGHT

TL = 740 to 790 mm; TV = 243 to 256 mm; HF = 92 to 99 mm; EAR = 43 to 45 mm. Weight: 3.1 to 3.8 kg.

DENTAL FORMULA: I 3/3, C 1/1, PM 4/4, M 2/2 = 40.

NATURAL HISTORY AND ECOLOGY: Information about the pygmy raccoon's habits was very scarce until a few years ago, when details about distribution, population ecology, genetics, diseases and parasites, diet, habitat, and space use were obtained (Cuarón et al., 2004). They tend to be more active during sunset and night. They climb trees with ease (Jones and Lawlor, 1965). These raccoons are omnivores. In Punta Chunchacab, they have been reported to eat crabs, ants, and lizards. They also feed on seeds of fruits and seeds from guamuchil (*Pithecellobium*), sapodilla (*Manilkara achras*), and grass leaves (*Panicum* sp.; Navarro and Suárez, 1989). Crabs represent more than 50% of the pygmy raccoon's diet, with fruits and insects also being very important (McFadden et al., 2006). Fecal analysis also showed that the relevance of the different food items varied importantly between seasons and sites (McFadden et al., 2006) and may change very significantly after major alterations in habitat availability or quality such as after Cozumel was hit by major hurricanes such as Emily and Wilma in 2005, when a decrease in the importance of crabs and plant material in the diet was observed and a proportional increase in the importance of vertebrates was recorded (D. Martínez-Godínez, unpubl. data). Data about weight, dentition, and signs of lactation of captured females in more than three years of intensive trapping efforts suggest that the majority of births should occur between November and January (García-Vasco, 2005; McFadden, 2004). But it is very likely that there is another birth peak in June–July, when nursing females also have been captured (Navarro and Suárez, 1989). It is likely to form small family groups; several individuals were seen together in a crop of palm (Jones and Lawlor, 1959, 1965). They are more frequently found in mangroves, particularly at the northwestern portion of Cozumel. They can also be found in lower densities in semi-evergreen and subdeciduous tropical forests and agricultural areas (Bautista, 2006; Copa-Alvaro, 2007; García-Vasco, 2005; McFadden, 2004; Navarro and Suárez, 1989). In Isla de La Pasion, which has an extension of 0.4 km^2, lives a resident population, which in July 1988 included around 20 individuals (Navarro and Suárez, 1989). Recent estimates showed that subpopulation sizes range between sites from 11 to 45 individuals and an average density of 22 ± 5.1 raccoons/km^2 (Copa-Alvaro, 2007; McFadden, 2004; McFadden et al., 2009). Density estimates also varied between sites and year of study, and ranged from 12.4 individuals/km^2 to 112 individuals /km^2 (Copa-Alvaro, 2007; McFadden, 2004). In some agricultural sites, as in El Cedral, it was considered a pest, so it was hunted and poisoned. On the Island of La Pasion, habitat disturbance by intense tourism can significantly affect the resident population (Navarro and Suárez, 1989). A conservative average home range of 67 ha has been estimated based on 8 radio-tracked individuals (McFadden et al., 2010). There are no natural predators for pygmy raccoons, but they are preyed on by feral dogs (Bautista, 2006). *P. pygmaeus* is particularly vulnerable to introduced pathogens and diseases such as mange, rabies, and canine distemper from exotic animals (Cuarón et al., 2004). The parasites *Eimeria nutalli*, *Placoconus lotoris*, *Capillaria procyonis*, and *Physaloptera* sp. have been identified in individuals of this carnivore. Also, it is confirmed that pygmy raccoons have been exposed to *Toxoplasma gondii* and to canine hepatitis, canine distemper, and feline panleukopenia viruses (McFadden et al., 2010; Mena, 2007).

DISTRIBUTION: This species is endemic to the island of Cozumel (1,989 km^2), located in the Caribbean Sea off the coast of Quintana Roo, in southern Mexico (Cuarón et al., 2004; Navarro and Suárez, 1989; Wozencraft, 1993).

VEGETATIONAL ASSOCIATIONS AND ELEVATION RAGE: *P. pygmaeus* is found in tropical subdeciduous and semi-evergreen forests, mangroves, and crops of palms. It is found from sea level up to 20 masl.

CONSERVATION STATUS: In Mexico *P. pygmaeus* is considered a species at risk of extinction due to its restricted distribution and the alteration of its habitat (Ceballos and Navarro, 1991; SEMARNAT, 2010). It is considered critically endangered by IUCN, with population trends decreasing (IUCN, 2010; Cuarón et al. 2004). Major threats to pygmy raccoons are introduced continental raccoons (genetic introgression), introduced predators, parasites and diseases from exotic animals, habitat fragmentation, hunting and capture as pets, and hurricanes (Copa-Alvaro, 2007; Cuarón et al., 2004, 2009).

Procyon pygmaeus. Photo: Christopher Gonzalez Baca.

Order Perissodactyla

Gerardo Ceballos

The Indians considered killing a tapir an act of great courage; the skin or parts of its body were offered in memory of its great son, as I myself have seen. They call it Tzimin and so they have called the horse.

— A.M. Tozzer, 1941

The order Perissodactyla is a little, diverse order consisting of 3 families and 17 species of tapirs, horses, donkeys, and rhinoceroses, which are distributed in Africa, Asia, Europe, and America (Wilson and Reeder, 2005). In Mexico, the order is exclusively represented by one family (Tapiridae) and a species of tapir. The order is characterized by the presence of an odd number of fingers — one, three, or five — covered with hooves in the extremities (Nowak, 1991). Although the variation in size is not as great as in the artiodactyls, it is quite impressive since the tapirs are the species of smaller size, weighing 225 kg, while the rhinoceroses, which are the largest, weigh up to 4,000 kg. No species has true horns, as in the artiodactyls, although the rhinoceros has "horns" on the face, which are actually fibrous dermal structures covered with keratin.

Perissodactyls are herbivores and their diet consists of grasses and herbs. In general, their teeth are highly specialized with extensive occlusal surfaces that allow them to crush their food. They have terrestrial habits, but tapirs spend considerable periods in the water. They are solitary or live in couples or herds. In the case of the zebras and other equidae, they form groups of thousands of individuals.

Most of the perissodactyls are in danger of extinction because of indiscriminate hunting and destruction of their habitat. In the case of some species like the Sumatran rhinoceros (*Dicerorhinus sumatrensis*) and the black rhino (*Diceros bicornis*), with populations only in the tens or hundreds of individuals, the situation is extremely critical.

(*Opposite*) Central American tapir (*Tapirus bairdii*). Tropical semi-green forest. Tuxtla Gutiérrez Zoo, Chiapas. Photo: Claudio Contreras Koob.

Family Tapiridae

The family Tapiridae comprises a single genus and four species. It has a very peculiar distribution since it is known in the Neotropics and the southeast of Asia (Wilson and Reeder, 2005). In Mexico, it is only represented by Baird's tapir (*Tapirus bairdii*), whose distribution has been severely affected by the destruction of its habitat and hunting. Currently, it is at risk of extinction.

Tapirus bairdii (Gill, 1865)

Baird's tapir

Iván Lira, Ignacio J. March, and Eduardo Naranjo

SUBSPECIES IN MEXICO

T. bairdii is a monotypic species.

DESCRIPTION: *Tapirus bairdii* has a robust body with a small tail and large head. Its nose is elongated into a large, short, and prehensile proboscis. The limbs are short and thick, with four fingers in the forefeet and three in the hind feet. The coat is

Tapirus bairdii. Tropical semi-green forest. Tuxtla Gutiérrez Zoo, Chiapas. Photo: Claudio Contreras Koob.

short; it is dark brown to black on most of the body, and pale on the chest, throat, and tips of the ears. The offspring are reddish-brown with a pattern of spots and white stripes (Leopold, 1959).

EXTERNAL MEASURES AND WEIGHT
TL = 1,930 to 2,020 mm; TV = 70 to 100 mm; HF = 372 to 380 mm; EAR = 130 to 140 mm.
Weight: 150 to 300 kg.

DENTAL FORMULA: I 3/3, C 1/1, PM 4/3, M 3/3 = 42.

NATURAL HISTORY AND ECOLOGY: Tapirs inhabit extensive forested areas (1,000 ha) with little disruption and with permanent bodies of water. They are excellent swimmers and the bodies of water are important elements of their habitat because they are used as refuges in case of danger and as sites of rest during the warmest hours. On land, tapirs circulate in a complex network of trails that allow them to move rapidly through the vegetation. They are strict herbivores; leaves, tender buds, fruits, flowers, and numerous crust species of plants constitute most of their diet; hence they act as important dispersers and/or predators of many plants. Their foraging and travel activities have a strong impact on the structure and dynamics of the vegetation in the areas where they inhabit. Mating can occur at any time of the year. The gestation period lasts from 390 to 400 days (approximately 13 months), after which a single young is born, rarely two. The offspring loses its mottled pattern at six months old, but stays with the mother at least one year. Females reach sexual maturity between two and three years of age and males at three years. They are usually solitary, although sometimes they can form small groups of two to five individuals during periods of estrus (Naranjo and Cruz, 1998). Its population density is generally less than 0.6 individual/km^2, and its home range varies from 1 km^2 to 4 km^2, although females with offspring use areas that are considerably smaller. They are very active during the first and last hours of the night, although they can occasionally move during the day in areas with little human activity. The jaguar and crocodiles are the main predators of the offspring and juveniles, while people hunt adults. The communication between congeners consists in acute vocalizations, similar to whistles. Tapirs' senses of smell and hearing are well developed, which usually enables them to flee from their natural enemies before being detected. Tapirs frequently defecate in water and spray their urine on plants and other objects, which may be linked to territorial marking and family communication. In Panama, mutualism with the coati (*Nasua narica*) has been observed since they feed from the tapir's ticks. In Mexico and Central America, tapirs are a common species hunted as food.

DISTRIBUTION: *T. bairdii* is distributed from the southeast of Mexico to the northwest of Colombia. Its historic distribution in Mexico included wet tropical forests from Veracruz to Oaxaca, Chiapas, and the Yucatan Peninsula. Currently, their distribution is severely restricted, and isolated populations are only found in Veracruz, Oaxaca, Chiapas, and Quintana Roo. It has been registered in the states of CA, CS, QR, OX, and VE.

VEGETATIONAL ASSOCIATIONS AND ELEVATIONAL RANGE: Tapirs inhabit tropical forests and subdeciduous tropical forests, cloud forests, and wetlands. In Mexico it is found from sea level to 2 500 masl. Most of the localities where it inhabits are below 600 masl.

CONSERVATION STATUS: *T. bairdii* is listed in appendix I of the Convention on International Trade in Endangered Species (CITES), considered a vulnerable species by the International Union for Conservation of Nature (IUCN), and catalogued as at risk of extinction by the U.S. Endangered Species Act. In Mexico it is considered at risk of extinction by Norma Oficial Mexicana (NOM-59-ECOL). The populations of tapir in Mexico are now restricted to the scarce wild areas in the humid tropics of the southeast of the country. The species has been exterminated in the states of Yucatan and Tabasco, and very few individuals survive in Veracruz. The main causes of its local extinction are the fragmentation or loss of habitat and unregulated hunting.

Pronghorn antelope (*Antilocapra americana*). Scrublands. Guerrero Negro, Baja California. Photo: Claudio Contresas Koob.

Mule deer (*Odocoileus hemionus*). Pine forest. San Pedro Martir National Park, Baja California. Photo: Gerardo Ceballos.

Order Artiodactyla

Gerardo Ceballos

In accordance with the testimony of the elders, the bison was very abundant in Monclova and Parras when the first colonizers arrived, probably half a century after the conquest. For some years, they killed large numbers for food, but suddenly they ceased to appear.

— Edward A. Palmer, 1881

The order Artiodactyla is one of the most diverse orders among mammals. There are 10 families and 220 species, which include pigs, peccaries, wild boars, hippos, camels, deer, and giraffes (Wilson and Reeder, 2005). They are distributed in the entire planet, with the exception of Australia, New Zealand, Antarctica, Madagascar, and other islands (Nowak, 1999b). In Mexico there are 10 species from 4 families: Antilocapridae (pronghorn), Bovidae (bison and bighorn sheep), Tayassuidae (peccaries), and Cervidae (deer). The main characteristic of the order is the presence of a pair—two or four—of toes. The only exceptions are the members of the genus *Tayassu* (Nowak, 1999b). In general, the legs are long and thin and the toes are fitted with hooves. The variation in sizes is impressive. The Tragulidae (mouse deer), which weigh from 1 kg to 8 kg, are the smallest species, while hippos can weigh up to 8,000 kg. Species of this order have glands in different parts of the body, which are especially developed in the ruminants and secrete important substances for their sex and social life (Ceballos and Miranda, 2000). The families of cattle, deer, and pronghorn have horns or antlers, which can be perennial or annual. In cattle such as the gaur (*Bos gaurus*) or the bison (*Bison bison*) the horns can develop in both sexes or only in the male; they are permanent, without ramifications, and are formed by a bone base with a corneum. In the family Antilocapridae, horns have one ramification and are exclusive to males. In this group the corneum changes annually. The family Cervidae has another kind of bony process known as antlers, which are shed and regrow each year; their growth might or might not ramify, and they often occur only in males. The pigs, tragulidae, and hippos have highly developed incisors and canines (Nowak, 1999b). The majority of the species in this order are herbivores. A few species such as the peccaries are omnivores, and feed on fruits, plants, and animals (Eisenberg, 1981; Sowls, 1997). The majority of the herbivore species are called ruminants because they have a very complex stomach that allows them to feed on plants that are poor in nutrients and hard to digest (Nowak, 1999b). The stomach is divided into four chambers: the rumen, the reticulum, the omasum, and the abomasum. The food is swallowed almost without chewing and enters the rumen, which maintains a huge amount of anaerobic bacteria and protozoa symbionts that ferment and degrade the cellulose. The fermented food is returned to the oral cavity, where it is chewed (rumination), and then sent to the omasum and the abomasum. In these organs digestion is carried out; the food is broken down further and is channelled to the duodenum, where nutrients are absorbed (Eisenberg, 1981; Nowak, 1999b). The majority of the species are terrestrial, but some, such as the sitatunga (*Tragelaphus spekii*), are adapted to survive in flooded sites, and others, such as the hippo (*Hippopotamus amphibius*), are definitely amphibious. Artiodactyls inhabit all kinds of environments, from pastures and tropical meadows to the wastelands of high mountains. They can be solitary or live in couples or herds of thousands of individuals; they can be active during the day, at twilight, and at night. The caribou (*Rangifer tarandus*) in Canada and the wildebeest (*Connochaetes taurinus*) in East Africa (Nowak, 1999b) undertake local movements or extensive migrations. In the African and Asian savannas there are groups formed by individuals of several species. In general, they are polygamous. During the breeding season encounters between males battling over the females are frequent. The dominant male controls the majority of females, which after a period of prolonged gestation give birth to one or two offspring. In pigs and wild boars the number of offspring is considerably higher, up to 15 (Sowls, 1997). The offspring are generally precocious, and can walk or run a few hours after birth. This precocity is an adaptation to avoid predators (Geist, 1998). The family has great economic importance.

Family Tayassuidae

The family Tayassuidae is a little, diverse family composed of only three species, which have a restricted distribution to the American continent from the south of the United States to Argentina (Wilson and Reeder, 2005). The peccary of Chaco (*Catagomus wagneri*) was not scientifically described until 1975 (Wetzel, 1977). Two species inhabit Mexico, the collared peccary (*Pecari tajacu*) of wide distribution and the white-lipped peccary or senso (*Tayassu pecari*), whose populations are currently found in only a few regions and is considered in danger of extinction.

Pecari tajacu (Linnaeus, 1758)

Collared peccary

Ignacio J. March and Salvador Mandujano

SUBSPECIES IN MEXICO

Pecari tajacu angulatus (Cope) 1889
Pecari tajacu crassus Merriam 1901
Pecari tajacu humeralis Merriam 1901
Pecari tajacu nanus Merriam 1901
Pecari tajacu nelsoni (Goldman) 1926
Pecari tajacu sonorensis (Mearns) 1897
Pecari tajacu yucatanensis Merriam 1901

DESCRIPTION: The collared peccary is an artiodactyl similar in size to a medium-sized dog. It is the smallest of the Tayassuidae family (Sowls, 1997).The body is robust, the tail is vestigial, and the head is large. The canines are widely developed, and the nose ends in a nasal disk. The legs are short and thin, and end in hooves; the

(*Right and opposite*) *Pecari tajacu*. Tropical dry forest. Chamela-Cuixmala Biosphere Reserve, Jalisco. Photos: Gerardo Ceballos

forefeet have four digits and the hind feet three, and in both cases only two fingers are functional. The coloration of adults varies from gray to black in the limbs and trunk; the belly and the tip of the ears are pale; it has a yellow or white stripe in the shape of a necklace on both sides of the neck. In the northern part of its distribution the coloration in the adults becomes more obscure in winter. The color of the piglets varies from yellow to light brown or reddish, with a darker stripe in the dorsal midline; after three months of age, the coloration of the offspring changes to that of an adult. Bristles, which are thicker and longer on the back than on the rest of the body, constitute the coat; it has an erectile mane along the dorsal line, starting in the head (Leopold, 1959). They have a musk gland in the lower part of the back. There is no significant sexual dimorphism; individuals in the United States and north of Mexico are larger than those in Central and South America (Sowls, 1997).

EXTERNAL MEASURES AND WEIGHT
TL = 800 to 980 mm; TV = 19 to 55 mm; HF = 170 to 200 mm; EAR = 70 to 100 mm.
Weight: 17 to 30 kg.

DENTAL FORMULA: I 2/3, C 1/1, PM 3/3, M 3/3 = 38.

NATURAL HISTORY AND ECOLOGY: In the wild, females possibly acquire sexual maturity in the first year of age, and remain fertile until death. The males mature between 10 and 11 months of age; the production of semen declines at 7 years of age. The males are sexually active throughout the year. Estrus lasts three to five days, occurring in cycles of 22 to 24 days throughout the year. The gestation period varies from 141 to 151 days, with 146 on average. The average size of the litter is two offspring, although there are deliveries of one or three offspring (Lochmiller et al., 1984, 1986; Mayer and Brandt, 1982). In captivity, the greatest registered longevity was 24 years and 8 months (Crandall, 1964), while in the wild the average life span is of about 7 years (Sowls, 1997). The peccary is an omnivorous animal, although mainly herbivorous. In arid areas of the north of its distribution, cacti predominate in its diet (Corn and Warren, 1985; Eddy, 1961; McCoy et al., 1990; Sowls, 1997). In the tropical forest it feeds on fruit, seeds of palms, roots and tubers, herbs, snails, and other small animals (Kiltie, 1981a, b, c). Commonly, it can cause considerable damage to

DISTRIBUTION: *Pecari tajacu* is distributed from the south of the United States to the north of Argentina and Uruguay (Sowls, 1997). In Mexico, they extend from the southern states and the Yucatan Peninsula to the north of the country along the coasts and the two cordilleras (western and eastern Sierra Madre) until they reach Texas by the Gulf, and Arizona, New Mexico, and Texas to the west (Reyna-Hurtado et al., 2008). The collared peccary is only absent from the highlands and large populated central areas of the center of the country from Edo in Mexico to the U.S. border. Recently it was recorded in Morelos state, which is defined as the northern record of the central area of the country (Mason-Romo et al., 2008). It has been registered in the states of AG, CA, CH, CL, CO, CS, DU, GJ, GR, HG, JA, MI, MO, NL, NY, OX, PU, QE, QR, SI, SL, SO, TA, TB, VE, YU, and ZA.

maize and other crops. The diet of the white-lipped peccary in the deciduous tropical forest of Jalisco is composed of 46% and 50% of roots, 43% and 39% of leaves, and 10% and 11% of fruits during the rainy and dry seasons, respectively (Martínez-Romero and Mandujano, 1995). Roots are the main constituents during the dry season, while in the rainy season leaves are. In other tropical deciduous forests it has also been observed that the consumption of roots of low nutritional quality increases with the reduction in the availability of leaves and fruits with the greatest nutritional quality (McCoy et al., 1990; Olmos, 1993). The peccary consumes fruits of *Opuntia excelsa*, *Ficus* spp., *Brosimum alicastrum*, *Spondias purpurea*, *Vitex mollis*, *Sideroxylon capiri*, and some legumes (Martínez-Romero and Mandujano, 1995). Plums (*Spondias*) are favored by chachalacas (*Ortalis poliocephala*), whose foraging activities increase the rate of decline in these fruits (Mandujano and Martinez-Romero, 2006). A similar phenomenon has been associated with troops of monkeys foraging for fruits in trees (Robinson and Eisenberg, 1985). In the United States, herds of up to 50 individuals have been reported, while in tropical areas smaller groups of up to 20 (but usually of 6 to 9) individuals are more common (Day, 1985; Ellisor and Harwell, 1969; Schweinsburg, 1971; Sowls, 1997). In the tropical forests of Jalisco, up to 12 individuals have been found in a herd, but groups of 4 peccaries are more common (Mandujano, 1999). Peccaries form relatively large herds (5 to 20 individuals) in tropical forests. Throughout the year adult animals and young animals form groups; between October and November small offspring have been observed, so it is very likely that births occur between July and August. In El Eden, a semi-deciduous forest of the Yucatan Peninsula, groups of 2 to 15 were observed (González-Marín et al., 2008). In the Calakmul region the largest group observed was made up of 9 individuals, while in the Lacandon Forest groups of 1 to 15 have been observed with a mean of 2.3 per group (Reyna-Hurtado, 2009). In Jalisco the population density is 1.2 ± 0.9 herds/km^2 with a density of individuals of 4.9 ± 1.6 peccaries/km^2 (Mandujano and Martinez-Romero, 2006). But in the semi-deciduous forests in that region, density seems to be larger (Ceballos and Miranda, 2002). These values are similar to those reported for other subspecies in tropical forests of Central and South America. In Peru 1.2 herds/km^2 (Kiltie and Terborgh, 1983) and in Panama 7 individuals/km^2 (Eisenberg and Thorington, 1973). In Venezuela, Panama, and Peru, 47%, 75%, and 87% of the herds, respectively, were formed by less than 5 individuals (Kiltie and Terborgh, 1983; Robinson and Eisenberg, 1985). In contrast, in Arizona, where the same subspecies of peccary is distributed as that in Chamela, the density is 11.8 ± 4.4 peccaries/km^2 and the average size of the herds fluctuates between 7 and 16 animals, reaching herds of up to 20 or 30 peccaries (Sowls, 1997). In the tropical forests, there is a subdivision of the herds into smaller groups (Robinson and Eisenberg, 1985), as well as in the arid areas of Texas (Green et al., 1984), where a herd of peccaries is subdivided into smaller groups where there is low availability of food resources. In contrast, the whole herd and, on some occasions, aggregations of different herds in the same place, occur in areas with high abundance of food. The area of activity of a herd varies according to the latitude and time of year, and may cover 1.6 km^2 to 5 km^2; these peccaries mark their territory with musk and defend it actively (Fragoso, 1998; McCoy et al., 1990; Robinson and Eisenberg, 1985). The dorsal gland is involved in marking the territory and in social interactions (Byers and Bekoff, 1981; Mayer and Brandt, 1982; Sowls, 1974, 1997). They are active during the day and night. Predators include the puma (*Puma concolor*), the jaguar (*Panthera onca*), the coyote (*Canis latrans*), the black bear (*Ursus americanus*), and the golden eagle (*Aquila chrysaetos*) (Sowls, 1997). Potentially, the ocelot (*Felis pardalis*), the crocodile, the boa (*Boa constrictor*), and large raptors may prey on the offspring. In most of Mexico, the main predators of peccary are the jaguar and the puma (Ceballos et al., 2007). In subperennial forests, the likelihood of peccaries being hunted by one of their predators decreases if individuals remain in a large herd. The collared peccary is intensely hunted both for sport and for their meat and skins to meet the livelihoods of rural people (Vickers, 1984). In South America, the hunting pressure is very high due to the large-scale trade of skin (Broad, 1984). In the south of the United States, sport hunting of this species generates considerable income (Day, 1985). In Mexico, the collared peccary is heavily hunted in the areas where it still survives. In some forests around the Calakmul Biosphere Reserve, the collared peccary was the most frequently hunted animal for one community (Weber, 2005) and among the top five prey species in three others (Escamilla et al., 2000).

VEGETATIONAL ASSOCIATIONS AND ELEVATIONAL RANGE: The collared peccary occurs in a wide spectrum of vegetation types, which include tropical perennial forests, subdeciduous and deciduous forests, thorny forests, xeric scrublands, pastures, oak forests, pine forests, cloud forests, and transformed areas or areas with secondary vegetation (Sowls, 1997). It has been found from sea level up to 3,000 m. In Mexico, most of the records are located at sea level and 800 masl.

CONSERVATION STATUS: The collared peccary is listed as a species of least concern by the IUCN (IUCN, 2010) but is not included in the endangered species list of Mexico (SEMARNAT, 2010). This species is still relatively common around large forested areas and where large forest fragments intersperse with croplands. Most of its populations are suffering heavy hunting pressure, especially around large protected areas such as Calakmul, El Triunfo, La Sepultura, Montes Azules, Sian Ka'an Biosphere, Chamela-Cuixmala, Zicuiran-Infiernillo Reserves (Ceballos and Oliva, 2005). The species is listed in appendix II of CITES. It disappeared from large areas in the center of Mexico, but in the areas where it remains it can be very abundant. This is the case of the coast of Jalisco, San Luis Potosí, and Tamaulipas (G. Ceballos, pers. comm.), and in areas of tropical forest such as Calakmul in Campeche, the Lacandon Forest, and El Ocote in Chiapas.

Tayassu pecari (Link, 1795)

White-lipped peccary

Rafael Reyna-Hurtado and Ignacio J. March

SUBSPECIES IN MEXICO

Tayassu pecari ringens Merriam, 1901

DESCRIPTION: *Tayassu pecari* is a medium-sized, pig-like ungulate with an elongated snout that has a disk of round cartilage at the tip. The extremities end up in hooves, with four digits in the forefeet and three in the hind feet; in both extremities only two fingers are functional. It has a vestigial tail, and in the posterior middle part of the back it has a gland that secrets a strong and white musk. It has four pairs of breasts; only the two posterior ones are functional. The pelage is made up of coarse bristles. The color of adults is black over almost all the body with the exception of a white line around the lips that extends into the cheeks (which gives the species its name) and some reddish areas around the neck. The color of neonates is a mix of red-brown, black, and cream with some stripes; by the third year of age, the color changes from reddish to black (Mayer and Wetzel, 1987). Adult white-lipped peccaries are between 930 mm and 1,300 mm in length. Some individuals measured in the Calakmul Biosphere Reserve ranged from 1,040 mm to 1,240 mm in length (Reyna-Hurtado,

2009). Body mass ranges from 25 kg to 42 kg along its distribution range. Twelve individuals weighed in Calakmul were between 25 kg and 33 kg with a mean of 28.1 kg (Reyna-Hurtado, 2009). The sexes are not distinct morphologically. Adults have well-developed canines; the upper canines are pointed downward, and with constant chewing they acquire a cutting-edge. While canines have little functionality for feeding, they play an important role in defense against predators and in the agonistic interactions between congeners.

EXTERNAL MEASURES AND WEIGHT
TL = 1,100 to 1,200 mm; TV = 50 mm; HF = 215-225; EAR = 78 to 85 mm.
Weight: 35 to 42 kg.

DENTAL FORMULA: I 2/3, C 1/1, PM 3/3, M 3/3 = 38.

NATURAL HISTORY AND ECOLOGY: The white-lipped peccary is predominantly diurnal. They form herds of up to 300 individuals or more. Smaller herds are seen in perturbed areas or where hunting pressure is high (March, 1993). In Mexico group sizes reported for Calakmul vary from 11 to 35 (Reyna-Hurtado, 2009) and from 5 to 60 for Montes Azules, Chiapas (Naranjo, 2002). They are very mobile and can travel up to 16 km daily (Reyna-Hurtado et al., 2008). The white-lipped peccaries have one of the largest home ranges of tropical ungulates in the world as they move in a range several times larger than that of any other ungulate in tropical forests. They require a large home range to fulfill all their requirements for survival. In the only estimate of home range for white-lipped peccaries in Mexico (at the Calakmul Biosphere Reserve), it was found that they moved over 77.2 km^2 and 121.8 km^2 (minimum convex polygon [MCP] for two groups of 25 and 18, respectively; Reyna-Hurtado, 2009). In the Peruvian Amazon forest the area of activity of a herd has been estimated to be 60 km^2 to 200 km^2 (Kiltie and Terborgh, 1976, 1983) and 109

Tayassu pecari. Savanna. Photo: Gerardo Ceballos.

km^2 to 200 km^2 (Fragoso, 1998, 2004) for the Brazilian Amazon. It is likely that during certain seasons, large herds separate into smaller groups depending on the distribution and abundance of available food (March, 1990); however, four groups in Calakmul showed great fidelity and remained cohesive for 18 months (Reyna-Hurtado, 2009; Reyna-Hurtado et al., 2008). White-lipped peccaries are omnivorous, although they feed mainly on fruits and seeds (Kiltie, 1981b). Reyna-Hurtado et al. (2008) found that fruit accounts for 81% of the diet of this species in Calakmul, Campeche. Some authors have proposed that, like other cranial characteristics, the closed adjustment of the canines is an adaptation to avoid dislocation of the lower jaw when these mammals bite hard fruits and seeds (Kiltie, 1981a). The sense of smell is very developed, allowing them to detect food under the ground as well as potential predators, including humans. They prefer areas with dense vegetation or well-conserved forests with abundant water bodies. They reproduce throughout the year, with greater frequency when there is a greater availability of food. In Mexico, births are more frequent between December and March (Reyna-Hurtado, 2009). The gestation period lasts 156 to 162 days (Roots, 1966), with an average of two offspring per pregnancy. They reach sexual maturity between 12 and 24 months of age. At the back of the body they have a musk gland that is used for social recognition among members of the groups (Sowls, 1997). Their predators include humans, jaguars, pumas, and possibly crocodiles. Since pre-Columbian times the white-lipped peccary has been an important source of animal protein in the livelihoods of many indigenous and rural groups of Latin America (March, 1987; Vickers, 1984). For several decades, this species has been subjected to a large-scale exploitation to meet the demands of the European and Japanese pelt markets (Broad, 1984). In Peru a certification process is underway for the extraction of several thousand peccary skins every year from the extensive communal forests of the Loreto Department that are managed under sustainable plans (Reyna-Hurtado et al., 2008).

DISTRIBUTION: The historic distribution of the white-lipped peccary covered most of the southeast of Mexico, almost all Central America, reaching the north part of the Ecuador, Colombia, and throughout the Amazon Basin to the northeast of Argentina, Bolivia, and Paraguay (Eisenberg, 1989; Hall, 1981; Sowls, 1997). In Mexico, it was originally distributed from southern Veracruz and east of Oaxaca, through Tabasco and low-lying areas of Chiapas, to the Yucatan Peninsula (Ceballos and Navarro, 1991; Leopold, 1959). It was determined recently, however, that the historical range of the species has been reduced by 84 % (Reyna-Hurtado et al., 2008). Currently, it only persists in some of the tropical forests of greater extension as the biosphere reserves of Calakmul in Campeche, Montes Azules in Chiapas, Chimalapas in Oaxaca, and Sian Ka'an in Quintana Roo. It has been registered in the states of CA, CS, OX, QR, TB, VE, and YU.

VEGETATIONAL ASSOCIATIONS AND ELEVATION RANGE: In Mexico, white-lipped peccaries mainly inhabit well-conserved perennial tropical forests and subperennial forests, and occasionally semi-dry tropical forests such as some areas in the Yucatan Peninsula (Álvarez del Toro, 1977; Dalquest, 1949; Hall and Dalquest, 1963; Leopold, 1959; March, 1990, 1991; Reyna-Hurtado, 2009). In South America, it also occurs in the dry savannas of Venezuela and in the xeric areas of Chaco in Paraguay (Sowls, 1997). It is found from sea level up to 800 m; however, the majority of the localities are situated at less than 325 masl (March, 1990). In South America, they have been also found at 1,900 masl on the eastern side of the Andes in Peru (Osgood, 1914).

CONSERVATION STATUS: The IUCN Red List (2010) raised this species' status from lower risk/least concern to near threatened in 2008 due to the rapid decline this species is experimenting (Reyna-Hurtado et al., 2008). In Mexico it was determined recently that the species has been extirpated in 84% of its historical range. More than 50 years ago, Leopold (1959) pointed out the reduction in numbers and range of this species in Mexico due to habitat loss and excessive hunting. So true were his predictions that nowadays white-lipped peccaries have disappeared from Veracruz, Tabasco, and Yucatan states, and isolated populations remain in Oaxaca and Quintana Roo, with the only stable populations in Chiapas and Campeche (March, 1993, 2005; Reyna-Hurtado et al., 2008). The white-lipped peccary is also listed in appendix II of CITES. In Mexico it is considered in danger of extinction (SEMARNAT, 2010). In Mexico, the largest populations are found in the Calakmul Biosphere Reserve in Campeche and the Montes Azules Biosphere Reserve in Chiapas. The Chimalapas Biosphere Reserve in Oaxaca and the Sian Ka'an Biosphere Reserve in Quintana Roo maintain some herds.

Family Cervidae

Of the 43 species that comprise the family Cervidae, the deer are the most conspicuous. They are distributed in all continents with the exception of Australia. In Mexico five species of three genera are represented. The elk or American deer (*Cervus elaphus*) disappeared from Mexico in the early 1900s, although it has been reintroduced in Coahuila. The white-tailed deer (*Odocoileus virginianus*) is distributed in almost all the country and is abundant in many regions. The distribution of the mule deer (*Odocoileus hemionus*) has been reduced considerably due to poaching. The two species of brockets (*Mazama americana* and *M. pandora*) have tropical distributions; and despite the fact that only one species was considered to be in Mexico, recently the existence of *M. pandora* was proved (Medellín et al., 1998).

Subfamily Cervinae

The subfamily Cervinae contains fifty-six species in ten genera, with one species (*Cervus canadensis*) known from Mexico.

Cervus canadensis Erxleben, 1777

Elk, Wapiti

Manuel Weber

SUBSPECIES IN MEXICO

Cervus elaphus merriami (Nelson, 1902)

For a long time the American elk was considered a separate species (*C. canadensis*) and only recently has it been considered a subspecies.

DESCRIPTION: The elk is the largest deer of Mexico. Close to the skull, the antlers are large and angle upward and backward, with two to seven branches. The hindquarters and the perineum are yellowish-white, while the rest of the body is reddish-brown. Adult males have a mane around the neck, from the bottom of the jaw to the chest and part of the back (Hall, 1981; Thomas and Toweill, 1982).

EXTERNAL MEASURES AND WEIGHT

TL = 2,039 to 2,972 mm; TV = 80 to 213 mm; HF = 464 to 660 mm.

Weight: 120 to 350 kg.

DENTAL FORMULA: I 0/3, C 1/1, PM 3/3, M 3/3 = 34.

NATURAL HISTORY AND ECOLOGY: Although the wapiti is one of the species of mammals better studied in the world (Clutton-Brock et al., 1982), very little is known about the natural history of this species in Mexico since it was extirpated at the beginning of the century. In fact, *Cervis elaphus merriami* disappeared before its taxonomic status could be verified completely. The information about its natural history is based on the available literature of habitats similar to those of its original geographical distribution in Mexico. The major habitats of the elk in Mexico are forests of oak and pine-oak, with interspersed natural grasslands (Leopold, 1959). Its diet is based on pasture, but it also consumes herbs, leaves, and fruits of some trees and shrubs (Altmann, 1978). It has very gregarious habits, forming mixed groups of different ages and sexes numbering more than 50 individuals (Thomas and Toweill, 1982). The males form groups (harems) of 2 to 10 females during the reproductive season, which in Durango, Mexico, occurs in November and December (M. Weber, pers. obs.). In general, females are monotocous and twins are rather rare (Altmann, 1978; Clutton-Brock et al., 1982). The gestation period lasts between 230 and 260 days, depending on the subspecies. Reproduction is seasonal, and is mainly controlled by photoperiods and therefore by latitude (Clutton-Brock and Albon, 1989). It is a very territorial deer; males mark and defend territories of up to 70 km^2 (Thomas and Toweill, 1982). Its main predators in America are pumas and wolves. Occasionally, coyotes and lynx prey on fawns (Altmann, 1978; Thomas and Toweill, 1982). Due to the overlap of eating habits with domestic cattle and sheep, competition is a severe problem in the United States (Thomas and Toweill, 1982); several populations are infected with brucellosis and tuberculosis, which represents a serious animal health problem. The elk is subject to intensive handling for animal production (farms) as well as for hunting (ranches) in northern Mexico (Weber, 1993).

VEGETATIONAL ASSOCIATIONS AND ELEVATION RANGE: Originally, *C. canadensis* was distributed almost exclusively in temperate forests of conifers. Worldwide it has been introduced to a wide variety of temperate and even tropical ecosystems. It is found from sea level to heights above 3,000 m. In Mexico, it has been reintroduced in mountainous areas with heights of more than 2,000 m.

CONSERVATION STATUS: *C. elaphus merriami*, the Mexican subspecies, is considered extinct. There have been several attempts to reintroduce *C. elaphus nelsoni* in Mexico (Ceballos and Navarro, 1991; Leopold, 1959). Currently, some populations have been reintroduced in Coahuila (G. Ceballos, pers. comm.), Chihuahua, Durango, and Sonora, whose total does not exceed 500 individuals, all on private ranches. In addition, 900 European elk (*C. elaphus elaphus*) were imported from New Zealand to Mexico to boost the number of this species in several states of the country. Without good management, there is the possibility of hybridization of these animals with the American elk (*C. elaphus nelsoni*).

DISTRIBUTION: The global distribution of the elk is very wide; it includes Europe, Asia, North Africa, and North America. It has been introduced to other continents, including Central and South America, Australia, and New Zealand (Nowak, 1999b). In North America, its historic distribution included a large portion of Canada and virtually throughout the United States (Hall, 1981; Thomas and Toweill, 1982). Its original distribution in Mexico (*C. canadensis merriami*) was in the sierras of Chihuahua and Sonora along the border with the United States (Hall, 1981). The wapiti of the Rocky Mountains (*C. canadensis nelsoni*) has been recently reintroduced in Coahuila, Chihuahua, Durango, and Sonora (G. Ceballos, pers. comm.; Leopold, 1959). There is a wild population of around 500 animals in Coahuila (G. Ceballos, pers. obs.). It has been recorded in the states of CH and SO (Anderson, 1972; Leopold, 1965).

Subfamily Odocoileinae

The subfamily Odocoileinae contains thirty-seven species in ten genera, with four species in two genera known from Mexico.

Mazama americana (Erxleben, 1777)

Red brocket

Manuel Weber

SUBSPECIES IN MEXICO

Mazama americana temama (Kerr, 1792)

The species *pandora* was originally described by Merriam (1901). Hershkovitz (1951, in Genoways and Jones, 1975), however, considered it a subspecies under the name of *M. gouazoubira pandora*. The same author concluded later that it was a subspecies of the brocket deer (*M. americana*). Currently, *M. pandora* is considered a separate species (Medellín et al., 1998).

DESCRIPTION: *Mazama americana* is the smallest deer in North America. Unlike other deer, the male's antlers do not branch and rarely are more than 12 cm long, thick at the root and thinner at the end, oblique at the top and curving backward, with the surface covered with grooves (Gaumer, 1917; Nowak, 1999b). Shedding of the antlers occurs each year. Both males and females are sexually mature at approximately one year of age (Branan and Marchinton, 1984). Its small size, slender body, and long tail distinguish it from other species. At a young age, both males and females have canines that eventually disappear completely. The ears are big but not very long, the eyes are small with a barely marked lachrymal. The coat is reddish-brown in the back; *M. pandora* is browner. The belly and inner part of the extremities are white; the tail is brown on top and white underneath (Gaumer, 1917). They do not have a metatarsal gland. The canines may or may not be present. The coloration of the offspring is similar to that of the adults but six to eight longitudinal white spots or specks extend over the whole length of the back and the coat is denser and longer (Bisbal, 1991).

EXTERNAL MEASURES AND WEIGHT
TL = 1,050 to 1,420 mm; TV = 110 to 130 mm; HF = 260 to 280 mm; EAR = 90 to 120 mm.
Weight: 17 to 65 kg.

Mazama americana. Tropical semi-green forest. Tuxtla Gutierrez Zoo, Chiapas. Photo: Fulvio Eccardi.

DENTAL FORMULA: I 0/3, C 0-1/1, PM 3/3, M 3/3 = 32–34.

NATURAL HISTORY AND ECOLOGY: The gestation period lasts seven to eight months. In captivity, the gestation period lasts approximately 225 days. One to two fawns are born in each delivery, weighing between 510 g and 567 g, with a mottled coat that changes at two or three months of age (Thomas, 1975). The breeding season shows very little seasonality; in Suriname births occur from March to October, during the maximum rainfall peak (Branan and Marchinton, 1987). There is contradictory information about the shedding of the antlers; some authors affirm that it occurs, while others do not (Roa and Lozada, 1989). Their diet is based on fruits, fungi, leaves, and flowers; 57 species of plants of 36 families have been identified in the diet of the brock deer of Suriname (Branan et al., 1985). Álvarez del Toro (1977) mentions that a high percentage of its diet comprises wild fruits such as fig (*Ficus* spp.). In captivity, the plants (shrubs and trees mainly) accepted by a female in Catemaco, Veracruz, included flameberry (*Urera caracasana*), fig (*Ficus yoponensis*), *Odontonema callystachyum*, and whiteroot (*Gouania lupuloides*; Arceo-Castro and Sánchez-Mantilla, 1992). This deer may be an important disperser of seeds, thus having an important role in maintaining the diversity of tropical forests. The species has suffered strong pressure from hunting, habitat destruction, and farming (Ojeda-Castillo, 1991). It shows diurnal and nocturnal habits. They are solitary and are very difficult to sight due to their shyness and habit of staying motionless when they perceive any danger (Eisenberg, 1989; Emmons and Feer, 1997). Apparently, their activity area is small, and does not exceed a kilometer in diameter (Hall and Dalquest, 1963). Densities of 1 deer per km^2 have been reported in Suriname (Branan and Marchinton, 1984). In the region of Los Tuxtlas, Veracruz, Bello (1993) found more traces of this ungulate in comparison with those of white-tailed deer and peccary, and therefore concluded that there was a medium abundance of this species in the region; a greater density occurs in the best-preserved areas, which are far from human settlements, with rugged topography, and from 30 degrees to 70 degrees where the dominant vegetation is the perennial high jungle. In Quintana Roo, a density of 8.5 individuals/km^2 has been estimated in environments with dense canopy (Quinto, 1994). Its main predators are the jaguar and the puma (Bisbal, 1991), although Leopold (1959) also considers the ocelot and big birds of prey such as the harpy eagle (*Harpia harpyja*).

VEGETATIONAL ASSOCIATIONS AND ELEVATION RANGE: This species is mainly found in tropical rainforests, evergreen subperennial forest, and cloud forest. It is found from sea level up to 1,200 m (Bello, 1993). In South America it can be found at 4,000 masl (Bisbal, 1991).

CONSERVATION STATUS: *M. americana* is not considered at risk in Mexico because it can survive well in second-growth vegetation and perturbed areas. A permit for its hunting is required. Ceballos and Navarro (1991) considered it a fragile species since it is heavily hunted in some regions such as Quintana Roo (Quinto, 1994). In this regard more detailed studies are needed to prevent the risk of overexploiting the species.

DISTRIBUTION: *M. americana* is found from Mexico to South America; its distribution limit is northern Argentina, south of Bolivia, Brazil, and Paraguay (Grubb, 1993). In Mexico, it is distributed from the south of Tamaulipas to the southeast of the country, covering the entire Yucatan Peninsula, and along the Pacific coast in Chiapas and part of Oaxaca. It has been registered in the states of CA, CH, OX, PU, QR, SL, TA, TB, VE, and YU.

Mazama pandora Merriam, 1901

Yucatan brown brocket

Rodrigo A. Medellín

SUBSPECIES IN MEXICO

M. pandora is a monotypic species.

DISTRIBUTION: *M. pandora* is an endemic species that occupies the Yucatan Peninsula from the tropical forest in the south to the xeric scrubland in the north. It is possible that in the future it will be reported in northern Guatemala and Belize. Most recent records are from the Calakmul Biosphere Reserve in Campeche. It has been recorded from the states of CA, QR, and YU.

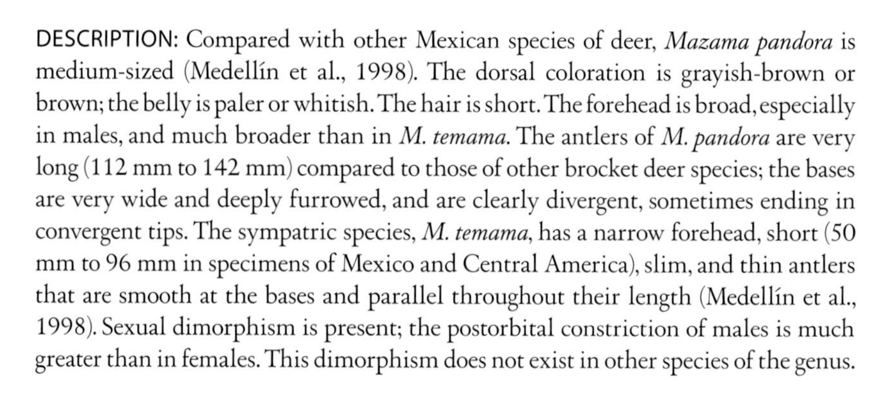

DESCRIPTION: Compared with other Mexican species of deer, *Mazama pandora* is medium-sized (Medellín et al., 1998). The dorsal coloration is grayish-brown or brown; the belly is paler or whitish. The hair is short. The forehead is broad, especially in males, and much broader than in *M. temama*. The antlers of *M. pandora* are very long (112 mm to 142 mm) compared to those of other brocket deer species; the bases are very wide and deeply furrowed, and are clearly divergent, sometimes ending in convergent tips. The sympatric species, *M. temama*, has a narrow forehead, short (50 mm to 96 mm in specimens of Mexico and Central America), slim, and thin antlers that are smooth at the bases and parallel throughout their length (Medellín et al., 1998). Sexual dimorphism is present; the postorbital constriction of males is much greater than in females. This dimorphism does not exist in other species of the genus.

EXTERNAL MEASURES AND WEIGHT

TL = 1,120 mm; TV = 75 mm; HF = 262 mm; EAR = 110 mm.
Weight: 21 kg.

DENTAL FORMULA: I 0/3, C 0/1, PM 3/3, M 3/3 = 32.

NATURAL HISTORY AND ECOLOGY: This deer coexists with two other species of deer in the south of the Yucatan Peninsula: the brocket deer (*M. temama*) and the white-tailed deer (*Odocoileus virginianus*). In the northern part of the peninsula only

Mazama pandora. Specimen in captivity. Mexico City. Photo: Gerardo Ceballos.

white-tailed deer and brown brockets are found. This species occupies the rainforests of Campeche and the scrublands and thorny forests of Yucatan, where they live in densities between 0.09 deer/km^2 and 1.5 deer/km^2 (Weber, 2005). The hearing bulla is larger in this species, thus suggesting a more sensitive ear than that in other brocket deer (Medellín et al., 1998).

VEGETATIONAL ASSOCIATIONS AND ELEVATION RANGE: *M. pandora* occupies all the major habitats of Yucatan, from the tropical rainforest to the xeric scrubland, including the deciduous tropical forest. It appears to be more abundant in less humid regions of the peninsula (Medellín et al., 1998). Detailed studies, however, indicate that the species seems to use habitats according to their availability (Reyna-Hurtado and Tanner, 2005; Weber, 2005). It has been reported from sea level to approximately 300 m in the Sierrita de Ticul, Yucatan. The brown brocket deer consumes plant matter of 20 species and the diet is dominated by fruit of a few species of trees. Fawns have been observed from March through October (Weber, 2005).

CONSERVATION STATUS: The International Union for Conservation of Nature (IUCN) lists the species as vulnerable due to a continuing population decline caused by habitat destruction. The Mexican list of species at risk does not include this species in any risk category. Given their relatively restricted distribution, further studies should focus on this species to evaluate its status.

Odocoileus hemionus (Rafinesque, 1817)

Mule deer

Manuel Weber and Carlos Galindo Leal

SUBSPECIES IN MEXICO
Odocoileus hemionus cerrocensis (Merriam, 1898)
Odocoileus hemionus crooki (Mearns, 1897)
Odocoileus hemionus fuliginatus (Cowan, 1933)
Odocoileus hemionus peninsulae (Lidekker, 1898)
Odocoileus hemionus sheldoni (Goldman, 1939)

DESCRIPTION: *Odocoileus hemionus* is a big and robust deer. Unlike in the white-tailed deer, the antlers branch in a dichotomous manner. The tail is scarcely covered by hair; its coloration is a homogeneous yellow with dark tips. The ears are very large. Adult males have a dark reddish or almost black forehead.

EXTERNAL MEASURES AND WEIGHT
TL = 1,000 to 1,800 mm; TV = 106 to 230 mm; HF = 325 to 590 mm; EAR = 115 to 250 mm.
Weight: 50 to 140 kg.

DENTAL FORMULA: I 0/3, C 0/1, PM 3/3, M 3/3 = 32.

NATURAL HISTORY AND ECOLOGY: Only few systematic studies and long-term studies on the ecology of the mule deer in Mexico exist (Galindo-Leal, 1993; Weber, 1993; Weber and Galindo-Leal, 1994). Mule deer are intermediate selectors of

Odocoileus hemionus. Pine forest. San Pedro Martir National Park, Baja California. Photo: Gerardo Ceballos.

food (browse and graze; Hofmann, 1989). Their diet varies with time and space, and comprises tender leaves, buds, fruits of trees and shrubs, various herbs, and green grass (Hervert and Krausman, 1986; Ordway and Krausman, 1986). In some localities of Mexico, a large proportion of their diet consists of oak (*Quercus* sp.), guasapol (*Ceanothus* sp.), mesquite (*Prosopis* sp.), herbs (*Olneya* sp., *Dysodia*), cactus (*Opuntia*), sotol (*Dasylirion*), yuccas (*Yucca*), and several grasses (*Mullenberghia* and *Bouteloua*) (Leopold, 1959). Even when this species is adapted to xeric climates, it requires water sources and populations are greatly influenced by this factor (Bowyer, 1986; Hervert and Krausman, 1986). There are no studies on the reproduction of the mule deer in Mexico. Reports on the mating season in Mexico range from December to February (Leopold, 1959). The males are physiologically ready to mate in December, but the majority of the females enter estrus during January and February (M. Weber, pers. obs.). Fawns are born in August and September after an average gestation period of 208 days (Leopold, 1959; M. Weber, pers. obs.). In productive habitats twins are commonly born (Wallmo, 1981). The fertility of females is directly proportional to the quality and quantity of available food in the pre-reproductive period (Leopold, 1959). This species is relatively more gregarious than the white-tailed deer (*O. virginianus*). Females with newborn offspring and fawns of the previous year form social units of two to eight individuals. Young males associate in groups of 4 to 10 individuals. The adult males are solitary. *O. hemionus* is a polygynous species. Males stay with a female until they mate and then they move, sometimes long distances, to find another female entering estrus. They are probably a species with "facultative territoriality" (Geist, 1981). The activity areas vary between 14 km^2 and 45 km^2 for males, and 2 km^2 and 18 km^2 for females (Ordway and Krausman, 1986). The puma is the main predator of the mule deer in Mexico, although coyotes and lynx can play an important role in regulating its populations in some locations (Krausman and Ables, 1981; Leopold, 1959). The competition with domestic livestock for food, water, and shade, as well as sport hunting and livelihood hunting, are decisive factors in the distribution and abundance of the mule deer in Mexico (Galindo-Leal, 1993; Leopold, 1959; Weber, 1993). *O. hemionus* has great economic importance, and is considered a prized hunting trophy; local inhabitants and indigenous people use it as a source of meat. It has been overexploited in Mexico.

DISTRIBUTION: *O. hemionus* is found throughout the west of the United States and Canada, to the limits with the Yukon and Northwestern Territories (Hall, 1981; Wallmo, 1981). In Mexico, its historic distribution included the entire Baja California Peninsula, Sonora, and deserts of Chihuahua and plateaus of the center, extending to Zacatecas, San Luis Potosí, and southwest of Tamaulipas (Leopold, 1959). Currently, it has disappeared from much of its historical distribution and its populations appear to be isolated in patches in several states of the country. The other subspecies have a fairly restricted distribution. Two subspecies occupy the Baja California Peninsula: one in the north (*O. hemionus peninsulae*) and the other in the south (*O. hemionus fuliginatus*). The subspecies *O. hemionus cerrocensis* lives on Isla Cedros (Merriam, 1898c) in the Pacific and *O. hemionus sheldoni* on Isla Tiburon in the Gulf of California. It has been recorded in the states of BC, BS, CH, CO, DU, NL, SL, SO, TA, and ZA.

VEGETATIONAL ASSOCIATIONS AND ELEVATION RANGE: The mule deer of the desert (*O. hemionus crooki*) prefers open habitats with little vegetation. It is mainly found in xeric bushes with creosote bush (*Larrea tridentata*), maguey (*Agave* spp.), cactus (*Opuntia* sp.), yuccas (*Yucca* sp.), ocotillo (*Fouqueria splendens*), and pastures of tobosa (*Hilaria mutica*) and blue grama (*Bouteloua gracilis*, *B. eriopoda*, *B. curtipendula*). It also frequents humid areas such as the "mogotes" of mesquite (*Prosopis* sp.) (Bowyer, 1986; Ordway and Krausman, 1986). The subspecies of the Baja California Peninsula inhabits xeric scrublands and forests of oak and pine (Galindo-Leal, 1993; Gallina, 1989; Gallina et al., 1991; Leopold, 1959; Perez-Gil, 1981). On Isla Tiburon, Sonora, *O. hemionus sheldoni* inhabits xeric scrublands that include trees and shrubs such as palo verde (*Cercidium floridum*), cuajiote (*Bursera microphylla*), ocotillo (*Fouqueria splendens*), *Jathropa cinerea*, *J. cuneata*, *Opuntia bigelovii*, and *Olneya tesota*. On Isla Cedros, the dominant vegetation is shrub or chaparral with species such as chamizo (*Adenostoma fasciculatum*), manzanita (*Arctostaphylos* spp.), (*Ceanothus* spp.), oak (*Quercus* spp.), and yucca (*Yucca whipplei*). In the high parts of the region there are patches of pine (*Pinus remorata*, *P. muricata*, *P. radiata*), and juniper (*Juniperus californica*). It is found from sea level up to 3,000 m.

CONSERVATION STATUS: At a subspecies level, *O. hemionus cerrocensis*, *O. hemionus fuliginatus*, *O. hemionus penninsulae*, and *O. hemionus sheldoni* are included in the lists of endangered species of the United States. In Mexico, the law protects only *O. hemionus cerrocensis*, *O. hemionus penninsulae*, and *O. hemionus sheldoni*. The subspecies with wider distribution in the country is *O. hemionus crooki*; hunting these deer is allowed under "special permits"; however, their populations are currently located in isolated patches (Galindo-Leal, 1993; Weber, 1994). This subspecies is in danger of disappearing in Coahuila, Nuevo Leon, Tamaulipas, Zacatecas, and San Luis Potosí. In the same way, the subspecies of Isla Cedros and Isla Tiburon are also in a precarious condition (Diaz-Castorena, 1989; Dietrich and Tijerina, 1988; Leopold, 1959; Perez-Gil, 1981). A small population was reintroduced a short while ago (1985) in Nuevo Leon and apparently it is stable and reproducing (Morrison et al., 1987).

Odocoileus virginianus (Zimmermann, 1780)

White-tailed deer

Carlos Galindo Leal and Manuel Weber

SUBSPECIES IN MEXICO

Odocoileus virginianus acapulcensis (Caton, 1877)
Odocoileus virginianus carminis Goldman and Kellogg, 1940
Odocoileus virginianus couesi (Coues and Yarrow, 1875)
Odocoileus virginianus mexicanus (Gmelin, 1788)
Odocoileus virginianus miquihuanensis Goldman and Kellogg, 1940
Odocoileus virginianus nelsoni Merriam, 1898
Odocoileus virginianus oaxacensis Goldman and Kellogg, 1940

Odocoileus virginianus sinaloae J.A. Allen, 1903
Odocoileus virginianus texanus (Mearns, 1898)
Odocoileus virginianus thomasi Merriam, 1898
Odocoileus virginianus toltecus (Saussure, 1860)
Odocoileus virginianus truei (Merriam, 1898)
Odocoileus virginianus veraecrucis Goldman and Kellogg, 1940
Odocoileus virginianus yucatanensis (Hays, 1872)
The white-tailed deer is the most polytypic species of deer in Mexico, with 14 subspecies in Mexico (Hall, 1981). It is very likely, however, that the several of these subspecies are artificial (Galindo-Leal and Weber 1998).

DISTRIBUTION: The white-tailed deer has the widest distribution of all deer in Mexico. Originally, it was found throughout the entire length and width of the national territory (Leopold, 1959). It has been recorded in the states of AG, BC, CA, CH, CL, CO, CS, DF, DU, GJ, GR, HG, JA, MI, MO, MX, NL, NY, OX, PU, QE, QR, SI, SL, SO, TA, TB, TL, VE, YU, and ZA.

DESCRIPTION: The tuft of white hairs on the base of the tail that bristles when it is excited or running away characterizes the white-tailed deer. *Odocoileus virginianus* is also distinguished by its stylized and fine body (Leopold, 1959). The neck is long and the head elongated. The legs are thin but strong. Only males have antlers, which are directed outside and ahead of the skull; the main emerging branch has 2 to 6 ramifications. The ears, of large size, are about 50 % of the total length of the head, although they are definitely smaller than those of *O. hemionus* (Leopold, 1959). The coloration is gray during the winter and reddish in summer, although not all subspecies change color. The fawns have a reddish mottled coat until three months old. The belly and inner side of the legs are white. The tail is brown or gray on top; long white hairs surround the edges and the bottom of the tail. The metatarsal gland measures between 15 mm and 30 mm in length. The lachrymal fossa is much less deep than in *O. hemionus.*

EXTERNAL MEASURES AND WEIGHT
TL = 1,000 to 2,400 mm; TV = 100 to 365 mm; HF = 279 to 538 mm; EAR = 140 to 229 mm.
Weight: 27 to 135 kg.

DENTAL FORMULA: I 0/3, C 0/1, PM 3/3, M 3/3 = 32.

NATURAL HISTORY AND ECOLOGY: The white-tailed deer is more active during the early hours of the morning and at dusk. Their activities are influenced by factors such as sex, age, reproductive time, habitat characteristics, availability of food, patterns of activity of the predators, and human activities. There are no reliable studies on the home range in Mexico. In Arizona, its habitats are similar to those in northern Mexico; the average home range of the white-tailed deer is 5.18 km^2 and 10.57 km^2 for females and males, respectively. The size of the home range varies greatly in relation to several factors. The most common social groups are females with their offspring. The males associate in mixed groups of age during the non-reproductive period and even shortly before the reproductive period, when they become solitary. In open habitats, the size of the groups tends to be larger than in closed habitats, where small groups predominate. The general pattern of dispersion of the white-tailed deer appears to be strongly biased on sex; the males are the ones who regularly disperse (Weber and Galindo-Leal, 1994). The white-tailed deer can reach population densities of up to more than 40 individuals per km^2. These densities have only been recorded in very productive places where there are no predators. In general the densities are less than 15 individuals per km^2 (Galindo-Leal 1992, 1993; Galindo-Leal et al., 1993, 1995; Mandujano and Gallina, 1993). The reproductive season of the white-tailed deer lasts about three months. In the north of Mexico, fawns are born in August and September. In general, females reach puberty in the first year of life or a little earlier, during the non-reproductive period. These females wait until the following reproductive season in January and February to mate. The majority of females are between 16 and 18 months (1.5 years) when they mate for the first time. The average weight at this age is 25 kg (Weber, 1992c, 1993; Weber and Galindo-Leal, 1992; Weber et al., 1995). The stages of increased energy demand in females are the last third of the pregnancy, delivery, and nursing. This phase of reproduction is synchronized with the highest quality and abundance of food. On average, the gestation period lasts between 196 and 205 days. In general, birth takes place in isolated and well-protected places, and once it starts it tends to be extremely fast. The whole process of a normal birth of twins takes less than 20 minutes. Deliveries of more than 40 minutes are considered difficult births (Weber and Galindo-Leal, 1992). In general, young females give birth to only one fawn. In the following births, two or even three offspring are born, if the area is productive. When born, the fawns tend to remain hidden in the vegetation and inactive during the first three or four weeks of age, except when the mothers visit to feed them. Twins are kept separate until four weeks of age (Galindo-Leal and Weber, 1998). The antlers of the white-tailed deer grow each year at the beginning of the rainy season (in Durango), approximately a week after the old antlers are shed (May and June). The size of the antlers is not an indicator of the age of the deer, but rather of the quality and quantity of food. The white-tailed deer communicates using its senses and through very sophisticated and specialized chemical communication. It possesses specialized exocrine glands in several parts of the body (tarsal, metatarsal, interdigital, prepucial, and lachrymal glands, as well as glands in the snout or nose), which play an important role in the specialized aspects of communication in this species, Their diet varies enormously depending on the season and region, and is perhaps the most notable characteristic of adaptability of this species in an environment in constant change. The white-tailed deer feeds on leaves, shoots, and fruits of a variety of shrubs and trees (Gallina, 1988, 1993a, b; Gallina et al., 1981).

Odocoileus virginianus. Tropical rainforest. Los Tuxtlas, Veracruz. Photo Gerardo Ceballos.

It consumes a large proportion of vegetal matter of "ligneous" origin from trees and shrubs. Their eating habits are selective. The digestive system of the white-tailed deer is adapted to its feeding habits; the reticulum and the abomasum are larger in relation to the size of the rumen and the omasum, and the salivary glands are more developed and better adapted to the consumption of plant material rich in dissolved sugars, tannins, and other chemical compounds of woody plants (Galindo-Leal and Weber, 1998). The white-tailed deer have relatively well-defined activity periods that vary seasonally and locally. They learn to select specific foods based on their preferences, palatability, and energy or protein intake; the search for these favorite foods modifies the diet constantly. In general, they show three peaks of foraging activity, which are more or less constant during the day; in the morning (05:00 hrs to 08:00 hrs), in the afternoon (17:00 hrs to 19:00 hrs), and at night and morning (22:00 hrs to 01:00 hrs; Galindo-Leal and Weber, 1998). The main predators of the white-tailed deer are the human and the puma, which prey on mainly adults and juveniles, and the coyote and the wild cat, which prey on fawns. In the north of Mexico, black bears and golden eagles occasionally prey on fawns or young deer. The white-tailed deer was the basic prey of the almost extinct Mexican wolf. Their presence and abundance can determine the success or failure in the future plans for reintroduction of this species in Mexico and in the Southwest of the United States. Currently, the uncontrolled populations of maroon dogs or ranch dogs appear to have a significant impact on the population, due to predation of newborn fawns during the reproductive season (Galindo-Leal and Weber, 1998). Many common parasites and diseases of domestic cattle can affect the white-tailed deer (Galindo-Leal and Weber, 1998).

VEGETATIONAL ASSOCIATIONS AND ELEVATION RANGE: The white-tailed deer is distributed in a huge variety of ecosystems, including tropical rainforests, coniferous forests, and semi-arid areas. Perhaps the only exception is the more marginal xeric climates of Sonora, Chihuahua, Durango, Coahuila, and Baja California, occupied by the mule deer. There are semi-arid areas of Sonora and Chihuahua where these species are sympatric (Galindo-Leal and Weber 1998; Leopold, 1965). The white-tailed deer is distributed from sea level to 2,800 m.

CONSERVATION STATUS: The white-tailed deer is the most adaptable and tolerant species to human activities (Leopold, 1959). It still persists in highly disturbed forests (Desierto de los Leones, Mexico City), agricultural (Ebano, Tamaulipas) and livestock (The Michilía, Durango) areas, and even in the surrounding areas of villages and towns of moderate size (Nuevo Laredo, Tamaulipas), as long as patches of suitable habitat provide food, water, and coverage in sufficient quantity and quality (Galindo-Leal and Weber, 1998). The two factors that have drastically affected their populations are the loss of habitat and indiscriminate hunting. The white-tailed deer is the main wild species of economic and hunting importance in Mexico (Leopold, 1959; Weber, 1993). In virtually any town, farming cooperative, ranch, lumberjack camp, or indigenous village this species is still found; locals hunt it for its meat. At times, its meat is sold commercially, and there are villagers dedicated to hunting and selling its meat and fur during several months of the year (Galindo-Leal and Weber, 1998). Coupled with this, sport hunting without the correct hunting permits and off-season has also a negative impact on the populations of the white-tailed deer. The pseudo sport hunters use prohibited methods such as "spotlighting" or shoot deer from vans on dirt roads (Leopold, 1959). Another strong threat is the translocation of subspecies of greater size in the distribution area of subspecies of smaller size. This activity, motivated by the obtaining of "trophies," may have serious consequences for the conservation status of the white-tailed deer (Galindo-Leal and Weber, 1998; Weber and Galindo-Leal, 1992).

Family Antilocapridae

The family Antilocapridae is endemic to North America with a single species (*Antilocapra americana*). It is distributed widely in Mexico, but in the early twentieth century it was almost wiped out. Currently, three isolated populations persist in the Baja California Peninsula, Sonora, and Chihuahua, numbering under 1,000 individuals.

Antilocapra americana (Ord, 1815)

Pronghorn antelope

Jorge Cancino and Rurik List

SUBSPECIES IN MEXICO
Antilocapra americana mexicana (Merriam, 1901)
Antilocapra americana peninsularis (Nelson 1912)
Antilocapra americana sonoriensis (Goldman, 1945)

DESCRIPTION: *Antilocapra americana* has contrasting colors, being white on the belly and inner part of the legs and reddish-brown or ash-colored on top. The hindquarters, stripes on the neck, lips, cheeks, and base of the horns are also white. It has a black mane. In general, males have a darker face and under the ears they have spots that resemble sideburns. The hair of the hindquarters is the longest on the body and can lift and/or flatten as a sign of alarm. The horns are present in both sexes; in females they are very small and without the prominence on front, while in the males they are bifurcated and are longer than the ears (O'Gara, 1990). An outer covering that combines keratin with hair forms the horns, which grow on a bone kernel and are shed and regrown every year. The longevity of the pronghorn is estimated to be from 9 to 16 years of age (O'Gara, 1978, 1990; Trainer et al., 1983).

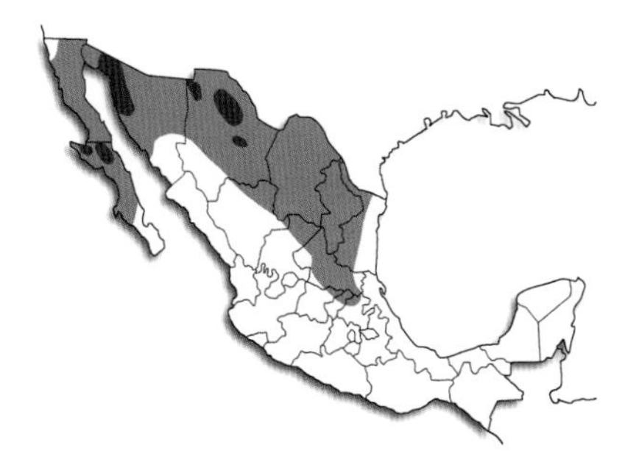

DISTRIBUTION: In Mexico *A. americana* is currently found in the Vizcaino desert, in the desert of Altar, on several ranches in Chihuahua (isolated), and in Coahuila (a reintroduced population). Until 1991 there were reports of some animals of the American subspecies reintroduced in San Luis Potosí. It is very abundant in the West of the United States of America and southern Canada. It has been recorded in the states of BC, BS, CO, DU, HG, NL, SI, SL, SO, TA, and ZA.

EXTERNAL MEASURES AND WEIGHT
TL = 1,245 to 2,062 mm; TV = 152 to 330 mm; HF = 362 to 521 mm; EAR = 140 to 229.
Weight: 50 to 56 kg.

DENTAL FORMULA: I 0/3, C 0/1, PM 3/3, M 3/3 = 32.

NATURAL HISTORY AND ECOLOGY: In general, pronghorn are found on large plains, in soft hills, between large mesas, and in broad riverbeds. They mainly consume herbaceous plants, shrubs, cacti, and grasses (Cancino, 1994; Palacio, 1991; Wright and de Vas, 1986). Habib and Peña (1982) reported a high consumption of herbs and *Artemisia ludoviciana* as the most important species in the diet of the pronghorn in Chihuahua. The desert pronghorn can survive long periods without drinking because the deserts of Vizcaino and Altar have no accessible and permanent freshwater sources. The pronghorn of the desert fill their water requirements through the consumption of dew, the moisture content of the plants, and probably metabolic water as well; some modifications of their behavior are also important (Yoakum, 1994). Females reach sexual maturity at 16 months. On average, gestation lasts 252 days. In the first birth, generally one offspring is born; from the second birth on, they give birth to two and occasionally three offspring. Sexual maturity in males is reported from the first year; however, young males are generally excluded from mating by dominant

Antilocapra americana. Grasslands. Janos Biosphere Reserve, Chihuahua. Photo: Gerardo Ceballos.

males (Kitchen and O'Gara, 1982). Their main predators are coyote (*Canis latrans*), puma (*Puma concolor*), bobcat (*Lynx rufus*), and golden eagle (*Aquila chrysaetos*) (O'Gara, 1978, 1990). There are 88 reports of parasites and diseases that include viruses, bacteria, protozoa, ticks, and other types of organisms. The impact at population level is unknown and there are no reports of epizootics (Yoakum and O'Gara, 1992). The main causes of mortality in Mexico are poaching, drought, and predation. Predation is increased by inadequate barbwire fencing, which reduces the pronghorn's mobility, speed, and escape routes (Meléndez et al., 2006; Valdés et al., 2006). They are gregarious. Their social structure is hierarchical with dominant systems of territorial mating (Kitchen, 1974). They are active during the day and at night, alternating resting and feeding periods. Adult males establish territories that are defended during the breeding season. Their territory is marked with urine, feces, secretions of interdigital and subauricular glands, and probably low-intensity vocalizations.

VEGETATIONAL ASSOCIATIONS AND ELEVATION RANGE: This species is generally found in places with xeric vegetation. The Mexican pronghorn (*Antilocapra mexicana americana*) is found between 1,220 m and 1,700 m (González-Romero and Lafón-Terrazas, 1993). The peninsular pronghorn (*A. americana peninsularis*), is found from sea level to 250 masl (Cancino et al., 1995). The Sonoran subspecies (*A. americana sonoriensis*) is found from 120 m to 1,200 m (Thompson-Olais, 1994).

CONSERVATION STATUS: *A. americana* in Mexico is considered in danger of extinction (SEMARNAT, 2010). The population is estimated between 1,000 and 1,500 individuals (Meléndez et al., 2006; Valdés et al., 2006). While predation of offspring by coyotes and poaching of adults have been the main reasons the remaining and reintroduced antelope have not increased, habitat fragmentation by conversion of grasslands to agriculture is becoming a major threat for the populations of Chihuahua (Valdés et al., 2006). More recently, the possibility of the persistence of the Sonora and Chihuahua populations in the border region is being reduced by the United States-Mexico border fence, which stops or limits the north-south movements and splits the already small populations (List, 2007). Globally, the species is considered as least concern in the Red List (IUCN, 2010). In the United States and Canada it is an important hunting species and most populations are stable (IUCN, 2010; O'Gara and Yoakum, 1992).

Family Bovidae

The family Bovidae is one of the most diverse families in the order; it has 137 species distributed on all continents with the exception of Australia. In Mexico it is represented by two species and two genera. The bighorn sheep (*Ovis canadensis*) is a species that was in danger of extinction but is now recovering. There is only one free-ranging bison (*Bison bison*) population in Mexico, which is considered endangered (List et al., 2007).

Subfamily Bovinae

The subfamily Bovinae contains a diverse group of ten genera with thirty extant species. Only one species is known from Mexico.

Bison bison (Linnaeus, 1758)

American bison

Jesús Pacheco R. and Rurik List

SUBSPECIES IN MEXICO
Bison bison bison Linnaeus, 1758

DESCRIPTION: *Bison bison* is the largest terrestrial mammal in the Americas. The coat of the head, neck, shoulders, and front legs is long, hirsute, and dark brown; the rest of the body is covered by short hair of lighter coloration (Nowak, 1999b). The head is large; the neck is short and has a prominent hump. The horns are short and bent, and an annual growth can be observed (Soper, 1941). The tail is short and hairy with a tuft of more dense and long hair on the tip. The eyes are placed in an anterior-laterally position (Meagher, 1986). They have sexual dimorphism. Males are relatively larger; the horns of females are shorter and thinner, the hump is smaller, and the neck is thinner (Nowak, 1999b).

EXTERNAL MEASURES AND WEIGHT
TL = 2,130 to 3,800 mm; TV = 300 to 910 mm; HF = 500 to 680 mm.
Weight: 318 to 907 kg.

DENTAL FORMULA: I 0/3, C 0/1, PM 3/3, M 3/3 = 32.

NATURAL HISTORY AND ECOLOGY: Bison are gregarious animals found in large herds varying with age, sex, season, and habitat. Females of all ages, males less than three years old, and old males form groups throughout the year. Although it is possible to find solitary individuals, it is common to observe them in herds (Fuller, 1960; McHugh, 1958). They are mainly diurnal, intercalating periods of feeding and resting, which are used to ruminate the eaten food. They mainly feed on grasses and shrubs. They move approximately 3 km daily; in the Great Plains of the United States they migrate hundreds of kilometers from the summer grounds to the wintering grounds (Banfield, 1974). The mating season takes place from late summer to late autumn (Dorn, 1995). Most births occur in April and June but can take place as late as October (Kirkpatrick et al., 1993). The gestation period lasts on average 262 to

DISTRIBUTION: Historically the bison has been distributed from Alaska and western Canada throughout the United States to northern Mexico. Currently, bison occupy < 1% of their original range and conservation herds a small fraction of that (Gates et al., 2010). There are isolated populations throughout their range, especially in parks, on private ranches, and in reserves (Meagher, 1986). Within historic times, the bison in Mexico ranged from western Sonora to eastern Coahuila across the plains of Chihuahua, Coahuila, and Durango (List et al., 2007). Today, the only free-ranging population moves between the north of the municipality of Janos, Chihuahua, in Mexico and the foothills of Hidalgo County in New Mexico, where it has been since at least the 1920s (List et al., 2007). The historical records include the states of CH, CO, DU, NL, SO, and ZA.

Bison bison. Grasslands. Janos Biosphere Reserve, Chihuahua. Photo: Gerardo Ceballos.

300 days, with one calf per delivery, which is born with an approximate weight of 15 kg to 25 kg (Dorn, 1995; Ruthberg, 1984). During parturition, the female isolates herself from the rest of the herd and gives birth where the grass is high enough to cover the calf; from that moment the mother defends it from any potential danger; in the males this type of behavior has not been observed (McHugh, 1958). Sexual maturity occurs between two and four years of age in males, but males usually do not reproduce until they are large enough to compete with other males for access to females (Wilson and Zittlau, 2004). There is a correlation between aggression and reproductive success, with frequent changes in the hierarchy of males during estrus, as dominant males get tired and temporarily abandon the herd (Lott, 1979). The wolf (*Canis lupus*) is the main predator of the bison. In some national parks in the United States bison can contribute to as much as 65 % of the winter diet of the wolf; however, this does not appear to have a severe negative impact in the bison population (Fuller, 1962). The only free-ranging population in Mexico fluctuates between 80 and 157 individuals with a natural age distribution (List et al., 2007). The fluctuation is due to organized hunting on a private ranch on New Mexico, where hunters take around 40 bison annually (adults, yearlings, and calves) (List and Solís, 2009).

VEGETATIONAL ASSOCIATIONS AND ELEVATION RANGE: The bison inhabits plains, grasslands, valleys, and, at times, forests (Reynolds et al., 1983). It is found from sea level to 3,900 masl (Beidelman, 1955; Fryxell, 1928). In Mexico the range includes the foothills and adjacent grasslands of the Sierra Madre Occidental of northwestern Chihuahua and extreme northeastern Sonora at an altitude range of 1,250 m to 2,550 m (List et al., 2007).

CONSERVATION STATUS: There are an estimated 500,000 to 600,000 plains bison in North America, but most are on private ranches and are managed as livestock; only around 20,500 are in conservation herds (Gates et al., 2010). In Mexico there is only one free-ranging bison population, which moves between Janos, Chihuahua, and Hidalgo, New Mexico. This population is considered endangered in Mexico (SEDESOL, 1994; SEMARNAT, 2002), but when they cross to the United States they are considered livestock and lack legal protection as wildlife (List et al., 2007). The population is subject to illegal (Mexico) and legal (United States) hunting; legal hunting takes most individuals (about 40 per year), but their numbers remained more or less stable throughout 2004-2009 (List and Solís, 2009). Other threats to this bison population are habitat encroachment through conversion of native grasslands in Mexico and the United States-Mexico border fence, which has divided roughly half of the area used by the bison to move into Mexico (List, 2007; List and Solís, 2009). The area used by the Janos bison is found within the recently created (2009) Janos Biosphere Reserve. A Bison Recovery Team for the bison in Mexico has been established and a Conservation Action Plan (Plan de Acción para la Conservación de Especies: *Bison bison*) has been drafted. As part of this plan, in November 2009, a group of genetically pure (without detectable cattle introgression) bison from Wind Cave National Park were translocated to a private property within the Janos Biosphere Reserve to start a breeding herd of pure bison to reintroduce new herds with ecological management in other areas of northern Mexico (Gates et al., 2010).

Subfamily Caprinae

The subfamily Caprinae contains thirty-five species in thirteen genera, with one species known from Mexico.

Ovis canadensis Shaw, 1804

Bighorn sheep

Rodrigo A. Medellín and Oscar Sánchez

SUBSPECIES IN MEXICO

Ovis canadensis cremnobates (Elliot, 1903)
Ovis canadensis mexicana (Merriam, 1901)
Ovis canadensis weemsi (Goldman, 1937)

DESCRIPTION: *Ovis canadensis* is the only native caprine of Mexico (Wilson and Reeder, 2005). There is a pronounced sexual dimorphism; males have massive and curved horns that form a spiral backward and outward, with the tips pointing forward or up in adult individuals. The horns of males may measure more than 500 mm in diameter and more than 660 mm in total length; in fact, in proportion, the bighorn sheep possesses one of the larger horns of all ruminants. The females have small and thin, slightly curved horns, directed backward and never forming a spiral. In general, the dorsal coloration is light brown to dark brown; the ventral coloration is pale or whitish, sometimes with yellowish tones. The rump and rear is paler with longer hair. Coloration varies according to the subspecies and distribution.

DISTRIBUTION: *O. canadensis* is found from Canada to Mexico. In Mexico it is found in the northern arid regions from the Baja California Peninsula to Sonora and Coahuila (Hall, 1981). It has been recorded in the states of BC, BS, CH, CO, and SO.

EXTERNAL MEASURES AND WEIGHT

TL = 1,240 to 1,950 mm; TV = 70 to 130 mm; HF = 315 to 440. EAR: 90 to 130 mm. Weight: 60 to 80 kg.

DENTAL FORMULA: I 0/3, C 1/1, PM 3/3, M 3/3 = 34.

NATURAL HISTORY AND ECOLOGY: The bighorn sheep in Mexico inhabits the dry mountains of the north and northwest, which have relatively dense plant cover that protects reproductive females and their offspring. Usually those areas have an adequate orientation for the detection of predators and for visual communication within each family group and between groups (Boyd et al., 1986). Other essential factors in the quality of the habitat of the bighorn sheep include the presence of short fodder (Ginnett and Douglas, 1982) with a continuous local distribution and sufficient availability throughout the year to supply the entire group, the proximity (1.5 km to 2.5 km; Elenowitz, 1984) of a permanent water body, and the presence of salt licks. Suitable habitat (in good condition and in continuous use) is fragmented; as a result, many family groups form independent herds, even in adjacent mountains. The population densities vary, but for the subspecies of the desert, including all Mexican populations, density is two to three sheep per km^2 (DeYoung et al., 2000; Lawson and Johnson, 1982). The diet of the bighorn sheep varies considerably, but in Mexico it basically consists of tender stems, leaves of shrubs, and pastures. They prefer fresh fodder of the short rainy season or growing in the vicinity of natural watering places or oases, but during a prolonged drought they depend basically on the dry pastures of the previous season, the food value of which is maximized by the efficient digestion of cellulose (Geist, 1971). For the desert subspecies the presence of surface water is of great importance (Turner and Weaver, 1980). Their activity is mainly diurnal and decreases toward noon (Lawson and Johnson, 1982). At dusk, they group at warm sites for protection from the wind, but usually close to escape routes (Wakelyn, 1984). The family groups include between two and nine animals that establish a defined social structure (Shackleton, 1985). Bighorn sheep are polygamous and, although males under two years of age can be physiologically mature, it is usually not until two and half years of age that they fight for the possession of females (DeYoung et al., 2000). Once the relations of dominance are established, the male mates with the females, which usually give birth to a single offspring (rarely twins) after nearly 175 days of pregnancy (Shackleton et al., 1984). Births are synchronized with climatic factors and availability of food (Lenarz, 1979); in Mexico births tend to cluster in the spring. Despite the lack of information, probably birth rates in females are between 8% in declining populations and up to 100% in expanding populations (Woodgerd, 1964). The proportion of males and females may be close to 3:10 in relatively stable populations (DeYoung et al., 2000).

VEGETATIONAL ASSOCIATIONS AND ELEVATION RANGE: The distribution area of the bighorn sheep in Mexico comprises the xeric scrubland habitat (Rzedowski, 1978). It is especially found in rugged terrain covered by this type of vegetation and basically in sites of interface between the xeric scrubland and arid pastures. It is found from sea level up to 1,800 m. The species has important populations in Baja California and Sonora, with one of the largest populations on Tiburon Island. After

Ovis canadensis. Scrubland, Hermosillo, Sonora. Photo: Claudio Contreras Koob.

being extirpated from other states, recently it was reported from Chihuahua (Pelz-Serrano et al., 2006) and a few animals have been reintroduced to Chihuahua and Coahuila (Medellín et al., 2005).

CONSERVATION STATUS: The Mexican federal list of species at risk includes *Ovis canadensis* (SEMARNAT, 2010) under the category of special protection. This indicates that despite the fact that the populations are not abundant, it is possible to exploit some of them for sport hunting under strict measures controlled by the authorities. Legally, less than 50 animals per season are allowed to be hunted. Mellink (1993b) provided a summary of historical aspects of the administration of this species in Mexico. The Mexican bighorn sheep populations are explicitly protected, since 1975, by the Convention on International Trade in Endangered Species (CITES) (1992), which includes the species in appendix II. Because of the great similarity in their habits, physiological traits, and subsistence, the introduction of domestic sheep (*Ovis aries*) and aoudad (*Ammotragus lervia*) in natural areas constitutes a high risk for the survival of the bighorn sheep. Both species are known to have caused the disappearance of entire populations of bighorn sheep.

Order Cetacea

Gerardo Ceballos, Sandra Pompa, and Pieter Folkens

The numbers [of gray whales], however, have been reduced to a point close to extinction. In recent years very few have occurred in the coast of California, but there is a possibility that if they are not bothered in their breeding sites, in the lagoons of Baja California, their populations can recover in time.

— G.M. Allen, 1942

The order Cetacea has long been organized as comprising three suborders — Archaeoceti (extinct archaic forms including the "terrestrial whales"); Mysticeti (whose extant forms have baleen hanging from the roof of the mouth instead of teeth); and Odontoceti (the toothed whales). The order contains 88 species, all of which are marine except for 4 species of freshwater dolphins (Hooker, 2009), one of which (*Lipotes vexillifer*) became extinct only recently.

These are aquatic mammals that inhabit all seas and oceans. Some species are found in large rivers in Asia and South America. They show great variation in size, with species ranging from the vaquita (*Phocoena sinus*) of the Gulf of California (25 kg) to the blue whale (*Balaenoptera musculus*) (145,150 kg), which is the largest mammal ever known. The body in cetaceans is specialized for aquatic activity; it is very long with the forelimbs modified into flippers. The end of the tail is flattened laterally into a pair of flukes. It is the main structure for locomotion, which is efficient and places some whales and dolphins among the fastest animals of the oceans. The skin is smooth, with little or no hair (Reynolds and Rommel, 1999).

Some cetaceans are capable of protracted dives of up to 2 hours and thousands of meters deep. There are a number of physiological adaptations that allow them to stay underwater for long periods; for example, they expel the air from their lungs before diving, the heart rate slows, and the respiratory center of the brain is relatively insensitive to changes in the concentration of carbon dioxide in the blood and tissues (Nowak, 1991).

In general, cetaceans are species with large areas of activity, regional seasonal movements, and annual migrations. For example, gray whales (*Eschrichtius robustus*) have an annual migration range of more than 15,000 km from their summer feeding areas in the Arctic to the coastal lagoons of Baja California, where they reproduce in the winter.

The populations of many whales have been seriously depleted by overexploitation to obtain products such as meat and oil. The population of the gray whale (*E. robustus*) in the northwest Pacific and the North Pacific right whale (*Eubalaena japonica*) are on the verge of extinction.

Cetacea has been going though higher taxonomic revision due to recent genetic discoveries. It appears that Cetacea is a clade within Artiodactyla with strong affinities to hippos. This has led to the establishment of the order Cetartiodactyla by some taxonomists and placing Cetacea as an Infraorder within Artiodactyla by others, but neither of these revisions have yet garnered universal acceptance.

(*Opposite*) Humpback whale (*Megaptera novaeangliae*). Bahia de Banderas, Nayarit. Photo: Sandra Pompa.

Suborder Mysticeti

The mysticetes (baleen whales) are represented by three families and four genera. Ten species of mysticetes live in the Northern Hemisphere, eight of which are known from Mexican territorial waters in the Pacific Ocean (Baja California). Six of them (the rorquals) are also known from the Gulf of Mexico.

Family Balaenidae

The little diverse family Balaenidae (right whales) is represented by three species, two of which are in the Northern Hemisphere, with one known from Mexican territorial waters in the Pacific Ocean (Baja California) (Mead and Brownell Jr., 2005). Right whales are characterized by very long baleen plates hanging from a narrow rostrum and a long, arching mouth line.

Eubalaena japonica (Lacépède, 1818)

North Pacific right whale

Mario A. Salinas Zacarías

SUBSPECIES IN MEXICO

E. japonica is a monotypic species.

Until only recently, this whale was considered the same species as the North Atlantic right whale (*Eubalaena glacialis*). Genetic analysis determined that *E. japonica* was distinct from *E. glacialis* and closer to the southern right whale (*E. australis*) than the North Atlantic species. All records of right whales from Mexico attributed to *E. glacialis* are actually *E. japonica.*

DESCRIPTION: The North Pacific right whale has not been recorded in Mexican waters for more than half a century, despite being easy to identify. It is almost completely very dark gray to black except for several small spots on the ventral part. The head is prominent, occupying one-third to one-fourth of the length of the body. A narrow

North Pacific right whale (*Eubalaena japonica*). Illustration: Pieter Folkens.

and extremely arched maxilla is characteristic of the genus. These animals lack gular furrows seen in all other mysticetes found in Mexican waters. The paired nostrils (blowholes) are sharply separated. The whale's blow when seen from an anterior or posterior view has the form of a V. Right whales lack a dorsal fin. They arch the body when initiating a dive, lifting their very broad flukes, which are slightly concave on the anterior edge with lobes separated by a deep notch. Both the clean arching back and shape of the flukes are diagnostic of the species. The pectoral fins are wide, angular, and slightly pointed at the tip. These whales have light, raised patches of skin about the head called callosities. The largest of them is located on the anterior part of the rostrum and maxilla. Smaller callosities are located on the lower jaw, around the eyes, and around the nostrils. The callosities are encrusted with crustaceans such as whale lice (Cyamidae), barnacles (*Balanomorpha*), and goose barnacles (*Lepadomorpha*). The coloration of these spots can be white, yellow, or orange, depending on the proportion of these crustaceans (Cummings, 1985; Leatherwood and Reeves, 1983).

DISTRIBUTION: This species of right whale dwells in the Northern Hemisphere, in cold and temperate coastal waters of the Pacific. It has rarely been reported in Mexico; its presence was documented by whalers of the nineteenth century, mainly by the reports of Captain Scammon (1874), who mentioned the capture of some animals between February and April as far south as the Bay of Sebastián Vizcaíno and Isla Cedros, Baja California, near the latitude 29 degrees north. In the 1960s a pair of whales was reported near Punta Abrejos, Baja California Sur (Leatherwood et al., 1988). There are several recent records off Baja California both along the Pacific coast and in the Gulf of California (Gendron et al., 1999: Urbán-R. et al., 2006).

EXTERNAL MEASURES AND WEIGHT

TL: 15.5 to 18 m.

Weight: up to 90,000 kg.

DENTITION

220 to 260 baleen plates on each side, up to 2.7 m long with very fine bristles.

NATURAL HISTORY AND ECOLOGY: North Pacific right whales undertake migration toward subarctic waters, where they feed during the summer, and then move toward warmer waters to reproduce; the offspring are born in winter (December to March). They have a single calf per year; gestation lasts 12 months and nursing 6 to 7 months. At birth, the calf measures around 3.6 m and grows rapidly in its first year, reaching a length of approximately 12 m within the first 18 months. Sexual maturity is reached at 10 years of age, when the whale has a length of 15 m in the case of males and 15.5 m in females (Omura et al., 1969). The whales feed near the surface, filtrating planktonic crustaceans such as copepods and euphausids that form large concentrations in the circumpolar waters. They often jump out of water, a behavior known as breaching. They are slow swimmers with relatively shallow dives. This slowness led to their being the first whales taken by humans in row boats. A thick layer of blubber (subcutaneous fat) makes them float when killed. They emit low sounds below 500 Hz, with an intensity of 172 dB to 187 dB and an average duration of 1.4 seconds. The sounds that they make are similar to mooing, and two types can be distinguished: low-frequency sounds (around 160 Hz) without significant changes in the frequency and complex sounds with marked changes in the frequencies and overtones varying from 30 Hz to 2,100 Hz (Cummings, 1985).

CONSERVATION STATUS: *Eubalaena japonica* is at risk of extinction because of the intense commercial takes conducted in the nineteenth and twentieth centuries. Despite absolute protection initiated in 1937 by the League of Nations and the IWC, a clandestine commercial operation conduced by Soviet whalers in the late twentieth century reduced the population to extremely critical levels. Small, isolated groups remain in the Bering Sea. No calf of the species has been sighted for decades. The genus *Eubalaena* is currently included in appendix I of CITES and it is considered endangered by the International Union for Conservation of Nature (IUCN) (IUCN, 2010). In Mexico, *E. japonica* is not included in the Mexican Official Regulation (SEMARNAT, 2010). And there is no specific legislation covering its occasional presence. Yet it is protected in Mexico, as all marine mammals inhabiting its waters are, in frameworks such as the ones in the Federal Law on Fisheries and the article 420-I of the Federal Penal Code (Diario Oficial de la Federación, 2009).

Family Balaenopteridae

The family Balaenopteridae is represented by two genera (*Balaenoptera* and *Megaptera*) and eight species, which have a cosmopolitan distribution (Wilson and Reeder, 1993). The family includes six species known from Mexican marine waters (*B. acutorostrata, B. borealis, B. edeni, B. physalus, B. musculus,* and *M. novaeangliae*). The smaller species weigh less than 10,000 kg and the largest, the blue whale, reaches 145,150 kg. Some species, including the blue whale, have been in danger of extinction due to commercial exploitation in the past. However, effective international protection measures have allowed some of these species to recover to relatively safe population levels. Members of the family are also known as "rorquals" or "finner whales."

Balaenoptera acutorostrata Lacépède, 1804

Minke whale

Luis Medrano González

SUBSPECIES IN MEXICO
Balaenoptera acutorostrata davidsoni (Omura, 1975)

DESCRIPTION: *Balaenoptera acutorostrata* is the smallest whale of the family of rorquals. Its appearance is thin and its head is very pointed, the feature from which its name is derived. It has a high, curved dorsal fin that is located in the third posterior part of the body; its pectoral fins are a little more than one-eighth of the total length of the animal. The dorsal coloration is black or gray ventrally and the ventral surface of the pectoral fins is white; it has a lighter coloration on the sides behind the pectoral fins. These whales have a white band on the dorsal side of the pectoral fins that helps identify them. The size and tone of this band is variable. At times, they have a pattern of bands that run from the back of the head, similar to those in the fin whale. They have 50 to 70 ventral furrows that may have a pink coloration, visible when they expand. The blow is low, inconspicuous, and silent (Leatherwood et al., 1983; Stewart and Leatherwood, 1985).

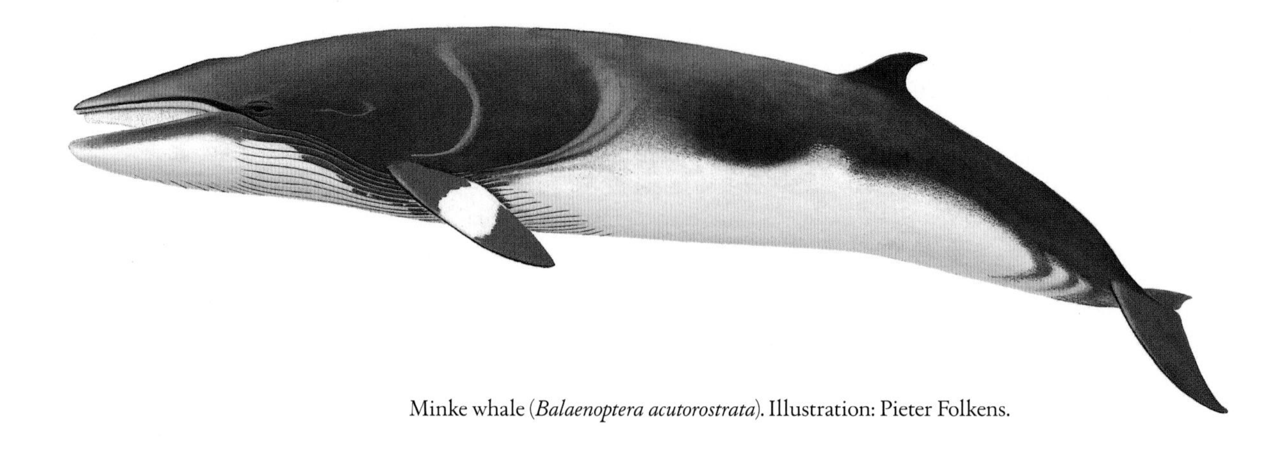

Minke whale (*Balaenoptera acutorostrata*). Illustration: Pieter Folkens.

EXTERNAL MEASURES AND WEIGHT
TL: adult males up to 9.8 m, females to 10.7 m.
Weight: up to 9,200 kg.

DENTITION
230 to 360 light yellowish baleen plates on each side, approximately 12 cm wide at the base and 20 cm long.

NATURAL HISTORY AND ECOLOGY: The migration cycle of minke whales varies considerably. Some populations travel thousands of kilometers, while others are essentially resident, traveling relatively short distances. Minke whales in the Gulf of California appear to be resident with little to no seasonal movements. Minke whales off the outer Pacific coast of Baja are believed to range north into the Southern California Bight in summer. Though rare, the minke whales off the eastern coast of Mexico are probably of the stock that travels to coasts of Maine and Eastern Canada. In the North Pacific they reproduce during the winter. Births occur between December and March. The gestation period lasts approximately 10 months and the size at birth is 2.4 m to 2.7 m in length. Females reach sexual maturity at 7 or 8 years of age when they measure 7.3 m. In the males, sexual maturity is reached at the same age with a length of 6.7 m to 7.0 m. Physical maturity is reached at 8.2 m in females and at 7.9 m in males. In the Northern Hemisphere they mainly feed on krill and fish such as cod of the family Gadidae, capelin, and herring. Individuals and groups of two or three are common, but they may form aggregations of several dozen in feeding events. They often come close to boats and can jump out of water in diverse ways. They have been held in captivity for short periods in Japan. Ectoparasites such as *Cyamus* and balaenopterae have been identified, as well as various endoparasites such as nematodes, trematodes, and cestodes. They are preyed on by killer whales (*Orcinus orca*) in all oceans.

CONSERVATION STATUS: *B. acutorostrata* has been traditionally captured since several centuries ago along the coasts of North America (west coast), Japan, and Norway. This species had commercial interest from the 1970s, when the populations of larger whales diminished and were placed under protection. Tens of thousands of individual minke whales have been caught for commercial purposes. After the moratorium established by the IWC in 1983, Japan and Norway continued hunting hundreds of these whales ostensibly for scientific purposes and have sought to allow commercial hunting. This has created a strong dispute among world conservationists. The actual number of these animals is subject to disagreement, but there is some consensus in the estimation of approximately 450,000 individuals for the whole world (Kock and Shimadzu, 1994; Leatherwood et al., 1983; Stewart and Leatherwood, 1985). As an alternative to hunting wild animals, in Japan in vitro fertilization has been conducted (Fukui et al., 1997). This species is found in appendix I of CITES. It is considered least concern by the IUCN (IUCN, 2010); it is subject to special protection in the Mexican Official Regulation (SEMARNAT, 2010) and its protection is included in the Federal Law on Fisheries and article 420-I of the Federal Penal Code (Diario Oficial de la Federación, 2009).

DISTRIBUTION: Minke whales are found in all oceans but less frequently in tropical regions (Leatherwood et al., 1983; Stewart and Leatherwood, 1985). There is controversy over the population structure of these animals, but studies on their anatomy and genetics agree on the division of three subspecies that correspond to the populations in the North Atlantic, the North Pacific, and the Southern Ocean (Amos and Dover, 1991; Hoelzel and Dover, 1991; Omura, 1975; Van Pijlen et al., 1991). There is a form of minke whales in the Southern Ocean called "dwarf" that is genetically similar to the Northwestern Pacific form (Wada et al., 1991). In the North Pacific, two subpopulations are recognized, one in the west and one in the east. The latter runs from the Sea of Chukchi and the coasts from Alaska to the Revillagigedo Islands. In Mexico, it can be found along the Pacific coast of Baja California and as far south as the Revillagigedo Islands, as well as in the Gulf of California (Stewart and Leatherwood, 1985; Tershy et al., 1990). In the Gulf of California, the minke can be found throughout the year (Tershy et al., 1990). The minke whale can be found in the Gulf of Mexico (Stewart and Leatherwood, 1985), but the author does not know of any records of the species in waters of the exclusive economic zone of Mexico. It has been registered in the areas of the GC, GM, NP, and TP.

Sei whale

Diane Gendron and Verónica Zavala Hernández

SUBSPECIES IN MEXICO

Balaenoptera borealis borealis (Lesson, 1828)

DESCRIPTION: *Balaenoptera borealis* shares characteristics with other species of the genus (Leatherwood et al., 1988). Its head is pointed, and from a lateral view curves with the ventral arch; the head makes up from 20% to 25% of the body length and this proportion increases with age. On the head is a pronounced medial crest, which distinguishes it from the tropical minke whale (three peaks), and medium-sized nostrils (Gambell, 1985). The dorsal fin is relatively large and is located about two-thirds back along the body; it measures from 3% to 4.5% of the length of animal (more than 60 cm). The dorsal fin is thin and curved, with the tip directed toward the back with which it forms an angle of about 45 degrees (Gambell, 1985). This feature alone is not enough for identification since this whale is similar to the Bryde's whale and to a lesser extent the fin whale. These species show variations in the shape (Leatherwood et al., 1988). The variation in the shape of the fin, coupled with the notches present, can serve for individual photo-identification (Schilling et al., 1992). The caudal and the pectoral fins are small and measure 25% and 9% of the body length, respectively (Gambell, 1985; Leatherwood et al., 1988). The color of the skin is dark gray with a bluish tint on both sides; scars (gray or white) made by lampreys and copepods give the skin a metal opaqueness (Leatherwood et al., 1988). *B. borealis* has 32 to 60 ventral folds, and its length constitutes a hallmark of the species since it is shorter than the other rorquals. The furrows end between the pectoral fins and the navel (Leatherwood et al., 1988).

DISTRIBUTION: *B. borealis* is distributed in all oceans, but it is found less frequently in polar areas (Gambell, 1985). Information about their distribution in the Pacific is scarce. During winter (December–March) this species is widely distributed between Punta Piedras Blancas, California, and the Revillagigedo Islands (Rice, 1974). In the Gulf of California there are only six records in the southern portion, two made in winter (Connally et al., 1986) and four between January 1993 and October 1995 (Gendron and Chávez Rosales, 1996). Recently, its presence in oceanic waters of the tropical Pacific has been documented in April, near the Galápagos, so that the range of the winter distribution of the population of the North Pacific may extend to southern Ecuador (Leatherwood et al., 1988). It less known in Mexico due to the confusion of its identification with the fin whale with which it shares the southern limit of its distribution (Leatherwood et al., 1988). Marking techniques show that this species moves from north to south along the west coast of North America, between southern California and Vancouver in Canada (Rice, 1974). It has been registered in the areas of the GC, NP, and TP.

EXTERNAL MEASURES AND WEIGHT

TL: 13.7 to 20 m. Females slightly larger than males.

Weight: up to 45,000 kg.

DENTITION

Its barbs are lengthy in comparison to those of other Balaenopterids (75 cm to 80 cm), with a grayish color and varying in number from 219 to 402.

NATURAL HISTORY AND ECOLOGY: The sei whale can be seen in groups of 6 to 12 individuals, but normally this number is greater in feeding areas (Gambell, 1985). In the southeast of the Gulf of California, animals were observed solitary or associated with other Balaenopterid whales (*B. edeni* and *B. physalus*; Gendron and Chávez Rosales, 1996). Its lifecycle is governed by seasonal migration patterns with mating peaks in winter. After a gestation period of 10.5 months, in the Northern Hemisphere; births occur in temperate waters of low latitudes between October and March and with a maximum in November (Gambell, 1985; Rice, 1974). The offspring are nursed for more than six months and are weaned to conclude the migration to higher latitudes, when they measure nearly 9 m in length (Gambell, 1985). In northern latitudes, they feed on zooplankton (copepods and euphausids), in addition to some fish (anchovy and sardine) and squid that are an important part of their diet (Rice, 1963). They can move at more than 20 knots, which ranks them among the fastest large whales (Leatherwood et al., 1988). Their breathing pattern is regular with 20 to 30-second intervals between each breath, thus facilitating an ability to remain immersed for 15 minutes or longer (Gambell, 1985).

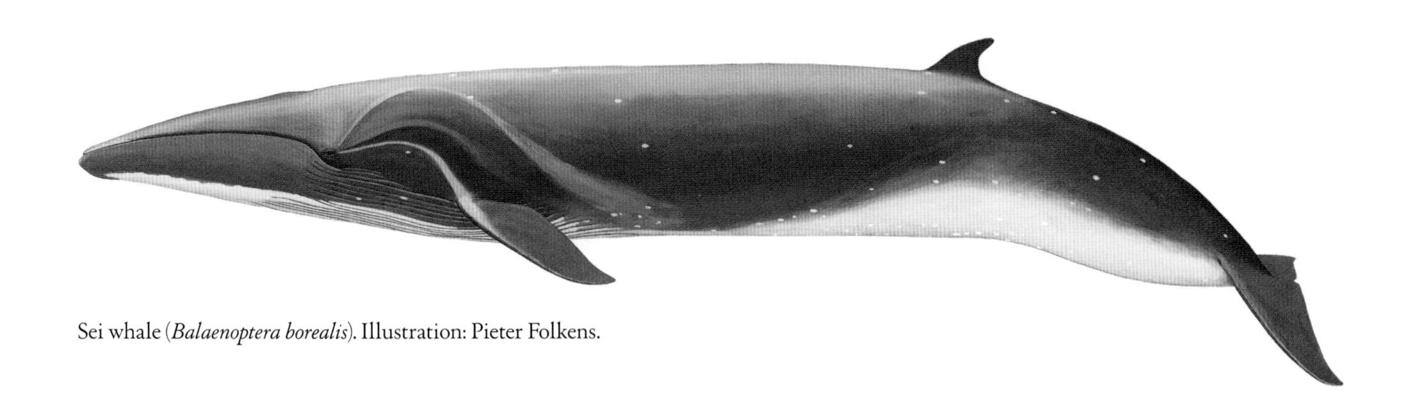

Sei whale (*Balaenoptera borealis*). Illustration: Pieter Folkens.

CONSERVATION STATUS: Sei whales began to be captured in the seventeenth century by Japanese whalers, but it only became a commercially attractive species after the blue whales and fin whales had decreased significantly. In the Southern Hemisphere the capture peak was in the 1960s, declining rapidly, until in 1977 the IWC gave it complete protection (Gambell, 1985). The scientific committee for the IWC estimates, for the Southern Hemisphere, a current population of 37,000 individuals, 14,000 in the North Pacific and few thousands in the North Atlantic (Allen, 1980). In Mexico its presence is considered rare (Aurioles Gamboa, 1993) and it is subject to special protection in the Mexican Official Regulation (SEMARNAT, 2010). Its protection is included in the Federal Law on Fisheries and article 420-I of the Federal Penal Code (Diario Oficial de la Federación, 2009). Today, the sei whale is listed as an endangered species in appendix I of CITES, referring to those in the area of Ecuador to Antarctica, but other regions are considered in appendix II. It is considered endangered by the IUCN (IUCN, 2010).

Balaenoptera edeni Anderson, 1879

Bryde's whale

Luis Medrano González and María de Jesús Vázquez Cuevas

SUBSPECIES IN MEXICO

B. edeni is a monotypic species.

Two types of Bryde's whales have been identified in the waters off Baja California that probably correspond to the coastal and oceanic shapes. This clade was then connected to the sei whale. A differentiated clade among these corresponds to the small

Bryde's whale (*Balaenoptera edeni*). Illustration: Pieter Folkens.

form of Bryde's whales in the Philippines. If this phylogeny is correct and if the holotype of *Balaenoptera edeni* actually corresponds to the small form of the Philippines, then it may be recognized that the Bryde's whale consists of two species. The name *B. edeni* would be assigned to the small form. In South Africa, Indochina, Japan, Baja California, and Brazil, a coastal form has been identified (Leatherwood et al., 1983).

DISTRIBUTION: Bryde's whales inhabit tropical and subtropical waters around the world; its normal distribution is limited by the isotherm of 20°C (Valdivia et al., 1981). Occasionally, they can be seen at high latitudes (Leatherwood et al., 1983). Concentrations are found in the waters off South Africa, West Africa, the Seychelles Islands, eastern Australia, the Fiji Islands, Midway, Bonin and Hawaii, the Gulf of Panama, the Galapagos Islands, Peru, Mexico, the Southeast of the United States, the Antilles, the Caribbean, and the Brazilian coast. In the Mexican Pacific, they are distributed within the Gulf of California from 26 degrees north along the coasts of Baja California to the mouth of the Gulf and along the continental coast, from the coast of Baja California to the Marias Islands, reaching the equatorial area. This species is also relatively abundant in the Gulf of Mexico and Caribbean Sea. Apparently, these cetaceans undertake small migrations from the coast to open sea and back, according to the fluctuations in prey abundance (Best, 1977; Gendron, 1993b; Jefferson and Schiro, 1995; Jefferson et al., 1992; Leatherwood et al., 1983; Tershy et al., 1990; Urbán-R. and Flores, 1996; Valdivia et al., 1981; Watson, 1981). It has been registered in the areas of the GC, GM, MC, NP, and TP.

DESCRIPTION: Bryde's whale is also called the tropical whale. Their color is dark smoky gray with faded sides and a light or white abdomen. In many individuals there is a gray region on each side in front of the dorsal fin. At sea, this species is easily confused with the sei whale (*B. borealis*). Bryde's whales differ from other members of the family by the presence of three prominent and parallel ridges that lie anterior of the nostrils on the dorsal surface of the rostrum. The dorsal fin, located on the third part of the body, measures from 45 cm to 50 cm in height. It is curved and pointed, and has an irregular rear edge that is useful for photo-identification. The pectoral fins are thin and sharp and measure approximately one-tenth of the total length of the body. The ventral-gular furrows number 40 to 50 with some reaching the navel.

EXTERNAL MEASURES AND WEIGHT

TL: 14 m to 15.6 m.
Weight: to 40,000 kg.

DENTITION

250 to 370 baleen plates (reaching a length of 42 cm and 24 cm wide) with long bristles. The plates are gray with darker fringes, with the exception of the most anterior, which are light gray and sometimes white.

NATURAL HISTORY AND ECOLOGY: The systematics of this species has been a source of controversy (Best, 1977; Dizon et al., 1996; Leatherwood et al., 1983). An analysis of the variation in the sequence of a fragment of mitochondrial DNA revealed a complex structure within the group of Bryde's and sei whales. In South Africa, there is a small coastal tropical Bryde's whale that probably corresponds to the holotype (Had et al., 1996) and to a larger ocean form (Best, 1977). The two forms have distinct barbs, marks on the body, diets, ovulation rates, and the seasonalities of reproduction. They do not undertake marked latitudinal migrations that define a feeding and a reproductive phase as do other mysticeti, but make longitudinal movements that are irregular and associated with changes in productivity in the ocean. They give birth almost all year round. In the temperate zones, they have annual cycles of latitudinal migrations, more similar to those of other mysticeti. It is estimated that, in the small coastal form, gestation lasts about 1 year, nursing 6 months, and the reproductive cycle 19 months. These animals reach sexual maturity when they measure a little more than 12 m (Best, 1977; Leatherwood et al., 1983). In the Gulf of California they can be found throughout the year without showing any seasonal regularity in their reproduction (Urbán-R. and Flores, 1996). Commonly, solitary individuals are observed, although these whales also form couples or groups of five to six individuals. At times, they have been seen in groups of 30 to 40 whales around sources of food (Leatherwood et al., 1983; Watson, 1981). In the Gulf of California they form feeding groups of more than 15 individuals that behave independently (Tershy et al., 1993). In the coastal areas they feed on epipelagic fish and offshore on mesopelagic fish and krill (Best, 1977). Their diet consists mainly of anchovies, herring, sardines, zooplankton, and squid. In the Gulf of California they mainly feed on fish and, in a lesser proportion, on krill and small lobsters (Best, 1977; Kawamura, 1977; Tershy et al., 1993; Urbán-R. and Flores 1996). It has been observed that they change their feeding pattern during El Niño phenomena (Ramírez, 1986). They mainly feed in the depths, trapping their food by descending to devour it. They may also swim horizontally near the surface lunging for prey. In both cases, this whale usually turns on its sides as fin and blue whales do (Leatherwood et al., 1983; Watson, 1981). They are deep divers and when they emerge to breathe they do so almost vertically, showing the tip of the rostrum. The blow is thin and tall (close to 4 m in height). They generally emerge 3 or 4 times in intervals of 10 to 15 seconds before diving 5 to 20 minutes, showing the caudal fin in very rare occasions. They are elusive; under water they often and drastically change the direction in which they are swimming. They may come close to ships and swim along with them for some time. The photo-identification of these animals is difficult in comparison to other whales, and there are few estimates of their current abundance and the identity of their populations. They have been observed forming aggregations of up to a few hundreds of individuals (Chávez, 1995; Leatherwood et al., 1983; Urbán-R. and Flores, 1996; Watson, 1981).

CONSERVATION STATUS: Bryde's whales were hunted for commercial purposes, but their record before the 1920s is scarce since the hunting effort for this species was low and it was confused with the fin and sei whales taken; hence, the information on the original population estimate and number of animals captured is incomplete. Currently, about 14,000 Bryde's whales live in the Northwestern Pacific, compared to 20,000 before commercial capture (Leatherwood et al., 1983). It is included in appendix I of CITES and is considered data deficient by the IUCN (IUCN, 2010). In Mexico it is subject to special protection in the Mexican Official Regulation (SEMARNAT, 2010) and as other marine mammals living in Mexican waters, it is protected by the Federal Law on Fisheries and article 420-I of the Federal Penal Code (Diario Oficial de la Federación, 2009).

Balaenoptera physalus (Linnaeus, 1758)

Fin whale

Lorenzo Rojas Bracho

SUBSPECIES IN MEXICO

Balaenoptera physalus physalus (Fischer, 1829)

DESCRIPTION: *Balaenoptera physalus* is the second largest rorqual whale (after the blue whale). *B. physalus* are larger in the Southern Hemisphere, reaching 23.8 m for the longest male, while females can exceeded 24.4 m. The body of the fin whale is the slimmest of all the members of the family. It has a hydrodynamic line and a marked flange that runs along the back (Leatherwood et al., 1988). The body is rounded anteriorly and compressed laterally posteriorly in the region of the caudal peduncle, which has a strong keel both dorsally and ventrally (Gambell, 1985). The dorsal fin is curved in its posterior margin. A distinctive feature is the gradual inclination of the anterior margin of the dorsal fin that is at a lower angle than in other rorquals. The span of the caudal fin (flukes) is one-third to one-fourth of the total length of the body (Tomilin, 1967). The pectoral fins (flippers) are small and lanceolate, and measure 12% of the body length (Nishiwaki, 1972). It has 50 to 100 ventral gular furrows (average 64) that start on the anterior edge of the jaw and run up at least to the navel (Nishiwaki, 1972; Watson, 1981). In adults, the head measures one-fourth of the total length of the body; dorsally, it has the form of a V and is less flat than in the blue whale (Gambell, 1985; Tomilin, 1967). The most distinctive feature of the fin whale is the asymmetrical coloration of the lower jaw and the barbs. The right side is white or grayish-white, while the left side is dark gray. The general coloration of the body is dark gray, slate to brownish at the back, lighter on the sides, and white on the ventral region. They have complex streaks and dorsal bands on the side and back of the head and between the pectoral fins, which are called saddles (chevrons). These are used by researchers for individual identification (Hall, 1981; Leatherwood et al., 1988; Nishiwaki, 1972; Tomilin, 1967).

EXTERNAL MEASURES AND WEIGHT

TL: females up to 24 m in the Northern Hemisphere, males to around 22 m.
Weight: up to 120,000 kg.

DENTITION

260 to 480 dark olive greenish to black baleen plates on each side of the mouth (up to 42 cm long and 24 cm wide) with lengthy and rigid fringes. The front third of the plates on the right side only are white or cream-colored.

NATURAL HISTORY AND ECOLOGY: The fin whale reproduces in temperate or subtropical waters during the winter (November to March) (Harrison, 1969; Slijper, 1979) with a peak of births between December and February (Gambell, 1985; Leatherwood et al., 1988). The gestation period lasts slightly more than 11 months and, in the Northern Hemisphere, the size at birth is 6 m (Haug, 1981; Laws, 1959). Lactation lasts about six months, and during this time the calves grow up from 6.0 m to 12.2 m. Mating peak occurs from January to February (Tomilin, 1967). First the multiparous females breed, followed in four to six weeks by the nulliparous (Harrison, 1969). On average, females give birth every 2.16 years (Slijper, 1979). Females usually have a resting period between pregnancies and generally give birth at intervals of three years. The size at which fin whales reach sexual maturity is 17.7 m for males and 18.3 m for females (Gambell, 1985). Due to population stress caused by the hunting of these whales, they may reach sexual maturity at an earlier age . It is estimated that it has declined from 10 years to 6 or 7 years since the mid-1930s (Gambell, 1985). Their diet is one of the most diverse within the mysticeti, and varies depending on the time and place (Kawamura, 1980). They mainly feed on planktonic crustaceans, but also on fish, and sometimes on cephalopods and mollusks (Nemoto and Kawamura, 1977). In the Northern Hemisphere, euphausids are a more important prey. Copepods are also an important part of the diet, in particular, *Calanus cristatus* (Kawamura, 1980, 1982), and fish such as sardines (Mizue, 1951), capelins and herring (Nemoto, 1959), and mackerel and anchovies (Kawamura, 1982). Although it is often found alone or in pairs, groups of up to 20 animals are common. In the feeding areas aggregations of 100 or more individuals have been reported (Gambell, 1985). They are fast whales that reach speeds exceeding 25 knots and can cover distances of 90 nautical miles per day. The tall blow of this whale measures up to 5 to 6 m high. The fin whale can submerge down to 230 m, which is deeper than blue whales and sei whales go. It can also breach, sending its entire body clear of the water in a leap (Leatherwood et al., 1988), but this is a rare behavior.

CONSERVATION STATUS: The fin whale, together with the blue whale, were primary targets of commercial mechanized hunting during the twentieth century, which eroded their populations throughout the world until the IWC banned their capture in 1976 (Rice, 1974). The population in the Northern Hemisphere, prior to its exploitation by the whaling industry, numbered 53,000 animals (Allen, 1980). At the time of receiving global protection, the population was estimated at 20,000 individu-

DISTRIBUTION: Fin whales are found in all oceans of the world. It conducts annual migrations between the feeding areas located in subpolar latitudes and the reproduction areas in temperate and subtropical latitudes (Mackintosh, 1965). In general, during the summer they are found in the Sea of Chukchi, near the Aleutian Islands (Bering Sea), in the Gulf of Alaska, and along the coast of California (Gambell, 1985). In the Northeastern Pacific, during the winter, fin whales are distributed from the south of the Bahia de Monterrey close to 35 degrees 30 minutes north toward the south to Cabo San Lucas (Leatherwood et al., 1988). Other records indicate that fin whales can be found throughout the year in the Gulf of California. It is possible that fin whales have a relatively isolated resident population in the Gulf of California (Aguayo et al., 1983; Gambell, 1957; Rojas-Bracho, 1984; Tershy et al., 1990, 1993). Studies on the differences between the vocalizations of fin whales of the Gulf of California and the Pacific and Atlantic support the assumption of a more or less isolated population (Thompson et al., 1992). Analysis on the variation of mitochondrial DNA supports the idea of a resident population of fin whales in the Gulf of California, but does not exclude the possibility that it is a faction of the population of the North Pacific with which there is continuous replacement (Bérubé et al., 1998). Nearly 300 fin whales live along the west coast of Baja California (Urbán-Ramírez, 1996). There are reports of the presence of offspring and feeding behavior around the major islands in the Gulf of California in the spring (Gendron, 1993). This suggests a local change of eating and reproductive habits with regard to other populations. It has been registered in the areas of the GC and NP.

Fin whale (*Balaenoptera physalus*). Illustration: Pieter Folkens.

als, that is, 38% of the pre-exploitation population (Gambell, 1985). Norwegian and Icelandic whalers have defied this ban on killing fin whales. Currently, this species is included in appendix I of CITES and the Mexican Official Regulation classifies it as special protection (SEMARNAT, 2010). It is considered endangered by the IUCN (IUCN, 2010); it is protected by the Federal Law on Fisheries and article 420-I of the Federal Penal Code (Diario Oficial de la Federación, 2009).

Balaenoptera musculus (Linnaeus, 1758)

Blue whale

Diane Gendron and Verónica Zavala Hernández

SUBSPECIES IN MEXICO
Balaenoptera musculus musculus (Linnaeus, 1758)

DESCRIPTION: The blue whale is distinguished by its large size, which is reflected both in its length and in the height of its blow of 6 m to 12 m produced when it

Blue whale (*Balaenoptera musculus*). Illustration: Pieter Folkens.

DISTRIBUTION: The blue whale is distributed in all oceans and is found in coastal and oceanic waters of the Baja California Peninsula and California. Its presence along the western coast of the Baja California Peninsula has been documented since the nineteenth century (Scammon, 1874), where it was intensively hunted by Norwegian whalers between 1924 and 1929, and during 1935 (Rice, 1974). Recently, the presence of the species in the Gulf of California has also been proved. In this area it has been mainly found during winter and spring, along the southwest coast of the Gulf (Gendron, 1990; Sears, 1987; Vidal et al., 1993), there are also observations in the region of the major islands during the summer (Tershy et al., 1990). Based on captures made in the North Pacific, blue whales are reported from the Aleutian Islands and the Gulf of Alaska to Baja California with an annual movement from north to south between the two areas (Rice, 1974), although the migration routes are not clearly defined. There is evidence that diet determines its activities in the waters of Baja California during the winter (Gendron, 1990). Their presence in the eastern tropical Pacific throughout the year indicates the possibility of a resident population (Berzín, 1978; Reilly and Thayer, 1990). Photo-identification has allowed the recognition of certain individuals that visit the waters of the Baja California Peninsula during the winter and the waters of California during the summer (Calambokidis et al., 1990). Despite its global distribution, there are few places in the world where they can be observed; the waters of the Peninsula of Baja California is one of them and represents the only known place in the northern Pacific where calf rearing takes place. It has been registered in the areas of the GC, NP, and PT.

emerges (Leatherwood et al., 1988). The proportionally smaller dorsal fin rises 25 cm to 50 cm and is positioned farther back on the body than in other members of the family. Its rostrum is U-shaped and wider than in other mysticetes. A very small dorsal fin along with a lighter overall coloration, a broad rostrum, and its massive body size are the main features identifying the blue whale. Its mottled light gray coloration varies and appears blue when seen just below the surface of the water. The pattern of pigmentation, with irregular areas along both sides, facilitates the recognition of individuals of this species by means of photo-identification (Sears et al., 1990). The triangular shape of the caudal fin (flukes), with a maximum of 6 m across, as well as the depth of the peduncle, are also features that distinguish the species (Leatherwood et al., 1988). The flukes are often raised prior to long feeding dives. The pectoral fins are long (2.5 m), representing around 15% of the total length (Leatherwood et al., 1988). The blue whale has from 55 to 88 ventral pleats that characterize the family Balenopteridae.

EXTERNAL MEASURES AND WEIGHT

TL: 20 to a maximum of 29.8 m (33.58 m in the Southern Hemisphere).
Weight: up to 180,000 kg (Southern Hemisphere).

DENTITION

260 to 400 black baleen plates (up to 42 cm long and 24 cm wide) with lengthy and rigid fringes. The plates are gray and the fringes are darker.

NATURAL HISTORY AND ECOLOGY: The blue whale is usually found alone, in pairs, and sometimes in trios, though larger numbers may be in an area during significant upwelling events. A feeding aggregation of an estimated 400 blue whales appeared in the late spring of 2012 off of northwest Baja. In the waters off the Baja California Peninsula, females accompanied by their offspring are commonly seen. The gestation period lasts between 10 and 11 months (Yochem and Leatherwood, 1985). Their diet is composed almost exclusively of euphausids (Gaskin, 1982); the species *Nyctiphanes simplexes* is an important prey in the Gulf of California (Gendron, 1990, 1992). The diving time varies between 10 and 30 minutes (Yochem and Leatherwood, 1985), with an average of 10 minutes for whales observed from 1988 to 1996 in the Gulf of California (D. Gendron, pers. obs.). Adult blue whales and their offspring are frequently observed in the waters of Baja California with ectoparasites such as barnacles and remoras (Rice, 1963).

CONSERVATION STATUS: Recent abundance estimates for the waters of California indicated approximately 2,800 individuals in the northeast Pacific population, out of a world population estimated to exceed 12,000 animals. Pre-exploitation populations exceeded 200,000 individuals. Due to the extreme sensitivity to boats, mainly those with outboard motors, tourism can be a potential threat. In 1966 the IWC protected this species (Yochem and Leatherwood, 1985); now it is listed in appendix I of CITES as endangered and as rare in Canada (Mansfield, 1985): IUCN considers it endangered (IUCN, 2010). It is subject to special protection in Mexico, which prohibits its capture, which is important because the waters of Baja California correspond to a priority area for the recovery of this species (SEMARNAT, 2010); it is protected by the Federal Law on Fisheries and article 420-I of the Federal Penal Code (Diario Oficial de la Federación, 2009).

Humpback whale

Paloma Ladrón de Guevara Porras

SUBSPECIES IN MEXICO

Megaptera novaeangliae is a monotypic species.

DESCRIPTION: Humpback whales have a robust body; with a dark gray to black coloration on the head, back, sides, and caudal peduncle, while the region of the throat, chest, and abdomen can have a white color that may extend to the flanks and to the fluke. The most distinctive external features are the presence of some tegumental protuberances on the head and jaw with large bumps on the edges. Each bump has a thick hair 1 cm to 3 cm long. They have 12 to 26 ventral-gular furrows. The pectoral fins are about 5 m long; the dorsal fin is less than 30 cm and varies from curved to rounded and is located on the last third of the body, usually over a hump; and a broad caudal fin (flukes) with a serrated posterior edge, a deep central notch, and a ventral surface with a variable combination of black and white coloration as well as numerous scars. These characteristics of the flukes can be used to identify individuals. The flukes are normally raised before a dive (Leatherwood et al., 1983; Tomilin, 1967).

EXTERNAL MEASURES AND WEIGHT

TL: males to about 15 m, females to 16 m.
Weight: 30,000 to 40,000 kg.

DENTITION

270 to 400 baleen plates on each side (up to 70 cm long and 30 cm at the base) with lengthy and rigid fringes. The plates are gray and the fringes are darker, with the exception of the most anterior, which are light gray and sometimes white.

NATURAL HISTORY AND ECOLOGY: Humpback whales are distributed along the coasts and around oceanic islands. During their migration, they travel away from the coast outside of the continental shelf (Winn and Reichley, 1985). In the Northern Hemisphere, they mainly feed on krill (*Euphausia*) and fish-like herring (*Clupea*), capelin (*Mallotus*), sand eel (*Ammodytes*), anchovy (*Engraulis*), mackerel (*Scomber*), and Arctic cod (*Boreogadus*) (Gaskin, 1985; Tomilin, 1967). Sexual maturity is reached between 4 and 6 years of age with an approximate body length of 10 m to 12 m. Females give birth during the winter after a gestation period of 11.5 months. Newborn

Humpback whale (*Megaptera novaeangliae*). Illustration: Pieter Folkens.

calves measure from 4.0 m to 4.5 m and are nursed for 11 months, reaching a length of 8 m to 9 m at the time of weaning (Leatherwood et al., 1983; Tomilin, 1967; Winn and Reichley, 1985). Different ectoparasites such as whale lice (Amphipoda) feed from the skin of the whale, while barnacles attach themselves to whales to feed on plankton (Tomilin, 1967; Winn and Reichley, 1985). With regard to their predators, there are reports of killer whales (Flórez-González et al., 1994; Whitehead and Glass, 1985); sharks usually attack only dead or weak animals (Winn and Reichley, 1985). During the feeding and reproduction seasons congregations of up to 24 whales have been be observed, but most commonly they are found alone or in groups of 3 to 4 individuals. In general they dive from 5 to 7 minutes, although dives of up to 30 minutes have been reported. Their blow is dense and measures about 2.5 m to 5 m high. Three feeding behaviors that vary regionally are recognized: (1) lunging: the whale approaches the food from the bottom, opening its mouth moments before reaching the surface and devouring its food; (2) tail slashing or stroke: the whale uses a rapid movement of its tail to produce a forward flow that concentrates the krill in front of the whale, which then moves forward devouring it; (3) bubble netting: using exhaled air the whale forms a spiral curtain of bubbles that surrounds the prey in the center of the ring through which the whale feeds (Gaskin, 1985; Winn and Reichley, 1985). The white undersides of the long flippers are sometimes used to "flash" prey, encouraging the fish to move into a more favorable position for capture. These various feeding techniques often come together in social feeding aggregations herding prey to the surface. Another collection of humpback whale behaviors is a wide variety of jumps, chin slaps, tail trows, and "pec slaps" (splashes made with the pectoral fins), especially during the breeding season. Humpback whales are considered the most acrobatic of the large whales (Tomilin, 1967). During the nursing season, agonistic displays and physical contact between males competing for the females occur and is commonly seen in the southwest Gulf of California. Another particular feature of the humpback whale is that it produces three types of sounds, denominated as singing, prey aggregation calls, and social sounds. Singing sounds are produced by solitary males and are structured in intricate patterns cast in a defined sequence; these occur in the subtropical mating areas. Social sounds are complex grunts, groans, and whistles. Prey aggregation calls are high-intensity cries around 400 to 600 Hz and can be heard in the feeding areas among feeding groups. These whales may have a hierarchical population structure that is a product of changes in the distribution associated with destinies of migration to which the humpback whales are faithful (Baker and Medrano-González, 1998). In Mexico, photo-identification and genetic studies have shown that the aggregations of the continental coast and the Revillagigedo Islands correspond to different population units (Medrano-González et al., 1995; Urbán-R. et al., 1994).

DISTRIBUTION: The humpback whale is a cosmopolitan species. There are populations in the North Atlantic, North Pacific, and Southern Ocean, each with a substructure. In the North Pacific they feed during the summer in the Bering Sea, Gulf of Alaska, the Chuckchi Sea of Okhotsk, Honshu Island in Japan, and off California from the Channel Islands to Cordell Bank. In the winter, during the breeding season, they concentrate in three different areas: the western North Pacific (Taiwan and Bonin Islands, Ryukyu, and Marianas), the central Pacific (Archipelago of Hawaii), and the eastern North Pacific (west coast of Mexico to Costa Rica and Revillagigedo Islands) (Johnson and Wolman, 1984). In the waters of the Mexican Pacific, two subregions of winter concentrations have been defined: (1) the west coast of Baja California Sur and Isabel Island, Tres Marias Islands to the Isthmus of Tehuantepec and (2) the Revillagigedo Archipelago. In addition, in the Gulf of California there is an aggregation present throughout the year, whose population identity has not been determined (Rice, 1974; Salinas et al., 1990; Urbán-R. and Aguayo, 1987b). In the Gulf of Mexico and Mexican Caribbean its presence is rare and it has been documented in Cuban and Dominican Republic waters and in the north coast of the Gulf, from the Florida Keys to Galveston, Texas (Jefferson et al., 1992; Weller and Schiro, 1996). It has been registered in the areas of the CS, GC, GM, NP, and TP.

CONSERVATION STATUS: Humpback whales have been captured in a traditional way in Greenland and the Antilles. They were commercially hunted from the nineteenth century until 1966, when the IWC banned their catch. It is estimated that in the North Pacific less than 1,000 individuals live in a population that originally exceeded 15,000 (Rice, 1974). In 1994 it was estimated that there were 3,350 individuals in the Mexican Pacific (Urbán-R. et al., 1994). Around the world, humpback whales appear to be recovering, but their coastal habits make them susceptible to pollution and activities such as tourism. This species is currently catalogued as endangered by the Endangered Species Act of 1973 of the United States; it is listed in appendix I of CITES and in the category of least concern by the IUCN (IUCN, 2010). In Mexico, it is subject to special protection (SEMARNAT, 2010) and it is protected by the Federal Law on Fisheries and article 420-I of the Federal Penal Code (Diario Oficial de la Federación, 2009). Tourism involving humpback whales has grown in recent years in Mexico and its practice is beginning to be regulated.

Family Eschrichtidae

The family Eschrichtidae is represented by a single extant species, the gray whale (*Eschrichtius robustus*), that at one time lived on both sides of the North Atlantic and North Pacific Oceans. The western Atlantic population did not reach the eastern shores of Mexico. The Atlantic populations were extinct by the early nineteenth century. (However, two vagrant gray whales have been identified in the Atlantic following an anomalous low ice year in the Arctic, one of which appeared in the Mediterranean Sea off Israel and the other off Namibia to become the first record of a gray whale in the Southern Hemisphere. Both of these whales were probably born in Mexico.) The eastern Pacific population lives along the west coast of North America from Baja California to the Bering and Chukchi Seas and was near extinction the early twentieth century. Currently, they have recovered due to the prohibition of hunting and the protection of their calving lagoon in Baja California. A population in the western Pacific is extremely endangered.

Eschrichtius robustus (Lilljeborg, 1861)

Gray whale

Silvia Regina Manzanilla Naim

SUBSPECIES IN MEXICO
E. robustus is a monotypic species.

DESCRIPTION: The coloration of the body of *Eschrichtius robustus* varies from a mottled light to dark gray with multiple spots produced by barnacles. Calves have dark gray smooth skin at birth. The dorsal view of the head is narrow and in the form of a V; it is covered with barnacles and cyamids; in a lateral view the lips have a convex curvature directed toward the tip of the face. The skin of the face has many cavities that contain small whiskers. Gray whales have two to three gular folds. They lack a dorsal fin; in its place is a small hump followed by a series of 6 to 12 protuberances covering the last third of the body toward the caudal peduncle. The fluke is wide, 3 m in size, and mottled. The pectoral fins have rounded edges and end in a tip. Females are slightly larger than males; it is estimated that males measure up to 11 m and

Gray whale (*Eschrichtius robustus*). Illustration: Pieter Folkens.

females up to 11.7 m when sexual maturity is reached, at around 5 and 11 years of age, respectively.

EXTERNAL MEASURES AND WEIGHT

TL: 12 to 15 m. Females slightly larger than males
Weight: up to 35,000 kg.

DENTITION

130 to 180 yellowish to cream-colored baleen plates on each side; 5 cm to 25 cm long, with coarse fringes.

DISTRIBUTION: *E. robustus* is a coastal species, which is distributed in the North Pacific Ocean. Three populations have been recognized: the one in the North Atlantic, extinct in the seventeenth century; the one in the Western Pacific (Korean), exploited until 1966 and currently near extinction; and the one in the Eastern Pacific (Californian), currently the only population that has recovered from the threat of extinction. The migration of this population ranges from the polar waters of the Bering Sea, where they feed during summer, to the coastal lagoons in Baja California in Mexico, where they reproduce during winter. The migration toward the south starts in November and ends in the coastal lagoons of Ojo de Liebre, Guerrero Negro, San Ignacio, and Bahía Magdalena, where the whales mate and breed. Some groups do not reach the Bering Sea and remain as residents in different latitudes of their migration toward the north. These groups have been reported in the Farallón Islands, Punta St. George, and the mouth of Klamath River in California, as well as near Oregon in United States and around Vancouver Island and Queen Charlotte Islands in British Columbia. The gray whales reach the High Gulf of California (Silber et al., 1994) and in cold years, such as 1989 and 1999, they can be seen as far south as Bahia de Banderas (L. Medrano González, pers. comm.). It has been registered in the areas of the GC and NP.

NATURAL HISTORY AND ECOLOGY: The migration of gray whales to the north starts in March and they maintain their coastal route at depths of no more than 183 m. Migration occurs in two stages separated by a period of eight weeks. During the first migration, adult males and females without calves generally compose the groups; in the second, females with offspring and juveniles integrate the groups (Leatherwood et al., 1983; Swartz, 1986). In spring they enter the Bering Sea through Step Unimak, separating into groups in different areas and distributing farther north as ice regresses. In late summer, when the masses of ice begin to form, the population moves southward toward Step Unimak (Leatherwood et al., 1983; Swartz, 1986). From Alaska, migrants pass Vancouver Island from November to late January (Darling, 1984). They travel along the coasts of Washington, Oregon, and Northern California, reaching the nursing lagoons in December (Braham, 1984; Dohl, 1979; Herzing and Mate, 1984; Swartz, 1986). In San Ignacio, the abundance of whales in courtship (without offspring) increases during a period of six weeks, from the last week of December to the second week of February (Swartz, 1986). Photo-identification studies reveal that 81% of the whales in courtship in Laguna San Ignacio remain in the lagoons for periods of less than a week (Jones, 1985). Its greatest feeding areas are located in the northern part of the Bering Sea and the southern part of the Sea of Chukchi, where they feed on amphipod shellfish that live on the sediments (Rice and Wolman, 1971). They have also been reported feeding in British Columbia (Oliver et al., 1983).

CONSERVATION STATUS: Around 1850 the intense exploitation within the nursing lagoons of Baja reduced significantly the population of the eastern Pacific gray whale. Hunting continued into the early twentieth century from coastal operations up and down its range and from mechanized pelagic whaling fleets involving only four nations—Japan, Norway, the Soviet Union, and the United States. By the 1930s, the gray whale was considered commercially extinct, prompting international cooperation and protection beginning after 1937, though some clandestine shore-based operations continued during World War II. In 1946, an international treaty protecting the population from commercial exploitation took effect, but permitted continued whaling "exclusively for local consumption by the aborigines." One significant step in the conservation of the gray whale was the Mexico declaration that the waters of the Laguna Ojo de Liebre be a refuge for the gray whale in 1972 (concurrent with the Marine Mammal Protection Act of the Unitied States). The eastern North Pacific population has recovered from estimated lows in the few hundreds to over 20,000 individuals. This may be near the carrying capacity of the whale's range in the absence of a warm Interglacial that would render the High Arctic ice-free for subsequent summers, such as the Holocene Roman Warm Period and the Eemian Interglacial when the population has been estimated to exceed 100,000 individuals. The western North Pacific population remains critically low. It is listed in appendix I of CITES and as least concern by the IUCN (IUCN, 2010). In Mexico it is subject to special protection (SEMARNAT, 2010) and it is protected by the Federal Law on Fisheries and article 420-I of the Federal Penal Code (Diario Oficial de la Federación, 2009).

Suborder Odontoceti

The Odontocetes (toothed whales) are represented by ten families, five of which occur in Mexican territorial waters. Within those families are 74 modern species, one of which is extinct, and 32 occur in Mexican waters. Odontocetes are much more diverse than the mysticetes, ranging from the diminutive porpoises to the large sperm whale. Odontocetes have one nostril (blowhole) compared with two in mysticetes (and all other mammals). The second nasal pathway in odontocetes is dedicated to providing air to a set of nasal sacs used for echolocation.

Family Physeteridae

There are three sperm whales within the genera *Kogia* and *Physeter* that inhabit tropical and temperate seas of the world. (*Kogia* is sometimes placed in its own family, Kogiidae.) All three species are found in Mexico. A population of the sperm whale (*Physeter macrocephalus*) is apparently resident in the Sea of Cortez.

Kogia breviceps (De Blainville, 1838)

Pygmy sperm whale

Mario A. Salinas Zacarías

SUBSPECIES IN MEXICO

Kogia breviceps is a monotypic species.

DESCRIPTION: Pygmy sperm whales are small whales and somewhat similar to sperm whales (*Physeter macrocephalus*). The triangular head is proportionally not as large as in the sperm whale at only 15% of the body length. The jaw terminates well before the tip of the snout. The single blowhole is located on top of the head. The pectoral fins are short and wide. The small, falcate dorsal fin reaches a height not greater than 20 cm and is located slightly behind the middle of the back. The fluke reaches a width of approximately 60 cm (one-fifth of the body length). *Kogia* have a semi-circular stripe behind the head known as a false operculum, that, with the general shape of the head, give it an appearance similar to that of a shark. Its coloration is dark gray on the back, sometimes referred to as coal or steel gray, which changes to light gray in the belly (Caldwell and Caldwell, 1989; Leatherwood et al., 1983).

EXTERNAL MEASURES AND WEIGHT

TL: up to 3.3 m.

Weight: 363 kg.

DENTAL FORMULA: 0/12–16.

NATURAL HISTORY AND ECOLOGY: Pygmy sperm whales have oceanic habits; thus they are rarely sighted and their biology is poorly understood. They are not gregarious and usually form groups of no more than three animals; groups of up to seven animals have been registered. According to sightings of calves at sea and the size of fetuses obtained from whales stranding, mating and delivery are estimated to happen between spring and autumn. Only one calf is delivered with a length from 1.0 m to

Pygmy sperm whale (*Kogi breviceps*). Illustration: Pieter Folkens

DISTRIBUTION: This species is found in tropical and temperate waters around the world. There are records in Australia, New Zealand, Mexico, the United States, Chile, Cuba, Puerto Rico, Uruguay, France, Germany, Great Britain, Japan, Hawaii, China, Indo-China, India, and South Africa (Caldwell and Caldwell, 1989; Leatherwood et al., 1983). In Mexico they have been observed in the Bay of La Paz, Baja California Sur, Bahia de Banderas, and Nayarit-Jalisco, and along the south coast of Jalisco; whales have been reported stranding in the Bay of La Paz (Baja California Sur), the Isle Holbox (Quintana Roo), and Veracruz coastline (Aurioles et al., 1993; Pérez-Sánchez et al., 1994; Salinas and Bourillón, 1988). The species has been registered in the areas of the CS, GC, GM, NP, and TP.

1.2 m and a weight of approximately 55 kg. Females reach sexual maturity when they measure 2.6 m to 2.8 m, and males from 2.7 m to 3.0 m. It is remarkable that the testicles in an animal of 3.0 m in length can measure up to 0.5 m. The gestation period lasts 9 to 11 months; lactation lasts about one year. Females may have offspring in consecutive years. Pygmy sperm whales mainly feed on squid of the genera *Sepioteuthis* and *Ommastrephes*, pelagic crustaceans, some fish, and, in some cases, shrimp from estuaries (*Penaeus californiensis*). They are slow swimmers that may dive great depths for up to 45 minutes. They emerge slowly, causing little disruption on the ocean's surface and taking barely noticeable breaths. They do not dive immediately after taking a breath, but keep near the surface for relatively long periods. Like large sperm whales, the back and dorsal fin are exposed on the water surface, and the caudal fin and peduncle hang downward. In this position, they look similar to a tree trunk in the water. They are timid and elusive when near boats. When they get frightened they expel feces, leaving behind a reddish blotch that has been interpreted as a strategy of escape similar to that made by cephalopods. Various researchers have suggested that this species performs seasonal movements, but no accurate data to set the pattern of these movements exist (Caldwell and Caldwell, 1989; Leatherwood et al., 1983).

CONSERVATION STATUS: From the 1970s, the interest in marine mammals and the effort of observation have increased and have allowed more frequent observations of this species. The number of individuals and females with calves stranding suggest that the numbers of this species are not as low as thought. Its abundance globally and in Mexican waters is unknown. It is believed that the species is in a stable situation since they are only caught occasionally in artisanal fisheries in Japan, the Lesser Antilles, Indonesia, Australia, and New Zealand. The pygmy sperm whale is listed in appendix II of CITES and is subject to special protection in the Mexican Official Regulation (SEMARNAT, 2010). It is protected by the Federal Law on Fisheries and article 420-I of the Federal Penal Code (Diario Oficial de la Federación, 2009). The IUCN (IUCN, 2010) conservation status is data deficient.

Kogia sima (Owen, 1866)

Dwarf sperm whale

David Aurioles Gamboa

SUBSPECIES IN MEXICO

K. sima is a monotypic species.

Dwarf sperm whale (*Kogia sima*). Illustration: Pieter Folkens

DESCRIPTION: *Kogia sima* has an external appearance very similar to that of *K. breviceps* with the most obvious difference being the size of the dorsal fin, which is larger and more falcate in *K. sima*. Its small mouth is located below and behind the melon. The false operculum stripe behind the head resembling a gill flap, the general coloration, and the form of the dorsal fin create the impression of a shark. The coloration is dark gray on the back and light on the belly. The head is prominent and pointed, and the blowhole is toward the left side, as in the other sperm whales.

EXTERNAL MEASURES AND WEIGHT
TL: up to 2.7 m.
Weight: 270 kg.

DENTAL FORMULA: 3/7–12.

NATURAL HISTORY AND ECOLOGY: The probability of observing the dwarf sperm whale at sea is remote because this small, timid species avoids boats, although it is possibly abundant in some regions (Aurioles et al., 1993). This species is well known in the Cerralvo Channel, along the east coast of Baja California Sur. The fishermen in that region confirmed reports that the whales shed red-colored feces when caught sleeping on the surface (Caldwell et al., 1971). It is more frequently sighted in pairs or alone than in groups. Sightings have been reported in the southwest Gulf of California during February to November. In this area, the species is probably an annual resident. Its diet includes cephalopods, mainly squid, fish, and pelagic and benthic shellfish (Nagorsen, 1985). The records of prey found in stomachs suggest that these whales feed at depths ranging from 250 m to 1,300 m (Fitch and Brownell, 1968; Ross, 1978). The species does not have a well-defined reproductive period, information available suggests births occur throughout the year. In both sexes, sexual maturity is reached at lengths of 2.0 m to 2.2 m (Ross, 1978). Two females stranded in the Bay of La Paz were pregnant, the first (October 1983) with a fetus of 591 mm, and the second (March 1985) with a fetus of 730 mm (Aurioles et al., 1993). The offspring at birth measure about 1 m in length (Nagorsen, 1985).

CONSERVATION STATUS: There is no information on the abundance of this cetacean in any part of the world, but by its cosmopolitan distribution in tropical and subtropical waters, and the frequency of strands, it could be considered abundant. Dwarf sperm whales do not suffer any important exploitation, although the species is occasionally used commercially in the Antilles and Japan (Nagorsen, 1985). There is no evidence of any threat on this species, which is listed in appendix II of CITES; it is subject to special protection in the Mexican Official Regulation (SEMARNAT, 2010), and is protected by the Federal Law on Fisheries and article 420-I of the Federal Penal Code (Diario Oficial de la Federación, 2009). It is considered data deficient by the IUCN (IUCN, 2010).

DISTRIBUTION: Details of the distribution of the dwarf sperm whale are not well known. It has been observed in tropical and temperate waters of the Atlantic, Pacific, and Indian Oceans. The largest number of strandings and records in the sea near the coast where the continental slope meets a narrower continental shelf, as well as the type of food found in their stomachs, suggest that *K. sima* is a more coastal species than *K. breviceps*. There are records that suggest that it is distributed in a large part of the Mexican Pacific (Aurioles et al., 1993; Rice, 1978; Salinas and Bourillón, 1988). It is also likely to be found in the Gulf of Mexico and Caribbean Sea (Jefferson et al., 1992; Leatherwood and Reeves, 1983; Nagorsen, 1985). A greater frequency of sightings made in the southwestern part of the Gulf of California could indicate the existence of regions with greater abundance of the species (Aurioles et al., 1993). It has been registered in the areas of the GC, NP, and TP.

Sperm whale

Diane Gendron and Verónica Zavala Hernández

SUBSPECIES IN MEXICO

Physeter macrocephalus is a monotypic species.

There has been controversy over the scientific name of the sperm whale. *P. catodon* was once common, but is no longer considered valid (Rice, 1989).

DESCRIPTION: The sperm whale is the largest of all toothed whales and is distinguished by its habits and characteristics (Scammon, 1874). Among the more conspicuous features, which facilitates its identification at sea, is the enormous head in the form of a box that occupies one-fourth to one-third of the total length of its body. This large forehead contains an organ that holds a large amount of oil once widely used by humans as a lubricant and in the production of cosmetics. These oils, called "spermaceti," drove a fierce commercial whaling industry during the nineteenth and twentieth centuries. This spermaceti organ is used by the whale for controlling echolocation sounds and possibly buoyancy. Another valued product of sperm whales is the ambergris (indigestible solid remains of squid in the intestine) used in perfumery. The blow slit is S-shaped and positioned on the left side of the forehead. The blow, measuring less than 2.4 m, is projected forward and to the left. This feature helps in identifying it and the direction it is moving from considerable distances. The dorsal fin is small, rounded, triangular, and located in the posterior third of the back. This is followed by a series of small peaks visible when the animal bends to dive. It raises the tail flukes at the initiation of a terminal (deep) dive. This allows photo-identification of individuals. The pectoral fins are small and have the shape of paddles with slightly rounded tips. The jaws are quite long and narrow. The body coloration is brownish-gray with lighter tones in the front of the head, on the belly, and around the mouth. The skin on the back of the head has a wrinkled appearance, which facilitates the identification of this species. The coloration of calves is gray, but the pigmentation increases in the course of the first weeks of life (Hoyt, 1984; Leatherwood and Reeves, 1983, 1988; Whitehead, 1990).

EXTERNAL MEASURES AND WEIGHT

TL: males up to 18.3 m; females to more than 11 m.

Weight: males to 57,000 kg; females to 24,000 kg.

Sperm whale (*Physeter macrocephalus*). Illustration: Pieter Folkens.

DENTAL FORMULA: 0/18–26.

NATURAL HISTORY AND ECOLOGY: Sperm whales are gregarious and travel in groups of 50 or more individuals. During reproduction herds of 50 to 150 animals are common (Leatherwood et al., 1988). Old males tend to be solitary, except in the breeding season when they join the herds composed primarily of females with their offspring. The rest of the year they can form exclusive groups of single, sexually inactive males or herds of mothers with offspring and young animals of both sexes (Leatherwood et al., 1983). In the Gulf of California groups of up to 50 animals composed of males or males and females with offspring have been observed (Gendron, 1993a). Mature males are found closer to the polar regions, while females and breeding males do not travel beyond the area limited by latitudes 40 degrees north and 40 degrees south. The general migratory pattern is toward the poles in spring and summer and toward tropical regions in autumn and winter (Leatherwood et al., 1983). Data obtained by the Japanese and Soviets in the North Pacific reveal substantial longitudinal dispersion (Rice, 1974). Births occur between the end of winter and summer after 14 to 15 months of pregnancy. Calves are born measuring 3.5 m to 4.5 m and weighing 1,000 kg. Lactation lasts 24 months, and the cycle between pregnancies has been estimated between 4 and 5 years. Females reach sexual maturity between 8 and 11 years of age with lengths of 8.3 m to 9.1 m. Males reach maturity after 10 years of age with lengths of 10 m to 12 m. Physical maturity is attained when they reach a size of 15.8 m and weight of 39,463 kg (Best, 1974; Best et al., 1984; Leatherwood et al., 1983, 1988; Rice, 1989). Sperm whales can dive at depths that exceed 1,000 m and remain submerged for periods of 1 hour or more; when they surface they spend a similar amount of time breathing up to 50 times before a new immersion (Leatherwood et al., 1988). They are long-lived animals that live between 45 and 60 years. Their diet consists mainly of meso- and bathypelagic cephalopods (Clarke, 1986; Kawakami, 1980). In the eastern tropical Pacific the giant squid (*Dosidicus gigas*) is the primary component of its diet (Clarke et al., 1993). It is suspected that in the Gulf of California it also feeds on this squid (Gendron 1993a; Vidal et al., 1993). They can be preyed on by killer whales (*Orcinus orca*) and false killer whales (*Pseudorca crassidens*) (Leatherwood et al., 1988; Palacios and Mate, 1996).

CONSERVATION STATUS: Historically, commercial whaling took a heavy toll on sperm whale populations around the globe throughout the nineteenth and twentieth centuries, until it was protected by decree of the ICW in 1979 (Leatherwood et al., 1983). The global abundance is estimated at 732,000 individuals, of which 410,000 live in the Southern Hemisphere (Hoyt, 1984). In Mexico there are no estimates of abundance but it is considered a rare species (Aurioles, 1993), but may be resident. The species is found in appendix II of CITES and is considered vulnerable by the IUCN (IUCN, 2010). In Mexico it is subjected to special protection (SEMARNAT, 2010) and it is protected by the Federal Law on Fisheries and article 420-I of the Federal Penal Code (Diario Oficial de la Federación, 2009). Some international organizations consider it endangered (Gosho et al., 1984).

DISTRIBUTION: Sperm whales are distributed in all oceans of the world; only the polar regions of both hemispheres are not frequented by this species. Their migratory routes are not as clear as in some whales given the complexity of the social structure and the migration pattern that segregates classes by sex and age. In the North Pacific they are frequently found south of latitude 40 degrees north, mainly in the winter; in central California they can be observed from November to April (Rice, 1974). There are some records of sperm whales from the nineteenth century, along the west coast of Baja California Sur, in the Gulf of California, and along the west coast of South Mexico (Leatherwood et al., 1983, 1988). In the Gulf of California their distribution is poorly known. In the early 1990s the substantial increase in the sightings of sperm whales, between the Bay of La Paz and Isla San Pedro Martir, was described referring to the movements of the giant squid *Dosidicus gigas* (Gendron, 1993a; Klett, 1981). In the Gulf of Mexico its distribution is not well known, but sperm whales are common in pelagic areas (Jefferson and Schiro, 1995; Jefferson et al. 1992); strandings on the shores of Veracruz and Quintana Roo have been reported (Fuentes and Aguayo, 1992; Pérez-Sánchez et al., 1994). It has been registered in the areas of the CS, GC, GM, NP, and TP.

Family Ziphiidae

Ziphiidae (beaked whales) comprises 19 species in five genera distributed across all the seas and oceans of the world (Wilson and Reeder, 1993). In Mexico, nine species of beaked whales represent four genera: *Ziphius, Berardius, Indopacetus*, and *Mesoplodon*. They are conspicuous cetaceans, whose distribution is not well known and are apparently not abundant in Mexican waters. It is interesting to note that *Mesoplodon peruvianus* was described in 1991.

Berardius bairdii Stejneger, 1883

Baird's beaked whale

Jorge Urbán Ramírez and David Aurioles Gamboa

SUBSPECIES IN MEXICO
B. bairdii is a monotypic species.

DESCRIPTION: The body of *Berardius bairdii* is cylindrical, robust, but not obese. In both sexes the jaw protrudes notoriously from the face and has two pairs of teeth, one pair of large and triangular teeth located in the apex of the jaw and one pair of smaller teeth placed behind a small diastema. Teeth erupt in both males and females. The face and jaw have the shape of a bottleneck delimited from the rest of the head by the melon. The coloration of immature animals is light gray. In mature individuals the coloration is dark slate gray that assumes a greenish-brown or brown sheen due to patches of diatoms. The top of the head can be lighter around the prominent melon. Scars are common about the entire body, presumably caused by intraspecific social encounters. Typically, it has white patches in the genital region, the navel, and the throat. The throat has V-shaped furrows that penetrate deep into the skin, and extend from the region of the maxillary symphysis to the articulation of the jaws. They regularly have a couple of large furrows and several small ones. The dorsal fin is small with an approximate height of 3% of the body length; its form varies from triangular to falcate, typically with a rounded tip. The size of the fluke is 26% to 30% of the body length. The pectoral fins are small and rounded at the tip (Balcomb, 1989; Leatherwood et al., 1983, 1988).

DISTRIBUTION: This species inhabits temperate waters of the North Pacific (Balcomb, 1989). In Mexico it is distributed along the western coast of the Baja California Peninsula and in the Gulf of California toward the north at least to the Isla of San Esteban (Urbán-R. and Jaramillo, 1992). Along the west coast of the peninsula two sightings have been made, one of 25 animals at 27 degrees north and another of 7 animals at 30 degrees north (Aguayo et al., 1988). In the Gulf of California 6 records have been made: a skull collected in Isla San Esteban, 2 strandings in Bay of La Paz (one and seven individuals, respectively), and 3 sightings in the Bay of La Paz, in the mouth of the Gulf, and in waters at 24 degrees 34 minutes north, 109 degrees 59 minutes west (Aurioles, 1992; Urbán-R. and Jaramillo, 1992; Vidal, 1990). It has been registered in the areas of the GC and NP.

EXTERNAL MEASURES AND WEIGHT
TL: males about 12 m; females to 12.8 m.
Weight: males more than 10,000 kg; females more than 11,000 kg.

DENTAL FORMULA: 0/4.

NATURAL HISTORY AND ECOLOGY: This whale has deep water pelagic habits. During summer and autumn it is found on both sides of the North Pacific, moving out to the open ocean in winter and spring (Aurioles, 1992). Most of the sightings in the coast of Japan suggest that the preferred depth, near the coast, is 1,000 m (Nishiwaki and Oguro, 1971). A lifetime of 80 years has been estimated (Balcomb, 1989), and in Mexico, individuals between 8 and 42 years of age have been found (Aurioles, 1992). It dives on average 25 to 30 minutes, reaching depths of up to 2,000 m. Its diet is

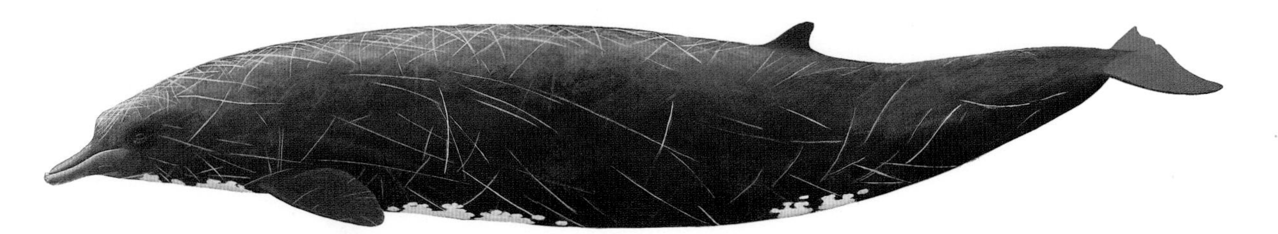

Baird's beaked whale (*Berardius bairdii*). Illustration: Pieter Folkens.

based on epibenthic and benthic organisms such as squid, rays, chimeras, octopus, and sea cucumbers, as well as pelagic organisms such as mackerel and sardine. They are gregarious animals and normally travel in groups of three or more individuals (Balcomb, 1989; Leatherwood et al., 1983, 1988). Females reach sexual maturity at 10 m in length and males at 9.5 m, which correspond to ages between 8 and 10 years. Physical maturity is reached at 20 years (Balcomb, 1989; Omura et al., 1955). Gestation lasts approximately 17 months. The peak of births occurs in March and April, and the maximum of mating occurs during October and November (Balcomb, 1989). At birth, these beaked whales measure 4.5 m; the lactation period is unknown (Kasuya, 1986; Leatherwood et al., 1982). The largest pair of teeth emerges at sexual maturity, suggesting that these teeth have a social function related to reproduction rather than feeding (Aurioles, 1992).

CONSERVATION STATUS: There is no information on the abundance of this species, except that it is not as scarce as previously thought. In Japan these animals are subject to exploitation (Kasuya, 1986). It is listed in appendix I of CITES and as data deficient by the IUCN (IUCN, 2010). It is subject to special protection in the Mexican Official Regulation (SEMARNAT, 2010) and it is protected by the Federal Law on Fisheries and article 420-I of the Federal Penal Code (Diario Oficial de la Federación, 2009).

Indopacetus pacificus (Longman, 1926)

Longman's beaked whale

Gerardo Ceballos

SUBSPECIES IN MEXICO

I. pacificus is a monotypic species.

DESCRIPTION: *Indopacetus pacificus* is one of the rarest cetaceans, known from 12 specimens and a smattering of sightings (Pitman, 2009). They are morphologically similar to both mesoplodon beaked whales and bottlenose whales. This has caused some taxonomic and geographic range confusion. Juveniles have short beaks like bottlenose whales; however, adult females have long beaks and an inconspicuous melon. The dorsal fins are large and triangular. Adult males have apparently a larger melon and two teeth located toward the front of the beak. The coloration is appar-

Longman's beaked whale (*Indopacetus pacificus*). Male (*top*) and female (*bottom*). Illustration: Pieter Folkens.

DISTRIBUTION: There are sightings and strandings of these whales at widespread locations in the tropical Pacific and Indian Oceans (Dalebout et al., 2003; Pitman, 2009). Specimens are from Australia, Somalia, South Africa, the Maldives, Kenya, the Philippines, Taiwan, and Japan. They are apparently common around the Maldives Archipelago (Anderson et al., 2006). The sightings from Mexico, which were mistakenly identified as *Hyperoodon* (bottlenose whales), are from the Guadalupe Islands and the Gulf of California (Gallo-Reynoso and Figueroa-Carranza, 1995; Urbán-R. et al., 1994). *I. pacificus* has been registered in the areas of the NP and TP.

ently variable. Juveniles have dark backs, turning light gray and then white. They have a light spot behind the eye. Females are grayish except for a brown head.

EXTERNAL MEASURES AND WEIGHT
TL: up to 8 m.
Weight: unknown.

DENTAL FORMULA: 0/2.

NATURAL HISTORY AND ECOLOGY: Longman's beaked whales apparently live in deep tropical and subtropical oceanic waters. Most sightings are from areas with surface water temperatures ranging from 21°C to 31°C. Almost nothing is known about its feeding habits, but it seems that these whales feed on squids (Pitman, 2009). They are apparently gregarious, forming groups of up to 100 whales, but most commonly around 20 animals. Their dives last up to 25 minutes.

CONSERVATION STATUS: *I. pacificus* is an extremely rare species. Little is known of its conservation status. It has been included in appendix II of CITES since 1979 and is listed as data deficient by the IUCN (IUCN, 2010). In Mexico it is not included in the list of protected species (SEMARNAT, 2010) but it is protected by the Federal Law on Fisheries and article 420-I of the Federal Penal Code (Diario Oficial de la Federación, 2009).

Mesoplodon carlhubbsi Moore, 1963

Hubbs' beaked whale

Gloria Eunice Panecatl Urquiza

SUBSPECIES IN MEXICO
M. carlhubbsi is a monotypic species.

DESCRIPTION: *Mesoplodon carlhubbsi* is relatively large for a mesoplodon. The head is small and the caudal peduncle is narrow. The middle part of the body is higher than wider, which gives an oval form in cross section. The melon descends to a prominent face. Secondary sexual characteristics are present, although the maximum size of males and females is similar. Adult males are characterized by a white high prominence located at the top of the head, forward of the single blowhole. This bulge is known as a white "beanie." The rostrum is also white. In males, the line of the mouth is interrupted in the central region by an area of the lower jaw where two teeth project. The top of the tooth is exposed in varying degrees. In females and juveniles these teeth remain within the jaw. A couple of ventral grooves are present in the throat area. The dorsal fin is located in the third posterior of the body directly above the anus. It is falcate, measuring 22 cm to 23 cm in adults. The pectoral fins are small in proportion to the body. The fluke has no median notch and the extremes are pointed. The rest of the body in males possesses a uniform color that goes from dark gray to black, while in females the sides and belly are lighter. Males show scars especially on their sides, measuring up to 2 m, often in parallel pairs, which are believed to be the result of intraspecific struggles. The circular scars are attributed to bites of sharks, lampreys, and ectoparasites such as barnacles. The scars are less extensive in females and juveniles (Leatherwood et al., 1983, 1988; Mead, 1989b; Mead et al., 1982).

EXTERNAL MEASURES AND WEIGHT

TL: up to 5.4 m.

Weight: 1,500 kg.

DENTAL FORMULA: 0/2. Teeth do not erupt in females.

NATURAL HISTORY AND ECOLOGY: We know very little of this whale's biology. It has pelagic habits; it is very elusive and inconspicuous on the surface of the water. Their diet is based on mesopelagic fish and squid. Several genera of squid (*Gonatus, Onychoteuthis, Histioteuthis,* and *Mastigoteuthis*) have been found in stomach contents, as well as traces of fish (*Chauliodus, Lampanictis, Poromitra, Icichthys,* and *Melamphaes*). This animal is not very gregarious, forming groups of 2 to 10 individuals. Births occur during the summer. The size of the offspring at birth is estimated at 2.5 m and physical maturity is reached at lengths of 5.0 m (Leatherwood et al., 1983, 1988; Mead, 1989a; Mead et al., 1982).

CONSERVATION STATUS: Due to few strandings and no positive identifications at sea, information regarding the distribution and abundance of this species is lacking. This whale is occasionally caught with harpoons in small fisheries in Japan. There is also a limited incidental catch. It is listed in appendix II of CITES since 1979 and as data deficient by the IUCN (IUCN, 2010). In Mexico it is not included in the list of protected species (SEMARNAT, 2010) and it is protected by the Federal Law on Fisheries and article 420-I of the Federal Penal Code (Diario Oficial de la Federación, 2009).

DISTRIBUTION: The precise distribution of this species is not known because records have been obtained from strandings, which do not constitute a complete overview of its distribution and abundance. It is believed to live in the temperate waters of the North Pacific near Japan, British Columbia, and California. Strandings registered farther north occurred in British Columbia in the latitude of 54 degrees north. The southernmost record was made in the eastern Pacific in the Isla San Clemente, California, at a latitude of less than 33 degrees north, but the southern boundary of its distribution is unknown. A skull was found north of New Zealand. These data suggest a relationship between the distribution of *M. carlhubbsi* with confluence of flows such as Oyashio and Kuroshio as well as the Alaska and California. The foregoing data and the strandings of this genus along Mexican coasts and Central American, between the Bay of Vizcaino and Ecuador, suggest that this species is in Mexican waters at least to the zone of influence of the flow of California (Leatherwood et al., 1983, 1988; Mead, 1989; Mead et al., 1982). It has been registered in the areas of the GC and NP.

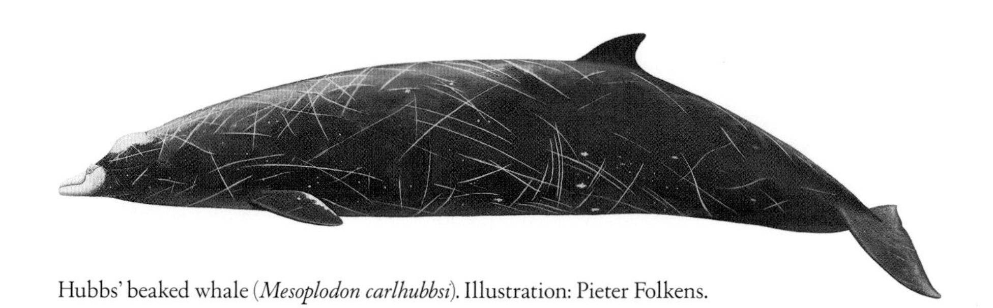

Hubbs' beaked whale (*Mesoplodon carlhubbsi*). Illustration: Pieter Folkens.

Blainville's beaked whale

Carlos Esquivel Macías

SUBSPECIES IN MEXICO

M. densirostris is a monotypic species.

DESCRIPTION: *Mesoplodon densirostris* is similar in shape and proportion as other members of the genus. It has a dorsal fin in the posterior third of the body, and only one pair of teeth located in the middle of the raised lower jaw. It is distinguished by having long teeth, 20 cm from the root, though little of it shows. and the ivory has the highest density known among mammals (Houston, 1990). The coloration is dark gray on the back with a lighter belly. It often appears greenish-brown due to diatoms on the skin. Adult males have white scars in the cephalic region, resulting from fights in which they use their teeth against other males (Raven, 1942). It is possible that regional variations of coloration exist. Scars from intraspecific social interactions are common on males, along with light oval scars caused by cookie cutter sharks.

DISTRIBUTION: This species is the most widely distributed of the genus *Mesoplodon*; it is found in tropical, subtropical, and warm temperate waters around the world, including Mexico. Its known distribution is based on strandings, particularly on oceanic islands. It is the only species of the genus that reaches the Equator to 35 degrees latitude in both hemispheres (Besharse, 1971; Casinos and Filella, 1981; Mead, 1989a; Moore, 1968; Pastene et al., 1989). Recently, a molecular identification allowed registration of the species for the first time in New Zealand (Dalebout et al., 1998). It has been registered in the areas of the GC, GM, NP, SC, and TP.

EXTERNAL MEASURES AND WEIGHT

TL: males at least 4.4 m; females at least 4.6 m.

Weight: males more than 800 kg; females at lest 1,000 kg.

DENTAL FORMULA: 0/2.

NATURAL HISTORY AND ECOLOGY: *M. densirostris* is a deep water species that can be found away from the coast. Until now, concentration areas have not been reported but there are some indications of this, as well as seasonal movements between the oceanic waters of eastern tropical Pacific and deep waters close to the continental coast of Mexico (Esquivel et al., 1993). They feed on squid such as *Todarodus* and *Octopoteuthis*, some pelagic fish such as *Cepula*, *Scopelogadus*, and *Lampanyctus*, and species of the families Ommastrephidae, Enoploteuthidae, and Neoteuthidae (Aguilar et al., 1982; Evans, 1987). The only known thing about their reproduction is that at birth the offspring measures between 1.9 m and 2.6 m, and that a female reaches sexual maturity at an approximate age of 9 years, judging by the bands in the dentine (Leatherwood et al., 1983; Mead, 1989a). They form groups of three to seven individ-

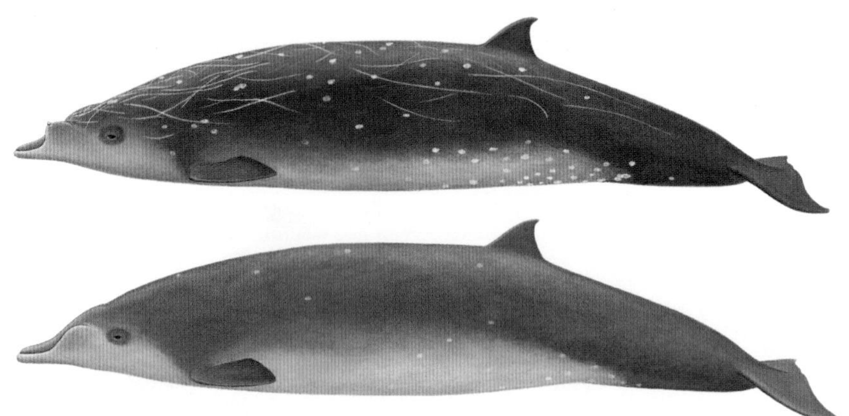

Blainville's beaked whale (*Mesoplodon densirostris*). Male (*top*) and female (*bottom*). Illustration: Pieter Folkens.

uals led by a male adult (Houston, 1990). When swimming, they do not expose the tail, and the blow is not visible but audible. Generally, they dive for 15 to 20 minutes, although they can remain submerged for 45 minutes (Watson, 1981).

CONSERVATION STATUS: They have never been captured intentionally (Ridgway, 1985). Blainville's beaked whales have no commercial or utilitarian value; they have been listed in appendix II of CITES and as data deficient by the IUCN (IUCN, 2010). Due to its apparent rarity the IUCN considers it to be in an indeterminate state (IUCN, 2010). In Mexico it is subject to special protection in the Mexican Official Regulation (SEMARNAT, 2010) and it is protected by the Federal Law on Fisheries and article 420-I of the Federal Penal Code (Diario Oficial de la Federación, 2009).

Mesoplodon europaeus (Gervais, 1855)

Gervais' beaked whale

Benjamín Morales Vela

SUBSPECIES IN MEXICO

M. europaeus is a monotypic species.

DESCRIPTION: Little is known about this whale, and what information that is available comes from stranded animals. Their skin is grayish-brown dorsally and lighter ventrally. The body has some lateral compression. Its head is small in relation to the body and has a well-defined and elongated face. The triangular dorsal fin is small. It is located in the posterior third of the body and is slightly falcate (Leatherwood et al., 1976). It has a tooth on each jaw, placed well behind the maxillary symphysis. Males often have a few linear scars on the upper body and genital region.

EXTERNAL MEASURES AND WEIGHT

TL: males to 4.5 m; females to 5.2 m.

Weight: about 1,200 to 1,800 kg.

DENTAL FORMULA: 0/2. Teeth do not erupt in females.

NATURAL HISTORY AND ECOLOGY: Little is known about the biology of *Mesoplodon europaeus*. Its length at birth is 2.1 m; sexual maturity is reached at sizes of 4.5 m to 4.8 m and its longevity is estimated at a minimum of 27 years (Leatherwood et al., 1983). Like other species of the genus *Mesoplodon*, it normally inhabits deep waters; it is inconspicuous and timid. When they are on the surface, killer whales can prey on them. Their diet is based on squid (Mead, 1989a).

DISTRIBUTION: The distribution of this animal is based solely on the registration of stranded animals. It is known only in Atlantic waters, showing preference for tropical and subtropical waters (Leatherwood et al., 1983). In America there are records from New York to the Caribbean (Leatherwood et al., 1983). One of the last strandings was recorded in Mauritania along the west coast of Africa (Robineau and Vely, 1993). In Mexico a stranded animal was registered in the coast of Campeche (Gallo and Pimienta, 1989). It has been registered in the area of the GM.

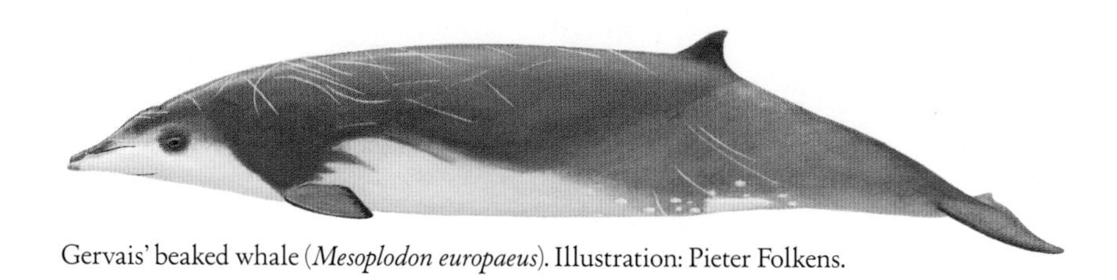

Gervais' beaked whale (*Mesoplodon europaeus*). Illustration: Pieter Folkens.

CONSERVATION STATUS: *M. europaeus* is a rare species, and little is known of its abundance (Mead, 1989). It is listed in appendix II of CITES and as data deficient by the IUCN (IUCN, 2010). It is subject to special protection in the Mexican Official Regulation (SEMARNAT, 2010) and it is protected by the Federal Law on Fisheries and article 420-I of the Federal Penal Code (Diario Oficial de la Federación, 2009).

Mesoplodon ginkgodens Nishiwaki and Kamiya, 1958

Ginkgo-toothed beaked whale

David Aurioles Gamboa

SUBSPECIES IN MEXICO
M. ginkgodens is a monotypic species.

DESCRIPTION: The Ginkgo-toothed beaked whale has a cylindrical body; the head ends in a sharp but moderate face compared with that of other beaked whales (Mead, 1989a). The color of adult animals is, in general, dark gray or almost black, but slightly lighter near the face. Adult males have small and light scattered spots on the back and the belly. The dorsal fin tends to be falcate, while the pectoral fins are small and very close to the body (Mead, 1989a). Males have a pair of teeth projecting out of a raised mouth line, but in *Mesoplodon ginkgodens* they do not project above the rostrum as in some of the other species of the genus. The teeth are the widest (in an anterior-posterior sense) registered in the genus. The tooth shape is similar to that of a Gingko (*Gingko biloba*) leaf, giving the species its name (Nishiwaki and Kamiya, 1958).

DISTRIBUTION: This species is found in temperate pelagic waters of the North Pacific. Only about 16 records exist, 8 for the eastern coast and 8 for the western coast (Mead, 1989a). In Mexico only one stranding occurred on the beach of Malarrimo in Vizcaíno Bay along the west coast of Baja California (Leatherwood et al., 1988). It has been registered in the area of the NP.

EXTERNAL MEASURES AND WEIGHT
TL: to 5 m for both sexes.
Weight: 1,500 kg.

DENTAL FORMULA: 0/2.

NATURAL HISTORY AND ECOLOGY: Little is known about this species due to the low number of collected specimens. It is believed that their habits are similar to those of other members of the genus; thus they are pelagic and feed mainly on squid in deep water, inconspicuous, and timid on the surface.

CONSERVATION STATUS: The total distribution and abundance are unknown. No intentional capture has been documented but, like other species of the genus, they

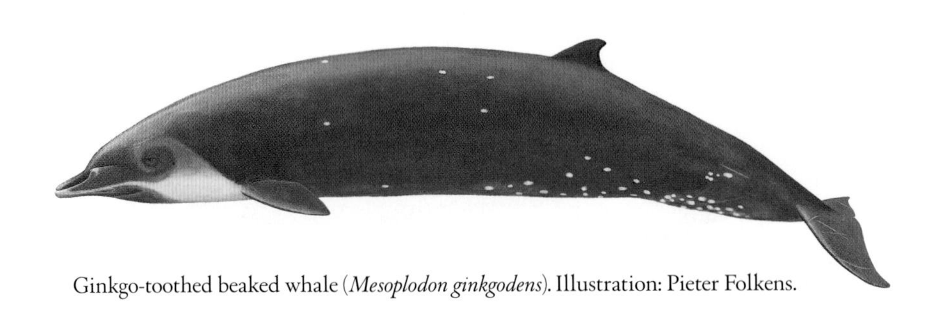

Ginkgo-toothed beaked whale (*Mesoplodon ginkgodens*). Illustration: Pieter Folkens.

may die intentionally or incidentally in small numbers in countries like Japan and Peru. The species is listed in appendix II of CITES and as data deficient by the IUCN (IUCN, 2010). It is subject to special protection in the Mexican Official Regulation (SEMARNAT, 2010) and it is protected by the Federal Law on Fisheries and article 420-I of the Federal Penal Code (Diario Oficial de la Federación, 2009).

Mesoplodon perrini Dalebout et al. 2002

Perrin's beaked whale

Luis Medrano González

SUBSPECIES IN MEXICO
M. perrini is a monotypic species.

DESCRIPTION: *Mesoplodon perrini* is the most recently described beaked whale. It is known only from the eastern tropical Pacific. Two colorations have been described that suggest a variation between classes of sex and age. One form is a combination of dark and light and the other is grayish-brown and more or less uniform in color. Both forms can be seen together. The light-dark form is larger than the brownish form. Both have a low triangular dorsal fin, with a wide base and slightly falcate. The head is small with a slight but distinctive melon. The face is long and similar to that of other mesoplodons. In the light-dark form the back is very dark from half of the back to the rear, covering the posterior flanks almost completely. Behind the head there is a white or cream area that extends dorsally for half the body, covering the sides of the thorax. The ventral part, including the jaw, is light colored. The head and face are grayish with an intermediate tone between the light and dark colors. The light coloration is similar to the saddle described in other cetaceans, and in young individuals it seems to form from the coalescence of the white spots. This form also has light-dark marks and various scars, similar to those found in other beaked whales. The brownish form is almost uniform and presents no marks. Everything suggests that this species shows sexual dimorphism. Males are the light-brown form, with slightly larger sizes, and the females and young are the brownish form (Pitman et al., 1987). The author spotted an individual of this species that measured approximately 4 m and had a dark brown with cream coloration that can be interpreted as an male.

DISTRIBUTION: Perrin's beaked while is known confidently only from stranded specimens on the southern coast of California between San Diego and Monterey. Its range is expected to be pelagic areas of the eastern tropical Pacific in water temperatures of at least 27°C. Possible sightings at sea have occurred near northern most Baja California, northwest of Ensenada. This species cannot be confidently identified at sea.

Perrin's beaked whale (*Mesoplodon perrini*). Male (*top*) and female (*bottom*). Illustration: Pieter Folkens.

EXTERNAL MEASURES AND WEIGHT
TL: 3.9 to 5.0 m.
Weight: approximately 1,500 kg.

DENTAL FORMULA: 0/2.

NATURAL HISTORY AND ECOLOGY: *M. perrini* has not been confidently identified at sea. Therefore, nothing is known about its habits, but its life history is expected to be similar to that of other members of the genus in the North Pacific.

CONSERVATION STATUS: *M. perrini* is considered data deficient by the IUCN (IUCN, 2010). The general laws and the Federal Law of Fisheries and article 254 bis of the Penal Code frame its protection in Mexican waters (SEPESCA, 1991). It is not included in the Mexican Official Regulation (SEMARNAT, 2010) but it is protected by the Federal Law on Fisheries and article 420-I of the Federal Penal Code (Diario Oficial de la Federación, 2009).

Mesoplodon peruvianus Reyes et al., 1991

Pygmy beaked whale

David Aurioles Gamboa and Jorge Urbán Ramírez

SUBSPECIES IN MEXICO
M. peruvianus is a monotypic species.

DESCRIPTION: The pygmy beaked whale is the smallest beaked whale. It was recently discovered in Mexico (4 specimens) and Peru (10 specimens), but its formal description was conducted in Peru (Reyes et al., 1991). The dorsal fin is small, triangular, and slightly concave in the anterior side and convex in the posterior. The coloration of two fresh stranded animals was dark gray in the back and slightly lighter in the belly. The color of the only recorded calf measured 1.59 m; coloration was brown in

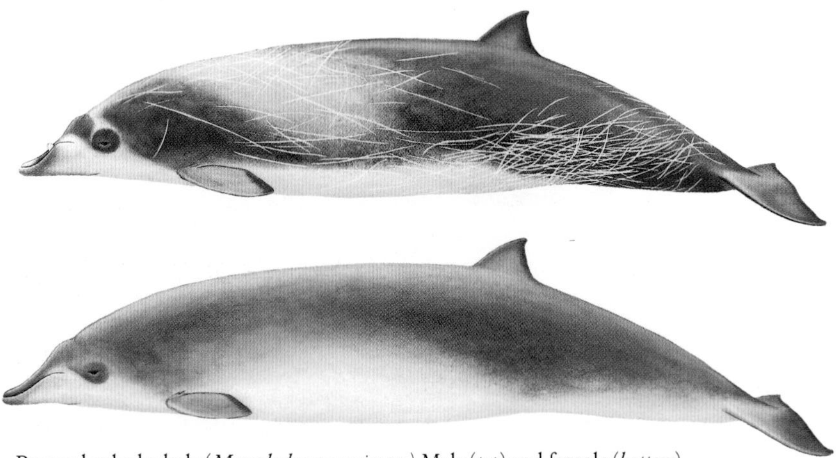

Pygmy beaked whale (*Mesoplodon peruvianus*). Male (*top*) and female (*bottom*). Illustration: Pieter Folkens.

the back and sides, while the belly was very light gray (Reyes et al., 1991). Like other species of the genus, it only has one pair of teeth that in general are larger in males and are exposed, while in the females and juveniles they do not project from the gum (Aurioles and Urbán, 1993; Mead, 1989a; Reyes et al., 1991). Externally, adult males and females are not different, except for the genital slits and the eruption of the maxillary teeth. In the skull, conspicuous differences exist that help distinguish each sex. The ossification and texture of some cranial bones vary between sexes, and in males there is a bump where the maxillary teeth are placed, as well as a couple of bone fins on the sides of the first third of the maxilla, which are unique among the cetaceans (Aurioles and Urbán, 1993).

DISTRIBUTION: Until 1997, 14 strandings and 8 sightings at sea were recorded. Four of the sightings were made in the Mexican Pacific between latitudes 10 degrees and 20 degrees north and longitudes 96 degrees and 105 degrees west, along the coast of Jalisco, Michoacán, Guerrero, and Oaxaca (J. Urbán, pers. comm.; Wade and Gerrodette, 1992). All records occurred in the eastern tropical Pacific. There are no records in the south of Peru or in northern Mexico, where there has been considerable research effort for a number of years, so these could be the limits of its distribution (Urbán and Aurioles, 1992). It has been registered in the areas of the GC and TP.

EXTERNAL MEASURES AND WEIGHT
TL: maximum known 3.90 m.
Weight: 800 to 1,000 kg.

DENTAL FORMULA: 0/4.

NATURAL HISTORY AND ECOLOGY: Little is known about *Mesoplodon peruvianus.* Like other beaked whales it has pelagic habits. The limited information available suggests that this animal feeds on fish of the orders Perciformes, Myctophiformes, and Ophidiformes (Reyes et al., 1991). The pair of teeth characteristic of the genus *Mesoplodon* and the scarce maxillary teeth in the species of the family Ziphiidae are not used for feeding but for courting females and/or combat between males. Parallel scars on the body from these bite wounds are frequent (Henyning, 1984). The projecting teeth in males emerge at sexual maturation (Aurioles, 1992).

CONSERVATION STATUS: Nothing is known about the abundance of this species. It is occasionally hunted by Peruvian fisheries (Reyes et al., 1991). It is included in appendix II of CITES, is subject to special protection in the Mexican Official Regulation (SEMARNAT, 2010), is protected by the Federal Law on Fisheries and article 420-I of the Federal Penal Code (Diario Oficial de la Federación, 2009) and is considered data deficient by the IUCN (IUCN, 2010).

Ziphius cavirostris G. Cuvier, 1823

Cuvier's beaked whale

Mario A. Salinas Zacarías and Daniel Castillo Lugardo

SUBSPECIES IN MEXICO
Z. cavirostris is a monotypic species.

DESCRIPTION: Cuvier's beaked whale has a long and robust body with a relatively small head in relation to the total length of the body. Females are slightly larger than males. The head in its anterior part is slightly concave; the angle of the front is not very pronounced and gradually decreases toward the tip of the muzzle, which is not well defined. This peculiar profile of the head and face is similar to that of a goose. The dorsal fin is gently falcate and is located in the third posterior part of the body. The nasal cavity is located at the top portion and back of the head. The blow is expelled slightly forward and slightly toward the left, and is usually low and inconspicuous.

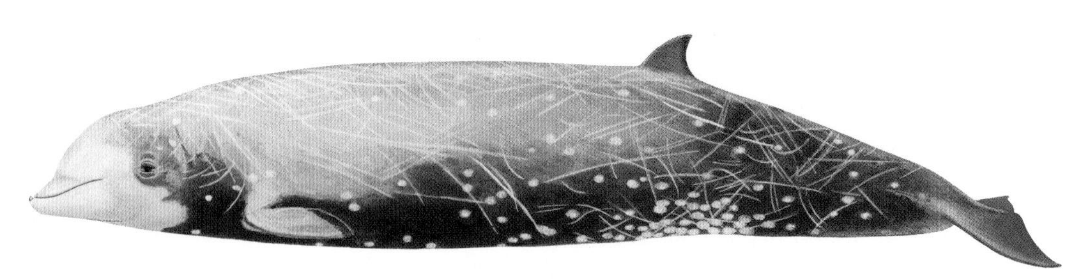

Cuvier's beaked whale (*Ziphius cavirostris*). Illustration: Pieter Folkens.

DISTRIBUTION: Currently, Cuvier's beaked whale is recognized as one of the cetaceans of wider distribution, inhabiting tropical and subpolar waters of the Pacific, Atlantic, and Indian Oceans in their east and west extremes (Mead, 1989a). In Mexico, it has been observed in all waters of the exclusive economic zone, along the western coast of the Baja California Peninsula, in the Gulf of California, especially in the Bay of La Paz, Bahia de Banderas, around the Socorro Island, in the Gulf of Mexico, and in the Caribbean Sea. Strandings have been reported on the shores of Baja California Sur, Michoacán, and Quintana Roo. It has been registered in the areas of the CS, GC, GM, NP, and TP.

The first breath after a long dive can be dense but is not visible at more than 1 km in appropriate conditions. The coloration patterns are variable; some animals can be dark gray to reddish-brown in the dorsal part and gradually fading toward the ventral region. A film of diatoms may give the skin a brownish or greenish-frown tinge. In juveniles, the coloration tends to be totally dark on both sides. The body is frequently covered with cream or white spots, mainly in the ventral region. The caudal peduncle is narrow and dark at its base. The coloration of the head in older males is paler than the rest of the body and can extend dorsally along the back. This appears to be a secondary sexual characteristic. Males have also a greater incidence of scars all over the body. Adult males have two teeth on the tip of the jaw that are visible externally (Leatherwood et al., 1983, 1988, Mead, 1989a).

EXTERNAL MEASURES AND WEIGHT
TL: up to 6.93 m.
Weight: males at least 2,600 kg; females to about 3,000 kg.

DENTAL FORMULA: 0/2.

NATURAL HISTORY AND ECOLOGY: *Ziphius cavirostris* has been observed in groups of 3 to 10 individuals and occasionally in groups up to 25. Solitary males are rarely sighted. These whales perform organized dives of at least 30 minutes; when they are immersing themselves in the water they show the fluke. They dive vertically and close to each other. Their food mainly consists of squid of the genera *Mesonychoteuthis*, *Teuthowenia*, and *Taonius*, fish (Gadiformes and Antheriniformes), and decapodous crustaceans. They are very cautious and do not approach boats. They have been observed jumping out of water. At birth they measure 2.7 m and reach sexual maturity at lengths of 5 m to 6 m. They can live up to 35 years. They may have epibionts such as *Xenobalanus* and *Conchoderma*, and endoparasites such as nematodes (*Anisakisy*, *Crassicauday*) and cestodes (*Phyllobothrium*) (Leatherwood et al., 1983, 1988; Mead, 1989a).

CONSERVATION STATUS: These whales have been exploited at a small scale in the fisheries of Japan and Isla San Vicente, and have been identified in the whale meat market of Korea. In New Zealand toxins such as DDT, DDE, and ADD have been detected in this species (Dalebout et al., 1998; Leatherwood et al., 1983, 1988; Mead, 1989a). Its abundance is unknown, and although it is rare it is more common than the other members of the family. It has been listed in appendix II of CITES and as least concern by the IUCN (IUCN, 2010). It is subject to special protection in the Mexican Official Regulation (SEMARNAT, 2010) and it is protected by the Federal Law on Fisheries and article 420-I of the Federal Penal Code (Diario Oficial de la Federación, 2009).

Family Delphinidae

The family Delphinidae includes 32 species distributed in all the seas and oceans of the world (Wilson and Reeder, 1993). In Mexico, it is represented by the thirteen genera *Delphinus, Feresa, Globicephala, Grampus, Lagenodelphis, Lagenorhynchus, Lissodelphis, Orcinus, Peponocephala, Pseudorca, Stenella, Steno,* and *Tursiops*, with 18 species. The range in size is highly variable, with very small species as the vaquita and ones of more than 9,000 kg such as the killer whales (*Orcinus orca*). The populations of most of the species are relatively stable, but the vaquita is extremely endangered.

Delphinus capensis Gray, 1828
Delphinus delphis Linnaeus, 1758

Short-beaked and long-beaked common dolphins

Carlos Esquivel Macías

SUBSPECIES IN MEXICO

D. delphis is a monotypic species. *D. capensis* is a monotypic species.
Taxonomy on common dolphins is complex and is still subject to investigation. Until recently the subspecies *D. delphis delphis* (Linnaeus, 1758; form of short face) and *D. delphis bairdii* (Dall, 1873; form of long face) were recognized, but recently they were both assigned a specific category (Evans, 1994; Heyning and Perrin, 1994). Since 1994 two species are recognized, *D. delphis* and *D. capensis*.

Long-beaked common dolphin (*Delphinus capensis*).

Short-beaked common dolphin (*Delphinus delphis*).
Illustrations: Pieter Folkens.

DISTRIBUTION: These species inhabit almost all tropical and temperate seas of the world. They are abundant in the North Pacific, especially along the west coast of Baja California, in the Gulf of California, and along the coast of Central America. In Mexico it is easily sighted in the islands of Baja California and the interior of the Gulf of California. They are common in the western portion of the Gulf of California and rare in the coast of Sinaloa and the Pacific of Nayarit. In the Mexican Southern Pacific it is possible to sight them at great distance from the shore (Evans, 1976; Perrin et al., 1985). The short-beaked dolphin is distributed in three populations: one in the north at latitude 32 degrees north, another between 28 and 30 degrees north, and another farther south at 15 degrees north. The long-beaked shape is found in coastal waters of northern Mexico at 20 degrees north, including the Gulf of California (Leatherwood et al., 1988). In the Mexican Atlantic it is less frequent than in the Pacific, and its presence in the Gulf of Mexico is doubtful (Caldwell, 1955; Jefferson and Schiro, 1995). These species have been registered in the areas of the CS, GC, NP, and TP.

DESCRIPTION: In common dolphins the length of the beak is used to distinguish the short-beaked (*Delphinus delphis*) and the long-beaked (*D. capensis*) common dolphins. This beak is sharply divided from the lower forehead by a deep groove; the length is intermediate between *Stenella longirostris* and *S. coeruleoalba*. The pointed dorsal fin varies from falciform to triangular, sloping backward and located in the middle of the back. The coloration of the common dolphin is distinctive; the back is almost black and the belly is white. Near the sides two more colors join, a white ochre patch on the chest and a light gray area that covers the caudal peduncle. This relationship of colors has four angles with different colors known as double cross-hatching, that is, a common back and ventral hatch and an anterior-posterior hatch (Evans, 1987; Perrin, 1972). Coloration and color patterns can vary even between short-beaked common dolphins, and is more evident between the two species. Between the face, pectoral fins, and sides, there are bands with variable numbers, intensity, and size that together with the body proportions, other details of coloration, and the size of the face are used to differentiate populations and even subspecies (Leatherwood et al., 1983, 1988). Recently, in California, Mexico, New Zealand, and France short-beaked dolphins with abnormal colorations have been sighted, which seem to have only the anterior-posterior hatch component (Perrin et al., 1995). *D. delphis* is more robust and considered an oceanic dolphin, while *D. capensis* is slimmer and more coastal.

EXTERNAL MEASURES AND WEIGHT
TL: 2.2 to 2.6 m.
Weight: 75 to 85 kg.

DENTAL FORMULA: 40–65/40–65.

NATURAL HISTORY AND ECOLOGY: Common dolphins are pelagic organisms, which rarely swim in freshwater (Walker, 1964). They are associated with fish populations that at times penetrate deep water (Hui, 1979). They change their habitat depending on the time of the year and undertake migrations in the Gulf of California. Males reach sexual maturity when they reach 2 m and females 1.9 m, at nearly 6 or 7 years of age (Evans, 1987). The gestation period lasts 10 to 11 months. When calves are born (in summer or spring), they measure 80 cm in length. Lactating females separate themselves from the school for five months while they nurse their offspring. They can give birth annually or in alternate years (Leatherwood et al., 1988). They feed during twilight on squid (Loliginidae, Sepiolidae, Sepiidae, and Gonatidae) and fish (*Micromesistius*, *Trisopterus*, *Merluccius*, *Trachurus*, and *Engraulis*) at depths of up to 280 m (Evans, 1987). In the eastern tropical Pacific, during the day, common dolphins are associated with tuna (*Thunnus albacares*) and thus are good indicators of shoals of fish, together with other tropical dolphins (Au and Pitman, 1986). At times, they share the food with the tuna and in groups numbering in the hundreds that are very active on the surface of the water. They can form groups of up to 15,000 individuals. These dolphins are preyed on by killer whales (*Orcinus orca*) and various sharks.

CONSERVATION STATUS: These two species are included in appendix II of CITES. Mexico regulates its release of the fences for tuna. Its mortality has not been determined in relation to the fishing activity in the Gulf of California, where these dolphins are used occasionally as bait in trawl nets. Mortality has been caused by explosives in some regions of the Gulf of California (Zavala-G. and Esquivel, 1990). The long-beaked common dolphin is considered data deficient and the short-beaked least concern by the IUCN (IUCN, 2010). The short- and long-beaked common dolphins are subject to special protection in the Mexican Official Regulation (SEMARNAT, 2010) and it is protected by the Federal Law on Fisheries and article 420-I of the Federal Penal Code (Diario Oficial de la Federación, 2009).

Pygmy killer whale

Carlos Esquivel Macías

SUBSPECIES IN MEXICO
F. attenuata is a monotypic species.

DESCRIPTION: *Feresa attenuata* is a small dolphin, closely related to pilot whales and killer whales. The dorsal fin is located on the middle of the back and it is hooked with a sharp edge. The head is rounded and lacks a beak. The tips of the pectoral fins are blunt. The body is torpedo-shaped. Its coloration is black on the back with dark gray on the sides and small white areas in the stomach and genitals. It is distinguished externally from the very similar *Peponocephala electra* by the shaape of the head.

EXTERNAL MEASURES AND WEIGHT
TL: males to 2.6 m; females to 2.5 m.
Weight: 150 to 170 kg.

DENTAL FORMULA: 11–9/11–13.

NATURAL HISTORY AND ECOLOGY: *F. atttenuata* is an exclusively pelagic species that can be found in groups of a few individuals up to 50. Pygmy killer whales are difficult to approach because they swim very fast without acrobatics (Leatherwood et al., 1988). They are aggressive animals (Pryor et al., 1965); they eat fish and other marine mammals. Many aspects of the natural history of this species are undocumented, but it is known that calves are born with an approximate length of 80 cm and that both sexes reach sexual maturity with a length of 2.2 m (Leatherwood et al., 1983).

CONSERVATION STATUS: Pygmy killer whales are seldom seen in nature and the size of their populations is unknown. A few are taken incidentally during tuna fishing (Leatherwood et al., 1983). *F. attenuata* is listed in appendix II of CITES, and considered data deficient by the IUCN (IUCN, 2010). It is subject to special protection in the Mexican Official Regulation (SEMARNAT, 2010) and it is protected by the Federal Law on Fisheries and article 420-I of the Federal Penal Code (Diario Oficial de la Federación, 2009).

DISTRIBUTION: The global distribution of *F. attenuata* is temperate pantropical (Leatherwood et al., 1988; Watson, 1981). In Mexico there are few records. There are no known areas of occurrence with particular abundance. It is more likely to be found in pelagic waters of the eastern tropical Pacific (Nishiwaki, 1972). It has been registered in the areas of the CS, GM, NP, and TP.

Pygmy killer whale (*Feresa attenuata*). Illustration: Pieter Folkens.

Short-finned pilot whale

Benjamín Morales Vela

SUBSPECIES IN MEXICO
G. macrorhynchus is a monotypic species.

DISTRIBUTION: *G. macrorhynchus* is a cosmopolitan species. It inhabits tropical and subtropical waters of the world. Its distribution in the Southern Hemisphere is not well known (Leatherwood et al., 1983). In the Northeast Pacific its distribution extends from Ecuador to the Gulf of Alaska. In Mexico, the species occurs in both coastal and oceanic waters (Hacker, 1992). It is relatively common in the central western Gulf of California. It has been registered in the areas of the CS, GC, GM, NP, and TP.

DESCRIPTION: The short-finned pilot whale is a large dolphin. The shape of its head is very characteristic—prominent and globular. Its dorsal fin is located in the first third of the body with a wide base and with its upper edge rounded and thick (Leatherwood et al., 1988). Its pectoral fins are thin and curved, with a length of up to one-sixth of its body (Leatherwood et al., 1976).

EXTERNAL MEASURES AND WEIGHT
TL: males to 5.3 m; females to 4.7m in the eastern North Pacific; larger elsewhere.
Weight: 1,400 kg to 4,000 kg.

DENTAL FORMULA: 7–7/12–12.

NATURAL HISTORY AND ECOLOGY: Pilot whales are social animals that can form pods of a few to herds of several hundred individuals. Commonly, they are found with other cetaceans such as bottlenose dolphins (Leatherwood et al., 1983). The duration of the reproduction period is not defined; apparently, females breed every three years. The gestation period lasts 15 to 16 months and calf rearing lasts more than a year (Leatherwood et al., 1988). They are born with a length of 1.4 m. Females reach sexual maturity at 3.0 m to 3.3 m and males at 4.2 m to 4.8 m. Males reach physical maturity at 5 m (Leatherwood et al., 1983). This species is strongly susceptible to stranding (Aurioles, 1993; Morales and Olivera, 1993). It feeds mainly on squid and other cephalopods (Hacker, 1992). They are probably preyed on by large sharks and killer whales (Leatherwood et al., 1988).

CONSERVATION STATUS: In the past century *Globicephala macrorhynchus* was regularly hunted in the waters of northwest Africa and the Antilles. Currently, in Japan it is captured for human consumption, and in other countries it is captured for public display. Its abundance in the eastern tropical Pacific is estimated at 60,000 animals (Leatherwood et al., 1983). No estimate of abundance exists for Mexico (Aurioles, 1993); it is listed in appendix II of CITES and considered data deficient by IUCN (IUCN, 2010). It is subject to special protection in the Mexican Official Regulation (SEMARNAT, 2010) and it is protected by the Federal Law on Fisheries and article 420-I of the Federal Penal Code (Diario Oficial de la Federación, 2009).

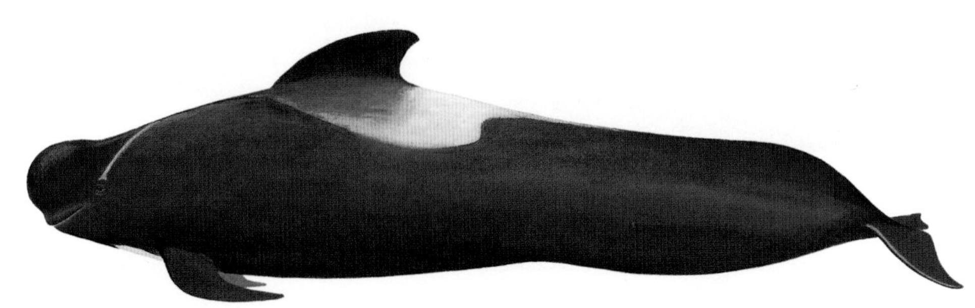

Pilot whale (*Globicephala macrorhynchus*). Illustration: Pieter Folkens.

Risso's dolphin

Luis Medrano Gónzález

SUBSPECIES IN MEXICO
Grampus griseus is a monotypic species.

DESCRIPTION: Risso's dolphin is one of the species of cetaceans on which little information is available. Its robust body is intermediate among the Delphinidae. At birth, they are totally gray; then they acquire a brownish appearance and grow to have a white belly and head, perhaps as a result of the accumulation of vacuoles of salt in the skin (Leatherwood et al., 1983; Viale, 1984). They have numerous scratches and marks on the body produced by the squid they feed on and from wounds (Leatherwood et al., 1983; Watson, 1981). They have a high dorsal fin, which is narrow and falcate and located in the middle of the body. The pectoral fins are prominent and pointed. Their head is rounded, similar to that of pilot whales (*Globicephala*), and the melon has a frontal rift visible at a short distance (Leatherwood et al., 1983).

DISTRIBUTION: Risso's dolphins can be found in warm and temperate regions of all the oceans, but prefer pelagic areas (Leatherwood et al., 1983). Gaps exist in their distribution in the northeastern Pacific along latitudes 21 degrees north and 42 degrees north (Leatherwood et al., 1980), but isolated groups, probably endogamic, have been observed along the Mexican coast at latitude 21 degrees north (Medrano et al., 2008). In Mexico, they have been reported in the Gulf of California, along the Pacific coast of Baja California, and along the continental Pacific coast (Esquivel et al., 1993). They have been reported in the northern Gulf of Mexico, where they show preference for waters on the continental slope with depths of 350 m to 975 m (Baumgartner, 1997; Jefferson and Schiro, 1995; Jefferson et al., 1992). It has been recorded in the areas of the CS, GC, GM, NP, and TP.

EXTERNAL MEASURES AND WEIGHT
TL: males to 3.83 m; females to 3.66 m.
Weight: 270 kg.

DENTAL FORMULA: 0/2 – 7.

NATURAL HISTORY AND ECOLOGY: Risso's dolphins are abundant in pelagic areas where they form schools of variable sizes from a few to several hundred individuals, which on average are composed of 10 to 20 animals (Leatherwood et al., 1980). They are known to feed on squid and occasionally on fish. They are commonly associated with other cetaceans such as pilot whales (*Globicephala*). Males are larger than females and reach sexual maturity at a length of 3 m. Sometimes they can hybridize with bottlenose dolphins (*Tursiops truncatus*) in wildlife and in captivity; they may live for several years in captivity (Leatherwood et al., 1980; Watson, 1981). They are less active in surface waters than other dolphins, although sometimes they make various jumps and splash with their fins on the water surface. It is uncommon for them to swim near boats.

CONSERVATION STATUS: Risso's dolphins are captured by fisheries in Canada, Japan, the Lesser Antilles, the Solomon Islands, and Indonesia. Around the world, they are

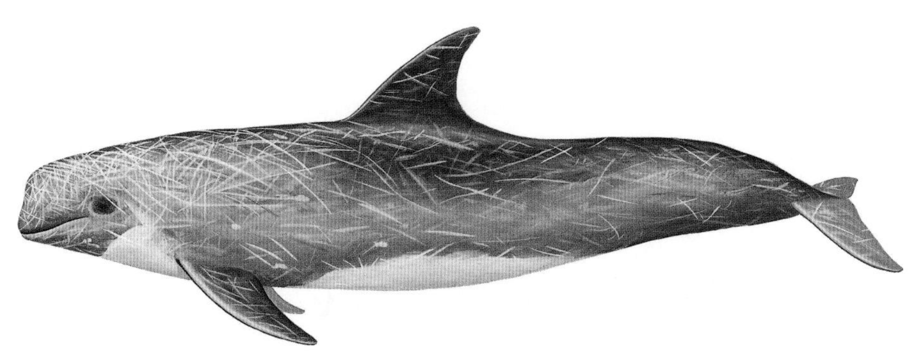

Risso's dolphin (*Grampus griseus*). Illustration: Pieter Folkens.

incidentally caught in various fishing nets. Although this mortality has not been adequately studied, it is considered minor (Leatherwood et al., 1983; Watson, 1981). The Risso's dolphin is not considered an endangered species (Leatherwood et al., 1983). It is listed in appendix II of CITES and as least concern by the IUCN (IUCN, 2010). In Mexico, it is subject to special protection in the Mexican Official Regulation (SEMARNAT, 2010) and it is protected by the Federal Law on Fisheries and article 420-I of the Federal Penal Code (Diario Oficial de la Federación, 2009).

Lagenodelphis hosei Fraser, 1956

Fraser's dolphin

Silvia Regina Manzanilla Naim

SUBSPECIES IN MEXICO

L. hosei is a monotypic species.

DESCRIPTION: The coloration in adult Fraser's dolphins (*Lagenodelphis hosei*) is very similar to that of the striped dolphin (*Stenella coeruleoalba*), with the exception of the dark band on the pectoral fins. Morphologically, it looks like *Lagenorhynchus* and *Delphinus*, accounting for its name, *Lagenodelphis*. Its body is robust, the face is short, and the pectoral fins are small in relation to the body. The dorsal fin is slightly falcate and ends in a tip (Leatherwood et al., 1983). Dorsally the dolphin is grayish-blue and ventrally it is white. The sides have a set of bands; the most conspicuous is a lateral flank stripe that extends from the eye to the flipper and anus. There is also a cream-colored stripe below and parallel to the black stripe. A couple of dark gray stripes

Fraser's dolphin (*Lagenodelphis hosei*); adult male (*top*); adult female (*bottom*). Illustrations: Pieter Folkens.

extend from the corner of the mouth to the anterior insertion of the pectoral fins. The lips and fins are dark gray to black (Leatherwood et al., 1983; Perrin et al., 1973).

EXTERNAL MEASURES AND WEIGHT
TL: females to 2.64 m; males to 2.7 m.
Weight: females to 160 kg; males to 210 kg.

DENTAL FORMULA: 34–44/34–44.

NATURAL HISTORY AND ECOLOGY: *L. hosei* is a poorly known cetacean. It can be found in groups mixed with *Peponocephala electra* in the eastern and western tropical Pacific and in the Gulf of Mexico. They travel in large groups of up to 500 individuals (Amano et al., 1996). Sexual dimorphism has been reported; males develop wider and darker stripes than females. A ventral hump has been observed near the caudal peduncle, similar to that in *Stenella longirostris*, which becomes more accentuated with sexual maturity (Amano et al., 1996). Through the few individuals examined, its diet is known to consist of deep-water fish, squid, and crustaceans (Leatherwood et al., 1983).

CONSERVATION STATUS: There is no information on migration movements or this dolphin's populations. It is not considered an endangered species or at risk of extinction. It has been included in appendix II of CITES and listed as least concern by the IUCN (IUCN, 2010). It is subject to special protection in the Mexican Official Regulation (SEMARNAT, 2010) and it is protected by the Federal Law on Fisheries and article 420-I of the Federal Penal Code (Diario Oficial de la Federación, 2009).

DISTRIBUTION: Fraser's dolphins have a wide tropical and pelagic distribution. It is more common in the Philippines and along the coasts of South Africa, Japan, Taiwan, and Malaysia (Amano et al., 1996). It is a rare species in the Gulf of Mexico and the Antilles. It is highly probable that it inhabits Mexican waters in the Gulf of Mexico since there are osteological records in the Florida Keys (Hersh and Odell, 1986) and sightings in the pelagic area of the Gulf (Jefferson and Schiro, 1995). There are two recorded sightings of the species in the vicinity of the Revillagigedo Islands, which constitute the most boreal record in the eastern Pacific (Aguayo and Sánchez, 1987). It has been registered in the areas of the GM and TP.

Lagenorhynchus obliquidens Gill, 1865

Pacific white-sided dolphin

Lourdes Flores Ochoa and Jorge Urbán Ramírez

SUBSPECIES IN MEXICO
L. obliquidens is a monotypic species.

DESCRIPTION: The white-sided dolphin is distinguished by its prominent falcate and bicolor dorsal fin, which is dark on the anterior part and pale gray on the posterior. The fin on young dolphins is less falcate. The body tapers and the face is small. The dominant pattern of coloration consists of a large lateral light gray patch and a smaller, darker light patch on either side of the caudal peduncle bounded by large black areas dorsally and laterally, and a thin dark line separating the lateral light gray from a white belly. Less obvious are two light stripes that contrast with the dark back and extend from the front of the head to the caudal peduncle. The mouth and lips are black, extending to the eyes. One of the white stripes extends from the side of the body to the front of the dorsal fin and continues diagonally until the anus. The pectoral fins are long and taper to a blunt tip. A dark line starts at the armpit and connects to the gray pectoral fin. Sometimes, the pectoral fins are like the dorsal fin, obscure on the front and light posteriorly. The fluke is dark and well defined with a notch in the center. Variations in the pattern of coloration have been observed,

Pacific white-sided dolphin (*Lagenorhynchus obliquidens*). Illustration: Pieter Folkens.

particularly in northern groups of the eastern North Pacific, including completely black or white individuals. The extension of this variation and its meaning in the separation of local forms is unknown. Males are larger and heavier than females.

DISTRIBUTION: Pacific white-sided dolphins inhabit temperate waters of the North Pacific from the Peninsula of Kamchatka and Isle Kodiak to the coast of Japan and Baja California. In Mexico, it is found in states of Baja California and Baja California Sur. These dolphins are found in the Gulf of California to the Isle of San Ildefonso, south of Bahía Concepción, and are frequently found in the Bahia de la Paz, mainly in winter and spring. There are no records farther south in the Baja California Peninsula. It seems that these dolphins spend winter and spring along the coast of California, and move offshore in summer and autumn. It has been registered in the areas of the GC and NP.

EXTERNAL MEASURES AND WEIGHT
TL: females to 2.36 m; males to 2.5 m.
Weight: females to 150 kg; males to 198 kg.

DENTAL FORMULA: 22–45/22–45.

NATURAL HISTORY AND ECOLOGY: These dolphins feed on a wide variety of fish and squid, mainly in the night, judging by telemetric studies. Squid, anchovies (*Engraulis mordax*), and hake (*Merluccius productus*) are most prevalent in analyzed stomach contents. In a sample reviewed by Heise (1996) on the coast of Canada, herring (*Clupea harengus*) and salmon (*Oncorhyncus*) predominated. White-sided dolphins are very active and sociable and generally form schools of 20 or more individuals; however, schools of hundreds of dolphins have been observed. They join groups of whales or other dolphins, including *Stenella coeruleoalba, Grampus griseus*, and *Orcinus orca*. They are sometimes sighted swimming or jumping along the waves made by boats. If they are trapped in fishing nets, they usually escape through holes or by jumping above the net. The white-sided dolphin jumps out of the water and spins in the air. It has been registered swimming at 15 miles per hour. The breeding season seems to occur in summer, from June to August, although offspring have been observed in early autumn. Calves are born after a gestation period of 12 months and calf rearing lasts 8 to 10 months; the interval between offspring is 4.6 years, on average.

CONSERVATION STATUS: The size of the populations of *Lagenorhynchus obliquidens* is unknown. It is caught in small numbers every year in areas of riparian fishing in Japan, but there is no organized fishery for this species. In some cases it has been captured for zoos or aquariums. They sometimes are caught accidentally in fishing nets in the North Pacific. It is listed in appendix II of CITES; in Mexico it is subject to special protection in the Mexican Official Regulation (SEMARNAT, 2010) and it is protected by the Federal Law on Fisheries and article 420-I of the Federal Penal Code (Diario Oficial de la Federación, 2009). The IUCN (IUCN, 2010) conservation status is least concern.

Northern right whale dolphin

Lorenzo Rojas Bracho

SUBSPECIES IN MEXICO
Lissodelphis borealis borealis

DESCRIPTION: The most distinctive features of this dolphin are the absence of a dorsal fin and its long, slender body. It is the only dolphin of the Northeastern Pacific with these features. Both pectoral fins and fluke are relatively small for any dolphin. The peduncle is very narrow and lacks a keel. Although the melon is not pronounced, the face is well defined with a short beak. The line of the mouth is elongated and straight. Most of the body is dark gray to black except for a middle ventral white stripe that extends from the thorax to the notch of the lobes of the caudal fin. The white coloration forms an area on the chest covering all the area between the pectoral fins. There is also a white spot in the anterior end of the jaw. The narrow and concave lobes of the caudal fin are light gray in the dorsal margins and white in the ventral margins (Leatherwood et al., 1983, 1988). Calves are a lighter gray.

EXTERNAL MEASURES AND WEIGHT
TL: males to 3.07 m; females about 20% shorter.
Weight: to 113 kg.

DENTAL FORMULA: 43 – 45/43 – 45.

NATURAL HISTORY AND ECOLOGY: These dolphins have gregarious habits and form herds of between 100 and 200 individuals; the largest school registered had up to 3,000 animals. Smaller groups are also common. These groups shift location according to the season. In autumn they move toward the south and the coast, whereas in the spring the movement is toward the north and away from the coast. In Baja California they have been reported from January to March. These movements are due to changes in temperature of water and abundance of their prey, in particular, the squid *Loligo opalescens*. The configuration of the herds is variable. Three types of arrangement have been reported: groups closely associated without subgroups and moving in the same direction; herds formed by subgroups of variable sizes, easily distinguishable, in which each group moves at their own pace; and a V formation similar to some birds and chorus lines, one next to the other. They are often associated with other species, particularly with *Lagenorhynchus obliquidens*. They are fast swimmers and can reach speeds of more than 33 km/h. They are born at a length of 80 cm to 100 cm. The size reached at sexual maturity is 2.2 m and 2.0 m for males and females, respectively. Apparently, the peak of births occurs in winter and spring. The species' gestation period, interval between births, and longevity are unknown (Leatherwood et al., 1983, 1988). Life span is approximately 42 years.

DISTRIBUTION: This dolphin is endemic to the North Pacific. It is found from Alaska to the Baja California Peninsula and from Japan to the Aleutian Islands. It prefers cold and temperate oceanic waters, between latitudes 50 degrees north and 30 degrees north. It can be found south of 30 degrees north only when the cold waters penetrate these latitudes. The temperatures in which it has been registered vary from 7.8°C to 18.9°C. It prefers deep water, although it can be found near the coast in the vicinity of submarine canyons (Leatherwood et al., 1983, 1988). It has been registered in the area of the NP.

Northern right-whale dolphin (*Lissodelphis borealis*). Illustration: Pieter Folkens.

CONSERVATION STATUS: Northern right whale dolphins are hunted in Japan and die incidentally in trawl and fence nets used by fisheries throughout the North Pacific. It is listed in appendix II of CITES and it is subject to special protection in the Mexican Official Regulation (SEMARNAT, 2010); it is protected by the Federal Law on Fisheries and article 420-I of the Federal Penal Code (Diario Oficial de la Federación, 2009). The IUCN (IUCN, 2010) conservation status is least concern.

Orcinus orca (Linnaeus, 1758)

Killer whale

Isabel Victoria Salas Rodarte

SUBSPECIES IN MEXICO

Orcinus orca is presently a monotypic species.

Up to nine eco-types of killer whales have been identified. At least three of these are likely to be recognized as separate species and the others may be subspecies.

DESCRIPTION: Killer whales are the largest members of the family Delphinidae. They have triangular dorsal fins that measure up to 2 m in males and 1 m in females and juveniles. The pectoral fins are big and oval. The head is blunt, with a conspicuous face. The dorsal coloration is black and the abdomen is white. A white post-ocular

Adult male (*upper*) and female (*lower*) killer whales (*Orcinus orca*). Illustration: Pieter Folkens.

patch and a grayish saddle in the back and under the dorsal fin on the flanks characterize them. The pectoral fins are black on both sides. Some killer whales of the Southern Hemisphere may have a brown coloration instead of black, and cream instead of white (Flores, 1991; Heyning and Dahlheim, 1988; Leatherwood et al., 1982, 1983).

EXTERNAL MEASURES AND WEIGHT
TL: adult males to 9 m; adult females to 7.9 m
Weight: males to at least 5,600 kg; females to at least 3,800 kg.

DENTAL FORMULA: 10–10/13–13.

NATURAL HISTORY AND ECOLOGY: Killer whales form matriarchal nomadic "pods" of 3 to 60 animals centered around a mature female. Pods can be segregated by age and sex. Their diet is very diverse and includes other marine mammals, from small and large mysticetes including blue whales *(Balaenoptera musculus)* to a variety of dolphins and pinnipeds, among many others. They also eat sea turtles, sea birds, different types of fish, and rays. They hunt in organized packs. They are resident in some areas, such as in British Columbia, throughout the year and conduct local movements related to food availability and climatic conditions. Others are transient, traveling long distances. Killer whales are polygamous. In the North Pacific they apparently mate from late spring to mid-summer. Gestation lasts 13 to 16 months, and calves are nursed for less than a year, but remain close to mothers for several years. Calves are born at a length of 2.0 m to 2.6 m and weigh 160 kg to 180 kg. The period between pregnancies exceeds two years (Balcomb, 1991).

CONSERVATION STATUS: Killer whales have no natural predators. They are captured for exhibition and for their oil and meat. They compete with fisheries and have been killed for this. In the North Atlantic and Southern Hemisphere, Norwegian and Soviet whalers have hunted them (Bigg et al., 1987; Leatherwood et al., 1983). This species is included in appendix II of CITES and is a species protected in the United States (Schouten, 1990). In Mexico, killer whales are subject to special protection (SEMARNAT, 2010) and are protected by the Federal Law on Fisheries and article 420-I of the Federal Penal Code (Diario Oficial de la Federacíon, 2009). The IUCN (IUCN, 2010) lists this species as data deficient.

DISTRIBUTION: Killer whales are cosmopolitan, but are more common in cold seas at high latitudes of both hemispheres (Dahlheim et al., 1982; Flores, 1991; Heyning and Dahlheim, 1988; Leatherwood et al., 1982, 1983). According to their ethology, movements, demography, and prey preference, two races, called resident and transient, are recognized (Bigg et al., 1987). These two forms are genetically distinct and allopatrically different, which demonstrates that resident populations are relatively isolated and locally adapted (Hoelzel, 1991). In Mexico, killer whales are not abundant but are regularly found in the Gulf of California and in the Mexican Pacific (Guerrero-Ruiz et al., 2006). Few records for the southeast of the Gulf of Mexico exist; thus it is considered rare in that area (Schmidly and Smith, 1981). Records of northern Gulf of Mexico indicate that the species has a regular presence, although its density is low (O'Sullivan and Mullin, 1997). It has been registered in the areas of the CS, GC, GM, NP, and TP.

Peponocephala electra (Gray, 1846)

Melon-headed whale

Daniel Castillo Lugardo and Mario A. Salinas Zacarías

SUBSPECIES IN MEXICO
P. electra is a monotypic species.

DESCRIPTION: Melon-headed whales have a fusiform and long body with a narrow caudal peduncle. The head is tapered and ends in a blunted tip with a muzzle slightly delimited by the jaw. The dorsal fin is high (40 cm), slightly falcate, with a rounded or relatively sharp tip. The pectoral fins are long and gently delineated, and comprise 20% of the total length of the body. Almost all the body is black; the ventral part is

Melon-headed whale (*Peponocephala electra*). Illustration: Pieter Folkens.

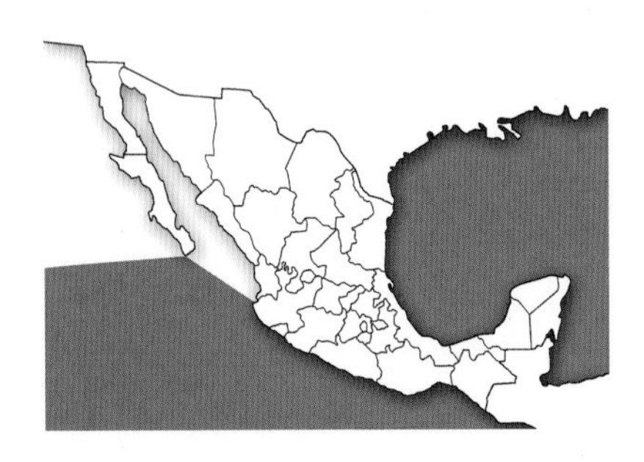

DISTRIBUTION: Concentration areas of *P. electra* are unknown. They are distributed in tropical and subtropical waters of the entire world. Due to their oceanic habits they are seldom sighted. They have been reported in oceanic waters from the Northeast Pacific to Hawaii; and in the mid Atlantic in the waters of Senegal, India, Malaysia, Indonesia, New Guinea, and northeastern Australia (Leatherwood et al., 1983). Whales stranding in Mexico have been reported in the southern tip of Baja California Sur, and they have been observed in pelagic waters of the Gulf of Mexico (Dullin et al., 1994; Jefferson and Schiro, 1995). It is believed that this species inhabits all oceanographic areas of the country. It has been registered in the areas of the CS, GC, GM, NP, and TP.

slightly lighter. It has white areas around the anus and the genital opening; a very slight white line around the lips; and a white spot in the form of an anchor between the gular region and the chest. They can be confused with other dolphins such as *Feresa attenuata* and *Pseudorca crassidens* but their length, the height and shape of the dorsal fin, and the coloration distinguish them from other species (Leatherwood et al., 1983).

EXTERNAL MEASURES AND WEIGHT
TL: males to 2.65 m; females to 2.75 m.
Weight: 160 to at least 210 kg.

DENTAL FORMULA: 22–26/20–24.

NATURAL HISTORY AND ECOLOGY: *Peponocephala electra* is a poorly known species of which only speculations on various aspects of its natural history can be made. It inhabits oceanic waters and is considered gregarious, forming herds of 20 to 30 individuals. In Japan groups of up to 500 individuals have been reported. It is commonly associated with Fraser's dolphin (*Lagenodelphis hosei*) in herds of 150 to 1,500 animals. It generally feeds on squid (*Loligo* and family Ommastrephidae) and epipelagic fish such as *Merluccius* (Best and Shaughnessy, 1981). Gestation lasts approximately 12 months and births occur throughout the year, although in the Southern Hemisphere they occur in July and August. The sexual maturity of females is reached at approximately 3.5 years at an average size of 2.1 m. Males reach maturity when they measure 2.2 m, at approximately 4 years of age. Migration patterns and local movements are poorly known.

CONSERVATION STATUS: The population size is unknown. There are artisanal fisheries in Japan where melon-headed whales are taken for human consumption or for bait. By-catch occurs in some small islands of the West Indies and there a small number are caught with yellow fin tuna (*Thunnus albacares*) in the eastern tropical Pacific. The species is included in appendix II of CITES and is subject to special protection in the Mexican Official Regulation (SEMARNAT, 2010); it is protected by the Federal Law on Fisheries and article 420-I of the Federal Penal Code (Diario Oficial de la Federación, 2009). The IUCN (2010) conservation status is least concern.

False killer whale

Isabel Victoria Salas Rodarte

SUBSPECIES IN MEXICO
P. crassidens is a monotypic species.

DESCRIPTION: False killer whales have a long and slender body. The head is smallish with a rounded face that overhangs the lower jaws. The long mouth line is slightly curved upward. The dorsal fin is high and falcate and is located in the middle of the back. The pectoral fins, with a remarkable curvature in the anterior margin in the form of an S, are a distinguishing feature of the species. The coloration is dark gray with lighter dark gray flanks that extend over the shoulders, and very light gray areas ventrally including a genital patch and a spot on the chest in the form of an anchor.

DISTRIBUTION: The false killer whale is widely distributed in temperate and tropical seas around the world. They prefer for pelagic waters, but can move to the coast. In Mexico, they are distributed throughout the Pacific and the Gulf of California. It is scarce in the Gulf of Mexico (Aguayo, 1986; Leatherwood et al., 1982, 1983; Schmidly and Smith, 1981; Urbán-R. and Aguayo, 1987a, b). It has been registered in the areas of the CS, GC, GM, NP, and TP.

EXTERNAL MEASURES AND WEIGHT
TL: males to 6 m; females to 5 m; calves 1.5–1.9 m.
Weight: 1,000 to 2,000 kg.

DENTAL FORMULA: varies in number 7 – 7 to 12 – 12.

NATURAL HISTORY AND ECOLOGY: False killer whales form herds of up to 500 individuals, but it is more common to find them in groups of a few dozen. It is unknown if groups have a structure according to age, sex, and season of the year or region, or if it shows seasonal movements. This species tends to strand in large groups. The offspring are born throughout the year, and sexual maturity is reached with lengths of 3.2 m to 3.8 m. They mainly feed on squid and pelagic fish, but they have been observed eating common and spotted dolphins freed from tuna nets, as well as attacking sperm whales and humpback whales (Flores-González et al., 1994; Leatherwood et al., 1983; Palacios and Mate, 1996; Perryman and Foster, 1980).

CONSERVATION STATUS: *Pseudorca crassidens* is occasionally taken in Japan and dies in small numbers at tuna fisheries in the Pacific. They are occasionally captured for public display. It is included in appendix II of CITES and is a species protected in the United States (Schouten, 1990). It is considered data deficient by the IUCN (IUCN, 2010). There is no information about its status or abundance in Mexico, and it is subject to special protection in the Mexican Official Regulation (SEMARNAT, 2010); as all marine mammals that inhabit Mexican seas, it is protected by the Federal Law on Fisheries and article 420-I of the Federal Penal Code (Diario Oficial de la Federación, 2009).

False killer whale (*Pseudorca crassidens*). Illustration: Pieter Folkens.

Stenella attenuata (Gray, 1846)

Pantropical spotted dolphin

Carlos Esquivel Macías

SUBSPECIES IN MEXICO
Stenella attenuata graffmani (Lonnberg, 1934)
Stenella attenuata attenuata (Perrin, 1975)

DISTRIBUTION: The taxonomy of the genus *Stenella* is complex. In the Pacific Ocean several management units have been differentiated based in their morphology, feeding habits, and distribution, according to the relationship between tuna and dolphins. A pelagic (*S. attenuata attenuata*) and a coastal (*S. attenuate graffmani*) form are recognized, but this is still provisional (Perrin, 1975). In the Atlantic Ocean, where the species has been studied much less, it is only known as *S. attenuata* and a subspecies has not been described or suggested. Spotted dolphins are found in all tropical and subtropical seas of the world, between Ecuador and 28 degrees north and south. It has pelagic and neritic habits, and therefore it can be found at any distance from the coast, especially in the eastern tropical Pacific. In the Mexican Atlantic it is not as frequently observed as along the Pacific coast. It is found in the Gulf of California to Topolobampo on the continental coast, and is completely absent along the Californian coastline and in the upper part of the Gulf of California. In the south, spotted dolphins can be found along the continental coast from the Baja California Peninsula to Peru (Perrin, 1975; Perrin et al., 1983). It has been registered in the areas of the CS, GC, GM, NP, and TP.

DISTRIBUTION: *Stenella attenuata* is slender and males have a moderate keel in the bottom of the caudal peduncle. Like other Delphinidae, it has a separation between the face and the frontal cephalic region (the melon), forming a marked fold; its face is narrow and long. The dorsal fin is in the middle of the back and has a hooked shape, characteristic of the family. The species is distinguished by having a coloration pattern of white spots in adulthood. Young dolphins are dark gray (Leatherwood et al., 1988; Nishiwaki, 1972). The abdomen is light gray with black spots that also occur in adulthood. The tip of the face is white and there is a thin stripe, almost black, from the face up to the eye, as well as another more tenuous stripe from the maxilla to the beginning of the pectoral fin. The intensity of the pattern varies within populations and in general is more intense in the coastal subspecies (*S. attenuata graffmani*) than in the oceanic subspecies (*S. attenuata attenuata*; Perrin, 1972).

EXTERNAL MEASURES AND WEIGHT
TL: 2 to 2.5 m.
Weight: up to 100 kg.

DENTAL FORMULA: 35–48/34–47.

NATURAL HISTORY AND ECOLOGY: Spotted dolphins are limited to tropical marine waters and travel with temperature changes of some ocean currents. When the water temperature is below 20°C, they move temporarily to warmer waters, but are not migratory. In pelagic localities they feed on squid of the families Ommastrephidae (*Dosidicus gigas*), Onychoteuthidae, and Enoploteuthidae, fish of the family Scombridae (*Auxis*), and smallwing flyingfish (*Oxyphoramphus micropterus*) (Perrin et al., 1973). Their diet is similar to that of the yellow fin tuna (*Thunnus albacares*), with which it associates commonly. This relationship is the best indicator of the presence of the largest shoal of tuna (Perrin, 1969a). It is a dolphin that competes with some coastal fisheries but the importance of this has not been evaluated (Zavala and Esquivel, 1991). Its main predators are other Odontoceti such as killer whales (*Orcinus orca*), short-finned pilot whales (*Globicephala macrorhynchus*), and false killer whales (*Pseudorca crassidens*; Perryman and Foster, 1980). Remoras and barnacles are common epibionts of this species (Perrin, 1969b). Nematodes, acantocephala, trematodes, and cestodes are common parasites. Calves are born 89 cm in length after a gestation period of 11.2 months. Slightly marked birth periods occur from February to March and from July to November. The offspring are breastfed until 29 months, and females rest 3 months before mating again (Kasuya et al., 1974). Sexual maturity is reached at 8.2 years, when females measure approximately 180 cm to 185 cm and males measure around 190 cm. Physical maturity occurs at 15 years old in females (195 cm) and 15 to 20 years in males (203 cm). Life span can reach 46 years. They can form subgroups within herds without separating completely. This means that there is a social organization where nursing mothers, groups of mothers with young calves, groups of subadults, and groups of adult males can be found (Pryor and Lang, 1980). It is the most abundant coastal dolphin in Mexico.

Pantropical spotted dolphin (*Stenella attenuata*). Illustration: Pieter Folkens.

CONSERVATION STATUS: This species of dolphin is the most affected by the fishing of tuna with seine nets, which started in the late 1950s. This activity was and still is particularly strong in the eastern tropical Pacific. It is estimated that 200,000 to 400,000 dolphins died annually during the 1960s and much of the 1970s when the incidental mortality was estimated. Today, the mortality of dolphins in the tuna fishery has been dramatically reduced and some recovery among the most affected populations has been observed (Compeán-Jiménez, 1993; Hall, 1994; Perrin et al., 1976). In Mexico they are used as bait for shark fishing and perhaps even as food for coastal fishermen (Zavala and Esquivel, 1991). The species is found in appendix II of CITES and is subject to special protection in the Mexican Official Regulation (SEMARNAT, 2010). In addition to its protection by the Federal Law on Fisheries and article 420-I of the Federal Penal Code (Diario Oficial de la Federación, 2009). The spotted dolphin is protected by the Official Mexican Regulation (001-CFSP-1993) on the fishing of tuna; in 1995 it established an allowed mortality of 1.5 dolphins per net catch. The IUCN (IUCN, 2010) conservation status is least concern.

Stenella clymene (Gray, 1846)

Clymene dolphin

Silvia Regina Manzanilla Naim

SUBSPECIES IN MEXICO

S. clymene is a monotypic species.

DESCRIPTION: The clymene dolphin has a pattern of coloration in three tones that distinguishes it from other species of the genus. The back is dark gray, almost black, and this coloration extends from the melon to the middle of the caudal peduncle. On the sides it has a gray tone that extends from the eyes to the peduncle and anus. The belly has a white coloration that extends from the jaw to the tail. It has a mottled pattern, especially where the gray and white colorations meet. Face color varies from completely dark in its dorsal part, white in the sides, with dark edges on the lips, to uniformly dark. The latter is less common and most of the individuals have a moustache or black stripe in the dorsal part of the face. The pectoral fins can be completely black or black ventrally and white dorsally, delineated by a dark color,

Clymene dolphin (*Stenella clymene*). Illustration: Pieter Folkens.

DISTRIBUTION: *S. clymene* is a little known species and rarely sighted. It inhabits deep waters from the isobath of 250 m up to 5,000 m or more in the tropical and subtropical Atlantic (Leatherwood et al., 1983). There are records in the northwestern coast of Africa, the central Atlantic waters, the southeastern United States, the northern and central Gulf of Mexico, and the Caribbean. The average temperature recorded at the areas where it has been sighted in the Gulf of Mexico varies from 22.8°C to 29.1°C (Jefferson and Schiro, 1995; Jefferson et al., 1992; Mullin et al., 1994). It has been registered in the areas of the CS and GM.

almost black. The pectoral fins are slightly smaller than in other species and end in a tip. The fluke has a well-defined notch, which is large and thin with pointed edges. The eyes are framed by a black coloration that extends as stripes toward the face (Perrin et al., 1981). It resembles *Stenella longirostris*, but it has a shorter and thicker face, and the dorsal fin is smaller. The dorsal fin is falcate, pointed, and located in the middle of the back. As in other dolphins, males are larger than females (Leatherwood et al., 1983).

EXTERNAL MEASURES AND WEIGHT

TL: adult females to 1.90 m; adult males to 1.97 m.

Weight: females to 75 kg; males to 80 kg.

DENTAL FORMULA: 38 – 49/38 – 49.

NATURAL HISTORY AND ECOLOGY: *Stenella clymene* is a pelagic species that travels in small groups of 2 to 100 individuals with an average of 42. Apparently, it feeds on fish and squid in the night. Some of the species found in the stomach contents of several individuals included *Brevortia tyrannus, Clupea harengus, Ceratoscopelus, Lampanyctus, Symbolophorus, Bregmaceros,* and unidentified squid. The diet of this species is similar to that of *S. longirostris* in the eastern Pacific (Jefferson et al., 1995). Its natural predators include pelagic sharks (Leatherwood et al., 1983). In the northern Gulf of Mexico it is sympatric with three other species (*S. longirostris, S. coeruleoalba,* and *S. attenuata*) of the same genus (Perrin et al., 1981). Information about their reproduction is unknown, but it has been suggested that the offspring measure approximately 77 cm at birth.

CONSERVATION STATUS: *S. clymene* is considered a rare but stable species because its mortality has not been associated with any fishery and it is not exploited directly or indirectly. It is listed in appendix II of CITES. It is subject to special protection in the Mexican Official Regulation (SEMARNAT, 2010) and it is protected by the Federal Law on Fisheries and article 420-I of the Federal Penal Code (Diario Oficial de la Federación, 2009). The IUCN (IUCN, 2010) conservation status is data deficient.

Striped dolphin

Carlos Esquivel Macías

SUBSPECIES IN MEXICO

S. coeruleoalba is a monotypic species.

DESCRIPTION: *Stenella coeruleoalba* is a robust dolphin with a clear separation between the face and the melon forming a fold. The dorsal fin is falciform and is located in the middle of the body. It is distinguished from other dolphins by the striped patterns of its coloration. The body is generally bluish-gray with stripes and blazes of various shades of gray. The first stripe is dark gray and runs from the eye to the beginning of the pectoral fin. The second is dark gray and thin, running from the eye and widening to the ventral-caudal region. The third is light gray, and originates in the melon, widening from the eye and bifurcating in the costal region and then extending to the caudal peduncle (Fraser and Noble, 1970; Leatherwood et al., 1983; Nishiwaki, 1972).

DISTRIBUTION: Striped dolphins inhabit all tropical, subtropical, and temperate oceans of the world, except the Indian Ocean, commonly between latitudes 20 degrees north and 20 degrees south. *S. coeruleoalba* is particularly abundant in the Mediterranean. In general, it has pelagic habits when it is infrequently sighted near the coasts. In Mexico it inhabits all oceans but it is only found in the south of the Gulf of California (Hubbs et al., 1973; Miyazaki, 1980, 1981). It has been registered in the areas of the CS, GC, GM, NP, and TP.

EXTERNAL MEASURES AND WEIGHT

TL: females to 2.4 m; males to 2.65 m.

Weight: females to 150 kg; males to 160 kg.

DENTAL FORMULA: 39 – 55/39 – 55.

NATURAL HISTORY AND ECOLOGY: These dolphins are pelagic but occasionally they are found near the coast, for example, in the western Pacific (Miyazaki, 1981). In Mexico its presence is barely known and there is no knowledge about its movements. Striped dolphins feed in equal parts on mesopelagic fish (family Myctophidae), squid (*Loligo*, *Symplectoteuthis*, and *Todarodus*), and shrimp (Miyazaki, 1980; Nishiwaki, 1972). In the Western Pacific the reproduction peaks occur in spring, winter, and perhaps late summer. The gestation period lasts 12 months and the offspring are born 1 m in length and with the complete pattern of coloration. Breastfeeding lasts 1.5 to 2 years. Sexual maturity is reached at 9 years old (females 212 cm in length, males 220 cm in length). Physical maturity occurs at 14 or 15 years of age (females 222 cm in length, males 236 cm in length). A complete reproductive cycle in females lasts 3 years. Life span can reach 57–58 years. Striped dolphins are preyed on by killer whales (*Orcinus orca*) and sharks of the family Carcharinidae. Very important causes of mortality are infections caused by nematodes, flukes, or cestodes (Dollfus, 1974). There is segregation by gender and age. Females in estrus sometimes separate from the rest of the group, and it is possible to find groups of young dolphins (Ka-

Striped dolphin (*Stenella coeruleoalba*). Illustration: Pieter Folkens.

suya, 1972; Miyazaki, 1980). They are gregarious and form schools of hundreds of thousands of individuals that are very active on the water surface (Leatherwood et al., 1988). They associate with shoals of tuna, which makes them good indicators of these fish, although in lesser degree than other species of the genus (Hubbs et al., 1973).

CONSERVATION STATUS: Striped dolphins are regularly hunted as food in New Guinea and Japan (Leatherwood et al., 1983; Miyazaki, 1981). In the Eastern Pacific there is an incidental mortality involving the tuna fishery though not well documented. The species is listed in appendix II of CITES and is subject to special protection in the Mexican Official Regulation (SEMARNAT, 2010). It is protected by the Federal Law on Fisheries and article 420-I of the Federal Penal Code (Diario Oficial de la Federación, 2009). The IUCN (IUCN, 2010) conservation status is least concern.

Stenella frontalis (G. Cuvier, 1829)

Atlantic spotted dolphin

Silvia Regina Manzanilla Naim

SUBSPECIES IN MEXICO

S. frontalis is a monotypic species.

Until recently, this species was called *S. plagiodon* (Cope, 1866).

DESCRIPTION: Morphologically *Stenella frontalis* resembles the bottlenose dolphin (*Tursiops truncatus*), but this dolphin is covered by a prominent mottled pattern. It is more robust than the Pacific spotted dolphin *(Stenella attenuata)*. In general, it has a grayish-blue coloration with the back darker than the abdomen, extending down on the sides and covering the caudal peduncle. A blue mottled pattern covers the dark back. Newborns have a lighter, more uniform color, similar to that in bottlenose

Atlantic spotted dolphin (*Stenella frontalis*). Physically mature adult (*top*); sexually mature adult (*bottom*). Illustrations: Pieter Folkens.

dolphins, and acquire the mottled pattern with age. The incidence of the spotted pattern decreases with the distance from the coast (Perrin et al., 1987). These dolphins have a high and falcate dorsal fin located in the middle of the back. The pectoral fins are curved and pointed. The fluke is wide and both sides end in a tip.

EXTERNAL MEASURES AND WEIGHT
TL: adult males to 2.26 m; adult females to 2.29 m.
Weight: males to 140 kg; females to 130 kg.

DENTAL FORMULA: 32–42/30–40.

NATURAL HISTORY AND ECOLOGY: Atlantic spotted dolphins travel in groups of up to 50 animals, but they can also form herds of several hundreds. Their diet is based on squid, small fish, and eels. Its natural predators are killer whales (*Orcinus orca*), tiger sharks (*Galeocerdo cuvieri*), bull sharks (*Carcharhinus been*), and hammer sharks (*Sphyrna*). Data about its history of life are scarce. It is estimated that females reach sexual maturity at approximately 9 years old, and that they reproduce for the first time at 10 years of age. The reproduction interval in females is three years on average, which is similar to what was described for *S. attenuata* (Herzing, 1997). The population inhabiting the Gulf of Mexico migrates to the coast at the end of spring and summer.

CONSERVATION STATUS: This species is subject to mortality by fishermen who use longlines to hunt sharks. It is common for fishermen to shoot the dolphins close to their ships and use them as bait (Leatherwood et al., 1983). It is listed in appendix II of CITES. In Mexico it is subject to special protection in the Mexican Official Regulation (SEMARNAT, 2010) and it is protected by the Federal Law on Fisheries and article 420-I of the Federal Penal Code (Diario Oficial de la Federación, 2009). The IUCN (IUCN, 2010) conservation status is data deficient.

DISTRIBUTION: This species is distributed in the tropical and semi-temperate waters of the Atlantic Ocean. In the North Atlantic, spotted dolphins have been reported from the coasts of New Jersey, in the United States, to the coast of England (Perrin et al., 1987). Farther south, they are found in the Bahama Islands, the Caribbean, and the Gulf of Mexico. In Mexico, they have been reported along the coast of Veracruz, Campeche, Quintana Roo, Yucatan, and coral atolls distributed in the Caribbean, in particular, Arrecife Alacranes and Cayo Arcas (Hugentobler and Gallo, 1986; Jefferson and Schiro, 1995; Pérez-Sánchez et al., 1994). It has been registered in the areas of the CS and GM.

Stenella longirostris (Gray, 1828)

Spinner dolphin

Carlos Esquivel Macías

SUBSPECIES IN MEXICO
S. longirostris longirostris (Gray, 1828)
S. longirostris orientalis Perrin, 1990
S. longirostris centroamericana Perrin, 1990

DESCRIPTION: *Stenella longirostris* is a slender dolphin that shows a clear demarcation between the melon and face, which is very long and conspicuous. The dorsal fin is located in the middle of the back and is completely triangular in both sexes, although its form varies between populations. In the male the dorsal fin tilts forward, giving the impression of being backward. Also in males of this species, there is a keel of connective tissue in the post-genital region. Its dorsal coloration is dark gray and the sides are slightly lighter, gradually varying until white in the genital and axillary regions. These two white areas converge ventrally in the eastern race, being one of the multiple geographical variations. In adulthood the edges of the white regions often spread (Evans, 1987; Perrin, 1972).

(*Top*) Pantropical spinner dolphin (*Stenella longirostris longirostris*); (*middle*) whitebelly spinner dolphin hybrid; (*bottom*) physically mature male eastern spinner dolphin (*Stenella longirostris orientalis*). Illustrations: Pieter Folkens.

DISTRIBUTION: The spinner dolphin inhabits deep waters of all tropical and subtropical oceans of the world, between latitudes 22 degrees north and 22 degrees south (Halley, 1978). In Mexico, they frequently come close to the coasts with deep waters in the southern part of the Gulf of California. Their presence in the northern Mexican Pacific has its most boreal limit in Bahía Magdalena (Perrin et al., 1985). In the eastern tropical Pacific, an eastern form has been distinguished from a white belly southern form and a northern white belly form. A study of genetic variation did not detect mitochondrial differences between schools, geographical regions, or the forms mentioned above, thus proving that in this whole area there is no reproductive segregation between populations (Dizon et al., 1991). It has been registered in the areas of the CS, GC, GM, NP, and TP.

EXTERNAL MEASURES AND WEIGHT

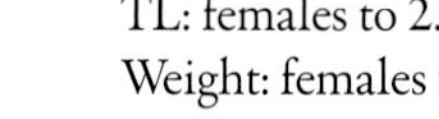

TL: females to 2.11; males to 2.35 m.
Weight: females to 65 kg; males to 78 kg.

DENTAL FORMULA: 45–65/45–65.

NATURAL HISTORY AND ECOLOGY: Spinner dolphins are pelagic and when they move toward the coast they do so following fish banks. Their movements have not been studied in Mexico, and it is unknown if they undertake migrations. Their diet is based on mesopelagic organisms such as squid of the families Ommastrephidae, Onychoteuthidae, and Enoploteuthidae, as well as fish of the families Myctophidae and Gonostomatidae. This diet is similar to that of *S. attenuata*, with which at times it forms herds associating with schools of yellow fin tuna (*Thunnus albacares*), thus creating a multispecies relationship (Hester et al., 1963; Perrin et al., 1973). It is one of the major indicators of yellow fin tuna, although sometimes it is associated with schools of smaller tuna; mortality is high due to tuna fisheries in the eastern tropical Pacific. Odontoceti, such as false killer whales, prey on them (*Pseudorca crassidens*) (Perryman and Foster, 1980). Calves are born with an average length of 77 cm after a gestation period of 11 months. Lactation lasts until two years of age; spinner dolphins reach sexual maturity when they measure 170 cm in the case of males and 165 cm in the case of females. The age of physical maturity is unknown (Leatherwood et al., 1988; Perrin et al., 1977). They form herds of hundreds of individuals, which have a variable composition and size. Only some small groups within the herds have long-term integrity. These dolphins show several exhibition behaviors; the spinning jump is the trademark jump for this species, and these jumps are associated with sounds. Each behavioral pattern is typical of the level of activity of the school. They show a daily pattern of behavior in which they feed at night and move in the mornings, resting at noon and moving again to seek food at dusk (Norris and Dohl, 1980).

CONSERVATION STATUS: Spinner dolphins are widely dispersed but some Pacific populations, such as the ones of northern and eastern pelagic areas, have been severely affected by the tropical tuna fishery, especially in the 1960s (Watson, 1981). Currently, the level of mortality has diminished, but the populations previously mentioned show signs of overexploitation in their reproductive parameters in comparison to the Atlantic populations (Perrin and Henderson, 1979; Perrin et al., 1977). There are about 1.8 million of these dolphins in the eastern Pacific (Evans, 1987). The species is included in appendix II of CITES. In Mexico there are regulations to curb the mortality rate in the fisheries that affect these dolphins, and the species is subject to special protection in the Mexican Official Regulation (SEMARNAT, 2010); it is protected by the Federal Law on Fisheries and article 420-I of the Federal Penal Code (Diario Oficial de la Federación, 2009). The IUCN (IUCN, 2010) conservation status is data deficient.

Steno bredanensis (G. Cuvier in Lesson, 1828)

Rough-toothed dolphin

Ivette Ruiz Boijseauneau

SUBSPECIES IN MEXICO

S. bredanensis is a monotypic species.

DESCRIPTION: Among the Delphinidae, *Steno bredanensis* is medium-sized. It is distinguished from other species by its face, which joins with the melon without a marked separation, giving the appearance of a long, almost conical head, seen from a lateral or dorsal view. The coloration is variable; the back varies from dark gray to gray purple, the sides can be white or pink, and the belly is white. At times they have yellowish spots in the sides and the ventral region. Some individuals have numerous linear and oval scars. The lips and tip of the snout are white, the pectoral fins are dark. The dorsal fin varies in form; it has a broad base, and is more falcate, slightly higher, and more triangular than in the bottlenose dolphin (*Tursiops truncatus*). These dolphins have a series of vertical grooves in the teeth, from which they get their name (Leatherwood et al., 1976, 1983; Nishiwaki, 1972). Watson (1981) believes that geographical forms or races, and even subspecies not yet described, probably exist.

DISTRIBUTION: The records on this dolphin are scarce but supposedly they are distributed in deep waters of tropical and temperate regions; they are rarely found in coastal waters. They have been sighted along the shores of the United States, Brazil, and Argentina, in Hawaii, in the Mediterranean Sea, along the shores of Holland, in the waters of South Africa, in the west of India, and along the coast of Japan (Nishiwaki, 1972; Richardson, 1973 in Miyazaki, 1980; Watson, 1981). It has been registered in the areas of the CS, GC, GM, NP, and TP.

EXTERNAL MEASURES AND WEIGHT

TL: 2.2 to 2.8 m

Weight: 150 kg.

DENTAL FORMULA: 20–27/20–27.

NATURAL HISTORY AND ECOLOGY: Groups of 50 animals have been reported, but it

Rough-toothed dolphin (*Steno bredanensis*). Illustration: Pieter Folkens.

is more common to find 10 to 20 individuals near the continental shelf. Sometimes they come near the bow of boats and occasionally jump out of the water. They often mix with other species such as the bottlenose dolphin (*Tursiops truncatus*), the spotted dolphin (*Stenella attenuata*), and the spinner dolphin (*Stenella longirostris*). The feeding habits of rough-toothed dolphins are little known, but they prey on various species of pelagic fish and squid. The diet of some individuals in the Gulf of Mexico includes common dolphinfish (*Coryphaena hippurus*) and octopus (*Termoctopus violaceus*). This species has been successfully trained in Japan and Hawaii (Leatherwood et al., 1983).

CONSERVATION STATUS: The species is not threatened. They are caught in coastal fisheries in Japan and occasionally die incidentally during tuna fishing (Leatherwood et al., 1983). The species is listed in appendix II of CITES; in Mexico it is subject to special protection in the Mexican Official Regulation (SEMARNAT, 2010). As all marine mammals of Mexico, the species it is protected by the Federal Law on Fisheries and article 420-I of the Federal Penal Code (Diario Oficial de la Federación, 2009). The IUCN (IUCN, 2010) conservation status is least concern.

Tursiops truncatus (Montagu, 1821)

Bottlenose dolphin

Araceli Mejía Olguín

SUBSPECIES IN MEXICO

Tursiops truncatus is a monotypic species.

The systematics of bottlenose dolphins is unclear and has described various species and subspecies. Currently, at least provisionally, it is considered a single species with various geographical forms that in general include coastal and oceanic types.

DESCRIPTION: Bottlenose dolphins are robust and medium in size with a triangular and falcate dorsal fin located in the central portion of the back (Leatherwood and Reeves, 1983; Watson, 1981). The coloration is highly variable; in general, the back is dark gray, although some animals can be black, brownish, purple gray, steel gray, bluish-gray, or slate gray. These tones change to light gray on the sides and the belly; the belly is pink. Compared to other dolphins, the face is well defined and wide; the melon has dark grooves that vary individually and geographically. The fluke slims abruptly in its tip, and the posterior part of the dorsal fin is thinner than that of the thoracic region, which is robust (Leatherwood and Reeves, 1983; Leatherwood et al., 1983; Nishiwaki, 1972; Walker, 1991).

EXTERNAL MEASURES AND WEIGHT

TL: males 2.45 to 4 m depending on population; females slightly smaller.

Weight: 260 to 500 kg.

DENTAL FORMULA: 20 – 26/18 – 24.

NATURAL HISTORY AND ECOLOGY: The coastal form herds with up to 50 individuals but most commonly includes 12 to 15 dolphins (Leatherwood et al., 1976, 1983). The oceanic form groups with hundreds of individuals (Leatherwood and Reeves, 1983). They have a lifespan of 25 to 30 years. Males mature at 10 to 13 years of age and females at 5 to 12 years (Sergeant et al., 1973). There are two peaks of reproduction, one in the spring and another in autumn, with births occurring in summer and even in winter (Nishiwaki, 1972; Salinas and Bourillón, 1988; Urbán-R., 1983). The courtship begins with a pre-copulatory bending of the body, forming an S. After mating, individuals maintain frequent friction in the genital area for 10 to 30 seconds (Tavolga and Essapian, 1957). The gestation period lasts 12 to 18 months. Calves are born within a social group. Other dolphins emit sounds around the pregnant mother; two adult females stay nearby during the delivery (Caldwell and Caldwell, 1972). Newborns measure 1 m and weigh 12 kg. Lactation lasts 18 months, but the offspring can remain close to their mothers until 2 or 3 years of age (Scott et al., 1990; Walker, 1991). They perform local movements that are related to changes in the tide and food (Shane et al., 1986). They consume from 8 kg to 15 kg of food per day. Their diet near the coast consists of a wide variety of fish, shellfish, and cephalopods, and in the pelagic area it varies regionally (Leatherwood and Reeves, 1982; Leatherwood et al., 1976, 1983). Generally, they form small groups of attack, although sometimes they hunt collectively, depending on the abundance of prey (Ballance, 1992; Irvine et al., 1981; Leatherwood, 1975). They commonly associate with shrimp trawlers and different types of nets, interaction regarded as competition with fisheries (Northridge, 1985). They often associate with rough-toothed dolphins (*Steno*), short-finned pilot whales (*Globicephala*), right whales (*Eubalaena*), gray whales (*Eschrichtius*), and humpback whales (*Megaptera*). The coastal and oceanic subspecies have different species of nematodes, cestodes, and flukes that parasitize the aerial cavities, lungs, abdominal cavities, and urogenital cavities (Walker, 1981). They have ectoparasites of the genus *Xenobalanus* in the tip of the dorsal fin (Wells et al., 1980). In the Gulf of Mexico they show recurring epizootics of Morbillivirus (Duignan et al., 1996). It is unknown whether in natural conditions predation by killer whales and sharks is important in their mortality (Wells et al., 1980).

CONSERVATION STATUS: Bottlenose dolphins adapt more easily to captivity than other dolphins, and their active behavior has allowed using them in parks and aquariums around the world. In New Zealand, tourism to observe this species in the wild is particularly prevalent. Mexico regulates its capture for captivity and export. In Mexico, as in other countries, because of commercial fishing operations, bottlenose dolphins are frequently killed, in addition to becoming trapped in fishing fleets and dying as a result (Northridge, 1985; Zavala and Esquivel, 1991). The species is listed in appendix II of CITES and is subject to special protection in the Mexican Official Regulation (SEMARNAT, 2010); it is protected by the Federal Law on Fisheries and article 420-I of the Federal Penal Code (Diario Oficial de la Federación, 2009). The IUCN (IUCN, 2010) conservation status is least concern.

DISTRIBUTION: Bottlenose dolphins inhabit coastal and oceanic areas of tropical, subtropical, and temperate seas from around the world, and are known in the Atlantic, Mediterranean, Indian, and Pacific Oceans, from the north of Japan and Southern California to New Zealand and Chile (Leatherwood et al., 1976, 1983). The genus is common in coastal and oceanic waters in the Mexican Pacific (Urbán-R., 1983), Gulf of Mexico, and Mexican Caribbean Sea (Fuentes and Aguayo, 1992; Jefferson and Schiro, 1995; Jefferson et al., 1992; Pérez-Sánchez et al., 1994). In the Mexican Pacific the population of coastal habits is distributed along the entire coast, including the Gulf of California. The oceanic population inhabits temperate seas of the north and has its southern boundary in Guadalupe Island. The other population lives along the west coast of Baja California to Bahía Magdalena and in the interior of the Gulf of California (Walker, 1981). It has been registered in the areas of the CS, GC, GM, NP, and TP.

Bottlenose dolphin (*Tursiops truncatus*). Offshore form (*top*) and near shore form (*bottom*). Illustration: Pieter Folkens.

Family Phocoenidae

The family Phocoenidae is represented by six species distributed in temperate and cold seas (Wilson and Reeder, 1993). In Mexico there are two species, one in the genus *Phocoena* and one in the genus *Phocoenoides*. Vaquita (*Phocoena sinus*) is a very small porpoise, endemic to the upper-most Gulf of California. It is in grave danger of extinction from fatalities due to by-catch from rolling hook and gill net fisheries, and the destruction of the estuarine system of the Colorado River.

Phocoena sinus Norris and McFarland, 1958

Vaquita

Silvia Regina Manzanilla Naim

SUBSPECIES IN MEXICO
P. sinus is a monotypic species.

DESCRIPTION: The vaquita is the smallest cetacean in the world. It is distinguished by its small size, an unusually high dorsal fin (compared with other porpoises), and its coloration, which varies from dark gray to black with a thick dark patch around the eyes and outlining the lips. In both areas, the dark color contrasts with the light gray coloration of the rest of the body. There is no discernible snout. Apparently, adult females are larger than males (Brownell et al., 1987).

EXTERNAL MEASURES AND WEIGHT
TL: males to 1.45 m; females to 1.50 m.
Weight: 45 to 50 kg.

DENTAL FORMULA: 16–22/17–20.

NATURAL HISTORY AND ECOLOGY: *Phocoena sinus* is a coastal species that moves in moderately shallow waters of the High Gulf of California (14 m to 56 m deep). Most of the sightings have been made in the vicinity of the Rocas Consag along the coast of San Felipe, but the localities where vaquitas have been collected from the beach and

Vaquita (*Phocoena sinus*). Illustration: Pieter Folkens.

dead in fishing nets include the coast of Sonora, the Gulf of Santa Clara, and Puerto Peñasco (Brownell, 1986). It is possible that the largest concentration of sightings in the vicinity of San Felipe, compared with the coast of Sonora, are due to high rates of productivity resulting from the greater speed of the tide, increased turbidity, mixture, upwelling, and exchange of water with the Central Gulf and hence, a slightly lower water temperature in summer (Silber, 1990b). They feed on benthic fish such as *Orthopristis reddingi* (bronze-striped grunt), *Bairdiella icistia* (Ronco croaker), and pelagic species such as *Cetengraulis mysticetus* (Pacific anchoveta), and squid of the species *Loliolopsis diomedea* (Fitch and Brownell, 1968). It is a poorly studied species, and it was not until 1985 that the first fresh specimens were collected (Brownell et al., 1987). It is likely that females reach sexual maturity at two or three years of age; breeding occurs once a year (S. Manzanilla, pers. obs.). A trend of early sexual maturation has been reported in the common porpoise (*P. phocoena*) as a response to the population pressure of incidental mortality suffered by fishing activities associated with herring in the North Atlantic (Read, 1990). The frequency of juvenile individuals caught in fishing nets suggests that adult females are traveling with the offspring of the previous year while bringing up the ones of the current year. This also occurs with the common porpoise of the North Atlantic (Read, 1990). Although there is no confirmed data on natural predation, shark fishermen of the area reported pieces of vaquitas in the stomachs of white and mako sharks (*Carcharodon*).

DISTRIBUTION: *P. sinus* is an endemic species to Mexico with a restricted distribution to the Gulf of California, in particular, to the Delta of Colorado River, located between latitudes 30 degrees north and 31 degrees north. The confirmed sightings of the vaquita are located in the north area of the Gulf of California, also called the High Gulf (Silber, 1990a), located between the states of Baja California and Sonora. Apparently, the vaquitas are concentrated along the coast of San Felipe in Baja California, in the Gulf of Santa Clara, and near Puerto Peñasco in Sonora (Brownell, 1986; Silber, 1990b; Wells et al., 1981). This species has been registered in the area of the GC.

CONSERVATION STATUS: Incidental mortality, as a result of the fishing activities associated with shark and totoaba (*Cynoscion macdonaldi*), has been one of the factors that placed the vaquita at risk of imminent extinction. For decades, it has been subjected to a constant degradation of its habitat due mainly to the reduction and modifications of the flow of freshwater from Colorado River (Álvarez-Borrego, 1983), which maintained an estuary rich in nutrients. Considering that the loss of these conditions has resulted in a hypersaline habitat, it is possible that the populations of potential prey of the vaquita and other marine mammals have been reduced (Barlow, 1986). Population size estimates are not very encouraging; Barlow (1986) estimated from 50 to 100 animals, whereas Silber (1990c) estimated that there were from 200 to 500 animals. Recently, Barlow et al. (1997) estimated a total population of 224 to 885 vaquitas that declines annually at a rate of about 18%. Rosel and Rojas-Bracho (1993) found no genetic variability in a portion of 400 bp of mitochondrial DNA, suggesting a severe bottleneck, inbreeding in a very small population, and/or founder effect. This is important because it implies that perhaps the population of vaquitas has never been numerous. Since 1978, this species was included in the list of vulnerable species of the IUCN (Brownell, 1983; Klinowska, 1991) as well as in the listing of Mexican vertebrates at risk of extinction (Villa-R., 1978). In 1979 it was listed as a species at risk of extinction in appendix I of CITES, as well as in the record of endangered species of the United States in 1985. In Mexico it is considered as an endemic species at risk of extinction (SEMARNAT, 2010) and it is protected by the Federal Law on Fisheries and article 420-I of the Federal Penal Code (Diario Oficial de la Federación, 2009); an official Mexican regulation (012-CFSP-1993) establishes measures for the specific protection of the vaquita and the totoaba. Furthermore, since June 1993, the area of the High Gulf, from the line that unites San Felipe with Puerto Peñasco, was decreed part of the El Pinacate Biosphere Reserve and Gran Desierto de Altar. This means that at the core zone, which includes the Delta of Colorado River in its flooding area, and the Ciénega of Santa Clara, any type of fishing activity for commercial, artisanal, or sport purposes is forbidden. The buffer zone includes the marine and terrestrial area surrounding the core zone from an imaginary line between Puerto Peñasco, Sonora, and San Felipe, Baja California. In this zone the use of trawl nets and meshes with a lumen of more than 13 cm is limited. The IUCN (IUCN, 2010) conservation status is critically endangered.

Dall's porpoise

Lorenzo Rojas Bracho

SUBSPECIES IN MEXICO

Phocoenoides dalli dalli

Two forms of *Phocoenoides* are recognized—*P. dalli dalli* and *P. dalli truei.* Marine mammalogists regard them as subspecies, though other taxonomists restrict the designations to "types" or "morphs."

DESCRIPTION: Dall's porpoise possesses an unusually robust body form for an odontocete, with the head, pectoral fins, and lobes of the flukes proportionally small compared with the rest of the body. There is no demarcation between the short snout and the melon, which is very small. Seen from the side, the head is triangular. The dorsal fin is triangular with a wide base and develops a forward cant in mature males. The caudal peduncle has a pronounced keel, both dorsally and ventrally, which is more extreme in physically mature males (Leatherwood et al., 1983, 1988). The flukes develop a forward cant in older males that appears as if the flukes were attached backwards with a straight anterior margin. Its coloration is atypical for a porpoise with a broad, highly contrasting white lateral field. There are two types of coloration patterns: *dalli* and *truei.* In general, the first type (*dalli*) is black with white patches in each side covering almost half of the body, which meet ventrally at the diaphragm. The white extends beyond the anus and a black stripe extends from the urogenital area to the navel. The dorsal fin is generally bicolor with white in the top. The margin of the lobes of the caudal fin is also white. In the second type (*truei*), the white patch extends anteriorly to the pectoral fins. Other patterns of coloration, including gray, gray chestnut, all black, or white, have been documented.

DISTRIBUTION: *Phocoenoides dalli* is an endemic species of the cool to cold temperate North Pacific, common between 32 degrees north and 62 degrees north. Generally, it prefers waters with depths greater than 170 m, more than 100 nautical miles off the coastline, but are common in near-coastal waters of Southeast Alaska. They appear to prefer water temperatures between 3°C and 18°C. They have been reported in the waters of Baja California, as far south as 28 degrees north, from March to May (Leatherwood et al., 1983, 1988; Morejohn, 1979). La Niña oceanic conditions with cooler ocean temperatures may influence this species' occurrence in waters off Baja California. It has been registered in the area of the NP.

EXTERNAL MEASURES AND WEIGHT

TL: males 2.2 to 2.4 m; females to 2.1 m; calves about 85 to 100 cm at birth.
Weight: males 170–200 kg; females up to 180 kg; calves around 11 kg.

DENTAL FORMULA: 21–38/21–38.

NATURAL HISTORY AND ECOLOGY: Generally, Dall's porpoises are found in groups of 2 to 12 individuals, but there are reports of groups of several thousand individuals in the open ocean far from shore, which are probably feeding aggregations (Jefferson, 1988). Unlike other odontocetes, large herds seem not to acquire specific configurations. The largest groups apparently are constituted by subgroups that form and break fluidly (Jefferson, 1993). Groups of 100 individuals swimming in a row have been reported (Norris and Prescott, 1961). Dall's porpoise is one of the more abundant species in the North Pacific and Bering Sea. The population size might be of up to 2.8 million animals (Jones et al., 1987). Even though they are found all year long, the changes in their abundance indicate certain movements in winter and summer, in the first case, moving to the south and near the coast, while in the second to the north and away from the coast (Leatherwood et al., 1988). They mainly feed on epibenthic squid and mesopelagic species, as well as on fish that form large schools such as the anchovetas (*Engraulis mordax*) and herring (*Clupea harengus*); occasionally, they feed on crustaceans (Jefferson, 1988; Morejohn, 1979). Apparently, they reproduce throughout the year (Morejohn, 1979) with two maximum peaks of births in spring and summer. There is no evidence of births from November to January (Jefferson, 1989). Males reach sexual maturity between 4 and 6 years of age when they measure from 180 cm to 186 cm. Females mature sexually between 3.5 and 4.5 years of age when they reach 174 cm to 177 cm. The gestation period lasts approximately 11 months (Jones et al., 1987; Leatherwood et al., 1983). Lactation lasts probably between two and four months (Jefferson, 1990). It is a long-lived dolphin. The maximum age reported is of 22 years, though 15 years is more common (Kasuya, 1978 in Jefferson, 1988). They are preyed on by killer whales and some sharks. As one of the fastest marine species it is able to evade most predators.

CONSERVATION STATUS: Dall's porpoises are taken incidentally in Japanese pelagic and coastal salmon fisheries in the Northeast Pacific (Leatherwood et al., 1983). This mortality has declined and currently results in low numbers taken annually. It is listed in appendix II of CITES and is subject to special protection in the Mexican Official Regulation (SEMARNAT, 2010); it is protected by the Federal Law on Fisheries and article 420-I of the Federal Penal Code (Diario Oficial de la Federación, 2009). The IUCN (IUCN, 2010) conservation status is least concern.

Dall's porpoise (*Phocoenoides dalli*); male (*top*), female (*middle*). calf (*bottom*). Illustration: Pieter Folkens.

Order Chiroptera

Gerardo Ceballos, Joaquín Arroyo Cabrales, and David Vazquez

> *Its grotesque figure, which in the eyes of ordinary people appears as a strange mixture of mouse and bird; the way they escape from light during day taking shelter in the darkness of dens and caves, and the way they launch to obtain their food in the murkiness of night, avoiding masterfully many barriers that prevent their passage, has given them a unique site in the imagination of men of all time.*
>
> — Bernardo Villa, 1967

Bats are the second most diverse order of mammals, a number exceeded only by rodent species. There are 1,116 species, grouped into 18 families (Wilson and Reeder, 2005). They are distributed almost everywhere in the world, with the exception of the coldest regions such as poles and summits of the highest mountains (Nowak, 1999a). In Mexico nine families are represented: Emballonuridae, Noctilionidae, Mormoopidae, Phyllostomidae, Natalidae, Thyropteridae, Vespertilionidae, Antrozoidae, and Molossidae, five of which are unique and endemic to America. The order is characterized by being the only mammals that actually fly, with most species using echolocation, a mechanism of obstacle detection through the transmission and reception of high-frequency sounds (Eisenberg, 1981). They have nocturnal or crepuscular habits, but if they are disturbed they are able to fly during the day. In general, they become less active during rainy nights or full moons, as rain interferes with echolocation and a full moon makes them more vulnerable to predation (Morrison, 1978b). Temporal and spatial activity is influenced by environmental factors such as food availability and climate (Eisenberg, 1981; Fleming, 1988). At higher-latitude regions, where they face severe winters, species hibernate, remaining several months in a state of torpor, where breathing and cardiac rhythm lower to minimum levels. This reduces energy usage, enabling them to survive on their fat reserves. Other species such as *Tadarida brasiliensis* and *Leptonycteris yerbabuenae* carry out seasonal migrations of hundreds or thousands of kilometers in search of better weather conditions, feeding sites, or breeding sites (Ceballos et al., 1997; Villa-R., 1967). Bats have a wide range of eating habits. There are species that feed on insects (insectivores), fruits (frugivores), vertebrates (carnivores), pollen and nectar (pollinivores/nectar-feeding), and blood (hematophages). Feeding patterns of some species vary with the seasonality of resources; they feed on insects or nectar and pollen in different seasons of the year (Gardner, 1977; Wilson, 1971). Because of their abundance, bats are a group that plays an important role in the structure and function of ecosystems; they annually consume thousands of tons of insects, as well as pollinate and disperse plants of arid and tropical regions. Hematophagus species can cause severe economic losses by transmitting paralytic rabies to cattle (Turner, 1975; Villa-R., 1967). The diversity of roosting sites is also remarkable; they include caves, wells, the foliage of trees and palms, abandoned mines, abandoned houses and buildings, bridges, and culverts (Eisenberg, 1981; Nowak, 1999a). Bats form the largest aggregations of mammals; some caves in northern Mexico and the southern United States are home to 20 million free-tailed bats (*Tadarida brasiliensis*). Reproduction patterns have a close relationship with food availability (Wilson, 1979). There are species that breed throughout the year (continuous polyestrus pattern); other species have two periods of high annual birth rate (bimodal polyestrus pattern) and some breed only once a year (monoestrus pattern).

(*Opposite top*) Greater fishing bat (*Noctilio leporinus*). Specimen in captivity. Costa Rica. Photo: Scott Altenbach.
(*Bottom*) Trumpeter bat (*Musonycteris harrisoni*). Tropical dry forest. Callejones, Colima. Photo: Marco Tschapka.

Family Emballonuridae

The family Emballonuridae comprises 13 genera and 47 species that have a wide distribution in tropical regions of America, Asia, Africa, and Australia (Wilson and Reeder, 2005). In Mexico, the family is represented by 9 species of 6 genera (*Balantiopteryx, Centronycteris, Diclidurus, Peropteryx, Rhynchonycteris,* and *Saccopteryx*).

Balantiopteryx io Thomas, 1904

Thomas' sac-winged bat

William López-Forment and Guadalupe Tellez-Giron

SUBSPECIES IN MEXICO

B. io is a monotypic species.

DESCRIPTION: *Balantiopteryx io* is a small bat. It can be distinguished from *B. plicata* by its smaller size, a forearm of less than 39 mm, and thinner wing and leg bones. As in *B. plicata*, the sack gland is located in the center of the propatagium. The calcaneus does not reach the knee; the uropatagium is covered with hair as far as the beginning of the tail and it exceeds half the dorsal surface. The hair color of the back is dark and ventrally is paler (Arroyo-Cabrales and Jones, 1988; Kirkpatrick et al., 1975; Villa-R., 1967).

DISTRIBUTION: Thomas' sac-winged bat is distributed in tropical lowlands, from central Veracruz and Oaxaca to Belize and Guatemala (Hall, 1981). It has been registered in the states of CS, OX, TB, and VE.

EXTERNAL MEASURES AND WEIGHT

TL = 52 to 54.5 mm; TV = 14 to 14.8 mm; HF = 6 to 6.3 mm; EAR = 12 to 12.4 mm; FA = 36 to 38 mm.

Weight: 3.7 to 5 g.

DENTAL FORMULA: I 1/3, C 1/1, PM 2/2, F 3/3 = 32.

NATURAL HISTORY AND ECOLOGY: Thomas' sac-winged bat mainly roosts in caves in the darkest sites near the entrance, but are sometimes found in the deepest sites; they also use gaps, cracks, or roofs; colonies of 500 to 1,000 individuals have been found (Hall and Dalquest, 1963). Other bat species such as *Saccopteryx bilineata, Pteronotus parnellii, Glossophaga soricina, Artibeus jamaicensis,* and *Desmodus rotundus* have been found in the same shelter (Hall and Dalquest, 1963; Kirkpatrick et al., 1975; Schaldach, 1964). Little is known about its natural history.

VEGETATIONAL ASSOCIATIONS AND ELEVATION RANGE: *B. io* is found in tropical rainforests, from sea level up to 1,500 m (Emmons and Feer, 1997; Hall, 1981; McCarthy, 1982a).

CONSERVATION STATUS: We do not know the status of the populations of *B. io.* It is not listed in any risk category (SEMARNAT, 2010).

Peters' sac-winged bat

William López-Forment and Guadalupe Tellez-Giron

SUBSPECIES IN MEXICO
Balantiopteryx plicata plicata Peters, 1867
Balantiopteryx plicata pallida Burt, 1948

DESCRIPTION: *Balantiopteryx plicata* is a small bat, but larger than *B. io*. It has big eyes, prominent nostrils, rounded ears, and thin legs. A distinctive feature is the sack gland located in the middle of the propatagium, which varies in color, size, and texture with age and time of year; it is very conspicuous in males and in females is rudimentary (Arroyo-Cabrales and Jones, 1988; López-Forment, 1981). The tail is half covered by the uropatagium. The hair color is bluish-gray on the back, the abdomen is paler (Villa-R., 1967).

EXTERNAL MEASURES AND WEIGHT
TL = 63 to 70 mm; TV = 12 to 21 mm, HF = 6 to 9 mm; EAR = 12 to 16 mm; FA = 39 to 42.5 mm.
Weight: 4.5 to 7.1 g.

DENTAL FORMULA: I 1/3, C 1/1, PM 2/2, F 3/3 = 32.

NATURAL HISTORY AND ECOLOGY: This bat lives in colonies of up to 2,000 individuals (Bradbury and Vehrencamp, 1976; Davis, 1944; López-Forment, 1981).

Balantiopteryx plicata. Tropical dry forest. Chamela-Cuixmala Biosphere Reserve, Jalisco. Photo: Gerardo Ceballos.

DISTRIBUTION: *B. plicata* is a tropical species that is distributed from southern Sonora and Baja California, on the Pacific side and from the north of Veracruz to Chiapas, reaching central and southern Mexico, in the Gulf side, to Costa Rica in Central America (López-Forment, 1981). It has been registered in the states of BS, CL, CS, GR, HG, JA, MI, MO, MX, NY, OX, PU, SI, SL, SO, TB, TL, VE, and ZA.

The colonies are mixed and vary in size seasonally. In Guerrero, the proportion of males is apparently higher during the breeding season, which occurs in the dry season. It roosts in caves, mines, cracks, holes in trees, and buildings. Individuals usually perch near the exit in places with plenty of light. Shelters are shared with other species such as *Artibeus jamaicensis*, *Glossophaga soricina*, *Peropteryx macrotis*, and *Desmodus rotundus* (Álvarez, 1968; Davis, 1944; López-Forment, 1981; Starret and Casebeer, 1968). It is an aerial opportunistic insectivore that feeds on a variety of insects such as hymenopterans, beetles, and hemipterans, of a size not exceeding 8 mm to 9 mm. They leave their shelters during twilight to feed in groups in areas that they guard for about 6 to 8 minutes (López-Forment, 1981). Females are slightly larger than males. Copulation occurs from late January to mid-February, and gestation lasts approximately four and a half months. The female delivers a single young in late June and July (López-Forment, 1981). Among their predators are the barn owl (*Tyto alba*), the pygmy skunk, the coati, and the domestic cat (López-Forment, 1981).

VEGETATIONAL ASSOCIATIONS AND ELEVATION RANGE: *B. plicata* is found in arid and semi-arid regions with strong seasonality, in scrublands and in deciduous and tropical rainforests (Dolan and Carter, 1979; López-Forment, 1981; Ramírez-Pulido et al., 1977; Villa-R., 1967). It has been found from sea level up to 1,500 m (López-Forment, 1981).

CONSERVATION STATUS: *B. plicata* is an abundant species, which persists in disturbed areas, and is even found in bridges (Reid, 1997). It is not at risk of extinction.

Centronycteris centralis Thomas, 1912

Thomas's shaggy bat

Erika Marce Santa and Lorena Morales Perez

SUBSPECIES IN MEXICO

C. centralis is a monotypic species.

Previously it was regarded as subspecies of *C. maximiliani*; however, Simmons and Handley (1998) assigned it as a separate species due to its distribution and its morphological skull and teeth differences.

DESCRIPTION: *Centronycteris centralis* is a small bat. The color of the back is gray brown, while the hairs that are near the eyes and the interfemoral membrane are reddish; in contrast with the dorsal hair, the abdomen is yellowish. Ears are long, sickle shaped, and pointed (Reid, 1997). The hair is long and smooth with an appearance similar to that of *Saccopteryx bilineata*, but lacks the white lines on the back. The tail is long and half covered by the uropatagium. It has no sacks in the wings and the glandular membrane originates from the metatarsus (Eisenberg, 1989; Hall, 1981; Villa-R., 1967). Unlike in *C. maximiliani* the basisphenoid fossa are relatively short and do not extend anteriorly in the pterygoid process. It has a distinctive notch in the lateral posterior edge of the palate, and in contrast to *C. maximiliani*, *C. centralis* lacks a dorsal lump at the base of the postorbital process; the nostrils are not embedded in the maxilla and extend anteriorly; and the upper canines are not inclined (Simmons and Handley, 1998).

Centronycteris centralis. Tropical rainforest. Photo: Gerardo Ceballos.

EXTERNAL MEASURES AND WEIGHT

LT = 60 to 93 mm; CV = 18 to 40 mm; O = 11 to 20; P = 7 to 9 mm; AN = 42 to 49 mm. Weight: 4 to 6 g.

DENTAL FORMULA: I 1/3, C 1/1, P 2/2, F 3/3 = 32.

NATURAL HISTORY AND ECOLOGY: *C. centralis* roosts in tree hollows. Their flight is slow and "maneuverable," which allows them to hunt insects among trees and in natural and man-made canopy gaps. Time of capture and sighting give evidence about their activity, which is believed to begin before darkness (Eisenberg, 1989; Emmons and Feer, 1997; Hall, 1963; Nowak, 1999a; Simmons and Handley, 1998). Very little is known about the reproductive biology of *C. centralis*; in Nicaragua and Costa Rica three females with an embryo, respectively, were collected during the month of May and in Ecuador a female with an embryo was collected in March (Nowak, 1999a; Simmons and Handley, 1998).

VEGETATIONAL ASSOCIATIONS AND ELEVATION RANGE: *C. centralis* is mainly found in tropical rainforests (Simmons and Handley, 1998; Villa-R., 1967). It is mainly distributed at low altitudes ranging from 0 to 300 masl (Eisenberg, 1989; Hall, 1981), but in Panama, Ecuador, and Peru it has been reported between 1,160 m and 1,450 m (Simmons and Handley, 1998).

CONSERVATION STATUS: *C. centralis* is classified as rare in the Mexican Official Regulation (SEMARNAT, 2010). In Mexico, the number of records is extremely low; hence, it is necessary to update the status of this bat's populations to ensure its conservation.

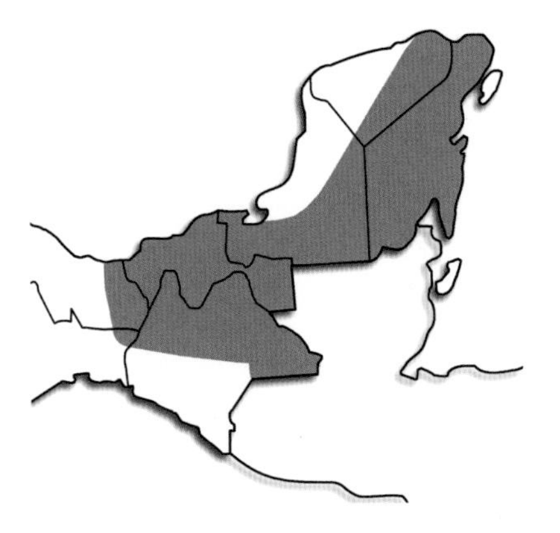

DISTRIBUTION: *C. centralis* is distributed from southern Veracruz (excluding the Yucatan Peninsula) south to Ecuador, and western and central Peru (Hall, 1981; Jones et al., 1977; Koopman, 1993; Simmons and Handley, 1998; Villa-R., 1966). It has been registered in the states of CA, CS, QR, TB, and VE.

Northern ghost bat

Gerardo Ceballos

SUBSPECIES IN MEXICO

Diclidurus albus virgo Thomas, 1903

Some authors (Goodwin, 1969; Ojasti and Linares, 1971) have considered this subspecies as a species, but this has not been generally accepted (Ceballos and Medellín, 1988; Wilson and Reeder, 2005).

DESCRIPTION: *Diclidurus albus* is of medium size with simple facial traits. It differs from other bat species in Mexico because it is white to grayish and because it has a glandular sac in the uropatagium. The uropatagium is large and triangular.

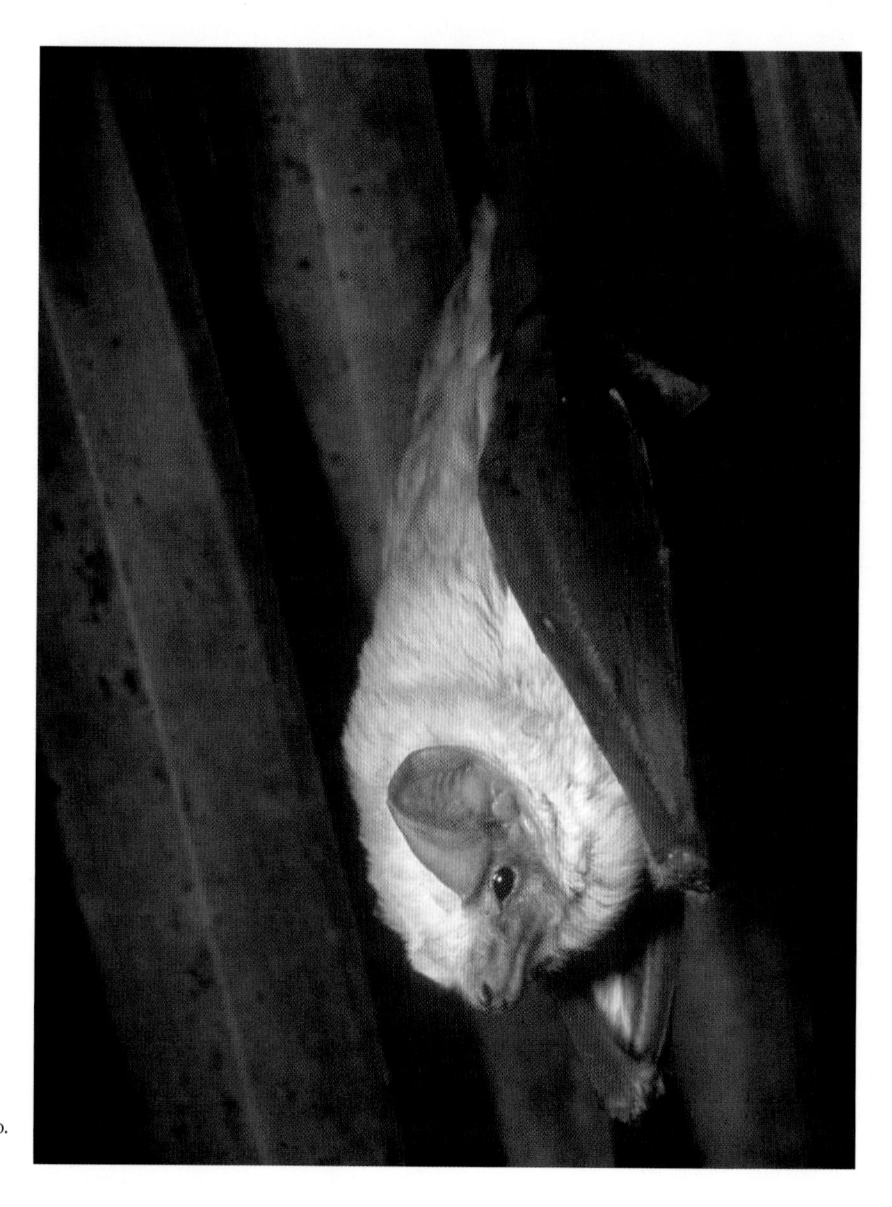

Diclidurus albus. Palm forest. Chamela-Cuixmala region, Jalisco. Photo: Gerardo Ceballos.

EXTERNAL MEASURES AND WEIGHT
TL = 86 to 103 mm; TV = 18 to 22 mm; HF = 10 to 12 mm; EAR = 16 to 24 mm; FA = 63 to 69 mm.
Weight: 17 to 24 g.

DENTAL FORMULA: I 1/3, C 1/1, PM 2/2, F 3/3 = 32.

NATURAL HISTORY AND ECOLOGY: *D. albus* is a species closely associated with tropical vegetation where palm trees are abundant (Ceballos and Medellín, 1988). During the day it roosts under the fronds of native species like the coquito palm (*Orbygnia cohune*) and chocho (*Astrocaryum mexicanum*); it has also been found in coconut palms (*Cocos nucifera*). Its diet is based on small insects, mostly moths. They are usually solitary, but in the breeding season, from November to June, it is possible to see up to four individuals in the same roosting site, usually separated (Ceballos and Medellín, 1988; Ceballos and Miranda, 1986). Apparently, they are monoestric; a single young is born between January and June (Ceballos and Medellín, 1988; Sanchez and Chavez, 1985). Apparently, on the coast of Jalisco, this species migrates regionally.

VEGETATIONAL ASSOCIATIONS AND ELEVATION RANGE: *D. albus* inhabits palm groves, deciduous and tropical rainforests, and disturbed sites with palms. It is found from sea level up to 1,500 m, but in Mexico it has only been collected up to altitudes of 200 masl.

CONSERVATION STATUS: In Mexico, *D. albus* is considered rare (Ceballos and Medellín, 1988). Apparently, it tolerates anthropogenic environmental changes, so it is possible that it is not in danger of extinction. In order to determine its dependence on primary forest, however, it is necessary to evaluate in detail its current distribution and habitat requirements. This information is essential to establish its conservation status.

DISTRIBUTION: *D. albus* is widely distributed in the Neotropical region, and it can be found from Mexico to Colombia and Venezuela in South America. In Mexico it is distributed in tropical regions on both sides, from Nayarit on the Pacific and Veracruz in the Gulf to Chiapas and Quintana Roo (Ceballos and Medellín, 1998; Sánchez-Hernández et al., 2002). It has been registered in the states of CL, CS, JA, MI, NY, OX, QR, and VE.

Peropteryx kappleri Peters, 1867

Greater dog-like bat

Ivan Castro-Arellano and Erika Marce Santa

SUBSPECIES IN MEXICO
Peropteryx kappleri kappleri Peters, 1867

DESCRIPTION: *Peropteryx kappleri* is the largest species of the genus. There are two coloration phases of the back, a dark brown and a light brown; the abdomen is lighter (Eisenberg, 1989; Villa-R., 1967). It is very similar to *P. macrotis*, although it can be distinguished by its larger ears and their black color (Medellín et al., 1997). The length of the skull varies from 16 mm to 17.8 mm (Villa-R., 1967).

EXTERNAL MEASURES AND WEIGHT
TL = 73.3 mm; TV = 13.71 mm, HF = 11.14 mm; EAR = 18.03 mm; FA = 59.58 mm.
Weight: 8.13 g.

DISTRIBUTION: *P. kappleri* ranges from southern Veracruz and Oaxaca, excluding the Yucatan Peninsula, to eastern Brazil and Peru (Koopman, 1993). It has been registered in the states of CA, CS, OX, TB, and VE.

DENTAL FORMULA: I 1/3, C 1/1, PM 2/2, F 3/3 = 32.

NATURAL HISTORY AND ECOLOGY: *P. kappleri* inhabits caves or cliffs of limestone, usually occupying the interstices between large rocks; they have been also found perched in hollow logs, close to the ground, forming compact groups of up to seven individuals (Eisenberg, 1989; Emmons and Feer, 1997; Medellín, 1993). It has been classified as an aerial insectivore because it usually uses open spaces in the upper canopy, rivers, and rarely farming areas (Emmons and Feer, 1997; Medellín, 1993). We are unaware of the details of this bat's reproduction; however, in *P. macrotis*, the number of embryos per female is one (Jones et al., 1973). The mating system is based in monogamous couples in which the male defends the female (Eisenberg, 1989).

VEGETATIONAL ASSOCIATIONS AND ELEVATION RANGE: The greater dog-like bat is mostly found in tropical rainforests (Eisenberg, 1989), from sea level up to 750 m (Hall, 1981).

CONSERVATION STATUS: Because of its dependence on primary jungle in Mexico, *P. kappleri* is considered rare. Details on its current situation are still unknown (SEMARNAT, 2010).

Peropteryx macrotis (Wagner, 1843)

Lesser dog-like bat

Erika Marce Santa and Ivan Castro-Arellano

SUBSPECIES IN MEXICO

Peropteryx macrotis macrotis (Wagner, 1843)

DISTRIBUTION: The lesser dog-like bat's range extends from the coast of Michoacan and east of Veracruz to Paraguay and south-eastern Brazil (Koopman, 1993; Sánchez et al., 1999). It has been registered in the states of CA, CS, GR, MI, OX, QR, TB, VE, and YU.

DESCRIPTION: *Peropteryx macrotis* is a small bat. The color of the body varies from dark brown to light gray and red. The hair is long with a solid color. The ears are dark and relatively large. It has a wing sack gland located in the propatagium. The skull is small and with an acute angle between the face and the head box. The upper incisors are very small and simple and the lower incisors are small and tricuspid (Eisenberg, 1989; Goodwin and Greenhall, 1961; Medellín et al., 2008; Villa-R., 1967). The dark-colored specimens are similar to *Saccopteryx bilineata*, but *P. macrotis* does not have white stripes on the back. It can be distinguished from *P. kappleri* by its smaller size (forearm length: 38 mm to 48 mm; Emmons and Feer, 1997; Villa-R., 1967).

EXTERNAL MEASURES AND WEIGHT

TL = 62.46 mm; TV = 14.17 mm, HF = 8.79 mm; EAR = 15.62 mm; FA = 48.29 mm. Weight: 5.61 g.

DENTAL FORMULA: I 1/3, C 1/1, PM 2/2, F 3/3 = 32.

NATURAL HISTORY AND ECOLOGY: *P. macrotis* inhabits caves or the interstices of limestone cliffs. They are usually found in the twilight zone, but they have been found in underground passages and in archaeological ruins perched on hollow logs (Emmons and Feer, 1997; Hatt and Villa-R., 1950; Kuns and Tashian, 1954; Villa-R., 1967). Females with embryos have been caught in April, and female infants

Peropteryx macrotis. Tropical rainforest. Photo: M.B. Fenton.

in August. The registered number of embryos is one (Jones et al., 1973). This species is insectivorous and feeds mainly on small beetles and dipterans during flight. They usually forage in mature rainforests but also in open spaces of plantations and near illuminated human settlements (Davis et al., 1964; Emmons and Feer, 1997; Goodwin and Greenhall, 1961). It has been captured along with individuals of *Balantiopteryx plicata*, *Glossophaga soricina*, and *Artibeus jamaicensis* (Ramírez-Pulido et al., 1977).

VEGETATIONAL ASSOCIATIONS AND ELEVATION RANGE: *P. macrotis* mostly inhabits tropical rainforests from 150 masl to 1 500 masl (Dalquest and Hall, 1949; Eisenberg, 1989; Hall, 1981).

CONSERVATION STATUS: *P. macrotis* is not included in any list of endangered species, but little is known of the status of its populations in Mexico (Ceballos et al., 2002).

Rhynchonycteris naso (Wied-Neuwied, 1820)

Proboscis bat

Heliot Zarza and Gerardo Ceballos

SUBSPECIES IN MEXICO

R. naso is a monotypic species.

DESCRIPTION: These are small bats with an elongated snout and with the nose extending beyond the tip of the lower lip, which gives them a pointed appearance (Villa-R., 1967). A distinctive character of the species is the lack of sack in the wing membrane, which is easily distinguishable from other taxa of the same family. On

DISTRIBUTION: *R. naso* is distributed throughout the tropical lowlands of southeastern Mexico, from northern Veracruz, Oaxaca, and the Yucatan Peninsula to northern Peru and central Brazil (Hall, 1981; Plumpton and Jones, 1992). It has been registered in the states of CA, CS, OX, QR, TB, and VE.

the back of the forearm there are five to seven tufts of yellow fur distributed at regular intervals (Medellín et al., 2008). The back is brown to gray, with two white or gray lines. The ventral side is brown-gray. The fur is bicolor with a dark base and lighter tip; it is soft and dense (Nowak, 1999a). The baculum is long and large (Brown et al., 1971). The brain is small in relation to the bat's body weight (Plumpton and Jones, 1992).

EXTERNAL MEASURES AND WEIGHT
TL = 49.0 to 59.18 mm; TV = 12 to 16 mm; HF = 7.18 to 7.45 mm; EAR = 11 to 15 mm; FA = 35 to 41 mm.
Weight: 2 to 4 g.

DENTAL FORMULA: I 1/3, C 1/1, PM 2/2, F 3/3 = 32.

NATURAL HISTORY AND ECOLOGY: These animals form colonies of 3 to 45 individuals, but generally include between 5 to 11 individuals; they roost in hollow logs, tree trunks, rocks, caves, or vegetation near riverbanks of slow motion. Each colony can occupy three to six roosting sites; individuals regularly move between different shelters depending on the needs of the colony (Nowak, 1999a). In general, they are not associated with other bats. When disturbed they fly aligned to another roosting site but after a while they return to the original one (Plumpton and Jones, 1992). The foraging area comprises approximately 1.1 ha. When foraging, they usually fly over bodies of water at a height of 3 m (Bradbury and Vehrencamp, 1977). Their diet consists of small insects such as mosquitoes, beetles, and moths (Nowak, 1999a). Females are able to reproduce at 18 months of age, and are likely to have a bimodal reproductive strategy. In Mexico and Central America, offspring are born in the dry season, shortly before the rains start; however, in South America another birth peak has been reported in the rainy season (Medellín, 1986). The population density has been estimated between 3 and 12 individuals/ha (Bradbury and Vehrencamp, 1976).

VEGETATIONAL ASSOCIATIONS AND ELEVATION RANGE: *Rhynchonycteris naso* is common in tropical rainforests, from sea level to 300 m (Plumpton and Jones, 1992).

CONSERVATION STATUS: Throughout its range, *R. naso* is a relatively common species. We do not know the impact of activities such as deforestation on its populations (Plumpton and Jones, 1992).

Rhynchonycteris naso. Tropical rain forest. Selva Lacandona, Chiapas. Photo: Gerardo Ceballos.

Greater sac-winged bat

Joaquin Arroyo-Cabrales

DISTRIBUTION: *S. bilineata* has Neotropical affinities; they are known from Mexico to Bolivia and southeastern Brazil. The subspecies *S. bilineata centralis* is distributed in low and warm lands of Mexico, from Jalisco in the Pacific slope, and Veracruz in the Gulf, to Chiapas. It has been registered in the states of CA, CL, CS, GR, JA, MI, OX, QR, TB, VE, and YU.

SUBSPECIES IN MEXICO

Saccopteryx bilineata centralis Thomas, 1904

Saccopteryx bilineata villai Álvarez and Gonzalez Ruiz, 2000

Until the mid-1960s, it was believed that the subspecies of *S. bilineata*, present in Mexico, was the typical *S. bilineata bilineata*. Álvarez (1968) noted that in Mexico these bat populations were the subspecies *S. bilineata centralis*. More recently, Polaco and Muñiz-Martinez (1987) analyzed a considerable number of specimens collected in Mexico, supporting the proposal of Álvarez (1968). Álvarez-Castañeda and Álvarez (1991) pointed to the possibility that the populations of *S. bilineata* in the Pacific Coast, which goes along Mexico and Central America, belong to a subspecies not mentioned, which was recently described by Álvarez and González-Ruíz (2000).

DESCRIPTION: *Saccopteryx bilineata* is a small bat that does not have a nose leaf. It has a sack in the glandular wing membrane near the forearm and elbow, which opens upward; males usually have more developed glandular sacks than females. It differs from another species of the genus in Mexico (*S. leptura*) because they are larger and dark brown, almost black, with two well-marked white lines that run on the back along its entire length and up to the uropatagium. There is no difference in color between males and females (Polaco and Muñiz-Martinez, 1987), but there is sexual dimorphism in some characters such as the sack gland.

EXTERNAL MEASURES AND WEIGHT

TL = 65 to 71 mm; TV = 10 to 21 mm, HF = 8 to 14 mm; EAR = 11 to 19 mm; FA = 37.7 to 45.2 mm.

Weight: 5.0 to 8.0 g.

DENTAL FORMULA: I 1/3, C 1/1, PM 2/2, F 3/3 = 32.

NATURAL HISTORY AND ECOLOGY: The largest number of specimens have been collected on trees in tropical forests; they are occasionally found in rolled banana leaves (*Musa* sp.), sewers, caves, and buildings (Villa-R., 1967). They are basically insectivorous; the groups are very social and consist of 1 male and several females (up to 8) that live in colonies of 40 to 50 individuals. Males actively defend their harem and territory. The roosting site is maintained for a long time, but the colonies move according to the distribution of insects. Their flight is slow, and they usually fly 1 or 2 m above the surface of waterways, roads, or canopy gaps, usually protected by the shade (Bradbury, 1983). Reproduction is highly synchronized with the rainy season; young are born from late May to July and early August. On average, females give birth to one young. The mothers, who do not leave their young even when foraging, carry out the care of the offspring; pregnant females have been found from February to May (Bradbury, 1983; Ceballos and Miranda, 1986; Hall and Dalquest, 1963; Vehrencamp and Bradbury, 1977).

Saccopteryx bilineata. Tropical semi-green forest. Chamela-Cuixmala Biosphere Reserve, Jalisco. Photo: Gerardo Ceballos.

VEGETATIONAL ASSOCIATIONS AND ELEVATION RANGE: *S. bilineata* mainly inhabits tropical rainforests, but is also found in deciduous or subdeciduous forests. It has been found from sea level up to 800 m, but most localities are located within 200 masl.

CONSERVATION STATUS: The greater sac-winged bat is not too abundant in most of its distribution. It is difficult to assess its current situation with the limited data available. It is therefore, necessary to carry out assessments to determine its vulnerability.

Saccopteryx leptura (Schreber, 1774)

Lesser sac-winged bat

Joaquin Arroyo-Cabrales

SUBSPECIES IN MEXICO

S. leptura is a monotypic species.

DESCRIPTION: *Saccopteryx leptura* is a small bat, without a nose leaf and with a glandular sack that opens upward in the wing membrane near the forearm and elbow. It differs from another species of the genus (*S. bilineata*) because it is smaller, is brown, and has a line on the back that is more dimly marked.

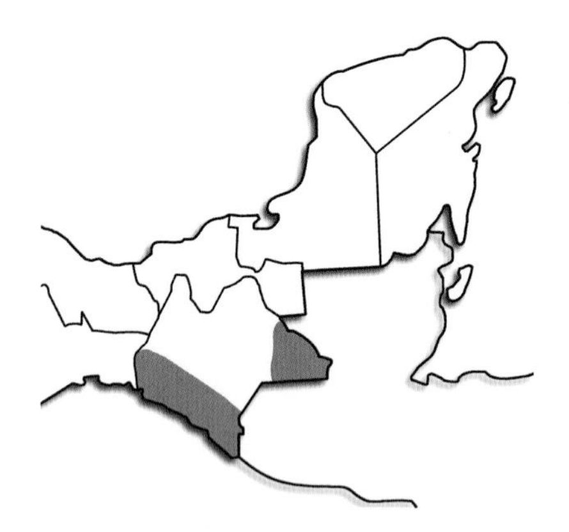

DISTRIBUTION: The distribution of *S. leptura* ranges from southern Mexico through Central America to Bolivia and southeastern Brazil. In Mexico, however, they are known only from three localities in Chiapas (Carter et al., 1966; R. Medellín, pers. comm.; Polaco et al., 1992). It has only been registered in the state of CS.

EXTERNAL MEASURES AND WEIGHT

TL = 55 to 62 mm; TV = 41 to 44 mm, HF = 6 to 9 mm; EAR = 11 to 15 mm; FA = 38.5 mm.
Weight: 3.5 g.

DENTAL FORMULA: I 1/3, C 1/1, PM 2/2, F 3/3 = 32.

NATURAL HISTORY AND ECOLOGY: Few studies relate to the biology of this species. We know more about their behavioral habits. Lesser sac-winged bats roost in not too confined places and do not defend them; in fact, they frequently change roosting sites. They fly in groups of up to nine individuals. Since they are monogamous, it is more common to find one female and one male (Bradbury and Vehrencamp, 1977). The young are born in May or November (Bradbury, 1983), and are carried by their mothers for 10 to 15 days, after which they are capable of flying; weaning occurs at two and a half weeks. They are insectivorous, and feed on small insects they catch during flight.

VEGETATIONAL ASSOCIATIONS AND ELEVATION RANGE: In Mexico, specimens have been collected in tropical deciduous forests and tropical rainforests at elevations of 35 m to 120 m. Elsewhere, it has been collected in tropical forests near streams or in swamps. In general, they are found in areas of less than 500 m of elevation.

CONSERVATION STATUS: Throughout its distribution, *S. leptura* is a rare species. To determine its current situation, further studies about its biology are required, especially about distribution and abundance. Its restricted distribution in tropical rainforests makes it necessary to emphasize the conservation of that habitat.

Family Phyllostomidae

The family Phyllostomidae is one of the most diverse and distinctive families of bats, endemic to the Americas and mainly found in tropical regions throughout the continent. It comprises 55 genera and 160 species (Wilson and Reeder, 2005). In Mexico there are 55 species of 37 genera (*Desmodus*, *Diaemus*, *Diphylla*, *Chrotopterus*, *Trachops*, *Vampyrum*, *Glyphonycteris*, *Lampronycteris*, *Lonchorhina*, *Lophostoma*, *Macrophyllum*, *Macrotus*, *Micronycteris*, *Mimon*, *Phylloderma*, *Phyllostomus*, *Tonatia*, *Trinycteris*, *Anoura*, *Choeroniscus*, *Choeronycteris*, *Glossophaga*, *Hylonycteris*, *Leptonycteris*, *Lichonycteris*, *Musonycteris*, *Artibeus*, *Carollia*, *Centurio*, *Chiroderma*, *Dermanura*, *Enchisthenes*, *Platyrrhinus*, *Sturnira*, *Uroderma*, *Vampyressa*, and *Vampyrodes*). The genus *Musonycteris* of the low deciduous forests of the Pacific coast of Mexico is endemic.

Subfamily Macrotinae

Macrotinae is a small subfamily containing only one genus (*Macrotus*) and two species, both of which are known from Mexico.

Macrotus californicus Baird, 1858

California leaf-nosed bat

Guadalupe Tellez-Giron and William López-Forment

SUBSPECIES IN MEXICO

M. californicus is a monotypic species.

Geographic and genetic variations found in populations where the genus *Macrotus* is distributed clearly indicate the presence of two species: *M. californicus* and *M. waterhousii* (Davis and Baker, 1974).

DISTRIBUTION: *M. californicus* is found from Southern California in the United States south to Sinaloa in Mexico; it occupies the entire Baja California Peninsula, Sonora, and the northern tip of the coastal region of Sinaloa (Davis and Baker, 1974; Jones et al., 1972). It has been registered in the states of BS, SI, SL, SO, and TA.

DESCRIPTION: *Macrotus californicus* is a medium-sized bat with an elongated face, a long and prominent nasal leaf, and large eyes. The ears are large and long and extend beyond the face when they are carried forward and are united by an inter-auricular membrane. The tragus is long and does not reach more than one-half of the ear. All over the body, the hair is dense and long, dark brown-gray, and lighter on the base of the back. The uropatagium is naked (Anderson, 1969a; Hall, 1981; Harris, 1991). *M. californicus* has a lighter coloration, and the ears and tympanic bulla are more developed than in *M. waterhousii* (Anderson and Nelson, 1965).

EXTERNAL MEASURES AND WEIGHT

TL = 85 to 103 mm; TV = 25 to 42 mm; HF = 11 to 18 mm; EAR = 30 to 35 mm; FA = 48 to 51 mm.

Weight: 9.5 to 14 g.

DENTAL FORMULA: I 2/2, C 1/1, PM 2/3, M 3/3 = 34.

NATURAL HISTORY AND ECOLOGY: *M. californicus* roosts in rocky sites, caves, mines, tunnels, and, occasionally, buildings or old structures. It perches on sites away from the entrance, where the temperature is lower than on the outside and with a humidity of more than 50%. These shelters are used for short periods; some groups are likely

Macrotus californicus. Photo: M.B. Fenton.

to migrate to hotter places during the winter (Harris, 1991). They are found in large groups. Females form maternity colonies in the summer, while males are associated in small groups in different roosting sites (Woloszyn and Woloszyn, 1982). Their nocturnal activity takes place one or two hours after sunset, and a second peak of activity occurs at approximately 22:00 hrs. These bats do not hibernate, but they can lower their body temperature for short periods (Harris, 1991). They feed mainly on the soft parts of a wide variety of nocturnal insects such as orthopterans, moths (Sphingidae, Noctuidae), beetles, and cycads that have an average size of 40 mm to 60 mm. Their prey is consumed within their refuge, where one can find the remains of the wings, legs, and other hard parts. They can occasionally consume fruits (Anderson, 1969a; Burt, 1938; Gardner, 1977; Vaughan, 1959). They hunt near the substrate and capture their prey close to the ground or on vegetation (Harris, 1991; Ross, 1967; Wilson, 1973a). Males do not reach sexual maturity until their second year of life. Reproduction takes place from September to November when copulation occurs. Females have delayed implantation; the embryo has a slow growth during the winter, followed by a rapid growth in March, ending in the birth of a young between May and June (Asdell, 1964); lactation lasts about a month (Cockrum, 1973; Woloszyn and Woloszyn, 1982).

VEGETATIONAL ASSOCIATIONS AND ELEVATION RANGE: *M. californicus* inhabits riparian forests and xeric scrublands (Harris, 1991; Woloszyn and Woloszyn, 1982); they are found from sea level up to 1,000 m (Findley and Jones 1965).

CONSERVATION STATUS: Although they are found in abundant groups, California leaf-nosed bats are very sensitive to disturbance of the sites where they live and for long periods do not return to places that have been disturbed (Harris, 1991). They are not considered in any risk category, but it is necessary to learn the current status of their populations.

Waterhouse's leaf-nosed bat

Xavier Lopez and Rodrigo A. Medellín

SUBSPECIES IN MEXICO
Macrotus waterhousii bulleri H. Allen, 1890
Macrotus waterhousii mexicanus Saussure, 1860

DESCRIPTION: *Macrotus waterhousii* is a medium-sized bat. The fur on the back varies from pale gray to dark brown, and on the abdomen varies from brownish to brown, usually being silver (Nowak, 1999a). Its ears are large and are united across the crown of the head by an inter-auricular membrane. The tragus is long and pointed; the uropatagium is wide and surrounds the tail, which extends beyond this membrane (Villa-R., 1967); the nose leaf is erect and lanceolate (7 mm; Anderson, 1969a). Chromosomal studies concluded that *Macrotus* is more closely related to Mormoopidae and Noctilionidae than to other phyllostomids (Baker, 1979). Davis and Baker (1974) separated *M. waterhousii* from *M. californicus* based on chromosomal characters. *Macrotus* can be distinguished from *Micronycteris* by its ears and longer tail.

DISTRIBUTION: *M. waterhousii* is distributed from Sonora on the Pacific side south to Guatemala. It is also found in some Caribbean islands (Hall, 1981; Ramírez-Pulido, 1988, 1993). It has been registered in the states of CH, CL, CS, DF, DU, GT, GR, JA, HG, MI, MO, MX, NY, OX, PU, QE, SI, SO, YU, and ZA.

EXTERNAL MEASURES AND WEIGHT
TL = 77 to 108 mm; TV = 25 to 42 mm; HF = 51 to 58 mm; EAR = 26 to 33 mm; FA = 44.7 to 58 mm.
Weight: 12 to 19 g.

DENTAL FORMULA: I 2/2, C 1/1, PM 2/3, M 3/3 = 34.

NATURAL HISTORY AND ECOLOGY: These bats can form colonies of hundreds of individuals. They are mainly cave dwellers, although they have been found in mines and abandoned buildings (Arita, 1993a, b). They do not require total darkness to establish their perching places, and have been found just 10 m from the partially illuminated entrances of caves and buildings (Baker et al., 1979; Nowak, 1999a). They feed on insects either in flight or on substrates; occasionally they eat fruits. Females have a delayed embryonic development (Baker et al., 1979). Ovulation, fertilization, and insemination occur in early autumn. Embryonic development is slow and gestation lasts eight months. Usually a single young is born per year, but females have been found with two embryos. These bats have been found in caves associated with *Desmodus*, *Glossophaga*, *Leptonycteris*, *Mormoops*, and *Pteronotus* (Barbour and Davis, 1969).

VEGETATIONAL ASSOCIATIONS AND ELEVATION RANGE: *M. waterhousii* inhabits tropical deciduous forests, thorny forests, pine-oak forests, tropical subdeciduous forests, and pastures, from sea level up to 1,400 m (Hall, 1981).

CONSERVATION STATUS: At the moment, the information available is inadequate and insufficient to determine the conservation status of *M. waterhousii*. The species seems to be as abundant in regions changed by humans as in areas under natural conditions.

Subfamily Micronycterinae

The subfamily Micronycterinae contains six genera and eighteen species, with four genera and five species known from Mexico.

Glyphonycteris sylvestris Thomas, 1896

Tricolored big-eared bat

Rodrigo A. Medellín

SUBSPECIES IN MEXICO
G. sylvestris is a monotypic species.

DISTRIBUTION: *G. sylvestris* is found in the Neotropical region from Mexico to southern Brazil (Hall, 1981; Koopman, 1982). In Mexico it inhabits the tropical Pacific Ocean side from Nayarit, and in the Gulf slope from Veracruz to southeastern Chiapas (Hall, 1981; Sanchez et al., 1999). It has been registered in the states of CL, CS, JA, MI, NY, OX, TB, and VE.

DESCRIPTION: *Glyphonycteris sylvestris* is a medium-sized bat. The ears are large and rounded. A thin but prominent lancet forms the nose leaf; it resembles a horseshoe whose sides are evident but its base is indistinguishable from the lower lip. The lower lip shows a furrow in the form of a V. The dorsal hair is long and wooly and has three bands of color clearly distinguishable: dark base, light center, dark tip. This pattern of coloration is unique in the genus (Emmons and Feer, 1990), and shared by the bats of the genera *Carollia* and *Chiroderma*, which can be distinguished by larger ears, the hairless forearm, and the upper lip's furrow in the form of a V. The hair of the belly is bicolor and lighter than that of the back. The tail is short and does not extend beyond the interfemoral membrane.

EXTERNAL MEASURES AND WEIGHT
TL = 53 to 73 mm; TV = 9 to 14 mm, HF = 11 to 12 mm; EAR = 20 to 22 mm; FA = 37 to 42 mm.
Weight: 7 to 11 g.

DENTAL FORMULA: I 2/2, C 1/1, PM 2/3, M 3/3 = 34.

NATURAL HISTORY AND ECOLOGY: Like other species of the genus, *G. sylvestris* probably feeds on insects and fruit (Goodwin and Greenhall, 1961). It roosts in hollow trees and natural caves, forming colonies of up to 75 bats (Goodwin and Greenhall, 1961), but generally smaller (Hall and Dalquest, 1963; Villa-R., 1967). The limited data on their reproduction suggest that they breed during the rainy season (Wilson, 1979).

VEGETATIONAL ASSOCIATIONS AND ELEVATION RANGE: *G. sylvestris* has been captured in the tropical rainforest (Handley, 1976) and other vegetation types, from sea level to 760 m (Hall and Dalquest, 1963).

CONSERVATION STATUS: *G. sylvestris* is not included in any official list of endangered species or in studies of rare or endangered species. Still, very little is known about this species. Given its wide distribution, it does not seem to face serious or immediate conservation problems.

Orange throated bat

Rodrigo A. Medellín

SUBSPECIES IN MEXICO
L. brachyotis is a monotypic species.

DESCRIPTION: *Lampronycteris brachyotis* is a medium-sized bat. The ears are relatively short and sharp, with a rounded lobe on the internal lateral edges. The nose leaf is small but conspicuous, with a thin and sharp lancet. The upper lip has two large pads forming a V (Medellín et al., 1985). The hair on the back is relatively short and dark brown with olive tones; specimens of northern distribution are orange-brown. In most adults, the abdomen is orange or reddish-yellow, which allows easy identification of this species. The tail is short and is fully included in the interfemoral membrane. The edge of the membrane is supported by a calcaneus, which has approximately the same length as the leg (Medellín et al., 1985).

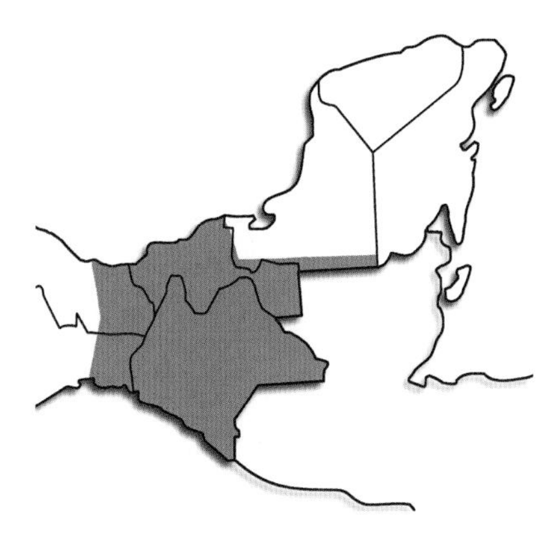

DISTRIBUTION: *L. brachyotis* inhabits the wetlands of the Neotropics from the Isthmus of Tehuantepec in Mexico to the Amazon basin in Brazil. In Mexico, it has been found in the southern tip of the Gulf basin from Veracruz to Chiapas (Hernández-Huerta et al., 2000; Medellín et al., 1985). It has been registered in the states of CA, CS, OX, and VE.

EXTERNAL MEASURES AND WEIGHT
TL = 57 to 75 mm; TV = 7 to 14 mm, HF = 11 to 14 mm; EAR = 12 to 19 mm; FA = 39 to 43 mm.
Weight: 9 to 15 g.

DENTAL FORMULA: I 2/2, C 1/1, PM 2/3, M 3/3 = 34.

NATURAL HISTORY AND ECOLOGY: The orange throated bat is predominantly insectivorous, but if insects are not very abundant, it can also consume fruit, nectar, and pollen (Bonaccorso, 1979; Medellín et al., 1985). It feeds on beetles, hymenopterans, dipterans, homopterans and arachnids, pollen of balsa (*Ochroma lagopus*), and unidentified fruits (Bonaccorso, 1979; Howell and Burch, 1974; Humphrey et al., 1983; Medellín et al., 1983). In general, they form small colonies of up to 10 individuals; howev-

Lampronycteris brachyotis. Photo: Tropical rainforests, La Tirimbina, Costa Rica. Photo: Gerardo Ceballos.

er, Medellín et al. (1983) studied a colony of at least 300 animals in Veracruz. Colonies roost in hollow trees, caves, mines, and archaeological ruins with little illumination. Apparently, they have a bimodal reproductive pattern; the first pulse of births coincides with the onset of the rainy season (Bonaccorso, 1979; Medellín et al., 1985).

VEGETATIONAL ASSOCIATIONS AND ELEVATION RANGE: In general, the orange throated bat has only been caught in the tropical rainforest from sea level to 525 m (Handley, 1976; Medellín et al. 1985).

CONSERVATION STATUS: *L. brachyotis* has not been included in any official list of endangered species. R. Medellín and H. Arita (pers. obs.) classify it as fragile, given its apparent dependence on the undisturbed tropical rainforest. In an analysis of local abundance and distribution, Arita (1993a, b) classifies it as rare because its distribution is relatively restricted and the local abundance is low.

Micronycteris microtis Miller, 1898

Common big-eared bat

Guadalupe Tellez-Giron and Gerardo Ceballos

SUBSPECIES IN MEXICO

Micronycteris microtis mexicana Miller, 1898

This species was known as *M. megalotis* (Simmons and Vose, 1998).

DESCRIPTION: *Micronycteris microtis* is a small bat. Its large and rounded ears, covered by hair on one-third of the inner edge, characterize it. An inter-auricular membrane that is moderately high with a slight cleft connects the ears; in other species of the genus the cleft is very deep. The eyes are small and the hair is long (approximately 10 mm in length). The fur color varies geographically, but in general is brown with white bases on the back and dark on the abdomen; in San Luis Potosí coloration varies from yellowish-gray to cinnamon brown (Dalquest, 1953a). The legs are long and thin, with the calcaneus relatively long. The tail is short, reaching the middle of the uropatagium, and only its tip is free (Dalquest, 1953a; Hall, 1981).

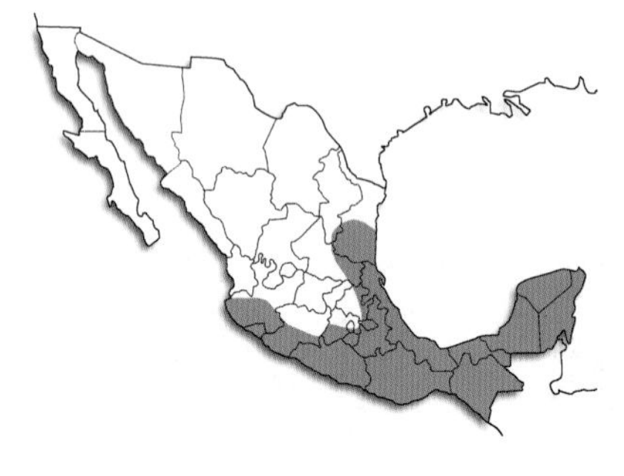

DISTRIBUTION: *M. microtis* is distributed from Mexico to Paraguay. In Mexico it is found from western Jalisco and Tamaulipas to the Yucatan Peninsula, including Cozumel Island. (Alonso-Mejía and Medellín, 1991; Álvarez, 1963a; Jones et al., 1973). It has been registered in the states of CL, CS, GR, HG, JA, MI, MO, MX, OX, PU, QR, SL, TA, TB, TL, VE, and YU.

EXTERNAL MEASURES AND WEIGHT

TL = 55 to 65 mm; TV = 11 to 17 mm, HF = 7 to 11 mm; EAR = 18 to 23 mm; FA = 32 to 38 mm.

Weight: 3.4 to 9.1 g.

DENTAL FORMULA: I 2/2, C 1/1, PM 2/3, M 3/3 = 34.

NATURAL HISTORY AND ECOLOGY: The common big-eared bat roosts in hollow trees and logs, but also occupies caves, culverts, tunnels, crevices between rocks, or buildings (Handley, 1976; LaVal and LaVal, 1980; Navarro, 1982). They are found solitary or in groups of up to 25 individuals (Fenton and Kunz, 1977; Tuttle, 1970). Its flight is not fast and it is considered a gleaner insectivore of substrate, whose foraging occurs at ground level or in foliage (Belwood, 1988; LaVal and LaVal, 1980). It can also consume fruit (Bonaccorso, 1979; Howell and Burch, 1974; Reis and Peracchi, 1987). The reproductive pattern is unimodal. In southern Mexico, however, the prolonged rainy season

Micronycteris microtis. Tropical rainforest. Photo: M.B. Fenton.

can cause two reproduction peaks. Pregnant females have been found in February, April, and May (Alonso-Mejía and Medellín, 1991; Sánchez-Hernández et al., 1985).

VEGETATIONAL ASSOCIATIONS AND ELEVATION RANGE: *M. microtis* inhabits tropical rainforests, deciduous and subdeciduous, but is also found in disturbed areas and in regions with secondary vegetation (Alonso-Mejía and Medellín, 1991). It has been found from sea level up to 3,400 m (Graham, 1983; Hall, 1981; Koopman, 1978; Navarro, 1982).

CONSERVATION STATUS: *M. microtis* is relatively common in undisturbed areas. The status of its populations is unknown. There are several populations in protected natural areas such as the Chamela-Cuixmala Biosphere Reserve, which ensures its long-term protection (Ceballos and Miranda, 2000).

Micronycteris schmidtorum Sanborn, 1935

Schmidt's big-eared bat

Rodrigo A. Medellín

SUBSPECIES IN MEXICO

M. schmidtorum is a monotypic species.

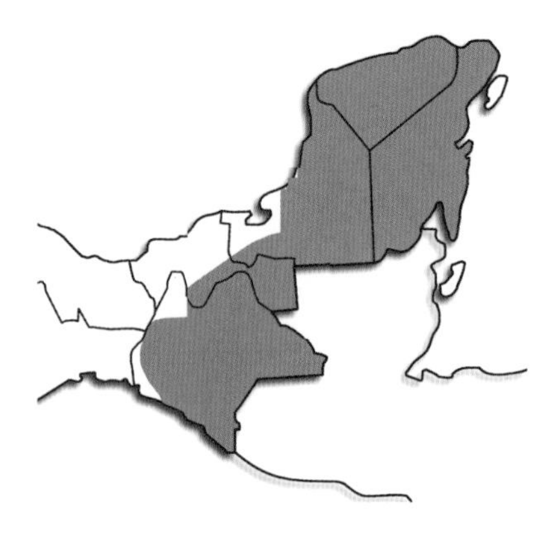

DISTRIBUTION: *M. schmidtorum* is distributed in the Neotropics from southern Mexico to Venezuela. In Mexico it has only been found in the Lacandon region and Cozumel Island. The Cozumel specimen, however, was a young individual whose classification as *M. schmidtorum* is in doubt (Jones and Lawlor, 1965; Jones et al., 1973). Two Yucatan specimens have been identified by Jones et al. (1973) as representatives of *M. megalotis mexicana*, not *M. schmidtorum*, as originally reported (Hatt and Villa-R., 1950). More recently, specimens from Campeche and mainland Quintana Roo have been reported by Escobedo-Cabrera et al. (2006), who also pointed out a record from the state of Yucatan. It has been registered in the states of CA, CS, QR, and YU.

DESCRIPTION: *Micronycteris schmidtorum* is a medium-sized bat with a delicate complexion. The ears are long and rounded, covered on the inner edge with long and thin hairs. The nasal leaf is well developed, thin, and pointed. On the head and joining the ears, there is a high membrane with a notable notch on the central part. The back is covered by long and soft brown hair with a white base. The belly is light gray, sometimes almost white; this feature separates it from other species of the genus. The tail tends to be longer than in other species of the same genus and does not reaches the edge of the uropatagium, which is relatively large.

EXTERNAL MEASURES AND WEIGHT
TL = 62 to 66 mm; TV = 15 to 17 mm, HF = 9 to 11 mm; EAR = 20 to 23 mm; FA = 33 to 36 mm.
Weight: 7 to 10 g.

DENTAL FORMULA: I 2/2, C 1/1, PM 2/3, M 3/3 = 34.

NATURAL HISTORY AND ECOLOGY: These bats are insectivorous but possibly include some fruit in their diet (Gardner, 1977). They roost in hollow trees and sometimes in human constructions. No information is available on their reproduction.

VEGETATIONAL ASSOCIATIONS AND ELEVATION RANGE: In Mexico, *M. schmidtorum* has only been caught in tropical rainforests. In other regions, it has been captured in deciduous forests and agricultural areas (Handley, 1976). It has been found at 600 masl. In Mexico, it has been captured at 120 masl.

CONSERVATION STATUS: *M. schmidtorum* is not listed in any official list of endangered species. R. Medellín and H. Arita (pers. obs.) classify it as endangered, given that the only Mexican population, inhabiting the rainforest in Chiapas and the Yucatan Peninsula, is currently facing serious problems due to habitat destruction.

Micronycteris schmidtorum. Photo: M.B. Fenton.

Niceforo's big-eared bat

Luis Escobedo and Livia Leon P.

SUBSPECIES IN MEXICO
T. nicefori is a monotypic species.
The genus was considered a subgenus of *Micronycteris* (Simmons and Voss, 1998).

DESCRIPTION: *Trinycteris nicefori* is a small-sized bat; the ears are medium-sized and have a triangular shape. The hair is grayish-brown, and the abdomen is lighter; the coat is tricolor, with a light band in the middle, so it can be confused with *Carollia brevicauda*. Sometimes, it has a tenuous gray line on the back. The face is medium-sized with small eyes. There is a commissure in form of V on the upper lip. The nasal leaf is small and somewhat thin. The calcaneus is shorter than the leg, and the tail is small and included in the first third of the uropatagium, with the exception of the tip (Reid, 1997; Sanborn, 1949). Most specimens have a faint, pale mid-dorsal stripe that is most evident on the lower back. The ventral margin of the nasal leaf is fused to the upper lip and lacks any demarcation between the horseshoe and the lip itself (Escobedo et al., 2006).

DISTRIBUTION: *T. nicefori* is distributed from southern Mexico and Belize to northern Colombia, Venezuela, Guyana, Brazil, and Peru (Koopman, 1993). Recent records infer an increment of its distribution heading northward (McCarthy, 1987). In Mexico, it was caught for the first time in April 1998 at the archaeological site of Yaxchilan, Chiapas (Escobedo-Morales et al., 2006). It has been registered in the state of CS.

EXTERNAL MEASURES AND WEIGHT
TL = 44 to 58 mm; TV = 8 to 15 mm, HF = 11 to 14 mm; EAR = 14 to 20 mm; FA = 35 to 40.
Weight: 7 to 11 g.

DENTAL FORMULA: I 2/2, C 1/1, PM 2/3, M 3/3 = 34.

NATURAL HISTORY AND ECOLOGY: Very little is known of the natural history of this species. It roosts in hollow logs and buildings. Apparently, it is more active one hour after sunset and one hour before sunrise (Reid, 1997). In a study of bat communities conducted in Costa Rica, it was only captured in harp traps placed in the canopy of the jungle (LaVal and Fitch, 1977). It has been collected along with *Sturnira lilium* and *Myotis keaysi*. It is possible that like other bats of sister genera such as *Micronycteris* or *Glyphonycteris*, it feeds on large cockroaches, orthopterans, and beetles; during the dry season it is likely to consume fruits (Bonacorso, 1979). The reproduction pattern is probably bimodal polyestric, with the first birth coinciding with the onset of rains (Wilson, 1979). The specimen caught in Yaxchilan had testicles in the scrotum (Escobedo et al., 2006).

VEGETATIONAL ASSOCIATIONS AND ELEVATION RANGE: Niceforo's big-eared bat is located in lowlands of the high jungle and the deciduous lowland (Reid, 1997). The specimen caught in Mexico was collected in high rainforest at an altitude of less than 100 m (Escobedo et al., 2006).

CONSERVATION STATUS: *T. nicefori* is a rare and local species of Central America; it is more common in South America (Reid, 1997). For Mexico, it should be considered threatened because it has only been registered in the region of the Lacandon Forest, which suffers severe deforestation problems. Yaxchilan is an archaeological protected area, located near the Montes Azules Biosphere Reserve, which protects thousands of acres of jungle in Chiapas (Escobedo et al., 2006).

Subfamily Desmodontinae

All three genera comprising the subfamily Desmodontinae are known from Mexico.

Desmodus rotundus (È. Geoffroy St.-Hilaire, 1810)

Common vampire bat

Gerardo Suzan A.

SUBSPECIES IN MEXICO

Desmodus rotundus murinus Wagner, 1840

DESCRIPTION: *Desmodus rotundus* is a medium-sized bat. The fur is thick and short, with a dark-gray dorsal coloration that can range from reddish to golden; the ventral part is lighter with a whitish tip; occasionally, it shows a lighter phase from the ventral region of the wing to the base of the ears. The ears are small, sharp, and separated. The forearms and legs have little hair. The thumb is well developed and has three well-marked pads (Greenhall et al., 1983; Villa-R., 1967). The skeleton has many specialized features adapting the bat for its feeding habits. The skull is very broad posteriorly and reduced anteriorly. Unlike other bats, it has a highly developed cerebral cortex and cerebellum. The face is reduced to withstand the large size of the upper incisors, which are particularly sharp. It is characterized by its quadruped locomotion and by the ability to take flight from the ground. The uropatagium has short and scarce hair. This bat has no tail (Villa-R., 1967).

EXTERNAL MEASURES AND WEIGHT

TL = 69 to 90 mm; TV = 0, HF = 13 to 20 mm; EAR = 15 to 20 mm; FA = 52 to 63 mm. Weight: 25 to 40 g.

Desmodus rotundus. Tropical dry forest. Sinaloa. Photo: Scott Altenbach.

DENTAL FORMULA: I 1/2, C 1/1, PM 1/2, F 1/1 = 20.

NATURAL HISTORY AND ECOLOGY: Colonies commonly have 20 to 100 individuals, but there have been reports of colonies of 500 to 5,000 individuals (Crespo et al., 1961). Females give birth to a single young. Gestation lasts seven months and vampire bats reproduce throughout the year. Their foraging area comprises 5 km to 8 km around their roosting site (Crespo et al., 1961); in certain regions they have been found at distances of 15 km to 20 km (Málaga-Alba, 1954). They leave their roosting site at dusk with a silent and low flight, usually near the ground. They roost in caves, wells, culverts, and dark buildings. They can be found in the interior of hollow trees such as the *Ceiba*, mahogany, *Fraxinus*, palm trees, oaks, cypresses, and *Sabine*. They are characterized by their diet that basically consists of the blood of different species of mammals such as cattle, horses, goats, sheep, pigs, and occasionally humans. The favorite biting sites are the neck, the base of the horns and ears, legs, vulva, anus, and tail. They can consume up to 20 ml of blood per individual per day; it takes about 40 minutes for them to eat (Greenhall, 1972). The natural predators of vampires are owls (*Tyto alba*, *Speotyto cunicularia*) and snakes (*Elaphe flavirufa*, *Constrictor constrictor*, *Bothrops atrox*). Villa-R. (1967) observed a carnivorous bat (*Chrotopterus auritus*) eating a vampire bat. They can transmit paralytic rabies, causing great economic losses in livestock (Hoare, 1972).

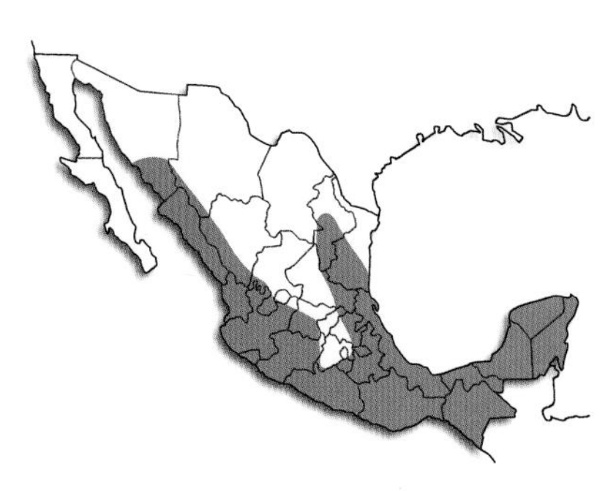

DISTRIBUTION: This species inhabits northern Sonora and Tamaulipas to Argentina (Villa-R., 1967). It has been registered in the states of CA, CH, CL, CS, DU, GR, HG, JA, MI, MO, MX, NL, NY, OX, PU, QE, QR, SI, SL, SO, TA, TB, TL, VE, YU, and ZA.

VEGETATIONAL ASSOCIATIONS AND ELEVATION RANGE: *D. rotundus* is restricted to tropical jungles in mature areas of secondary vegetation, crops, and pastures. In Peru it has been registered at 3,800 masl (Málaga-Alba et al., 1971). In Mexico it has been registered up to 2,300 masl (Villa-R., 1967).

CONSERVATION STATUS: The frequent increase of livestock and destruction of the tropics has favored the distribution of the vampire bat in Mexico. It is considered a public and animal health problem because of the possibility of transmission of paralytic rabies.

Diaemus youngi (Jentink, 1893)

White-winged vampire bat

William López-Forment and Guadalupe Tellez-Giron

SUBSPECIES IN MEXICO

D. youngi is a monotypic species.

DESCRIPTION: *Diaemus youngi* is of medium size, and is very similar to *Desmodus rotundus*. It has long and pointed ears that are more separated than those of *Desmodus*, the lower lip is vertically corrugated, and it has big eyes and a short face. The interfemoral membrane is narrow with hair. The calcaneus and forearm also have some hair. The tips of the wing membrane, between the second and third fingers, are white. The presence of two oral glands and the long thumb without pads are the most distinctive features among the three species of vampire bats (Eisenberg, 1989; Goodwin and Greenhall, 1961; Villa-R., 1967).

Diaemus youngi. Specimen in captivity. Photo: Scott Altenbach.

EXTERNAL MEASURES AND WEIGHT
TL = 78 to 87 mm; TV = 0, HF = 16 to 20.7 mm; EAR = 16 to 19 mm; FA = 50 to 54.7 mm.
Weight: 32 to 39 g.

DENTAL FORMULA: I 1/2, C 1/1, PM 1/2, F 2/1 = 20.

NATURAL HISTORY AND ECOLOGY: *D. youngi* lives in wet and open areas. It roosts in caves and hollow trees in colonies of more than 30 individuals (Goodwin and Greenhall, 1961). Within their shelters, they are located in illuminated sites near species such as *Saccopteryx bilineata* and *Desmodus rotundus*, and in association with other bat species such as *Peropteryx macrotis*, *Chrotopterus auritus*, *Carollia perspicillata*, *Sturnira lilium*, *Artibeus lituratus*, and *Platyrrhinus helleri* (Trajano, 1984). Like *Desmodus rotundus* it feeds on blood of vertebrates, preferably birds and occasionally livestock (Goodwin and Greenhall, 1961). They are polyestric; more information about their reproduction is still unavailable (Carter, 1970; Schmidt, 1988).

DISTRIBUTION: In Mexico, *D. youngi* is distributed from southern Tamaulipas to northern Argentina (Eisenberg, 1989; Goodwin and Greenhall, 1961). It has been registered in the states of CS, TA, TB, and VE.

VEGETATIONAL ASSOCIATIONS AND ELEVATION RANGE: White-winged vampire bats have been collected in tropical rainforests, as well as in dry tropical forests at 500 masl to 1,500 masl (Eisenberg, 1989; Emmons and Feer, 1999).

CONSERVATION STATUS: Mexican populations have been considered fragile (Arita, 1993a, b). In the Mexican Official Regulation, however, it is not included in any risk category (SEMARNAT, 2010).

Hairy-legged vampire bat

Guadalupe Tellez-Giron and William López-Forment

SUBSPECIES IN MEXICO

Diphylla ecaudata centralis Thomas, 1903

DESCRIPTION: *Diphylla ecaudata* is a medium-sized bat. Externally, it is very similar to *Desmodus*, but is recognizable by its short, rounded ears, shorter thumbs without apparent pads, and large and shiny eyes. The calcaneus is short; the legs have abundant hair that reaches the base of the nails. The hair color is brown with a light base, and is longer and softer than that of *Desmodus* or *Diaemus* (Schmidt, 1978; Villa-R., 1967).

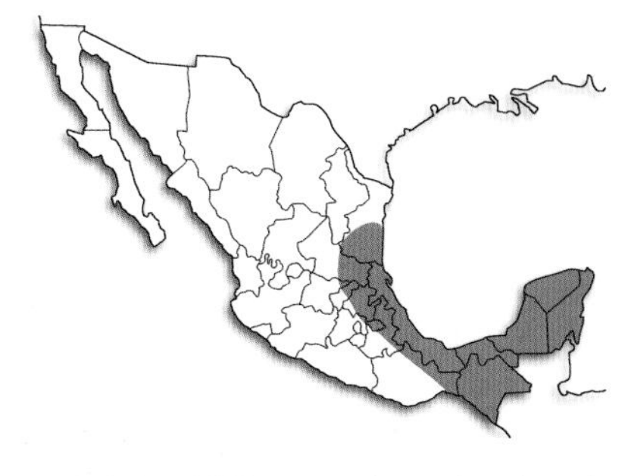

DISTRIBUTION: *D. ecaudata* is distributed from southern Texas in the United States, through the Gulf of Mexico, to the southeastern part of the Mexico, including the Yucatan Peninsula, reaching Central and South America (Greenhall et al., 1984). It has been registered in the states of CA, CS, HG, OX, PU, QE, QR, SL, TA, VE, and YU.

EXTERNAL MEASURES AND WEIGHT

TL = 75 to 96 mm; TV = 0, HF = 11 to 16 mm; EAR = 12 to 21 mm; FA = 50 to 56 mm. Weight: 24 to 43 g.

DENTAL FORMULA: I 2/2, C 1/1, PM 1/2, F 2/2 = 26.

NATURAL HISTORY AND ECOLOGY: These bats live in caves and abandoned mines with a temperature of 23°C to 24°C. In rare cases they have been registered in hollow trees. They form groups of up to 25 individuals in association with other bat species (Dalquest, 1955; Felten, 1956; Villa-R., 1967). It feeds on blood but has a preference almost exclusively for birds (Hoyt and Altenbach, 1981; Villa-R., 1967). Reproduction data indicate an aseasonal polyestry (Greenhall et al., 1984). They show altruistic behavior similar to that of the common vampire bats in feeding unfed individuals (Elizalde-Arellano et al., 2005).

VEGETATIONAL ASSOCIATIONS AND ELEVATION RANGE: *D. ecaudata* is found exclusively in tropical forests (Greenhall et al., 1984). It is found from sea level up to 1,200 m (Dalquest and Hall, 1947).

CONSERVATION STATUS: *D. ecaudata* is a common species, which is not within any category of extinction.

Diphylla ecaudata. Tropical semi-green forest. Calakmul Biosphere Reserve, Campeche. Photo: Gerardo Ceballos.

Subfamily Vampyrinae

The subfamily Vampyrinae contains four genera, three of which (*Chrotopterus, Trachops,* and *Vampyrum*) are known from Mexico.

Chrotopterus auritus (Peters, 1856)

Big-eared woolly bat

Rodrigo A. Medellín

SUBSPECIES IN MEXICO

Chrotopterus auritus auritus (Peters, 1856)

DESCRIPTION: *Chrotopterus auritus* is a big bat, one of the largest in the Americas. The ears are long and rounded; the nasal leaf is well developed, with a blunt lancet and a well-developed horseshoe. The hair is dark gray, long, and woolly. Given its size and the presence of the nasal leaf, it could only be confused with two species: *Vampyrum spectrum* or *Phylloderma stenops*. Both species have shorter hair and less developed ears; in addition, *V. spectrum* is larger.

Chrotopterus auritus, feeding on another bat. Tropical rainforest. Montes Azules Biosphere Reserve, Chiapas. Photo: Edmundo Huerta.

EXTERNAL MEASURES AND WEIGHT
TL = 103 to 122 mm; TV = 7 to 17 mm; HF = 21 to 28 mm; O = 40 to 48 mm; FA = 78.7 to 86.5 mm.
Weight: 66.8 to 96.1 g.

DENTAL FORMULA: I 2/1, C 1/1, PM 2/3, M 3/3 = 32.

NATURAL HISTORY AND ECOLOGY: This bat is almost exclusively carnivorous. It feeds on small birds, lizards, mice, bats, insects, and some large shrews. Their prey weigh between 10 g and 35 g, with a maximum of 70 g (Medellín, 1988). It roosts in hollow trees, mines, ruins, and other abandoned buildings, or in deep caves and cenotes, usually in family groups. In Costa Rica, the foraging area of a single individual was less than 4 ha (Medellín, 1989). Available data suggest monoestry; reproduction varies geographically depending on the climate cycle; in general, young are born at the beginning of the rainy season (Wilson, 1979). In Mexico, pregnant females have been found in April and breastfeeding in July (Medellín, 1989).

VEGETATIONAL ASSOCIATIONS AND ELEVATION RANGE: In Mexico, *C. auritus* inhabits tropical rainforests, cloud forests, and semi-wet scrublands of the Yucatan Peninsula, from sea level up to 2,000 m (Medellín, 1989).

CONSERVATION STATUS: *C. auritus* is not included in any official list, but it has been classified as threatened (R. Medellín and H. Arita, pers. comm.). Arita (1993a, b) classifies it as locally rare but with a wide distribution. Since it is a species associated with vegetation types that currently face high rates of destruction, while carnivorous animals are living at low densities, it is clear that this species faces conservation problems. If adequate extensions of pristine habitat are preserved, there would be no need for additional measures to ensure the permanence of this species.

DISTRIBUTION: This species inhabits humid tropical regions of America. In Mexico, it inhabits the southern part of the coastal plain of the Gulf of Mexico in Veracruz from Jesus Carranza to the Yucatan Peninsula and the southern tip of the Pacific coast of Chiapas. It has been registered in the states of CA, CS, OX, QR, TB, VE, and YU.

Trachops cirrhosus (Spix, 1823)

Fringe-lipped bat

Xavier Lopez and Rodrigo A. Medellín

SUBSPECIES OF MEXICO
Trachops cirrhosus coffini Goldman, 1925

DESCRIPTION: This bat has a robust body and long, wide, and rounded ears. It has many elongated warts around the mouth. Its soft and relatively long fur extends down to the middle of the forearm, being chestnut brown on the back and cinnamon brown on the abdomen, the wing membranes and the uropatagium are brown, the tip of the wings can be whitish. The well-developed nasal leaf has jagged lateral distal margins. The tail is almost fully included in the uropatagium and protrudes just a little bit. The claws are robust.

EXTERNAL MEASURES AND WEIGHT
TL = 72 to 95 mm; TV = 10 to 20 mm, HF = 16 to 22 mm; EAR = 30 to 40 mm; FA = 56.2 to 64 mm.
Weight: 28 to 45 g.

DENTAL FORMULA: I 2/1, C 1/1, PM 2/3, M 3/3 = 32.

DISTRIBUTION: *T. cirrhosus* is distributed from southern Mexico to Bolivia and southern Brazil (Hall, 1981; Linares, 1986; Peracchi et al., 1982; Ramírez-Pulido et al., 1986; Ramírez-Pulido and Castro-Campillo, 1993). It has been found in the states of CA, CS, OX, QR, VE, and YU.

NATURAL HISTORY AND ECOLOGY: *Trachops cirrhosus* is a carnivorous species and occasionally insectivorous (Pine and Anderson, 1979). It feeds on small vertebrates, particularly frogs, which are located by the chants of mating males (Findley, 1993; Tuttle and Ryan, 1981); it also consumes small mammals and birds (Emmons and Feer, 1999). Its feeding behavior is very specialized. It responds differently from the audible and, perhaps, not audible chants of frogs and toads, discriminating poisonous species. The submandibular salivary glands, unique in *Trachops* and *Megaderma*, produce a substance that protects the oral cavity and digestive tract from chemicals that can be toxic (Phillips et al., 1987; Studholme et al., 1986). Presumably males have harems, occupying one roosting site, and thus limiting aggressive interactions between males (Tuttle and Stevenson, 1982). It prefers caves, although it has been found roosting in houses, culverts, and abandoned buildings (Emmons and Feer, 1999; Handley, 1966). If other species share its roosting site, these bats segregate themselves (Arita, 1993a, b). There are not many data about their reproduction; however, pregnant females have been captured in Panama almost throughout the year (Bonaccorso, 1979; Honeycutt, et al., 1980; Tuttle, 1970).

VEGETATIONAL ASSOCIATIONS AND ELEVATION RANGE: This species inhabits gallery forests, deciduous forests, plantations, and wet jungles (Emmons and Feer, 1999; Linares, 1986). It is found from sea level to 330 m (McCarthy, 1987).

CONSERVATION STATUS: The information available at this time is inadequate and insufficient to determine the conservation status of *T. cirrhosus*; however, its dependence on the forest and water bodies makes it vulnerable (Medellín and Redford, 1992).

Vampyrum spectrum (Linnaeus, 1758)

Great false vampire bat

Rodrigo A. Medellín

DISTRIBUTION: In the Neotropics, *V. spectrum* is found from Mexico to Brazil, Peru, and Bolivia. In Mexico there are records on the coast of southern Veracruz, Campeche, Oaxaca (in owl pellets), and several in the Lacandon Forest in Chiapas (Alfaro et al., 2005; Hall, 1981; Hernández-Huerta et al., 2000; Medellín et al., 1998). It has been recorded in the states of CA, CS, OX, and VE.

SUBSPECIES IN MEXICO

V. spectrum is a monotypic species.

DESCRIPTION: *Vampyrum spectrum* is the largest bat of the Americas. When extended forward, the ears reach the top of the face and are somewhat sharp. Its nasal leaf is well developed, and its face is somewhat stretched. The eyes are small. This bat has no tail and the uropatagium is broad. The hair is relatively short and its coloration varies from light to dark reddish-brown. The large size of these bats allows observers to easily distinguish them from any other species in the continent.

EXTERNAL MEASURES AND WEIGHT

TL = 125 to 158 mm, HF = 32 to 38 mm; EAR = 39 to 49 mm; FA = 98 to 110 mm. Weight: 126 to 190 g.

DENTAL FORMULA: I 2/2, C 1/1, PM 2/3, M 3/3 = 34.

NATURAL HISTORY AND ECOLOGY: Great false vampire bats are carnivores. Their large size allows them to hunt mice, birds, and other bats. Apparently, they capture birds with a strong body odor and that form colonies such as the *ani* or *pijuy* (*Crotophaga* sp.). Prey size ranges from 20 g to 150 g (Vehrencamp et al., 1977). They mainly roost in family groups in hollow trees. They carry out feeding flights that last between 1 and 4.5 hours; adults hunt to provide their young who are waiting in the roosting site (Vehrencamp et al., 1977). They fly along streams and rivers, taking advantage of the absence of vegetation. There is not too much information on the reproductive cycle. They give birth to a single young apparently at the end of the dry season or the beginning of the rainy season (Navarro and Wilson, 1982). Barn owls (*Tyto alba*) are predators of this large bat (Alfaro et al., 2005).

VEGETATIONAL ASSOCIATIONS AND ELEVATION RANGE: This species is mainly found in tropical rainforests, but also occupies parts of tropical deciduous forest and swamps. It has been found from sea level up to 1,650 m. In Mexico it has only been found below 150 m.

CONSERVATION STATUS: *V. spectrum* has not been included in any official list. Arita (1993a, b) classified it as locally rare, but widely distributed. R. Medellín and H. Arita (pers. obs.) classified it as endangered, given its large size, habits, and reduced habitat.

Vampyrum spectrum. Tropical rainforest. Photo: Scott Altenbach.

Subfamily Phyllostominae

Phyllostominae contains twenty-six genera divided among three tribes: Phyllostomini, Glossophagini, and Stenodermatini.

Tribe Phyllostomini

Lonchorhina aurita Tomes, 1863

Tomes's sword-nosed bat

Rodrigo A. Medellín

SUBSPECIES IN MEXICO
Lonchorhina aurita aurita Tomes, 1863.

DESCRIPTION: *Lonchorhina aurita* is a medium-sized bat of a delicate complexion, with long and sharp ears. Its nasal leaf is the most developed among the Mexican bats, sometimes reaching more than 20 mm in length (Arita, 1990). The eyes are

Lonchorhina aurita. Tropical rainforest.
Photo: M.B. Fenton.

small. The fur is dark brown to black and very long and soft. The wing membranes are black. The tail is long and reaches the edge of the broad uropatagium. The legs are also long.

EXTERNAL MEASURES AND WEIGHT
TL = 53 to 67 mm; TV = 42 to 65 mm, HF = 13 to 15 mm; EAR = 19 to 35 mm; FA = 47 to 57 mm.
Weight: 10 to 22 g.

DENTAL FORMULA: I 2/2, C 1/1, PM 2/3, M 3/3 = 34.

NATURAL HISTORY AND ECOLOGY: Tomes's sword-nosed bats are insectivorous, but there is a report of an individual who had eaten fruit (Nelson, 1965). It roosts inside caves generally prone to flooding, deep tunnels, and culverts in roads; they can be found in small colonies of 10s to large groups of 500 animals. The roosting sites are usually occupied by other bat species, so they have been classified as "integrationist" (Arita, 1993a). Their flight is slow, paused, and agile, and their sonar is fine and sensitive (Lassieur and Wilson, 1989). Females are pregnant during the dry season and give birth to a single young. The young are born at the beginning of the rainy season (Lassieur and Wilson, 1989).

VEGETATIONAL ASSOCIATIONS AND ELEVATION RANGE: In Mexico *L. aurita* has been recorded in the tropical and subtropical forest. In other regions of Central and South America, it has been found in the semi-dry deciduous forest and agricultural areas (Handley, 1976). It is found from sea level up to 1,537 m.

DISTRIBUTION: In the Neotropics *L. aurita* is distributed from southern Mexico to Brazil and Peru. In Mexico, it has only been found in the most humid parts of the southern coastal plain of the Gulf and south of the Yucatan Peninsula. It has been registered in the states of CS, OX, QR, TB, and VE (Polaco et al., 1992).

CONSERVATION STATUS: *L. aurita* is not included in any official list of endangered species. Throughout the Neotropics it is locally abundant and widely distributed (Arita, 1993b). It has been classified as fragile (R. Medellín and H. Arita, pers. obs.), but it is necessary to further assess its current situation.

Lophostoma brasiliense Peters, 1867

Pygmy round-eared bat

Rodrigo A. Medellín

SUBSPECIES IN MEXICO

L. brasiliense is a monotypic species.

Before it was assigned to the current genus, it was included in *Tonatia* (Lee et al., 2002).

DESCRIPTION: *Lophostoma brasiliense* is a small-sized bat. The nasal leaf is well formed, but the lower element (horseshoe) is fused with the upper lip. This feature distinguishes it from *Micronycteris microtis*, where the horseshoe of the nasal leaf is well formed and is differentiated from the upper lip. The ears are large and rounded with hair only on the base of the inner edge. The upper lip has several tiny warts, and the eyes are small. This bat is dark brown to black, with fine hair of average length. The tail is short and is fully included in the uropatagium.

Lophostoma brasiliense. Tropical rainforest.
Photo: M.B. Fenton

DISTRIBUTION: This species occupies in tropical America down to Peru and Brazil. In Mexico, it has only been found in the wettest portion of the Gulf coastal plain in Veracruz, Campeche, Chiapas, and the south of the Yucatan Peninsula (Briones-Salas and Santos Moreno, 2002; Hall, 1981; Hernández-Huerta et al., 2000). It has been registered in the states of CA, CS, OX, QR, and VE.

EXTERNAL MEASURES AND WEIGHT
TL = 58 to 65 mm; TV = 7 to 11 mm, HF = 9 to 13 mm; EAR = 24 to 26 mm; FA = 33 to 36 mm.
Weight: 9 to 11 g.

DENTAL FORMULA: I 2/1, C 1/1, PM 2/3, M 3/3 = 32.

NATURAL HISTORY AND ECOLOGY: This bat is a gleaner insectivore of substrate and may be partly frugivorous (Gardner, 1977). There is a lot of information about its biology. Apparently, it can roost in the interior of hollow trees and in termite nests (Handley, 1976). They give birth to a single young; birth is likely occur at the end of the dry season or the beginning of the rainy season (Wilson, 1979).

VEGETATIONAL ASSOCIATIONS AND ELEVATION RANGE: This species has been found in tropical rainforests, deciduous forests, and disturbed areas, from sea level to 500 m.

CONSERVATION STATUS: *L. brasiliense* has not been included in any official list. R. Medellín and H. Arita (pers. obs.) classify it as fragile, but given the little information available on this species, its current status is unclear.

Lophostoma evotis (Davis and Carter, 1978)

Davis's round-eared bat

Rodrigo A. Medellín

SUBSPECIES IN MEXICO
L. evotis is a monotypic species.
Until 1978 it was considered a subspecies of *Tonatia silvicola.*

DISTRIBUTION: The species is endemic to Mesoamerica; it is found from southern Mexico to Honduras. In Mexican territory, it has been found only in the wettest portion of the Gulf coastal plain and in the south of the Yucatan Peninsula (Medellín and Arita, 1989). It has been registered in the states of CA, CS, QR, TB, and VE.

DESCRIPTION: *Lophostoma evotis* is a medium-sized bat. The ears are long and rounded. The nasal leaf is well developed, with the lower element (horseshoe) fused with the upper lip. The coloration of the head is uniform, which distinguishes it from *Tonatia saurophila.* The body is uniform dark gray with little contrast between the back and the abdomen. The hair is fine and of average length. The tail is short and is fully included in the uropatagium.

EXTERNAL MEASURES AND WEIGHT
TL = 79 to 91 mm; TV = 14 to 18 mm; HF = 13 to 16 mm; EAR = 33 to 35 mm; FA = 47 to 53 mm.
Weight: 20 g.

DENTAL FORMULA: I 2/1, C 1/1, PM 2/3, M 3/3 = 32.

NATURAL HISTORY AND ECOLOGY: Davis's round-eared bats are partly frugivorous and insectivorous of substrate; they capture their prey on foliage and other surfaces. The stomach contents of four animals of Quintana Roo contained remains of beetles and orthopterans. They give birth to a single young and apparently delivery takes place at the end of the dry season or the beginning of the rains.

Lophostoma evotis. Tropical rainforest.
Photo: M.B. Fenton.

VEGETATIONAL ASSOCIATIONS AND ELEVATION RANGE: *L. evotis* has been caught in tropical forest and disturbed rainforest, from sea level to 200 m.

CONSERVATION STATUS: *L. evotis* has not been included in any official list. Arita (1993a, b) classifies it as locally rare and with restricted distribution. R. Medellín and H. Arita (pers. obs.) classify it as threatened due to its association with habitats that are being degraded and its restricted distribution, but given how little is known of this species, its status is unclear.

Macrophyllum macrophyllum (Schinz, 1821)

Long-legged bat

Rodrigo A. Medellín

SUBSPECIES IN MEXICO
M. macrophyllum is a monotypic species.

DESCRIPTION: *Macrophyllum macrophyllum* is a small bat of delicate complexion with big and sharp ears, small eyes, and a very wide and high nasal leaf that totally dominates its features. It has long gray and black hair, which is thin and rough. The tail is long and is fully included in the uropatagium. The membranes are dark and the edge of the uropatagium has tactile papillae that are easily recognizable. The legs are long and highly developed.

EXTERNAL MEASURES AND WEIGHT
TL = 40 to 62 mm; TV = 38 to 49 mm, HF = 13 to 16 mm; EAR = 17 to 20 mm; FA = 33 to 40 mm.
Weight: 6 to 11 g.

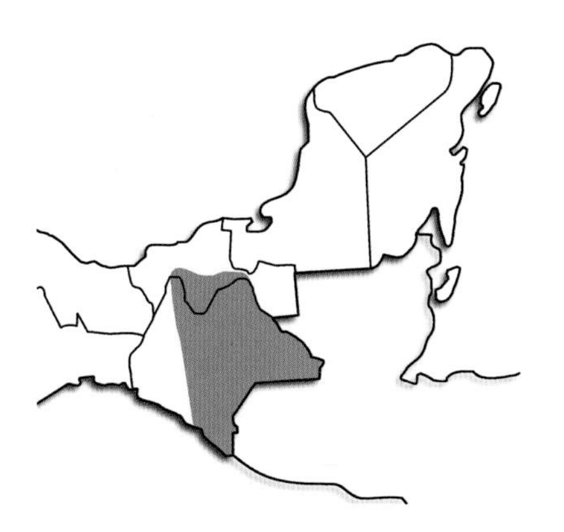

DISTRIBUTION: This species occupies a large area from southern Mexico to southern Brazil and northern Argentina (Harrison, 1975). In Mexico it has been only found in the most humid and tropical Gulf coastal plain. It has been recorded in the states of CS and TB.

DENTAL FORMULA: I 2/2, C 1/1, PM 3/3, F 2/2 = 32.

NATURAL HISTORY AND ECOLOGY: *M. macrophyllum* feeds on insects, dipterans, and lepidopterans. They form small groups of up to 10 individuals; they roost inside caves, culverts, and trees (Linares, 1966; Seymour and Dickermann, 1982). They fly along the course of rivers and streams in dense forests, where most of the specimens have been captured. It has been suggested, by the peculiar anatomy of the legs, that these bats catch some of their prey from the water surface (Gardner, 1977). They give birth to a single young and have a bimodal polyestric reproductive cycle; the first delivery occurs at the beginning of the rainy season and the second in the middle of the rainy season (Harrison, 1975).

VEGETATIONAL ASSOCIATIONS AND ELEVATION RANGE: The few Mexican specimens were caught in the tropical rainforest. In Venezuela, it has been caught in conditions of high humidity in various types of forests (Handley, 1976). It has been found from sea level up to approximately 500 m.

CONSERVATION STATUS: *M. macrophyllum* is not included on any official list of wildlife protection. R. Medellín and H. Arita (pers. obs.) classify it as fragile, but there is insufficient information available to define its current situation. Arita (1993a, b) classifies it as locally rare and widely distributed.

Mimon cozumelae Goldman, 1914

Cozumelan golden bat

Rodrigo A. Medellín

DISTRIBUTION: *M. cozumelae* is found in wet lowland tropical forests. It is distributed from southeastern Veracruz in Mexico, down to the Atlantic coast of southeastern Brazil, including the Yucatan Peninsula, Central America, and the coastal region of northern Colombia, Venezuela, and Guyana (Eisenberg, 1989). It has been registered in the states of CA, CS, OX, QR, TB, VE, and YU.

SUBSPECIES IN MEXICO

M. cozumelae is a monotypic species.

There are doubts about the taxonomic status of *M. cozumelae*, as it has been considered a subspecies of *M. bennettii* (Koopman, 1993). Some authors have regarded it as separate species, however, because their integration has not been fully verified; in addition, genetic studies show that both species have a different chromosome number (Baker et al., 1981; Jones and Carter, 1976).

DESCRIPTION: *Mimon cozumelae* is a medium- to large-sized bat. The nasal leaf is large and in the form of a lancet, and has low cuts on the sides. It lacks hair and has a simple horseshoe at the base. The fur is dark brown with a paler base; the latter color is dominant at the base of the neck and in some individuals it extends to the shoulders (Eisenberg, 1989; Villa-R., 1967). The fur on the back is long (7 mm) and on the abdomen is short (5 mm). The interfemoral membrane is broad and extends to the base of the legs. The tail is long, almost the size of the femur, and it is fully included in the interfemoral membranes. It differs from *M. crenulatum* by the absence of hairs on the nasal leaf, by the lack of white lines on the back, and by its greater size and coloration (*M. crenulatum* is almost black). The skull is short, with the cranial cavity located above the rostrum and a small tympanic bulla. The upper incisors are not permanent, which is a characteristic of the subfamily to which it belongs (Phillips, 1971; Phillips et al., 1977).

EXTERNAL MEASURES AND WEIGHT
TL = 85 to 95 mm; TV = 20 to 25 mm, HF = 32 to 36 mm; EAR = 36 to 38 mm; FA = 53.1 to 59 mm.
Weight: 35 g.

DENTAL FORMULA: I 2/1, C 1/1, PM 2/2, F 3/3 = 30.

NATURAL HISTORY AND ECOLOGY: This is a foliage gleaner bat; although some individuals have been observed feeding on fruits and insects, they mostly eat small vertebrates such as lizards and birds, so are considered carnivores (Fenton, 1992). The carnivore hypothesis is supported by the odor of the roosting sites, similar to that of the pellets of prey birds (Hall and Dalquest, 1963; Villa-R., 1967; Whitaker and Findley, 1980). They form small colonies of up to 10 individuals, usually in caves, mines, and hollow logs (LaVal, 1977). This species reproduces once a year during the rainy season (Wilson, 1979). In southeastern Mexico, pregnant females, with embryos at the last stages, have been collected in April and May (Hall and Dalquest, 1963). Moreover, young and young females have been reported during the rainy season in various Central American countries (Rick, 1968; Valdéz and LaVal, 1971).

Mimon cozumelae. Tropical rainforest. Photo: M.B. Fenton.

VEGETATIONAL ASSOCIATIONS AND ELEVATION RANGE: In Mexico, the distribution of this species is limited to tropical rainforests, tropical subdeciduous forests, and tropical deciduous forests. It is distributed from sea level up to 600 m.

CONSERVATION STATUS: *M. cozumelae* is not included in any list of endangered species, although it has been classified as locally rare species but widely distributed (Arita, 1993a, b).

Mimon crenulatum (È. Geoffroy St.-Hilaire, 1810)

Striped hairy-nosed bat

Rodrigo A. Medellín

SUBSPECIES IN MEXICO
Mimon crenulatum keenani Handley, 1960

DESCRIPTION: *Mimon crenulatum* is a medium-sized bat with a fragile complexion. The ears are very long and sharp and, at the base, they have yellow hair. The nasal leaf is long and relatively thin with an irregular edge and with fine hairs on both sides. The back is blackish with a whitish-yellow line along the back from the neck to the hip. The abdomen is gray with very long and soft hair. The tail is long but only reaches the center of the uropatagium, which is broad. The long-haired, serrated nasal leaf distinguishes it from other bats; *Lonchorhina aurita* also has a long nasal leaf, but its edges are smooth and hairless. In addition, *L. aurita* lacks the whitish-yellow line on the back.

Mimon crenulatum. Tropical rainforest. Montes Azules Biosphere Reserve, Chiapas. Photo: Rodrigo Medellín.

EXTERNAL MEASURES AND WEIGHT
TL = 80 to 91 mm; TV = 21 to 27 mm, HF = 10 to 14 mm; EAR = 24 to 28 mm; FA = 48 to 52 mm.
Weight: 12 to 16 g.

DENTAL FORMULA: I 2/1, C 1/1, PM 2/2, F 3/3 = 30.

NATURAL HISTORY AND ECOLOGY: *M. crenulatum* is an insectivorous bat that feeds on beetles, lepidopterans, dipterans, hemipterans, and some small vertebrates (Humphrey et al., 1983; Whitaker and Findley, 1980). It probably includes some fruits in its diet. It roosts inside hollow trees and sometimes in buildings. The colonies are very small (less than 10 individuals) (Eisenberg, 1989; Timm et al., 1989). Breeding season takes place once a year; the little information available suggests that the offspring are born at the end of the dry season or beginning of the rainy season.

VEGETATIONAL ASSOCIATIONS AND ELEVATION RANGE: Mexican specimens were captured in the rainforest. In Venezuela it has been captured in deciduous forests and agricultural areas (Handley, 1976). It has been found from sea level to about 600 m.

CONSERVATION STATUS: *M. crenulatum* is not included in any official list, but as the only Mexican population inhabits the rainforest of Chiapas and Campeche, it should be considered endangered. It has been considered locally abundant and widely distributed (Arita, 1993a, b), but in Mexican territory it is considered rare (Medellín, 1993).

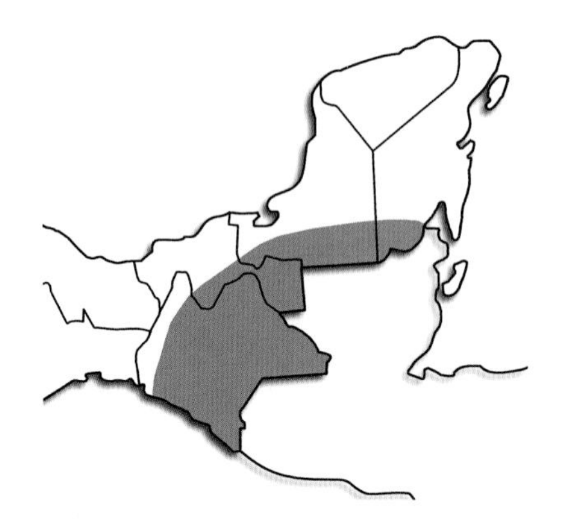

DISTRIBUTION: This species is found in the Neotropics from the southern tip of Mexico to Brazil and Peru. In Mexican territory it has been found only in the Lacandon region and south of the Yucatan Peninsula. It has been registered in the states of CA, CS, and QR.

Pale-faced bat

Rodrigo A. Medellín

DISTRIBUTION: This species inhabits the Neotropics from southern Mexico to Brazil and Bolivia (Eisenberg, 1989; Koopman, 1982). In Mexico, it has only been registered in the wettest parts of Chiapas. It has been registered in the state of CS.

SUBSPECIES IN MEXICO
P. stenops is a monotypic species.

DESCRIPTION: *Phylloderma stenops* is a large and robust bat. The ears are short; the nasal leaf is well developed with a relatively short and wide lancet and a well-defined horseshoe. The eyes are large and black. The canines are relatively short for its size. Males have a prominent gular gland. Coloration varies from pearl gray to reddish; the hair is short and sparse. The tips of the wings are white; the tail is short and fully included in the broad uropatagium.

EXTERNAL MEASURES AND WEIGHT
TL = 105 to 115 mm; TV = 12 to 24 mm; HF = 19 to 25 mm; EAR = 25 to 32 mm; FA = 66 mm to 73 mm.
Weight: 41 to 80 g.

DENTAL FORMULA: I 2/2, C 1/1, PM 2/3, M 3/3 = 34.

NATURAL HISTORY AND ECOLOGY: There is very little information on this bat. Apparently, they are insectivorous and frugivorous. They have been observed on a wasp nest feeding on larvae and pupae (Jeanne, 1970). An individual excreted large seeds of an Annonaceae, and in captivity they were fed bananas (LaVal, 1977). Its roosting sites are unknown. Mating season occurs once a year, apparently at the end of the dry season.

VEGETATIONAL ASSOCIATIONS AND ELEVATION RANGE: The only specimens that have been caught in Mexico inhabited the tropical rainforest. In South America, it has also been caught in tropical deciduous forests, swamps, and agricultural areas, from sea level up to approximately 600 m.

CONSERVATION STATUS: *P. stenops* has not been included in any official list of endangered species. R. Medellín and H. Arita (pers. obs.) classified it as threatened since the Mexican specimens inhabit regions that are currently facing conservation problems.

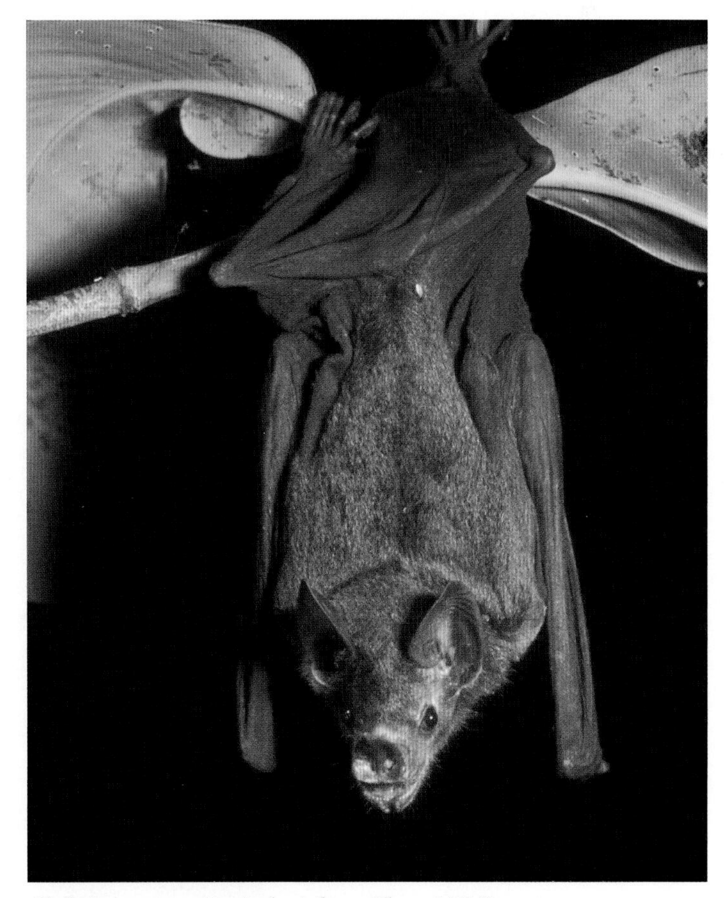

Phylloderma stenops. Tropical rainforest. Photo: M.B. Fenton.

Pale spear-nosed bat

Rodrigo A. Medellín

SUBSPECIES IN MEXICO

Phyllostomus discolor verrucosus Elliot, 1905

DISTRIBUTION: This species is found in the Neotropical region from southern Mexico to Paraguay and Argentina (Eisenberg, 1989; Hall, 1981; Koopman, 1982). In Mexico, it has been caught in warm and low areas of southern Veracruz, Oaxaca, Chiapas, and Tabasco. It has been registered in the states of CS, OX, TB, and VE.

DESCRIPTION: *Phyllostomus discolor* is a medium-sized bat. Its ears are short with rounded tips; the nasal leaf is small but conspicuous. The eyes are large. Males have a gland in the upper chest that produces a whitish discharge. The hair is short and reddish-brown (lighter in the abdomen) and sometimes very dark. Some animals have portions of gray hair. The short tail is included in the uropatagium, which is relatively narrow.

EXTERNAL MEASURES AND WEIGHT

LT = 90 to 102 mm; CV = 9 to 17 mm, P = 16 to 20 mm; O = 17 to 22 mm; AN = 60 to 66 mm.

Weight: 30 to 45 g.

DENTAL FORMULA: I 2/2, C 1/1, PM 2/2, F 3/3 = 32.

NATURAL HISTORY AND ECOLOGY: Pale spear-nosed bats feed on insects such as beetles, hymenopterans, dipterans, and lepidopterans; flowers and nectar from various plants, like *Ceiba*, *Bauhinia*, *Hymenaea*, *Parkia*, *Crescentia*, *Ochroma*, *Pseudobombax*, and, *Manilkara*; and fruits of trees such as *Spondias*, *Ficus*, *Diospyros*, *Manilkara*, *Piper*, *Acnistus*, and *Musa* (Gardner, 1977). Their diverse diet makes them important pollinators. They roost inside hollow trees and caves in groups of up to 25 animals organized in polygynous units of 1 to 12 females per male (Nowak, 1999a). The information on reproduction suggests that there is no defined reproductive cycle, but in some regions it suggests monoestry. In Costa Rica, births occur at the beginning of the rainy season (Wilson, 1979).

Phyllostomus discolor. Tropical rainforest. Photo: M.B. Fenton.

VEGETATIONAL ASSOCIATIONS AND ELEVATION RANGE: This species inhabits tropical rainforests and deciduous forests. In Venezuela, it has been reported in tropical rainforests, thorny forests, deciduous and cloud forests, and agricultural areas (Handley, 1976). They have been captured from sea level up to 1,160 m (Handley, 1976), but in general the sites were below 600 m in elevation.

CONSERVATION STATUS: *P. discolor* has not been included in any list of endangered species. In Mexico, it has a restricted distribution but does not appear to be rare.

Tonatia saurophila Koopman and Williams, 1951

Greater round-eared bat

Rodrigo A. Medellín

SUBSPECIES IN MEXICO

T. saurophila is a monotypic species.

DESCRIPTION: *Tonatia saurophila* is a bat of average size. The ears are long and rounded. The nasal leaf is well developed; however, the horseshoe is merged with the upper lip and there is no discernible edge. The hair is of average length and coloration is generally grayish-brown with a whitish strip at the top of the head. The tail is of average size and is fully included in the uropatagium.

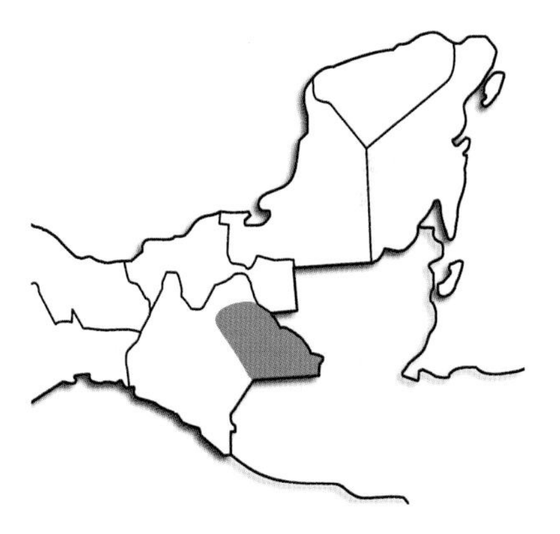

DISTRIBUTION: *T. saurophila* is distributed from southern Mexico to Brazil and Paraguay. In Mexico, it has only been caught in the Lacandon Forest, Chiapas. It has been registered in the state of CS.

EXTERNAL MEASURES AND WEIGHT

TL = 93 to 102 mm; TV = 13 to 20 mm; HF = 13 to 18 mm; EAR = 29 to 35 mm; FA = 55 to 60 mm.

Weight: 23 to 29 g.

DENTAL FORMULA: I 2/1, C 1/1, PM 2/3, M 3/3 = 32.

NATURAL HISTORY AND ECOLOGY: Very little information about the biology of the greater round-eared bat exists. They are insectivorous and frugivorous animals that feed on orthopterans, coleopterans, lepidopterans, homopterans, dipterans, and some vertebrates (Humphrey et al., 1983; Whitaker and Findley, 1980). Apparently, they catch insects on the substrate (vegetation, soil), as other species of this group do. They roost inside hollow trees, sometimes alone and sometimes in the company of other species of bats (Eisenberg, 1989; Handley, 1976). They form small colonies. Apparently, their reproductive cycle is bimodal polyestric, with births at the beginning of the rainy season and in the middle of the wet season (Medellín, 1986; Wilson, 1979).

VEGETATIONAL ASSOCIATIONS AND ELEVATION RANGE: *T. saurophila* is mainly associated with tropical rainforests. In South America, it is also found in thorny scrublands and agricultural areas, from sea level to at least 700 m.

CONSERVATION STATUS: *T. saurophila* is not included in any official list of endangered species. It has been classified as endangered due to the fact that the only known Mexican population is in the Lacandon Forest, which is a region currently facing a strong process of deterioration and destruction (R. Medellín and H. Arita, pers. obs.).

Anoura geoffroyi Gray, 1838

Geoffroy's hairy-legged long-tongued bat

Jorge Ortega R. and Hector T. Arita

DISTRIBUTION: In the Pacific slope, *A. geoffroyi* is distributed from Sonora, and in the Gulf of Mexico from Tamaulipas, including the Altiplane, to the southeast of Peru and Brazil (Eisenberg, 1989; Emmons and Feer, 1999). It has been registered in the states of CL, CS, DU, DF, GR, HG, JA, MI, MO, MX, NY, OX, PU, QE, QR, SI, SL, SO, TA, TL, VE, and ZA.

SUBSPECIES IN MEXICO
Anoura geoffroyi lasiopyga (Peters, 1868)

DESCRIPTION: *Anoura geoffroyi* is a medium-sized glossophaginae bat. The face is long with small ears and a simple nasal leaf. The uropatagium is virtually invisible because it is very narrow and completely covered with hair. The species lacks a tail and lower incisors; these traits differentiate it from other species of the genus *Glossophaga*. The hair color (brown-gray) is uniform all over the body (Linares, 1987).

EXTERNAL MEASURES AND WEIGHT
LT = 60 to 70 mm; CV = 0, P = 10 to 12 mm; OR = 7 to 9 mm; AN = 40 to 47 mm.
Weight: 15 to 20 g.

DENTAL FORMULA: I 2/0, C 1/1, PM 3/3, M 3/3 = 32.

NATURAL HISTORY AND ECOLOGY: The feeding behavior of Geoffroy's hairy-legged long-tongued bat depends on the availability and abundance of food, which includes pollen, nectar of flowers, and insects (Gardner, 1977). Of the 69 specimens collected in various states of Mexico, 50% contained grains of pollen of some species of the family Compositae and the other half only had remains of insects (Álvarez and Gonzalez-Q., 1970). In some areas of South America this species is closely related to the pollination of plants such as *Eperua falcata*, which only opens its flowers at night (Eisenberg, 1989). Their reproductive period is unusual, as it coincides with the end of the rainy season. Each female produces a single young per year (Wilson, 1979). It only roosts inside caves and tunnels, forming small groups of less than 100 individuals, but sometimes with up to 300 individuals (Álvarez and Ramirez-Pulido, 1972; Hoffman et al., 1986; Ortega and Alarcón-D., 2008). The proportion of sexes in the colonies varies seasonally (Willig, 1983; Wilson, 1979). It shares its roosting sites with very few species (Arita, 1993a, b).

VEGETATIONAL ASSOCIATIONS AND ELEVATION RANGE: This species inhabits tropical deciduous forests, tropical subdeciduous forests, dry scrublands, pine (*Pinus*) and oak (*Quercus*) forests, and mountain cloud forests. Most of the records in Mexico occur at about 1,000 m, but it is found from sea level up to 3,000 m (Matson and Baker, 1986). In Peru it is found from 700 m to 3,600 m (Graham, 1983).

CONSERVATION STATUS: This species is widely distributed in Mexico and in some areas it is abundant. Yet, data about the current status of their populations and their conservation are scarce.

Godman's long-tailed bat

Hector T. Arita

SUBSPECIES IN MEXICO

C. godmani is a monotypic species.

DESCRIPTION: *Choeroniscus godmani* is a small-sized bat. Like other glossophagines it has small ears, a small nasal leaf in form of an equilateral triangle, and an elongated face and tongue. The color of the back is dark brown in males and more gray in females (Hall, 1981). The uropatagium is broad and contains the whole tail, which is short and does not reach the edge of the uropatagium. It is very similar to *Hylonycteris underwoodi*, and it is not possible to distinguish the two species without examining the skull. Most specimens of *H. underwoodi* have three colored bands on the back, while *C. godmani* only has two.

DISTRIBUTION: In the Pacific slope *C. godmani* is found from southern Sinaloa; in the Gulf it is found from Veracruz to the northern parts of South America as Colombia, Venezuela and the Guyanas (Eisenberg, 1989). It has been recorded in the states of CS, GR, JA, NY, OX, SI, and VE.

EXTERNAL MEASURES AND WEIGHT

TL = 53 to 55 mm; TV = 6 to 11 mm, HF = 7 to 11 mm; EAR = 11 to 13 mm; FA = 31 to 35 mm.
Weight: 5 to 8 g.

DENTAL FORMULA: I 2/0, C 1/1, PM 2/3, M 3/3 = 30.

NATURAL HISTORY AND ECOLOGY: Godman's long-tailed bat's roosting habits are unknown. Another species of the same genus (*C. minor*) has been found forming a group of two males and six females, roosting beneath a fallen trunk on a stream of water (Nowak, 1999a; Tuttle, 1976). It feeds on nectar, pollen, and insects (Gardner, 1977); births occur at the beginning of the rainy season (Wilson, 1979). In Mexico, a pregnant female was found in July in Sinaloa and a nursing female in May in Oaxaca.

VEGETATIONAL ASSOCIATIONS AND ELEVATION RANGE: In Mexico, *C. godmani* seems to be more common in tropical deciduous forests of the Pacific slope, although it has been reported in more humid tropical forests. In Venezuela is associated with tropical rainforests (Eisenberg, 1989). In Mexico, it is usually found in areas below 500 m. In Venezuela it has been found from 350 m (Eisenberg, 1989).

CONSERVATION STATUS: *C. godmani* is not included in any official list of endangered species. It can be considered a rare species, however, by virtue of both its restricted range and its low numbers (Arita, 1993a, b).

Mexican long-tongued bat

Jorge Ortega R. and Hector T. Arita

DISTRIBUTION: The distribution of *C. mexicana* extends from the southwestern United States, northern and central Mexico, including the Baja California Peninsula and the Marias Islands, to El Salvador and Honduras (Arroyo-Cabrales et al., 1987). It has been reported in the states of AG, BC, BS, CH, CL, CO, DF, DU, GJ, GR, HG, JA, MI, MO, MX, NL, NY, OX, PU, QE, SI, SO, SL, TA, TL, VE, and ZA.

SUBSPECIES IN MEXICO

C. mexicana is a monotypic species.

DESCRIPTION: *Choeronycteris mexicana* is a medium-sized bat. The color of its back hair ranges from gray to brown, while the abdomen is lighter. The snout is very long and the tongue is elongated and extendible. The tail is short and extends up to one-third of the uropatagium, which is naked (Arroyo-Cabrales et al., 1987). Its face is shorter and wider than that of *Musonycteris*, and the teeth are less separated (Phillips, 1971).

EXTERNAL MEASURES AND WEIGHT

TL = 81 to 103 mm = TV = 6 to 12 mm; HF = 10 to 13 mm; EAR = 3 to 7 mm; FA = 39 to 45 mm

Weight: 20 g.

DENTAL FORMULA: I 2/0, C 1/1, PM 2/3, M 3/3 = 30.

NATURAL HISTORY AND ECOLOGY: *C. mexicana* feeds on nectar and pollen (Gardner, 1977; Villa-R., 1967). Stomach contents of 16 specimens collected in central Mexico (Álvarez and Gonzalez-Q., 1970) showed a large percentage of pollen grains of pitaya (*Lemaireocereus*), cazahuate (*Ipomoea*), maguey (*Agave*), and garambullo (*Myrtillocactus*). They also consume nectar and pollen from *Pseudobombax ellipticum* (Eguiarte et al., 1987). It roosts at the entrances of caves and abandoned mines, forming small groups; they have also been collected in basements of houses and hollow trees (Watkins et al., 1972). It is monoestrous (Wilson, 1979). In Mexico, pregnant females have been found mainly in February and March, but in Jalisco they have been reported in September (Watkins et al., 1972). Generally, females have a single young and occasionally two (Arroyo-Cabrales et al., 1987). It is suspected that, like other bats of the genus *Leptonycteris*, it undertakes latitudinal migrations (Hayward and Cockrum, 1971).

Choeronycteris mexicana. Scrubland. Tolcayuca, Hidalgo. Photo: Carlos Galindo.

VEGETATIONAL ASSOCIATIONS AND ELEVATION RANGE: *C. mexicana* has been collected in different habitats. In Mexico, there are records in tropical deciduous forests, tropical subdeciduous forests, thorny forests, xeric scrublands and coniferous and oak forests. It has been found from 300 m to 3,600 m (Arroyo-Cabrales et al., 1987).

CONSERVATION STATUS: The current state of the populations of *C. mexicana* is unknown. Although it has a wide distribution in Mesoamerica, it is only locally abundant (Arroyo Cabrales et al., 1987; Villa-R., 1967).

Commissari's long-tongued bat

Hector T. Arita

SUBSPECIES IN MEXICO
Glossophaga commissarisi commissarisi Gardner, 1962
Glossophaga commissarisi hespera Webster and Jones, 1982

DESCRIPTION: *Glossophaga commissarisi* is a small-sized phyllostomid bat. As in other species of the genus, the ears are short, the face is long, and the nasal leaf is comparatively small. The uropatagium is wide and the tail is short, completely covered by the interfemoral membrane. It is very similar to *Glossophaga soricina*, but it can be distinguished by its smaller size, darker color, a longer nasal leaf (approximately 8 mm, measured in fresh specimens), larger lower incisors, and non-procumbent upper incisors (Gardner, 1962; Jones et al., 1972).

DISTRIBUTION: *G. commissarisi* includes three subspecies with separated distributions. *G. commissarisi hespera* is located on the Pacific side of Mexico, from Sinaloa to Colima (Webster and Jones, 1982d). *G. commissarisi commissarisi* is distributed from southern Mexico (Veracruz, Oaxaca, and Chiapas) to Panama (Eisenberg, 1989). Finally, *G. commissarisi bakeri* is found in the Amazon in Brazil, Colombia, Ecuador, and Peru (Webster and Jones, 1987). It has been registered in the states of CL, CS, DU, JA, NY, OX, SI, and VE.

EXTERNAL MEASURES AND WEIGHT
TL = 42 to 61 mm; TV = 4 to 10 mm; HF = 10 to 12 mm.; EAR = 12 to 15 mm; FA = 31 to 37 mm.
Weight: 8 to 12 g.

DENTAL FORMULA: I 2/2, C 1/1, PM 2/3, M 3/3 = 32.

NATURAL HISTORY AND ECOLOGY: Despite being a relatively common species, very little is known about the natural history of *G. commissarisi*. Most specimens have been captured on streams of forested areas (Gardner, 1962; Watkins et al., 1972) and the only information about shelters is the report of Jones et al. (1972), who captured five individuals in a cave in Sinaloa. The only empirical information about their feeding habits is the discovery of traces of lepidopterans, fruit, and pollen of *Musa* and *Mucuna* in the stomach of two individuals of Costa Rica (Howell and Burch, 1974). Wilson (1979) suggests that this species has two annual peaks of reproductive activity.

VEGETATIONAL ASSOCIATIONS AND ELEVATION RANGE: This species has been collected in a variety of habitats, from savannahs and thorny forests to tropical rainforests, passing through tropical deciduous forests, tropical subdeciduous forests, and cloud forests. It has been found from sea level up to 2,000 m. Most localities have intermediate altitudes (between 500 m and 1,500 m).

CONSERVATION STATUS: *G. commissarisi* is not included in any official list of endangered species. Its range is wide and in some localities it is comparatively abundant. In addition, it has been frequently found in disturbed areas such as artificial savannas and orchards. All this suggests that this species does not requires urgent protective measures.

Gray long-tongued bat

Hector T. Arita

SUBSPECIES IN MEXICO
G. leachii is a monotypic species.
It was known as a subspecies of *G. soricina.*

DESCRIPTION: *Glossophaga leachii* is a small-sized bat. It has short ears, relatively large eyes, a small nasal leaf in form of an equilateral triangle, and an elongated face of less extent than that of other specialized glossophagines. The dorsal fur is cinnamon brown (Webster, 1984). The uropatagium is large; the tail is short and does not reach the edge of the interfemoral membrane. In North America, it is the largest species of the genus. It differs from *G. soricina* and *G. morenoi* by the non-procumbent upper incisors, and from *G. commissarisi* by having larger lower incisors (Webster, 1984).

DISTRIBUTION: *G. leachii* is an endemic species of Mesoamerica. It is distributed from the south of Colima and Jalisco, along the Pacific slope, to Costa Rica (González-Ruíz and Villalpando, 1997; González-Ruíz et al., 2000; Webster, 1984). It has been registered in the states of CL, CS, GR, JA, MI, MO, MX, TL, and VE.

EXTERNAL MEASURES AND WEIGHT
TL = 47 to 64 mm; TV = 4 to 10 mm, HF = 10 to 12 mm; EAR = 12 to 15 mm; FA = 35 to 38 mm.
Weight: 11 g.

DENTAL FORMULA: I 2/2, C 1/1, PM 2/3, M 3/3 = 34

NATURAL HISTORY AND ECOLOGY: The gray long-tongued bat roosts inside caves, abandoned buildings, and culverts. The only information about its feeding behavior is an observation of an animal visiting the flowers of *Pseudobombax ellipticum* in Morelos (Eguiarte et al., 1987). It is likely that, like other species of the same genus, it bases its diet on a combination of nectar, pollen, fruits, and insects. Pregnant females have been found with one embryo from February to November, and lactating females have been found in February, March, June, and November (Webster, 1983).

VEGETATIONAL ASSOCIATIONS AND ELEVATION RANGE: This species is mainly found in tropical deciduous forests or subdeciduous forests. It has also been found in the transition of pine-oak forests. It has been collected from sea level up to 2,400 m (Davis, 1944). Most localities have intermediate altitudes (between 500 m and 1,500 m).

CONSERVATION STATUS: *G. leachii* is not included in any official list of endangered species. Due to its apparent local rarity, and because it is endemic to Mesoamerica, it is necessary to conduct studies on its situation.

Western long-tongued bat

Hector T. Arita

SUBSPECIES IN MEXICO

Glossophaga morenoi mexicana Webster and Jones, 1980
Glossophaga morenoi morenoi Martinez and Villa, 1938
Before the revision of Gardner (1986) this species was known as *G. mexicana.*

DESCRIPTION: *Glossophaga morenoi* is a small bat of the family Phyllostomidae. It can be recognized by the elongated face and tongue and the reduced nose leaf shaped like an equilateral triangle. The fur has two dorsal bands clearly defined, with the bases clear and tips dark (Webster and Jones, 1985). The species *Glossophaga* are difficult to separate and sometimes there is a need to examine the skull to achieve a proper identification. *G. morenoi* can be distinguished from other species by its procumbent upper incisors (projected forward) and lower incisors and small spaces between them.

DISTRIBUTION: *G. morenoi* is an endemic species of Mexico. It is distributed in the lowlands of the Pacific slope from Chiapas and Michoacan, and in the depression of the Rio Balsas (Chávez and Ceballos, 1998; González-Ruíz et al., 2000; Webster and Jones, 1985). It has been registered in the states of CL, CS, GU, MI, MO, MX, OX, and PU.

EXTERNAL MEASURES AND WEIGHT

TL = 52 to 58 mm; TV = 5 to 11 mm, HF = 10 to 11 mm; EAR = 13 to 15 mm; FA = 32 to 37 mm.
Weight: 7 to 10 g.

DENTAL FORMULA: I 2/2, C 1/1, PM 2/3, M 3/3 = 34.

NATURAL HISTORY AND ECOLOGY: The information on this bat is scarce and it is difficult to interpret it because of the complexity of the taxonomy of the genus; it has been summarized by Webster and Jones (1985). A good number of reports about the ecology of *G. soricina* published before the reports of Webster and Jones (1980) most probably refer to *G. morenoi.* This bat roosts inside caves, tunnels, and old buildings. Apparently, there is no information about their feeding habits, but their diet is probably based on a combination of nectar, pollen, fruits, and insects. The only information on its reproductive cycle is about a pregnant female found in March and a breastfeeding female in April.

VEGETATIONAL ASSOCIATIONS AND ELEVATION RANGE: *G. morenoi* is mainly found in tropical deciduous forests and subdeciduous forests, but it has also been found in thorny forests and forests of pine-oak. Of the *Glossophaga* species, this is the species that has greater affinity for arid sites. It is found from sea level up to 1,500 m, although most locations are located almost at sea level or lower than 300 m (Webster and Jones 1985).

CONSERVATION STATUS: *G. morenoi* is not included in any official list of endangered species. It is an endemic species of Mexico and, judging by its relative scarcity in collections, it can be assumed that it is rather rare. Studies are needed in order to know more about their natural history and to determine the actual status of their populations.

Pallas long-tongued bat

Jorge Uribe and Hector T. Arita

SUBSPECIES IN MEXICO

Glossophaga soricina handleyi Webster and Jones, 1980
Glossophaga soricina mutica Merriam, 1898

DISTRIBUTION: *G. soricina* is distributed from Mexico to South America; it is considered a tropical species. In Mexico, its distribution virtually includes all the Neotropical region, from Chihuahua, Sonora in the west, and Tamaulipas in the east, it follows both coastal slopes reaching the Eje Volcanico Transversal and hence the Yucatan Peninsula. The subspecies (*G. s. mutica*) is a resident of the Islas Marias (Álvarez, 1991; Hall, 1981). It has been reported in the states of CH, CL, CS, DF, DU, GR, JA, MI, MX, NY, OX, PU, QE, QR, SL, SO, TA, YU, and ZA.

DESCRIPTION: For the genus, *Glossophaga soricina* is a medium-sized bat. It has a long snout fitted with a nasal leaf and a very long, tubular, and protractile tongue provided with filiform papillae. The face has a length similar to that of the cranial box, with a moderate decline of the face toward the skull; the ears are small and rounded, the interfemoral membrane is wide, and the postpalatal crest is relatively uniform in height (Álvarez et al. 1991; Webster, 1983). They have relatively small wings with an average wingspan of 25 cm (Lemke, 1984). The length of the forearm is on average 60% of the total length (Álvarez et al., 1991). The color of the back varies from dark brown to light reddish-brown; ventrally it is toffee colored (Álvarez et al., 1991; Eisenberg, 1989; Nowak, 1999a). The incisors are well developed; the lower incisors are relatively longer than the upper incisors, and dental cavities have been reported. Females are generally larger than males, so there is sexual dimorphism. The chromosome number is diploid, with 32 chromosomes (Álvarez et al., 1991).

EXTERNAL MEASURES AND WEIGHT

TL = 49 to 64 mm; TV = 5 to 10 mm, HF = 7 to 11 mm; EAR = 9 to 15 mm; FA = 32 to 39 mm.
Weight: 9 to 10.5 g.

DENTAL FORMULA: I 2/2, C 1/1, PM 2/3, M 3/3 = 34.

NATURAL HISTORY AND ECOLOGY: These bats roost in a variety of sites, including caves, abandoned mines, tunnels, hollow trees, culverts on roads, buildings, and bridges, among others (Álvarez, 1968; Nowak, 1999a; Tuttle, 1976). They can associate with 30 species of bats (Ramírez et al., 1984). Males and females form colonies, but females with their offspring form maternity colonies in certain periods of the year. The number of individuals per colony is variable and may be greater than 2,000 individuals of both sexes (Eisenberg, 1989; Hall and Dalquest, 1963). In comparison with larger species of the same area, its home range is reduced; this suggests a probable relationship between body size and home range (Álvarez et al., 1991). They do not show a specific pattern in flight routes and habitats change temporarily (Bonaccorso, 1979). It feeds on insects, fruits, pollen, nectar, and floral parts (Gardner, 1977). They are specialists in feeding on nectar and pollen as they have a long and extendible tongue and their hair has divergent scales to catch pollen; they have a very specialized physiology to digest the pollen and nectar so they are considered important pollinators of some plants, especially shrubs and trees (Eisenberg, 1989; Lemke, 1984). They forage either by hovering at a flower or by hanging or sprawling on a flower using their thumbs (Lemke, 1984). In Mexico, they have been observed feeding on pollen above 800 m and rarely below this elevation (Álvarez and González-Q., 1970). They have a bimodal pattern of activity, with peaks of activity just after dark and before sunrise (LaVal, 1970). They also have unimodal patterns or even polymodal ones, visiting flowers every one or two hours (Ramírez et al., 1984). The peak of feeding activity occurs between the first and fourth hour after dusk; subsequently activity declines as a result of the gradual reduction of nectar in the plant, reaching the lowest rate of foraging when the amount of nectar produced by the plant is below 50% (Lemke, 1984). Some bats defend small territories, guarding the food and foraging line with well-defined routes (Lemke, 1984). The size of the foraging sites with agave ranges between 3 m^2 and 10 m^2 (Lemke, 1984). They experience polyestry, and therefore females may be pregnant every month (Wilson, 1979). They have an average of two to three offspring per year (Eisenberg, 1989). Usually, each female gives birth to a single young, which is fed with milk until two months old; young are able to fly between 25 and 28 days old (Álvarez et al., 1991). The young are carried by the females in a lateral position (Kleiman and Davis, 1979). Longevity is variable, but Jones (1982) reported a captive specimen that lived 10 years.

VEGETATIONAL ASSOCIATIONS AND ELEVATION RANGE: Although these bats prefer open and humid areas, they can be found in all types of tropical vegetation; it is also a common inhabitant of scrub temperate zones. Due to their enormous plasticity, they can tolerate damaged areas or crops (Álvarez et al., 1991; Eisenberg, 1989; Lemke, 1984). They have been found from sea level up to 2,600 m (Álvarez et al., 1991). Álvarez (1968) mentions that this species is more commonly found in low-lying areas.

CONSERVATION STATUS: *G. soricina* is not included in any official list of endangered species. It is an abundant species widely distributed; it survives in disturbed areas, and so it is not considered threatened.

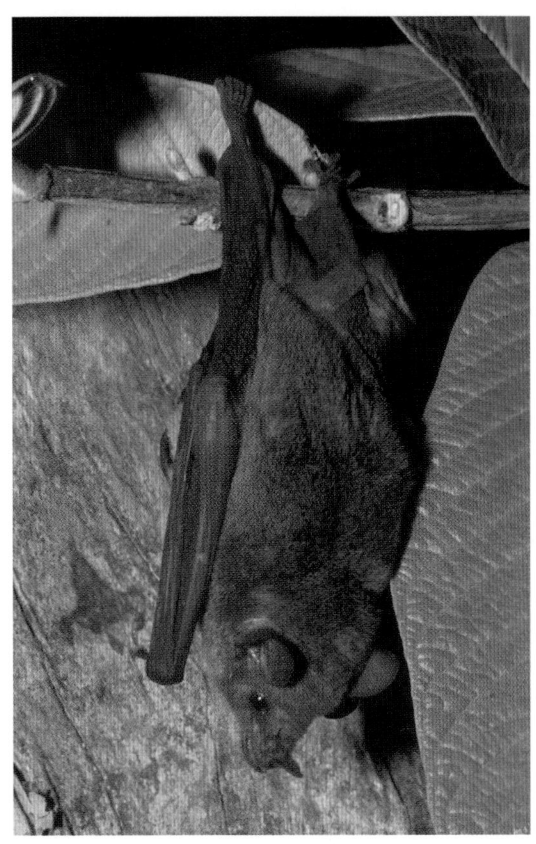

Glossophaga soricina. Oak forest. Pinal de Amoles, Querétaro. Photo: Gerardo Ceballos.

Hylonycteris underwoodi Thomas, 1903

Underwood's long-tongued bat

Hector T. Arita

SUBSPECIES IN MEXICO

Hylonycteris underwoodi minor Phillips and Jones, 1971
Hylonycteris underwoodi underwoodi Thomas, 1903
Álvarez and Álvarez-Castañeda (1991b) proposed to classify the two subspecies at a higher level. This change has not been adopted, however (Arita and Ceballos, 1997; Cervantes et al., 1996; Wilson and Reeder, 1993).

DESCRIPTION: *Hylonycteris underwoodi* is a small-sized bat. Females are larger than males (Phillips and Jones, 1971). As in other glossophagines, the interfemoral membrane is broad and the tail is small. Externally it is similar to bats of the genus *Glossophaga*, but differs from them by its smaller size, darker color (dark chocolate), more delicate face, the absence of lower incisors, and the three-banded pattern of its dorsal fur (although some individuals have only two bands). The latter also distinguishes *Choeroniscus godmani.*

DISTRIBUTION: *H. underwoodi* is an endemic species of Mesoamerica. In the Pacific slope it is distributed from Nayarit, and in the Gulf from Veracruz to Panama (Chavez and Ceballos, 1998; Emmons and Feer, 1999; Hall, 1981). It has been registered in the states of CL, CS, GR, JA, MX, NY, OX, TB, and VE.

EXTERNAL MEASURES AND WEIGHT
TL = 50 to 60 mm; TV = 7 to 10 mm, HF = 7 to 11 mm; EAR = 9 to 13 mm; FA = 32 to 38 mm.
Weight: 6 to 9 g.

DENTAL FORMULA: I 2/0, C 1/1, PM 2/3, M 3/3 = 30.

NATURAL HISTORY AND ECOLOGY: Underwood's long-tongued bats form small groups of 2 to 10 individuals, roosting in superficial caves (Hall and Dalquest, 1963), under bridges, in culverts, and in hollow logs (LaVal, 1977). They mainly feed on nectar and pollen and supplement their diet with insects such as moths (Howell and Burch, 1974; Sánchez-Casas and Álvarez, 1997). They have a bimodal pattern of reproduction (Wilson, 1979). In Costa Rica, pregnant females have been found from January to April and from August to November.

VEGETATIONAL ASSOCIATIONS AND ELEVATION RANGE: The subspecies *H. underwoodi underwoodi* is mainly found in tropical rainforests, forests of oak (*Quercus*) and pine, and cloud forests. The subspecies *H. underwoodi minor* is typical of tropical deciduous forests and the transition of these to the forest of oak (*Quercus*). It is distributed from approximately 100 masl to 2,100 masl. Most localities have intermediate altitudes (between 500 m and 2,000 m).

CONSERVATION STATUS: *H. underwoodi* is not included in any official list of endangered species. Its relative scarcity in collections, however, indicates that it is not a common species and its range is very restricted (Arita, 1993a, b). Therefore, it is necessary to monitor its populations and establish accurately its current situation.

Hylonycteris underwoodi. Tropical rainforest.
Photo: M.B. Fenton.

Mexican long-nosed bat

Hector T. Arita

SUBSPECIES IN MEXICO
L. nivalis is a monotypic species.
Due to the perplexity of the natural history of the genus *Leptonycteris*, many references, previous to the revision of Davis and Carter (1962) on *L. nivalis*, actually relate to *L. yerbabuenae*.

DESCRIPTION: *Leptonycteris nivalis* is the largest species of Mexican glossophagines. As in other nectar-feeding bats, the ears and nasal leaf are small and the face and tongue are elongated. The overall color is grizzly, grayer than in *L. yerbabuenae*, and the dorsal hair is longer. The interfemoral membrane is reduced to a very narrow band covered with abundant hair on the edge. The tail consists of three tiny vertebrae not appreciated externally. It differs from *L. yerbabuenae* by its grayer and longer dorsal hair, its narrowest and hairy uropatagium, and its longer wings (Arita and Humphrey, 1988; Davis and Carter, 1962).

EXTERNAL MEASURES AND WEIGHT
TL=76 to 85 mm; TV = 0, HF = 13 to 19 mm; EAR = 17 to 19 mm; FA = 53 to 59 mm.
Weight: 18 to 30 g.

DENTAL FORMULA: I 2/2, C 1/1, PM 2/3, F 2/2 = 30.

NATURAL HISTORY AND ECOLOGY: *L. nivalis* mainly roosts in caves, tunnels, and abandoned mines (Hensley and Wilkins, 1988). In the southern part of its distribution, colonies are generally small, with less than 500 individuals. In contrast, northern colonies of up to 10,600 individuals have been found (Easterla, 1972b). This species is migratory, at least in the northern part of its distribution (Easterla, 1972b). It is possible that southern populations perform altitudinal migrations.

Leptonycteris nivalis. Specimen in captivity.
Photo: Scott Altenbach.

DISTRIBUTION: *L. nivalis* is virtually an endemic species of Mexico (Arita and Humphrey, 1988). Outside of Mexico, it is only found in two small regions of the United States: in Hidalgo County in New Mexico and the National Park Big Bend in Texas. In Mexico, it is mainly distributed in intermediate elevations along the Sierra Madre Oriental and the Eje Neovolcanico. The supposed specimens of *L. nivalis* of Guatemala (Hall, 1981) are probably *L. yerbabuenae* (Arita and Humphrey, 1988). It has been registered in the states of CO, DF, DU, GJ, HG, JA, MI, MO, MX, NY, NL, PU, QE, SI, SL, TA, VE, and ZA.

Very little is known about their diet. Most reports on the diet actually refer to *L. yerbabuenae*. *L. nivalis* feeds on nectar and pollen of some maguey (*Agave*; Carter and Jones, 1978; Davis, 1974) and other plants, including Convolvulaceas (*Ipomoea*), Bombacaceas (*Ceiba*), and cacti such as garambullo (*Myrtillocactus geometrizans*; Álvarez and González-Q., 1970). Its pattern of reproduction is poorly understood. In Texas, nursing females were found in June and July, but pregnant females have never been found, suggesting that mating and births occur in Mexico during the spring and females then reach the caves in Texas to form maternity colonies (Davis, 1974; Easterla, 1972b). In Mexico, pregnant females have been found in March and April, and nursing females and juveniles in July (Wilson et al., 1985), thus coinciding with the cycle observed in Texas. In Morelos a pregnant female and a lactating female were found in January, however, suggesting that the pattern of reproduction may be different in the southern distribution of this species. They could be the pollinators of plants of economic importance such as wild populations of the giant agave (*Agave salmiana*), which is a plant pollinated by bats (Martinez del Rio and Eguiarte, 1987).

VEGETATIONAL ASSOCIATIONS AND ELEVATION RANGE: This species is mainly found in xeric environments, pine-oak forests, and the transition between the latter and tropical deciduous forests (Arita, 1991). Most localities are between 1,000 m and 2,200 m (Arita, 1991). Of the known locations, only 5 are below 1,000 m. Two records indicate that this bat can reach high altitudes: Saussure (1860) described the type locality of *L. nivalis* as "the snow line of the Pico de Orizaba," while Koestner (1941) found a colony of Mexican long-nosed bats at a height of 3,780 m at Cerro Potosi, Nuevo Leon.

CONSERVATION STATUS: *L. nivalis* has been classified as an endangered species on the list of endangered species in the United States and Mexico (SEMARNAT, 2010; Shull, 1988). Apparently, populations in Mexico have declined in recent years (Wilson et al., 1985), although further studies are needed to understand their situation. This species is almost endemic to Mexico and is not abundant in the sites where it is located. It is also a migratory species. All these factors contribute to classify it as an obvious candidate for future studies aimed at ensuring its conservation.

Leptonycteris yerbabuenae Martínez and Villa, 1940

Lesser long-nosed bat

Hector T. Arita

SUBSPECIES IN MEXICO

L. yerbabuenae is a monotypic species

DESCRIPTION: *Leptonycteris yerbabuenae* is a medium-sized bat. It has small ears, an elongated face, and a small nasal leaf. The hair on the back is short and light brown. These bats have no visible external tail and the uropatagium is reduced to a narrow membrane without a fringe of hair on the edge. It differs from *L. nivalis* by its cranial characteristics, its smaller size, its fur (the fur of *L. nivalis* is more gray and long), and by having shorter wings, particularly the last phalanx of the third finger (Arita and Humphrey, 1988).

Leptonycteris yerbabuenae. Arid scrubland. Photo: Scott Altenbach.

EXTERNAL MEASURES AND WEIGHT
TL = 75 to 85 mm; TV = 0, HF = 12 to 15 mm; EAR = 14 to 17 mm; FA = 47 to 56 mm. Weight: 15 to 25 g.

DENTAL FORMULA: I 2/2, C 1/1, PM 2/3, F 2/2 = 30.

NATURAL HISTORY AND ECOLOGY: Lesser long-nosed bats roost in caves and abandoned mines, forming colonies of up to 100,000 (Hayward and Cockrum, 1971; Tuttle, 1976). Northern populations are migratory and travel every year from the southwestern United States to Mexico, probably following a "nectar corridor" formed by different species of plants that provide food along the migratory route (Ceballos et al., 1997; Cockrum, 1991). It mainly feeds on nectar and pollen of tropical and subtropical plants such as *Pseudobombax* and *Ceiba*, maguey-related species (*Agave* spp., *Manfredo brachystachya*), and various cacti (*Carnegiea gigantea*, *Pachycereus pringlei*, *Stenocereus thurberi*; Álvarez and González-Q., 1970; Eguiarte and Burquez, 1987; Eguiarte et al., 1987; Hayward and Cockrum, 1971; Howell, 1974; Quiroz et al., 1986). They also feed on fruit of the cacti and occasionally on insects (Gardner, 1977). In northern Mexico and the southwestern United States, they reproduce at the end of the spring and early summer, forming large maternity colonies (Cockrum, 1991; Hayward and Cockrum, 1971). In the south, their reproduction pattern has not been studied, but it does not appear to coincide with that of the northern populations (Wilson, 1979). Evidence of a winter migration has been found; populations move from the coast of Jalisco to the Baja California Peninsula and Sonora to reproduce (Ceballos et al., 1998). A close relationship between lesser long-nosed bats and some plants they pollinate has been proved (Ceballos et al., 1997; Eguiarte and Burquez, 1987; Howell and Burch, 1974). Eguiarte and Burquez (1988) speculated that the population decline of these bats could have caused the decline of 75% in fertility of *Manfreda brachystachya*, a cactus pollinated by bats. It has been suggested that lesser long-nosed bats are important pollinators of wild populations of certain plants of commercial importance, particularly the variegated Caribbean agave (*Agave angustifolia*) and the blue agave (*Agave tequilana*; Arita, 1991, Arita and Wilson, 1987). Recent research has shown, however, that cultivated populations of *A. tequilana* have very little connection with the wild, so the importance of bats might not be so great, at least in the case of the maguey (Nabham and Fleming, 1992).

DISTRIBUTION: *L. yerbabuenae* is distributed from the southwestern United States (Arizona and New Mexico) and northern Mexico to El Salvador. Another population is found in the northern tip of South America, in Colombia and Venezuela (Arita and Humphrey, 1988; Eisenberg, 1989). In Mexico, it is mainly distributed in the dry parts of the tropics and subtropics. It is located along the Pacific slope from Sonora to Chiapas, including the Rio Balsas. In the slope of the Gulf of Mexico it is distributed from southern Tamaulipas to southern Veracruz. It has also been found in some locations of the Mexican plateau. It has been registered in the states of BS, CH, CL, CS, DF, DU, GJ, HG, JA, MI, MO, MX, NY, NL, OX, PU, QE, SI, SO, SL, TA, VE, and ZA.

VEGETATIONAL ASSOCIATIONS AND ELEVATION RANGE: This species mainly inhabits tropical deciduous forests, although it has been collected in tropical subdeciduous forests, thorny forests, pine-oak forests, cloud forests, and xeric scrublands (Arita, 1991; Handley, 1976). It is found from sea level up to 2,400 m. Most locations are located within 1,700 m. Unlike *L. nivalis*, it has been collected at sea level (Arita, 1991).

CONSERVATION STATUS: This species is included in the list of endangered species in the United States and Mexico (SEMARNAT, 2010; Shull, 1988) and is considered vulnerable by the International Union for Conservation of Nature (IUCN) (2010). In Mexico, some populations have declined in recent years (Eguiarte and Burquez, 1988; Wilson et al., 1985). Cockrum and Petryszyn (1991) recently questioned the vulnerability of this species, however, suggesting that its inclusion in the list of endangered species was premature because available data do not indicate any reduction of its populations.

Lichonycteris obscura Thomas, 1895

Dark long-tongued bat

Hector T. Arita

SUBSPECIES IN MEXICO

L. obscura is a monotypic species.

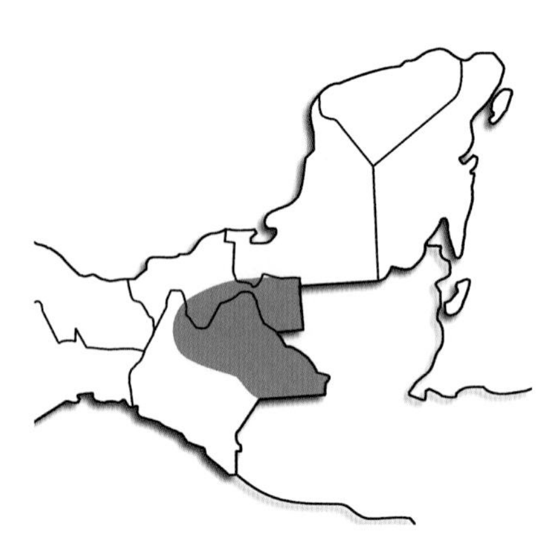

DISTRIBUTION: In Mexico, *L. obscura* is known only in the proximity of Pichucalco and in the Lacandon Forest, Chiapas. It is distributed from this town to Bolivia, passing through forested areas of Central America, northern South America, the Amazon basin, and the jungles of Peru and Bolivia. It has been registered in the states of CS and TB.

DESCRIPTION: *Lichonycteris obscura* is a small-sized bat. Like other species of the same family, it has an elongated snout and tongue and a very small nasal leaf in form of an equilateral triangle. It differs from other glossophagines by its tail and wide interfemoral membrane and by its three chocolate brown bands of fur.

EXTERNAL MEASURES AND WEIGHT

TL = 50 to 55 mm; TV = 6 to 10 mm, HF = 7 to 11 mm; EAR = 10 to 13 mm; FA = 30 to 34 mm.

Weight: 6 to 8 g.

DENTAL FORMULA: I 2/0, C 1/1, PM 2/3, F 2/2 = 26.

NATURAL HISTORY AND ECOLOGY: The roosting sites used by this species are unknown. According to its taxonomy and morphology, it is assumed to be polynivorous, but the only information about its diet is that of Álvarez-Castañeda and Álvarez (1991), who reported pollen of *Lonchocharpus* in the stomach of a specimen collected in Chiapas. The few data available on its reproduction indicate the presence of juveniles during the dry season (beginning of the year in Mexico and Central America).

VEGETATIONAL ASSOCIATIONS AND ELEVATION RANGE: *L. obscura* has been found almost exclusively in the tropical rainforest. It has also been collected in orchards, but always near mature rainforest. Almost all localities are located below 500 masl. Known localities in Mexico are found below 200 m.

CONSERVATION STATUS: *L. obscura* is not included in any official list, but its scarcity in collections indicates that it is not locally abundant but has a wide distribution (Emmons and Feer, 1999). In Mexico, its distribution is considered restricted, making it vulnerable to extirpation (Ceballos, 2007).

Musonycteris harrisoni Schaldach and McLaughlin, 1960

Colima long-nosed bat

Hector T. Arita

SUBSPECIES IN MEXICO

M. harrisoni is a monotypic species.

It is closely related to *Choeronycteris*; in fact, some authors included it in that genus (Hall, 1981), but most experts recognize the genus *Musonycteris* (Koopman, 1993; Webster et al., 1982).

DESCRIPTION: *Musonycteris harrisoni* is a medium-sized bat. It has a broad interfemoral membrane and a small tail. It differs from other polynivorous bats by the extremely elongated face and tongue, comparable only to that of *Platalina genovensium*, a bat exclusive to South America.

EXTERNAL MEASURES AND WEIGHT

TL = 75 to 85 mm; TV = 6 to 10 mm, HF = 10 to 14 mm; EAR = 13 to 17 mm; AN = 40 to 43 mm.

Weight: 7 to 13 g.

DENTAL FORMULA: I 2/0, C 1/1, PM 2/3, M 3/3 = 30.

NATURAL HISTORY AND ECOLOGY: Apparently, these bats roost in culverts and caves not too deeply, forming groups of a few individuals. Their feeding habits are unknown, but the extremely elongated face suggests that this species is one of the most specialized polynivores. It feeds on nectar and pollen from plants like Ceiba (*Ceiba pentandra*), *Cordia alliodora*, and *Ipomoea* (Sanchez and Álvarez, 1997). The generic name means "bat from the banana plantations," and it has been suggested that *Musonycteris* feeds on pollen from the flowers of the banana (Villa-R., 1967), but there is no empirical evidence to support this assumption. The only data on reproduction are two pregnant females collected in Colima in September (Wilson, 1979). Sánchez-Hernández (1978) reported the presence of juveniles and breastfeeding females in Jalisco, but did not mention the date of collection.

Musonycteris harrisoni. Tropical dry forest. Callejones, Colima. Photo: Marko Tschapka.

DISTRIBUTION: *M. harrisoni* is an endemic species of Mexico, restricted to areas covered with scrublands and dry lowlands of western Mexico. It has been found in localities along the Pacific coast and the Balsas depression. It has been registered in the states of CL, GR, JA, MI, MO, and MX.

VEGETATIONAL ASSOCIATIONS AND ELEVATION RANGE: *M. harrisoni* is known only in areas that were originally covered with tropical deciduous forests or xerophytes near these forests. It has been collected in banana plantations (*Musa* sp.) adjacent to this type of vegetation. It is distributed from sea level up to 1,700 m (Álvarez and Sánchez-C., 1997b; Ramírez-Pulido and Martínez Vázquez, 2007). Most of the localities are sited within 500 masl.

CONSERVATION STATUS: *M. harrisoni* is not included in any official list of endangered species. Its extreme rarity and local low density, its restricted range, and its habitat specificity suggest that this species is seriously threatened by severe disruption of its habitat (Ceballos and Miranda, 2000). The genus *Musonycteris* is endemic to Mexico; for this reason, this bat deserves special attention in conservation programs. There is a population in the Chamela-Cuixmala Biosphere Reserve (Ceballos and Miranda, 2000).

Tribe Stenodermatini

Artibeus hirsutus Andersen, 1906

Hairy fruit-eating bat

Guadalupe Tellez-Giron

SUBSPECIES IN MEXICO

A. hirsutus is a monotypic species.

DESCRIPTION: *Artibeus hirsutus* is a medium-sized bat, similar to but smaller than *A. jamaicensis*. It has no tail and its main feature is the presence of a fringe of hair on the brink of the interfemoral membrane that extends to the tibia. Facial lines are not visible or very weak when present. The length of the face is slightly longer than half of the cranial box; internal incisors are bifid and slightly larger than the external ones. The hair on the body is brown and on the abdomen is lighter (Anderson, 1960a; Hall, 1981; Webster and Jones, 1983).

DISTRIBUTION: *A. hirsutus* is an endemic species of Mexico, distributed in the Pacific slope, from southern Sonora to the southern Guerrero, penetrating the lowlands of the Eje Volcanico Transversal to the state of Morelos (Villa-R., 1967). It has been registered in the states of CL, GR, JA, MI, MO, MX, NY, SI, and SO.

EXTERNAL MEASURES AND WEIGHT

TL = 73.5 to 95 mm; TV = 0, HF = 12 to 18 mm; EAR = 17 to 22 mm; FA = 52 to 58 mm.

Weight: 32 to 47 g.

DENTAL FORMULA: I 2/2, C 1/1, PM 2/2, F 2/2 = 28.

NATURAL HISTORY AND ECOLOGY: These fruit bats prefer undisturbed sites with tropical vegetation. It roosts in caves, abandoned mines, and sometimes houses. It often shares its roosting site with other species such as *A. jamaicensis*, *A. intermedius*, *Balantiopteryx plicata*, *Leptonycteris nivalis*, *Anoura geoffroyi*, *Glossophaga soricina*, *Chiroderma salvini*, and *Sturnira lilium* (Anderson, 1960; Lukens and Davis, 1957). Its mating season takes place from February through September; during those months pregnant, lactating females and males with testicles in the scrotum have been collected (Watkins et al., 1972; Wilson, 1979).

VEGETATIONAL ASSOCIATIONS AND ELEVATION RANGE: This species inhabits tropical deciduous forests and occasionally fruit crops (Davis and Russell, 1953; Webster and Jones, 1983). Its altitudinal range goes from sea level up to 2,575 m (Webster and Jones, 1983).

CONSERVATION STATUS: *A. hirsutus* is a rare species, and very little is known about its populations. The information available is inadequate to determine its current situation.

Artibeus jamaicensis Leach, 1821

Jamaican fruit-eating bat

Jorge Ortega R. and Gabriela Steers

SUBSPECIES IN MEXICO

Artibeus jamaicensis paulus Davis, 1970
Artibeus jamaicensis richardsoni J.A. Allen, 1908
Artibeus jamaicensis triomylus Handley, 1966
Artibeus jamaicensis yucatanicus J.A. Allen, 1904

DESCRIPTION: *Artibeus jamaicensis* is a large-sized bat. There is significant morphological variation throughout its range, with smaller specimens in the north and larger ones in the south (Handley, 1987). They have an erect nasal leaf of 4 mm to 6 mm in length and a series of warts in the form of V in the upper lip (Silva-Taboada, 1979). It has no tail and the uropatagium is notched. The fur is dark brown, dense, but not too long, from 7 mm to 12 mm on the back and 4 mm to 11 mm on the abdomen. It has two facial lines of light white (Dalquest et al., 1952). The hair on the legs, forearms, and uropatagium is short. The skull is large and robust, with a flat and wide face; the postorbital processes are poorly developed, but the sagittal crests and occipitotemporalis are conspicuous (Davis, 1970b; Handley, 1987). In the subspecies *A. jamaicensis triomylus* the third lower molar is evident, whereas in the other subspecies this tooth is absent or reduced.

Artibeus jamaicensis. Cloud forest. El Jabalí, Colima. Photo: Gerardo Ceballos.

EXTERNAL MEASURES AND WEIGHT

TL = 65 to 94 mm; TV = 0, HF = 10 to 18 mm; EAR = 20 to 24 mm; FA = 54 to 61 mm.
Weight: 45 g.

DENTAL FORMULA: I 2/2, C 1/1, PM 2/2, F 3/3 = 32.

NATURAL HISTORY AND ECOLOGY: The Jamaican fruit-eating bat specializes in wild figs (*Ficus*); it also consumes fruits of other plants

DISTRIBUTION: *A. jamaicensis* is distributed in tropical regions, from the coastal slope of the states of Sinaloa and Tamaulipas in Mexico to the north of Bolivia and Argentina (Dalquest, 1953; Eisenberg, 1989). It has been registered in the states of CA, CL, CS, DU, GR, HG, JA, MI, MO, MX, NY, OX, PU, QE, QR, SI, SL, TA, TB, VE, YU, and ZA.

such as the breadnut (*Brosimum alicastrum*), the plum (*Spondias purpurea*), the pomarrosa (*Syzigium jambos*), the sapodilla (*Manilkara zapota*), the chancarro (*Cecropia obtusifolia*), the canelilla (*Quararibea funebris*), the hierba santa (*Piper sanctum*), and the ceiba (*Ceiba pentandra*) (August, 1981; Gardner, 1977; Tuttle, 1968). It complements its diet with insects, pollen, nectar, and leaves (Fleming et al., 1972; Kunz and Diaz, 1995). It can visit a fruit tree up to 15 times a night, indicating a good memory to locate and return to the same site (Fleming et al., 1977; Handley et al., 1991). The fruits to be consumed are brought to the roosting site, which might be at a considerable distance (175 m on average); this is usually a tree or a cave. These sites are easy to detect because under the foraging site are remnants of fruits, seeds, and newly sprouted seedlings (Goodwin, 1970; Kunz and Díaz, 1995; Vazquez-Yañes et al., 1975). Consuming only the flesh of the fruits of *Ficus* increases the level of sodium and decreases the level of potassium in the urine (Studier et al., 1983). It is an abundant bat of the Neotropics; a population of nearly 200 individuals per km^2 in Barro Colorado Island has been estimated (Handley et al., 1991). It roosts in a variety of places, including caves, foliage, holes in trunks, abandoned buildings, and tunnels (Goodwin and Greenhall, 1961). Occasionally, males roost alone, forming shelters in palms by cutting the midrib of the leaves and causing them to bend and provide a secure protection (Foster and Timm, 1976). Their main predators are owls (*Tyto alba*), opossums, falcons (*Falco rufigularis*), and some snakes (Eisenberg, 1989; Handley et al., 1991; Kunz et al., 1983; Morrison, 1978b). Its activity is significantly reduced during nights of full moon (Bonaccorso, 1979; Morrison, 1978a, b). *A. jamaicencis* has a harem-like social structure in which groups of adult females roost with a male over a long period. Their mating system is polygenic; the bats that live in hollow trunks at Barro Colorado defend the resources available and the individuals that roost in caves defend the females (Kunz et al., 1983; Morrison, 1979; Morrison and Morrison, 1981). Their reproductive cycle is bimodal polyestrous. They usually give birth to a single young after a gestation period of 2.5 months. Females are receptive during the postpartum and breastfeeding, so they can have up to two offspring per year. The birth peaks are strongly influenced by the availability of resources in the region, and delays in the embryonic development can occur during times of increased scarcity (Fleming, 1970; Handley et al., 1991; Willig, 1985).

VEGETATIONAL ASSOCIATIONS AND ELEVATION RANGE: This species is frequently found in the lowlands of the coast associated with tropical subdeciduous forests and thorny forests, but it has also been found in tropical rainforests, cloud forests, and secondary vegetation (Eisenberg, 1989). It is distributed from sea level up to 2,230 m (Handley, 1976; Naranjo and Espinoza, 2001).

CONSERVATION STATUS: *A. jamaicensis* is one of the most abundant species of bats in the tropics, including disturbed areas, so it is not endangered.

Great fruit-eating bat

Gabriela Steers and Jose Juan Flores M.

DISTRIBUTION: *A. lituratus* is found in the tropical lowlands from southern Sonora and Tamaulipas to the Yucatan Peninsula and Chiapas southward to South America (Davis, 1984; Redford and Eisenberg, 1992). It has been registered in the states of CA, CH, CL, CS, JA, GR, MX, NA, NL, OX, PU, QR, SI, SO, TA, TB, TL, VE, and YU.

SUBSPECIES IN MEXICO
Artibeus lituratus palmarum J.A. Allen and Chapman, 1897

DESCRIPTION: *Artibeus lituratus* is the largest species among the members of the genus *Artibeus*. It has an erect and well-developed nasal leaf and has no tail. The color of the coat is dark brown. The lower limbs are brown with little hair; the uropatagium has more hair (Lukens and Davis, 1957). The facial markings are white and well defined, extending from the nasal leaf toward the top of the head within the limits of the ears. Additionally, this bat has a tenuous white infraorbital line (Lukens and Davis, 1957). The skull is large and robust. It has a postorbital process and a prominent preorbital process united by a stripe that crosses the frontal region in an oblique manner. The sagittal crest is prominent, while the superorbital and preorbital peaks are well developed and have an angular position, forming a lateral extension in the nasofrontal depression (Davis, 1970b; Lukens and Davis, 1957).

EXTERNAL MEASURES AND WEIGHT
TL = 81 to 102 mm; TV = 0, HF = 16 to 26 mm; EAR = 10 to 27 mm; FA = 65.2 to 74.3 mm.
Weight: 48 to 80 g.

DENTAL FORMULA: I 2/2, C 1/1, PM 2/2, F 2/2 = 28.

NATURAL HISTORY AND ECOLOGY: *A. lituratus* is mainly frugivorous; its diet includes a wide variety of fruits, flowers, and leaves. In some sites of Central America the volume of insects in its diet is high (25%), so they are an important resource for this species (Fleming et al., 1972). Among the plants these bats consume are *Acroconia, Anacardium, Brosimum alicastrum* (Ramon), *Cecropia obtusifolia* (chancarro), *Cordia, Eugenia, Ficus* (higuillo), *Mangifera* (mango), *Musa* (banana), *Persea*, and *Spondias purpurea* (plum). Even when the presence of blood in the stomach content has been reported, this type of food is considered atypical (Gardner, 1977). It has a polyestric bimodal reproductive pattern with two peaks, in March and July. In the northern part of its distribution, females have a single young per year, while in the southern part they can have two offspring per year (Wilson, 1979). In July and April, the largest number of lactating females has been collected in southeastern Mexico (Ramírez-Pulido et al., 1983). They have gregarious habits and form colonies of approximately 25 individuals (Dalquest and Walton, 1970; Eisenberg, 1989). It occupies a wide range of roosting sites, including caves, tunnels, tree trunks, abandoned buildings, bridges, and the leaves of some palm species (Dalquest and Walton, 1970).

VEGETATIONAL ASSOCIATIONS AND ELEVATION RANGE: This species inhabits a wide variety of tropical vegetation, including tropical deciduous forests, tropical rainforests, scrublands, riparian vegetation, mangroves, and holm oak woods. They are also abundant in acahuales and other disturbed areas, including gardens. It is distributed from sea level up to approximately 1,080 m (Davis, 1984).

CONSERVATION STATUS: *A. lituratus* is a relatively common bat in the tropical regions of Mexico. Its density is apparently not affected by anthropogenic disturbances.

DISTRIBUTION: In Mexico, *C. perspicillata* is distributed in the tropical regions from Tamaulipas and Oaxaca to the Yucatan Peninsula. In Central and South America it has been found in southern Bolivia, Brazil, and Paraguay (Cloutier and Thomas, 1992). It has been registered in the states of CA, CS, OX, PU, QR, SL, TA, TB, VE, and YU.

Common short-tailed bat

Guadalupe Tellez-Giron and Oscar Sanchez

SUBSPECIES IN MEXICO

C. perspicillata is a monotypic species.

Pine (1972) indicated that there are three subspecies: *C. carollia azteca*, *C. perspicillata perspicillata* and *C. perspicillata tricolor*. McLellan (1984) concluded, however, that there are not enough features to recognize subspecies; thus, it is considered a monotypic species.

DESCRIPTION: *Carollia perspicillata* is a robust small to medium bat. It is the largest of the three species of the genus in Mexico. The main feature that identifies it is the length of the forearm, which is larger than 43 mm (in other species it is shorter). The hair color varies from brown to grayish-cinnamon brown. The hair of the intrascapular region has three bands; the base and tip are dark and the middle part is lighter. The interfemoral membrane is naked, and the tail is integrated into this membrane. The face is short and the nasal leaf is small and triangular. The skull does not have a complete zygomatic arch, the jaw in an aerial view is shaped like a V, and the internal incisors are covered by the external incisors; the third molars are larger than in other species (Cloutier and Thomas, 1992; Hall, 1981; Pine, 1972).

Carollia perspicillata. Tropical rainforest. Photo: M.B. Fenton.

EXTERNAL MEASURES AND WEIGHT

TL = 66 to 95 mm; TV = 11 to 14 mm, HF = 8 to 19 mm; EAR = 17 to 22 mm; FA = 43 to 46 mm.

Weight: 18 to 23.5 g.

DENTAL FORMULA: I 2/2, C 1/1, PM 2/2, F 3/3 = 32

NATURAL HISTORY AND ECOLOGY: The common short-tailed bat is relatively abundant. Captures usually occur almost at ground level. It is a generalist frugivore that feeds on at least 50 species of plants, fruits, and flowers (Bonaccorso, 1979; Fleming, 1988; Fleming et al., 1972). The fruits it consumes have a high protein value and are low in fiber. In the dry season their food is complemented by pollen and nectar. It disperses up to 2,500 seeds of the fruits consumed each night (Fleming, 1988). They also consume insects, mainly in the dry season (Fleming et al., 1972). It has gregarious habits, forming groups of 10 to 100 individuals; it roosts in caves, trunks, tunnels, and rock crevices and under the leaves of trees and constructions. Males are territorial and form harems with several adult females and juveniles; adult males and subadults without territory form groups and occupy sites away from the harem. Only between 12% and 17% of the colony form harems (Fleming, 1988). Foraging occurs in areas located up to 2 km from its daytime roosting site; night shelters are only used to consume food (Bonnacorso and Gush, 1987; Fleming and Heithaus, 1986; Heithaus and Fleming, 1978). In Central America, the pattern of reproduction is bimodal; the first period occurs between June and August, which coincides with the greater presence of fruit; the second period occurs when flowers are most abundant in the dry season (Fleming et al., 1972; Heithaus et al., 1975; Williams, 1986; Wilson, 1973b). Its predators include snakes (*Boa constrictor* and *Trimorphodon biscutatus*), owls (*Tyto alba*), and some mammals such as opossums, kinkajou, and the false vampire (*Vampyrum spectrum*) (Fleming, 1988).

VEGETATIONAL ASSOCIATIONS AND ELEVATION RANGE: *C. perspicillata* is found in tropical rainforests and dry deciduous forests. It is very common in secondary vegetation (DosReis and Guillaument, 1983; Fleming, 1988; LaVal and Fitch, 1977; Pine 1972). In Mexico, it is found from sea level up to 1,000 m. In South America it has been recorded up to 2,150 m (Koopman, 1978; McLellan, 1984; Pine 1972).

CONSERVATION STATUS: *C. perspicillata* is an abundant species even in very disturbed regions; it is not within any category of extinction.

Carollia sowelli Baker et al., 2002

Sowell's short-tailed bat

Guadalupe Tellez-Giron

SUBSPECIES IN MEXICO

C. sowelli is a monotypic species.

This species was known as *C. subrufa*. Baker et al. recently acknowledged, however, that the populations of Mexico and northern Central America are different species (Baker et al., 2002).

DESCRIPTION: *Carollia sowelli* is a small-sized bat. It is medium-sized between *C. perspicillata* and *C. subrufa*. It is distinguished by the following characteristics: the forearm has hair but not as abundant as in *C. perspicillata* and measures less than 41 mm;

Carollia sowelli. Dry forest. Photo: M.B. Fenton.

the hair of the body is long, dense, and smooth; the color varies both in the abdomen and in the back from cinnamon brown, dark gray, to dusky brown. An important feature is the four bands of color in the hair of the dorsal region; the basal and subterminal bands are obscure, and the second band and the tip are lighter. The face is more elongated than in *C. subrufa* with rounded warts in front of the lower lip. The skull is robust and less globular, with an incomplete zygomatic arch (Álvarez and Álvarez-Castañeda, 1991b; Pine, 1972; Sánchez-Hernández and Romero-Almaraz, 1995).

DISTRIBUTION: In the Mexican Gulf, *C. sowelli* is distributed from northern Veracruz, and in the Pacific side, in the states of Chiapas and Oaxaca. Its distribution extends into Costa Rica (Baker et al., 2002). It has been registered in the states of CA, CS, HG, OX, PU, QR, SL, TB, and VE.

EXTERNAL MEASURES AND WEIGHT
TL = 60 to 79 mm; TV = 7 to 14 mm, HF = 12 to 14 mm; EAR = 16 to 21 mm; FA = 37 to 42 mm.
Weight: 12 to 21 g.

DENTAL FORMULA: I 2/2, C 1/1, PM 2/2, F 3/3 = 32.

NATURAL HISTORY AND ECOLOGY: *C. sowelli* feeds primarily on wild fruits like figs (*Ficus* sp.) and possibly some insects trapped below the canopy of the jungle (Gardner, 1977; Handley, 1966; Pine, 1972). It roosts in caves, crevices of rocks, and houses and beneath the leaves of bananas (*Musa* sp.). It has been found with other bat species, including *Pteronotus davyi*, *P. parnellii*, *Mormoops megalophyla*, *Desmodus rotundus*, *Natalus stramineus*, *Glossophaga soricina*, and *Myotis* sp. (Pine, 1972). Males and females form separate colonies (Pine, 1972). There are records of pregnant and lactating females from March to July (Pine, 1972). Their reproductive pattern is bimodal (Wilson, 1979).

VEGETATIONAL ASSOCIATIONS AND ELEVATION RANGE: Sowell's short-tailed bats are often found in tropical rainforests, although they have also been collected in tropical deciduous forests (LaVal, 1970; Pine, 1972). It is found from sea level up to 2,400 m (Pine, 1972).

CONSERVATION STATUS: Very little is known about the populations of this species. It is not in any risk category.

Carollia subrufa (Hahn, 1905)

Gray short-tailed bat

Guadalupe Tellez-Giron

SUBSPECIES IN MEXICO
C. subrufa is a monotypic species.

DESCRIPTION: *Carollia subrufa* is a small-sized bat. It is the smallest of the three species of the genus in Mexico. Its distinctive features are the length of the forearm, which is less than 41 mm; the hair, which is short, scraggly, and thick; and its color, which varies from reddish-brown to dark brown in the intrascapular region. Also, the basal and intermediate bands are not clearly distinguished, thus giving the appearance of two bands. The forearm is almost naked, although some individuals may have a little hair. The face is short and the central chin has a wart with a double row of smaller warts in form of a V. The skull is robust, with the zygomatic arch incomplete. The palate has half of the row of teeth divergent, and in the jaw (occlusal view) the external incisors are clearly visible (Cloutier and Thomas, 1992; Owen, 1987; Pine, 1972; Ramírez-Pulido et al., 1977; Villa-R., 1967).

DISTRIBUTION: The distribution of this species includes western Mexico on the Pacific slope from Jalisco to Chiapas, extending possibly to Costa Rica and Panama in Central America (Eisenberg, 1989; McLellan, 1984; Pine, 1972). It has been registered in the states of CL, CS, GR, JA, MI, and OX.

EXTERNAL MEASURES AND WEIGHT
TL = 56 to 78 mm; TV = 5 to 15 mm, HF = 10 to 15 mm; EAR = 12 to 19 mm; FA = 34 to 40 mm.
Weight: 18 g.

DENTAL FORMULA: I 2/2, C 1/1, PM 2/2, F 3/3 = 32.

NATURAL HISTORY AND ECOLOGY: The gray short-tailed bat feeds primarily on sapodilla (*Mastichodendron* sp.), figs (*Ficus* sp.), nectar, and sometimes insects (Gardner, 1977). It roosts in caves, culverts, hollow trees, and abandoned houses or buildings in groups of several individuals (Ceballos and Miranda, 2000; Ramírez-Pulido et al., 1977). Its activity begins at dusk (Sanchez-Hernandez, 1984). Its pattern of reproduction is bimodal polyestric, with two peaks of births in the months of April–May and July–August. Females give birth to a single young (Wilson, 1979).

VEGETATIONAL ASSOCIATIONS AND ELEVATION RANGE: *C. subrufa* is frequently found in tropical dry forests, but also in tropical humid forests, in oak and pine forests, and in orchards and fruit crops (Álvarez and Álvarez-Castañeda, 1991; Pine, 1972; Ramírez-Pulido et al., 1977). It is found from sea level up to 1,200 m (Pine, 1972).

CONSERVATION STATUS: *C. subrufa* is a relatively common species that is not within any category of risk.

Centurio senex Gray, 1842

Wrinkle-faced bat

Guadalupe Tellez-Giron and Oscar Sanchez

SUBSPECIES IN MEXICO
Centurio senex senex Gray, 1842

DESCRIPTION: *Centurio senex* is a relatively small-sized bat. It has a distinctive face, without a nasal leaf and with very round and big eyes; the jaw is projected forward (conspicuous prognathism) and it has prominent folds. The ears are short with two lobes that are part of the tragus. The hair color is variable, but generally is light brown and lighter in the belly. Each shoulder has patches of white hair. The skin below the neck is loose and is used to cover the face when resting. The wing membranes have very obvious transversal bands that are not present in any other species of Mexican bat (Goodwin and Greenhall, 1961; Villa-R., 1967).

EXTERNAL MEASURES AND WEIGHT
TL = 58 to 65 mm; TV = 0, HF = 9 to 14 mm; EAR = 14 to 17 mm; FA = 41 to 45 mm.
Weight: 17 to 23 g.

DENTAL FORMULA: I 2/2, C 1/1, PM 2/2, F 2/2 = 28.

Centurio senex. Mangrove forest. Chamela-Cuixmala Biosphere Reserve, Jalisco. Photo: Gerardo Ceballos.

DISTRIBUTION: Wrinkle-faced bats are found in tropical regions from Sinaloa and Tamaulipas, including the Yucatan Peninsula, to the west of Venezuela (Nowak, 1999a). It has been registered in the states of CA, CL, CS, DU, GR, HG, JA, MI, NY, OX, QR, SI, TA, VE, and YU.

NATURAL HISTORY AND ECOLOGY: This species mainly roosts under the leaves of trees (Jones et al., 1971). Their main food is fleshy and juicy fruit. Little is known about the species of plants used for food. In Mexico, it feeds on fruit of "azulillo" (*Vitex molix*; Ramírez-Pulido and López-Forment, 1979). Occasionally, it has been observed in groups of up to eight individuals in the site where they feed (Ramírez-Pulido and López-Forment, 1979). We have little information about their reproduction, but they probably are polyestric; pregnant females have been found from January to August and young between February and August (Gardner, 1977; Jones et al., 1971; Nowak, 1999a; Wilson, 1979).

VEGETATIONAL ASSOCIATIONS AND ELEVATION RANGE: This species has been collected in several types of vegetation, from dry and wet tropical forests, up to sites with xeric vegetation (Watkins et al., 1972). It has been found from sea level up to 2,230 m (Jones et al., 1971; Naranjo and Espinosa, 2001).

CONSERVATION STATUS: *C. senex* is a relatively rare species. Very little is known about its populations, but it has not been included in any risk category.

Chiroderma salvini Dobson, 1878

Salvin's big-eyed bat

Jesus Pacheco R.

SUBSPECIES IN MEXICO

Chiroderma salvini salvini Dobson, 1878
Chiroderma salvini scopaeum Handley, 1966

DESCRIPTION: *Chiroderma salvini* is a medium-sized bat. The color of the body is brown, and it has four white lines clearly marked on its face, which distinguish it from *C. villosum* (Nowak, 1999a). Two of the lines extend from the posterior base of the nasal leaf to the top of the head, and the other two from the base of the upper lip to the base of the ears. It has a light-colored band on the back, from the shoulders up to the interfemoral membrane (Ceballos and Miranda, 1986). The face is relatively short. The ears are rounded with a yellowish margin and tragus. The nasal leaf and the interfemoral membranes are well developed; the interfemoral membrane along with the tibia and two-thirds of the proximal forearm are covered with hair. It has no tail or nose bones (Ceballos and Miranda, 1986; Nowak, 1999a).

DISTRIBUTION: *C. salvini* is distributed from southern Chihuahua and the central-west of Mexico to northern South America (Jones and Carter, 1979). It has been registered in the states of CH, CL, CS, DU, GR, HG, JA, MI, MX, NY, OX, PU, SI, VE, and ZA.

EXTERNAL MEASURES AND WEIGHT

TL = 67 to 77 mm; TV = 0, P = 12 to 15 mm; EAR = 15 to 19 mm; FA = 43.9 to 51.5 mm. Weight: 27.2 to 29.1 g.

DENTAL FORMULA: I 2/2, C 1/1, PM 2/2, F 2/2 = 28.

NATURAL HISTORY AND ECOLOGY: Salvin's big-eyed bat roosts in the foliage and hollow trees, in small groups or solitary (Emmons and Feer, 1997). Pregnant and lactating females have been reported from January to May and from July to December (Wilson, 1979). It reproduces twice a year, having a bimodal polyestric pattern. They are active two or three hours after nightfall (Ceballos and Miranda, 1986). Their food consists mainly of fruit (Gardner, 1977). Its stomach has a very elongated fundal blind gut and a very long small intestine, indicating that its diet is mainly based on plant material (Pacheco and Salazar, 1990).

VEGETATIONAL ASSOCIATIONS AND ELEVATION RANGE: This species mainly inhabits tropical deciduous forests and adjacent areas with secondary vegetation. It is found from sea level up to 1,722 m (Eisenberg, 1989).

CONSERVATION STATUS: The status of this species is undetermined. Despite their widespread distribution in western Mexico, they are not very abundant. The destruction of their habitat is an important factor that probably endangers their existence. Studies are needed to assess the status of its populations, as well as studies on its current geographical distribution.

Chiroderma salvini. Tropical rainforest. Photo: Bernal Rodriguez.

Hairy big-eyed bat

Jesus Pacheco R.

SUBSPECIES IN MEXICO
Chiroderma villosum jesupi J.A. Allen, 1900

DISTRIBUTION: *C. villosum* has a wide distribution in Central and South America; it is found from southern Mexico to western Brazil (Eisenberg, 1989; Jones and Carter, 1976; Jones and Carter, 1979; Nowak, 1999a). The distribution limit on the Pacific side is the region of the Isthmus of Tehuantepec in Oaxaca, and on the basin of the Gulf of Mexico are Veracruz and Hidalgo. It has been recorded in the states of CA, CS, HG, QR, OX, and VE.

DESCRIPTION: *Chiroderma villosum* is a medium-sized bat. The facial lines are tenuous and in some individuals are absent (Nowak, 1999a). They have no tail. The fur is light brown, soft, and long; the hair color of the back does not contrast with the ventral part (Eisenberg, 1989). The skull and teeth are robust; its molars are modified for chewing fruit, and thus they have lost the stylar cusps and are flattened and thick. The eyes are large and the snout is short and wide (Emmons and Feer, 1999). The genus is characterized by the absence of nasal bones (Nowak, 1999). It differs from *C. salvini* by being smaller and having more tenuous or absent facial lines (Hall, 1981).

EXTERNAL MEASURES AND WEIGHT
TL = 65 to 82 mm; TV = 0, HF = 10 to 16 mm; EAR = 14 to 20 mm; FA = 42.7 to 50.3 mm.
Weight: 20 to 35 g.

DENTAL FORMULA: I 2/2, C 1/1, PM 2/2, F 2/2 = 28.

NATURAL HISTORY AND ECOLOGY: The hairy big-eyed bat roosts in hollow trunks. It feeds on fruits, especially on figs, and occasionally insects (Emmons and Feer, 1999; Gardner, 1977). Pregnant and lactating females have been reported in March–May, July, and December (Medellín et al., 1986; Wilson, 1979). Judging by the gastric characteristics, its stomach has one of the largest specializations of the species of the tribe Stenodermatini (Gardner, 1977).

Chiroderma villosum. Tropical rainforest.
Photo: M.B. Fenton.

VEGETATIONAL ASSOCIATIONS AND ELEVATION RANGE: Hairy big-eyed bats inhabit tropical rainforests, including open areas and disturbed areas of secondary vegetation (Emmons and Feer, 1999; Medellín et al., 1986; Nowak, 1999a). It is found from sea level up to 500 m (Eisenberg, 1989).

CONSERVATION STATUS: *C. villosum* is a rare bat in the country, although in Chiapas it appears to be abundant (Álvarez-Castañeda and Álvarez, 1991). The data available show that its presence does not accurately indicate its abundance. It is therefore necessary to establish the status of its populations.

Dermanura azteca (Andersen, 1906)

Aztec fruit-eating bat

Gerardo Lopez Ortega and Maribel Ayala

SUBSPECIES IN MEXICO
Dermanura azteca azteca (Andersen, 1906)
Dermanura azteca minor (Davis, 1969)

DESCRIPTION: *Dermanura azteca* is one of the largest species of the genus *Dermanura.* It has a short face and a well-developed nasal leaf. It has no tail and its interfemoral membrane is narrow and covered with hair. Its back fur varies from dark brown to light brown. It differs from other species of the genus that live in Mexico by a skull larger than 21 mm; the row of teeth in the jaw of more than 7 mm; the forearm of no less than 41 m;, and its altitude range.

EXTERNAL MEASURES AND WEIGHT
TL = 58 to 73 mm; TV = 0, HF = 12 to 13 mm; EAR = 13 to 15 mm; FA = 35 to 41 mm. Weight: 18 to 24 g.

DENTAL FORMULA: I 2/2, C 1/1, PM 2/2, F 2/2 = 28.

NATURAL HISTORY AND ECOLOGY: The Aztec fruit-eating bat roosts inside caves, abandoned mines, and small wells, and between tree branches. It forms small colonies. The digestive tract is typical of a frugivore. It feeds on tejocotes, capulin, and cones of *Cupressus* and *Juniperus*, as well as insects (Ceballos and Galindo, 1984). Apparently, it breeds in the spring and summer. There are records of males with testicles in the scrotum in October and November, and pregnant females in March, April, and July. Young have been recorded between April and June.

VEGETATIONAL ASSOCIATIONS AND ELEVATION RANGE: This species is associated with temperate forests of oak, pine, fir, cloud forest, and disturbed sites such as banana orchards. It is distributed from 1,400 masl to 3,300 masl. Most records are located around 2,000 m.

CONSERVATION STATUS: This species is relatively common and survives in disturbed areas; therefore, it is not considered at risk.

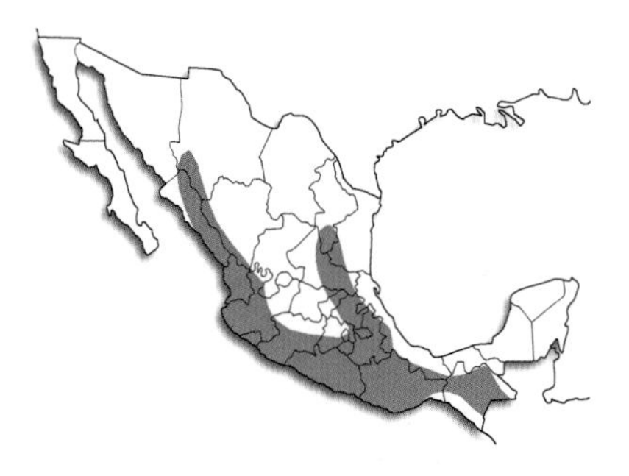

DISTRIBUTION: Three isolated areas of distribution are known (split distribution), two of them in Mexico. The first one includes mountainous areas surrounding the Mexican Altiplane and toward the highlands of Eje Volcanico Transversal to the Sierra Madre del Sur in Oaxaca (Davis, 1969; Hall, 1981; González-Christen et al., 2002). The other area covers the mountainous regions of Chiapas to Honduras, and the third covers the mountains from Costa Rica to the west of Panama. It has been registered in the states of CS, CL, DU, GR, HG, JA, MI, MO, MX, NL, NY, OX, QE, SI, SL, TA, TL, and VE.

Dwarf fruit-eating bat

Gerardo Lopez Ortega and Maribel Ayala

SUBSPECIES IN MEXICO
Dermanura phaeotis phaeotis (Miller, 1902)
Dermanura phaeotis nana (Andersen, 1906)
Dermanura phaeotis palate (Davis, 1970)

DESCRIPTION: *Dermanura phaeotis* is a small-sized bat with dark or gray thin and silky hair, an arched skull, and a short face (Davis, 1958); the interfemoral membrane is broad and hairless (Dalquest, 1953c). The four facial lines and the edges of the ears are light-colored (Davis, 1970a). It differs from *D. azteca* and *D. tolteca* by the broad and hairless interfemoral membrane and by the pale yellow facial lines and the margins of the ears (Lukens and Davis, 1957). It differs from *D. watsoni* by having two small molars in the jaw (Davis, 1958).

EXTERNAL MEASURES AND WEIGHT
TL = 51 to 60 mm; TV = 0, HF = 9 to 12 mm; EAR = 14 to 18 mm; FA = 35 to 42 mm. Weight: 8 to 16 g.

DENTAL FORMULA: I 2/2, C 1/1, PM 2/2, F 2/2 = 28.

Dermanura phaeotis. Tropical dry forest. Chamela-Cuixmala Biosphere Reserve, Jalisco. Photo: Gerardo Ceballos.

NATURAL HISTORY AND ECOLOGY: Dwarf fruit-eating bats form small colonies. They roost in caves and in the underside of large leaves; some alter the shape of the leaf to build their diurnal shelters (Davis, 1970a, b; Timm, 1987; Villa-R., 1967). Like in other stenodermatini, the digestive tract is typical of a frugivore (Forman et al., 1979). Their diet is based on fruits, but they also consume pollen and insects (Fleming et al., 1979; Heithaus et al., 1975; Timm, 1987). It has a polyestric bimodal reproductive pattern with females pregnant from January to April and from June to August (Davis, 1970a, b; Timm et al., 1989). Males with testicles in the scrotum are present during summer and winter, while younger individuals are reported for July (Davis, 1970a, b; Lukens and Davis, 1957; Timm, 1987).

VEGETATIONAL ASSOCIATIONS AND ELEVATION RANGE: This species is associated with cloud forests, coniferous forests, tropical subdeciduous forests, tropical deciduous forests, tropical rainforests, scrublands, pastures, and banana plantations. It is distributed from sea level up to 1,350 m. Most localities are found at heights of 400 m and 600 m (Davis, 1970a, b; Lukens and Davis, 1957; Ramírez-Pulido and López-Forment, 1979; Sanchez-Hernandez, 1978; Villa-R., 1967; Watkins et al., 1977).

CONSERVATION STATUS: *D. phaeotis* is not considered at risk because it is abundant in natural and disturbed regions (Ceballos and Miranda, 1986, 2000).

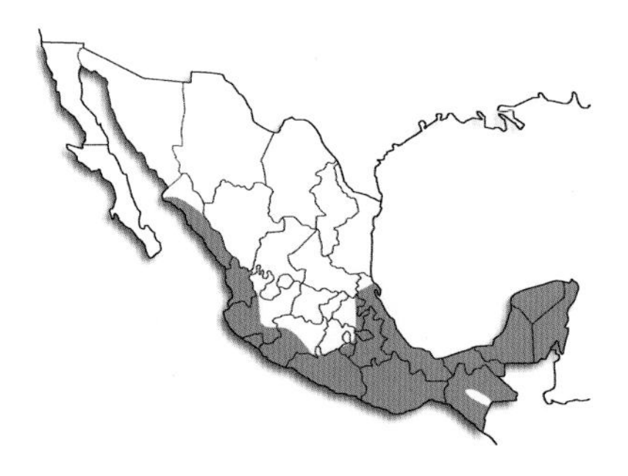

DISTRIBUTION: The dwarf fruit-eating bat's distribution covers both sides of the country; on the Pacific side it is found from southern Sinaloa to Costa Rica, and on the Gulf of Mexico in the middle part of Veracruz, the Yucatan Peninsula, and the Caribbean to Brazil and Peru (Davis, 1970a; Timm, 1985). It has been registered in the states of CA, CL, CS, GR, JA, MI, NY, OX, QR, SI, TA, TB, VE, and YU.

Dermanura tolteca (Saussure, 1860)

Toltec fruit-eating bat

Gerardo Lopez Ortega and Maribel Ayala

SUBSPECIES IN MEXICO

Dermanura tolteca tolteca (Saussure, 1860)
Dermanura tolteca hespera (Davis, 1969)

DESCRIPTION: *Dermanura tolteca* is one of the smallest species of the genus *Dermanura*. The fur is dark gray, with tenuous or absent white lines in the supraorbital region. The interfemoral membrane is deeply notched and has hair around the uropatagium (Davis, 1969). It differs from *D. azteca* by having a smaller skull, row of teeth in the jaw, and forearm and because it inhabits in lowlands. It differs from *D. phaeotis* and *D. watsoni* because these species do not have the notched uropatagium with hair (Davis, 1969; Hall, 1981; Webster and Jones, 1982).

EXTERNAL MEASURES AND WEIGHT

TL = 51 to 63 mm; TV = 0, HF = 8 to 12 mm; EAR = 15 to 19 mm; FA = 39 to 42 mm. Weight: 15 to 20 g.

DENTAL FORMULA: I 2/2, C 1/1, PM 2/2, F 2/2 = 28.

NATURAL HISTORY AND ECOLOGY: Toltec fruit-eating bats form small groups that roost in caves, buildings, and the underside of large leaves (Tuttle, 1976; Webster and Jones, 1982). They cut some lateral midribs of wide leaves to build their shelters in the form of a tent (Timm, 1987). The digestive tract is typical of a frugivore (Forman et al., 1979). It feeds on fruits of plants like *Ficus*, *Solanum*, *Cecropia*, and *Piper* (Dinerstein,

DISTRIBUTION: This species is distributed along the coastal plain of the Gulf and Pacific, from the southern part of Nuevo Leon and northern Sinaloa to the northwest of Colombia and Ecuador (Webster and Jones, 1982a). It has been registered in the states of CH, CL, CS, DU, GR, HG, JA, MI, MO, MX, NL, NY, OX, PU, QE, SI, SL, TA, VE, and ZA.

1986). It has a bimodal polyestric pattern of reproduction. Breastfeeding is seasonal and reproductive periods coincide with peaks of abundance of fruits (Dinerstein, 1986; Timm et al., 1989; Wilson, 1973b). Pregnant females and infants have been recorded from January to September and juveniles have been reported in August and September (Álvarez and Álvarez-Castañeda, 1991b; Carter et al., 1966; Davis, 1970a; Watkins et al., 1972; Wilson, 1979).

VEGETATIONAL ASSOCIATIONS AND ELEVATION RANGE: This species is associated with xeric scrublands and tropical deciduous and subdeciduous forests. It has also been found in temperate forests, including the cloud forest and the pine-oak forest. There are records in disturbed vegetation such as banana and coffee plantations (Álvarez and Álvarez-Castañeda, 1991b; Anderson, 1972; Iñiguez-Dávalos, 1993; Watkins et al., 1972). It is found from sea level up to 2,200 m (León Paniagua and Romo-Vázquez, 1991; Watkins et al., 1972). Most records are below 1,500 m.

CONSERVATION STATUS: Although its current situation is not completely known, *D. tolteca* is not considered at risk of extinction. It is relatively abundant in some regions, including disturbed areas.

Dermanura watsoni (Thomas, 1901)

Watson fruit-eating bat

Barbara Vargas Miranda

SUBSPECIES IN MEXICO
D. watsoni is a monotypic species.

DESCRIPTION: *Dermanura watsoni* is one of the smallest bats of the genus. The color of the back fur varies from brown to bright orange, while the ventral part is lighter. It has two facial lines and a white line on the edge of the ear (Hall, 1981). The interfemoral membrane is wide and naked. The lateral part of the frontal bone in the orbital area is significantly inflated (Davis, 1970a; Hall, 1981).

DISTRIBUTION: *D. watsoni* is distributed from southern Veracruz and Oaxaca to Panama. Apparently, it is absent in the tropical dry regions of Central America (Davis, 1970a; Hall, 1981). It has been registered in the states of CA, CS, OX, TB, and VE.

EXTERNAL MEASURES AND WEIGHT
TL = 51 to 61 mm; TV = 0, HF = 7 to 11 mm; EAR = 13 to 19 mm; FA = 36 to 41 mm. Weight: 9 to 15 g.

DENTAL FORMULA: I 2/2, C 1/1, PM 3/3, F 2/2 = 32.

NATURAL HISTORY AND ECOLOGY: Pregnant females have been collected in Chiapas in February, May, July, August, and November, which suggests the possibility of asynchronous polyestry and the possibility of three periods of birth (Álvarez and Álvarez-Castañeda, 1991b). Females collected in February, March, April, August, and November contained one embryo (Davis, 1970a). It basically feeds on fruits, and it has been called a specialized frugivore of *Ficus* (Medellín, 1993).

VEGETATIONAL ASSOCIATIONS AND ELEVATION RANGE: *D. watsoni* inhabits high jungle, median jungle and wetlands, from 150 m to 1,524 m (Álvarez and Álvarez-Castañeda, 1991b; Davis, 1970a; Hall, 1981).

Dermanura watsoni. Tropical dry forest. Calakmul Biosphere Reserve, Campeche. Photo: Gerardo Ceballos.

CONSERVATION STATUS: *D. watsoni* is not included in the list of endangered species (SEMARNAT, 2010), but it is a rare bat along its distribution (Davis and Russell, 1954b).

Enchisthenes hartii (Thomas, 1892)

Hart's little fruit bat

Joaquin Arroyo-Cabrales

SUBSPECIES IN MEXICO

E. hartii is a monotypic species (Arroyo-Cabrales and Owen, 1998).

Owen (1987) conducted a phylogenetic analysis of the species belonging to the tribe Stenodermatini (subfamily Phyllostominae), which included *Enchisthenes* (= *Artibeus*) *hartii*. He did not reach a definitive conclusion as to the phylogenetic relationships of these species, and therefore suggested that this species should be provisionally included within the genus *Dermanura*. Recently, it has been suggested, based on molecular studies, that *Enchisthenes* should be recognized as a valid genus (Van Den Bussche et al., 1993).

DISTRIBUTION: *E. hartii* is distributed in the highlands of the coastal slopes of Mexico, from central Jalisco and northeastern Tamaulipas to Chiapas. Their distribution generally reaches to Bolivia in South America; there is an isolated record in Tucson, Arizona (Hall, 1981). It has been registered in the states of CL, CS, GR, JA, MI, MX, OX, PU, TA, and VE.

DESCRIPTION: *Enchisthenes hartii* is a medium-sized bat. The third upper molar is well developed. The hair is almost black, thus it is easily distinguishable from other species of the genus. It has a couple of lighter lines that extend from the base of the nasal leaf to the inside base of the ears, and another pair from the commissure of the mouth to the external base of the ears. It has no tail (Villa-R., 1967).

EXTERNAL MEASURES AND WEIGHT
TL = 59 to 66 mm; TV = 0, HF = 10 to 13 mm; EAR = 12 to 16 mm; FA = 36 to 41 mm.
Weight: 14 to 18 mm.

DENTAL FORMULA: I 2/2, C 1/1, PM 2/2, F 3/3 = 32.

NATURAL HISTORY AND ECOLOGY: In Mexico, most individuals have been collected in mist nets placed across rivers or streams in coniferous and pine-oak forests, and in areas with tropical deciduous forests near fruit trees, amates (*Ficus* sp.), fig, macahuite, and zapotillo (Ramirez-Pulido and López-Forment, 1979; Villa-R., 1967). Ninety percent of the individuals collected in Venezuela are from wet low-montane forests (Handley, 1976). They are frugivores (Gardner, 1979); after cutting a fruit they perch on a branch to eat, using their forearms to handle and turn the fruit (De la Torre, 1955). Their reproductive activity is unknown; a nursing female was captured in Michoacan in April (Sánchez-Hernández et al., 1985), another was collected in Chiapas in July (Baker et al., 1971), and a pregnant female was caught in August, also in Chiapas (Álvarez and Álvarez-Castañeda, 1991b). In Costa Rica it reproduces all year (Gardner et al., 1970).

VEGETATIONAL ASSOCIATIONS AND ELEVATION RANGE: Hart's little fruit bat has been collected in coniferous forests, cloud forests, and tropical deciduous forests. It is distributed from sea level to nearly 2,000 m. In Mexico it has been collected above 600 m (McCarthy and Bitar, 1983).

CONSERVATION STATUS: The specimens collected in Mexico are mostly isolated records, so we need more fieldwork to determine the conservation status of this species. In Mexico this is a rare species, but not in some localities of South America (Arroyo-Cabrales and Owen, 1998).

Platyrrhinus helleri (Peters, 1866)

Heller's broad-nosed bat

Guadalupe Tellez-Giron

SUBSPECIES IN MEXICO
P. helleri is a monotypic species.
The name *Vampyrops helleri* has been widely used to designate this species, but Gardner and Ferrell (1990) stated that the correct name for the genus is *Platyrrhinus*.

DESCRIPTION: *Platyrrhinus helleri* is a medium-sized bat. It is easily distinguishable from other species, such as *Uroderma* and *Vampyrodes*, by being smaller, having darker hair, the presence of two pairs of white lines on its face and a white line that runs from the head to the middle of the back, and the notched interfemoral membrane covered with hair and a lighter fringe of hair on the edge. In addition, three-quarters of the forearm is covered with hair. The color of the back hair ranges from dark to intense brown; the abdomen is lighter (Davis et al., 1964; Hall, 1981).

DISTRIBUTION: Heller's broad-nosed bat's distribution extends in the tropical region from eastern and southeastern Mexico, excluding the northern part of the Yucatan Peninsula, into the heart of Bolivia and Brazil (Eisenberg, 1991; Hall, 1981). It has been registered in the states of CS, OX, TB, and VE.

EXTERNAL MEASURES AND WEIGHT
TL = 50 to 66 mm; TV = 0, HF = 9 to 14 mm; EAR = 15 to 19 mm; AN = 35 to 41 mm. Weight: 13 to 18 g.

DENTAL FORMULA: I 2/2, C 1/1, PM 2/2, F 3/3 = 32.

NATURAL HISTORY AND ECOLOGY: Heller's broad-nosed bat is common in wet habitats such as tropical rainforests and deciduous forests near streams and rivers (Davis et al., 1964). They roost inside caves or tunnels in the foliage, and in hollow branches, buildings, or fallen branches of palms (Tuttle, 1976). They are commonly captured in the passages of rivers, between the canopy of the jungle, or near banana plantations. Their main food is the fruit of the fig tree (*Ficus* sp.; Bonacorso, 1979). Its activity begins an hour after nightfall (Fenton and Kunz, 1977). Its pattern of reproduction is bimodal polyestric, with the first peak during the second half of the dry season and the second peak early in the first half of the rainy season (Fleming et al., 1972).

VEGETATIONAL ASSOCIATIONS AND ELEVATION RANGE: Heller's broad-nosed bat is mainly found in tropical rainforests, deciduous forests, and disturbed areas (Ferrell and Wilson, 1991). It is known as an inhabitant of lowlands, but in America it has been found up to 1,500 masl (Handley, 1976).

CONSERVATION STATUS: *P. helleri* is a rare species in Mexico; little is known about the status of its populations, thus it is difficult to determine precisely its conservation status. It is not considered in immediate danger.

Sturnira hondurensis Anthony, 1924

Highland yellow-shouldered bat

Guadalupe Tellez-Giron

SUBSPECIES IN MEXICO
Sturnira hondurensis ludovici Anthony, 1924
Sturnira hondurensis occidentalis Jones and Phillips, 1964

DESCRIPTION: *Sturnira hondurensis* is similar to *S. lilium*, but larger. The uropatagium is vestigial or very small and covered with dense hair. It has no tail. The color of the back hair is brown with dark gray and the abdomen is lighter. It has ochre-colored stains on the shoulders resembling epaulettes. The last phalanx of the third finger measures less than 15 mm, the lower incisors are bilobed, and the forearm is larger than 42 mm (Álvarez and Ramírez-Pulido, 1980; Baker and Phillips, 1965a; Davis, 1984).

EXTERNAL MEASURES AND WEIGHT
TL = 66 to 71 mm; TV = 0, HF = 10 to 14 mm; EAR = 14-17 mm; FA = 42 to 47 mm. Weight: 19 to 24 g.

DENTAL FORMULA: I 2/2, C 1/1, PM 2/2, F 3/3 = 32.

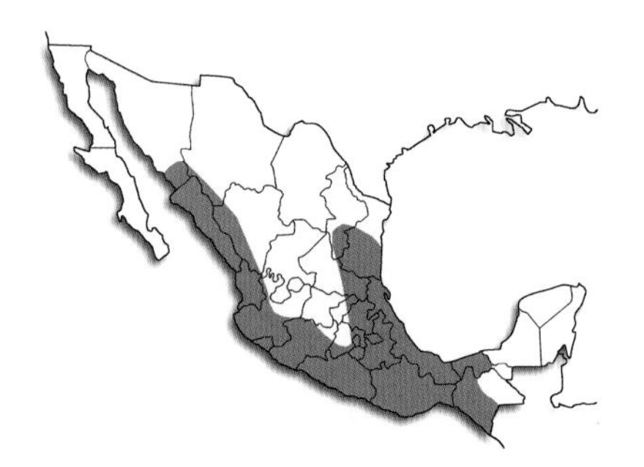

DISTRIBUTION: *S. hondurensis* is distributed on the Atlantic side, from southern Tamaulipas, and on the Pacific slope from southern Sinaloa, to Guyana, Ecuador, and Venezuela (Eisenberg, 1989; Polaco and Muñiz-Martinez, 1987). It has been registered in the states of CL, CO, GR, JA, MI, MX, OX, SL, TA, and VE.

NATURAL HISTORY AND ECOLOGY: *S. hondurensis* is frequently found in very humid environments. It is usually collected on streams or bodies of water and canyons. It is commonly found on the margins of tropical deciduous forests and pine-oak forests (Baker and Womochel, 1966; Webb et al., 1980). Its activity begins after dusk. It is believed to practice sexual segregation (Ramírez-Pulido et al., 1977). It feeds mainly on fruit; its reproductive pattern is bimodal polyestric; pregnant females have been found in the months of April, July, August, and November (Gardner, 1977; Watkins et al., 1972; Wilson, 1979).

VEGETATIONAL ASSOCIATIONS AND ELEVATION RANGE: *S. hondurensis* inhabits temperate forests of pine and oak, tropical deciduous forests, rainforests, and crops of bananas and coffee (Baker and Phillips, 1965a; Baker et al., 1971; Carter and Jones, 1978; LaVal, 1972). It has been found from sea level up to 2,240 m. But it is abundant in localities between 500 m and 1,200 m (Eisenberg, 1989; Ramírez-Pulido et al., 1977).

CONSERVATION STATUS: *S. hondurensis* is a common species, which is not in any risk category.

Sturnira lilium (È. Geoffroy St.-Hilaire, 1810)

Yellow-shouldered fruit bat

Guadalupe Tellez-Giron and Miguel Amin

SUBSPECIES IN MEXICO

Sturnira lilium parvidens Goldman, 1917

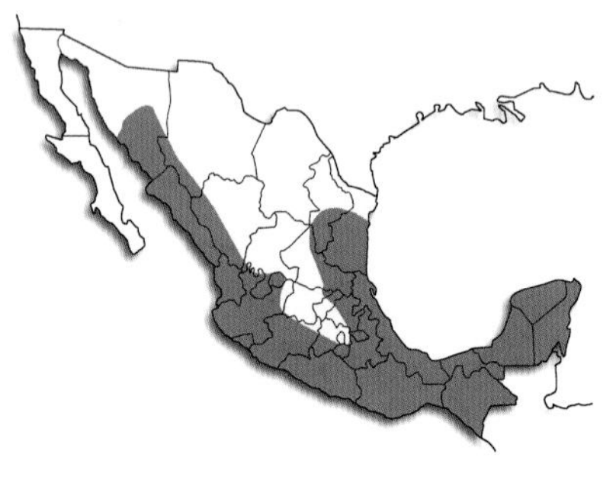

DISTRIBUTION: In Mexico, *S. lilium* is distributed throughout the Neotropical region from Sonora in the Pacific slope and Tamaulipas in the Gulf slope, to the Yucatan Peninsula and Chiapas. Its distribution reaches northern Argentina and Paraguay (Gannon et al., 1989; Ramírez-Pulido et al., 1983). It has been registered in the states of CA, CH, CL, CS, DU, GR, HG, JA, MI, MO, MX, NY, OX, PU, QE, QR, SI, SL, SO, TA, TB, VE, YU, and ZA.

DESCRIPTION: *Sturnira lilium* is a medium-sized bat. It is characterized by having a very small interfemoral membrane and a vestigial or absent tail; this region is covered with hair that reaches the legs. The calcaneus is very low or absent. Throughout the body, the hair is abundant and dense and its color varies according to the sex and distribution, but generally ranges from dark gray to dark red. The head, neck, and shoulders are more yellowish; the abdomen is paler. A striking feature in most individuals, especially males, is the presence of reddish or yellowish stains on the shoulders resembling epaulettes. The forearm measures from 37 mm to 45 mm, the last phalanx of the third finger is larger than 15 mm and the lower incisors are trilobed (Davis, 1984; Gannon et al., 1989; Husson, 1978).

EXTERNAL MEASURES AND WEIGHT

TL = 62 to 65 mm; TV = 0, HF = 10 to 16 mm; EAR = 11 to 18 mm; FA = 36 to 45 mm. Weight: 18 to 19 g.

DENTAL FORMULA: I 2/2, C 1/1, PM 2/2, F 3/3 = 32.

NATURAL HISTORY AND ECOLOGY: This species inhabits the most humid tropical forests and open areas. It is commonly collected on rivers (Handley, 1976). Their roosting sites are diverse and include caves, tunnels, hollow trees, and buildings

(Goodwin and Greenhall, 1961; Handley, 1976). It feeds on a wide variety of fruits, pollen, and insects; among the fruits they eat are bananas (*Musa* sp.), wild figs (*Ficus* sp.), guarumo (*Cecropia* sp.), and piperacillin (*Piper* spp.; Gardner, 1977; Heithaus et al., 1974; Howell and Burch, 1974). Its pattern of reproduction is a continuous polyestry, with three peaks of reproductive activity, from January to March, from July to September, and from November to December (Jones et al., 1973; Sanchez-Hernandez et al. 1986; Willig, 1985; Wilson, 1979).

VEGETATIONAL ASSOCIATIONS AND ELEVATION RANGE: The yellow-shouldered fruit bat is found in wet and dry tropical forests and in fruit crops (Nowak, 1999a). It is found from sea level up to 2,000 m (Eisenberg, 1989).

CONSERVATION STATUS: *S. lilium* is an abundant species, which is not found in any risk category.

Sturnira lilium. Cloud forest. El Jabalí, Colima. Photo: Gerardo Ceballos.

Uroderma bilobatum Peters, 1866

White-lined tent-making bat

Jesus Pacheco R.

SUBSPECIES IN MEXICO

Uroderma bilobatum davisi R.J. Baker and McDaniel, 1972
Uroderma bilobatum molaris Davis, 1968

DESCRIPTION: *Uroderma bilobatum* is a medium-sized bat. Males and females are almost identical in size. The facial lines and the line on the back are very evident; the first ones contrast sharply with the dark color of the head, which differs from *U. magnirostrum* (Davis, 1968). The nasal leaf is simple but well developed. The margin of the ears is yellow (Davis, 1968; Eisenberg, 1989). The eyes are large. The dorsal profile of the skull shows a depression in the front, a feature that sets it apart from *U. magnirostrum* (Baker and Clark, 1987; Davis, 1968).

EXTERNAL MEASURES AND WEIGHT

TV = 54 to 61 mm; TV = 0, HF = 9 to 13 mm; EAR = 12 to 18 mm; FA = 38 to 46 mm. Weight: 13 to 21 g.

DENTAL FORMULA: I 2/2, C 1/1, PM 2/2, F 3/3 = 32.

NATURAL HISTORY AND ECOLOGY: *U. bilobatum* roosts surrounded by the foliage or under banana leaves (*Musa* sp.) and palms (*Scheelea rostrata*, *Cocos nucifera*, *Asterogyne* sp., *Prichardia pacifica*, *Livistona chinensis*, *Sabal mauitiiformis*, and *S. glaucescens*). They modify the leaves of these plants to form their own shelters, or "tents" (Baker and Clark, 1987; Foster and Timm, 1976). Two different styles of these shelters have

DISTRIBUTION: In Mexico the distribution of the white-lined tent-making bat is restricted to lowlands in the Pacific slope (Oaxaca and Chiapas) and the Gulf (Veracruz, Tabasco, and Campeche); it continues to South America (Baker and Clark, 1987; Davis, 1968; Dowler and Engstrom, 1988; Jones et al., 1977; Koopman 1993). It has been registered in the states of CS, GR, MI, and OX.

Uroderma bilobatum. Tropical rainforest. Photo: M.B. Fenton.

been recorded, one for plants with big and wide leaves and another for species with long and pinnate leaves (Timm, 1987). In general, they form their shelters through a number of bites, whether on the midrib or on the side ribs, causing both sides of the leaves to collapse down around the midrib (Eisenberg, 1989; Nowak, 1999a; Timm, 1987). The incisions made are always in the form of an inverted V (Timm, 1987). During the day, individuals rest in their tents but remain vigilant and fly immediately when disturbed (Baker and Clark, 1987). Individuals can be found solitary or in groups of 2 to 59 bats of both sexes (Eisenberg, 1989; Kunz, 1982a). They are mainly frugivorous, but can also eat insects, pollen, and nectar (Eisenberg, 1989; Flemming et al., 1972; Gardner, 1977). They reproduce twice a year with a single young born per breeding (Wilson, 1979). The gestation period probably lasts four to five months. In Chiapas and Quintana Roo, breastfeeding females have been found in April (Medellín et al., 1986). It is likely to have a bimodal pattern of activity, with two peaks of activity after dusk (Erkert, 1982).

VEGETATIONAL ASSOCIATIONS AND ELEVATION RANGE: This species is closely associated with tropical rainforests, but also inhabits tropical subdeciduous forests (Eisenberg, 1989; Timm, 1987). It is found from sea level up to 1,800 m (Baker and Clark, 1987; Davis, 1968); but most records in Mexico are below 300 m.

CONSERVATION STATUS: *U. bilobatum* is a bat with a restricted distribution. The records indicate that it is a scarce species; little is known about its biology. It faces serious conservation problems because it lives in vegetation with high rates of deforestation. Its permanence in the tropics of Mexico involves the conservation of these habitats.

Uroderma magnirostrum Davis, 1968

Brown tent-making bat

Jesus Pacheco R.

SUBSPECIES IN MEXICO

U. magnirostrum is a monotypic species.

DESCRIPTION: *Uroderma magnirostrum* is a medium-sized bat. It characteristically has a rugged face and a skull profile, from the crown to tip of the snout, straight and gradually tilted. The four facial lines are weakly marked or absent, being less evident than in *U. bilobatum* (Davis, 1968; Eisenberg, 1989). It has a nasal leaf and lacks an external tail. The color of the body is usually light brown and the edge of the ears is yellowish. The edge of the wings is translucent (Emmons and Feer, 1997). The dorsal surface of the uropatagium has hair beyond the knee (Davis, 1968).

EXTERNAL MEASURES AND WEIGHT
TL = 58 to 65 mm; TV = 0, HF = 9 to 12 mm; EAR = 13 to 17 mm; FA = 36 to 47 mm. Weight: 12 to 21 g.

DENTAL FORMULA: I 2/2, C 1/1, PM 2/2, F 3/3 = 32.

NATURAL HISTORY AND ECOLOGY: Very little information is available on the biology of this species. Their feeding habits are unknown (Gardner, 1977). The gastric anatomy reveals a preference for fruit, but the diet could also include insects and pollen (Emmons and Feer, 1997; Gardner, 1977; Pacheco and Salazar, 1990). They probably act as seed dispersers. Females make up groups when nursing; males and females perch in separate locations during this season (Eisenberg, 1989). They modify palm leaves to build their shelters, which are used as daytime roosts and serve as protection against sun, rain, and predators (Foster and Timm, 1976; Timm, 1987). It has been reported that they only use palm leaves of *Astrocaryum murumuru* (Timm, 1987), as follows: without cutting the midrib, these bats cut about two-thirds of the palm fronds using a variety of bites, causing the leaf to collapse or fold down on both sides of the midrib in the form of an inverted V. The bats perch on both sides of the leaves and the midrib (Timm, 1987).

DISTRIBUTION: The geographic range of this species is restricted to the Pacific slope of Michoacan to Chiapas, continuing its distribution to Central America to the central part of Brazil (Eisenberg, 1989; Jones and Carter, 1979). It has been recorded in the states of CS, GR, MI, and OX.

VEGETATIONAL ASSOCIATIONS AND ELEVATION RANGE: This species inhabits tropical deciduous forests and open areas in this type of vegetation (Emmons and Feer, 1997; Nowak 1999a; Pacheco and Salazar, 1990). It seems to be more abundant in arid environments than *Uroderma bilobatum* (Eisenberg, 1989). It is found from sea level up to 1,000 m; however, most individuals have been captured below 500 m (Eisenberg, 1989, Ramírez-Pulido and López-Forment, 1979).

CONSERVATION STATUS: The records indicate that throughout Mexico *U. magnirostrum* is a rare bat. Although it has been caught in disturbed areas, it is likely that deforestation is causing its populations to decrease. Studies are needed to determine accurately its true abundance and establish appropriate measures to protect them (Jones, 1976).

Vampyressa thyone Thomas, 1909

Northern little yellow-eared bat

Guadalupe Tellez-Giron

SUBSPECIES IN MEXICO
V. thyone is a monotypic species.

DESCRIPTION: *Vampyressa thyone* is a small-sized bat. The face is short and wide, with two tenuous facial lines that extend up between the ears. The nasal leaf is long and narrow; the tragus and the base of the ears have a yellow edge. The edge of the uropatagium is naked, with few hairs on the center. The upper incisors have different sizes; the interiors are bifid and twice as big as the externals. The color of the back varies from grayish to reddish-brown, with the abdomen slightly grayer (Goodwin, 1963; Lewis and Wilson, 1987; Linares, 1986).

DISTRIBUTION: *V. thyone* is distributed from southern Mexico to southeastern Brazil and southwestern Peru and Ecuador (Eisenberg, 1989; Lewis and Wilson, 1987). In Mexico it is only found in warm-humid regions of the southeastern parts of the country (Ramírez-Pulido et al., 1983). It has been registered in the states of CA, CS, OX, and VE.

Vampyressa thyone. Tropical rainforest. Photo: M.B. Fenton.

EXTERNAL MEASURES AND WEIGHT
TL = 47 to 52 mm; TV = 0, HF = 9 to 10 mm; EAR = 14 to 15 mm; FA = 29 to 33.2 mm.
Weight: 10 g.

DENTAL FORMULA: I 2/2, C 1/1, PM 2/2, F 2/2 = 28.

NATURAL HISTORY AND ECOLOGY: This species is found in very wet habitats, for example, on streams in gallery forests. They have been collected between 3 masl and 12 masl (Bonaccorso, 1979; Handley, 1966). They build their daytime shelters in leaves of trees and shrubs such as Araceae (*Philodendron* sp.). These leaves are partially cut in the base, so that the leaf surface collapses inward, forming a pyramidal shape. The shelters are occupied only once (Timm, 1984). It is a frugivorous bat that specializes in eating wild figs (*Ficus* sp.). In Costa Rica, it also eats fruit of the genus *Acnistus* (Bonaccorso, 1979; Howell and Burch, 1974; Navarro, 1982). The peak of activity occurs at the first two hours after dusk (Bonaccorso, 1979). The reproductive pattern is bimodal polyestric, with more births at the end of the dry season and in the middle of the rainy season, and pregnant females from March to July and from November to March; lactating females have been observed in March, April, May, July, and August (Bonaccorso, 1979; Fleming et al., 1972; Wilson, 1979).

VEGETATIONAL ASSOCIATIONS AND ELEVATION RANGE: This species is only found in tropical rainforests, generally below 500 masl; however, in Chiapas it has been captured at 2,000 m and in Venezuela above 1,500 m (Arnold and Schonewald, 1972; Eisenberg, 1989).

CONSERVATION STATUS: In Mexico this species is scarce. It has been collected in a few sites and on each occasion only a few individuals were captured; little is known about their populations.

Vampyrodes caraccioli (Thomas, 1889)

Great stripe-faced bat

Guadalupe Tellez-Giron

SUBSPECIES IN MEXICO
Vampyrodes caraccioli major G.M. Allen, 1908
Jones and Carter (1976) consider *V. caraccioli major* as a subspecies of Mexico. Hall (1981) and Starret and Casebeer (1968) consider it as another species.

DESCRIPTION: *Vampyrodes caraccioli* is a medium-sized bat, easily recognizable by the two pairs of wide white lines on its face and the thin white line on the back. The hair color is a slight variation between grayish-brown and cinnamon brown; females have a paler coloration than males. Although *Platyrrhinus helleri* has similar characteristics, the lines on the face and back, and the darker and larger hair, set it apart from *V. caraccioli*. *V. caraccioli* has a broad and well-developed nasal leaf in form of a U. The forearm and limbs are covered with hair that reaches the nails. The interfemoral membrane is narrow and in form of a U. The edge of the ears is intense yellow (Eisenberg, 1989; Emmons and Feer, 1997; Villa-R., 1967).

DISTRIBUTION: *V. caraccioli* is distributed from southern Veracruz and Chiapas to northwestern Brazil and Peru (Eisenberg, 1989). It has been registered in the states of CS, OX, and VE.

EXTERNAL MEASURES AND WEIGHT

TL = 73 to 80 mm; TV = 0, HF = 12 to 17 mm; EAR = 20 to 25 mm; FA = 45 to 57 mm. Weight: 28 to 40 g.

DENTAL FORMULA: I 2/2, C 1/1, PM 2/2, F 2/3 = 30.

NATURAL HISTORY AND ECOLOGY: These are fruit bats that live under the leaves of palms and other plant species, forming small groups of two to four individuals (Coates-Estrada and Estrada, 1986; Lay, 1962). They have always been collected on streams or rivers. Data on reproduction indicate a bimodal pattern. In Chiapas lactating females were captured in April, May, September, and October (Álvarez and Álvarez, 1991; Medellín, 1986; Wilson, 1979).

VEGETATIONAL ASSOCIATIONS AND ELEVATION RANGE: This species is strongly associated with undisturbed evergreen and mature tropical rainforests (Carter et al., 1966; Hall and Dalquest, 1963). It is frequently found below 500 m; sometimes they have been collected at 1,000 m (Koopman, 1978).

CONSERVATION STATUS: *V. caraccioli* is a rare species found only in undisturbed areas. In order to determine its current status, it is important to know if this is a specialized species in undisturbed habitats.

Vampyrodes caraccioli. Tropical rainforest. Montes Azules Biosphere Reserve, Chiapas. Photo: Rodrigo Medellín.

Family Mormoopidae

The family Mormoopidae is a little diverse family of bats, which is formed by two genera and eight species; their distribution is restricted to arid and tropical regions of the Americas (Wilson and Reeder, 2005). In Mexico, both genera are represented, *Mormoops* and *Pteronotus*, with one and four species, respectively.

Mormoops megalophylla (Peters, 1864)

Ghost-faced bat

Luis Ignacio Iñiguez Davalos

SUBSPECIES IN MEXICO
Mormoops megalophylla megalophylla (Peters, 1864)

DESCRIPTION: *Mormoops megalophylla* is a medium-sized bat. The face has complex folds of skin that gives it a strange look. The chin has two large concave shelves connected to a series of pleats under and on its sides the upper lip is complexly ornamented, with intricate folds and many warty tubers (Villa-R., 1967). The nostrils are separated and resemble short tubes. The ears are short and rounded, connected by a big pleat to the forehead, and they are deeply notched on the centerline. This edge

Mormoops megalophylla. Tropical dry forests. Merida, Yucatan. Photo: Jens Rydell.

that connects the ears is diagnostic and separates this species from other mormoopids. The interior of the ear extends forward, forming a fold on the commissure of the mouth; the eyes are medium-sized and surrounded by the tunnel that forms the ears, which are very open. The general appearance of the head is almost spherical. The fur is soft, long, and limp, and of variable coloration that ranges from chocolate or reddish to gray. The uropatagium is longer than the legs, with the tail included in it, but the tail emerges through the center of the membrane.

DISTRIBUTION: *M. megalophylla* is distributed from southern Texas to Peru. It has colonized Venezuela and some Caribbean islands (Hall, 1981). It has been registered in the states of AG, BS, CA, CH, CL, CO, CS, DF, DU, GJ, GR, HG, JA, MI, MO, MX, NL, NY, OX, PU, QE, QR, SI, SL, SO, TA, TB, TL, VE, YU, and ZA.

EXTERNAL MEASURES AND WEIGHT
TL = 68 to 107 mm; TV = 18 to 31 mm; HF = 7 to 17 mm; EAR = 10 to 18 mm; FA = 45 to 61 mm.
Weight: 10 to 20 g.

DENTAL FORMULA: I 2/2, C 1/1, PM 2/3, M 3/3 = 34.

NATURAL HISTORY AND ECOLOGY: *M. megalophylla* is an insectivorous bat that often forages over bodies of water at low altitude. It roosts in caves in a gregarious manner, with concentrations of individuals that vary seasonally. The caves used by this species are generally characterized by relatively high humidity and relatively high and constant temperature (Ceballos and Miranda, 1986; Emmons and Feer, 1990). This bat does not do well at low temperatures for extended periods and never enters torpor. The colonies vary from a few to several thousand individuals, often sharing the shelter with other mormoopidae and phyllostomidae species, though separated spatially (Arita, 1993a, b). When foraging, they leave the cave in groups. They seem to be nomadic and undertake extensive migrations. They are more common in tropical dry areas than in wet ones. Although information on reproduction is sparse and scattered, mating season seems to occur at the end of winter and births coincide with the beginning of the rainy season.

VEGETATIONAL ASSOCIATIONS AND ELEVATION RANGE: *M. megalophylla* is commonly found in the tropical deciduous forest and thorny forest, but has also been caught in tropical rainforests, cloud forests, and arid and semi-arid regions. It is distributed from sea level up to 2,230 m (Naranjo and Espinoza, 2001).

CONSERVATION STATUS: Locally, *M. megalophylla* is a scarce species, although sometimes it can form large concentrations. It is rare to find them in large numbers. It is not considered at risk of extinction.

Pteronotus davyi Gray, 1838

Davy's naked-backed bat

Arturo Jimenez-Guzman and Gerardo Ceballos

SUBSPECIES IN MEXICO
Pteronotus davyi fulvus (Thomas, 1892)

DESCRIPTION: *Pteronotus davyi* is very similar to but smaller than *P. gymnonotus*; both are easily distinguishable from all bat species in Mexico because the wings cover the back, giving them the appearance of being naked. There are two coloration

DISTRIBUTION: *P. davyi* has a wide distribution in tropical regions of the Americas, from southern Sonora and Tamaulipas in Mexico to Brazil. It has been recorded in the states of CA, CL, CS, DU, GR, HG, JA, MI, MO, MX, NL, NY, TA, TB, OX, PU, QR, SI, SL, SO, VE, YU, and ZA.

phases: a reddish light one and a dark brown one. The hairless back is dark brown. The tail comes out from the middle part of the uropatagium.

EXTERNAL MEASURES AND WEIGHT
TL = 63 to 87 mm; TV = 15 to 29 mm, HF = 8 to 19 mm; EAR = 6 to 16 mm; FA = 40 to 47 mm.
Weight: 7 to 8 g.

DENTAL FORMULA: I 2/2, C 1/1, PM 2/3, M 3/3 = 34.

NATURAL HISTORY AND ECOLOGY: Davy's naked-backed bats are gregarious and form colonies of hundreds of thousands of individuals (Ceballos and Miranda, 2000). They find refuge in caves, wells, tunnels, and mines that are dark, very wet, and hot. They share their shelters with other species of the same family, including *P. personatus*, *P. parnellii*, and *Mormoops megalophylla*. They become active shortly after sunset, before the majority of other bat species. They carry out their activities in small groups, and follow well-defined travel routes. They feed exclusively on insects, particularly moths (Lepidoptera) caught at low altitude. They are monoestric, and their mating season takes place during the dry season and the beginning of the rains, between February and August. Usually a single young is born.

Pteronotus davyi. Tropical dry forest. Chamela-Cuixmala Biosphere Reserve, Jalisco. Photo: Gerardo Ceballos.

VEGETATIONAL ASSOCIATIONS AND ELEVATION RANGE: *P. davyi* is mainly found in tropical biomes such as tropical deciduous forests, tropical rainforests, dry scrublands, and disturbed environments; from sea level up to 1,500 m; most locations are below 600 masl.

CONSERVATION STATUS: These bats are common in tropical regions of the Americas in general, and the Pacific coast of Mexico in particular. At present, they do not face risks endangering their survival.

Pteronotus gymnonotus (Wagner, 1843)

Big naked-backed bat

Jorge Ortega R. and Hector T. Arita

SUBSPECIES IN MEXICO

P. gymnonotus is a monotypic species.

DESCRIPTION: *Pteronotus gymnonotus* is similar to *P. davyi*; in both species the coat is dark brown or black; the wings are joined on the middle part of the back, giving the bats the appearance of being naked, although hair is present below the wing membrane. The basic difference between the two species is that *P. gymnonotus* is substantially larger. Like other mormoopids it lacks a nasal leaf and has ornamented lips with folds and warts. The tail is included in the uropatagium, extending up about half of the membrane, and only the distal portion is free. Echolocation calls were reported for a Mexican specimen (Ibañez et al., 2000).

EXTERNAL MEASURES AND WEIGHT

TL = 81 to 96 mm; TV = 21 to 24 mm; HF = 9 to 13 mm; EAR = 10 to 18 mm; FA = 49 to 56 mm.

Weight: 12 to 16 g.

DENTAL FORMULA: I 2/2, C 1/1, PM 2/3, M 3/3 = 34.

NATURAL HISTORY AND ECOLOGY: *P. gymnonotus* is an aerial insectivorous species; it catches insects on the fly (Eisenberg, 1989). It mainly roosts in caves, which are shared with other species of the same genus and with several phyllostomids (Villa-R., 1967). Few data about their reproductive patterns as well as number of offspring are known, and they are supposed to have only a single young per year and a single annual reproductive period. In Mexico, a pregnant female was recorded for May (Ibañez et al., 2000); a nursing female was reported for Costa Rica in August (Dolan and Carter, 1979).

DISTRIBUTION: The big naked-backed bat's range extends from Veracruz to Brazil. For Mexico there are only three records, one for the Cueva de la Laguna Encantada in the state of Veracruz, one for Cueva de las Sardinas, Tabasco (Ibañez et al., 2000), and one for the Yaxchilan archaeological site in Chiapas. It has been recorded in the states of CS, TB, and VE.

VEGETATIONAL ASSOCIATIONS AND ELEVATION RANGE: In Mexico the big naked-backed bat has been recorded in tropical rainforests, although in Venezuela it prefers dry semi-deciduous forests (Eisenberg, 1989). It has been collected in places lower than 400 m.

CONSERVATION STATUS: The current status of *P. gymnonotus* is unknown, but for Mexico, due to the low number of records, it could be considered as rare.

Parnelli's mustached bat

Jorge Ortega R.

SUBSPECIES IN MEXICO

Pteronotus parnellii mesoamericanus Smith, 1972

Pteronotus parnellii mexicanus (Miller, 1902)

DESCRIPTION: *Pteronotus parnellii* is a medium-sized bat, although it is the largest species in its genus (Eisenberg, 1989; Fleming, 1988; Herd, 1983; Villa-R., 1967). The upper lip has a series of papillae and small warts on the bottom edge and tactile hairs on the sides. The nostrils are fused and expanded on the upper lip, forming a bulge at the base of the nose (Herd, 1983). The face has small sebaceous glands at the base of the eyelids (Dalquest and Werner, 1954). The ears are large and lanceolate. One-third of the tail is included in the uropatagium, which is wide and lacks hair. It has two stages of coloration: light gray and brown. Specimens of the north are lighter and smaller than those of the south. Hair is usually white at the base and colored at the tip (Herd, 1983; Smith, 1972). The skull is large and robust and moderate in length; the zygomatic arches are slightly extended. It lacks postorbital processes and sagittal ridges (Herd, 1983; Villa-R., 1967).

Pteronotus parnellii. Tropical dry forest. Chamela-Cuixmala Biosphere Reserve, Jalisco. Photo: Gerardo Ceballos.

EXTERNAL MEASURES AND WEIGHT
TL = 70.4 to 71.7 mm; TV = 19 to 26 mm; HF = 14.62 to 14.68 mm; EAR = 14 to 28 mm; FA = 59.83 to 59.3 mm.
Weight: 19.6 to 24.2 g.

DENTAL FORMULA: I 2/2, C 1/1, PM 2/3, M 3/3 = 34.

NATURAL HISTORY AND ECOLOGY: Parnell's mustached bat is basically insectivorous (Fleming et al., 1972; Nowak, 1999a). It has been reported that a colony of 600,000 individuals might consume 1,900 kg to 3,000 kg of insects per night. The average foraging distance is 3.5 km with respect to a roost site; the bats return to the cave after six hours, on average. (Bateman and Vaughan, 1974; Dalquest, 1954). Their echolocation is based on a series of long-term emissions with a low modular frequency (Smith, 1972). It usually perches in caves, preferring internal chambers with higher humidity and temperature (Álvarez, 1963a). It can form large colonies of up to 800,000 individuals. Sometimes they roost along with other species within the caves (Dalquest, 1953a; Medellín and López-Forment, 1986; Villa-R., 1967). Mating season occurs in December; a single young is born between June and July, although pregnant females have been captured in March and April (Álvarez and Álvarez-Castañeda, 1991; Cockrum, 1955). It has a monoestric pattern (Fleming et al., 1972; LaVal and Fitch, 1977). The presence of rabies virus has been reported for this bat (Malaga-Alba and Villa-R., 1957).

VEGETATIONAL ASSOCIATIONS AND ELEVATION RANGE: *P. parnellii* has been collected in a large number of habitats, including tropical deciduous forests, tropical rainforests, cloud forests, thorny forests, swamps, grasslands, and secondary vegetation (Eisenberg, 1989; Handley, 1976). It is distributed from sea level up to 3,000 m (Smith, 1972).

CONSERVATION STATUS: The current status of Parnell's mustached bat is unknown; however, it is one of the most abundant bats, which survives even in disturbed areas, so it is not considered at risk.

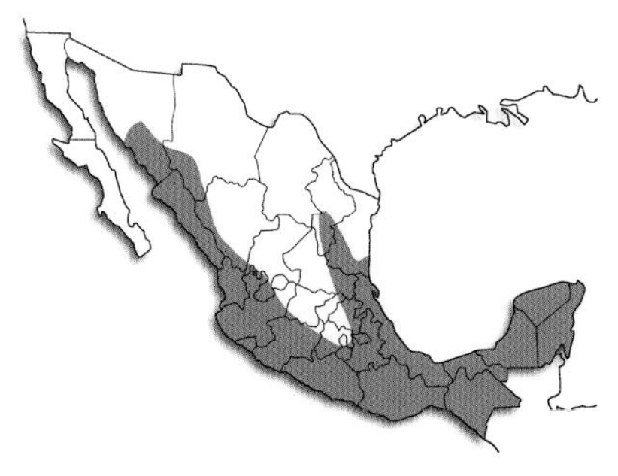

DISTRIBUTION: In Mexico *P. parnellii* is distributed in the entire Neotropical region from Sonora in the Pacific slope and in Tamaulipas in the Gulf Slope throughout the Yucatan Peninsula and Chiapas. Its distribution reaches northern Argentina and Paraguay (Gannon et al., 1989; Jiménez-G. and Zúniga, 1992; Ramírez-Pulido et al., 1983). It has been registered in the states of CA, CH, CL, CO, CS, DU, GJ, GR, HG, JA, MI, MO, MX, NL, NY, OX, PU, QE, QR, SI, SL, SO, TA, TB, VE, YU, and ZA.

Pteronotus personatus (Wagner, 1843)

Wagner's mustached bat

Jorge Ortega R.

SUBSPECIES IN MEXICO
Pteronotus personatus psilotis (Dobson, 1878)

DESCRIPTION: *Pteronotus personatus* has extended and ornamented lips with folds and warts, fitted with tactile vibrissae (Ceballos and Miranda, 1986; Villa-R., 1967). The hair color varies from light brown to dark brown; it sometimes has red and orange tones (Smith, 1972). The ears are small and lanceolate. One-third of the tail is included within the uropatagium, which lacks hair. *P. personatus* is similar in shape, color, and dentition to *P. parnellii*; size is the distinguishing feature between the two species (Eisenberg, 1989). In addition, *P. personatus* has a tragus with a secondary sheet remarkably developed, which is not present in *P. parnellii* (Smith, 1972). The skull is small and robust with a short face; no sagittal crest or postorbital processes are present (Ceballos and Miranda, 1986).

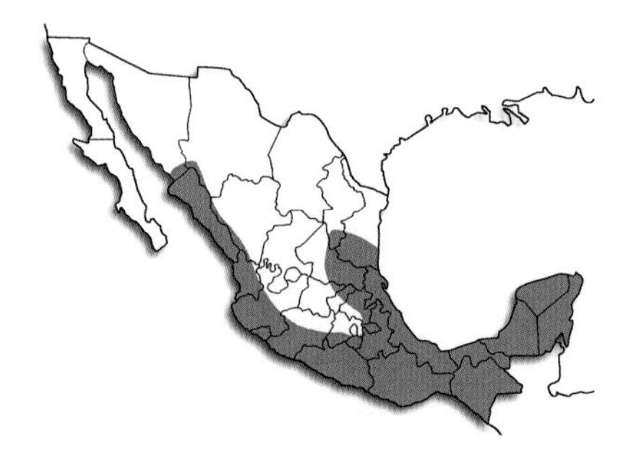

DISTRIBUTION: *P. personatus* is distributed along the Mexican coastal plains, from Sonora and Tamaulipas to the northwest of Brazil (Eisenberg, 1989). It has been registered in the states of CA, CL, CS, GJ, GR, HG, JA, MI, MO, NY, OX, QE, QR, SI, SL, SO, TA, TB, and VE.

EXTERNAL MEASURES AND WEIGHT

TL = 49 to 55 mm; TV = 10 to 18 mm; HF = 12.7 to 13.5 mm; EAR = 6 to 16 mm; FA = 40.8 to 47.4 mm.

Weight: 8 g.

DENTAL FORMULA: I 2/2, C 1/1, PM 2/3, M 3/3 = 34.

NATURAL HISTORY AND ECOLOGY: *P. personatus* has insectivorous habits (Novick, 1965). It uses the first four hours of the night (on average) to go out and forage, returning immediately to its roosting site (Bateman and Vaughan, 1974). Echolocation is based on a series of short-term emissions that gradually change in frequency of modulation (Smith, 1972). It perches in caves in colonies of hundreds of individuals, although not in dense clusters (Villa-R., 1967). It has been reported to roost along with other bats (Dalquest and Hall, 1949). Little is known about their reproductive habits, but mating has been reported in December and the presence of offspring during July to September; hence, it is thought to have a monoestric pattern (Cockrum, 1955). Individuals infected with rabies have been reported (Malaga-Alba and Villa-R., 1957).

VEGETATIONAL ASSOCIATIONS AND ELEVATION RANGE: *P. personatus* is frequently associated with tropical and subdeciduous forests, although it has also been collected in grassland and secondary vegetation (Eisenberg, 1989). It is distributed from sea level to 400 m (Handley, 1976).

CONSERVATION STATUS: The current status of the populations of Wagner's mustached bat is unknown, but due to the fact that it can form colonies of hundreds of individuals, it is not considered at risk.

Pteronotus personatus. Tropical dry forest. Chamela-Cuixmala Biosphere Reserve, Jalisco. Photo: Gerardo Ceballos.

Family Noctilionidae

The family Noctilionidae is endemic to tropical regions of the Americas. It consists of a single genus (*Noctilio*) and two species, *N. leporinus* and *N. albiventris*, which have a wide distribution from southern Mexico to northern Argentina (Wilson and Reeder, 2005).

Noctilio albiventris Desmarest, 1818

Lesser bulldog bat

Ivan Castro-Arellano and Jorge Uribe

SUBSPECIES IN MEXICO
Noctilio albiventris minor Osgood, 1910

DESCRIPTION: *Noctilio albiventris* is the smallest bat of the two species of the genus. The fur is short and its coloration has a wide geographical variation, which ranges from reddish-brown to dark brown on the back and yellowish-brown to a bright white cream on the abdomen. It has a tenuous dorsal stripe running from the interscapular region up to the posterior part of the body (Davis, 1976). The snout, nose, and lips are large and, in general, the face gives the appearance of a bulldog. The ears are separated, thin, and sharp; the tail is longer than half of the femur and its tip is free on the dorsal surface of the interfemoral membrane. The legs are long and robust, with claws not as developed as in *N. leporinus* (Hood and Pitochelli,

Noctilio albiventris. Photo: M.B. Fenton.

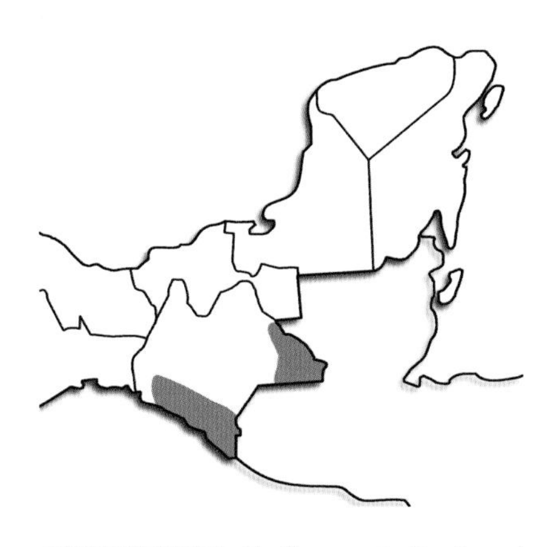

DISTRIBUTION: *N. albiventris* is distributed from southern Mexico and Central America to Brazil and northern Argentina. In Mexico, it has only been recorded in southernmost Chiapas (Hood and Pitocchelli, 1983; Medellín et al., 2008; Polaco, 1987). It has been registered in the state of CS.

1983). The braincase is high, the mastoid bone is prominent, and the sagittal crest is well developed in males (Davis, 1976; Hood and Pitochelli, 1983). It differs from *N. leporinus* by its smaller size (40 g; forearm < 70 mm; Davis, 1973, 1976).

EXTERNAL MEASURES AND WEIGHT
TL = 65 to 68 mm; TV = 13 to 16 mm, HF = 16 to 19 mm; EAR = 22 to 24 mm; FA = 70 mm.
Weight: 27 to 35 g.

DENTAL FORMULA: I 2/1, C 1/1, PM 1/2, F 3/3 = 28.

NATURAL HISTORY AND ECOLOGY: These bats live in rainforests near water bodies. They roost in hollow trees and buildings; they have been found in large groups along with *Molossus coibensis* and other species (Davis et al., 1964; Hood and Pitochelli, 1983; Tuttle, 1970). Their activity patterns have two peaks, one at dusk until 19:00 hours and another after midnight (Brown, 1968). It feeds in small groups near rivers and streams (Davis et al., 1964). The reproductive season begins in late November and December; gestation lasts the spring months and the young are born from April to August (Hood and Pitochelli, 1983). The litter consists of a single young (Hood and Pitochelli, 1983). They are mainly insectivorous, but also consume fish (Davis et al., 1964; Fleming et al., 1972; Howell and Burch, 1974). Analyses of stomach content found beetles, dipterans, hemipterans, homopterans, hymenopterans, lepidopterans, and orthopterans. They feed on the disticids (Coleoptera), which are beetles that develop most of their lifecycle in water. The consumption of fruits of *Brosimum*, *Ceiba*, and *Morus* has been reported (Hood and Pitochelli, 1983; Hooper and Brown, 1968; Howell and Burch, 1974).

VEGETATIONAL ASSOCIATIONS AND ELEVATION RANGE: *N. albiventris* is mainly found in mature tropical jungle near rivers, streams, and swamps (Davis, 1976; Davis et al., 1964; Dickerman et al., 1981). It inhabits elevations from sea level up to 1,100 m (Hood and Pitochelli, 1983), but in Mexico it has only been recorded at sea level.

CONSERVATION STATUS: The subspecies that inhabits Mexico is considered rare (SEDESOL, 1994; SEMARNAT, 2002), mainly because of its limited geographical distribution.

Noctilio leporinus (Linnaeus, 1758)

Greater bulldog bat

Jorge Uribe and Ivan Castro-Arellano

SUBSPECIES IN MEXICO
Noctilio leporinus mastivus (Vahl, 1797)

DESCRIPTION: Both species of the genus *Noctilio* are similar in their cranial and external morphological characters, but *N. leporinus* is larger. It has a short snout with the upper lips hanging, wide, and divided by a fold of skin just below the nostrils (Ceballos and Miranda, 1986; Findley, 1993; Hood and Jones, 1984; Nowak, 1999a; Silva-Taboada, 1979). The chin has highly developed fleshy edges that give the bat the appearance of a bulldog. The ears are separated, narrow, and pointy; the tragus is pinnate and lobed with serrated projections in the form of fingers. The tail is long and extends a third of the length of the interfemoral membrane, and the third terminal is free. The wings are narrow and very long, measuring up to two and a half times the total length. The hind legs are robust with highly developed thumbs and conspicuous claws; calcaneum bones are extremely long and developed (Ceballos and Miranda, 1986). The calcar and tibia are laterally flattened so they offer little resistance to water (Hill and Smith, 1992). The coat is extremely short and dorsal coloration varies considerably (Hood and Jones, 1984). In males, the coloration of the back varies from light brown to dark or intense orange and the abdomen is lighter; in females, the color of back varies from light brown to gray. All colorations have a dorsal light-colored midline from the shoulders to the base of the tail. The wings and tail are almost furless and brown (Ceballos and Miranda, 1986; Hood and Jones, 1984; Nowak, 1999a). Apparently, in the Mexican subspecies variations of the coloration exist: dark reddish for the coastal plain of the Gulf and light yellow for the Pacific coastal plain (Álvarez-Castañeda and Álvarez, 1991). There is evidence of geographic variation in size; the largest individuals are found in the northern and southern parts of its distribution. Males are generally larger than females (Hood and Jones, 1984; Nowak, 1999a). The skull lacks a postorbital process; males show a more developed sagittal crest than females; the premaxilla has palatal and nasal branches merged together, and a parallel row of maxillar teeth, the tympanic bulla is relatively small (Ceballos and Miranda, 1986; Hood and Jones, 1984).

Noctilio leporinus. Mangrove forest. Chamela-Cuixmala Biosphere Reserve, Jalisco. Photo: Gerardo Ceballos.

EXTERNAL MEASURES AND WEIGHT
TL = 98 to 132 mm; TV = 25 to 28 mm; HF = 25 to 34 mm; RAC = 28 to 30 mm; FA = 70 to 92 mm.
Weight: 50 to 90 g.

DENTAL FORMULA: I 2/1, C 1/1, PM 1/2, F 3/3 = 28.

NATURAL HISTORY AND ECOLOGY: Greater bulldog bats are common in the coastal plains. Usually, they are observed and captured near water bodies and quiet streams, but they can be found in large rivers, estuaries, bays, and coastal lagoons (Hood and Jones, 1984). They roost in hollow trees, caves, and cracks in rocks and have also been found in buildings (Carter et al., 1966; Ceballos and Miranda, 1986; Hill and Smith, 1992; Nowak, 1999a). In Trinidad and Tobago, colonies perched in tree holes and red mangrove (*Rhizophora mangle*; Goodwin and Greenhall, 1961). Roosting sites are characterized by a strong fish smell (Hill and Smith, 1992; Hood and Jones, 1984).

DISTRIBUTION: *N. leporinus* is distributed from Sinaloa, Veracruz, and the Yucatan Peninsula in Mexico to Argentina in South America, but its distribution is discontinuous and is virtually restricted to coastal regions (Aguilar-Cervantes and Álvarez-Solorzano, 1991; Ceballos and Miranda, 1986; Hood and Jones, 1984; Nowak, 1999a). It has been registered in the states of CA, CS, CL, GR, JA, MI, NY, OX, QR, SI, TB, VE, and YU.

They may be solitary or form colonies of dozens of individuals (Ceballos and Miranda, 1986). There is, however, much variation in colony size, for example, a colony in Yucatan had only three adult males, seven females, and six young (Jones et al., 1973). They become active after dusk. Their flight pattern is not particularly fast, and occurs at low altitude with zigzag movements (Coates-Estrada and Estrada, 1986); near bodies of water their wings almost touch the surface (Ceballos and Miranda, 1986; Hill and Smith, 1992). They feed mainly on fish and occasionally on aquatic or winged insects such as ants (Hemiptera), crickets (Orthoptera), and beetles (Coleoptera) (Aguilar-Cervantes and Álvarez-Solorzano, 1991; Altenbach, 1989; Carter et al., 1966; Findley, 1993; Fleming et al., 1972; Hill and Smith, 1992; Hood and Jones, 1984; Nowak, 1999a; Silva-Taboada, 1979). There are references to the consumption of fruits and the presence of plant material in stomach contents (Aguilar-Cervantes and Álvarez-Solorzano, 1991; Álvarez del Toro, 1977; Villa-R., 1967). It has been hypothesized that piscivory evolved from capturing swimming or floating insects (Altenbach, 1989; Hill and Smith, 1992). This hypothesis is supported by observations on the conduct of foraging on insects, which are captured as accurately as fish (Altenbach, 1989). They use echolocation to detect disturbances on the water surface, which indicate the presence of small fish (Hill and Smith, 1992). Neither smell nor vision seems to be needed during captures (Jones and Hood, 1984). They feed at dusk and at night (Nowak, 1999a); apparently, they are active throughout the night (Jones and Hood, 1984). Once the fish is located, they submerge their legs, armed with hook-like claws, to catch their prey and subsequently bend their knees; the food is transferred directly to the mouth (Altenbach, 1989). They can feed on the fly or perch and consume it (Altenbach, 1989). Under laboratory conditions, in a single night, this species is able to capture up to 40 fish (Hill and Smith, 1992; Nowak, 1999a). Although Hill and Smith (1992) reported that the greater bulldog bat only consumes freshwater fish, Nowak (1999a) mentions that they sometimes fish at sea. The chewing apparatus is supplemented with small sacks on the cheeks that temporarily store poorly chewed food while the bat continues capturing prey (Aguilar-Cervantes and Álvarez-Solorzano, 1991; Hood and Jones, 1984). The digestive tract is anatomically and histochemically adapted to piscivory, so it is very efficient in the absorption of nutrients. There is little information regarding their general physiology and thermoregulation; average body temperature is 35.6°C (Silva-Taboada, 1979). Females are mono-ovulatory and produce a single offspring. The reproduction season usually starts in November and December and young are born between April and June (Jones and Hood, 1984). Yet, there seems to be a second reproductive peak, and in Chamela, this species seems to be reproducing throughout the year (Ceballos and Miranda, 2000).

VEGETATIONAL ASSOCIATIONS AND ELEVATION RANGE: *N. leporinus* has been found in mature forests and secondary vegetation near rivers, lakes, and lagoons (Coates-Estrada and Estrada, 1986). In Chamela, Jalisco, it is distributed in riparian vegetation, mangroves, and palm forests, mainly in the coquito palm (*Orbygnia guacuyule*). In the region of the Panama Canal, it has been captured more frequently in riparian vegetation areas. In the archaeological site of Yaxchilan, Chiapas, the species was found associated with palm trees and high tropical forests (Álvarez and Álvarez-Castañeda, 1991; Fleming et al., 1972). It inhabits elevations from sea level to 300 m (Álvarez and Álvarez-Castaneda, 1991).

CONSERVATION STATUS: *N. leporinus* is not included in any list of endangered species. Throughout its distribution range, it is relatively common, so it is not considered to have conservational problems. Deforestation and pollution of water bodies where it feeds, however, are problems that affect its populations.

Family Thyropteridae

The family Thyropteridae comprises a single genus and two species. It is endemic to the humid tropical regions throughout the Americas. The two species are characterized by a disk-shaped structure in the extremities used to hold on to the leaves used as roosting sites. In Mexico there is only one species.

Thyroptera tricolor Spix, 1823

Spix's disk-winged bat

Guadalupe Tellez-Giron

SUBSPECIES IN MEXICO
Thyroptera tricolor albiventer (Tomes, 1856)

DESCRIPTION: *Thyroptera tricolor* is a very small bat that has an adhesive disk in each limb, a feature that distinguishes it from all other species of bats in Mexico. The disks at the base of its thumbs are larger than those of the legs (Villa-R., 1967). The face is short, with well-separated and funnel-shaped ears, with a small tragus and a small basal lobe. The interfemoral membrane has a fringe of hair from the edge to the leg. The calcaneus has two cartilaginous projections directed toward the edge of the interfemoral membrane. The tail is long and is projected beyond the edge of the uropatagium. The hair is long and dense; the back is dark brown to red and the abdomen is light-colored to pale yellow (Wilson and Findley, 1977).

DISTRIBUTION: The distribution of *T. tricolor* extends from Veracruz and Chiapas in southern Mexico to Ecuador and Brazil (Eisenberg, 1989; Koopman, 1982; Villa-R., 1967; Wilson and Findley, 1977). It has been recorded in the states of CS and VE.

EXTERNAL MEASURES AND WEIGHT
TL = 34 to 52 mm; TV = 24 to 27 mm, HF = 3 to 4 mm; EAR = 11 to 12 mm; FA = 32 to 38 mm.
Weight: 3 to 5 g.

DENTAL FORMULA: I 2/3, C 1/1, PM 3/3, M 3/3 = 38.

NATURAL HISTORY AND ECOLOGY: Spix's disk-winged bats form small colonies of two to nine individuals; they roost inside rolled leaves of Musacea (*Heliconia* sp. and *Musa* sp.) and Marantaceae (*Calathea* sp.), found in foliage gaps in tropical rainforests. They require leaves with an opening of 50 mm to 100 mm. Members of the colonies remain together in their shelter during the day (Villa-R., 1963; Wilson and Findley, 1977). They are insectivorous and capture their prey in flight; one individual consumes 0.8 g of insects per night (Eisenberg, 1989; Wilson and Findley, 1977). Data on reproduction are scarce. In Chiapas, males with testicles in the scrotum and abdomen have been found in November, and a female with an embryo in May (Álvarez and Ramírez-Pulido, 1972; Gardner, 1963).

VEGETATIONAL ASSOCIATIONS AND ELEVATION RANGE: *T. tricolor* has only been recorded in tropical rainforests. In Mexico, it occupies the lowlands, from sea level to 140 m (Álvarez and Ramírez-Pulido, 1972; Villa-R., 1963).

CONSERVATION STATUS: This species is rare in Mexico, and it has a very restricted distribution, so it is important to study and obtain more data about its populations. It is not considered endangered.

Family Natalidae

The family Natalidae is represented by only one genus and five species. It is endemic to the tropical lowlands of America, and its distribution extends from Mexico to northern Argentina and the Caribbean islands.

Natalus lanatus Tejedor, 2005

Woolly funnel-eared bat

David Vazquez and Joaquin Arroyo

SUBSPECIES IN MEXICO
N. lanatus is a monotypic species.

DISTRIBUTION: *N. lanatus* is an endemic species of Mexico, distributed in the Pacific slope and western region of the Sierra Madre Occidental, from southwest of Chihuahua and northern Durango to middle Guerrero, penetrating the lowlands of the Eje Volcanico Transversal in Orizaba, Veracruz, to northern lowlands of Veracruz (Tejedor, 2005). It has been registered in the states of CH, DU, GR, JA, NY, SI, and VE.

DESCRIPTION: *Natalus lanatus* is a small bat with a fragile structure and is light and small like *N. mexicanus*. The ears are funnel-shaped and are relatively large with a thick tragus. In dorsal view, the pinna in the ears has a straight medial margin and a broader apex. The dorsal surface of the pinna lacks pleats or rarely has one pleat and woollier pelage. The legs are significantly shorter than the forearms. The ventral fur is bicolor; dorsal and ventral hairs are always darker at the base than at the tips. The pelage is dense, woolly, and dull, grayish to ochraceous in color. The dorsal pelage is often tricolor, with hair shafts having a dark base, a lighter middle part, and a tip slightly darker than the middle part (in some individuals the tips are the same color as the middle of the hair and the dorsal pelage looks bicolored). Overall the dorsal pelage color varies from gray (mouse gray) to ochraceous (tawny olive). The ventral pelage color also varies from gray to ochraceous but is noticeably bicolored with hair tips lighter than bases. The feet are markedly hairy, with the long hair tufts covering the distal end of each digit. A fringe of hair is present along the edge of the Europatagium (Tejedor, 2005).

EXTERNAL MEASURES AND WEIGHT
TL = 90 mm; TV = 47 to 55 mm, HF = 8 to 10 mm; EAR = 13 to 16 mm; FA = 35 to 38 mm.
Weight: 5 to 6.5 g.

DENTAL FORMULA: I 2/3, C 1/1, PM 3/3, M 3/3 = 38.

NATURAL HISTORY AND ECOLOGY: This insectivorous species was described only recently (2005). *N. lanatus* remained undetected because of the wide morphological variation of *N. mexicanus*. They are likely sympatric species. With respect to ecology, *N. lanatus* differs considerably in climatic regimes. In both latitudinal and altitudinal distributions *N. lanatus* is included within the range of *N. mexicanus*, and, as has been suggested for other Natalidae (Tejedor et al., 2004), its distribution is probably more limited by the availability of roosts than vegetation type.

VEGETATIONAL ASSOCIATIONS AND ELEVATION RANGE: *N. lanatus* has been reported in dry subtropical with marked seasonal variations in temperature and precipitation (e.g., La Bufa, Chihuahua; Anderson, 1972) to continuously moist montane tropical forest (Orizaba, Veracruz; Hall and Dalquest, 1963). This species has been collected at altitudes that range from lowlands to middle elevations (43 m to 1,200 m).

CONSERVATION STATUS: *N. lanatus* is probably a relatively abundant species that is not included in any risk category, but the precise status of this species is undetermined and needs more research, beginning with the correct identification of the individuals at present named *N. stramineus* or *N. mexicanus.*

Natalus mexicanus Miller, 1902

Mexican funnel-eared bat

Guadalupe Tellez-Giron and Gerardo Ceballos

SUBSPECIES IN MEXICO

N. mexicanus is a monotypic species.

The species was known as *N. stramineus saturatus*, but Tejedor (2006) proposed to change the specific name for the entire *Natalus* genus from Mexico to Panama. Recently, this proposal was supported by Lim (2009). *N. stramineus* is an endemic species of the Lesser Antilles (Tejedor, 2006).

DISTRIBUTION: This species is widely distributed in the Americas, from northern Mexico to Brazil. In Mexico, it is distributed on the Pacific slope from western Sonora including the southern tip of Baja California Sur to Chiapas, and on the Gulf slope from central Nuevo Leon to the Yucatan Peninsula (Goodwin, 1959; Hall, 1981; Villa-R., 1967). It has been registered in the states of BS, CA, CH, CL, CS, DF, DU, GR, HG, JA, MI, MO, MX, NL, NY, OX, PU, QR, SI, SL, SO, TA, TB, VE, YU, and ZA.

DESCRIPTION: *Natalus mexicanus* is a small bat with a very fragile structure. The ears are funnel-shaped and are relatively large with a thick tragus. The legs are long and thin. The tail is fully integrated into the interfemoral membrane and is longer than the length of the body and head. The third phalanx of the third finger is cartilaginous, even in adults. A distinctive feature in males is the presence of a rounded gland on the forehead. The hair color varies from gray to cinnamon yellow or dark reddish. The long hair is soft and silky. There is a slight difference in size between subspecies (Goodwin, 1959b; Villa-R., 1967; Woloszyn and Woloszyn, 1982).

EXTERNAL MEASURES AND WEIGHT

TL = 85 to 98 mm; TV = 47 to 55 mm, HF = 8 to 10 mm; EAR = 13 to 15 mm; FA = 35 to 46 mm.

Weight: 3 to 6 g.

DENTAL FORMULA: I 2/3, C 1/1, PM 3/3, M 3/3 = 38.

NATURAL HISTORY AND ECOLOGY: These insectivorous bats emerge from their shelters to feed before nightfall (Woloszyn and Woloszyn, 1982). They roost inside caves or abandoned mines and on the roofs of houses with a relative high humidity, at sites of greatest darkness. They form colonies of up to 10,000 individuals. They carry out local migrations every day (Jones et al., 1965). During the breeding season, which extends from March until July, females form maternity colonies and males form groups of up to 300 individuals in adjacent roosting sites (Medellín and López-Forment, 1986). One of its most frequent predators is the pine snake (*Pituophis* spp.; Jones et al., 1965). They are generally associated with other species of bats such as *Glossophaga soricina, Desmodus rotundus, Pteronotus parnellii, P. personatus, Carollia perspicillata, Balantiopteryx plicata,* and *Macrotus waterhousii* (Hall and Dalquest, 1963; Jones et al., 1972; Medellín and López-Forment, 1986).

Natalus mexicanus. Tropical dry forest. Chamela-Cuixmala Biosphere Reserve, Jalisco. Photo: Gerardo Ceballos.

VEGETATIONAL ASSOCIATIONS AND ELEVATION RANGE: Mexican funnel-eared bats are mainly found in dry and seasonal tropical forests, rainforests, shrublands, and secondary vegetation (Eisenberg, 1989; Webb et al., 1980). It has been collected from sea level up to 2,190 m (Watkins et al., 1972).

CONSERVATION STATUS: *N. mexicanus* is a relatively abundant species that is not included in any risk category.

Family Molossidae

The family Molossidae is relatively diverse; it comprises 12 genera and 80 species. Its wide distribution includes the tropical and subtropical regions of the world. In Mexico 19 species of 6 genera are known (*Cynomops, Eumops, Molossus, Nyctinomops, Promops*, and *Tadarida*).

Subfamily Molossinae

The subfamily Molossinae is comprised of fifteen genera and many species, with five genera and eighteen species known from Mexico.

Cynomops mexicanus (Jones and Genoways, 1967)

Mexican dog-faced bat

Luis Ignacio Iñiguez Davalos

SUBSPECIES IN MEXICO
C. mexicanus is a monotypic species.
This species was considered a subspecies of *Molossops greenhalli*. Morphometric and genetic evidence has shown, however, that they are different species and that the correct name of the genus is *Cynomops* (Peters et al., 2002).

DESCRIPTION: *Cynomops mexicanus* is a medium-sized bat. The muzzle is broad with the lips lacking vertical folds; its rounded ears are not interconnected and are folded lengthwise. The hair is short and shiny, the dorsal coloration is dark brown to reddish brown, with the base of the hairs lighter to white; ventrally it is grayish-brown; the nose, ears, feet, tail, and wing membranes are dark brown to black. It has a thin band of hair spread along the external side of the forearm to the base of the proximal carpal bones. The tail measures approximately 40% to 50% of the length of the head and body, protruding less than half of its length.

DISTRIBUTION: *C. mexicanus* is an endemic species of Mexico, distributed along the Pacific coastal plain from Durango to Oaxaca (Muñiz-Martinez et al., 2003; Peters et al., 2002). It has been registered in the states of CL, DU, GR, JA, NY, and OX.

EXTERNAL MEASURES AND WEIGHT
TL = 94 to 148 mm; TV = 28 to 37 mm; HF = 8 to 10 mm; EAR = 13.8 to 18 mm; FA = 33.3 to 38.2 mm.
Weight: 10.8 to 18.9 g.

DENTAL FORMULA: I 2/2, C 1/1, P 2/3, M 3/3 = 34.

NATURAL HISTORY AND ECOLOGY: This bat is insectivorous. It forages around and over ponds and marshes or in gaps within the forest. It roosts in holes of trees, branches, and poles, as well as in buildings. It forms colonies of 50 to 75 individuals.

VEGETATIONAL ASSOCIATIONS AND ELEVATION RANGE: *C. mexicanus* is found in deciduous rainforests, perennial rainforests, gallery forests, grasslands, forests of pine-oak, and cloud forests. It is distributed from sea level to 1,600 masl.

CONSERVATION STATUS: *C. mexicanus* is a species with a very wide distribution that is not considered at risk of extinction.

Eumops auripendulus (Shaw, 1800)

Shaws mastiff bat

Hector T. Arita

SUBSPECIES IN MEXICO
Eumops auripendulus auripendulus (Shaw, 1800)

DISTRIBUTION: *E. auripendulus* is distributed from southern Mexico to Brazil, Paraguay, and Argentina (Eger, 1977; Redford and Eisenberg, 1992). In Mexico, it is known in only four localities in the lowlands of the country. It has been registered in the states of CS, OX, QR, and TB.

DESCRIPTION: *Eumops auripendulus* is a large bat. Cranial measures in males are larger than in females, but externally they look the same (Eger, 1977). As in other molossids, the tail is thick with a free end that protrudes beyond the uropatagium. The dorsal pelage is blackish brown to dark brown; in Mexican specimens it is almost black. It has large and rounded ears, joined in the frontal region; unlike in other species, the tragus is pointed.

EXTERNAL MEASURES AND WEIGHT
TL = 80 mm; TV = 43 to 54 mm; HF = 12 to 18 mm; EAR = 19 to 25 mm; FA = 55 to 68 mm.
Weight: 23 to 35 g.

DENTAL FORMULA: I 1/2, C 1/1, PM 2/2, M 3/3 = 30.

NATURAL HISTORY AND ECOLOGY: The natural history of *E. auripendulus* is almost unknown. It has been found in both dry and humid tropical forests (Eisenberg, 1989; Redford and Eisenberg, 1992). In Venezuela, it roosts in small groups in the roofs of houses and hollows of trees (Linares, 1987). Although there is no information on its diet, it might consume insects.

VEGETATIONAL ASSOCIATIONS AND ELEVATION RANGE: In Mexico, *E. auripendulus* inhabits the perennial rainforest, from sea level up to 1,000 m.

CONSERVATION STATUS: *E. auripendulus* is not included in any official list of endangered species. In South America, it is relatively common and has a wide distribution (Arita, 1993a, b). In Mexico, it is rare, but its scarcity in collections may be due to the difficulty in catching it. As it is associated with forested areas, it might be threatened by deforestation in the southern part of the country.

Eumops ferox (Gundlach, 1862)

Wagner's mastiff bat

Jorge Ortega-R.

SUBSPECIES IN MEXICO
It is a monotypic species.

DESCRIPTION: *Eumops ferox* is a large-sized bat. It has a wide and triangular tragus; the ears are short, round, and wide, and are joined in the forehead by a longitudinal keel-shaped structure projected to the front (Eger, 1977; Silva-Taboada, 1979). The hair is short; the dorsal region is dark with white bands at the base and there is a lighter tonality on the abdomen (Eger, 1977; Eisenberg, 1989). The uropatagium is short and wide and, as in all molossids, the tail is free (Eger, 1977; Silva-Taboada, 1979). It is one of the largest species of the genus; it is distinguished from other species by the lack of lacrimal edge and by having a well-developed and semi-oval basisphenoid (Eger, 1977; Villa-R., 1967). It only has a couple of curved incisors; the molars have a typical pattern in the form of a W.

DISTRIBUTION: *E. ferox* is distributed from Mexico to Central America and the islands of Jamaica and Cuba (Silva-Taboada, 1979). In Mexico it has been registered from the south of Jalisco and north of Veracruz, spreading centrally and south to the Yucatan Peninsula (González-Ruiz and Villalpando, 1997; Hall, 1981). It has been registered in the states of CL, CS, HG, JA, MI, MO, VE, and YU.

EXTERNAL MEASURES AND WEIGHT

TL = 80 to 85 mm; TV = 40 to 54 mm; HF = 10 to 15 mm; EAR = 17 to 23 mm; FA = 57 to 64 mm.
Weight: 30 to 45 g.

DENTAL FORMULA: I 1/2, C 1/1, PM 2/2, M 3/3 = 30.

NATURAL HISTORY AND ECOLOGY: This species is crepuscular. Its diet consists of insects caught at great heights, although it also has been collected foraging in the vicinity of water bodies (Eisenberg, 1989; Jones et al., 1973). For the populations of the coast of Jalisco, the mating season occurs during the months of March and April (Nuñez et al., 1981), while for the populations of the Yucatan Peninsula it occurs from April to June (Bowles et al., 1990). Females give birth to a single young in mid-June and the nursing period lasts from five to six weeks (Bowles et al., 1990). The populations of this species roost in shelters located in holes of trees, small cliffs, abandoned constructions, and the cracks of cliffs (Emmons and Feer, 1997). They form colonies of adults and juveniles without segregation of sexes and share their shelters with other species of bats (Silva-Taboada, 1979). *E. ferox* exhibits a marked geographical variation manifested with polymorphisms and genetic differences along its entire distribution (Eger, 1977; Wagner et al., 1974).

VEGETATIONAL ASSOCIATIONS AND ELEVATION RANGE: In Mexico *E. ferox* mainly inhabits deciduous and subdeciduous rainforests. It has also been collected in an ecotone of jungle-oak forest (Sanchez-Hernandez, 1978). It is distributed from sea level to 900 masl.

CONSERVATION STATUS: The current status of this species' populations is unknown, but apparently it is relatively common. Because of its foraging and roosting habits, however, it has been classified as a rare species (Bowles et al., 1990; Emmons and Feer, 1997; Jones et al., 1973).

Eumops ferox. Scrubland. Photo: Scott Altenbach.

Eumops hansae Sanborn, 1932

Hansa mastiff bat

Rodrigo A. Medellín

SUBSPECIES IN MEXICO
E. hansae is a monotypic species.

DESCRIPTION: *Eumops hansae* is a medium-sized bat. The ears are long in comparison with those of other molossids, reaching the tip of the face and connecting on the top of the head. The tragus is comparatively wide and square. Unlike other species in this family, the upper lip lacks wrinkles and possesses in the center, just under the nostrils, a group of short and rigid whiskers. The dorsal pelage is very dark, almost black, and ventrally it has lighter underfur.

DISTRIBUTION: The species is distributed from the southeast of Mexico to Brazil and Paraguay (Eger, 1977). In Mexico, it has only been registered in two locations in the state of Chiapas: in the south of the Lacandon Forest and in the coastal region of Mapastepec (Álvarez and Álvarez-Castañeda, 1990; Medellín et al., 1992). It has been registered in the state of CS.

EXTERNAL MEASURES AND WEIGHT
TL = 100 to 110 mm; TV = 25 to 28 mm; HF = 8 to 12 mm; EAR = 17 to 22 mm; FA = 37 to 41 mm.
Weight: 16 to 21 g.

DENTAL FORMULA: I 1/2, C 1/1, PM 2/2, M 3/3 = 30.

NATURAL HISTORY AND ECOLOGY: Very little is known about the biology of this species. They are insectivores that catch their prey in flight. They have been found roosting in the interior of hollow trees (Handley, 1976). The single specimen of the Lacandon Forest was found outside a house, near the ground (Medellín et al., 1992).

VEGETATIONAL ASSOCIATIONS AND ELEVATION RANGE: In Mexico *E. hansae* inhabits evergreen tropical forests. In South America, it has been found in thorny scrublands and in agricultural areas. It has been registered from sea level up to 700 masl (Handley, 1976; Medellín et al., 1992).

CONSERVATION STATUS: *E. hansae* is a rare species; currently only three specimens have been caught in Mexico. It is not included in any official list of endangered species; R. Medellín and H. Arita (pers. obs.) classify it as endangered, due to the fact that the only three known specimens come from regions that currently face strong alterations. Arita (1993a, b) classifies it as rare and with a restricted distribution and therefore among the most vulnerable species.

Eumops nanus (Miller, 1900)

Dwarf bonneted bat

Hector T. Arita

SUBSPECIES IN MEXICO
Eumops nanus (Miller, 1900)
Some authors considered *nanus* a subspecies of *E. bonariensis* (Eger, 1977; Jones et al., 1988; Koopman, 1993).

DESCRIPTION: *Eumops nanus* is the smallest species of the genus (Bowles et al., 1990; Eger, 1977). The large and rounded ears, joined in the front, characterize it. The tragus is large and rounded. The dorsal pelage is reddish-brown or dark brown. The tail has a free end that projects beyond the edge of the uropatagium. It is distinguished from other species of the same genus by its smaller size and from *E. hansae* by its less dark coloration. It is distinguished from *Nyctinomops* and *Tadarida brasiliensis* because these species have deep channels in the upper lip.

DISTRIBUTION: *E. nanus* is distributed from southern Veracruz and the Yucatan Peninsula to Buenos Aires, Argentina (Eger, 1977; Redford and Eisenberg, 1992). Subfossil material has been found in the Gruta de Loltun, Yucatan (Arroyo-Cabrales and Álvarez, 1990). It has been registered in the states of QR, TB, VE, and YU.

EXTERNAL MEASURES AND WEIGHT
TL = 42 mm; TV = 30 to 31 mm; HF = 6 to 11 mm; EAR = 18 to 19 mm; FA = 37 to 49 mm.
Weight: 6 to 14 g.

DENTAL FORMULA: I 1/2, C 1/1, PM 2/2, M 3/3 = 30.

NATURAL HISTORY AND ECOLOGY: The roosting sites used by this bat are unknown; they might include slopes and cliffs. It has been found under roofs (Bowles et al., 1990; Redford and Eisenberg, 1992) and in a thatch roof (Lay, 1962). Apparently, they show high fidelity to their roosting site. In a study conducted in Merida, Yucatan, several individuals were captured repeatedly in the same roosting site, the tiles of a house (Bowles et al., 1990). It is an insectivorous species. The stomach contents of four individuals captured in Yucatan contained, in order of abundance, remnants of lepidopterans, beetles, and hemipterans (Bowles et al., 1990). In the Yucatan Peninsula, reproduction takes place in the spring. Pregnant females have been captured in March, May, and June, and nursing females in June, July, and August (Birney et al., 1974; Bowles et al., 1990; Engstrom et al., 1989).

VEGETATIONAL ASSOCIATIONS AND ELEVATION RANGE: In Mexico, *E. nanus* is known in tropical perennial forests and tropical subdeciduous forests. It seems to be more abundant in dry tropical forests (Bowles et al., 1990; Eisenberg, 1989). In Yucatan, it has been found in urban and suburban environments in the city of Merida (Bowles et al., 1990). Most of the locations were found near sea level. In Mexico, all localities are located below 40 masl. In Venezuela, the species has affinity for places of "extreme low altitude" (Eisenberg, 1989).

CONSERVATION STATUS: *E. nanus* is not included in any official list of endangered species. It was considered a very rare species in Mexico because until very recently only three specimens had been caught (Eger, 1977); however, it is very abundant in Yucatan (Birney et al., 1974; Bowles et al., 1990). Its ability to inhabit urban areas and use human constructions as roosting sites, suggests that it is less affected than other species by the loss of natural habitat.

Eumops nanus. Tropical dry forest. Mérida, Yucatán. Photo: Jens Rydell.

Western mastiff bat

Elizabeth E. Aragon

SUBSPECIES IN MEXICO
Eumops perotis californicus (Merriam, 1890)

DESCRIPTION: *Eumops perotis* is a large-sized bat. Sexual dimorphism is present since males are larger than females in several morphologic characters (Eger, 1977). Additionally, males have a gland in the middle ventral part of the body, but its function is unknown (Aragón, unpublished data; Villa-R., 1967). A hallmark of the species is the large and not erect ears, which stand out from the nose, almost hiding the eyes, and are joined together in the middle line of the head. As in other molossids, the lips are smooth throughout the muzzle, without wrinkles. The dorsal pelage is dark brown (chocolate) and the fur is lighter on the abdomen.

EXTERNAL MEASURES AND WEIGHT
TL = 135 to 188 mm; TV = 53.3 to 68.9 mm; HF = 14 to 19 mm; EAR = 34.3 to 47.2 mm; FA = 72 to 80 mm.
Weight : 68 to 70 g.

Eumops perotis. Scrubland. Photo: Scott Altenbach.

DENTAL FORMULA: I 1/2, C 1/1, PM 2/2, M 3/3 = 30.

NATURAL HISTORY AND ECOLOGY: In Mexico this species apparently migrates. It has only been captured from September to November and from January to May (E. Aragon, pers. obs.; Sánchez, 1984). Some populations of Texas are residents (Ohlendorft, 1972). It is a rapid flyer that starts its activity at evening. It forms small colonies of usually less than 100 members. It roosts in cracks, hollow trees, caves, cracks of rocks, tunnels, and bridges (Álvarez and Polaco, 1980; Cockrum, 1960; Sánchez, 1984). It inhabits shelters of not less than 3 m in height from which to begin its flight because its large size and long, thin wings prevent it from taking off from the ground. They share roosting sites with other bats, including *Desmodus rotundus*, *Balantiopteryx plicata*, *Myotis occultus*, and *Tadarida brasiliensis* (E. Aragon, pers. obs.; Sánchez et al., 1993). They are insectivorous. In Mexico, fecal samples show remnants of locusts (Orthoptera), spiders (Aranae), and bees (E. Aragon, pers. obs.; Sanchez, 1984). It also feeds on moths, grasshoppers, dragonflies, leaf insects, beetles, and cicadas (Easterla and Whittaker, 1972; Ross, 1967; Schmidly, 1991). Its pattern of reproduction in Mexico has not been established, but pregnant females have been captured in January and a nursing female in September (Baker, 1956; Jiménez-G. and Zuñiga-R., 1992; Sánchez, 1984). Resident populations of Texas begin their reproduction in the spring with an estimated gestation period of 18 to 19 days. In general, females give birth to one young in June and July (Schmidly, 1991). Both females and males use the same shelter throughout the year, even during the time of birth and growth of the offspring, a situation rare in bats (Schmidly, 1991). They are preyed on by owls (*Tyto alba*), peregrine hawks (*Falco peregrinus*), American kestrels (*Falco sparverius*), and red-tailed hawks (*Buteo jamaicensis*) (Sánchez et al., 1993; Schmidly, 1991).

DISTRIBUTION: *E. perotis* is distributed from the southwestern United States to South America. In Mexico, it is known in the Pacific slope and the Altiplane (Álvarez and Polaco, 1980; Dalquest and Roth, 1970; Jiménez-G. and Zuñiga-R., 1992; Jones, et al., 1972, 1977, 1988; Hall, 1981; Matson and Baker, 1986; Ramírez-P., et al., 1983; Sánchez, 1984; Sánchez et al., 1993; Villa-R., 1967). It has been registered in the states of CO, DF, DU, HG, JA, MI, MO, NL, SI, SO, TA, and ZA.

VEGETATIONAL ASSOCIATIONS AND ELEVATION RANGE: *E. perotis* inhabits arid and semi-arid vegetation, xeric scrublands, and vegetation with succulent and rosette scrub. It is found from 850 masl to 2,240 masl (Sánchez et al., 1993).

CONSERVATION STATUS: *E. perotis* is not included in the Mexican Official Regulation of endangered species (SEDESOL, 1994; SEMARNAT, 2002). It is considered as a candidate for conservation in the United States (Bogan et al., 1996). In Mexico, it should be included in the list of endangered species since it is migratory, its shelters are scarce, and its colonies have a punctual distribution.

Eumops underwoodi Goodwin, 1940

Underwood's mastiff bat

Luis Ignacio Iñiguez Davalos

SUBSPECIES IN MEXICO

Eumops underwoodi underwoodi Goodwin, 1940
Eumops underwoodi sonoriensis Benson, 1947

DESCRIPTION: *Eumops underwoodi* is a large bat with large and flattened ears united in the base by the upper edge of the front; the ears go beyond the nostrils when extended forward. The muzzle is sharp, and the upper lip has vertical folds. The nostrils have elevated edges that join together and form a small central keel. The color of the back

Eumops underwoodi. Scrubland. Photo: Scott Altenbach.

varies from bright brown ochre to dark brown with a whitish base. The wings are very long and narrow (Eger, 1977).

EXTERNAL MEASURES AND WEIGHT

TL = 150 to 162 mm; TV = 48 to 52 mm; HF = 17 to 19 mm; EAR = 31.5 to 33 mm; FA = 65.3 to 75.3 mm.
Weight: 40 to 68 g.

DENTAL FORMULA: I 1/2, C 1/1, PM 2/2, M 3/3 = 30.

NATURAL HISTORY AND ECOLOGY: This bat is insectivorous; it feeds on prey, mainly small coleopterans, caught at high altitudes. Their peak of activity starts late at night. They fly in very straight courses, making audible shrieks. They roost inside hollow trees, rocky walls with cracks, and building roofs. Because of the form of their wings, it is difficult for them to take off from the ground, so their shelters are high and they initiate flight by falling. Births occur at the beginning of the rains.

VEGETATIONAL ASSOCIATIONS AND ELEVATION RANGE: *E. underwoodi* inhabits the deciduous rainforest, forest of pine-oak, and perennial rainforest. It is found from sea level to 1,960 masl, predominately in low elevations.

CONSERVATION STATUS: *E. underwoodi* is a rare species. There is not sufficient information to determine with precision its current status. It is possible that its rarity is due to its habits of flying at high altitudes and to the difficult access to its shelters.

DISTRIBUTION: This species is distributed from Arizona, on the side of the Pacific, to Guatemala, El Salvador, and Honduras. In Mexico, it is found from Sonora to Chiapas on the coastal plain and the mountainous systems. It has been registered in the states of CH, CL, CS, JA, MI, MO, MX, OX, SO, and TB.

Molossus alvarezi González-Ruiz, Ramírez-Pulido & Arroyo-Cabrales, 2011

Álvarez's mastiff bat

Joaquín Arroyo-Cabrales

SUBSPECIES IN MEXICO

M. alvarezi is a monotypic species.

DESCRIPTION: *Molossus alvarezi* is a medium-sized *Molossus*. The dorsal region is very pale yellowish gray-brown, with black over the middle, forming an indistinct dorsal band; the sides are yellowish gray, slightly varied with blackish-tipped hairs. The ventral region is clear white, the fur lead-colored at the base. The skull is relatively narrow.

DISTRIBUTION: The species is known from the Yucatan Peninsula, including the states of Yucatan and Quintana Roo, Mexico, and Belize. It is recorded from the states of QR and YU.

EXTERNAL MEASURES AND WEIGHT

TL = 120 mm; TV = 41 mm, HF = 11 mm; EAR = 14 mm, FA = 42.7 to 47.4 mm.
Weight: 23.2 g.

DENTAL FORMULA: I 1/1, C 1/1, PM 1/2, M 3/3 = 26.

NATURAL HISTORY AND ECOLOGY: *M. alvarezi* is an aerial insectivore. This species can be found in evergreen and dry deciduous forests, pastures, and populated areas. It roosts in caves and houses, often in large groups. A long-term study in Yucatan, Mexico, found this species to be the most commonly encountered molossid in the region. Individuals are most active during the first two hours after sunset and again before dawn. The diet consists mainly of moths, with some beetles and other insects. In Yucatan, pregnant females have been recorded from March to June. These bats may be found in rural and urban areas.

VEGETATIONAL ASSOCIATIONS AND ELEVATION RANGE: *M. alvarezi* inhabits deciduous tropical woodlands.

CONSERVATION STATUS: This species is uncommon to locally common, but never as abundant as *M. rufus*. It is found in protected areas.

Molossus aztecus Saussure, 1860

Aztec mastiff bat

Mery Santos G. and Ivan Castro-Arellano

SUBSPECIES IN MEXICO

M. aztecus is a monotypic species.

It has been considered a subspecies of *M. molossus*, as *M. molossus aztecus*; with such classification, it has been reported in different states of the country (Hall, 1981). Recent studies have shown that it is a different species (Dolan, 1989).

DESCRIPTION: *Molossus aztecus* is one of the smallest bats within the genus. Like *M. molossus* and *M. sinaloae*, it has a white band on the dorsal hair. Although *M. aztecus*

DISTRIBUTION: *M. aztecus* is found from the center of Mexico to possibly South America. Dolan (1989) raised the possibility that this species is endemic to high and cold regions from Mexico to Costa Rica and Panama. Eisenberg (1989) mentions that its range extends from Sinaloa to Tamaulipas, in Mexico, to Panama and Venezuela. In Mexico it is known in the Altiplane, and the Sierra Madre Occidental, Sierra Madre Oriental, and Sierra Madre Sur. It has been registered in the states of CO, CS, GR, JA, MX, OX, SI, SL, and TA.

and *M. molossus* are similar in appearance, the membranes, nose, and ears of *M. aztecus* are very dark. The dorsal pelage is chocolate brown and the white band on the base of the hairs of the back is less conspicuous. The forearm is shorter, the skull bigger, the face wider, and the incisors flatter and broader. In terms of genetic characteristics, type of coat, and configuration of the skull, *M. aztecus* is more similar to *M. coibensis* than to *M. molossus* (Dolan, 1989).

EXTERNAL MEASURES AND WEIGHT
TL = 84 to 92 mm; TV = 33 to 33 mm; HF = 10 to 11 mm; EAR = 13 to 13 mm; FA = 34 to 35 mm.
Weight: 12 to 16 g.

DENTAL FORMULA: I 1/1, C 1/1, PM 1/2, M 3/3 = 26.

NATURAL HISTORY AND ECOLOGY: This bat feeds on insects. It inhabits hollow trees and is strongly associated with humid habitats (Handley, 1976). Females of *M. aztecus* and *M. sinaloae* are the only ones of the entire genus that may be pregnant while nursing other young (Dolan, 1989).

VEGETATIONAL ASSOCIATIONS AND ELEVATION RANGE: This bat inhabits temperate oak forests and pine-oak forests. It is generally found at altitudes higher than 1,500 masl (Dolan, 1989).

CONSERVATION STATUS: The current status of the populations of *M. aztecus* is unknown, but due to its broad range of distribution it might not be threatened.

Molossus coibensis J.A. Allen, 1904

Coiban mastiff bat

Hector T. Arita

SUBSPECIES IN MEXICO
M. coibensis is a monotypic species.
It has been classified as a subspecies of *Molossus molossus* (Hall, 1981), but Dolan (1989) classified it as a different species. The Mexican specimens, examined by Gardner (1966), which were used for the description of the subspecies *lambi*, were assigned to *M. coibensis* by Dolan (1989).

DESCRIPTION: *Molossus coibensis* is a small-sized molossid, the smallest in Mexico. The general color of the body is very dark, almost black, although in some individuals reddish colors are present. The individual hairs of the back are very short (2 mm to 3 mm) and each one has a black tip and a lighter base of cream to white color approximately one-quarter of its length. It is distinguished from other species of the genus by the combination of size and color; *M. rufus* has a similar coloration but is much bigger (length of forearm > 47 mm). *M. molossus* and *M. sinaloae* have longer dorsal hair and a larger basal band (Dolan, 1989).

EXTERNAL MEASURES AND WEIGHT

TL = 86 to 97 mm; TV = 28 to 37 mm; HF = 8 to 10 mm; EAR = 13 to 15 mm; FA = 32 to 37 mm.
Weight: 12 to 16 g.

DENTAL FORMULA: I 1/1, C 1/1, PM 1/2, M 3/3 = 26.

NATURAL HISTORY AND ECOLOGY: Very little is known about the biology of this species. Most of the known specimens have been captured in nets placed on flows of water. In Chiapas, an individual of this species was seen emerging from a hole in the top of an amate (*Ficus* sp., Gardner, 1966). The details of its diet are unknown, but it is likely to feed on hard-shelled insects as other species do. In Chiapas, pregnant females with a single embryo have been found in March and June (Gardner, 1966).

VEGETATIONAL ASSOCIATIONS AND ELEVATION RANGE: This bat inhabits dry tropical areas of the Pacific. The vegetation in the Mexican localities mentioned by Gardner (1966) corresponds to that of tropical subdeciduous forests. It is distributed from sea level up to 350 m, although apparently in South America some localities reach 800 masl (Dolan, 1989).

CONSERVATION STATUS: *M. coibensis* is not included in any official list of endangered species. Its total area of distribution is unknown, so its real situation cannot be established. Its distribution in Mexico is restricted to a small area of Chiapas, although judging by the data of Gardner (1966), local populations could be comparatively numerous.

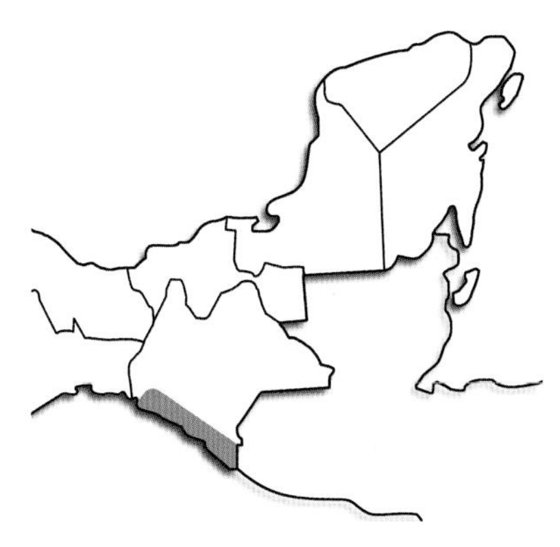

DISTRIBUTION: The known distribution of the species includes the Pacific slope from southern Chiapas to Panama (Dolan, 1989). It is known to exist in South America, but it is impossible to define its distribution without more specimens (Dolan, 1989). Due to the recent taxonomic change of *M. coibensis*, it is likely that a number of specimens, reported in the literature as *M. molossus*, correspond in fact to this species, and that its range is wider than believed. It has been registered in the state of CS.

Molossus molossus (Pallas, 1766)

Pallas mastiff bat

Mery Santos G. and Ivan Castro-Arellano

SUBSPECIES IN MEXICO

Molossus molossus lambi Gardner, 1966

DESCRIPTION: *Molossus molossus* is a medium-sized bat. The dorsal pelage varies from grayish-brown to very dark brown; the abdomen is pale brown (Eisenberg, 1989). It is characterized by the thick and relatively free tail, narrow and long wings, short and velvety coat, wide ears that project over the eyes, and a practically naked face (Muñoz, 1995). *M. molossus* and *M. aztecus* are extremely similar; however, the membranes, nose, and ears of *M. aztecus* are very dark, the dorsal pelage is brown, and the white band of the base of the hair is less conspicuous. *M. molossus* bats of Central America are candy brown, noticeably paler, and more opaque than *M. molossus* bats of the Minor Antilles, which are ebony (Dolan, 1989).

EXTERNAL MEASURES AND WEIGHT

TL = 99 to 101 mm; TV = 34 to 36 mm; HF = 9 to 10 mm; EAR = 12 to 13 mm; FA = 38 to 39 mm.
Weight: 13 to 15 g.

DENTAL FORMULA: I 1/1, C 1/1, PM 1/2, M 3/3 = 26.

NATURAL HISTORY AND ECOLOGY: *M. molossus* is an insectivorous bat that feeds on flying insects, including moths, beetles, and flying ants, among others. It often forages in the vicinity of streams and roosts in caves, hollow trees, or human constructions. It has been found sharing roosting sites with *M. rufus* and forming colonies of hundreds of animals (Eisenberg, 1989; Emmons and Feer, 1997). Reproduction takes place at the beginning of the rainy season, and females form maternity colonies in which to raise the newborns. The offspring form compact groups while their mothers emerge from the cave to feed; on their return each mother identifies her young by its distinctive sounds. The offspring begin to thermoregulate at 20 days of age; maternal care stops at 60 days of age, and the size of the forearm reaches the adult size at 65 days of age (Eisenberg, 1989).

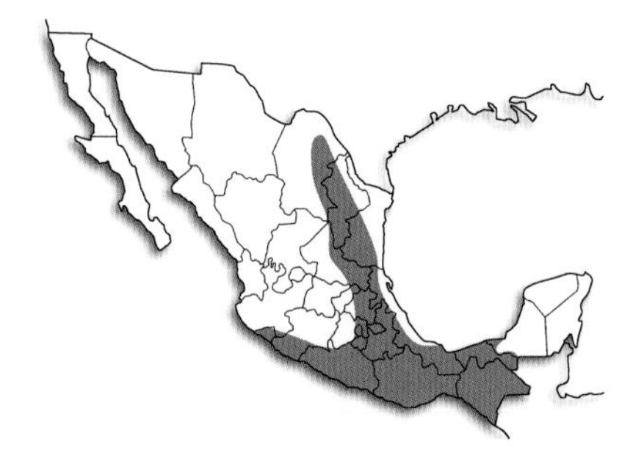

DISTRIBUTION: The range of distribution of this species is very broad; it is known from Mexico to Uruguay and Argentina (Dolan, 1989; Eisenberg, 1989; Hall, 1981; Muñiz-Martinez et al., 2003; Sánchez-Hernandez et al., 1999). In Mexico, it is known in the lowlands of the Pacific, from Sinaloa to Chiapas. It has been registered in the states of CL, CO, CS, DU, JA, MI, MX, OX, SI, SL, and TA.

VEGETATIONAL ASSOCIATIONS AND ELEVATION RANGE: Together with *M. rufus*, *M. pretiosus*, and *M. coibensis*, this species prefers open areas such as savannahs, pastures, open and dry areas with shrubs, or zones of cacti, thorny bushes, and other similar environments (Dolan, 1989). It is found from sea level to 2,000 masl (Muñoz, 1995).

CONSERVATION STATUS: The current status of the populations of *M. molossus* is not known, but due to its broad range of distribution it might not be threatened.

Molossus rufus E. Geoffroy, 1805

Black mastiff bat

Mery Santos G. and Ivan Castro-Arellano

SUBSPECIES IN MEXICO
Molossus rufus nigricans Miller, 1902

DESCRIPTION: *Molossus rufus* is the largest species of the genus. These bats are characterized by having a thick and relatively free tail, long and narrow wings, short and velvety coat, wide ears projected over the eyes, and a relatively naked face. The dorsal pelage is dark brown, reddish, or blackish; ventrally it is slightly paler; the uropatagium is naked, the nose is very long and truncated, extending 6 mm or 7 mm behind anterior edge of the ears, the skull is robust with a high sagittal crest and a well-developed lamboid crest (Muñoz, 1995). *M. rufus* is also known as *M. ater* (Dolan, 1989; Emmons and Feer, 1997).

DISTRIBUTION: The distribution range of this species extends from Sinaloa, in the west, and Nuevo Leon in northeastern Mexico, to Argentina (Chavez and Ceballos, 1998; Eisenberg, 1989; Hernández-Huerta et al., 2000). It has been registered in the states of CA, CL, CO, CS, DF, DU, GR, HG, JA, MI, MO, MX, NY, OX, PU, QE, QR, SI, SL, TA, TB, VE, and YU.

EXTERNAL MEASURES AND WEIGHT
TL = 120 to 125 mm; TV = 44 to 46 mm; HF = 13 to 14 mm; EAR = 16 to 17 mm; FA = 47 to 49 mm.
Weight: 27 to 31 g.

DENTAL FORMULA: I 1/1, C 1/1, PM 1/2, M 3/3 = 26.

NATURAL HISTORY AND ECOLOGY: *M. rufus* is an insectivorous bat that feeds mainly on water scavenger beetles and leaf beetles; its diet also includes flies, long-horned

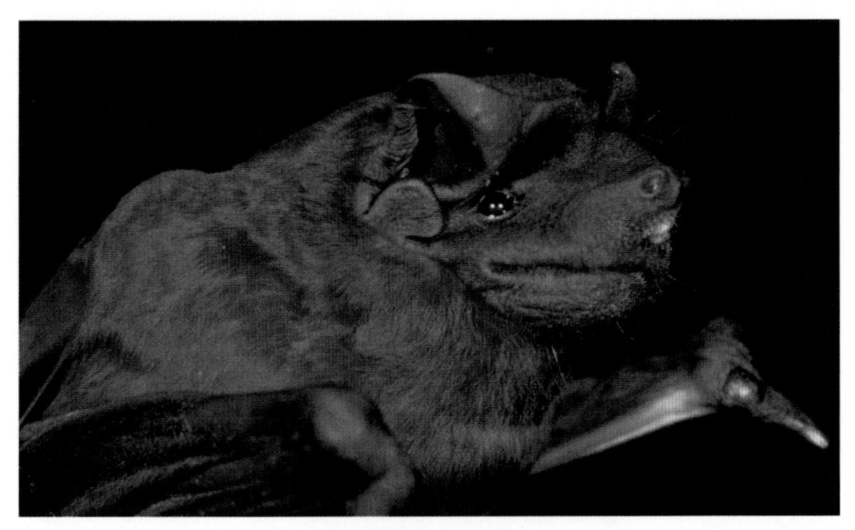

Molossus rufus. Tropical dry forest. Merida, Yucatan. Photo: Jens Rydell.

beetles and sawyer beetles, diving beetles, and hemipterans (Fenton et al., 1998). They go out to forage at dusk for approximately one hour before returning; only a few leave the cave to forage again. It is found in rainforests and many types of dry environments, in towns and cities. The consumption of massive quantities of insects is beneficial to humans (Eisenberg, 1989; Emmons and Feer, 1997). It roosts in hollow trees and human constructions. In general, it prefers wet areas. It forms colonies of more than 50 animals (Goodwin and Greenhall, 1961). It is found in caves along with *M. molossus*.

VEGETATIONAL ASSOCIATIONS AND ELEVATION RANGE: This species is very tolerant and inhabits the deciduous and perennial rainforest, scrubland, oak forest, and disturbed vegetation (Handley, 1976). It is found from sea level to 2,000 masl (Muñoz, 1985).

CONSERVATION STATUS: The current status of the populations of *M. rufus* is unknown, but due to its great tolerance and wide distribution, it is presumed that this species is not at risk of extinction.

Molossus sinaloae J.A. Allen, 1906

Allen's mastiff bat

Mery Santos G.

SUBSPECIES IN MEXICO

M. sinaloae is a monotypic species.

DESCRIPTION: *Molossus sinaloae* is the second largest species of the genus. In broad terms, the thick and relatively free tail, long and narrow wings, short and velvety

Molossus sinaloae. Tropical dry forest. Merida, Yucatan. Photo: Jens Rydell.

coat, wide ears projected over the eyes, and a practically naked face characterize it. In particular, the dorsal pelage is dark brown with a slight reddish tint and ventrally the fur is paler. The dorsal pelage is opaque, bicolor (the hairs have a white band in the base, extending two-thirds of the length), long (6.5 mm), soft, and limp. The pelage extends to the interfemoral membrane. The basisphenoid concavity is long and deep, the edge of the palate has no medial projection, the upper anterior premolar is relatively big, the sagittal and occipital crests are low and smooth, and the upper dental series are almost parallel (Dolan, 1989; Eisenberg, 1989; Muñoz, 1995). Until recently, *M. sinaloae* was known as *M. trinitatus* (Eisenberg, 1989; Muñoz, 1995).

EXTERNAL MEASURES AND WEIGHT
TL = 48 to 115 mm; TV = 38 to 54 mm; HF = 10 to 13 mm; EAR = 12 to 17 mm; FA = 46 to 49 mm.
Weight: 14 to 28 g.

DENTAL FORMULA: I 1/1, C1/1, PM 1/2, M 3/3 = 26.

NATURAL HISTORY AND ECOLOGY: *M. sinaloae* feeds on flying, soft-bodied insects, probably because of its narrow skull. Together with *M. aztecus*, females of this species are the only ones of the genus that can be pregnant while nursing. The gestation period lasts 90 days (Dolan, 1989).

VEGETATIONAL ASSOCIATIONS AND ELEVATION RANGE: This species is tolerant of many environmental conditions. It inhabits deciduous rainforests and tropical precloud forests (Dolan, 1989; Handley, 1976). It is distributed from sea level to 2,000 masl (Muñoz, 1995).

CONSERVATION STATUS: This species has a wide distribution and a great tolerance to anthropogenic disturbances; therefore, it is not threatened.

DISTRIBUTION: *M. sinaloae* is distributed from northeastern Mexico to northern South America (Dolan, 1989; Eisenberg, 1989). In Mexico, it is found from the north of Sinaloa, reaching the Sierra Madre del Sur in Jalisco; then, it continues on the Eje Volcanico Transversal to the Gulf of Mexico and the Yucatan Peninsula (Dolan, 1989; González-Ruíz et al., 2000). It has been registered in the states of CL, GR, JA, MI, MO, MX, PU, SI, and YU.

Peale's free-tailed bat

Hector T. Arita

SUBSPECIES IN MEXICO

N. aurispinosus is a monotypic species.

It is also known as *Tadarida aurispinosa*, but the arrangement of Freeman (1981a) is widely accepted (Koopman, 1993).

DESCRIPTION: *Nyctinomops aurispinosus* is a medium-sized bat. As in other species of the genus, it possesses large ears united on their bases. The dorsal coloration is dark brown, although some individuals have reddish or grayish colors. Ventrally it is lighter. The tail is thick and has a free end that stands out from the edge of the uropatagium. It is distinguished from other species of the genus by certain features of the skull and by its intermediate size, being smaller than *N. macrotis* but larger than *N. femorosaccus* and *N. laticaudatus*.

EXTERNAL MEASURES AND WEIGHT

TL = 120 to 140 mm; TV = 45 to 55 mm; HF = 9 to 11 mm: EAR = 18 to 22 mm; FA = 47 to 53 mm.

Weight: 17 to 23 g.

DENTAL FORMULA: I 1/2, C 1/1, PM 2/2, M 3/3 = 30.

NATURAL HISTORY AND ECOLOGY: Very little is known of the natural history of this species. Apparently, it roosts in caves, forming small groups (Arita, 1993a, b; Carter and Davis, 1961), although it has also been found in human constructions (Taddei and Garutti, 1981). Its diet is unknown, but it possibly feeds on moths (Lepidoptera) and other soft-bodied insects, as other species of the genus do (Freeman, 1979). The only data on reproduction in Mexico comes from a nursing female caught in July in Zacatecas (Álvarez, 1963a).

VEGETATIONAL ASSOCIATIONS AND ELEVATION RANGE: In Mexico, *N. aurispinosus* has been found in dry tropical areas, mainly in the deciduous rainforest, thorny forest, and xeric scrubland. This species has been found from sea level up to 3,150 m (Jones and Arroyo-C., 1990), but in Mexico it is only known in areas with altitudes of less than 1,100 m (Álvarez and Aviña, 1964; Jones and Arroyo-C., 1990).

CONSERVATION STATUS: *N. aurispinosus* is not included in any official list of endangered species. In Mexico, it is known in very few locations, and it is not very abundant, which would suggest rarity, but its real situation is unknown.

DISTRIBUTION: *N. aurispinosus* is found in the Neotropics, from Mexico to Venezuela, Bolivia, and Brazil (Jones and Arroyo-C., 1990). In Mexico, it is distributed on the side of the Pacific from the south of Sonora, and on the side of the Gulf of Mexico, from the south of Tamaulipas to Oaxaca. It has been registered in the states of CL, JA, MI, MO, NY, OX, PU, SI, SL, SO, TA, and ZA.

Pocketed free-tailed bat

Hector T. Arita

SUBSPECIES IN MEXICO

N. femorosaccus is a monotypic species.

Until the revision of Freeman (1981a), this bat was known as *Tadarida femorosacca.*

DESCRIPTION: *Nyctinomops femorosaccus* is a small-sized bat, and one of the smallest within the genus *Nyctinomops*. Females are slightly smaller than males, at least in measures of the skull (Kumirai and Jones, 1990). The ears are relatively large, united in their bases, and slightly extended beyond the snout when folded forward. The dorsal pelage is brown, but the general appearance is more grayish than in other species of the genus. As in other molossids, the tail has a free end that extends beyond the edge of the uropatagium. It has a membranous sack on the uropatagium, between the femur and the tibia, which gives this bat its name. It is slightly larger than *N. laticaudatus* and clearly smaller than *N. macrotis* and *N. aurispinosus*.

EXTERNAL MEASURES AND WEIGHT

TL = 100 to 110 mm; TV = 31 to 40 mm; HF = 10 to 11 mm; EAR = 22 to 24 mm; FA = 45 to 48 mm.

Weight: 11 to 18 g.

DENTAL FORMULA: I 1/2, C 1/1, PM 2/2, M 3/3 = 30.

Nyctinomops femorosaccus. Scrubland. Photo: Scott Altenbach.

NATURAL HISTORY AND ECOLOGY: *N. femorosaccus* mainly roosts in rocky cliffs, sometimes sharing them with other molossids, but not in direct contact with them. (Barbour and Davis, 1969; Easterla, 1970, 1973). It also roosts in small caves, in buildings, and under the tiles of roofs (Jones et al., 1972; Matson and Baker, 1986). Their colonies are formed by a few dozens of individuals that roost closely to each other. Apparently, individuals emerge from the shelters late at night (Kumirai and Jones, 1990). They are insectivorous, and base their diet on large moths (macrolepidopterans), but also consume other flying insects, including Hymenoptera, dipterans, and coleopterans (Easterla and Whitaker, 1972; Freeman, 1979; 1981a). In the stomach contents of some individuals were the remains of non-flying insects, grasshoppers, and locusts (Orthoptera), which were probably caught in the roosting site (Easterla and Whitaker, 1972; Kumirai and Jones, 1990). They reproduce during the summer, giving birth to a single young. Mating takes place at the end of spring. The offspring are born in late June and July, and are nursed during August and September (Easterla, 1973; Jones et al., 1972; Kumirai and Jones, 1990). Predators of this bat include snakes and owls (Jones et al., 1972; Kumirai and Jones, 1990).

VEGETATIONAL ASSOCIATIONS AND ELEVATION RANGE: *N. femorosaccus* is mainly found in dry tropical and subtropical areas, covered with deciduous tropical forest, thorny forest, or xeric scrubland. Occasionally, it penetrates the transition area toward the pine-oak forest (Álvarez-Castañeda, 1991; Watkins et al., 1972). This species has been collected from sea level to 2,200 masl (Kumirai and Jones, 1990; Watkins et al., 1972).

CONSERVATION STATUS: The conservation status of this species is unknown. The area of distribution of this species is restricted. The relative scarcity of specimens in scientific collections would suggest that, at local level, it is a rare species. As some intensive studies have shown, however, in certain localities it is relatively abundant (Easterla, 1973).

DISTRIBUTION: *N. femorosaccus* is an almost endemic species of Mexico. In the United States, it is only found in the southern portion of California, Arizona, New Mexico, and Texas. In Mexico, it is mainly found in the coastal plain of the Pacific, from Sonora to the basin of the Balsas River in the states of Mexico, Morelos, and Puebla. It has also been reported in the north of Coahuila (Easterla, 1970) and in Monterrey, Nuevo Leon (Jiménez-G., 1968). It has been registered in the states of BC, BS, CL, CO, DU, JA, MI, MO, MX, NL, PU, SI, and SO.

Nyctinomops laticaudatus (È. Geoffroy St.-Hilaire, 1805)

Broad-tailed bat

Hector T. Arita

SUBSPECIES IN MEXICO

Nyctinomops laticaudatus ferrugineus (Goodwin, 1954)
Nyctinomops laticaudatus yucatanicus Miller, 1902
Some authors use the name *Tadarida laticaudata* for this species (Jones et al., 1988), but the majority has adopted the settlement of Freeman (1981a), who named it *Nyctinomops.*

DESCRIPTION: *Nyctinomops laticaudatus* is a medium-sized bat, but the smallest within the genus. Males are slightly larger than females (Silva-Taboada, 1979). It possesses large ears united in their bases. The dorsal pelage is brown, with lighter underfur. The tail is thick with an end that extends beyond the uropatagium. It is distinguished from other species by its smaller size. It can be differentiated from *N. femorosaccus*, the species most similar in size, by having the first upper molar not square and shorter ears (Hall, 1981; Kumirai and Jones, 1990).

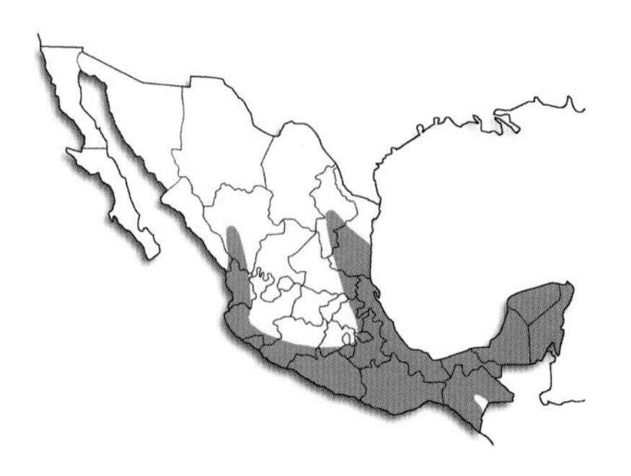

DISTRIBUTION: This species has a wide distribution; it is found from Mexico to Paraguay and northern Argentina (Redford and Eisenberg, 1992). It has also been found in Cuba (Silva-Taboada, 1979). In Mexico it is distributed on the Pacific slope from Durango, and on the Gulf of Mexico, from Tamaulipas to the south, including the Yucatan Peninsula (Hall, 1981; Muñiz-Martinez et al., 2003). It has been registered in the states of CA, CL, CS, DU, GR, JA, NY, OX, QR, SL, TA, VE, and YU.

EXTERNAL MEASURES AND WEIGHT
TL = 90 to 116 mm; TV = 35 to 48 mm; HF = 10 to 12 mm; EAR = 18 to 21 mm; FA = 41 to 47 mm.
Weight: 10 to 16 g.

DENTAL FORMULA: I 1/2, C 1/1, PM 2/2, M 3/3 = 30.

NATURAL HISTORY AND ECOLOGY: We know very little about the biology of this species. It roosts in vertical cracks in caves or cliffs near bodies of water, forming groups of a few tens of individuals (Álvarez, 1963a; Linares, 1987). In Yucatan, it has also been found in constructions and inside Mayan ruins, forming groups of up to 1,000 individuals (Bowles et al., 1990; Jones et al., 1973). In Cuba, it has been found roosting in the foliage of a palm (*Copernicia vepertilionum*), along with other molossids, and in human constructions (Silva-Taboada, 1979). There is scarce data on the diet of this species, but it is likely to consume soft insects, mainly lepidopterans (Freeman, 1979, 1981a, b). In Cuba, traces of coleopterans and lepidopterans have been found in the digestive tract (Silva-Taboada, 1979). In Tamaulipas, pregnant females (each with a single embryo) and lactating females have been found in June (Álvarez, 1963a). In Cuba and Yucatan, pregnant females have been reported in April, May, and June, and nursing females in July and August (Bowles et al., 1990; Jones et al., 1973; Silva-Taboada, 1979). These data indicate a monoestric pattern synchronized with the rainy season. Their predators include owls. Remain of this bat has been found in regurgitation pellets of owls from the cave El Abra, Tamaulipas, and the cave Lol-Tun, Yucatan (Álvarez, 1963a; Arroyo-Cabrales and Álvarez, 1990). They are relatively abundant in fossil and subfossil material of caves and cenotes (Álvarez, 1976; Arroyo-C. and Álvarez, 1990; Dalquest and Roth, 1970).

VEGETATIONAL ASSOCIATIONS AND ELEVATION RANGE: In Mexico, *N. laticaudatus* has been found in tropical areas, mainly in tropical deciduous forests and tropical subdeciduous forests. It has also been collected in thorny forests and in tropical perennial forests. It is known only in altitudes lower than 1,000 masl (Eisenberg, 1989).

CONSERVATION STATUS: The current status of the populations of *N. laticaudatus* is unknown. The subspecies *N. laticaudatus yucatanicus* is relatively abundant and is frequently found in sites inhabited by humans (Bowles et al., 1990; Jones et al., 1973; Silva-Taboada, 1979). Judging by the number of collected individuals, the subspecies *N. laticaudatus ferrugineus* is more rare. The species as a whole does not require immediate actions of conservation.

Nyctinomops macrotis (Gray, 1840)

Big free-tailed bat

Hector T. Arita

SUBSPECIES IN MEXICO
N. macrotis is a monotypic species.
It is also known as *Tadarida macrotis*. Freeman (1981a) proposed to include it in the genus *Nyctinomops*, which is generally accepted.

DESCRIPTION: *Nyctinomops macrotis* is a relatively large bat, the largest within the genus. The males are slightly larger than females (Silva-Taboada, 1979). The ears are large, are united in their bases, and extend beyond the snout when folded forward. The dorsal pelage is brown, varying from reddish to blackish, with dark tips. The tail is thick and long, and protrudes from the edge of the uropatagium. It is easily distinguishable from other species of the genus by its large size and from other molossids by the wrinkles on the upper lip.

DISTRIBUTION: In North America, *N. macrotis* is distributed from Colorado and Utah to Chiapas, Mexico City, and the center of Veracruz (Freeman, 1981a; Hall, 1981; Martínez-Coronel and López-Vidal, 1997). There are no records in Central America. In South America, it is found from Venezuela to northern Argentina (Eisenberg, 1989; Redford and Eisenberg, 1992). It has also been found in Cuba, Jamaica, and La Espanola (Ton, 1989). It has been registered in the states of BS, CH, CL, CS, DF, DU, GR, JA, MI, MX, NL, PU, QR, SI, SO, TA, VE, and ZA.

EXTERNAL MEASURES AND WEIGHT
TL = 120 to 139 mm; TV = 46 to 62 mm; HF = 10 to 11 mm; EAR = 25 to 32 mm; FA = 54 to 63 mm.
Weight: 17 to 24 g.

DENTAL FORMULA: I 1/2, C 1/1, PM 2/2, M 3/3 = 30.

NATURAL HISTORY AND ECOLOGY: The wing morphology of *N. macrotis* shows a rapid and straight flight. A flight of more than 40 km/h has been estimated (Findley et al., 1972). The sounds of its echolocalization system are of very low frequency (less than 20 kHz) and therefore audible by the human being (Fenton and Bell, 1981). It mainly roosts in the cracks of cliffs, although it also uses buildings, caves, and holes in the trees (Easterla, 1973; Silva-Taboada, 1979). In Chihuahua, a colony of about 20 to 30 individuals, roosting in a crack of the vertical walls of a canyon of approximately 450 m high, was found (Easterla, 1972). The colony was found at about 15 m high from the floor of the canyon. In Mexico City, this species has been observed roosting

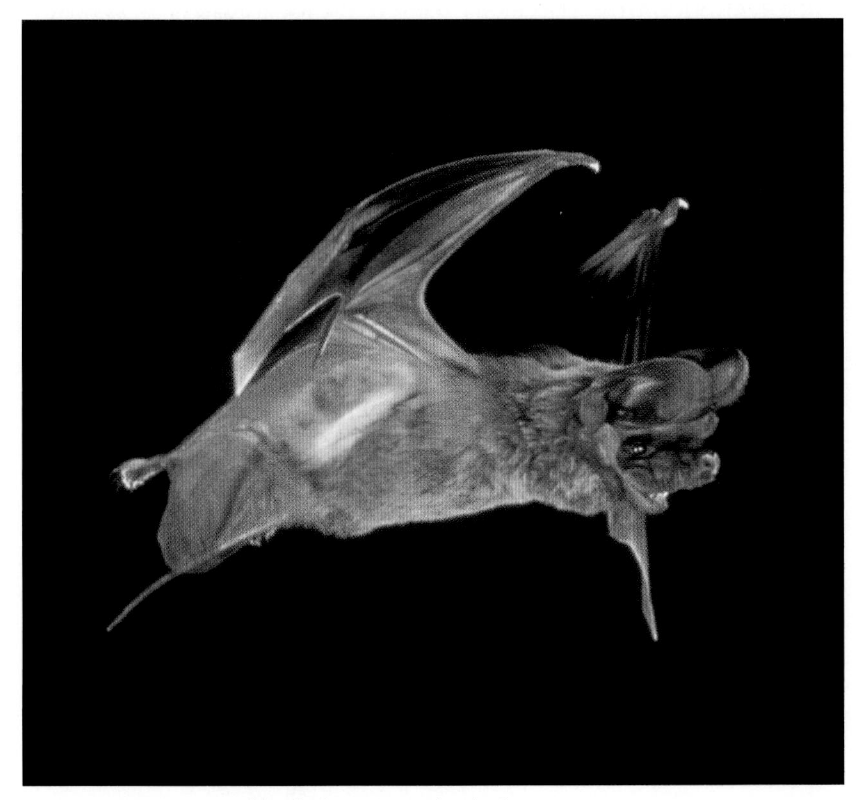

Nyctinomops macrotis. Scrubland. Photo: Scott Altenbach.

in tall buildings, in conditions that resemble their natural refuges in cliffs (Sánchez-Hernández et al., 1989). In Cuba, it is often found solitary or forming groups of two to four individuals inside holes of the caves (Silva-Taboada, 1979). They show migratory movements, at least in the northern part of their distribution (Easterla, 1973; Milner et al., 1990). Most of the specimens collected during the summer, in the north of Mexico and in the southwest of the United States, have been nursing or pregnant females (Easterla, 1973). This suggests that females migrate to the north during the summer, forming maternity colonies, as Brazilian free-tailed bats (*Tadarida brasiliensis*) do. Their diet consists mainly of lepidopterans, but also includes other insects such as grasshoppers (Gryllidae), locusts (Tettigoniidae), and flying ants (Formicidae; Easterla and Whitaker, 1972; Freeman, 1979, 1981 a, b). Females give birth to a single young per year. In the north of its distribution, the breeding season takes place at the end of spring and early summer. Pregnant females have been found from April to July, and nursing females from June to September (Easterla, 1973). Data from Silva-Taboada (1979), for this species in Cuba, coincide with this pattern of reproduction. The reproductive pattern for the populations of the center of Mexico and South America is unknown.

VEGETATIONAL ASSOCIATIONS AND ELEVATION RANGE: In North America, *N. macrotis* is mainly distributed in arid areas covered with xeric scrubland or deciduous tropical vegetation, and is occasionally found in upland pine forests (Milner et al., 1990). In South America and in the West Indies, it has also been found in more humid sites with tropical perennial vegetation. This species has been collected in sites from sea level to 2,600 masl (Milner et al., 1990). In Mexico, it has been found in Mexico City, at an altitude of 2,300 masl (Sánchez-Hernández et al., 1989), although most localities are below 1,800 masl.

CONSERVATION STATUS: The current status of the populations of *N. macrotis* is unknown. Their relative scarcity and their migratory movements across international borders could hamper the design of a conservation strategy for this species. Moreover, the fact that this bat can take advantage of construction sites in urban areas such as Mexico City (Sánchez-Hernández et al., 1989) would suggest that it is not very susceptible to severe environmental changes.

Promops centralis Thomas, 1915

Big crested mastiff bat

Guadalupe Tellez-Giron

SUBSPECIES IN MEXICO

Promops centralis centralis Thomas, 1915

DESCRIPTION: *Promops centralis* is a large bat. The face is short; the ears are rounded and short and do not reach half the face; the tragus is semi-circular. The sagittal crest is strong and pronounced, and the palate is deeply vaulted. The hair is short, and no more than 5 mm long on the back. The dorsal coloration varies from dark brown to reddish-brown, with white underfur; the abdomen is lighter (Eisenberg, 1989; Goodwin, 1969).

EXTERNAL MEASURES AND WEIGHT
TL = 127 to 161 mm; TV = 47 to 60 mm; HF = 9 to 13 mm; FA = 49 to 55 mm.
Weight: 19 to 27 g.

DENTAL FORMULA: I 1/2, C 1/1, PM 2/2, M 3/3 = 30.

NATURAL HISTORY AND ECOLOGY: *P. centralis* mainly roosts under the bark of trees, in hollow trees, or under the leaves of palms, forming small groups of up to six individuals. They have been captured in the gaps of the forests (Eisenberg, 1989; Emmons and Feer, 1997; Freeman, 1981a, b; Watkins et al., 1972). It is a rare species (Ceballos and Miranda, 2000). It has been captured over rivers with shallow water in the dry season (Bowles et al., 1990). There are no known data on their diet. Freeman (1981a) mentions that its skull is more fragile, and that it catches softer insects than those caught by Molossus. Mating season takes place from March to July (Birney et al., 1974; Bowles et al., 1990; Sanchez-Hernandez et al., 1985).

VEGETATIONAL ASSOCIATIONS AND ELEVATION RANGE: *P. centralis* has been found in tropical perennial forests, deciduous rainforests, xeric scrublands, and pine-oak forests (Eisenberg, 1989; Emmons and Feer, 1997; Goodwin, 1969; Watkins et al., 1972). It is distributed from sea level to 1,050 masl (Sanchez-Hernandez et al., 1985).

CONSERVATION STATUS: *P. centralis* is a rare species in Mexico. The status of its populations is not known, but due to its affinity for undisturbed environments, it is likely to face conservation problems.

DISTRIBUTION: *P. centralis* is distributed from Mexico to northern Argentina. In Mexico it is found in the lowlands from Jalisco to Chiapas, and from Puebla to the Yucatan Peninsula (Birney, et al., 1972; Eisenberg, 1989; Hall, 1981; Urbano et al., 1987; Vidal Lopez and Martinez Coronel, 1999; Watkins et al., 1972). It has been registered in the states of CL, CS, CL, GR, JA, MI, OX, PU, and YU.

Subfamily Tadarinae

The subfamily Tadarinae has one genus of ten species with one species known from Mexico.

Tadarida brasiliensis (È. Geoffroy St.-Hilaire, 1824)

Mexican free-tailed bat

Hector T. Arita and Jorge Ortega R.

SUBSPECIES IN MEXICO
Tadarida brasiliensis intermedia Shamel, 1931
Tadarida brasiliensis mexicana (Saussure, 1860)

DESCRIPTION: *Tadarida brasiliensis* is a small-sized molossid. On average, females are larger than males (Wilkins, 1989). The tail has a free end that protrudes beyond the edge of the uropatagium. The ears are rounded and long, but do not extend beyond the snout. The dorsal pelage is relatively short, brown, and without notorious bands, although the base of the hairs tends to be lighter than the tip. The wings are elongat-

Tadarida brasiliensis. Scrubland. Photo: Scott Altenbach.

ed and narrow, with a high wing load (0.16 g/cm^2), typical of species with rapid flight in open spaces (Vaughan, 1966). Both sexes have a gular gland, which is much more developed in mature males. It produces a very characteristic musky odor. The upper lip is marked by a series of profound furrows or wrinkles, a feature that distinguishes it from other Mexican molossids, with the exception of the genus *Nyctinomops* (the larger fifth finger distinguishes it from this species).

EXTERNAL MEASURES AND WEIGHT

TL = 46 to 65 mm; TV = 29 to 42 mm; HF = 9 to 11 mm; EAR = 14 to 19 mm; FA = 36 to 46 mm.
Weight: 11 to 15 g.

DENTAL FORMULA: I 1/3, C 1/1, PM 2/2, M 3/3 = 32.

NATURAL HISTORY AND ECOLOGY: The subspecies *T. brasiliensis mexicana* is migratory and conducts seasonal movements of up to 1,840 km (Cockrum, 1969; Glass, 1982; Villa-R. and Cockrum, 1962). The general pattern of migration is that individuals spend autumn and winter in the south (Mexico) and move to the north (the Southwest of the United States) during the spring and summer. Cockrum (1969) identified four separate migrant groups. The first two only carry out local movements and are found in California, western Arizona, and Nevada. The third group migrates annually from eastern Arizona and New Mexico to the south along the Sierra Madre Occidental. The fourth group moves from Texas toward the center of Mexico, following the Sierra Madre Oriental. Apparently, the populations of Baja California Sur and Chiapas do not migrate (H. Arita, pers. obs.; Woloszyn and Woloszyn, 1982).

They mainly roost in caves, forming groups of a few hundred to several millions of individuals, with a density of up to 1,800 adults/m^2 or 5,000 offspring/m^2 (Davis et al., 1962; McCracken, 1986). They also use hollow trees, cellars, stadiums, and other high buildings (Jennings, 1958; Sánchez-Hernández et al., 1989; Wilkins, 1989). It is a segregationist species that shares its roosting sites with very few species (Arita, 1993a, b). In the north of its distribution, the subspecies *T. brasiliensis mexicana* forms maternity colonies of up to 20 million individuals, made up of females, offspring, young, and very few males. They are insectivores that feed mainly on moths (Lepidoptera; Wilkins, 1989). They leave their roosting sites at dusk, forming on occasion large "columns" that can take several hours to leave the shelter. Some individuals may move up to 50 km from their refuge to the foraging area (Davis et al., 1962). It has been observed flying at more than 40 km/hr at an altitude of more than 3,000 m above the ground (Williams et al., 1973). In Mexico, the mating season probably occurs during the spring (Davis et al., 1962). Females migrate to the north at the beginning of summer and form maternity colonies in the Southwest of the United States. The offspring are born in late June or early July, and every female gives birth to only a single young. Females are able to find and recognize their own offspring from among the millions that form the colony, apparently using auditory and odor stimuli (McCracken, 1984). The large colonies attract a large variety of predators. Among the most common are several birds, like red-tailed hawks (*Buteo jamaicensis*), owls (*Bubo virginianus*), and barn owls (*Tyto alba*); medium-sized mammals such as skunks, raccoons, and opossums, and several types of snakes (*Elaphe*, *Masticophis*, *Micrurus*; Davis et al., 1962; Wilkins, 1989). Close to the entrance of the caves, it is common to observe birds of prey flying in circles, attacking the emerging bats. The guano of *T. brasiliensis*, which accumulates in large quantities in the caves occupied by these bats, is important to the local economy of some areas (it is used as fertilizer; Villa-R., 1967). It has been proposed that the species contributes to agriculture because it feeds on tons of insects that otherwise would be pests (Villa-R., 1967). The presence of heavy metals and organochlorine compounds has been documented in the tissues and guano of this species (Clark et al., 1975; Geluso et al., 1976). In particular, organochlorine compounds from insecticides are concentrated in fatty tissue that these bats accumulate in preparation for their long migrations. When the bats consume these reserves during migration, the concentration of pesticides can reach lethal levels (Geluso et al., 1976; McCracken, 1986).

DISTRIBUTION: The distribution of *T. brasiliensis* ranges from 40 degrees north latitude, in the center of the United States, to 40 degrees south latitude in Chile and Argentina (Hall, 1981; Redford and Eisenberg, 1992). Apparently, it is not found in the Amazon Basin. In Mexico, it is distributed throughout the country, except in the lowlands of the southeast and in the Yucatan Peninsula. It has been registered in the states of AG, BC, BS, CH, CL, CO, CS, DF, DU, GJ, GR, HG, JA, MI, MO, MX, NL, OX, PU, QE, SL, SI, SO, TA, TL, VE, and ZA.

VEGETATIONAL ASSOCIATIONS AND ELEVATION RANGE: In Mexico, *T. brasiliensis* is mainly found in xeric scrublands of the Altiplane. It has also been observed in tropical forests and dry forests of pine-oak, but always close to arid areas. It has been found from sea level to 3,200 masl, although most locations are between 1,000 masl and 2,000 masl.

CONSERVATION STATUS: *T. brasiliensis* is not included in any official list of endangered species, although drastic reductions of its populations have been documented. In the cave Eagle Creek of Arizona, the colony collapsed from 25 million bats in 1963 to 30,000 individuals in 1969, which represents a decrease of 99.9% of the original population (Cockrum, 1970). In Mexico, several caves that in the 1950s and 1960s contained colonies of millions of these bats have been developed as tourist sites or have been burned and dynamited. Of course, these sites do not shelter any bat (McCracken, 1986; P. Robertson, pers. comm.). Despite being a relatively abundant species, *T. brasiliensis* could be threatened with extinction because it is particularly vulnerable due to its migratory habits, because it forms enormous colonies in few sites, and because of their susceptibility to pesticides. The huge colonies of these animals are a unique natural phenomenon that must be protected.

Family Vespertilionidae

The family Vespertilionidae is the most diverse family of bats in the world; it comprises 35 genera and 318 species (Wilson and Reeder, 1993). It is distributed in tropical and temperate regions, and is the dominant group in temperate zones. In Mexico there are 44 species of 11 genera (*Corynorhinus, Eptesicus, Euderma, Idionycteris, Lasionycteris, Lasiurus, Myotis, Nycticeius, Parastrellus, Perimyotis,* and *Rhogeessa*). The genera *Antrozous* and *Bauerus*, previously included in this family, have been recently placed in another family (Simmons, 1998).

Subfamily Myotinae

Although the subfamily Myotine is comprised of only one genus, it includes more than a hundred species. The taxonomy of many of those species remains unresolved. Nineteen species are known from Mexico.

Myotis albescens (È. Geoffroy St.-Hilaire, 1806)

Silver-tipped myotis

Rodrigo A. Medellín

SUBSPECIES IN MEXICO
M. albescens is a monotypic species.

DISTRIBUTION: *M. albescens* inhabits the tropical rainforests of southeastern Mexico, from the Isthmus of Tehuantepec across Central and South America to southern Mexico and Argentina. It occupies the jungles of the Coastal Plain from southern Veracruz to Tabasco and Chiapas (Eisenberg, 1989; Hall, 1981; LaVal, 1973a; Ramírez-Pulido et al., 1986). It has been registered in the states of CS, OX, TB, and VE.

DESCRIPTION: *Myotis albescens* is a small-sized bat. The face is simple and lacks ornamentation. The eyes are black; the ears are small and relatively short. *M. albescens* is slightly larger than *M. keaysi* or *M. nigricans*. The hair color is black, but the tips of the hairs are white, giving this bat a frosty appearance. The abdomen is slightly lighter. The wing membranes and uropatagium have a very distinctive white appearance on the Mexican specimens. The edge of the uropatagium has a fringe of hair usually visible only with a magnifying glass.

EXTERNAL MEASURES AND WEIGHT
TL = 76 to 85 mm; TV = 25 to 36 mm, HF = 8 to 11 mm; EAR = 32 to 38 mm; FA = 33 to 36 mm.
Weight: 5 to 8 g.

DENTAL FORMULA: I 2/3, C 1/1, PM 3/3, M 3/3 = 38.

NATURAL HISTORY AND ECOLOGY: There is much information on this species. They are insectivorous and capture their prey in flight. In the stomach contents of an individual caught in Bolivia, remains of lepidopterans and arachnids were found (Aguirre, 1994). They roost in hollow tree trunks, in caves, and in buildings, forming small groups. Most of the Mexican specimens and those captured in other countries have been caught in nets placed on streams in the jungle. They seem to be associated with watercourses (Handley, 1976; R. Medellín, pers. comm.). Subadults have been caught in Chiapas in April, and a female caught in January contained an embryo (Medellín et al., 1986). Apparently, its reproductive pattern is bimodal (Myers, 1977).

VEGETATIONAL ASSOCIATIONS AND ELEVATION RANGE: In Mexico this species has only been caught in tropical rainforests (high evergreen forests). In other countries, it has been found in flooded savannah, tropical forests, and other tropical deciduous habitats (Eisenberg, 1989; Handley, 1976; Laval, 1973a). In Mexico it has been found from sea level up to 350 m. In South America, it has been caught at altitudes of up to 500 masl, and only three specimens have been captured at an altitude of 1,500 m (Handley, 1976; LaVal, 1973a).

CONSERVATION STATUS: The status of this species is undetermined; relatively few specimens are known in Mexico. All of them were collected in high rainforests. Because streams are among the first habitats deforested with the entry of humans to tropical regions, and given its apparent dependence of these habitats, *M. albescens* could face a threat not fully recognized. It is included in the Mexican Official Regulation (SEDESOL, 1994; SEMARNAT, 2002) as a rare species.

Myotis auriculus Baker and Stains, 1955

Southwestern myotis

Jorge Uribe and Hector T. Arita

SUBSPECIES IN MEXICO

Myotis auriculus apache Hoffmeister and Krutzsch, 1955
Myotis auriculus auriculus R.H. Baker and Stains, 1955

Myotis auriculus. Temperate forest. Photo: M.B. Fenton.

DESCRIPTION: Among the species in its genus, *Myotis auriculus* is a medium-sized bat. The length of its ears distinguishes it from other *Myotis* of North America (Warner, 1982). There is no sexual dimorphism in size (Williams and Findley, 1979). It is distinguished from other species of the same genus by the presence of a relatively more rounded skull and a sagittal crest and the lack of a distinctive stripe or fringe of microscopic hairs on the back of the margin of the uropatagium (Hall, 1981; Warner, 1982). The dorsal fur color varies from light brown to ochre and light brown-olive with dark underfur. The abdomen coloration is slightly different, but it is generally classified as buffy (Warner, 1982). The wing membrane is brown and has no stripes or fringes of microscopic hairs on the back of the interfemoral membrane. Coloration of the ears varies from dark brown to light pale brown (Hall, 1981; Warner, 1982).

EXTERNAL MEASURES AND WEIGHT

TL = 86 to 97 mm; TV = 39 to 45 cm, HF = 8 to 10 mm; EAR = 18 to 22 cm; FA = 35.7 to 40.4 mm.
Weight: 4 to 6 g.

DENTAL FORMULA: I 2/3, C 1/1, PM 3/3, M 3/3 = 38.

NATURAL HISTORY AND ECOLOGY: The southwestern myotis is a bat with flight patterns that are relatively quick and

DISTRIBUTION: *M. auriculus* is only distributed in North America, from the central region of Mexico to the southwestern region of the United States, where it reaches its more boreal limit. In Mexico, it has been reported from Sonora and Coahuila to Zacatecas, Jalisco, southern and northern Tamaulipas, and Veracruz (Hall, 1981). It has been recorded in the states of CH, CL, CO, DU, JA, NL, NY, SI, SO, TA, VE, and ZA.

straight; its average speed has been estimated at 12.9 km/hr at distances of 30 m (Hayward and Davis, 1964). It is considered insectivorous, and its main prey items are moths (Lepidoptera) between 30 mm and 40 mm in size (Fenton and Bell, 1979). Males consume significantly more butterflies than females. They can make small "landings" in the short substrate while capturing prey (Fenton and Bell, 1979). They use a variety of places to take refuge, with caves as their main shelters; however, they are often found in hollow trees, cracks in rocks, and man-made structures (Nowak, 1999a; Warner, 1982). Periods of flight are commonly altered with rest periods, when captured prey are consumed. These bats begin their activity between 1.5 and 2 hours after sunset at temperatures between 11°C and 19°C. They have a trimodal pattern of activity, with periods of peak activity between 30 and 89 minutes after sunset and minimum periods of 120 to 140 minutes and 180 to 209 minutes after sunset (Cockrum and Cross, 1964). Females make up maternity colonies so there is segregation of sexes (Hall, 1981; Nowak, 1999a). Breeding usually starts in June or early July. In captivity this species has lived up to four years (Warner, 1982).

VEGETATIONAL ASSOCIATIONS AND ELEVATION RANGE: *M. auriculus* inhabits arid xeric scrublands and forests of pine in the United States (Barbour and Davis, 1969; Nowak, 1999a; Warner and Czaplewski, 1981). The altitudinal range varies from 366 m to 2,250 m (Warner, 1982).

CONSERVATION STATUS: *M. auriculus* is not included in any official list of endangered species. Apparently, it does not face any risk of extinction.

Myotis californicus (Audubon and Bachman, 1842)

California myotis

Luis Ignacio Iñiguez Davalos

SUBSPECIES IN MEXICO

Myotis californicus californicus (Audubon and Bachman, 1842)
Myotis californicus mexicanus (Saussure, 1860)
Myotis californicus stephensi Dalquest, 1946

DISTRIBUTION: *M. californicus* is distributed from western Canada and the United States to Guatemala. In Mexico, it is found in northern Guerrero, Oaxaca, and Chiapas. It has been registered in the states of AG, BC, BS, CH, CL, CO, CS, DF, DU, GJ, GR, HG, JA, MI, MO, MX, NL, NY, OX, PU, QE, SI, SL, SO, TA, TL, VE, and ZA.

DESCRIPTION: *Myotis californicus* is a small-sized bat with a triangular head, wide ears, and a pointed nose. It has no nasal leaf. The eyes are small and the long ears exceed the tip of the snout when stretched forward. The hair is long, thick, thin, and brightly colored, varying from brown (forests) to light cream (deserts); there is a marked contrast between the light tips and dark underfur. The abdomen is usually paler than the back. The hair is short and ends at half the dorsal side of the uropatagium and a little less on the ventral side. The tail is longer than the legs, and has a keeled-shaped extension.

EXTERNAL MEASURES AND WEIGHT

TL = 70 to 87 mm; TV = 28 to 41 mm, HF = 6 to 8 mm; EAR = 11 to 16 mm; FA = 27 to 36 mm.
Weight: 2 to 5 g.

DENTAL FORMULA: I 2/3, C 1/1, PM 3/3, M 3/3 = 38.

Myotis californicus. Scrubland. Photo: Scott Altenbach.

NATURAL HISTORY AND ECOLOGY: *M. californicus* is an insectivore. It is found in places with abundant water. Its activity starts with the twilight. It hunts at low altitudes, in areas within or near vegetation. Its flight is markedly erratic. It roosts in dry leaves, mines, tree holes, loose rocks, buildings, bridges, and cracks; it sometimes uses caves. It is found solitary or in small colonies. In the United States, it is a nomadic species that hibernates in the north of its distribution; however, there are no records of such behavior in Mexico. Sexes are separate almost year round; mating occurs in autumn, births at the end of the dry season, and lactation with the beginning of rains.

VEGETATIONAL ASSOCIATIONS AND ELEVATION RANGE: The California myotis inhabits temperate forests of conifers and hardwoods, pine, pine-oak, pine-oak-fir, arid and semi-arid, thorny forests, and subtropical mountain forests. It is distributed from sea level up to 3,000 m.

CONSERVATION STATUS: Apparently, *M. californicus* is a common species throughout its geographical distribution. It is not considered endangered.

Myotis carteri LaVal, 1973

Carter's myotis

Don E. Wilson

SUBSPECIES IN MEXICO

M. carteri is a monotypic species.

M. carteri was originally described as a subspecies of *M. nigricans*. The species level was proposed by Bogan (1978) and accepted in subsequent publications (Hall, 1981;

Jones et al., 1988). Without presenting new arguments, Koopman (1993) proposed to reinstate *M. carteri* under *M. nigricans*. We follow the classification of Bogan (1978), since this is the last formal taxonomic review on this group.

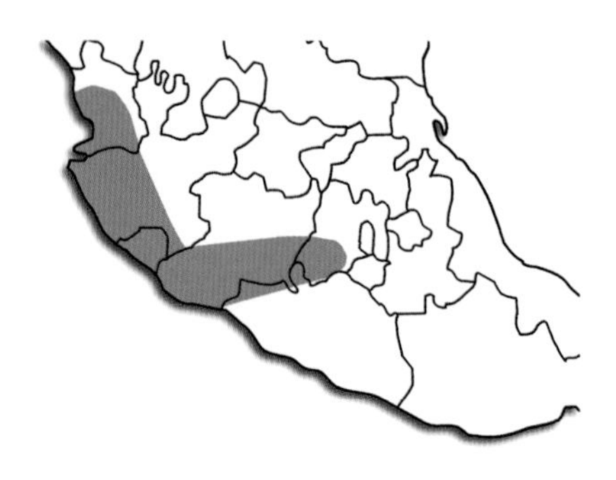

DISTRIBUTION: *M. carteri* is an endemic species of Mexico. It is distributed in the Pacific coast from Nayarit and Michoacan to the Balsas basin to the southern state of Mexico (Chavez and Ceballos, 2002; Laval, 1973a; Polaco and Muñiz-Martinez, 1987). It has been reported in the states of CL, JA, MI, MX, and NY.

DESCRIPTION: *Myotis carteri* is one of the smallest species of bats in Mexico. The dorsal coloration is dark brown to almost black, although some individuals are light brown. It differs from *M. findleyi* by its larger size and darker coloration.

EXTERNAL MEASURES AND WEIGHT
TL = 68 to 82 mm; TV = 28 to 39 mm, HF = 6 to 9 mm; EAR = 10 to 13 mm; FA = 32.9 to 36.2 mm.
Weight: 3 to 4 g.

DENTAL FORMULA: I 2/3, C 1/1, PM 3/3, M 3/3 = 38.

NATURAL HISTORY AND ECOLOGY: Most specimens have been collected in nets placed on water currents. Their roosting sites are unknown, but it is likely that, like *M. nigricans*, it uses holes in trees, old buildings, and caves. It is insectivorous, but details of its diet are unknown. Its pattern of reproduction has not been studied.

VEGETATIONAL ASSOCIATIONS AND ELEVATION RANGE: *M. carteri* is found in tropical deciduous forests, xeric scrublands, and mountain cloud forests. It is mainly distributed in the lowlands of the Pacific slope. There are, however, records up to 1,500 m.

CONSERVATION STATUS: *M. carteri* is not included in any official list of endangered species. Because it is an endemic species, restricted to a small area of the Pacific coast, however, it is important to research and establish accurately the real state of its populations.

Myotis elegans Hall, 1962

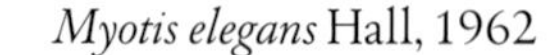

Elegant myotis

Hector T. Arita

DISTRIBUTION: *M. elegans* is an endemic species of Mesoamerica. It is distributed in the slope of the Gulf of Mexico from the east of San Luis Potosí and south of the Yucatan Peninsula to Costa Rica. In the Mexican Pacific it is known from two localities in Chiapas (Hall, 1981; LaVal, 1973a). It has been registered in the states of CA, CS, SL, TB, and VE.

SUBSPECIES IN MEXICO
M. elegans is a monotypic species.

DESCRIPTION: *Myotis elegans* is a very small bat. The overall color of the back is dark; the hair has brown or cinnamon tips and darker underfur. The wing membranes are brown or dark brown. It differs from *M. keaysi* by having two-thirds of the uropatagium hairless. It is very similar to *M. nigricans*, but can be distinguished by the dorsal pelage (more contrasting bands) and the less inflated shape of the skull (LaVal, 1973a).

EXTERNAL MEASURES AND WEIGHT
TL = 39 to 45 mm; TV = 31 to 35 mm, HF = 6 to 7 mm; EAR = 11 to 13 mm; FA = 32 to 34 mm.
Weight: 3 to 5 g.

DENTAL FORMULA: I 2/3, C 1/1, PM 3/3, M 3/3 = 38.

NATURAL HISTORY AND ECOLOGY: The natural history of this species is virtually unknown. Apparently, all specimens have been caught in nets placed in the forest or on water currents; thus, we do not know what kind of roosting sites they use. The only data on reproduction come from a female with an embryo captured in February in Campeche (Jones et al., 1973). Like other close species, *M. elegans* probably feeds on insects that it captures in flight.

VEGETATIONAL ASSOCIATIONS AND ELEVATION RANGE: *M. elegans* inhabits primarily tropical rainforests and tropical subdeciduous forests. In Central America, it has also been found in deciduous forests and scrublands (LaVal, 1973a). All localities in Mexico are below 200 m. In Honduras, it has been captured at an altitude of 750 m (LaVal, 1973a).

CONSERVATION STATUS: *M. elegans* is not included in any official list of endangered species. Because it is an endemic species of Mesoamerica and is rare it should be considered a candidate for any of these lists. Only a detailed study of its populations would determine its actual situation.

Myotis evotis (H. Allen, 1864)

Long-eared myotis

Hector T. Arita

SUBSPECIES IN MEXICO

Myotis evotis evotis (H. Allen, 1864)

The only specimen of this species known in Mexico is *M. micronyx* (Nelson and Goldman, 1909). Miller and Allen (1928) considered *micronyx* a synonym of *evotis*. However, the Mexican specimens showed notable cranial differences from other specimens of *M. evotis* (Genoways and Jones, 1969), and the habitat in which it was collected is atypical for that species. M. milleri was the sister species but is now considered a synonym (Manning and Jones, 1989; Reduker et al., 1983).

DISTRIBUTION: This species is widely distributed in the western United States and Canada, from British Columbia, Alberta, and Saskatchewan to California, New Mexico, and Arizona. In Mexico it is only known in Comondu, Baja California Sur. It has been seen in the state of BS.

DESCRIPTION: *Myotis evotis* is a small-sized bat, but among the species of its genus it is of medium size. Of the species of the genus *Myotis*, this is the species with the longest ears in proportion to the body. The general coloration is brown, the uropatagium is wide, and the tail extends to the edge of the interfemoral membrane. *M. evotis* can be distinguished from other species of the genus by their long ears, which are dark, almost black.

EXTERNAL MEASURES AND WEIGHT

TL = 42 to 56 mm; TV = 36 mm, HF = 8 to 10 mm; EAR = 22 to 25 mm; FA = 35 to 41 mm.

Weight: 8 g.

DENTAL FORMULA: I 2/3, C 1/1, PM 3/3, M 3/3 = 38.

NATURAL HISTORY AND ECOLOGY: The information on the natural history of *M. evotis* comes from studies in the United States. This species usually forms groups of a few dozen individuals in old buildings, wood constructions, holes in trees, caves, and fissures in the rocks (Barbour and Davis, 1969; Manning and Jones, 1989). Reproduction takes place in the summer, and each female produces only one offspring. It is an insectivore and captures its prey when it is resting on the vegetation. It feeds primarily on moths (Lepidoptera), beetles (Coleoptera), flies (Diptera), and other insects (Warner, 1985). At the sites where it is sympatric with *M. auriculus*, *M. evotis* specializes in feeding on beetles, while *M. auriculus* feeds on moths (Black, 1974; Husar, 1976).

VEGETATIONAL ASSOCIATIONS AND ELEVATION RANGE: In the United States *M. evotis* is mainly found in coniferous forests or in temperate rainforests on the West Coast, although it has also been found in dry scrublands (Barbour and Davis, 1969; Manning and Jones, 1989). The specimen of Mexico was collected in xeric scrublands, in conditions atypical for this species. In the United States, it is found from sea level to nearly 3,000 m (Barbour and Davis, 1969; Manning and Jones, 1989). The only locality in Mexico is 400 masl.

CONSERVATION STATUS: *M. evotis* is not included in any official list of endangered species. In the United States and Canada, it has a wide distribution and is relatively abundant. The specimen of Comondu was captured in 1905 and since then it has not been found in Mexico again. To establish its conservation status in Mexico it is necessary to make a revision of the taxonomic group of species of the genus *Myotis* with long ears and a study on its possible presence in the country.

Myotis findleyi Bogan, 1978

Findley's myotis

Don E. Wilson

SUBSPECIES IN MEXICO
M. findleyi is a monotypic species.

DESCRIPTION: *Myotis findleyi* is a small-sized bat. Along with *M. planiceps*, it is the smallest bat of the genus *Myotis* in Mexico. Like other members of the genus, the dorsal pelage is dark brown, almost black, although some individuals are brown. It differs from other species of the genus by its small size and because its distribution is restricted to the Marias Islands.

DISTRIBUTION: *M. findleyi* is an endemic species of Mexico, restricted to the Islas Marias near the coast of Nayarit (Bogan, 1978; Wilson, 1991). It has been registered in the state of NY.

EXTERNAL MEASURES AND WEIGHT
TL = 71 mm; TV = 26 mm, HF = 6.6 mm; EAR = 12 mm; FA = 29 to 33 mm.
Weight: 2 to 3 g.

DENTAL FORMULA: I 2/3, C 1/1, PM 3/3, M 3/3 = 38.

NATURAL HISTORY AND ECOLOGY: The roosting sites these bats use are unknown. All specimens have been caught in nets placed on small streams. They feed on insects, but the details of their diet are unknown. The pattern of reproduction has not been studied, but apparently their breeding season begins in April (Wilson, 1991).

VEGETATIONAL ASSOCIATIONS AND ELEVATION RANGE: *M. findleyi* is only known in tropical deciduous forests. It is found from sea level up to 600 m.

CONSERVATION STATUS: *M. findleyi* is not included in any official list of endangered species. Because it is an endemic of an island, however, it would be important to undertake studies on its current situation.

Myotis fortidens Miller and Allen 1928

Cinnamon myotis

Cuauhtemoc Chávez Tovar and Gerardo Ceballos

SUBSPECIES IN MEXICO

Myotis fortidens fortidens Miller and G.M. Allen, 1928
Myotis fortidens sonoriensis Findley and Jones, 1967

DESCRIPTION: *Myotis fortidens* is a small-sized bat. In general, the dorsal pelage is yellow with very dark underfur; the ventral region is lighter (Ceballos and Miranda, 1986), although they can have a reddish-golden coloration and yellow abdomen (Polaco and Muñiz-Martínez, 1987). The facial features are simple, and the long tail is included in the uropatagium. The uropatagium is shaped like a V, is highly developed, and lacks both dorsal and ventral hairs. The skull is delicate, with separated incisors and a well-developed zygomatic arch (Villa-R., 1967).

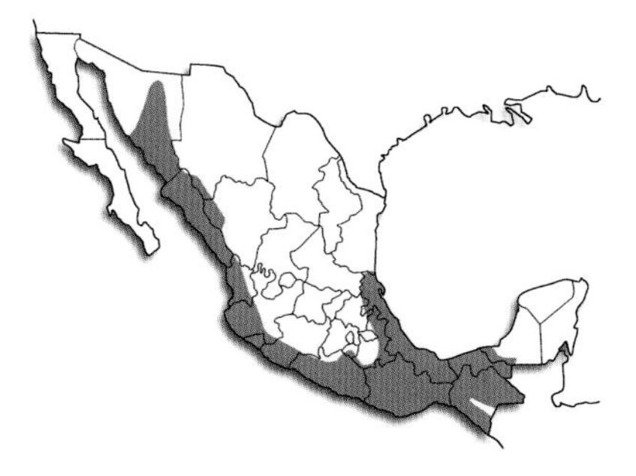

DISTRIBUTION: On the side of the Pacific, *M. fortidens* is distributed from northern Sonora, and on the Gulf side, from southern Tamaulipas, to Guatemala (Dickerman et al., 1981; Hall, 1981). It has been recorded in the states of CL, CS, GR, JA, NY, MI, OX, SI, SO, TB, and VE.

EXTERNAL MEASURES AND WEIGHT

TL = 80 to 96 mm; TV = 31 to 40 mm, HF = 7 to 11 mm; EAR = 10 to 14 mm; FA = 33 to 38 mm.
Weight: 5 to 8 g.

DENTAL FORMULA: I 2/3, C 1/1, PM 3/3, M 3/2 = 36.

NATURAL HISTORY AND ECOLOGY: *M. fortidens* mainly roosts in crevices, tunnels, caves, and abandoned buildings, as well as under the bark of trees; for example, in Chajul, Chiapas, it was collected under the leaves of Platanillo (*Heliconia* sp.) in the rainforest (Medellín et al., 1986). They are active during the early hours of the night and at dawn. Their food consists of small insects, preferably dipterans, formosan termites, or Isoptera, which are caught in flight (Nowak, 1999a). It breeds only once a year, coinciding with the rainy season; females have a single young per birth (Ceballos and Miranda, 1986). Pregnant females have been collected in April and May (Dickerman et al., 1981, Garrido, 1980); while in August and September males with testicles in the scrotum have been found (Álvarez and Álvarez-Castañeda, 1991; Garrido, 1980).

VEGETATIONAL ASSOCIATIONS AND ELEVATION RANGE: *M. fortidens* mainly inhabits tropical deciduous forests and areas of secondary vegetation adjacent to forests. It has also been collected in tropical rainforests and in crops. It is distributed from sea level to 870 m.

CONSERVATION STATUS: *M. fortidens* is a scarce species, almost endemic to Mexico. We do not know its conservation status, but because of its wide distribution, it probably is out of danger.

Hairy-legged myotis

Jorge Ortega R. and Hector T. Arita

SUBSPECIES IN MEXICO

Myotis keaysi pilosatibialis LaVal, 1973

DISTRIBUTION: The distribution of the hairy-legged myotis extends from southern Tamaulipas on the Gulf Coast to the east of Venezuela and Trinidad (LaVal, 1973a). It has been registered in the states of CA, CS, OX, PU, QR, SL, TA, TB, VE, and YU.

DESCRIPTION: *Myotis keaysi* is a small-sized bat with long hair (5 mm to 6 mm) that extends to the tibia. The dorsal coloration is black or dark brown. Each hair is bicolor; the base is dark brown and the tips are yellow. The abdomen has lighter tones (Eisenberg, 1989). Externally, it differs from other species of the same genus by the pelage and the dorsal surface of the uropatagium, which is covered, to the middle of the tibia, by hair visible with the naked eye.

EXTERNAL MEASURES AND WEIGHT

TL = 36 to 47 mm; TV = 33 to 41 mm, HF = 6 to 10 mm; EAR = 9 to 13 mm; FA = 31 to 41 mm.
Weight: 5 g.

DENTAL FORMULA: I 2/3, C 1/1, PM 3/3, M 3/3 = 38.

NATURAL HISTORY AND ECOLOGY: The hairy-legged myotis forms groups of two to eight individuals that roost in cracks or small cavities of the caves (Arita, 1992; Medellín and López-Forment, 1986). In Yucatan, numerous colonies of more than 500 individuals have been found on the roof of flooded caves (Arita, 1992; Villa-R., 1967). In one group the proportion of females to males was 23:1, which probably indicates that as in other species of bats there is segregation of sexes. Although this species is found throughout the year, apparently, the population number decreases during the dry season. Reproduction occurs between April and May; each female usually has a single young (Villa-R., 1967).

VEGETATIONAL ASSOCIATIONS AND ELEVATION RANGE: *M. keaysi* has been recorded in a variety of habitats, which often are very different between close populations (LaVal, 1973a). In Mexico, it has been collected in forest of conifers and oak, tropical deciduous forest, tropical subdeciduous forest, tropical rainforest ,and cloud forest. It is found from sea level up to 2,500 m.

CONSERVATION STATUS: The current status of *M. keaysi* is unknown, but in the tropical forests it is a common species and it is widely distributed; therefore, it has not been listed as threatened with extinction.

Western small-footed myotis

Hector Godinez Álvarez

SUBSPECIES IN MEXICO

Myotis ciliolabrum melanorhinus (Merriam, 1886)

Previously, it was considered a subspecies of *M. leibii*. Van Zyll de Jong (1984), after a morphometric analysis of the skull, proposed to separate *M. ciliolabrum* as a species of *M. leibii* with two subspecies: *M. ciliolabrum ciliolabrum* and *M. ciliolabrum melanorhinus*.

DISTRIBUTION: *M. melanorhinus* is distributed from Canada to the north and central portion of Mexico, in the states of Zacatecas and Michoacan (Hall, 1981; Matson and Baker, 1986; Nowak, 1999a). It has been registered in the states of BC, CH, CL, CO, DU, MI, NL, NY, SO, and ZA.

DESCRIPTION: *Myotis melanorhinus* is a medium-sized bat similar to *M. leibii*. Its upper incisors are smaller, and the premolars and molars are larger (Van Zyll de Jong, 1984). The hair is long and usually glossy. The dorsal pelage varies from blond to brown. The face and ears are black. Extended forward, the large ears reach the tip of the nose or slightly exceed it. The skull is flat with relatively large teeth. The legs are small and the calcar has a pronounced keel-shaped extension (Hall, 1981).

EXTERNAL MEASURES AND WEIGHT

TL = 73 to 81 mm; TV = 29 to 41 mm, HF = 9 to 10 mm; EAR = 10 to 13 mm; FA = 32 to 35 mm.

Weight: 2 to 4 g.

DENTAL FORMULA: I 2/3, C 1/1, PM 3/3, M 3/3 = 38.

NATURAL HISTORY AND ECOLOGY: Information on Mexican populations is scarce. There is some information on populations of United States and Canada. It is characterized by hibernating in caves for a five-month period, which begins in late November and ends in early April (Banfield, 1974). The bats form small groups of both sexes, located near the entrance of the caves. These sites are exposed to cold winds, so this species is probably able to withstand low temperatures (Kunz, 1982a, b). In the sum-

Myotis melanorhinus. Temperate forest. Photo: Scott Altenbach.

mer, females form small maternity colonies located in horizontal or vertical cracks. The number of offspring per litter varies from 1 to 2; weight of young males is 1.9 g to 3.8 g and of females is 1.6 g to 3.9 g (Tuttle and Heaney, 1974). It is estimated to live up to 12 years (Kunz, 1982a, b). It feeds primarily on flying insects captured with the uropatagium or snout. Foraging takes place mainly in clear or open sites; it can also be found in small bodies of water or wells, fluttering near vegetation (Findley, 1987).

VEGETATIONAL ASSOCIATIONS AND ELEVATION RANGE: *M. melanorhinus* is often found in forests of pine-oak (Findley, 1987). It is found from 540 m to 3,100 m (Hall, 1981; Matson and Baker, 1986; Villa-R., 1967; Wilson et al., 1985).

CONSERVATION STATUS: The conservation status of *M. melanorhinus* is unknown. In Mexico the records are scarce and represent the southern limit of its distribution; therefore, it is necessary to know more about its natural history and current status of its populations.

Myotis nigricans (Schinz, 1821)

Black myotis

Don E. Wilson

DISTRIBUTION: *M. nigricans* is distributed from the south of Tamaulipas on the Gulf of Mexico side to the north of Argentina and Paraguay (LaVal, 1973a; Redford and Eisenberg, 1992). The distribution in Mexico is confusing because until the revision of LaVal (1973a) the name *M. nigricans* was used to designate specimens of other species (*M. elegans* and *M. keaysi*). The discrepancies between the list of states presented here and the one of Ramírez-Pulido et al. represent the changes in the taxonomy. The records of Guanajuato and Queretaro are doubtful because they are not backed by specimens (Ramírez-Pulido et al., 1986). The subspecies *nigricans* is distributed in Mexico from the south of Tamaulipas to Chiapas, whereas the subspecies *extremus* is known in the region of Huehuetan, Chiapas (Bogan, 1978; LaVal, 1973a). It has been registered in the states of CS, HG, OX, PU, SL, TA, and VE.

SUBSPECIES IN MEXICO

Myotis nigricans extremus Miller and G.M. Allen, 1928
Myotis nigricans nigricans (Schinz, 1821)
Bogan (1978) recognized the validity of *M. nigricans extremus* and proposed a species level for *M. carteri*, originally described as subspecies of *M. nigricans*. Recently, Koopman (1993) included *M. carteri* as subspecies of *M. nigricans*. This arrangement, however, is not based on new information and is simply an opinion not followed here.

DESCRIPTION: *Myotis nigricans* is a small-sized bat. Along with *M. findleyi* and *M. planiceps* it is one of the smallest species in the genus in Mexico. The dorsal pelage is dark brown to almost black, although some individuals are brown. The individual hairs are dark all along the pelage, with lighter tips in some individuals. The ears are relatively short and have rounded edges. The tragus is elongated but with a blunt tip. The wing membranes and uropatagium are dark. The tail is long and reaches the edge of the uropatagium.

EXTERNAL MEASURES AND WEIGHT

TL = 39 to 52 mm; TV = 28 to 39 mm; HF = 6 to 9 mm; EAR = 10 to 13 mm; FA = 31 to 36 mm.
Weight: 3 to 4 g.

DENTAL FORMULA: I 2/3, C 1/1, PM 3/3, M 3/3 = 38.

NATURAL HISTORY AND ECOLOGY: This bat roosts in the attics of buildings, hollow logs, and caves (Hall and Dalquest, 1963; Wilson, 1983b). Some colonies of this species are formed by more females than males (Hall and Dalquest, 1963; Wilson, 1983b). The pattern of reproduction has been studied in detail in Panama (Wilson and Findley, 1970), but we know very little about the populations in Mexico.

In Panama, three annual peaks of births have been observed (February, April–May, August), correlated with fluctuations in the availability of food. It appears that the populations in Mexico have a pattern of reproduction more similar to that of the species of temperate zones, with a single annual reproductive peak (Wilson and Findley, 1971). This species is insectivorous, although the details of its diet are unknown. The colonies of this bat attract predators that include opossums (*Didelphis*), cats (*Felis domesticus*), and snakes. Individuals from this species can live up to seven years in the wild (Wilson, 1983b).

VEGETATIONAL ASSOCIATIONS AND ELEVATION RANGE: *M. nigricans* is found in tropical deciduous forests, tropical subdeciduous forests, tropical rainforests, and cloud forests. In Mexico, it is found from sea level to 2,500 masl, although most locations are below 1,500 m. In other places it has been found up to 3,150 m (Wilson, 1983b).

CONSERVATION STATUS: *M. nigricans* is not included in any official list of endangered species. The subspecies *M. nigricans nigricans* has a broad distribution and is relatively abundant in some locations. The subspecies *M. nigricans extremus* has an extremely restricted distribution and its status is unknown. A new evaluation on the taxonomic state and conservation status of this subspecies is necessary.

Myotis occultus Hollister, 1909

Arizona bat

Jorge Ortega R. and Hector T. Arita

SUBSPECIES IN MEXICO

M. occultus is a monotypic species.

It was considered a subspecies of *M. lucifugus* (Wilson and Reeder, 2005).

DESCRIPTION: This vespertilionid is small. The pelage is long and silky; the coloration varies from black and brown to golden yellow (Fenton and Barclay, 1980). *Myotis occultus* is distinguished from *M. lucifugus* by the presence of a highly developed sagittal crest and large skull and teeth (Findley and Jones, 1967). The interfemoral membrane is wide and includes the tail, except for its distal portion. The tragus is medium in size with indented edges.

DISTRIBUTION: *M. occultus* is distributed from the arid portions of the southwestern United States to the half portion of central Mexico. It has been registered in the states of CH, DF, MX, and ZA.

EXTERNAL MEASURES AND WEIGHT

TL = 41 to 54 mm; TV = 29 to 41 mm, HF = 7 to 11 mm; EAR = 10 to 13 mm; FA = 33 to 41 mm.

Weight: 7 g.

DENTAL FORMULA: I 2/3, C 1/1, PM 3/3, M 3/3 = 38.

NATURAL HISTORY AND ECOLOGY: Virtually everything that is known about the natural history of this species comes from studies conducted in Canada and the United States (Fenton and Barclay, 1980). It begins its activities of foraging during the twilight, with a second period of activity during the early hours of the night. It is insectivorous and captures its prey on bodies of water (Humphrey, 1982).

Myotis occultus. Photo: M.B. Fenton.

Reproduction takes place during the summer. The reproductive rate for adult females is very high; 90% of them become pregnant. The gestation period lasts between 50 and 60 days; each female has a single young and occasionally two. It roosts forming two different groups, one formed females and their offspring and one with males and nulliparous females. Their presence in the roosting sites depends mainly on the temperature and relative humidity. Individuals roosting in old buildings have been reported (Humphrey and Cope, 1976; Sánchez et al., 1989). In the United States, this species migrates to more temperate zones in the winter (Humphrey, 1982).

VEGETATIONAL ASSOCIATIONS AND ELEVATION RANGE: In the United States *M. occultus* has been recorded in arid and temperate habitats. In Mexico it has been found in arid grasslands, xeric scrublands, and oak and pine forests in a range of altitudes from 2,250 m to 2,700 m.

CONSERVATION STATUS: In the United States, the indiscriminate use of insecticides has been a major cause in the decline of this species' populations (Fenton and Barclay, 1980). For Mexico there are no detailed reports on the size of the populations and their current situation, although its limited distribution indicates that the species is rare and very restricted.

Myotis peninsularis Miller, 1898

Peninsular myotis

Gerardo Ceballos and Eric Mellink

SUBSPECIES IN MEXICO

M. peninsularis is a monotypic species.

This species has been considered a subspecies of *M. velifer* (Hall and Kelson, 1959;

Villa-R., 1967). There is a consensus to classify it as species (Hall, 1981; Hayward, 1970).

DESCRIPTION: *Myotis peninsularis* is very similar in appearance to *M. velifer*, but smaller (Hall, 1981; Villa-R., 1967). It is relatively large within the species of the genus. It is characterized by having simple facial features, without ornamentation. The tail is long and is included in the uropatagium. The uropatagium, which is in the form of a V, is well developed and lacks hair both dorsally and ventrally. The dorsal coloration is dark brown; however, in some specimens it is yellowish brown (Woloszyn and Woloszyn, 1982).

DISTRIBUTION: *M. peninsularis* is an endemic species of Mexico, which is distributed exclusively in the south of the Baja California Peninsula, in Baja California Sur. It has been registered in the state of BS.

EXTERNAL MEASURES AND WEIGHT
TL = 77 to 94 mm; TV = 36 to 46 mm; HF = 7 to 10 mm; EAR = 15 mm; AN = 37 to 43 mm.
Weight: 4 to 6 g.

DENTAL FORMULA: I 2/3, C 1/1, PM 3/3, M 3/3 = 38.

NATURAL HISTORY AND ECOLOGY: Little is known about the biology of *M. peninsularis*. It is a gregarious species that roosts in caves, forming groups of up to thousands of individuals in caves in the "Cerro Martir" in the Sierra of the Laguna in Baja California Sur (Jones et al., 1965; Woloszyn and Woloszyn, 1982). It roosts in caves, tunnels, sewers, churches, and abandoned buildings. It eats small insects hunted in flight. It reproduces in the summer; females give birth to a single young in late June and July (Woloszyn and Woloszyn, 1982). It is not known if it hibernates. This species is sympatric with *M. volans*, *Eptesicus fuscus*, and *Corynorhynus townsendii*.

VEGETATIONAL ASSOCIATIONS AND ELEVATION RANGE: This species is found in all types of vegetation, from xeric scrublands and low rainforests to forests of oak and pine-oak in the Sierra of the Laguna, from sea level to 2,200 m (Woloszyn and Woloszyn, 1982).

CONSERVATION STATUS: *M. peninsularis* is a species that should be considered fragile because of its restricted distribution. It can occupy disrupted areas and inhabits villages and suburban areas, however. The La Laguna Biosphere Reserve protects a large portion of its range.

Myotis planiceps Baker, 1955

Flat-headed myotis

Arturo Jimenez-Guzman

SUBSPECIES IN MEXICO
M. planiceps is a monotypic species.

DESCRIPTION: Within its genus, *Myotis planiceps* is a very small bat. As in other *Myotis*, the face has no ornamentation, the ears are relatively long and naked, and the tail is included in the uropatagium. It has a long coat (9.9 mm), with a dark base and brown tip (Baker, 1955, 1956).

EXTERNAL MEASURES AND WEIGHT
TL = 51 to 76 mm; TV = 25; HF = 8; EAR = 10; FA = 26.5 to 27.5 mm.
Weight: 7 g.

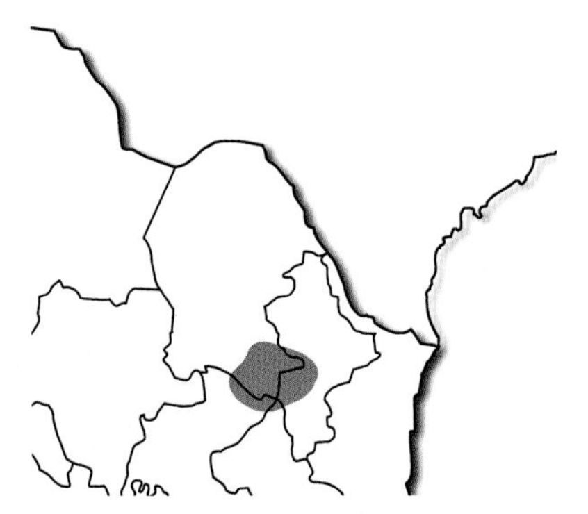

DISTRIBUTION: *M. planiceps* is an endemic species of Mexico, inhabiting a small region between the states of Coahuila, Zacatecas, and Nuevo León (Arroyo Cabrales et al., 2005). It has the most restricted distribution of all species of bats in North America. It has been registered in the states of CO, NL, and ZA.

DENTAL FORMULA: I 2/3, C 1/1, PM 3/3, M 3/3 = 38.

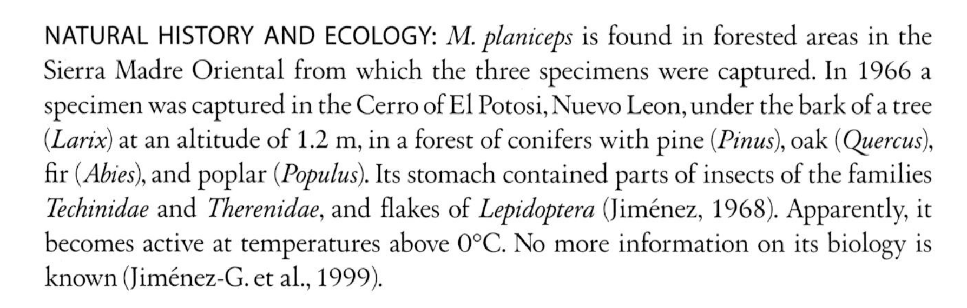

NATURAL HISTORY AND ECOLOGY: *M. planiceps* is found in forested areas in the Sierra Madre Oriental from which the three specimens were captured. In 1966 a specimen was captured in the Cerro of El Potosi, Nuevo Leon, under the bark of a tree (*Larix*) at an altitude of 1.2 m, in a forest of conifers with pine (*Pinus*), oak (*Quercus*), fir (*Abies*), and poplar (*Populus*). Its stomach contained parts of insects of the families *Techinidae* and *Therenidae*, and flakes of *Lepidoptera* (Jiménez, 1968). Apparently, it becomes active at temperatures above 0°C. No more information on its biology is known (Jiménez-G. et al., 1999).

VEGETATIONAL ASSOCIATIONS AND ELEVATION RANGE: This species inhabits coniferous forests and mixed forests (*Abies* sp., *Pinus* sp., *Quercus* sp.), from 1,000 masl to 3,500 masl.

CONSERVATION STATUS: *M. planiceps* is a scarce species; only three specimens are known. It inhabits very disturbed regions, and thus is not consider endangered (Ceballos and Navarro, 1991; SEDESOL, 1994; SEMARNAT, 2002). The International Union for Conservation of Nature (IUCN) (1996) believed it to be extinct. Despite this, it is possible that the species is not extinct, but with a reduced population. It is necessary to carry out an evaluation to determine if there is a population and establish appropriate measures for its long-term conservation.

Myotis planiceps. Pine forest. Arteaga, Coahuila. Photo: Bernal Rodríguez.

Fringed myotis

David Vazquez Ruiz

SUBSPECIES IN MEXICO

Myotis thysanodes aztecus Miller and G.M. Allen, 1928
Myotis thysanodes thysanodes Miller, 1897

DESCRIPTION: *Myotis thysanodes* is a small-sized bat. It has well-developed ears, and is distinguished from other species of the genus by a well-developed fringe at the rear edge of the uropatagium. It has a unique dental characteristic (the absence of cusps in molars 1 and 2). This dental simplification is not observed in any other species of the genus *Myotis* (O'Farrell and Studier, 1980). The coloration of the coat varies from yellowish-brown tones to dark olive and is lighter on the abdomen. (Hall, 1981). This bat displays sexual dimorphism; females are larger with longer heads, bodies, and forearms (O'Farrell and Studier, 1980).

DISTRIBUTION: This species is distributed in the central region and Southwest of the United States. The most northern record is located in British Columbia, Canada. In Mexico, the species is distributed from the desert regions of the north of the country in Sonora, Durango, and San Luis Potosí, including the highland regions of the Eje Volcanico Transversal, to the tropical forests of Veracruz and humid tropics of Chiapas (Hall, 1981; Moreno-V., 1999). It has been registered in the states of BC, CH, CL, CO, CS, DU, HG, JA, MI, MX, NL, OX, PU, QE, SL, SO, TA, VE, and ZA.

EXTERNAL MEASURES AND WEIGHT

TL = 43 to 59 mm; TV = 34 to 45 mm; HF = 9 to 10 mm; EAR = 16 to 20 mm; FA = 40 to 47 mm.
Weight: 7.5 g.

DENTAL FORMULA: I 2/3, C 1/1, PM 3/3, M 3/3 = 38.

NATURAL HISTORY AND ECOLOGY: This species inhabits caves, mine tunnels, and construction sites. Migratory phenomena have been observed for the species, but the magnitude of the movements is unknown. O'Farrell and Studier (1973) inferred that migration should be of short distance and at low altitude, especially in the southern regions of the United States, where bats can be active in winter. Mating season occurs in the middle of spring; pregnant females have been found in the early summer in Chihuahua (Barbour and Davis, 1969). The gestation period lasts between 50 and

Myotis thysanodes. Scrubland. Photo: Scott Altenbach.

60 days; the female gives birth to only a single young (O'Farrell and Studier, 1980). During the development and growth of the young, a number of "nursing" females guard and care young bats during the night to protect them (O'Farrell and Studier, 1973). Like most vespertilionids, its food consists mainly of insects. It has a peculiar feeding pattern, since it performs a slow but highly maneuverable flight on the canopy, feeding especially on beetles and moths (Black, 1974).

VEGETATIONAL ASSOCIATIONS AND ELEVATION RANGE: This species is found in areas with xeric vegetation, pastures, forests of conifers, forests of oak, cloud forests, and tropical subdeciduous forests (O'Farrell and Studier, 1980). It occurs from 1,200 masl to 2,850 masl (O'Farrell and Studier, 1980).

CONSERVATION STATUS: While in the United States this species is abundant and is not found in any list of endangered species (O'Farrell and Studier, 1980), in Mexico its conservation status is unknown. The scarce records of *M. thysanodes* reflect the necessity to intensify the research on its biology, geographic distribution, and conservation status.

Myotis velifer (J.A. Allen, 1890)

Cave myotis

Alondra Castro-Campillo, Elsa Gonzalez, Ulises Aguilera, and Jose Ramirez-Pulido

SUBSPECIES IN MEXICO
Myotis velifer incauta (J.A. Allen, 1896)
Myotis velifer velifer (J.A. Allen, 1890)

DESCRIPTION: *Myotis velifer* is comparable in size to *M. thysanodes* and *M. occultus*. *M. velifer* is the largest among the other Mexican species of the genus. Females have significantly larger forearms and maxillary teeth series (Williams and Findley, 1979). The forearm of M. velifer is the longest of the genus in North America, and varies from 35 mm to 46 mm. The dorsal coloration varies from light brown or sepia to dark brown, the ventral region is a chamois creamy color. The hair is long, slightly silky, and bicolor. The ears have a lobe in the base of the internal edge and extend beyond the tip of the nose. The tragus is slim with the anterior edge almost straight and the distal edge ending in a blunt tip; its length is close to half the length of the ear. The interfemoral membrane surrounds the tail, which does not protrude beyond the posterior edge. The legs are robust and large. In adult bats, the sagittal crest is well developed. Molars are robust; their width compared with the palatine is greater than in any other American members of the genus *Myotis* (Allen, 1890, 1906; Álvarez, 1963a; Álvarez and Polaco, 1984; Anderson, 1972; Baker, 1956; Bogan and Williams, 1970; Fitch et al., 1981; Hall, 1981; Hayward, 1970; Martinez and Villa Ramirez, 1938; Miller and Allen, 1928; Ramírez-Pulido, 1969b; Villa-R., 1967). It is distinguished from *M. occultus* by the presence of a sharp sagittal crest and because the molars are wider, the hair is shorter and less silky; *M. thysanodes* lacks the typical hair of the interfemoral membrane and from both species because it has a wider face and shorter cranial box (Fitch et al., 1981).

Myotis velifer. Pine forest. Photo: Scott Altenbach.

EXTERNAL MEASURES AND WEIGHT
TL = 80 to 109 mm; TV = 35 to 55 mm, HF = 7 to 12 mm; EAR = 13 to 16.6 mm; FA = 36.5 to 47 mm.
Weight: 6 to 11 g.

DENTAL FORMULA: I 2/3, C 1/1, PM 3/3, M 3/3 = 38.

NATURAL HISTORY AND ECOLOGY: This cave bat forms large colonies whose size and function vary with the season: from 600 to 5,000 individuals in a colony of reproduction and up to 15,000 individuals in a maternity colony. While females form the big maternity colonies, males remain in small groups of less than 100 individuals. It has been found in mines, caves, cracks, and the roofs of old churches and abandoned houses (Fitch et al., 1981; Hayward, 1970; Matson and Baker, 1986; Villa-R., 1967). It feeds on insects; its diet fluctuates according to the time of year and the habitat, and includes especially microlepidopterans and coleoptera (Álvarez and Polaco, 1984; Davis and Russell, 1953; Fitch et al., 1981; Hayward, 1970). It only reproduces once a year; spermatogenesis occurs from late summer to early autumn; mating occurs in autumn, probably extending to winter. Females give birth to only one young after a gestation period that ranges from approximately 60 to 70 days (Ceballos and Galindo, 1984). The births begin in late June and end in early July. Males do not reach sexual maturity until the second year of life (Fitch et al., 1981; Hayward, 1970; Jones et al., 1970; Kunz, 1973). *M. velifer* has been found associated with *M. yumanensis*, *M. californicus*, *M. volans*, and *Tadarida brasiliensis* (Álvarez and Ramírez-Pulido, 1972; Fitch et al., 1981; Villa-R., 1967).

DISTRIBUTION: *M. velifer* is found from Kansas in the United States to northeastern Guatemala. In Mexico, two species are found: *M. velifer incauta* in the north and northeast, and *M. velifer velifer* distributed in the rest of the country (Fitch et al., 1981; Hall, 1981; Miller and Allen, 1928; Villa-R., 1960, 1967). It has been registered in the states of AG, BS, CH, CL, CO, CS, DF, DU, GJ, GR, HG, JA, MI, MO, MX, NL, OX, PU, QE, SI, SL, SO, TA, TB, TL, VE, and ZA.

VEGETATIONAL ASSOCIATIONS AND ELEVATION RANGE: *M. velifer* inhabits a variety of vegetation types, including deciduous rainforests, thorny rainforests, xeric scrublands, oak forests, pine forests, fir forests, and disturbed areas (Baker and Greer,

1962; Baker and Webb, 1966; Davis, 1944; Drake, 1958; Jones and Webster, 1977; León Paniagua and Romo-Velázquez, 1993; Redell, 1981; Urbano-V. et al., 1987; Webb and Baker, 1969; Wilson, 1985; Wilson et al., 1985). It has been found from sea level to 3,300 m (Jones et al., 1970; Villa-R., 1967; Watkins et al., 1972).

CONSERVATION STATUS: Within the chiropterans, *M. velifer* is perhaps one of the species with better prospects of survival in terms of availability of roosting sites since it is very tolerant to various environmental conditions (Sánchez et al., 1989). Furthermore, due to its wide distribution it does not face immediate conservation problems, except when occupying caves and caverns, which are very often visited and altered.

Myotis vivesi Menegaux, 1901

Fish-eating myotis

Rafael Avila Flores and Rodrigo A. Medellín

SUBSPECIES IN MEXICO

M. vivesi is a monotypic species.

DESCRIPTION: This bat is much bigger than the rest of the American species of the genus *Myotis*. It is distinguished from the rest of them by the remarkably developed legs (the length ranges between 21.2 mm and 24.8 mm, approximately the same of that of the tibia); the claws are long and laterally compressed; the plagiopatagium is remarkably narrow at knee level and has hemorrhagic nodules between the forearm and the fifth finger; the teeth cusps are taller and thinner; and the second premolar (upper and inferior) is taller than the first. The dorsal pelage is dark buffy or pale cinnamon, and the fur is whitish on the abdomen (Baker and Patton, 1967; Hall, 1981; Menegaux, 1901; Miller, 1906; Miller and Allen, 1928; Villa-R., 1967). Some authors consider that these morphologic differences, along with the bat's physiologic metabolism of water and salt (Carpenter, 1968), are enough to locate *Myotis vivesi* in a different genus, *Pizonyx* (Bradshaw, 1963; Burt, 1932b; Cockrum and Miller, 1906; Miller and Allen, 1928; Reeder and Norris, 1954; Villa-R., 1967). Some palearctic species of the genus *Myotis* have some degree of piscivory and similar morphologic adaptations, however (Patten and Findley, 1970); thus, according to these observations, and taking into account the set of shared characteristics with other members of the genus (Baker and Patton, 1967; Patten and Findley, 1970), other authors have chosen to keep this species in the genus *Myotis* (Hall, 1981; Patten and Findley, 1970; Wilson and Reeder, 1993).

EXTERNAL MEASURES AND WEIGHT

TL = 138 to 163 mm; TV = 50 to 74.8 mm; HF = 22 to 24 mm; EAR = 20 to 26 mm; FA = 53.7 to 63 mm.
Weight: 27.5 g.

DENTAL FORMULA: I 2/3, C 1/1, PM 3/3, M 3/3 = 38.

NATURAL HISTORY AND ECOLOGY: The fish-eating myotis is the only American member of the family Vespertilionidae that feeds on vertebrates, especially small sea fish. These, along with some crustaceans, compose most of its diet, which can be

Myotis vivesi. Scrubland. Isla Partida, Baja California. Photo: Gerardo Ceballos.

complemented with insects and occasionally algae (Burt, 1932b; Patten and Findley, 1970; Reeder and Norris, 1954; Villa-R., 1967). Its feeding activity begins at dusk, and apparently, ends at dawn (Burt, 1932b; Reeder and Norris, 1954; Villa-R., 1979). Its flight has been described as a circular pattern close to the water, and once it detects a prey near the surface, it extends its legs (and sometimes part of the tail and uropatagium) to try and catch the prey (Altenbach, 1989; Brown and Berry, 1981; Burt, 1932b; Reeder and Norris, 1954). Its predators include barn owls (*Tyto alba*), which may prey on this species as main food item (R. Medellín, pers. obs.; E. Velarde, pers. comm.; Villa-R., 1979), as well as the loggerhead shrike (*Lanius ludovicianus*), ring-billed gulls (*Larus delawarensis*), yellow-footed gulls (*Larus livens*), ravens (*Corvus corax*), peregrine falcons (*Falco peregrinus*), osprey (*Pandion haliaetus*), and rats (*Rattus norvegicus*) (Burt, 1932b; Velarde and Medellín, 1981; Villa-R., 1979). During the day it roosts in loose rocks, sometimes sharing the space with petrels (genus *Oceanodroma*; Burt, 1932b; Velarde and Medellín, 1981; Villa-R., 1967). Reproductive activity (gestation, delivery, and nursing) takes place during April and June. They give birth to one young per pregnancy (Burt, 1932b; Cockrum, 1955; Reeder and Norris, 1954). Individuals can live up to 10 years (Orr, 1965).

VEGETATIONAL ASSOCIATIONS AND ELEVATION RANGE: This bat does not seem to be directly associated with any particular type of vegetation since all its activities take place in coastal areas that every so often have cactus and dispersed bushes (e.g., cardon cactus; Burt, 1932b). Xeric shrubs often cover the continental areas that surround the coasts inhabited by these bats. It has only been observed in areas close to sea level.

CONSERVATION STATUS: Because this species is restricted to an area relatively reduced with dispersed and peculiar habitats, the Mexican government has included this endemic species of the northeastern part of the country in the category of rare (SEDESOL, 1994). Its largest population, located in Isla Partida, has been estimated at between 12,000 and 15,000 individuals (Villa-R., 1979).

DISTRIBUTION: *M. vivesi* is an endemic species of Mexico that dwells in a large part of the coastal region (mainly islands) of the Gulf of California, from the Bay of San Jorge to the coast of Guaymas, in Sonora, and from the middle of Baja California Norte (Islas Encantadas) to the Bay of Paz and Punta Coyotes, in Baja California Sur. Furthermore, it has been collected in areas of the Bay Sebastian Vizcaino and the Bay Tortuga in the Pacific coast (Hall, 1981; Reeder and Norris, 1954, Villa-R., 1979). It has been registered in the states of BC, BS, and SO.

Long-legged myotis

Jorge Uribe and Hector T. Arita

DISTRIBUTION: This species is distributed only in North America, from the Eje Neovolcanico Transverso in Mexico to northeastern Canada, crossing through the central and western United States. In Mexico, it is found along the Baja California Peninsula and in the northern regions of Chihuahua, Coahuila, and Durango; its most northern distribution comprises the entire Eje Neovolcanico Transverso from Jalisco to Veracruz (Hall, 1981; Ramírez Pulido et al., 1980). It has been registered in the states of BC, BS, CH, CL, CO, DF, DU, JA, MI, MO, MX, PU, SI, SO, TL, and VE.

SUBSPECIES IN MEXICO

Myotis volans amotus Miller, 1914
Myotis volans interior Miller, 1914
Myotis volans volans (H. Allen, 1866)

DESCRIPTION: *Myotis volans* is a small-sized bat, but is large compared to the other species of the genus. The ears are small and rounded and do not exceed the tip of the nose when extended forward; the tragus is short with a small lobe in its basal part; the legs are also small. This species is characterized by hair on the plagiopatagium, in addition to a very small face with the front and occiput abruptly elevated. The coloration of the coat varies from reddish-brown to dark brown; the coloration of the belly is variable, although usually it is cinnamon brown or grayish-brown (Ceballos and Galindo, 1984; Miller and Allen, 1928). Some populations of this species show local adaptations of coloration. There is some sexual dimorphism; the females have a larger forearm (Warner and Czaplewski, 1984). Morphologically, the three subspecies show a slight geographic variation, with the subspecies of the peninsula of Baja California (*M. volans volans*) being significantly smaller than the other two subspecies present in Mexico.

Myotis volans. Pine forest. Janos Biosphere Reserve, Chihuahua. Photo: Juan Cruzado.

EXTERNAL MEASURES AND WEIGHT

TL = 83 to 106 mm; TV = 32 to 49 mm; HF = 5 to 9 mm; EAR = 11 to 14 mm; FA = 35 to 41 mm.
Weight: 6 g.

DENTAL FORMULA: I 2/3, C 1/1, PM 3/3, M 3/3 = 38.

NATURAL HISTORY AND ECOLOGY: This bat shows a rapid and direct flight, covering relatively long distances flying through, around, below, or above the canopy of the forests while pursuing its prey (Fenton and Bell, 1979). It is considered insectivorous. Its main prey includes butterflies (Lepidoptera), termites (Isoptera), flies (Diptera), bedbugs (Hemiptera), grasshoppers (Orthoptera), wasps (Hymenoptera), and small beetles (Coleoptera) (Black, 1974; Jones et al., 1973). Based on comments made by Grinnell (1918), it is believed that this bat uses the same routes of foraging. Unlike other species of the genus, *M. volans* is opportunistic, capturing any available prey (Bell, 1980). They are gregarious, and usually change habitat within seasons; they do not form colonies as numerous as other species of the genus (Ceballos and Galindo, 1984; Hoffmeister, 1970). They use a variety of roosting sites, including abandoned buildings, cracks in trees, and holes between the crust and trunk and also in caves and tunnels; these last roosting sites are, apparently, occupied only to hibernate (Ceballos and Galindo, 1984; Dalquest and Ramage, 1940; Quay, 1948; Schowalter, 1980). They show a peak of activity in the first three or four hours after nightfall (Cockrum and

Cross, 1964). Furthermore, they are active at temperatures ranging between 12°C and 18°C, so they are considered tolerant to temperatures. Records of rabies in this species are only known for North America (Warner and Czaplewski, 1984). They reproduce in late spring. Each female gives birth to a single young, and apparently segregation of sexes exists in differentiated areas (Ceballos and Galindo, 1984).

VEGETATIONAL ASSOCIATIONS AND ELEVATION RANGE: *M. volans* inhabits forests of conifers and oak, tropical deciduous forests, and xeric scrublands (Ceballos and Galindo, 1984). The interval altitudinal can vary from 60 masl to 3,770 masl, but usually is found between 2,000 masl and 3,000 masl (Warner and Czaplewski, 1984).

CONSERVATION STATUS: *M. volans* is not listed in any official list of endangered species. Bats with residues of DDT, DTT, and DDD in the brain and skeleton have been reported in Oregon; however, there is not enough information to assume that these chemicals are affecting local populations (Warner and Czaplewski, 1984).

Myotis yumanensis (H. Allen, 1864)

Yuma myotis

Barbara Vargas Miranda and Miguel Angel Briones S.

SUBSPECIES IN MEXICO

Myotis yumanensis lambi Beson, 1947
Myotis yumanensis lutosus Miller and G.M. Allen, 1928
Myotis yumanensis sociabilis Grinnell, 1914
Myotis yumanensis yumanensis (H. Allen, 1864)

DISTRIBUTION: *M. yumanensis* is distributed from Canada to Mexico, in the north and center of the country (Hall, 1981). It has been registered in states of AG, BC, BS, CH, CL, CO, DU, GJ, HG, JA, MI, MO, MX, PU, QE, SI, SL, SO, and ZA.

DESCRIPTION: *Myotis yumanensis* is a relatively small-sized bat. The skull has an abrupt elevation on the front (Hall, 1981). The ears of *M. yumanensis* are of average length; unlike *M. occultus*, the tragus is bigger and the contour is semicircular. The pelage is abundant, of medium length, and in general bicolor (Villa-R., 1967). Dorsal coloration is opaque, whereas the coloring of the belly is grayish or whitish (Álvarez et al., 1994).

EXTERNAL MEASURES AND WEIGHT

TL = 80 to 86 mm; TV = 36 to 39 mm; HF = 5 to 7 mm; EAR = 13.5 to 15.5 mm; FA = 33.8 to 36.1 mm.
Weight: 5.0 to 6.1 g.

DENTAL FORMULA: I 2/3, C 1/1, PM 3/3, M 3/3 = 38.

NATURAL HISTORY AND ECOLOGY: *M. yumanensis* has nocturnal habits and its diet consists basically of insects. It roosts in caves and abandoned houses and at night it uses different roosting sites. Data on its reproduction relate to the subspecies *M. yumanensis lutosus*, with lactating females or with one embryo collected in May, June, and August; and males with testicles in the scrotum in June. Females form maternity colonies. They form colonies of hundreds or thousands of adult individuals (Ceballos and Galindo, 1984). Only a study of karyotype has been made of the species (Bickham, 1979a).

VEGETATIONAL ASSOCIATIONS AND ELEVATION RANGE: This species is distributed in arid deserts and xeric scrubland; however, it has also been collected in grasslands, forests of pine and oak, and tropical regions with tropical deciduous forests and tropical subdeciduous forests. It inhabits altitudes from sea level to 3,000 m.

CONSERVATION STATUS: *M. yumanensis* is not classified within any level of importance for conservation.

Subfamily Vespertilioninae

Thirty-nine genera comprise the diverse subfamily Vespertilioninae, ten of which and twenty-five species are known from Mexico.

Corynorhinus mexicanus G.M. Allen, 1916

Mexican big-eared bat

Ricardo Lopez-Wilchis

SUBSPECIES IN MEXICO
C. mexicanus is a monotypic species.
The genus is also known as *Plecotus* (Hall, 1981).

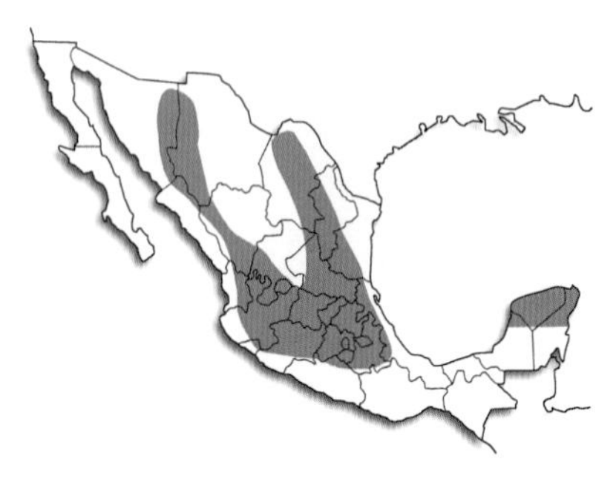

DISTRIBUTION: *C. mexicanus* is an endemic species of Mexico; it is distributed in the highest and most humid parts of the Sierra Madre Oriental, the Eje Volcanico Transversal, and the Sierra Madre Occidental. An isolated population lives in the Yucatan Peninsula (Hall, 1981, Koopman, 1974). It has been registered in the states of AG, CH, CO, CL, DF, GT, HG, JA, MO, MX, MI, NL, OX, PU, QT, QR, SL, SI, SO, TA, TL, VE, YU, and ZA.

DESCRIPTION: *Corynorhinus mexicanus* is the smallest species of the genus. Its large ears, exceeding 33 mm in length, and the atrium with transverse ribs of varying length characterize it. The tragus is large (> 13 mm) and has a prominent basal lobe. There is sexual dimorphism; males are larger than females. The back is dark brown without a marked contrast between the base and the tip of the hair. The color at the base of the rear atrium is lighter; the hair on the base of the ventral region is dark brown and on the top is light brown or cream. The upper incisors generally have an accessory unicusp (Handley, 1959; Tumlison, 1992).

EXTERNAL MEASURES AND WEIGHT
TL = 90 to 112 mm; TV = 41 to 51 mm, HF = 9 to 13 mm; EAR = 30 to 36 mm; FA= 39 to 47 mm.
Weight: 6 to 12 g.

DENTAL FORMULA: I 2/3, C 1/1, PM 2/3, M 3/3 = 36.

NATURAL HISTORY AND ECOLOGY: In the roosting sites, this species' population numbers vary throughout the year; individuals can be found isolated, in pairs, or in large colonies formed by approximately 1,000 individuals. They are temporarily segregated by sex, resulting in four population types: (1) maternal populations, formed by females and their offspring; (2) male populations in spring; (3) transitional summer-autumn populations, where most individuals are males, and (4) winter populations formed by both sexes. They are excellent flyers, agile and capable of carrying out delicate movements that allow them to have a great maneuverability to capture their food among the dense forests. Their diet is based almost exclusively on microlepidopterans. They perform a very precise selection of their roosting sites, preferring sites that are thermally stable and not too cold, with high humidity and little air currents. They maintain a daily torpor as a pattern of thermoregulation, which allows them to lower their metabolic rate and body temperature 2 degrees above the substrate on which they roost. Mating takes place in late autumn; females store the sperm for four months and fertilization takes place in winter. The birth peak occurs in early spring and lasts about two weeks. Breastfeeding spans nearly eight weeks and weaning takes place in early summer (Galvan Leon et al., 1999; López-Wilchis, 1989). Two species of insects and six species of mites have been reported as ectoparasites (López-Wilchis, 1989; Morales-Malacara and López-Wilichis, 1990, 1998). Young and males seem to be more susceptible to parasitism (López-Wilchis, 1989). Twelve species of bats have been associated with this species, including nine vespertilionids, two phyllostomids, and one molossid (López-Wilchis, 1998).

VEGETATIONAL ASSOCIATIONS AND ELEVATION RANGE: This species commonly inhabits temperate rainforests of pine-oak and conifers, but can also be found in transition areas of this type of vegetation. It commonly roosts in constructions such as tunnels, mines, basements, and abandoned houses (Handley, 1959). It has been found from sea level up to 2,400 m, although most records are located in localities above 1,800 m (Morales-Malacara and López-Wilches, 1990; Tumlison, 1992).

CONSERVATION STATUS: *C. mexicanus* is a rare species (Morales-Malacara and López-Wilches, 1990; Tumlison, 1992). Apparently, their populations are declining due to the deterioration of their habitat and the constant disruption of their roosting sites. It is possible that the species is threatened.

Corynorhinus mexicanus. Pine forest. La Marquesa, state of Mexico. Photo: Gerardo Ceballos.

Townsend's big-eared bat

Ricardo Lopez-Wilchis

SUBSPECIES IN MEXICO

Corynorhinus townsendii australis Handley, 1955

Corynorhinus townsendii pallescens Miller, 1897

DISTRIBUTION: This species has the widest geographic range within its genus. Its distribution covers a small portion of southwestern Canada, it extends widely throughout the United States and penetrates to Mexico mainly by the central and northern mountain ranges, reaching the Isthmus of Tehuantepec (Ramírez-Pulido and Castro-C., 1990; Ramírez Pulido et al., 1986; Sánchez-Hernández et al., 2002). It has been registered in the states of AG, BC, BS, CH, CL, CO, DF, DU, GJ, GR, HG, JA, MI, MO, MX, OX, QE, SL, SO, TA, VE, and ZA.

DESCRIPTION: *Corynorhinus townsendii* is a medium-sized bat with small and dark eyes. The face is short and the large ears exceed 33 mm in length; in the atrium they have transverse ribs that vary in length. The tragus is also large (> 13 mm) and has a prominent basal lobe. The measures and weights of males are generally higher than those of females. The base of the back hair is gray and the tip varies from dark brown to dark reddish-brown. In the ventral region, the base is gray and the tip is yellowish. It can be distinguished from *C. mexicanus* because the fur on the back is lighter and shows a marked contrast between the base and tip. The upper incisor is simple (Handley, 1959; Kunz and Martin, 1982).

EXTERNAL MEASURES AND WEIGHT

TL = 90 to 112 mm; TV = 35 to 54 mm, HF = 9 to 13 mm; EAR = 30 to 39 mm; FA = 39 to 45 mm.

Weight: 5 to 13 g.

DENTAL FORMULA: I 2/3, C 1/1, PM 2/3, M 3/3 = 36.

NATURAL HISTORY AND ECOLOGY: Unlike the vast information available on the subspecies that lives in the north part of the country, little is known about this species in Mexico. Individuals have been usually captured solitary in wet and cold caves, even when the site is dry. It is also commonly found in mines, tunnels, and aban-

Corynorhinus townsendii. Photo: Scott Altenbach.

doned buildings. They maintain torpor as a daily pattern of thermoregulation. It is possible that the population of northern Mexico undergoes a true hibernation. For northern subspecies, a reproductive pattern similar to that of *C. mexicanus* has been reported, which suggests that the individuals living in Mexico could follow this pattern as well, including sperm storage. In central Mexico, the presence of a large colony made up by lactating females and males with abdominal testicles was reported in August (Villa-R., 1967). They are good flyers that feed mainly on microlepidopterans. It has been reported to share roosting sites with *C. mexicanus* (Muñiz-Martinez and Polaco, 1996).

VEGETATIONAL ASSOCIATIONS AND ELEVATION RANGE: Townsend's big-eared bat inhabits a wide variety of vegetation types such as tropical deciduous forests and subdeciduous forests, oak forests, juniper forests, thorny forests, xeric scrublands, pastures, disturbed vegetation, and crops. It has been found from sea level up to 2,300 m, although most records are located below 1,000 m. Throughout its distribution, it has been located in deciduous forests and coniferous forests, but most of the records come from the arid central plateau and northern parts of the country, and from the dry valleys of the transitional zone of the Eje Volcanico Transversal (Handley, 1959; Kunz and Martin, 1982).

CONSERVATION STATUS: *C. townsendii* is a rare species; only few reports of its populations in Mexico exist (Handley, 1959; Villa-R., 1967). Like other bat species, population numbers appear to be decreasing due to the deterioration of the foraging areas by intense deforestation of forests and widespread vandalism that occurs in their shelters.

Eptesicus brasiliensis (Desmarest, 1819)

Brazilian brown bat

Rodrigo A. Medellín

SUBSPECIES IN MEXICO

Eptesicus brasiliensis andinus J.A. Allen, 1914

Eptesicus brasiliensis propinquus 1953

There is no consensus on the taxonomic status of this species. Some authors believe that the Mexican specimens are actually *E. andinus*, a monotypic species (Davis, 1966; Hall, 1981), while others believe that *E. andinus* is a subspecies of *E. brasiliensis* (Koopman, 1978, 1993; Timm et al., 1989).

DESCRIPTION: *Eptesicus brasiliensis* is a small-sized bat. There are a couple of subcutaneous glands on both sides of the face between the nose and eyes. The eyes are small and the face is almost hairless, except for the front and the back portion of the face almost reaching the nose. The ears are short and elongated, with rounded tips. The hair on the back is very long (5 mm to 9 mm) and thick, reddish-brown to dark black. The abdomen has almost the same color as the back, but the hair is shorter. The membranes are black. The tail is long and is fully included in the uropatagium, reaching its end.

Eptesicus brasiliensis. Cloud forest. El Jabalí, Colima. Photo: Gerardo Ceballos.

EXTERNAL MEASURES AND WEIGHT

TL = 92 to 114 mm; TV = 33 to 46 mm; HF = 9 to 12 mm; EAR = 12 to 17 mm; FA = 39 to 47 mm.
Weight: 8 to 13 g.

DENTAL FORMULA: I 2/3, C 1/1, PM 1/2, F 3/3 = 32.

NATURAL HISTORY AND ECOLOGY: The information on the biology of this species is scarce. They feed on insects they catch in flight, but there is no specific information about their diet. They roost in hollow trees or buildings and houses (Handley, 1976). Apparently, their reproductive cycle shows polyestry.

VEGETATIONAL ASSOCIATIONS AND ELEVATION RANGE: In Mexico, this species has been captured in temperate forests. In other countries, it has been found in savannas, swamps, and rainforests (Handley, 1976). In Mexico, it has been found between 900 masl and 2,000 masl. In Venezuela, it has been captured in low-lying areas, almost at sea level (Handley, 1976).

CONSERVATION STATUS: The status of this species is undetermined. There are few specimens and the vast majority of them came from the cloud forest, vegetation currently facing severe conservation problems; this vegetation was never abundant in Mexico and a large proportion of its original extent has been converted into man-made environments. Although they are not in any official list of endangered species they are likely vulnerable.

DISTRIBUTION: *E. brasiliensis* inhabits forested wet regions of medium altitude, from Nayarit and Veracruz to the south of South America. In Mexico, it occupies middle elevations of the Sierra Madre Occidental, Sierra Madre Oriental, and Chiapas (Hall, 1981; Koopman, 1982; R. Medellín, pers. obs.; Téllez-G. et al., 1997). It has been registered in the states of CL, CS, GR, NY, and VE.

Argentine brown bat

Guadalupe Tellez-Giron

DISTRIBUTION: In Mexico, *E. furinalis* is distributed from Jalisco on the Pacific coast, and from southern Tamaulipas on the Gulf side to Chiapas, the southeast, and the Yucatan Peninsula. There is a population in the Eje Neovolcanico. Its distribution limit in South America is northern Argentina (Eisenberg, 1989; Sánchez-Hernández et al., 2002; Villa-R., 1967). It has been recorded in the states of CA, CL, CS, GR, JA, MI, MO, MX, OX, QR, SL, TA, TB, VE, and YU.

SUBSPECIES IN MEXICO

Eptesicus furinalis gaumeri (J.A. Allen, 1897)

DESCRIPTION: *Eptesicus furinalis* is the smallest of the three species of the genus *Eptesicus* of Mexico. Externally, it looks like *E. brasiliensis* and, because of its size, it could be confused with some species of *Myotis*. The tragus is short and rounded. The distinguishing characteristics between *Eptesicus* and *Myotis* are found in the skull and teeth; the molars are small, and the teeth row of the jaw measures between 5.4 mm and 6.0 mm; the sagittal and occipital crests are little developed. The upper incisors are triangular; the internal incisive is bifid and the external is almost the same size as the internal, while in *E. brasiliensis* both are rounded. The color of the back hair is dark brown with reddish-golden tips; the abdomen is lighter and can be whitish (Davis, 1965a, b, 1966; Eisenberg, 1989; Villa-R., 1967).

EXTERNAL MEASURES AND WEIGHT

TL = 96 to 105 mm; TV = 30 to 39 mm; HF = 7 to 11 mm; EAR = 12 to 14 mm; FA = 36 to 40 mm.

Weight: 7 to 9 g.

DENTAL FORMULA: I 2/3, C 1/1, PM 1/2, F 3/3 = 32.

NATURAL HISTORY AND ECOLOGY: The Argentine brown bat's activity starts in the first hours after nightfall. Apparently, it is strongly associated with sites where there is surface water (streams, pits, ponds, and cenotes) and undisturbed forests (Hall and Dalquest, 1963; Hollander and Jones, 1988; Jones, 1964; Navarro, 1982). Its roosting sites include hollow trees, rock crevices, cenotes, and caves; it forms colonies of 12 to 100 individuals (Coates-Estrada and Estrada, 1986; Navarro, 1982). Occasionally, they share roosting sites with *Artibeus jamaicensis*, *A. lituratus*, *Dermanura toltecus*, *Sturnira lilium*, *Desmodus rotundus*, *Rhogeessa tumida*, and *Cynomops mexicanus* (Hollander and Jones, 1988; Jones and Dunningan, 1965). They have a slow and erratic flight, usually in irregular circles at a height of between 6 m and 9 m; they feed exclusively on insects (Dalquest, 1953a). Data on reproduction are scarce. In Yucatan, Quintana Roo, and Michoacan, males with testicles in the scrotum and pregnant females with two embryos have been reported in May (Birney et al., 1974; Sánchez-Hernández et al., 1985). Hollander and Jones (1988) found a female with a two-day-old young in August.

Eptesicus furinalis. Tropical dry forest. Yucatan. Photo: Jens Rydell.

VEGETATIONAL ASSOCIATIONS AND ELEVATION RANGE: *E. furinalis* is frequently collected in pine-oak forests, but is also found in low deciduous forests, riparian forests, and temperate forests (Jones, 1964b;

Polaco and Muñiz-Martinez, 1987). It occupies the lowlands, from sea level up to 1,000 m (Davis, 1965a).

CONSERVATION STATUS: *E. furinalis* is a relatively common species. There is no information on its conservation status.

Eptesicus fuscus (Palisot de Beauvois, 1796)

Big brown bat

Guadalupe Tellez-Giron

SUBSPECIES IN MEXICO
Eptesicus fuscus miradorensis (H. Allen, 1866)
Eptesicus fuscus pallidus Young, 1908
Eptesicus fuscus peninsulae (Thomas, 1898)

DESCRIPTION: *Eptesicus fuscus* is an insectivorous bat of medium size. Females are relatively larger than males. The body is robust; the face and nose are wide with fleshy lips. The ears are short and rounded; folded forward they hardly reach the nostrils. The tail tip projects slightly beyond the uropatagium. It possesses a cartilaginous calcar that articulates with the calacaneum, and has a keel-shaped extension (Medellín et al., 1983). The dorsal hair is soft and shiny with an oily appearance and measures less than 10 mm. One-quarter of the interfemoral membrane is covered with hair. Pelage color varies between subspecies, but in general, dorsally, it ranges from pinkish tans to rich chocolates, with longer hair with brilliant ends; the ventral fur is lighter, being near pinkish to olive buff. The face, ears, wing membranes, and tail are more obscure and contrast with the body that is lighter (Davis, 1965a, b).

Eptesicus fuscus. Pine forest. Ajusco, Distrito Federal. Photo: Claudio Contreras Koob.

EXTERNAL MEASURES AND WEIGHT
TL = 87 to 138 mm; TV = 34 to 57 mm, HF = 8 to 14 mm; EAR = 10 to 20 mm; FA = 39 to 54 mm.
Weight: 11 to 23 g.

DENTAL FORMULA: I 2/3, C 1/1, PM 1/2, F 3/3 = 32.

NATURAL HISTORY AND ECOLOGY: Big brown bats are more abundant in forests of conifers. They roost in barns, houses, churches, hollow trees, and caves (Kurta et al., 1989). Males are solitary during summer. After hibernation, adult females form maternity colonies of 5 to 700 individuals. In mountainous regions, males were found at higher altitudes than females (Davis et al., 1986; Fenton and Kunz, 1977; Kurta et al., 1989). Its activity begins two hours after dusk. It feeds an average of 100 minutes per night, mainly on beetles. Capture begins at about 50 m of height and descends to 10 m to 15 m; young forage farther down than adults (Freeman, 1981b; Kunz, 1973). Hibernation begins approximately in November, but females begin to store fat one month before hibernating. Males begin their torpor after females. Hibernating sites are cold and dry, and with air currents (Nowak, 1999a). Their predators are usually birds, hawks, rats, weasels, and domestic cats (Black, 1976; Nowak, 1999a). The reproductive pattern is unimodal and occurs from May to July, with an average of one young per breeding, and rarely two. In the lowlands, births occur before May (Barbour and Davis, 1969; Wimsatt, 1945). This species is host to ectoparasites such as mites, nematodes, and endoparasites such as flukes. They are susceptible to the rabies virus (Nadin-Davis et al., 2010; Whitaker, 1973).

VEGETATIONAL ASSOCIATIONS AND ELEVATION RANGE: *E. fuscus* is found in scrublands, grasslands, temperate forests of pine and oak, and tropical deciduous forests. It is, however, most common in regions of temperate forests (Nowak, 1999a; Villa-R., 1967). It is found from 500 masl to 3,466 masl. In South America it has been found at intermediate elevations (Eisenberg, 1989; Handley, 1976; Moreno-Valdez, 1998; Villa-R., 1967).

CONSERVATION STATUS: This bat is abundant in Mexico and is considered out of danger. Two factors affecting its populations are the frequent control and disruption of its colonies, as well as the pollution by pesticides, as high concentrations of these chemicals in milk, embryos, and adult tissues can cause death (Clark, 1981; Clark et al., 1975).

DISTRIBUTION: This species has a wide distribution, ranging from Alberta and Quebec in Canada to the northwest of Colombia and Venezuela. In Mexico, it occupies almost the entire territory except the Yucatan Peninsula (Davis, 1966; Hall, 1981). It has been registered in the states of AG, BC, BS, CH, CL, CO, CS, DF, DU, GJ, GR, HG, JA, MI, MO, MX, NL, NY, OX, PU, QE, SI, SL, SO, TA, TB, TL, VE, and ZA.

Euderma maculatum (J.A. Allen, 1891)

Spotted bat

Livia Leon P.

SUBSPECIES IN MEXICO

E. maculatum is a monotypic species.

Frost and Timm (1992) conducted a phylogenetic analysis of the tribe Plecotini and suggested that the genus *Idionycteris* should join *Euderma*, thus having two species, *E. maculatum* and *E. phyllote*. This arrangement has not been widely accepted (Koopman, 1993).

DESCRIPTION: This species of bat is relatively large and is very easy to distinguish by its pelage color, with the back almost black with well-defined white spots, a blackish band around the neck, and a light abdomen with dark underfur. The ears are very long, 45 mm to 50 mm from the notch to the tip when they are fully stretched, and at the base there is a tuft of white hair. The membranes of the wings, ears, and uropatagium are thin, foldable, and pink (Easterla, 1973; Handley, 1959).

EXTERNAL MEASURES AND WEIGHT

TL = 107 to 115 mm; TV = 48 mm, HF = 12 mm; EAR = 37 to 47 mm; FA = 48 to 51 mm.

Weight: 16 to 20 g.

DENTAL FORMULA: I 2/3, C 1/1, PM 2/2, F 3/3 = 34.

Euderma maculatum. Scrubland. Photo: Scott Altenbach.

DISTRIBUTION: *E. maculatum* is found from British Columbia, in Canada (Woodsworth et al., 1981) and the northern United States (Montana and Wyoming) to Queretaro, in central Mexico (Hall, 1981). In Mexico, only five specimens have been recorded, three in Durango (Álvarez and Polaco, 1984; Gardner, 1965) and two in Queretaro (León-Paniagua et al., 1990; Schmidly and Martin, 1973). One constant component of its habitat is the presence of rocks, cliffs or large canyons (Easterla, 1973). It has been registered in the states of DU and QE.

NATURAL HISTORY AND ECOLOGY: *Euderma maculatum* roosts in crevices of cliffs and rocks. Apparently, it does not use trees as roosting sites (Easterla, 1970). It is a solitary species, except possibly during hibernation. Apparently, it is a late nocturnal flyer; of 54 specimens captured in Texas, 49 were caught after midnight (Easterla, 1973). It feeds on insects, primarily moths (97%;other insects, 0.2%) (Easterla, 1973). It does not catch its prey on the substrate or foliage; it forages alone, approximately 10 m to 30 m above the substrate (Leonard and Fenton, 1983; Woodsworth et al., 1981). It avoids contact with other echolocating bats. This species is considered a fast flyer. A considerable number of specimens have been collected in open areas, where permanent streams or ponds exist near the cliffs. Very little is known about its reproduction; apparently, it happens at the end of June to mid-August. Easterla (1973) captured a pregnant female in June and lactating females in June and August. A lactating female was collected in August in Queretaro (León-Paniagua et al., 1990). The scarce evidence indicates that it only gives birth to a single young every year (Watkins, 1977). Ectoparasites of the genera *Cryptonyssus* sp. and *Ornithodorus* sp. and the species *Basilia rondanni* have been collected (Whitaker and Easterla, 1975). In New Mexico evidence of rabies has been found (Constantine, 1961b). These bats are preyed on by kestrels (*Falco sparverius*) and other birds (Black, 1976).

VEGETATIONAL ASSOCIATIONS AND ELEVATION RANGE: *E. maculatum* inhabits xeric scrublands and mixed chaparrals of oaks, pines, and cedars. In the United States, it has also been found in coniferous forest (Easterla, 1973). In Mexico, it is only found in desert-like and semi desert-like areas (Gardner, 1965; León-Paniagua et al., 1990; Schmidly and Martin, 1973). In Mexico, it has been found between 1,820 masl and 2,438 masl, but in the United States it has been registered from 57 m to 2,438 m.

CONSERVATION STATUS: The current situation of *E. maculatum* in Mexico is unknown (Watkins, 1977). It has been classified as a rare species by the U.S. Wildlife Service.

Allen's big-eared bat

Jorge Ortega R. and Hector T. Arita

SUBSPECIES IN MEXICO

Idionycteris phyllotis phyllotis (G.M. Allen, 1916)

Until recently, this species was considered as monotypic; but subspecies variation between the populations of the western United States and Mexico has been found (Tumilson, 1993).

DESCRIPTION: *Idionycteris phyllotis* is a medium-sized bat. It is characterized by large ears (4 mm to 5 mm wide and 30 mm in length), a conspicuous tragus, and a pair of lobes projected on the base of the front of the ears to the most prominent part of the muzzle (Barbour and Davis, 1969). The dorsal coat is very dense, the underfur is black, and the tips of the hair are lighter, giving a general appearance of a light brown color (Czaplewski, 1983).

EXTERNAL MEASURES AND WEIGHT

TL = 103 to 118 mm; TV = 46 to 55 mm, HF = 9 to 12 mm; EAR = 34 to 43 mm; FA = 45 mm.

Weight: 8 to 16 g.

DENTAL FORMULA: I 2/3, C 1/1, PM 2/3, M 3/3 = 36.

NATURAL HISTORY AND ECOLOGY: The main roosting sites of *I. phyllotis* are gaps between rocks. They form clusters of a few individuals during the breeding season,

DISTRIBUTION: The distribution of this species is restricted to mountainous regions of the southwestern United States to central Mexico. It has been reported in 15 locations in Mexico, which are distributed in 11 states of central-northern Mexico. It has been recorded in the states of CH, CO, DF, DU, HG, JA, MX, NL, OX, PU, QE, SL, and TA.

Idionycteris phyllotis. Pine forest. Arteaga, Coahuila. Photo: Bernal Rodriguez.

which takes place from June to August; females usually give birth to a single young (Czaplewski, 1983). It is an opportunistic insectivore (Czaplewski, 1983). Its activity period starts one or two hours after sunset (Barbour and Davis, 1969).

VEGETATIONAL ASSOCIATIONS AND ELEVATION RANGE: *I. phyllotis* mainly inhabits arid scrublands dominated by xerophytes; it has also been found in areas of pine-oak and fir forests. In Mexico, the altitudinal interval varies from 855 m to 3,225 m (Ceballos and Galindo, 1984).

CONSERVATION STATUS: There are no data on the current status of this species, nor on the population size in the country; it appears to be a very rare species. This probably makes it vulnerable to extinction.

Lasionycteris noctivagans (Le Conte, 1831)

Silver-haired bat

Jorge Ortega R. and Hector T. Arita

SUBSPECIES IN MEXICO

L. noctivagans is a monotypic species.

DESCRIPTION: Silver-haired bats are medium-sized bats. They are characterized by the darker hair on the head; the body pelage combines dark brown hair with some silver, which gives this bat the appearance to which it owes its name. The interfemoral membrane is wide and triangular with abundant hair on the back. The ears are short, rounded, and exposed.

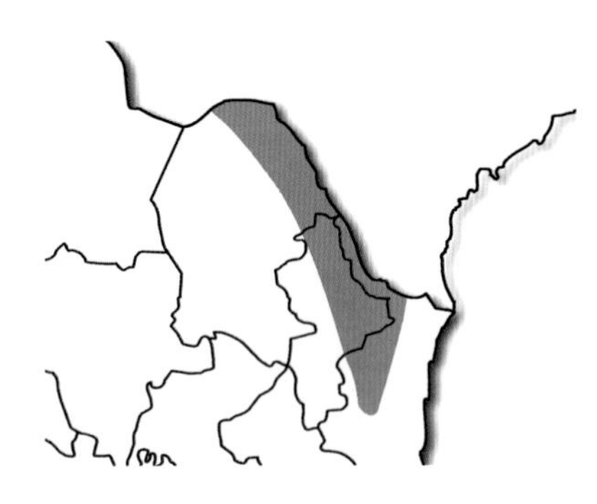

DISTRIBUTION: In Mexico *L. noctivagans* is known only in two localities in the Sierra de San Carlos (Arriaga-Flores, 2010; Schmidly and Hendricks, 1984). It is distributed from the south of the Pacific coast to Alaska and Canada, along virtually all the coastline of United States, with the exception of the Florida peninsula, to northeastern Mexico. It has been recorded in the state of TA.

EXTERNAL MEASURES AND WEIGHT

TL = 92 to 115 mm; TV = 38 to 50 mm, HF = 7 to 9 mm; EAR = 8 to 10 mm; FA = 37 to 44 mm.
Weight: 8 to 11 g.

DENTAL FORMULA: I 2/3, C 1/1, PM 2/3, M 3/3 = 36.

NATURAL HISTORY AND ECOLOGY: *Lasionycteris noctivagans* mainly roosts in hollow trees. It is considered an opportunist insectivore. It has two activity periods, which occur two to three hours, and seven to eight hours, after sunset (Kunz, 1982a, b). Due to its migratory behavior, it has been reported in the north in spring–summer and in the south in autumn–winter (Barbour and Davis, 1969). The gestation period occurs in May and June, and offspring are born after approximately 50 to 60 days (Kunz, 1982a). Breastfeeding lasts 36 days; each female usually gives birth to only a single young (Thomas, 1992).

VEGETATIONAL ASSOCIATIONS AND ELEVATION RANGE: The typical habitats of this species are temperate forests combined with deciduous conifer forests. It is found in areas with water bodies, streams, or ponds. There is a close relationship between *L. noctivagans* and temperate deciduous forests of the middle portion of the western United States (Thomas, 1992). The only record of this species in Mexico was made in a coniferous forest at a height of 646 m (Schmidly and Hendricks, 1984; Yates et al., 1976).

Lasionycteris noctivagans. Photo: Scott Altenbach.

CONSERVATION STATUS: *L. noctivagans* is a species with a marginal distribution, making it vulnerable to extinction. Deforestation and forest management practices reduce the availability of roosting sites.

Lasiurus blossevillii (Lesson and Garnot, 1826)

Western red bat

Juan Carlos Morales, Saul Aguilar, and Livia Leon P.

SUBSPECIES IN MEXICO

Lasiurus blossevillii teliotis (H. Allen, 1891).

Based on significant morphological and genetic differences Baker et al. (1988) noted that eastern and western populations of *Lasiurus*, classified as *borealis*, were different species, and that the name *L. borealis* should apply to eastern populations (*L. blossevillii borealis*). As the first synonym for western populations is *Vespertilio blossevillii*, the trinominal combination of this species in Mexico is *L. blossevillii teliotis*. Morales and Bickham (1995) supported the existence of *L. blossevillii*, but consider *L. blossevillii teliotis* and *L. blossevillii frantzi* synonymous. More studies on the geographical distribution and variation of the species, as well as a systematic review of the group, are needed (McCarthy, 1987).

DISTRIBUTION: *L. blossevillii* is found in some parts of the southwestern United States and Central America (Findley et al., 1975; Genoways and Baker, 1988). In Mexico, it is found in the western part of the country, including a large part of the Central Highland (Anderson, 1972; Bogan and Williams, 1970; Genoways and Baker, 1988; Jones et al., 1988; Schmidly, 1991; Schmidly and Hendricks, 1984). It has been registered in the states of BC, BS, CH, CL, CO, CS, GR, HG, JA, MI, MX, NL, NY, OX, QE, SL, SO, and TA.

DESCRIPTION: *Lasiurus blossevillii* is of medium size, and is very similar to the eastern red bat (*Lasiurus borealis borealis*). It has short and rounded ears, and the tail is relatively long. The coat ranges from dark reddish to brown; the hair lacks white tips, giving the bat the appearance of *L. borealis*. The third posterior part of the interfemoral membrane is naked or only scattered. *L. blossevillii* is slightly smaller than *L. borealis*, and its cranial measures are significantly smaller.

EXTERNAL MEASURES AND WEIGHT
TL = 103 mm; TV = 49 mm, HF = 10 mm; EAR = 13 mm; FA = 50 to 57 mm.
Weight: 7 to 12 g.

DENTAL FORMULA: I 1/3, C 1/1, PM 2/2, F 3/3 = 32.

NATURAL HISTORY AND ECOLOGY: The western red bat's feeding habits and reproductive pattern are poorly documented. They roost in the foliage of trees. There are records of pregnant females in mid-year and they possibly lactate during the autumn (Schmidly, 1991; Schmidly and Hendricks, 1984).

VEGETATIONAL ASSOCIATIONS AND ELEVATION RANGE: *L. blossevillii* has been captured in riparian areas, cotton crops, walnuts, and pine-oak forests; it is associated with multistratum xeric scrublands and thorny and deciduous forests (Schmidly, 1991; Schmidly and Hendricks, 1984), It is found between 400 masl and 800 masl (Schmidly, 1991; Schmidly and Hendricks, 1984).

CONSERVATION STATUS: *L. blossevillii* is a relatively common species; however, the status of its populations is unknown.

Lasiurus blossevillii. Temperate deciduous forest. Photo: Scott Altenbach.

Eastern red bat

Juan Carlos Morales, Saul Aguilar, and Livia Leon P.

SUBSPECIES IN MEXICO

L. borealis is a monotypic species.

Baker et al. (1988) consider it as monotypic; however, Koopman et al. (1957), Handley (1960), and Silva-Taboada (1979) recognize it as polytypical, with eight species and subspecies, thus recognizing all Antillean *Lasiurus* as a subspecies of *borealis*. Handley (1960) recognizes two subspecies: *L. borealis borealis* and *L. borealis teliotis* (*L. b. ornatus*). Baker et al. (1988) includes *degelidus*, *minor*, *pfeifferi*, and *blossevilli* in the subgenus *Lasiurus* (Morales and Bickham, 1995). Wilson and Reeder (1993) consider it necessary to conduct a review of this taxonomic group.

DESCRIPTION: The color of *Lasiurus borealis* distinguishes it; it ranges from bright red to pale red in the upper parts of the body; males are generally brighter than females. In low-lying areas the pelage is lighter, which suggests sexual dimorphism. In the back of the shoulders is a pale white patch (Barbour and Davis, 1969; Hall, 1981; Schmidly, 1991; Schmidly and Hendricks, 1984). The ears are wide, short, and rounded (7 mm to 11 mm); carried forward, they do not exceed the edge of the snout. The distal end of the tragus, at half the height of the ear, is triangular. The wing membranes are very dark, ornate, and grate shaped. The interfemoral membrane is completely covered with dense hair in its dorsal part. Throughout the body, the hair is very dense and long. The claws are black (Hall and Jones, 1961; Schmidly, 1991). *L. borealis* resembles *L. seminolus*, but the coloration is different in the two species (Schmidly, 1991; Schmidly and Hendricks, 1984).

EXTERNAL MEASURES AND WEIGHT

TL = 106 mm; TV = 45 mm, HF = 8 mm; EAR = 13 mm.

Weight: 7 to 13 g.

DENTAL FORMULA: I 1/3, C 1/1, PM 2/2, F 3/3 = 32.

NATURAL HISTORY AND ECOLOGY: Eastern red bats are solitary except for small family groups consisting of an adult female and her offspring; they roost in trees and shrubs; they never use caves, mines, or similar sites. They are migratory and typically follow a specific feeding territory at the crown of trees; they often hunt near the lamps in the streets, catching insects such as beetles (Banfield, 1974; Barbour and Davis, 1969; Bogan and Williams, 1970; Hall and Jones, 1961; Handley, 1960; Mumford, 1973; Schmidly, 1991; Whitaker and Mumford, 1972a). They become active when the temperature ranges between 13°C and 20°C (Davis and Lidicker, 1956; Jones, 1965; Lewis, 1940). Females give birth from 1 to 4 pups between May and July, after a gestation period of 80 to 90 days. Each young weighs about 15 g at birth and is completely hairless. After three to four weeks it is able to fly and is weaned at five or six weeks of age (Cockrum, 1955; Constantine, 1966a, b; Jackson, 1961; MacClure, 1942; Mumford, 1973; Whitaker and Mumford, 1972). Parasites reported for this species are mites, fleas, bedbugs, protozoa, and helminthes (Jackson, 1961; Lowery, 1974; Whitaker and Wilson, 1974).

VEGETATIONAL ASSOCIATIONS AND ELEVATION RANGE: *L. borealis* is associated with multistratum of xeric scrublands, thorny forests, and deciduous forests

DISTRIBUTION: *L. borealis* is distributed from central and southern Canada to Chile and Argentina. It is also found in the Caribbean islands (Baker and Genoways, 1978; Banfield, 1974; Barbour and Davis, 1969; Bogan and Williams, 1970; Shump and Shump, 1982; Wilson and Reeder, 1993). In Mexico, it is located in the northern part of Chihuahua, Coahuila, Nuevo Leon, and Tamaulipas (Bogan and Williams, 1970; Hall, 1981; Hall and Jones, 1961; Jones et al. 1988; Schmidly, 1991; Villa-R., 1967). It has been registered in the states of CH, CO, NL, and TA.

(Schmidly and Hendricks, 1984). It has been captured between 400 masl and 1,600 masl (Moreno-Valdéz, 1998; Schmidly and Hendricks, 1984).

CONSERVATION STATUS: We do not know the current status of this species in Mexico. It is considered out of danger.

Lasiurus cinereus (Palisot de Beauvois, 1796)

Hoary bat

Juan Carlos Morales, Saul Aguilar, and Livia Leon P.

SUBSPECIES IN MEXICO

Lasiurus cinereus cinereus (Palisot of Beauvois, 1796)

DESCRIPTION: This bat is easy to recognize because of its large size and conspicuous coloration. The coloring is a mixed red-gray with a heavy white tinge, giving these bats a frosty appearance. *Lasiurus cinereus* differs from other members of the genus by its skeletal characteristics: the humerus is relatively short, the forearm is long, and the first and third fingers are short. The skull is long, wide, and heavy; the face is broad and possesses wide nasal openings.

Lasiurus cinereus. Pine forest. Ajusco, Distrito Federal. Photo: Hector Arita.

EXTERNAL MEASURES AND WEIGHT
TL = 123 to 138 mm; TV = 47 to 61 mm; HF = 8 to 13 mm; EAR = 13 to 19 mm; FA = 50 to 57 mm.
Weight: 20 to 35 g.

DENTAL FORMULA: I 1/3, C 1/1, PM 2/2, F 3/3 = 32.

NATURAL HISTORY AND ECOLOGY: The hoary bat typically inhabits the foliage of trees, hanging on the tips or edges of the branches. They are migratory and exhibit an interesting seasonal distribution. In summer, females move to give birth and take care of their offspring. Males, however, move to different places, usually in mountainous areas. The migratory patterns of females precede the movements of males in contrast to the patterns found in migratory birds. Breeding takes place in the winter, before migration. Males tend to have a territorial area prepared for the arrival of females and do not make a very long journey to the north. Births occur from mid-May to July; females give birth from two to four pups (Barbour and Davis, 1969; Constantine, 1966a, b; Cowan and Guiguet, 1965; Jackson, 1961; Jones, 1965; Lowery, 1974; Schmidly, 1991; Schmidly and Hendricks, 1984; Whitaker and Mumford, 1972a). Their activity begins with sunset. They cover large areas and sometimes forage at considerable distances from their daytime roosting sites; they prefer moths as prey items, from which they only swallow the thorax and abdomen (Barbour and Davis, 1969; Schmidly, 1991; Schmidly and Hendricks, 1984). Various parasites have been reported for this species: mites, helminths, and protozoa (Jackson, 1961; Radovsky, 1967; Tromba, 1954; Whitaker, 1973; Whitaker and Wilson, 1974).

VEGETATIONAL ASSOCIATIONS AND ELEVATION RANGE: *L. cinereus* has been collected in temperate forests of pine-oak and in the lowlands of deciduous forests (Álvarez and Ramírez-Pulido, 1972; Ceballos and Galindo, 1984; Polaco and Muñiz-Martínez, 1987; Villa-R., 1967). It has been found at altitudes ranging from 500 m to 1,900 m (Álvarez and Ramírez-Pulido, 1972; Álvarez-Castañeda, 1991; Moreno-Valdez, 1998; Polaco and Muñiz-Martínez, 1987; Villa-R., 1967).

CONSERVATION STATUS: In the United States *L. cinereus* is considered endangered. In Mexico, it has no immediate problems of conservation.

DISTRIBUTION: This species has a wide distribution from southern Canada to Chile and Argentina (Barbour and Davis, 1969; Cabrera, 1958; Hall, 1981; Shump and Shump, 1982; Wilson and Reeder, 1993). On the Pacific slope, it is distributed throughout northern Mexico, including the Baja California Peninsula, to Oaxaca, and on the slope of the Gulf, it has been registered in Veracruz (Álvarez-Castañeda, 1991; Hall, 1981; Ramírez-Pulido et al., 1983). It has been registered in the states of BC, BS, CH, CL, CO, DF, DU, GJ, JA, MI, MO, MX, NL, NY, OX, QE, SI, SL, SO, TA, and VE.

Lasiurus ega (Gervais, 1856)

Southern yellow bat

Juan Carlos Morales, Saul Aguilar, and Livia Leon P.

SUBSPECIES IN MEXICO
Lasiurus ega panamensis (Thomas, 1901)
Handley (1960) and Hall and Jones (1961) recognized *L. ega xanthinus* (*Dasypterus ega xanthinus*) and *L. ega panamensis* (*Dasypterus ega panamensis*). However, according to genetic studies, Baker et al. (1988), recommended that these subspecies were elevated to a species level (*L. xanthinus* and *L. ega panamensis*).

DESCRIPTION: *Lasiurus ega* is a medium-sized bat. The pelage on the interfemoral membrane characterizes it. Ears are short, rounded, and wide. The tail is long and is

Lasiurus ega. Tropical dry forest.
Photo: M.B. Fenton.

fully included in the uropatagium. The calcar is long and does not have a keel-shaped extension. It can be confused with *L. intermedius*, although this species is larger. The coat is yellow with dark underfur; the abdomen is lighter. The skull is short and wide and has a sagittal crest (Hall, 1981; Hall and Jones, 1961).

DISTRIBUTION: *L. ega* is distributed from southern California, Arizona, New Mexico, and Texas, and from eastern and south-central Mexico, to South America (Baker et al. 1988; Eisenberg, 1989; Jones et al., 1988; Ramírez-Pulido and Lopez-W., 1983; Schmidly, 1991; Spencer et al., 1988). It has been registered in the states of CA, CS, GR, HG, JA, MO, NL, OX, PU, QE, TA, VE, and YU.

EXTERNAL MEASURES AND WEIGHT

TL = 115 mm; TV = 45 to 58 mm, HF = 8 mm; EAR = 13 mm; FA = 43 to 52 mm. Weight: 15 g.

DENTAL FORMULA: I 1/3, C 1/1, PM 2/2, F 3/3 = 32.

NATURAL HISTORY AND ECOLOGY: Little is known about the biology of this species. It is associated with tropical habitats, especially high and medium jungles below 1,000 m in altitude (Davis and Lidicker, 1956). Like all lasiurids, it roosts under tree branches and leaves of palm trees and buildings. It has also been collected in a pasture with small areas of ocotillo, organ pipe cactus, and mesquite (Cockrum, 1963). It is found solitary or in small groups. Its activity begins two hours after dusk (Baker, 1956), although most prey are caught before midnight (Kurta and Lehr, 1995). They are insectivorous. Females carry from two to four embryos in early June (Ceballos and Galindo, 1984; Cockrum, 1982; Schmidly, 1991).

VEGETATIONAL ASSOCIATIONS AND ELEVATION RANGE: *L. ega* is mainly associated with tropical habitats, especially in medium and high jungles. It has been found at altitudes of less than 1,000 m.

CONSERVATION STATUS: The southern yellow bat's current situation is unknown, but by virtue of its relative abundance and wide geographical distribution, it is considered out of danger.

Lasiurus intermedius H. Allen, 1862

Northern yellow bat

Juan Carlos Morales, Saul Aguilar, and Livia Leon P.

SUBSPECIES IN MEXICO

Lasiurus intermedius intermedius H. Allen, 1862

L. intermedius and *L. ega* are more closely interrelated, as karyotype study suggests (Morales and Bickham, 1995).

DESCRIPTION: *Lasiurus intermedius* gets its name from its yellow-orange to yellow-brown fur. Its membranes are brown. The ears are pointed and short, the wings and face are relatively long. It has white patches on the shoulders and wrists. Only half of the interfemoral membrane is covered with hair. The calcar has a slight, keel-shaped

Lasiurus intermedius. Tropical dry forests, Merida, Yucatan. Photo: Jens Rydell.

DISTRIBUTION: *L. intermedius* is found from the eastern United States, on the shores of Virginia, to Central America (Hellebuyck et al., 1985; Koopman, 1965; Silva-Taboada, 1976, 1979; Schmidly, 1991; Webster, 1980; Wilson and Reeder, 1993). In Mexico, it is distributed on both sides; in the east, it is found all along the Gulf coast to the western Yucatan Peninsula. And it occurs from the southern part of Sinaloa to northern Oaxaca (Baker and Dickerman, 1956; Carter et al., 1966; Carter and Jones, 1978; Gardner, 1962; Jones, 1964; Loomis and Jones, 1964; Villa-R., 1967). It has been recorded in states of CL, CS, DF, GR, HG, JA, MI, MX, NL, NY, OX, QR, SI, TA, VE, and YU.

extension. The sagittal crest is pronounced (Hall and Jones, 1961; Schmidly, 1991; Webster et al., 1980). It differs from *L. ega* by its larger body.

EXTERNAL MEASURES AND WEIGHT
TL = 121 to 164 mm; TV = 51 to 77 mm, HF = 8 to 13 mm; EAR = 15 to 19 mm; FA = 45 to 52 mm.
Weight 3 to 17 g.

DENTAL FORMULA: I 1/3, C 1/1, PM 2/2, F 3/3 = 32.

NATURAL HISTORY AND ECOLOGY: *L. intermedius* is solitary. It roosts under trees and in buildings. In winter sexual segregation exists; in spring and summer females form maternity colonies. Reproduction occurs in autumn and winter (Barbour and Davis, 1969; Hall and Jones, 1961); births occur in May and June (Davis, 1960; Lowery, 1974). Females carry three to four embryos and deliver two or three offspring (Schmidly, 1991). They forage in open areas; their prey items include flies, beetles, ants, and mosquitoes (Sherman, 1939).

VEGETATIONAL ASSOCIATIONS AND ELEVATION RANGE: The northern yellow bat is commonly found under palm trees, in grasslands and forest areas of conifers, and in low pine-oak forests. It has also been found in crops (Baker and Dickerman, 1956; Carter and Jones, 1978; Carter et al., 1966; Jones, 1964). It can be found from sea level up to 3,000 m.

CONSERVATION STATUS: This species is not considered in any risk category.

Lasiurus xanthinus (Thomas, 1897)

Western yellow bat

Saul Aguilar, Livia Leon P., and Juan Carlos Morales

SUBSPECIES IN MEXICO
L. xanthinus is a monotypic species.
This species was considered a subspecies of *L. ega* until Baker et al. (1988), based on cytogenetic studies, elevated it to a species level. Prior to this work, chromosomal studies of the genus showed significant differences in the sex chromosomes and coloration of *L. ega* and *L. xanthinus* (Baker and Patton, 1967; Baker et al., 1971; Bickham, 1979b, 1987).

DESCRIPTION: *Lasiurus xanthinus* is a small-sized bat (Hall and Jones, 1961). In general, it is larger than *L. ega*. Aguilar-Morales (pers. obs.) found that of 12 measures of the skull, only the width of the box and the width of the cranial postorbital constriction are smaller than in *L. ega* (less than 8.35 mm and 4.71 mm, respectively). The dorsal pelage (bright yellow with black) is very similar to that of *L. ega*. The hair of the interfemoral membrane is shiny yellow, contrasting with the posterior part of the back, and the hair is longer than in other species. The color of the abdomen is light brown or orange (Handley, 1960). The face is not black as in *L. ega*. Ears are short, wide, without hair, and light brown. The legs are long and light brown colored, as well as the claw of the first finger, which contrasts with the blond color of the body.

Lasiurus xanthinus. Scrubland. Río Bravo, Acuña, Coahuila. Photo: Juan Cruzado.

EXTERNAL MEASURES AND WEIGHT
TL = 115 to 120 mm; TV = 40 to 50 mm; HF = 8 to 10 mm; EAR = 15 to 19 mm; FA = 42 to 47 mm.
Weight: 16 g.

DENTAL FORMULA: I 1/3, C 1/1, PM 2/2, F 3/3 = 32.

NATURAL HISTORY AND ECOLOGY: Little is known of the biology of *L. xanthinus*, which was considered part of *L. ega*; environmental characteristics of both species have been confused. Like all lasiurids, it roosts under branches of some trees and in buildings (Ceballos and Galindo, 1984; Cockrum, 1982; Schmidly, 1991). It is solitary. Its period of activity begins before dusk, although available literature cites that the activity of this species takes place several hours after dusk (Baker, 1956; Dalquest, 1953b; Jones et al., 1965). They fly rapidly along streams and ponds. In Baja California Sur, only females were found (Aguilar-Morales, pers. comm.). It has been found along with *L. borealis*, *Mormoops megallophyla*, *Myotis californicus*, *Eptesicus fuscus*, *Antrozous pallidus*, *Pipistrellus hesperus*, and *Choeronycteris mexicana*. In Baja California Sur, municipality of Los Cabos, Ejido of Miraflores, 11 pregnant females were captured in June 1995, most of which had 2 embryos and only 4 had 3. Ectoparasites of the families Macronyssidae and Myobiidae were found. Most ectoparasites were found on the body and very few on the wings (S. Aguilar-Morales, pers. obs., pers. comm.)

VEGETATIONAL ASSOCIATIONS AND ELEVATION RANGE: *L. xanthinus* is associated with semi-arid xeric scrubland vegetation (Baker and Patton, 1967). This bat has been collected in lowlands with *Yucca*, *Agave*, *Larrea*, *Acacia*, and *Opuntia* (Greer, 1960). *L. xanthinus* is considered a lowland species (Handley, 1976; Koopman, 1978; Woloszyn and Woloszyn, 1982); however, it can be found above 2,300 m (Genoways and Jones, 1968b).

DISTRIBUTION: *L. xanthinus* is found from the southeastern United States (California and Arizona), the eastern slope of Mexico and the Altiplane to Guerrero, Chiapas, although its exact limits are unknown; recent studies suggest the existence of contact zones in the distribution of *L. ega* and *L. xanthinus* in the states of Puebla, Morelos, Michoacan, Guerrero, and Chiapas (Hall, 1981; Koopman, 1993). It has been registered in the states of BC, BS, CL, CO, DF, DU, GR, HG, JA, MI, MO, NL, NY, PU, QE, SI, SL, SO, and ZA.

CONSERVATION STATUS: The current status of the populations of *L. xanthinus* is unknown. Further information on its distribution and ecology is necessary in order to define its conservation status.

Nycticeius humeralis (Rafinesque, 1818)

Evening bat

Arturo Moreno Valdez

SUBSPECIES IN MEXICO
Nycticeius humeralis humeralis Rafinesque, 1818
Nycticeius humeralis mexicanus Davis, 1944

DISTRIBUTION: *N. humeralis* is distributed from northern Tuxpan, Veracruz, to the north of Coahuila and almost all the states of the eastern United States, to the south of Ontario, Canada. It has been registered in the states of CO, NL, SL, TA, and VE.

DESCRIPTION: *Nycticeius humeralis* is a relatively small-sized bat. Sexual dimorphism exists since females are slightly larger than males. The dorsal pelage is dark brown and the fur is lighter on the abdomen. The skull is massive, short, broad, and low. *N. humeralis* is similar to *Eptesicus fuscus* but smaller in size; it also resembles some members of the genus *Myotis*, differing from them by the short and blunt tragus. In addition, it possesses very remarkable submaxillary glands that give it a slightly musky odor. The wings are short (260 mm to 280 mm wingspan) and narrow; the ears are small and dark; the tail protrudes less than 1 mm beyond the edge of the uropatagium, which is covered with hair at its basal portion.

EXTERNAL MEASURES AND WEIGHT
TL = 83 to 102 mm; TV = 33 to 41 mm; HF = 8 to 10.8 mm; EAR =11 to 14 mm; FA = 33 to 35 mm.
Weight: 6.4 to 13 g.

DENTAL FORMULA: I 1/3, C 1/1, PM 1/2, M 3/3 = 30.

NATURAL HISTORY AND ECOLOGY: Evening bats form small groups, mainly in hollow trees or under bark. It has also been observed in houses, in hay (*Tillandsia* sp.), under palms, and rarely in caves. Females congregate by the hundreds in maternity colonies (caves) during the reproductive period. In the northeast of Mexico, *N. humeralis* is common throughout the year, and usually bats are more abundant in some localities with gallery forests comprising sabinos (*Taxodium mucronatum*) and sycamore (*Platanus* sp.). It has rarely been captured in forests of pine and oak. It is believed that the populations in the north of its distribution are migratory; however, in Mexico and Texas the bat has not been observed in the winter, but in Florida it has. These bats have two foraging periods, one after dusk and another before dawn. As for other species of bats, little is known about its diet, but in some analyzed excreta remains of beetles, bedbugs, moths, and cicadas have been found. Mating occurs in autumn; usually twins are born between May and June, and are capable of carrying out their first flight at three weeks of age. Before leaving the maternity colony, females rub the faces of their offspring with an exudate produced in the submaxillary glands. After hunting, the mothers return and seek their offspring, locating them by their distinctive odor. Among its predators are snakes, raccoons, and domestic cats.

VEGETATIONAL ASSOCIATIONS AND ELEVATION RANGE: *N. humeralis* inhabits gallery forests, xeric scrublands, thorny forests of conifers, and rainforests. It is found from sea level in the Mar de la Pesca, Tamaulipas, to 914 masl in the Hacienda de la Mariposa, Coahuila; most localities are below 500 masl.

CONSERVATION STATUS: Not much is known about this bat's conservation status, but probably some animals are being affected by the destruction of the gallery forest by the construction of dams along the main rivers of the northeast of the country.

Parastrellus hesperus (H. Allen, 1864)

Canyon bat

Livia Leon P.

SUBSPECIES IN MEXICO

Parastrellus hesperus hesperus (H. Allen, 1864)
Parastrellus hesperus maximus Hatfield, 1936

The interpretation of the geographic variation in *P. hesperus* has been vastly controversial. Four subspecies were previously considered for Mexico (Cockrum, 1960; Hall and Kelson, 1959; Villa-R., 1967). According to the latest taxonomic revision of this species (Findley and Traut, 1970), in Mexico only two subspecies are distributed: *P. hesperus hesperus* (including *P. hesperus australis*), which is located in the west (including the Baja California Peninsula) to the southeast to Guerrero, Michoacan, and Morelos; and *P. hesperus maximus* (including *P. hesperus potosinus*), which is found from Hidalgo and Queretaro to the north, to Chihuahua and Coahuila (Jones, 1988).

DESCRIPTION: *Parastrellus hesperus* is one of the smallest species of bats in Mexico. The dorsal pelage is bright gray or yellowish and lighter on the abdomen, the base of the hairs on both sides is more obscure. The ears are short and rounded and, like the facial mask, the wings and the interfemoral membrane are dark. The tragus is curved forward. The tail is fully included in the uropatagium and the calcar has a keel-shaped extension. As in almost all the vespertilionids, males are slightly smaller than females of the same age. It differs from other American vespertilionids by size and from *Perimyotis subflavus* by having the skull almost straight in dorsal profile, whereas *P. subflavus* has a concave skull in the same profile, large ears, tricolored dorsal pelage, and long intermingled and bright hair (Dalquest, 1953b; Hall and Dalquest, 1950; Hall, 1981, Hayward and Cross, 1979).

Parastrellus hesperus. Scrubland. Photo: Scott Altenbach.

DISTRIBUTION: *P. hesperus* is found in western North America, from Washington to the center of Mexico through the central plateau to Hidalgo and Queretaro, and on the side of the Pacific to Guerrero (Jones et al., 1988; Ramírez-Pulido and Castro-Campillo, 1990, 1994). It has been registered in the states of BC, BS, CH, CO, DU, GR, HG, JA, MI, MO, MX, NL, QE, SI, SL, SO, TA, and ZA.

EXTERNAL MEASURES AND WEIGHT
TL = 84 to 95 mm; TV = 34 to 44 mm; HF = 7 to 10 mm; EAR = 11 to 14 mm; FA = 29 to 32 mm.
Weight: 3 to 6 g.

DENTAL FORMULA: I 2/3, C 1/1, PM 2/2, M 3/3= 34.

NATURAL HISTORY AND ECOLOGY: *P. hesperus* is one of the most abundant bats in arid areas of Mexico. Members of this species spend the day in cracks of crags and rocks (Twente, 1955), and thus they are rarely found within their shelters (Hayward and Cross, 1979). They are the first to emerge during the twilight. Their activity during the hours preceding the twilight during the autumn and winter has been well documented (Cockrum and Cross, 1964; Cross, 1965; Jones et al., 1965; Mumford et al., 1964; O'Farrel et al., 1967). It flies at a distance of 4 m to 10 m over the vegetation; its flight is slow and erratic. It is commonly observed flying over ponds or along the surface of water in search of food. They feed on insects (moths, beetles, and mosquitoes) hunted in flight (Ross, 1967). Its feeding habits make it a key pest controller. It reproduces in the late spring and early summer; most of the deliveries occur in the months of June and July (Asdell, 1964; Schmidly, 1977b). Two offspring are born after a gestation period of 90 days. O'Farrel (1972) described in detail the birth of this species; it can take between 45 and 120 minutes. Females occasionally form maternity colonies (Schmidly, 1977b). Possibly males are attracted by vocalizations made by females to carry out the copulation. Females held in captivity called wild males (O'Farrel, 1972). These are solitary bats, although sometimes they can be found forming colonies of a few animals. With regard to the longevity, only a record of a specimen recaptured after five years exists (Hayward and Cross, 1979). Its main predators are owls, some falcons, and sometimes other bats (Hayward and Cross, 1979; Orr, 1950). Ectoparasites, including mites, fleas, hemipterans, and dipterans, have been collected.

VEGETATIONAL ASSOCIATIONS AND ELEVATION RANGE: *P. hesperus* is restricted to desert and semi-desert environments with xeric vegetation, coniferous forest, oak, juniper, and riparian vegetation adjacent to arid areas (Hayward and Cross, 1979). It appears that the temperature is not a limiting factor in their distribution (O'Farrel and Bradley, 1970). The altitudinal range in which it is distributed goes from sea level (Anthony, 1925; Grinnel, 1933; Hayward and Cross, 1979; Jones, et al., 1965) to 2,800 masl (Burt, 1934).

CONSERVATION STATUS: There is no information on the situation of this species in Mexico, but their eating habits may be affected by an excessive use of insecticides.

Perimyotis subflavus (F. Cuvier, 1832)

Eastern pipistrelle bat

Osiris Gaona and Rodrigo A. Medellín

SUBSPECIES IN MEXICO
Perimyotis subflavus clarus R.H. Baker, 1954
Perimyotis subflavus veraecrucis (Ward, 1891)

DESCRIPTION: *Perimyotis subflavus* is distinguished by its tricolor pelage, dark at the base, whitish or yellowish in the middle, and dark in the tips (Barbour and Davis, 1969; Nason, 1948). It is characterized by its dorsal color that goes from a faint yellow-orange to a strong reddish-brown, and a faint yellow-orange to a dark mahogany on the abdomen (Davis, 1959). There are melanic individuals in which the typical colors of the bands are not so marked (Osgood, 1936; Trapido and Crowe, 1942).

DISTRIBUTION: *P. subflavus* is distributed from central Canada and United States to Guatemala and Honduras. Its distribution in Mexico comprises the southern part of Tamaulipas along a stripe from northeastern Puebla and all the coastal plain of the Gulf, as well as the state of Tamaulipas, the Yucatan Peninsula, and Campeche (Hall, 1981). It has been registered in the states of CA, CO, CS, HG, PU, TA, TB, TL, VE, and YU.

EXTERNAL MEASURES AND WEIGHT
TL=77 to 89 mm; TV=34 to 41 mm; HF=7.3 to 9.9 mm; EAR = 12.4 to 14.1 mm; FA = 31.4 to 34.1 mm.
Weight: 4.6 to 7.9 g.

DENTAL FORMULA: I 2/3, C 1/1, PM 2/2, M 3/3 = 34.

NATURAL HISTORY AND ECOLOGY: The roosting sites for hibernation and maternity are found in different locations (Griffin, 1934, 1936; Guthrie, 1933). During winter females are separate from the males (Griffin, 1940) and hibernate in caves, mines, and other structures made by man. They are often found associated with other species, including *Myotis occultus* and *Eptesicus fuscus* (Davis, 1964; Fenton, 1970; Folk, 1940). They select the deepest part of the caves where the environmental temperature is relatively constant (Hall, 1962; Hitchcock, 1949). Populations with sexual proportions biased toward males have been found during winter; this is attributed to the high survival range of males (Davis, 1959) and the difference of selection made by females and males (Jones and Pagels, 1968). It has a strong loyalty to its wintering sites (LaVal and LaVal, 1980). In spring they disperse from their wintering sites and females migrate to maternity caves. Maternity colonies are found more frequently in barns and in other structures made by man (Hoying, 1983; Jones and Suttkus, 1973). They also use trees, caves, and crevices in the rocks. During the period of maternity

Perimyotis subflavus. Photo: M.B. Fenton.

sexes are segregated and males are solitary. They can fly tens of miles in search of their wintering sites. Information on its longevity is that of a female captured in Illinois 14.8 years after having been marked (Walley and Jarvis, 1972). Higher mortality occurs in juveniles (Davis and Hitchcock, 1965). Males have a higher probability of survival. They perform migratory movements (Baker, 1978). A flaw in the reserves of fat is perhaps the cause of mortality in subadults during hibernation (Davis, 1966). It is likely that mortality is higher during the pre-flight period, when young flyers are unable to return to their refuge and are not found by their mothers (Hoying, 1983). Its predators include other mammals like *Microtus ochrogaster* and *Lasiurus cinereus*, the leopard frog (*Rana pipiens*), and birds of prey. High losses of the population occur with floods in caves and snowstorms. They commonly hunt their prey in places where there is water and in the edges of forests (Schmidly et al., 1977). They are slow and their flight is erratic and agitated while foraging (Hoying, 1983). They feed in the first hours of sunset. Their prey usually consist of small insects from 4 mm to 10 mm long, including beetles, cicada, mosquitoes, formicids, and moths (Ross, 1967; Whitaker, 1972). The incidence of infections of rabies is relatively low (Whitaker and Miller, 1974). Copulation and insemination of females occur in autumn (LaVal and LaVal, 1980) and ovulation in spring (Guthrie, 1933). The sperm is deposited in the uterus of the females who are in hibernation and stay there until ovulation in spring. There are only two implants in each uterine horn. The period of gestation, measured from implantation to birth, is 44 days after which 2 offspring generally are born (Wimsatt, 1945). Newborn twins weigh up to 58% of the weight of the mother; these means that the mass of the offspring constitute approximately one-third of the maternal mass. The period of births goes from early June to July, although this is variable. Young are born without hair and pink, with eyes closed (Lane, 1946). The newborns are capable of creating sounds that help their mother identify them. Young begin to fly at three weeks of age and acquire their flying and foraging skills a week later. They have a rapid postnatal growth, which is determined by changes in the forearm length; this happens during the first 18 days after the birth (Hoying, 1983). The complete closure of the epiphyses hollow in the fourth metacarpal happens at approximately 45 days of age (Hoying, 1983).

VEGETATIONAL ASSOCIATIONS AND ELEVATION RANGE: *P. subflavus* mainly inhabits tropical environments like shrubland, deciduous forest, and rainforest. Occasionally it is located in forests of oak and pine. It has been found from 400 masl to 1,200 masl (Hall, 1981).

CONSERVATION STATUS: *P. subflavus* is a relatively common species that is not considered at risk of extinction (IUCN, 1996; SEDESOL, 1994; SEMARNAT, 2002).

Rhogeessa aeneus Goodwin, 1958

Yucatan yellow bat

Joaquin Arroyo-Cabrales and Robert J. Baker

SUBSPECIES IN MEXICO

R. aeneus is a monotypic species.

It has been grouped in the complex *tumida-parvula* by Genoways and Baker (1996). It

Rhogeessa aeneus. Tropical dry forest. Merida, Yucatan. Photo: M.B. Fenton.

was considered part of *R. tumida*, but it was recently moved to a species level (Audet et al., 1993).

DESCRIPTION: *Rhogeessa aeneus* is a small-sized bat. Males have glands in the ears; females do not (sexual dimorphism) (Audet et al., 1993). It differs from *R. tumida* by its number of chromosomes (2n = 32 in *aeneus* versus 2n = 34 in *tumida*) and from *R. parvula* by its more reddish coloration (Goodwin, 1958).

EXTERNAL MEASURES AND WEIGHT
TL = 63 to 79 mm; TV = 26 to 36 mm; HF = 4 to 6 mm; EAR = 11 to 14 mm; FA = 25.7 to 29.2 mm.
Weight= 3 to 5 g.

DENTAL FORMULA: I 1/3, C 1/1, PM 1/2, M 3/3 = 30.

NATURAL HISTORY AND ECOLOGY: Little is known about the biology of this species. Two females, each with two embryos, were collected in May (Birney et al., 1974). The sounds made during the flight (echolocalization), are short and broadband (Audet et al., 1993). These bats have been collected in or near houses (Bowles et al., 1990).

VEGETATIONAL ASSOCIATIONS AND ELEVATION RANGE: *R. aeneus* inhabits tropical forests, especially rainforests, subdeciduous forests, and deciduous forests. It is known in localities below 50 masl.

CONSERVATION STATUS: The lack of a better understanding of the populations makes it hard to define the conservation status of *R. aeneus*. Its distribution restricted to the Yucatan Peninsula makes it vulnerable.

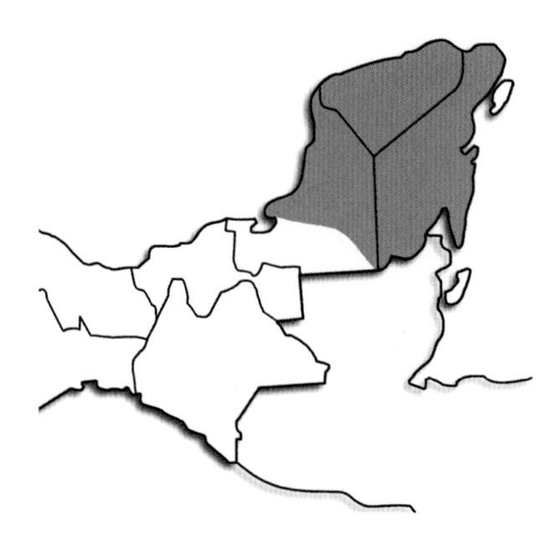

DISTRIBUTION: This species has Neotropical affinities and is distributed in the Yucatan Peninsula, Peten (Guatemala), and Belize. It has been registered in the states of CA, QR, and YU.

Rhogeessa alleni Thomas, 1892

Allen's yellow bat

Joaquin Arroyo-Cabrales and Robert J. Baker

SUBSPECIES IN MEXICO

R. alleni is a monotypic species.

Miller (1906) described a new genus of *Baeodon* bat, as the typical species of *R. alleni*, classified since then as *B. alleni*. In the taxonomic revision of the genus *Rhogeessa*, LaVal (1973b) confirms the differences between *Baeodon* and *Rhogeessa* at the subgenera level, and this is the taxonomic assignment currently being followed.

DISTRIBUTION: *R. alleni* is an endemic species of Mexico. Twenty-seven specimens from various localities of mountainous areas in western Mexico are known. It is found from the center of Oaxaca to the north of Jalisco; there is a record in the Mexican Altiplane (Santa Maria del Rio, San Luis Potosí; Polaco et al. 1992). It has been registered in the states of JA, MI, MO, OX, PU, SL, and ZA.

DESCRIPTION: *Rhogeessa alleni* is a small-sized bat; however, it is large within its genus and is distinguished by the third molar that is reduced. The hair is short and the ears are blackish (Álvarez and Aviña, 1965).

EXTERNAL MEASURES AND WEIGHT

TL = 81 to 90 mm; TV = 34 to 42 mm; HF = 6 to 7 mm; EAR = 14 to 15 mm; FA = 32.4 to 34.2 mm.

Weight: 5.8 to 8.0 g.

DENTAL FORMULA: I 1/3, C 1/1, PM 1/2, M 3/3 = 30.

NATURAL HISTORY AND ECOLOGY: There are limited data on the biology of this species (Matson and Baker, 1986; Watkins et al., 1972). Females have been collected in March, August, and November; no reproductive activity was observed (LaVal, 1973b; Sanchez-Hernandez et al., 1993). Four of the 27 specimens had tooth malformations (Polaco et al., 1992; Ramírez-Pulido and Müdespacher, 1987b); this could express the presence of a deleterious gene in the species.

VEGETATIONAL ASSOCIATIONS AND ELEVATION RANGE: *R. alleni* has been collected frequently in tropical deciduous forests, and also in thorny forests, coniferous forests, and xeric scrublands. It is known in localities that go from 125 masl to 1,990 masl; most of them are above 1,000 masl.

CONSERVATION STATUS: Only 27 specimens have been collected; therefore, much research is necessary to determine the conservation status of *R. alleni*. It is considered a rare species.

Rhogeessa genowaysi Baker, 1984

Genoways's yellow bat

Joaquin Arroyo-Cabrales and Robert J. Baker

SUBSPECIES IN MEXICO

R. genowaysi is a monotypic species.

Baker (1984) allocated this species to the complex *tumida-parvula* within the subgenus *Rhogeessa*.

DESCRIPTION: *Rhogeessa genowaysi* is a small-sized bat. It is distinguished from *R. tumida* only by its number of chromosomes (2n = 42 in *R. genowaysi* and 2n = 34 in *R. tumida*; Baker, 1984). Another useful feature in the recognition of this species is the size of the ears. which are less than 12 mm, although at times they are similar to those in *R. tumida*, ranging from 11 mm to 12 mm.

EXTERNAL MEASURES AND WEIGHT
TL = 37 to 50 mm; TV = 27 to 33 mm; HF = 5 to 8 mm; EAR = 10 to 12 mm; FA = 27.8 to 30.5 mm.
Weight: 3 to 5 g.

DENTAL FORMULA: I 1/3, C 1/1, PM 1/2, M 3/3 = 30.

NATURAL HISTORY AND ECOLOGY: Little is known about the habitats and habits of this species, with the exception of the collections made among trees in secondary tropical forests (Baker, 1984). In both locations where it inhabits, it has been collected together with the species *R. tumida*.

VEGETATIONAL ASSOCIATIONS AND ELEVATION RANGE: *R. genowaysi* has been collected in secondary bushes of the tropical subdeciduous forest. The typical location is a few meters above sea level.

CONSERVATION STATUS: The few specimens collected come from two localities very close together. This restricted distribution suggests that it is a rare species. In addition, the collection of this bat has only been made among trees; this indicates the need for the forest where the localities are to be protected to ensure the survival of the populations of this species. It is likely to be threatened with extinction.

DISTRIBUTION: This species is endemic to Chiapas. It is only known in two locations in the vicinity of Huixtla, separated by only a few kilometers. It has only been registered in the state of CS.

Rhogeessa gracilis Miller, 1897

Slender yellow bat

Joaquin Arroyo-Cabrales and Robert J. Baker

SUBSPECIES IN MEXICO
R. gracilis is a monotypic species.
LaVal (1973b) ranks this species within the subgenus *Rhogeessa* (H. Allen, 1866).

DESCRIPTION: *Rhogeessa gracilis* is a small-sized bat with long ears (nearly 18 mm), which makes it different from other species of the genus (Jones, 1977). Hairs with three defined bands constitute the dorsal pelage. The lingual cingulum surface is smooth and regular.

EXTERNAL MEASURES AND WEIGHT
TL = 84 to 89 mm; TV = 36 to 43 mm; HF = 6 to 8 mm; EAR = 17.5 to 18 mm; FA = 32.7 to 33.5.
Weight: 3 to 4 g.

DENTAL FORMULA: I 1/3, C 1/1, PM 1/2, M 3/3 = 30.

DISTRIBUTION: This species is Neotropical and endemic to Mexico. Sixteen specimens from 8 localities of mountainous areas of western Mexico are known; these come from central Oaxaca to northern Jalisco. It has been registered in the states of JA, MO, OX, and PU.

NATURAL HISTORY AND ECOLOGY: The biology of this species has not been studied. Some specimens have been collected in forests of pine and pine-oak, others in arid areas (Jones, 1977), and seven in very arid tropical forest (Sanchez-Hernandez et al., 1993). One female was captured in Jalisco in May with an embryo 17 mm long; two subadult females were captured in June (Jones, 1977).

VEGETATIONAL ASSOCIATIONS AND ELEVATION RANGE: The vegetation where *R. gracilis* has been collected is frequently conifer forests, but it has also been collected in xeric scrublands and deciduous rainforests. It has been found from 600 masl to 2,000 masl, although the majority of locations are above 1,000 masl.

CONSERVATION STATUS: Only six specimens have been collected. While this could indicate the fragility of the species, more studies are needed to determine the conservation status of the populations.

Rhogeessa mira LaVal, 1973

Least yellow bat

Joaquin Arroyo-Cabrales and Robert J. Baker

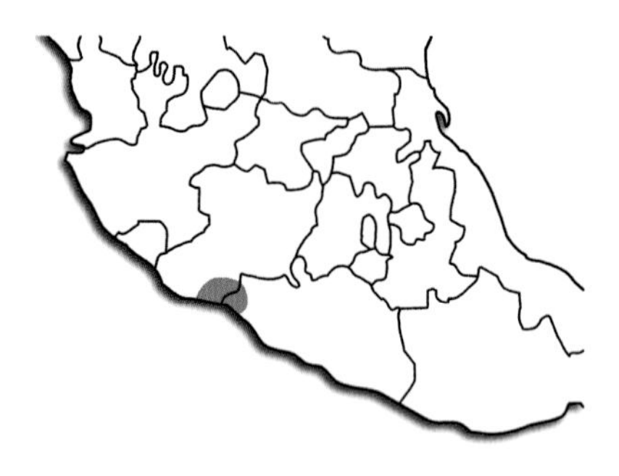

DISTRIBUTION: This species is endemic to the state. Only about 30 specimens from 4 locations of the area around the village El Infiernillo are known (Hall, 1981; Villalpando-R. and Arroyo-Cabrales, 1996). It has been registered in the state of MI.

SUBSPECIES IN MEXICO
R. mira is a monotypic species.
It was assigned to the subgenus *Rhogeessa* (H. Allen, 1866) by LaVal (1973b), while Baker et al. (1985) assigned it to the complex *tumida-parvula*.

DESCRIPTION: *Rhogeessa mira* is the smallest bat in Mexico. Its dorsal pelage is yellowish brown and the ears are obscure, contrasting sharply with the hair. The cingulum of the canine has soft edges. It differs from other species of the genus by its smaller size.

EXTERNAL MEASURES AND WEIGHT
TL =64 to 72 mm; TV = 26 to 33 mm; HF = 5 to 6 mm; EAR = 10 to 12 mm; FA = 25.0 to 27.8 mm.
Weight: 3 g.

DENTAL FORMULA: I 1/3, C 1/1, PM 1/2, M 3/3 = 30.

NATURAL HISTORY AND ECOLOGY: Little is known about the habitat and habits of this species, with the exception that it inhabits semi-arid areas dominated by mesquites (*Prosopis* sp.) and cacti (Álvarez and Aviña, 1965). In one of the localities (7 km north of El Infiernillo) it was found next to two other species of the genus, *R. alleni* and *R. parvula*, suggesting a distribution of food resources (ecological segregation), possibly based on the size of the prey (LaVal, 1973b).

VEGETATIONAL ASSOCIATIONS AND ELEVATION RANGE: The dominant vegetation in the town type is the xeric scrub. The altitude of only three of the towns are known, located at 125 masl and 340 masl.

CONSERVATION STATUS: The few specimens collected come from three locations very close together; more studies are needed to determine the conservation status of *R. mira.* Due to increasing environmental degradation in the Balsas River Basin, it should be considered fragile or vulnerable and the region where it is distributed should be protected.

Rhogeessa parvula H. Allen, 1866

Little yellow bat

Joaquin Arroyo-Cabrales and Robert J. Baker

SUBSPECIES IN MEXICO

Rhogeessa parvula major Goodwin, 1958
Rhogeessa parvula parvula H. Allen, 1866

This species was assigned to the subgenus *Rhogeessa* by LaVal (1973b) and to the complex *tumida-parvula* by Baker et al. (1985). Wilson (1991) noted the possibility that the populations in the north of the country in reality belong to a subspecies different to which they were currently assigned, *R. parvula major.*

DESCRIPTION: *Rhogeessa parvula* is a small bat (Ceballos and Miranda, 1986). It has short ears and a relatively hairy uropatagium, at least halfway between the knee and the leg; this feature distinguishes it from *R. tumida*, a sympatric species (LaVal, 1973b).

Rhogeessa parvula. Tropical dry forest. Chamela-Cuixmala Biosphere Reserve, Jalisco. Photo: Gerardo Ceballos.

EXTERNAL MEASURES AND WEIGHT
TL = 61 to 79 mm; TV = 21 to 32 mm; HF = 4 to 7 mm; EAR = 11 to 14 mm; FA = 26.2 to 29.8 mm.
Weight: 5 to 8 g.

DENTAL FORMULA: I 1/3, C 1/1, PM 1/2, M 3/3 = 30.

NATURAL HISTORY AND ECOLOGY: These bats are rapid flyers, and have been collected in caves, hollow trees, and cracks in rocks and on streams with running water (Villa-R., 1967); it has also been observed emerging from palm leaves and from a roof of a house. In the Marias Islands, they are especially visible because they fly before dusk in significant numbers (Wilson, 1991). Pregnant females have been collected at the end of February and in early June; there have been females with infants at the end of April and in early July; lactating females have been found from June to September. Females can carry one or two embryos (LaVal, 1973b).

DISTRIBUTION: This species is endemic to Mexico. The subspecies *R. parvula minor* is distributed across the coastal plains of the Pacific, in most of the west of the country, from Sonora to Oaxaca, including Durango, Zacatecas, and Morelos. The nominal subspecies is located in the Three Marias Islands. It has been registered in the states of CL, DU, GR, JA, MI, MO, MX, NY, OX, SI, SO, and ZA.

VEGETATIONAL ASSOCIATIONS AND ELEVATION RANGE: *R. parvula* has been collected mainly in thorny forests and tropical subdeciduous forests. They are found in localities ranging from sea level to 1,480 m; from Nayarit to the north, most of the localities are below 600 m.

CONSERVATION STATUS: A large number of specimens of this species has been collected, in particular, the population inhabiting the Marias Islands; however, in the majority of the cases it has been captured in islands; that is why its biology is poorly understood, and more studies are needed to determine the status of conservation of the species. The areas where it is distributed should be protected, in particular, the thorny and deciduous forests, because this will ensure the permanence of the environmental conditions where this species lives. This species inhabits the Chamela-Cuixmala Biosphere Reserve in Jalisco (Ceballos and Miranda, 2000).

Rhogeessa tumida H. Allen, 1866

Central American yellow bat

Joaquin Arroyo-Cabrales and Robert J. Baker

SUBSPECIES IN MEXICO
R. tumida is a monotypic species.
It was assigned to the subgenus Rhogeessa by LaVal (1973b) and to the complex *tumida-parvula* by Baker et al. (1985).

DESCRIPTION: *Rhogeessa tumida* is a small-sized bat with short ears and with hair on the base of the uropatagium. The dorsal pelage is usually dark with two bands; the lingual cingulum of the upper canine has cusps, which distinguishes it from other species of the genus in which the canine is smooth.

EXTERNAL MEASURES AND WEIGHT
TL = 70 to 86 mm; TV = 27 to 35 mm; HF = 6.4 to 8 mm; EAR = to 12.5 14.2 mm; FA = 28.3 to 31.5 mm.
Weight: 3 to 5 g.

DENTAL FORMULA: I 1/3, C 1/1, PM 1/2, M 3/3 = 30.

NATURAL HISTORY AND ECOLOGY: *R. tumida* has been collected with mist nets placed on wells and streams; they feed on insects and roost in tree branches (Hall and Dalquest, 1963). Based on data from the majority of the available specimens, LaVal (1973b) suggested that pregnancy, birth, and breastfeeding occur from mid-February to mid-July; some specimens had more than one embryo. Their reproductive cycle is monoestral synchronous; they reach sexual maturity at the end of their first year and females give birth twice a year (Baker et al., 1985). A specimen from Veracruz was positive for rabies (Villa-R., 1967).

DISTRIBUTION: This species has Neotropical affinities, and is distributed from northeastern Mexico, through Central America, to the south of Brazil, Bolivia, and Ecuador. In Mexico, it is distributed on the coastal slope of the Gulf and the south of the country, except in the Yucatan Peninsula. It has been registered in the states of CS, HG, OX, QE, SL, TA, TB, and VE.

VEGETATIONAL ASSOCIATIONS AND ELEVATION RANGE: *R. tumida* has been collected in tropical forests, in particular from subdeciduous forests. It is distributed from sea level to 1,800 masl, although most locations are lower than 500 masl.

CONSERVATION STATUS: This species has a variation in the number of chromosomes (Baker et al., 1985), which has led to it being distinguished from a similar species (*R. genowaysi*), as well as *R. aeneus*, which was previously known as subspecies of *R. tumida*, being classified as a separate species. In addition, there are other genetic variants that might classify it as a different species; the taxonomy of the group has not been solved completely. The conservation status of *R. tumida* is difficult to define and will require more studies in order to have a thorough knowledge on the systematic of these bats. In his analysis on Neotropical bats, Arita (1993a, b) considers this species as locally rare but widely distributed.

Rhogeessa tumida. Tropical dry forest. Ticul, Yucatan. Photo: Héctor T. Arita.

Family Antrozoidae

The family Antrozoidae is represented by two genera (*Antrozous* and *Bauerus*), which until recently were considered part of the family Vespertilionidae (Simmons, 1998). It is an endemic family of North America, with a distribution that runs from the south of the United States to Costa Rica.

Antrozous pallidus (Le Conte, 1856)

Pallid bat

Jorge Ortega R.

SUBSPECIES IN MEXICO
Antrozous pallidus minor Miller, 1902
Antrozous pallidus packardi Martin and Schmidly, 1982
Antrozous pallidus pallidus (Le Conte, 1856)

DISTRIBUTION: *A. pallidus* is distributed from British Columbia in Canada to the highlands of Mexico. Its distribution includes the western states of the United States, northern states of Mexico, and central plateau of Mexico (González-Ruiz and Villalpando-R., 1997; Martin and Schmidly, 1982; Hermanson and O'Shea, 1983). It has been registered in the states of BC, BS, CO, CH, DU, HG, JA, MI, NL, QE, SL, SO, TA, and ZA.

DESCRIPTION: *Antrozous pallidus* is a relatively large bat. The snout is wide and short and is equipped with a series of sebaceous glands that secrete a musk-like smell (Nowak, 1999a). The ears are big and prominent with a lanceolate tragus that extends beyond the first half of the pinna (Hermanson and O'Shea, 1983). Its dorsal pelage is brown and the fur is almost white on the abdomen. The coloration varies according to a geographic gradient, with the northern and subspecies of Baja California more obscure than those in the central portion of Mexico (Martin and Schmidly, 1982). The tail is fully included in the uropatagium, which extends beyond the lower limbs. *A. pallidus* differs from *Bauerus dubiaquercus*, by its larger size and lighter color; it also has a clear division between the incisors that is not visible in *B. dubiaquercus* (Van Gelder, 1959b).

EXTERNAL MEASURES AND WEIGHT
TL = 92 to 135 mm; TV = 35 to 53 mm; HF = 11 to 16 mm; EAR = 21 to 37 mm; FA = 45 to 60 mm.
Weight: 13 to 28 g.

DENTAL FORMULA: I 1/2, C 1/1, PM 1/2, M 3/3 = 28.

NATURAL HISTORY AND ECOLOGY: *A. pallidus* consumes mainly insects of large size (20 mm to 70 mm), caught on the ground or on the low parts of vegetation. It feeds on beetles, grasshoppers, and scorpions. It also captures some vertebrates such as small lizards. It has been found feeding on the inflorescences of *Agave*, so it is considered an eventual nectarivore (Barbour and Davis, 1969; Black, 1974; Herrera et al., 1993). Its activity starts at night and its foraging area comprises approximately 3 km^2 (Bell, 1982). Apparently, it returns several times during the night to its roosting site to mark its place and to protect its partner (Vaughan and O'Shea, 1976). Mating occurs from October to December; the sperm stay in the female until the beginning of spring. Births occur from May to July. The offspring are born after nine months of pregnancy and are weaned six weeks after birth (Cockrum, 1955). This species is seasonal monoestric, and have one young per birth, but there are reports of up to two offspring for multiparous females (Davis, 1969; Orr, 1954). They are gregarious and form colonies of up to 200 individuals. During the summer mainly females and

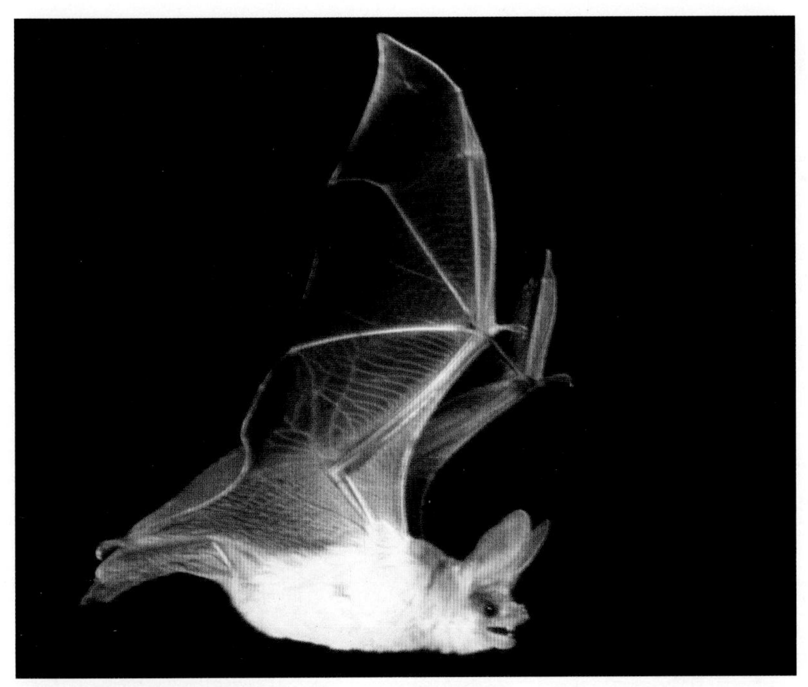

Antrozous pallidus. Scrubland. Photo: Scott Altenbach.

young form certain colonies, although in winter these bats form mixed groups that are inactive and presumably hibernate (O'Shea and Vaughan, 1977; Twente, 1955). The main roosting sites are caves, tunnels, cracks, and constructions. Occasionally, they share roosting sites with other species as *Myotis velifer* and *Tadarida brasiliensis* (Hermanson and O'Shea, 1982). Its main predators are birds of prey (Baker, 1953; López-Forment and Urbano, 1977).

VEGETATIONAL ASSOCIATIONS AND ELEVATION RANGE: *A. pallidus* is common in areas of xeric vegetation and grasslands. In the United States it has been collected in habitats with karstic or calcareous rocks (Hermanson and O'Shea, 1983). It is common in lowlands (500 masl), although it has been collected in sites up to 2,440 masl (Black, 1974).

CONSERVATION STATUS: The current status of the pallid bat's populations is unknown, although it is a relatively common species of arid areas of Mexico.

Bauerus dubiaquercus (Van Gelder, 1959)

Van Gelder's bat

Julio Juarez-G.

SUBSPECIES IN MEXICO

B. dubiaquercus is a monotypic species.

Bauerus dubiaquercus. Tropical rainforest. Photo: M.B. Fenton.

DISTRIBUTION: *B. dubiaquercus* has been reported in only a few localities. In Mexico isolated records exist in several states of the Pacific, from Nayarit to Chiapas (Briones, 1998; Hall, 1981; Hernández-Huerta et al., 2000). In Central America it is known in Costa Rica, Honduras, Guatemala, and Belize. It has been registered in the states of CA, CS, GR, JA, NY, OX, QR, and VE.

DESCRIPTION: *Bauerus dubiaquercus* is a medium-sized bat with a bloated and slightly dark face. The ears are long with the apex slightly rounded. The tail is long and is fully included in the interfemoral membrane. The calcar is well developed and has a small, keel-shaped extension. The pelage is thick, fine, and medium in size. The dorsal coloration is clay and the belly is lighter.

EXTERNAL MEASURES AND WEIGHT
TL = 110 to 120 mm; TV = 45 to 54 mm; HF = 10.5 to 12 mm; EAR = 22 to 24 mm; FA = 50.5 to 55.8 mm.
Weight: 15 g.

DENTAL FORMULA: I 1/3, C 1/1, PM 1/2, M 3/3 = 30.

NATURAL HISTORY AND ECOLOGY: This species has solitary habits. Its roosting sites are unknown. It has been speculated that its feeding strategy is exclusively aerial. Data on reproduction suggest that it only gives birth to one young. Reproduction is carried out between May and July.

VEGETATIONAL ASSOCIATIONS AND ELEVATION RANGE: *B. dubiaquercus* has been found in the high jungle, median subdeciduous jungle, forest of pine-oak, tropical jungle, and cloud forest. It is distributed from sea level to 2,350 masl.

CONSERVATION STATUS: *B. dubiaquercus* is a rare species and its conservation status is unknown. It is not included in the official list of species at risk (SEDESOL, 1994; SEMARNAT, 2002).

Appendix A

Tracks of Some Mammal Species of Mexico

Marcelo Aranda

In this section we present illustrations of the tracks of medium-size and large mammals from Mexico. This section can be complemented by information in Aranda and March (1987) and Emmons and Feer (1997).

Tracks are not shown to scale. See species accounts for size of hind foot.

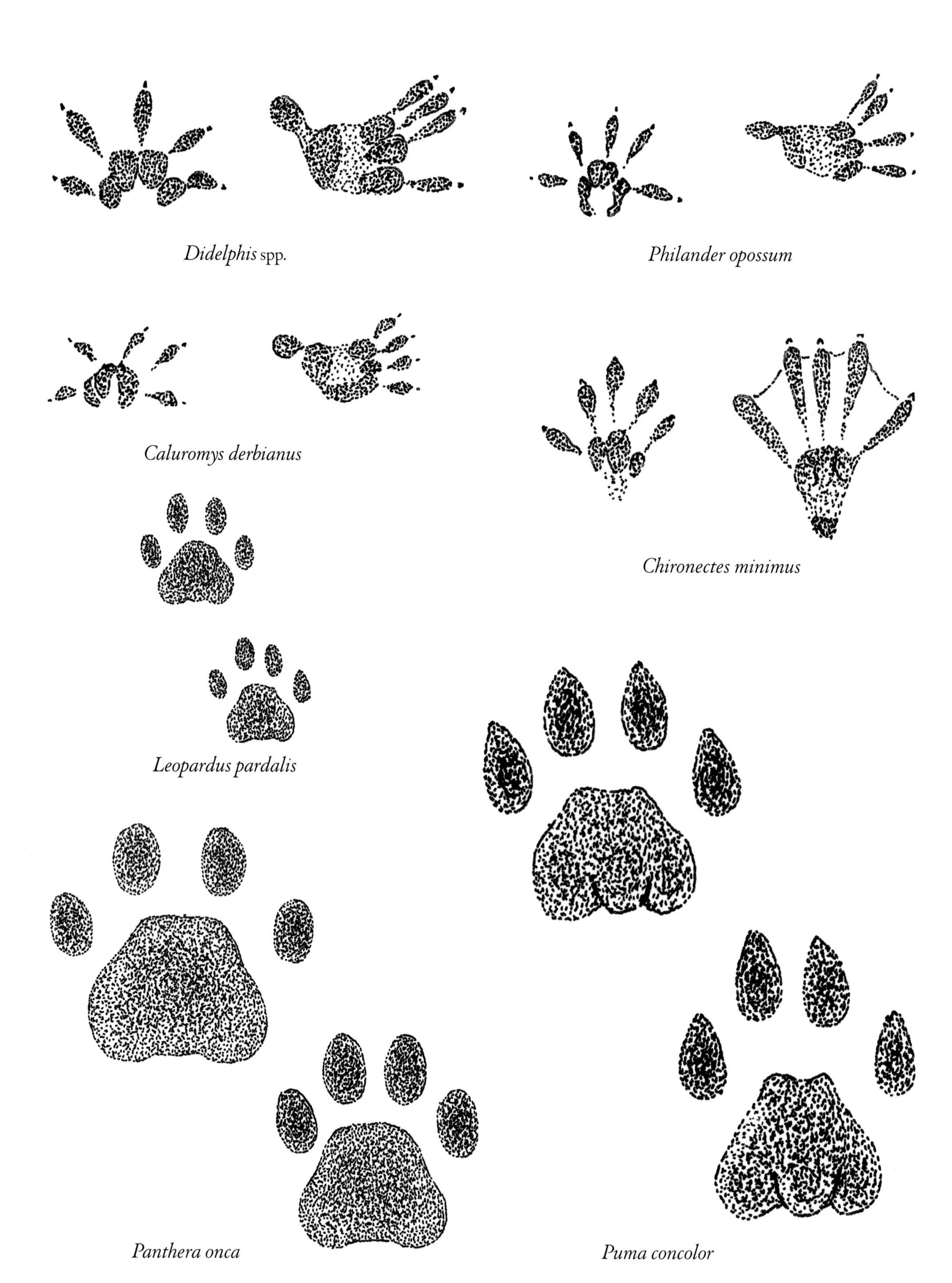
Didelphis spp.
Philander opossum
Caluromys derbianus
Chironectes minimus
Leopardus pardalis
Panthera onca
Puma concolor

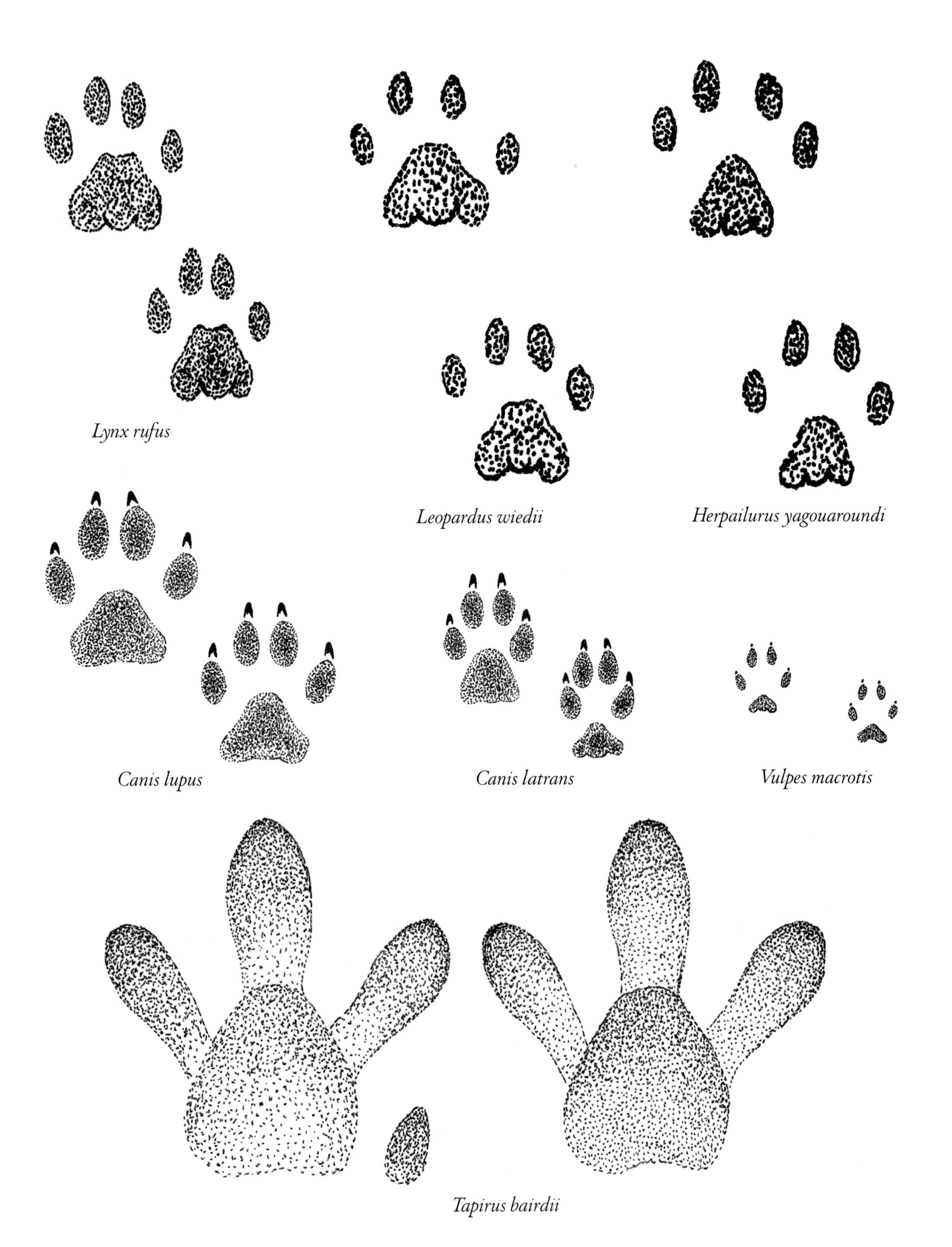
Lynx rufus
Leopardus wiedii
Herpailurus yagouaroundi
Canis lupus
Canis latrans
Vulpes macrotis
Tapirus bairdii

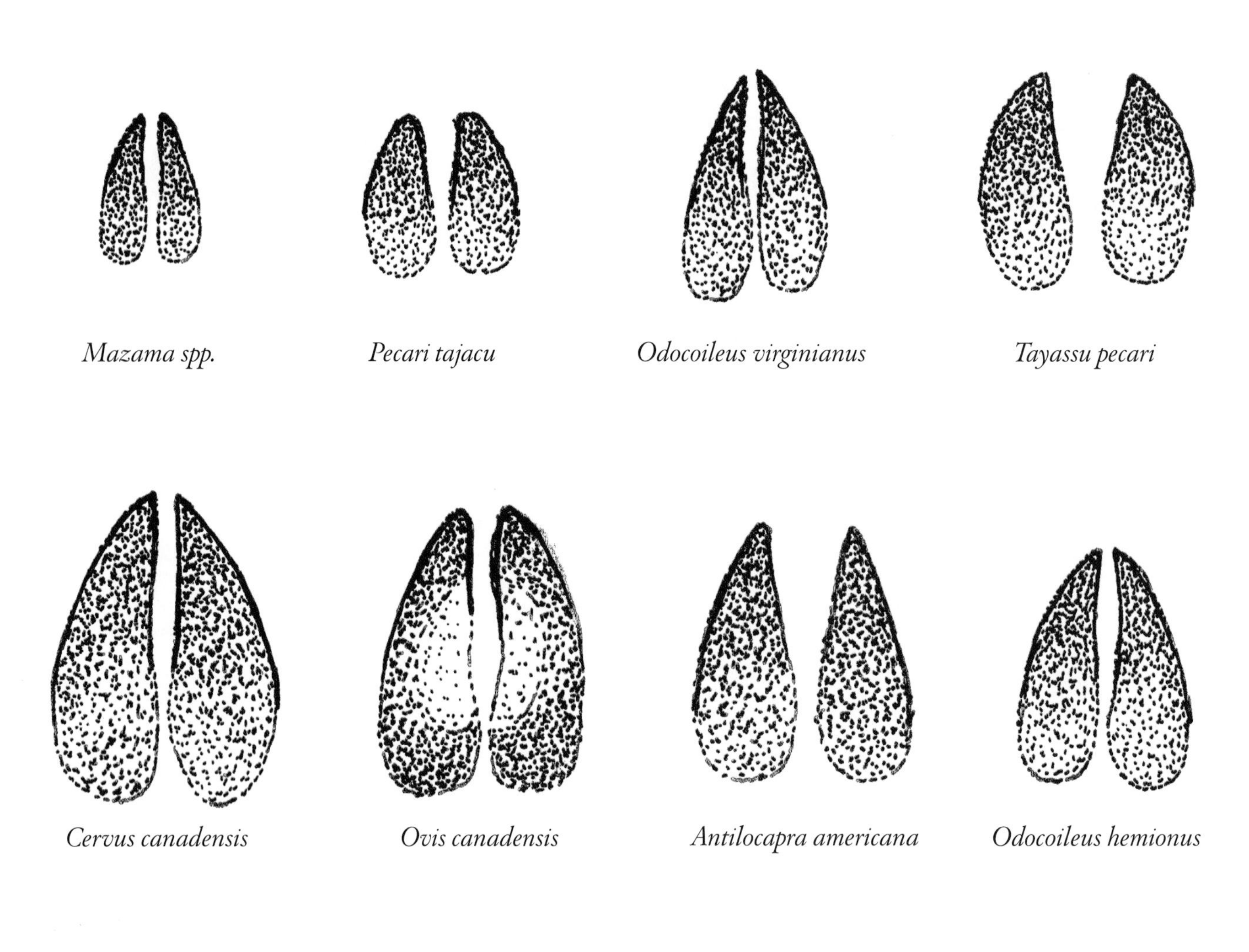
Mazama spp.
Pecari tajacu
Odocoileus virginianus
Tayassu pecari
Cervus canadensis
Ovis canadensis
Antilocapra americana
Odocoileus hemionus

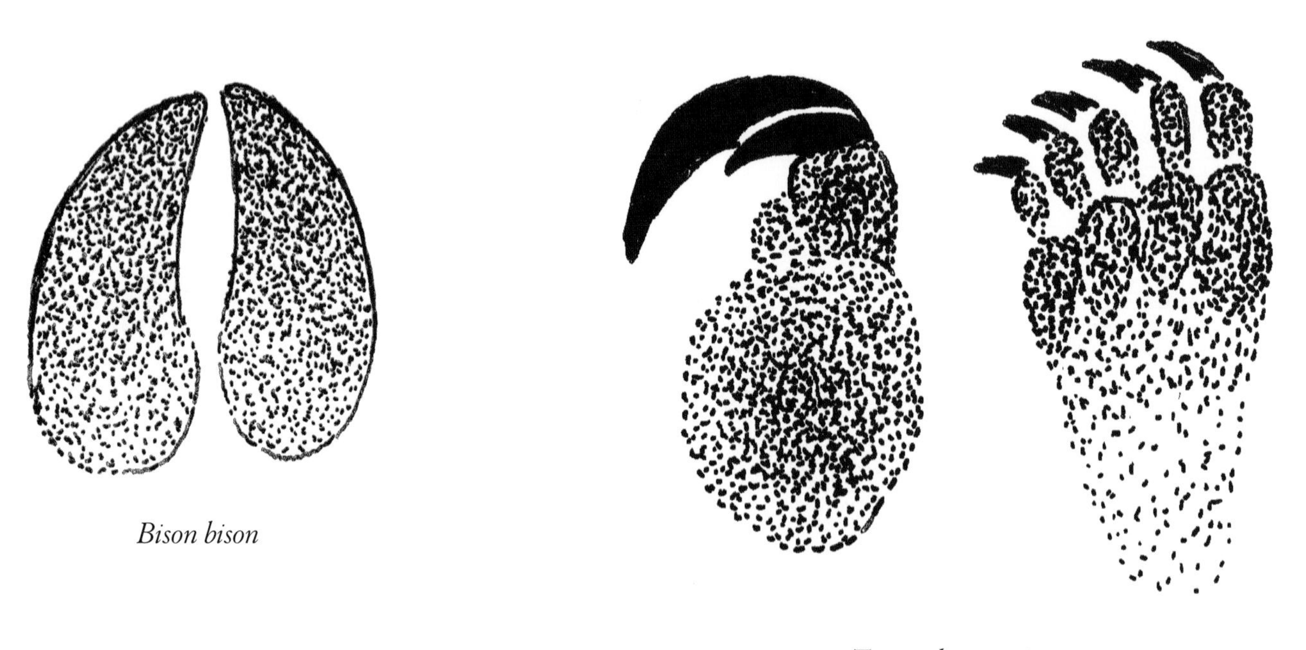
Bison bison
Tamandua mexicana

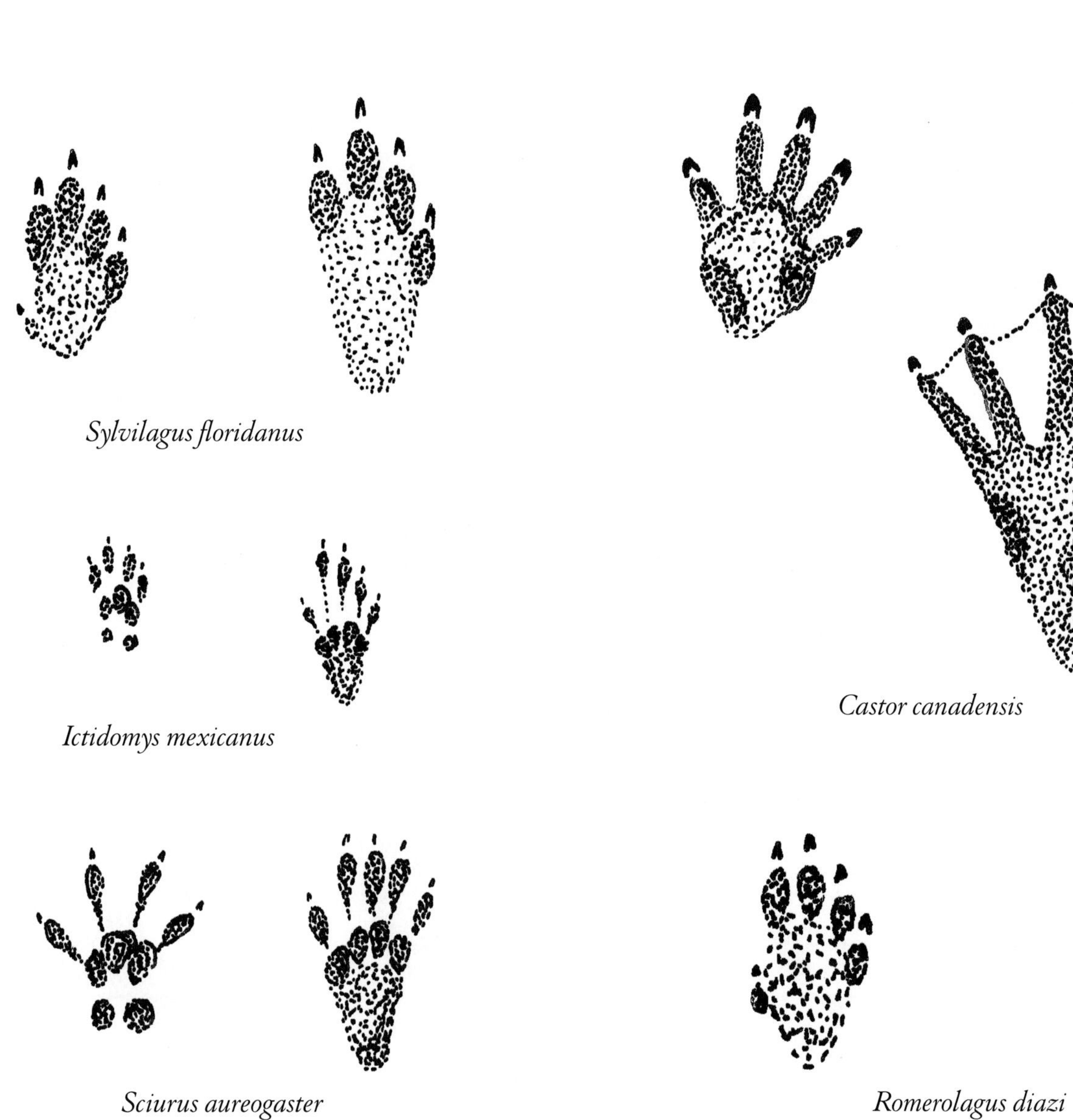

Sylvilagus floridanus

Castor canadensis

Ictidomys mexicanus

Sciurus aureogaster

Romerolagus diazi

Dasypus novemcinctus

Otospermophilus variegatus

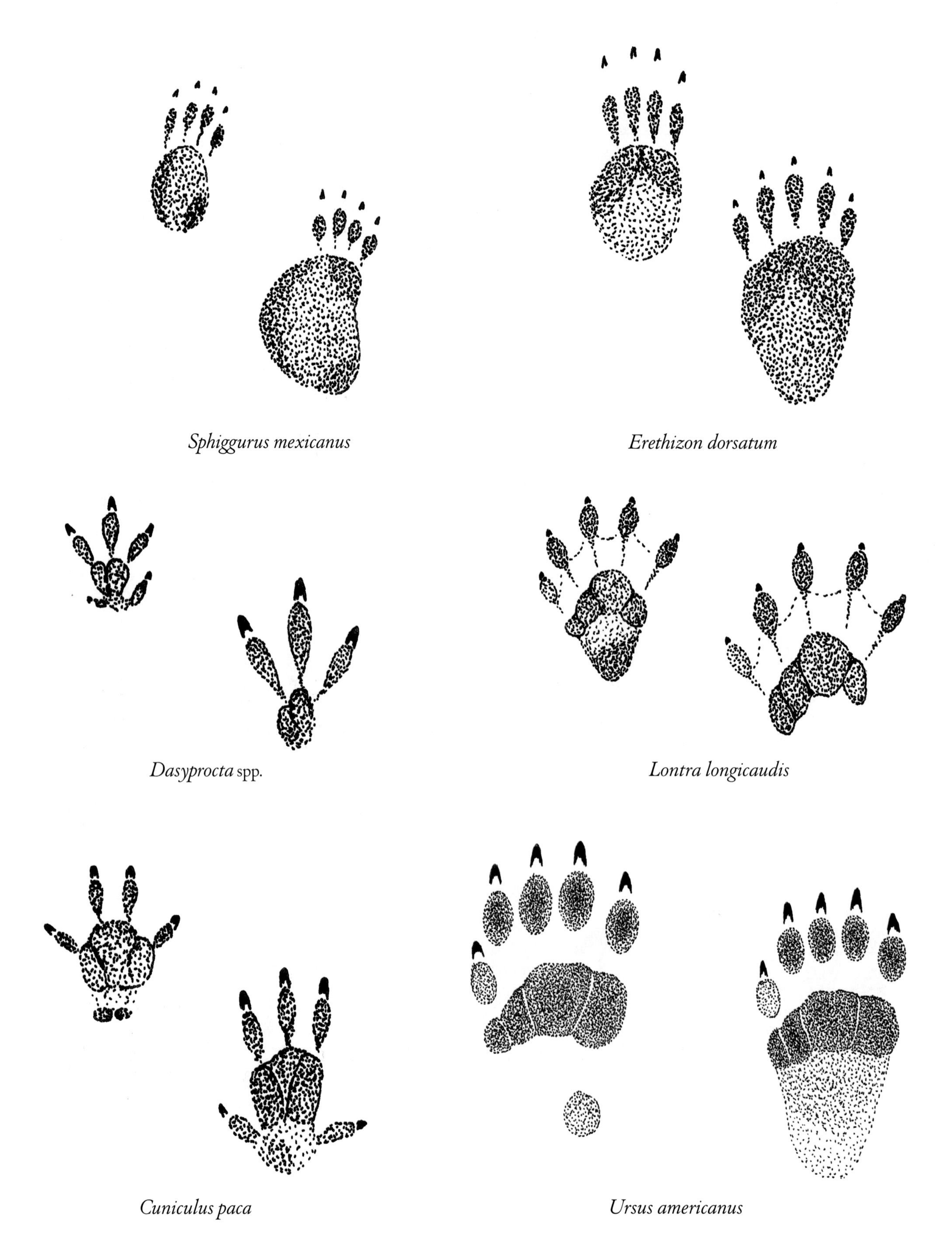
Sphiggurus mexicanus
Erethizon dorsatum
Dasyprocta spp.
Lontra longicaudis
Cuniculus paca
Ursus americanus

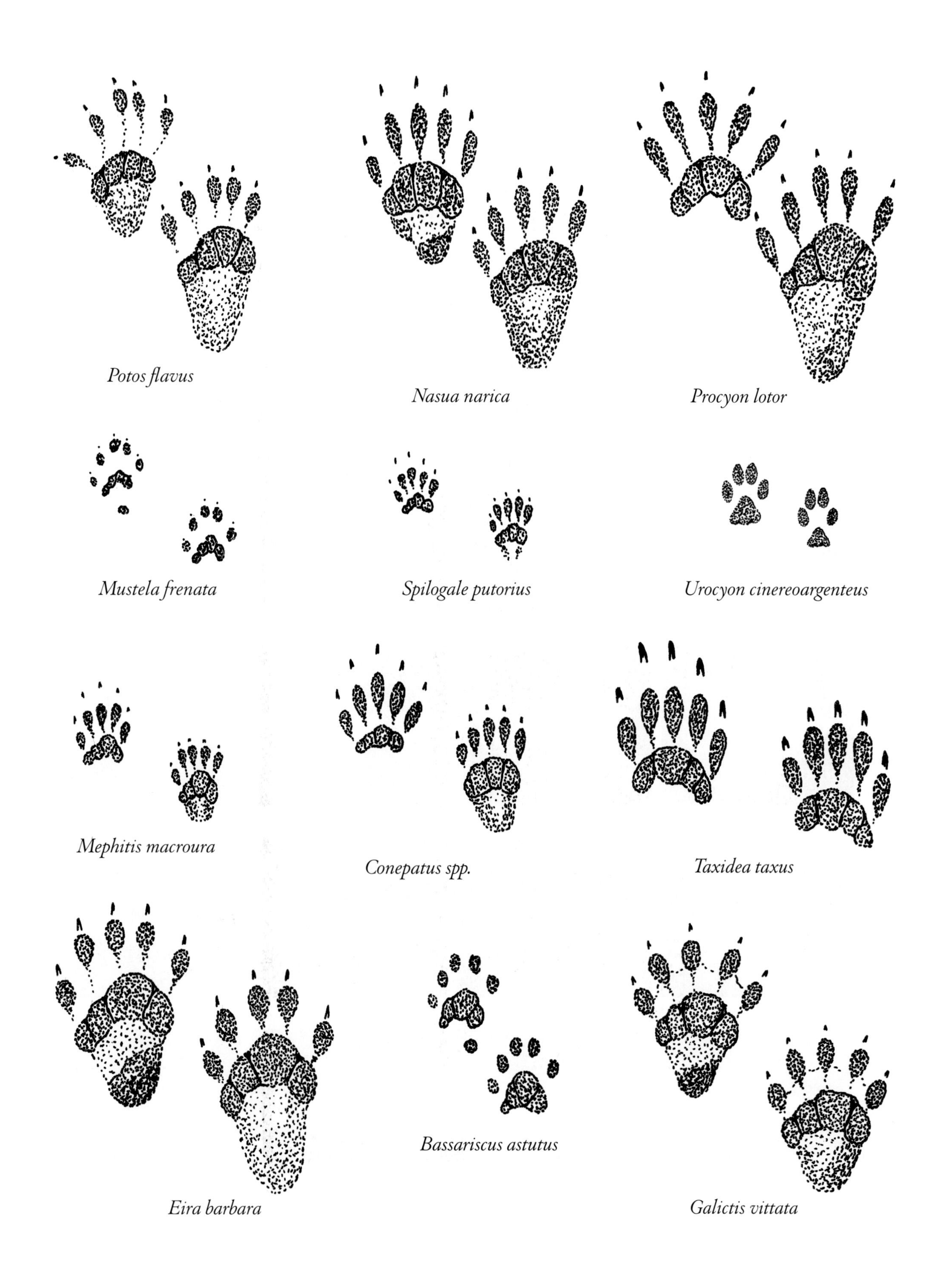
Potos flavus
Nasua narica
Procyon lotor
Mustela frenata
Spilogale putorius
Urocyon cinereoargenteus
Mephitis macroura
Conepatus spp.
Taxidea taxus
Eira barbara
Bassariscus astutus
Galictis vittata

Appendix B

Skulls of Mammal Genera from Mexico

Gerardo Ceballos, Ana Isabel Bieler, and Gisselle Oliva

In this section we present a photographic catalogue of most of the mammal genera from Mexico. The section can be complemented using the work of Goodwin (1969) and Ceballos and Miranda (1986, 2000). For each genus dorsal, ventral, and lateral views of the skull and the mandible are presented, although for some genera the mandible is missing. The number next to the species name indicates a scale in centimeters. Ana Isabel Bieler took all the photos. The specimens for the photos were requested from either the National Mammals Collection of the Instituto de Biologia or the Museo de Zoologia of the Facultad de Ciencias, from the Universidad Nacional Autonoma de Mexico.

ORDER DIDELPHIMORPHIA

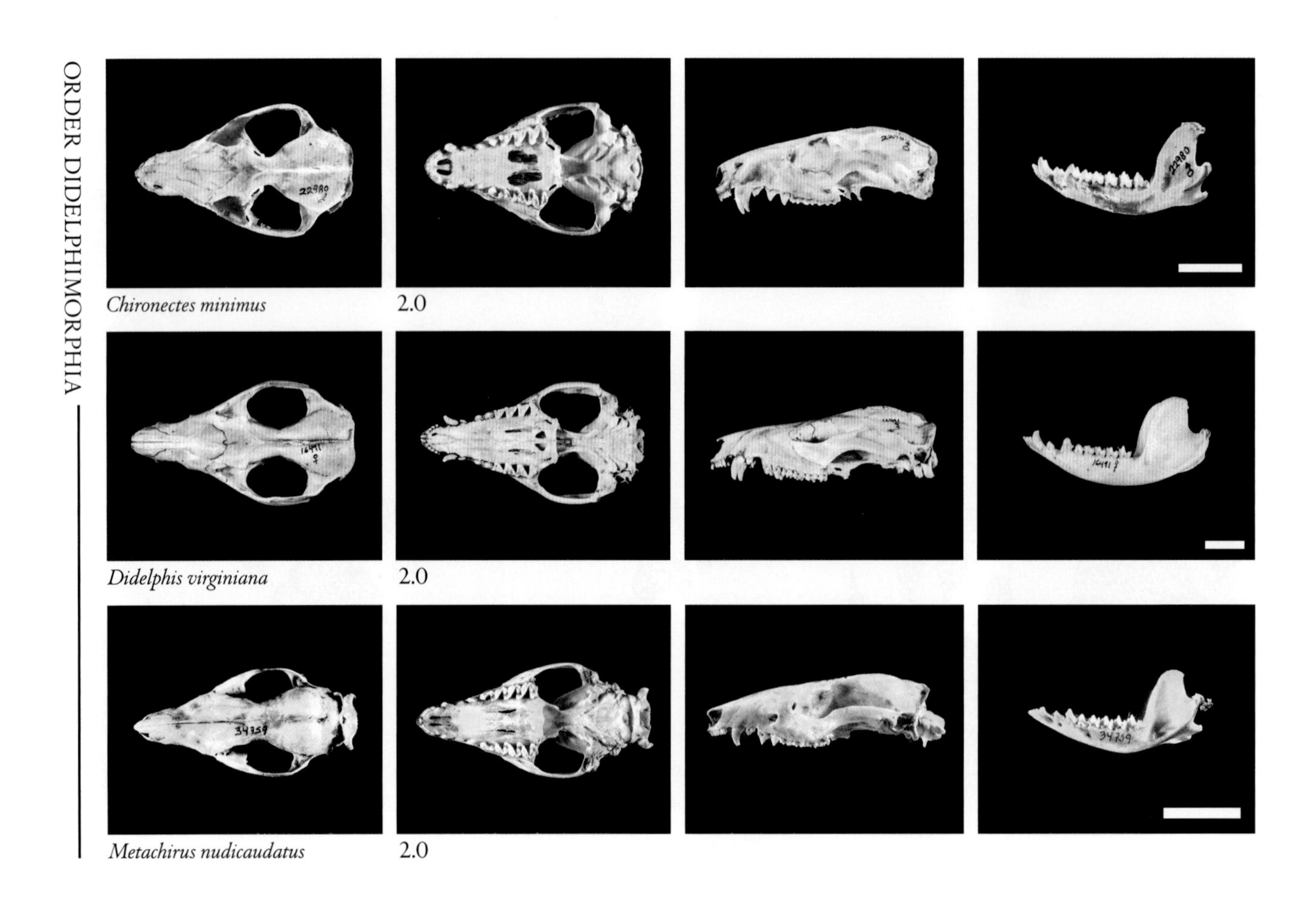

Chironectes minimus 2.0

Didelphis virginiana 2.0

Metachirus nudicaudatus 2.0

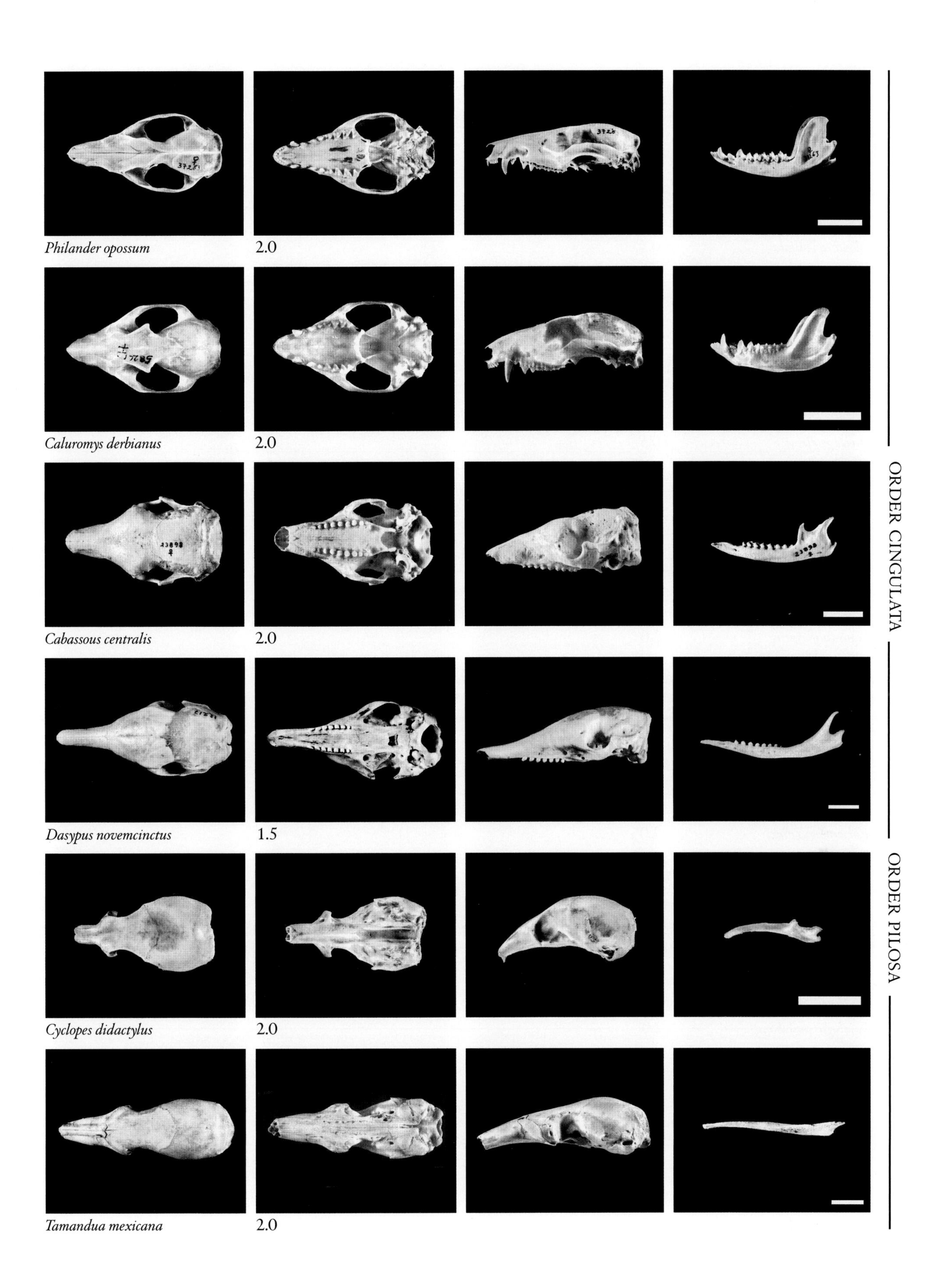

Philander opossum 2.0

Caluromys derbianus 2.0

Cabassous centralis 2.0

Dasypus novemcinctus 1.5

Cyclopes didactylus 2.0

Tamandua mexicana 2.0

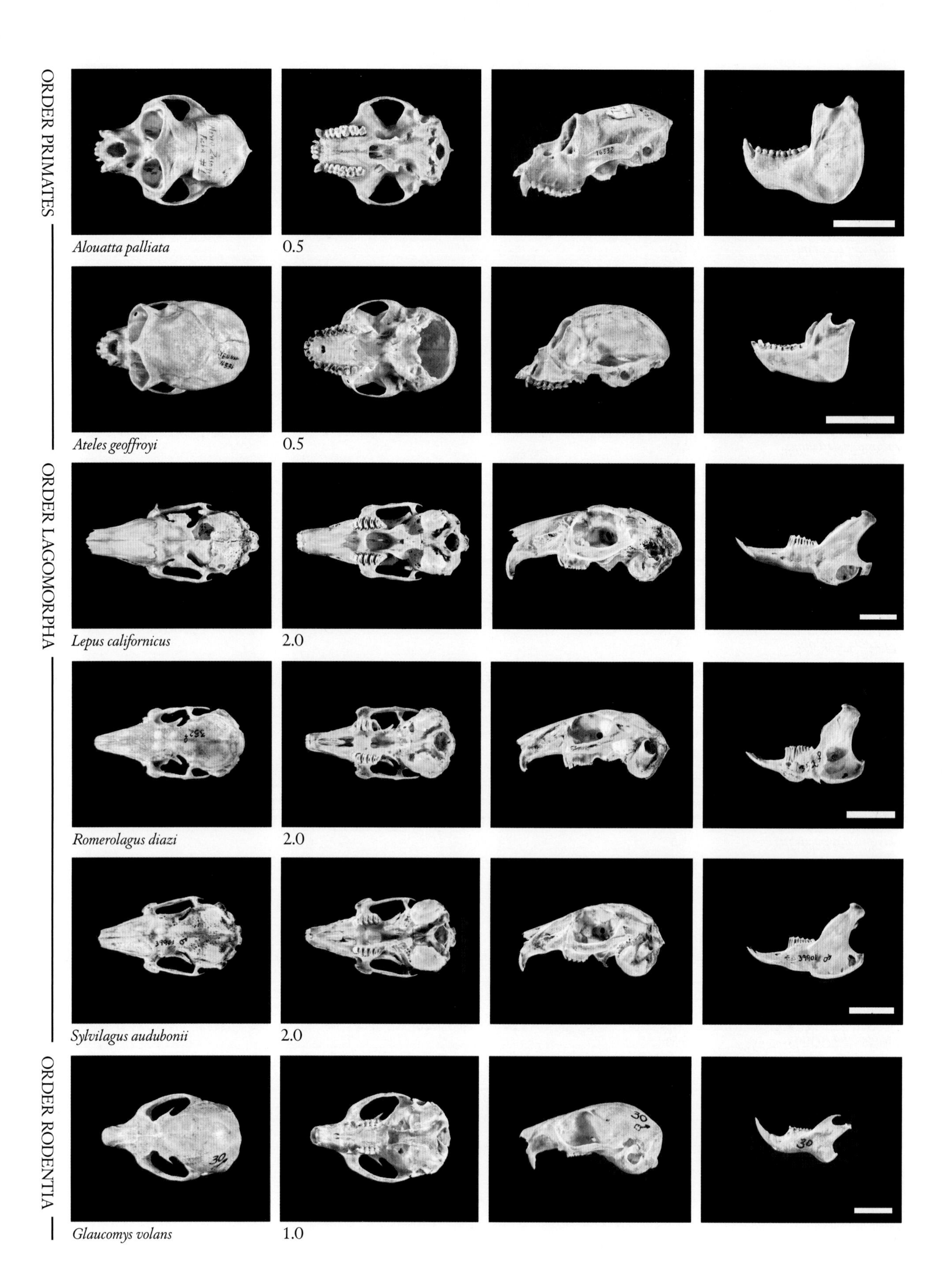

Alouatta palliata 0.5

Ateles geoffroyi 0.5

Lepus californicus 2.0

Romerolagus diazi 2.0

Sylvilagus audubonii 2.0

Glaucomys volans 1.0

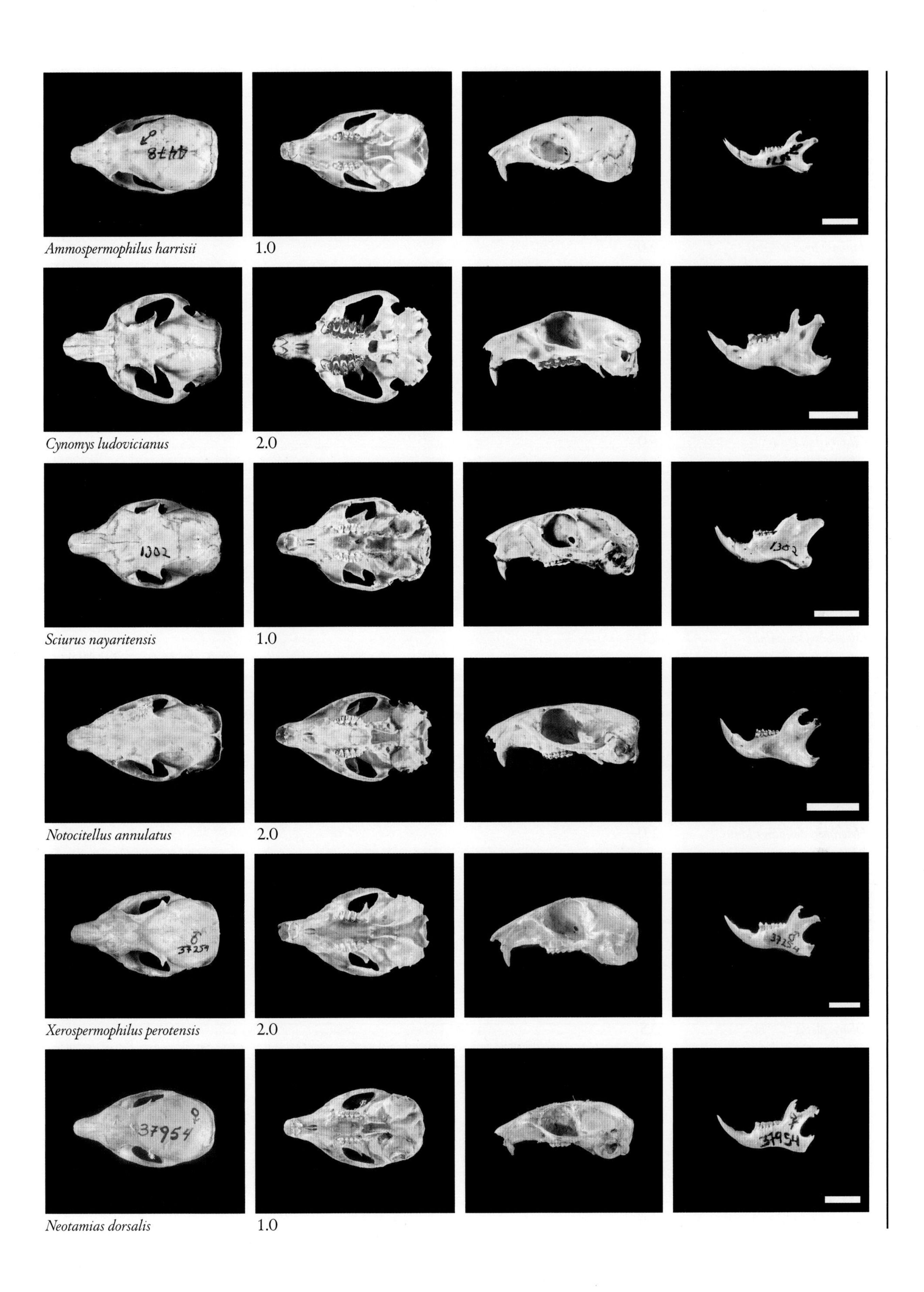

Ammospermophilus harrisii 1.0

Cynomys ludovicianus 2.0

Sciurus nayaritensis 1.0

Notocitellus annulatus 2.0

Xerospermophilus perotensis 2.0

Neotamias dorsalis 1.0

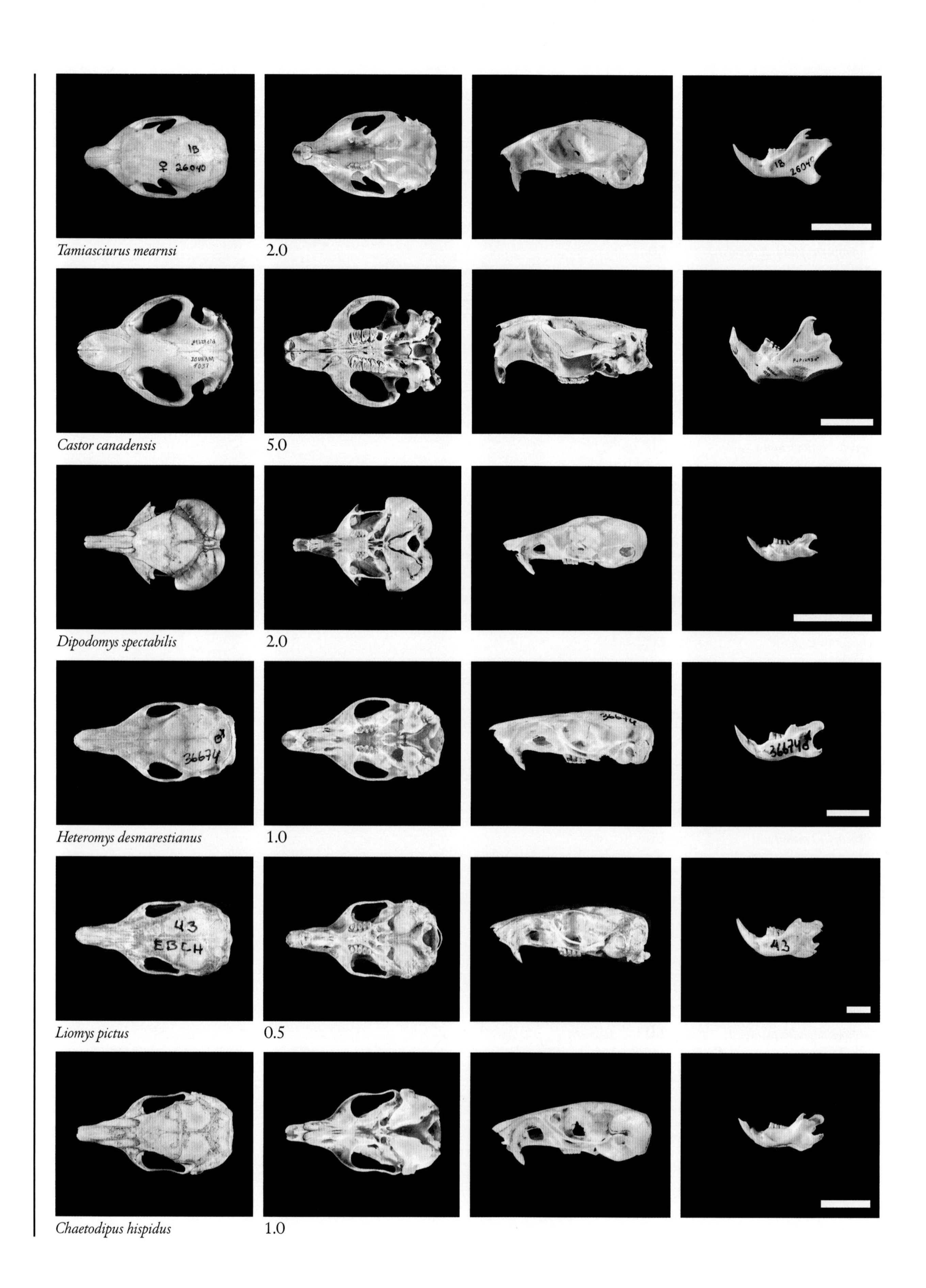

Tamiasciurus mearnsi 2.0

Castor canadensis 5.0

Dipodomys spectabilis 2.0

Heteromys desmarestianus 1.0

Liomys pictus 0.5

Chaetodipus hispidus 1.0

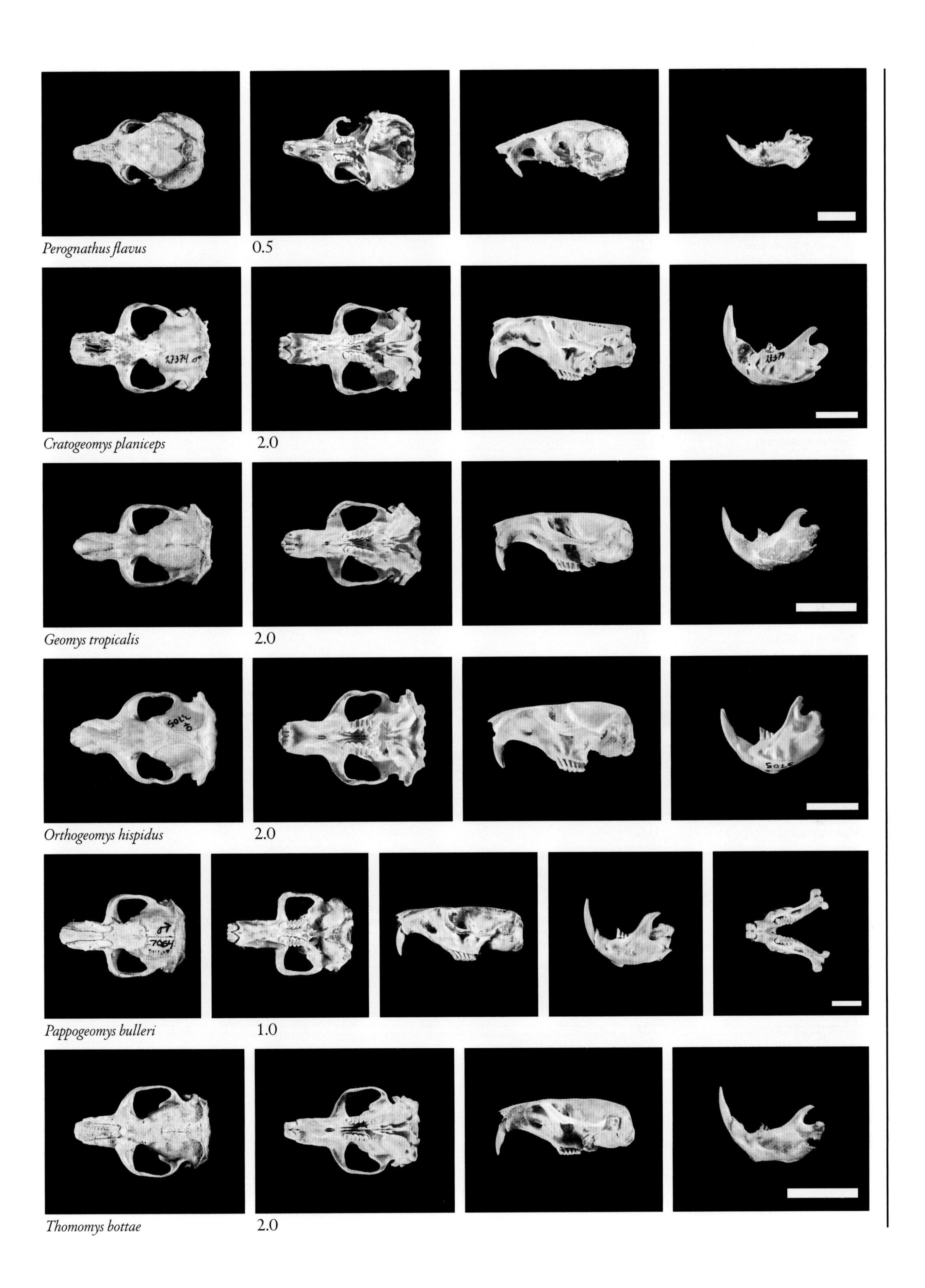

Perognathus flavus 0.5

Cratogeomys planiceps 2.0

Geomys tropicalis 2.0

Orthogeomys hispidus 2.0

Pappogeomys bulleri 1.0

Thomomys bottae 2.0

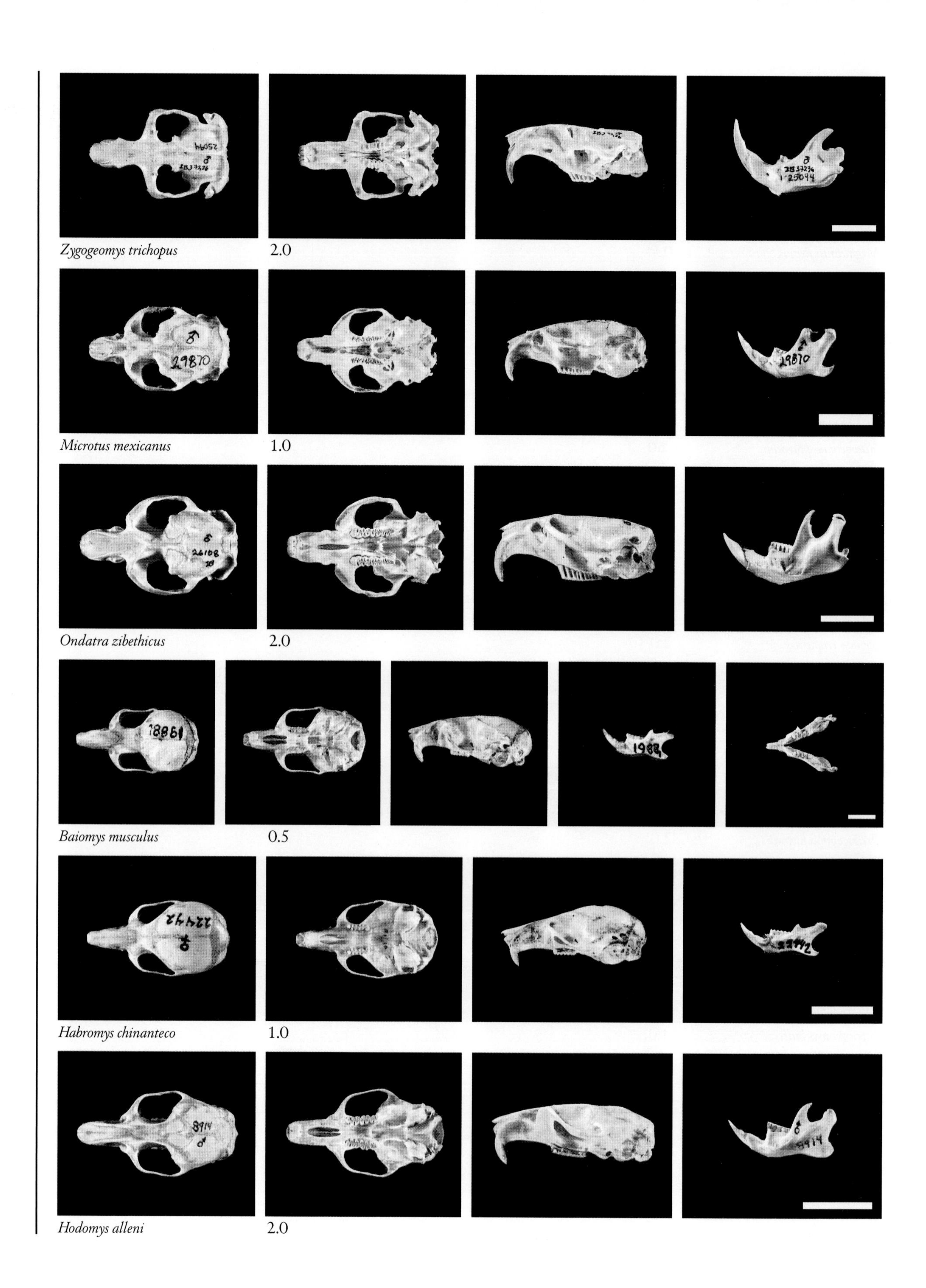

Zygogeomys trichopus 2.0

Microtus mexicanus 1.0

Ondatra zibethicus 2.0

Baiomys musculus 0.5

Habromys chinanteco 1.0

Hodomys alleni 2.0

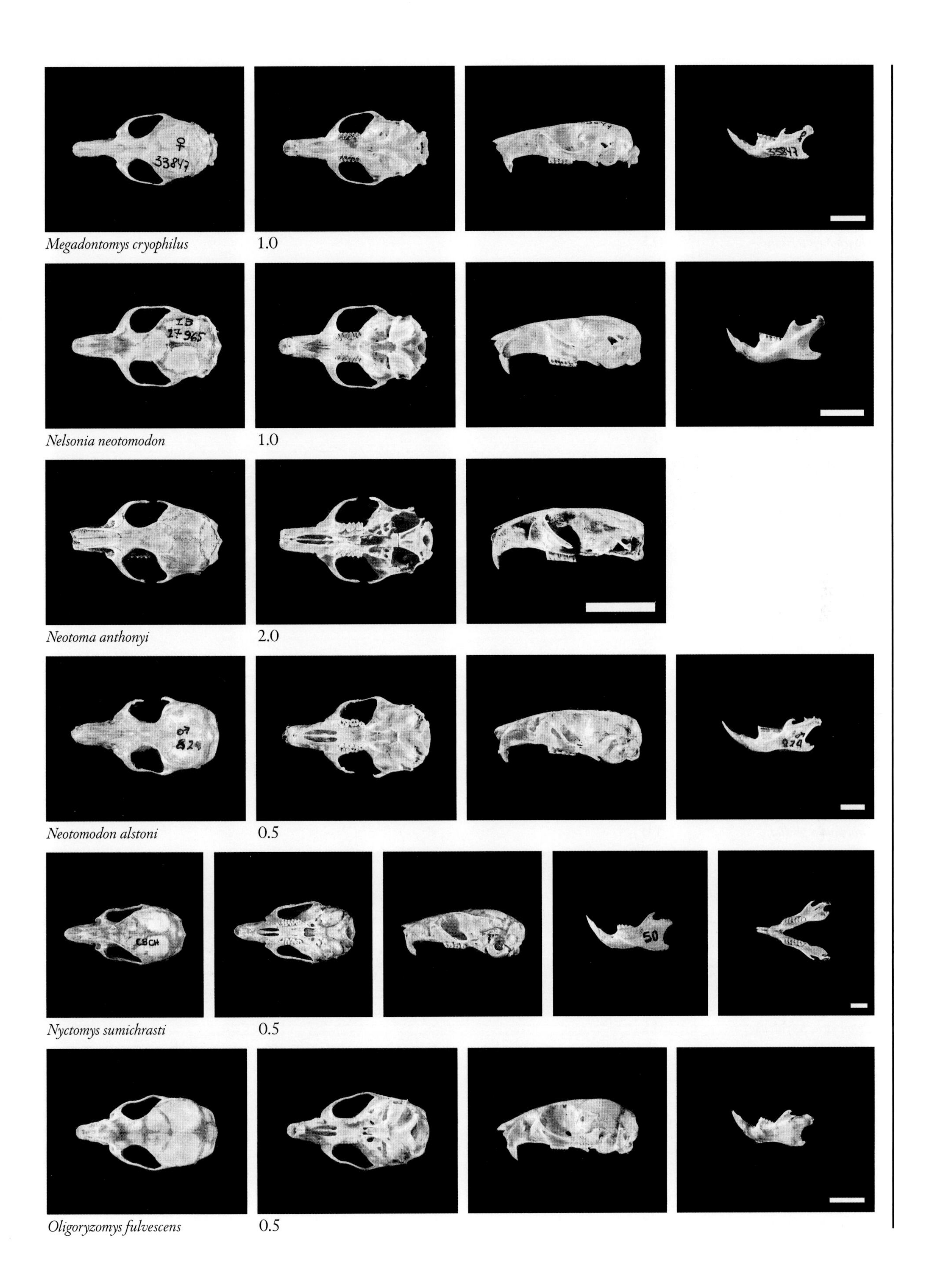

Megadontomys cryophilus 1.0

Nelsonia neotomodon 1.0

Neotoma anthonyi 2.0

Neotomodon alstoni 0.5

Nyctomys sumichrasti 0.5

Oligoryzomys fulvescens 0.5

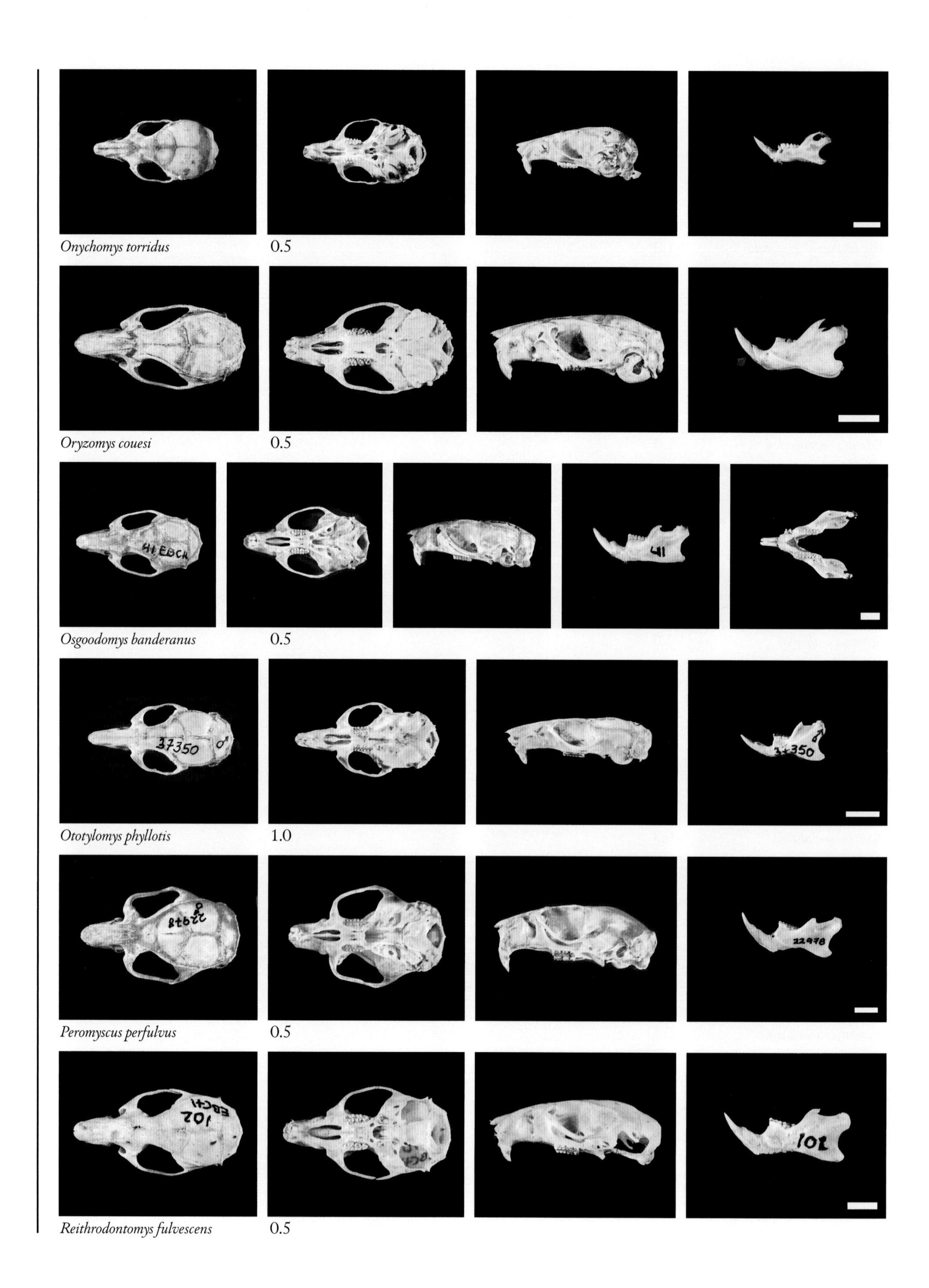

Onychomys torridus 0.5

Oryzomys couesi 0.5

Osgoodomys banderanus 0.5

Ototylomys phyllotis 1.0

Peromyscus perfulvus 0.5

Reithrodontomys fulvescens 0.5

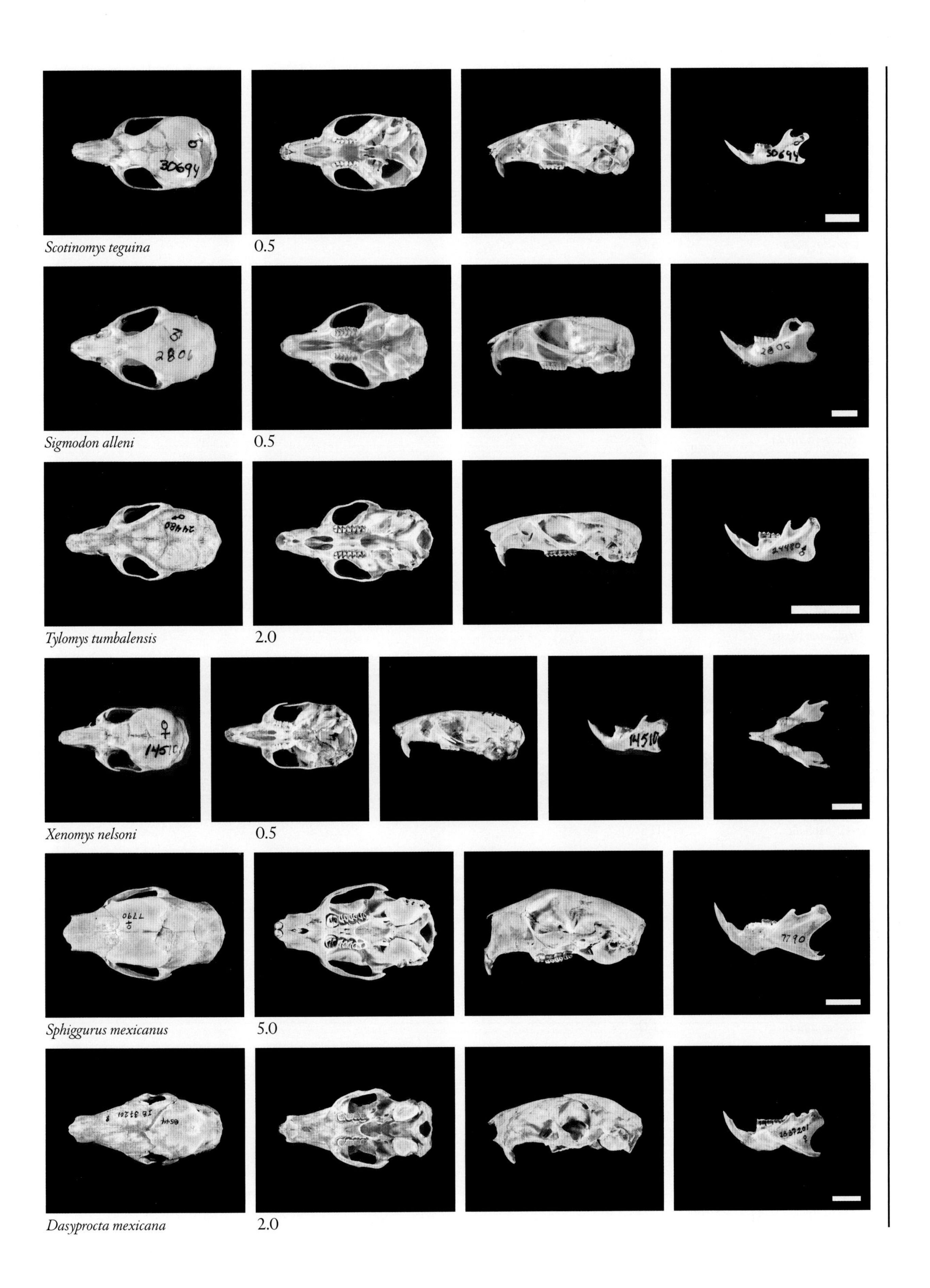

Scotinomys teguina 0.5

Sigmodon alleni 0.5

Tylomys tumbalensis 2.0

Xenomys nelsoni 0.5

Sphiggurus mexicanus 5.0

Dasyprocta mexicana 2.0

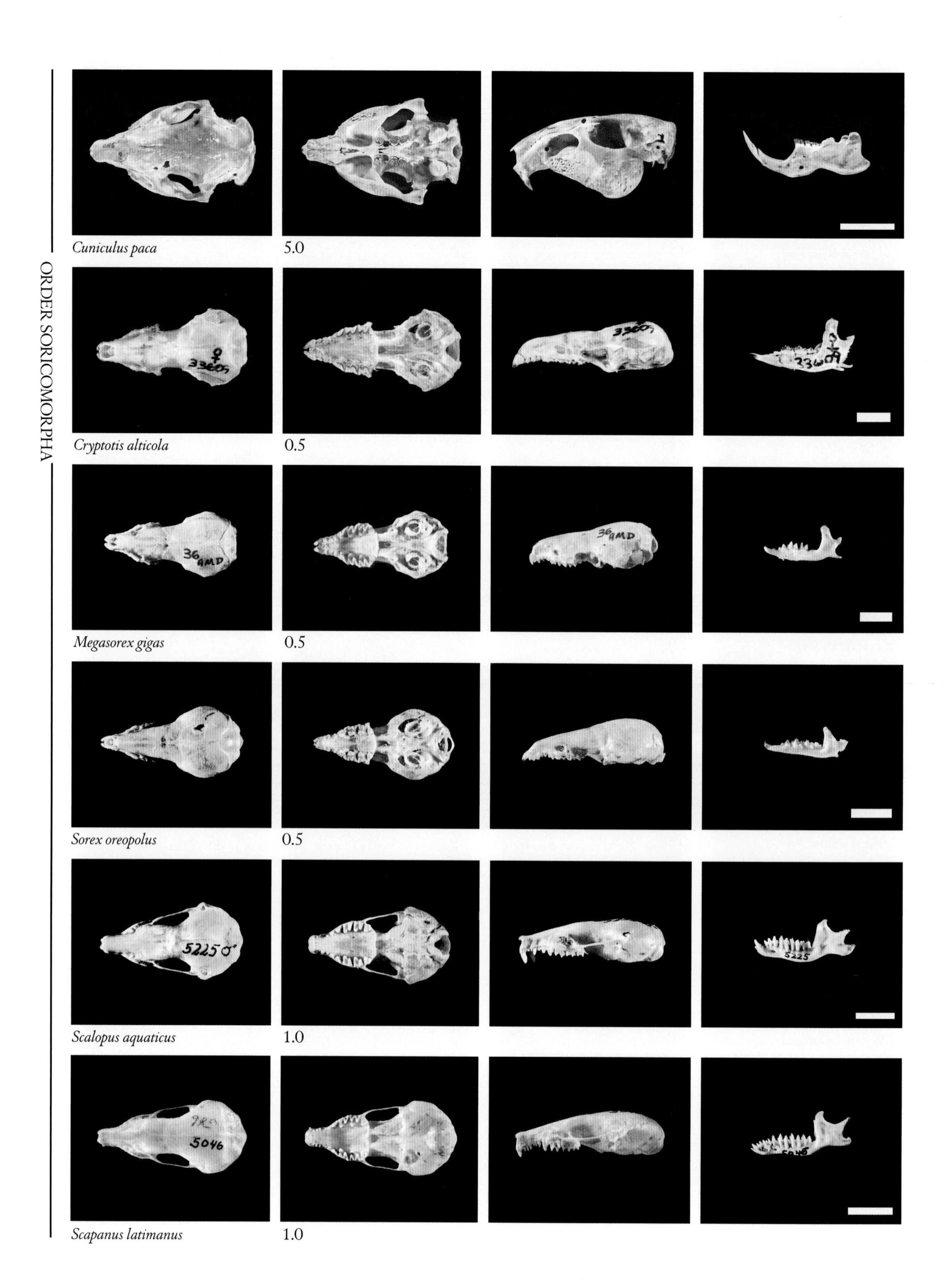

Cuniculus paca 5.0

Cryptotis alticola 0.5

Megasorex gigas 0.5

Sorex oreopolus 0.5

Scalopus aquaticus 1.0

Scapanus latimanus 1.0

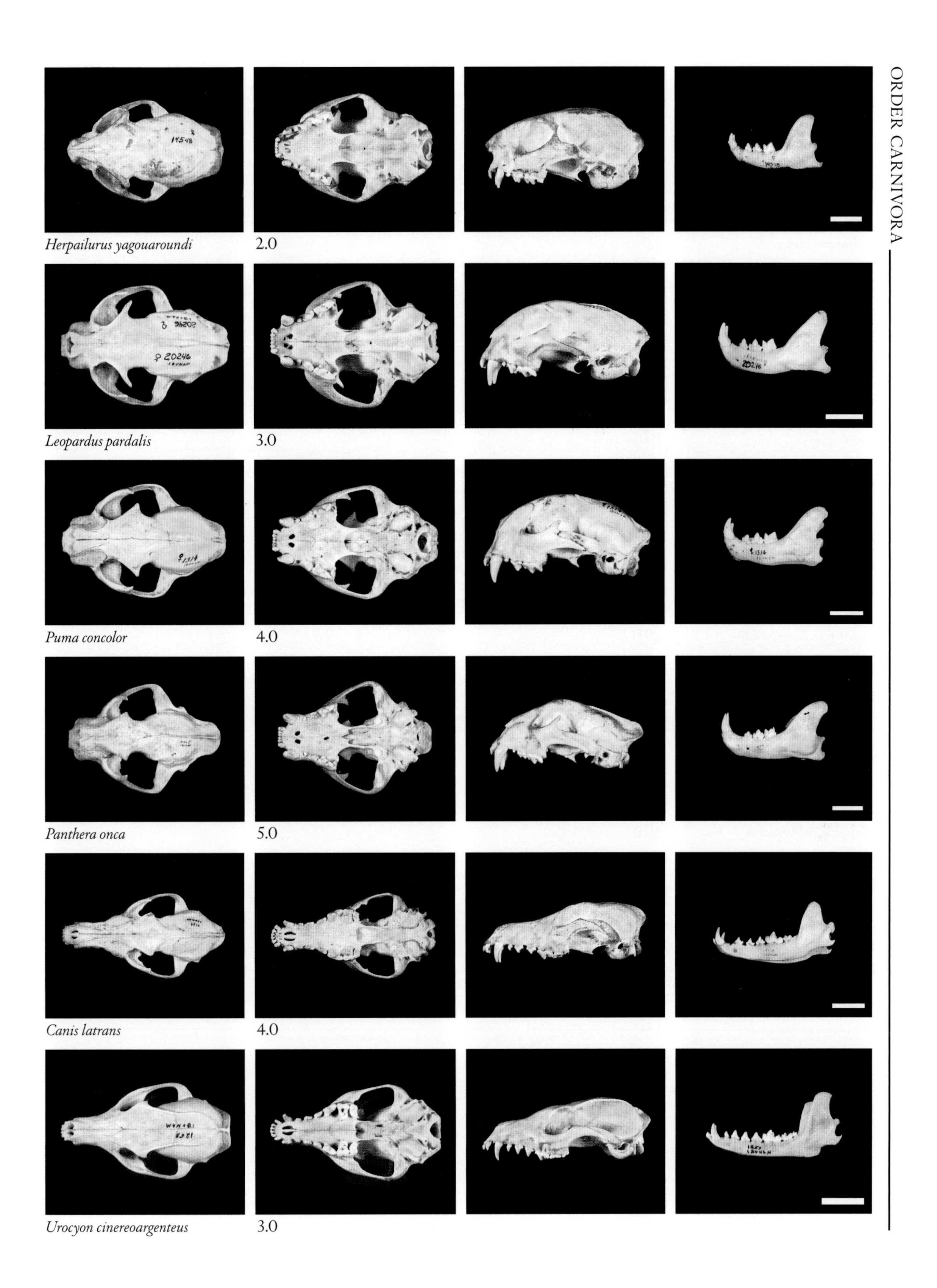

Herpailurus yagouaroundi 2.0

Leopardus pardalis 3.0

Puma concolor 4.0

Panthera onca 5.0

Canis latrans 4.0

Urocyon cinereoargenteus 3.0

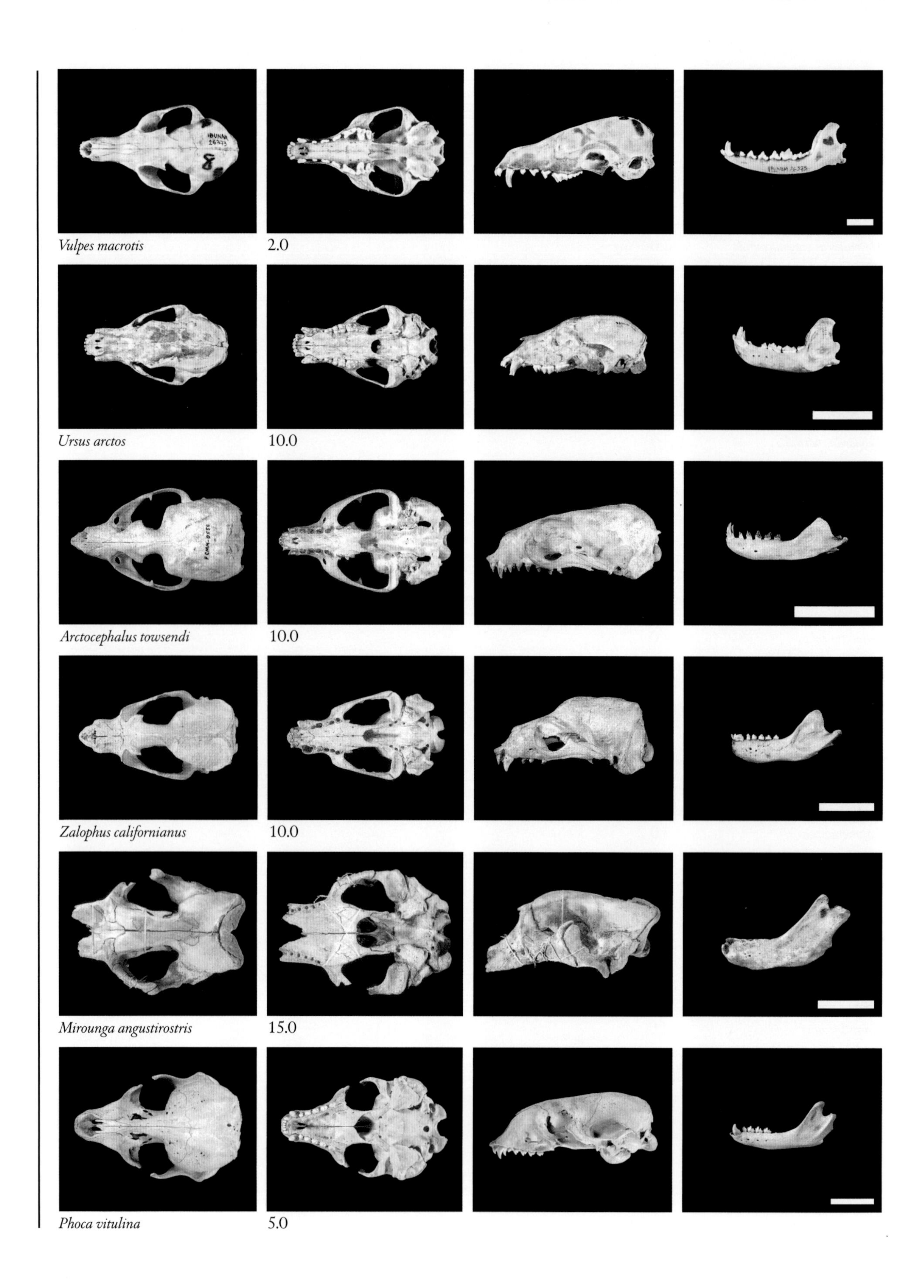

Vulpes macrotis 2.0

Ursus arctos 10.0

Arctocephalus towsendi 10.0

Zalophus californianus 10.0

Mirounga angustirostris 15.0

Phoca vitulina 5.0

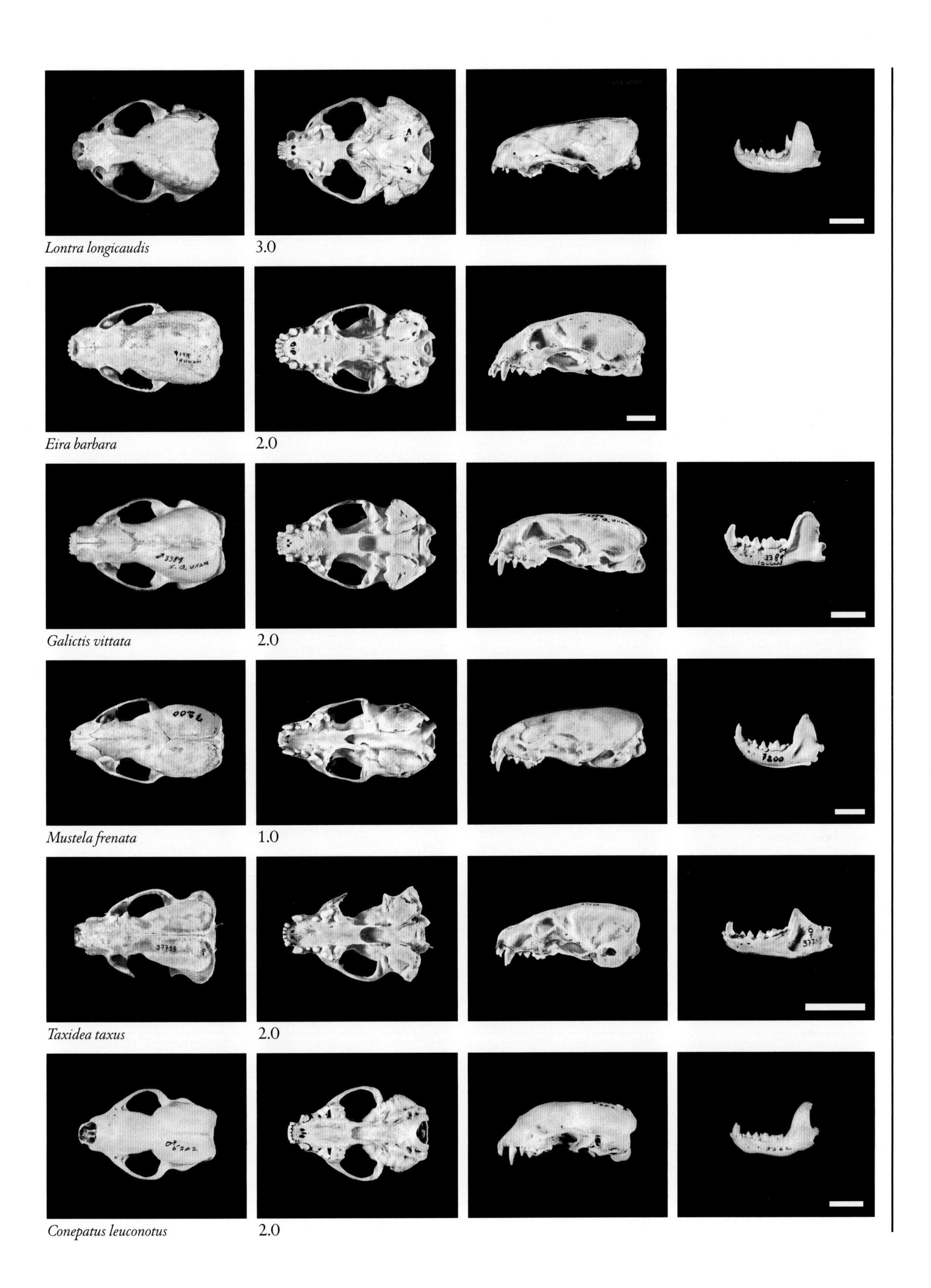
Lontra longicaudis 3.0

Eira barbara 2.0

Galictis vittata 2.0

Mustela frenata 1.0

Taxidea taxus 2.0

Conepatus leuconotus 2.0

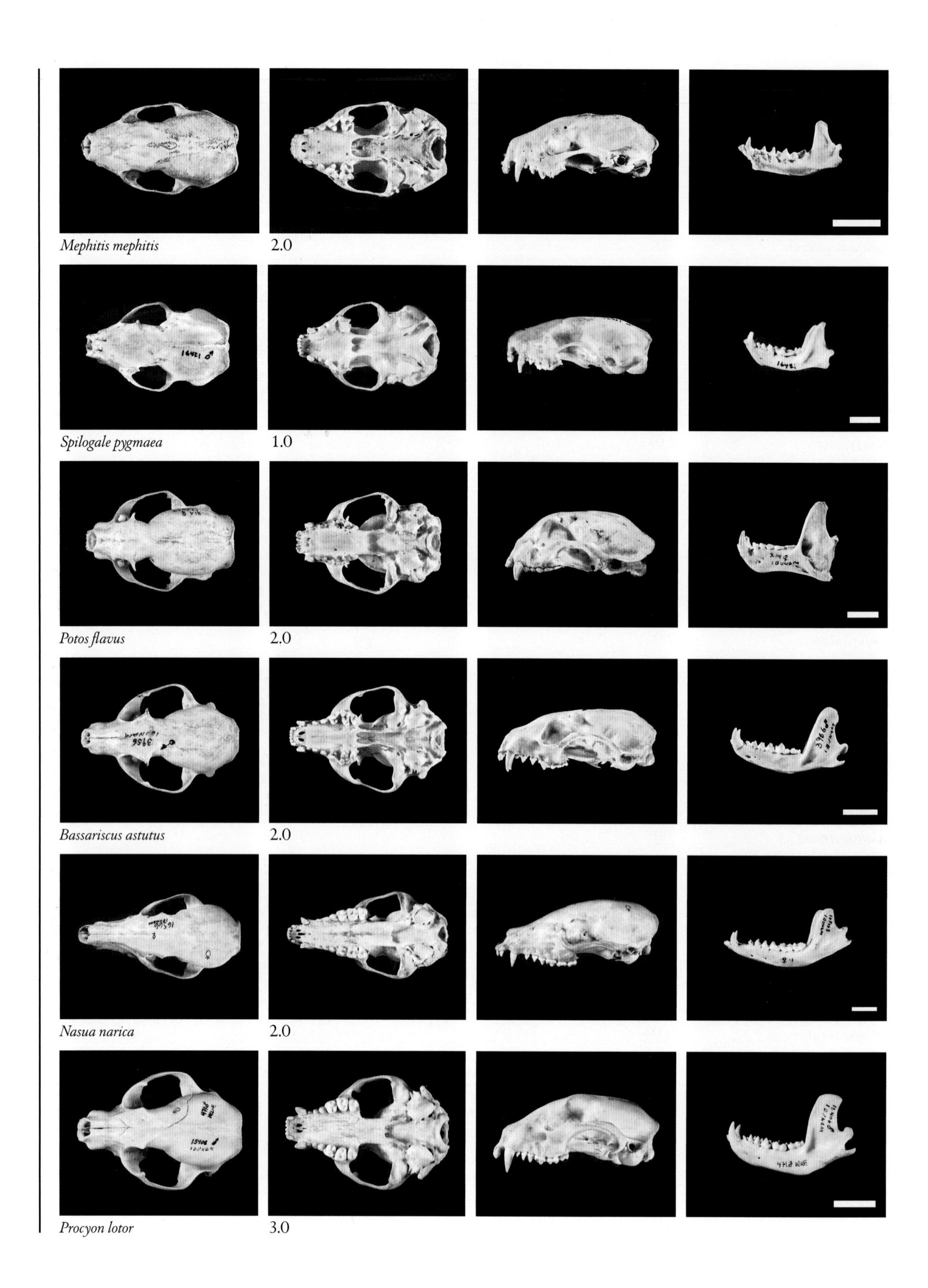

Mephitis mephitis 2.0

Spilogale pygmaea 1.0

Potos flavus 2.0

Bassariscus astutus 2.0

Nasua narica 2.0

Procyon lotor 3.0

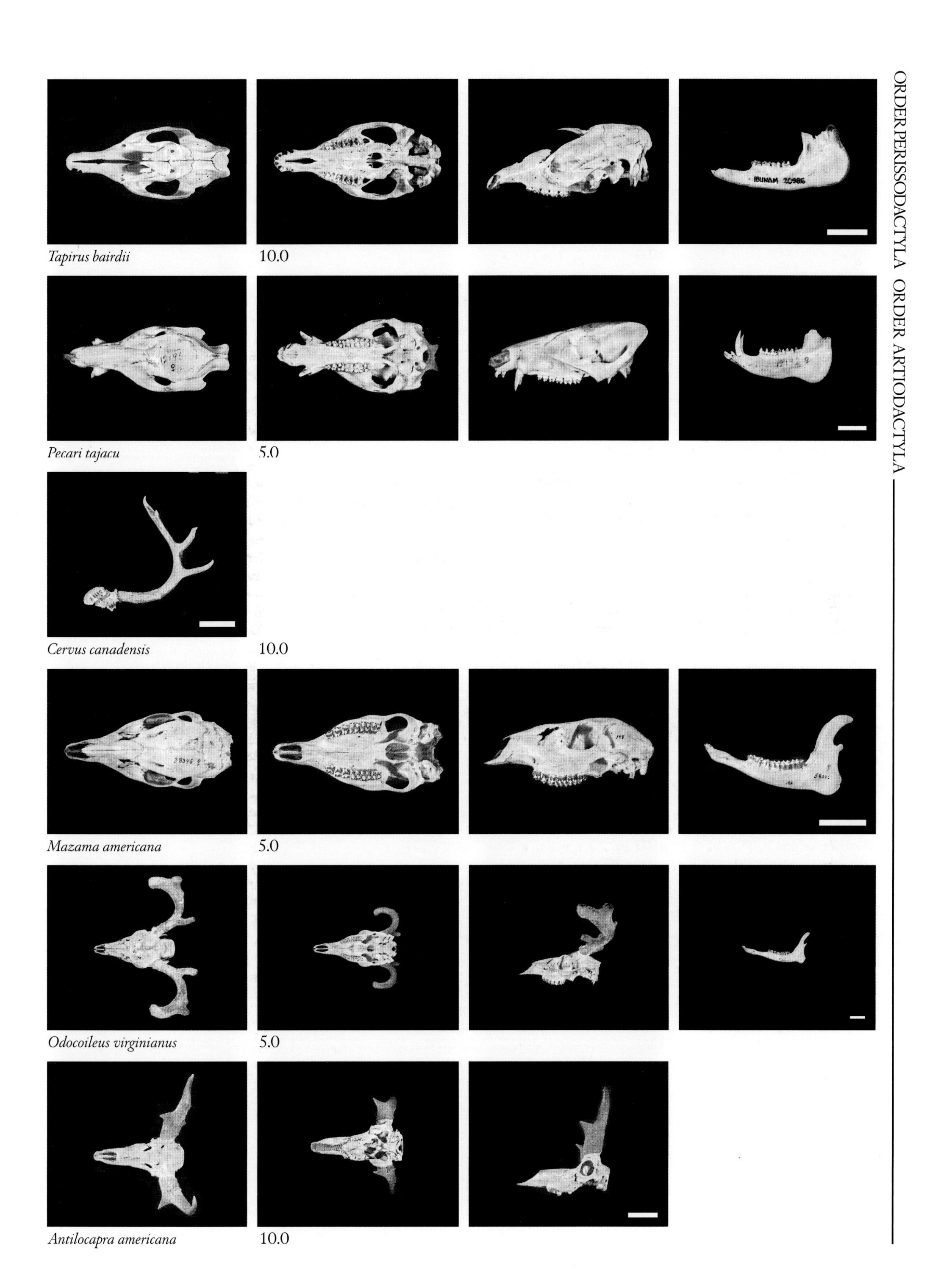

Tapirus bairdii 10.0

Pecari tajacu 5.0

Cervus canadensis 10.0

Mazama americana 5.0

Odocoileus virginianus 5.0

Antilocapra americana 10.0

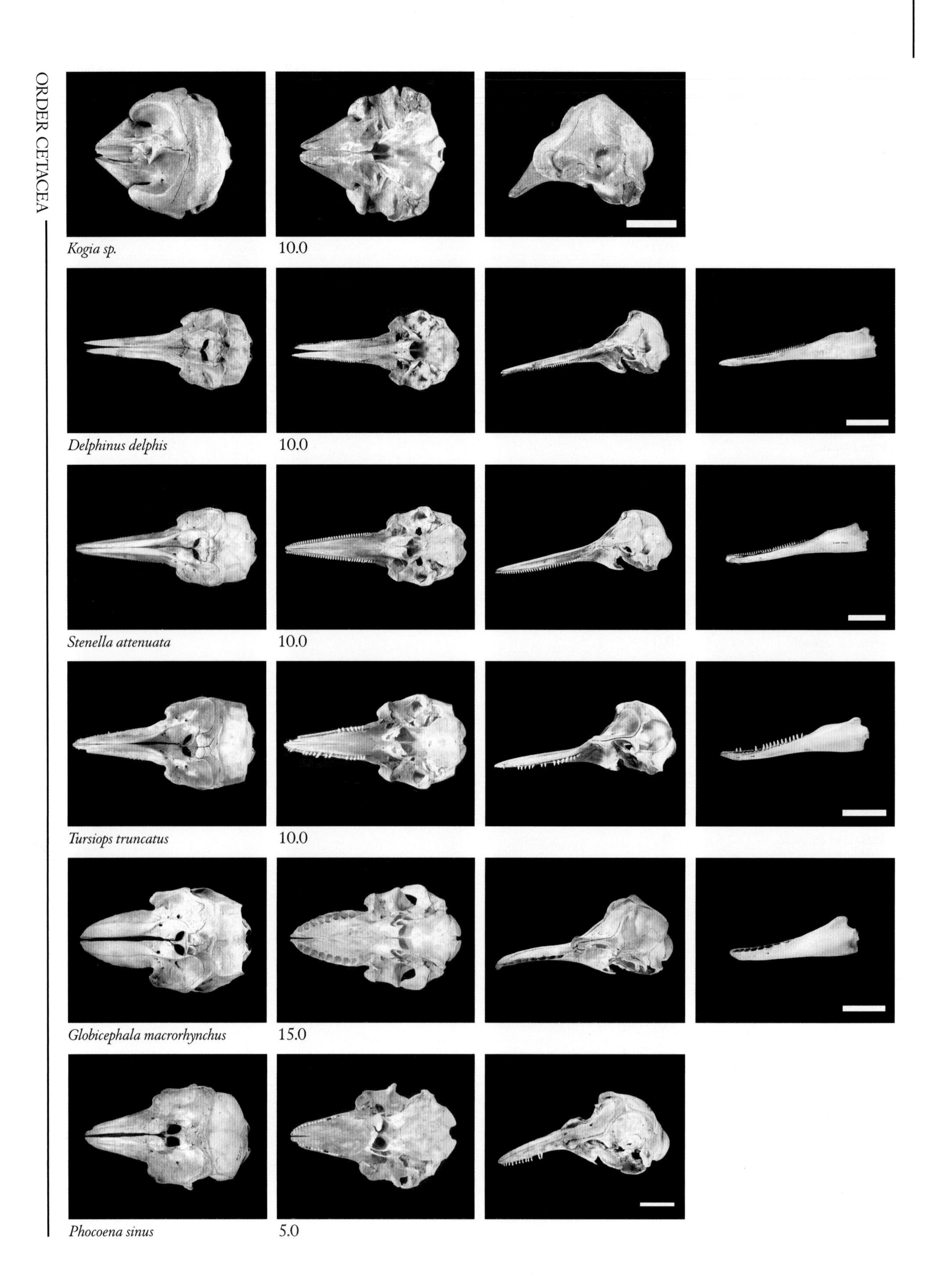
Kogia sp.
10.0
Delphinus delphis
10.0
Stenella attenuata
10.0
Tursiops truncatus
10.0
Globicephala macrorhynchus
15.0
Phocoena sinus
5.0

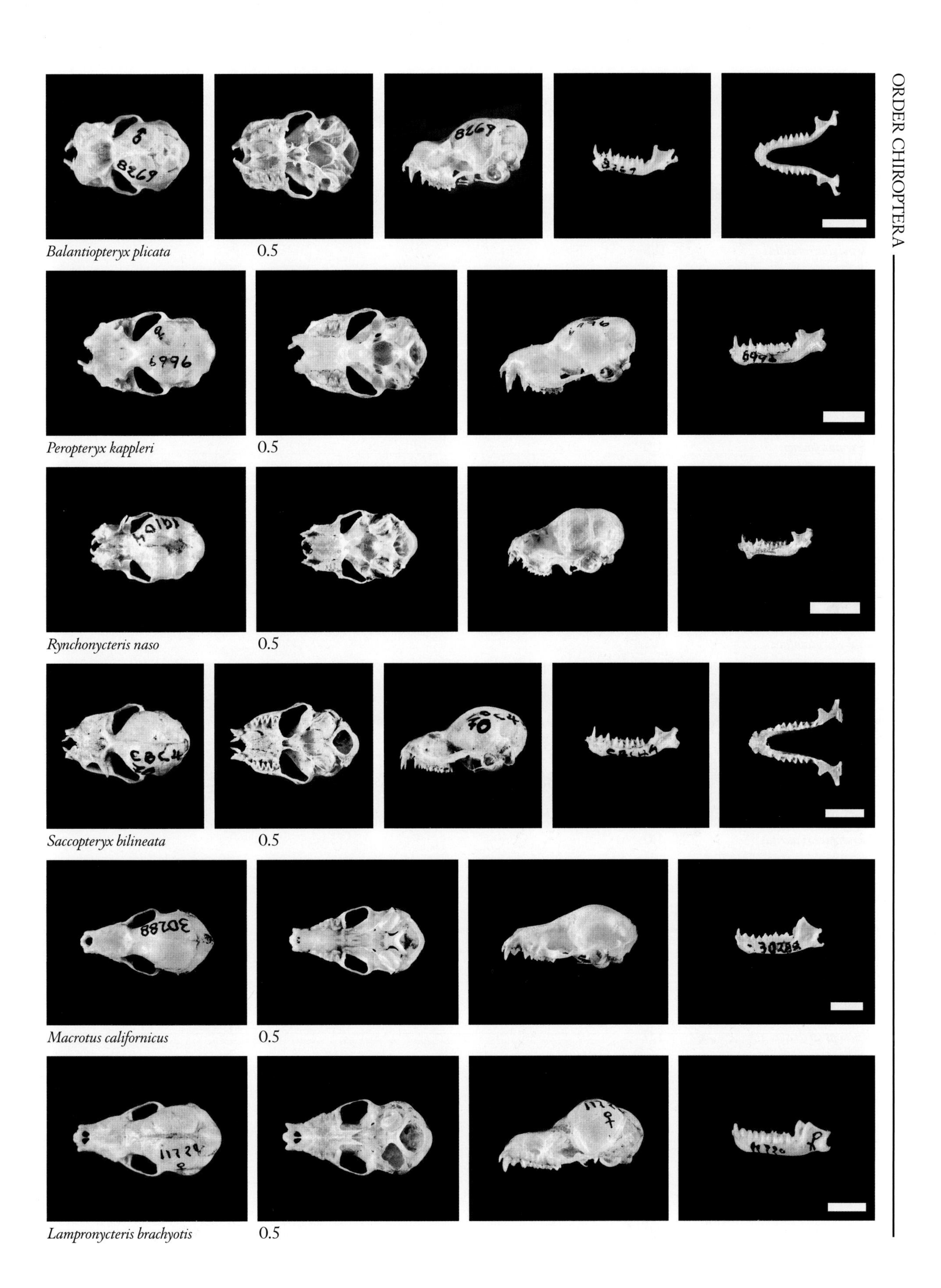

Balantiopteryx plicata 0.5

Peropteryx kappleri 0.5

Rynchonycteris naso 0.5

Saccopteryx bilineata 0.5

Macrotus californicus 0.5

Lampronycteris brachyotis 0.5

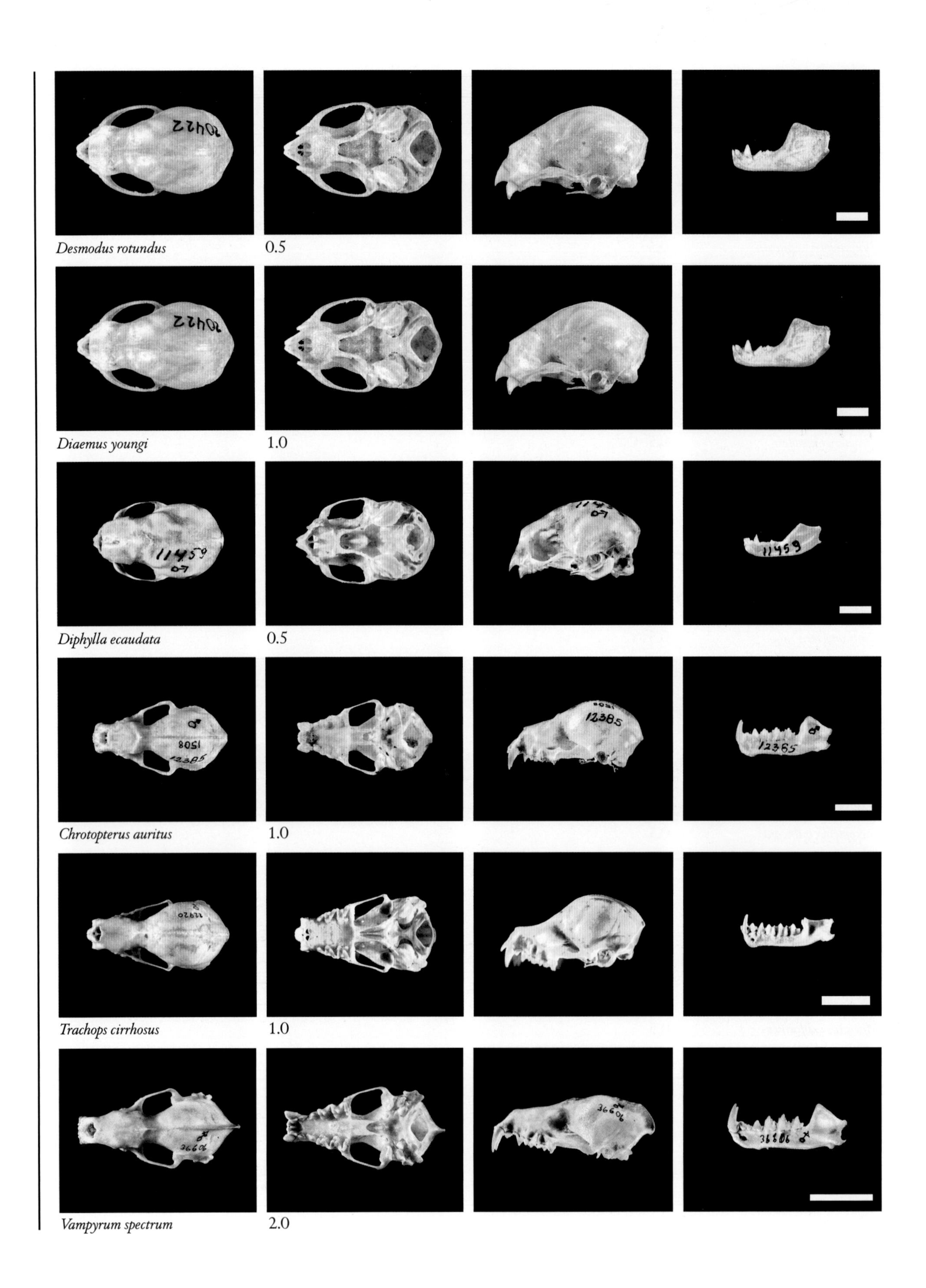

Desmodus rotundus 0.5

Diaemus youngi 1.0

Diphylla ecaudata 0.5

Chrotopterus auritus 1.0

Trachops cirrhosus 1.0

Vampyrum spectrum 2.0

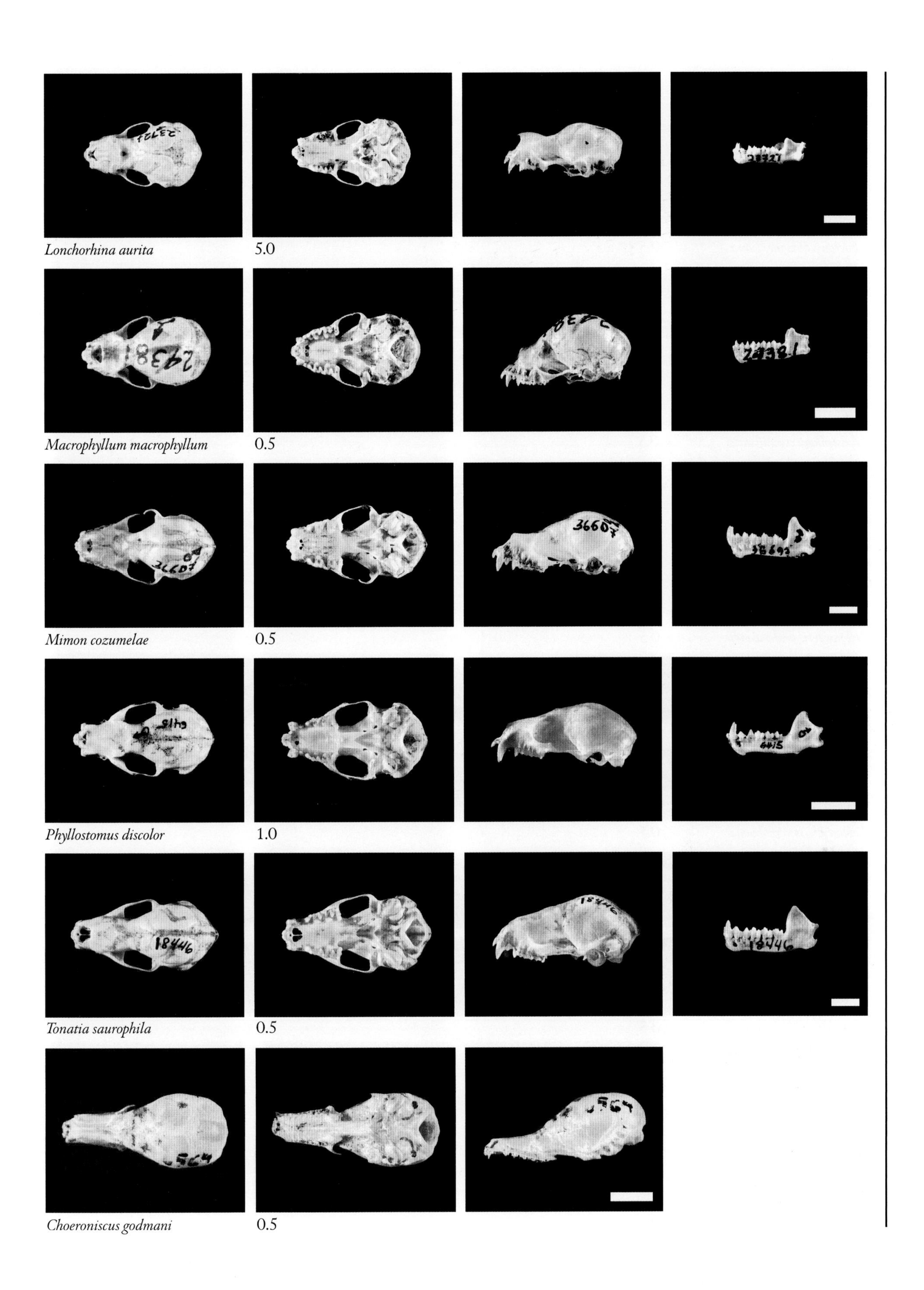

Lonchorhina aurita 5.0

Macrophyllum macrophyllum 0.5

Mimon cozumelae 0.5

Phyllostomus discolor 1.0

Tonatia saurophila 0.5

Choeroniscus godmani 0.5

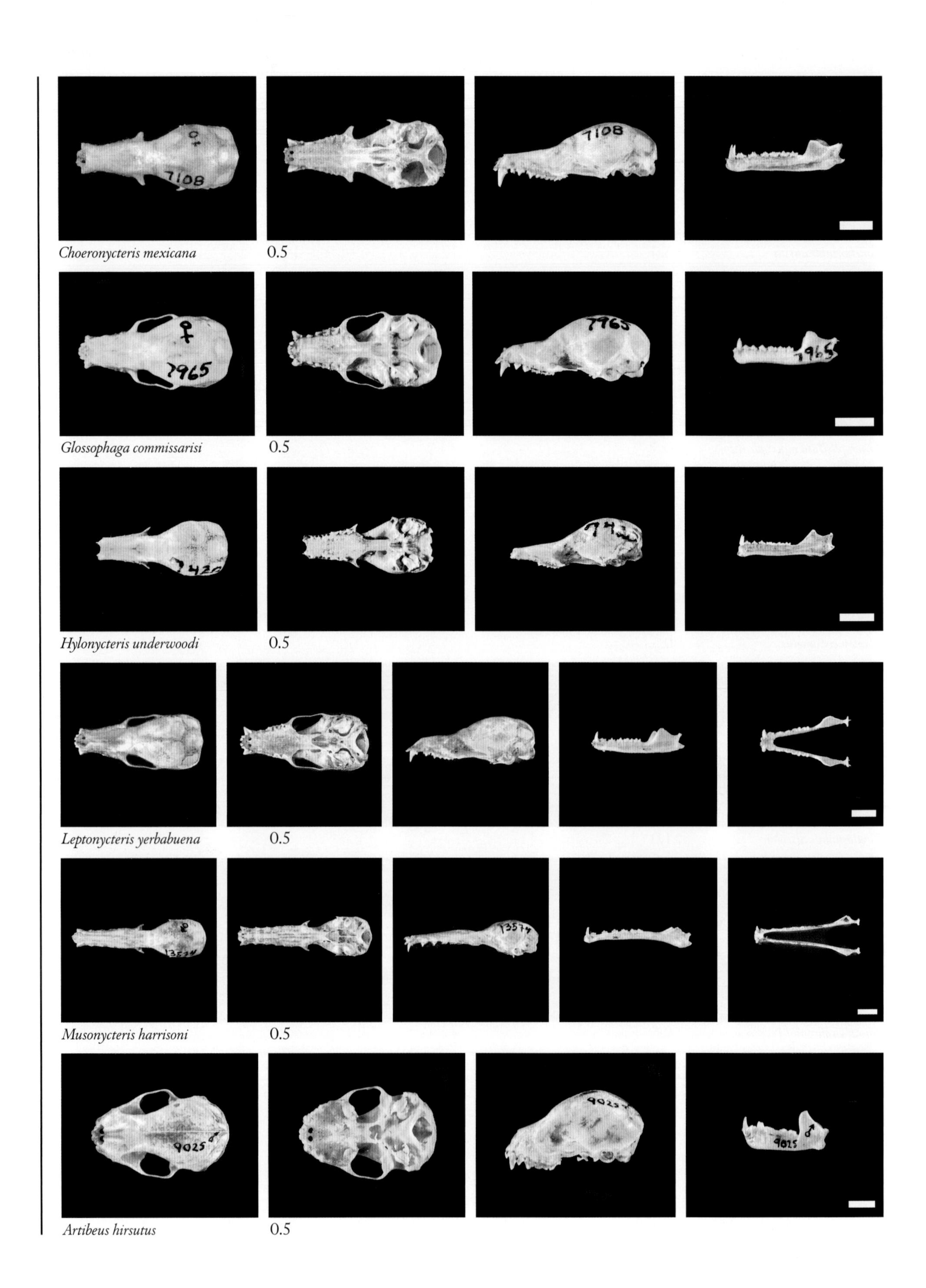

Choeronycteris mexicana 0.5

Glossophaga commissarisi 0.5

Hylonycteris underwoodi 0.5

Leptonycteris yerbabuena 0.5

Musonycteris harrisoni 0.5

Artibeus hirsutus 0.5

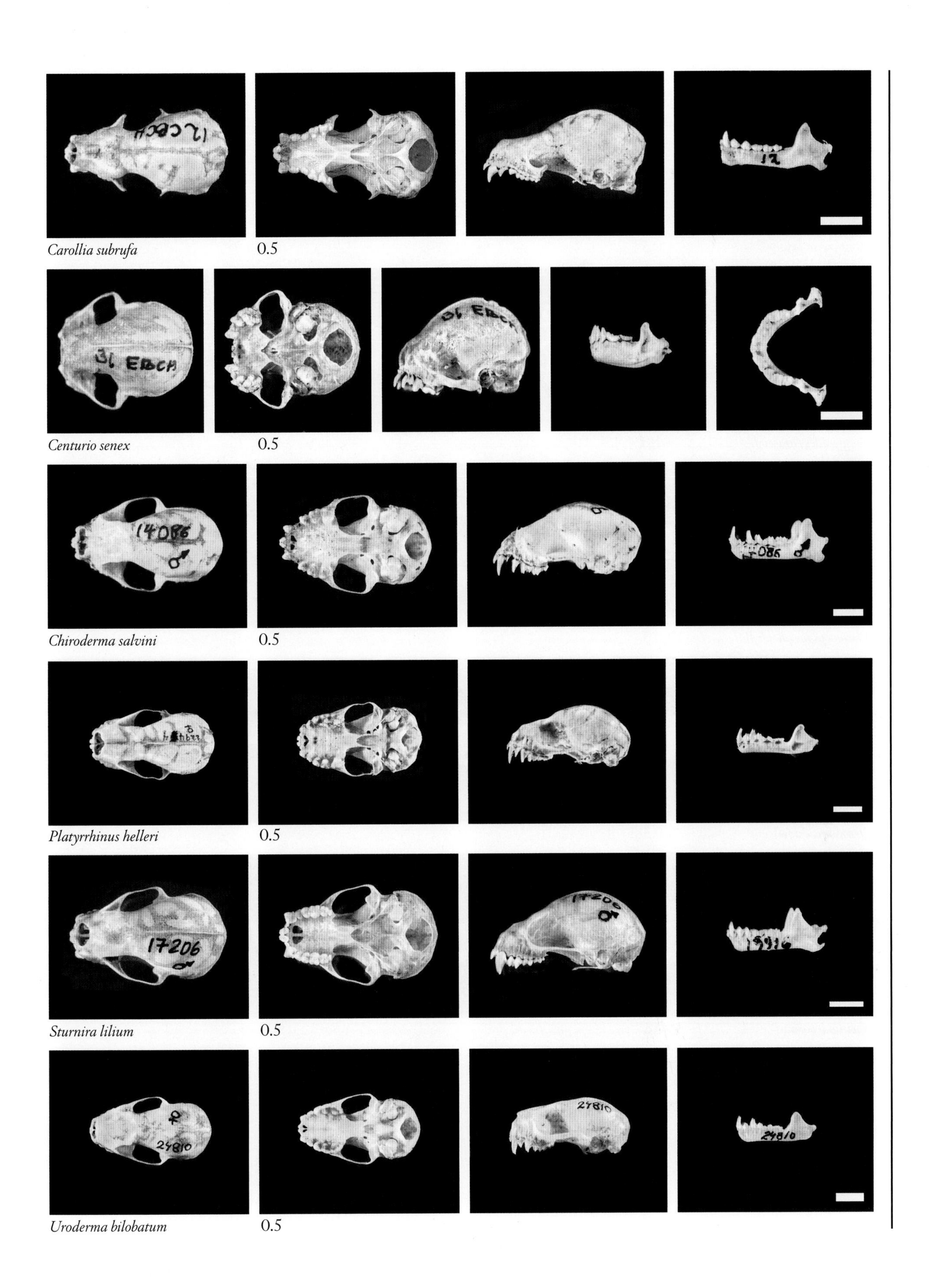
Carollia subrufa
0.5
Centurio senex
0.5
Chiroderma salvini
0.5
Platyrrhinus helleri
0.5
Sturnira lilium
0.5
Uroderma bilobatum
0.5

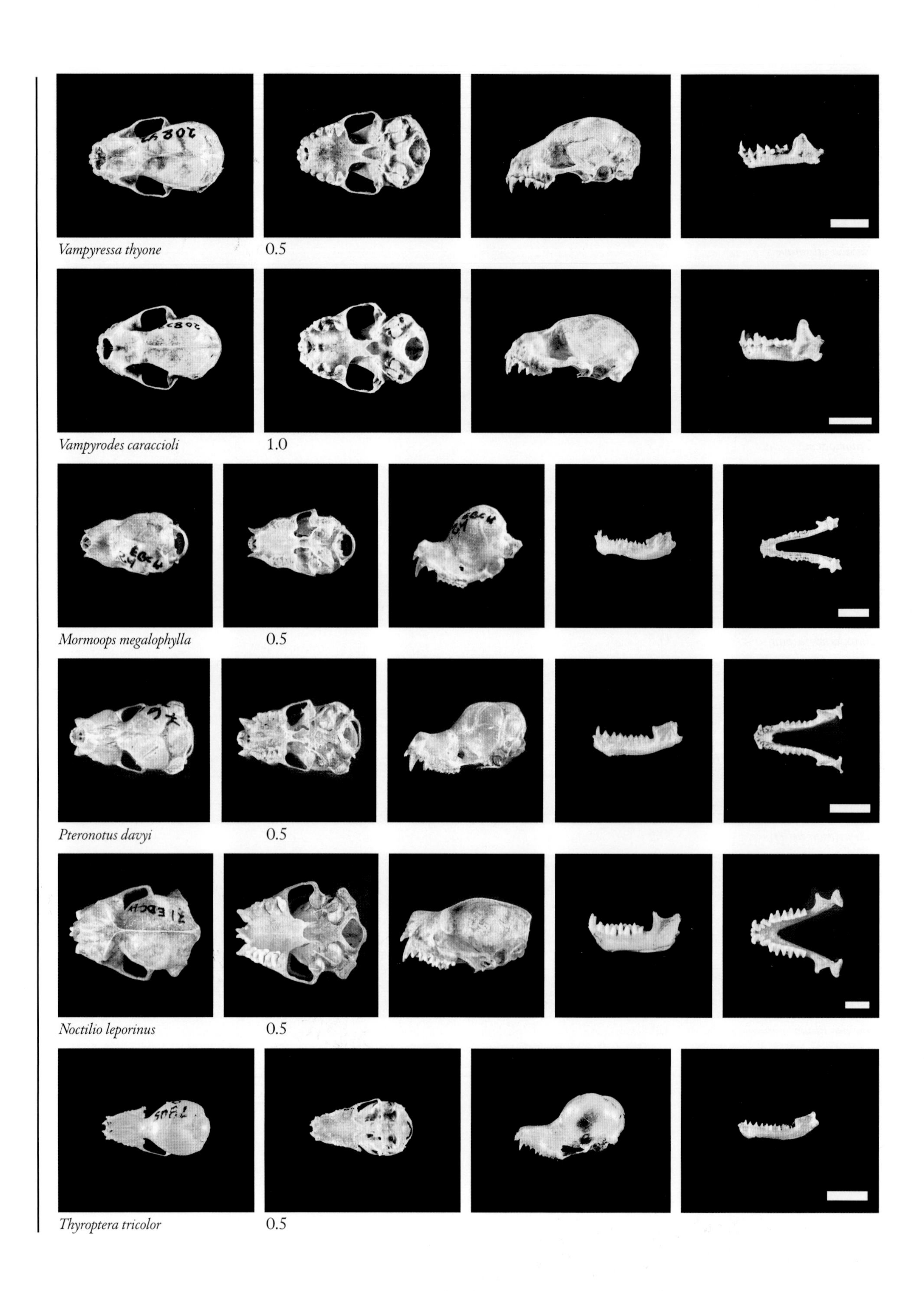

Vampyressa thyone 0.5

Vampyrodes caraccioli 1.0

Mormoops megalophylla 0.5

Pteronotus davyi 0.5

Noctilio leporinus 0.5

Thyroptera tricolor 0.5

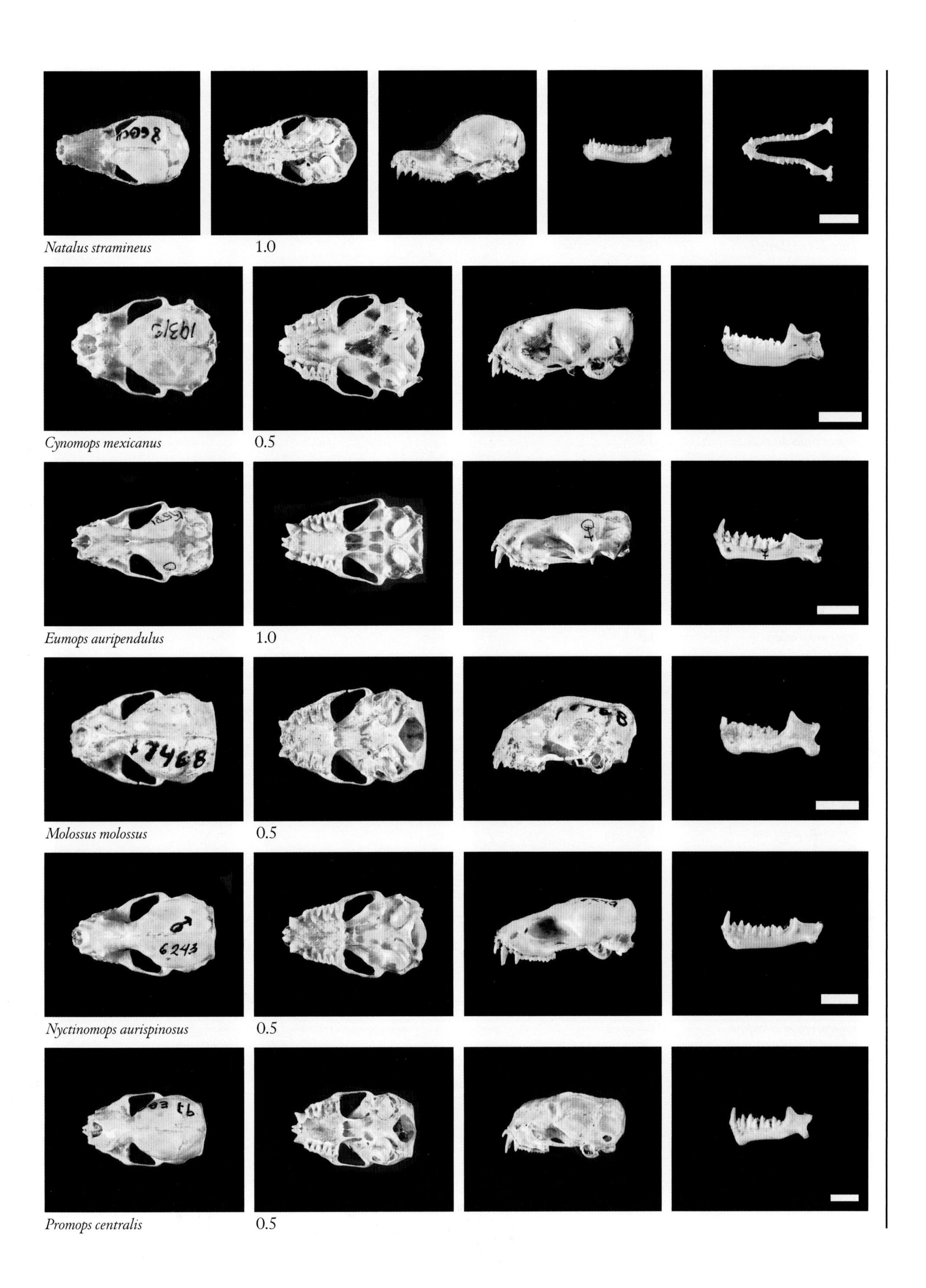

Natalus stramineus 1.0

Cynomops mexicanus 0.5

Eumops auripendulus 1.0

Molossus molossus 0.5

Nyctinomops aurispinosus 0.5

Promops centralis 0.5

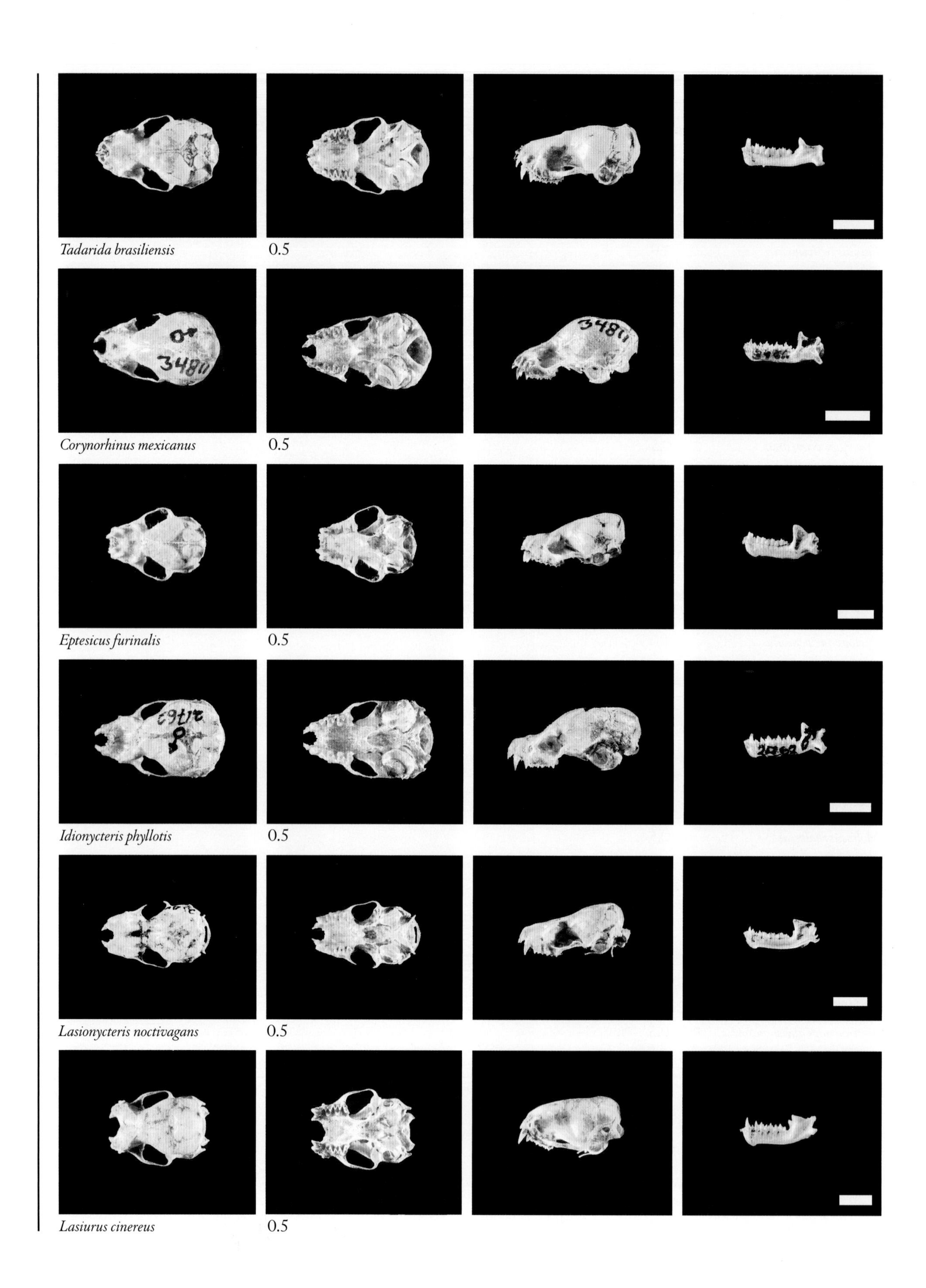
Tadarida brasiliensis
0.5
Corynorhinus mexicanus
0.5
Eptesicus furinalis
0.5
Idionycteris phyllotis
0.5
Lasionycteris noctivagans
0.5
Lasiurus cinereus
0.5

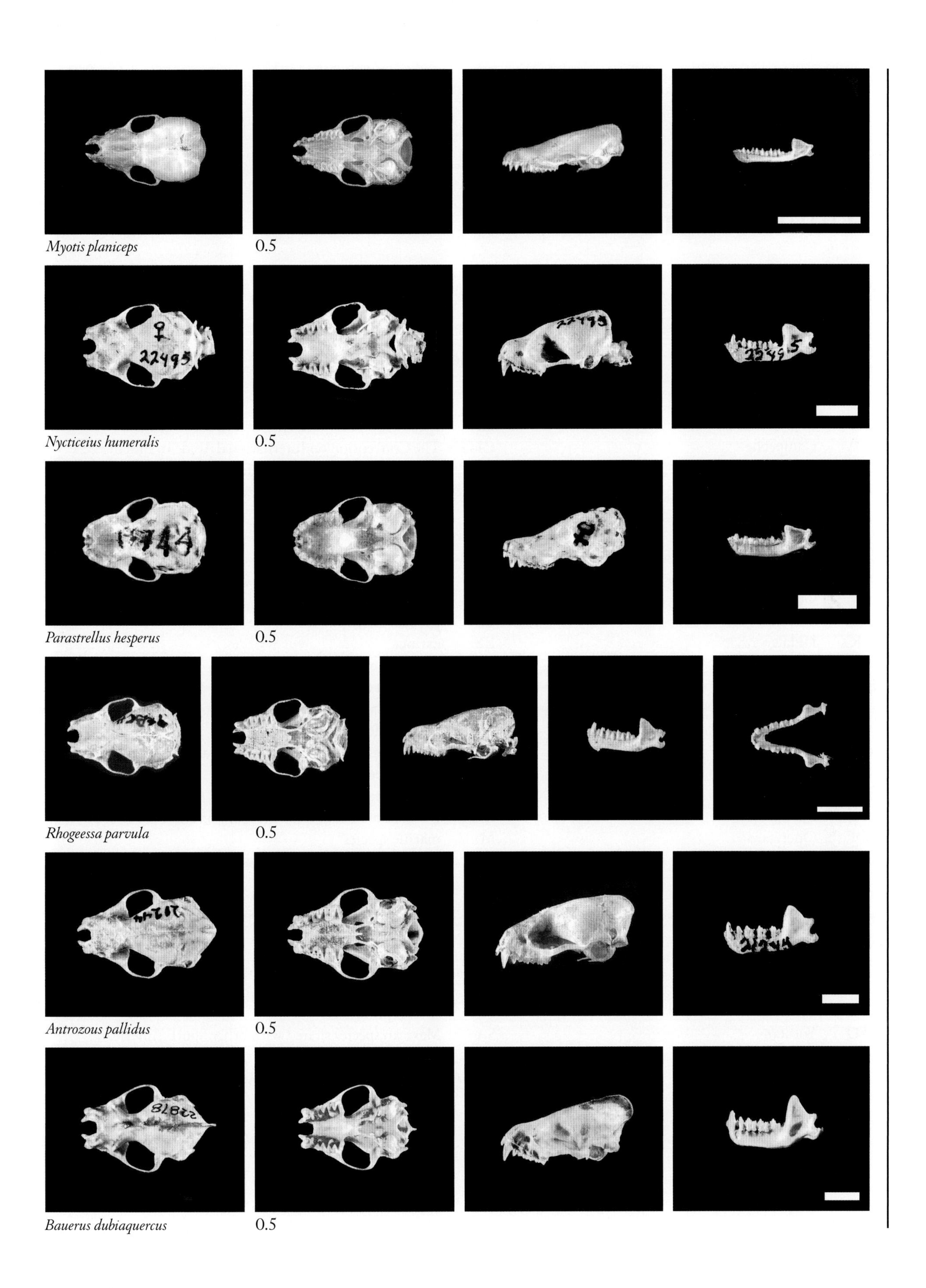
Myotis planiceps
0.5
Nycticeius humeralis
0.5
Parastrellus hesperus
0.5
Rhogeessa parvula
0.5
Antrozous pallidus
0.5
Bauerus dubiaquercus
0.5

Appendix C

Abbreviations

Mexican States

AG Aguascalientes
BC Baja California
BS Baja California Sur
CA Campeche
CH Chihuahua
CL Colima
CO Coahuila
CS Chiapas
DF Distrito Federal
DU Durango
GJ Guanajuato
GR Guerrero
HG Hidalgo
JA Jalisco
MI Michoacán
MO Morelos
MX Estado de México
NY Nayarit
NL Nuevo Leon
OX Oaxaca
PU Puebla
QE Querétaro
QR Quintana Roo
SI Sinaloa
SL San Luis Potosí
SO Sonora
TA Tamaulipas
TB Tabasco
TL Tlaxcala
VE Veracruz
YU Yucatan
ZA Zacatecas

Other Abbreviations

AM American continent
C continental species
CI species found in both islands and the continent
CITES The Convention on International Trade in Endangered Species of Wild Fauna and Flora
CONANP Comision Nacional de Areas Naturales Protegidas
CS Caribbean Sea
CR critically endangered
E endangered
EN endangered
ET extirpated
EC? probably extirpated
EX extinct
EX? probably extinct or extirpated
EW extinct in the wild
GC Gulf of California region
GM Gulf of Mexico
HD habitat destruction or habitat degradation
I insular species
IE introduced species
IUCN International Union for the Conservation of Nature
M marine species
MA Mexico and Central America
MX Mexico
NA North America
NT near threatened
NP Northern Pacific
OE overexploitation
PO pollution
PR special protection
SA Mexican species shared with other South American countries
SEMARNAT Secretaria del Medio Ambiente y Recursos Naturales
T threatened
TP Tropical Pacific
VU threatened or vulnerable

References

Ackerman, B.B., F.G. Lindzey, and T.P. Hemker. 1984. Cougar food habits in southern Utah. Journal of Wildlife Management, 48:147-155.

Acosta y Lara, E.F. 1951. Notas ecológicas sobre algunos quirópteros de Brasil. Comunicaciones Zoológicas del Museo de Montevideo, 3:1-2.

Adler, G.H. 1996. Habitat relations of two endemic species of highland forest rodents in Taiwan. Zoological Studies, 35:105-110.

Agenbroad, L.D. 1984. New World mammoth distribution. Pp. 90-108, in: Quaternary Extinctions: A Prehistoric Revolution (P.S. Martin and R.G. Klein, eds.). University of Arizona Press, Tucson.

Agnew, W., D.W. Uresk, and R.M. Hansen. 1986. Flora and fauna associated with prairie dog colonies and adjacent ungrazed mixed-grass prairie in western South Dakota. Journal of Range Management, 39:135-139.

Aguayo, L.A. 1986. Perspectivas de la investigación de los mamíferos marinos en México. P. 32, in: Primer Simposio Nacional sobre el Desarrollo Histórico de las Investigaciones Oceanográficas en México. México.

Aguayo, L.A. 1989. Aprovechamiento de los mamíferos marinos en América Latina. Pp. 500-531, in: Memorias del 2° Simposio Internacional sobre Vida Silvestre (Wildlife Society de México, Secretaría de Desarrollo Urbano y Ecología). Talleres Gráficos de la Nación, México.

Aguayo, L.A. 1990. Perspectivas de la investigación de los mamíferos marinos en México. P. 12, in: Primer Simposio Nacional sobre el Desarrollo Histórico de las Investigaciones Oceanográficas en México. México.

Aguayo, L.A., and R.T. Sánchez. 1987. Sighting records of Fraser's dolphin in the Mexican Pacific waters. Scientific Reports of the Whales Research Institute, 38:187-188.

Aguayo, L.A., L. Findley, L. Rojas-Bracho, and O. Vidal. 1983. The population of fin whales in the Gulf of California, Mexico. Fifth Biennial Conference on the Biology of Marine Mammals. Boston.

Aguayo, L.A., D. Aurioles G., J. Urbán R., M. Salinas Z., O. Vidal, and L.T. Findley. 1988. Beaked whales in Mexican waters. Document SC/40/SM/23, International Whaling Commission, Washington, DC.

Aguilar, A., L.L. Jover, and J. Nadal. 1982. A note on the organochlorine contamination in a Blainville's beaked whale, *Mesoplodon densirostris* (De Blainville, 1817) from the Mediterranean Sea. Publicaciones del Departamiento de Zoología, Universidad de Barcelona, 7:85-90.

Aguilar-Cervantes, A., and T. Álvarez-Solorzano. 1991. Notas sobre la dieta alimentaria del murciélago pescador. Anales de la Escuela Nacional de Ciencias Biológicas, Instituto Politécnico Nacional, México, 35:123-127.

Aguilar-Morales, S. (in prep.). Análisis taxonómico, distribucional y parasitológico de *Lasiurus ega* y *Lasiurus xanthinus* en México. Tesis de maestría en ciencias (biología animal), Facultad de Ciencias, UNAM, México.

Aguirre, A.E., and M. Ulloa. 1982. Mohos que se desarrollan en el estiércol de algunos ratones silvestres de México. Boletin de la Sociedad Mexicana de Micología, 17:55-66.

Aguirre, G., and E. Fey. 1981. Estudio preliminar del tepezcuintle (Agouti paca nelsoni Goldman, 1913) en la selva Lacandona, Chiapas. Pp. 45-54, in: Estudios Ecológicos en el Neotrópico Mexicano (P.R. Castillo, ed.). Instituto de Ecología, A.C., México.

Aguirre, L.F. 1994. Estructura y ecología de las comunidades de murciélagos de la Sabana Espíritu (Beni, Bolivia). Tesis de licenciatura, Universidad Mayor de San Andrés, La Paz, Bolivia.

Alfaro, A.M., J.L. García-García, and A. Santos-Moreno. 2005. The false vampire bat Vampyrum spectrum in Oaxaca, Mexico. Bat Research News, 46:145-146.

Aliaga-Rossel, E., R.S. Moreno, R.W. Kays, and J. Giacalone. 2006. Ocelot (*Leopardus pardalis*) predation on agouti (*Dasyprocta punctata*). Biotropica, 38:691-694.

Allen, G.M. 1914. A new bat from Mexico. Proceedings of the Biological Society of Washington, 27:109-111.

Allen, H. 1866. Notes on the Vespertilionidae of tropical America. Proceedings of the Academy of Natural Sciences of Philadelphia, 18:279-288.

Allen, J.A. 1881. List of mammals collected by Dr. Edward Palmer in northeastern Mexico, with field notes by the collector. Bulletin of the Museum of Comparative Zoology, 8:183-189.

Allen, J.A. 1887. The West Indian seal (*Monachus tropicalis* Gray). Bulletin of the American Museum of Natural History, 1:1-34.

Allen, J.A. 1889. Notes on a collection of mammals from southern Mexico, with descriptions of a new species of the genera *Sciurus, Tamias,* and *Sigmodon.* Bulletin of the American Museum of Natural History, 2:165-181.

Allen, J.A. 1890. Notes on collections of mammals made in central and southern Mexico by Dr. Audley C. Buller, with descriptions of new species of the genera Vespertilio, Sciurus, and Lepus. Bulletin of the American Museum of Natural History, 3:175-194.

Allen, J.A. 1893a. Description of the new species of opossum from the Isthmus of Tehuantepec, Mexico. Bulletin of the American Museum of Natural History, 5:235-236.

Allen, J.A. 1893b. On a collection of mammals from the San Pedro Martir region of Lower California, with notes on other species, particularly of the genus Sitomys. Bulletin of the American Museum of Natural History, 5:181-202.

Allen, J.A. 1894. Remarks on specimens of *Chilonycterys rubiginosus* from western Mexico, and on the color phases of *Pteronotus davyi* Gray. Bulletin of the American Museum of Natural History, 6:317-332.

Allen, J.A. 1895. On a collection of mammals from Arizona and Mexico, by Mr. W.W. Price, with field notes by the collector. Bulletin of the American Museum of Natural History, 7:193-258.

Allen, J.A. 1896. Description of new North American mammals. Bulletin of the American Museum of Natural History, 8:233-240.

Allen, J.A. 1897. Further notes on mammals collected in Mexico by Dr. Audley C. Buller, with descriptions of new species. Bulletin of the American Museum of Natural History, 9:47-58.

Allen, J.A. 1898. Descriptions of new mammals from western Mexico and Lower California. Bulletin of the American Museum of Natural History, 10:143-158.

Allen, J.A. 1899. Description of five new American rodents. Bulletin of the American Museum of Natural History, 12:11-17.

Allen, J.A. 1901a. A preliminary study of the North American opossums of the genus *Didelphis.* Bulletin of the American Museum of Natural History, 14:149-188.

Allen, J.A. 1901b. Descriptions of two new opossums of the genus *Metachirus.* Bulletin of the American Museum of Natural History, 14:213-218.

Allen, J.A. 1903. List of mammals collected by Mr. J. H. Batty in New Mexico and Durango, with descriptions of new species and subspecies. Bulletin of the American Museum of Natural History, 19:587-612.

Allen, J.A. 1906. Mammals from the states of Sinaloa and Jalisco, Mexico, collected by J.H. Batty during 1904 and 1905. Bulletin of the American Museum of Natural History, 34:191-262.

Allen, J.A. 1916. The Neotropical weasels. Bulletin of the American Museum of Natural History, 35:86-111.

Allen, J.A., and F.M. Chapman. 1897a. On a collection of mammals from Jalapa and Las Vigas, state of Veracruz, Mexico. Bulletin of the American Museum of Natural History, 9:197-208.

Allen, J.A., and F.M. Chapman. 1897b. On mammals from Yucatan with descriptions of a new species. Bulletin of the American Museum of Natural History, 9:1-12.

Allen, K.R. 1980. Conservation and Management of Whales. University of Washington Press, Seattle.
Allen, L., R. Engeman, and H. Kupra. 1996. Evaluation of three relative abundance indices for assessing dingo populations. Wildlife Research, 23:197-206.
Almazán-Catalán, J.A., A. Taboada-Salgado, C. Sánchez-Hernández, Ma. de L. Romero-Almaraz, Y.Q. Jiménez-Salmerón, and E. Guerrero Ibarra. 2009. Registros de murciélagos para el estado de Guerrero, México. Acta Zoologica Mexicana (n.s.), 25:177-185.
Alonso-Mejía, A., and R.A. Medellín, 1991. *Micronycteris megalotis*. Mammalian Species, 376:1-6.
Alonso-Mejía, A., and R.A. Medellín, 1992. *Marmosa mexicana*. Mammalian Species, 421:1-4.
Alston, E.R. 1876. On the genus Dasyprocta, with description of a new species. Proceedings of the Zoological Society of London, 1876:347-352.
Alston, E.R. 1879-1882. Biology Central-Americana: Mammalia. Taylor and Francis, London.
Alston, E.R. 1968. Records of opossum and kit fox from Zacatecas. Journal of Mammalogy, 49:318.
Altenbach, J.S. 1989. Prey capture by the fishing bats *Noctilio leporinus* and *Myotis vivesi*. Journal of Mammalogy, 70:421-424.
Altmann, J. 1978. Elk. Pp. 205-211, in: Big Game of North America: Ecology and Management (J.L. Schmidt and D.L. Gilbert, eds.). Stackpole Books, Harrisburg, PA.
Altmann, S.A. 1959. Field observations on a howling monkey society. Journal of Mammalogy, 40:317-330.
Altmann, S.A. 1966. Vocal communication in howling monkeys (7.5 i.p.s. tape). Library of Natural History Sounds, Laboratory of Ornithology, Cornell University, Ithaca, NY.
Altmann, S.A. 1968. Primates. Pp. 466-522, in: Animal Communication (T.S. Sebeok, ed.). Indiana University Press, Bloomington.
Álvarez, C., A.L. Aguayo, and J. Mujica. 1988. Observaciones sobre el manatí *Trichechus manatus*, en la región media del Usumacinta, Tabasco. Pp. 617-624, in: Ecología y Conservación del Delta de los Ríos Usumacinta y Grijalva. Memorias. Instituto nacional sobre Recursos Bióticos (INIREB)-División Regional Tabasco, México.
Álvarez, J., M.R. Willig, J.K. Jones Jr., and W.D. Webster. 1991. *Glossophaga soricina*. Mammalian Species, 379:1-7.
Álvarez, T. 1958. Roedores colectados en el Territorio de la Baja California. Acta Zoológica Mexicana, 2:1-7.
Álvarez, T. 1960. Sinopsis de las especies mexicanas del genero *Dipodomys*. Revista de la Sociedad Mexicana de Historia Natural, 21:391-424.
Álvarez, T. 1961a. Sinopsis de las ardillas arbóreas del género *Sciurus* en México (Mammalia. Sciuridae). Anales de la Escuela Nacional de Ciencias Biológicas, México, 10:123-148.
Álvarez, T. 1961b. Taxonomic status of some mice of the *Peromyscus boylii* group in eastern Mexico, with description of a new subspecies. University of Kansas Publications, Museum of Natural History, 14:111-120.
Álvarez, T. 1963a. The recent mammals of Tamaulipas, Mexico. University of Kansas Publications, Museum of Natural History, 14:363-473.
Álvarez, T. 1963b. Restos de mamíferos encontrados en una cueva del Valle Nacional, Oaxaca, México. Revista de Biología Tropical, 11:57-61.
Álvarez, T. 1964. Nota sobre restos óseos de mamíferos del Reciente, encontrados cerca de Tepeapulco, Hidalgo, México. Publicación del Departamento de Prehistoria, Instituto Nacional de Antropología e Historia, 15:1-15.
Álvarez, T. 1968. Notas sobre una colección de mamíferos de la región costera del Río Balsas entre Michoacán y Guerrero. Revista de la Sociedad Mexicana de Historia Natural, 29:21-35.
Álvarez, T. 1976. Restos óseos rescatados del Cenote Sagrado de Chichén-Itzá, Yucatán. Cuadernos de Trabajo, Departamento de Prehistoria, Instituto Nacional de Antropología e Historia, México, 15:13-39.
Álvarez, T. 1982. Restos de mamíferos recientes y pleistocénicos procedentes de las grutas de Loltún, Yucatán, México. Pp. 7-35, in: Restos de Moluscos y Mamíferos Cuaternarios Procedentes de Loltún, Yucatán (T. Álvarez and O.J. Polaco, eds.). Instituto Nacional de Antropología e Historia, México.
Álvarez, T., and S.T. Álvarez-Castañeda. 1990. Cuatro nuevos registros de murciélagos (Chiroptera) del estado de Chiapas. Anales de la Escuela Nacional de Ciencias Biológicas, 55:157-161.
Álvarez, T., and S.T. Álvarez-Castañeda. 1991a. Análisis de la fauna de roedores del área de El Cedral, San Luis Potosí, México. Anales del Insituto de Biología, UNAM, México, Serie Zoología, 62:169-180.
Álvarez, T., and S.T. Álvarez-Castañeda. 1991b. Notas sobre el estado taxonómico de *Pteronotus davyi* en Chiapas y de Hylonycteris en México (Mammalia: Chiroptera). Anales de la Escuela Nacional de Ciencias Biológicas, Instituto Politécnico Nacional, México, 34:223-229.
Álvarez, T., and C. Aviña. 1963. Notas acerca de algunas especies mexicanas de ardillas del género *Sciurus* (Rodentia: Sciuridae). Revista de la Sociedad Mexicana de Historia Natural, 24:33-39.
Álvarez, T., and C.E. Aviña. 1965. *Baedon alleni, Rhogeessa tumida major* and *R. p. parvula* newly reported for Michoacan, with notes on the qualitative differentiation of the two Rhogeessas. The Southwestern Naturalist, 10:75-76.
Álvarez, T., and L. González-Q. 1970. Análisis polínico del contenido gástrico de murciélagos Glossophaginae de México. Anales de la Escuela Nacional de Ciencias Biológicas, Instituto Politécnico Nacional, México, 18:137-165.
Álvarez, T., and N. González-Ruíz. 2000. Variación geográfica de *Saccopteryx bilineata* (Chiroptera: Emballonuridae) en México, con descripción de una nueva subespecie. Anales de la Escuela Nacional de Ciencias Biológicas, Instituto Politécnico Nacional, México, 46:305-316.
Álvarez, T., and N. González-Ruíz. 2001. Nuevos registros de Notiosorex crawfordi (Insectivora: Soricidae) para México. Acta Zoológica Mexicana (n.s.), 84:175-177.
Álvarez, T., and J.J. Hernández-Chávez. 1990. Cuatro nuevos registros del ratón de campo *Peromyscus* (Rodentia: Muridae) en el Estado de México, México. Anales de la Escuela Nacional de Ciencias Biológicas, Instituto Politécnico Nacional, México, 33:163-173.
Álvarez, T., and F. Lachica. 1974. Zoogeografía de los vertebrados de México. Pp. 221-295, in: El Escenario Geográfico (A. Flores-Díaz, T. González-Quintero, and F. Lachica, eds.). Secretaria de Educación Pública, Instituto Nacional de Antropología e Historia, México.
Álvarez, T., and O. Polaco. 1980. Nuevos registros de murciélagos para el Estado de Hidalgo, México. Anales de la Escuela Nacional de Ciencias Biológicas, Instituto Politécnico Nacional, México, 23:135-141.
Álvarez, T., and O. Polaco. 1984. Estudio de los mamíferos capturados en La Michilía, sureste de Durango, México. Anales de la Escuela Nacional de Ciencias Biológicas, Instituto Politécnico Nacional, México, 28:99-148.
Álvarez, T., and J. Ramírez-Pulido. 1968. Descripción de una nueva subespecie de *Spermophilus adocetus* (Rodentia, Sciuridae) de Michoacán, México y estado taxonómico de la S. a. arceliae (Villa, B. 1942). Revista de la Sociedad Mexicana de Historia Natural, 29:181-189.
Álvarez, T., and J. Ramírez-Pulido. 1972. Notas acerca de murciélagos mexicanos. Anales de la Escuela Nacional de Ciencias Biológicas, Instituto Politécnico Nacional, México, 19:167-178.
Álvarez, T., and N. Sánchez-Casas. 1997a. Contribución al conocimiento de los mamíferos, excepto Chiroptera y Rodentia, de Michoacán, México. Anales de la Escuela Nacional de Ciencias Biológicas, Instituto Politécnico Nacional, México, 42:47-74.
Álvarez, T., and N. Sánchez-C. 1997b. Notas sobre la alimentación de *Musonycteris* y *Choeroniscus* (Mammalia: Phyllostomidae), en México. Revista Mexicana de Mastozoología, 2:113-115.
Álvarez, T., S.T. Álvarez, and J.C. López-Vidal. 1994. Claves para Murciélagos Mexicanos. Escuela Nacional de Ciencias Biológicas, Instituto Politécnico Nacional, México.

Álvarez, T., S.T. Álvarez-Castañeda, and M. González-Escamilla. 1997. Localidades típicas de mamíferos terrestres de México. Centro de Investigaciones Biológicas del Noroeste, S. C. y Escuela Nacional de Ciencias Biológicas, México.

Álvarez, T., J. Arroyo-Cabrales, and M. González. 1987. Mamíferos (excepto Chiroptera) de la costa de Michoacán, México. Anales de la Escuela Nacional de Ciencias Biológicas, Instituto Politécnico Nacional, México, 31:13-62.

Álvarez, T., P. Domínguez, and J. Arroyo-Cabrales. 1984. Mamíferos de La Angostura, región central de Chiapas. Cuaderno de Trabajo 24, Departamento de Prehistoria, Instituto Nacional de Antropología e Historia, México.

Álvarez-Borrego, S. 1983. Gulf of California. Pp. 427-449, in: Estuaries and Enclosed Seas (B.H. Ketchum, ed.). Elsevier, New York.

Álvarez-Castañeda, S.T. 1991. Nuevos registros de murciélagos (Orden Chiroptera) para los Estados de México y Chiapas. Anales de la Escuela Nacional de Ciencias Biológicas, Instituto Politécnico Nacional, México, 34:215-222.

Álvarez-Castañeda, S.T. 1996. Los Mamíferos del Estado de Morelos. Centro de Investigaciones Biológicas del Noroeste, La Paz, Baja California Sur.

Álvarez-Castañeda, S.T. 2010. Phylogenetic structure of the *Thomomys bottae-umbrinus* complex in North America. Molecular Phylogenetics and Evolution, 54:671-679.

Álvarez-Castañeda, S.T., and T. Álvarez. 1991. Los Murciélagos de Chiapas. Instituto Politécnico Nacional, México.

Álvarez-Castañeda, S.T., and P. Cortes Calva. 1999. Familia Muridae. Pp. 445-468, in: Mamíferos del Noroeste de México (S.T. Álvarez-Castañeda and J.L. Patton, eds.). Centro de Investigaciones Biológicas del Noroeste, S.C., La Paz, México.

Álvarez-Castañeda, S.T., and W. López-Forment, C. 1995. Datos sobre los mamíferos del área aledaña a Palpan, Morelos, México. Anales del Instituto de Biología, Universidad Nacional Autónoma de México, 66:123-133.

Álvarez-Castañeda, S.T., and A. Ortega-Rubio. 2003. Current status of rodents on islands in the Gulf of California. Biological Conservation, 109:157-163.

Álvarez-Castañeda, S.T., and J.L. Patton. 1999. Mamíferos del Noroeste de México. Vol. 1. Centro de Investigaciones Biológicas del Noroeste, La Paz, Baja California Sur.

Álvarez-Castañeda, S.T., and J.L. Patton. 2000. Mamíferos del Noroeste de México. Vol. 2. Centro de Investigaciones Biológicas del Noroeste, La Paz, Baja California Sur.

Álvarez-Castañeda, S. T., and E. Ríos. 2010. Revision of *Chaetodipus arenarius* (Rodentia: Heteromyidae). Zoological Journal of the Linnean Society, 1-16.

Álvarez-Castañeda, S.T., P. Cortés-Calva, and C. Gómez-Machorro. 1998. Peronyscus caniceps. Mammalian Species, 602:1-3.

Álvarez-Romero, J.G., R.A. Medellín, A. Oliveras de Ita, H. Gómez de Silva, and O. Sánchez. 2008. Animales exóticos en México: una amenaza para la biodiversidad. CONABIO, Instituto de Ecología-UNAM, SEMARNAT, México D.F.

Álvarez del Toro, M. 1952. Los Animales Silvestres de Chiapas. Gobierno del Estado, Tuxtla Gutierrez, Chiapas.

Álvarez del Toro, M. 1966. A note on the breeding of Baird's tapir at Tuxtla Gutierrez Zoo. International Zoo Yearbook, 6:196-197.

Álvarez del Toro, M. 1977. Los Mamíferos de Chiapas. Universidad Autónoma de Chiapas, Tuxtla Gutiérrez, Chiapas, México.

Álvarez del Toro, M. 1991. Los Mamíferos de Chiapas. Reimpresión. Instituto de Historia Natural de Chiapas, Gobierno del Estado, Tuxtla Gutiérrez, Chiapas.

Alves-Costa, C.P., G.A.B. Fonseca, and C. Christofaro. 2004. Variation in the diet of the brown-nosed coati (*Nasua nasua*) in southeastern Brazil. Journal of Mammalogy, 85:478-482.

Amano, M., N. Miyazaki, and F. Yanagisawa. 1996. Life history of Fraser's dolphin, *Lagenodelphis hosei*, based on a school captured off the Pacific coast of Japan. Marine Mammal Science, 12:199-214.

Amman, B.R., and R.D. Bradley. 2004. Molecular evolution in *Baiomys* (Rodentia: Sigmodontidae): Evidence for a genetic subdivision in *B. musculus*. Journal of Mammalogy, 85:162-166.

Ammes, N. 1982. Mexican wolf recovery plan. U.S. Fish and Wildlife Service. Albuquerque, NM.

Amos, B., and G.A. Dover. 1991. The use of satellite DNA sequences in determining population differentiation in the Minke whale. Reports of the International Whaling Commission (special issue), 13:235-244.

Amstrup, S.C., and J.J. Beecham. 1976. Activity patterns of radiocollared black bears in Idaho. Journal of Wildlife Management, 40:340-348.

Anciaux de Faveaux. 1971. Catalogue des Acarines parasites et commensaux des Chiropteres Working Document, Institute Royal Scientifique Nationale de Belgique (Bruxelles), 7:201-327.

Andelt, W.F. 1985. Behavioral ecology of coyotes in South Texas. Wildlife Monographs, 94:1-45.

Andersen, K. 1906. Brief diagnoses of a new genus and ten new forms of stenodermatous bats. Annals and Magazine of Natural History (series 7), 18:419-423.

Anderson, R.C., R. Clark, P.T. Madsen, C. Johnson, J. Kiszka, and O. Breysse. 2006. Observations of Longman's beaked whale (*Indopacetus pacificus*) in the western Indian Ocean. Aquatic Mammals, 32:223-231.

Anderson, S. 1956. Extensions of known ranges of Mexican bats. University of Kansas Publications, Museum of Natural History, 9:347-351.

Anderson, S. 1960a. Neotropical bats from western Mexico. University of Kansas Publications, Museum of Natural History, 14:1-8.

Anderson, S. 1960b. The baculum of microtine rodents. University of Kansas Publications, Museum of Natural History, 12:181-216.

Anderson, S. 1962. Tree squirrels (*Sciurus colliaei* group) of western Mexico. American Museum Novitates, 2093:1-13.

Anderson, S. 1964. The systematic status of *Perognathus artus* and *Perognathus goldmani* (Rodentia). American Museum of Natural History, 2184:1-27.

Anderson, S. 1969a. *Macrotus waterhousii*. Mammalian Species, 1:1-4.

Anderson, S. 1969b. Taxonomic status of the woodrat, *Neotoma albigula*, in southern Chihuahua, Mexico. University of Kansas Publications, Museum of Natural History, 51:25-50.

Anderson, S. 1972. Mammals of Chihuahua, taxonomy and distribution. Bulletin of the American Museum of Natural History, 148:149-410.

Anderson, S., and A.S. Gaunt. 1962. A classification of the whitesided jackrabbits of Mexico. American Museum Novitates, 2088:1-16.

Anderson, S., and S. Hadary. 1965. A kit fox from Zacatecas. Journal of Mammalogy, 46:343.

Anderson, S., and J.P. Hubbard. 1971. Notes on geographic variation of *Microtus pennsylvanicus* (Mammalia, Rodentia) in New Mexico and Chihuahua. American Museum Novitates, 2460:1-8.

Anderson, S., and J.K. Jones Jr. 1960. Records of harvest mice *Reithrodontomys*, from Central America, with description of a new subspecies from Nicaragua. University of Kansas Publications, Museum of Natural History, 9:521-529.

Anderson, S., and C.A. Long. 1961. Small mammals in pellets of barn owls from Miñaca, Chihuahua. American Museum Novitates, 2052:1-4.

Anderson, S., and C.E. Nelson, 1960. Birds and mammals in barn owl pellets from near Laguna, Chihuahua, Mexico. The Southwestern Naturalist, 5:99-101.

Anderson, S., and C.E. Nelson. 1965. A systematic revision of *Macrotus* (Chiroptera). American Museum Novitates, 2212:1-39.

Andrews, A.P. 1998. El comercio marítimo de los mayas del Posclásico. Arqueología Mexicana, 6(33):16-23.

Angermann, R., J.E.C. Flux, J.A. Chapman, and A.T. Smith. 1990. Lagomorph classification. Pp. 7-13, in: Rabbits, Hares and Pikas: Status Survey and Conservation Action Plan (J.A. Chapman and J.E.C. Flux, eds.). IUCN and Natural Resources, Gland, Switzerland.

Ankel-Simons, F. 1999. Primate Anatomy: An Introduction. Academic Press, San Diego, CA.
Anonymous. 1981. Manual de campo para la determinación de las especies presentes en las capturas comerciales de atún. Secretaría de Pesca, México.
Anonymous. 1984. Marine Mammals Protection Act of 1972. Annual Report 1983/1984. U.S. Department of Commerce, Washington, DC.
Anonymous. 1989. Found and lost: The rare Florida mastiff bat. Bats, 7:11.
Anthony, A.W. 1925. Expedition to Guadalupe Island, Mexico, in 1922: The birds and mammals. Proceedings of the California Academy of Sciences (4th series), 14:277-322.
Anthony, H.E. 1932. A new genus of rodents from Yucatan. American Museum Novitates, 586:1-13.
Antonelis, G.A., Jr., and C.H. Fiscus. 1980. The pinnipeds of the California current. CalCOFI Report, 21:68-77.
Antonelis, G.A., C.H. Fiscus, and R.L. De Long. 1984. Spring and summer prey of California sea lions, *Zalophus californianus* at San Miguel Island, California 1978-1979. Fishery Bulletin, 82:67-76.
Aplin, K.P., and M. Archer. 1987. Recent advances in marsupial systematics with a new syncretic classification. Pp. xv–lxi, in: Possums and Opossums: Studies in Evolution (M. Archerm, ed.). Royal Zoological Society of New South Wales, Sydney.
Aragón, E.E., and C. Baudoin. 1990. Algunos aspectos reproductivos de dos especies de ardillas del género *Spermophilus* (Rodentia: Sciuridae) en una zona de simpatría del Desierto Chihuahuense. Acta Zoológica Mexicana (n.s.), 36:1-25.
Aragón, E.E, N. Millán, and C. Baudoin. 1993. Ciclos de actividad y organización espacial de las ardillas *Spermophilus spilosoma* y *S. mexicanus* (Rodentia: Sciuridae) en el Desierto Chihuahuense. Pp. 273-287, in: Avances de la Mastozoología en México (R.A. Medellín and G. Ceballos, eds.). Publicaciones Especiales No. 1, Asociación Mexicana de Mastozoología, México.
Araiza, M. 2001. Determinación de sitios potenciales para la reintroducción del lobo gris mexicano. Tesis de maestría, Universidad Nacional de Costa Rica, Heredia, Costa Rica.
Araiza, M., L. Carrillo, R. List, C.A. López González, E. Martínez Meyer, P.G. Martínez-Gutiérrez, O. Moctezuma, N.E. Sánchez-Morales, and J Servín. 2012. Consensus criteria for potential areas for wolf reintroduction in Mexico. Conservation Biology, 26:630-637.
Aranda, M. 1981. Rastros de mamíferos silvestres de México. Instituto Nacional de Investigaciones sobre Recursos Bióticos, Xalapa, Veracruz.
Aranda, M. 1991. Wild mammal skin trade in Chiapas, México. Pp. 175-177, in: Neotropical Wildlife Use and Conservation (J.G. Robinson and K.H. Redford, eds.). University of Chicago Press, Chicago.
Aranda, M. 1992. El jaguar (Panthera onca) en la Reserva Calakmul, México: morfometría, hábitos alimentarios y densidad de población. Pp. 235-274, in: Memorias del Simposio Felinos de Venezuela: Biología, Ecología y Conservación. Caracas, Venezuela.
Aranda, M. 1993. Hábitos alimentarios del jaguar (Panthera onca) en la Reserva de la Biósfera de Calakmul, Campeche. Pp. 231-238, in: Avances en el Estudio de los Mamíferos Mexicanos (R.A. Medellín and G. Ceballos, eds.). Publicaciones Especiales No. 1, Asociación Mexicana de Mastozoología, A.C., México.
Aranda, M. 1994. Importancia de los pecaries (Tayassu spp.) en la alimentación del jaguar (Panthera onca). Acta Zoologica, 62:11-22.
Aranda, M. 1998. Densidad y estructura de una población del jaguar (Panthera onca) en la Reserva de la Biosfera Calakmul, Campeche, Mexico. Acta Zoologica Mexicana, 75:199-201.
Aranda, M. 2000. Huellas y otros rastros de los mamíferos grandes y medianos de México. Instituto de Ecología, CONABIO, México.
Aranda, M., J.E. Escobedo, and C. Pozo. 1997. Registros recientes de *Otonyctomys hatti* (Rodentia: Muridae) en Quintana Roo, México. Acta Zoológica Mexicana, 72:63-65.
Aranda, M., N. López Rivera, and L. López de Buen. 1995. Habitos alimentarios del coyote (Canis latrans) en la Sierra del Ajusco, Mexico. Acta Zoológica Mexicana (n.s.), 65:89-99.
Aranda, M., and I. March. 1987. Guía de los Mamíferos Silvestres de Chiapas. Instituto Nacional de Investigaciones sobre Recursos Bióticos, Xalapa, Veracruz, México.
Aranda, M.S., C. Martínez del Río, L.C. Colmenero, and V.M. Magallón. 1979. Los mamíferos de la Sierra del Ajusco. Comisión Coordinadora para el Desarrollo Agropecuario, Departamento del Distrito Federal, México.
Aranda, M., O. Rosas, J.J. Ríos, and N. García. 2002. Análisis comparativo de la alimentación del gato montés (*Lynx rufus*) en dos diferentes ambientes de México. Acta Zoológica Mexicana (n.s.), 87:99-109.
Arceo-Castro, G., and R. Sánchez-Mantilla. 1992. Especies vegetales consumidas por una hembra de temazate (Mazama americana) en condiciones de cautiverio en el predio de Pipiapan, Mpo. de Catemaco, Ver. Pp. 174-188, in: Memorias X Cabrera Simposio sobre Fauna Silvestre Gral. M.V.Z. Manuel Valtierra. UNAM, México.
Archer, M, and J. Kirsch. 2006. The evolution and classification of marsupials. Pp. 1-18, in: Marsupials (P.J. Armati, C.R. Dickman, and I.D. Hume, eds.). Cambridge University Press, Cambridge, MA.
Arellano, E., and D.S. Rogers. 1994. *Reithrodontomys tenuirostris*. Mammalian Species, 477:1-3.
Arita, H.T. 1988. Revisión taxonómica de los murciélagos magueyeros del género *Leptonycteris* (Chiroptera: Phyllostomatidae). Acta Zoológica Mexicana (n.s.), 29:1-60.
Arita, H.T. 1990. Nose leaf morphology ecological correlates in phyllostomid bats. Journal of Mammalogy, 71:36-47.
Arita, H.T. 1991. Spatial segregation in long-nosed bats, *Leptonycteris nivalis* and *Leptonycteris curasoae*, in Mexico. Journal of Mammalogy, 72:706-714.
Arita, H.T. 1992. Ecology and conservation of cave bat communities in Yucatan, Mexico. PhD dissertation, University of Florida, Gainesville.
Arita, H.T. 1993a. Riqueza de especies de la mastofauna de México. Pp. 109-125, in: Avances en el Estudio de los Mamíferos de México (R.A. Medellín and G. Ceballos, eds.). Asociación Mexicana de Mastozoología, A.C.
Arita, H.T. 1993b. Conservation biology of the cave bats of Mexico. Journal of Mammalogy, 74:693-702.
Arita, H.T. 1993c. Rarity in Neotropical bats: Correlation with phylogeny, diet, and body mass. Ecological Applications, 3:506-517.
Arita, H.T. 1997. Species composition and morphological structure of the bat fauna of Yucatan, Mexico. Journal of Animal Ecology, 66:83-97.
Arita, H.T., and G. Ceballos. 1997. Los mamíferos de México, distribución y estado de conservación. Revista Mexicana de Mastozoología, 2:33-71.
Arita, H.T., and F. Figueroa. 1999. Geographic patterns of body-mass diversity of Mexican mammals. Oikos, 85:310-319.
Arita, H.T., and S.R. Humphrey. 1988. Revisión taxonómica de los murciélagos magueyeros del género *Leptonycteris* (Chiroptera: Phyllostomidae). Acta Zoológica Mexicana (n.s.), 29:1-60.
Arita, H.T., and J. Ortega. 1998. The Middle-American bat fauna: Conservation in the Neotropical-Nearctic border. Pp. 295-308, in: Bat Biology and Conservation (T.H. Kunz and P.A. Racey, eds.). Smithsonian Institution Press, Washington, DC.
Arita, H.T., and P. Rodríguez. 2002. Geographic range, turnover rate, and the scaling of species diversity. Ecography, 25:541-553.
Arita, H.T., and D.E. Wilson. 1987. Long-nosed bats and agaves: The tequila connection. Bats, 5:3-5.
Arita, H.T., J.G. Robinson, and K.H. Redford. 1990. Rarity in Neotropical forest mammals and its ecological correlates. Conservation Biology, 4:181-192.
Arita, H.T., F. Figueroa, A. Frisch, P. Rodríguez, and K. Santos-del Prado. 1997. Geographical range sizes and the conservation of Mexican mammals. Conservation Biology, 11:92-100.
Armstrong, D. M. 1972. Distribution of mammals in Colorado. Monographs of the Museum Natural History, University of Kansas, 3:1-415.

Armstrong, D.M., and J.K. Jones Jr. 1971. Mammals from the Mexican state of Sinaloa. I. Marsupialia, Insectivora, Edentata, Lagomorpha. Journal of Mammalogy, 52:747-757.

Armstrong, D.M., and J.K. Jones Jr. 1972. *Megasorex gigas*. Mammalian Species, 16:1-3.

Armstrong, D.M., J.K. Jones Jr., and E.C. Birney. 1972. Mammals from the Mexican state of Sinaloa. III. Carnivora and Artiodactila. Journal of Mammalogy, 53:48-61.

Arnaud, G., and E. Troyo. 1995. Populations of *Peromyscus pseudocrinitus* in Coronados Island, Gulf of California, Mexico. Peromyscus Newsletter, 20:20-21.

Arnold, J.R., and J. Schonewald. 1972. Notes on the distribution of some bats in southern Mexico. The Wasmann Journal of Biology, 30:171-174.

Arnold, M.L., L.W. Robbins, R.K. Chesser, and J.C. Patton. 1983. Phylogenetic relationships among six species of *Reithrodontomys*. Journal of Mammalogy, 64:128-132.

Arnould, J.P.Y. 2002. Southern fur seals: *Arctocephalus* spp. Pp. 190-201, in: Encyclopedia of Marine Mammals (W.F. Perrin, B. Würsig, and J.G.M. Thewissen, eds.). Academic Press, London.

Arriaga-Flores, J.C. 2010. Segundo registro de *Lasionycteris noctivagans* (Chiroptera, Vespertilinidae) en Mexico. Revista Mexicana de Mastozoologia, 14:46-50.

Arriaga Weiss, S., and W. Contreras Sánchez. 1993. El manati (*Trichechus manatus*) en Tabasco: informe técnico. Villahermosa, Tabasco, Mexico, Universidad Juárez Autónoma de Tabasco División Académica de Ciencias Biológicas.

Arroyo-Cabrales, J., and T. Álvarez. 1990. Restos óseos de murciélagos procedentes de las excavaciones en las grutas de Lol-Tún. Instituto Nacional de Antropología e Historia, México.

Arroyo-Cabrales, J., and J.K. Jones Jr. 1988. Balantiopteryx oi and Baliantopteryx infusca. Mammalian Species, 313:1-2.

Arroyo-Cabrales, J., and R.D. Owen. 1998. Intraspecific variation and phenetic affinities of *Dermanura hartii*, with a comment on use of the generic name *Enchisthenes*. Pp. 67-81, in: Volume Honoring the Late Dr. J. Knox Jones, Jr. (H.H. Genoways and R.J. Baker, eds.). Texas Tech University Press, Lubbock.

Arroyo-Cabrales, J., E.K.V. Kalko, R.K. LaVal, J.E. Maldonado, R.A. Medellín, O.J. Polaco, and B. Rodríguez-Herrera. 2005. Rediscovery of the Mexican flat-headed bat *Myotis planiceps* (Vespertilionidae). Acta Chiropterologica, 7:309-314.

Arroyo-Cabrales, J., R.R. Hollander, and J.K. Jones Jr. 1987. *Choeronycteris mexicana*. Mammalian Species, 291:1-5.

Arroyo-Cabrales, J., O.J. Polaco, and E. Johnson. 2002. La mastofauna del Cuaternario tardío de México. Pp. 103-123, in: Avances en los Estudios Paleomastozoológicos en México (M. Montellano-Ballesteros and J. Arroyo-Cabrales, eds.). Colección Científica 443, Instituto Nacional de Antropología e Historia, México.

Arroyo-Cabrales, J., O. Polaco, D.E. Wilson, and A.L. Gadner. 2008. Nuevos registros de murciélagos para el Estado de Nayarit, México. Revista Mexicana de Mastozoología, 12:141-162.

Asdell, S.A. 1964. Patterns of Mammalian Reproduction. Cornell University Press, Ithaca, NY.

Atramentowicz, M. 1986. Dynamique de population chez trois marsupiaux didelphidés de Guyane. Biotropica, 18:136-149.

Au, D.W.K., and R.L. Pitman. 1986. Sea bird interactions with dolphins and tuna in the eastern tropical Pacific. The Condor, 88:304-317.

Au, D.W.K., W.L. Perryman, and W.F. Perrin. 1977. Dolphin distribution and the relationship to environmental features in the Eastern Tropical Pacific. National Oceanic Atmospheric Administration, Southwest Fisheries Center, Administrative report LJ-79-43. La Jolla, CA.

Audet, D., M.D. Engstrom, and M.B. Fenton. 1993. Morphology, karyology, and echolocation calls of *Rhogeessa* (Chiroptera, Vespertilionidae) from the Yucatan Peninsula. Journal of Mammalogy, 74:498-502.

August, P.V. 1981. Fig fruit consumption by *Artibeus jamaicensis* in the Llanos of Venezuela. Biotropica, 13:70-76.

Auguston, G.F. 1944. A new mouse flea, *Pleochaetoides bullisi*, n.gen., sp., from Texas. Journal of Parasitology, 30:366-368.

Aurioles, G.D. 1982. Contribución al conocimiento de la conducta migratoria del lobo marino de California, Zalophus californianus. Tesis de licenciatura en biología marina, Universidad Autónoma de Baja California Sur, La Paz, Baja California Sur.

Aurioles, G.D. 1988. Behavioral ecology of California sea lions in the Gulf of California. PhD dissertation, University of California, Santa Cruz.

Aurioles, G.D. 1992. On a mass stranding of Baird's beaked whales in the Gulf of California. Mexico. California Fish and Game, 78:116-123.

Aurioles, G.D. 1993. Biodiversidad y estado actual de los mamíferos marinos en México. Revista de la Sociedad Mexicana de Historia Natural, 44:397-412.

Aurioles, G.D., and J. Llinas. 1987. Western gulls as possible predator of California sea lion pups. The Condor, 89:923-924.

Aurioles, G.D., and F. Trillmich. 2008. *Arctocephalus galapagoensis*. In: IUCN 2010. IUCN Red List of Threatened Species. Version 2010.4. www.iucnredlist.org.

Aurioles, G.D., and J.R. Urbán. 1993. Sexual dimorphism in the skull of the pigmy beaked whale, *Mesoplodon peruvianus*. Revista de Investigación Científica, Número Especial SOMEMMA, Universidad Autónoma de Baja California Sur, 1:39-52.

Aurioles, G.D., and A. Zavala-González. 1994. Algunos factores ecológicos que determinan la distribución y abundancia del lobo marino *Zalophus californianus*, en el Golfo de California. Ciencias Marinas, 20(4):535-553.

Aurioles, G.D., J.R. Urbán, and P.L. Enríquez. 1993a. Dwarf sperm whale strandings and sightings on the southwest coast of the Gulf of California, México. In: X Biennial Conference on Biology of Marine Mammals (S.H. Ridgway and R. Harrison, eds.). Academic Press, New York.

Aurioles, G.D., J. Urbán Ramírez, and B. Morales Vela. 1993b. Programa nacional de investigación sobre mamíferos marinos. Pp. 139-159, in: Biodiversidad marina y costera de México (S.I. Salazar-Vallejo and N.E. González, eds.). Comisión Nacional para el conocimiento y uso de la Biodiversidad, Centro de Investigaciones de Quintana Roo, México.

Aurioles, G.D., C. Fox, F. Sinsel, and G. Tanos. 1984. Prey of California sea lions (*Zalophus californianus*) in the Bay of La Paz, Baja California Sur, Mexico. Journal of Mammalogy, 65:519-521.

Aurioles, G.D., J.P. Gallo R., E. Muñoz L,. and J. Egido V. 1989. El delfín de costados blancos *Lagenorhynchus obliquidens* Gill, 1865 (Cetacea: Delphinidae), residente estacional en el suroeste del Golfo de California, México. Anales del Instituto de Biología, UNAM, México, Serie Zoología, 60(3):459-472.

Aurioles, G.D., F. Sinsel, C. Fox, E. Alvarado, and O. Maravilla. 1983. Winter migration of subadult male California sea lions (*Zalophus californianus*) in the southern part of Baja California. Journal of Mammalogy, 64:513-518.

Aurioles-Gamboa, D., and A. Zavala-González. 1994. Ecological factors that determine distribution and abundance of the California sea lion *Zalophus californianus* in the Gulf of California. Ciencias Marinas, 20:535-553.

Aurioles-Gamboa, D., B.J. Le Boeuf, and L.T. Findley. 1993. Registros de pinnípedos poco comúnes para el Golfo de California. Revista de la Investigación Científica de la Universidad Autónoma de Baja California Sur, 1 (No. Especial de la SOMEMMA), 13-19.

Aurioles-Gamboa, D., Y. Schramm, and S. Mesnick. 2004. Galapagos fur seals, *Arctocephalus galapagoensis*, in Mexico. Latin American Journal of Aquatic Mammals 3:77-80.

Ávila Foucat, S., and L. Saad Alvarado. 1998. Valuación de la ballena gris (*Eschrichtius robustus*) y la ballena jorobada (*Megaptera novaeangliae*) en México. Pp. 123-143, in: Aspectos Económicos sobre la Biodiversidad de México (H. Benítez, E. Vega López, A. Peña Jiménez, and S. Ávila Foucat, eds.). CONABIO-Semarnap, México.

Avise, J.C., M.H. Smith, and R.K. Selander. 1974. Biochemical polymorphism and systematics in the genus *Peromyscus*. VI. The *boylii* species group. Journal of Mammalogy, 55:751-763.
Avise, J.C., M.H. Smith, and R.K. Selander. 1979. Biochemical polymorphism and systematics in the genus *Peromyscus*. VII. Geographic differentiation in members of the *truei* and *maniculatus* species groups. Journal of Mammalogy, 60:177-192.
Avise, J.C., M.H. Smith, R.K. Selander, T.E. Lawlor, and P.R. Ramsey. 1974. Biochemical polymorphism and systematics in the genus *Peromyscus*. V. Insular and mainland species of the subgenus *Haplomylomys*. Systematic Zoology, 23:226-238.
Ayala-Castañares, A. 1982. Las ciencias del mar y el desarrollo de México. Ciencia y Desarrollo, 6:6-28.
Ayres, J.M., D. de Magalhaes, E. de Sousa, and J.K. Barreiros. 1991. On the track of the road: Changes in subsistence hunting in a Brazilian Amazonian village. Pp. 82-92, in: Neotropical Wildlife Use and Conservation (J.G. Robinson and K.H. Redford, eds.). University of Chicago Press, Chicago.
Baar, S.L., E.D. Fleharty, and M.F. Artman. 1974. Utilization of deep burrows and nests by cotton rats in west-central Kansas. The Southwestern Naturalist, 19:440-444.
Baca, M.J. del. 1982. Estudio ecológico de la subcomunidad de roedores en el Ajusco, D.F. Tesis de licenciatura, Escuela Nacional de Ciencias Biológicas, Instituto Politécnico Nacional, México.
Baccus, J.F. 1978. Notes on distribution of some mammals from Coahuila. The Southwestern Naturalist, 23:706-708.
Bachelard, G. 1942. El agua y los sueños. Ensayo sobre la imaginación de la materia. Fondo de Cultura Económica. México.
Bailey, T.N. 1971. Biology of striped skunk on a southwestern Lake Erie marsh. The American Midland Naturalist, 85:196-207.
Bailey, T.N. 1974. Social organization in a bobcat population. Journal of Wildlife Management, 38:435-446.
Bailey, V. 1900. Revision of American voles of the genus *Microtus*. North American Fauna, 17:1-88.
Bailey, V. 1905. Biological survey of Texas. North American Fauna, 25:1-222.
Bailey, V. 1931 (= 1932). Mammals of New Mexico. North American Fauna, 53:1-412.
Baillie, J., and B. Groombridge. 1996. 1996 IUCN Red List of Threatened Animals. IUCN, Gland, Switzerland.
Baird, A.B., M.R. Marchán-Rivadeneira, S.G. Pérez, and R.J. Baker. 2012. Morphological analysis and description of two new species of *Rhogeessa* (Chiroptera: Vespertilionidae) from the Neotropics. Occasional Papers of the Museum, Texas Tech University, 307:1-32.
Baird, R.W., and P.J. Stacey. 1990. Status of the Northern right whale dolphin *Lissodelphis borealis* in Canada. The Canadian Field Naturalist, 105:243-250.
Baird, S.F. 1857. Part I. General report upon the zoology of the several Pacific Railroad routes. Reports on explorations and surveys to ascertain the most practicable and economic route for a railroad from the Mississippi River to the Pacific Ocean. Vol. 8. Beverly Tucker, Printer, Washington, DC.
Baird, S.F. 1858. Part II. Reports of explorations and surveys, to ascertain the most practicable and economical route for a railroad from the Mississippi River to the Pacific Ocean. Beverley Tucker, Printer, Washington, DC.
Baker, C.L. 1971. Geologic reconnaissance of the eastern Cordillera of Mexico. Geological Society of America Special Paper, 131:1-83.
Baker, C.S., and L. Medrano-González. 2002. World-wide distribution and diversity of humpback whale mitochondrial DNA lineages. In: Molecular and Cell Biology of Marine Mammals (C.J. Pfeiffer, ed.). Krieger Publishing Co., Malabar, FL.
Baker, J.G. 1986. Patterns of plant invasions in North America. Pp. 44-57, in: Ecology of Biological Invasions of North America and Hawaii (H.A. Mooney and J.A. Drake, eds.). Springer-Verlag, New York.
Baker, R.H. 1951a. Mammals from Tamaulipas, Mexico. University of Kansas Publications, Museum of Natural History, 5:207-218.
Baker, R.H. 1951b. Two new moles (genus *Scalopus*) from Mexico and Texas. University of Kansas Publications, Museum of Natural History, 5:17-24.
Baker, R.H. 1953. Mammals from owl pellets taken in Coahuila, Mexico. Transactions Kansas Academy of Sciences, 56:253-254.
Baker, R.H. 1954. The silky pocket mouse (*Perognathus flavus*) of Mexico. University of Kansas Publications, Museum of Natural History, 7:339-347.
Baker, R.H. 1955. A new species of bat (genus *Myotis*) from Coahuila, Mexico. Proceedings of the Biological Society of Washington, 68:165-166.
Baker, R.H. 1956. Mammals of Coahuila, Mexico. University of Kansas Publications, Museum of Natural History, 9:125-335.
Baker, R.H. 1960. Mammals of the Guadiana Lava field, Durango, Mexico. Publications of the Museum of Michigan State University, Biological Series, 1:305-327.
Baker, R.H. 1962. Additional records of *Notiosorex crawfordi* from Mexico. Journal of Mammalogy, 43:283.
Baker, R.H. 1966. Further notes on the mammals of Durango, Mexico. Journal of Mammalogy, 47:344-345.
Baker, R.H. 1968. Records of opossum and kit fox from Zacatecas. Journal of Mammalogy, 49:318.
Baker, R.H. 1969a. Cotton rats of the *Sigmodon fulviventer* group. University of Kansas Publications, Museum of Natural History, 51:177-232.
Baker, R.H. 1969b. Records of the badger from Mexico. The Southwestern Naturalist, 14:251-252.
Baker, R.H. 1977. Mammals of the Chihuahuan desert region: Future prospects. Pp. 221-225, in: Transactions of the Symposium on the Biological Resources of the Chihuahuan Desert Region United States and Mexico (R.H. Wauer and D.H. Riskind, eds.). U.S. Department of the Interior, National Park Service Transactions and Proceedings Series, 3, Washington, DC.
Baker, R.H. 1991. The classification of Neotropical mammals--a historical résumé. Pp. 7-32, in: Latin American Mammalogy: History, Biodiversity, and Conservation (M.A. Mares and D.J. Schmidly, eds.). University of Oklahoma Press, Norman.
Baker, R.H., and M.W. Baker. 1975. Montane habitat use by the spotted skunk (*Spilogale putorius*) in Mexico. Journal of Mammalogy, 56:671-673.
Baker, R.H., and R.W. Dickerman. 1956. Daytime roost of the yellow bat in Veracruz. Journal of Mammalogy, 37:443.
Baker, R.H., and J.K. Greer. 1960. Notes on Oaxaca mammals. Journal of Mammalogy, 41:413-415.
Baker, R.H., and J.K. Greer. 1962. Mammals of the Mexican State of Durango. Publications of the Museum, Michigan State University, Biological Series, 2:29-159.
Baker, R.H., and M.K. Petersen. 1965. Notes on a climbing rat, Tylomys from Oaxaca, Mexico. Journal of Mammalogy, 494-495.
Baker, R.H., and C.J. Phillips. 1965a. Mammals from El Nevado de Colima, Mexico. Journal of Mammalogy, 46:691-693.
Baker, R.H., and C.J. Phillips. 1965b. *Peromyscus ochraventer* in San Luis Potosí. Journal of Mammalogy, 46:337-338.
Baker, R.H., and C. Sánchez-Hernández. 1973. Observaciones sobre el zorrillo pigmeo manchado, *Spilogale pygmaea*. Anales del Instituto de Biología, UNAM, México, Serie Zoología, 44:61-64.
Baker, R.H., and K.A. Shump Jr. 1978a. *Sigmodon fulviventer*. Mammalian Species, 94:1-4.
Baker R.H., and K.A. Shump Jr. 1978b. *Sigmodon ochrognathus*. Mammalian Species, 97:1-2
Baker, R.H., and H.J. Stains. 1955. A new long-eared Myotis (*Myotis evotis*) from northeastern Mexico. University of Kansas Publications, Museum of Natural History, 9:81-84.
Baker, R.H., and B. Villa-Ramírez. 1953. Mamíferos registrados por primera vez en el Estado de Hidalgo. Revista de la Sociedad Mexicana de Historia Natural, 14:93-108.

Baker, R.H., and B. Villa-Ramírez. 1960. Distribución geográfica y poblaciones actuales del lobo gris en México. Anales del Instituto de Biología, UNAM, México, Serie Zoología, 30:369-374.

Baker, R.H., and R.G. Webb. 1966. Notas acerca de los anfibios, reptiles y mamíferos de la Pesca, Tamaulipas. Revista de la Sociedad Mexicana de Historia Natural, 27:179-190.

Baker, R.H., and R.G. Webb. 1976. *Thamnophis elegans* captures Sorex emarginatus. Herpetological Review, 7:112.

Baker, R.H., and D. Womochel. 1966. Mammals from southern Oaxaca. The Southwestern Naturalist, 11:306.

Baker, R.H., R.G. Webb, and P. Dalbyn. 1967. Notes on reptiles and mammals from southern Zacatecas. The American Midland Naturalist, 77:223-226.

Baker, R.H., R.G. Webb, and E. Stern. 1971. Amphibians, reptiles and mammals from north-central Chiapas. Anales del Instituto de Biología, UNAM, México, Serie Zoología, 42:77-85.

Baker, R.H., M.W. Baker, J.D. Johnson, and E.G. Webb. 1981. New records of mammals and reptiles from northwestern Zacatecas, México. The Southwestern Naturalist, 25:568-569.

Baker, R.J. 1967. Karyotypes of bats of the family Phyllostomidae and their taxonomic implications. The Southwestern Naturalist, 12:407-428.

Baker, R.J. 1973. Comparative cytogenetics of the New World leafnosed bats (Phyllostomidae). Periodicum Biologorum, 75:37-45.

Baker, R.J. 1979. Karyology. Pp. 107-176, in: Biology of Bats of the New World Family Phyllostomatidae. Part III. (R.J. Baker, J.K. Jones Jr., and D.C. Carter, eds.). Special Publications of the Museum, Texas Tech University, 16:1-441.

Baker, R.J. 1984. A sympatric cryptic species of mammal: A new species of *Rhogeessa* (Chiroptera: Vespertilionidae). Systematic Zoology, 33:178-183.

Baker, R.J., and C.L. Clark. 1987. *Uroderma bilabatum*. Mammalian Species, 279:1-4.

Baker, R.J., and T.C. Hsu. 1970. Further studies of the sex-chromosome system of the American leaf-nosed bats (Chiroptera, Phyllostomidae). Cytogenetics, 9:131-138.

Baker, R.J., and J.K. Jones Jr. 1972. *Tadarida aurispinosa* in Sonora, Mexico. The Southwestern Naturalist, 17:308-309.

Baker, R.J., J.K. Jones Jr., and D.C. Carter (eds.). 1979. Biology of bats of the New World family Phyllostomatidae. Part III. Special Publications of the Museum, Texas Tech University, 16:1-441.

Baker, R.J., and R.G. Jordan. 1970. Chromosomal studies of some Neotropical bats of the families Emballonuridae, Noctilionidae, Natalidae and Vespertilionidae. Caryologia, 4:595-604.

Baker, R.J., and J.T. Mascarello. 1969. Karyotypic analyses of the genus *Neotoma* (Cricetidae: Rodentia). Cytogenetics, 8:187-198.

Baker, R.J., and J.L. Patton. 1967. Karyotypes and karyotypic variation of North American vespertilionid bats. Journal of Mammalogy, 48:270-286.

Baker, R.J., and M.C. Ward. 1967. Distribution of bats in southeastern Arkansas. Journal of Mammalogy, 48:130-132.

Baker, R.J., and S.L. Williams. 1974. *Geomys tropicalis*. Mammalian Species, 35:1-4.

Baker, R.J., J.W. Bickham, and M.L. Arnold. 1985. Chromosomal evolution in Rhogeessa (Chiroptera: Vespertilionidae): Possible speciation by centric fusions. Evolution, 39:233-243.

Baker, R.J., C.G. Dunn, and K. Nelson. 1988. Allozymic study of the relationships of Phylloderma and four species of Phyllostomus. Occasional Papers of the Museum, Texas Tech University, 125:1-14.

Baker, R.J., A.L. Gardner, and J.L. Patton. 1972. Chromosomal polymorphism in the phyllostomatid bat, Mimon crenulatum (Geoffroy-St. Hilaire, 1810). Experientia, 28:969-970.

Baker, R.J., H.H. Genoways, and P.A. Seyfarth. 1981. Results of the Alcoa Foundation-Suriname expedition. VI. Additional chromosomal data for bats (Mammalia: Chiroptera) from Suriname. Annals of Carnegie Museum, 50:333-344.

Baker, R.J., T. Mollhagen, and G. Lopez. 1971. Notes on *Lasiurus ega*. Journal of Mammalogy, 52:849-853.

Baker, R.J., S. Solari, and F.G. Hoffmann. 2002. A new Central American species from the *Carollia brevicauda* complex. Occasional Papers of the Museum, Texas Tech University, 217:i + 1-11.

Baker, R.J., M.B. O'Neill, and L.R. McAliley. 2003. A new species of desert shrew, *Notiosorex*, based on nuclear and mitochondrial sequence data. Occasional Papers of the Museum, Texas Tech University, 22:1-12.

Baker, R.J., J.C. Patton, H.H. Genoways, and J.W. Bickham. 1988. Genetic studies of *Lasiurus* (Chiroptera:Vespertilionidae). Occasional Papers of the Museum, Texas Tech University, 17:1-15.

Baker, R.J., C.A. Porter, J.C. Patton, and R.A. Van Den Bussche. 2000. Systematics of bats of the family Phyllostomidae based on RAG2 DNA sequences. Occasional Papers of the Museum, Texas Tech University, 202:1-16.

Baker, R.R. 1978. The Evolutionary Ecology of Animal Migration. Holmes and Meier, New York.

Balcomb, K.C., III. 1989. Baird's beaked whale *Berardius bairdii* Stejneger, 1883; Arnoux's beaked whale *Berardius arnuxii* Dubernoy, 1851. Pp. 188-261, in: Handbook of Marine Mammals (S.H. Ridway and S.R. Harrison, eds.). Vol. 4. Academic Press, Orlando, FL.

Balcomb, K.C., III. 1991. Kith and kin of the killer whale. Pacific Discovery, California Academy of Sciences, 44:8-17.

Baldwin, J.D., and J.I. Baldwin. 1972. Population density and use of space in howling monkeys (*Alouatta villosa*) in southwestern Panama. Primates, 18:161-184.

Baldwin, J.D., and J.I. Baldwin. 1974. Warum brüllen Brülaffen? Umschau, 22:712-713.

Ballance, L.T. 1992. Habitat use patterns and dolphin ranges of the bottlenose dolphin in the Gulf of California, Mexico. Marine Mammal Science, 8:262-274.

Banbury, J.L., and G.S. Spicer. 2007. Molecular systematics of chipmunks (*Neotamias*) inferred by mitochondrial control region sequences. Journal of Mammalian Evolution, 14:149-162.

Banfield, A.W.F. 1974. The Mammals of Canada. University of Toronto Press, Toronto.

Banks, R.C. 1964a. Birds and mammals of the voyage of "La Gringa." Transactions of the San Diego Society of Natural History, 13:177-184.

Banks, R.C. 1964b. Range extensions for three bats in Baja California, Mexico. Journal of Mammalogy, 45:489.

Banks, R.C. 1965. Some information from barn owl pellets. Auk, 82:506.

Banks, R.C. 1967. The *Peromyscus guardia-interparietalis* complex. Journal of Mammalogy, 48:210-218.

Banks, R.C., and R.L. Brownell. 1969. Taxonomy of the common dolphins of the eastern Pacific Ocean. Journal of Mammalogy, 50:262-271.

Barbour, H.L. 1974. A north temperate bat community: Structure and prey populations. Journal of Mammalogy, 55:138-157.

Barbour, R.W., and W.H. Davis. 1969. Bats of America. University Press of Kentucky. Lexington.

Barcénas, H.V., Y. Rubio-Rocha, E. Nájera-Solís, L.D. López, and R.A. Medellín. 2009. Ampliación de la distribución de tres carnívoros en el noroeste de México. Revista Mexicana de Mastozoología, 13:115-122.

Barkalow, F.S., and J.B. Funderburg. 1960. Probable breeding and additional records of the seminole bats in North Carolina. Journal of Mammalogy, 41:394-395.

Barlow, J. 1986. Factors affecting the recovery of *Phocoena sinus*, the vaquita or Gulf of California harbor porpoise. Southwest Fisheries Center Administrative Report LJ-86<H>37.

Barlow, J., and J. Calambokidis. 1995. Abundance of blue and humpback whales in California: A comparison of mark-recapture and line-transect estimates. P. 8, in: Eleventh Biennial Conference on the Biology of Marine Mammals. Orlando, FL.

Barlow, J., T. Gerrodette, and G. Silber. 1997. First estimates of vaquita abundance. Marine Mammal Science, 13:44-58.

Barrera, A.M. 1952. Algunas observaciones sobre las ratas que constituyen plaga en los cocotales de la Costa Chica de Guerrero. Fitófilo, 6:5-10.

Barrera, A.M. 1953. Sinopsis de los sifonápteros de la cuenca de México. Anales de la Escuela Nacional de Ciencias Biológicas, Instituto Politécnico Nacional, México, 7:155-245.

Barrera, A.M. 1954. Notas sobre sifonápteros VII. Lista de especies colectadas en el Municipio de Huitzilac, Morelos y descipción de *Peromyscopsylla zempoalensis*. Ciencia México, 14:87-90.

Barrera, A.M. 1955a. Las especies mexicanas del género Pulex Linnaeus (Siph. Pulicid.). Anales de la Escuela Nacional de Ciencias Biológicas, Instituto Politécnico Nacional, México, 8:219-236.

Barrera, A.M. 1955b. Notas sobre sifonápteros. I. Algunas especies mexicanas; consideraciones sobre su distribución geográfica. Revista de la Sociedad Mexicana de Entomología, 1-2:83-98.

Barreto, G.R., and O.E. Hernández. 1988. Aspectos bioecológicos de los Báquiros (Tayassu tajacu y T. pecari) en el Estado de Cojedes: estudio comparativo. Tesis de licenciatura, Escuela de Biología, Universidad Central de Venezuela, Caracas.

Bartig, J.L., T.L. Best, and S.L. Burt. 1993. *Tamias bulleri*. Mammalian Species, 438:1-4.

Bassols, B.I. 1981. Catálogo de los ácaros Mesostigmata de mamíferos de México. Anales de la Escuela Nacional de Ciencias Biológicas, Instituto Politécnico Nacional, México, 24:9-49.

Bateman, G.C., and T.A. Vaughan. 1974. Nightly activities of mormoopid bats. Journal of Mammalogy, 55:45-65.

Baudoin, C., and E.E. Aragón. 1991. Caractéristiques des déplacements et des domaines vitaux de deux espèces sympatriques de Spermophiles du Bolson de Mapimi (Mexique): étude par la méthode de capture-recapture et par radiopistage. Pp. 153-161, in: Le Rongeur et l' Espace (M. Le Berre and L. Le Guelte, eds.). R. Chabaud, Paris.

Baudoin, C., G. Arnaud, N. Millán, E.E. Aragón, and A. Demouron. 1991. Etude expérimentale de l' adaptabilité de deux espèces sympatriques de Spermophiles aux variations d'enviromment pendant la période hivernale. Conséquences possibles sur leur répartition dans le désert Chihuahua (Durango, México). Pp. 91-99, in: Le Rongeur et l' Espace (M. Le Berre and L. Le Guelte, eds.). R. Chabaud, Paris.

Baumgardner, G.D. 1991. *Dipodomys compactus*. Mammalian Species, 369:1-4.

Baumgardner, G.D., and M.L. Kennedy. 1994. Patterns of interspecific morphometric variation in kangaroo rats (genus Dipodomys). Journal of Mammalogy, 75:203-211.

Baumgardner, G.D., and D.J. Schmidly. 1981. Systematics of the southern races of two species of kangaroo rats (*Dipodomys compactus* and *Dipodomys ordii*). Occasional Papers of the Museum, Texas Tech University, 73:1-27.

Baumgarther, L.L. 1939. Fox squirrel dens. Journal of Mammalogy, 20:456-465.

Baumgartner, M.F. 1997. The distribution of Risso's dolphin (*Grampus griseus*) with respect to the physiography of the northern Gulf of Mexico. Marine Mammal Science, 13(4):614-638.

Bautista, S. M. 2006. Distribución, abundancia y dieta de perros y gatos ferales en la Isla Cozumel. Master's thesis, Instituto de Ecología, A.C., Xalapa, México.

Bautista-Romero, J., H. Reyes Bonilla, D.B. Lluch-Cota, and S.E. Lluch-Cota. 1994. Aspectos generales de la fauna marina. Pp. 247-275, in: La Isla Socorro, Reserva de la Biósfera Archipiélago de Revillagigedo (A. Ortega Rubio and A. Castellanos Vera, eds.). Publicaciones Especiales No. 8. CIBNOR, La Paz, Baja California Sur.

Beasom, S.L., and R.A. Moore. 1977. Bobcat food habit response to a change in prey abundance. The Southwestern Naturalist, 21:451-457.

Beck, C.A., and D.J. Forrester. 1988. Helminths of the Florida manatee, *Trichechus manatus latirostris*, with a discussion and summary of the parasites of sirenians. Journal of Parasitology, 74:628-637.

Beck, T.D.I. 1991. Black bears in west-central Colorado. Technological Publication, 39:39-91. Colorado Division of Wildlife, Fort Collins.

Beidelman, R.G. 1955. An altitudinal record for bison in northern Colorado. Journal of Mammalogy, 36:470-471.

Beier, P. 1993. Determining minimum habitat areas and habitat corridors for cougars. Conservation Biology, 7:94-108.

Bekoff, M., and M.C. Wells. 1980. The social ecology of coyotes. Scientific American, 242:130-148.

Bell, D.J., J. Hoth, A. Velázquez, F.J. Romero, L. León, and M. Aranda. 1985. A survey of the distribution of the volcano rabbit *Romerolagus diazi*: An endangered Mexican endemic. Dodo, Journal of the Jersey Wildlife Preservation Trust, 22:42-48.

Bell, D.M., M.J. Hamilton, C.W. Edwards, L.E. Wiggins, R. Muñiz-Martínez, R.E. Strauss, R.D. Bradley, and R.J. Baker. 2001. Patterns of karyotypic megaevolution in *Reithrodontomys*: Evidence from a cytochrome-b phylogenetic hypothesis. Journal of Mammalogy, 82:81-91.

Bell, G.P. 1980. Habitat use and response to patches of prey by desert insectivorous bats. Canadian Journal of Zoology, 58:1876-1883.

Bell, G.P. 1982. Behavioral and ecological aspects of gleaning by a desert insectivorous bat, *Antrozous pallidus* (Chiroptera: Vespertilionidae). Behavioral Ecology and Sociobiology, 10:217-223.

Bello, J. 1993. Situación actual del Orden Artiodactyla en la región de los Tuxtlas, Veracruz. Tesis de licenciatura, Facultad de Biología, Universidad Veracruzana, Xalapa, Veracruz.

Beltran, E. 1941. Examen protozoólogico de dos especies de murciélagos (*Myotis velifer* y *Tadarida brasiliensis*) de Tepoztlán, Morelos. Revista de la Sociedad Mexicana de Historia Natural, 2:273-277.

Belwood, J.J. 1988. The influence of bat predation on calling behavior in Neotropical forest katydis (Insecta: Orthoptera: Tettigonidae). PhD dissertation, University of Florida, Gainesville.

Benavides, T., and J. Villarreal. 1994. Bosquejo histórico del manejo y administración del venado cola blanca texano (*Odocoileus virginianus texanus*) en el Noreste de México. Pp. 5-14, in: IV Simposio sobre Venados de México. Universidad Nacional Autónoma de México, México.

Bendel, P.R., and J.E. Gates. 1987. Home range and microhabitat partitioning of the southern flying squirrel (*Glaucomys volans*). Journal of Mammalogy, 68:243-255.

Bennett, E.T. 1833. Characters of new species of mammalia from California. Proceedings of the Zoological Society of London, 39-42.

Benson, D.L., and F.R. Gehlbach. 1979. Ecological and taxonomic notes on rice rat (*Oryzomys couesi*) in Texas. Journal of Mammalogy, 60:225-228.

Benson, S.B. 1938. Notes on kit foxes (*Vulpes macrotis*) from Mexico. Proceedings of the Biological Society of Washington, 51:17-24.

Benson, S.B. 1939. Description and records of harvest mice (*Reithrodontomys*) from Mexico. Proceedings of the Biological Society of Washington, 52:147-150.

Benson, S.B. 1947. Description of a subspecies of *Myotis yumanensis* from Baja California, Mexico. Proceedings of the Biological Society of Washington, 60:45-46.

Benson, S.B. 1953. A record of the porcupine (*Erethizon dorsatum*) from Sonora, Mexico. Journal of Mammalogy, 34:511-512.

Benton, M., and D. Harper. 1997. Basic Palaeontology. Addison Wesley Longman, Essex, England.

Berg, William E. 1981. Ecology of bobcats in northern Minnesota. Pp. 55-61, in: Bobcat Research Conference: Proceedings (L.G. Blum and P.C. Escherich, eds.), 16-18 October 1979, Front Royal, VA. NWF Science and Technical Series No. 6. National Wildlife Federation, Washington, DC.

Bernal, J.A. 1978. Estado actual del castor, *Castor canadensis* V. Bailey 1913, en el estado de Nuevo León, México. Tesis de licenciatura, Facultad de Ciencias Biológicas, Universidad Autónoma de Nuevo León, Monterrey.

Bernardi, G., S.R. Fain, J.P. Gallo-Reynoso, A.L. Figueroa-Carranza, and B.J. Le Boeuf. 1998. Genetic variability in Guadalupe fur seals. Journal of Heredity, 89:301-305.

Bernardz, J.C., and J.A. Cook. 1984. Distribution and numbers of the white-sided jackrabbit (*Lepus callotis gaillardi*) in New Mexico. The Southwestern Naturalist, 29:358-360.

Berry, D.L., and R.J. Baker. 1972. Chromosomes of pocket gophers of the genus *Pappogeomys*, subgenus *Cratogeomys*. Journal of Mammalogy, 53:303-309.

Bertram, G.C.L.. and C.K.R. Bertram. 1973. The modern Sirenia: Their distribution and status. Biological Journal of the Linnean Society 5:297-338.

Bérubé, M., A. Aguilar, D. Dendanto, F. Larsen, G. Notobartolo Di Sciara, R. Sears, J. Sigurjónsson, J. Urbán-R, and P.J. Palsboll. 1998. Population genetic structure of North Atlantic, Mediterranean Sea and Sea of Cortez fin whales, *Balaenoptera physalus* (Linnaeus 1758): Analysis of mitochondrial and nuclear loci. Molecular Ecology, 7:585-599.

Berzín, A.A. 1978. Whale distribution in tropical eastern Pacific waters. Reports of the International Whaling Commission, 28:173-177.

Besharse, J.C. 1971. Maturity and sexual dimorphism in the skull, mandible and teeth of the beaked whale, *Mesoplodon densirostris*. Journal of Mammalogy, 52:297-315.

Best, P.B. 1974. The biology of the sperm whale as it relates to stock management. Pp. 257-293, in: The Whale Problem, A status Report (W.E. Schevill, ed.). Harvard University Press, Cambridge, MA.

Best, P.B. 1977. Two allopatric forms of Bryde's whale off South Africa. Report of the International Whale Commission (special issue), 10-38.

Best, P.B., and P.D. Shaughnessy. 1981. First record of the melon-head whale, *Peponocephala electra*, from South Africa. Annals of the South African Museum, 83:33-47.

Best, P.B., P.A.S. Canham, and N. MacLeod. 1984. Patterns of reproduction in sperm whale, *Physeter macrocephalus*. Reports of the International Whaling Commission (special issue), 16:51-79.

Best, R.C. 1979. Food and feeding habitats of wild and captive Sirenia. Mammal Review, 11:3-29.

Best, R.C., and A.Y. Harada. 1985. Food habits of the silky anteater (*Cyclopes didactylus*) in the Central Amazon. Journal of Mammalogy, 66:780-781.

Best, T.L. 1973. Ecological separation of three genera of pocket gophers (Geomyidae). Ecology, 54:1311-1319.

Best, T.L. 1978. Variation in kangaroo rats (genus *Dipodomys*) of the *heermanni* group in Baja California, Mexico. Journal of Mammalogy, 59:160-175.

Best, T.L. 1981. Bacular variation in kangaroo rat (Genus *Dipodomys*) of the *heermanni* group in Baja California, Mexico. The Southwestern Naturalist, 25:529-534.

Best, T.L. 1983. Morphologic variation in the San Quintin kangaroo rat (*Dipodomys gravipes* Huey 1925). The American Midland Naturalist, 109:409-413.

Best, T.L. 1988a. *Dipodomys nelsoni*. Mammalian Species, 326:1-4.

Best, T.L. 1988b. *Dipodomys spectabilis*. Mammalian Species, 311:1-10.

Best, T.L. 1992. *Dipodomys venustus*. Mammalian Species, 403:1-4.

Best, T.L. 1993. Patterns of morphology and morphometric variation in heteromyid rodents. Pp. 197-230, in: Biology of the Heteromyidae (H.H. Genoways and J.H. Brown, eds.). Special Publication 10, American Society of Mammalogists.

Best, T.L. 1995a. *Spermophilus annulatus*. Mammalian Species, 508:1-4.

Best, T.L. 1995b. *Sciurus alleni*. Mammalian Species, 501:1-4.

Best, T.L., and G. Ceballos. 1995. *Spermophilus perotensis*. Mammalian Species, 507:1-3.

Best, T.L., and L.L. Janecek. 1992. Allozymic and morphologic variation among *Dipodomys insularis*, *Dipodomys nitratoides*, and two populations of *Dipodomys merriami* (Rodentia: Heteromyidae). The Southwestern Naturalist, 37:1-8.

Best, T.L., and J.A. Lackey. 1985. *Dipodomys gravipes*. Mammalian Species, 236:1-4.

Best, T.L., and J.A. Lackey. 1992a. *Chaetodipus artus*. Mammalian Species, 418:1-3.

Best, T.L., and J.A. Lackey. 1992b. *Chaetodipus pernix*. Mammalian Species, 420:1-3.

Best, T.L., and H.H. Thomas. 1991a. *Dipodomys insularis*. Mammalian Species, 374:1-3.

Best, T.L., and H.H. Thomas. 1991b. *Spermophilus madrensis*. Mammalian Species, 378:1-2.

Best, T.L., S.L. Burt, and J.L. Bartig. 1993. *Tamias durangae*. Mammalian Species, 437:1-4.

Best, T.L., C. Intress, and K.D. Shull. 1988. Mound structure in the three taxa of Mexican kangaroo (*Dipodomys spectabilis cratodon*, *D. s. zygomaticus* and *D. nelsoni*). The American Midland Naturalist, 119:216-220.

Best, T.L., H.A. Ruiz-Piña, and L. León-Paniagua. 1995. *Sciurus yucatanensis*. Mammalian Species, 506:1-4.

Best, T.L., R.K. Chesser, D.A. McCullough, and G.D. Baumgardner. 1996. Genic and morphometric variation in kangaroo rats, genus *Dipodomys*, from coastal California. Journal of Mammalogy, 77:785-800.

Best, T.L., C.L. Lewis, K. Caesar, and A.S. Titus. 1990a. *Ammospermophilus interpres*. Mammalian Species, 365:1-6.

Best, T.L., R.M. Sullivan, J.A. Cook, and T.L. Yates. 1986. Chromosomal, genic, and morphological variation in the agile kangaroo rat, *Dipodomys agilis* (Rodentia: Heteromyidae). Systematic Zoology, 35: 311-324.

Best, T.L., A.S. Titus, K. Caesar, and C.L. Lewis. 1990b. *Ammospermophilus harrisii*. Mammalian Species, 366:1-7.

Best, T.L., K. Caesar, A.S. Titus, and C.L. Lewis. 1990c. *Ammosperphilus insularis*. Mammalian Species, 364:1-4.

Bickham, J.W. 1979a. Banded karyotypes of 11 species of American bats (genus *Myotis*). Cytologia, 44:789-797.

Bickham, J.W. 1979b. Chromosomal variation and evolutionary relationships of vespertilionid bats. Journal of Mammalogy, 60:350-363.

Bickham, J.W. 1987. Chromosomal variation among seven species of lasiurine bats (Chiroptera:Vespertilionidae). Journal of Mammalogy, 68:837-842.

Bigg, M.A. 1981. Harbour seal—*Phoca vitulina* and *P. largha*. Pp. 1-27, in: Handbook of Marine Mammals (S.H. Ridgway and R.J. Harrison, eds.). Vol. 2. Academic Press, London.

Bigg, M.A., G.M. Ellis, J.K.B. Ford, and K.C. Balcomb. 1987. Killer Whales. Phantom Press, British Columbia, Canada.

Bigler, W.J. 1966. A marking harness for the collared peccary. Journal of Wildlife Management, 30:213-214.

Bigler, W.J., and J.H. Jenkins. 1975. Population characteristics of *Peromyscus gossypinus* and *Sigmodon hispidus* in tropical hammocks of South Florida. Journal of Mammalogy, 56:633-644.

Bigler, W.J., E.H. Lebetkin, P.M. Cumbie, A.G. Cook, and W.G. Meginnis. 1977. Response of a cotton rat population to increased density. The American Midland Naturalist, 97:10-17.

Birney, E.C. 1973. Systematics of three species of woodrats (genus *Neotoma*) in central North America. University of Kansas Publications, Museum of Natural History, 58:1-173.

Birney, E.C., and J.K. Jones Jr. 1972. Woodrats (genus *Neotoma*) of Sinaloa, Mexico. Transactions of the Kansas Academy of Science, 74:197-211.

Birney, E.C., J.B. Bowles, R.M. Timm, and S.L. Williams. 1974. Mammalian distributional records in Yucatan and Quintana Roo, with comments on reproduction, structure, and status of peninsular populations. Occasional Papers of the Bell Museum of Natural History, University of Minnesota, 13:1-25.

Bisbal, F.J. 1991. Biología y habitat del venado matacan. Pp. 67-75, in: Memorias Simposio El Venado en Venezuela: Conservación, Manejo Aspectos Biológicos y Legales. Fudeci-Profauna-Fedecave, Caracas, Venezuela.

Bissonette, J.A. 1982. Collared peccary. Pp. 841-850, in: Wild Mammals of North America: Biology, Management, Economics (J.A. Chapman and G.A. Feldhamer, eds.). Johns Hopkins University Press, Baltimore.

Black, H. 1974. A north temperate bat community: Structure and prey populations. Journal of Mammalogy, 55:138-157.

Black, H.L. 1976. American kestrel predation on the bats *Eptesicus fuscus*, *Euderma maculatum* and *Tadarida brasiliensis*. The Southwestern Naturalist, 21:250-251.

Blair, W. F. 1940. A contribution to the ecology and faunal relationships of the mammals of the Davis Mountain region, southwestern Texas. Miscellaneous Publications of the Museum of Zoology, University of Michigan, 46:1-39.

Blair, W.F. 1943. Ecological distribution of mammals in the Tularosa Basin, New Mexico. Contributions of the Laboratory of Vertebrate Biology, University of Michigan, 20:1-24.

Blair, W.F. 1952. Mammals of the Tamaulipan biotic providence in Texas. Texas Journal of Science, 4:230-250.

Blair, W.F., and C.E. Miller Jr. 1949. The mammals of the Sierra Vieja region, southwestern Texas, with remarks on the biogeographic position of the region. Texas Journal of Science, 2:93-117.

Blanco, S., G. Ceballos, C. Galindo, M. Maass, R. Patrón, A. Pescador, and A. Suárez. 1981. Ecología de la Estación Experimental Zoquiapan: descripción general, vegetación y fauna. Cuadernos Universitarios No. 2. Universidad Autónoma de Chapingo, México.

Bloedel, P. 1955a. Hunting methods of fish eating bats, particularly *Noctilio leporinus*. Journal of Mammalogy, 36:390-399.

Bloedel, P. 1955b. Observations on life histories of Panama bats. Journal of Mammalogy, 36:232-235.

Bodmer, R.E., T.G. Fang and L. Moya. 1988. Estudio y manejo de los pecaríes (*Tayassu tajacu* y *T. pecari*) en la Amazonia peruana, Matero. Universidad Nacional de la Amazonía Peruana, Lima, 2:18-25.

Bogan, M.A. 1978. A new subspecies of Myotis from the Islas Tres Marías, Nayarit, Mexico, with comments on variation in *Myotis nigricans*. Journal of Mammalogy, 59:519-530.

Bogan, M.A., and C. Jones. 1975. Observations on *Lepus callotis* in New Mexico. Proceedings of the Biological Society of Washington, 88:45-49.

Bogan, M.A., and P. Melhop. 1983. Systematic relationships of gray wolves in North America. Occasional Papers of the Museum of Southwestern Biology, 1:1-21.

Bogan, M.A., and D.F. Williams. 1970. Additional records of some Chihuahuan bats. The Southwestern Naturalist, 15:131-134.

Bogan, M., T.J. O'Shea, and L. Ellison. 1996. Diversity and conservation of bats in North America. Endangered Species Update, 14:1-4.

Boitani, L., and S. Bartoli. 1982. Simon and Schuster´s Guide to Mammals. Simon and Schuster, New York.

Bolaños, J., E. Naranjo, G. Escalona, and C. Lorenzo. 2006. *Eumops underwoodi* (Chiroptera: Molossidae) en Campeche. Revista Mexicana de Mastozoología, 10:75-79.

Bolin, I. 1981. Male parental behavior in black howler monkeys (*Alouatta palliata pigra*) in Belize and Guatemala. Primates, 22:349-360.

Bolles, K. 1980. Variation and evolution in alarm vocalization of *Ammospermophilus*, antelope squirrel. American Zoologist, 20:275.

Bolles, K. 1988. Evolution and variation of antipredator vocalizations of antelope squirrels, *Ammospermophilus* (Rodentia: Sciuridae). Zeitschrift für Säugetierkunde, 53:129-147.

Bonaccorso, F.J. 1979. Foraging and reproductive ecology in a Panamanian bat community. Bulletin of the Florida State Museum of Biological Science, 24:359-408.

Bonacorso, F.J., and T.J. Gush. 1987. An experimental study of the feeding behavior and foraging strategies of phyllostomid fruit bats. Journal of Animal Ecology, 56:907-920.

Bonilla, C., E. Cisneros, and V. Sánchez-Cordero. 1992. First record of the Mexican big-eared bat *Idionycteris phyllotis* (Vespertilionidae) in the state of Oaxaca, Mexico. The Southwestern Naturalist, 37:429-430.

Bonner, E.N. 1979. Harbour (Common) seal. Pp. 58-62, in: Mammals in the Seas (Food and Agriculture Organization of the United Nations, Rome, ed.). Vol. 2. Pinniped species summaries and report on sirenians. United Nations FAO, Rome.

Booth, E.S. 1957. Mammals collected in Mexico from 1951 to 1956 by the Walla Walla College Museum of Natural History. Walla Walla College Publications, Department of Biological Sciences, 20:1-19.

Borell, A.E., and M.D. Bryant. 1942. Mammals of the Big Bend area of Texas. University of California Publications in Zoology, 48:1-61.

Botello, F., P. Illoldi, M. Linaje, G. Monroy, and V. Sánchez-Cordero. 2005. Nuevos registros del tepezcuintle (Agouti paca) para el norte del estado de Oaxaca, México. Revista Mexicana de Biodiversidad, 76:103-105.

Bourillón, L., A. Cantú, F. Eccardi, E. Lira, J. Ramírez, E. Velarde, and A. Zavala. 1988. Islas del Golfo de California. Secretaría de Gobernación-Universidad Nacional Autónoma de México, México.

Bowen, W.D. 1981. Variation in coyote social organization: The influence of prey size. Canadian Journal of Zoology 59:639-652.

Bowen, W.D., C.A. Beck, and D.A. Austin. 2002. Pinniped ecology. In: Encyclopedia of Marine Mammals (W.F. Perrin, B. Würsig, and J.G.M. Thewissen, eds.). Academic Press, London.

Bowers, J.H. 1974. Genetic compatibility of *Peromyscus maniculatus* and *Peromyscus melanotis*, as indicated by breeding studies and morphometrics. Journal of Mammalogy, 55:720-737.

Bowles, J.B., P.D. Heideman, and K.R. Erickson. 1990. Observations on six species of free-tailed bats (Molossidae) from Yucatan, Mexico. The Southwestern Naturalist, 35:151-157.

Bowyer, R.T. 1986. Habitat selection by southern mule deer. Journal of California Fish and Game, 72:153-169.

Bowyer R.T., J.W.J. Testa, and J.B. Faro. 1995. Habitat selection and home ranges of river otters in a marine environment: Effects of the Exxon Valdez Oil SIPI. Journal of Mammalogy 76:1-11.

Boyd, R.J., A.Y. Cooperrider, P.C. Lent, and J.A. Bailey. 1986. Ungulates. Pp. 569-564, in: Inventory and Monitoring of Wildlife Habitat (A.Y. Cooperrider, R.J. Boyd, and H.S. Stuart, eds.). U.S. Department of the Interior, Bureau of Land Management, U.S. Government Printing Office, Washington, DC.

Bradbury, J.W. 1983. *Saccopteryx bilineata* (Murciélago de Saco, sac-wing bat). Pp. 488-489, in: Costa Rican Natural History (D.H. Janzen, ed.). University of Chicago Press, Chicago.

Bradbury, J.W., and L.H. Vehrencamp. 1976. Social organization and foraging in emballonurid bats. I. Field studies. Behavioral Ecology and Sociobiology, 1:337-381.

Bradbury, J.W., and S L. Vehrencamp. 1977. Social organization and foraging in emballonurid bats. IV. Parental investment patterns. Behavioral Ecology and Sociobiology, 2:19-29.

Bradford, D.F. 1974. Water stress of free-living *Peromyscus truei*. Ecology, 55:1407-1414.

Bradley, R. 1983. The pre-Columbian exploitation of the manatee in Mesoamerica. Papers in Anthropology. University of Oklahoma, Norman.

Bradley, R.D., and D.J. Schmidly. 1987. The glans penes and bacula in Latin American taxa of the *Peromyscus boylii* group. Journal of Mammalogy, 68:595-616.

Bradley R.D., D.D. Henson, and N.D. Durish. 2008. Re-evaluation of the geographic distribution and phylogeography of the *Sigmodon hispidus* complex based on mitochondrial DNA sequences. The Southwestern Naturalist, 53:301-310.

Bradley, R.D., D.J. Schmidly, and C.W. Kirpatrick. 1996. The relationships of *Peromyscus sagax* to the P. boylii and P. truei species groups in Mexico based on morphometric, karyotypic, and allozymic data. Pp. 95-106, in: Contributions in Mammalogy: A Memorial Volume Honoring Dr. J. K. Jones, Jr. (H.H. Genoways and R.J. Baker, eds.). Museum of Texas Tech University, Lubbock.

Bradley, R D., D.J. Schmidly, and R.D. Owen. 1989. Variation in the glan penes and bacula among Latin American populations of the *Peromyscus boylii* species complex. Journal of Mammalogy, 70:712-725.

Bradley, R.D., D.J. Schmidly, and R.D. Owen. 1990. Variation in the glans penes and bacula among Latin American populations of the *Peromyscus aztecus*. Occasional Papers of the Museum, Texas Tech University, 135:1-15.

Bradley, R.D., I. Tiemann-Boege, C.W. Kilpatrick, and D.J. Schmidly. 2000. Taxonomic status of *Peromyscus boylii sacarensis*: Inferences from DNA sequences of the mitochondrial cytochrome-b gene. Journal of Mammalogy, 81:875-88.

Bradley, R.D., F. Mendez-Harclerode, M.J. Hamilton, and G. Ceballos. 2004. A new species of Reithrodontomys from Guerrero, Mexico. Occasional Papers of the Museum, Texas Tech University, 231:1-12.

Bradley, R.D., N.D. Durish, D.S. Rogers, J.R. Miller, M.D. Engstrom, and C.W. Kilpatrick. 2007. Towards a molecular phylogeny for *Peromyscus*: Evidence from mitochondrial cytochrome-b sequences. Journal of Mammalogy, 88:1146-1159.

Bradley, W. G. 1967. Home range, activity patterns, and ecology of the antelope ground squirrel in southern Nevada. The Southwestern Naturalist, 12:231-252.

Bradley, W.G., and E.L. Cockrum. 1968. A new subspecies of the meadow vole (*Microtus pennsylvanicus*) from northwestern Chihuahua, Mexico. American Museum Novitates, 2325:1-7.

Bradley, W.G., and R.A. Mauer. 1971. Reproduction and food habits of Merriam's kangaroo rat, *Dipodomys merriami*. Journal of Mammalogy, 52:497-507.

Bradshaw, G.V., and B.J. Hayward. 1960. Mammal skulls recovered from owl pellets in Sonora, Mexico. Journal of Mammalogy, 41:282-283.

Braham, H.W. 1984. Distribution and migration of gray whales in Alaska. Pp. 249-266, in: The Gray Whale *Eschrichtius robustus* (M.L. Jones, S.L. Swartz, and J.S. Leatherwood, eds.). Academic Press, Orlando, FL.

Branan, W.V., and R.L. Marchinton. 1984. Biology of red brocket deer in Suriname with emphasis on management potential. The Royal Society of New Zealand Bulletin, 22:41-42.

Branan, W.V., and R.L. Marchinton. 1987. Reproductive ecology of white-tailed and red brocket deer in Suriname. Pp. 344-351, in: Biology and Management of the Cervidae (C.M. Wemmer, ed.). Smithsonian Institution Press, Washington, DC.

Branan, W.V., M. Werkhoven, and R.L. Marchinton. 1985. Food habits of brocket and white-tailed deer in Suriname. Journal of Wildlife Management, 49:972-976.

Brand, L.R., and R.E. Ryckman. 1968. Laboratory life histories of *Peromyscus eremicus* and *Peromyscus interparietalis*. Journal of Mammalogy, 49:495-501.

Brand, L.R., and R.E. Ryckman. 1969. Biosystematics of *Peromyscus eremicus, P. guardia*, and *P. interparietalis*. Journal of Mammalogy, 50:501-513.

Brashear, W.A., R.C. Dowler, and G. Ceballos. 2010. Climbing as an escape behavior in the American hog-nosed skunk, *Conepatus leuconotus*. Western North American Naturalist 70:258-260.

Braun, J.K. 1988. Systematics and biogeography of the southern flying squirrel *Glaucomys volans*. Journal of Mammalogy, 69:422-426.

Brennan, J.M., and H.C. Dalmat. 1960. Chiggers of Guatemala (Acarina: Trombiculidae). Annals of the Entomological Society of America, 53:183-191.

Brennan, J.M., and E.K. Jons. 1959. *Pseudoschoengastia* and four new Neotropical species of the genus (Acarina: Trombiculidae). Journal of Parasitology, 45:421-429.

Briese, L.A., and M.H. Smith. 1974. Seasonal abundance and movement of nine species of small mammals. Journal of Mammalogy, 55:615-629.

Brinton, D.G. 1882 [1970]. American Hero-Myths: A Study in the Native Religions of the Western Continent. Johnson Reprint Corporation, New York.

Briones-Salas, M.A. 1998. First record of *Bauerus dubiaquercus* (Vespertilionidae) in Oaxaca, Mexico. The Southwestern Naturalist, 43:495-496.

Briones-Salas, M.A. 1999. Los mamíferos de la región Sierra Norte de Oaxaca, México. Reporte de CONABIO.

Briones-Salas, M., and A. Santos-Moreno. 2002. First record of *Tonatia brasiliense* (Chiroptera: Phyllostomidae) in Oaxaca, Mexico. The Southwestern Naturalist, 47:137-138.

Britt, L. 1972. Some aspects of the ecology of *Perognathus flavus, Dipodomys ordii* and *Dipodomys merriami*. PhD dissertation, University of New Mexico, Albuquerque.

Broad, S. 1984. The peccary skin trade. Traffic Bulletin, 6:27-28.

Brower, L.P., B.E. Horner, M.A. Marty, C.M. Moffitt, and B. Villa-Ramírez. 1985. Mice (*Peromyscus maniculatus, P. spicilegus* and *Microtus mexicanus*) as predators of overwintering monarch butterflies (*Danaus plexippus*) in Mexico. Biotropica, 17:89-99.

Brown, D.E. (ed.). 1983. The Wolf in the Southwest. University of Arizona Press, Tucson.

Brown, D.E. 1984. Arizona's Tree Squirrels. Arizona Fish and Game Department, Phoenix.

Brown, D.E. 1985. The Grizzly in the Southwest. University of Oklahoma Press, Norman.

Brown, H.L. 1946. Rodent activity in a mixed prairie near Hays, Kansas. Transactions of the Kansas Academy of Science, 48:448-456.

Brown, J.H. 1968. Activity patterns of some Neotropical bats. Journal of Mammalogy, 49:754-757.

Brown, J.H. 1971. Mammals on mountaintops: Nonequilibrium insular biogeography. The American Naturalist, 105:467-478.

Brown, J.H., and B.A. Harney. 1993. Population and community ecology of heteromyid rodents in temperate habitats. Pp. 618-651, in: Biology of the Heteromyidae (H.H. Genoways and J.H. Brown, eds.). Special Publication 10, American Society of Mammalogists.

Brown, J.H., and R.C. Lasiewski. 1972. Metabolism of weasels: The costs of being long and thin. Ecology, 53:939-943.

Brown, J.H., and G.A. Lieberman. 1973. Resource utilization and coexistence of seed-eating desert rodents in sand dune habitats. Ecology, 54:788-797.

Brown, J.H., and M.V. Lomolino. 1999. Biogeography. 2nd ed. Sinauer Associates, Sunderland.

Brown, J.H., and P.F. Nicoletto. 1991. Spatial scaling of species composition: Body masses of North American land mammals. The American Naturalist, 138:1478-1512.

Brown, J.H., and C.F. Welser. 1968. Serum albumin polymorphisms in natural and laboratory populations of *Peromyscus*. Journal of Mammalogy, 49:420-426.

Brown, P., and R. Berry. 1981. Echolocation and foraging behavior in the Mexican fishing bat, *Pizonyx vivesi*. Abstracts of the 12th Annual North American Symposium on Bat Research. Cornell University, Ithaca, NY.

Brown, R.E., H.H. Genoways, and J.K. Jones Jr. 1971. Bacula of some Neotropical bats. Mammalia, 35:456-464.

Brownell, R.L., Jr. 1983. *Phocoena sinus*. Mammalian Species, 198:1-3.

Brownell, R.L. 1986. Distribution of the vaquita, *Phocoena sinus*, in Mexican waters. Marine Mammal Science, 2(4):299-305.

Brownell, R.L., Jr., W.A. Walker, and K.A. Forney. 1996. Pacific white sided dolphin, *Langenorhynchus obliquidens* Gill, 1865. In: Handbook of Marine Mammals (S.H. Ridgway and R. Harrison, eds.). Vol. 6. Academic Press, London.

Brownell, R.L., L.T. Findley, O. Vidal, A. Robles, and N.S. Manzanilla. 1987. External morphology and pigmentation of the vaquita, *Phocoena sinus* (Cetacea: Mammalia). Marine Mammal Science, 3(1):22-30.

Bryant, M.D. 1945. Phylogeny of Nearctic Sciuridae. The American Midland Naturalist, 33:257-390.

Bryant, W.E. 1889. Provisional description of some Mexican mammals. Proceedings of the California Academy of Sciences (series 2), 2:25-27.

Brylski, P. 1990a. Cactus mouse. Pp. 222-223, in: California's Wildlife, vol. 3, Mammals (D.C. Zeiner, W.F. Laudenslayer, K.E. Mayer, and M. White, eds.). California Department of Fish and Game, Sacramento.

Brylski, P. 1990b. California pocket mouse. Pp. 182-183, in: California's Wildlife, vol. 3, Mammals (D.C. Zeiner, W.F. Laudenslayer, K.E. Mayer, and M. White, eds.). California Department of Fish and Game, Sacramento.

Brylski, P. 1990c. California vole, *Microtus californicus*. Pp. 260-261, in: California's Wildlife, vol. 3, Mammals (D.C. Zeiner, W.F. Laudenslayer, K.E. Mayer, and M. White, eds.). California Department of Fish and Game, Sacramento.

Brylski, P. 1990d. Desert kangaroo rat *Dipodomys deserti*. Pp. 210-211, in: California's Wildlife, vol. 3, Mammals (D.C. Zeiner, W.F. Laudenslayer, K.E. Mayer, and M. White, eds.). California Department of Fish and Game, Sacramento.

Brylski, P. 1990e. Desert woodrat. *Neotoma lepida*. In: California's Wildlife, vol. 3, Mammals (D.C. Zeiner, W.F. Laudenslayer, K.E. Mayer, and M. White, eds.). California Department of Fish and Game, Sacramento.

Brylski, P. 1990f. Dusky-footed woodrat. Pp. 291-292, in: California's Wildlife, vol. 3, Mammals (D.C. Zeiner, W.F. Laudenslayer, K.E. Mayer, and M. White, eds.). California Department of Fish and Game, Sacramento.

Brylski, P. 1990g. Little pocket mouse. Pp. 184-186, in: California's Wildlife, vol. 3, Mammals (D.C. Zeiner, W.F. Laudenslayer, K.E. Mayer, and M. White, eds.). California Department of Fish and Game, Sacramento.

Buchanan, G.D. 1957. Variation in litter size of nine-banded armadillos. Journal of Mammalogy, 38:529.

Bucher, J.E., and H.I. Fritz. 1977. Behavior and maintenance of the woolly opossum (*Caluromys*) in captivity. Laboratory Animal Science, 27:1007-1012.

Bucher, J.E., and R.S. Hoffman. 1980. *Caluromys derbianus*. Mammalian Species, 140:1-4.

Buergelt, C.D., R.K. Bonde, C.A. Beck, and T.J. O'Shea. 1984. Pathological findings in manatees in Florida. Journal of the American Veterinary Association 11:1331-1334.

Burnie, D. 2001. Animal. Dorling Kindersley, London.

Burt, W.H. 1932a. Descriptions of heretofore unknown mammals from islands in the Gulf of California, Mexico. Transactions of the San Diego Society of Natural History, 7:161-182.

Burt, W.H. 1932b. The fish-eating habits of *Pizonyx vivesi* (Menegaux). Journal of Mammalogy, 13:363-365.

Burt, W.H. 1934. The mammals of southern Nevada. Transactions of the San Diego Society of Natural History, 7:375-428.

Burt, W.H. 1938. Faunal relationships and geographic distribution of mammals in Sonora, Mexico. Miscellaneous Publications of the Museum of Zoology, University of Michigan, 39:1-77.

Burt, W.H. 1960. Bacula of North American mammals. Miscellaneous Publications of the Museum of Zoology, University of Michigan, 113:1-76.

Burt, W.H. 1961. Some effects of volcan Paricutin on vertebrates. Occasional Papers of the Museum of Zoology, University of Michigan, 620:1-24.

Burt, W.H., and R.P. Grossenheider. 1964. A Field Guide to the Mammals. Houghton Mifflin Co., Boston.

Burt, W.H., and R.P. Grossenheider. 1976. A Field Guide to the Mammals of North America and North of Mexico. Houghton Mifflin Co., Boston.

Burt, W.H., and E.T. Hooper. 1941. Notes on mammals from Sonora and Chihuahua, Mexico. Occasional Papers of the Museum of Zoology, University of Michigan, 430:1-7.

Burt, W.H., and R.A. Stirton. 1961. The mammals of El Salvador. Miscellaneous Publications of the Museum of Zoology, University of Michigan, 117:1-69.

Burton, A.M., and G. Ceballos. 2006. Northern-most record of the collared anteater (*Tamandua mexicana*) from the Pacific slope of Mexico. Revista Mexicana de Mastozoología, 10:67-70.

Burton, J.A., and B. Pearson. 1987. Rare Mammals of the World. The Stephan Greene Press, Boston.

Bustamante, A. 2008. Densidad y uso de hábitat por el ocelote (*Leopardus pardalis*), puma (*Puma concolor*) y jaguar (Panthera onca) en el sureste del área de amortiguamiento del Parque Nacional Corcovado, Península de Osa, Costa Rica. Tesis de maestría, Instituto Internacional en Conservación y Manejo de Vida Silvestre. Heredia, Costa Rica.

Byers, J.A. 1983. Social interactions of juvenile collared peccaries, *Tayassu tajacu* (Mammalia: Artiodactyla). Journal of Zoology, 201:83-96.

Byers, J.A., and M. Bekoff. 1981. Social, spacing and cooperative behavior of the collared peccary, *Tayassu tajacu*. Journal of Mammalogy, 62:76-785.

Caballero, D.J. 1970. Descripción de *Brachylaemus* (*Brachylaemus*) *bravoae* sp. nov. (Trematoda: Digenea), de roedores del estado de Jalisco, México. Anales del Instituto de Biología, UNAM, México Serie Zoología, 41:39-44.

Caballero, E. 1948. Mamíferos de Nuevo León y consideraciones sobre las especies del género Filaria Miller, 1787. Revista de la Sociedad Mexicana de Historia Natural, 9:257-261.

Caballero y Caballero, E. 1944. Una nueva especie del género Litomosoides y consideraciones acerca de los caracteres sistemáticos de las especies de este género. Anales del Instituto de Biología, UNAM, México. Seria Zoológica, 15:383-388.

Cabrera, A. 1957-1961. Catálogo de los mamíferos de América del Sur. Revista del Museo Argentino de Ciencias Naturales, "Bernardo Rivadavia," 4:1-732.

Cabrera, A., and J. Yepes. 1940. Mamíferos Sudamericanos. Compañía Argentina de Editores, Buenos Aires.

Cabrera, A., and J. Yepes. 1960. Mamíferos sudamericanos. 2nd ed. Ediar, Buenos Aires, 1:1-187.

Cabrera, H., S.T. Álvarez-Castañeda, N. Gonzalez-Ruiz, and J.P. Gallo-Reynoso. 2007. Distribution and natural history of Schmidly's deermouse (*Peromyscus schmidlyi*). The Southwestern Naturalist, 52:620-623.

Cahalane, V.H. 1939. Mammals of the Chiricahua Mountains, Cochise Country, Arizona. Journal of Mammalogy, 20:418-440.

Cain, G.D., and E.E. Studier. 1974. Parasitic helminths of bats from the southeastern United States and Mexico. Proceedings of the Helminthological Society of Washington, 41:113-114.

Caire, W. 1978. The distribution and zoogeography of the mammals of Sonora, Mexico. PhD dissertation, University of New Mexico, Albuquerque.

Caire, W. 1997. Annotated checklist of the recent land mammals of Sonora, Mexico. Special Publication of the Museum of Southwestern Biology, 3:69-80.

Caire, W., J.E. Vaughan, and V.E. Diersing. 1978. First record of *Sorex arizonae* (Insectivora: Soricidae) from Mexico. The Southwestern Naturalist, 25:532-533.

Calambokidis, J., G.H. Steiger, J.C. Cubbage, K.C. Balcomb, C. Ewald, S. Kruse, R. Wells, and R. Sears. 1990. Sightings and movements of blue whales off central California 1986-88 from photo-identification of individuals. Pp. 343-348, in: Individual recognition of Cetaceans: Use of photo-identification and other techniques to estimate population parameters (P.S. Hammond, S.A. Mizroch, and G.P. Donovan, eds.). Special Issue 12, Reports of the International Whaling Commission.

Caldecott, J.O., M.D. Jenkins, T.H. Johnson, and B. Groombridge. 1996. Priorities for conserving global species richness and endemism. Biodiversity and Conservation, 5:699-727.

Caldwell, D.K. 1955. Notes on the spotted dolphin *Stenella plagiodon*, and the first record of the common dolphin, *Delphinus delphis*, in the Gulf of Mexico. Journal of Mammalogy, 36:467-470.

Caldwell, D.K., and M.C. Caldwell. 1972. The World of the Bottlenose Dolphin. Lippincott, Philadelphia.

Caldwell, D.K., and M.C. Caldwell. 1985. Manatees *Thrichechus manatus* Linnaeus, 1758; Trichechus senegalensis Link, 1795; and *Trichechus inunguis* (Natterer, 1883). Pp. 33-66, in: Handbook of Marine Mammals (S.H. Ridgway and R. Harrison, eds.). Vol. 3. Academic Press, New York.

Caldwell, D.K., and M.C. Caldwell. 1989. Pygmy sperm whale *Kogia breviceps* (de Blaiville, 1838); Dwarf sperm whale *Kogia simus* Owen, 1866. Pp. 235-260, in: Handbook of Marine Mammals (S.H. Ridgway and R. Harrison, eds.). Vol. 4. Academic Press, New York.

Caldwell, D.K., H. Neuhauser, M.C. Caldwell, and H.W. Coolidge. 1971. Recent records of marine mammals from the coasts of Georgia and South Carolina. Cetology, 5:1-12.

Calhoun, J.B. 1945. Daily activity rhythms of the rodents, *Microtus ochrogaster* and *Sigmodon hispidus hispidus*. Ecology, 26:251-273.

Callahan, J.R. 1975. Status of the peninsula chipmunk. Journal of Mammalogy, 56:266-269.

Callahan, J.R. 1976. Systematics and biogeography of the *Eutamias obscurus* complex (Rodentia: Sciuridae). PhD dissertation, University of Arizona, Tucson.

Callahan, J.R. 1977. Diagnosis of *Eutamias obscurus* (Rodentia, Sciuridae). Journal of Mammalogy, 58:188-201.

Callahan, J.R. 1980. Taxonomic status of *Eutamias bulleri*. The Southwestern Naturalist, 25:1-8.

Callahan, J.R., and R. Davis. 1976. Desert chipmunks. The Southwestern Naturalist, 21:127-130.

Callahan, J.R., and R. Davis. 1977. A new subspecies of the cliff chipmunk from coastal Sonora, Mexico. The Southwestern Naturalist, 22:67-75.

Camacho, V. 1940. Los metoritos del Valle de México. Revista de la Sociedad Mexicana de Historia Natural, 1:109-118.

Camacho-Cruz, A., M. González-Espinosa, J.H.D. Wolf, and B.H.J. De Jong. 2000. Germination and survival of tree species in disturbed forests of the highlands of Chiapas, Mexico. Canadian Journal of Botany, 78: 1309-1318.

Cameron, G.N. 1977. Experimental species removal: Demographic responses by *Sigmodon hispidus* and *Reithrodontomys fulvescens*. Journal of Mammalogy, 58:488-506.

Cameron, G.N., and S.R. Spencer. 1981. *Sigmodon hispidus*. Mammalian Species, 158:1-9.

Cameron, G.N., E.D. Fleharty, and B.A. Carnes. 1979a. Daily movement patterns of *Sigmodon hispidus*. The Southwestern Naturalist, 24:63-70.

Cameron, G.N., W.B. Kincaid, and B.W. Carnes. 1979b. Experimental species removal: Variation in movements of *Sigmodon hispidus* and *Reithrodontomys fulvescens*. Journal of Mammalogy, 60:195-197.

Cancino, J. 1994. Food habits of the peninsular pronghorn. 16th Biennial Pronghorn Workshop. Emporia, KS.

Cancino, J., P. Miller, J. Bernal Stoopen, and J. Lewis (eds.). 1995. Population and Habitat Assessment for the Peninsular Pronghorn (*Antilocapra americana peninsularis*). IUCN/SSC Conservation Breeding Specialist Group, Apple Valley, MT.

Cant, J.G.H. 1977. A census of the agouti (Dasyprocta punctata) in seasonally dry forest at Tikal, Guatemala, with some comments on strip censusing. Journal of Mammalogy, 58:688-690.

Cant, J.G.H. 1986. Locomotion and feeding postures of spider and howling monkeys: Field study and evolutionary interpretation. Folia Primatologica, 46:1-14.

Capella, J.J., L. Flórez, P. Falk, and D. Palacios. 2002. Regular appearance of otariid pinnipeds along the Colombian Pacific coast. Aquatic Mammals, 28:67-72.

Carbajal Villareal, S. 2005. Ámbito hogareño y patrón de actividad del Margay Leopardus wiedii (Schinz, 1821) en la Reserva de la Biosfera "El Cielo" Tamaulipas, México. Tesis de maestría en Ciencias en Biología, Instituto Tecnológico de Ciudad Victoria, División de estudios de Posgrado en Investigación, Tamaulipas, México.

Carbyn, L.N. 1987. Grey wolf and red wolf. Pp. 358-376, in: Wild Furbearer Management in North America (M. Novak, G.A. Baker, M.E. Obbard, and B. Malloch, eds.). Ontario Trappers Association, Ministry of Natural Resources, Ontario, Canada.

Carbyn, L.N., and M. Killaby. 1989. Status of the swift fox in Saskatchewan. Blue Jay, 47:41-52.

Carey, A.B. 1982. The ecology of red foxes, gray foxes and rabies in the eastern United States. Wildlife Society Bulletin, 10:18-26.

Carleton, M.D. 1973. A survey of gross stomach morphology in New World Cricetinae (Rodentia, Muroidea) with comments on functional interpretations. Miscellaneous Publications of the Museum of Zoology, University of Michigan, 146:1-43.

Carleton, M.D. 1977. Interrelationships of populations of the *Peromyscus boylii* species group (Rodentia, Muroidea) in western Mexico. Occasional Papers of the Museum of Zoology, University of Michigan, 675:1-47.

Carleton, M.D. 1979. Taxonomic status and relationships of *Peromyscus boylii* from El Salvador. Journal of Mammalogy, 60:280-296.

Carleton, M.D. 1980. Phylogenetic relationships in neotominepromyscine rodents (Muroidea) and a reappraisal of the dichotomy within New World Cricetinae. Miscellaneous Publications of the Museum of Zoology, University of Michigan, 157:1-146.

Carleton, M.D. 1985. Macroanatomy. Pp. 116-175, in: Biology of New World Microtus (R.H. Tamarin, ed.). Special Publication 8, American Society of Mammalogists.

Carleton, M.D. 1989. Systematics and evolution. Pp. 7-141, in: Advances in the Study of *Peromyscus* (Rodentia) (G.L. Kirkland and J.N. Layne, eds.). Texas Tech University Press, Lubbock.

Carleton, M.D., and J. Arroyo-Cabrales. 2009. Review of the *Oryzomys couesi* complex (Rodentia: Cricetidae: Sigmodontinae) in western Mexico. Bulletin of the American Museum of Natural History, 331: 94-127.

Carleton, M.D., and D.G. Huckaby. 1975. A new species of *Peromyscus* from Guatemala. Journal of Mammalogy, 56:444-451.

Carleton, M.D., R.D. Fisher, and A.L. Gardner. 1999. Identification and distribution of cotton rats, genus *Sigmodon* (Muridae: Sigmodontinae), of Nayarit, Mexico. Proceedings of the Biological Society of Washington, 112:813-856.

Carleton, M.D., O. Sánchez, and G. Urbano Vidales. 2002. A new species of *Habromys* (Muridae: Neotominae) from Mexico, with generic review of species definitions and remarks on diversity patterns among Mesoamerican small mammals restricted to humid montane forests. Proceedings of the Biological Society of Washington, 115:488-533.

Carleton, M.D., D.E. Wilson, A.L. Gardner, and M.A. Bogan. 1982. Distribution and systematics of *Peromyscus* (Mammalia: Rodentia) of Nayarit, Mexico. Smithsonian Contributions in Zoology, 352:1-46.

Carpenter, C.R. 1934. A field study of the behavior and social relations of howling monkeys. Comparative Psychological Monographs, 10:1-168.

Carpenter, C.R. 1935. Behavior of red spider monkeys (*Ateles geoffroyi*) in Panama. Journal of Mammalogy, 16:171-180.

Carpenter, C.R. 1965. The howlers of Barro Colorado Island. Pp. 250-291, in: Primate Behavior (I. de Vore, ed.). Holt, Rinehart and Winston, New York.

Carpenter, R.E. 1968. Salt and water metabolism in the marine fish-eating bat, *Pizonyx vivesi*. Comparative Biochemistry and Physiology, 24:951-964.

Carraway, L.N. 2005. Comparative osteology of the glenoid fossa and condyloid process in North American Soricinae shrews (Soricomorpha: Soricidae). Pp. 75-83, in: Contribuciones Mastozoológicas en Homenaje a Bernardo Villa (V. Sánchez-Cordero and R.A. Medellín, eds.). Instituto de Biología e Instituto de Ecología, Universidad Nacional Autónoma de México, México City, México.

Carraway, L.N. 2007. Shrews (Eulypotyphla: Soricidae) of Mexico. Monographs of the Western North American Naturalist, 3:1-91.

Carraway, N.L., and R.M. Timm. 2000. Revision of the extant taxa of the genus *Notiosorex* (Mammalia: Insectivora: Soricidae). Proceedings of the Biological Society of Washington, 113:302-318.

Carreira-J., C.A, A.M. Jansen, M.P. Deane, and H.L. Lenzi. 1996. Histopathological study of experimental and natural infections by *Trypanosoma cruzi* in *Didelphis marsupialis*. Memorias do Instituto Oswaldo Cruz, 91:609-618.

Carrera, J. 1985. Manejo de un hato de venado cola blanca (*Odocoileus virginianus texanus*) en el noreste de Coahuila. Pp. 756-761, in: I Simposium Internacional de Fauna Silvestre. Wildlife Society de México, A.C. y Sedue, México.

Carrera, R., W. Ballard, P. Gipson, B.T. Kelly, P.R. Krausman, M.C. Wallace, C. Villalobos, and D.B. Wester. 2008. Comparison of Mexican wolf and coyote diets in Arizona and New Mexico. Journal of Wildlife Management, 72:376-381.

Carroll, D.S., and R.D. Bradley. 2005. Systematics of the genus *Sigmodon*: DNA sequences from beta-fibrinogen and cytochrome b. The Southwestern Naturalist, 50:342-349.

Carroll, C., M.K. Phillips, C.A. Lopez-Gonzalez, and N.H. Schumaker. 2006. Defining recovery goals and strategies for endangered species: The wolf as a case study. Bioscience, 56:25-37.

Carroll, D.S., L.L. Peppers, and R.D. Bradley. 2005. Molecular systematics and phylogeography of the *Sigmodon hispidus* species group. Pp. 87-100, in: Contribuciones Mastozoológicas en Homenaje a Bernardo Villa (V. Sánchez-Cordero and R.A. Medellín, eds.). Instituto de Biología e Instituto de Ecología, UNAM, CONABIO, México.

Carrera, J. 1990. Survival in Mexico today. Proceedings of the Arizona Wolf Symposium 90.

Carter, D.C. 1970. Chiropteran reproduction. Pp. 233-246, in: About Bats (B.H. Slaughter and D.W. Walton, eds.). Southern Methodist University Press, Dallas, TX.

Carter, D.C., and W.B. Davis. 1961. *Tadarida aurispinosa* (Peale) (Chiroptera: Molossidae) in North America. Proceedings of the Biological Society of Washington, 74:161-165.

Carter, D.C., and P.G. Dolan. 1978. Catalogue of type specimens of Neotropical bats in selected European museums. Special Publications of the Museum, Texas Tech University, 15:1-136.

Carter, D.C., and J.K. Jones Jr. 1978. Bats from the Mexican state of Hidalgo. Occasional Papers of the Museum, Texas Tech University, 54:1-12.

Carter, D.C., R.H. Pine, and W.B. Davis. 1966. Notes on Middle American bats. The Southwestern Naturalist, 11:488-499.

Casariego-Madorell, M.A. 2004. Abundancia relativa y hábitos alimentarios de la nutria de rio (*Lontra longicaudis annectens*) en la costa de Oaxaca, México. Tesis de maestría, Posgrado en Ciencias Biologicas, Facultad de Ciencias, UNAM, México.

Casariego-Madorell, A.M., R. List, and G. Ceballos. 2008. Population size and the diet of the river otter (*Lontra longicaudis annectens*) on the Oaxaca coast, Mexico. Acta Zoológica Mexicana, 24:179-199.

Cascelli de Azevedo, F.C. 2008. Food habits and livestock depredation of sympatric jaguars and pumas in the Iguaçu National Park area, South Brazil. Biotropica, 40:494-500.

Case T.J., M.L. Cody, and E. Ezcurra (eds.). 2002. A New Island Biogeography of the Sea of Cortés. Oxford University Press, New York.

Casinos, A., and S. Filella. 1981. Notes on cetaceans of the Iberian coasts. IV. A specimen of *Mesoplodon densirostris* (Cetacea, Hyperoodontidae) stranded on the Spanish Mediterranean littoral. Säugethierkundliche Mitteilungen, 29:61-67.

Castro Arellano, I., and T. Lacher Jr. 2005. A new record and altitudinal extensions for the Cielo Biosphere Reserve mammals, Tamaulipas, Mexico. Revista Mexicana de Mastozoología, 9:150-155.

Castro-Campillo, A., and J. Ramírez-Pulido. 2000. Systematics of the smooth-toothed pocket gopher, *Thomomys umbrinus*, in the Mexican Transvolcanic Belt. American Museum Novitates, 3297:1-30.

Cavalcanti, S.M.C. 2008. Predator-prey relationships and spatial ecology of jaguars in the southern Pantanal, Brazil: Implications for conservation and management. PhD dissertation, Utah State University, Logan.

Ceballos, G. 1985. The importance of riparian habitats for the conservation of endangered mammals in Mexico. Pp. 96-100, in: Riparian Ecosystems and Their Management: Reconciling Conflicting Uses. Proceedings of the First North American Riparian Conference, U.S. Department of Agriculture, Forest Service General Technical Report RM-120, EUA.

Ceballos, G. 1989. Population and community ecology of small mammals in tropical deciduous forests in western Mexico. PhD dissertation, University of Arizona, Tucson.

Ceballos, G. 1990. Comparative natural history of small mammals from tropical forests in western Mexico. Journal of Mammalogy, 71:263-266.

Ceballos, G. 1993. La extinción de especies. Revista Ciencias (special number), 7:5-10.

Ceballos, G. 1999. Áreas prioritarias para la conservación de los mamíferos de México. Biodiversitas, 27:1-8.

Ceballos, G. 2007. Conservation priorities for mammals in megadiverse Mexico: The efficiency of reserve networks. Ecological Applications, 17:569-578.

Ceballos, G., and J.H. Brown. 1995. Global patterns of mammalian diversity, endemism, and endangerment. Conservation Biology, 9:559-568.

Ceballos, G., and F. Eccardi. 2003. Animales de México en Peligro de Extinción. Fundación IUSA, México.

Ceballos, G., and P.R. Ehrlich. 2002. Mammal population losses and the extinction crisis. Science, 296:904-907.

Ceballos, G., and C. Galindo. 1983. *Glaucomys volans goldmani* (Rodentia: Sciuridae) in Central Mexico. The Southwestern Naturalist, 28:375-376.

Ceballos, G., and C. Galindo. 1984. Mamíferos Silvestres de la Cuenca de México. Edit. Limusa, México.

Ceballos, G., and A. García. 1995. Conserving Neotropical biodiversity: The role of dry forests in western Mexico. Conservation Biology, 9:1349-1353.

Ceballos, G., and R. Medellín. 1988. *Diclidurus albus*. Mammalian Species, 316:1-4.

Ceballos, G., and E. Mellink. 1990. Distribución y estatus de los perros llaneros (*Cynomys mexicanus* and *Cynomys ludovicianus*) en México. Pp. 327-344, in: Areas Naturales Protegidas y Especies en Peligro de Extinción (J.L. Camarillo and F. Rivera, eds.). Escuela Nacional de Estudios Profesionales Iztacala, Universidad Nacional Autónoma de México.

Ceballos, G., and A. Miranda. 1985. Notes on the biology of Mexican flying squirrels (*Glaucomys volans*) (Rodentia: Sciuridae). The Southwestern Naturalist, 30:449-450.

Ceballos, G., and A. Miranda. 1986. Los Mamíferos de Chamela, Jalisco. Manual de Campo, Instituto de Biología, Universidad Nacional Autónoma de México, México.

Ceballos, G., and A. Miranda. 2000. Guía de Campo de los Mamíferos de la Costa de Jalisco, México. Fundación Ecológica Cuixmala, A.C., México.

Ceballos, G., and D. Navarro. 1991. Diversity and conservation of Mexican mammals. Pp. 167-198, in: Topics in Latin American Mammalogy: History, Biodiversity, and Education (M.A. Mares and D.J. Schmidly, eds.). University of Oklahoma Press, Norman.

Ceballos, G., and G. Oliva. 2005. Los mamíferos silvestres de México. CONABIO-UNAM.

Ceballos, G., and J. Pacheco. 2000. Los perros llaneros de Chihuahua: Importancia biológica y conservación. Biodiversitas, 31:1-5.

Ceballos, G., and P. Rodríguez. 1993. Diversidad y conservación de los mamíferos de México: II. Patrones de endemicidad. Pp. 87-108, in: Avances en el Estudio de los Mamíferos de México (R.A. Medellín and G. Ceballos, eds.). Publicaciones Especiales vol. 1, Asociación Mexicana de Mastozoología, México.

Ceballos, G., and J. Simonetti. 2002. Diversidad y conservación de los mamíferos Neotropicales. CONABIO-UNAM, México.

Ceballos, G., and D.E. Wilson. 1985. *Cynomys mexicanus*. Mammalian Species, 248:1-3.

Ceballos, G., J. Arroyo-Cabrales, R.A. Medellín, and Y. Domínguez-Castellanos. 2005. Lista actualizada de los mamíferos de México. Revista Mexicana de Mastozoología, 9:21-71.

Ceballos, G., J. Arroyo-Cabrales, and R.A. Medellín. 2002. The mammals of Mexico: Composition, distribution, and status. Occasional Papers of the Museum, Texas Tech University, 218:1-27.

Ceballos, G., J. Arroyo-Cabrales, and E. Ponce. 2010. The effect of Pleistocene environmental changes on the distribution and community structure of mammals in Mexico. Quaternary Research, 73:464-473.

Ceballos, G., E. Mellink, and L.R. Hanebury. 1993. Distribution and conservation status of prairie dogs *Cynomys mexicanus* and *Cynomys ludovicianus* in Mexico. Biological Conservation, 63:105-112.

Ceballos, G., J. Pacheco, and R. List. 1999. Influence of prairie dogs (*Cynomys ludovicianus*) on habitat heterogeneity and mammalian diversity in Mexico. Journal of Arid Environments, 41:161-172.

Ceballos, G., P. Rodríguez, and R. Medellín. 1998. Assessing conservation priorities in megadiverse Mexico: Mammalian diversity, endemicity, and endangerment. Ecological Applications, 8:8-17.

Ceballos, G., B. Vieyra, and J. Ramírez Pulido. 1998. A recent record of the Volcano rabbit (*Romerolagus diazi*) from the Nevado de Toluca, State of Mexico. Revista Mexicana de Mastozoología, 3:149-150.

Ceballos, G., H. Zarza, and M. Steele. 2002. *Xenomys nelsoni*. Mammalian Species, 704:1-3.

Ceballos, G., C. Chávez, R. List, and H. Zarza. 2007. Conservación y manejo del jaguar en México: estudios de caso y perspectivas. CONABIO-UNAM-Alianza WWF Telcel., México D.F.

Ceballos, G., T.H. Fleming, C. Chávez, and J. Nassar. 1997. Population dynamics of *Leptonycteris curasoae* (Chiroptera, Phyllostomidae) in Jalisco, Mexico. Journal of Mammalogy, 78:1220-1230.

Ceballos, G., S. Pompa, E. Espinoza, and A. García. 2010. Extralimital distribution of Galapagos (*Zalophus wollebaeki*) and northern (*Eumetopias jubatus*) sea lions in Mexico. Aquatic Mammals, 36:188-194.

Ceballos, G., P. Manzano, F.M. Mendez-Harclerode, M.L. Haynie, D.H. Walker, and R.D. Bradley. 2010. Geographic distribution, genetic diversity, and conservation status of the southern flying squirrel (*Glaucomys volans*) in Mexico. Occasional Papers of the Museum, Texas Tech University, 299:1-15.

Ceballos, G., C. Chávez, R. List, H. Zarza, and R. Medellín (eds.). 2011. Jaguar Conservation and Management in Mexico: Case Studies and Perspectives. Alianza WWF-Telcel/Universidad Nacional Autonóma de México, México D.F.

Ceballos, G., C. Chávez, H. Zarza, and C. Manterola. 2005. Ecología y conservación del jaguar en la región de Calakmul. Biodiversitas, 62:1-7.

Ceballos, G., P.R. Ehrlich, J. Soberón, I. Salazar, and J.P. Fay. 2005. Global mammal conservation: What must we manage? Science, 309:603-607.

Ceballos, G., A. Davidson, R. List, J. Pacheco, P. Manzano-Fischer, and G. Santos. 2010. Rapid collapse of a grassland system and its large scale ecological and conservation implications. PLoS ONE 5(1).

Ceballos, G., R. List, G. Garduño, M.J. Muñozcano Quintanar, R. López Cano, and E. Collado (eds.). 2009. La diversidad biológica del Estado de México. Estudio de Estado. UNAM-Instituto de Ecología-FES Iztacala-CONABIO-Gobierno del Estado de México, Toluca, Estado de México.

Cervantes, F.A. 1993a. Conejos y liebres silvestres de México. Ciencia y Desarrollo, 110:18-28.

Cervantes, F.A. 1993b. *Lepus flavigularis*. Mammalian Species, 423:1-3.

Cervantes, F.A., and L. Guevara. 2010. Rediscovery of the critically endangered Nelson's small-eared shrew (*Cryptotis nelsoni*), endemic to Volcán San Martín, eastern Mexico. Mammalian Biology, 75:451-454.

Cervantes, F.A., and M. Hortelano. 1991. Mamíferos pequeños de la Estación Biológica "El Morro de la Mancha," Veracruz, México. Anales del Instituto de Biología, UNAM, México, Serie Zoología, 62:129-136.

Cervantes, F.A., and L. Yepes Mulia. 1995. Species richness of mammals from the vicinity of Salina Sruz, coastal Oaxaca, México. Anales del Instituto de Biologia, UNAM, México, Serie Zoologia, 66:113-122.

Cervantes, F.A., J. Loredo, and J. Vargas. 2002. Abundance of sympatric skunks (Mustelidae: Carnivora) in Oaxaca, Mexico. Journal of Tropical Biology, 18:463-469.

Cervantes, F.A., C. Lorenzo, and F.X. González. 2004. The Omiltemi rabbit is not extinct. Mammalian Biology, 69:61-64.

Cervantes, F.A., C. Lorenzo, and R.S. Hoffmann. 1990. *Romerolagus diazi*. Mammalian Species, 360:1-7.

Cervantes, F.A., J. Martínez, and R.M. González. 1994. Karyotypes of the Mexican tropical voles *Microtus quasiater* and *M. umbrosus* (Arvicolinae: Muridae). Acta Theriologica, 39:373-377.

Cervantes, F.A., G. Matamoros-Trejo, and I. Martínez-Mateos. 1995. Mamíferos silvestres de la unidad de evaluación y monitoreo de la biodiversidad "Ing. Luis Macías Arellano," San Cayetano, Estado de México. Anales del Instituto de Biología, UNAM, México, Serie Zoología, 66(2):233-239.

Cervantes, F.A., J. Ramírez-P., and A. Castro-C. 1994. Diversidad taxonómica de los mamíferos de México. Anales del Instituto de Biología, UNAM, México, Serie Zoología, 65:177-190.

Cervantes, F., J.N. Ramirez-Vite, and S. Ramirez-Vite. 2002. Mamíferos pequeños de los alrededores del Poblado de Tlanchinol, Hidalgo. Anales del Instituto de Biología, UNAM, México, Serie Zoología, 73(2):225-237.

Cervantes, F.A., C. Lorenzo, J. Vargas, and T. Holmes. 1992. *Sylvilagus cunicularius*. Mammalian Species, 412:1-4.

Cervantes, F., J.N. Ramirez-Vite, S. Ramirez-Vite, and C. Ballesteros. 2004. New records of mammals from Hidalgo and Guerrero, Mexico. The Southwestern Naturalist, 49:122-124.

Cervantes, F.A., V.J. Sosa, J. Martínez, R.M. González, and R.C. Dowler. 1993. *Pappogeomys tylorhinus*. Mammalian Species, 433:1-4.

Cervantes, F.A., B. Villa, C. Lorenzo, J. Vargas, I. Villaseñor, and J. López. 1999. Búsqueda de poblaciones supervivientes de la liebre endémica Lepus flavigularis. Comisión Nacional para el Conocimiento y Uso de la Biodiversidad, México.

Cervantes-Reza, F.A. 1979. El conejo de los volcanes *Romerolagus diazi* (Mammalia: Lagomorpha), especie mexicana seriamente amenazada. Pp. 359-368, in: Memorias de la II Reunión Iberoamericana sobre Conservación y Zoología de Vertebrados. Universidad Hispanoamericana de Cáceres, Cáceres, España.

Cervantes-Reza, F.A. 1980. Principales Características Biológicas del Conejo de los Volcanes *Romerolagus diazi*, Ferrari-Pérez, 1893 (Mammalia: Lagomorpha). Tesis de licenciatura, Facultad de Ciencias, Universidad Nacional Autónoma de México, México.

Cervantes-Reza, F.A. 1981. Some predators of the zacatuche (*Romerolagus diazi*). Journal of Mammalogy, 62:850-851.

Cervantes-Reza, F.A. 1982. Observaciones sobre la reproducción del zacatuche o teporingo *Romerolagus diazi* (Mammalia: Lagomorpha). Doñana Acta Vertebrata, 9:416-420.

Cervantes-Reza, F.A., and W. López-Forment. 1981. Observations on the sexual behavior, gestation period, and young of captive Mexican volcano rabbits, *Romerolagus diazi*. Journal of Mammalogy, 62:634-635.

CFE (Compañía Federal de Electricidad). 1991. Manifiesto de impacto ambiental, "Línea de transmisión Emiliano Zapata IIPalenque." Reporte de la Residencia General de Construcción de Líneas de Transmisión y Subestaciones del Sureste. Coordinación de Proyectos de Transmisión y Transformación, CFE, México.

Challenger, A. 1998. Utilización y Conservación de los ecosistemas terrestres de México. Pasado, presente y futuro. CONABIO, UNAM y Agrupación Sierra Madre, México.

Chapman, B.R., and R.L. Packard. 1974. An ecological study of Merriam's pocket mouse in southeastern Texas. The Southwestern Naturalist, 19:281-291.

Chapman, F.M. 1936. White-lipped peccary. Natural History, 38:408-413.

Chapman, J.A. 1971. Organ weights and sexual dimorphism of the brush rabbit. Journal of Mammalogy, 52:453-455.

Chapman, J.A. 1974. *Sylvilagus bachmani*. Mammalian Species, 34:1-4.

Chapman, J.A., and G. Ceballos. 1990. The cottontails. Pp. 95-110, in: Rabbits, Hares, and Pikas: Status Survey and Conservation Action Plan (J.A. Chapman and J.E.C. Flux, eds.). IUCN, Gland, Switzerland.

Chapman, J.A., and G.A. Feldhamer. 1982. Wild Mammals of North America: Biology, Management, Economics. Johns Hopkins University Press, Baltimore.

Chapman, J.A., and R.P. Morgan. 1973. Systematic status of the cottontail complex in western Maryland and nearby West Virginia. Wildlife Monograph, 56:1-73.

Chapman, J.A., and G.L. Wilmer. 1978. *Sylvilagus audubonii*. Mammalian Species, 106:1-4.

Chapman, J.A., J.G. Hockman, and W.R. Edwards. 1982. Cottontails. Pp. 83-123, in: Wild Mammals of North America: Biology, Management, Economics (J.A. Chapman and G.A. Feldhamer, eds.). Johns Hopkins University Press, Baltimore.

Chapman, J.A., J.G. Hockman, and C.M.M. Ojeda. 1980. *Sylvilagus floridanus*. Mammalian Species, 136:1-8.

Chapman, J.A., J.E. C. Flux, A.T. Smith, E.J. Bell, G. Ceballos, K.R. Dixon, F.C. Dobler, N.A. Formozov, R.K. Ghose, W.L.R. Oliver, T. Robinson, E. Schnider, S.N. Stuart, K. Sujimurua, and Z. Changlin. 1990. Conservation action needed for rabbits, hares and pikas. Pp. 154-168, in: Rabbits, Hares, and Pikas: Status Survey and Conservation Action Plan (J.A. Chapman and J.E.C. Flux, eds.). IUCN and Natural Resources, Gland, Switzerland.

Charles-Dominique, P., M. Atramentowicz, M. Charles-Dominique, H. Gerard, A. Hladik, C.M. Hladik, and M.F. Prevost. 1981. Les mammifères frugivores arboricoles nocturnes d'e une forêt guyanaise: Interrelations plants-animaux. Revué Ecologia (Terre et Vie), 35:341-345.

Chavero, A. 1884-1889 [1974]. Historia antigua y de la conquista. In: México a través de los siglos (V. Riva Palacio, ed.). Editorial Cumbre, México, 1974.

Chávez, C., and G. Ceballos. 1998. Diversidad y conservación de los mamíferos del Estado de México. Revista Mexicana de Mastozoología, 3:113-134.

Chávez, C., and G. Ceballos. 2001. Diversidad y abundancia de murciélagos en selvas secas de estacionalidad contrastante en el oeste de México. Revista Mexicana de Mastozoología, 5:27-44.

Chávez, C., and G. Ceballos. 2002. New records of tropical dry forest mammals from the state of México. Revista Mexicana de Mastozoología, 6:90-98.

Chavez, C., M. Amin, and G. Ceballos (in review). Habitat use, prey use, and coexistence of jaguars (*Panthera onca*) and pumas (*Puma concolor*) in tropical forests of southern Mexico. Biological Conservation.

Chávez, G., and S. Zaragoza. 2009. Riqueza de mamíferos del Parque Nacional Barranca del Cupatitzio, Michoacán, México. Revista Mexicana de Biodiversidad, 80:95-104.

Chávez, J.C. 1993. Dinámica poblacional y uso de hábitat por roedores en un matorral de palo loco (Senecio praecox). Tesis licenciatura, Facultad de Ciencias, Universidad Nacional Autónoma de México, México.

Chávez, J.C. 1993b. Los roedores silvestres del Pedregal. Oikos, 21:4.

Chávez, C., 2006. Ecología poblacional y conservación del jaguar en la Reserva de la Biosfera Calakmul, Campeche, México. Tesis de maestría, Universidad Nacional Autónoma de México.

Chávez, J.C., and G. Ceballos. 1994. Historia natural comparada de los pequeños mamíferos de la Reserva El Pedregal. Pp. 229-238, in: Reserva Ecológica "El Pedregal" de San Angel: Ecología, Historia Natural y Manejo (A. Rojo, ed.). Universidad Nacional Autónoma de Mexico, México.

Chávez, C., and G. Ceballos. 1998. Diversidad y estado de conservación de los mamíferos del Estado de México. Revista Mexicana de Mastozoología, 3:113-134.

Chávez, L.G. 2005. A recent record of *Leopardus pardalis* (Linnaeus, 1758) from Michoacan, Mexico. Revista Mexicana de Mastozoología, 9:123-127.

Chávez, S.R. 1995. Estimación poblacional del rorcual tropical Balaenoptera edeni (Anderson, 1878) en la Bahía de La Paz, B.C.S., México. Tesis de maestría, Universidad Autónoma de Baja California Sur. La Paz, Baja California Sur, México.

Chávez, T.C. 1988. Ecología y dinámica poblacional de la comunidad de roedores de la Sierra del Ajusco, México. Tesis doctoral, Universidad Nacional Autónoma de México.

Chávez, T.C., and L. Espinosa. 1993. Ecología de roedores del Estado de Hidalgo. Pp. 433-471, in: Investigaciones Recientes sobre Flora y Fauna de Hidalgo (M. Villavicencio and Y. Marmolejo, eds.). Centro de Investigaciones Biológicas, Universidad Autónoma del Estado de Hidalgo, Pachuca, Hidalgo.

Chávez, C., G. Ceballos, R.A. Medellín, and H. Zarza. 2007. Primer censo nacional del jaguar. Pp. 133-142, in: Conservación y manejo del jaguar en México: estudios de caso y perspectivas (G. Ceballos, C. Chávez, R. List, and H. Zarza, eds.). CONABIO-UNAM-Alianza WWF-Telcel., México D.F.

Chávez, T.C., and R. Gallardo V. 1993. Demografía y reproducción de *Neotomodon alstoni alstoni* en la Sierra del Ajusco, México. Pp. 317-331, in: Avances en el Estudio de los Mamíferos de México (R.A. Medellín and G. Ceballos, eds.). Publicaciones Especiales No. 1, Asociación Mexicana de Mastozoología, México.

Chebez, J.C. 1994. Los que se van. Especies argentinas en peligro. Editorial Albatros SACI, Buenos Aires.

Chiasson, B. 1957. The dentition of the Alaskan fur seal. Journal of Mammalogy, 38(3):310-319.

Chinchilla, F.A. 1997. La dieta del jaguar (*Panthera onca*), el puma (*Felis concolor*) y el manigordo (*Felis pardalis*) en el Parque Nacional Corcovado, Costa Rica. Revista de Biologia Tropical, 45:1223-1229.

Chipman, R.K. 1965. Age determination of the cotton rat (*Sigmodon hispidus*). Tulane Studies in Zoology, 12:19-38.

Chivers, D.J. 1969. On the daily behaviour and spacing of howling monkey groups. Folia Primatologica, 10:48-102.

Choate, J.R. 1969. Taxonomic status of the shrew, *Notiosorex* (*Xenosorex*) *phillipsii* Schaldach, 1926 (Mammalia: Insectivora). Proceedings of the Biological Society of Washington, 82:469-476.

Choate, J.R. 1970. Systematics and zoogeography of Middle American shrews of the genus *Cryptotis*. University of Kansas Publications, Museum of Natural History, 19:195-317.

Choate, J.R. 1973. *Cryptotis mexicana*. Mammalian Species, 28:1-3.

Choate, J.R., and E.D. Fleharty. 1974. *Cryptotis goodwini*. Mammalian Species, 44:1-3.

Choate, W., and J.K. Jones Jr. 1970. Additional notes on reproduction in the Mexican vole, *Microtus mexicanus*. The Southwestern Naturalist, 14:356-358.

Christen, D.M. 1985. Seasonal tenancy of artificial nest structures of residence in an unexplored fox squirrel population. The American Midland Naturalist, 115:209-215.

Churchfield, S., V.A. Nesterenko, and E.A. Shvarts. 1999. Food niche overlap and ecological separation amongst six species of coexisting forest shrews (Insectivora: Soricidae) in the Russian Far East. Journal of Zoology (London), 248:349-359.

Cimé, P.J.A., J.B. Chablé, J.E. Sosa, and S.F. Hernández. 2006. Quirópteros y pequeños roedores de la Reserva de la Biosfera, Ría Celestún, Yucatán, México. Acta Zoológica Mexicana (n.s.), 22:127-131.

CITES (Convention on International Trade in Endangered Species). 1982. Identification Manual, vol. 1, Mammalia. IUCN, Gland, Switzerland.

CITES. 1992. Appendices I, II and III to the Convention on International Trade in Endangered Species of Wild Fauna and Flora. Publication Unit, U.S. Department of the Interior, Fish and Wildlife Service, Washington, DC.

CITES. 2010. Appendices I, II and III to the Convention on International Trade in Endangered Species of Wild Fauna and Flora. Publication Unit, U.S. Department of the Interior, Fish and Wildlife Service, Washington, DC.

Claire, W. 1976. Phenetic relationships of pocket mice in the subgenus *Chaetodipus* (Rodentia: Heteromyidae). Journal of Mammalogy, 57:375-378.

Clark, D.R., Jr. 1981. Bats and environmental contaminants: A review. United States Department of the Interior, Fish and Wildlife Service, Special Scientific Report-Wildlife, 235:1-27.

Clark, D.R., Jr., C.O. Martin, and D.M. Swineford. 1975. Organochlorine insecticide residue in the free-tailed bat (*Tadarida brasiliensis*) at Bracken Cave, Texas. Journal of Mammalogy, 56:429-443.

Clark, T.W. 1989. Conservation biology of the black-footed ferret *Mustela nigripes*. Wildlife Preservation Trust Special Scientific Report No. 3.

Clark, W.H. 1983. Nuevo registro de la musaraña desertícola *Notiosorex crawfordi* (Coues) del desierto central de Baja California. Anales del Instituto de Biología, UNAM, México, Serie Zoología, 3:439-441.

Clarke, M.R. 1986. Cephalopods in the diet of odontocetes. Pp. 281-322, in: Research on Dolphins (M. Bryden and R. Harrison, eds.). Clarendon Press, Oxford.

Clarke, M.R., O. Paliza, and A. Aguayo. 1993. Riesgo para la recuperación de la existencia de cachalotes en el Pacífico sureste debido al desarrollo de la pesca de la pota. Boletin de Lima, 85:73-78.

Clavijero, F.X. 1780 [1980]. Historia Antigua de México. 7th ed. Editorial Porrúa, México.

Clavijero, F.X. 1789 [1970]. Historia de la Antigua o Baja California. Editorial Porrúa, México.

Clemente, A.M. 1994. Hábitos alimenticios de dos especies de didélfidos (*Didelphis marsupialis* y *Philander opossum*) en dos hábitats diferentes de la Reserva de la Biosfera Montes Azules, Selva Lacandona, Chiapas, México. Tesis de licenciatura, Escuela de Biología, Instituto de Ciencias y Artes de Chiapas, Universidad Autónoma de Chiapas, Tuxtla Gutiérrez, Chiapas.

Clemente-Sánchez, F. 1984. Utilización de la vegetación nativa en la alimentación del venado cola blanca (*Odocoileus virginianus*) en el Estado de Aguascalientes. Tesis de maestría, Colegio de Posgraduados de Chapingo, Chapingo, México.

Cloutier, D., and D.W. Thomas. 1992. *Carollia perspicillata*. Mammalian Species, 417:1-9.

Clutton-Brock, T., and S.D. Albon. 1989. Red Deer in the Highlands. Blackwell, London.

Clutton-Brock, T., G.B. Corbett, and M. Hills. 1976. A review of the family Canidae with a classification with a numerical method. Bulletin of the British Museum, Zoology, 29:119-199.

Clutton-Brock, T., F. Guinness, and S.D. Albon. 1982. Red Deer: Behavior and Ecology of Two Sexes. University of Chicago Press, Chicago.

Coates-Estrada, R., and A. Estrada. 1985. Occurrence of the whitebat, *Diclidurus virgo* (Chiroptera: Emballonuridae) in the region "Los Tuxtlas," Veracruz. The Southwestern Naturalist, 30:332-323.

Coates-Estrada, R., and A. Estrada. 1986. Manual de Identificación de Campo de los Mamíferos de la Estación de Biología "Los Tuxtlas." Instituto de Biología, Universidad Nacional Autónoma de México.

Cockrum, E.L. 1952. Mammals of Kansas. University of Kansas Publications, Museum of Natural History, 7:1-303.

Cockrum, E.L. 1955. Reproduction in North American bats. Transactions of the Kansas Academy of Sciences, 58:487-511.

Cockrum, E.L. 1960a. Distribution, habitat and habits of the mastiff bat Eumops perotis in North America. Journal of the Arizona Academy of Science, 1:79-84.

Cockrum, E.L. 1960b. The Recent Mammals of Arizona: Their Taxonomy and Distribution. University of Arizona Press, Tucson.

Cockrum, E.L. 1969. Migration in the guano bat, University of Kansas Publications, Museum of Natural History, 51:303-336.

Cockrum, E.L. 1970. Insecticides and guano bats. Ecology, 51:761-762.

Cockrum, E.L. 1973. Additional longevity records for American bats. Journal of Arizona Academy of Sciences, 8:108-110.

Cockrum, E.L. 1982. Mammals of the Southwest. University of Arizona Press, Tucson.

Cockrum, E.L. 1991. Seasonal distribution of northwestern populations of the long-nosed bats, *Leptonycteris sanborni* family Phyllostomidae. Anales del Instituto de Biología, UNAM, México, Serie Zoología, 62:181-202.

Cockrum, E.L., and G.R. Bradshaw. 1959. *Pteronotus davyi* in northwestern Mexico. Journal of Mammalogy, 40:442.

Cockrum, E.L., and G.R. Bradshaw. 1963. Notes on mammals from Sonora, Mexico. American Museum Novitates, 2138:1-9.

Cockrum, E.L., and S.P. Cross. 1964. Time of bat activity over water holes. Journal of Mammalogy, 45:635-636.

Cockrum, E.L., and Y. Petryszyn. 1991. The long-nosed bat, *Leptonycteris*: An endangered species in the Southwest? Occasional Papers of the Museum, Texas Tech University, 142:1-32.

Coelho, A.M., C. Bramblett, L. Quick, and S. Bramblett. 1976. Resource availability and population density in primates: A sociobioenergetic analysis of the energy budgets of Guatemalan howler and spider monkeys. Primates, 17:63-80.

Collet, S.F., C. Sánchez-Hernández, K.A. Shump Jr. and W.R. Teska. 1975. Algunas características poblacionales demográficas de pequeños mamíferos en dos hábitats mexicanos. Anales del Instituto de Biología, UNAM, México, Serie Zoología, 46:101-124.

Collins, L.R. 1973. Monotremes and Marsupials. Smithsonian Institution Press, Washington, DC.

Colmenero, R.L.C. 1984. Nuevos Registros del Manatí (*Trichechus manatus*) en el Sureste de México. Anales Instituto de Biología UNAM, México, Serie Zoología, 1:243-254.

Colmenero, R.L.C. 1991. Propuesta de un Plan de Manejo para la Población del Manatí Trichechus manatus de México. Anales Instituto de Biología, UNAM, México, Serie Zoología, 62(2):203-218.

Colmenero, R.L.C., and E.Z. Hoz. 1986. Distribución de los Manatíes, Situación y su Conservación en México. Anales del Instituto de Biología, UNAM, México, Serie Zoología, 56(3): 955-1020.

Colmenero, L.C., and E.B. Zarate. 1990. Distribution, status and conservation of the West Indian manatee in Quintana Roo, Mexico. Biological Conservation, 52:27-35.

Colmenero-Rolon, L.C. 1986. Aspectos de la ecología y comportamiento de una colonia de manatíes (*Trichechus manatus*) en el municipio de Emiliano Zapata, Tabasco. Anales del Instituto de Biología, UNAM, México, Serie Zoología, 2:589-602.

Colón, C. 1492-1493 [1995]. Diario del primer viaje. Diputación Provincial de Granada. Granada.

Compeán-Jiménez, G. 1993. Aprovechamiento del atún y protección del delfín. Pp. 129-138, in: Biodiversidad Marina y Costera de México (S.I. Salazar-Vallejo and N.E. González, eds.). CONABIO y Ciqro, México.

Conaway, C.H. 1958. Maintenance, reproduction, and growth of the least shrew in captivity. Journal of Mammalogy, 39:507-512.

Connally, D.C., S. Leatherwood, G. James, and B. Winning. 1986. A note on vessel surveys for whales in the Sea of Cortez, January through April 1983-1985 and on the establishment of a data reporting center for the area. Report of the International Whaling Commission, Document SC/37/025, Washington, DC.

Conner, D.A. 1985. Analysis of the vocal repertoire of adult pikas: Ecological and evolutionary perspectives. Animal Behaviour, 33:124-134.

Conroy, C.J., E.A. Hadly, and C.J. Bell. 2001. Dating the origin of New World voles with multiple rates and calibration dates. Society for the Study of Evolution, Knoxville, TN.

Conroy, V.C.J., Y. Hortelano, F.A. Cervantes, and J.A. Cook. 2001. The phylogenetic position of southern relictual species of Microtus (Muridae: Rodentia) in North America. Mammalian Biology, 66:332-344.

Constantine, D.G. 1946. A record of *Dasypterus ega xanthinus* from Palm Springs, California. Bulletin of the Southern California Academy of Science, 65:107.

Constantine, D.G. 1958. Ecological observations on lasiurine bats in Georgia. Journal of Mammalogy, 36:64-70.

Constantine, D.G. 1959. *Pteronotus davyi* in northwestern Mexico. Journal of Mammalogy, 40:442.

Constantine, D.G. 1961a. Gestation in the spotted skunk. Journal of Mammalogy, 42:421-422.

Constantine, D.G. 1961b. Spotted bat and big-free tailed bat in northern New Mexico. The Southwestern Naturalist, 6:92-97.

Constantine, D.G. 1966a. Ecological observations on lasiurine bats in Iowa. Journal of Mammalogy, 47:34-41.

Constantine, D.G. 1966b. New bat locality records from Oaxaca, Arizona and Colorado. Journal of Mammalogy, 47:125-126.

Constantine, D.G. 1979a. An updated list of rabies-infected bats in North America. Journal of Wildlife Diseases, 15:347-348.

Constantine, D.G. 1979b. Rabies in *Myotis thysanodes, Lasiurus ega, Euderma maculatum* and *Eumops perotis* in California. Journal of Wildlife Diseases, 15:343-345.

Conway, M.C. 1976. A rare hare. New Mexico Wildlife, 21:21-23.

Conway, M., and C.G. Schmitt. 1978. Record of the Arizona shrew (*Sorex arizonae*) from New Mexico. Journal of Mammalogy, 59:631.

Cook, J.A. 1986. The mammals of the Animas Mountains and adjacent areas, Hidalgo Co., New Mexico. Occasional Papers of the Museum of Southwestern Biology, 4:1-45.

Copa-Alvaro, M.E. 2007. Efectos de los huracanes Emily y Wilma en la abundancia de mamíferos medianos en la Isla Cozumel, México. Tesis de maestría en ciencias, Universidad Nacional Autónoma de México, Ciudad de México.

Corbet, G.B. 1982. The occurrence and significance of a pectoral mane in rabbits and hares. Journal of Zoology (London), 198:541-546.

Corbet, G.B., and J.E. Hill. 1991. A World List of Mammalian Species. 3rd ed. Natural History Museum Publications, Oxford University Press, Oxford.

Cordero, G.A., and R.A. Nicolas. 1987. Feeding habits of the opossum (*Didelphis marsupialis*) in northern Venezuela. Fieldiana: Zoology (n.s.), 39:125-131.

Corn, J.L., and R.J. Warren. 1985. Seasonal food habits of the collared peccary in south Texas. Journal of Mammalogy, 66:155-159.

Cornely, J.E., and R.J. Baker. 1986. *Neotoma mexicana*. Mammalian Species, 262:1-7.

Cornely, J.E., D.J. Schmidly, H.H. Genoways, and R.J. Baker. 1981. Mice of the genus *Peromyscus* in Guadalupe Mountains National Park, Texas. Occasional Papers of the Museum, Texas Tech University, 74:1-35.

Corona-M., E., M. Montellano-Ballesteros, and J. Arroyo-Cabrales. 2008. A concise history of Mexican paleomammalogy. Arquivos do Museu Nacional, Rio de Janeiro, 66:179-189.

Cortés-Calva, P., and S. Álvarez-Castañeda. 1997. Estimación y número de camada de *Chaetodipus arenarius* (Rodentia:Heteromydae) en Baja California Sur, México. Revista de Biología Tropical (Costa Rica), 44-45:301-304

Cotera, M. 1996. Untersuchungen zur ökologischen Anpassung des Wüstenfuchses Vulpes macrotis zinseri B. in Nuevo León, Mexiko. Tesis para obtener el grado de doctor en filosofía de la Universidad Ludwig-Maximilians-Universität München, Alemania.

Coues, E. 1867. The quadrupeds of Arizona. The American Naturalist, 1:281-292, 351-363, 393-400, 531-541.

Coues, E. 1874. Synopsis of the Muridae of North America. Proceedings of the Academy of Natural Sciences of Philadelphia, 3:173-196.

Coues, E. 1877. Fur-bearing animals: A monograph of North American Mustelidae. U.S. Geological Survey of the Territories, Miscellaneous Publications. 8:1-348.

Cowan, I. McT. 1940. Distribution and variation in the native sheep of North America. The American Midland Naturalist, 24:505-580.

Cowan, I. McT., and C.J. Guiguet. 1965. The mammals of British Columbia. British Columbia Provincial Museum, 11:1-141.

Cowlishaw, G., and R. Dunbar. 2000. Primate Conservation Biology. University of Chicago Press, Chicago.

Cox, J.R., A. De Alba-Avila, R.W. Rice, and J.N. Cox. 1993. Biological and physical factors influencing *Acacia constricta* and *Prosopis velutina* establishment in the Sonoran desert. Journal of Range Management, 46:43-48.

Crabb, W.D. 1941. Food habits of the prairie spotted skunk in southeastern Iowa. Journal of Mammalogy, 22:349-664.

Crabb, W.D. 1944. Growth, development and seasonal weights of spotted skunks. Journal of Mammalogy, 25:213-221.

Crabb, W.D. 1948. The ecology and management of the prairie spotted skunks in Iowa. Ecological Monographs, 18:201-232.

Craighead, F.C., Jr. 1979. Track of the Grizzly. Sierra Books, San Francisco.

Craighead, J.J., J. Varney, and F.C. Craighead Jr. 1974. A population analysis of the Yellowstone grizzly bears. Montana Forest and Conservation Experimental Station Bulletin, 40:1-20.

Crandall, L.S. 1964. The Management of Wild Mammals in Captivity. University of Chicago Press, Chicago.

Crawshaw, P.G. 1995. Comparative ecology of ocelot (*Felis pardalis*) and jaguar (Panthera onca) in a protected subtropical forest in Brazil and Argentina. PhD dissertation, University of Florida, Gainesville.

Crawshaw, P.G., and H.B. Quigley. 1989. Notes on ocelot movements and activity in the Pantanal region, Brazil. Biotropica, 21:377-379.

Crawshaw, P.G., and H.B. Quigley. 1991. Jaguar spacing, activity and habitat use in a seasonally flooded environment in Brazil. Journal of Zoology (London), 223:357-370.

Crespo, J.A., J.M. Vanella, B.J. Blood, and J.M. de Carlo. 1961. Observaciones ecológicas del Vampiro *Desmodus r. rotundus* (Geoffroy) en el noreste de Córdoba. Revista del Museo Argentino de Ciencias Naturales, "Bernardino Rivadavia," 6:131-160.

Cross, S.P. 1965. Roosting habits of *Pipistrellus hesperus*. Journal of Mammalogy, 46:270-279.

Crossin, R.S., O.H. Soule, R.G. Webb, and R.H. Baker. 1973. Biotic relationships in the Cañon del Río Mezquital, Durango, Mexico. The Southwestern Naturalist, 18:187-200.

Crowe, D.M. 1975. Aspects of age, growth and reproduction of bobcats from Wyoming. Journal of Mammalogy, 56:177-198.

Cruz-L., L.E., C. Lorenzo, L. Soto, E. Naranjo, and N. Ramírez-M. 2004. Diversidad de mamíferos en cafetales y selva mediana de las cañadas de la Selva Lacandona, Chiapas, México. Acta Zoológica Mexicana (n.s.), 20:63-81.

Csuti, B. 1980. Type specimens of recent mammals in the Museum of Vertebrate Zoology, University of California, Berkeley. University of California Publications in Zoology, 114:1-75.

Cuarón, A.D. 1987. Hand-rearing a Mexican anteater *Tamandua mexicana* at Tuxtla Gutiérrez Zoo. International Zoo Yearbook, 26:255-260.

Cuarón, A.D. 1991. Conservación de los primates y sus hábitats en el sur de México. Tesis de maestría en Manejo de Vida Silvestre, Universidad Nacional, San José, Costa Rica.

Cuarón, A.D. 2000. A global perspective on habitat disturbance and tropical rainforest mammals. Conservation Biology, 14:1574-1579.

Cuarón, A.D., I.J. March, and P.M. Rockstroh. 1989. A second armadillo (*Cabassous centralis*) for the faunas of Guatemala and Mexico. Journal of Mammalogy, 70:870-871.

Cuarón, A.D., M.A. Martínez-Morales, K.W. McFadden, D. Valenzuela, and M.E. Gompper. 2004. The status of dwarf carnivores on Cozumel Island, Mexico. Biodiversity and Conservation, 13:317-331.

Cuarón, A., D. Valenzuela-Galván, D. García-Vasco, M.E. Copa, S. Bautista, H. Mena, D. Martínez-Godínez, C. González-Baca, L.A. Bojórquez-Tapia, L. Barraza, P.C. De Grammont, F. Galindo-Maldonado, M. A. Martínez-Morales, E. Vázquez-Domínguez, E. Andresen, J. Benítez-Malvido, D. Pérez-Salicrup, K.W. McFadden, and M.E. Gompper. 2009. Conservation of the endemic dwarf carnivores of Cozumel Island, Mexico. Small Carnivore Conservation, 41:15-21.

Culvert, M., W.E., Johnson, J. Pecon-Slattery, and S.J. O'Brien. 2000. Genomic ancestry of the American puma (*Puma concolor*). The American Genetic Association, 91:186-197.

Cummings, W.C. 1985. Right whales. *Eubalaena glacialis* (Muller, 1776) and *Eubalaena australis* (Desmoulins, 1822). Pp. 275-304, in: Handbook of Marine Mammals, vol. 3, The Sirenians and Baleen Whales (S.H. Ridgway and R. Harrison, eds.). Academic Press, New York.

Curds, T. 1993. Distribución geográfica de las dos especies de mono zaraguate que habitan en Guatemala: *Alouatta palliata* y *Alouatta pigra*. Pp. 317-329, in: Estudios Primatológicos en México (A. Estrada, E. Rodríguez-Luna, R. López-Wilchis, and R. Coates-Estrada, eds.). Vol. 1. Universidad Veracruzana, Xalapa, Veracruz, México.

Currier, M.J.P. 1983. Felis concolor. Mammalian Species. 20:1-17.

Cypher, B.L., and K.A. Spencer. 1998. Competitive interactions between coyotes and San Joaquin kit foxes. Journal of Mammalogy, 79:204-214.

Czaplewski, N.J. 1983. *Idionycteris phyllotis*. Mammalian Species, 208:1-4.

D'Agrosa, C., C.E. Lennert-Cody, and O. Vidal. 2000. Vaquita bycatch in Mexico's artisanal gillnet fisheries: Driving a small population to extinction. Conservation Biology, 14:1110-1119.

Dahlheim, M.E. 1981. A review of the biology and exploitation of the killer whale, *Orcinus orca*, with comments on recent sightings from Antarctica. Reports of the International Whaling Commission, 31:541-546.

Dahlheim, M.E., S. Leatherwood, and W.F. Perrin. 1982. Distribution of killer whales in the warm temperate and tropical eastern Pacific. Reports of the International Whaling Commission. 32:647-653.

Daily, G.C. 1997. Countryside biogeography and the provision of ecosystem services. Pp. 104-113, in: Nature and Human Society: The Quest for a Sustainable World (P.H. Raven, ed.). National Research Council, National Academy Press, Washington, DC.

Daily, G.C., G. Ceballos, J. Pacheco, G. Suzán, and A. Sánchez-Azofeifa. 2003. Countryside biogeography of Neotropical mammals: Conservation opportunities in agricultural landscapes of Costa Rica. Conservation Biology, 17:1814.

Dalby, P.L., and R.H. Baker. 1967. Crawford's desert shrew first reported from Nuevo Leon. The Southwestern Naturalist, 12:195-196.

Dalebout, M.L., A. van Helden, K. van Waerebeek, and C.S. Baker. 1998. Molecular genetic identification of Southern Hemisphere beaked whales (Cetacea: Ziphiidae). Molecular Ecology, 7:687-694.

Dalebout, M.L., J.G. Mead, C.S. Baker, A.L. Van Helden, and A.N. Baker. 2002. A new species of beaked whale *Mesoplodon perrini* sp. n. (Cetacea: Ziphiidae) discovered through phylogenetic analyses of mitochondrial DNA sequences. Marine Mammal Science, 18:577-608.

Dalebout, M.L., G.J.B. Ross, C.S. Baker, R.C. Anderson, P.B. Best, V.G. Cockcroft, H.L. Hinsz, V. Peddemors, and R.L. Pitman. 2003. Appearance, distribution and genetic distinctiveness of Longman's beaked whale, *Indopacetus pacificus*. Marine Mammal Science, 19:421-461.

Dalquest, W.W. 1948. Mammals of Washington. University of Kansas Publications, Museum of Natural History, 2:1-444.

Dalquest, W.W. 1949. The white-lipped peccary in the state of Veracruz, Mexico. Anales del Instituto de Biología, UNAM, México, Series Zoología, 20:411-413.

Dalquest, W.W. 1950. Records of mammals from the Mexican State of San Luis Potosi. Occasional Papers of the Museum of Zoology, Louisiana State University, 23:1-15.

Dalquest, W.W. 1951. Six new mammals from the state of San Luis Potosi, Mexico. Journal of the Washington Academy of Sciences, 41:361-364.

Dalquest, W.W. 1953a. Mammals of the Mexican state of San Luis Potosi, Mexico. Louisiana State University Studies Biological Series, 1:1-229.

Dalquest, W.W. 1953b. Mexican bats of the genus *Artibeus*. Proceedings of the Biological Society of Washington, 66:61-66.

Dalquest, W.W. 1953c. Aves y mamíferos del estado de Morelos. Revista de la Sociedad Mexicana de Historia Natural, 14:77-147.

Dalquest, W.W. 1954. Neeting bats in tropical Mexico. Transactions of the Kansas Academy of Science, 57:1-10.

Dalquest, W.W. 1955. Natural history of the vampire bats of eastern Mexico. The American Midland Naturalist, 53:79-87.

Dalquest, W.W. 1957. American bats of the genus *Mimon*. Proceedings of the Biological Society of Washington, 70:45-47.

Dalquest, W.W., and E.R. Hall. 1947. Geographic range of the hairylegged vampire in eastern Mexico. Transactions of the Kansas Academy of Science, 50:315-317.

Dalquest, W.W., and E.R. Hall. 1949. Five bats to the known fauna of Mexico. Journal of Mammalogy, 30:434-437.

Dalquest, W.W., and M.C. Ramage. 1946. Notes on the long-legged bat (*Myotis volans*) at Old Fort Tejon and vicinity, California. Journal of Mammalogy, 27:60-63.

Dalquest, W.W., and E. Roth. 1970. Late Pleistocene mammals from a cave in Tamaulipas, Mexico. The Southwestern Naturalist, 15:217-230.

Dalquest, W.W., and D.W. Walton. 1970. Diurnal retreats of bats. Pp. 162-187, in: About Bats (B.H. Slaughter and D.W. Walton, eds.). Southern Methodist University Press, Dallas, TX.

Dalquest, W.W., and H.J. Werner. 1954. Histological aspects of the faces of North American bats. Journal of Mammalogy, 35:147-160.

Dalquest, W.W., H.J. Werner, and J.H. Roberts. 1952. The facial glands of a fruit-eating bat, *Artibeus jamaicensis* Leach. Journal of Mammalogy, 33:102-103.

Daly, M., M.I. Wilson, and P.R. Behrends. 1984. Breeding of captive kangaroo rats, *Dipodomys merriami* and *Dipodomys microps*. Journal of Mammalogy, 65:338-341.

Daly, M., M.I. Wilson, P.R. Beherends, and L.F. Jacobs. 1990. Characteristics of kangaroo rats, *Dipodomys merriami*, associated with differential predation risk. Animal Behaviour, 40:380-389.

Darling, J.D. 1984. Gray whales off Vancouver Island, British Columbia. Pp. 267-288, in: The Gray Whale *Eschrichtius robustus* (M.L. Jones, S.L. Swartz, and J.S. Leatherwood, eds.). Academic Press, Orlando, FL.

Davidow-Henry, B.R., J. Knox Jones Jr., and R.R. Hollander. 1989. *Cratogeomys castanops*. Mammalian Species, 338:1-6.

Davies, J.L. 1963. The antitropical factor in cetacean speciation. Evolution, 17:107-116.

Davis, B.L., and R J. Baker. 1974. Morphometrics, evolution, and cytotaxonomy of mainland bats of the genus *Macrotus* (Chiroptera: Phyllostomidae). Systematic Zoology, 23:26-39.

Davis, B.L., S.L. Williams, and G. López. 1971. Chromosomal studies of Geomys. Journal of Mammalogy, 52:617-620.

Davis, J.A. 1978. A classification of the otters. In: Otters, Proceedings of the First Working Meeting of the Otter Specialist Group (N. Duplaix, ed.). IUCN, Morges, Switzerland.

Davis, R. 1969. Growth and development of young pallid bats, *Antrozous pallidus*. Journal of Mammalogy, 50:729-736.

Davis, R.B., C.F. Herreid, and H.L. Short. 1962. Mexican free-tailed bat in Texas. Ecological Monographs, 32:311-346.

Davis, W.B. 1944. Notes on Mexican mammals. Journal of Mammalogy, 25:370-403.

Davis, W.B. 1955. A new four-toed anteater from Mexico. Journal of Mammalogy, 36:557-559.

Davis, W.B. 1957. Notes on the Mexican shrew *Megasorex gigas* (Merriam). The Southwestern Naturalist, 2:174-175.

Davis, W.B. 1958. Review of Mexican bats of the *Artibeus* "*cinereus*" complex. Proceedings of the Biological Society of Washington, 71:163-166.

Davis, W.B. 1960. The mammals of Texas. Game and Fish Commission, 41:1-252.

Davis, W.B. 1965a. Review of South American bats of the genus *Eptesicus*. The Southwestern Naturalist, 11:245-274.

Davis, W.B. 1965b. Review of the *Eptesicus brasiliensis* complex in Middle America with the description of a new subspecies from Costa Rica. Journal of Mammalogy, 46:229-240.

Davis, W.B. 1968. Review of the genus *Uroderma* (Chiroptera). Journal of Mammalogy, 49:676-698.

Davis, W.B. 1969. A review of the small fruit bats (genus *Artibeus*) of Middle America. Part I. The Southwestern Naturalist, 14:15-29.

Davis, W.B. 1970a. A review of the small fruit bats (genus *Artibeus*) of Middle America. Part II. The Southwestern Naturalist, 14:389-402.

Davis, W.B. 1970b. The large fruit bats (genus *Artibeus*) of Middle America with a review of the *Artibeus jamaicensis* complex. Journal of Mammalogy, 51:105-122.

Davis, W.B. 1970c. A review of the small fruit bats (genus *Artibeus*) of Middle America. Part II. The Southwestern Naturalist, 14:389-402.

Davis, W.B. 1973. Geographical variation in the fishing bat, *Noctilio leporinus*. Journal of Mammalogy, 54:862-874.

Davis, W.B. 1974. The mammals of Texas. Bulletin of the Texas Parks and Wildlife Department, 41:1-294.

Davis, W.B. 1976. Geographical variation in the lesser noctilio, *Noctilio albiventris* (Chiroptera). Journal of Mammalogy, 57:687-707.

Davis, W.B. 1982. Biogeography of the bats of South America. Pp. 273-302, in: Mammalian Biology in South America (M.A. Mares and H.H. Genoways, eds.). Special Publication Series, Pymatuning Laboratory of Ecology, University of Pittsburgh, 6:1-539.

Davis, W.B. 1984. Review of the large fruit-eating bats of the *Artibeus "lituratus"* complex (Chiroptera: Phyllostomidae) in Middle America. Occasional Papers of the Museum, Texas Tech University, 93:1-16.

Davis, W.B. 1993. Order Chiroptera. Pp. 137-241, in: Mammal Species of the World: A Taxonomic and Geographic Reference (D.E. Wilson and D.M. Reeder, eds.). Smithsonian Institution Press, Washington, DC.

Davis, W.B., and D.C. Carter. 1962. Review of the genus *Leptonycteris* (Mammalia: Chiroptera). Proceedings of the Biological Society of Washington, 75:193-198.

Davis, W.B., and DC. Carter. 1978. A review of the round-eared bats of the *Tonatia silvicola* complex, with descriptions of three new taxa. Occasional Papers of the Museum, Texas Tech University, 53:1-12.

Davis, W.B., and L.A. Follansbee. 1945. The Mexican volcano mouse, *Neotomodon*. Journal of Mammalogy, 26:401-411.

Davis, W.B., and A.L. Gardner. 2007. Genus *Eptesicus* Rafinesque, 1820. Pp. 440-450, in: Mammals of South America, vol. 1, Marsupials, Xenarthrans, Shrews, and Bats (A.L. Gardner, ed.). University of Chicago Press, Chicago.

Davis, W.B., and W.Z. Lidicker Jr. 1956. Winter range of the red bat, *Lasiurus borealis*. Journal of Mammalogy, 37:280-281.

Davis, W.B., and P.W. Lukens Jr. 1958. Mammals of the Mexican state of Guerrero, exclusive of Chiroptera and Rodentia. Journal of Mammalogy, 39:347-367.

Davis, W.B., and R.J. Russell Jr. 1953. Aves y mamíferos del estado de Morelos. Revista de la Sociedad Mexicana de Historia Natural, 14:77-147.

Davis, W.B., and R.J. Russell Jr. 1954a. Mammals from the Mexican state of Morelos. Journal of Mammalogy, 53:63-80.

Davis, W.B., and R.J. Russell Jr. 1954b. Bats of the Mexican state of Morelos. Journal of Mammalogy, 35:63-80.

Davis, W.B., D.C. Carter, and R.H. Pine. 1964. Note-worthy records of Mexican and Central American bats. Journal of Mammalogy, 45:375-387.

Davis, W.H. 1959a. Taxonomy of the eastern pipistrel. Journal of Mammalogy, 40:521-531.

Davis, W.H. 1959b. Disproportionate sex ratios in hibernating bats. Journal of Mammalogy, 40:16-19.

Davis, W.H. 1964. Winter awakening patterns in the bats *Myotis lucifugus* and *Pipistrellus subflavus*. Journal of Mammalogy, 45:645-647.

Davis, W.H. 1966. Population dynamics of the bat *Pipistrellus subflavus*. Journal of Mammalogy, 47:383-396.

Davis, W.H., and H.B. Hitchcock. 1965. Biology and migration of the bat, *Myotis lucifugus*, in New England. Journal of Mammalogy, 46:296-313.

Dawson, G.A., and J.W. Lang. 1973. The functional significance of nest building by a Neotropical rodent (*Sigmodon hispidus*). The American Midland Naturalist, 89:503-509.

Day, B.N., J.H. Egoscue, and A.M. Woodbury. 1956. Ord kangaroo rat in captivity. Science, 124:485-486.

Day, G.I. 1985. Javelina. Arizona Fish and Game Department, Phoenix.

Day, G.I., and W.K. Carrel. 1986. Aging Javelina by Tetracycline Labeling of Teeth. Arizona Fish and Game Department, Tucson.

De la Lanza Espino, G.1991. Oceanografía de los mares mexicanos. AGT Editor, México.

De la Riva Hernández, G. 1989. La Mastofauna en Aguascalientes (Zona Semiárida). Universidad Autónoma de Aguascalientes, Aguascalientes.

De la Torre, A.J., and G. De la Riva. 2009. Food habits of pumas (Puma concolor) in a semiarid region of central Mexico. Mastozoología Neotropical, 16:211-216.

De la Torre, J.A., C. Muench, and M.C. Arteaga. 2009. Nuevos registros de grisón (*Galictis vittata*) para la selva Lacandona, Chiapas, México. Revista Mexicana de Mastozoología, 13:109-114.

De la Torre, L. 1955. Bats from Guerrero, Jalisco and Oaxaca, Mexico. Fieldiana Zoology, 37:695-701.

De Poorter, M., and W. Van der Loo. 1981. Report of the breeding and behavior of the volcano rabbit at the Antwerp Zoo. Pp. 956-972, in: Proceedings of the World Lagomorph Conference (K. Myers and C.D. MacInnes, eds.). University of Guelph, Guelph, Ontario.

Decker, D.M. 1991. Systematics of the coatis, genus *Nasua* (Mammalia: Procyonidae). Proceedings of the Biological Society of Washington, 104:370-386.

Dekker, D. 1974. On the natural history of manatees (*Trichechus manatus manatus*) from Surinam for the Amsterdam Zoo. Journal of Aquatic Mammals, 2(2):1-3.

Del Barco, M. 1757 [1973]. Historia Natural y Crónica de la Antigua California (Adiciones y Correcciones a la noticia de Miguel Venegas). Instituto de Investigaciones Históricas, Serie de Historiadores y cronistas de Indias, Universidad Nacional Autónoma de México, México.

Delgadillo Villalobos, J.A., B.R. McKinney, F. Heredia-Pineda, and S. Gibertisern. 2005. Nest record of Sorex milleri from Maderas del Carmen, México. The Southwestern Naturalist, 50:94-95.

Delibes, M., and F. Hiraldo. 1987. Food and habits of the bobcat in two habitats of the southern Chihuahua desert. The Southwestern Naturalist, 32:457-461.

Delibes, M., L. Hernández, and F. Hiraldo. 1985. Datos preliminares de la ecología del coyote y gato montés en el sur del Desierto de Chihuahua. Pp. 1018-1032, in: Primer Simposio Internacional de Fauna Silvestre. Wildlife Society de México, A.C., México.

Delibes, M., L. Hernández, and F. Hiraldo. 1989. Comparative food habits of three carnivores in western Sierra Madre, Mexico. Zeitschrift für Säugetierkunder, 54:107-110.

Demastes J.W, T.A. Spradling, M.S. Hafner, D.J. Hafner, and D.L. Reed. 2003. Systematics and phylogeography of pocket gophers in the genera *Cratogeomys* and *Pappogeomys*. Molecular Phylogenetics and Evolution, 22:144-154.

Denyes, H.A. 1956. Natural terrestrial communities of Brewster County, Texas, with special reference to the distribution of the mammals. The American Midland Naturalist, 55:289-320.

Desmastes, J.W., A.L. Butt, M.S. Hafner, and J.E. Light. 2003. Systematics of a rare species of pocket gopher, *Pappogeomys alcorni*. Journal of Mammalogy, 84:753-761.

Deswbury, D.A. 1979. Copulatory behavior of four Mexican species of *Peromyscus*. Journal of Mammalogy, 60:844-846.

Deutsch, C.J., D.E. Crocker, D.P. Costa, and B.J. Le Boeuf. 1994. Sex- and age-related variation in reproductive effort of northern elephant seals. Pp. 169-210, in: Elephant Seal (B.J. Le Boeuf and R. Laws, eds.). University of California Press, Los Angeles.

Deutsch, L.A. 1983. An encounter between bush dog (*Speothos venaticus*) and paca (Agouti paca). Journal of Mammalogy, 64:532-533.

De Villa Meza., A., E. Martínez, and C. López. 2002. Ocelot (*Leopardus pardalis*) food habits in a tropical deciduous forest of Jalisco, Mexico. The American Midland Naturalist, 148:146-154.

DeWalt, T.S., P.D. Sudman, M.S. Hafner, and S.K. David. 1993. Phylogenetic relationships of pocket gophers (*Cratogeomys* and *Pappogeomys*) based on mitochondrial DNA cytochrome b sequences. Molecular Phylogeny and Evolution, 2:193-204.

Díaz de Leon, J. 1905. Catálogo de los Mamíferos de la República Mexicana. Imp. Ricardo Rodriguez R., Aguascalientes.

Díaz-Camacho, S.P., F. Delgado-Vargas, K. Willms, M.C. De la Cruz-Otero, J.G. Rendón-Maldonado, L. Robert, S. Antuna, and Y. Nawa. 2010. Intrahepatic growth and maturation of *Gnathostoma turgidum* in the natural definitive oposum host, *Didelphis virginiana*. Parasitology International, 59(3):338-343.

Díaz del Castillo, B. 1521. Historia Verdadera de la Conquista de la Nueva España. Fernández Editores, México.

Díaz-Castorena, M.A. 1989. Distribución histórica y actual del venado bura en el estado de Zacatecas. Pp. 108-110, in: Memorias del III Simposio del Venado en México, Universidad Nacional Autónoma de México, México.

Di Bitetti, M., A. Paviolo, and C. De Angelo. 2006. Density, habitat use and activity patterns of ocelots (*Leopardus pardalis*) in the Atlantic forest of Misiones, Argentina. Journal of Zoology, 270:153-163.

Dice, L.R. 1937. Mammals of the San Carlos Mountains and vicinity. University of Michigan Studies, Science Series, 12:246-268.

Dice, L.R., and P.M. Blossom. 1937. Studies of mammalian ecology in southwestern North America with special attention to the colors of desert mammals. Carnegie Institute, Washington Publication, 485:1-129.

Dickermann, R.W. 1962. *Erethizon dorsatum* from Coahuila, Mexico. Journal of Mammalogy, 43:108.

Dickermann, R.W., K.F. Koopman, and C. Seymour. 1981. Notes on bats from the Pacific lowlands of Guatemala. Journal of Mammalogy, 62:406-411.

Dierauf, L., and F. Gulland. 2001. Marine Mammal Medicine. CRC Press, Boca Raton, FL.

Diersing, V.E. 1976. An analysis of *Peromyscus difficilis* from the Mexican-United States boundary area. Proceedings of the Biological Society of Washington, 89:451-466.

Diersing, V.E. 1980. Systematics of the flying squirrels *Glaucomys volans* (Linnaeus) from Mexico, Guatemala and Honduras. The Southwestern Naturalist, 25:157-172.

Diersing, V.E. 1981. Systematic status of *Sylvilagus brasiliensis* and *S. insonus* from North America. Journal of Mammalogy, 62:539-556.

Diersing, V.E., and D.F. Hoffmeister. 1977. Revision of the shrews *Sorex merriami* and description of the new species of the subgenus *Sorex*. Journal of Mammalogy, 53:321-333.

Diersing, V.E., and D.E. Wilson. 1980. Distribution and systematics of the rabbits (*Sylvilagus*) of West-Central Mexico. Smithsonian Contributions to Zoology, 297:1-34.

Dietrich, P.U., and G. Tijerina. 1988. Situación actual y perspectivas de la población del venado bura en el estado de Nuevo León y experiencia con el centro reproductivo de venado de la Facultad de Ciencias Forestales (UANL) Linares, Nuevo León. Pp. 18-32, in: II Simposio sobre el Venado en México. Universidad Nacional Autónoma de México, México.

Dillon, A., and M.J. Kelly. 2008. Ocelot home range, overlap and density: Comparing radio telemetry with camera trapping. Journal of Zoology, 275:391-398.

Dinerstein, E. 1985. First records of *Lasiurus castaneus* and *Antrozous dubiaquercus* from Costa Rica. Journal of Mammalogy, 66:411-412.

Dinerstein, E. 1986. Reproductive ecology of fruit bats and the seasonality of fruit production in a Costa Rican cloud forest. Biotropica, 18:307-318.

Dirzo, R., and M.C. García. 1992. Rates of deforestation in Los Tuxtlas, a Neotropical area in southeast Mexico. Conservation Biology, 6:84-90.

Disney, R.H.L. 1968. Observations on a zoonosis: *Leischmaniasis* in British Honduras. Journal of Applied Ecology, 5:1-59.

Dizon, A.E., S.O. Southern, and W.F. Perrin. 1991. Molecular analysis of mtDNA types in exploited populations of spinner dolphins (*Stenella longirostris*). Reports of the International Whaling Commission (special issue), 13:183-202.

Dizon, A.E., C.A. Lux, R.G. LeDuc, J.R. Urbán, M. Henshaw, C.S. Baker, F. Cipriano, and F.R.L. Brownell Jr. 1996. Molecular phylogeny of the Bryde's/Sei whale complex: Separate species status for the Pygmy Bryde's form. Reporte a la Comisión Ballenera Internacional SC/48/027.

Doan-Crider, D.L. 1995a. Population's characteristics and home range dynamics of the black bear in northern Coahuila, Mexico. M.S. thesis. Texas A&M University-Kingsville.

Doan-Crider, D.L. 1995b. Food habits of the Mexican black bear in Big Bend National Park, Texas and Serranías del Burro, Coahuila, Mexico. Report to Big Bend National Park.

Doan-Crider, D.L. 2003. Movements and spaciotemporal variation in relation to food productivity and distribution, and population dynamics of the Mexican black bear in the Serranías del Burro, Coahuila, Mexico. PhD dissertation, Texas A&M University and Texas A&M University<N>Kingsville.

Doan-Crider, D.L., and E.C. Hellgren. 1996. Population characteristics and winter ecology of black bears in Coahuila, Mexico. Journal of Wildlife Management, 60:398-407.

Dohl, T.P. 1979. Evidence for increasing offshore migration of the California gray whale, *Eschrichtius robustus*, in Southern California, 1975 through 1978. Abstracts, Third Biennial Conference of the Biology of Marine Mammals. Seattle, WA.

Dolan, P.G. 1989. Systematics of Middle American mastiff bats of the genus *Molossus*. Special Publications of the Museum, Texas Tech University, 29:1-71.

Dolan, P.G., and D.C. Carter. 1977. *Glaucomys volans*. Mammalian Species, 78:1-6.

Dolan, P.G., and D.C. Carter. 1979. Distributional notes and records for Middle America Chiroptera. Journal of Mammalogy, 60:644-649.

Dollfus, R.P. 1974. Pholeter (Trematoda, Digenea) from an intestinal cyst of *Stenella coeruleoalba* Meyen, 1833 (Odontoceti Delphinidae): Comments on the family Pholeteridae, R.Ph. Dollfus, 1939. List of helminthes identified to date in *Stenella coeruleoalba* Meyen. Investigations on Cetacea, 5:331-338.

Domínguez, C.Y. 2000. Estructura y contenido de madrigueras de Liomys pictus en selva mediana subperennifolia, de la estación de Biología Chamela, Jalisco. Tesis de licenciatura, Fes Iztacala, UNAM, Edo. de México.

Domínguez-Castellanos, Y., and G. Ceballos. 2005. Un registro notable del tigrillo (*Leopardus wiedii*) en la Reserva de la Biosfera Chamela-Cuixmala, Jalisco. Revista Mexicana de Mastozoología, 9:146-149.

Domning, D.P. 1994. Paleontology and evolution of sirenians: Status of knowledge and research needs. Proceedings of the 1st International Manatee and Dugong Research Conference. Gainesville, FL.

Domning, D., and V. Buffrenil. 1991. Hydrostasis in the Sirenia: Quantitative data and functional interpretations. Marine Mammal Science, 7(4):331-368.

Domning, D., and L. Hayek. 1986. Interspecific and intraspecific morphological variation in manatees (Sirenia: *Trichechus*). Marine Mammal Science, 2(2):87-144.

Donkin, R.A. 1985. The peccary, with observations on the introduction of pigs to the New World. Transactions of the American Philosophical Society, 75:152.

Dooley, M. 1988. Abstract of the results for the Oxford University Expedition to the Islas Marias, Mexico, 1987. Lagomorph Newsletter, 7:4-5.

Dorn, C.G. 1995. Application of reproductive technologies in North American bison (Bison bison). Theriogenology, 43:13-20.

DosReis, N.R., and J.L. Guillaumet. 1983. Les chauves-souris frugivores de la région de Manaus et leur rôle dans la dissemination des espèces végétales. Terre et Vie, 38:147-169.

Douglas, C.L. 1969. Comparative ecology of pinyon mice and deer mice in Mesa Verde National Park, Colorado. University of Kansas Publications, Museum of Natural History, 18:421-504.

Dowler, R.C., and M.D. Engstrom. 1988. Distributional records of mammals from the southwestern Yucatan Peninsula of Mexico. Annals of Carnegie Museum, 57:159-166.

Dowler, R.C., and H.H. Genoways. 1978. *Liomys irroratus*. Mammalian Species, 82:1-6.

Drabek, C.M. 1970. Ethoecology of the round-tailed ground squirrel, *Spermophilus tereticaudus*. PhD dissertation, University of Arizona, Tucson.

Dragoo, J.W., and R.L. Honeycutt. 1997. Systematics of mustelid-like carnivores. Journal of Mammalogy, 78:426-443.

Dragoo, J.W., R.L. Honeycutt, and D.J. Schmidly. 2003. Taxonomic status of white-backed hog-nosed skunks, genus *Conepatus* (Carnivora: Mephitidae). Journal of Mammalogy, 84:159-176.

Dragoo, J.W., R.D. Bradley, R.L. Honeycutt, and J.W. Templeton. 1993. Phylogenetic relationships among the skunks: A molecular perspective. Journal of Mammalian Evolution, 1:255-267.

Dragoo, J., J. Choate, T. Yates, and T. O'Farrell. 1990. Evolutionary and taxonomic relationships among North American arid-land foxes. Journal of Mammalogy, 71:318-332.

Drake, J.J. 1958. The brush mouse *Peromyscus boylii* in southern Durango. Publications of the Museum, Michigan State University, Biological Series, 1:97-132.

Drickamer, L.C., and J. Bernstein. 1972. Growth in two subspecies of *Peromyscus maniculatus*. Journal of Mammalogy, 53:228-231.

Drickamer, L.C., and B.M. Vestal. 1973. Patterns of reproduction in a laboratory colony of *Peromyscus*. Journal of Mammalogy, 54:523-528.

Ducommun, M.A., F. Jeanmaire-Besacon, and P. Vogel. 1994. Shield morphology of curly overhair in 22 genera of *Soricidae* (Insectivora, Mammalia). Revue Suisse de Zoologie, 101:623-643.

Dugelby, B., D. Foreman, R. List, B. Miller, J. Humphrey, M. Seidman, and R. Howard. 2001. Rewilding the Sky Islands Region of the Southwest. Pp. 65-81, in: Large Mammal Restoration: Ecological and Sociological Challenges in the 21st Century (D.S. Maehr, R.F. Noss, and J.L. Larkin, eds.). Island Press, Washington, DC.

Duignan, P.J., C. House, D.K. Odell, R.S. Wells, L.J. Hansen, M.T. Walsh, D.J. St. Aubin, B. Rima, and J.R. Geraci. 1996. Morbillivirus infection in bottlenose dolphins: Evidence for recurrent epizootics in the western Atlantic and Gulf of Mexico. Marine Mammal Science, 12(4):499-515.

Duke, K.L. 1944. The breeding season of two species of *Dipodomys*. Journal of Mammalogy, 25:155-160.

Duke, K.L. 1957. Reproduction in Perognathus. Journal of Mammalogy, 38:207-210.

Dullin, K.D., T.A. Jefferson, L.J. Hansen, and W. Hoggard. 1994. First sightings of melon-headed whales (*Peponocephala electra*) in the Gulf of Mexico. Marine Mammal Science, 10(3):342-348.

Dunford, C.J. 1977a. Behavioural limitation of round-tailed ground squirrel density. Ecology, 58:1254-1268.

Dunford, C.J. 1977b. Social system of round-tailed ground squirrels. Animal Behaviour, 25:885-906.

Dunford, C., and R. Davis. 1975. Cliff chipmunk vocalizations and their relevance to the taxonomy of coastal Sonora chipmunks. Journal of Mammalogy, 56:207-212.

Dunn, J.P., J.A. Chapman, and R.E. Marsh. 1982. Jackrabbits: *Lepus californicus* and allies. Pp. 124-145, in: Wild Mammals of North America: Biology, Management, and Economics (J.A. Chapman and G.A. Feldhamer, eds.). Johns Hopkins University Press, Baltimore.

Durand, J. 1950 [1983]. Ocaso de Sirenas, Esplendor de Manatíes. 2nd ed. Fondo de Cultura Económica, México.

Durrel, G., and J. Mallison. 1968. The volcano rabbit or teporingo (*Romerolagus diazi*). Pp. 29-36, in: Fifth Annual Report. Jersey Wildlife Preservation Trust, Jersey, England.

Easterla, D.A. 1970. First records of the spotted bat in Texas and notes on its natural history. The American Midland Naturalist, 83:306-308.

Easterla, D.A. 1972a. A diurnal colony of big freetail bats, *Tadarida macrotis* (Gray), in Chihuahua, Mexico. The American Midland Naturalist, 88:468-470.

Easterla, D.A. 1972b. Status of *Leptonycteris nivalis* (Phyllostomatidae) in Big Bend National Park, Texas. The Southwestern Naturalist, 17:287-292.

Easterla, D.A. 1973. Ecology of the 18 species of Chiroptera at Big Bend National Park, Texas. Part II. Northwestern Missouri State University Studies, 34:54-165.

Easterla, D.A., and J. Baccus. 1973. A collection of bats from the Fronteriza Mountains, Coahuila, Mexico. The Southwestern Naturalist, 17:424-427.

Easterla, D.A., and J.O. Whittaker. 1972. Food habits of some bats from Big Bend National Park, Texas. Journal of Mammalogy, 53:887-890.

Eddy, T.A. 1961. Foods and feeding patterns of the collared peccary in southern Arizona. Journal of Wildlife Management, 25:248-257.

Edwards, C.W., and R.D. Bradley. 2002. Molecular systematics and historical phylogeography of the *Neotoma mexicana* species group. Journal of Mammalogy, 81:20-30.

Edwards, C.W., C.F. Fulhorst, and R.D. Bradley. 2001. Molecular phylogenetics of the *Neotoma albigula* species group: Further evidence of a paraphyletic assemblage. Journal of Mammalogy, 82:267-279.

Edwards, R.L. 1946. Some notes on the life history of the Mexican ground squirrel in Texas. Journal of Mammalogy, 27:105-121.

Eger, J.L. 1977. Systematics of the genus *Eumops* (Chiroptera: Molossidae). Life Sciences Contributions of Royal Ontario Museum, 110:1-69.

Egoscue, H.J. 1956. Preliminary studies of kit fox in Utah. Journal of Mammalogy, 37:351-357.

Egoscue, H.J. 1957. The desert woodrat: A laboratory colony. Journal of Mammalogy, 38:472-481.

Egoscue, H.J. 1962. Ecology and life history of the kit fox in Tooele County, Utah. Ecology, 43:481-497.

Egoscue, H.J. 1964. Ecological notes and laboratory life history of the canyon mouse. Journal of Mammalogy, 45:387-396.

Egoscue, H.J. 1979. Vulpes velox. Mammalian Species, 122:1-5.

Eguiarte, L.E., and A. Búrquez. 1987. Reproductive ecology of *Manfreda brachystachya*, an iteroparous species of Agavaceae. The Southwestern Naturalist, 32:169-178.

Eguiarte, L.E., and A. Búrquez. 1988. Reducción en la fecundidad en *Manfreda brachystachya* (Cav.) Rose, una agavacea polinizada por murciélagos: los riesgos de la especialización en la polinización. Boletín Sociedad Botánica de México, 48:147-149.

Eguiarte, L.E., C. Martínez del Río, and H.T. Arita. 1987. El néctar y el polen como recursos: el papel ecológico de los visitantes a las flores de Pseudobombax ellipticum (H.B.K.) Dugand. Biotropica, 19:74-82.

Ehrlich, P., and G. Ceballos. 1997. Población y medio ambiente: ¿Qué nos espera? Ciencia, 48(4):19-30.

Eidemiller, B.J. 1980. Influence on competition on habitat use by desert omnivores. American Zoology, 20:953.

Eiler, J.H., W.G. Wathen, and M.R. Pelton. 1989. Reproduction in black bears in the southern Appalachian mountains. Journal of Wildlife Management, 53:353-360.

Eisenberg. J.F., and K.H. Redford. 1999. Mammals of the Neotropics. Vol. 3. The Central Neotropics: Ecuador, Perú, Bolivia, Brasil. University of Chicago Press, Chicago.

Eizirik E., J.H. Kim, M. Menotti-Raymond, P.G. Crawshaw, S.J. O'Brien, and W.E. Johnson. 2001. Phylogeography, population history and conservation genetics of jaguars (Panthera onca, Mammalia, Felidae). Molecular Ecology, 10:67-79.

Eisenberg, J.F. 1963. The behavior of heteromyid rodents. University of California Publications in Zoology, 69:1-102.

Eisenberg, J.F. 1967. A comparative study on rodent ethology with emphasis on evolution of social behavior I. Proceedings of the United States National Museum, 122:1-51.

Eisenberg, J.F. 1976. Communication mechanisms and social integration in the black spider monkey, *Ateles fusciceps robustus*, and related species. Smithsonian Contributions to Zoology, 213:1-108.

Eisenberg, J.F. 1981. The Mammalian Radiations. University of Chicago Press, Chicago.

Eisenberg, J.F. 1989. Mammals of the Neotropics. Vol. 1. The Northern Neotropics: Panama, Colombia, Venezuela, Suriname, French Guiana. University of Chicago Press, Chicago.

Eisenberg, J.F. 1993. Ontogeny. Pp. 479-490, in: Biology of the Heteromyidae (H.H. Genoways and J.H. Brown, eds.). Special Publication 10, American Society of Mammalogists.

Eisenberg, J.F., and D.E. Isaac. 1963. The reproduction of heteromyid rodents in captivity. Journal of Mammalogy, 44:61-67.

Eisenberg, J.F., and R. Kuehn. 1966. The behavior of *Ateles geoffroyi* and related species. Smithsonian Miscellaneous Collection, 151:1-63.

Eisenberg, J.F., and R.W. Thorington Jr. 1973. A preliminary analysis of a Neotropical mammal fauna. Biotropica, 5:150-161.

Eisenberg, J.F., M.A. O'Connell, and P.V. August. 1979. Density, productivity, and distribution of mammals in two Venezuelan habitats. Pp. 187-207, in: Vertebrate Ecology in the Northern Neotropics (J.F. Eisenberg, ed.). Smithsonian Institute Press, Washington, DC. El Libro del Consejo. 4th ed. Universidad Nacional Autónoma de México, México. 1984.

Eizirik, E., J. Kim, M. Menotti-Raymond, P.G. Crawshaw Jr., S.J. O'Brien, and W.E. Johnson. 2002. Phylogeography, population history and conservation genetics of jaguars (*Panthera onca*, Mammalia, Felidae). Molecular Ecology, 10:65-79.

Elder, F.B., and M.R. Lee. 1985. The chromosomes of *Sigmodon ochrognatus* and *S. fulviventer* suggest a realignment of Sigmodon species group. Journal of Mammalogy, 66:511-518.

Eldredge, N. 2002. Life on Earth: An Encyclopedia of Biodiversity, Ecology and Evolution. ABC-CLIO, Santa Barbara, CA.

Elenowitz, A. 1984. Group dynamics and habitat use of transplanted desert bighorn sheep in the Peloncillo Mountains, New Mexico. Desert Bighorn Council, 28:1-8.

Ellerman, J.R. 1940. The families and genera of living rodents. British Museum (Natural History), 1:1-689; 2:1-690.

Elliot, D.G. 1903a. A list of a collection of Mexican mammals with a description of some apparently new forms. Field Columbia Museum, Publication 71, Zoological Series, 3:141-149.

Elliot, D.G. 1903b. A list of mammals collected by Edmund Heller, in the San Pedro Martir and Hanson Laguna mountains and the accompanying coast regions of Lower California, with descriptions of apparently new species. Field Columbia Museum, Publication 79, Zoological Series, 3:199-232.

Elliot, D.G. 1904. The land and sea mammals of Middle America and the West Indies. Field Columbia Museum, Publication 95, Zoological Series, 4:411-850.

Elliot, D.G. 1905a. A catalogue of the collection of mammals in the Field Columbia Museum. Field Columbia Museum, Publication 95, Zoological Series, 8:viii + 1-694.

Elliot, D.G. 1905b. A check list of mammals of the North American continent, the West Indies and the neighboring seas. Field Columbia Museum. Publication 105, Zoological Series, 6:1-701.

Elliot, D.G. 1905c. Descriptions of apparently new species and subspecies of mammals from Mexico and San Domingo. Proceedings of the Biological Society of Washington, 18:233-236.

Elliot, D.G. 1917. A check-list of mammals of the North American continent, the West Indies, and the neighboring seas. Supp. American Museum of Natural History, IV.

Ellisor, J.E., and W.F. Harwell, 1969. Mobility and home range of collared peccary in southern Texas. Journal of Wildlife Management, 33:425-427.

Elizalde-Arellano, C., J.C. López-Vidal, J. Arroyo-Cabrales, R.A. Medellín, and J.W. Laundré. 2007. Food sharing behavior in the hairy-legged vampire bat *Dyphylla ecaudata*. Acta Chiropterologica, 9:314-319.

Emerson, K.C. 1971. New records of *Anoplura* from Mexico. Journal of Kansas Entomology Society, 44:374-377.

Emmons, L.H. 1987a. Comparative feeding ecology of felids in a Neotropical forest. Behavioral Ecology and Sociobiology, 20:271-283.

Emmons, L.H. 1987b. Ecological considerations on the farming of game animals: Capybaras yes, pacas no. Vida Silvestre Neotropical, 1:54-55.

Emmons, L.H. 1988. A field study of ocelots *(Felis pardalis)* in Peru. Revue d'Ecologie (Terre Vie), 43:133-157.

Emmons, L.H. 1990. Neotropical Rainforest Mammals: A Field Guide. University of Chicago Press, Chicago.

Emmons, L.H. 1997. Neotropical Rainforest Mammals. 2nd ed. University of Chicago Press, Chicago.

Emmons, L.H., and F. Feer. 1990. Neotropical Rainforest Mammals. A Field Guide. University of Chicago Press, Chicago.

Emmons, L.H., and F. Feer. 1997. Neotropical Rainforest Animals: A Field Guide. 2nd ed. University of Chicago Press, Chicago.

Enders, R.K. 1930. Banana stowaways again. Science, 71:438-439.

Enders, R.K. 1935. Mammalian life histories from Barro Colorado Island, Panama. Bulletin of the Museum of Comparative Zoology, 78:385-502.

Engstrom, M.D. 1984. Chromosomal, genic, and morphological variation in the *Oryzomys melanotis* species group. PhD dissertation, Texas A&M University, College Station.

Engstrom, M.D. 1988. Distributional records of mammals from the southwestern Yucatan Peninsula of Mexico. Annals of Carnegie Museum, 57:159-166.

Engstrom, M.D., and J.W. Bickham. 1983. Karyotype of *Nelsonia neotomodon*, with notes on the primitive karyotype of peromyscine rodents. Journal of Mammalogy, 64:685-688.

Engstrom, M.D., and D.E. Wilson. 1981. Systematics of *Antrozous dubiaquercus* (Chiroptera: Vespertilionidae), with comments on the status of Bauerus Van Gelder. Annals of Carnegie Museum, 50:371-383.

Engstrom, M.D., H.H. Genoways, and P.K. Tucker. 1987. Morphological variation, karyology, and systematic relationships of *Heteromys gaumeri* (Rodentia: Heteromyidae). Pp. 289-303, in: Studies in Neotropical Mammalogy: Essays in Honor of Phillip Herhkovitz (B.D. Patterson and R.M. Timm, eds.). Fieldiana-Zoology, n.s. 39, Field Museum of Natural History, Chicago.

Engstrom, M.D., T.E. Lee, and D.E. Wilson. 1987. *Bauerus dubiaquercus*. Mammalian Species, 282:1-3.

Engstrom, M.D., F.A. Reid, and B.K. Lim. 1993. New records of two mammals from Guatemala. The Southwestern Naturalist, 1:80-82.

Engstrom, M.D., O. Sánchez-H., and G. Urbano-V. 1992. Distribution, geographic variation, and systematic relationships within *Nelsonia* (Rodentia: Sigmodontinae). Proceedings of the Biological Society of Washington, 105:867-881.

Engstrom, M.D., C.A. Schmidt, J.C. Morales, and R.C. Dowler. 1989. Records of mammals from Isla Cozumel, Quintana Roo, Mexico. The Southwestern Naturalist, 34:413-449.

Enríquez Vázquez, P., R. Mariaca Méndez, O.G. Retana Guiascón, and E.J. Naranjo Piñera. 2006. Medicinal use of wild fauna in los Altos de Chiapas, Mexico. Interciencia, 37:491-499.

Erickson, A.B. 1949. Summer populations and movements of the cotton rat and other rodents on the Savannah River Refuge. Journal of Mammalogy, 30:133-140.

Erickson, A.B., and J. Nellor. 1964. Breeding biology of the black bear. Part 1. Pp. 4-45, in: The Black Bear in Michigan (A.W. Erickson, J. Nellor, and G.A. Petrides, eds.). Michigan State Agriculture Experimental Research Station Bulletin 4, East Lansing.

Erickson, A.B., and H.I. Scudder. 1947. The raccoon as a predator of turtles. Journal of Mammalogy, 28:406-407.

Erkert, H.G. 1982. Ecological aspects of bat activity rhythms. Pp. 201-242, in: Ecology of Bats (T.H. Kunz, ed.). Plenum Press, New York.

Ernest, K.A., and M.A. Mares. 1987. *Spermophilus tereticaudus*. Mammalian Species, 274:1-9.

Errington, P.L. 1963. Muskrat Populations. Iowa State University. Ames.

Escalante, T., D. Espinosa, and J.J. Morrrone. 2003. Using parsimony analysis of endemicity to analyze the distribution of Mexican land mammals. The Southwestern Naturalist, 48:563-578.

Escalante, P.P., A.G. Navarro, and A.T. Peterson. 1993. A geographic, ecological, and historical analysis of land bird diversity in Mexico. Pp. 281-307, in: Biological Diversity of Mexico (T.P. Ramamoorthy, R. Bye, A. Lot, and J. Fa, eds.). Oxford University Press, New York.

Escobedo-Cabrera, E., L. León-Paniagua, and J. Arroyo-Cabrales. 2006. Geographic distribution and some taxonomic comments on *Micronycteris schmidtorum* Sanborn (Chiroptera: Phyllostomidae) in Mexico. Caribbean Journal of Science, 42:129-135.

Escobedo-Morales L.A., L. León-Paniagua, J. Arroyo-Cabrales, and O.J. Polaco. 2005. Diversidad y abundancia de los mamíferos de Yaxchilán, municipio de Ocosingo, Chiapas. Pp. 283-298, in: Contribuciones mastozoológicas en homenaje a Bernardo Villa (V. Sánchez-Cordero and R.A. Medellín, eds.). Instituto de Biología, Instituto de Ecología, UNAM, CONABIO. México.

Escobedo-Morales, L.E., L. León-Paniagua, J. Arroyo-Cabrales, and F. Greenaway. 2006. Distributional records for mammals from Chiapas, Mexico. The Southwestern Naturalist, 51:269-271.

Eshelman, B.D., and G.N. Cameron. 1987. *Baiomys taylori*. Mammalian Species, 285:1-7.

Esher, R.J., J.L. Wolfe, and J.N. Layne. 1978. Swimming behavior of rice rats (*Oryzomys palustris*) and cotton rats (*Sigmodon hispidus*). Journal of Mammalogy, 59:551-558.

Espinoza, M.E., H. Nuñez, P. Gonzalez, R. Luna, M.A. Altamirano, E. Cruz, G. Cartas, and C. Guichard. 1999. Listado preliminar de los vertebrados terrestres de la Reserva de la Biosfera El Triunfo, Chiapas. Publicación Especial, Instituto de Historia Natural, 1:1-38.

Espinoza Medinilla, E., A.A. Dadda, and E.A. Cruz. 1998. Mamíferos de la Reserva de la Biosfera El Triunfo, Chiapas. Revista Mexicana de Mastozoología, 3:79-94.

Espinoza Medinilla, E., E. Cruz, H. Kramsky, and I. Sánchez. 2003. Mastofauna de la Reserva de la Biosfera "La Encrucijada," Chiapas. Revista Mexicana de Mastozoología, 7:5-19.

Espinoza Medinilla, E., E. Cruz, I. Lira, and I. Sánchez. 2004. Mamíferos de la Reserva de la Biosfera "La Sepultura," Chiapas, México. Revista Biología Tropical, 52:249-259.

Espinosa, T.J. 1982. Los quirópteros del estado de Aguascalientes. Pp. 74-97, in: Estudio Taxonómico Ecológico de la Flora y Fauna del Estado de Aguascalientes. Universidad Autónoma de Aguascalientes, Aguascalientes.

Esquivel, L. 1998. El posible significado mítico de la pintura rupestre de Baja California Sur. Pp. 25-40, in: Historia Comparativa de las Religiones (H.K. Kocyba and Y. González Torres, eds.). Eduvem, Instituto Nacional de Antropología e Historia, México.

Esquivel, C., L. Sarti, et al. 1993. Primera observacion directa documentada sobre la depredacion de la tortuga marina *Lepidochelys olivacea* por *Orcinus orca*. Cuadernos Mexicanos de Zoologia, 1:96-98.

Estes, J.A. 1980. *Enhydra lutris*. Mammalian Species, 133:1-8.

Estes, J.A. 1990. Growth and equilibrium in sea otter populations. Journal of Animal Ecology, 59:385-401.

Esteva, M., F. Cervantes Reza, S.V. Brant, and J. Cook. 2010. Molecular phylogeny of long-tailed shrews (genus *Sorex*) from México and Guatemala. Zootaxa, 2615:47-65.

Estrada, A. 1984. Resource use by howler monkeys (*Alouatta palliata*) in the rain forest of Los Tuxtlas, Veracruz, Mexico. International Journal of Primatology, 5:105-131.

Estrada, A., and R. Coates-Estrada. 1984a. Fruit-eating and seed dispersal by howling monkeys (*Alouatta palliata*) in the tropical rain forest of Los Tuxtlas, Mexico. American Journal of Primatology, 6:77-91.

Estrada, A., and R. Coates-Estrada. 1984b. Some observations on the present distribution and conservation of *Alouatta* and *Ateles* in southern Mexico. American Journal of Primatology, 7:133-137.

Estrada, A., and R. Coates-E. 1985. A preliminary study of resource overlap between howling monkeys (*Alouatta palliata*) and other arboreal mammals in the tropical rainforest of Los Tuxtlas, Mexico. American Journal of Primatology, 9:27-37.

Estrada, A., and R. Coates-Estrada. 1986. Manual de los Mamíferos de la Estación de Biología Tropical Los Tuxtlas. Instituto de Biología, Universidad Nacional Autónoma de México, México.

Estrada, A., and R. Coates-Estrada. 1988. Estudios de primatología de campo en México. Pp. 61-87, in: Tendencias Actuales de la Primatología (J. Martínez, ed.). Universidad Autónoma Metropolitana-Iztapalapa, México.

Estrada, A., and R. Coates-Estrada. 1989. La destrucción de la selva y la conservación de los primates silvestres de México (Alouatta y Ateles). Pp. 211-233, in: Primatología en México: Comportamiento, Ecología, Aprovechamiento y Conservación de Primates (A. Estrada, R. López-Wilchis, and R. Coates-Estrada, eds.). Universidad Autónoma Metropolitana, México.

Estrada, A., S.J. Solano, T. Ortiz Martínez, and R. Coates-Estrada. 1999. Feeding and general activity patterns of a howler monkey (*Alouatta palliata*) troop living in a forest fragment at Los Tuxtlas, Mexico. American Journal of Primatology, 48:167-183.

Evans, P.G.H. 1987. The Natural History of Whales and Dolphins. Facts on File, New York.

Evans, W.E. 1976. Distribution and differentiation of the stocks of *Delphinus delphis* Linnaeus, in the northeastern Pacific. ACMRR/MM/SC/18.

Evans, W.E. 1994. Common dolphin, white-bellied porpoise *Delphinus delphis* Linnaeus, 1758. Pp. 191-224, in: Handbook of Marine Mammals (S.H. Ridgeway and R. Harrison, eds.). Vol. 5. Academic Press, London.

Ewer, R.F. 1973. The Carnivores. Cornell University Press, Ithaca, NY.

Ewer, R.F. 1977. The Carnivores. 2nd ed. Cornell University Press, Ithaca, NY.

Ezcurra, E., and S. Gallina. 1981. Biology and population dynamics of white-tailed deer in northwestern Mexico, Pp. 77-108, in: Deer Biology, Habitat Requirements and Management in Western North America (P.F. Folliott and S. Gallina, eds.). Instituto de Ecología, A.C., Mexico.

Fa, J.E. 1989. Conservation-motivated analysis of mammalian biogeography in the Trans-Mexican Neovolcanic Belt. National Geographic Research, 5:296-316.

Fa, J.E., and D.J. Bell. 1990. The volcano rabbit *Romerolagus diazi*. Pp. 143-146, in: Rabbits, Hares, and Pikas: Status Survey and Conservation Action Plan (J. Chapman and J.E.C. Flux, eds.). IUCN, Gland, Switzerland.

Fa, J. E., and L.M. Morales. 1993. Patterns of mammalian diversity in Mexico. Pp. 319-361, in: Biological Diversity of Mexico (T.P. Ramamoorthy, R. Bye, A. Lot, and J. Fa, eds.). Oxford University Press, New York.

Fa, J.E., and L.M. Morales. 1998. Patrones de diversidad de mamíferos de México. Pp. 315-352, in: Diversidad Biológica de México. Orígenes y Distribución (T.P. Ramamoorthy, R. Bye, A. Lot, and J. Fa, eds.). Instituto de Biología, UNAM, México.

Fa, J.E., F.J. Romero, and J. López-Paniagua. 1992. Habitat use by parapatric rabbits in a Mexican high-altitude grassland system. Journal of Applied Ecology, 29:357-370.

Fa, J.E., J. López-Paniagua, F.J. Romero, J.L. Gómez, and J.C. López. 1990. Influence of habitat characteristics on small mammals in Mexican high-altitude grassland. Journal of Zoology (London), 221:275-292.

Fair, J., 1990. The Great American Bear. North Word Press, Inc., Minocque, WI.

Faller, J.C., C. Chávez, S. Johnson, and G. Ceballos. 2007. Densidad y tamaño de la población de jaguar en el noreste de la Península de Yucatán. Pp. 111-122, in: Conservación y Manejo del Jaguar en México: Estudios de Caso y Perspectivas (G. Ceballos, C. Chávez, R. List, and H. Zarza, eds.). CONABIO-Alianza WWF-Telcel-Universidad Nacional Autónoma de México, México D.F.

Faller-Menéndez, J.C., T. Urquiza-Haas, C. Chávez, S. Johnson, and G. Ceballos. 2005. Registro de mamíferos en la reserva privada el Zapotal, en el noreste de la península de Yucatán. Revista Mexicana de Mastozoología, 9:128-140.

Feagle, F.R. 1998. Primate Adaptation and Evolution. Academic Press, San Diego.

Fedigan, L.M. 1990. Vertebrate predation in *Cebus capucinus*: Meat eating in a Neotropical monkey. Folia Primatologica, 54:196-205.

Fedigan, L.M., and M.J. Baxter. 1984. Sex differences and social organization in free-ranging spider monkeys (*Ateles geoffroyi*). Primates, 25:279-294.

Fedriani, J.M., T.K. Fuller, R.M. Sauvajot, and E. York. 2000. Competition and intraguild predation among three sympatric carnivores. Oecologia (Berlin), 125:258-270.

Felder, D.L, D.K. Camp, and J.W. Tunnell. 2009. An introduction to Gulf of Mexico biodiversity assessment. Pp. 1-13, in: Gulf of Mexico Origin, Waters, and Biota (D.L. Felder and D.K. Camp, eds.). Vol. 1, Biodiversity. Texas A&M University Press, College Station.

Felten, H. 1956. Fledermäuse (Mammalia, Chiroptera) aus El Salvador, Teil IV. Senckenbergiana Biologica, 37:341-367.

Fenton, M.B. 1970. Population studies of *Myotis lucifugus* (Chiroptera: Vespertilionidae) in Ontario. Life Sciences Contribution, Royal Ontario Museum, 77:1-34.

Fenton, M.B. 1985. Communication in the Chiroptera. Indiana University Press, Bloomington.

Fenton, M.B. 1992. Bats. Facts on File, New York.

Fenton, M.B., and R.M. Barclay. 1980. *Myotis lucifugus*. Mammalian Species, 142:1-8.

Fenton, M.B., and G.P. Bell. 1979. Echolocation and feeding in four species of Myotis (Chiroptera). Canadian Journal of Zoology, 57:1271-1277.

Fenton, M.B. and, G.P. Bell. 1981. Recognition of species of insectivorous bats by their echolocation calls. Journal of Mammalogy, 62:233-243.

Fenton, M.B., and T.H. Kunz. 1977. Movements and behavior. Pp. 351-364, in: Biology of the Bats of the New World Family Phyllostomatidae, Parts I–II (R.J. Baker, J.K. Jones, and D.C. Carter, eds.). Special Publications of the Museum, Texas Tech University, 13:1-364.

Fenton, M.B., I.L. Rautenbach., J. Rydell, H.T. Arita, and J. Ortega. 1998. Emergence, echolocation, diet and foraging behavior of *Molossus ater* (Chiroptera: Molossidae). Biotropica, 30:314-320.

Fenton, M.B., L. Acharya, D. Audet, M.B. Hickey, M.K. Obrist, D.M. Syme, and B. Adkins. 1992. Phyllostomid bats (Chiroptera: Phyllostomidae) as indicator of habitat disruption in the Neotropics. Biotropica, 24:440-446.

Fenton, M.B., I.L. Rautenbach, J. Rydell, H.T.W. Arita, J. Ortega, S. Bouchard, M.D. Hovorka, B.K. Lim, E. Odgren, C.V. Portfors-Yeomans, W. Scully, D.M. Syme, and M.J. Vonhof. 1998. Emergence, echolocation, diet and foraging of *Molossus ater*. Biotropica, 30:314-320.

Ferrari-Pérez, F. 1886. Catalogue of animals collected by the Geographical and Exploring Commission of the Republic of Mexico. Proceedings of the United States National Museum, 9:125-199.

Ferrell, C.S., and D.E. Wilson. 1990. *Platyrrhinus helleri*. Mammalian Species, 373:1-5.

Ferrusquía-Villafranca, I. 1977. Distribution of Cenozoic vertebrate faunas in Middle America and problems of migration between North and South America. Pp. 193-321, in: Conexiones terrestres entre Norte y Sudamérica (I. Ferrusquía-Villafranca, ed.). Boletín del Instituto de Biología, Universidad Nacional Autónoma de México, 101, México.

Ferrusquía-Villafranca, I. 1993. Geology of Mexico: A synopsis. Pp. 3-107, in: Biological Diversity of Mexico: Origins and Distribution (T.P. Ramamoorthy, R. Bye, A. Lot, and J. Fa, eds.). Oxford University Press, New York.

Findley, J.S. 1953. Pleistocene soricidae from San Josecito cave, Nuevo Leon, Mexico. University of Kansas Publications, Museum of Natural History, 5:633-639.

Findley, J.S. 1955a. Speciation of the wandering shrew. University of Kansas Publications, Museum of Natural History, 9:1-68.

Findley, J.S. 1955b. Taxonomy and distribution of some American shrews. University of Kansas Publications, Museum of Natural History, 7:613-618.

Findley, J.S. 1965. Shrews from Hermit Cave, Guadalupe Mountains, New Mexico. Journal of Mammalogy, 46:206-210.

Findley, J.S. 1967. A black population of the Goldman pocket mouse. The Southwestern Naturalist, 12:191-192.

Findley, J.S. 1969. Biogeography of southwestern boreal and desert mammals. Pp. 113-128, in: Contributions of Mammalogy, A Volume Honouring Professor E. Raymond Hall (J.K. Jones Jr., ed.). University of Kansas Museum Natural History, Miscellaneous Publications, No. 51, Lawrence.

Findley, J.S. 1972. Phenetic relationships among bats of the genus *Myotis*. Systematic Zoology, 21:31-52.

Findley, J.S. 1987. The Natural History of New Mexican Mammals. New Mexico Natural History Series. University of New Mexico Press, Albuquerque.

Findley, J.S. 1993. Bats, a Community Perspective. Cambridge University Press, Cambridge.

Findley, J.S., and W. Caire. 1977. The status of mammals in the northern region of the Chihuahuan desert. Pp. 127-139, in: Transactions of the Symposium on the Biological Resources of the Chihuahuan Desert Region United States and Mexico (R.H. Wauer and D.H. Riskind, eds.). U.S. Department of the Interior, National Park Service Transactions and Proceedings Series, No. 3, Washington, DC.

Findley, J.S., and C.J. Jones. 1960. Geographic variation in the yellow-nosed cotton rat. Journal of Mammalogy, 41:462-469.

Findley, J.S., and C.J. Jones. 1965. Northernmost records of some Neotropical bat genera. Journal of Mammalogy, 46:330-331.

Findley, J.S., and C. Jones, 1967. Taxonomic relationships of bats of the species Myotis fortidens, *M. lucifugus* and *M. occulatus*. Journal of Mammalogy, 48:429-444.

Findley, J.S., and G.L. Traut. 1970. Geographic variation in *Pipistrellus hesperus*. Journal of Mammalogy, 51:741-765.

Findley, J.S., E.H. Studier, and D.E. Wilson. 1972. Morphologic properties of bat wings. Journal of Mammalogy, 53:429-444.

Findley, J.S., A.H. Harris, D.E. Wilson, and C. Jones. 1975. Mammals of New Mexico. University of New Mexico Press, Albuquerque.

Finley, R.B. 1958. The wood rats of Colorado: Distribution and ecology. University of Kansas Publications, Museum of Natural History, 10:213-252.

Fisher, H.I. 1941. Notes on shrews of the genus *Notiosorex*. Journal of Mammalogy, 22:263-269.

Fisher, R.D., and M.A. Bogan. 1977. Distributional notes on *Notiosorex* and *Megasorex* in western Mexico. Proceedings of the Biological Society of Washington, 90:826-828.

Fisler, G.F. 1965. Adaptations and speciation in harvest mice of the San Francisco Bay region. Journal of Mammalogy, 48:549-556.

Fisler, G.F. 1971. Age structure and sex ratio in populations of *Reithrodontomys*. Journal of Mammalogy, 52:653-662.

Fitch, J.E., and R.L. Brownell. 1968. Fish otoliths in cetacean stomachs and their importance in interpreting feeding habits. Journal of Fisheries Research Board of Canada, 25:2561-2574.

Fitch, J.H., K.A. Shump, and A.U. Shump. 1981. *Myotis velifer*. Mammalian Species, 149:1-5.

Flake, L.D. 1974. Reproduction of four rodent species in a shortgrass prairie in Colorado. Journal of Mammalogy, 55:213-216.

Fleharty, E.D., and M.A. Mares. 1973. Habitat preference and spatial relations of *Sigmodon hispidus* on a remnant prairie in west-central Kansas. The Southwestern Naturalist, 18:21-29.

Fleharty, E.D., and L.E. Olson. 1969. Summer food habits of *Microtus ochrogaster* and *Sigmodon hispidus*. Journal of Mammalogy, 50:475-486.

Fleischer, L.A. 1987. The distribution, abundance and population characteristics of the Guadalupe fur seals *Arctocephalus townsendi* (Merriam, 1897). Master's thesis, University of Washington, Washington, DC.

Fleming, T.H. 1970. *Artibeus jamaicensis*: Delayed embryonic development in a Neotropical bat. Science, 171:402-404.

Fleming, T.H. 1972. Aspects of the population dynamics of three species of opossums in the Panama Canal Zone. Journal of Mammalogy, 53:619-623.

Fleming, T.H. 1973. The reproductive cycles of three species of opossums and other mammals in the Panama Canal Zone. Journal of Mammalogy, 54:439-455.

Fleming, T.H. 1974. The population ecology of two species of Costa Rican heteromyid rodents. Ecology, 55:493-510.
Fleming, T.H. 1979. Life-history strategies. Pp. 1-61, in: Ecology of Small Mammals (D.M. Stoddart, ed.). Chapman and Hall, New York.
Fleming, T.H. 1983. *Liomys salvini*. Pp. 475-477, in: Costa Rican Natural History (D.H. Janzen, ed.). University of Chicago Press, Chicago.
Fleming, T.H. 1988. The Short-tailed Fruit Bat: A Study in Plant-Animal Interactions. University of Chicago Press, Chicago.
Fleming, T.H., and G.J. Brown. 1975. An experimental analysis of seed hoarding and burrowing behavior in two species of Costa Rican heteromyid rodents. Journal of Mammalogy, 56:301-315.
Fleming, T.H., and E.R. Heithaus. 1986. Seasonal foraging behavior of the frugivorous bat *Carollia perspicillata*. Journal of Mammalogy, 67:660-671.
Fleming, T.H., E.R. Heithaus, and W.B. Sawyer. 1977. An experimental analysis of the food location behavior of frugivorous bats. Ecology, 58:619-627.
Fleming, T.H., E.T. Hooper, and D.E. Wilson. 1972. Three Central American bat communities: Structure, reproductive cycles, and movement patterns. Ecology, 53:655-670.
Fleming, T.H., R.A. Nuñez, and L.S.L. Sternberg. 1993. Nectar corridors and the diet of migratory nectarivorous bats. Oecologia (Berlin), 94:72-75.
Flores, O.L.M. 1991. Observaciones de conducta de una orca en cautiverio en la Ciudad de México. Tesis de licenciatura, Facultad de Ciencias, Universidad Nacional Autónoma de México, México.
Flores-González, L., J.J. Capella, and H.C. Rosenbaun. 1994. Attack of killer whales (*Orcinus orca*) on humpback whales (*Megaptera novaeangliae*) on a South American Pacific breeding ground. Marine Mammal Science, 10:218-222.
Flores-González, L., J. Capella, G. Bravo, and H. Rosenbaum. 1994. Ataque de orcas, *Orcinus orca*, y orcas falsas, *Pseudorca crassidens*, sobre ballenas jorobadas en el Pacífico Colombiano. P. 91, in: IX Seminario Nacional de Ciencias y Tecnologías del Mar y Congreso Latinoamericano en Ciencias del Mar. Medellín, Colombia.
Flores-Villela, O. 1993. *Herpetofauna mexicana*. Carnegie Museum Natural History, Special Publication.
Flores-Villela, O., and P. Gerez. 1988. Conservación en México: Síntesis sobre Vertebrados Terrestres, Vegetación y Uso del Suelo. Instituto Nacional para la Conservación de los Recursos Bióticos, Xalapa, Veracruz.
Flux, J.E.C., and R. Angermann. 1990. The hares and jackrabbits. Pp. 61-94, in: Rabbits, Hares, and Pikas: Status Survey and Conservation Action Plan (J.A. Chapman and J.E.C. Flux, eds.). IUCN, Gland, Switzerland.
Flyger, V., and J.E. Gates. 1982a. Fox and gray squirrels. Pp. 209-229, in: Wild Mammals of North America: Biology, Management, Economics (J.A. Chapman and G.A. Feldhamer, eds.). Johns Hopkins University Press, Baltimore.
Flyger, V. and J.E. Gates. 1982b. Pine squirrels. Pp. 230-238, in: Wild Mammals of North America: Biology, Management, Economics (J.A. Chapman and G.A. Feldhamer, eds.). Johns Hopkins University Press, Baltimore.
Folk, G.E., Jr. 1940. Shift population among hibernating bats. Journal of Mammology, 21:306-315.
Folkens, P. A., R. Reeves, B. S. Stewart, P. J. Clapham, and J. A. Powell 2002. National Audubon Society Guide to Marine Mammas of the World. Alfred A. Knopf, New York.
Folliott, P.F., and S. Gallina (eds.). 1981. Deer Biology, Habitat Requirements and Management in Western North America. Instituto de Ecología, A.C., México.
Fonseca, G.A.B., and M.C.M. Kierulff. 1988. Biology and natural history of Brazilian Atlantic forest small mammals. Bulletin of the Florida State Museum, Biology Series, 34(3-4):99-152.
Forbes, R. 1962. Notes on food of silky pocket mice. Journal of Mammalogy, 43:278-279.
Ford, B., 1981. Black Bear: The Spirit of the Wilderness. Houghton Mifflin, Boston.
Ford, L.S., and R.S. Hoffmann. 1988. *Potos flavus*. Mammalian Species, 321:1-9.
Forester-Turley, P., S. MacDonald, and C. Mason (eds.). 1990. Otters. An Action Plan for Their Conservation. Gland, Switzerland.
Forman, G.L. 1971. Histochemical differences in gastric mucus of bats. Journal of Mammalogy, 52:191-193.
Forman, G.L., and C.J. Phillips. 1988. Histological variation in the proximal colon of heteromyid and cricetid rodents. Journal of Mammalogy, 69:144-149.
Forman, G.L., C.J. Phillips, and C.S. Rouk. 1979. Alimentary tract. Pp. 205-219, in: Biology of Bats of the New World Family Phyllostomidae, Part III (R.J. Baker, J.K. Jones Jr., and D.C. Carter, eds.). Special Publications of the Museum, Texas Tech University, 16:1-411.
Foster, M.S., and R.M. Timm. 1976. Tent-making by *Artibeus jamaicensis* (Chiroptera: Phyllostomatidae) with comments on plants used by bats for tents. Biotropica, 8:265-269.
Foster, N.S., and H.S. Hygnstrom. 1990. Prairie Dogs and Their Ecosystem. University of Nebraska–Lincoln.
Fox, M.W. 1971. Behavior of Wolves, Dogs and Related Canids. Harper and Row, New York.
Fox, M.W. 1975. The Wild Canids: Their Systematics, Behavioral Ecology and Evolution. Van Nostrand Reinhold, New York.
Fragoso, J. M. 1991. The effect of hunting on tapirs in Belize. In: Neotropical Wildlife Use and Conservation (J.G. Robinson and K.H. Redford, eds.). University of Chicago Press, Chicago.
Fragoso, J.M.V. 1998. Home range and movement patterns of white-lipped peccary (*Tayassu pecari*) herds in the northern Brazilian Amazon. Biotropica, 30:458-469.
Fraser, F.C., and B.A. Noble. 1970. Variation of pigmentation pattern in Meyen's dolphin *Stenella coeruleoalba* (Meyen). Pp. 147-163, in: Investigations in Cetacea (F. Dilleri, ed.). Berne, Switzerland.
Freeman, P.W. 1979. Specialized insectivory: Beetle-eating and moth-eating molossid bats. Journal of Mammalogy, 60:467-479.
Freeman, P.W. 1981a. A multivariate study of the family Molossidae (Mammalia: Chiroptera): Morphology, ecology, and evolution. Fieldiana Zoology (n.s.), 7:1-173.
Freeman, P.W. 1981b. Correspondence of food habits and morphology in insectivorous bats. Journal of Mammalogy, 62:166-173.
Freese, C.H. 1976. Censusing *Alouatta palliata, Ateles geoffroyi* and *Cebus capucinus* in the Costa Rican dry forest. Pp. 4-9, in: Neotropical Primates: Field Studies and Conservation (R.W. Thorington Jr., and P.G. Heltne, eds.). National Academy of Sciences, Washington, DC.
Fritts, S.H., and J.A. Sealander. 1978. Reproductive biology and population characteristics of bobcats (Lynx rufus) in Arkansas. Journal of Mammalogy, 59:347-353.
Fritzell, E.K. 1978. Aspects of raccoon (*Procyon lotor*) social organization. Canadian Journal of Zoology, 56:260-271.
Frost, D.R., and R.M. Timm. 1992. Phylogeny of Plecotini bats (Chiroptera: Vespertilionidae): Summary of the evidence and proposal of a logically consistency taxonomy. American Museum Novitates, 3034:1-16.
Fryxell, F.M. 1928. The former range of the bison in the Rocky Mountains. Journal of Mammalogy, 9:129-139.
Fuentes, I.A., and A.L. Aguayo. 1992. Distribución de cetáceos en el Golfo y Caribe mexicano. XVII Reunión Internacional para el Estudio de los Mamíferos Marinos. La Paz, Baja California.
Fukui, Y., T. Mogoe, H. Ishikawa, and S. Ohsumi. 1997. In vitro fertilization of in vitro matured minke whale (*Balaenoptera acutorostrata*) follicular oocytes. Marine Mammal Science, 13:395-404.
Fuller, B., M.R. Lee, and L.R. Maxson. 1984. Albumin evolution in *Peromyscus* and *Sigmodon*. Journal of Mammalogy, 65:466-473.
Fuller, T.K., and L.B. Keith. 1980. Wolf population and prey relationships in northeastern Alberta. Journal of Wildlife Management, 44:538-601.

Fuller, W.A. 1960. Behavior and organization of the wild bison of Buffalo National Park, Canada. Arctic, 13:3-19.
Fuller, W.A. 1962. The biology and management of the bison of Wood Buffalo National Park. Canadian Wildlife Service, Wildlife Management Bulletin Series 1, 16:1-52.
Furman, D.P. 1955. *Steptolaelaps* (Acarina: Laelaptidae) a new genus of mites parasitic on Neotropical rodents. Journal of Parasitology, 41:519-525.
Gadner, A.L. 1965. New bat records from the Mexican state of Durango. Proceedings of the Western Foundation of Vertebrate Zoology, 1:101-106.
Gadner, A.L., and J.L. Patton. 1976. Karyotypic variation in oryzomyine rodents (Cricetinae) with comments on chromosomal evolution in the Neotropical cricetine complex. Occasional Papers of the Museum of Zoology, Louisiana State University, 49:1-48.
Gaines, M.S. 1985. Genetics. Pp. 845-883, in: Biology of New World Microtus (R.H. Tamarin, ed.). Special Publication 8, American Society of Mammalogists.
Galef, B.G., R.A. Mittermeier, and R.C. Bailey. 1976. Predation by the Tayra (Eira barbara). Journal of Mammalogy, 52:461-465.
Galina, P., S. Álvarez-Cárdenas, A. González-Romero, and S. Gallina. 1988. Mastofauna. Pp. 209-228, in: La Sierra de La Laguna de Baja California Sur (L. Arriaga and A. Ortega, eds.). Centro de Investigaciones Biológicas de Baja California Sur, A.C., Publicación No. 1:1-237.
Galindo, J.R., M. de la Rosa, A. González, L. Snook, and J.H. Shaw. 1985. Manejo Forestal y el Venado Cola Blanca (*Odocoileus virginianus*) en Macuiltianguis, Oaxaca, México. Pp. 512-529, in: Primer Simposium Internacional de Fauna Silvestre. Wildlife Society México-SEDUE, México.
Galindo-Leal, C. 1992. Overestimation of deer densities in Michilia Biosphere Reserve. The Southwestern Naturalist, 37:209-212.
Galindo-Leal, C. 1993. Densidades poblacionales de los venados cola blanca, cola negra y bura en Norteamérica. Pp. 371-391, in: Avances en el Estudio de los Mamíferos de México (R.A. Medellín and G. Ceballos, eds.). Publicaciones Especiales, Vol. 1, Asociación Mexicana de Mastozoología, A.C., México.
Galindo-Leal, C. 1999. La gran región de Calakmul: Prioridades biológicas de conservación y propuesta de modificación de la Reserva de la Biosfera. Reporte Final a World Widlife Fund--México, México D.F.
Galindo-Leal, C., and M. Weber. 1994. Translocation of deer subspecies: Reproductive implications. Wildlife Society Bulletin, 22:117-120.
Galindo-Leal, C., and M. Weber. 1998. El Venado de la Sierra Madre Occidental: Ecología, Conservación y Manejo. Edicusa-CONABIO, México.
Galindo-Leal, C., A. Morales G., and M. Weber. 1993. Distribution and abundance of Coues deer and cattle in Michilia Biosphere Reserve. The Southwestern Naturalist, 38:127-135.
Galindo-Leal, C., A. Morales, and M. Weber. 1995. Utilización de hábitat, abundancia y dispersión del venado de Coues: un experimento seminatural. Pp. 315-332, in: Ecología y Manejo del Venado Cola Blanca en México y Costa Rica (C. Vaughan and M. Rodriguez, eds.). Editorial de la Universidad Nacional, Heredia, Costa Rica.
Galliez, M., M. De Souza L., T. Lopes Queiroz, and F.A. Dos Santos F. 2009. Ecology of the water opossum *Chironectes minimus* in Atlantic forest streams of southeastern Brazil. Journal of Mammalogy, 90:93-103.
Gallina, S. 1981. Contribución al conocimiento de los hábitos alimenticios del tepezcuintle (*Agouti paca* Lin) en Lacanjá-Chansayab, Chiapas. Pp. 57-67, in: Estudios Ecológicos en el Neotrópico Mexicano (P.R. Castillo, ed.). Instituto de Ecología, A.C., México.
Gallina, S. 1988. Importancia del injerto (*Phoradendron* spp.) para el venado. The Southwestern Naturalist, 33:21-25.
Gallina, S. 1989. El habitat del venado bura en la Sierra de La Laguna, BCS. Pp. 450-463, in: Memorias del VI Simposio sobre Fauna Silvestre. Universidad Nacional Autónoma de México, México.
Gallina, S. 1990. El Venado cola blanca y su hábitat en La Michilía, Durango. Tesis doctoral. Facultad de Ciencias, Universidad Nacional Autónoma de México, México.
Gallina, S. 1993a. Biomasa disponible y capacidad de carga para el venado y el ganado en la Reserva La Michilia, Durango. Pp. 437-453, in: Avances en el Estudio de los Mamíferos de México (R.A. Medellín and G. Ceballos, eds.). Publicaciones Especiales, Vol. 1, Asociación Mexicana de Mastozoología, A.C., México.
Gallina, S. 1993b. White-tailed deer and cattle diets at La Michilia, Durango, Mexico. Journal of Range Management, 46:487-492.
Gallina, S., P. Galina-Tessaro, and S. Álvarez-Cardenas. 1991. Mule deer density and pattern distribution in the pine-oak forest at the Sierra de la Laguna in Baja California Sur. Ethology, Ecology and Evolution, 3:27-33.
Gallina, S., M.E. Maury, and V. Serrano. 1978. Hábitos alimenticios del venado cola blanca (*Odocoileus virginianus* Rafinesque) en la Reserva La Michilía, Estado de Durango. Pp. 47-108, in: Reservas de la Biósfera en el Estado de Durango (G. Halffter, ed.). Instituto de Ecología, México.
Gallina, S., M.E. Maury, and V. Serrano. 1981. Food habits of whitetailed deer. Pp. 133-148, in: Deer Biology, Habitat Requirements and Management in Western North America (P.F. Folliott and S. Gallina, eds.). Instituto de Ecología, A.C., México.
Gallo, J.P. 1989. Distribución y estado actual de la nutria o perro de agua (Lutra longicaudus annectens Major 1897) en la Sierra Madre del Sur, México. Tesis de maestría, Facultad de Ciencias, Universidad Nacional Autónoma de México, México.
Gallo, J.P. 1991. The status and distribution of river otters (*Lutra longirostris annectens* Major, 1897) in Mexico. Habitat. Proceedings of the 5th International Otter Colloquium. Hankensbüttel, Germany.
Gallo, J.P., and D. Aurioles. 1984. Distribución y estado actual de la población de la foca común (*Phoca vitulina richardsi*) Gray, 1864, en Baja California, México. Anales del Instituto de Biología, UNAM, México, Serie Zoología, 55:323-332.
Gallo, J.P., and O.A. Ortega. 1986. First record of *Zalophus californianus* in Acapulco, Mexico. Marine Mammal Science, 2:158.
Gallo, J.P., and F. Pimienta. 1989. Primr registro del zifio de las Antillas (*Mesoplodon europaeus* Gervais, 1855) (Cetacea: Ziphiidae) en México. Anales del Instituto de Biología, UNAM, México, Serie Zoología, 60:267-278.
Gallo, J.R. 1983. Notas sobre la distribución del manatí (*Trichechus manatus*) en las costas de Quintana Roo. Anales del Instituto de Biologia, UNAM, México, Serie Zoologia, 53:443-448.
Gallo-Reynoso, J.P. 1983. Notas sobre la distribución del manatí (*Trichechus manatus*) en las Costas de Quintana Roo. Anales del Instituto de Biología, UNAM, México, Serie Zoología, 53(1):443-448.
Gallo-Reynoso, J.P. 1997. Situación y distribución de las nutrias en México, con énfasis en *Lontra longicaudis annectens* Major, 1897. Revista Mexicana de Mastozoología, 2:10-32.
Gallo-Reynoso, J.P., and A.L. Figueroa-Carranza. 1995. Occurrence of bottlenose whales in the waters of Isla Guadalupe, Mexico. Marine Mammal Science, 11:573-575.
Gallo-Reynoso, J.P., and G.B. Rathbun. 1997. Status of sea otters (*Enhydra lutris*) in Mexico. Marine Mammal Science, 13:332-340.
Gallo Reynoso, J.P., and L. Rojas Bracho. 1985. Nombres científicos y comunes de los mamíferos marinos de México. Anales del Instituto de Biología, UNAM, México, Serie Zoología, 3:1043-1056.
Gallo-Reynoso, J.P., G. Suarez-Gracida, H. Cabrera-Santiago, E. Coria-Galindo, J. Egido-Villarreal, and L.C. Ortiz. 2002. Status of beavers (Castor canadensis frondator) in Rio Bavispe, Sonora, Mexico. The Southwestern Naturalist, 47:501-504.
Gallo-Reynoso, J.P., T. Van Devender, A.L. Reina-Guerrero, J. Egido-Villarreal, and E. Pfeiler. 2008. Probable occurrence of a brown bear (Ursus arctos) in Sonora, Mexico, in 1976. The Southwestern Naturalist, 53(2):256-260.
Gambell, R. 1985. Fin whale, *Balaenoptera physalus* (Linnaeus, 1758). In: Handbook of Marine Mammals, vol. 3, The Sirenians and Baleen Whales (S.H. Ridgway and R. Harrison, eds.). Academic Press, Orlando, FL.

Gannon, M.R., M.R. Willig, and J.K. Jones Jr. 1989. *Sturnira lilium*. Mammalian Species, 333:1-5.
Gaona, S., and G. López. 1991. Conejos y liebres endémicos de México. Cemanáhuac, 3:12-13.
Gaona, S., A. González-Christen, and R. López-Wilchis. 2003. Síntesis del conocimiento de los mamíferos silvestres del Estado de Veracruz, México. Revista de la Sociedad Mexicana de Historia Natural (series 3), 1:91-123.
García, L.C., and R. Monroy. 1985. Estimación de la población de venado cola blanca (*Odocoileus virginianus*) en la selva baja caducifolia del sureste del Estado de Morelos. Pp. 68-80, in: III Simposio sobre Fauna Silvestre, Universidad Nacional Autónoma de México, Linares, N.L., México.
García, R.M.C. 1992. Conducta territorial del lobo marino *Zalophus californianus* en la Lobera Los Cantiles, Isla Ángel de la Guarda, Golfo de California, México. Tesis de licenciatura en biología, Facultad de Ciencias, Universidad Nacional Autónoma de México, México.
García-G., J.L. 2007. Estructura poblacional del murciélago Dermanura tolteca (Saussure, 1860), en el municipio de Santiago Comaltepec, Oaxaca. Tesis de maestría, IPN.
García-García, J.L., A.M. Alfaro-E., and A. Santos-Moreno. 2006. Registros notables de los murciélagos en el estado de Oaxaca, México. Revista Mexicana de Mastozoología, 10:88-91.
García-Rodríguez, A.I., B.W. Bowen, D. Domning, A.A. Mignucci-Giannoni, M. Marmontel, R.A. Montoya-Ospina, B. Morales-Vela, B.M. Rudin, R.K. Bonde, and P.M. McGuire. 1998. Phylogeography of the West Indian manatee (*Trichechus manatus*): How many populations and how many taxa? Molecular Ecology, 7:1137-1149.
García-Rodríguez, F., and D. Aurioles-Gamboa. 2004. Spatial and temporal variations in the diet of the California sea lion (*Zalophus californianus*) in the Gulf of California, México. Fishery Bulletin, 102:47-62.
García-Vasco, D. 2005. Distribución, abundancia y aspectos poblacionales del mapache enano *(Procyon pygmaeus)*, un carnívoro insular endémico. Advanced bachelor of science (licenciatura) thesis, Universidad Veracruzana, Xalapa.
Gardner, A.L. 1962a. A new bat genus *Glossophaga* from Mexico. Contributions in Science, Los Angeles County Museum, 54:1-7.
Gardner, A.L. 1962b. Bat records from the Mexican states of Colima and Nayarit. Journal of Mammalogy, 43:102-103.
Gardner, A.L. 1965. New bat records from the Mexican state of Durango. Proceedings of the Western Foundation of Vertebrate Zoology, 1:101-106.
Gardner, A.L. 1966. A new subspecies of the Aztec mastiff bat, *Molossus aztecus* Sussure, from southern Mexico. Los Angeles Country Museum Contributions in Science, 111:1-5.
Gardner, A.L. 1973. The systematics of the genus *Didelphis* (Marsupialia: Didelphidae) in North and Middle America. Special Publications of the Museum, Texas Tech University, 4:1-81.
Gardner, A.L. 1977. Feeding habits. Pp. 293-350, in: Biology of Bats of the New World Family Phyllostomatidae, Part II (R.J. Baker, J.K. Jones Jr., and D.C. Carter, eds.). Special Publications of the Museum, Texas Tech University, 13:1-364.
Gardner, A.L. 1986. The taxonomic status of *Glossophaga morenoi* Martinez and Villa, 1938 (Mammalia: Chiroptera: Phyllostomidae). Proceedings of the Biological Society of Washington, 99:489-492.
Gardner, A.L. 1993. Order Didelphiomorphia. In: Mammal Species of the World: A Taxonomic and Geographic Reference (D.E. Wilson and D.M. Reeder, eds.). 2nd ed. Smithsonian Institution Press, Washington, DC.
Gardner, A.L. (ed.). 2008. Mammals of South America: Xenarthras, Shrews, and Bats. University of Chicago Press, Chicago.
Gardner, A.L., and G.K. Creighton. 1989. A new generic name for Tate's (1933) *Microtarsus* group of South American mouse opossums (Marsupialia: Didelphidae). Proceedings of the Biological Society of Washington, 102:3-7.
Gardner, A.L., and J.L. Patton. 1976. Karyotypic variation in oryzomyine rodents (Cricetinae) with comments on chromosomal evolution in the Neotropical cricetine complex. Occasional Papers of the Museum of Zoology, Louisiana State University, 49:1-48.
Gardner, A.L., R.K. LaVal, and D.E. Wilson. 1970. The distributional status of some Costa Rican bats. Journal of Mammalogy, 51:712-729.
Garrido, R. 1980. La distribución geográfica de los murciélagos de la Costa de Michoacán, México. Tesis de licenciatura en biología, Facultad de Ciencias, Universidad Nacional Autónoma de México, México.
Garrison, T.E., and T.L. Best. 1990. *Dipodomys ordii*. Mammalian Species, 353:1-10.
Garshelis, D.L. 1992. Mark-recapture density estimation for animals with large home ranges. Pp. 1098-1111, in: Wildlife 2001: Populations (D.R. McCullough and R.H. Barrett, eds.). Elsevier Applied Science, New York.
Garshelis, D.L. 2009. Family Ursidae (Bears). Pp. 448-497, in: Handbook of the Mammals of the World (D.E. Wilson and R.A. Mittermeier, eds.). Vol. 1. Carnivores. Lynx Edicions. Barcelona, Spain.
Gaskin, D.E. 1985. The Ecology of Whales and Dolphins. Heinemann Educational Books, Auckland.
Gates, C.C., C.H. Freese, P.J.P. Gogan, and M. Kotzman (eds.). 2010. American Bison: Status Survey and Conservation Guidelines 2010. IUCN, Gland, Switzerland.
Gaumer, G. 1913. Monografía sobre el Lagomys diazi Ferrari-Pérez. Dirección General Agrícola, Departamento de Exploración Biológica, Serie Zoología, No. 4. México.
Gaumer, G.F. 1917. Monografía de los Mamíferos de Yucatán. Departamento de Talleres Gráficos de la Secretaría de Fomento, México.
Geist, V. 1971. Mountain Sheep: A Study in Behavior and Evolution. University of Chicago Press, Chicago.
Geist, V. 1981. Behavior: Adaptive strategies in mule deer. Pp. 157-233, in: Mule and Black-tailed Deer of North America: Ecology and Management. (O.C. Wallmo, ed.). Wildlife Management Institute, University of Nebraska Press, Lincoln.
Geist, V. 1998. Deer of the World: Their Evolution, Behavior and Ecology. Stackpole Books, Mechanicsburg, PA.
Geluso, K.N., J.S. Altenbach, and D.E. Wilson. 1976. Bat mortality: Pesticide poisoning and migratory stress. Science, 194:184-186.
Gendron, D. 1990. Relación entre la abundancia de eufáusidos y de ballenas azules (*Balaenoptera musculus*) en el Golfo de California. Tesis de maestría, CICIMAR-IPN, La Paz, Baja California Sur.
Gendron, D. 1992. Population structure of daytime surface swarms of *Nyctiphanes simplex* (Crustacea: Euphausiacea) in the Gulf of California, Mexico. Marine Ecology Progress, 87:1-6.
Gendron, D. 1993a. El cachalote (*Physeter macrocephalus*) en el Golfo de California: avistamientos recientes. XVIII Reunión Internacional para el Estudio de los Mamíferos Marinos. La Paz, Baja California Sur.
Gendron, D. 1993b. Índice de avistamientos y distribución del género Balaenoptera en el Golfo de California, México durante febrero, marzo y abril 1988. Revista de Investigación Científica (número especial SOMEMMA), 1:21-30.
Gendron, D., and S. Chávez Rosales. 1996. Recent sei whale (*Balaenoptera borealis*) sightings in the Gulf of California, Mexico. Aquatic Mammals, 22:127-130.
Gendron, D., and J. Urbán-R. 1993. Evidence of feeding by humpback whales (*Megaptera novaeangliae*) in the Baja California breeding ground, Mexico. Marine Mammal Science, 9:76-81.
Gendron, D., S. Lanham, and M. Carwardine. 1999. North Pacific right whale (*Eubalaena glacialis*) sighting south of Baja California, Mexico. Aquatic Mammals, 25:31-34.
Genoways, H.H. 1971. A new species of spiny pocket mouse (genus *Liomys*) from Jalisco, Mexico. University of Kansas Publications, Museum of Natural History, 5:1-17.

Genoways, H.H. 1973. Systematics and evolutionary relationships of spiny pocket mice, genus *Liomys*. Special Publications of the Museum, Texas Tech University, 5:1-368.

Genoways, H.H., and R.J. Baker. 1988. *Lasiurus blossevillii* (Chiroptera: vespertilionidae) in Texas. Texas Journal of Science, 40:1-13.

Genoways, H.H., and R.J. Baker. 1996. A new species of the genus *Rhogeessa*, with comments on geographic distribution and speciation in the genus. Pp. 83-87, in: Contributions in Mammalogy: A Memorial Volume Honoring Dr. J. K. Jones, Jr. (H.H. Genoways and R.J. Baker, eds.). Museum of Texas Tech University.

Genoways, H.H., and E.C. Birney. 1974. *Neotoma alleni*. Mammalian Species, 41:1-4.

Genoways, H.H., and J.H. Brown. 1993. Biology of the Heteromyidae. Special Publication of the American Society of Mammalogists, 10:1-719.

Genoways, H.H., and J.K. Jones Jr. 1967. Notes on the distribution and variation in the Mexican big-eared bat, *Plecotus phyllotis*. The Southwestern Naturalist, 12:477-480.

Genoways, H.H., and J.K. Jones Jr. 1968a. Notes on spotted skunks (genus *Spilogale*) from western Mexico. Anales del Instituto de Biología, UNAM, México, Serie Zoología, 39:123-132.

Genoways, H.H., and J.K. Jones Jr. 1968b. A new mouse of the genus *Nelsonia* from southern Jalisco, Mexico. Proceedings of the Biological Society of Washington, 81:97-100.

Genoways, H.H., and J.K. Jones Jr. 1968c. Notes on bats from the Mexican state of Zacatecas. University of Kansas Publications, Museum of Natural History, 49:743-745.

Genoways, H.H., and J.K. Jones Jr. 1969a. Notes on pocket gophers from Jalisco, Mexico, with descriptions of two new subspecies. Journal of Mammalogy, 50:748-755.

Genoways, H.H., and J.K. Jones Jr. 1969b. Taxonomic status of certain long eared bats (genus *Myotis*) from the southwestern United States and Mexico. The Southwestern Naturalist, 14:1-13.

Genoways, H.H., and J.K. Jones Jr. 1971. Systematics of southern banner-tailed kangaroo rats of the *Dipodomys phillipsii* group. Journal of Mammalogy, 52:265-287.

Genoways, H.H., and J.K. Jones Jr. 1972. Variation and ecology in a local population of the vesper mouse (*Nyctomys sumichrasti*). Occasional Papers of the Museum, Texas Tech University, 9:1-22.

Genoways, H.H., and J.K. Jones Jr. 1973. Notes on some mammals from Jalisco, Mexico. Occasional Papers of the Museum, Texas Tech University, 9:1-22.

Genoways, H.H., and J.K. Jones Jr. 1975. Annotated checklist of mammals of the Yucatan Peninsula, Mexico. IV. Carnivora, Sirenia, Perissodactyla, Artiodactyla. Occasional Papers of the Museum, Texas Tech University, 26:1-22.

Genoways, H.H., R.M. Timm, and M.D. Engstrom. 2005. Natural history and karyology of the Yucatán vesper mouse, *Otonyctomys hatti*. Pp. 215-220, in: Contribuciones Mastozoológicas en Homenaje a Bernardo Villa (V. Sánchez-Cordero and R.A. Medellín, eds.). Instituto de Biología, UNAM, Instituto de Ecología, UNAM, CONABIO, México.

Genoways, H.H., S.L. Williams, and J.A. Groen. 1981. Results of the Alcoa Foundation-Suriname expeditions. V. Noteworthy records of Surinamese mammals. Annals of Carnegie Museum, 50:319-332.

Geoffroy, É. 1810. Sur le Phyllostomes et les Mégadermes. Annals of Museum of Natural History, 15:157-198.

George, L. 1986. Baja Islands Project: Faunal surveys and management recommendations. Department of Biology, University of New Mexico, Albuquerque.

George, S.B. 1988. Systematics, historical biogeography, and evolution of the genus *Sorex*. Journal of Mammalogy, 69:443-461.

Gese, E.M., O.J. Rongstand, and W. R. Mytton. 1988. Relationship between coyote group size and diet in southeastern Colorado. Journal of Wildlife Management, 52:639-652.

Getz, L.L. 1985. Habitats. Pp. 286-309, in: Biology of New World Microtus (R.H. Tamarin, ed.). Special Publication No. 8, American Society of Mammalogists.

Gewalt, W. 1968. Kleine Beobachtungen an selteneren Beuteltieren in Berliner Zoo. V. Zwergbeutelratte (Marmosa Mexicana Merriam 1897). Zoologische Garten, Berlin, 35:288-303.

Giles, R.H., Jr. 1978. Wildlife Management. Freeman, San Francisco.

Gilmore, R.M. 1947. Report on a collection of mammal bones from archeological cave-sites in Coahuila, Mexico. Journal of Mammalogy, 28:147-165.

Gilmore, R.M. 1978. Right whale. Pp. 62-69, in: Marine Mammals (D. Haley, ed.). Pacific Search Press, Seattle.

Ginnett, T.F., and C.L. Douglas. 1982. Food habits of feral burros and desert bighorn sheep in Death Valley National Monument. Desert Bighorn Council Transactions, 26:81-87.

Ginsberg, J.R., and D.W. Macdonald. 1990. Foxes, wolves, jackals and dogs: An action plan for the conservation of canids. IUCN, Gland, Switzerland.

Gittleman, J. 1989. Carnivore group living: Comparative trends. Pp. 183-207, in: Carnivore Behavior, Ecology and Evolution (J. L. Gittleman, ed.). Vol. 1. Cornell University Press, Ithaca, NY.

Gittleman, J. L. 1996. Carnivore Behavior, Ecology, and Evolution. Cornell University Press, Ithaca, NY.

Glander, K.E. 1975. Habitat description and resource utilization: A preliminary report on mantled howling monkey ecology. Pp. 37-57, in: Socioecology and Psychology of Primates (R. Tuttle, ed.). Mouton, The Hague.

Glander, K.E. 1978. Drinking from arboreal water sources by mantled howler monkeys (*Alouatta palliata* Gray). Folia Primatologica, 29:206-217.

Glanz, W.E. 1982. The terrestrial mammal fauna of Barro Colorado Island: Censuses and long-term changes. In: The Ecology of a Tropical Forest: Seasonal Rhythms in a Tropical Ecosystem (E.G. Leigh Jr., A.S. Rand, and D.M. Windsor, eds.). Smithsonian Institution Press, Washington, DC.

Glanz, W.E. 1984. Food and habitat use by two sympatric Sciurus in Central America. Journal of Mammalogy, 65:342-347.

Glanz, W.E. 1990. Neotropical mammal densities: How unusual is the community on Barro Colorado Island, Panama? In: Four Neotropical Rainforests (A.H. Gentry, ed.). Yale University Press, New Haven.

Glass, B.P. 1982. Seasonal movements of Mexican free-tailed bats *Tadarida brasiliensis mexicana* banded in the Great Plains. The Southwestern Naturalist, 27:127-133.

Glazier, D.S. 1980. Ecological shifts and the evolution of geographical restricted species of North American Peromyscus (mice). Journal of Biogeography, 7:63-83.

Glendinning, J.I. 1992. Range extension for the diminutive woodrat *Nelsonia neotomodon*, in the Mexican Transvolcanic range. The Southwestern Naturalist, 37:92-93.

Godin, A.J. 1982. Striped and hooded skunks. Pp. 674-687, in: Wild Mammals of North America. Biology, Management, Ecology. (J.A. Chapman and G.A. Felhamer, eds.). Johns Hopkins University Press, Baltimore.

Goertz, J.W. 1964. The influence of habitat quality upon density of cotton rat populations. Ecological Monographs, 34:359-381.

Goertz, J.W. 1965. Reproductive variation in cotton rats. The American Midland Naturalist, 74:329-340.

Goldman, E.A. 1904. Descriptions of five new mammals from Mexico. Proceedings of the Biological Society of Washington, 17:79-82.

Goldman, E.A. 1905. Twelve new wood rats of the genus Neotoma from Mexico. Proceedings of the Biological Society of Washington, 22:139-142.

Goldman, E.A. 1910. Revision of the wood rats of the genus *Neotoma*. North American Fauna, 31:1-124.

Goldman, E.A. 1911. Revision of the spiny pocket mice (genera *Heteromys* and *Liomys*). North American Fauna, 34:1-70.

Goldman, E.A. 1915. Five new mammals from Mexico and Arizona. Proceedings of the Biological Society of Washington, 28:133-138.
Goldman, E.A. 1918. The rice rats of North America. North American Fauna, 43:1-100.
Goldman, E. A. 1932. Review of wood rats of *Neotoma lepida* group. Journal of Mammalogy, 13:59-67.
Goldman, E.A. 1938a. A new woodrat of the genus *Hodomys*. Journal of the Washington Academy of Sciences, 28:498-499.
Goldman, E.A. 1938b. Three new races of *Microtus mexicanus*. Journal of Mammalogy, 19:493-495.
Goldman, E.A. 1939. Review of the pocket gopher of the genus *Platygeomys*. Journal of Mammalogy, 20:87-93.
Goldman, E.A. 1943. The races of the ocelot and margay in Middle America. Journal of Mammalogy, 24:372-385.
Goldman, E.A. 1944. Classification of wolves. Part 2. Pp. 389-694, in: The Wolves of North America (S.P. Young and E.A. Goldman, eds.). American Wildlife Institute, Washington, DC.
Goldman, E.A. 1951. Biological investigations in Mexico. Smithsonian Miscellaneous Collections, 115:1-176.
Goldman, E.A., and R.T. Moore. 1946. The biotic provinces of Mexico. Journal of Mammalogy, 26:347-360.
Golley, F.B. 1962. Mammals of Georgia: A Study of Their Distribution and Functional Role in the Ecosystem. University of Georgia Press, Athens.
Gómez, J.J.L. 1990. Ecología de Poblaciones de Pequeños Mamíferos en el Volcán Pelado, D. F. Tesis de licenciatura, Facultad de Ciencias, Universidad Nacional Autónoma de México.
Gompper, M.E. 1995. *Nasua narica*. Mammalian Species, 487:1-10.
González, F.X. 1992. Comparación cromosómica entre el conejo zacatuche Romerolagus diazi y la liebre torda *Lepus callotis* (Mammalia: Lagomorpha). Tesis de licenciatura, Facultad de Ciencias, Universidad Nacional Autónoma de México, México.
González, M.1984. Estudio taxonómico de algunos nematodos parásitos de roedores y lagomorfos de México. Tesis de licenciatura, Facultad de Ciencias, Universidad Nacional Autónoma de México, México.
González, S.F.N. 1982. Estudio preliminar sobre el cacomixtle Bassariscus astutus flavus, Rhoads (1894), en el municipio de Agualeguas, Nuevo León, México. Tesis de licenciatura, Universidad Autónoma de Nuevo León, Nuevo León, México.
González-Bocanegra, K., E.I. Romero-Berny, M.C. Escobar-Ocampo, and Y. García-Del Valle. 2011. Wildlife use by rural communities in the Catazaja-La Libertad Wetlands, Chiapas, Mexico. Ra Ximhai, 7:219-230.
González-Christen, A., S. Gaona, and G. López-Ortega. 2002. Registros adicionales de mamíferos para el Estado de Veracruz, México. Vertebrata Mexicana, 11:9-16.
González-Romero, A. 1980. Roedores plaga en las zonas agrícolas del Distrito Federal. Instituto de Ecología y Museo de Historia Natural de la Ciudad de México, México.
González-Romero, A., and A. Lafón-Terrazas. 1993. Distribución y estado actual del berrendo (*Antilocapra americana*) en México. Pp. 411-419, in: Avances en el Estudio de los Mamíferos de México (R.A. Medellín and G. Ceballos, eds.). Publicaciones especiales 1, Asociación Mexicana de Mastozoología, México.
González-Ruíz, N., and J. Villalpando-R. 1997. Primer registro de murciélagos y segundo de *Myotis auriculus apache* (Mammalia: Chiroptera) para Michoacán, México. Vertebrata Mexicana, 4:13-16.
González-Ruíz, N., J. Navarro-Frías, and T. Álvarez. 2000. Notas sobre algunos nuevos registros de murciélagos del Estado de México, México. Vertebrata Mexicana, 9:1-6.
González-Ruiz, N., J. Ramírez-Pulido, and J. Arroyo-Cabrales. 2011. A new species of mastiff bat (Chiroptera: Molossidae: *Molossus*) from Mexico. Mammalian Biology, 76:461-469.
González-Suárez, M., R. Flatz, D. Aurioles-Gamboa, P.W. Hedrick, and R. Gerber. 2009. Isolation by distance among California sea lion populations in Mexico: Redefining management stocks. Molecular Ecology, 18:1088-1099.
González-Zamora1, A., et al. 2011. The northern naked-tailed armadillo in the Lacandona rainforest, Mexico: New records and potential threats. Revista Mexicana de Biodiversidad, 82:581-586.
Goodwin, G.G. 1932. Three new *Reithrodontomys* and two new *Peromyscus* from Guatemala. American Museum Novitates, 506:1-5.
Goodwin, G.G. 1934. Mammals collected by A. W. Anthony in Guatemala, 1924-1928. Bulletin of the American Museum of Natural History, 68:1-60.
Goodwin, G.G. 1946. Mammals of Costa Rica. Bulletin of the American Museum of Natural History, 87:271-473.
Goodwin, G.G. 1954a. A new short-tailed shrew and a new freetailed bat from Tamaulipas, Mexico. American Museum Novitates, 1670:1-3.
Goodwin, G.G. 1954b. Mammals from Mexico collected by Marian Martin for the American Museum of Natural History. American Museum Novitates, 1689:1-16.
Goodwin, G.G. 1956. A preliminary report on the mammals collected by Thomas MacDougall in southwestern Oaxaca, Mexico. American Museum Novitates, 1757:1-15.
Goodwin, G.G. 1958. Bats of the genus *Rhogeessa*. American Museum Novitates, 1923:1-17.
Goodwin, G.G. 1959a. Description of some new mammals. American Museum Novitates, 1967:1-8.
Goodwin, G.G. 1959b. Bats of the subgenus *Natalus*. American Museum Novitates, 1977:1-22.
Goodwin, G.G. 1961. Flying squirrels (*Glaucomys volans*) of Middle America. American Museum Novitates, 2059:1-22.
Goodwin, G.G. 1964. A new species and a new subspecies of *Peromyscus* from Oaxaca, Mexico. American Museum Novitates, 2183:1-8.
Goodwin, G.G. 1966. A new species of vole (genus *Microtus*) from Oaxaca, Mexico. American Museum Novitates, 2243:1-4.
Goodwin, G.G. 1969. Mammals from the state of Oaxaca, Mexico, in the American Museum of Natural History. Bulletin of the American Museum of Natural History, 141:1-270.
Goodwin, G.G., and A.M. Greenhall. 1961. A review of the bats of Trinidad and Tobago. Bulletin of the American Museum of Natural History, 122:187-302.
Goodwin, H.A., and C.W. Holloway. 1972. Red Data Book, vol. 1, Mammalia. IUCN and Natural Resources.
Goodwin, R.E. 1970. The ecology of Jamaican bats. Journal of Mammalogy, 51:571-579.
Gordon, D.G., and A. Baldridge. 1991. Gray Whales. Monterey Bay Aquarium, Monterey, CA.
Gori, M., G.M. Carpaneto, and P. Ottino. 2003. Spatial distribution and diet of the Neotropical otter *Lontra longicaudis* in the Ibera lake (northern Argentina). Acta Theriologica, 48:495-504.
Gosho, M.E., D.W. Rice, and J.M. Breiwick. 1984. The sperm whale *Physeter macrocephalus*. Marine Fisheries Review, 46:54-64.
Grafodatskii, A.S., D.V. Ternovsky, A.A. Isaenko, and S.I. Radzhabli. 1977. Constitutive heterochromatin and DNA content in some mustelids (Mustelidae, Carnivora). Genetika, 13:2123-2128.
Graham, G.L. 1983. Changes in bat species diversity along an elevational gradient up the Peruvian Andes. Journal of Mammalogy, 64:559-571.
Graham, R.W., and J. Mead. 1987. Environmental fluctuations and evolution of mammalian faunas during the last deglaciation in North America. Pp. 183-220, in: North America and Adjacent Oceans during the Last Deglaciation (W.F. Ruddiman, ed.). Geological Society of America, Boulder, CO.
Granados, H. 1980. El conejo de los volcanes. *Romerolagus diazi*. Naturaleza, 2:161-166.
Granados, H. 1981. Basic information on the volcano rabbit. Pp. 940-948, in: Proceedings of the World Lagomorph Conference (K. Myers and C.D. MacInnes, eds.). University of Guelph, Guelph, Ontario.

Granados, H., and J. Hoth. 1989. Estudios sobre la biología del ratón de los volcanes (*N. a. alstoni*). XVI. Comparación de la capacidad reproductora de hembras silvestres en el laboratoriosiguiendo 3 sistemas de apareamiento. Archivos de Investigación Médica (México), 20:95-105.

Granados, H., and J. Luis. 1987. Estudios sobre la biología del ratón de los volcanes (*N. a. alstoni*). XI. Investigación comparativa sobre la reproducción de hembras silvestres en el laboratorio apareadas durante uno y dos ciclos estrales. Archivos de Investigación Médica (México), 18:111-118.

Granados, H., and J. Ramírez. 1986. Estudios sobre la biología del ratón de los volcanes (*N. a. alstoni*). IX. Crecimiento de animales silvestres en el laboratorio. Archivos de Investigación Médica, México, 17:285-297.

Granados, H., R. Zulbarán, and D. Juárez. 1980. Estudios sobre la biología del conejo de los volcanes. III. Presencia de un triángulo de pelo amarillo dorado en la nuca. Memorias del VIII Congreso Latinoamericano de Zoología. Mérida, Venezuela.

Gray, J.A., Jr. 1943. Rodent populations in the sagebrush desert of the Yakima Valley, Washington. Journal of Mammalogy, 24:191-193.

Gray, J. E. 1846. On the British Cetacea. Annual Magazine of Natural History, 17:82-85.

Gray, J.E. 1869. Catalogue of Carnivorous, Pachydermatous, and Edentata, Mammalia in the British Museum. British Museum (Natural History), London.

Green, G.E., W.E. Grant, and E. Davis. 1984. Variability of observed group sizes within collared peccary herds. Journal of Wildlife Management, 48:244-248.

Gregorin, R., G.L. Capusso, and V.R. Furtado. 2008. Geographic distribution and morphological variation in *Mimon bennettii* (Chiroptera, Phyllostomidae). Iheringia, Série Zoologia, 98:404-411.

Greenbaum, I.F., and R.J. Baker, 1976. Evolutionary relationships in *Macrotus* (Mammalia: Chiroptera): Biochemical variation and karyology. Systematic Zoology, 25:15-25.

Greenhall, A.M. 1968. Notes on the behavior of the false vampire bat. Journal of Mammalogy, 49:337-340.

Greenhall, A.M. 1972. The biting and feeding habits of the vampire bat, *Desmodus rotundus*. Journal of Zoology (London), 168:451-461.

Greenhall, A.M., U. Schmidt, and G. Joermann. 1984. *Diphylla ecaudata*. Mammalian Species, 227:1-3.

Greenhall, A.M., G. Joermann, U. Schmidt, and M.R. Seidel. 1983. *Desmodus rotundus*. Mammalian Species, 202:1-6.

Greer, J.K. 1960. Southern yellow bat from Durango, Mexico. Journal of Mammalogy, 41:511.

Greer, J.K., and M. Greer. 1970. Record of the pygmy spotted skunk (*Spilogale pygmaea*) from Colima, Mexico. Journal of Mammalogy, 51:629-630.

Grenot, C. 1983. Fauna del Bolsón de Mapimí. Ecología y conservación de los vertebrados. Departamento de Zonas Áridas, Universidad Autónoma de Chapingo, Chapingo, Estado de México.

Grenot, C., and V. Serrano. 1981. Ecological organization of small mammal communities of the Bolson de Mapimí, Mexico. Pp. 89-100, in: Ecology of the Chihuahuan Desert (R. Barbault and G. Halffter, eds.). Instituto de Ecología, Mexico.

Grenot, C., and V. Serrano. 1982. Distribution spatiale et structure des communautés de petits vertébres du désert de Chihuahua. Comptés Rendus Société Biogéographie, 58:159-192.

Griffin, D.R. 1934. Marking bats. Journal of Mammalogy, 15:202-207.

Griffin, D.R. 1936. Bat banding. Journal of Mammalogy, 17:235-239.

Griffin, D.R. 1940. Notes on the life histories of New England cave bats. Journal of Mammalogy, 21:181-187.

Grigione, M., A. Caso, R. List, and C. López-González. 2001. Status and conservation of endangered cats along the U.S.-Mexico border. Endangered Species Update, 18:129-132.

Grigione, M.M., P. Beier, R.A. Hopkins, D. Neal, W.D. Padley, C.M. Schonewald, and M.L. Johnson. 2002. Ecological and allometric determinants of home-range size for mountain lions (*Puma concolor*). Animal Conservation, 5:317-324.

Grinnell, H.W. 1918. A synopsis of the bats of California. University of California Publications in Zoology, 17:223-404.

Grinnell, J. 1933. Review of the recent mammal fauna of California. University of California Publications in Zoology, 40:71-234.

Grinnell, J., and J. Dixon. 1918. Natural history of the ground squirrels of California. Monthly Bulletin of the California State Commission in Horticulture, 7:597-708.

Grinnell, J., J.S. Dixon, and J.M. Lindsdale. 1937. Fur-bearing Mammals of California: Their Natural History, Systematic Status and Relations to Man. University of California Press, Berkeley.

Groombridge, B., and M.D. Jenkins. 2002. World Atlas of Biodiversity: Earth's Living Resources in the 21st Century. University of California Press, Berkeley.

Groves, C. 2001. Primate Taxonomy. Smithsonian Institution Press, Washington, DC.

Grubb, P. 1993. Order Artiodactyla. Pp. 377-414, in: Mammal Species of the World (D.E. Wilson and D.M. Reeder, eds.). Smithsonian Institution Press, Washington, DC.

Güereña, G.L., M. Uribe-Alcocer, and F.A. Cervantes. 1982. Estudio cromosómico del conejo tropical (*Sylvilagus brasiliensis*). Mammalian Chromosome Newsletter, 23:157-161.

Guerrero, J. A., E. De Luna, and D. González. 2004. Taxonomic status of *Artibeus jamaicensis* inferred from molecular and morphometric data. Journal of Mammalogy, 85:866-874.

Guerrero, S., and F.A. Cervantes. 2003. Lista comentada de los mamíferos terrestres del Estado de Jalisco, México. Acta Zoológica Mexicana (n.s.), 89:93-110.

Guerrero, V. S. Mastofauna de Jalisco. www.acude.udg.mx/jalisciencia/diagnostico/biotico/faunajalisco/mamdiag.pdf.

Guerrero Ruiz, M., J. Urbán R., and L. Rojas Bracho. 2006. Las ballenas del Golfo de California. Instituto Nacional de Ecología (INE-SEMARNAT), México D.F.

Guevara-Chumacero, L., R. López-Wilchis, and V. Sánchez-Cordero. 2001. 105 años de investigación mastozoológica en México (1890-1995): una revisión de sus enfoques y tendencias. Acta Zoológica Mexicana (n.s.), 83:35-72.

Guggisberg, C.A.W. 1975. Wild Cats of the World. Taplinger Publishing Co., New York.

Guthrie, M.J. 1933a. Notes on the seasonal movements and habits of some cave bats. Journal of Mammalogy, 14:1-19.

Guthrie, M.J. 1933b. The reproductive cycles of some cave bats. Journal of Mammalogy, 14:199-216.

Gutiérrez, E., S.A. Jansa, and R.S. Voss. 2010. Molecular systematics of mouse opossums (Didelphidae: *Marmosa*): Assessing species limits using mitochondrial DNA sequences, with comments on phylogenetic relationships and biogeography. American Museum Novitates, 3692:1-22.

Habib, P.R., and N.J. Peña. 1982. Hábitos alimenticios del berrendo (*Antilocapra americana*) en la región central de Chihuahua. Boletín Pastizales CE "La Campana" INIP-SARH, (13):6.

Hacker, S.E. 1992. Stomach contents of four short-finned pilot whales (*Globicephala macrorhynchus*) from the southern California bight. Marine Mammal Science, 8:76-81.

Hafner, D.J. 1984. Evolutionary relationships of the Nearctic Sciuridae. Pp. 3-23, in: The Biology of Ground-dwelling Squirrels: Annual Cycles, Behavioral Ecology, and Sociality (J.O. Murie and G.R. Michener, eds.). University of Nebraska Press, Lincoln.

Hafner, D.J., and K.N. Geluso. 1983. Systematic relationships and historical biogeography of desert pocket gophers, *Geomys arenarius*. Journal of Mammalogy, 64:405-413.

Hafner, D.J., M.S. Hafner, G.L. Hasty, T.A. Spradling, and J.W. Demastes. 2008. Evolutionary relationships of pocket gophers (*Cratogeomys castanops* species group) of the Mexican Altiplano. Journal of Mammalogy, 89:190-208.

Hafner, D. J., B.R. Riddle, and S.T. Álvarez-Castañeda. 2001. Evolutionary relationships of white-footed mice (*Peromyscus*) on islands in the Sea of Cortez, Mexico. Journal of Mammalogy, 82:775-790.

Hafner, M.S. 1979. Evolutionary relationships and systematics of the Geomyidae (Mammalia: Rodentia). Unpublished PhD dissertation, University of California, Berkeley.

Hafner, M.S., and L.J. Barkley. 1984. Genetics and natural history of a relictual pocket gopher, *Zygogeomys* (Rodentia, Geomydae). Journal of Mammalogy, 65:474-479.

Hafner, M.S., and J.C. Hafner. 1982. Structure of surface mounds of *Zygogeomys* (Rodentia: Geomyidae). Journal of Mammalogy, 63:536-538.

Hafner, M.S., and D.J. Hafner. 1987. Geographic distribution of two Costa Rican species of *Orthogeomys*, with comments on dorsal pelage markings in the Geomyidae. The Southwestern Naturalist, 32:5-11.

Hafner, M.S., D.J. Hafner, J.W. Demastes, G.L. Hasty, J.E. Light, and A. Spradling. 2009. Evolutionary relationships of pocket gophers of the genus Pappogeomys (Rodentia: Geomyidae). Journal of Mammalogy, 90:47.

Hafner, M.S., A.R. Gates, V.L. Mathis, J.W. Demastes, and D.J. Hafner. 2011. Redescription of the pocket gopher *Thomomys atrovarius* from the Pacific coast of mainland Mexico. Journal of Mammalogy, 92:1367-1382.

Hafner, M.S., J.E. Light, D.J. Hafner, S.V. Brant, T.A. Spradling, and J.W. Demastes. 2005. Cryctic species in the Mexican pocket gopher *Cratogeomys merriami*. Journal of Mammalogy, 86:1095-1108.

Hafner, M.S., J.C. Hafner, J.L. Patton, and M.F. Smith. 1987. Macrogeographic patterns of the genetic differentiation in the pocket gopher *Thomomys umbrinus*. Systematics Zoology, 36:18-34.

Hafner, M.S., T.A. Spradling, J.E. Light, D.S. Hafner, and J. R. Demboski. 2004. Systematic revision of pocket gophers of the *Cratogeomys gymnurus* species group. Journal of Mammalogy, 85:1170-1183.

Hafner, D.J., J.E. Light, M.S. Hafner, E. Reddington, D.S. Rogers, and B.R. Riddle. 2007. Basal clades and molecular systematics of Heteromyid rodents. Journal of Mammalogy, 88:1129-1145.

Hafner, D.J., M.S. Hafner, G.L. Hasty, T.A. Spradling, and J.W. Demastes. 2008. Evolutionary relationships of pocket gophers (*Cratogeomys castanops* species group) of the Mexican Altiplano. Journal of Mammalogy, 89:190-208.

Hafner, M.S., J.E. Light, D.J. Hafner, S.V. Brant, T.A. Spradling, and J.W. Demastes. 2005. Cryctic species in the Mexican pocket gopher *Cratogeomys merriami*. Journal of Mammalogy, 86:1095-1108.

Hafner, D. J., B.R. Riddle, and S.T. Álvarez-Castañeda. 2001. Evolutionary relationships of white-footed mice (*Peromyscus*) on islands in the Sea of Cortez, Mexico. Journal of Mammalogy, 82:775-790.

Haiduk, M.W., J.W. Bickham, and D.J. Schmidly. 1979. Karyotypes of six species of *Oryzomys* from Mexico and Central America. Journal of Mammalogy, 60:610-615.

Haines, H. 1961. Seasonal changes in the reproductive organs of the cotton rats. Texas Journal of Science, 13:219-230.

Haines, A.M., J. Janecka, M.E. Tewes, and L. Grassman Jr. 2006. The importance of private lands for ocelot (*Leopardus pardalis*) conservation in the United States. Oryx, 40:90-94.

Hall, D.O. 1974. Reducing a maturity bias in estimating populations: An example with cotton rats (*Sigmodon hispidus*). Journal of Mammalogy, 55:477-480.

Hall, E.R. 1936. Mustelid mammals from the Pleistocene of North America. With systematic notes on some recent members of the genera *Mustela, Taxidea*, and *Mephitis*. Carnegie Institution, Washington Publications, 473:1-119.

Hall, E.R. 1946. Mammals of Nevada. University of California Press, Berkeley.

Hall, E.R. 1951a. American weasels. University of Kansas Publications, Museum of Natural History, 4:1-466.

Hall, E.R. 1951b. Mammals obtained by Dr. Curt von Wedel from the barrier beach of Tamaulipas, Mexico. University of Kansas Publications, Museum of Natural History, 5:33-47.

Hall, E.R. 1962. A new bat (*Myotis*) from Mexico. University of Kansas Publications, Museum of Natural History, 14:163-164.

Hall, E.R. 1968. Variation in blackish deer mouse *Peromyscus furvus*. Anales del Instituto de Biología, UNAM, México, Serie Zoología, 39:149-159.

Hall, E.R. 1981. The Mammals of North America. 2 vols. John Wiley and Sons, New York.

Hall, E.R. 1984. Geographic variation among brown and grizzly bears (*Ursus arctos*) in North America. University of Kansas Publications, Museum of Natural History, 13:1-16.

Hall, E.R., and T. Álvarez. 1961. A new species of the mouse (*Peromyscus*) from northwestern Veracruz, Mexico. Proceedings of the Biological Society of Washington, 74:202-205.

Hall, E.R., and E.L. Cockrum. 1953. A synopsis of the North American microtine rodents. University of Kansas Publications, Museum of Natural History, 141:1-270.

Hall, E.R., and W.W. Dalquest. 1950. A synopsis of the American bats of the genus *Pipistrellus*. University of Kansas Publications, Museum of Natural History, 1:591-602.

Hall, E.R., and W.W. Dalquest. 1963. The mammals of Veracruz. University of Kansas Publications, Museum of Natural History, 14:165-362.

Hall, E.R., and H.H. Genoways. 1970. Taxonomy of the *Neotoma albigula*-group of wood rats in central Mexico. Journal of Mammalogy, 51:504-516.

Hall, E.R., and J.K Jones Jr. 1961. North American yellow bats, *Dasypterus*, and a list of named kinds of the genus *Lasiurus* Gray. University of Kansas Publications, Museum of Natural History, 14:73-98.

Hall, E.R., and K.R. Kelson. 1959. The Mammals of North America. 2 vols. The Ronald Press Co., New York.

Hall, E.R., and K.R. Kelson. 1980. The Mammals of North America. John Wiley and Sons, New York.

Hall, E.R., and B. Villa-R. 1949a. An annotated check list of the mammals of Michoacan, Mexico. University of Kansas Publications, Museum of Natural History, 1:431-472.

Hall, E.R., and B. Villa-R. 1949b. A new harvest mouse from Michoacan, Mexico. Proceedings of the Biological Society of Washington, 62:163-164.

Hall, E.R., and B. Villa-R. 1950. Lista anotada de los mamíferos de Michoacán, México. Anales del Instituto de Biología, UNAM, México, Serie Zoología, 21:159-214.

Hall, M. 1994. Reducción de la mortalidad incidental de delfines en la pesquería del atún. P. 50, in: XIX Reunión internacional para el estudio de los mamíferos marinos. La Paz, Baja California Sur.

Hallett, J.G. 1982. Habitat selection and the community matrix of a desert small-mammals fauna. Ecology, 63:1400-1410.

Halley, D. (ed.). 1978. Marine Mammals of Eastern North Pacific and Arctic Waters. Pacific Search Press, Seattle.

Halloran, A.F. 1942. A surface nest and the young of *Sigmodon* in Texas. Journal of Mammalogy, 23:91.

Halls, L.K. 1978. The white-tailed deer, Pp. 43-65, in: Big Game of North America, Ecology and Management (J.L. Schmidt and D.L. Gilbert, eds.). Wildlife Management Institute. Stackpole Books, Harrisburg. PA.

Halls, L.K. (ed.) 1984. White-tailed Deer: Ecology and Management. Stackpole Books, Harrisburg, PA.

Hamilton, D.A. 1982. Ecology of the bobcat in Missouri. Master's thesis, University of Missouri.

Hamilton, W.J., Jr. 1944. The biology of the little short-tailed shrew, *Cryptotis parva*. Journal of Mammalogy, 25:1-7.

Hamilton, W.J., Jr., and J.O. Whitaker Jr., 1979. Mammals of the Eastern United States. Cornell University Press, Ithaca, NY.

Handley, C.O., Jr. 1959. A revision of American bats of the genera *Euderma* and *Plecotus*. Proceedings of the United States Natural Museum, 110:95-246.

Handley, C.O., Jr. 1960. Descriptions of new bats from Panama. Proceedings of the United States Natural Museum, 112:459-479.

Handley, C.O., Jr. 1966. Checklist of the mammals of Panama. Pp. 753-795, in: Ectoparasites of Panama (R.L. Wenzel and V.J. Tipton, eds.). Field Museum of Natural History, Chicago.

Handley, C.O., Jr. 1976. Mammals of the Smithsonian Venezuelan Project. Brigham Young University Science Bulletin, Biological Series, 20:1-90.

Handley, C.O., Jr. 1987. New species of mammals from northern South America: Fruit-eating bats, genus *Artibeus* Leach. Pp. 163-172, in: Studies in Neotropical Mammalogy: Essays in Honor of Philip Hershkovitz (B.D. Patterson and R.M. Timm, eds.). Fieldiana-Zoology, n.s. 39, Field Museum of Natural History, Chicago.

Handley, C.O., Jr., D.E. Wilson, and A.L. Gardner. 1991. Demography and natural history of the common fruit bat *Artibeus jamaicensis*, on Barro Colorado Island, Panama. Smithsonian Contributions to Zoology, 511:11-573.

Hanken, J., and D.B. Wake. 1998. Biology of tiny animals: Systematics of the minute salamanders (Thorius: Plethodontidae) from Veracruz and Puebla, Mexico, with descriptions of five new species. Copeia, 312-345.

Hanna, G.D. 1925. Expedition to Guadalupe Island, Mexico, in 1922. Proceedings of the California Academy of Sciences, 14:217-275.

Hansen, K. 2007. Bobcat: Master of Survival. Oxford University Press, New York.

Hansen, L.P., C.M. Nixon, and S.P. Havera. 1986. Recapture rates and length of residence in an unexploited fox squirrel population. The American Midland Naturalist, 115:209-215.

Hanson, J.D., J.L. Indorf, V.J. Swier, and R.D. Bradley. 2010. Molecular divergence within the *Oryzomys palustris* complex: Evidence for multiple species. Journal of Mammalogy, 91:336-347.

Harris, A.H. 1974. *Myotis yumanensis* in the interior southwestern North America, with comments on *Myotis lucifigus*. Journal of Mammalogy, 55:589-607.

Harris, A.H. 1977. Wisconsin age environments in the northern Chihuahuan desert: Evidence from higher vertebrates. Pp. 23-52, in: Transactions of the Symposium on the Biological Resources of the Chihuahuan Desert Region, United States and Mexico (R.H. Wauer and D.H. Riskind, eds.). U.S. National Park Service Transactions and Proceedings Series, No. 3, U.S. National Park Service, Washington, DC.

Harris, C.E., and M.K. Petersen. 1979. Olfactory discrimination in *Sigmodon* and its possible relevance to habitat segregation. Occasional Papers Zoologicas of M. K. Peters, 2:1-16.

Harris, D., D.S. Rogers, and J. Sullivan. 2000. Phylogeography of *Peromyscus furvus* (Rodentia; Muridae) based on cytochrome b sequence data. Molecular Ecology, 9:2129-2135,

Harris, J. 1990. California chipmunk. Pp. 114-114, in: California's Wildlife, Mammals (D.C. Zeiner, W.F. Laudenslayer, K.E. Mayer, and M. White, eds.). California Department of Fish and Game, Sacramento.

Harris, J. 1991. California leaf-nosed bat. Pp. 34-35, in: California's Wildlife, Mammals (D.C. Zeiner, W.F. Laudenslayer, K.E. Mayer, and M. White, eds.). California Department of Fish and Game, Sacramento.

Harrison, D.L. 1975. *Macrophyllum macrophyllum*. Mammalian Species, 62:1-3.

Harrison, R.J. 1969. Reproduction and reproductive organs. Pp. 253-348, in: The Biology of Marine Mammals (H.T. Andersen, ed.). Academic Press, Orlando, FL.

Hart, E.B. 1992. *Tamias dorsalis*. Mammalian Species, 399:1-6.

Hartman, D.S. 1979. Ecology and behavior of the manatee (*Trichechus manatus*) in Florida. American Society of Mammalogists Special Publication No. 5.

Hartman, G.D., and T.L. Yates. 1985. *Scapanus orarious*. Mammalian Species, 253:1-5.

Harvey, M.J. 1976. Home range movements and diel activity of the eastern mole, *Scalopus aquaticus*. The American Midland Naturalist, 95:436-445.

Harvey, T.E., and C. Polite. 1990. Merriam's chipmunk. Pp. 112-113, in: California's Wildlife, Mammals (D.C. Zeiner, W.F. Laudenslayer, K.E. Mayer, and M. White, eds.). California Department of Fish and Game, Sacramento.

Hasbrouck, J.J., W.R. Clark, and R.D. Andrews. 1992. Factors associated with raccoon mortality in Iowa. Journal of Wildlife Management, 56:693-699.

Hass, Chr. C. 2002. Home-range dynamics of white-nosed coatis in southern Arizona. Journal of Mammalogy, 83:934-946.

Hass, C.C., and D. Valenzuela. 2002. Anti-predator benefits of group living in white-nosed coatis (*Nasua narica*). Behavioral Ecology and Sociobiology, 51:570-578.

Hatt, R. 1934. The American Museum Congo Expedition manatee and other recent manatees. Bulletin of the American Museum of Natural History, 66:533-566.

Hatt, R.T. 1938a. Feeding habits of the least shrew. Journal of Mammalogy, 19:247-248.

Hatt, R.T. 1938b. Notes concerning mammals collected in Yucatan. Journal of Mammalogy, 19:333-337.

Hatt, R.T. 1953. The mammals. In: Faunal and Archeological Researches in Yucatan Caves (R. Hatt, H.I. Fisher, D.A. Langebartel, and G.W. Brainard, eds.). Cranbrook Institute Science Bulletin, 33:1-119.

Hatt, R.T., and B. Villa-R. 1950. Observaciones sobre algunos mamíferos de Yucatán y Quintana Roo. Anales del Instituto de Biología, UNAM, México, Serie Zoología, 21:215-240.

Haug, T. 1981. On some reproduction parameters in fin whales *Balaenoptera physalus* (L) caught off Norway. Reports of the International Whaling Commission, 31:373-378.

HaysSen, V., F. Miranda, and B. Pasch. 2012. *Cyclopes didactylus*. Mammalian Species, 44:51-58.

Hayward, B.J. 1970. The natural history of the cave bat *Myotis velifer*. Western New Mexico University Research in Science, 1:1-74.

Hayward, B.J., and E.L. Cockrum. 1971. The natural history of the western long-nosed bat *Leptonycteris sanborni*. Western New Mexico University Research in Science, 1:75-123.

Hayward, B.J., and S.P. Cross. 1979. The natural history of *Pipistrellus hesperus* (Chiroptera: Vespertilionidae). Western New Mexico University Research in Science, 13:1-36.

Hayward, B.J., and R. Davis. 1964. Flight speeds in western bats. Journal of Mammalogy, 45:236-242.

Heaney, L.R., and E.C. Birney. 1977. Distribution and natural history notes on some mammals from Puebla, Mexico. The Southwestern Naturalist, 21:543-559.

Heath, D.R., G.A. Heidt, D.A. Sangey, and V.R. MacDaniel. 1983. Arkansas range extensions of the seminole bat (*Lasiurus seminolus*) and eastern big eared bat (*Plecotus rafinesquii*) and additional county records for the hoary bat (*Lasiurus cinereus*), silver-haired bat (*Lasionycteris noctivagans*) and evening bat (*Nycticesius humeralis*) Arkansas. Academic Science Proceedings, 37:90-91.

Hedgal, P.L., A.L. Ward, A.M. Johnson, and H.P. Tietjen. 1965. Notes on the life history of the Mexican pocket gopher (*Cratogeomys castanops*). Journal of Mammalogy, 46:334-335.

Heidt, G.A. 1970. The least weasel, *Mustela nivalis* L., developmental biology in comparison to other North American Mustela. Michigan State University, Publication Museum (Biology Series), 4:227-282.

Heilbrun, R.D., N.J. Silvy, M.E. Tewes, and M.J. Peterson. 2003. Using automatically triggered cameras to individually identify bobcats. Wildlife Society Bulletin, 31:748-755.

Heinsohn, G.E., A.V. Spain, and P.K. Anderson. 1976. Populations of dugongs (Marnmalia: Sirenia): Aerial survey over the inshore waters of tropical Australia. Biological Conservation, 9:21-23.

Heise, K. 1996. Life history and population parameters of Pacific white-sided dolphins (*Lagenorhynchus obliquidens*). Trabajo Inédito. SC/48/SM45. International Whaling Commission, Washington, DC.

Heithaus, E.R., and T.H. Fleming. 1978. Foraging movements of frugivorous bat, *Carollia perspicillata* (Phyllostomidae). Ecological Monographs, 48:127-143.

Heithaus, E.R., T.H. Fleming, and P.A. Opler. 1975. Foraging patterns and resource utilization in seven species of bats in a seasonal tropical forest. Ecology, 56:841-854.

Helgen, K.M., and D.E. Wilson. 2005. A systematic and zoogeographic overview of the raccoons of Mexico and Central America. Pp. 221-236, in: Contribuciones Mastozoológicas en Homenaje a Bernado Villa (V. Sánchez-Cordero and R.A. Medellín, eds.). Instituto de Biología e Instituto de Ecología, UNAM, Mexico City.

Helgen, K.M., F.R. Cole, L.E. Helgen, and D.E. Wilson. 2009. Generic revision in the Holarticground squirrel genus *Spermophilus*. Journal of Mammalogy, 90:270-305.

Hellebuyck, V., J.R. Tamsit, and J.G. Hartman. 1985. Records of bats new to El Salvador. Journal of Mammalogy, 66:783-788.

Hellgren, E.C. 1993. Status, distribution, and summer food habits of black bear in Big Bend National Park. The Southwestern Naturalist, 38:77-80.

Hellgren, E.C. 1998. Physiology of hibernation in bears, *Ursus*. International Association for Bear Research and Management, 10:467-477.

Hellgren, E.C., R.L. Lochmiller, and W.E. Grant, 1985. Pregnancy diagnosis in the collared peccary by ultrasonic amplitude depth analysis. Journal of Wildlife Management, 49:71-73.

Helm, J.D., III. 1975. Reproductive behavior of *Ototylomys* (Cricetidae). Journal of Mammalogy, 56:575-590.

Helm, J.D., III, C. Sánchez-Hernández, and R.H. Baker. 1974. Observaciones sobre los ratones de las marismas *Peromyscus perfulvus* Osgood (Rodentia, Cricetidae). Anales del Instituto de Biología, UNAM, México, Serie Zoología, 45:141-146.

Hemker, T.P., F.G. Lindzey, and B.B. Ackerman. 1984. Population characteristics and movement patterns of cougars in southern Utah. Journal of Wildlife Management, 48:1275-1284.

Henderson, D.A. 1975. Whalers on the coasts of Baja California: Opening the peninsula to the outside world. Geoscience and Man, 12(June 20):49-56.

Henderson, D.A. 1979. Taxonomic status of the southwestern stocks of spinner dolphin *Stenella longirostris* and spotted dolphin *S. attenuata*. Reports of the International Whaling Commission, 29:175-183.

Hennings, D., and R.S. Hoffmann. 1977. A review of the *Sorex vagrans* species complex from western North America. University of Kansas Publications, Museum of Natural History, 68:1-35.

Hensley, A.P., and K.T. Wilkins. 1988. *Leptonycteris nivalis*. Mammalian Species, 307:1-4.

Henson, D.D., and R.D. Bradley. 2009. Molecular systematics of the genus *Sigmodon*: Results from mitochondrial and nuclear gene sequences. Canadian Journal of Zoology, 87:211-220.

Herd, R. 1983. *Pteronotus parnellii*. Mammalian Species, 209:1-5.

Hermann, J.A. 1950. The mammals of the Stockson Plateau of northeastern Terrell County, Texas. Texas Journal of Science, 2:368-393.

Hermanson, W., and T.J. O'Shea. 1983. *Antrozous pallidus*. Mammalian Species, 213:1-8.

Hernández, C.J.J. 1990. Taxonomía y distribución del género *Peromyscus* (Rodentia: Cricetidae) en el Estado de México, México. Tesis de licenciatura, Escuela Nacional de Ciencias Biológicas, Instituto Politécnico Nacional, México.

Hernández, F. 1959. Historia Natural de Nueva España. Vol. II. Universidad Nacional Autónoma de México.

Hernández, L., and A. Lafón. 1986. Inventario y análisis del lobo mexicano (*Canis lupus baileyi*, Nelson y Goldman). VII Ciclo de conferencias sobre fauna silvestre y recreación de áreas naturales, Simposio sobre lobo mexicano. Saltillo, Coahuila.

Hernández, M.L.A., R. Gálvez, M. Díaz, and C.M. Cruz. 2008. Nuevas localidades en la distribución de murciélagos filostóminos (Chiroptera: Phyllostomidae), en Chiapas, México. Revista Mexicana de Mastozoología, 12:163-169.

Hernández-Camacho, C.J., D. Aurioles-Gamboa, and L.R. Gerber. 2008. Age-specific birth rates of California sea lions (*Zalophus californianus*) in the Gulf of California, Mexico. Marine Mammals Science, 24:664-676.

Hernández-Cardona, A., L.A. Lago-Torres, L.I. González, J.C. Fallet-Menéndez, and Y. Pereyra-Arellano. 2007. Registro del tlacuachin (Tlacuatzin canescens) en el área de conservación El Zapotal, en el noreste del estado de Yucatán. Revista Mexicana de Mastozoología, 11:85-90.

Hernández-Chávez, J.J. 1990. Taxonomía y distribución del género *Peromyscus* (Rodentia: Cricetidae) en el Estado de México, México. Tesis de licenciatura, Escuela Nacional de Ciencias Biológicas, Instituto Politécnico Nacional, México D.F.

Hernández-Chávez, J.J. 1997. La alimentación de Tyto alba en la Ciénaga de Chapala, Michoacán, México. Pp. 157-174, in: Homenaje al Profesor Ticul Álvarez (J. Arroyo Cabrales and O.J. Polaco, eds.). Instituto Nacional de Antropología e Historia, Colección Científica, México D.F.

Hernández-Huerta, A. 1992. Los carnívoros y sus perspectivas de conservación en las áreas protegidas de México. Acta Zoológica Mexicana (n.s.), 54:1-23.

Hernández-Huerta, A., V.J. Sosa, J.M. Aranda, and J. Bello. 2000. Noteworthy records of small mammals from the Calakmul biosphere reserve in the Yucatan Peninsula, Mexico. The Southwestern Naturalist, 45:340-344.

Hernández-Meza, B. 2000. Caracterización espacial y contenido de las madrigueras de *Liomys pictus* en una selva baja de Jalisco. Tesis de licenciatura, Fes Iztacala, UNAM, Edo. de México.

Herrera, A.L. 1890. Notas acerca de los vertebrados del Valle de México. La Naturaleza 2a. Serie, 1:299-342.

Herrera, A.L. 1892. Quirópteros de México. La Naturaleza, 5:218-226, 298-299.

Herrera, A.L. 1897. Primates, carnívoros e insectívoros de México. Anales del Museo Nacional de México, Primera época, 4:63-70.

Herrera, L., and S. Urdaneta-Morales. 1992. Didelphis marsupialis: A primary reservoir of Trypanosoma cruzi in urban areas of Caracas, Venezuela. Annals of Tropical Medicine and Parasitology, 86(6):607-612.

Herrera, M.L.G., T.H. Fleming, and J.S. Findley. 1993. Geographic variation in carbon composition of the pallid bat, *Antrozous pallidus*, and its dietary implications. Journal of Mammalogy, 74:601-606.

Herrera, M.L.G., T.H. Fleming, and L.S. Sternberg. 1998. Trophic relationships in a Neotropical bat community: A preliminary study using carbon and nitrogen isotopic signatures. Tropical Ecology, 1:23-29.

Herrera-Pérez, G. 1993. Estudio comparativo de las morfometrías del venado cola blanca *Odocoileus virginianus texanus* en dos ranchos del Noreste de México. Tesis de licenciatura, Facultad de Ciencias Biológicas, Universidad Autónoma de Nuevo León, Monterrey, Nuevo León, México.

Herrero, S. 1985. Bear Attacks: Their Causes and Avoidance. Lyons and Burford, New York.

Herrero, S., and D. Hamer. 1977. Courtship and copulation of a pair of grizzly bears, with comments on reproductive plasticity and strategy. Journal of Mammalogy, 58:441-444.

Hersh, S.L., and D.K. Odell. 1986. Mass stranding of Fraser's dolphin, *Lagenodelphis hosei*, in the western North Atlantic. Marine Mammal Science, 2:73-76.

Hershkovitz, P. 1950. Mammals of northern Colombia. Preliminary report no. 6: Rabbits (Leporidae), with notes on the classification and distribution of the South American forms. Proceedings of the United States National Museum, 100:327-375.

Hershkovitz, P. 1951. Mammals from British Honduras, Mexico, Jamaica and Haiti. Fieldiana Zoology, 31:547-569.

Hershkovitz, P. 1971. A new rice rat of Mexico. Journal of Mammalogy, 52:700-709.

Hershkovitz, P. 1976. Comments on generic names of four-eyed opossums (family Didelphidae). Proceedings of the Biological Society of Washington, 89:295-304.

Hershkovitz, P. 1992. The South America genus *Gracilinanus* Gardner and Creighton, 1989 (Marmosidae, Marsupialia): A taxonomic review with notes on general morphology and relationships. Fieldiana Zoology (n.s.), 70:1-56.

Hervert, J., and P. Krausman. 1986. Desert mule deer use of water developments in Arizona. Journal of Wildlife Management, 50:670-676.

Herzing, D.L. 1997. The life history of free-ranging Atlantic spotted dolphins (Stenella frontalis): Age classes, color phases and female reproduction. Marine Mammal Science, 13:576-595.

Herzing, D.L., and B.R. Mate. 1984. Gray whale migrations along the Oregon coast, 1978-81. Pp. 289-308, in: The Gray Whale *Eschrichtius robustus* (M.L. Jones, S.L. Swartz, and J.S. Leatherwood, eds.). Academic Press, Orlando, FL.

Hester, F.J., J.R. Hunter, and R.R. Whitney. 1963. Jump and spinning behavior in the spinner porpoise. Journal of Mammalogy, 44:587-588.

Hewitt, D., and D. Doan-Crider. 2008. Metapopulations, food and people: Bear management in northern Mexico. P. 372, in: Wildlife Science: Linking Ecological Theory and Management Applications (T.E. Fulbright and D.G. Hewitt, eds.). CRC Press, Boca Raton, FL.

Heyning, J.E. 1984. Functional morphology involved in intraspecific fighting of the beaked whale, *Mesoplodon carlhubbsi*. Canadian Journal of Zoology, 62:1645-1654.

Heyning, J.E. 1989. Cuvier's beaked whale *Ziphius cavirostris* G. Cuvier, 1823. Pp. 289-308, in: Handbook of Marine Mammals, vol. 4, River Dolphins and the Larger Toothed Whales (S.H. Ridgway and R. Harrison, eds.). Academic Press, New York.

Heyning, J.E., and M.E. Dahlheim. 1988. *Orcinus orca*. Mammalian Species, 304:1-9.

Heyning, J.E., and W.F. Perrin. 1994. Evidence for two species of common dolphins (genus *Delphinus*) from the eastern North Pacific. Contributions in Science, Los Angeles Country Museum, 442:1-35.

Hickman, G.C. 1977. Burrow system structure of *Pappogeomys castanops* (Geomyidae) in Lubbock County, Texas. The American Midland Naturalist, 97:50-58.

Hill, J., and J.D. Smith. 1992. Bats: A Natural History. University of Texas Press, Austin.

Hilton-Taylor, C. 2000. 2000 IUCN Red List of Threatened Species. IUCN, Glanz, Switzerland.

Hinesley, L.L. 1979. Systematics and distribution of two chromosome forms in the southern grasshopper mouse, genus *Onychomys*. Journal of Mammalogy, 60:117-128.

Hisaw, F.L. 1923a. Observations on the burrowing habits of the moles. Journal of Mammalogy, 4:79-88.

Hisaw, F.L. 1923b. Feeding habits of moles. Journal of Mammalogy, 23:9-20.

Hitchcock, H. B. 1949. Hibernation of bats in southeastern Ontario and adjacent Quebec. Canadian Field-Naturalist, 63:47-59.

Hladik, A., and C.M. Hladik. 1969. Rapports trophiques entre végetation et primates dans la forêt de Barro Colorado (Panama). La Terre et la Vie, 23:25-117.

Hoare, C.A. 1972. The Trypanosomes of Mammals: A Zoological Monograph. Blackwell Scientific Publications, Oxford.

Hobbs, D.E. 1980. The effects of habitat sound properties on alarm calling behavior in two species of tree squirrels. PhD dissertation, University of Arizona, Tucson.

Hoelzel, A.R. 1991. Analysis of regional mitochondrial DNA variation in the killer whale: Implications for cetacean conservation. Reports of the International Whaling Commission (special issue), 13:225-233.

Hoelzel, A.R., and G.A. Dover. 1991. Mitochondrial D-loop DNA variation within and between populations of the minke whale (*Balaenoptera acutorostrata*). Reports of the International Whaling Commission (special issue), 13:171-181.

Hoffmann, A. 1990. Los Trombiculidos de México (Acarida: Trombiculidae). Publicaciones especiales 2, Instituto de Biología, Universidad Nacional Autónoma de México, México.

Hoffmann, A. 1993. Las Colecciones de Artrópodos de A. Hoffmann. Serie Cuadernos No. 19, Instituto de Biología, Universidad Nacional Autónoma de México, México.

Hoffmann, A., and I. Bassols de B. 1970. Ácaros de la familia Spelaeorhynchidae. Revista Latinoamericana de Microbiología, 12:145-149.

Hoffmann, A., I. Bassols de B., and C. Méndez. 1972. Nuevos hallazgos de ácaros en México. Revista de la Sociedad Mexicana de Historia Natural, 33:151-159.

Hoffmann, A., J.G. Palacios-V., and J.B. Morales M. 1986. Manual de Bioespeleología. Facultad de Ciencias, Universidad Nacional Autónoma de México, México.

Hoffmann, R.S, and J.K. Jones Jr. 1970. Influence of late-glacial and postglacial events on the distribution of recent mammals on the Northern Great Plains. In: Pleistocene and Recent Environments of the Central Great Plains (W. Dort Jr., and J.K. Jones Jr., eds.). University of Kansas Press, Lawrence.

Hoffmann, R.S., and J.W. Koeppl. 1985. Zoogeography. Pp. 84-115, in: Biology of New World Microtus (R.H. Tamarin, ed.). Special Publication 8, American Society of Mammalogists.

Hoffmann, R.S., and A.T Smith. 2005. Lagomorpha. Pp. 185-211, in: Mammals of the World (D. Wilson and D.A. Reeder, eds.). 3rd ed. Johns Hopkins University Press, Baltimore.

Hoffmann, R.S., C.G. Anderson, R.W. Thoringthon Jr., and L.R. Heaney. 1993. Family Sciuridae. Pp. 419-466, in: Mammal Species of the World (D.E. Wilson and D.M. Reeder, eds.). Smithsonian Institution Press, Washington, DC.

Hoffmeister, D.F. 1951. A taxonomic and evolutionary study of the piñon mouse, *Peromyscus truei*. Illinois Biological Monographs, 21:1-104.

Hoffmeister, D.F. 1963. The yellow-nosed cotton rat, *Sigmodon ochrognathus*, in Arizona. The American Midland Naturalist, 70:429-441.

Hoffmeister, D.F. 1970. The seasonal distribution of bats in Arizona: A case for improving mammalian range maps. The Southwestern Naturalist, 15:11-22.

Hoffmeister, D.F. 1974. The taxonomic status of *Perognathus penicillatus minumus* Burt. The Southwestern Naturalist, 19:213-214.

Hoffmeister, D.F. 1981. *Peromyscus truei*. Mammalian Species, 161:1-5.

Hoffmeister, D.F. 1986. Mammals of Arizona. University of Arizona Press and the Arizona Fish and Game Department, Tucson.

Hoffmeister, D.F., and L. De la Torre. 1961. Geographic variation in the mouse *Peromyscus difficilis*. Journal of Mammalogy, 42:1-13.

Hoffmeister, D.F., and V.E. Diersing. 1973. The taxonomic status of *Peromyscus merriami goldmani* Osgood, 1904. The Southwestern Naturalist, 18:354-357.

Hoffmeister, D.F., and M.R. Lee. 1963. The status of the sibling species *Peromyscus merriami* and *Peromyscus eremicus*. Journal of Mammalogy, 44:201-213.

Hoffmeister, D.F., and M.R. Lee. 1967. Revision of the pocket mice, *Perognathus penicillatus*. Journal of Mammalogy, 4883:361-380.

Hoffmeister, R.G., and D.F. Hoffmeister. 1991. The hyoid in North American squirrels, Sciuridae, with remarks on associated musculature. Anales del Instituto de Biologia, UNAM, México, Serie Zoología, 62:219-234.

Hofmann, R.R. 1989. Evolutionary steps of ecophysiological adaptation and diversification of ruminants: A comparative view of their digestive system. Oecologia, 3:443-457.

Hollander, R.R., and J.K. Jones Jr. 1988. Northernmost record of the tropical brown bat, *Eptesicus furinalis*. The Southwestern Naturalist, 33:100.

Hollister, N. 1909. Two new bats from the southwestern United States. Proceedings of the Biological Society of Washington, 22:43-44.

Homan, J.A., and H.H. Genoways. 1978. An analysis of hair structure and its phylogenetic implications among heteromyid rodents. Journal of Mammalogy, 59:740-760.

Honacki, J.H., K.E. Kiman, and J.K. Koeppl. 1982. Mammalian Species of the World: A Taxonomic and Geographic Reference. Allen Press, Lawrence, KS.

Honeycutt, R.L., and V.M. Sarich. 1987. Albumin evolution and subfamilial relationships among New World leaf-nosed bats (family Phyllostomidae). Journal of Mammalogy, 68:508-517.

Honeycutt, R.L., and S.L. Williams. 1982. Genic differentiation in pocket gophers of the genus *Pappogeomys*, with comments on intergeneric relationships in the subfamily Geomyinae. Journal of Mammalogy, 63:208-217.

Honeycutt, R.L., J.B. Baker, and H.H. Genoways. 1980. Results of the Alcoa foundation-Suriname expeditions. III. Chromosomal data for bats (Mammalia: Chiroptera) from Suriname. Annals of Carnegie Museum, 49:237-250.

Honeycutt, R.L., M.P. Moulton, J.R. Roppe, and L. Fifield. 1981. The influence of topography and vegetation on the distribution of small mammals in southwestern Utah. The Southwestern Naturalist, 26:295-300.

Hood, C.S., and R.J. Baker. 1986. G- and C-banding chromosomal studies of bats in the family Emballonuridae. Journal of Mammalogy, 67:705-711.

Hood, C.S., and J.K. Jones Jr. 1984. *Noctilio leporinus*. Mammalian Species, 216:1-7.

Hood, C.S., and J. Pitochelli. 1983. *Noctilio albiventris*. Mammalian Species, 197:1-5.

Hood, C.S., L.W. Robbins, R.J. Baker, and H.S. Shellhamer. 1984. Chromosomal studies and evolutionary relationships of an endangered species, *Reithrodontomys raviventris*. Journal of Mammalogy, 65:655-667.

Hoofer, S.R., and R.A. Van Den Bussche. 2003. Molecular phylogenetics of the Chiropteran family Vespertilionidae. Acta Chiropterologica, 5(suppl.):1-63

Hoofer, S.R., R.A. Van Den Bussche, and I. HoràcĐk. 2006. Generic status of the American pipistrelles (Vespertilionidae) with description of a new genus. Journal of Mammalogy, 87:981-992.

Hoogesteijn R., A, Hoogesteijn, and E. Mondolfi. 1993. Jaguar predation and conservation: Cattle mortality caused by felines on three ranches in the Venezuelan Llanos. Symposia of the Zoological Society of London, 65:391-407.

Hoogland, J.L. 1995. The Black-tailed Prairie Dog: Social Life of a Burrowing Mammal. University of Chicago Press, Chicago.

Hoogland, J.L. 1996. *Cynomys ludovicianus*. Mammalian Species, 535:1-10.

Hooker, Sascha K. 2009. Encyclopedia of Marine Mammals (W.F. Perrin, B. Wursig, and J.G.M. Thewissen, eds.). 2nd ed. Elsevier/Academic Press, Amsterdam.

Hooper, E.T. 1941. Mammals of the lava fields and adjoining areas in Valencia County, New Mexico. Miscellaneous Publications of the Museum of Zoology, University of Michigan, 51:1-47.

Hooper, E.T. 1946. Two genera of pocket gophers should be congeneric. Journal of Mammalogy, 27:397-399.

Hooper, E.T. 1947. Notes on Mexican mammals. Journal of Mammalogy, 28:40-57.

Hooper, E.T. 1950. Descriptions of two sub-species of harvest mice (genus *Reithrodontomys*) from Mexico. Proceedings of the Biological Society of Washington, 63:167-170.

Hooper, E.T. 1952a. A systematic review of the harvest mice (genus *Reithrodontomys*) of Latin America. Miscellaneous Publications of the Museum of Zoology, University of Michigan, 77:1-255.

Hooper, E.T. 1952b. Notes on the pygmy mouse (*Baiomys*), with a description of a new subspecies from Mexico. Journal of Mammalogy, 33:90-97.

Hooper, E.T. 1952c. Records of the flying squirrels (*Glaucomys volans*) in Mexico. Journal of Mammalogy, 33:109-110.

Hooper, E.T. 1952d. Notes on mammals of western Mexico. Occasional Papers of the Museum of Zoology, University of Michigan, 565:1-26.

Hooper, E.T. 1953. Notes on mammals of Tamaulipas, Mexico. Occasional Papers of the Museum of Zoology, University of Michigan, 544:1-12.

Hooper, E.T. 1954. A synopsis of the cricetine rodent genus *Nelsonia*. Occasional Papers of the Museum of Zoology, University of Michigan, 558:1-12.

Hooper, E.T. 1955. Notes on mammals of western Mexico. Occasional Papers of the Museum of Zoology, University of Michigan, 565:1-26.

Hooper, E.T. 1957. Records of Mexican mammals. Occasional Papers of the Museum of Zoology, University of Michigan, 586:1-9.

Hooper, E.T. 1958. The male phallus in mice of the genus *Peromyscus*. Miscellaneous Publications of the Museum of Zoology, University of Michigan, 105:1-24.

Hopper, E.T. 1961. Notes on mammals of western and southern Mexico. Journal of Mammalogy, 42:120-122.

Hooper, E.T. 1964. A synopsis of the cricetine rodent genus *Nelsonia*. Occasional Papers, Museum of Zoology, University of Michigan, 558:1-12.

Hooper, E.T. 1968a. Classification. Pp. 27-70, in: Biology of *Peromyscus* (Rodentia) (J. King, ed.). American Special Publications, No. 2.

Hooper, E.T. 1968b. Habitats and food of amphibious mice of the genus *Rheomys*. Journal of Mammalogy, 49:550-553.

Hooper, E.T. 1972. A synopsis of the rodent genus *Scotinomys*. Occasional Papers of the Museum of Zoology, University of Michigan, 665:1-32.

Hooper, E.T., and J.H. Brown. 1968. Foraging and breeding in two sympatric species of Neotropical bats, genus *Noctilio*. Journal of Mammalogy, 49:310-312.

Hooper, E.T., and M.D. Carleton. 1976. Reproduction, growth, and development in two contiguously allopatric rodent species, genus *Scotinomys*. Miscellaneous Publications of the Museum of Zoology, University of Michigan, 151:1-52.

Hooper, E.T., and G.G. Musser. 1964. Notes on classification on the rodent genus *Peromyscus*. Occasional Papers of the Museum of Zoology, University of Michigan, 635:1-13.

Horáček, I., and V. Hanák. 1985/1986. Generic status of *Pipistrellus savii* and comments on classification of the genus *Pipistrellus* (Chiroptera, Vespertilionidae). Myotis, 23-24:9-16.

Horner, B.E. 1954. Arboreal adaptations of *Peromyscus* with special references to use of the tail. Contribution of the Laboratory of Vertebrate Biology, University of Michigan, 61:1-85.

Hornocker, M. 1969. Winter territoriality in mountain lions. Journal of Wildlife Management, 33:457-464.

Hornocker, M. 1970. An analysis of mountain lion predation upon mule deer and elk in the Idaho primitive areas. Wildlife Monographs, 21:39.

Hortelano, M.Y., and F.A. Cervantes. 1989. Variación del tamaño de camada del ratón meteorito (*Microtus mexicanus*). Anales del Instituto de Biología, UNAM, México, Serie Zoología, 60:211-222.

Horváth, A., and D.A. Navarrete-Gutiérrez. 1997. Ampliación del área de distribución de *Peromyscus zarhynchus* Merriam, 1898 (Rodentia: Muridae). Revista Mexicana de Mastozoología, 2:122-125.

Horváth, A., I. March, and J. Wolf. 2001. Rodent diversity and land use in Montebello, Chiapas, Mexico. Studies on Neotropical Fauna and Environment, 36:169-176.

Horváth, A., R. Vidal López, O. Pérez Macías, C. Chávez Gloria, Y. Aguirre Bonifaz, D. Gallegos Castillo, M. Ramírez Lozano, E. Sánchez Vázquez, and E. Espinoza Medinilla. 2008. Mamíferos de los Parques Nacionales Lagunas de Montebello y Palenque, Chiapas. El Colegio de la Frontera Sur, Unidad San Cristóbal de las Casas, Informe Final SNIB-CONABIO, Proyecto BK047, México.

Horwich, R.H. 1983. Species status of the black howler monkey, *Alouatta pigra*, of Belize. Primates, 24:288-289.

Horwich, R.H. 1989. The geographic distribution of the black howler monkey (*Alouatta pigra*) in Central America and efforts to conserve it in Belize. Pp. 191-201, in: Primatología en México: Comportamiento, Ecología, Aprovechamiento y Conservación de Primates (A. Estrada, R. López-Wilchis, and R. Coates-Estrada, eds.). Universidad Autónoma Metropolitana, México.

Horwich, R.H., and E.D. Johnson. 1986. Geographic distribution and status of the black howler (*Alouatta pigra*) in Central America. Primates, 27:53-62.

Horwich, R.H., and J. Lyon. 1987. Development of the Community Baboon Sanctuary in Belize: An experiment in grass roots conservation. Primate Conservation, 8:32-34.

Hoth, J., A. Velázquez, F.J. Romero, L. León, M. Aranda, and D.J. Bell. 1987. The volcano rabbit—a shrinking distribution and a threatened habitat. Oryx, 21:85-91.

Houseal, T.W., Y.F. Greenbaum, D.J. Schmidly, S.A. Smith, and K.M. Davis. 1987. Karyotypic variation in *Peromyscus boylii* from Mexico. Journal of Mammalogy, 68:281-296.

Houston, J. 1990. Status of Blainville's beaked whale, *Mesoplodon densirostris*, in Canada. Canadian Field Naturalist, 104:117-120.

Howard, W.E., and L.G. Ingles. 1951. Outline for an ecological life history for pocket gophers and other fossorial mammals. Ecology, 32:537-544.

Howard, W.E., and R.E. Marsh. 1982. Spotted and hog-nosed skunks (*Spilogale putorius* and allies). Pp. 664-673, in: Wild Mammals of North America: Biology, Management, Economics (J.A. Chapman and G.A. Feldhamer, eds.). Johns Hopkins University Press, Baltimore.

Howell, A.H. 1914. Revision of the American harvest mice (genus *Reithrodontomys*). North American Fauna, 36:1-97.

Howell, A.H. 1922. Diagnoses of seven new chipmunks of the genus *Eutamias*, with a list of American species. Journal of Mammalogy, 3:178-185.

Howell, A.H. 1929. Revision of the American chipmunks (genera *Tamias* and *Eutamias*). North American Fauna, 52:1-157.

Howell, A.H. 1938. Revision of the North American ground squirrels, with a classification of the North American Sciuridae. North American Fauna, 56:1-256.

Howell, D.J. 1974. Bats and pollen: Physiological aspects of the syndrome of chiropterophily. Comparative Biochemistry and Physiology, Part A. Comparative Physiology, 48:263-276.

Howell, D.J., and D. Burch. 1974. Food habits of some Costa Rican bats. Revista de Biología Tropical, 21:281-294.

Hoying, K. 1983. Growth and development of the eastern pipistrelle bat, *Pipistrellus subflavus*. Unpublished master's thesis, Boston University.

Hoyt, E. 1984. The Whale Watcher's Handbook. Madison Press Book, Toronto.

Hoyt, R.A., and J.S. Altenbach. 1981. Observations on *Diphylla ecaudata* in captivity. Journal of Mammalogy, 62:215-216.

Hsu, T.C., and R.A. Mead. 1969. Mechanisms of chromosomal changes in mammalian speciation. Pp. 8-17, in: Comparative Cytogenetics (K. Benirschke, ed.). Springer-Verlag, New York.

Hsu, T.C., R.J. Baker, and T. Utakoyi. 1968. The multiple sex chromosome system of American leaf nosed bats (Chiroptera, phyllostomatidae). Cytogenetics, 7:27-38.

Hubbard, C.A. 1958. Mexican jungle and desert fleas with three new descriptions. Entomological News, 69:161-166.

Hubbard, J.P. 1972. Hooded skunk on Monollon Plateau, New Mexico. The Southwestern Naturalist, 16:458.

Hubbs, C. 1956. Back from oblivion. Pacific Discovery, 9:14-21.

Hubbs, C.L., W.F. Perrin, and K.C. Balcomb. 1973. *Stenella coeruleoalba* in the eastern and central tropical Pacific. Journal of Mammalogy, 54:549-552.

Huckaby, D.G. 1980. Species limits in the *Peromyscus mexicanus* group (Mammalia: Rodentia: Muroidea). Contributions from the Sciences, Natural History Museum, Los Angeles Country, 326:1-24.

Hudson, J.W. 1964. Temperature regulation in the round-tailed ground squirrel, *Citellus tereticaudus*. Annals of the Academy of Sciences, Fennica, Serie A (IV), 71:219-233.

Hudson, J.W., and D.R. Deavers. 1973. Metabolism, pulmocutaneous water loss and respiration of eight species of ground squirrels from different environments. Comparative Biochemistry and Physiology, 45:69-100.

Huey, L.M. 1925. Two new kangaroo rats of the genus *Dipodomys* from Lower California. Proceedings of the Biological Society of Washington, 38:83-88.

Huey, L.M. 1940. A new coastal form of brush rabbit from the vicinity of San Quintin, Lower California, Mexico. Transactions of the San Diego Society of Natural History, 9:221-224.

Huey, L.M. 1960. Comments on the pocket mouse, *Perognathus fallax*, with descriptions of two new races from Baja California, Mexico. Transactions of the San Diego Society of Natural History, 12:407-408.

Huey, L.M. 1964. The mammals of Baja California, Mexico. Transactions of the San Diego Society of Natural History, 13:85-165.

Hugentobler, H., and J.P. Gallo 1986. Un registro de la estenela moteada del Atlántico (*Stenella plagiodon* Cope, 1866) (Cetacea: Delphinidae) del estado de Campeche, Mexico. Anales del Instituto de Biología, UNAM, México, Serie Zoología, 56:1039-1042.

Hui, C.A. 1979. Undersea topography and distribution of dolphins of the genus *Delphinus* in Southern California. Journal of Mammalogy, 60:521-527.

Hulley, J.I. 1976. Maintenance and breeding of captive jaguarondis *Felis yagouaroundi* at Chester zoo and Toronto. International Zoo Year Book, 16:120-122.

Humphrey, S.R. 1982. Bats (Vespertilionidae and Molossidae). Pp. 52-70, in: Wild Mammals of North America, Biology, Management, Economics (J.A. Chapman and G.A. Feldhamer, eds.). Johns Hopkins University Press, Baltimore.

Humphrey, S.R., and J.B. Cope. 1976. Population ecology of the little brown bat, *Myotis lucifugus*, in Indiana and north-central Kentucky. Special Publications, American Society of Mammalogists, 4:1-81.

Humphrey, S.R., F.J. Bonaccorso, and T.L. Zinn. 1983. Guild structure of surface-gleaning bats in Panama. Ecology, 62:284-294.

Hunsaker, D., II. 1977. Ecology of New World marsupials. Pp. 95-156, in: The Biology of Marsupials (D. Hunsaker II, ed.). Academic Press, New York.

Hunsaker, D., II, and D. Shupe. 1977. Behavior of New World marsupials. Pp. 279-347, in: The Biology of Marsupials (D. Hunsaker II, ed.). Academic Press, New York.

Husar, S.L. 1976. Behavioral character displacement: Evidence for food partitioning in insectivorous bats. Journal of Mammalogy, 57:331-338.

Husson, A.M. 1978. The Mammals of Suriname. E.J. Brill, Leiden.

Hutterer, R. 1980. A record of Goodwin's shrew, *Cryptotis goodwini*, from Mexico. Mammalia, 44:413.

Hutterer, R. 1993. Order Insectivora. Pp. 69-130, in: Mammal Species of the World (D.E. Wilson and D.M. Reeder, eds.). Smithsonian Institution Press, Washington, DC.

Hutterer, R. 2005. Order Soricomorpha. Pp. 220-311, in: Mammal Species of the World (D.E. Wilson and D.M. Reeder, eds.). Johns Hopkins University Press, Baltimore.

Hwang, Y.T., and S. Larivière. 2001. *Mephitis macroura*. Mammalian Species, 686:1-3.

Ibañez, C., R. Lopez-Wilchis, J. Juste B., and M.A. León-Galván. 2000. Echolocation calls and a noteworthy record of *Pteronotus gymnonotus* (Chiroptera, Mormoopidae) from Tabasco, México. The Southwestern Naturalist, 45:345-347.

Ibarra Cerdeña, C.N., V. Sánchez Cordero, P. Ibarra López, and E. Iñiguez. 2007. Noteworthy records of *Musonycteris harrisoni* and *Tlacuatzin canescens* pollinating a columnar cactus in west-central Mexico. International Journal of Zoology Research, 3:223-226.

ICZN (International Commission on Zoological Nomenclature).1998. Opinion 1894. Regnum Animale. Ed. 2 (M.J. Brisson, 1762): rejected for nomenclatural purposes, with the conservation of the mammalian generic names for *Philander* (Marsupialia), *Pteropus* (Chiroptera), *Glis*, *Cuniculus* and *Hydrochoerus* (Rodentia), *Meles*, *Lutra* and *Hyaena* (Carnivora), *Tapirus* (Perissodactyla), *Tragulus* and *Giraffa* (Artiodactyla). Bulletin of Zoological Nomenclature, 55:64-71.

Ingles, L.G. 1958. Notas acerca de los mamíferos mexicanos. Anales del Instituto de Biología, UNAM, México, Serie Zoología, 29:379-408.

Ingles, L.G. 1965. Mammals of the Pacific States. Stanford University Press, Stanford, CA.

Iñiguez-Dávalos, L.I. 1993. Patrones ecológicos en la comunidad de murciélagos de Manantlán. Pp. 355-370, in: Avances en el Estudio de los Mamíferos de México (R.A. Medellín and G. Ceballos, eds.). Asociación Mexicana de Mastozoología, A.C., México.

Iñiguez-Dávalos, L.I., and E. Santana C. 1993. Patrones de distribución y riqueza de especies de los mamíferos del occidente de México. Pp. 65-86, in: Avances en el Estudio de los Mamíferos de México (R. A. Medellín and G. Ceballos, eds.). Publicaciones Especiales No. 1, Asociación Mexicana de Mastozoología, México.

Innes, D.G.L. 1978. A reexamination of litter sizes in some North American microtines. Canadian Journal of Zoology, 56:1488-1496.

International Whaling Commission. 2001. Appendix 3. Classification of the Order Cetacea (whales, dolphins and porpoises). Journal of Cetacean Research and Management, 3(1):v–xii.

Iriarte, J.A., W.L. Franklin, W.E. Johnson, and K.H. Redford. 1990. Biographic variation of food habits and body size of the America puma. Oecologia, 85:185-190.

Irvine, A.B. 1983. Manatee metabolism and its influence on distribution in Florida. Biological Conservation, 25:315-334.

Irvine, A.B., M.D. Scott, R.S. Wells, and J.H. Kaufmann. 1981. Movements and activities of the Atlantic bottlenose dolphin, *Tursiops truncatus*, near Sarasota, Florida. Fishery Bulletin, 79:671-687.

Irvine, A.B., R.C. Neal, R.T. Cardeilhac, J.A. Popp, F.H. Whiter, and R.C. Jenkins. 1980. Clinical observations on captive and free-ranging West Indian manatees, *Trichechus manatus*. Aquatic Mammals, 8:2-10.

Irwin, D.W., and R.J. Baker. 1967. Additional records of bats from Arizona and Sinaloa. The Southwestern Naturalist, 12:195.

IUCN (International Union for Conservation of Nature). 1988. Red List of Threatened Animals. Gland, Switzerland.

IUCN. 1990. Red List of Threatened Animals. World Conservation Monitoring Centre, Cambridge, UK.

IUCN. 1994. Red List of Threatened Animals. World Conservation Monitoring Centre, Cambridge, UK.

IUCN. 1996. 1996 IUCN Red List of Threatened Animals. Gland, Switzerland.

IUCN. 2003. Sustainable Livelihoods. Media Brief for the World Parks Congress. Gland, Switzerland.

IUCN. 2009. IUCN Red List of Threatened Species. Versión 2009.1. Revisado el 7 de junio de 2009 de: www.iucnredlist.org.

IUCN 2010. IUCN Red List of Threatened Species. Version 2010.1. www.iucnredlist.org.

Iudica, C.A. 2000. Systematic revision of the Neotropical fruit bats of the genus *Sturnira*: A molecular and morphological approach. PhD dissertation, University of Florida, Miami.

Izawa, K. 1976. Group sizes and composition in the upper Amazon Basin. Primates, 16:295-316.

Jablonski, D. 1995. Extinction in the fossil record. Pp. 25-44, in: Extinction Rates (J.H. Lawton and R.M. May, eds.). Oxford University Press, Oxford.

Jackson, H.H.T. 1914. New moles of the genus *Scalopus*. Proceedings of the Biological Society of Washington, 27:19-22.

Jackson, H.H.T. 1915. A review of the American moles. North American Fauna, 38:1-100.

Jackson, H.H.T. 1925. Preliminary descriptions of seven shrews of the genus *Sorex*. Proceedings of the Biological Society of Washington, 38:127-130.

Jackson, H.H.T. 1928. A taxonomic review of the American longtailed shrews (genera *Sorex* and *Microsorex*). North American Fauna, 51:1-238.

Jackson, H.H.T. 1947. A new shrew (genus *Sorex*) from Coahuila. Proceedings of the Biological Society of Washington, 60:131-132.

Jackson, H.H.T. 1961. Mammals of Wisconsin. University of Wisconsin Press, Madison.

Jaeger, E.C. 1961. Desert Wildlife. Stanford University Press, Stanford, CA.

Jameson, E.W., Jr. 1999. Host-ectoparasite relationships among North American chipmunks. Acta Theriologica, 44:225-231.

Jameson, E.W., and H.J. Peeters. 1988. California Mammals. California Natural History Guides, No. 52. University of California Press, Berkeley.

Janecek, L. 1990. Genic variation in the *Peromyscus truei* group (Rodentia: Cricetidae). Journal of Mammalogy, 71:301-308.

Janzen, D.H. 1970. Altruism by coatis in the face of predation by boa constrictor. Journal of Mammalogy, 51:387-389.

Janzen, D.H. 1981. Digestive seed predation by a Costa Rican Baird's tapir. Biotropica (suppl.): 59-63.

Janzen, D.H. 1982a. Wild plant acceptability to a captive Costa Rican Baird's tapir. Brenesia, 19-20:99-128.

Janzen, D.H. 1982b. Seeds in tapir dung in Santa Rosa National Park, Costa Rica. Brenesia, 19-20:129-135.

Janzen, D.H. 1982c. Seed removal from fallen Guanacaste fruits (*Enterolobium ciclocarpum*) by spiny pocket mice (*Liomys salvini*). Brenesia, 19-20:425-429.

Janzen, D.H. 1983. *Tapirus bairdii*. Pp. 496-497, in: Costa Rican Natural History (D.H. Janzen, ed.). University of Chicago Press, Chicago.

Janzen, D.H., and D.E. Wilson. 1983. Mammals. Pp. 426-442, in: Costa Rican Natural History (D.H. Janzen, ed.). University of Chicago Press, Chicago.

Jau-Mexia, N., O.J. Polaco, and J. Arroyo-Cabrales. 2000. New mammals for the Pleistocene of Zacatecas, Mexico. Current Research in the Pleistocene, 17:124-125.

Jeanne, R.L. 1970. Note on a bat (*Phylloderma stenops*) preying upon the brood of a social wasp. Journal of Mammalogy, 51:624-625.

Jefferson, T.A. 1988. *Phocoenoides dalli*. Mammalian Species, 319:1-7.

Jefferson, T.A. 1989a. Status of the Dall's porpoise, *Phocoenoides dalli*, in Canada. Canadian Field Naturalist, 104:112-116.

Jefferson, T.A. 1989b. Calving seasonality of Dall's porpoise in the eastern North Pacific. Marine Mammal Science, 5:196-200.

Jefferson, T.A., and S. Leatherwood. 1994. *Lagenodelphis hosei*. Mammalian Species, 470:1-5.

Jefferson, T.A., and A.J. Schiro. 1995. Registros históricos de cetáceos de la costa del Golfo de México. P. 39, in: XX Reunión Internacional para el Estudio de los Mamíferos Marinos. La Paz, Baja California Sur.

Jefferson, T.A., S. Leatherwood, and M.A. Weber. 1993. FAO Species Identification Guides. Marine Mammals of the World. United Nations Environmental Program, Rome.

Jefferson, T.A., D.K. Odell, and K.T. Prunier. 1995. Notes on the biology of the clymene dolphin (*Stenella clymene*) in the northern Gulf of Mexico. Marine Mammal Science, 11:564-572.

Jefferson, T.A., S. Leatherwood, L.K.M. Shoda, and R.L. Pitman. 1992. Marine Mammals of the Gulf of Mexico. A Field Guide for Aerial and Shipboard Observers. Texas A&M University, Galveston.

Jenkins, S.H., and P.E. Busher. 1979. *Castor canadensis*. Mammalian Species, 120:1-8.

Jennings, W.L. 1958. The ecological distribution of bats in Florida. PhD dissertation, University of Florida, Gainesville.

Jentink, F.A. 1892. Catalogue Sustematique des Mamiferes (Singes, Carnivores, Ruminants, Pachydermes, Sirenes et Cetaces). Tome XI Museum D'Histoire Naturelle des Pays-Bas. E.J. Brill, Leiden.

Jiménez, J.J. 1971. Comparative post-natal growth in five species of the genus *Sigmodon*: I. External morphological character relationships. Revista de Biología Tropical, 19:33-148.

Jiménez-Almaráz, M.T. 1991. Los mamíferos del Parque Ecológico Estatal de Omiltemi, Municipio de Chilpancingo, Guerrero. Tesis de licenciatura, Facultad de Ciencias, Universidad Nacional Autónoma de México, México.

Jiménez-Almaraz, M.T., J.J. Gómez, and L.L. Paniagua. 1993. Mamíferos. Pp. 503-549, in: Historia Natural del Parque Ecológico Estatal Omiltemí, Chilpancingo, Guerrero, México (I. Luna Vega and J. Llorente-Bousquets, eds.). CONABIO-UNAM, México.

Jiménez-Guzman, A., 1966. Mammals of Nuevo Leon, Mexico. Master's thesis, University of Kansas, Lawrence.

Jiménez-Guzman, A., 1968. Nuevos registros de murciélagos para Nuevo León México. Anales del Instituto de Biología, UNAM, México, Serie Zoología, 39:133-144.

Jiménez-Guzman, A., and J.H., López-S. 1992. Estado actual de la zorra del desierto *Vulpes velox zinseri*, en el Ejido El Tokio, Galeana, Nuevo León, México. Publicaciones Biológicas, Facultad de Ciencias Biológicas, Universidad Autónoma de Nuevo León, 6:53-60.

Jiménez-Guzman, A., and M.A. Zúñiga-R. 1992. Nuevos registros de mamíferos para Nuevo León, México. Publicaciones Biológicas, Facultad de Ciencias Biológicas, Universidad Autónoma de Nuevo León, 6:189-191.

Jiménez-Guzman, A., M.A. Zúñiga-R., and J.A. Niño. 1999. Mamíferos de Nuevo León. Universidad Autónoma de Nuevo León, Monterrey.

Jiménez-Huerta, J. 1992. Distribución y abundancia del recurso alimenticio en un fragmento de selva alta perennifolia y su uso por Ateles y Alouatta en el Ejido Magallanes (Municipio de Soteapan, Veracruz). Tesis de licenciatura, Universidad Veracruzana, Xalapa, Veracruz.

Johnson, D.W., and D.M. Armstrong. 1987. *Peromyscus crinitus*. Mammalian Species, 287:1-8.

Johnson, E.W., and R.K. Selander. 1971. Protein variation and systematics in kangaroo rats (genus Dipodomys). Systematic Zoology, 20:377-405.

Johnson, J.H., and A.A. Wolman. 1984. The humpback whale, *Megaptera novaeangliae*. Pp. 30-37, in: The Status of Endangered Whales (Jeffrey M. Breiwick and Howard W. Braham, eds.). Marine Fisheries Review, 46(4). NOAA/NMFS.

Johnson, M.L., and S. Johnson. 1982. Voles, Microtus species. Pp. 326-354, in: Wild Mammals of North America. Biology, Management, Economics (J.A. Chapman and G.A. Feldhamer, eds.). Johns Hopkins University Press, Baltimore.

Johnson, W.E., and R.K. Selander. 1971. Protein variation and systematics in kangaroo rats (genus *Dipodomys*). Systematic Zoology, 20:377-405.

Johnson, W.E., E. Eizirik, and G.M. Lento. 2001. The control, exploitation, and conservation of carnivores. Pp. 197-219, in: Carnivore Conservation (J.L. Gittleman, S.M. Funk, D. Macdonald, and R.K. Wayne, eds.). Conservation Biology Series 5. Cambridge University Press, Cambridge.

Johnston, R.F. 1956. Breeding of the Ord kangaroo rat (*Dipodomys ordii*) in southern New Mexico. The Southwestern Naturalist, 1:190-193.

Jolley, T.W., R.L. Honeycutt, and R.D. Bradley. 2000. Phylogenetic relationships of pocket gophers (genus *Geomys*) based on the mitochondrial 12S rRNA gene. Journal of Mammalogy, 81:1025-1034.

Jones, C. 1965. Ecological distribution and activity period of bats of the Mogollan mountains area of New Mexico and adjacent Arizona. Tulane Studies in Zoology, 12:93-100.

Jones, C. 1976. Economics and conservation. Part I. Pp. 133-145, in: Biology of Bats of the New World Family Phyllostomatidae (R.J. Baker, J.K. Jones Jr., and D.C. Carter, eds.). Special Publications of the Museum, Texas Tech University, 10:1-218.

Jones, C., and J. Pagels. 1968. Notes on a population of *Pipistrellus subflavus* in southern Louisiana. Journal of Mammalogy, 49:134-139.

Jones, C., and R.D. Suttkus. 1973. Colony structure and organization of *Pipistrellus subflavus* in southern Louisiana. Journal of Mammalogy, 54:962-968.

Jones, C.A., and C.N. Baxter. 2004. *Thomomys bottae*. Mammalian Species, 742:1-14.

Jones, G.S., and J.D. Webster. 1977. Notes on distribution, habitat and abundance of some mammals of Zacatecas, México. Anales del Instituto de Biología, UNAM, México, Serie Zoología, 47:75-84.

Jones, J.H., and N.S. Smith. 1979. Bobcat (Lynx rufus) density and prey selection in central Arizona, USA. Journal of Wildlife Management, 43:666-672.

Jones, J.K., and H.H. Genoways. 1975. *Dipodomys phillipsii*. Mammalian Species, 51:1-3.

Jones, J.K., and H.H. Genoways. 1978. *Neotoma phenax*. Mammalian Species, 108:1-3.

Jones, J.K., and T.L. Yates. 1983. Review of the white-footed mice, genus *Peromyscus*, of Nicaragua. Occasional Papers of the Museum, Texas Tech University, 82:1-15.

Jones, J.K., Jr. 1963. Additional records of mammals from Durango, Mexico. Transactions of the Kansas Academy of Science, 66:750-753.

Jones, J.K., Jr. 1964a. A new subspecies of harvest mouse *Reithrodontomys gracilis*, from isla del Carmen, Campeche. Proceedings of the Biological Society of Washington, 77:123-124.

Jones, J.K., Jr. 1964b. Bats from western and southern Mexico. Transactions of the Kansas Academy of Science, 67:509-516.

Jones, J.K., Jr. 1966a. Recent records of the shrew *Megasorex gigas* (Merriam, 1897), from western Mexico. The American Midland Naturalist, 75:249-250.

Jones, J.K., Jr. 1966b. Bats from Guatemala. University of Kansas Publications, Museum of Natural History, 16:439-472.

Jones, J.K., Jr. 1977. *Rhogeessa gracilis*. Mammalian Species, 76:1-2.

Jones, J.K., Jr., and J. Arroyo-C. 1990. *Nyctinomops aurispinosus*. Mammalian Species, 350:1-3.

Jones, J.K., Jr., and D.C. Carter. 1976. Annotated checklist, with keys to subfamilies and genera. Part I. Pp. 7-38, in: Biology of Bats of the New World Family Phyllostomatidae (R.J. Baker, J.K. Jones Jr., and D.C. Carter, eds.). Special Publications of the Museum, Texas Tech University, 10:1-218.

Jones, J.K., Jr., and D.C. Carter. 1979. Systematic and distributional notes. Part III. Pp. 7-11, in: Biology of Bats of the New World Family Phyllostomatidae (R.J. Baker, J.K. Jones Jr., and D.C. Carter, eds.). Special Publications of the Museum, Texas Tech University, 16:1-441.

Jones, J.K., Jr., and P.B. Dunningan. 1965. *Molossops greenhalli* and other bats from Guerrero and Oaxaca, Mexico. Transactions of the Kansas Academy of Sciences, 68:461-462.

Jones, J.K., Jr., and H.H. Genoways. 1967a. Distribution of the porcupine, *Erethizon dorsatum*, in Mexico. Mammalia, 32:709-711.

Jones, J.K., Jr., and H.H. Genoways. 1967b. Notes on the Oaxacan vole, *Microtus oaxacensis* Goodwin, 1966. Journal of Mammalogy, 48:320-321.

Jones, J.K., Jr., and H.H. Genoways. 1969. Holotypes of recent mammals in the Museum of Natural History, The University of Kansas. Pp. 129-146, in: Contributions in Mammalogy (J.K. Jones Jr., ed.). University of Kansas Publications, Museum of Natural History, 51:1-428.

Jones, J.K., Jr., and H.H. Genoways. 1970. Harvest mice (genus *Reithrodontomys*) of Nicaragua. Occasional Papers, Western Foundation for Vertebrate Zoology, 2:1-16.

Jones, J.K., Jr., and J.A. Homan. 1974. *Hylonycteris underwoodi*. Mammalian Species, 32:1-2.

Jones, J.K., Jr., and T.E. Lawlor. 1965. Mammals from isla Cozumel, Mexico, with description of a new species of harvest mouse. University of Kansas Publications, Museum of Natural History, 16:409-419.

Jones, J.K., Jr., T. Álvarez, and M.R. Lee. 1962. Noteworthy mammals from Sinaloa, Mexico. University of Kansas Publications, Museum of Natural History, 14:145-159.

Jones, J.K., Jr., J. Arroyo-Cabrales, and R.D. Owen. 1988. Revised checklist of bats (Chiroptera) of Mexico and Central America. Occasional Papers of the Museum, Texas Tech University, 120:1-34.

Jones, J.K., Jr., D.C. Carter, and W.D. Webster. 1983. Records of mammals from Hidalgo, Mexico. The Southwestern Naturalist, 28:378-380.

Jones, J.K., Jr., J.R. Choate, and A. Cadena. 1972. Mammals from the Mexican state of Sinaloa. II. Chiroptera. University of Kansas Publications, Museum of Natural History, 6:1-29.

Jones, J.K., Jr., H. Genoways, and T.E. Lawlor. 1974a. Annotated checklist of mammals of the Yucatan Peninsula, Mexico. II. Rodentia. Occasional Papers of the Museum, Texas Tech University, 22:1-24.

Jones, J.K., Jr., H.H. Genoways, and J.D. Smith. 1974b. Annotated checklist of mammals of the Yucatan Peninsula, Mexico. III: Marsupialia, Insectivora, Primates, Edentata, Lagomorpha. Occasional Papers of the Museum, Texas Tech University, 23:1-12.

Jones, J.K., Jr., H.H. Genoways, and L.C. Watkins. 1970. Bats of the genus *Myotis* from western Mexico with a key to species. Transactions of the Kansas Academy of Science, 73:409-418.

Jones, J.K., Jr., J.D. Smith, and T. Álvarez. 1965. Notes on bats from the cape region of Baja California. Transactions of the San Diego Society of Natural History, 14:53-56.

Jones, J.K., Jr., J.D. Smith, and H.H. Genoways. 1973. Annotated checklist of mammals of the Yucatan Peninsula, Mexico. I. Chiroptera. Occasional Papers of the Museum, Texas Tech University, 13:1-31.

Jones, J.K., Jr., J.D. Smith, and R.W. Turner. 1971. Noteworthy records of bats from Nicaragua, with a checklist of the chiropteran fauna of the country. University of Kansas Publications, Museum of Natural History, 2:1-35.

Jones, J.K., Jr., P. Swannepoel, and D.C. Carter, 1977. Annotated checklist of the bats of Mexico and Central America. Occasional Papers of the Museum, Texas Tech University, 47:1-35.

Jones, J.K., Jr., R.P. Lampe., C.A. Spernath, and T.H. Kunz. 1973. Notes on the distribution and natural history of bats in southern Montana. Occasional Papers of the Museum, Texas Tech University, 15:1-12.

Jones, J.K., Jr., R.S. Hoffmann, D.W. Rice, C. Jones, R.J. Baker, and M. Engstrom. 1992. Revised checklist of North American mammals north of Mexico, 1991. Occasional Papers of the Museum, Texas Tech University, 146:1-23.

Jones, L.L., G.C. Bouchet, and B.J. Turnock. 1987. Comprehensive report of the incidental take, biology, and status of the Dall's porpoise. International North Pacific Fisheries Commission Document, No. 3156.

Jones, M.L. 1982. Longevity of captive mammals. Zoologische Garten, 52:113-28.

Jones, M.L. 1985. Photographic identification study of gray whale reproduction, distribution, and duration of station in San Ignacio lagoon, and inter-lagoon movements in Baja California Sur, Mexico. Abstracts, Sixth Biennial Conference on the Biology of Marine Mammals. Vancouver, Canada.

Jones, T. 1993. The social systems of heteromyid rodents. Pp. 575-595, in: Biology of the Heteromyidae (H.H. Genoways and J.H. Brown, eds.). American Society of Mammalogists.

Jones, W.T. 1984. Natal philopatry in banner-tailed kangaroo rats. Behavioral Ecology and Sociobiology, 15:151-155.

Jones, W.T. 1985. Body size and life-history variables in heteromyids. Journal of Mammalogy, 66:128-132.

Jones, W.T. 1993. The social systems of Heteromyd rodents. Pp. 575-595, in: Biology of the Heteromyidae (H.H. Genoways and J.H. Brown, eds.). Special Publication 10, American Society of Mammalogists.

Jonkel, C.J., and I. McT. Cowan. 1971. The black bear in the sprucefir forest. Wildlife Monographs, 27:57.

Joule, J., and G.N. Cameron. 1974. Field estimation of demographic parameters: Influence of *Sigmodon hispidus* population structure. Journal of Mammalogy, 55:309-318.

Joule, J., and G.N. Cameron. 1975. Species removal studies and dispersal strategies of sympatric *Sigmodon hispidus* and *Reithrodontomys fulvescens* populations. Journal of Mammalogy, 56:378-396.

Juárez, J. 1992. Distribución altitudinal de los roedores de la Sierra de Atoyac de Álvarez, Guerrero. Tesis de licenciatura, Facultad de Ciencias, Universidad Nacional Autónoma de México. México.

Juárez, J., T. Jiménez A., and D. Navarro L. 1988. Additional records of *Bauerus dubiaquercus* (Chiroptera:Vespertilionidae) in Mexico. Journal of Mammalogy, 3:385.

Judd, F.W. 1969. Distributional notes for some mammals from western Texas and eastern New Mexico. Texas Journal of Science, 22:381-383.

Juelson, T.C. 1970. A study of the ecology and ethology of the rock squirrel, *Spermophilus variegatus* (Erxleben) in northern Utah. PhD dissertation, University of Utah, Salt Lake City.

Julian-LaFerriére, D. 1991. Organisation du peuplement de marsupiaux en Guyane Française. Revue d'Ecologie. La Terre et la Vie, 46:125-144.

Julian-LaFerriére, D. 1993. Radio-tracking observations on ranging and foraging patterns by kinkajous (*Potos flavus*) in French Guiana. Journal of Tropical Ecology, 9:19-32.

Junge, A.J., and R.S. Hoffmann. 1981. An annotated key to the longtailed shrews (genus *Sorex*) of the United States and Canada, with notes on Middle American Sorex. University of Kansas Publications, Museum of Natural History, 94:1-48.

Karasov, W.H. 1983. Water flux and water requirement in freeliving antelope ground squirrels *Ammospermophilus leucurus*. Physiological Zoology, 56:94-105.

Karasov, W.H. 1985. Nutrient constraints in the feeding ecology of an omnivore in a seasonal environment. Oecologia, 66:280-290.

Kasuya, T. 1972. Growth and reproduction of *Stenella coeruleoalba* based on the age determination by means of dentinal growth layers. Scientific Reports of the Whales Research Institute, 24:57-79.

Kasuya, T. 1986. Distribution and behavior of Baird's beaked whales of the Pacific coast of Japan. Scientific Reports of the Whales Research Institute, 37:61-83.

Kasuya, T., N. Miyazaki, and W.H. Dawbin. 1974. Growth and reproduction of *Stenella attenuata* on the Pacific coast of Japan. Scientific Reports of the Whales Research Institute, 26:157-226.

Kaufmann, J.H. 1962. Ecology and social behavior of the coati, *Nasua narica* on Barro Colorado Island, Panama. University of California Publications, Zoology, 60:95-222.

Kaufmann, J.H. 1982. Raccoon and allies. Pp. 578-585, in: Wild Mammals of North America (J.A. Chapman and G.A. Feldhamer, eds.). Johns Hopkins University Press, Baltimore.

Kaufmann, J.H. 1983. Nasua narica (Pizote, Coati). Pp. 478-480, in: Costa Rican Natural History (D.H. Janzen, ed.). University of Chicago Press, Chicago.

Kaufmann, J.H. 1987. Ringtail and coati. Pp. 501-508, in: Wild Furbearer Management and Conservation in North America (M. Novak, J. Baker, M.E. Obbard, and B. Malloch, eds.). Ministry of Natural Resources, Ontario, Canada.

Kaufmann, J.H., and A. Kaufmann. 1965. Observations on the behavior of tayras and grisons. Zeitschrift für Säugetierkunde, 30:146-155.

Kaufmann, J.H., D.V. Lanning, and S.E. Poole. 1976. Current status and distribution of the coati in the United States. Journal of Mammalogy, 57:621-637.

Kawakami, T. 1980. A review of sperm whale food. Scientific Reports of the Whales Research Institute, 32:199-218.

Kawamichi, T. 1981. Vocalizations of Ochotona as a taxonomic character. Pp. 324-339, in: Proceedings of the World Lagomorph Conference (K. Myers and C.D. MacInnes, eds.). University of Guelph, Guelph, Ontario, Canada.

Kawamura, A. 1977. On the food of Bryde's whales caught in the South Pacific and Indian Oceans. Scientific Reports of the Whales Research Institute, 29:49-58.

Kawamura, A. 1980. A review of food of balaenopterid whales. Scientific Reports of the Whales Research Institute, 32:155-197.

Kawamura, A.1982. Food habits and prey distribution of three rorcual species in the North Pacific Ocean. Scientific Reports of the Whales Research Institute, 34:59-92.

Kays, R., and J. Gittleman. 2001. The social organization of the kinkajou *Potos flavus* (Procyonidae). Journal of Zoology, 253:491-504.

Keith, J.O. 1965. The Abert squirrel and its dependence on Ponderosa pine. Ecology, 46:150-163.

Keller, B.L. 1985. Reproductive patterns. Pp. 725-778, in: Biology of the New World Microtus (R.H. Tamarin, ed.). Special Publication 8, American Society of Mammalogists.

Kellogg, L.L., and E.A. Goldman. 1944. Review of the spider monkey. Proceedings of the United States National Museum, 96:1-45.

Kelly, J.F. 2000. Stable isotopes of carbon and nitrogen in the study of avian and mammalian trophic ecology. Canadian Journal of Zoology, 78:1-27.

Kelson, K.R. 1952. Comments on the taxonomy and geographic distribution of some North American woodrats (genus *Neotoma*). University of Kansas Publications, Museum of Natural History, 5:233-242.

Kelt, D.A., K. Rogovin, G. Shenbrot, and J.H. Brown. 1999. Patterns in the structure of Asian and North American desert small mammal communities. Journal of Biogeography, 26:825-841.

Kenagy, G.J., and G.A. Bartholomew. 1979. Effects of day length and endogenous control on the annual reproductive cycle of the antelope ground squirrel *Ammospermophilus leucurus*. Journal of Comparative Physiology, 130:131-136.

Kenagy, G.J., and G.A. Bartholomew. 1985. Seasonal reproductive patterns in five coexisting California desert rodent species. Ecological Monographs, 55:371-397.

Kennedy, M.L., and G.D. Schnell. 1978. Geographic variation and sexual dimorphism in Ord's kangaroo rat, *Dipodomys ordii*. Journal of Mammalogy, 59:45-59.

Kennedy, M.L., T.L. Best, and M.J. Harvey. 1984a. Bats of Colima, Mexico. Mammalia, 48:397-408.

Kennedy, M.L., P.K. Kennedy, and G.D. Baumgadner. 1984b. First record of the seminole bat (*Lasiurus seminolus*) in Tennessee. Journal of the Tennessee Academy of Science. 58:89-90.

Kennedy, M.L., P.K. Price, and O.S. Fuller. 1977. Flight speeds of five species of Neotropical bats. The Southwestern Naturalist, 22:401-404.

Kenney, R.D. 2002. North Atlantic, North Pacific, and Southern right whales. In: Encyclopedia of Marine Mammals (W.F. Perrin, B. Würsig, and J.G.M. Thewissen, eds.). Academic Press, London.

Kenney, R.D., G.P. Scott, T.J. Thompson, and H.E. Winn. 1997. Estimates of prey consumption and trophic impacts of cetaceans in the USA Northeast continental shelf ecosystem. Journal of Northwest Atlantic Fisheries Science, 22:155-171.

Kenyon, K.W. 1969. The sea otter in the eastern Pacific Ocean. North American Fauna, 68:1-352.

Kenyon, K.W. 1977. Caribbean monk seal extinct. Journal of Mammalogy, 58:97-98.

Kerr, R. 1792. The Animal Kingdom, or Zoological System, of the Celebrated Sir Charles Linnaeus. Class I. Mammalia. London.

Kight, J. 1962. An ecological study of the bobcat *Lynx rufus* (Schreber) in west-central South Carolina. Master's thesis, University of Georgia, Athens.

Kilgore, D.L., Jr. 1970. The effect of northward dispersal on growth rate of young, size of young at birth, and litter size in *Sigmodon hispidus*. The American Midland Naturalist, 84:510-520.

Kilpatrick, C.W., and E.G. Zimmerman. 1975. Genetic variation and systematics of four species of mice of the *Peromyscus boylii* species group. Systematic Zoology, 24:143-162.

Kilpatrick, C.W., and E.G. Zimmerman. 1976a. Hemoglobin polymorphism in the encinal mouse, *Peromyscus pectoralis*. Biochemistry and Genetics, 14:137-143.

Kilpatrick, C.W., and E.G. Zimmerman. 1976b. Biochemical variation and systematics of *Peromyscus pectoralis*. Journal of Mammalogy, 57:506-521.

Kiltie, R.A. 1981a. The function of interlocking canines in rain forest peccaries (Tayassuidae). Journal of Mammalogy, 62:459-469.

Kiltie, R.A. 1981b. Stomach contents of rain forest peccaries (*Tayassu tajacu* and *T. pecari*). Biotropica, 13:234-236.

Kiltie, R.A. 1981c. Distribution of palm fruits on a rain forest floor: Why do whitelipped peccaries forage near objects? Biotropica, 13:141-145.

Kiltie, R.A., and J. Terborgh. 1976. Ecology and behavior of rain forest peccaries in southern Peru. National Geographic Research Report, 873-882.

Kiltie, R.A., and J. Terborgh. 1983. Observations on the behavior of rain forest peccaries in Peru: Why do whitelipped peccaries form herds? Zeitschrift für Tierpsychology, 62:241-255.

Kim, K.C. 1966. The species of *Enderleinellus* (Anophura, Hoplopleuridae) parasitic on the Sciurini and Tamiasciurini. Journal of Parasitology, 52:988-1024.

Kincaid, W.B. 1975. Species removal studies: III Niche dynamics and competition in *Sigmodon hispidus* and *Reithrodontomys fulvescens*. Master's thesis, University of Houston, Texas.

King, C. 1989. Natural History of Weasels and Stoats. Comstock Publishing Associates, Ithaca, NY.

King, J.A. 1955. Social behavior, social organization, and population dynamics in black-tailed prairie dog towns in Black Hills in South Dakota. Contributions Laboratory of Vertebrate Biology, No. 67, University of Michigan, Ann Arbor.

King, J.E. 1953. The monk seals (genus *Monachus*). Bulletin of the British Museum. (Natural History) Zoology, 3:201-256.

King, J.E. 1964. Seals of the World. British Museum of Natural History, London.

King, J.E. 1983. Seals of the World. Oxford University Press, Oxford.

Kingston, N., B. Villa-R., and W. López-Forment. 1971. New host and locality records for species of the genera *Periglischrus* and *Cameronista* (Acarina: Spinturnicidae) on bats from Mexico. Journal of Parasitology, 57:927-928.

Kinlaw, A. 1995. *Spilogale putorius*. Mammalian Species, 511:1-7.

Kirkpatrick, C.M. 1955. The testis of the fox squirrel in relation to age and season. American Journal of Anatomy, 97:229-256.

Kirkpatrick, R.D., and L.K. Sowls, 1962. Age determination of the collared peccary by tooth-replacement pattern. Journal of Wildlife Management, 26:214-217.

Kirkpatrick, R.D., A.M. Cartwright, J.C. Brier, and E.J. Spicka. 1975. Additional mammal records from Belize. Mammalia, 39:330-331.

Kirkaptrick, J.F., D.F. Gudermuth, R.L. Flagan, J.C. McCarthy, and B.L. Lasley. 1993. Remote monitoring of ovulation and pregnancy of Yellowstone bison. Journal of Wildlife Management, 57:407-412.

Kirsch, J.A.W., and J.H. Calaby. 1977. The species of living marsupials: An annotated list. Pp. 9-26, in: The Biology of Marsupials (B. Stonehouse and D. Gilmore, eds.). University Park Press, Baltimore.

Kitchen, D.W. 1974. Social behavior and ecology of the pronghorn. Wildlife Monographs, 38:1-196.

Kitchen, D.W., and B.W. O'Gara. 1982. Pronghorn (*Antilocapra americana*). Pp. 960-971, in: Wild Mammals of North America: Biology, Management, Economics (J.A. Chapman and G.A. Feldhamer, eds.). Johns Hopkins University Press, Baltimore.

Kitchings, J.T., and J.D. Story. 1979. Home range and diet of bobcats in eastern Tennessee. Proceedings of the Bobcat Research Conference, National Wildlife Federation Science and Technology Service, 6:47-52.

Kleiman, D.C., and T.M. Davis. 1979. Ontogeny and maternal care. Pp. 387-402, in: Biology of Bats of the New World Family Phyllostomatidae, Part III (R.J. Baker, J.K. Jones Jr., and D.C. Carter, eds.). Special Publications of the Museum, Texas Tech University, 16:1-441.

Kleiman, D.G., J.K. Eisenberg, and E. Maliniak. 1979. Reproductive parameters and productivity of caviomorph rodents. Pp. 153-181, in: Vertebrate Ecology in the Northern Neotropics (J.K. Eisenberg, ed.). Academic Press, New York.

Klein, L.L. 1971. Observations on copulation and seasonal reproduction of two species of spider monkeys, *Ateles belzebuth* and *Ateles geoffroyi*. Folia Primatologica, 15:233-248.

Klett, A. 1981. Estado actual de la pesquería del calamar gigante en el estado de Baja California Sur. Serie Científica Pesca, 21:11-28.

Klinowska, M. 1991. Dolphins, Porpoises and Whales of the World. The IUCN Red Databook. IUCN, Gland, Switzerland.

Knaus, R.M., N. Kinler, and R. Linscombe. 1983. Estimating river otter populations: The feasibility of 65Zn to label faeces. Wildlife Society Bulletin, 11:375-377.

Knipe, T. 1957. The javelina in Arizona. Wildlife Bulletin, Arizona Fish and Game Department, 2:96.

Knobloch, I.W. 1942. Notes on a collection of mammals from Sierra Madre of Chihuahua, Mexico. Journal of Mammalogy, 23:297-298.
Kock, K.H., and Y. Shimadzu. 1994. Tropic relationships and trends in population size and reproductive parameters in Antarctic high-level predators. Pp. 287-312, in: Southern Ocean Ecology: The Biomass Perspective (S.Z. El-Sayed, ed.). Cambridge University Press, Cambridge.
Koehler, G.M., and M.G. Hornocker. 1989. Influences of seasons on bobcats in Idaho. Journal of Wildlife Management, 53:197-202.
Koestner, E.J. 1941. An annotated list of mammals collected in Nuevo Leon, Mexico, in 1938. The Great Basin Naturalist, 2:9-15.
Koestner, E.J. 1944. Populations of small mammals on Cerro Potosi, Nuevo Leon, Mexico. Journal of Mammalogy, 25:284-289.
Koford, C.B. 1958. Prairie dogs, whitefaces, and blue grama. Wildlife Monographs, 3:6-78.
Koford, C.B. 1969. The last of the Mexican grizzly bear. International Union for Conservation of Nature and Natural Resources Bulletin.
Kohs, G.M., and C.M. Clifford. 1966. Three new species of *Ixodes* from Mexico and descriptions of the male of *I. auritulus auritulus* Newmann, *I. conepati* Cooley and Kohls, and *I. lasallei* Méndez and Ortiz (Acarina: Ixodidae). Journal of Parasitology, 52:810-820.
Konecny, M.J. 1989. Movement patterns and food habits of four sympatric carnivore species in Belize, Central America. Pp. 243-264, in: Advances in Neotropical Mammalogy (K.H. Redford and J.F. Eisenberg, eds.). Sandhill Crane Press, Gainesville, FL.
Konstant, W., R.A. Mittermeier, and S.D. Nash. 1985. Spider monkeys in captivity and in the wild. Primate Conservation, 5:82-109.
Koop, B.F., R.J. Baker, and J.T. Mascarello. 1985. Cladistical analysis of chromosomal evolution within the genus *Neotoma*. Occasional Papers of the Museum, Texas Tech University, 96:1-9.
Koopman, K.F. 1956. Bats from San Luis Potosi with a new record for *Balantiopteryx plicata*. Journal of Mammalogy, 37:547-548.
Koopman, K.F. 1959. The zoogeographical limits of the West Indies. Journal of Mammalogy, 40:236-240.
Koopman, K.F. 1965. A northern record of yellow bat. Journal of Mammalogy, 46:695.
Koopman, K.F. 1971. The systematics and historical status of the Florida *Eumops* (Chiroptera: Molossidae). American Museum Novitates, 2478:1-6.
Koopman, K.F. 1974. Eastern limits of *Plecotus* in Mexico. Journal of Mammalogy, 58:872-873.
Koopman, K.F. 1978. Zoogeography of Peruvian bats with special emphasis on the role of the Andes. American Museum Novitates, 2651:1-33.
Koopman, K.F. 1982. Biogeography of the bats of South America. Pp. 273-302, in: Mammalian Biology in South America (M.A. Mares and H.H. Genoways, eds.). Special Publications Series, Pymatuning Laboratory of Ecology, University of Pittsburgh.
Koopman, K.F. 1989. A review and analysis of the bats of the West Indies. Pp. 635-644, in: Biogeography of the West Indies: Past, Present, and Future (C.A. Woods, ed.). Sandhill Crane Press, Gainesville, FL.
Koopman, K.F. 1993. Chiroptera. Pp. 137-241, in: Mammal Species of the World. A Taxonomic and Geographic Reference (D.E. Wilson and D.M. Reeder, eds.). Smithsonian Institution Press, Washington, DC.
Koopman, K.F. 1994. Chiroptera: Systematics. Pp. 70-110, in: Handbook of Zoology, vol. VIII, Mammalia (J. Niethammer, H. Schliemann, and D. Starck, eds.). Walter de Gruyter, Berlin.
Koopman, K.F., and P.S. Martin. 1959. Subfossil mammals from the Gomez Farias Region and the Tropical Gradient of eastern Mexico. Journal of Mammalogy, 40:1-12.
Koopman, K.F., M.K. Hecht, and E. Ledecky-Janecek. 1957. Notes on the mammals of the Bahamas with special reference to the bats. Journal of Mammalogy, 38:164-174.
Koprowski, J.L. 1994. *Sciurus niger*. Mammalian Species, 479:1-9.
Koprowski, J.L., J.L. Roseberry, and W.D. Klimstra. 1988. Longevity records for the fox squirrel. Journal of Mammalogy, 69:383-384.
Korschgen, L.J. 1981. Foods of fox and gray squirrels in Missouri. Journal of Wildlife Management, 45:260-266.
Korschgen, L.J., and H.B. Stuart. 1972. Twenty years of avian predator-small mammal relationships in Missouri. Journal of Wildlife Management, 36:269-282.
Kortlucke, S.M. 1973. Morphological variation in the kinkajou, *Potos flavus* (Mammalia: Procyonidae) in Middle America. University of Kansas Publications, Museum of Natural History, 17:1-36.
Kratter, A.W. 1991. First nesting record for Williamson's sapsucker (*Sphyrapicus hyroides*) in Baja California, Mexico, and comment on the biogeography of the fauna of the Sierra San Pedro Martir. The Southwestern Naturalist, 36:247-250.
Krausman, P., and E. Ables. 1981. Ecology of Carmen mountains white-tailed deer. U.S. National Park Service, Scientific Monograph Series, 15:1-114.
Krumbiegel, V.I. 1942. Die sa¨ugetiere der Su¨damerika-expeditionen Prof. Dr. Kriegs. Zoologischer Anzeiger, 139:81-96.
Kuban, F.J., and G. Schwartz. 1985. Nectar as a diet item of ring-tailed cat. The Southwestern Naturalist, 30:311-312.
Kumirai, A., and J.K. Jones Jr. 1990. *Nyctinomops femorosaccus*. Mammalian Species, 349:1-5.
Kuns, M.L., and R.E. Tashian. 1954. Notes on mammals from northern Chiapas, Mexico. Journal of Mammalogy, 35:100-103.
Kunz, T.H. 1973. Population studies of the cave bat (*Myotis velifer*): Reproduction, growth and development. University of Kansas Publications, Museum of Natural History, 15:1-43.
Kunz, T.H. 1982a. Ecology of Bats. Plenum Press, New York.
Kunz, T.H. 1982b. *Lasionycteris noctivagans*. Mammalian Species, 172:1-5.
Kunz, T.H. 1982c. Roosting ecology. Pp. 1-55, in: Ecology of Bats (T.H. Kunz, ed.). Plenum Press, New York.
Kunz, T.H., and C.A. Díaz. 1995. Folivory in fruit-eating bats, with new evidence from *Artibeus jamaicensis* (Chiroptera: Phyllostomidae). Biotropica, 27:106-120.
Kunz, T.H., and R.A. Martin. 1982. *Plecotus townsendii*. Mammalian Species, 175:1-6.
Kunz, T.H., P.V. August, and C.D. Burnett. 1983. Harem social organization in cave roosting *Artibeus jamaicensis* (Chiroptera: Phyllostomidae). Biotropica, 15:133-138.
Kurta, A., and G.C. Lehr. 1995. *Lasiurus ega*. Mammalian Species, 515:1-7.
Kurtén, B., and E. Anderson. 1980. Pleistocene Mammals of North America. Columbia University Press, New York.
Lackey, J.A. 1967. Biosystematics of heermanni group kangaroo rats in Southern California. Transactions of the San Diego Society of Natural History, 14:313-344.
Lackey, J.A. 1970. Distributional records of bats from Veracruz. Journal of Mammalogy, 51:384-385.
Lackey, J.A. 1991. *Chaetodipus arenarius*. Mammalian Species, 384:1-4.
Lackey, J.A., and T.L. Best. 1992. *Chaetodipus goldmani*. Mammalian Species, 419:1-5.
Lackey, J.A., D.G. Huckaby, and B.G. Ormiston. 1985. *Peromyscus leucopus*. Mammalian Species, 247:1-10.
Lamont, M.M., J.T. Vida, J.T. Harvey, S. Jeffries, R. Brown, H.H. Huber, R. DeLong, and W.K. Thomas. 1996. Genetic substructure of the Pacific harbor seal (*Phoca vitulina richardsi*) off Washington, Oregon and California. Marine Mammal Science, 12:402-413.
Landa, Fr. D. 1560 [1973]. Relación de las cosas de Yucatán. 12th ed. Editorial Porrúa. México.
Lane, H.K. 1946. Notes on *Pipistrellus subflavus subflavus* (F. Cuvier) during the season of parturition. Proceedings of the Pennsylvania Academy of Science, 20:57-61.
Lao, M. 1995. Las sirenas. Historia de un símbolo. Ediciones Era, México.
Larivière, S. 1999. Lontra provocax. Mammalian Species, 610:1-4.
Larivière, S., and L.R. Walton. 1997. Lynx rufus. Mammalian Species, 563:1-8.

Larson, S.E. 1997. Taxonomic re-evaluation of the jaguar. Zoo Biology, 16:107-120.

Lassieur, S., and D.E. Wilson. 1989. *Lonchorhina aurita*. Mammalian Species, 347:1-4.

Laurie, E.M.O. 1953. Rodents from British Honduras, Mexico, Trinidad, Haiti and Jamaica collected by Mr. I.T. Sanderson. Annals and Magazine of Natural History, 6:382-394.

LaVal, R.K. 1970. Banding returns and activity periods of some Costa Rican bats. The Southwestern Naturalist, 15:1-10.

LaVal, R.K. 1972. Distributional records and band recoveries of bats from Puebla, Mexico. The Southwestern Naturalist, 16:449-451.

LaVal, R.K. 1973a. A revision of the Neotropical bats of the genus *Myotis*. Natural History Museum of Los Angeles County Science Bulletin, 15:1-54.

LaVal, R.K. 1973b. Systematics of the genus *Rhogeessa* (Chiroptera: Vespertilionidae). University of Kansas Publications, Museum of Natural History, 19:1-47.

LaVal, R.K. 1977. Notes on some Costa Rican bats. Brenesia, 10:77-83.

LaVal, R.K., and H.S. Fitch. 1977. Structure, movements and reproduction in three Costa Rican bat communities. University of Kansas Publications, Museum of Natural History, 69:1-28.

Laval, R.K., and M.L. Laval. 1980. Prey selection by a Neotropical foliage-gleaning bat, *Micronycteris megalotis*. Journal of Mammalogy, 61:327-330.

Lawhead, D.N. 1977. Home range, density, and habitat preference of the bobcat on the Three Bar Wildlife Area, Arizona. Arizona Cooperative Wildlife Research Unit Quarterly Report, 27:7-8.

Lawhead, D.N. 1984. Bobcat Lynx rufus home range density and habitat preference in south-central Arizona USA. The Southwestern Naturalist, 29:105.

Lawlor, T.E. 1965. The Yucatan deer mouse, *Peromyscus yucatanicus*. University of Kansas Publications. Museum of Natural History, 16:421-438.

Lawlor, T.E. 1969. A systematic study of the rodent genus *Ototylomys*. Journal of Mammalogy, 50:29-42.

Lawlor, T.E. 1971a. Distribution and relationships of six species of *Peromyscus* in Baja California and Sonora, Mexico. Occasional Papers of the Museum of Zoology, University of Michigan, 661:1-22.

Lawlor, T.E. 1971b. Evolution of *Peromyscus* on northern islands in the Gulf of California, Mexico. Transactions of the San Diego Society of Natural History, 16:91-124.

Lawlor, T.E. 1982a. *Ototylomys phyllotis*. Mammalian Species, 181:1-3.

Lawlor, T.E. 1982b. The evolution of body size in mammals: Evidence from insular populations in Mexico. The American Naturalist, 119:54-72.

Lawlor, T.E. 1983. The mammals. Pp. 269-285, in: Island Biogeography in the Sea of Cortez (T.J. Case and M.L. Cody, eds.). University of California Press, Berkeley.

Laws, R.M. 1959. On the breeding season of Southern Hemisphere fin whales, *Balaenoptera physalus* (Linn). Norsk Hvalfangst-Tiden, 48:329-351.

Lawson, B., and R. Johnson. 1982. Mountain sheep. Pp. 1036-1055, in: Wild Mammals of North America: Biology, Management, Economics (J.A. Chapman and G.A. Feldhamer, eds.). Johns Hopkins University Press, Baltimore.

Lay, D.M. 1962. Seis mamíferos nuevos para la fauna de México. Anales del Instituto de Biología, UNAM, México, Serie Zoología, 33:373-377.

Layne, J.N. 1967. Lagomorphs. In: Recent Mammals of the World, Synopsis of Families (S. Anderson and J.K. Jones Jr., eds.). Ronald Press, New York.

Layne, J.N. 1974. Ecology of small mammals in flatwoods habitat in north-central Florida, with emphasis on the cotton rat (*Sigmodon hispidus*). American Museum Novitates, 2544:1-48.

Layne, J.N., and D. Glover. 1977. Home range of the armadillo in Florida. Journal of Mammalogy, 58:411-413.

Lazcano, M.A., and J.M. Packard. 1989. The occurrence of manatees (*Trichechus manatus)* in Tamaulipas, Mexico. Marine Mammal Science, 5(2):202-205.

Le Boeuf, B.J. 1994. Variation in the diving pattern of northern elephant seal with age, mass, sex, and reproductive condition. Pp. 237-252, in: Reproductive Success (T. H. Clutton-Brock, ed.). University of Chicago Press, Chicago.

Le Boeuf, B.J., and R.M. Laws. 1994. Elephant seals: An introduction to the genus. Pp. 1-26, in: Elephant Seal (B.J. Le Boeuf and R. Laws, eds.). University of California Press, Los Angeles.

Le Boeuf, B.J., and R.S. Peterson. 1969. Social status and mating activity in elephant seals. Science, 163:91-93.

Le Boeuf, B.J., and J. Reiter. 1988. Life time reproductive success in northern elephant seals. Pp. 344-362, in: Reproductive Success (T. H. Clutton-Brock, ed.). University of Chicago Press, Chicago.

Le Boeuf, B.J., Y. Naito, A.C. Huntley, and T. Asaga. 1989. Prolonged, continuous, deep diving by northern elephant seals. Canadian Journal of Zoology, 67:2514-2519.

Le Boeuf, B.J., D. Aurioles G., R. Condit, C. Fox, R. Gisiner, R. Romero, and F. Sinsel. 1983. Size and distribution of the California sea lion population in Mexico. Proceedings of the California Academy of Sciences, 43:77-85.

Le Conte, J.L. 1852. *Geomys hispidus*. Proceedings of the Academy of Natural Sciences of Philadelphia, 6:58.

Le Conte, J.L. 1853. Description of three new species of American arvicolae, with remarks upon some other American rodents. Proceedings of the Academy of Natural Sciences of Philadelphia, 6:404-415.

Le Count, A.L. 1982. Characteristics of a central Arizona black bear population. Journal of Wildlife Management, 46:861-868.

Leatherwood, S. 1975. Some observations of feeding behavior of bottlenose dolphin *Tursiops truncatus* in northern Gulf of Mexico and (*Tursiops* cf. *T. gilli*) of Southern California and Nayarit, Mexico. Marine Fisheries Review, 37:10-16.

Leatherwood, S., and R.R. Reeves. 1983. The Sierra Club Handbook of Whales and Dolphins. Sierra Club Books, San Francisco.

Leatherwood, S., and W.A. Walker. 1979. The northern right whale dolphin *Lissodelphis borealis* peale in the eastern North Pacific. Pp. 85-141, in: Behavior of Marine Animals, vol. 3, Cetaceans (H.E. Winn and B.L. Olla, eds.). Plenum Press, New York.

Leatherwood, S., D.K. Caldwell, and H.E. Winn. 1976. Whales, dolphins, and porpoises of the western North Atlantic: A guide to their identification. National Oceanic and Atmospheric Administration Technical Report, NMFS CIRC-396, Seattle.

Leatherwood, S., L. Harrington-Coulombe, and C.L. Hubbs. 1978. Relict survival of the sea otter in Central California and evidence of its recent redispersal south of Point Conception. Bulletin of the Southern California Academy of Science, 77:109-115.

Leatherwood, S., R.R. Reeves, and L. Foster. 1983. Whales and Dolphins. Sierra Club Books, San Francisco.

Leatherwood, S., R.R. Reeves, W.F. Perrin, and W.E. Evans. 1982. Whales, dolphins and porpoises of the eastern North Pacific and adjacent arctic waters: A guide to their identification. NOAA Technical Report NMFS-CIRC/444, La Jolla, CA.

Leatherwood, S., W.F. Perrin, V.L. Kirby, C.L. Hubbs, and M. Dahlheim. 1980. Distribution and movements of Risso's dolphin, *Grampus griseus*, in the eastern North Pacific. Fishery Bulletin, 77:951-962.

Leatherwood, S., R.R. Reeves, W.F. Perrin, W.E. Evans, and L. Hobbs. 1988. Ballenas, delfines y marsopas del Pacífico Nororiental y de las Aguas Árticas Adyacentes: una guía para su identificación. Informe Especial N° 6, Comisión Interamericana del Atún Tropical.

LeConte, J., 1856. Observations on the North American species of bats. Proceedings of the Academy of Natural Sciences of Philadelphia, 7:431-438.

Lechleitner, R.R. 1958. Certain aspects of behavior of the black-tailed jack rabbit. The American Midland Naturalist, 60:145-155.

Lechuga, G., and A. Núñez G. 1992. Los mamíferos silvestres de los valles Los Reyes y Tocumbo, Michoacán, México. Pp. 258-270, in: X Simposio sobre Fauna Silvestre Gral. MV. Manuel Cabrera Valtierra (M. de los A. Roa Riol and T. Estrada A., eds.). Universidad Nacional Autónoma de México, Gobierno del Estado de Guerrero y Asociación de Zoológicos, Criaderos y Acuarios de la República Mexicana, A.C., México.

Lee, D.S., J.D Funderburg Jr., and M.K. Clark. 1982. A distributional survey of North Carolina mammals. Occasional Papers, North Carolina Biological Survey, 10:1-70.

Lee, H.K., and R.J. Baker. 1987. Cladistical analysis of chromosomal evolution in pocket gophers of the *Cratogeomys castanops* complex (Rodentia: Geomyidae). Occasional Papers of the Museum, Texas Tech University, 114:1-15.

Lee, M.R., and F.F.B. Elder. 1977. Karyotypes of eight species of Mexican rodents (Muridae). Journal of Mammalogy, 58:479-487.

Lee, M.R., and W.S. Modi. 1983. Chromosomes of *Spilogale pygmaea* and *S. putorius leucoparia*. Journal of Mammalogy, 64:493-495.

Lee, R.M., and D.J. Schmidly. 1977. A new species of Peromyscus (Rodentia: Muridae) from Coahuila, Mexico. Journal of Mammalogy, 58:263-268.

Lee, T.E., Jr. 1987. Distributional records of *Lasiurus seminolus* (Chiroptera: Vespertilionidae). Texas Journal of Science, 39:193.

Lee, T.E., Jr., and M.D. Engstrom. 1991. Genetic variation in the silky pocket mouse (*Perognathus flavus*) in Texas and New Mexico. Journal of Mammalogy, 72:273-285.

Lee, T.E., Jr., B. Riddle, and P.L. Lee. 1996. Speciation in the desert pocket mouse (*Chaetodipus penicillatus* Woodhouse). Journal of Mammalogy, 77:58-68.

Lee, T.E., Jr., S.R. Hoofer, and R.A. Van Den Bussche. 2002. Molecular phylogenetics and taxonomic revision of the genus *Tonatia* (Chiroptera: Phyllostomidae). Journal of Mammalogy, 83:49-57.

Lembeck, M. 1978. Bobcat study, San Diego County, California. California Fish and Game Department, Sacramento. Project E-W-2, Study IV, Job 1.7.

Lemke, T.O. 1984. Foraging ecology of the long-nosed bat, *Glossophaga soricina*, with respect to resource availability. Ecology, 65:538-548.

Lenarz, M.S. 1979. Social structure and reproductive strategy in desert bighorn sheep (*Ovis canadensis mexicana*). Journal of Mammalogy, 60:671-678.

Lento, G.M., R.E. Hickson, G.K. Chambers, and D. Penny. 1995. Use of spectral analysis to test hypotheses on the origin of pinnipeds. Molecular Biology and Evolution, 12(1):28-52.

León, L., A.L. Martínez, M.G. Torres, E.M. Figueroa, A.H. Flores, L. Garduño, B.M. González, M. Mayorga, A. Mata, E.A. Pérez, L. Ríos, M.S. Valencia, E.V. Contreras, and V. Villavicencio. 1990. Estudio faunístico preliminar de la zona de Ocuilan y sus alrededores, Estado de México y Morelos. Biología de Campo, Departamento de Biología, Facultad de Ciencias, UNAM, México D.F.

León-Galván, M.A., R. López-Wilchis, O. Hernández-Pérez, E. Arenas-Ríos, and A. Rosado. 2005. Male reproductive cycle of the Mexican big-eared bat, *Corynorhinus mexicanus*, (Chiroptera, Vespertilionidae). The Southwestern Naturalist, 50:453-460.

León Paniagua., L. 1994. XXVI. Algunos aspectos de la taxonomía mastozoológica en México: historia, problemática y alternativas. Pp. 485-504, in: Taxonomía biológica (J. Llorente B. and I. Luna, eds.). Universidad Nacional Autónoma de México-Fondo de Cultura Económica, México.

León Paniagua, L., and E. Romo-Vázquez. 1991. Catálogo de Mamíferos (Vertebrata: Mammalia). Serie Catálogos del Museo de Zoología "Alfonso L. Herrera." Facultad de Ciencias, Universidad Nacional Autónoma de México, México.

León Paniagua, L., and E. Romo-Vázquez. 1993. Mastofauna de la Sierra de Taxco, Guerrero. Pp. 55-64, in: Avances en el Estudio de los Mamíferos de México (R.A. Medellín and G. Ceballos, eds.). Publicaciones Especiales 1, Asociación Mexicana de Mastozoología, México.

Leon-Paniagua, L., A. Navarro-Siguenza, B. Hernandez-Banos, and J.C. Morales. 2007. Diversification of the arboreal mice of the genus *Habromys* (Rodentia: Cricetidae: Neotominae) in the Mesoamerican highlands. Molecular Phylogenetics and Evolution, 42:653-664.

León-Paniagua, L., E. Romo-Vázquez, J.C. Morales, D.J. Schmidly, and D. Navarro-López. 1990. Noteworthy records of mammals from the state of Queretaro, Mexico. The Southwestern Naturalist, 35:231-235.

Leonard, M.L., and M.B. Fenton. 1983. Habitat use by spotted bats (*Euderma maculatum*, Chiroptera: Vespertilionidae): Roosting and foraging behavior. Canadian Journal of Zoology, 61:1487-1491.

Leonard, M.L., and M.B. Fenton. 1984. Echolocation calls of *Euderma maculatum* (Vespertilionidae): Use in orientation and communication. Journal of Mammalogy, 65:122-126.

León-Galván, M.A., T. Fonseca, R. López-Wilchis, and A. Rosado. 1999. Prolonged storage of spermatozoa in the genital tract of female Mexican big-eared bat (*Corynorhinus mexicanus*): The role of lipid peroxidation. Canadian Journal of Zoology, 77:1-6.

Leopold, A.S. 1959. Wildlife of Mexico: The Game Birds and Mammals. University of California Press, Berkeley.

Leopold, A.S. 1965. Fauna Silvestre de México. Aves y Mamíferos de Caza. Instituto Mexicano de Recursos Naturales Renovables, México.

Leopold, A.S. 1967. Grizzlies of the Sierra del Nido. Pacific Discovery, 20:30-32.

Leopold, A.S. 1977. Fauna Silvestre de México. Reimpresión. Editorial Pax, México.

Leopold, A.S. 1982. Fauna Silvestre de México. Aves y Mamíferos de Caza. Reimpresión. Editorial Pax, México.

Levenson, H. 1990. Sexual size dimorphism in chipmunks. Journal of Mammalogy, 71:161-170.

Levenson, H., R.S. Hoffman, C.F. Nadler, L. Detsch, and S.D. Freeman. 1985. Systematics of the Holarctic chipmunks (Tamias). Journal of Mammalogy, 66:219-242.

Levi-Strauss, C. 1964 [1986]. Mitológicas. Lo crudo y lo cocido. Fondo de Cultura Económica. México.

Lewis, J.B. 1940. Mammals of Amelia Country, Virginia. Journal of Mammalogy, 21:422-428.

Lewis, S.E., and D.E. Wilson. 1987. *Vampyressa pusilla*. Mammalian Species, 292:1-5.

Lidicker, W.Z., Jr. 1960. An analysis of intraspecific variation in the kangaroo rat *Dipodomys merriami*. University of California, Publications in Zoology, 67:125-218.

Lim, B.K. 1993. Cladistic reappraisal of Neotropical stenodermatine bat phylogeny. Cladistics, 9:147-165.

Lim, B.K. 2009. Review of the origins and biogeography of bats in South America. Chiroptera Neotropical, 15:391-410.

Lim, B.K., W.A. Pedro, and F.C. Passos. 2003. Differentiation and species status of the Neotropical yellow-eared bats *Vampyressa pusilla* and *V. thyone* (Phyllostomidae) with a molecular phylogeny and review of the genus. Acta Chiropterologica, 5:15-29.

Linares, O.J. 1966. Notas acerca de *Macrophyllum macrophyllum* (Wied) (Chiroptera). Memorias de la Sociedad de Ciencias Naturales La Salle, 26:53-61.

Linares, O.J. 1986. Murciélagos de Venezuela. Cuadernos Lagoven, Caracas, Venezuela.

Lindsay, G.E. 1962. The Belvedere expedition to the Gulf of Baja California. Transactions of the San Diego Society of Natural History, 13:1-44.

Lindsay, G.E. 1964. Sea of Cortez expedition of the California Academy of Sciences. Proceedings of the California Academy of Sciences, 30:11-19.

Lindsay, S.L. 1981. Taxonomic and biogeographic relationships of Baja California chickarees (*Tamiasciurus*). Journal of Mammalogy, 62:673-682.

Lindsdale, J.M. 1946. The California Ground Squirrel. University of California Press. Berkeley.

Lindstedt, S.L. 1980. Energetics and water economy of the smallest desert mammal. Physiological Zoology, 53:82-97.

Lindstedt, S.L., J.F. Hakanson, D.J. Wells, S.D. Swain, H. Hoppeler, and V. Navarro. 1991. Running energetics in the pronghorn antelope. Nature, 353:748-750.

Lindzey, F.G. 1982. Badger Taxidea taxus. Pp. 653-663, in: Wild Mammals of North America: Biology, Management, Economics (J.A. Chapman and G.A. Feldhamer, eds.). Johns Hopkins University Press, Baltimore.

Linsdale, J.M. 1946. The California Ground Squirrel. University of California Press, Berkeley.

Linzey, A.V., and J.N. Layne. 1969. Comparative morphology of the male reproductive tract in the rodent genus *Peromyscus* (Muridae). American Museum Novitates, 2355:1-47.

Linzey, A.V., and J.N. Layne. 1974. Comparative morphology of spermatozoa of the rodent genus *Peromyscus* (Muridae). American Museum Novitates, 2532:1-20.

Lira, I. 2007. Nuevo registro de *Balaenoptera musculus* Linnaeus, 1758 (Mysticeti: Balaenopteridae) para la costa de Oaxaca, México. Revista Mexicana de Mastozoología, 11:69-72.

List, R. 1997. Ecology of the kit fox (*Vulpes macrotis*) and coyote (*Canis latrans*) and the conservation of the prairie dog ecosystem in northern Mexico. PhD dissertation, University of Oxford.

List, R. 2007. The impacts of the border fence on wild mammals. Pp. 77-86, in: A Barrier to Our Shared Environment: The Border Wall between Mexico and the United States (A. Cordova and C.A. de la Parra, eds.). Instituto Nacional de Ecologıa-El Colegio de la Frontera Norte, México D.F.

List, R., and B. Cypher. 2004. Pp. 106-109, in: Status Survey and Conservation Action Plan. Canids: Foxes, Wolves, Jackals and Dogs (C. Sillero-Zubiri, M. Hoffmann, and D.W. Macdonald, eds.). IUCN/Species Survival Commission Canid Specialist Group, Gland, Switzerland.

List, R., and D.W. Macdonald. 1998. Carnivora and their larger mammalian prey: Species inventory and abundance in the Janos-Nuevo Casas Grandes prairie dog complex. Revista Mexicana de Mastozoología, 3:95-112.

List, R., and D.W. Macdonald. 2003. Home range and habitat use of the kit fox (*Vulpes macrotis*) in a prairie dog (*Cynomys ludovicianus*) complex. Journal of Zoology, 259:1-5.

List, R., and V. Solís. 2008. Diagnóstico sobre el estado actual de la población de Bisonte americano (*Bison bison*) en la frontera entre México y Estados Unidos y recomendaciones para su conservación y manejo. Informe técnico no publicado, presentado al Instituto Nacional de Ecología-SEMARNAT.

List, R., G. Ceballos, and J. Pacheco. 1999. Distribution and conservation status of the North American porcupine (Erethizon dorsatum) in Mexico. The Southwestern Naturalist, 44:400-404.

List, R., P. Manzano-Fischer, and D.W. Macdonald. 2003. Coyote and kit fox diets in a prairie dog complex in Mexico. Pp. 183-188, in: The Swift Fox: Ecology and Conservation of Swift Foxes in a Changing World (M. Sovada and L. Carbyn, eds.). Canadian Plains Research Center, University of Regina, Canada.

List, R., G. Ceballos, C, Curtin, P.J.P. Gogan, J. Pacheco, and J. Truett. 2007. Historic distribution and challenges to bison recovery in the northern Chihuahuan desert. Conservation Biology, 21(6):1487-1494.

List, R., O. Pergmans, J. Pacheco, J. Cruzado, and G. Ceballos. 2010. Genetic divergence on the Chihuahuan meadow vole (Microtus pennsylvanicus chihuahuensis) and conservation implications of marginal population extinctions. Journal of Mammalogy, 91(5):1093-1101.

List, R., J. Pacheco, E. Ponce, R. Sierra-Corona, and G. Ceballos. 2010. The Janos Biosphere Reserve, northern Mexico. International Journal of Wilderness, 16:35-41.

Litvaitis, J.A., J.A., Sherburne, and J.A. Bissonette. 1986. Bobcat habitat use and home range size in relation to prey density. Journal of Wildlife Management, 50:110-117.

Litvaitis, J.A., C.L. Stevens, and W.W. Mautz. 1984. Age, sex and weight of bobcat in relation to winter diet. Journal of Wildlife Management, 48:632-635.

Livorell, B., P. Goaut, and C. Baudoin. 1993. A comparative study of social behaviour of two sympatric ground squirrels (*Spermophilus spilosoma* and *S. mexicanus*). Ethology, 93:236-246.

Lluch, B.D. 1969. El Lobo Marino de California *Zalophus californianus*, (Lesson, 1828) Allen, 1880: Observaciones sobre su Ecología y Explotación. Instituto Mexicano de Recursos Naturales Renovables, México.

Lochmiller, R.L., E.C. Hellgren, and W.E. Grant. 1984. Selected aspects of collared peccary (*Dicotyles tajacu*) reproductive biology in a captive Texas herd. Zoo Biology, 3:145-149.

Lochmiller, R.L., E.C. Hellgren, and W.E. Grant. 1986. Reproductive responses to nutritional stress in adult female collared peccaries. Journal of Wildlife Management, 50:295-300.

Lochmiller, R.L., E.C. Hellgren, R.M. Robinson, and W.E. Grant. 1984. Techniques for collecting blood from collared peccaries *Dicotyles tajacu* (L.). Journal of Wildlife Diseases, 20:47-50.

Lockard, R.B., and J.S. Lockard. 1971. Seed preference and buried seed retrieval of *Dipodomys deserti*. Journal of Mammalogy, 52:219-221.

Long, C.A. 1962. Records of reproduction for harvest mice. Journal of Mammalogy, 43:103-104.

Long, C.A. 1972. Taxonomic revision of the North American badger, *Taxidea taxus*. Journal of Mammalogy, 53:725-759.

Long, C.A 1973. *Taxidea taxus*. Mammalian Species, 26:1-4.

Loomis, R.B. 1969. Chiggers (Acarina: Trombiculidae) from vertebrates of the Yucatan Peninsula, Mexico. University of Kansas Publications, Museum of Natural History, 50:1-30.

Loomis, R.B. 1971. The genus *Euschoengastoides* (Acarina: Trombiculidae) from North America. Journal of Parasitology, 57:689-707.

Loomis, R.B., and J.K. Jones Jr. 1964. The northern yellow bat in Sinaloa, Mexico. Bulletin of the California Academy of Science, 63:32.

Loomis, R.B., and J.P. Webb 1971. A new intranasal chigger of the subgenus *Criticula*, genus *Microtrombicula* (Acarina: Trombiculidae) from Texas. Bulletin of the California Academy of Science, 70:1032-103.

López, C., and D.F. García. 2006. Murciélagos de la Sierra Tarahumara, Chihuahua, México. Acta Zoológica Mexicana (n.s.), 22:109-135.

López Austin, A. 1990. Los mitos del tlacuache. Caminos de la mitología mesoamericana. Alianza Editorial Mexicana, México.

López-Forment, C.W. 1968. Aspectos biológicos de la tuza *Cratogeomys tylorhinus tylorhinus* (Rodentia: Geomyidae) del Valle de México. Tesis de licenciatura, Universidad Nacional Autónoma de México, México.

López-Forment, C.W. 1981. Algunos aspectos ecológicos del murciélago *Balantiopteryx plicata plicata* Peters, 1867 (Chiroptera: Emballonuridae) en México. Anales del Instituto de Biología, UNAM, México, Serie Zoología, 50:673-699.

López-Forment, C.W. 1997. Algunas notas faunísticas del estudio de regurgitaciones de lechuza Tyto alba, en el sur del Valle de México. Pp. 175-181, in: Homenaje al Profesor Ticul Álvarez (L. Arroyo Cabrales and O. J. Polace, eds.). Museo Nacional de Antropología e Historia, Colección Científica, México D.F.

López-Forment, C.W., and F. Cervantes-Reza. 1981. Preliminary observations on the ecology of *Romerolagus diazi* in Mexico. Pp. 949-955, in: Proceedings of the World Lagomorphs Conference (K. Myers and C.D. MacInnes, eds.). University of Guelph, Guelph, Ontario.

López-Forment, C.W., and G. Urbano. 1977. Restos de pequeños mamíferos recuperados en regurgitaciones de lechuza, Tyto alba, en México. Anales del Instituto de Biología, UNAM, México, Serie Zoología, 48:231-242.

López-F., C.W., and G. Urbano-V. 1979. Historia natural del zorrillo manchado pigmeo, *Spilogale pygmaea*, con la descripción de una nueva subespecie. Anales del Instituto de Biología, UNAM, México, Serie Zoología, 50:721-728.

López-Forment, C.W., I.E. Lira, and C. Müdespacher. 1996. Mamíferos: Su biodiversidad en las islas mexicanas. AGT Editor, México D.F.

López-González, C.A. 2003. Murciélagos (Chiroptera) del Estado de Durango, México: composición, distribución y estado de conservación. Vertebrata Mexicana, 13:15-23.
López-González, C., and S.J. Presley. 2001. Taxonomic status of *Molossus bondae* J.A. Allen, 1904 (Chiroptera: Molossidae), with description of a new subspecies. Journal of Mammalogy, 82:760-774.
López-González, C., and L. Torres-Morales. 2004. Use of abandoned mines by two species of long-eared bats, genus *Corynorhinus* (Chiroptera: Vespertilionidae) in Durango, Mexico. Journal of Mammalogy, 85:989-994.
López O.E., and J. Ramírez-Pulido. 1999. VI. La zoología en México. Contribuciones, estado actual y perspectivas. Pp. 212-254, in: Las ciencias naturales en México (H. Aréchiga and C. Beyer, eds.). Fondo de Cultura Económica, México.
López-Vidal, J.C., and T. Álvarez. 1993. Biología de la rata montera *Neotoma mexicana*, en la Michilía, Durango, México. Pp. 185-195, in: Avances en el Estudio de los Mamíferos de México (R.A. Medellín and G. Ceballos, eds.). Publicaciones Especiales, Asociación Mexicana de Mastozoología, México.
López-Wilchis, R. 1989. Biología de *Plecotus mexicanus* (Chiroptera: Vespertilionidae) en el Estado de Tlaxcala, México. Tesis doctoral, Universidad Nacional Autónoma de México, México.
López-Wilchis, R. 1999. Murciélagos asociados a una colonia de *Corynorhinus mexicanus* (Chiroptera: Vespertilionidae). Vertebrata Mexicana, 5:9-16.
López-Wilchis, R., and J. López Jardines. 1995. Bases de datos para colecciones mastozoológicas. Ciencia, 46:298-308.
López-Wilchis, R., and J. López Jardines. 1998. Los Mamíferos de México Depositados en Colecciones de Estados Unidos y Canadá. Vol. 1. Universidad Autónoma Metropolitana Unidad Iztapalapa, México.
López-Wilchis, R., and J. López Jardines. 1999. Los Mamíferos de México Depositados en Colecciones de Estados Unidos y Canadá. Vol. 2. Universidad Autónoma Metropolitana Unidad Iztapalapa, México.
López-Wilchis, R., and J. López Jardines. 2000. Los Mamíferos de México Depositados en Colecciones de Estados Unidos y Canadá. Vol. 3. Universidad Autónoma Metropolitana Unidad Iztapalapa, México.
López-Wilchis, R., S. Gaona, and G. López-Ortega. 1992. Algunas consideraciones sobre los mamíferos terrestres de importancia cinegética. Ciencia, 43:245-260.
López-Wilchis, R., G. López Ortega, and S. Gaona. 1992. Mapa de zonas de importancia de mamíferos terrestres raros, amenazados y en peligro de extinción. In: Regionalización Mastofaunística. IV.8.9. Sección Naturaleza, Subsección Biogeografía. Atlas Nacional de México. Instituto de Geografía, UNAM e INEGI. México D.F.
Lorenzo, A. 1986. Xochicalco en la leyenda de los soles. Grupo editorial Miguel Angel Porrúa. México.
Lorenzo, C., F.A. Cervantes, and M.A. Aguilar. 1993. The chromosomes of some Mexican cottontail rabbits of the genus *Sylvilagus*. Pp. 129-136, in: Avances en el Estudio de los Mamíferos de México (R. Medellín and G. Ceballos, eds.). Publicaciones Especiales, Asociación Mexicana de Mastozoología, México.
Lott, D.F. 1979. Dominance relations and breeding rate in mature male American bison. Zeitschrift für Tierpsychology, 49:418-432.
Lotze, J.H., and S. Anderson. 1979. *Procyon lotor*. Mammalian Species, 119:1-8.
Lowery, G.H., Jr. 1974. The Mammals of Louisiana and its Adjacent Waters. Louisiana State University Press, Baton Rouge.
Lowery, G.H., and W.B. Davis. 1942. A revision of the fox squirrel of the lower Mississippi Valley and Texas. Occasional Papers of the Museum of Zoology, Louisiana State University, 9:153-172.
Lowry, M.S., and O.M. Maravilla-Chávez. 2005. Recent abundance of California sea lions in western Baja California, Mexico and the United States. Pp. 485-498, in: Proceedings of the 6th California Islands Symposium (D.K. Garcelon and C.A. Schwenn, eds.). National Park Service Technical Publication CHIS-05-01. Institute for Wildlife Studies, Arcata.
Lowry, M.S., B.S. Stewart, C.B. Heath, P.K. Yochem, and J.M. Francis. 1991. Seasonal and annual variability in the diet of California sea lions, *Zalophus californianus* at San Nicolas Island, California, 1981-86. Fishery Bulletin, 89:331-336.
Lubin, D.Y. 1983. *Tamandua mexicana* (oso jaceta, hormiguero, tamandua, banded anteater, lesser anteater). Pp. 494-496, in: Costa Rican Natural History (D.H. Janzen, ed.). University of Chicago Press, Chicago.
Lubin, D.Y., and G.G. Montgomery. 1981. Defenses of Nasutitermes termites (Isoptera, Termitidae) against Tamandua anteaters (edentata, Myrmecophagidae). Biotropica, 13:66-76.
Lubin, D.Y., G.G. Montgomery, and O.P. Young. 1977. Food resources of anteaters (Edentata: Myrmecophagidae). I. A year's census of arboreal nests of ants and termites on Barro Colorado Island, Panama Canal Zone. Biotropica, 9:26-34.
Lucas, J.L., and R.B. Loomis. 1968. The genus *Hexidionis* (Acarina, Trombiculidae) with the description of a new species from western Mexico. Bulletin of the Southern California Academy of Sciences, 67:233-264.
Ludlow, M.E., and M.E. Sunquist. 1987. Ecology and behavior of ocelots in Venezuela. National Geographic Research, 3:447-461.
Luis, J., and H. Granados 1990. Estudios sobre la biología del ratón de los volcanes (*Neotoma alstoni alstoni*) XXI: Capacidad reproductora de hembras silvestres en 15 apareamientos sucesivos. Archivos de Investigación Médica (México), 21:51-56
Luis, J., T. Arenas, G. López, and H. Granados. 1993. Estudios sobre la biología del ratón de los volcanes (*Neotoma alstoni alstoni*) XXVIII Existencia de estro postparto. Revista de la Facultad de Medicina, Universidad Nacional Autónoma de México, 36:41-44.
Lukens, D.W., Jr., and W.B. Davis. 1957. Bats of the Mexican state of Guerrero. Journal of Mammalogy, 38:1-14.
Luna, S.H., and G.C. López. 2006. Abundance and food habits of cougars and bobcats in the Sierra de San Luis, Sonora, México. USDA Forest Service Proceedings RMRS-P-36.
Luna Vega, I., and J.B. Llorente (eds.) 1993. Historia Natural del Parque Ecológico Estatal Omiltemi, Chilpancingo, Guerrero, México. CONABIO-UNAM, México D.F.
Lunaschi, L. 2002. Redescripción y comentarios taxonómicos sobre Ochoterenatrema lamda (*Digenea: Lecithodendriinae*), parásito de quirópteros en México. Anales del Instituto de Biología, UNAM, México, Serie Zoología, 73:11-18.
Lundy, W.E. 1954. Howlers. Natural History, 63:128-133.
Lux, C.A., A.S. Costa, and A.E. Dizon. 1996. Mitochondrial DNA population structure of the Pacific white-sided dolphin: Evolutionarily significant units and management units. Trabajo SC/48/SM6 presentado al Comité Científico de la International Whaling Comisión, Seattle.
Lyon, M.W., and W.H. Osgood. 1909. Catalogue of the type specimens of mammals in the United States National Museum, including the Biological Survey Collection. Smithsonian Institute Bulletin, United States Natural Museum, 62:1-325.
Maass, J.P., P. Balvanera, A. Castillo, G.C. Daily, H.A. Mooney, P. Ehrlich, M. Quesada, A. Miranda, V.J. Jaramillo, F. García-Oliva, A. Martínez-Yrizar, H. Cotler, J. López-Blanco, A. Pérez-Jiménez, A. Búrquez, C. Tinoco, G. Ceballos, L. Barraza, R. Ayala, and J. Sarukhán. 2005. Ecosystem services delivered by tropical dry forests: A case study from the Pacific coast of Mexico. Ecology and Society, 10(1). www.ecologyandsociety.org/viewarticle.php?id=1219.
Mab, D.M. 1975. Aspects of age, growth and reproduction of bobcats from Wyoming. Journal of Mammalogy, 56:177-198.
Macdonald, D. 2001. The New Encyclopedia of Mammals. Oxford University Press, Oxford.
Macdonald, D.W. 2006. The Encyclopedia of Mammals. Oxford University Press, Oxford.
MacClure, H.E. 1942. Summer activities of bats (genus *Lasiurus*) in Iowa. Journal of Mammalogy, 23:430-434.

McFadden, K.W., R.N. Sambrotto, R.A. Medellín, and M.E. Gomper. 2006 Feeding habits of endangered pygmy raccoons (*Procyon pygmaeus*) based on stable isotope and fecal analyses. Journal of Mammalogy, 87:501-509.

Macías-Sánchez, S.S., and M. Aranda. 1999. Análisis de la alimentación de la nutria *Lontra longicaudis* (Mammalia: Carnivora) en el sector del río Los Pescados, Veracruz, México. Acta Zoologica Mexicana (n.s.), 76:49-57.

Mackintosh, N.A. 1965. The Stock of Whales. Fishing News, London.

MacMahon, J.A. 1985. Deserts. National Audubon Society Nature Guides, New York.

Macmillen, R.E. 1964. Population ecology, water relations, and social behavior of a Southern California semidesert rodent fauna. University of California Publications in Zoology, 71:1-59.

Macswiney G.M.C., J. Sosa, and C. Selem-Salas. 2003. Ampliación en la distribución de *Eumops underwoodi* Goodwin, 1940 (Chiroptera: Molossidae) en la península de Yucatán, México. Revista Mexicana de Mastozoología, 7:55-57.

Macswiney G.M.C., A.B. Bolívar C., F.M. Clarke, and P.A. Racey. 2006. Nuevos registros de *Pteronotus personatus* y *Cynomops mexicanus* (Chiroptera) en el estado de Yucatán, México. Revista Mexicana de Mastozoología, 10:102-109.

MacSwiney, M.C., S. Hernández-Betancourt, and R. Avila-Flores. 2009. *Otonyctomys hatti*. Mammalian Species, 825:1-5.

Machado-Allison, C.E., and A. Barrera. 1964. Notas sobre *Megamblyopinus*, *Amblyopinus* y *Amblyopinodes*. Revista de la Sociedad Mexicana de Historia Natural, 5:173-191.

Madison, D.M. 1978. Movement indicators of reproductive events among female meadow voles as revealed by radiotelemetry. Journal of Mammalogy, 59:835-843.

Maehr, D.S., E. Darrell, and J.C. Roof. 1991. Social ecology of Florida panthers. National Geographic Research and Exploration, 7:414-431.

Maehr, D.S., E.C. Hellgren, R.L. Bingham, and D.L. Doan-Crider. 2001. Body mass of American black bears from Florida and México. The Southwestern Naturalist, 46:129-133.

Maffei, L., and A.J. Noss. 2008. How small is too small? Camera trap survey areas and density estimates for ocelots in the Bolivian Chaco. Biotropica, 40:71-75.

Maffei, L., A.J. Noss, E. Cuellar, and D. Rumiz. 2005. Ocelot (*Felis pardalis*) population densities, activity, and ranging behaviour in the dry forests of eastern Bolivia: Data from camera trapping. Journal of Tropical Ecology, 21:349-353.

Mailliard, J. 1924. A new mouse (*Peromyscus slevini*) from the Gulf of California. Proceedings of the California Academy of Sciences, 12:1219-1222.

Málaga-Alba, A. 1954. Vampire bats as a carrier of rabies. American Journal of Public Health, 44:909-918.

Málaga-Alba, A., and B. Villa-R. 1957. Algunas notas acerca de la distribución de los murciélagos de América del Norte, relacionados con el problema de la rabia. Anales del Instituto de Biología, UNAM, México, Serie Zoologíca, 27:529-569.

Málaga-Alba, A., H. Samame, and S. Gonzales. 1971. Constatación de un nido natural de rabia en El Alto Ucayali-Departamento de Loreto. Boletín de Divulgación No. 4, Centro de Investigaciones del Instituto Veterinario, Lima, Perú.

Malcolm, J.R. 1990. Estimation of mammalian densities in continuous forest north of Manaus. Pp. 339-357, in: Four Neotropical Rain Forests (A.H. Gentry, ed.). Yale University Press, New Haven, CT.

Maldonado, J.E., M. Cotera, E. Geffen, and R.K. Wayne. 1997. Relationships of the endangered Mexican kit fox (*Vulpes macrotis zinseri*) to North American arid-land foxes based on mitochondrial DNA sequence data. The Southwestern Naturalist, 42:460-470.

Maldonado, J.E., F. Orta Dávila, B.S. Stewart, E. Geffen, and R.K. Wayne. 1995. Intraspecific genetic differentiation in California sea lions (*Zalophus californianus*) from Southern California and the Gulf of California. Marine Mammal Science, 11(1):46-58.

Mandujano, S. 1992. Estimaciones de la densidad del venado cola blanca (*Odocoileus virginianus*) en un bosque tropical caducifolio de Jalisco. Tesis de maestría, Facultad de Ciencias, Universidad Nacional Autónoma de México, México.

Mandujano, S. 1997. Densidad poblacional de la ardilla gris del Pacífico (*Sciurus colliaei*) en un bosque tropical caducifolio de Jalisco. Revista Mexicana de Mastozoología, 2:90-96.

Mandujano, S. 1999. Variation in herd size of collared peccaries in a Mexican tropical forest. The Southwestern Naturalist, 44:199-204.

Mandujano, S., and S. Gallina. 1992. Tendencia poblacional del venado cola blanca en un bosque tropical caducifolio de Jalisco. Pp. 299-305, in: X Simposio sobre Fauna Silvestre Gral. M.V. Manuel Cabrera Valtierra. Universidad Nacional Autónoma de México, México.

Mandujano, S., and S. Gallina. 1993. Densidad del venado cola blanca basada en conteos en transectos en un bosque tropical de Jalisco. Acta Zoológica Mexicana (n.s.), 56:1-37.

Mandujano, S., and G. Hernández. 1990. Análisis de los factores ambientales que influyen sobre el nivel poblacional del venado cola blanca (*Odocoileus virginianus*) en el Parque Desierto de los Leones, D.F. Pp. 351-364, in: Áreas Naturales Protegidas en México y Especies en Peligro de Extinción (J.L. Camarillo and F. Rivera, eds.). Escuela Nacional de Estudios Profesionales-Iztacala, Universidad Nacional Autónoma de México, México.

Mandujano. S., and L.E. Martinez-Romero, 1997. Fruit fall caused by chachalacas (*Ortalis poliocephala*) on red mombin trees (*Spondias purpurea*): Impact on terrestrial fruit consumers, especially the white-tailed deer (*Odocoileus virginianus*). Studies on Neotropical Fauna and Environment, 32:1-3.

Mandujano, S., and L.E. Martínez-Romero. 2006. Pecarí de collar, *Pecari tajacu*. Pp. 411-413, in: Historia Natural de Chamela (F. Noguera, J. Vega, and R. Ayala, eds.).Universidad Nacional Autónoma de México, México D.F.

Mandujano, S., and V. Rico-Gray. 1991. Hunting use and knowledge of the biology of the white-tailed deer (*Odocoileus virginianus* Hays) by the Maya of central Yucatan, Mexico. Journal of Ethnobiology, 11:175-183.

Manning, R.W., and J.K. Jones Jr. 1989. *Myotis evotis*. Mammalian Species, 329:1-5.

Mansfield, A.W. 1985. Status of the blue whales, *Balaenoptera musculus*, in Canada. The Canadian Field-Naturalist, 99:417-420.

Manzano, P. 1993. Distribución geográfica y selección de hábitat de la ardilla voladora (*Glaucomys volans*) en México. Tesis de licenciatura, Facultad de Ciencias, Universidad Nacional Autónoma de México, México.

Manzano Fischer, P., R. List, and G. Ceballos. 1999. Grassland birds in prairie-dog towns in northwestern Chihuahua, Mexico. Studies in Avian Biology, 19:263-271.

March, I.J. 1987. Los lacandones de México y su relación con los mamíferos silvestres: Un estudio etnozoológico. Biótica, 12:43-56.

March, I.J. 1990. Evaluación de hábitat y situación actual del pecarí de labios blancos *Tayassu pecari* en México. Tesis de maestría, Programa Regional en Manejo de Vida Silvestre para Mesoamérica y El Caribe, Universidad Nacional, Heredia, Costa Rica.

March, I.J. 1991. Monographie des WeiBbartpekaris (*Tayassu pecari*). Bongo, Berlín, 18:151-170.

March, I.J. 1999. El Tapir: Un mamífero en peligro de extinción en México. Centro de Estudios para la Conservación de los Recursos Naturales, A.C., San Cristóbal de las Casas, Chiapas, México.

March, I.J. 2005. Tayassu tajacu. Pp. 522-524, in: Los Mamiferos Silvestres de Mexico (G. Ceballos and G. Oliva, eds.). Fondo de Cultura-CONABIO, México D.F.

March, I.J., and M. Aranda. 1992. Mamíferos de la Selva Lacandona. Pp. 197-216, in: Reserva de la Biosfera Montes Azules, Selva Lacandona: Investigación para su Conservación (M.A. Sánchez-V. and M.A. Ramos, eds.). Publicaciones Ocasionales Ecosfera, México.

Marine Mammal Commission. 1993. Annual Report of the Marine Mammal Commission, Calendar Year 1992. Report to the Congress, Washington, DC.

Marineros, L., and G.F. Martínez. 1988. Mamíferos Silvestres de Honduras. Asociación Hondureña de Ecología, Tegucigalpa.

Marmontel, M., S.R. Humphrey, and T.J. O'Shea. 1997. Population viability analysis of the Florida manatee (Trichechus manatus latrirostris), 1976-1991. Conservation Biology, 11:467-481.

Márquez-Huitzil, R. 1994. Distribución geográfica y ecológica de Geomys tropicalis, una tuza endémica de México. Tesis de licenciatura, Facultad de Ciencias, Universidad Nacional Autónoma de México, México.

Marsh, H., and L.W. Lefebvre. 1994. Sirenian status and conservation efforts. Aquatic Mammals, 20:155-170.

Marshall, L.G. 1978. Chironectes minimus. Mammalian Species, 109:1-6.

Marshall, L.G., J.A. Case, and M.O. Woodbourne. 1990. Phylogenetic relationships of the families of marsupials. Pp. 433-505, in: Current Mammalogy (H.H. Genoways, ed.). Vol. 2. Plenum Press, New York.

Martín del Campo, R. 1941. Ensayo de interpretación del libro undécimo de la historia de Sahagún, III. Mamíferos. Anales del Instituto de Biología, UNAM, México, Serie Zoologíca, 12:489-506.

Martin, C.O., and D.J. Schmidly. 1982. Taxonomic review of the pallid bat, Antrozous pallidus (Le Conte). Special Publications of the Museum, Texas Tech University, 18:1-48.

Martin, F.E., and S. Álvarez. 1982. Crecimiento y desarrollo en el laboratorio de Neotomodon alstoni (Rodentia:Cricetidae). Anales de la Escuela Nacional de Ciencias Biológicas, 26:55-84.

Martin, P.S. 1955. Zonal distribution of vertebrates in a Mexican cloud forest. The American Naturalist, 89:347-361.

Martin, P.S. 1958. A biogeography of reptiles and amphibians in the Gomez Farias region, Tamaulipas, Mexico. Miscellaneous Publications of the Museum of Zoology, University of Michigan, 101:1-102.

Martin, P.S. 1960. Southwestern animal communities in the late Pleistocene. Unpublished manuscript, Department of Geosciences, University of Arizona, Tucson.

Martin, P.S., and B.E. Harrell. 1957. The Pleistocene history of temperate biotas in Mexico and Eastern United States. Ecology, 38(3):472-480.

Martin, P.S., and R.G. Klein (eds.). 1984. Quaternary Extinctions: A Prehistoric Revolution. University of Arizona Press, Tucson.

Martin, R.A. 1987. Notes on the classification and evolution of some North American fossil Microtus (Mammalia: Rodentia). Journal of Vertebrate Paleontology, 7:270-283.

Martínez, M. 1979. Catálogo de nombres vulgares y científicos de las plantas mexicanas. FCE. México.

Martínez, L., and B. Villa Ramírez. 1938. Contribuciones al conocimiento de los murciélagos de México. Anales del Instituto de Biología, UNAM, México, Serie Zoologíca, 9:339-360.

Martínez-Coronel, M., and R. Vidal-López. 1999. Nota de distribución de dos murciélagos molósidos en Chiapas, México. Vertebrata Mexicana, 4:17-19.

Martínez-Coronel, M., J. Ramírez-Pulido, and T. Álvarez. 1991. Variación intrapoblacional de *Peromyscus melanotis* (Rodentia: Muridae) en el Eje Volcánico Transverso, México. Acta Zoológica Mexicana (n.s.), 47:1-51.

Martínez del R., C., and L.E. Eguiarte. 1987. Bird visitation to Agave salmiana: Comparisons among hummingbirds and perching birds. The Condor, 89:357-363.

Martinez-Gallardo, R., and V. Sanchez-Cordero. 1993. Dietary value of fruits and seeds to spiny pocket mice, *Heteromys desmarestianus* (Heteromyidae). Journal of Mammalogy, 74:436-442.

Martínez-Gallardo, R., and V. Sánchez-Cordero. 1997. Historia natural de algunas especies de mamíferos terrestres. Pp. 591-610, in: Historia Natural de Los Tuxtlas (E. González-Soriano, R. Dirzo, and R. C. Vogt, eds.). Instituto de Biología, CONABIO e Instituto de Ecología, México.

Martínez Gracida, M. 1891. Flora y Fauna del Estado Libre y Soberano de Oaxaca, Recopiladas. Imprenta del Estado, Oaxaca.

Martínez-Meyer, E., M. Martínez-Morales, and J. Sosa-Escalante. 1998. First record of *Potos flavus* (Carnivora: Procyonidae) from Cozumel Island, Quintana Roo, Mexico. The Southwestern Naturalist, 43(1):101-102.

Martínez-Morales, M.A., and A.D. Cuarón. 1999. Boa constrictor, an introduced predator threatening the endemic fauna on Cozumel Island, Mexico. Biodiversity and Conservation, 8:957-963.

Martínez-Romero, L.E., and S. Mandujano. 1995. Hábitos alimentarios del pecarí de collar (*Pecari tacaju*) en un bosque tropical caducifolio de Jalisco, México. Acta Zoologica Mexicana, 64:1-20.

Martínez-Vazquez, J. 1987. Estudio sobre la variación estacional de la dieta del zacatuche o teporingo *Romerolagus diazi* (Mammalia: Lagomorpha). Tesis de licenciatura, Facultad de Ciencias, Universidad Nacional Autónoma de México, México.

Martínez Kú, D.H., G. Escalona Segura, and J.A. Vargas Contreras. 2007. Primer registro del zorrillo manchado del Sur *Spilogale angustifrons* Howell 1902 para el estado de Campeche, México. Acta Zoológica Mexicana (n.s.), 23:175-177.

Mascarello, J.T. 1978. Chromosomal, biochemical, mensural, penile, and cranial variation in desert woodrats (*Neotoma lepida*). Journal of Mammalogy, 59:477-495.

Mascarello, J.T., and K. Boyes. 1980. C- and G-banded chromosomes of *Ammosperphilus insularis* (Rodentia: Sciuridae). Journal of Mammalogy, 61:714-716.

Mascarello, J.T., and T.C. Hsu. 1976. Chromosome evolution in woodrats, genus *Neotoma* (Rodentia: Cricetidae). Evolution, 30:152-169.

Mascarello, J.T., J.W. Warner, and R.J. Baker. 1974. A chromosome banding analysis of the mechanisms involved in the karyological divergence of *Neotoma phenax* (Merriam, 1903) and *Neotoma micropus* Baird, 1855. Journal of Mammalogy, 55:831-834.

Mason, C., and S.M. MacDonald. 1986. Otters: Ecology and Conservation. Cambridge University Press, London.

Mason-Romo, E., E.P. Villa-Mendoza, G.A. Rendón, and D. Valenzuela. 2008. Primer registro del Pecari de collar (*Pecari tajacu*) en el estado de Morelos. Revista Mexicana de Mastozoología, 12:170-175.

Mass, J., R. Patron, A. Suarez, S. Blanco, G. Ceballos, C. Galindo, and A. Pescador. 1981. Ecología de la Estación Experimental Zoquiapan (Descripción General, Vegetación y Fauna). Universidad Autónoma de Chapingo, Dirección de Difusión Cultural, Departamento de Bosques, Chapingo, Estado de México.

Masterson, L. 2006. Living with Bears. PixyJack Press, Masonville, CO.

Mate, B.R. 1977. Aerial censusing of pinnipeds in the eastern Pacific for assessment of population numbers, migratory distribution, rookery stability effort, and recruitment. Report to the U.S. Marine Mammal Commission NTIS No. PB265859, Springfield, VA.

Matocha, K. 1968. A study of certain aspects of the reproduction, growth and development of the Mexican ground squirrel (*Citellus mexicanus*) in southern Texas. Master's thesis, Texas A&M University, Kingsville.

Matocq, M.D. 2002. Morphological and molecular analysis of a contact zone in the *Neotoma fuscipes* species complex. Journal of Mammalogy, 83:866-883.

Matson, J.D. 1977. Records of mammals from Zacatecas, Mexico. Journal of Mammalogy, 58:110.

Matson, J.O. 1980. The status of banner-tailed kangaroo rats, genus *Dipodomys*, from central Mexico. Journal of Mammalogy, 61:563-566.

Matson, J.O., and R.H. Baker. 1986. Mammals of Zacatecas. Special Publications of the Museum, Texas Tech University, 24:1-88.

Matson, J.O., and D.P. Christian. 1977. A laboratory study of seed caching in two species of *Liomys* (Heteromyidae). Journal of Mammalogy, 58:670-671.

Matson, J.O., and D.R. Patten. 1975. Notes on some bats from the state of Zacatecas, Mexico. Natural History Museum of Los Angeles Country, Contributions in Science, 263:1-12.

Matsuzaki, T., M. Kamiya, and H. Suzuki. 1985. Gestation period of the laboratory reared volcano rabbit (*Romerolagus diazi*). Experimentation Animale, 34:63-66.

Matsuzaki, T., M. Saito, and H. Kamiya. 1982. Breeding and rearing of the volcano rabbit (*Romerolagus diazi*) in captivity. Experimentation Animale, 31:185-188.

Matthey, R. 1973. The chromosome formulae of eutherian mammals. Pp. 531-616, in: Cytotaxonomy and Vertebrate Evolution (A.B. Chiarelli and E. Campanna, eds.). Academic Press, London.

Maxwell, M.H., and L.N. Brown. 1968. Ecological distribution of rodents on the high plains of eastern Wyoming. The Southwestern Naturalist, 13:143-158.

May, L.A. 1976. Fauna de vertebrados del gran Desierto, Sonora, México. Anales del Instituto de Biología, UNAM, México, Serie Zoología, 47:143-182.

Mayer, J.J., and P.N. Brandt. 1982. Identity, distribution and natural history of the peccaries, Tayassuidae. Pp. 433-455, in: Mammalian Biology in South America (M.A. Mares and H.H. Genoways, eds.). Special Publication Series 6, Pymatuning Laboratory of Ecology, University of Pittsburgh, Linesville.

Mayer, J.J., and R.M. Wetzel. 1987. *Tayassu pecari*. Mammalian Species. 293:1-7.

Mayr, E. 1978. Origin and history of some terms in systematics and evolutionary biology. Systematic Zoology, 27:83-88.

Mazzoti, L. 1946. Hallazgo del *Trypanosoma vespertilionis* en murciélagos mexicanos. Revista de la Sociedad Mexicana de Historia Natural, 7:49-50.

McBee, K., and R.J. Baker. 1982. *Dasypus novemcinctus*. Mammalian Species, 162:1-9.

McBride, R. 1980. The Mexican wolf (*Canis lupus baileyi*): A historical review and observations on its status and distribution. U.S. Fish and Wildlife Service, Progress Report, Albuquerque, NM.

McCann, C. 1975. A study of the genus *Berardius* Duvernoy. Scientific Reports of the Whales Research Institute, 27:111-137.

McCarthy, T.J. 1982a. Bat records from the Caribbean lowlands of the El Peten, Guatemala. Journal of Mammalogy, 63:683-685.

McCarthy, T.J. 1982b. *Chironectes*, *Cyclopes*, *Cabassous* and probably *Cebus* in southern Belize. Mammalia, 46:397-400.

McCarthy, T.J. 1987. Distributional records of bats from the Caribbean lowlands of Belize and adjacent Guatemala and Mexico. Pp. 137-162, in: Studies in Neotropical Mammalogy (B.D. Patterson and R.M. Timm, eds.). Fieldiana-Zoology, N.S. 39, Field Museum of Natural History, Chicago.

McCarthy, T.J., and N.A. Bitar. 1983. New bat records (*Enchisthenes* and *Myotis*) from the Guatemalan central highlands. Journal of Mammalogy, 64:526-527.

McCarty, R. 1978. *Onychomys leucogaster*. Mammalian Species, 87:1-6.

McClearn, D. 1992. Locomotion, posture and feeding behavior of kinkajous, coatis, and raccoons. Journal of Mammalogy, 73:245-261.

McClenaghan, L.R., Jr. 1977. Genic variability, morphological variation and reproduction in central and marginal populations of *Sigmodon hispidus*. Unpublished PhD dissertation, University of Kansas, Lawrence.

McCord, C.M. 1974. Selection of winter habitat by bobcats (*Lynx rufus*) on the Quabbin Reservation, Massachusetts. Journal of Mammalogy, 55:428-437.

McCord, C.M., and J.E. Cardoza. 1982. Bobcat and lynx (*Felis rufus* and *Felis lynx*). Pp. 728-766, in: Wild Mammals of North America, Biology Management, Economics (J.A. Chapman and G.A. Feldhamer, eds.). Johns Hopkins University Press, Baltimore.

McCoy, M.B., C. Vaughan, M. Rodríguez, and D. Kitchen, 1990. Seasonal movement, home range, activity and diet of collared peccaries (*Tayassu tajacu*) in Costa Rican dry forest. Vida Silvestre Neotropical, 2:6-90.

McCracken, G.F. 1984. Communal nursing in Mexican free-tailed bat maternity colonies. Science, 223:1090-1091.

McCracken, G.F. 1986. Why are we losing our Mexican free-tailed bats? Bats Newsletter (Bat Conservation International), 3:1-4.

McDaniel, J., 1979. Report from Florida. Pp. 34-37, in: The Black Bear in Modern North America (D. Burk, ed.). Boone and Crockett Club and Amwell Press, Clinton, NJ.

McDonough, M.M., L.K. Ammerman, R.M. Timm, H.H. Genoways, P.A. Larsen, and R.J. Baker. 2008. Speciation within bonneted bats (genus *Eumops*): The complexity of morphological, mitochondrial, and nuclear data sets in systematics. Journal of Mammalogy, 89:1306-1315.

McFadden, K.W. 2004. The ecology, evolution and natural history of the endangered carnivores of Cozumel Island, Mexico. PhD dissertation, Columbia University.

McFadden, K.W., M.E. Gompper, D. Valenzuela, and J.C. Morales. 2008. Evolutionary history of the critically endangered Cozumel dwarf carnivores inferred from mitochondrial DNA analyses. Journal of Zoology, 276:176-186.

McFadden, K.W., D. García-Vasco, A.D. Cuarón, D. Valenzuela-Galván, R.A. Medellín, and M.E. Gompper. 2009. Vulnerable island carnivores: The endangered endemic dwarf procyonids from Cozumel Island. Biodiversity Conservation Online.

McFadden, K.W., D. García-Vasco, A. Cuarón-Orozco, D. Valenzuela-Galván, R.A. Medellín, and M. E. Gompper. 2010. Vulnerable island carnivores: The endangered endemic dwarf procyonids from Cozumel Island. Biodiversity and Conservation, 19:491-502.

McFadden, K.W., R.N. Sambrotto, R.A. Medellín, and M.E. Gompper. 2006. Feeding habits of endangered pygmy raccoons (*Procyon pygmaeus*) based on stable isotope and fecal analyses. Journal of Mammalogy, 87:501-509.

McFadden, K.W., S.E. Wade, E.J. Dubovi, and M.E. Gompper. 2005. A serology and fecal parasitology survey of the critically endangered pygmy raccoon (*Procyon pygmaeus*). Journal of Wildlife Diseases, 41:615-617.

McGee, W.J. 1971 [1980]. Los Seris. Sonora, México. Instituto Nacional Indigenista, México.

McGhee, M.E., and H.H. Genoways. 1978. *Liomys pictus*. Mammalian Species, 83:1-5.

McGrew, J.C. 1979. Vulpes macrotis. Mammalian Species, 123:1-6.

McHugh, T. 1958. Social behavior of American buffalo (*Bison bison bison*). Zoologica, 43:1-40.

McKenna, M.C., and S.K. Bell. 1997. Classification of Mammals above the Species Level. Columbia University Press, New York.

McKillop, H.I. 1985. Prehistoric exploitation of the manatee in the Maya and circum-Caribbean areas. World Archaeology, 16:337-353.

Mckinney, B.R., and M.T. Pittman. 2000. Habitat, diet, home range, and seasonal movement of resident and relocated black bears in west Texas. Texas Parks and Wildlife Department, Project WER57-STATE, Austin.

McLean, L.M., S.T. McCay, and M.J. Lovallo. 2005. Influence of age, sex and time of year on diet of bobcat (*Lynx rufus*) in Pennsylvania. The American Midland Naturalist, 153:450-453.

McLellan, L.J. 1984. A morphometric analysis of Carollia (Chiroptera, Phyllostomidae). American Museum Novitates, 2791:1-35.

McLellan, M.E. 1926. Expedition to the Revillagigedo Islands, Mexico, in 1925. VI. The birds and mammals. Proceedings of the California Academy of Science, 15:279-322.

McLellan, M.E. 1927. Pizonyx vivesi on Isla Partida, Gulf of California. Journal of Mammalogy, 8:243.

McManus, J.J. 1974. *Didelphis virginiana*. Mammalian Species, 40:1-6.

McNab, B.K. 1978. The comparative energetic of Neotropical marsupials. Journal of Comparative Physiology, 125:115-128.

McNab, B.K. 1985. Energetics, population biology, and distribution of xenarthrans, living and extinct. Pp. 219-232, in: The Evolution and Ecology of Armadillos, Sloths, and Vermilinguas (G.G. Montgomery, ed.). Smithsonian Institution Press, Washington, DC.

McPherson, A.B. 1985. A biogeographical analysis of factors influencing the distribution of Costa Rican rodents. Brenesia, 23:97-273.

McPherson, A.B., R. Zeledón, and S. Shelton. 1985. Comments on the status of *Metachirus nudicaudatus dentatus* (Goldman, 1912) in Costa Rica. Brenesia, 24:375-377.

McWhirter, D.W., P.L. Dalby, and J.H. Asher Jr. 1974. A new mutant in the yellow-bellied cotton rat. Journal of Heredity, 65:316-319.

Mead, I.J., and T.R. Van Devender. 1981. Late Holocene diet of *Bassariscus astutus* in the Grand Canyon, Arizona. Journal of Mammalogy, 62:439-442.

Mead, J.G. 1989a. Beaked whales of the genus *Mesoplodon*. Pp. 349-413, in: Handbook of Marine Mammals, vol. 4, River Dolphins and the Larger Toothed Whales (S.H. Ridgway and R.L. Harrison, eds.). Academic Press, London.

Mead, J.G. 1989b. Bottlenose whales *Hyperoodon ampullatus* (Forster, 1770) and *Hyperoodon planifrons* Flower, 1882. Pp. 321-349, in: Handbook of Marine Mammals, vol. 4, River Dolphins and the Larger Toothed Whales (S.H. Ridgway and R. Harrison, eds.). Academic Press, London.

Mead, J.G., and R.L. Brownell Jr. 2005. Order Cetacea. Pp. 723-743, in: Mammal Species of the World: A Taxonomic and Geographic Reference (D.E. Wilson and D.M. Reeder, eds.). 3rd ed. Johns Hopkins University Press, Baltimore.

Mead, J.G., W.A. Walker, and W.J. Houck. 1982. Biological observations on *Mesoplodon carlhubbsi* (Cetacea: Ziphiidae). Smithsonian Contributions to Zoology, 1-40.

Mead, R.A. 1968. Reproduction in western forms of the spotted skunk (genus *Spilogale*). Journal of Mammalogy, 48:606-616.

Meadows, A. 1991. Burrows and burrowing animals: An overview. Symposia of the Zoological Society of London, 63:1-13.

Meagher, M. 1973. The bison of Yellowstone National Park. Natural Park Service Science Monographs, 1:1-161.

Meagher, M. 1986. *Bison bison*. Mammalian Species, 266:1-8.

Mearns, E.A. 1890. Descriptions of supposed new species and subspecies of mammals from Arizona. Bulletin of the American Museum of Natural History, 2:277-307.

Mearns, E.A. 1895. Preliminary description of a new subgenus and six new species and subspecies of hares from the Mexican border of the United States. Proceedings of the United States National Museum, 18:551-565.

Mearns, E.A. 1896. Preliminary diagnosis of new mammals from the Mexican border of the United States. Proceedings of the United States National Museum, 18:443-447.

Mearns, E.A. 1907. Mammals of the Mexican Boundary of the United States. Bulletin 56, United States National Museum, Washington, DC.

Mech, L.D. 1970. The Wolf: The Ecology and Behavior of an Endangered Species. Natural History Press, Garden City, NY.

Mech, L.D. 1974. *Canis lupus*. Mammalian Species, 32:6.

Mech, L.D., and L. Boitani. 2004. Canis lupus. Pp. 124-129, in: Canids: Foxes, Wolves, Jackals and Dogs. Status Survey and Conservation Action Plan (C. Sillero-Zubin, M. Hoffmann, and D. MacDonald, eds.). IUCN World Conservation Union, Gland, Switzerland.

Mech, L.D., and Boitani, L. (eds.). 2004. Wolves: Behavior, Ecology and Conservation. University of Chicago Press, Chicago.

Medellín, R.A. 1983. Tonatia bidens and Mimon crenulatum in Chiapas, Mexico. Journal of Mammalogy, 64:150.

Medellín, R.A. 1986. Murciélagos de Chajul. Tesis de licenciatura, Facultad de Ciencias, Universidad Nacional Autónoma de México, México.

Medellín, R.A. 1988. Prey of *Chrotopterus auritus*, with notes on feeding behavior. Journal of Mammalogy, 69:841-844.

Medellín, R. A. 1989. *Chrotopterus auritus*. Mammalian Species, 343:1-5.

Medellín, R.A. 1991. Ecomorfología del cráneo de cinco didélfidos: tendencias, divergencias e implicaciones. Anales del Instituto de Biología, UNAM, México, Serie Zoología, 62:269-286.

Medellín, R.A. 1992. Community, ecology and conservation of mammals in a Mayan tropical rainforest and abandoned agricultural fields. PhD dissertation, University of Florida, Gainesville.

Medellín, R.A. 1993. Estructura y diversidad de una comunidad de murciélagos en el trópico húmedo mexicano. Pp. 333-354, in: Avances en el Estudio de los Mamíferos de México (R.A, Medellín and G. Ceballos, eds.). Asociación Mexicana de Mastozoología, Publicaciones Especiales.

Medellín, R.A. 1994. Seed dispersal of *Cecropia obtusifolia* by two species of opossums in the Selva Lacandona, Chiapas, Mexico. Biotropica, 26:400-407.

Medellín, R.A., and H.T. Arita. 1989. *Tonatia evotis* and *Tonatia silvicola*. Mammalian Species, 334:1-5.

Medellín, R.A., and H.V. Bárcenas. 2010. Estimación de la densidad y dieta del lince (Lynx rufus) en seis localidades de México. Informe final SNIBCONABIO proyectos No. ES003 y ES009. México D.F.

Medellín, R.A.. and G. Ceballos (eds.) 1993. Avances en el Estudio de los Mamíferos de México. Publicaciones Especiales 1, Asociación Mexicana de Matozoología, A.C., México.

Medellín, R.A., and W. López-Forment. 1986. Las cuevas: un recurso compartido. Anales del Instituto de Biología, UNAM, México, Serie Zoología, 56:1027-1034.

Medellín, R.A., and K.H. Redford. 1992. The role of mammals in Neotropical forest-savanna boundaries. Pp. 519-548, in: Nature and Dynamics of Forest-Savanna Boundaries (P.A. Furley, J. Proctor, and J.A. Ratter, eds.). Chapman and Hall, London.

Medellín, R.A., H.T. Arita, and O. Sánchez. 1997. Identificación de los Murciélagos de México, Clave de Campo. Publicaciones Especiales No. 2, Asociación Mexicana de Mastozoología, A.C., México.

Medellín R., G. Ceballos, and H. Zarza. 1998. *Spilogale pygmaea*. Mammalian Species, 600:1-3.

Medellín, R.A., M. Equihua, and M.A. Almin. 2000. Bat diversity and abundance as indicators of disturbance in Neotropical rainforests. Conservation Biology, 14(6):1666-1675.

Medellín, R.A., A.L. Gardner, and J.M. Aranda. 1998. The taxonomic status of the Yucatan brown brocket, *Mazama pandora* (Mammalia: Cervidae). Proceedings of the Biological Society of Washington, 111:1-14.

Medellín, R.A., D.E. Wilson, and D. Navarro. 1985. *Micronycteris brachyotis*. Mammalian Species, 251:1-4.

Medellín, R.A., G. Cancino Z., A. Clemente M., and R. Guerrero V. 1992. Noteworthy records of three mammals from Mexico. The Southwestern Naturalist, 37:427-429.

Medellín, R.A., D. Navarro, W.B. Davis, and V.J. Romero. 1983. Notes on the biology of *Micronycteris brachyotis* (Dobson) (Chiroptera), in southern Veracruz, Mexico. Brenesia, 21:7-11.

Medellín, R.A., G. Urbano-Vidales, O. Sánchez-Herrera, G. Téllez-Girón, and H. Arita. 1986. Notas sobre murciélagos del este de Chiapas. The Southwestern Naturalist, 4:532-535.

Medellín, R.A., C. Manterola, M. Valdéz, D.G. Hewitt, D. Doan-Crider, and T.E. Fulbright. 2005. History, ecology, and conservation of the pronghorn antelope, bighorn sheep, and black bear in Mexico. In: Biodiversity, Ecosystems and Conservation in Northern Mexico (J.L. Cartron, G. Ceballos, and R. Felger, eds.). Oxford University Press, Oxford.

Medellín, R.A., C. Equihua, C. Chetkiewics, A. Rabinowitz, P. Crawshaw, K. Redford, J.G. Robinson, E. Sanderson, and A. Taber (eds.). 2002. El jaguar en el nuevo milenio. Fondo de Cultura Económica, Universidad Nacional Autónoma de México y Wildlife Conservation Society, México D.F.

Medina T., J.G., and J.A. De la Cruz C. 1976. Ecología y control del perrito de la pradera mexicano (*Cynomys mexicanus merriam*). Monografía Técnico Científica, 2(5):365-418.

Medrano-González, L., C.S. Baker, M.R. Robles-Saavedra, et al. 2001. Transoceanic population genetic structure of humpback whales in North and South Pacific. Memoirs of the Queensland Museum, 47:465-479.

Medrano-González, L., A. Aguayo-Lobo, J. Urbán-Ramírez, and C.S. Baker. 1995. Diversity and distribution of mitochondrial DNA lineages among humpback whales, *Megaptera novaeangliae*, in the Mexican Pacific Ocean. Canadian Journal of Zoology, 73(9):1735-1743.

Medrano González, L., H. Rosales-Nanduca, M.J. Vázquez-Cuevas, J. Urbán-Ramírez, et al.. 2008. Diversidad, composiciones comunitarias y estructuras poblacionales de la mastofauna marina en el Pacífico mexicano y aguas adyacentes. Pp. 469-492, in: Avances en el Estudio de los Mamíferos de México II (C. Lorenzo, E. Espinoza, and J. Ortega, eds.). Asociación Mexicana de Mastozoología, A.C., San Cristóbal de las Casas.
Meier, T.P. 1983. Relative brain size within North American sciuridae. Journal of Mammalogy, 64:246.
Meléndez, C., R. Paredes, and M. Valdés. 2006. Poblaciones del Berrendo en Sonora. Pp. 71-82, in: El Berrendo en México, Acciones de Conservación 2006 (M. Valdés, E. de La Cruz, E. Peters, and E. Pallares, eds.). INE-SEMARNAT, México.
Meliani, P. 1991. Manual sobre los rumiantes silvestres de México. Tesis de licenciatura, Facultad de Medicina Veterinaria y Zootecnia, Universidad Nacional Autónoma de México, México.
Mellink, E. 1986. Some ecological characteristics of three dry farming systems in the San Luis Potosi Plateau, Mexico. PhD dissertation, University of Arizona, Tucson.
Mellink, E. 1992a. Status de los Heterómidos y Cricétidos Endémicos del Estado de Baja California. Comunicaciones Académicas, Serie Ecología, CICESE, Ensenada, B.C., México.
Mellink, E. 1992b. The status of *Neotoma anthonyi* (Rodentia, Muridae, Cricetidae) of Todos los Santos Island, Baja California, Mexico. Bulletin of the Southern California Academy of Sciences, 91:137-140.
Mellink, E. 1993a. Biological conservation of Isla de Cedros, Baja California, Mexico: Assessing multiple threats. Biodiversity and Conservation, 2:62-69.
Mellink, E. 1993b. The president spoke. Pp. 201-220, in: Counting Sheep, 20 Ways of Seeing Desert Bighorn (G.P. Nabhan, ed.). University of Arizona Press, Tucson.
Mellink, E. 1995. Status of the muskrat in the Valle de Mexicali and Delta del Río Colorado, Mexico. California Fish and Game, 8:33-38.
Mellink, E. 1998. Ampliación de la distribución del tlacuache (*Didelphis virginiana*) en Baja California. Revista Mexicana de Mastozoología, 3:148.
Mellink, E., and J. Contreras. 1993. Western gray squirrels in Baja California. California Fish and Game, 79:169-170.
Mellink, E., and J. Luévano. 1998. Status of beavers (*Castor canadiensis*) in Valle de Mexicali, Mexico. Bulletin of the Southern California Academy of Sciences, 97:115-120.
Mellink, E., and H. Madrigal. 1993. Ecology of Mexican prairie dogs, *Cynomys mexicanus,* in El Manantial, northeastern Mexico. Journal of Mammalogy, 74:631-635.
Mellink, E., and P.S. Martin. 2001. Mortality of cattle on a desert range: Paleobiological implications. Journal of Arid Environments, 49:671-675.
Mellink, E., J.R. Aguirre R., and E. García Moya. 1986. Utilización de la Fauna Silvestre en el Altiplano Potosino-Zacatecano. Colegio de Postgraduados, Chapingo, México.
Mellink, E., G. Ceballos, and J. Luevano. 2002. Population demise and extinction threat of the Angel de la Guarda deer mouse (*Peromyscus guardia*). Biological Conservation, 108:107-111.
Mellink, E., A. Orozco-Meyer, B. Contreras, and M. González-Jaramillo. 2002. Observations on nesting seabirds and insular rodents in the Middle Sea of Cortés in 1999 and 2000. Bulletin of the Southern California Academy of Sciences, 101:28-35.
Melquist, W.E., and A.E. Dronkert. 1987. River otter. Pp. 627-641, in: Wild Furbearer Management and Conservation in North America (M. Novak, J.A. Baker, M.E. Obbard, and B. Malloch, eds.). Ontario Ministry of Natural Resources and the Ontario Trappers Association, Toronto, Ontario, Canada.
Melquist, W.E., and M.G. Hornocker. 1983. Ecology of river otters in west-central Idaho. Wildlife Monographs, 83:1-60.
Mena, H. 2007. Presencia de Leptospira spp. y moquillo canino en poblaciones de perros y carnívoros silvestres en la Isla Cozumel. Tesis de maestría en ciencias, Universidad Nacional Autónoma de México, Ciudad de México, México.
Mendel, F.C. 1976. Postural and locomotor behavior of *Alouatta palliata* on various substrates. Folia Primatologica, 26:36-53.
Mendes, S.L. 1985. Uso do espaço, padrões de atividades diárias e organização social de *Alouatta fusca* (Primates, Cebidae) em Caratinga-MG. Tesis de maestría, Universidad de Brasilia, Brasilia.
Mendoza D., M.A. 1997. Efecto de la adición de alimento en la dinámica poblacional y estructura de comunidades de pequeños roedores en un bosque tropical caducifolio. Tesis de maestría (ecología y ciencias ambientales). Facultad de Ciencias, UNAM, México.
Menegaux, M.A. 1901. Description d'une variété et d'une espèce nouvelles de chiroptères rapportées du Mexique par M. Diguet. Bulletin du Muséum d'Histoire Naturelle, Paris, Serie, 7:321-327.
Menu, H. 1984. Révision du statut de *Pipistrellus subflavus* (F. Cuvier, 1832). Proposition d'un taxon générique nouveau: *Perimyotis* nov. gen. = Revision of the status of *Pipistrellus subflavus* (F. Cuvier, 1832). A new generic name is proposed: *Perimyotis* gen. nov. Mammalia, 48:409-416.
Mercure, A., K. Ralls, K.P. Koepfli, and R.K. Wayne. 1993. Genetic subdivisions among small canids: Mitochondrial DNA differentiation of swift, kit, and Arctic foxes. Evolution, 47:1313-1328.
Meritt, D.A. 1985. Naked-tailed armadillos, *Cabassous* sp. Pp. 389-391, in: The Evolution and Ecology of Armadillos, Sloths, and Vermilinguas (G.G. Montgomery, ed.). Smithsonian Institution Press, Washington, DC.
Merriam, C.H. 1887. Description of a new species of wood rat from Cerros Island, off Lower California (*Neotoma bryanti* sp. nov.). The American Naturalist, 21:191-193.
Merriam, C.H. 1890. Annotated list of mammals of the San Francisco mountain plateau and desert of the Little Colorado in Arizona, with notes on their vertical distribution, and description of new species. Pp. 43-123, in: Results of a biological survey of the San Francisco mountain region and desert of the Little Colorado, Arizona. North American Fauna, 3:1-136.
Merriam, C.H. 1892. Description of nine new mammals collected by E.W. Nelson in the states of Colima and Jalisco, Mexico. Proceedings of the Biological Society of Washington, 7:164-174.
Merriam, C.H. 1893. Descriptions of eight new ground squirrels of the genera *Spermophilus* and *Tamias* from California, Texas, and Mexico. Proceedings of the Biological Society of Washington, 8:129-138.
Merriam, C.H. 1894. A new subfamily of murine rodents--the Neotominae—with description of a new genus and species and a synopsis of the known forms. Proceedings of the Academy of Natural Sciences of Philadelphia, 14:225-252.
Merriam, C.H. 1895. Monographic revision of the pocket gopher family Geomyidae (exclusive of the species of *Thomomys*). North American Fauna, 8:1-258.
Merriam, C.H. 1896a. *Romerolagus nelsoni*, a new genus and species of rabbit from Mt. Popocatepetl, Mexico. Proceedings of the Biological Society of Washington, 10:169-174.
Merriam, C.H. 1896b. Synopsis of weasels of North America. North American Fauna, 11:1-44.
Merriam, C.H. 1897a. Descriptions of five new shrews from Mexico, Guatemala, and Colombia. Proceedings of the Biological Society of Washington, 11:227-230.
Merriam, C.H. 1897b. *Romerolagus nelsoni*. Naturaleza, 2:525-528.
Merriam, C.H. 1898a. Descriptions of twenty new species and a new subgenus of *Peromyscus* from Mexico and Guatemala. Proceedings of the Biological Society of Washington, 12:115-125.
Merriam, C.H. 1898b. Descriptions of two new subgenera and three new species of *Microtus* from Mexico and Guatemala. Proceedings of the Biological Society of Washington, 12:105-108.

Merriam, C.H. 1898c. Mammals of Tres Marias Islands off western Mexico. Proceedings of the Biological Society of Washington, 12:13-19.

Merriam, C.H. 1901. Descriptions of 23 new harvest mice (genus *Reithrodontomys*). Proceedings of the Washington Academy of Science, 3:547-558.

Merriam, C.H. 1902a. Five new mammals from Mexico. Proceedings of the Biological Society of Washington, 15:67-69.

Merriam, C.H. 1902b. Twenty new pocket mice (*Heteromys* and *Liomys*) from Mexico. Proceedings of the Biological Society of Washington, 15:41-50.

Merriam, C.H. 1903. Four new mammals, including a new genus (*Teanopus*), from Mexico. Proceedings of the Biological Society of Washington, 16:79-82.

Merriam, C.H. 1907. Descriptions of ten new kangaroo rats. Proceedings of the Biological Society of Washington, 20:75-80.

Merriam, C.H. 1918. Review of the grizzly and big brown bears of North America (genus *Ursus*), with description of a new genus *Vetularctos*. North American Fauna, 41:1-136.

Merriam, M.D. 1895. Synopsis of the American shrews of the genus Sorex. North American Fauna, 10:57-124.

Merritt, D.A., Jr. 1975. The lesser anteater, *Tamandua tetradactyla*, in captivity. International Zoo Yearbook, 15:41-45.

Merritt, J.F. 1978. *Peromyscus californicus*. Mammalian Species, 85:1-6.

Meserve, P.L. 1974. Ecological relationships of two sympatric woodrats in California coastal sage scrub community. Journal of Mammalogy, 55:442-447.

Meserve, R.L. 1976a. Three-dimensional home ranges of cricetid rodents. Journal of Mammalogy, 58:549-558.

Meserve, R.L. 1976b. Food relationships of a rodent fauna in a California coastal sage scrub community. Journal of Mammalogy, 57:300-319.

Meserve, P.L. 1977. Three-dimensional home ranges of cricetid rodents. Journal of Mammalogy, 58:549-558.

Mesnick, S.L., M.C. García Rivas, B.J. Le Boeuf, and S.M. Peterson. 1998. Northern elephant seals in the Gulf of California, Mexico. Marine Mammal Science, 14:171-178.

Messick, J.P. 1987. North American badger. Pp. 587-597, in: Wild Furbearer Management and Conservation in North America (M. Novak, G.A. Baker, M.E. Obbard, and B. Malloch, eds.). Ontario Trappers Association, Ministry of Natural Resources, Ontario, Canada.

Messick, J.P., and M.G. Hornocker. 1981. Ecology of the badger in southwestern Idaho. Wildlife Monographs, 76:1-53.

Messing, J.H. 1986. A late Pleistocene-Holocene fauna from Chihuahua, Mexico. The Southwestern Naturalist, 31:277-288.

Meyer, B.J., and R.K. Meyer. 1944. Growth and reproduction of the cotton rat, *Sigmodon hispidus hispidus*, under laboratory conditions. Journal of Mammalogy, 25:107-129.

Michalski, F., P.G. Crawshaw, T.G. de Oliveira, and M.E. Fabian. 2006. Notes on home range and habitat use of three small carnivore species in a disturbed vegetation mosaic of southeastern Brazil. Mammalia, 70:52-57.

Michener, G.R. 1984. Age, sex and species differences in the annual cycles of ground-dwelling sciurids: Implications for sociality. Pp. 79-107, in: The Biology of Ground-dwelling Squirrels (J.O. Murie and G.O. Michener, eds.). University of Nebraska Press, Lincoln.

Millar, J.S. 1989. Reproduction and development. Pp. 169-232, in: Advances in the Study of *Peromyscus* (Rodentia) (G.L. Kirkland and J.N. Layne, eds.). Texas Tech University Press, Lubbock.

Miller, A.H., and R.C. Stebbins. 1964. The Lives of Desert Animals in Joshua Tree National Monument. University of California Press, Berkeley.

Miller, R.S. 1964. Ecology and distribution of pocket gophers (Geomyidae) in Colorado. Ecology, 45:256-272.

Miller, B., G. Ceballos, and R. Reading. 1994. The prairie dog and biotic diversity. Conservation Biology, 8:677-681.

Miller, B., R.P. Reading, and S. Forrest. 1996. Prairie Night: Black-footed Ferrets and the Recovery of Endangered Species. Smithsonian Institution Press, Washington, DC.

Miller, B., R. Reading, J. Hoogland, T. Clark, G. Ceballos, R. List, S. Forrest, L. Hanebury, P. Manzano, J. Pacheco, and D. Uresk. 2000. The role of prairie dogs as keystone species: A response to Stapp. Conservation Biology, 14:318-321.

Miller, D.S., and D.W. Speake. 1978a. Prey utilization by bobcats on quail plantations in southern Alabama. Proceedings of the Annual Conference of the Southeastern Association of Fish and Wildlife Agencies, 32:100-111.

Miller, D.S., and D.W. Speake. 1978b. Status of the bobcat: An endangered species? Pp. 145-153, in: Proceedings of the Rare and Endangered Wildlife Symposium (R. Odom and L. Landers, eds.). Athens, GA.

Miller G.S., Jr. 1900. Three new bats from the island of Curacao. Proceedings of the Biological Society of Washington, 13:123-127.

Miller, G.S., Jr 1902. Note on the *Chilonycteris davyi fulvus* of Thomas. Proceedings of the Biological Society of Washington, 15:155.

Miller, G.S., Jr. 1906. Twelve new genera of bats. Proceedings of the Biological Society of Washington, 19:83-86.

Miller, G.S., Jr. 1912. List of North American land mammals in the United States National Museum, 1911. Bulletin of the United States National Museum, 79:1-455.

Miller, G.S. Jr. 1924. List of North American recent mammals 1923. Bulletin of the United States National Museum, 128:1-673.

Miller, G.S., Jr., and G.M. Allen. 1928. The American bats of the genera *Myotis* and *Pizonyx*. Bulletin of the United States National Museum, 144:1-218.

Miller, G.S., Jr., and R. Kellogg. 1955. List of North American recent mammals. Bulletin of the United States National Museum, 205:1-954.

Miller, G.S., Jr., and J.A. Rehn. 1901. Systematic results of the study of North American land mammals to the close of the year 1900. Proceedings of the Boston Society of Natural History, 30:1-352.

Miller, W., and O. Carranza-Castañeda. 2002. Importance of Mexico´s late Tertiary mammalian faunas. Pp. 83-102, in: Avances en los Estudios Paleomastozoológicos en México (M. Montellano-Ballesteros and J. Arroyo-Cabrales, eds.). Colección Científica, Instituto Nacional de Antropología e Historia, México.

Milner, J., C. Jones, and J.K. Jones Jr. 1990. *Nyctinomops macrotis*. Mammalian Species, 351:1-4.

Milton, K. 1980. The Foraging Strategy of Howler Monkeys: A Study in Primate Economics. Columbia University Press, New York.

Milton, K. 1981. Estimates of reproductive parameters for freeranging *Ateles geoffroyi*. Primates, 22:574-579.

Minasian, S., K. Balcomb III, and L. Foster. 1984. The World´s Whales. Smithsonian Institution, Washington, DC.

Minta, S.C., K.A. Minta, and D.F. Lott. 1992. Hunting associations between badgers (*Taxidea taxus*) and coyotes (*Canis latrans*). Journal of Mammalogy, 73:814-820.

Miranda, J.M.D., I.P. Bernardi, K.C. Abreu, and F.C. Passos. 2005. Predation of *Alouatta guariba clamitans* Cabrera (Primates, Atelidae) by *Leopardus pardalis* (Linnaeus) (Carnivora, Felidae). Revista Brasileira do Zoologia, 22:793-795.

Mitchell, E. 1968. Northeast Pacific stranding distribution and seasonality of Cuvier's beaked whale, *Ziphius cavirostris*. Canadian Journal of Zoology, 46:265-279.

Mitchell, E. 1975. Porpoise, Dolphin and Small Whale Fisheries of the World: Status and Problems. IUCN, Morges, Switzerland.

Mitchell, H. 1964a. Comparative Nutrition of Man and Domestic Animals. Vol. 2. Academic Press, New York.

Mitchell, H. 1964b. Investigations of the cave atmosphere of a Mexican bat colony. Journal of Mammalogy, 45:468-577.

Mittermeier, R.A. 1978. Locomotion and posture in *Ateles geoffroyi* and *Ateles paniscus*. Folia primatologica, 30:161-193.

Mittermeier, R.A., and A.F. Coimbra-Filho. 1981. Systematics: Species and subspecies. Pp. 29-109, in: Ecology and Behavior of Neotropical Primates (A.F. Coimbra-Filho and R.A. Mittermeier, eds.). Vol. 1. Academia Brasileña de Ciencias, Rio de Janeiro, Brazil.

Mittermeier, R.A., and A.F. Coimbra-Filho. 1988. Systematics: Species and subspecies — an update. Pp. 13-75, in: Ecology and Behavior of Neotropical Primates (R.A. Mittermeier, A.B. Rylands, A.F. Coimbra-Filho, and G.A.B. da Fonseca, eds.). Vol. 2. World Wildlife Foundation, Washington, DC.

Mittermeier, R.A., and C. Goettsch. 1992. La importancia de la diversidad biológica de México. Pp. 57-62, in: México ante los Retos de la Biodiversidad (R. Dirzo and J. Sarukhán, eds.). Conabia, México.

Mittermeier, R.A., N. Myers, and C.G. Mittermeier. 1999. Biodiversidad Amenazada. Las ecorregiones terrestres prioritarias del mundo. CEMEX, Conservation International y Agrupación Sierra Madre, México D.F.

Mittermeier, R.A., P. Robles, and C. Goettsch. 1997. Megadiversidad. Los países biológicamente más ricos del mundo. CEMEX, México.

Miyazaki, N. 1980. Preliminary note on age determination and growth of the rough-toothed dolphin, *Steno bredanensis* of the Pacific Coast of Japan. Reports of the International Whaling Commission (special issue), 3:171-178.

Miyazaki, N. 1981. An outline of the biological studies on *Stenella coeruleoalba.* Studies on the Levels of Organochlorine Compounds and Heavy Metals in the Marine Organisms (T. Fujiyama. ed.). Okinawa, University of Ryukyus, 1-5.

Mizue, K. 1951. Food of whales (in adjacent waters of Japan). Scientific Reports of the Whales Research Institute, 5:81-90.

Moctezuma, B.J., and M. Serrato (eds.). 1988. Islas del Golfo de California. Secretaría de Gobernación y Universidad Nacional Autónoma de México, México.

Mody, W.S., and M.R. Lee. 1984. Systematic implications of chromosomal banding analyses of populations of *Peromyscus truei* (Rodentia: Muridae). Proceedings of the Biological Society of Washington, 97:116-123.

Moehrenschlager, A., B. Cypher, K. Ralls, M.A. Sovada, and R. List. 2004. Comparative ecology and conservation priorities of swift and kit foxes. Pp. 181-194, in: The Biology and Conservation of Wild Canids (D.W. Macdonald, C. Sillero-Zubiri, and J. Ginsberg, eds.). Oxford University Press, Oxford.

Mollohan, C.M. 1987. Black bear habitat use in northern Arizona. Final Perf. Rep., Proj. W-78-R, Work Plan 4, Job 19. Arizona Fish and Game Department, Research Branch, Phoenix.

Mondolfi, E. 1986. Notes on the biology and status of the small wild cats in Venezuela. Pp. 125-146, in: Cats of the World: Biology, Conservation, and Management (S.D. Miller and D.D. Everett, eds.). National Wildlife Federation, Washington, DC.

Mondolfi, E., and G. Medina P. 1957. Contribución al conocimiento del "Perrito de Agua" (*Chironectes minimus* Zimmermann, 1811). Memorias de la Sociedad de Ciencias Naturales La Salle, 17:140-155.

Mones, A. 1968. Restos óseos de mamíferos contenidos en regurgitaciones de lechuza del estado de Oaxaca, México. Anales del Instituto de Biología, UNAM, México, Serie Zoología, 39:169-172.

Monson, G. 1943. Food habits of the banner-tailed kangaroo rat in Arizona. Journal of Wildlife Management, 7:98-102.

Monson, G. 1972. Unique Birds and Mammals of the Coronado National Forest. U.S. Forest Service, Washington, DC.

Montellano-Ballesteros, M., and J. Arroyo-Cabrales (eds.). 2002. Avances en los studios paleomastozoológicos en México. Colección Científica, Instituto Nacional de Antropología e Historia, 443:1-248.

Monterrubio, R.T. 1995. Biología y distribución de la tuza queretana *Cratogeomys neglectus* (Rodentia: Geomyidae) en Pinal de Amoles, Querétaro. Tesis de licenciatura, Universidad Autónoma Metropolitana, Unidad Xochimilco, México.

Monterrubio, T., J.W. Demastes, L. León-Paniagua, and M.S. Hafner. 2000. Systematics relationships of the endangered Queretaro pocket gopher (*Cratogeomys neglectus*). The Southwestern Naturalist, 45:249-252.

Montgomery, G.G. 1983. *Cyclopes didactylus* (tapacara, serafín de platanar, silky anteater). Pp. 461-463, in: Costa Rican Natural History (D.H. Janzen, ed.). University of Chicago Press, Chicago.

Montgomery, G.G. 1985a. Impact of vermilinguas (Cyclopes, Tamandua: Xenarthra = Edentata) on arboreal ant populations. Pp. 351-363, in: The Evolution and Ecology of Armadillos, Sloths, and Vermilinguas (G.G. Montgomery, ed.). Smithsonian Institution Press, Washington, DC.

Montgomery, G.G. 1985b. Movements, foraging and food habits of the four extant species of Neotropical vermilinguas (Mammalia: Myrmecophagidae). Pp. 365-377, in: The Evolution and Ecology of Armadillos, Sloths, and Vermilinguas (G.G. Montgomery, ed.) Smithsonian Institution Press, Washington, DC.

Moore, J.C. 1968. Relationships among the living genera of beaked whales. Field Zoology, 53:209-298.

Mora, J.M., and I. Moreira. 1984. Mamíferos de Costa Rica. Editorial Universidad Estatal A Distancia, San José, Costa Rica.

Morales, A. 1985. Análisis cuantitativo de las dietas de ganado vacuno y venado cola blanca en La Michilía, Durango. Tesis de licenciatura, Facultad de Ciencias, Universidad Nacional Autónoma de México, México.

Morales, A., M. Weber, and C. Galindo-Leal. 1989. Factores que afectan las estimaciones de abundancia del venado cola blanca por métodos indirectos. Pp. 92-104, in: III Simposio sobre Venados en México. Universidad Nacional Autónoma de México, Linares, Nuevo León.

Morales, J.C. 1992. Molecular Systematics of the Three Bats (Genus *Lasiurus*). Texas A&M University.

Morales, J.C., and J.W. Bickham. 1995. Molecular systematics of the genus *Lasiurus* (Chiroptera: Vespertilionidae) based on restriction site maps of the mitochondrial ribosomal genes. Journal of Mammalogy, 76:730-749.

Morales, J.C., and M.D. Engstrom. 1989. Morphological variation in the painted spiny pocket mouse, *Liomys pictus* (family Heteromyidae), from Colima and southern Jalisco, Mexico. Life Sciences Occasional Paper, 38:1-16.

Morales, V.B. 1990. Parámetros reproductivos del lobo marino en la Isla Ángel de la Guarda, Golfo de California, México. Tesis de maestría, Facultad de Ciencias, Universidad Nacional Autónoma de México, México.

Morales, V.B., and D. Olivera G. 1993. Varamiento de calderones *Globicephala macrorhynchus* (Cetácea: Delphinidae) en la isla de Cozumel, Quintana Roo, México. Anales del Instituto de Biología, UNAM, México, Serie Zoología, 64:177-180.

Morales-Malacara, J.B., and R. López-W. 1990. Epizoic fauna of *Corynorhinus mexicanus* (Chiroptera:Vespertilionidae) in Tlaxcala, Mexico. Journal of Medical Entomology, 27:440-445.

Morales-Malacara, J.B., and R. López-Wilchis. 1998. New species of the genus *Spinturnix* (Acari: Mesostigmata: Spinturnicidae) on *Corynorhinus mexicanus* (Chiroptera: Vespertilionidae) in central Mexico. Journal of Medical Entomology, 35:543-550.

Morales Vela, B., and L. Medrano González. 1997. Variación genética del manatí (*Trichechus manatus*), en el sureste de México y monitoreo con radiotransmisores en Quintana Roo. Reporte inédito, CONABIO, México.

Morales Vela, B., and L.D. Olivera Gómez. 1995. Distribución y abundancia del manatí *Trichechus manatus manatus* en las costas de Belice y Bahía de Chetumal, México. XX Reunión Internacional para el Estudio de los Mamíferos Marinos, La Paz, Baja California Sur.

Morales-Vela, B., D. Olivera-Gómez, J.E. Reynolds III, and G.B. Rathbun. 2000. Distribution and habitat use by manatees (*Trichechus manatus manatus*) in Belize and Chetumal Bay, Mexico. Biological Conservation, 95:67-75.

Moreira, J., R. Garcia, R. McNab, T. Dubón, F. Córdova, and M. Córdova. 2007. Densidad de ocelotes (*Leopardus pardalis*) en la parte este del Parque Nacional Mirador Rió Azul, Guatemala. Sociedad para la Conservación de la Vida Silvestre (WCS-Guatemala). Informe Técnico.

Morejohn, G.V. 1979. The natural history of Dall's porpoise in the North Pacific Ocean. Pp. 45-83, in: Behavior of Marine Animals, vol. 3, Cetaceans (H.E. Winn and B.L. Olla, eds.). Plenum Press, New York.

Moreno, A. 1987. Determinación y distribución de los mamíferos nativos del Cañón del Huajuco. Santiago, Nuevo León, México. Tesis de licenciatura, Facultad de Ciencias Biológicas, Universidad Autónoma de Nuevo León. Nuevo León, México.

Moreno, R., and A. Bustamante. 2007. Estatus del jaguar, otros felinos y sus presas en el Alto Chagres, utilizando cámaras trampa. Reporte técnico. Sociedad Mastozoológica de Panamá.

Moreno, R., and A. Bustamante. 2009. Datos ecológicos del ocelote (*Leopardus pardalis*) para el área de Cana, Parque Nacional Darien, Panamá, utilizando el método de Cámaras trampa. Tecnociencia, 11:91-102.

Moreno, R., and J. Giacalone. 2006. Ecological data obtained from latrine use by ocelots (Leopardus pardalis) on Barro Colorado Island, Panama. Tecnociencia, 8:7-21.

Moreno, R., and J. Giacalone. 2008. El comportamiento de ocelotes (*Leopardus pardalis*) en letrinas. Mesoamericana, 12:36.

Moreno, R., R. Kays, and R. Samudio Jr. 2006. Competitive release in the diet of ocelots (*Leopardus pardalis*) and puma (*Puma concolor*) after jaguar (*Panthera onca*) decline. Journal of Mammalogy, 87:808-816.

Moreno-Valdez, A. 1996. First record for the kinkajou, *Potos flavus* (Carnivora: Procyonidae) in Tamaulipas, Mexico. The Southwestern Naturalist, 41:457-458

Moreno-Valdez, A. 1998. Mamíferos nativos del Cañón del Huajuco. Santiago, Nuevo León, México. Revista Mexicana de Mastozoologia, 3:5-25.

Moreno-Valdez, A. 1999. The fringed *Myotis thysanodes* (Chiroptera: Vespertilionidae), in Tamaulipas, Mexico. The Southwestern Naturalist, 33(3).

Moreno-Valdez, A., P.A. Lavín-M., and O.M. Hinojosa-F. 1997. El Tepezcuintle, *Agouti paca* (Rodentia: Agoutidae), en Tamaulipas, México. Revista Mexicana de Mastozoología, 2:129-131.

Morrell, S.H. 1972. Life history of the San Joaquin kit fox. California Fish and Game Journal, 58:162-174.

Morris, D. 1965. The Mammals: A Guide to Living Species. Harper and Row, New York.

Morrison, B., B. Muller-Using, and M. Cotera. 1992. The translocation of mule deer in Nuevo Leon, México. Pp. 75-79, in: Global Trends in Wildlife Management (B. Bobek, ed.). Vol. 2. Jagelonian University Press, Krakow, Poland.

Morrison, D.W. 1978a. Foraging ecology and energetics of the frugivorous bat *Artibeus jamaicensis*. Ecology, 59:716-723.

Morrison, D.W. 1978b. Lunar phobia in a Neotropical fruit bat, *Artibeus jamaicensis* (Chiroptera: Phyllostomidae). Animal Behavior, 26:852-856.

Morrison, D.W. 1979. Apparent male defense of tree hollows in the fruit bat, *Artibeus jamaicensis*. Journal of Mammalogy, 60:11-15.

Morrison, D.W., and S.H. Morrison. 1981. Economics of harem maintenance by a Neotropical bat. Ecology, 62:864-866.

Morrison, M.L., I.C. Timoss, and K.A. With. 1987. Development and testing linear regression models predicting bird-habitat relationships. Journal of Wildlife Management, 51:247-253.

Morton, S.R. 1979. Diversity of desert-dwelling mammals: A comparison of Australia and North America. Journal of Mammalogy, 60:253-264.

Moulton, M P., J.R. Choate, S.J. Bissell, and R.A. Nicholson. 1981. Associations of small mammals on the central high plains of eastern Colorado. The Southwestern Naturalist, 26:53-57.

Müller-Schwarze, D., and L. Sun. 2003. The Beaver: Natural History of a Wetlands Engineer. Cornell University Press, Ithaca, NY.

Müllerried, K.G. 1957. La Geología de Chiapas. Gobierno del Estado de Chiapas, Tuxtla Gutiérrez.

Mullin, K.D., L.V. Higgins, T.A. Jefferson, and L.J. Hansen. 1994. Sightings of clymene dolphin (*Stenella clymene*) in the Gulf of México. Marine Mammal Science, 10:464-470.

Mumford, R.E. 1973. Natural history of the red bat (*Lasiurus borealis*) in Indiana. Periodicum Biologorum, 75:155-158.

Mumford, R.E., L.L. Oakley, and D.A. Zimmerman. 1964. June bat records from Guadalupe Canyon, New Mexico. The Southwestern Naturalist, 12:163-171.

Muñiz, M. R. 2001. Vertebrados terrestres de San Juan de Camarones, Durango. Informe final a CONABIO, Instituto Politécnico Nacional. www.conabio.gob.mx/institucion/proyectos/resultados/InfR008.pdf.

Muñiz-Martínez, R., and J. Arroyo-Cabrales. 1996. El registro más norteño de la rata enana Nelsonia neotomodon (Rodentia: Muridae). Vertebrata Mexicana, 2:13-16.

Muñiz-Martínez, R., and O.J. Polaco. 1996. Nuevos registros de simpatría de dos especies del género *Corynorhinus* (Chiroptera: Vespertilionidae) en México. Vertebrata Mexicana, 1:13-16.

Muñiz-Martínez, R., C. López-González, J. Arroyo-Cabrales, and M. Ortiz Gómez. 2003. Noteworthy records of free-tailed bats (Chiroptera: Molossidae) from Durango, Mexico. The Southwestern Naturalist, 48:138-144.

Muñoz, J. 1995. Clave de murciélagos vivientes en Colombia. Universidad de Antioquia, Editorial Ciudad, Medellín.

Murie, A. 1981. The Grizzlies of Mount McKinley. University of Washington Press, Seattle.

Musser, G.G. 1964. Notes on geographic distribution, habitat and taxonomy of some Mexican mammals. Occasional Papers of the Museum of Zoology, University of Michigan, 636:1-22.

Musser, G.G. 1968. A systematic study of the Mexican and Guatemalan gray squirrel, *Sciurus aureogaster* F. Cuvier (Rodentia: Sciuridae). Miscellaneous Publications of the Museum of Zoology, University of Michigan, 137:1-112.

Musser, G.G. 1969. Notes on *Peromyscus* (Muridae) of Mexico and Central America. American Museum Novitates, 2357:1-23.

Musser, G.G. 1971. *Peromyscus allophylus* Osgood: *A synonym of Peromyscus gymnotis* Thomas (Rodentia, Muroidea). American Museum Novitates, 2453:1-10.

Musser, G.G., and M.D. Carleton. 1993. Familia Muridae. Pp. 501-755, in: Mammal Species of the World: A Taxonomic and Geographic Reference (D.E. Wilson and D.M. Reeder, eds.). 2nd ed. Smithsonian Institution Press, Washington, DC.

Musser, G.G., and M.D. Carleton. 2005. Superfamily Muroidea. Pp. 894-1531, in: Mammal Species of the World: A Taxonomic and Geographic Reference (D.E. Wilson and D.M. Reeder, eds.). Vol. 2. 3rd ed. Johns Hopkins University Press, Baltimore,

Muul, Y. 1974. Geographic variation in nesting habitats of *Glaucomys volans*. Journal of Mammalogy, 55:840-844.

Myers, P. 1977. Patterns of reproduction of four species of vespertilionid bats in Paraguay. University of California, Publications in Zoology, 107:1-41.

Myers, P. 1978. Sexual dimorphism in size of vespertilionid bats. The American Naturalist, 112:701-711.

Myers, P., and R.M. Wetzel. 1983. Systematics and zoogeography of the bats of the Chaco Boreal. Miscellaneous Publications of the Museum of Zoology, University of Michigan, 165:1-59.

Nabham, G.P., and T.H. Fleming. 1992. The conservation of mutualisms. Species, 19:32-34.

Nader, L.A. 1978. Kangaroo rats: Intraspecific variation in *Dipodomys spectabilis* Merriam and Dipodomys deserti Stephens. Illinois Biological Monographs, 49:1-116.

Nadin-Davis, S.A., Y.Q. Feng, D. Mousse, A.I. Wandeler, and S. Aris-Brosou. 2010. Spatial and temporal dynamics of rabies virus variants in big brown bat populations across Canada: Footprints of an emerging zoonosis. Molecular Ecology, 19:2120-2136.

Nagorsen, D. 1985. *Kogia simus*. Mammalian Species, 239:1-6.
Napier, J.R. 1976. Evolutionary aspects of primate locomotion. American Journal of Physical Anthropology, 27:333-342.
Napier, J.R., and P.H. Napier. 1967. A Handbook of Living Primates. Academic Press, New York.
Napier, P.H. 1976. Catalogue of Primates in the British Museum (Natural History), Part I: Families Callitrichidae and Cebidae. British Museum of Natural History, London.
Naranjo, E.J., and E. Cruz. 1998. Ecología del tapir (*Tapirus bairdii*) en la Reserva de la Biosfera La Sepultura, Chiapas, México. Acta Zoologica Mexicana (n.s.), 73:111-125.
Naranjo, E.J., and E. Espinosa. 2001. Los mamíferos de la Reserva Ecológica Huitepec, Chiapas, México. Revista Mexicana de Mastozoología, 5:58-67.
Nash, D.J., and R.N. Seaman. 1977. *Sciurus aberti*. Mammalian Species, 80:1-5.
Nason, E.S. 1948. Morphology of hair of eastern North American bats. The American Midland Naturalist, 39:345-361.
Navarro L., D. 1982. Mamíferos de la Estación de Biología Tropical "Los Tuxtla," Veracruz. Tesis de licenciatura, Facultad de Ciencias, Universidad Nacional Autónoma de México, México.
Navarro L., D. 1985. Status and distribution of the ocelot (*Felis pardalis*) in South Texas. Master's thesis, Texas A&M University, Kingsville.
Navarro L., D., and M. Suárez. 1989. A survey of the pygmy raccoon (*Procyon pygmaeus*) of Cozumel, Mexico. Mammalia, 53:458-461.
Navarro L.D., and D.E. Wilson. 1982. *Vampyrum spectrum*. Mammalian Species, 184:1-4.
Navarro L., D., T. Jiménez, and J. Juárez. 1991. Los mamíferos de Quintana Roo. Pp. 371-450, in: Diversidad Biológica en la Reserva de la Biosfera de Sian Ka'an, Quintana Roo, México (D. Navarro and J.G. Robinson, eds.). Centro de Investigaciones de Quintana Roo, Chetumal, Q.R.
Neal, B.J. 1965a. Seasonal changes in body weights, fat deposition, adrenal glands and temperatures of *Citellus tereticaudus* and *Citellus harrisii* (Rodentia). The Southwestern Naturalist, 10:156-166.
Neal, B.J. 1965b. Reproductive habits of the round tailed and Harris antelope ground squirrels. Journal of Mammalogy, 46:200-206.
Neal, B.J. 1965c. Growth and development of the round tailed and Harris antelope ground squirrels. The American Midland Naturalist, 73:479-489.
Negus, N.C., E. Gould, and R.K. Chipman. 1961. Ecology of the rice rat *Oryzomys palustris* (Harlan) on Breton Island, Gulf of Mexico, with a critique of the social stress theory. Tulane Studies in Zoology, 8:95-123.
Nelson, C.E. 1965. *Lonchorhina aurita* and other bats from Costa Rica. Texas Journal of Science, 17:303-306.
Nelson, E.W. 1898. What is *Sciurus variegatus* Erxleben? Science (n.s.), 8:897-898.
Nelson, E.W. 1899a. Mammals of the Tres Marias Islands. North American Fauna, 14:15-19.
Nelson, E.W. 1899b. Natural History of the Tres Marias Islands, Mexico: General account of the islands, with reports on mammals and birds. North American Fauna, 14:1-97.
Nelson, E.W. 1899c. Revision of the squirrels of Mexico and Central America. Proceedings of the Washington Academy of Sciences, 1:15-110.
Nelson, E.W. 1901. Descriptions of two new squirrels from Mexico. Proceedings of the Biological Society of Washington, 14:131-132.
Nelson, E.W. 1904a. Description of seven new rabbits from Mexico. Proceedings of the Biological Society of Washington, 17:103-110.
Nelson, E.W. 1904b. Descriptions of new squirrels from Mexico. Proceedings of the Biological Society of Washington, 17:147-150.
Nelson, E.W. 1907. Descriptions of new North American rabbits. Proceedings of the Biological Society of Washington, 20:81-84.
Nelson, E.W. 1909. The rabbits of North America. North American Fauna, 29:1-314.
Nelson, E.W. 1922. Lower California and its natural resources. National Academy of Sciences, First Memoir, 16:1-94.
Nelson, E.W. 1930. Wild Animals of North America: Intimate Studies of Big and Little Creatures of the Mammal Kingdom. National Geographic Society, Washington, DC.
Nelson, E.W., and E.A. Goldman. 1909. Eleven new mammals from Lower California. Proceedings of the Biological Society of Washington, 22:23-28.
Nelson, E.W., and E.A. Goldman 1931a. New carnivores and rodents from Mexico. Journal of Mammalogy, 12:302-306.
Nelson, E.W., and E.A. Goldman. 1931b. Six new white-footed mice (*Peromyscus maniculatus* group), from islands off the Pacific Coast. Journal of the Washington Academy of Science, 21:530-535.
Nelson, E.W., and E.A. Goldman. 1932. A new white-footed mouse from Lower California, Mexico. Transactions of the San Diego Society of Natural History, 7:51-52.
Nelson, E.W., and E.A. Goldman. 1933. Revision of the jaguars. Journal of Mammalogy, 14:221-240.
Nelson, E.W., and E.A. Goldman. 1934. Pocket gophers of the genus *Thomomys* of Mexican main lands and bordering territory. Journal of Mammalogy, 15:105-124.
Nelson, J. E. 1965. Movements of Australian flying foxes (Pteropodidae: Megachiroptera). Australian Journal of Zoology, 13:53-73.
Nemoto, T. 1959. Food of baleen whales with reference to whale movements. The Scientific Reports of the Whales Research Institute, Tokyo, 14:149-290.
Nemoto, T., and A. Kawamura, 1977. Characteristics of food habits and distribution of baleen whales with special reference to the abundance of North Pacific sei and Bryde's whales. Reports of the International Whaling Commission (special issue), 1:80-87.
Neville, M.K.. K.E. Glander, F. Braza, and A.B. Rylands. 1988. The howling monkeys, genus *Alouatta*. Pp. 349-453, in: Ecology and Behavior of Neotropical Primates (R.A. Mittermeier, A.B. Rylands, A.F. Coimbra-Filho, and G.A.B. da Fonseca, eds.). World Wildlife Fund, Washington, DC.
Newby T.C. 1973. Observations on the breeding behavior of the harbor seal in the state of Washington. Journal of Mammalogy, 54:540-543.
Newcomer, M.W., and D.D. DeFarcy. 1985. White-faced capuchin (*Cebus capucinus*) predation on a nestling coati (*Nasua narica*). Journal of Mammalogy, 66:185-186.
Nicholson, W.S., E.P. Hill, and D. Briggs. 1985. Denning, pup rearing and dispersal in the gray fox in east-central Alabama. Journal of Wildlife Management, 49:33-37.
Niethamer, J. 1964. Contribution a la connaissance des mammiferes terrestres de l´ile Indefatigable (Santa Cruz), Galapagos. Résultats de l´expedition Allemagne aux Galapagos 1962/1963. Mammalia, 28:593-606.
Nilsson, M.A., U. Arnason, P.B.S. Spencerb, and A. Janke. 2004. Marsupial relationships and a timeline for marsupial radiation in South Gondwana. Gene, 340:189-196.
Nishiwaki, M. 1972. General biology: Cetacea: Pp. 3-204, in: Mammals of the Sea: Biology and Medicine (S.H. Ridgway, ed.). Charles C. Thomas Publisher, Springfield, IL.
Nishiwaki, M., and T. Kamiya. 1958. A beaked whale Mesoplodon stranded at Oiso Beach Japan. Scientific Reports of the Whales Research Institute, 13:53-83.
Nishiwaki, M., and Oguro N. 1971. Baird's beaked whales caught on the coast of Japan in recent years. Scientific Reports of the Whales Research Institute, 7:1-22.
Nishiwaki, M., T. Kasuya, K. Kureha, and N. Oguro. 1972. Further comments on *Mesoplodon ginkgodens*. Scientific Reports of the Whales Research Institute, 24:43-56.
Nixon, C.M., and L.P. Hansen. 1987. Managing forest to maintain populations of gray and fox squirrels. Illinois Department of Conservation Technical Bulletin, 5:1-35.
Nixon, C.M., and M.W. McClain. 1969. Squirrel population decline following a late spring frost. Journal of Wildlife Management, 33:353-357.

Nixon, C.M., D.M. Worley, and M.W. McClain. 1968. Food habits of squirrels in Southeast Ohio. Journal of Wildlife Management, 32:294-305.

Nolasco, A.L., I. Lira, and G. Ceballos. 2007. Ampliacíon en la distribución histórica del tapir (*Tapirus bairdii*) en el Pacífico Mexicano. Revista Mexicana de Mastozoología, 11:91-94.

Norris, K.S., and T.P. Dohl. 1980. Behavior of the Hawaiian spinner dolphin, Stenella longirostris. Fishery Bulletin, 77:821-849.

Norris, K.S., and W.N. McFarland. 1958. A new harbor porpoise of the genus Phocoena from the Gulf of California. Journal of Mammalogy, 39:22-39.

Norris, K.S., and J.H. Prescott. 1961. Observations on Pacific cetaceans of Californian and Mexican waters. University of California Publications in Zoology, 63:291-402.

Northridge, S.P. 1985. Estudio mundial de las interacciones entre los mamíferos marinos y las pesquerías. FAO. Información Pesquera, 251:1-234.

Novick, A. 1963. Orientation in Neotropical bats. II. Phyllostomatidae and Desmodontidae. Journal of Mammalogy, 44:44-56.

Novick, P. 1965. Echolocation of flying insects by the bat *Chilonycteris psilotis.* Biological Bulletin, 128:297-314.

Novick, P., and J.R. Valsnys. 1964. Echolocation of flying insects by the bat *Chilonecyeris parnellii.* Biological Bulletin, 127:478-488.

Nowak, M.R. 1881. A perspective on the taxonomy of wolves in North America. Pp. 10-19, in: Wolves in Canada and Alaska: Their Status, Biology, and Management (L.N. Carbyn, ed.). Canadian Wildlife Service, Ottawa.

Nowak, R.M. 1991. Walker´s Mammals of the World. 5th ed. Johns Hopkins University Press, Baltimore.

Nowak, R.M. 1999a. Walker's Bats of the World. 6th ed. Johns Hopkins University Press, Baltimore.

Nowak, R.M. 1999b. Walker's Mammals of the World. 6th ed. Johns Hopkins University Press, Baltimore.

Nowak, R.M., and J. Paradiso. 1983. Walker´s Mammals of the World. Johns Hopkins University Press, Baltimore.

Nowell, K., and P. Jackson (comps. and eds.). 1996. Wild Cats: Status Survey and Action Plan. IUCN, Gland, Switzerland.

Núñez, G.A., and G. Pastrana H. 1990. Los Roedores Michoacanos. Departamento de Biología, Coordinación de la Investigación Científica, Universidad Michoacana de San Nicolás de Hidalgo, Morelia, Michoacán.

Núñez, G.A., C.B.T. Chávez, and C.H. Sánchez. 1981. Mamíferos silvestres de la región de El Tuito, Jalisco, México. Anales del Instituto de Biología, UNAM, México, Serie Zoología, 55:647-668.

Núñez, R., B. Miller, and F. Lindzey. 2000. Food habits of jaguars and pumas in Jalisco, Mexico. Journal of Zoology, 252:373-379.

Nuñez-Perez, R., E. Corona-Corona, J. Torres-Villanueva, C. Anguiano-Méndez, M. Tornez, I. Solorio, and A. Torres. 2011. Nuevos Registros del Oso Hormiguero, *Tamandua mexicana*, en el Occidente de México. Edentata, 12:58-62.

O'Connell, M.A. 1979. Ecology of didelphid marsupials from northern Venezuela. Pp. 73-87, in: Vertebrate Ecology in the Northern Neotropics (J.F. Eisenberg, ed.). The National Zoological Park, Smithsonian Institution Press, Washington DC.

O'Farrel, M.J. 1972. Notes on parturition and behavior in *Pipistrellus hesperus* in the laboratory. Occasional Papers of the Biology Society of Nevada, 32:1-3.

O'Farrel, M.J., and W.G. Bradley. 1970. Activity patterns of bats over a desert spring. Journal of Mammalogy, 51:18-26.

O'Farrel, M.J., and W.A. Clark. 1984. Notes on the white-tailed antelope squirrel, *Ammospermphilus leucurus*, and the pinyon mouse, *Peromyscus truei*, in north-central Nevada. The Great Basin Naturalist, 44:428-430.

O'Farrel, M.J., and B.W. Miller. 1972. Pipistrelle bats attracted to vocalizing females and to a blacklight insect trap. The American Midland Naturalist, 88:462-463.

O'Farrel, M.J., W.G. Bradley, and G.W. Jones. 1967. Fall and winter bat activity at a desert spring in southern Nevada. The Southwestern Naturalist, 12:163-171.

O'Farrell, E.R., and E.H. Studier. 1973. Reproduction, growth and development in *Myotis thysanodes* and *Myotis lucifugus* (Chiroptera: Vespertilionidae). Ecology, 54:18-30.

O'Farrell, E.R., and E.H. Studier. 1980. *Myotis thysanodes.* Mammalian Species, 137:1-5.

O'Farrell, T.P. 1987. Kit fox. Pp. 423-431, in: Wild Furbearer Management and Conservation in North America (M. Novak, G.A. Baker, M.E. Obbard, and B. Malloch, eds.). Ontario Trappers Association, Ministry of Natural Resources, Ontario, Canada.

O'Gara, B. 1978. *Antilocapra americana.* Mammalian Species, 90:1-7.

O'Gara, B.W. 1990. The pronghorn (*Antilocapra americana*). Pp. 231-264, in: Horns, Pronghorns and Antlers: Evolution, Morphology, Physiology and Social Significance (G.A. Bubenik and A. B. Bubenik, eds.). Springer-Verlag, New York.

O'Gara, B.W., and J.D. Yoakum. 1992. Pronghorn Management Guide. Pronghorn: An Antelope Workshop. Rock Springs, WY.

O'Shea, T.J., and T.A. Vaughan. 1977. Nocturnal and seasonal activities of the pallid bat, *Antrozous pallidus.* Journal of Mammalogy, 58:269-284.

O'Sullivan, S., and K.D. Mullin. 1997. Killer whales (*Orcinus orca*) in the northern Gulf of Mexico. Marine Mammal Science, 13:141-147.

Ockenfels, R.A. 1994. Factors affecting adult pronghorn mortality rates in central Arizona. Arizona Fish and Game Department, Phoenix.

Ochoa-Gaona, S., and M. González-Espinosa. 2000. Land use and deforestation in the highlands of Chiapas, Mexico. Applied Geography, 20:17-42.

Odell, D.K. 1974. Seasonal occurrence of the northern elephant seal, *Mirounga angustirostris*, on San Nicolas Island, California, Journal of Mammology, 55:81-95.

Oestner, E.J. 1944. Populations of small mammals on Cerro Potosi, Nuevo Leon, Mexico. Journal of Mammology, 25:284-289.

Ohlendorft, H.M. 1972. Observations on a colony of *Eumops perotis* (Molossidae). The Southwestern Naturalist, 17:297-300.

Ojasti, J., and O. Linares. 1971. Adiciones a la fauna de murciélagos de Venezuela, con notas sobre las especies del género *Diclidurus* (Chiroptera). Acta Biológica Venezuelica, 7:421-441.

Ojeda, R.A., P.G. Blendinger, and R. Brandl. 2000. Mammals in South American drylands: Faunal similarity and trophic structure. Global Ecology and Biogeography, 9:115-123.

Ojeda-Castillo, M.M. 1991. Las especies del género Mazama en Venezuela, sus estudios bioecológicos. Pp. 159-163, in: El Venado en Venezuela: Conservación, Manejo Aspectos Biológicos y Legales. FUDECI-PROFAUNA-FEDECAVE, Caracas, Venezuela.

Olin, G., and D. Thompson. 1982. Mammals of the Southwest Deserts. Rush Press, San Diego.

Oliveira, T.G. 1998. *Leopardus wiedii.* Mammalian Species, 579:1-6.

Oliver, J.S., P.N. Slattery, M.A. Silbertein, and E.F. O'Connor. 1983. Gray whale feeding on dense ampeliscid amphipod communities near Bamfield, British Columbia. Canadian Journal of Zoology, 62:41-49.

Olivera, M., J. Ramírez P., and S.L Williams. 1986. Reproducción de *Peromyscus* (*Neotomodon*) alstoni (Mammalia: Muridae) en condiciones de laboratorio. Acta Zoologica Mexicana, 16:1-27.

Olmos, F. 1993. Diet of sympatric Brazilian caatinga peccaries (*Tayassu tajacu* and *T. pecari*). Journal of Tropical Ecology, 9:255-258.

Olsen, R.W. 1968. Gestation period in *Neotoma mexicana*. Journal of Mammalogy, 49:533-534.

Omura, H. 1975. Osteological study of the minke whale from the Antarctic. Scientific Reports of the Whale Research Institute, 27:1-36.

Omura, H., K. Fujino, and S. Kimura. 1955. Beaked whale *Berardius bairdii* of Japan, with notes on *Ziphius cavirostris.* Scientific Reports of the Whales Research Institute, 10:89-132.

Omura, H., S. Ohsumi, T. Nemoto, K. Nasu, and T. Kasuya. 1969. Black right whales in the North Pacific. Scientific Reports of the Whales Research Institute, 21:1-78.

Oporto, S., M.G. Hidalgo, and J. Bello. 2008. Efecto de un periodo de inundación sobre la abundancia de seis especies de phillostomidos en Villahermosa Tabasco. Semana de Divulgación y Video Científico.

Orbell, M. 1995. The Illustrated Encyclopedia of Māori Myth and Legend. Canterbury University Press, Christchurch.

OrduÑa Trejo, C., A. Castro, and J. Ramirez Pulido. 1999-2000. Mammals from the Tarsacan Plateau, Michoacan, Mexico. Revista Mexicana de Mastozoologia, 4:53-68.

Ordway, L., and P. Krausman. 1986. Habitat use by desert mule deer. Journal of Wildlife Management, 50:677-683.

Orozco-Lugo, C.L., D. Valenzuela-Galván., L.B. Vázquez, A.J. Rhodes, A. de León-Ibarra, A. Hernández, M.E. Copa-Alvaro, L.G. Avila-Torres Agatón, and M. de La Peña-Domene. Velvety fruit-eating bat (*Enchistenes hartii*; Phyllostomidae) in Morelos, Mexico. The Southwestern Naturalist, 53:517-520.

Orozco-Meyer, A., and B. Morales-Vela. 1998. Distribution and abundance of the river otter (*Lutra longicaudis annectens* Major, 1897), in the Rio Hondo, Quintana Roo, México. XXIII Reunión Internacional para el Estudio de los Mamíferos Marinos. 20-24 de Abril. Xcaret, Quintana Roo, México.

Orr, R.T. 1940. The rabbits of California. Occasional Papers, California Academy of Sciences, 19:1-227.

Orr, R.T. 1942. Observations on the growth of young brush rabbits. Journal of Mammalogy, 23:298-302.

Orr, R.T. 1950. Notes on the seasonal occurrence of red bats in San Francisco. Journal of Mammalogy, 31:457-458.

Orr, R.T. 1954. Natural history of the pallid bat, *Antrozous pallidus* (Le Conte). Proceedings of the California Academy of Sciences, 28:165-246.

Orr, R.T. 1963. An extension of the range of Merriam's desert shrew. Journal of Mammalogy, 44:424.

Orr, R.T. 1965. Longevity in *Pizonyx vivesi*. Journal of Mammalogy, 46:497.

Orr, R.T., J. Stonewald, and K.W. Kenyon. 1970. The California sea lion skull growth: Comparison of two populations. Proceedings of the California Academy of Sciences, 37:381-394.

Ortega, J., and I. Alarcón-D. 2008. *Anoura geoffroyi*. Mammalian Species, 818:1-7.

Ortega, C., and G. Massieu. 1963. Aminoácidos libres del encéfalo y del hígado de diversos géneros y especies de murciélagos. Anales del Instituto de Biología, UNAM, México, Serie Zoología, 34:27-34.

Ortega, J., and H.T. Arita. 1999. Structure and social dynamics of harem groups in *Artibeus jamaicensis* (Chiroptera: Pyhllostomidae). Journal of Mammalogy, 80:1173-1185.

Ortega, J., and H.T. Arita. 2000. Defensive behavior of females by dominant males of *Artibeus jamaicensis* (Chiroptera: Phyllostomidae). Ethology, 106:395-407.

Ortega Ortiz, J.G. 2000. Diversidad de cetáceos en el sur del Golfo de México. XXV Reunión internacional para el estudio de los mamíferos marinos. La Paz, Baja California Sur.

Osgood, W.H. 1900. Revision of the pocket mice of the genus *Perognathus*. North American Fauna, 18:1-73.

Osgood, W.H. 1904. Thirty new mice of the genus *Peromyscus* from Mexico and Guatemala. Proceedings of the Biological Society of Washington, 17:55-77.

Osgood, W.H. 1909. A revision of the mice of the American genus *Peromyscus*. North American Fauna, 28:1-285.

Osgood, W.H. 1914. Mammals of an expedition across northern Peru. Field Museum of Natural History Zoological Series, 10:143-185.

Osgood, F.L. 1936. *Melanistic pipistrelles*. Journal of Mammalogy, 17:64.

Osgood, W.H. 1938. A new woodrat from Mexico. Field Museum of Natural History Zoological Series, 20:475-476.

Osgood, W.H. 1945. Two new rodents from Mexico. Journal of Mammalogy, 26:299-301.

Osgood, W.H., and C.H. Merriam. 1909. Revision of the mice American genus *Peromyscus*. North American Fauna, 28:1-285.

Overall, K.L., 1980. Coatis, tapirs and ticks: A case of mammalian interspecific grooming. Biotropica, 12:158.

Owen, J.G., and R.S. Hoffmann. 1983. *Sorex ornatus*. Mammalian Species, 212:1-5.

Owen, R.D. 1987. Phylogenetic analyses of the bat subfamily Stenodermatinae (Mammalia: Chiroptera). Special Publications of the Museum, Texas Tech University, 26:1-65.

Owen, R.D. 1988. Phenetic analyses of the bat subfamily Stenodermatinae (Chiroptera Phyllostomidae). Journal of Mammalogy, 69:795-810.

Pacheco, J., and L. Salazar. 1990. Anatomía gruesa y descriptiva del aparato digestivo de los quirópteros frugívoros de la Costa Chica de Guerrero, México. Tesis profesional., Escuela Nacional de Estudios Profesionales, Zaragoza, Universidad Nacional Autónoma de México, México.

Pacheco, J., G. Ceballos, and R. List. 1999. Los mamíferos de la región de Janos-Casas Grandes, Chihuahua, México. Revista Mexicana de Mastozoología, 4:71-85.

Pacheco, J., G. Ceballos, and R. List. 2000. Mamíferos de la región de Janos-Nuevo Casas Grandes, Chihuahua, México. Revista Mexicana de Mastozoología, 4:73-87.

Pacheco, J., G. Ceballos, and R. List. 2002. Reintroducción del hurón de patas negras en las praderas de Janos, Chihuahua. Biodiversitas, 6:1-5.

Pacheco, J., G. Ceballos, G. Daily, P.R. Ehrlich, G. Suzan, B. Rodríguez, and E. Marcé. 2005. Diversidad, historia natural y conservación de los mamíferos de la región de San Vito de Coto Brus, Costa Rica. Revista de Biología Tropical, 54:219-240.

Pacheco L.F., and J.A. Simonetti. 2000. Genetic structure of a mimosoid tree deprived of its seed disperser, the spider monkey. Conservation Biology, 14:1766-1775.

Packard, R.L. 1956. The tree squirrels of Kansas: Ecology and economic importance. University of Kansas Publications, Museum of Natural History, 11:1-67.

Packard, R.L. 1960. Speciation and evolution of the pygmy mice, genus Baiomys. University of Kansas Publications, Museum of Natural History, 9:579-670.

Packard, R.L. 1968. An ecological study of the fulvous harvest mouse in eastern Texas. The American Midland Naturalist, 79:68-88.

Packard, R.L. 1977. Mammals of the southern Chihuahuan desert: An inventory. Pp. 141-153, in: Transactions of the Symposium on the Biological Resources of the Chihuahuan Desert Region United States and Mexico (R.H. Wauer and D.H. Riskind, eds.). U.S. Department of the Interior, National Park Service Transactions and Proceedings Series, 3:1-658.

Packard, R.L., and J.B. Montgomery. 1978. *Baiomys musculus*. Mammalian Species, 102:1-3.

Packard, J.M., G.B. Rathbun, and D.P. Doming. 1984. Sea cows and manatees. Pp. 292-295, in: The Encyclopedia of Mammals (D. Macdonald, ed.). Facts on File, New York.

Padilla, V.A. 1990. Aspectos biológicos de la foca común (*Phoca vitulina richardsi* Grey, 1864) en la costa occidental de Baja California. Tesis de licenciatura, Facultad de Ciencias, Universidad Nacional Autónoma de México, México.

Padilla García, U., and R. Pineda López. 1997. Vertebrados del Estado de Querétaro. FOMES, Universidad Autónoma de Querétaro, Querétaro.

Paintiff, J.A., and D.E. Anderson. 1980. Breeding the margay at New Orleans Zoo. International Zoo Year Book, 20:223-224.

Palacio, J. 1991. Composición botánica de la dieta del berrendo (*Antilocapra americana*) y uso del hábitat en primavera y verano. Tesis profesional, Universidad Autónoma de Aguascalientes, Aguascalientes, Aguascalientes.

Palacios, D.M., and B.R. Mate. 1996. Attack by false killer whales (Pseudorca crassidens) on sperm whales (*Physeter macrocephalus*) in the Galapagos Islands. Marine Mammal Science, 12:582-587.

Palmer, F.G. 1937. Geographic variation in the mole *Scapanus latimanus*. Journal of Mammalogy, 18:280-314.

Pardini R., and E. Trajano. 1999. Use of shelters by the Neotropical river otter (*Lontra longicaudis*) in Atlantic forest stream, southeastern Brazil. Journal of Mammalogy, 80(2):600-610.

Parera, A. 1996. Las nutrias verdaderas de la Argentina. Boletín Técnico de la Fundación Vida Silvestre Argentina. Buenos Aires, Argentina.

Parker, R.L. 1975. Rabies in skunks. Pp. 41-50, in: The Natural History of Rabies (G.M. Baer, ed.). Academic Press, New York.

Pastene, L.A., K. Numachi, M. Jofre, M. Acevedo, and G. Joyce. 1989. First record of the Blainvelle's beaked whale, *Mesoplodon densirostris* Blainville, 1871 (Cetacea, Ziphiidae) in the eastern South Pacific. Marine Mammal Science, 6:82-84.

Patten, D.R., and L.T. Findley. 1970. Observations and records of Myotis (*Pizonyx vivesi*) Menegaux (Chiroptera: Vespertilionidae). Contributions in Science, Los Angeles Country Museum of Natural History, 183:1-9.

Patterson, B.D. 1980. A new subspecies of *eutamias-quadrivittatus* (Rodentia, Sciuridae) from the Organ Mountains, New Mexico. Journal of Mammalogy, 61:455-464.

Patton, J.C., and R.J. Baker. 1978. Chromosomal homology and evolution of phyllostomatoid bats. Systematic Zoology, 27:449-462.

Patton, J.L. 1967. Chromosome studies of certain pocket mice, genus *Perognathus* (Rodentia: Heteromyidae). Journal of Mammalogy, 48:27-37.

Patton, J.L. 1969. Karyotypic variation in the pocket mouse, *Perognathus penicillatus* Woodhouse (*Rodentia-Heteromyidae*). Caryologia, 22:351-358.

Patton, J. L. 1970. Karyotypes of five species of pocket mouse, *Perognathus* (Rodentia: Heteromyidae), and a summary of chromosome data for the genus. Mammalian Chromosome Newsletter, 11:3-8.

Patton, J.L. 1973. An analysis of natural hybridization between the pocket gophers *Thomomys bottae* and *Thomomys umbrinus*, in Arizona. Journal of Mammalogy, 54:561-584.

Patton, J.L. 1993a. Familia Heteromidae. Pp. 477-486, in: Mammal Species of the World, 2nd ed. (D.E. Wilson and D.M. Reeder, eds.). Smithsonian Institution Press, Washington, DC.

Patton, J.L. 1993b. Family Geomidae. Pp. 469-486, in: Mammal Species of the World (D.E. Wilson and D.M. Reeder, eds.). Smithsonian Institution Press, Washington, DC.

Patton, J.L. 1999. Familia Geomyidae. Pp. 321-350, in: Mamìferos Del Noroeste Mexicano (S.T Álvarez-Castañeda and J.L. Patton, eds.). Centro de Investigaciones Biológicas del Noroeste, SC.

Patton, J.L. 2005. Family Geomyidae. Pp. 859-871, in: Mammal Species of the World: A Taxonomic and Geographic Reference (D.E. Wilson and D.A. Reeder, eds.). Vol. 2. Johns Hopkins University Press, Baltimore.

Patton, J.L., and P.V. Brylski. 1987. Pocket gophers in alfalfa fields: Causes and consequences of habitat-related body size variation. The American Naturalist, 130:493-506.

Patton, J.L., and R.E. Dingman. 1968. Chromosome studies of pocket gophers, genus *Thomomys*. I: The specific status of *Thomomys umbrinus* (Richardson) in Arizona. Journal of Mammalogy, 49:1-13.

Patton, J.L., and J.K. Jones. 1972. First records of *Perognathus baileyi* from Sinaloa, Mexico. Journal of Mammalogy, 53:371-372.

Patton, J.L., and D.S. Rogers. 1993. Cytogenetics. Pp. 236-258, in: Biology of the Heteromyidae (H.H. Genoways and J.H. Brown, eds.). Special Publication, American Society of Mammalogists.

Patton, J.L., and M.F. Smith. 1981. Molecular evolution in *Thomomys*: Phylectic systematics, paraphyly, and rates of evolution. Journal of Mammalogy, 62:493-500.

Patton, J.L., and M.F. Smith. 1990. The evolutionary dynamics of the pocket gopher *Thomomys bottae*, with emphasis on California populations. University of California Publications in Zoology, 123:1-161.

Patton, J.L., and O.H. Soulé. 1967. Natural hybridization in pocket mice, genus *Perognathus*. Mammalian Chromosome Newsletter, 8:236-264.

Patton, J.L., D.G. Huckaby, and S.T. Álvarez-Castañeda. 2007. The evolutionary history and a systematic revision of woodrats of the *Neotoma lepida* group. University of California Publications in Zoology, 135.

Patton, J.L., S.W. Sherwood, and S.Y. Yang. 1981. Biochemical systematics of chaetodipine pocket mice, genus *Perognathus*. Journal of Mammalogy, 62:477-492.

Paulson, D.D. 1988. Chaetodipus baileyi. Mammalian Species, 297:1-5.

Peltz-Serrano, K., E. Ponce-Guevara, R. Sierra-Corona, R. List, and G. Ceballos. 2006. Recent records of desert bighorn sheep (*Ovis canadensis mexicana*) in eastern Sonora and northwestern Chihuahua, Mexico. The Southwestern Naturalist, 51:450-454.

Pembleton, E.F., and S.L. Williams. 1978. *Geomys pinetis*. Mammalian Species, 86:1-3.

Peña-Escalante, J.E., S. Hernández, and A. Segovia. 2001. *Chiroderma villosum* (Chiroptera: Phyllostomidae) en el estado de Yucatán, México. Revista Mexicana de Mastozoología, 5:68-71.

Pence, D.B., and H.H. Genoways. 1974. *Neolabidophrus yucatanensis* gen. et. sp. n. and a new record for *Dermacarus ornatus*, 1967 (Acarina: Glycyphagidae) from *Heteromys gaumeri* Allen and Chapman, 1877. Gaumer´s spiny pocket mouse (Rodentia: Heteromyidae). Journal of Parasitology, 60:712-715.

Penney, D.F., and E.G. Zimmerman. 1976. Genic divergence and local population differentiation by random drift in the pocket gopher genus Geomys. Evolution, 30:473-483.

Peppers, L.L., and R.D. Bradley. 2000. Cryptic species in *Sigmodon hispidus*: Evidence from DNA sequences. Journal of Mammalogy, 81:332-343.

Peppers, L.L., D.S. Carroll, and R.D. Bradley. 2002. Molecular systematic of the genus *Sigmodon* (Rodentia: Muridae): Evidence from mitochondrial cytochrome-b gene. Journal of Mammalogy, 83:396-407.

Peracchi, A.L., and S.T. Alburquerque. 1976. Sobre os habitos almentares de *Choropterus auritus australis* Thomas 1905 (Mammalia, Chiroptera, Phyllostomidae). Revista Brasileira de Biología, 36:176-184.

Peracchi, A.L., S.T. Albuquerque, and S.D.L. Raimundo. 1982. Contribuição ao conhecimento dos hábitos alimentares de *Trachops cirrhosus* (Spix, 1823) (Mammalia: Chiroptera: Phyllostomidae). Arquivos da Universidade Federal Rural do Rio de Janeiro, 5(1):1-5.

Pérez, S.A. 1978. Observaciones sobre la morfología, alimentación y reproducción de *Liomys pictus* (Rodentia: Heteromyidae). Tesis de licenciatura, Facultad de Ciencias, UNAM, México

Pérez-Gil, R. 1981. A preliminary study of the deer from Cedros Island, Baja California, Mexico. Master's thesis, School of Natural Resources, University of Michigan, Ann Arbor.

Pérez-Lustre, M., R.G. Contreras, and A. Santos-Moreno. 2006. Mamíferos del bosque mesófilo de montaña del Municipio de San Felipe Usila, Tuxtepec, Oaxaca, México. Revista Mexicana de Mastozoología, 10:29-40.

Pérez-Sánchez, C.E., A.A. Ortega, O.S. Gordillo, H.V. Alafita, L.G. Medrano, and C.M. Esquivel. 1994. Proyecto "Mamíferos marinos de la costa de Veracruz, México." P. 32, in: XIX Reunión Internacional para el Estudio de los Mamíferos Marinos. La Paz, Baja California Sur.

Perrin, W.F. 1969a. Using porpoises to catch tuna. World Fishing, 18:4.

Perrin, W.F. 1969b. The barnacle *Conchoderma auritum*, on a porpoise (*Stenella graffmani*). Journal of Mammalogy, 50:150-151.

Perrin, W.F. 1972. Color patterns of spinner porpoises (*Stenella* cf. *S. longirostris*) of the eastern Pacific and Hawaii, with comments on dolphin pigmentation. Fishery Bulletin, 70:983-1003.

Perrin, W.F. 1975. Distribution and differentiation of population of dolphins of the genus *Stenella* in the eastern tropical Pacific. Journal of the Fishery Research Board Canadia, 32:1059-1067.

Perrin, W.F., and J.R. Henderson. 1979. Growth and Reproductive Rates in Two Populations of Spinner Dolphins, *Stenella longirostris*, with Different Histories of Exploitation. South West Fisheries Center, Administrative Reports, LJ-79–29.

Perrin, W.F., J.M. Coe, and J.R. Zweifel. 1976. Growth and reproduction of the spotted porpoise *Stenella attenuata*, in the offshore eastern tropical Pacific. Fishery Bulletin, 74:229-269.

Perrin, W.F., D.B. Holts, and R.B. Miller. 1977. Growth and reproduction of the eastern spinner dolphin, a geographical form of *Stenella longirostris* in the eastern tropical Pacific. Fishery Bulletin, 75:725-750.

Perrin, W.F., M.D. Scott., G.J. Walker, and V.L. Cass. 1985. Review of geographical stocks of tropical dolphins (*Stenella* spp. and *Delphinus delphis*) in the eastern tropical Pacific. National Oceanic and Atmospheric Administration, National Marine Fishery Services. Technical Reports 28.

Perrin, W.F., E.D. Mitchell, J.G. Mead, D.K. Caldwell, and P.J.H. van Bree. 1981. *Stenella clymene*, a rediscovered tropical dolphin of the Atlantic. Journal of Mammalogy, 62:538-598.

Perrin, W.F., M.D. Scott, G.J. Walker, F.M. Ralston, and D.K.W. Au. 1983. Distribution of four dolphins (*Stenella* spp. and *Delphinus delphis*) in the eastern tropical Pacific with annotated catalog of data sources. National Atmospheric Administration, South West Fishery Center Technical Report No. 38:65.

Perrin, W.F., P.B. Best, W.H. Dawbin, K.C. Balcomb, R. Gambell, and G.J.B. Ross. 1973. Rediscovery of Fraser's dolphin *Lagenodelphis hosei*. Nature, 241:345-350.

Perrin, W.F., E.D. Mitchell, J.G. Mead, D.K. Caldwell, M.C. Caldwell, P.J.H. van Bree, and W.H. Dawbin. 1987. Revision of the spotted dolphin *Stenella* spp. Marine Mammal Science, 3:99-170.

Perrin, W.F., W.A. Armstrong, A.N. Baker, J. Barlow, S.R. Benson, A.S. Collet, J.M. Cotton, D.M. Everhart, T.D. Farley, R.M. Mellon, S.K. Miller, V. Philbrick, J.L. Quan, and H.R. Lira Rodríguez. 1995. An anomalously pigmented form of the shortbeaked common dolphin (*Delphinus delphis*) from the southwestern Pacific, eastern Pacific and eastern Atlantic. Marine Mammal Science, 11:240-247.

Perry, H.R. 1982. Muskrats. Pp. 282-325, in: Wild Mammals of North America: Biology, Management, Economics (J.A. Chapman and G.A. Feldhamer, eds.). Johns Hopkins University Press, Baltimore.

Perryman, W.L., and T.C. Foster. 1980. Preliminary report on predation by small whales, mainly the false killer whale (*Pseudorca crassidens*), on dolphins (*Stenella* spp. and *Delphinus delphis*) in the eastern tropical Pacific. National Oceanic Atmospheric Administration. South West Fishery Center, La Jolla, CA.

Peters, S.L., B.K. Lim, and M.D. Engstrom. 2002. Systematics of dog-faced bats (*Cynomys*) based on molecular and morphometric data. Journal of Mammalogy, 83:1097-1110.

Petersen, J. 1980. A comparison of small mammal populations sampled by pit-fall and live traps in Durango, Mexico. The Southwestern Naturalist, 25:122-124.

Petersen, M.K. 1968. Electrophoretic blood-serum patterns in selected species of *Peromyscus*. The American Midland Naturalist, 79:130-148.

Petersen, M.K. 1972. Female *Peromyscus pectoralis* fosters litter of *Reithrodontomys megalotis*. The Southwestern Naturalist, 17:301-302.

Petersen, M.K. 1973. Interactions between the cotton rats, *Sigmodon fulviventer* and *S. hispidus*. The American Midland Naturalist, 90:319-333.

Petersen, M.K. 1976. Noteworthy range extensions of some mammals in Durango, Mexico. The Southwestern Naturalist, 21:139-142.

Petersen, M.K. 1978. Rodent Ecology and Natural History Observations on the Mammals of Atotonilco de Campa, Durango, México. Carter Press, Inc., Ames, Iowa.

Petersen, M.K. 1979a. A laboratory test of species exclusion in *Sigmodon fulviventer* and *Sigmodon hispidus*. Iowa State, Journal of Research, 53:201-304.

Petersen, M.K. 1979b. A temporal comparison of owl pellet contents with small mammal population levels in Durango, Mexico. Centzontle, 2:2-21.

Petersen, M.K. 1980. The taxonomic status of a dusky-colored population of *Peromyscus* at Atotonilco, Durango, Mexico. Occasional Papers Zoology, Michael K. Peterson, 5:1-19.

Petersen, M.K. 1993. Dietary overlap and livestock forage relationships in two species of *Sigmodon* from Durango, México. Pp. 289-300, in: Avances en el Estudio de los Mamíferos de México (R.A. Medellín and G. Ceballos, eds.). Publicaciones Especiales, vol. 1, Asociación Mexicana de Mastozoología, A.C., México.

Petersen, M.K., and M.J. Helland. 1978. Behavioral interactions in *Sigmodon fulviventer* and *Sigmodon hispidus*. Journal of Mammalogy, 59:118-124.

Petersen, M.K., and M.K. Petersen. 1979. A temporal comparison of owl pellet contents with small mammal population levels in Durango, Mexico. Occasional Papers Zoology, Michael K. Petersen, 4:1-27.

Peterson, A.T., J. Soberón, and V. Sánchez-Cordero. 1999. Conservatism of ecological niches in evolutionary time. Science, 285:1265-1267.

Peterson, R.L. 1966. Notes on the Yucatan vesper rat, *Otonyctomys hatti*, with a new record, the first from British Honduras. Canadian Journal of Zoology, 44:281-284.

Peterson, R.L., and P. Kirmse. 1969. Notes on *Vampyrum spectrum*, the false vampire bat, in Panama. Canadian Journal of Zoology, 47:140-142.

Peterson, R.S., and G.A. Bartholomew. 1967. The Natural History and Behavior of the California Sea Lion. Special Publication 1, American Society of Mammalogists.

Peterson, T.C., K. P. Gallo, J. Lawrimore, T.W. Owen, A. Huang, and D.A. Mckittrick. 1999. Global rural temperature trends. Geophysical Research Letters, 26(3):329-332.

Pfeiffer, C.J., and G.H. Gass. 1963. Note on the longevity and habits of captive *Cryptotis parva*. Journal of Mammalogy, 44:427-428.

Phillips, C.J. 1971. The dentition of Glossophagine bats: Development, morphological characteristics, variation, pathology, and evolution. University of Kansas Publications, Museum of Natural History, 54:1-138.

Phillips, C.J., and J.K. Jones Jr. 1968. Dental abnormalities in North American bats. I. Emballonuridae, Noctilionidae, and Chilonycteridae. Transactions of the Kansas Academy of Science, 71:509-520.

Phillips, C.J., and J.K. Jones Jr. 1971. A new subspecies of the longnosed bats, *Hylonycteris underwoodi*, from Mexico. Journal of Mammalogy, 52:77-80.

Phillips, C.J., G.W. Grimes, and G.L. Forman. 1977. Oral biology. Part II. Pp. 121-246, in: Biology of the Bats of the New World: Family Phillostomidae (R.J. Baker, J.K. Jones Jr., and D.C. Carter, eds.). Special Publications of the Museum, Texas Tech University, 13:1-364.

Phillips, C., B. Tandler, and C.A. Pinkstaff. 1987. Unique salivary glands in two genera of tropical *Michrochiroptera* bats: An example of evolutionary convergence in histology and histochemistry. Journal of Mammalogy, 68:235-242.

Piaggio, A., and G.S. Spicer. 2001. Molecular phylogeny of the chipmunks inferred from mitochondrial cytochrome b and cytochrome oxidase II gene sequences. Molecular Phylogeny and Evolution, 20:335-350.

Piaggio, A.J., E.W. Valdez, M.A. Bogan, and G.S. Spicer. 2002. Systematics of *Myotis occultus* (Chiroptera: Vespertilionidae) inferred from sequences of two mitochondrial genes. Journal of Mammalogy, 83:386-395.

Pine, R.H. 1972. The bats of the genus *Carollia*. Technical monograph. The Texas Agricultural Experiment Station, 8:1-125.

Pine, R.H. 1973. Anatomical and nomenclatural notes on opossums. Proceedings of the Biological Society of Washington, 86:391-402.

Pine, R.H., and J.E. Anderson. 1979. Notes on stomach contents in *Trachops cirrhosus* (Chiroptera: Phyllostomatidae). Mammalia, 43:568-570.

Pinkham, C.F.A. 1973. The evolutionary significance of locomotor patterns in the Mexican spiny pocket mouse, *Liomys irroratus*. Journal of Mammalogy, 54:742-746.

Pitman, R. 2009. Indo-Pacific beaked whale, *Indopacetus pacificus*. Pp. 600-602, in: Encyclopedia of Marine Mammals (W. Perrin, B. Würsig, and J. Thewissen, eds.). Academic Press, Burlington, MA.

Pitman, R.L., A.L. Aguayo, and J.R. Urbán. 1987. Observations of an unidentified beaked whale (*Mesoplodon* sp.) in the eastern tropical Pacific. Marine Mammal Science, 3:345-352.

Pizzimenti, J.J. 1975. Evolution of the prairie dog genus *Cynomys*. University of Kansas Publications, Museum of Natural History, 39:1-73.

Plumpton, D.L., and J.K. Jones. 1992. *Rhynchonycteris naso*. Mammalian Species, 413:1-5.

Pocock, R.I. 1939. The races of the jaguar (*Panthera onca*). Novitates Zoologicae, 41:406-422.

Poelker, R.J., and H.O. Hartwell. 1973. Black Bear of Washington. Biological Bulletin 18, Washington State Fish and Game Department, Seattle.

Poglayen-Neuwall, I. 1975. Copulatory behavior, gestation and parturition of the tayra (*Eira barbara* L., 1758). Zeitschrift für Säugertierkunde, 40:176-189.

Poglayen-Neuwall, I., and E.D. Toweill. 1988. Mammalian species *Bassariscus astutus*. The American Society of Mammalogists, 327:1-8.

Polaco, O.J. 1987. First record of *Noctilio albiventris* (Chiroptera: Noctilionidae) in Mexico. The Southwestern Naturalist, 32:508-509.

Polaco, O.J., and J. Arroyo-Cabrales. 2001. El ambiente durante el poblamiento de América. Arqueología Mexicana, 9:30-34.

Polaco, O.J., and R. Muñiz-Martínez. 1987. Los murciélagos de la costa de Michoacán, México. Anales de la Escuela Nacional de Ciencias Biológicas, Universidad Nacional Autónoma de México, 31:63-89.

Polaco, O.J., J. Arroyo-Cabrales, and J.K. Jones Jr. 1992. Noteworthy records of some bats from Mexico. Texas Journal of Science, 44:331-338.

Polaco, O.J., J. Arroyo-Cabrales, E. Corona-M. and J.G. Oliva-López. 2001. The American mastodon *Mammut americanum* in Mexico. Pp. 237-242, in: La Terra degli Elefanti (G. Cavarretta, P. Gioia, M. Mussi, and M.R. Palombo, eds.). Consiglio Nazionale delle Ricerche, Rome.

Polechla, P.J. 1990. Action plan for North American otters In: Otters: An Action Plan for Their Conservation (P. Foster-Turley, S, MacDonald, and D. Mason, eds.). IUCN/SSC Otter Specialist Group, Gland, Switzerland.

Polechla, P.J., P. Gallo, and F. Tovar. 1987. Distribution, occupied habitat and status of the Neotropical river otter (*Lutra longicaudis annectens*) in the southern portions of Sierra Madre del Sur, Mexico. 68 Reunión de la American Society of Mammalogists. Clemson University, Clemson, SC.

Pompa, S., P.R. Ehlich, and G. Ceballos. 2011. Global distribution and conservation of marine mammals. Proceedings of the National Academy of Sciences, 108:13600-13605.

Ponce, E., J. Pacheco, R. List, and G. Ceballos. In press. Threatened North American porcupine (*Erethizon dorsatum*) populations at its southern distribution limit in Mexico. Journal of Mammalogy.

Ponce-Ulloa, H.E., and J.E. Llorente Bousquets. 1993. Distribución de Siphonaptera (Arthropoda, Insecta) en la Sierra de Atoyac de Álvarez, Guerrero, México. Instituto de Biología, Universidad Nacional Autónoma de México, Publicaciones Especiales, 11:1-77.

Pontrelli, M.J. 1968. Mating behavior of the black-tailed jackrabbit. Journal of Mammalogy, 49:785-786.

Poole, A.J., and V.S. Schantz. 1942. Catalog of the type specimens of mammals in the United States National Museum, including the Biological Surveys Collection. Smithsonian Institution, Bulletin United States Natural Museum, 178:1-705.

Porras-Peters, H., D. Aurioles-Gamboa, V.H. Cruz-Escalona, and P.L. Koch. 2008. Position, breadth and trophic overlap of sea lions (*Zalophus californianus*) in the Gulf of California, Mexico. Marine Mammal Science, 24:554-576.

Pozo de la Tijera, C., and J.E. Escobedo-C. 1999. Mamíferos terrestres de la Reserva de la Biosfera de Sian Ka'an, Quintana Roo, México. Revista Biología Tropical, 47:251-262.

Pozos, M.E.D. 2006. Educación ambiental para la conservación de los murciélagos. Una experiencia en la Esc. Primaria "Ignacio Manuel Altamirano," Buena Vista, Mpio. Emiliano Zapata, Veracruz, México. Tesis de licenciatura, Universidad Veracruzana.

Pressey, R.L., C.J. Humprey, C.J. Margules, C.R. Vane-Wright, and P.H. Williams. 1993. Beyond opportunism: Key principles for systematic reserve selection. Trends in Ecology and Evolution, 8:124-128.

Price, M.V., and J.H. Brown. 1983. Patterns of morphology and resource use in North American desert rodent communities. In: Biology of Desert Rodents (O.J. Reichman and J.H. Brown, eds.). Great Basin Naturalist Memoirs No. 7.

Prieto, B.M. 1988. Hábitos alimenticios de tres especies de roedores cricétidos. Tesis de maestría, Facultad de Ciencias, Universidad Nacional Autónoma de México, México.

Prieto Bosch, M., and V. Sánchez-Cordero. 1993. Sistemas de información geográficos: un caso de estudio en Veracruz. Pp. 455-464, in: Avances en el estudio de los mamíferos de México (R. A. Medellín and G. Ceballos, eds.). Asociación Mexicana de Mastozoología, A.C., Publicaciones Especiales, 1:1-464.

Pryor, K., and L. Lang. 1980. Social behavior and school structure in pelagic porpoises (*Stenella attenuata* and *Stenella longirostris*) during purse seining for tuna. National Oceanic Atmospheric Administration, National Marine Fisheries Service-South West Fisheries Center, La Jolla, Administrative Report No. LJ-80<H>11c.

Pryor, T., K. Pryor, and K.S. Norris. 1965. Observations on a pygmy killer whale (*Feresa attenuata* Gray, 1875) from Hawaii. Journal of Mammalogy, 46:450-460.

Puig, H., R. Bracho, and V. Sosa. 1983. Composición florística y estructura del bosque mesófilo en Gómez Farías, Tamaulipas. México. Biótica, 8:339-359.

Pumo, D.E., E.Z. Goldin, B. Elliot, C.J. Phillips, and H.H. Genoways. Mitochondrial DNA polymorphism in three Antillean island populations of the fruit bat, *Artibeus jamaicensis*. Molecular Biology and Evolution, 5:79-89.

Quadros, J., and E.L.A. Monteiro-Filho. 2000. Fruit occurrence in the diet of the Neotropical otter, *Lontra longicaudis*, in southern Brazilian Atlantic forest and its implications for seed dispersion. Journal of Neotropical Mammalogy, 7:33-36.

Quay, W.B. 1948. Notes on some bats from Nebraska and Wyoming. Journal of Mammalogy, 29:181-182.

Quintanilla, J.B., R.G. Ramírez, and J. Aranda. 1989. Determinación de la composición botánica de la dieta seleccionada por el venado cola blanca (*Odocoileus virginianus texanus*) en el Municipio de Anahuac, Nuevo León. Pp. 41-45, in: III Simposio sobre el Venado en México. Universidad Nacional Autónoma de México, Linares, Nuevo León.

Quinto, F. 1994. Avances para el manejo de venados en selvas tropicales del Sureste de México. Pp. 45-52, in: IV Simposio sobre Venados de México. Universidad Nacional Autónoma de México, Nuevo Laredo, Tamaulipas.

Quiring, D.P., and C.F. Harlan. 1953. On the anatomy of the manatee. Journal of Mammalogy, 34:192-203.

Quiroz, D.L., M.S. Xelhuantzi, and M.C. Zamora. 1986. Análisis palinológico del contenido gastrointestinal de los murciélagos Glossophaga soricina y Leptonycteris yerbabuenae de las Grutas de Juxtlahuaca, Guerrero. Instituto Nacional de Antropología e Historia, México.

Rabinowitz, A.R., and B.G. Nottingham. 1986. Ecology and behavior of the jaguar (*Panthera onca*) in Belize, Central America. Journal of Zoology (London), 210:149-159.

Raddell, J.R. 1971. A preliminary bibliography of Mexican cave biology with a checklist of published records. Association for Mexican Cave Studies Bulletin, 3:1-184.

Raddell, J.R. 1981. A review of the cavernicole fauna of Mexico, Guatemala and Belize. Texas Memorial Museum, University of Texas Bulletin, 27:327.

Radowsky, F.J. 1967. The Macronyssidae and Laelapidae (Acarina: Mesostigmata) parasitic on bats. University of California Publications in Entomology, 46:1-237.

Ralls, K., and P.J. White. 1995. Predation on San Joaquin kit foxes by larger canids. Journal of Mammalogy, 76:723-729.

Ralston, G.L., and W.H. Clark. 1971. Occurrence of *Mustela frenata* in northern Baja California, Mexico. The Southwestern Naturalist, 16:209.

Ramírez, N., C. Sobrevilla, N.X. de Erlich, and T. Ruiz-Zapata. 1984. Floral biology and breeding system of Bauhinia benthamiana Taub. (Leguminosae), a bat pollinated tree in Venezuelan "Llanos." American Journal of Botany, 71:273-280.

Ramírez, P.A. 1986. Distribución y alimentación de la ballena Bryde durante el fenómeno "El Niño" 1982-1983. CPPS, Boletín ERFEN, 17:20-27.

Ramírez, R.J. 1984. Revisión sistemática del grupo hispidus del género *Sigmodon* (Rodentia: Muridae) en la República Mexicana. Tesis de licenciatura, Facultad de Ciencias, Universidad Nacional Autónoma de México, México.

Ramírez Bravo, E. 2012. New records of the Mexican hairy porcupine (*Coendou mexicanus*) and Tamandua (*Tamandua mexicana*) in Puebla, Central Mexico. The Western North American Naturalist, 72:93-95.

Ramírez-Pulido, J. 1969a. Contribución al estudio de los mamíferos del Parque Nacional "Lagunas de Zempoala," Morelos, México. Anales del Instituto de Biología, UNAM, México, Serie Zoología, 40:253-290.

Ramírez-Pulido, J. 1969b. Nuevos registros de murciélagos para el Estado de Morelos, México. Anales del Instituto de Biología, UNAM, México, Serie Zoología, 40:253-290.

Ramírez-Pulido, J., and M.A. Armella. 1987. Activity patterns of Neotropical bats (Chiroptera: Phyllostomidae) in Guerrero, Mexico. The Southwestern Naturalist, 32:363-370.

Ramírez-Pulido, J., and M.C. Britton. 1981. An historical synthesis of Mexican mammalian taxonomy. Proceedings of the Biological Society of Washington, 94:1-17.

Ramírez-Pulido, J., and A. Castro-C. 1990. Bibliografía reciente de los mamíferos de México. 1983/1988. Universidad Autonóma Metropolitana-Iztapalapa, México.

Ramírez-Pulido, J., and A. Castro-Campillo. 1993. Diversidad mastozoológica en México. Pp. 413-427, in: Diversidad biológica en México (R. Gío-Argáez and E. López-Ochoterena, eds.). Revista de la Sociedad Mexicana de Historia Natural (volumen especial), 44:1-427.

Ramírez-Pulido, J., and A. Castro-C. 1994. Bibliografía reciente de los mamíferos de México. 1989-1993. Universidad Autónoma Metropolitana, Unidad Iztapalapa, Mexico.

Ramírez-Pulido, J., and N. González-Ruiz. 2006. Las colecciones de mamíferos de México: Origen y destino. Pp. 73-110, in: Colecciones Mastozoólogicas de México (C. Lorenzo, E. Espinoza Medinilla, M. Briones, and F. A. Cervantes, eds.). AMMAC, México D.F.

Ramírez-Pulido, J., and W. López-Forment. 1976. Daños de la ardilla arborícola (*Sciurus aureogaster*) en los cocoteros de la Costa Grande de Guerrero, México. Anales del Instituto de Biologia, UNAM, México, Serie Zoología, 48:67-74.

Ramírez-Pulido, J., and W. López-Forment. 1979. Additional records of some Mexican bats. The Southwestern Naturalist, 24:541-544.

Ramírez-Pulido, J., and J. Martínez Vázquez. 2007. Diversidad de los mamíferos de la Reserva de la Biosfera Tehuacan-Cuicatlán, Puebla-Oaxaca. México. Institución. Informe Final. SNIB-CONABIO proyecto No. BK022. México D.F. www.conabio.gob.mx/institucion/proyectos/resultados/Inf%20BK022.pdf.

Ramírez-Pulido, J., and C. Müdespacher. 1987a. Estado actual y perspectivas del conocimiento de los mamíferos de México. Ciencia, 38:49-67.

Ramírez-Pulido, J., and C. Müdespacher. 1987b. Fórmulas dentarias anormales en algunos murciélagos mexicanos. Acta Zoológica Mexicana (n.s.), 23:1-54.

Ramírez-Pulido, J., and A.R. Phillips. 1968. Primer registro de comadreja (Mustela) en Quintana Roo, Mexico. Anales del Instituto de Biología, UNAM, México, Serie Zoología, 39:145-148.

Ramírez-Pulido, J., and C. Sánchez-Hernández. 1971. *Tylomys nudicaudus* from the Mexican states of Puebla and Guerrero. Journal of Mammalogy, 52:481.

Ramírez-Pulido, J., and C. Sánchez-Hernández. 1972. Regurgitaciones de Lechuza, procedentes de la cueva del Cañón del Zopilote, Guerrero, México. Revista de la Sociedad Mexicana de Historia Natural, 23:107-112.

Ramírez-Pulido, J., M.A. Armella, and A. Castro-Campillo. 1993. Reproductive patterns of three Neotropical bats (Chiroptera: Phyllostomidae) in Guerrero, Mexico. The Southwestern Naturalist, 38:24-29.

Ramírez-Pulido, J., J. Arroyo Cabrales, and A. Castro-C. 2005. Estado actual y relación nomenclatural de los mamiferos terrestres de México. Acta Zoologica Mexicana (n.s.), 21:21-82.

Ramírez-Pulido, J., A. Castro-Campillo, and U. Aguilera. 1995. Sinopsis de los mamíferos del Estado de México. Revista de la Sociedad Mexicana de Historia Natural, 46:205-246.

Ramírez-Pulido, J., G. Ceballos, and J.L. Williams. 1980. A noteworthy record of the long-legged Myotis from central Mexico. The Southwestern Naturalist, 25:124.

Ramírez-Pulido J., A. Martínez, and G. Urbano. 1977. Mamiferos de la Costa Grande de Guerrero, México. Anales del Instituto de Biologia, UNAM, México, Serie Zoología, 48:243-292.

Ramírez-Pulido, J., M. Martínez-Coronel, and A. Castro-Campillo. En prensa. Variación No-Geográfica de *Peromyscus furvus* (Rodentia, Muridae). INAH. Vol. esp. dedicado al M. en C. Ticul Álvarez.

Ramírez-Pulido, J., D.F. Ran, and A. Castro-Campillo. 1994. Análisis multivariado estatal de los mamíferos mexicanos con una modificación al Algoritmo de Peters. Revista de la Sociedad Mexicana de Historia Natural, 45:61-74.

Ramírez-Pulido, J., M.C. Britton, A. Perdomo, and A. Castro. 1986. Guía de los Mamíferos de México. Referencias Hasta 1983. Universidad Autónoma Metropolitana, Iztapalapa, México.

Ramírez-Pulido, J., A. Castro-Campillo, M.A. Almeida, and A. Salame-Méndez. 2000. Bibliografía reciente de los mamíferos de México 1994-2000. Universidad Autónoma Metropolitana, México.

Ramírez-Pulido, J., A. Castro-Campillo, J. Arroyo-Cabrales, and F. Cervantes. 1996. An updated checklist of the Mexican native mammals. Occasional Papers of the Museum, Texas Tech University, 158:1-62.

Ramírez-Pulido, J., R. López W., C. Müdespacher, and I. Lira. 1982. Catálogo de los Mamíferos Terrestres Nativos de México. Universidad Autónoma Metropolitana, Iztapalapa, Trillas, México.

Ramírez-Pulido, J., R. Lopez-Vilchis, C. Müdespacher, and I.E. Lira. 1983. Lista y Bibliografía Reciente de los Mamíferos de México. Universidad Autónoma Metropolitana, Unidad Iztapalapa, Edit. Contraste, México.

Ramsey, P.R., and L.A. Briese. 1971. Effects of inmigrants on the spatial structure of a small mammal community. Acta Theriologica, 16:191-202.

Randall, J.A. 1986. Preference for estrous female urine by male Kangaroo rats (*Dipodomys spectabilis*). Journal of Mammalogy, 67:736-739.

Randall, J.A. 1989. Neighbor recognition in a solitary desert rodent (*Dipodomys merriami*). Ethology, 81:123-133.

Randall, J.A. 1991a. Mating strategies of a nocturnal desert rodent (*Dipodomys spectabilis*). Behavioural Ecology and Sociobiology, 28:215-220.

Randall, J.A. 1991b. Sandbathing to establish familiarity in the Merriam's kangaroo rat, *Dipodomys merriami*. Animal Behaviour, 41:267-275.

Randall, J.A. 1993. Behavioural adaptations of desert rodents (Heteromyidae). Animal Behaviour, 45:263-287.

Randi, E., N. Mucci, F. Claro-Hergueta, A. Bonnet, and E.J.P. Douzery. 2001. A mitochondrial DNA control region phylogeny of the Cervinae: Speciation in *Cervus* and implications for conservation. Animal Conservation, 4:1-11.

Randolph, P.A., et al. 1977. Energy costs of reproduction in the cotton rat, *Sigmodon hispidus*. Ecology, 58:31-45.

Rangel, M.G., and E. Mellink. 1993. Historia natural de la rata magueyera en el Altiplano Mexicano. Pp. 173-183, in: Avances en el Estudio de los Mamíferos de México (R.A. Medellín and G. Ceballos, eds.). Asociación Mexicana de Mastozoología, A.C., México.

Raun, G.G. 1960. Barn owl pellets and small mammal populations near Mathis, Texas, in 1956 and 1959. The Southwestern Naturalist, 5:194-200.

Raven, H.C. 1942. On the structure of *Mesopodon densirostris*, a rare beaked whale. Bulletin of the American Museum of Natural History, 80:23-50.

Ray, J., and M. Sunquist. 2001. Trophic relations in a community of African rainforest carnivores. Oecologia, 127:395-408.

Read, A.J. 1990. Estimation of body condition in harbour porpoises, *Phocoena phocoena*. Canadian Journal of Zoology, 68:1962-1966.

Ream, R., and V. Burkanov. 2005. Trends in abundance of Steller sea lions and northern fur seals across the North Pacific Ocean. PICES XIV Annual Meeting, Vladivostock, Russia.

Reddell, J. 1981. A review of the cavernicole fauna of Mexico, Guatemala and Belize. Texas Memorial Museum Bulletin, 27:327.

Reddell, J.R., and W.R. Elliot. 1973. A checklist of the cave fauna of Mexico. IV. Additional records from the Sierra del Abra, Tamaulipas and San Luis Potosi. Association for Mexican Cave Studies, 5:171-180.

Redford. K.H., and F.J. Eisenberg. 1989. Mammals of the Neotropics. Vol. 1. University of Chicago Press, Chicago.

Redford, K.H., and J.F. Eisenberg. 1992. Mammals of the Neotropics. Vol. 2. University of Chicago Press, Chicago.

Redford, K.H., and J.G. Robinson. 1987. The game of choice: Patterns of indian and colonist hunting in the Neotropics. American Anthropologist, 89:650-667.

Redford, K.K. 1985. Food habits of armadillos (Xenarthra: Dasypodidae). Pp. 429-437, in: The Evolution and Ecology of Armadillos, Sloths, and Vermilinguas (G.G. Montgomery, ed.). Smithsonian Institution Press, Washington, DC.

Redondo, R.A.F., L.P.S. Brina, R.F. Silva, A.D. Ditchfield, and F.R. Santos. 2008. Molecular systematics of the genus *Artibeus* (Chiroptera: Phyllostomidae). Molecular Phylogenetics and Evolution, 49:44-58.

Reduker, D.W., T.L. Yates, and I.F. Greenbaum. 1983. Evolutionary affinities among southwestern long-eared Myotis (Chiroptera: Vespertilionidae). Journal of Mammalogy, 64:666-677.

Reed, J.E., W.B. Ballard, P.S. Gipson, B.T. Kelly, P.R. Krausman, M.C. Wallace, and D.B. Webster. 2008. Diets of free-ranging Mexican gray wolves in Arizona and New Mexico. Wildlife Society Bulletin, 34:1127-1133.

Reeder, W.G., and K.S. Norris. 1954. Distribution, type locality, and habits of the fish-eating bat, *Pizonyx vivesi*. Journal of Mammalogy, 35:81-87.

Reeves, R.R., B.S. Stewart, and S. Leatherwood. 1992. The Sierra Club Handbook of Seals and Sirenians. Sierra Club Books, San Francisco.

Reeves, R.R., S. Leatherwood, S.A. Karl, and E.R. Yohe. 1985. Whaling results at Akutan (1912-1939) and Port Hobron (1926-37), Alaska. Reports of the International Whaling Commission, 35:441-457.

Reeves, R., B. Stewart, P. Clapham, and J. Powell. 2002. National Audubon Society Guide for Marine Mammals of the World. Alfred A. Knopf, New York.

Rehn, J.A. 1904. A study of the bats of the genus *Dermonotus* (*Pteronotus* auct.). Proceedings of the Academy of Natural Sciences of Philadelphia, 56:250-256.

Reich, L.M. 1981. *Microtus pennsylvanicus*. Mammalian Species, 159:1-8.

Reichman, O.J. 1975. Relation of desert rodent diets to available resources. Journal of Mammalogy, 56:731-745.

Reichman, O.J., and D. Oberstein. 1977. Selection of seed distribution types by *Dipodomys merriami* and *Perognathus amplus*. Ecology, 58:636-643.

Reichman, O.J., and M.V. Price. 1993. Ecological aspects of heteromyid foraging. Pp. 539-573, in: Biology of the Heteromyidae (H.H. Genoways and J.H. Brown, eds.). Special Publication, American Society of Mammalogists.

Reichman, O.J., and S.C. Smith. 1990. Burrows and burrowing behavior by mammals. Current Mammalogy, 2:197-244.

Reichman, O.J., and K.M. Van de Graaff. 1973. Seasonal activity and reproductive patterns of five species of Sonoran desert rodents. The American Midland Naturalist, 90:118-126.

Reichman, O.J., D.T. Wicklow, and C. Rebar. 1985. Ecological and mycological characteristics of caches in the mounds of D*ipodomys spectabilis*. Journal of Mammalogy, 66:643-651.

Reid, F.A. 1997. A Field Guide to the Mammals of Central America and Southern Mexico. Oxford University Press, New York.

Reijnders, P., S. Brasseur, J. van der Toorn, P. van der Wolf, I. Boyd, J. Harwood, D. Lavigne, and L. Lowry. 1993. Seals, Fur Seals, Sea Lions, and Walrus: Status Survey and Conservation Action Plan. IUCN, Gland, Switzerland.

Reilly, S.B., and V.G. Thayer. 1990. Blue whale (*Balaenoptera musculus*) distribution in the eastern tropical Pacific. Marine Mammal Science, 6:265-277.

Reilly, S.B., J.L. Bannister, P.B. Best, M. Brown, R.L. Brownell Jr., D.S. Butterworth, P.J. Clapham, J. Cooke, G.P. Donovan, J. Urbán, and A.N. Zerbini. 2008. *Eubalaena glacialis*. In: IUCN Red List of Threatened Species. Version 2010.4. www.iucnredlist.org.

Reinhardt, J. 1873. Et Bidrag til Kudnskab on Aberne: Mexico og Centralamerika. Videnskabernes Meddelingen Naturhistorik Forening Kjöbenhavn, 4(ser. 3):150-158.

Reis, N.R., and A.L. Peracchi. 1987. Qurópteros de regiao de Manaus, Amazonas, Brasil (Mammalia, Chiroptera). Boletín de Museum Parense Emilio Goeldi, Sere Zoologia, 3:161-182.

Rennert, P.D., and C.W. Kilpatrick. 1986. Biochemical systematics of populations of *Peromyscus boylii*. I. Populations from east-central Mexico with low fundamental numbers. Journal of Mammalogy, 67:481-488.

Rennert, P.D., and C.W. Kilpatrick. 1987. Biochemical systematics of *Peromyscus boylii*. II. Chromosomally variable population from eastern and southern Mexico. Journal of Mammalogy, 68:799-811.

Repenning, C.A., R. Peterson, and C. Hubbs. 1971. Contribution to the systematics of the southern fur seal, with particular reference to the Juan Fernandez and Guadalupe species. In: Antarctica Pinnipedia (W.H. Burt, ed.). Antarctica Researcher Series, 18.

Revees, R.R., S.B. Stewart, and S. Leatherwood. 1992. The Sierra Club Handbook of Seals and Sirenians. Sierra Club Handbooks, San Francisco.

Reyes, J.C., J.G. Mead, and K.V. Waerebeek. 1991. A new species of beaked whale *Mesoplodon peruvianus* (Cetacea: Ziphiidae) from Peru. Marine Mammals Science, 7:1-24.

Reyes-Martínez, Y.P.M. 2001. Estudio poblacional de *Peromyscus zarhynchus* (Rodentia: Muridae) en el Parque Nacional Lagos de Montebello, Chiapas. Tesis de licenciatura, Universidad de Guadalajara, Jalisco, México.

Reyes-Osorio, S. 1981. Condición actual de la población de la población del venado bura en la isla Tiburón, Sonora. Pp. 104-108, in: Reunión sobre Necesidades de Investigación sobre Fauna Silvestre y Pastizales en el Norte de México y Suroeste de Estados Unidos. U.S. Forest Service, General Technical Report WO-365, Rio Rico, AZ.

Reyna-Hurtado, R. 2009. Conservation status of the white-lipped peccary (*Tayassu pecari*) outside the Calakmul Biosphere Reserve in Campeche, Mexico: A synthesis. Tropical Conservation Science, 2:159-172.

Reyna-Hurtado, R., and G.W. Tanner. 2005. Habitat preferences of ungulates in hunted and nonhunted areas in the Calakmul Forest, Campeche, Mexico. Biotropica, 37:676-685.

Reyna-Hurtado, R., A. Taber, M. Altrichter, J. Fragoso, A. Keuroghlian, and H. Beck. 2008. *Tayassu pecari*. In: IUCN (International Union for Conservation of Nature). 2010. IUCN Red List of Threatened Species. Version 2010.4. International Union for Conservation of Nature. www.iucnredlist.org.

Reynolds, D.G., and J.J. Beecham. 1980. Home range activities and reproduction of black bears in west-central Idaho. International Conference for Bear Research and Management, 4:181-190.

Reynolds, H.G. 1958. The ecology of the Merriam kangaroo rat (*Dipodomys merriami* Mearns, 1890) on the grazing lands of southern Arizona. Ecological Monographs, 28:111-127.

Reynolds, H.G., and H.S. Haskell. 1949. Life history notes on Price and Bailey pocket mice of southern Arizona. Journal of Mammalogy, 30:150-156.

Reynolds, H.G., and F. Turkowski. 1972. Reproductive variations in the round-tailed ground squirrel as related to winter rainfall. Journal of Mammalogy, 53:893-898.

Reynolds, H.W., R.D. Glaholt, and A.W.L. Hawley. 1983. Bison. In: Wild Mammals of North America: Biology, Management, Economics (J.A. Chapman and G.A. Feldhamer, eds.). Johns Hopkins University Press, Baltimore.

Reynolds, J.E., and R. Rommel. 1999. Biology of Marine Mammals. Smithsonian Institution Press, Washington, DC.

Rice, D.W. 1963. Progress report on biological studies of the larger cetacea in the waters off California. Norsk Hvalfangst-Tidende, 52:181-187.

Rice, D.W. 1974. Whale and whale research in the eastern North Pacific. Pp. 170-195, in: The Whale Problem, A Status Report (W.E. Schevill, ed.). Harvard University Press, Cambridge, MA.

Rice, D.W. 1978. Sperm whale. Pp. 83-87, in: Marine Mammals of Eastern North Pacific and Arctic Waters (D. Haley, ed.). Pacific Search Press, Seattle.

Rice, D.W. 1989. Sperm whale *Physeter macrocephalus* Linnaeus, 1758. Pp. 177-233, in: Handbook of Marine Mammals, vol. 4, River Dolphins and the Larger Toothed Whales (S.H. Ridway and R. Harrison, eds.). Academic Press, London.

Rice, D.W. 1998. Marine Mammals of the World: Systematics and Distribution. Special Publication 4, Society for Marine Mammalogy.

Rice, D.W., and A.A. Wolman. 1971. Life history and ecology of the gray whale, *Eschrichtius robustus*. Special Publication 3, American Society of Mammalogists.

Richard, A. 1970. A comparative study of the activity patterns and behavior of *Alouatta villosa* and *Ateles geoffroyi*. Folia Primatologica, 12:241-263.

Rick, A.M. 1965. *Otonyctomys hatti* in Guatemala. Journal of Mammalogy, 46:335-336.

Rick, A.M. 1968. Notes on bats from Tikal, Guatemala. Journal of Mammalogy, 49:516-520.

Rickart, E.A. 1977. Reproduction, growth and development in two species of cloud forest *Peromyscus* from southern Mexico. University of Kansas Publications, Museum of Natural History, 67:1-22.

Rickart, E.A., and P.B. Robertson. 1985. *Peromyscus melanocarpus*. Mammalian Species, 241:1-3.

Rico-Gray, V., and E.S. Watts. 1989. Estado actual del hábitat ocupado por Ateles y Alouatta en la Península de Yucatán. Pp. 176-190, in: Primatología en México: Comportamiento, Ecología, Aprovechamiento y Conservación de Primates (A. Estrada, R. López-Wilchis, and R. Coates-Estrada, eds.). UAM-Iztapalapa, México D.F.

Riddle, B.R. 1995. Molecular biogeography in the pocket mice (*Perognathus* and *Chaetodipus*) and grasshopper mice (*Onychomys*): The late Cenozoic development of a North American aridlands rodent guild. Journal of Mammalogy, 76:283-301.

Riddle, B.R., and R.L. Honeycutt. 1990. Historical biogeography in North American arid regions: An approach using mitochondrial-DNA phylogeny in grasshopper mice (genus *Onychomys*). Evolution, 44:1-15.

Riddle, B.R., D.J. Hafner, and L.F. Alexander. 2000a. Comparative phylogeography of Bailey's pocket mouse (*Chaetodipus baileyi*) and the *Peromyscus eremicus* species group: Historical vicariance of the Baja California Peninsular desert. Molecular Phylogenetics and Evolution, 17:161-172.

Riddle, B.R., D.J. Hafner, and L.F. Alexander. 2000b. Phylogeography and systematics of *Peromyscus eremicus* species group and historical biogeography of North American warm regional deserts. Molecular Phylogenetics and Evolution, 17:145-160.

Riddle, B.R., D.J. Hafner, L.F. Alexander, and J.R. Jaeger. 2000. Cryptic vicariance in the historical assembly of a Baja California peninsular desert biota. Proceedings of the National Academy of Sciences, 97(10):14438-14443.

Riechers, P.A. 2004. Análisis mastofaunística de la zona sujeta a conservación ecológica Laguna de Bélgica, Chiapas, México. Anales del Instituto de Biología, UNAM, México, Serie Zoología, 75:363-382.

Riechers, P.A., and R. Vidal-L. 2009. Registros de *Choeronycteris mexicana* (Chiroptera: Phyllostomidae) en Chiapas. Revista Mexicana de Biodiversidad, 80:879-882.

Riedman, M. 1990. The Pinnipeds: Seals, Sea Lions, and Walruses. University of California Press, Berkeley.

Rios, E., and S.T. Álvarez Castañeda. 2007. Environmental responses to altitudinal gradients and subspecific validity in pocket gophers (*Thomomys bottae*). Journal of Mammalogy, 88:926-934.

Ripple, W.J., and R.L. Beschta. 2003. Wolf reintroduction, predation risk, and cottonwood recovery in Yellowstone National Park. Forest Ecology and Management, 184:299-313.

Ripple, J., and D. Perrine. 1999. Manatees and Dugongs of the World. Voyageur Press.

Rizo, A.A. 2008. Descripción y análisis de los pulsos de ecolocalización de 14 especies de murciélagos insectívoros aéreos del Estado de Morelos. Tesis de maestría, Instituto de Ecología, A.C., Xalapa, Veracruz.

Rizo-Díaz, B.L.E. 1990. Análisis de algunos aspectos físicos y biológicos de los varamientos de cetáceos en la Bahía de la Paz, B.C.S., México. Tesis profesional, Facultad de Ciencias, Universidad Nacional Autónoma de México, México.

Roa, M.A., and J. Lozada. 1989. Temazates (*Mazama* spp.). Memorias III Simposio sobre Venados en México. Linares, N.L. 29-35.

Robbins, L.W., and R.J. Baker. 1981. An assessment of the nature of rearrangements in eighteen species of *Peromyscus* (Rodentia: Cricetidae). Cytogenetics and Cell Genetics, 31:194-202.

Robbins, L.W., and V.M. Sarich. 1988. Evolutionary relationships in the family Emballonuridae (Chiroptera). Journal of Mammalogy, 69:1-13.

Roberts, M.W., and J.L. Wolfe. 1974. Social influences on susceptibility to predation in cotton rats. Journal of Mammalogy, 55:869-872.

Robertson, P.B. 1975. Reproduction and community structure of rodents over a transect in southern Mexico. PhD dissertation, University of Kansas.

Robertson, P.B., and E.A. Rickart. 1975. *Cryptotis magna*. Mammalian Species, 61:1-2.

Robertson, P.B., and G.G. Musser. 1976. A new species of *Peromyscus* (Rodentia: Cricetidae), and a new specimen of Peromyscus simulatus from southern Mexico, with comments on their ecology. University of Kansas Publications, Museum of Natural History, 47:1-8.

Robineau, D., and M. Vely. 1993. Stranding of a specimen of Gervais' beaked whale (*Mesoplodon europaeus*) on the coast of West Africa (Mauritania). Marine Mammal Science, 9:438-440.

Robinson, J.G., and J.F. Eisenberg. 1985. Group size and foraging habits of the collared peccary *Tayassu tajacu*. Journal of Mammalogy, 66:153-155.

Robinson, J.G., and K.H. Redford. 1986. Body size and population density of Neotropical forest mammals. The American Naturalist, 128:665-680.

Robinson, R.G., and K.H. Redford. 1991. Neotropical Wildlife Use and Conservation. University of Chicago Press, Chicago.

Robles Gil, P., G. Ceballos, and F. Eccardi. 1993. Diversidad de fauna mexicana. CEMEX, Monterrey, México.

Rodrígues, A., and K. Gaston. 2002. Optimisation in reserve selection procedures--why not? Biological Conservation, 107:123-129.

Rodríguez, E., F. García, and D. Canales. 1993. Translocación del mono aullador (*Alouatta palliata*): una alternativa conservacionista. Pp. 129-178, in: Estudios Primatológicos en México (A. Estrada, E. Rodríguez-Luna, R. López-Wilchis, and R. Coates-Estrada, eds.). Vol. 1. Universidad Veracruzana. Xalapa, Veracruz, México.

Rodríguez, P., J. Soberón, and H.T. Arita. 2003. El componente beta de la diversidad de mamíferos de México. Acta Zoologica Mexicana (n.s.), 89:1-19.

Rodríguez-Herrera, B., R.A. Medellín, and R.M. Timm. 2007. Neotropical Tent-roosting Bats Murciélagos. INBio, Costa Rica.

Rodríguez-Jaramillo, M. del C., and D. Gendron. 1996. Report of a sea otter, *Enhydra lutris*, off the coast of Isla Magdalena, Baja California Sur, Mexico. Marine Mammal Science, 12:153-156.

Rodríguez-Martínez, L., J. Vázquez, and A. Bautista. 2007. Primer registro del gato montés (Lynx rufus) en el Parque Nacional La Malinche, Tlaxcala, México. Revista Mexicana de Mastozoología, 11:80-84.

Rodríguez Vela, H. 1999. *Notiosorex crawfordi* (Coues, 1877) en el matorral desértico de Nuevo León, México. Vertebrata Mexicana, 5:5-8.

Rogers, A.R., and H. Harpending. 1992. Population growth makes waves in the distribution of pairwise genetic differences. Molecular Biology and Evolution, 9(3):552-569.

Rogers, D.S. 1989. Evolutionary implications of chromosomal variation among spiny pocket mice, genus *Heteromys* (order Rodentia). The Southwestern Naturalist, 34:85-100.

Rogers, D.S. 1990. Genic evolution, historical biogeography, and systematic relationships among spiny pocket mice (subfamily Heteromyinae). Journal of Mammalogy, 71:668-685.

Rogers, D.S., and M.D. Engstrom. 1992. Evolutionary implications of allozymic variation in tropical *Peromyscus* of the *mexicanus* species group. Journal of Mammalogy, 73:55-69.

Rogers, D.S., and E.J. Heske. 1984. Chromosomal evolution of the brown mice, genus *Scotinomys* (Rodentia: Cricetidae). Genetica, 63:221-228.

Rogers, D.S., and J.W. Rogers. 1992. *Heteromys nelsoni*. Mammalian Species, 397:1-2.

Rogers, D.S., and D.J. Schmidly. 1982. Systematics of spiny pocket mice (Genus *Heteromys*) of the *Desmarestianus* species group from Mexico and northern Central America. Journal of Mammalogy, 63:375-386.

Rogers, D.S., E.J. Heske, and D.A. Good. 1983. Karyotype and a range extension of *Reithrodontomys tenuirostris*. The Southwestern Naturalist, 21:372-374.

Rogers, D.S., C.C. Funk, J.R. Miller, and M.D. Engstrom. 2007. Molecular phylogenetic relationships among crested-tailed mice (genus *Habromys*). Journal of Mammalian Evolution, 14:37-55.

Rogers, D.S., I.F. Greebaum, S.J. Gunn, and M.D. Engstrom. 1984. Cytosystematic value of chromosomal inversion data in the genus *Peromyscus* (Rodentia: Cricetidae). Journal of Mammalogy, 65:457-465.

Rogers, L.L. 1977. Social relationships, movements, and population dynamics of black bears in north-eastern Minnesota. PhD dissertation, University of Minnesota, Minneapolis.

Rogers, L.L. 1987. Effects of food supply and kinship on social behavior of black bears in northeastern Minnesota. Wildlife Monographs.

Rogovin, K., G. Shenbrot, and A. Surov. 1991. Analysis of spatial organization of a desert rodent community in Bolson de Mapimi, Mexico. Journal of Mammalogy, 72:347-359.

Rojas, M.E. 1984. Descripción del microhabitat de cinco especies de ratones en la Sierra del Ajusco. Tesis de licenciatura, Facultad de Ciencias, Universidad Nacional Autónoma de México, México.

Rojas, P. 1951. Estudio biológico del conejo de los volcanes (Género *Romerolagus*) (Mammlia: Lagomorpha). Tesis de licenciatura. Facultad de Ciencias, Universidad Nacional Autónoma de México, México.

Rojas-Bracho, L. 1984. Presencia y Distribución del Rorcual Común (*Balaenoptera physalus* (Linnaeus, 1758) (Cetacea: Balaenopteridae) en el Golfo de California, México. Tesis Profesional, Facultad de Ciencias, Universidad Nacional Autónoma de México, México.

Rojas-Martínez, A.E., and C. Sánchez-Hernández. 1988. El ratón de cola corta *Microtus mexicanus mexicanus* depredado por la garza chapulinera. Anales del Instituto de Biología, UNAM, México, Seria Zoología, 58:941-942.

Rojas-Martínez, A.E., and A. Valiente-Banuet. 1996. Análisis comparativo de la quiropterofauna del Valle de Tehuacán-Cuicatlán, Puebla-Oaxaca. Acta Zoológica Mexicana (n.s.), 67:1-23.

Rolley, R.E. 1983. Behavior and population dynamics of bobcats in Oklahoma. PhD dissertation, Oklahoma State University, Stillwater.

Rolley, R.E. 1985. Dynamics of a harvested bobcat population in Oklahoma. Journal of Wildlife Management, 49:283-292.

Rolley, R.E. 1999. Wild furbearer management and conservation in North America. Section IV: Species Biology, Management and Conservation. Chapter 50.

Romero, A.M.L. 1993. Biología de Liomys pictus. Tesis doctoral, Facultad de Ciencias, Universidad Nacional Autónoma de México, México.

Romero, F.J. 1993. Análisis de habitat para fauna silvestre, una propuesta para su estudio. Pp. 371-380, in: Memorias del XI Simposio Nacional y I Simposio Internacional de Fauna Silvestre. División de Educación Contínua, Facultad de Medicina Veterinaria y Zootecnia, Universidad Nacional Autónoma de México, México.

Romero, F.J. 1994. Patrones de uso de hábitat de *Neotomodon alstoni* y *Peromyscus difficilis* en un bosque templado del Eje Neovolcánico Transversal. Segundo Congreso Nacional de Mastozoología, Guadalajara, Jalisco. Asociación Mexicana de Mastozoología, A.C.

Romero, F.J., and J. López-Paniagua. 1991. Estudio sobre la descripción de los patrones de uso de hábitat de *Romerolagus diazi* en el Volcán Pelado. Pp. 36-37, in: I Congreso Nacional de Mastozoología. Asociación Mexicana de Mastozoología, A.C., Xalapa, Veracruz, México.

Romero, F.J., J. López-Paniagua, and G. Campos-R. 1987. Estudios preliminares sobre la descripción del hábitat de *Romerolagus diazi* en una de sus áreas de distribución. Pp. 96-97, in: Memorias del V Simposio sobre Fauna Silvestre. División de Educación Contínua, Facultad de Medicina Veterinaria y Zootecnia, Universidad Nacional Autónoma de México, México.

Romero, F.J., A. Velázquez, and L. León. 1991. Fragmentación del hábitat del zacatuche (*Romerolagus diazi* Ferrari-Pérez, 1893). Pp. 98-104, in: Memorias del IX Simposio sobre Fauna Silvestre "General M.V. Manuel Cabrera Valtierra." División de Educación Continua, Facultad de Medicina Veterinaria y Zootecnia, Universidad Nacional Autónoma de México, México.

Romero, M.E.R. 1998. La navegación maya. Arqueología Mexicana, 6(33): 6-15.

Romero-Nájera, I. 2004. Distribución, abundancia y uso de hábitat de Boa constrictor introducida a la Isla Cozumel. Tesis de maestría en ciencias, Universidad Nacional Autónoma de México, Morelia, México.

Romero-R., F. 1987. Análisis de la alimentación del lince (Lynx rufus escuinapae) en el Volcán Pelado, Ajusco, Distrito Federal, México. Tesis de licenciatura, Faculatad de Ciencias, Universidad Nacional Autónoma de México, México.

Romero-R., F. 1993. Análisis de la alimentación del lince (*Lynx rufus* escuinapae) de en el centro de México. Avances en el Estudio de los Mamíferos de México, Publicacion Especial de la Asociación Mexicana de Mastozoología, A.C., 1 (en Prensa), México.

Rommel, S., and J. Reynolds. 2000. Diaphragm structure and function in the Florida manatee (*Trichechus manatus latirostris*). The Anatomical Record, 259(1):41-51.

Romo, M. 1987. Dinámica de la población del venado cola blanca (*Odocoileus virginianus*) en la Sierra San Blas de Pabellón del Estado de Aguascalientes. Tesis de licenciatura, Universidad Autónoma de Aguascalientes, Aguascalientes, México.

Romo, M., and S. Gallina. 1988. Estudio de la población del venado cola blanca (*Odocoileus virginianus*) en la Sierra San Blas de Pabellón del Estado de Aguascalientes. Pp. 8-17, in: II Simposio sobre el Venado en México, Universidad Nacional Autónoma de México, México.

Romo, V.E. 1993. Distribución altitudinal de los roedores al noreste del estado de Querétaro. Tesis de licenciatura, Facultad de Ciencias, Universidad Nacional Autónoma de México, México.

Romo-Vazquez, E., L. Leon, and O. Sanchez. 2005. A new species of *Habromys* (Rodentia: Sigmodontinae) from Mexico. Proceedings of the Biological Society of Washington, 118:605-611.

Roosmalen, M.G.M. van. 1980. Habitat preferences, diet, feeding strategy and social organization of the black spider monkey (*Ateles paniscus paniscus* L.) in Surinam. Unpublished PhD dissertation, Agricultural University of Wageningen, Wageningen.

Roosmalen, M.G.M. van, and L.L. Klein. 1988. The spider monkeys, genus Ateles. Pp. 455-537, in: Ecology and Behavior of Neotropical Primates (R.A. Mittermeier, A.B. Rylands, A.F. Coimbra-Filho, and G.A.B. da Fonseca, eds.). Vol. 2. World Wildlife Fund, Washington, DC.

Roots, C.G. 1966. Notes on the breeding of white-lipped peccaries. International Zoo Yearbook, 6:198-199.

Rosas, A. 1990. Comportamiento del venado en cautiverio. Tesis de licenciatura, Facultad de Ciencias, Universidad Nacional Autónoma de México, México.

Rosas-Becerril, P. 1992. Patrones reproductivos del venado cola blanca (*Odocoileus virginianus couesi*) durante su estación reproductiva en la Reserva de la Biosfera La Michilía, Durango, México. Tesis de licenciatura, Escuela Nacional de Estudios Profesionales-Iztacala, Universidad Nacional Autónoma de México, México.

Rosas-Rosas, O.C. 2006. Ecological status and conservation of jaguars in northeastern Sonora, Mexico. PhD dissertation, New Mexico State University, Las Cruces.

Rosatte, R.C. 1987. Striped, spotted, hooded, and hog-nosed skunk. Pp. 598-613, in: Wild Furbearer Management and Conservation in North America, North Bay (M. Novak, J.A. Baker, M.E. Obbard, and B. Mallouch, eds.). Ontario Ministry of Natural Resources, Toronto.

Rosel, P.E., and L. Rojas-Bracho. 1993. Genetic variation, or lack thereof, in the vaquita. P. 13, in: 10th Biennial Conference on the Biology of Marine Mammals. Galveston, TX.

Ross, A. 1967. Ecological aspects of the food habits of insectivorous bats. Proceedings of the Western Foundation of Vertebrate Zoology, 1:205-264.

Ross, G.J.B. 1978. Records of pygmy and dwarf sperm whales genus *Kogia*, from southern Africa, with biological notes and some comparisons. Annals of the Cape Province Museum of Natural History, 11:259-327.

Rossi, R.V., R.S. Voss, and D.P. Lunde. 2010. A revision of the didelphid marsupial genus *Marmosa*. Part 1, The species in Tate's "Mexicana" and "mitis" sections and other closely related forms. Bulletin of the American Museum of Natural History, 334:1-83.

Roth, E.L. 1976. A new species of pocket mouse (*Perognathus*: *Heteromyidae*) from the Cape Region of Baja California Sur, Mexico. Journal of Mammalogy, 57: 562–566.

Rovirosa, J.N. 1885. Apuntes para la zoología de Tabasco, vertebrados observados en el territorio de Macuspas. La Naturaleza, 7:345-389.

Ruddiman, W.F. 1987. Northern oceans. Pp. 137-154, in: North America and Adjacent Oceans during the Last Deglaciation (W.F. Ruddiman and H.E. Wright Jr., eds.). The Geology of North America, vol. K-3, Geological Society of America.

Ruedas, L.A. 1998. Systematics of Sylvilagus Gray, 1867 (Lagomorpha: Leporidae) from southwestern North America. Journal of Mammalogy, 79:1355-1378.

Ruedas, L.A., and J. Salazar-Bravo. 2007. Morphological and chromosomal taxonomic assessment of *Sylvilagus brasiliensis gabbi* (Leporidae). Mammalia, 71:63–69.

Ruiz, G.F., and F.E. Rocha-V. 2009. Mamíferos del Ejido El Durazno, Municipio de Coyuca de Catalán, Guerrero. www.sites.google.com/site/vertebradosdeguerrero/Home/proyectos/mamiferos-del-ejido-el-durazno-municipio-de-coyuca-de-catalan-guerrero.

Ruiz, R.G. 1983. *Mimon crenulatum keenani* (Chiroptera: Phyllostomidae) from Belize. The Southwestern Naturalist, 28:374.

Ruiz-García, M., E. Payán, A. Murillo, and D. Álvarez. 2006. DNA microsatellite characterization of the jaguar (*Panthera onca*) in Colombia. Genes and Genetic Systems, 81:115-127.

Ruiz-Piña, H.A. 1994. Variación geográfica y sistemática de *Sciurus yucatanensis* (Rodentia: Sciuridae). Tesis de maestría en ciencias, Universidad Nacional Autonóma de México, México.

Russell, A.L., R.A. Medellín, and G.F. Mc Cracken. 2005. Genetic variation and migration in the Mexican free-tailed bat (*Tadarida brasiliensis mexicana*). Molecular Ecology, 14:2207-2222.

Russell, J.K. 1981. Exclusion of adult male coatis from social groups: Protection from predation? Journal of Mammalogy, 62:206-208.

Russell, J.K. 1982. Timing of reproduction by coatis (*Nasua narica*) in relation to fluctuations in food resources. Pp. 413-431, in: The Ecology of a Tropical Forest: Seasonal Rhythms in a Tropical Ecosystem (E.G. Leigh Jr., A.S. Rand, and D.M. Windsor, eds.). Smithsonian Institution Press, Washington, DC.

Russell, R.J. 1953. Four new pocket gophers of the genus *Cratogeomys* from Jalisco, Mexico. University of Kansas Publications, Museum of Natural History, 5:535-542.

Russell, R.J. 1957. A new species of pocket gopher (*Pappogeomys*) from Jalisco, Mexico. University of Kansas Publications, Museum of Natural History, 9:357-361.

Russell, R.J. 1968a. Evolution and classification of the pocket gophers of the subfamily Geomyinae. University of Kansas Publications, Museum of Natural History, 16:473-579.

Russell, R.J. 1968b. Revision of the pocket gophers of the genus *Pappogeomys*. University of Kansas Publications, Museum of Natural History, 16:581-776.

Russell, R.J., and A. Alcorn. 1957. *Reithrodontomys burti* in Sinaloa. Journal of Mammalogy, 38:417-418.

Ruthberg, A.T. 1984. Birth synchrony in American bison (*Bison bison*): Response to predation or season? Journal of Mammalogy, 65:418-423.

Rzedowski, J. 1978. Vegetación de México. Limusa, México.

Rzedowski, J. 1981. Vegetación de Mexico. Limusa, Mexico.

Rzedowski, J. 1983. Vegetación de México. Limusa, México.

Rzedowski, J. 1998. Diversidad y orígenes de la flora fanerogámica de México. Pp. 129-145, in: Diversidad biológica de México: Orígenes y distribución (T.P. Ramamoorthy, R. Bye, A. Lot, and J. Fa, eds.). Instituto de Biología, UNAM, México.

Rzedowski, J., and G.C. Rzedowski. 1981. Flora Fanerogámica del Valle de México. Vol. I. Cia. Editorial Continental, S. A. de C.V. (CECSA), México.

Sacchi-Santos, L.H., P.H. Oh, S. Loredano, R. Mondín-Machado, and I.B. Moreno. 1992. Primeiras observacoes sobre a ocorrencia de *Steno bredanensis* no litoral norte do estado do Rio Grande do sul, Brasil. In: Memorias de la Quinta Reunión de Especialistas en Mamíferos Acuáticos de América del Sur. Buenos Aires, Argentina.

Sáenz, J. 1994. Ecología del pizote (*Nasua narica*) y su papel como dispersador de semillas en el bosque seco tropical, Costa Rica. Tesis de maestría en manejo de vida silvestre, Universidad Nacional, San José, Costa Rica.

Sahagún, Fr. B. ca. 1580 [1992, 8th ed.]. Historia general de las cosas de Nueva España. Editorial Porrúa, México.

Salazar, S. 2002. Lobo marino y lobo peletero. Pp. 267-290, in: Reserva Marina de Galápagos, Línea de Base de la Biodiversidad (E. Danulat and G.J. Edgar, eds.). Fundación Charles Darwin/Servicio del Parque Nacional Galápagos, Santa Cruz, Galápagos, Ecuador .

Saldaña, V.R. 2008. Comparación de la diversidad de murciélagos filostómidos en fragmentos de bosque mesófilo de montaña y cafetales de sombra, del centro de Veracruz. Tesis de maestría, Instituto de Ecología, A.C., Xalapa, Veracruz.

Salinas, M., and P. Ladrón de Guevara. 1993. Riqueza y diversidad de mamíferos marinos. Pp. 85-93, in: Biología y Problemática de los Vertebrados en México (O. Flores-V. and A. Navarro-S., eds.). Ciencias (special number), 7:1-110.

Salinas, Z.M., C. Álvarez F., P. Ladrón de Guevara P., and A. Aguayo L. 1990. La importancia de la fotoidentificación en el estudio de los cetáceos en México. La Ballena jorobada (*Megaptera novaeangliae*) un ejemplo. In: Resúmenes IV Reunión de Trabajo de Especialistas en Mamíferos Acuáticos de América del Sur. Valdivia, Chile.

Salinas Z., M., and L.F. Bourillón. 1988. Taxonomía, diversidad y distribución de los cetáceos de la Bahía de Banderas, México. Tesis profesional, Facultad de Ciencias, Universidad Nacional Autónoma de México, México.

Sanborn, C.C. 1949. Bats of the genus *Micronycteris* and its subgenera. Fieldiana Zoology, 31:215-233.
Sánchez, H.O. 1985. Los mamíferos en las culturas antiguas de México. Zacatuche, 1(2):2-12.
Sánchez, H.O. 1993. Análisis de algunas tendencias ecogeográficas del género *Reithrodontomys* (Rodentia: Muridae) en México. Pp. 25-44, in: Avances en el Estudio de los Mamíferos de México (R.A. Medellín and G. Ceballos, eds.). Vol. 1. Publicaciones Especiales, Asociación Mexicana de Mastozoología, A.C., México.
Sánchez, H.O., and G. Magaña Cota. 2008. Murciélagos de Guanajuato: perspectiva histórica y actualización de su conocimiento. Acta Universitaria (Universidad de Guanajuato), 3:27-39.
Sánchez, H.O., G. López-O., and R. López-Wilchis. 1989. Murciélagos de la ciudad de México y sus alrededores. Pp. 141-165, in: Ecología Urbana (R. Gío-Argaéz, I. Hernández-R., and E. Sáinz-H., eds.). Universidad Nacional Autónoma de México, Sedue, SEP, Sociedad Mexicana de Historia Natural, Universidad Autónoma Metropolitana, México.
Sánchez, H.O., G. Téllez-Girón, R.A. Medellín, and G. Urbano-V. 1986. New records of mammals from Quintana Roo, Mexico. Mammalia, 50:275-278.
Sánchez, H.O., J. Ramírez-Pulido, U. Aguilera-Reyes, and O. Monroy-Vilchis. 2002. Felid records from the state of Mexico, Mexico. Mammalia, 66:289-294.
Sánchez, H.O., R. Medellín, A. Aldama, B. Goettsch, J. Soberón, and M. Tambutti. 2007. Método de evaluación del riesgo de extinción de las especies silvestres en México (MER). Secretaría de Medio Ambiente y Recursos Naturales, México D.F.
Sánchez, R.V.H. 1987. Observaciones sobre el comportamiento durante el periodo reproductivo del lobo marino común *Zalophus californianus* en la Lobera Morro de Santo Domingo, B.C. México. Tesis de licenciatura en biología, Facultad de Ciencias, Universidad Nacional Autónoma de México.
Sánchez-Casas, N., and T. Álvarez. 1997. Notas sobre la dieta de *Hylonycteris* (Chiroptera: Phyllostomatidae) en México. Vertebrata Mexicana, 3:9-12.
Sánchez-Cordero, V. 1993. Estudio poblacional de la rata espinosa (*Heteromys desmarestianus*) en una selva húmeda de Veracruz, México. Pp. 301-316, in: Avances en el Estudio de los Mamíferos de México (R.A. Medellín and G. Ceballos, eds.). Vol. 1. Asociacion Mexicana de Mastozoología, A.C., Publicaciones Especiales.
Sánchez-Cordero, V. 2001. Elevational gradients of diversity for bats and rodents in Oaxaca, Mexico. Global Ecology and Biogeography, 10:63-76.
Sánchez-Cordero, V., and T.H. Fleming. 1993. Ecology of tropical heteromyids. Pp. 596-617, in: Biology of the Heteromyidae (H.H. Genoways and J.H. Brown, eds.). Special Publication 10, American Society of Mammalogists.
Sánchez-Cordero, V., and R.A. Valadez. 1996. Hábitat y distribución del género Oryzomys (Rodentia: Cricetidae). Anales del Instituto de Biología, UNAM, México, Serie Zoología, 59:99-112.
Sánchez-Cordero, V., and B. Villa-R. 1988. Variación morfométrica en *Peromyscus spicilegus*. Anales del Instituto de Biología, UNAM, México, Serie Zoología, 58:819-836.
Sánchez-Cordero, V., P. Illoldi-Rangel, M. Linaje, S. Sarkar, and A. Townsend Peterson. 2005. Deforestation and extant distributions of Mexican endemic mammals. Biological Conservation, 126:465-473.
Sánchez-Hernandez, C. 1978. Registro de murciélagos para el estado de Jalisco, México. Anales del Instituto de Biología, UNAM, México, Serie Zoología, 1:155-158.
Sánchez-Hernandez, C. 1984. Los murciélagos de la Estación de Investigación, Experimentación y Difusión "Chamela," Jalisco, México. II Reunión Iberoamericana de Conservación y Zoología de Vertebrados, 385-398.
Sánchez-Hernandez, C., and C. Chávez. 1985. Observaciones del murciélago de cápsula Diclidurus virgo Thomas. II Reunión Iberoamericana de Conservación y Zoología de Vertebrados, 1:411-416.
Sánchez-Hernandez, C., and Ma. de L. Romero A. 1995. Murciélagos de Tabasco y Campeche: una propuesta para su conservación. Instituto de Biología, Universidad Nacional Autónoma de México, México.
Sánchez-Hernandez, C., Ma. de L. Romero A., and G.A. Nuñez. 1992. El oso hormiguero *Tamandua mexicana* en la costa del Estado de Michoacán. The Southwestern Naturalist, 37:88-89.
Sánchez-Hernandez, C., C. Chávez, E. Ceballos, and M. A. Gurrola. 1985. Notes on the distribution and reproduction of bats from coastal regions in Michoacan, Mexico. Journal of Mammalogy, 66:549-553.
Sánchez-Hernandez,.C., M. de L. Romero A., R. Vargas Y., and G. Gaviño de la Torre. 1993. Noteworthy records of some bats from Morelos, Mexico. Bat Research News, 34:1-2.
Sánchez-Hernandez,. C., Ma. de L. Romero-Almaraz, R.D. Owen, A. Núñez-Garduño, and R. López-Wilchis. 1999. Noteworthy records of mammals from Michoacán, México. The Southwestern Naturalist, 44:231-235.
Sánchez-Hernandez, C., M.L. Romero, H. Colín, and C. García. 2001. Mamíferos de cuatro áreas con diferente grado de alteración en el sureste de México. Acta Zoológica Mexicana (n.s.), 84:35-48.
Sánchez-Hernández, C., M.L. Romero-Almaraz, H. Colín-Martínez, and C. García-Estrada. 2001. Mamíferos de cuatro áreas con diferente grado de alteración en el sureste de México. Acta Zoológica Mexicana (n.s.), 84:35-48.
Sánchez-Hernández, C., M.L. Romero-Almaraz, and G.D. Schnell. 2005. New species of *Sturnira* (Chiroptera: Phyllostomidae) from northern South America. Journal of Mammalogy, 86:866-872.
Sánchez-Hernández, C., M. de L. Romero-Almaraz, G.D. Schnell, M.L. Kennedy, T.L. Best, R.D. Owen, and C. López-González. 2002. Bats of Colima, Mexico: New records, geographic distribution, and reproductive condition. Occasional Papers, Oklahoma Museum of Natural History, 12:1-23.
Sánchez-Herrera, O. 1996. Una técnica para capturar mamíferos pequeños sobre árboles, evitando daños forestales. Vertebrata Mexicana, 1:17-23.
Sanderson, G.C. 1983. Procyon lotor (Mapache, Raccoon). Pp. 485-488, in: Costa Rican Natural History (D.H. Janzen, ed.). University of Chicago Press, Chicago.
Sanderson, G.C. 1987. Raccoon. Pp. 487-499, in: Wild Furbearer Management and Conservation North America (M. Novak, J. Baker, M.E. Obbard, and B. Malloch, eds.). Ministry of Natural Resources, Ontario, Canada.
Sanderson, E.W., C.L.B. Chetkiewicz, R.A. Medellín, A. Rabinowitz, K.H. Redford, J.G. Robinson, and A.B. Taber. 2002. Un análisis geográfico del estado de conservación y distribución de los jaguares a través de su area de distribución. Pp. 551-600, in: El Jaguar en el Nuevo Milenio (R.A. Medellín, C. Equihua, C.L.B. Chetkiewicz, P.G. Crawshaw, A. Rabinowitz, K.H. Redford, J.G. Robinson, E.W. Sanderson, and A. Taber, eds.). Ediciones Cientificas Universitarias, México.
Sanderson, E. W., K.H. Redford, B. Weber, K. Aune, D. Baldes, J. Berger, D. Carter, C. Curtin, J. Derr, S. Dobrott, E. Fearn, C. Fleener, S. Forrest, C. Gerlach, C.C. Gates, J. Gross, P. Gogan, S. Grassel, J.A. Hilty, M. Jensen, K. Kunkel, D. Lammers, R. List, K. Minkowski, T. Olson, C. Pague, P.B. Robertson, and B. Stephensont. 2008. The ecological future of the North American bison: Conceiving long-term, large-scale conservation of wildlife. Conservation Biology, 22:252-266.
Santos-Moreno, A., M. Briones-Salas, G. González-Pérez, and T. de J. Ortiz. 2003. *Rheomys mexicanus* (Rodentia, Muridae) and *Lontra longicaudis annectens* (Carnivora, Mustelidae) in Sierra Norte de Oaxaca, Mexico. The Southwestern Naturalist, 48:312-313.
Santos del Prado, K. 1996. Diversidad y conservación de mamíferos en México: un enfoque taxonómico y filogenético. Tesis de licenciatura, Facultad de Ciencias, UNAM.
SARH (Secretaria de Agricultura y Recursos Hidráulicos). 1994. Calendario Cinegético, 1994-1995. México D.F.

Sarmiento-Aguilar, R. 1999. Estudio poblacional de tres especies de roedores (Rodentia: Muridae) en el Parque Nacional Lagos de Montebello, Chiapas. Tesis de licenciatura, Escuela de Biología, Universidad de Ciencias y Artes de Chiapas, Tuxtla Gutiérrez, Chiapas, México.

Sarti, L., L. Flores, and A. Aguayo. 1991. Una nota sobre alimentación de Orcinus orca. XVI Reunión Internacional para el Estudio de los Mamíferos Marinos. Nuevo Vallarta, Bahía de Banderas, Nayarit, México.

Saussure, H. De. 1993. Voyage aux Antilles et au Mexique 1854-1856 présenté par Louis de Roguin et Claude Weber. Editions Olizane, Ginebra, Switzerland.

Saussure, M.H. 1860. Note sur quelques mammiferes du Mexique. Revue et Magazine de Zoologie, Paris (series 2), 12:3-11.

Saussure, M.H. 1861. Note compléntaire sur quelques mamiféres du Mexique. Revue et Magazine de Zoologie, Paris (series 2), 13:3.

Scammon, C.M. 1874. The Marine Mammals of the North Western Coast of North America, Described and Illustrated: Together with an Account of the American Fishery. John H. Carmany and Co. San Francisco. Reprinted (1968) with a new introduction by Victor B. Scheffer. Dover Publications, Inc., New York.

Schaldach, W.J., Jr. 1960. *Xenomys nelsoni* Merriam, sus relaciones y sus hábitos. Revista de la Sociedad Mexicana de Historia Natural, 21:425-434.

Schaldach, W.J., Jr. 1964. Notas breves sobre algunos mamíferos del Sur de México. Anales del Instituto de Biología, UNAM, México, Serie Zoología, 35:129-137.

Schaldach, W.J., Jr. 1966. New forms of mammals from southern Oaxaca, Mexico, with notes on some mammals of the coastal range. Säugethierkundliche Mitteilungen,14:286-297.

Schaldach, W.J., Jr. and C.A. McLaughlin. 1960. A new genus and species of glossophagine bat from Colima, Mexico. Los Angeles Country, Museum Contributions in Science, 37:1-8.

Schaller, G.B., and P.G. Crawshaw. 1980. Movement patterns of jaguar. Biotropica, 12:161-168.

Schantz, V.S. 1948a. A new badger from Mexico-United States boundary. Proceedings of the Biological Society of Washington, 61:175-176.

Schantz, V.S. 1948b. Extension of range of *Taxidea taxus sonoriensis*. Journal of Mammalogy, 29:75.

Schantz, V.S. 1949. Three new races of badgers (Taxidea) from southwestern United States. Journal of Mammalogy, 30:301-305.

Scheffer, V.B. 1958. Seals, Sea Lions, and Walruses; A Review of the Pinnipedia. Stanford University Press, Stanford.

Schilling, M.R., I. Seipt, M. Weinrich, S.E. Frohock, A.E. Kuhlberg, and P.J. Clapham. 1992. Behavior of individually-identified sei whales *Balaenoptera borealis* during an episodic influx into the southern Gulf of Marine in 1986. Fishery Bulletin, 90:749-755.

Schlitter, D.A. 1973. *Notiosorex crawfordi evotis* from Nayarit. The Southwestern Naturalist, 17:423.

Schmidly, D.J. 1972. Geographic variation in the white-ankled mouse *Peromyscus pectoralis*. The Southwestern Naturalist, 17:113-138.

Schmidly, D.J. 1973. Geographic variation and taxonomy of *Peromyscus boylii* from Mexico and the southwestern United States. Journal of Mammalogy, 54:111-130.

Schmidly, D.J. 1974. *Peromyscus pectoralis*. Mammalian Species, 49:1-3.

Schmidly, D.J. 1977a. Factors governing the distribution of mammals in the Chihuahuan desert region. Pp. 163-192, in: Transactions of the Symposium on the Biological Resources of the Chihuahuan Desert Region United States and Mexico (R.H. Wauer and D.H. Riskind, eds.). U.S. Department of the Interior, National Park Service Transactions and Proceedings Series, 3:1-658.

Schmidly, D.J. 1977b. The Mammals of Trans-Pecos Texas. Texas A&M University Press, College Station.

Schmidly, D.J. 1983. Texas Mammals East of the Balcones Fault Zone. Texas A&M University Press, College Station.

Schmidly, D.J. 1991. The Bats of Texas. Texas A&M University Press, College Station.

Schmidly, D.J., and R.D. Bradley. 1995. Morphological variation in the Sinaloan mouse *Peromyscus simulus*. Revista Mexicana de Mastozoología, 1:44-58.

Schmidly, D.J., and W.B. Davis. 2004. The Mammals of Texas. 6th ed. University of Texas Press, Austin.

Schmidly, D.J., and F.S. Hendricks. 1976. Systematics of the southern races of Ord's kangaroo rat, *Dipodomys ordii*. Bulletin of the Southern California Academy of Science, 75:225-237.

Schmidly, D.J., and F.S. Hendricks. 1984. Mammals of the San Carlos Mountains of Tamaulipas, Mexico. Pp. 15-69, in: Contributions in Mammalogy in Honor of Robert L. Packard (R.E. Martin and B.R. Chapman, eds.). Special Publications of the Museum, Texas Tech University, 22:1-234.

Schmidly, D.J., and C.O. Martin. 1973. Notes on bats from the Mexican state of Queretaro. Bulletin of the Southern California Academy of Science, 72:90-92.

Schmidly, D.J., and G.L. Schroeter. 1974. Karyotypic variation in *Peromyscus boylii* (Rodentia: Cricetidae) from Mexico and corresponding taxonomic implications. Systematic Zoology, 23:333-342.

Schmidly, D.J., and D.M. Smith. 1981. Marine Mammals of the Southeastern United States Coast and the Gulf of Mexico. U.S. Department of the Interior, Washington, DC.

Schmidly, D.J., R.D. Bradley, and P.S. Cato. 1988. Morphometric differentiation and taxonomy of three chromosomally characterized groups of *Peromyscus boylii* from east-central Mexico. Journal of Mammalogy, 69:462-480.

Schmidly, D.J., K.T. Wilkins, and J.N. Derr. 1993a. Biogeography. Pp. 319-356, in: Biology of Heteromyidae (H.G. Genoways and J.H. Brown, eds.). Special Publication, American Society of Mammalogists.

Schmidly, D.J., K.T. Wilkins, and J.N. Derr. 1993b. Taxonomy. Pp. 38-196, in: Biology of the Heteromyidae (H.H. Genoways and J.H. Brown, eds.). Special Publication 10, American Society of Mammalogists.

Schmidly, D.J., M.R. Lee, W.S. Modi, and E.G. Zimmerman. 1985. Systematics and notes on the biology of *Peromyscus hooperi*. Occasional Papers of the Museum, Texas Tech University, 97:1-40.

Schmidly, D.J., K.T. Wilkins, R.L. Honeycutt, and B.C. Weynand. 1977. The bats of east Texas. Texas Journal of Science, 28:127-143.

Schmidt, C. 1988. Reproduction. Pp. 99-109, in: Natural History of Vampire Bats (A.M. Greenhall and U. Schmidt, eds.). CRC Press, Boca Raton, FL.

Schmidt, C.A., and M.D. Engstrom. 1994. Genic variation and systematics of rice rats (*Oryzomys palustris* species group) in south Texas and northeastern Tamaulipas, Mexico. Journal of Mammalogy, 75:914-928.

Schmidt, C.A., M.D. Engstrom, and H.H. Genoways. 1989. *Heteromys gaumeri*. Mammalian Species, 345:1-4.

Schmidt, U. 1978. Vampirfledermäuse. Neue Brehem-Bücherei, Ziemsen Verlag, Wittenberg Lutherstadt.

Schmidt-Nielsen, K., and B. Schmidt-Nielsen. 1952. Water metabolism of desert mammals. Physiological Reviews, 32:135-166.

Schneider, L.K. 1977. Marsupial chromosomes, cell cycles, and cytogenetics. In: The Biology of Marsupials (D. Hunsaker II, ed.). Academic Press, New York.

Schnell, J.H. 1968. The limiting effects of natural predation on experimental cotton rat populations. Journal of Wildlife Management, 32:698-711.

Schobinger, J. 1997. Arte prehistórico de América. Consejo Nacional para la Cultura y las Artes, México.

Schön-Ybarra, M.A. 1984. Locomotion and postures of red howlers in a deciduous forest-savanna interface. American Journal of Physical Anthropology, 63:65-76.

Schouten, K. 1990. Checklist of CITES Fauna y Flora. Secretarial of the Convention on International Trade in Endangered Species of Wild Fauna and Flora. Lausanne, Switzerland.

Schowalter, D.R. 1980. Swarming, reproduction, and early hibernation of *Myotis lucifugus* and *Miotys volans* in Alberta, Canada. Journal of Mammalogy, 61:350-354.

Schramm, P. 1961. Copulation and gestation in the pocket gopher. Journal of Mammalogy, 42:167-170.

Schramm, Y., S.L. Mesnick, J. de la Rosa, D.M. Palacios, M.S. Lowry, D. Aurioles-Gamboa, H.M. Snell, and S. Escorza-Treviño. 2009. Phylogeography of California and Galapagos sea lions and present population structure within the California sea lion. Marine Biology, 156:1375-1387.

Schreiber, A., R. Wirth, M. Riffel, and H. Van Rompaey. 1989. Weasels, Civets, Mongooses, and Their Relatives: An Action Plan for the Conservation of Mustelids and Viverids. IUCN, Gland, Switzerland.

Schug, M.D., H.S. Vessey, and A.I. Koriytko. 1991. Longevity and survival in a population of white-footed mice (*Peromyscus leucopus*). Journal of Mammalogy, 72:360-366.

Schultz, T.A., F.J. Radovski, and P.D. Budwiser. 1970. First insular record of *Notiosorex crawfordi*, with notes on other mammals on San Martin Island, Baja California, Mexico. Journal of Mammalogy, 51:148-150.

Schwanz, L.E. 2006. Annual cycle of activity, reproduction, and body mass in Mexican ground squirrels (*Spermophilus mexicanus*). Journal of Mammalogy, 87: 1086-1095.

Schwartz, C.W., and E.R. Schuartz. 1959. The Wild Mammals of Missouri. University of Missouri Press and Missouri Conservation, Columbia.

Schweinsburg, R.E. 1971. Home range, movements and herd integrity of the collared peccary. Journal of Wildlife Management, 35: 455-460.

Scott, D.M., W.L. Perryman, and P. Hammond. 1985. Obtención de información sobre la reproducción de delfines, magnitud y estructura de los grupos según fotografias aéreas, in: Memorias de la X Reunión Internacional sobre el Estudio de los Mamíferos Marinos (Sociedad Mexicana para el Estudio de los Mamíferos Marinos). 24 al 27 marzo. La Paz, Baja California Sur, México.

Scott, B.M.V., and D.M. Shackleton. 1982. A preliminary study of the social organization of the Vancouver Island wolf. Pp. 12-25, in: Wolves of the World: Perspectives of Behavior, Ecology and Conservation (F.H. Harrington and P.C. Paquet, eds.). Noyes Publications, Park Ridge, NJ.

Scott, M.D., R.S. Wells, and A.B. Irvine. 1990. A long-term study of bottlenose dolphin on the west coast of Florida. Pp. 235-244, in: The Bottlenose Dolphin (S. Leatherwood and R.R. Reeves, eds.). Academic Press, San Diego.

Scott-M, K.M. 1984. Taxonomía y relación con los cultivos de roedores y lagomorfos en el ejido Tokio Galeana, Nuevo León, México. Tesis profesional, Facultad de Ciencias Biológicas, Universidad Autónoma de Nuevo León, Monterrey, Nuevo León, México.

Sears, R. 1987. The photographic identification of individual blue whales (*Balaenoptera musculus*) in the Sea of Cortez. Cetus, 1:14-17.

Sears, R., J.M. Williamson, F.W. Wenzel, M. Bérubé, D. Gendron, and P. Jones. 1990. Photographic identification of the blue whale (*Balaenoptera musculus*) in the Gulf of St. Lawrence, Canada. Pp. 335-342, in: Individual Recognition of Cetaceans: Use of Photo-Identification and Other Techniques to Estimate Population Parameters (P.S. Hammond, S.A. Mizroch, and G.P. Donovan, eds.). Special Issue 12, Reports of the International Whaling Commission.

Secretaría de Desarrollo Social. 1991. Gaceta Ecológica. Mayo.

Secretarías de Desarrollo Social, de Agricultura y Recursos Hidraúlicos, de Relaciones Exteriores y de Turismo. 1994. Calendario Cinegético, agosto 1994–abril 1995.

SEDESOL (Secretaría de Desarrollo Social). 1994. Norma Oficial Mexicana. NOM-059-ECOL-1994. Que determina las especies y subespecies de flora y fauna silvestres terrestres y acuáticas, en peligro de extinción, amenazadas, raras y sujetas a protección especial y que establece especificaciones para suprotección. Diario Oficial de la Federación, 488(10):2-60.

SEDUE (Secretaria de Desarrollo Urbano y Ecología). 1991a. Acuerdo por el que se establecen los criterios ecológicos CT-CERN-001–91 que determinan las especies raras, amenazadas, en peligro de extinción o sujetas a protección especial y sus endemismos, de la flora y la fauna terrestres y acuáticas en la República Mexicana. Gaceta Ecológica, 3(15):2-27.

SEDUE. 1991b. Calendario Cinegético. México.

Seidensticker, J.C., M.G. Hornocker, W.V. Wiles, and J.P. Messick. 1973. Mountain lion social organization in the Idaho Primitive Area. Wildlife Monographs, 35:1-60.

Sekulic, R. 1983. The effect of female call on male howling in red howler monkeys (*Alouatta seniculus*). International Journal of Primatology, 4:291-305.

Selander, R.K., R.F. Johnston, B.J. Wilks, and G.G. Raun. 1962. Vertebrates of the barrier islands of Tamaulipas, Mexico. University of Kansas Publications, Museum of Natural History, 12:309-345.

SEMARNAT (Secretaría de Medio Ambiente y Recursos Naturales). 2009. Norma Oficial Mexicana NOM-059-ECOL-2009, Protección ambiental-Especies nativas de México de flora y fauna silvestres — Categorías de riesgo y especificaciones para su inclusión, exclusión o cambio-Lista de especies en riesgo. Diario Oficial de la Federacion, México D.F.

SEMARNAT. 2010. Norma Oficial Mexicana NOM-059-ECOL-2001, Protección ambiental-Especies nativas de México de flora y fauna silvestres--Categorías de riesgo y especificaciones para su inclusión, exclusión o cambio-Lista de especies en riesgo. Diario Oficial de la Federacion, México D.F.

SEPESCA. 1991. Estudio de Manifestación de Impacto Ambiental del parque camaronicola en la zona de Mar Muerto, Municipio de Arriaga, Chiapas. Secretaría de Pesca. Subsecretaria de Fomento y Desarrollo Pesquero. Dirección General de Acuacultura. México.

Sergeant, D.E., D.K. Caldwell, and M.C. Caldwell. 1973. Age, growth and maturity of bottlenose dolphin (*Tursiops truncatus*) from northeast Florida. Journal of the Fisheries Research Board of Canada, 30:1009-1011.

Serrano, V. 1982. Hábitos alimenticios de las principales especies de roedores del Bolson de Mapimí (Reserva de la Biosfera de Mapimí, Durango). Pp. 873-879, in: Zoología Neotropical. Actas del VIII Congreso Latinoamericano de Zoología (P.J. Salinas, ed.). Caracas, Venezuela.

Serrano, V. 1987. Las comunidades de roedores desertícolas del Bolsón de Mapimí. Acta Zoológica Mexicana (n.s.), 20:1-22.

Servin, M.J.I. 1986. Estudio de la Recuperación del Lobo Mexicano Canis lupus baileyi en el Estado de Durango. II Etapa, Instituto de Ecología, Xalapa.

Servin, J. 1991. Algunos aspectos de la conducta social del lobo mexicano (*Canis lupus baileyi*). Acta Zoologica Mexicana (n.s.), 44:1-35.

Servin, J., and C. Huxley. 1991. La dieta del coyote en un bosque de encino-pino de la Sierra Madre Occidental de Durango, México. Acta Zoologica Mexicana, 44:1-26.

Servin, J., E. Chacón, and R. Rodríguez-Mazzini. 1994. ¿Evidencia de respuesta númerica en una población de Peromyscus truei a la abundancia de frutos de Juniperus deppeana? Segundo Congreso Nacional de Mastozoología, 16 al 19 de marzo de 1994, Guadalajara, Jalisco, Asociación Mexicana de Mastozoología, A.C.

Setzer, H.W. 1949. Subspeciation in the kangaroo rat, *Dipodomys ordii*. University of Kansas Publications, Museum of Natural History, 1:473-573.

Severinghaus, W.D., and D.F. Hoffmeister. 1978. Qualitative cranial characters distinguishing *Sigmodon hispidus* and *Sigmodon arizonae* and the distribution of these two species in northern Mexico. Journal of Mammalogy, 59:868-870.

Seymour, C., and R.W. Dickermann. 1982. Observations on the longlegged bat, *Macrophyllum macrophyllum*, in Guatemala. Journal of Mammalogy, 63:530-532.

Seymour, K.L. 1989. *Panthera onca*. Mammalian Species, 340:1-9.

Shackleton, D.M. 1985. *Ovis canadensis*. Mammalian Species, 230:1-9.

Shackleton, D.M., R.G. Peterson, J. Haywood, and A. Bottrell. 1984. Gestation period in *Ovis canadensis*. Journal of Mammalogy, 65:337-338.

Shane, S.H., R.S. Wells, and B. Würsig. 1986. Ecology, behavior and social organization of the bottlenose dolphin: A review. Marine Mammal Science, 2:34-63.

Sharp, J., and D. Uresk. 1990. Ecological review of black-tailed prairie dogs and associated species in western South Dakota. The Great Basin Naturalist, 50:339-345.

Shaughnessy, P.D., and R.H. Fay. 1977. A review of the taxonomy and nomenclature of North Pacific harbour seals. Journal of Zoology (London), 182:385-419.

Sheldon, J.W. 1992. Wild Dogs: The Natural History of the Nondomestic Canidae. Academic Press, San Diego.

Sherbrooke, W.C. 1976. Differential acceptance of toxic jojoba seed (*Simmondsia chinensis*) by four Sonoran desert heteromyid rodents. Ecology, 57:596-602.

Sherman, H.B. 1939. Note on the food of some Florida bats. Journal of Mammalogy, 20:103-104.

Shull, A.M. 1988. Endangered and threatened wildlife and plants: Determination of endangered status for two long-nosed bats. Federal Register, 53:38456-38460.

Shump. K.A., Jr. 1975. Temperature of spiny pocket mice (*Liomys pictus*). Transactions of the Kansas Academy of Science, 78:171-172.

Shump, K.A., Jr. 1978. Ecological importance of nest construction in the hispidus cotton rat (*Sigmodon hispidus*). The American Midland Naturalist, 100:103-115.

Shump, K.A., Jr., and R.H. Baker. 1978. *Sigmodon leucotis*. Mammalian Species, 96:1-2.

Shump, K.A., Jr., and A.U. Shump. 1982. *Lasiurus cinereus*. Mammalian Species, 185:1-5.

Signorini, I. 1979. Los Huaves de San Mateo del Mar, Oaxaca. Instituto Nacional Indigenista, México.

Silber, G.K. 1990a. Distributional relations of cetaceans in the northern Gulf of California with special reference to the vaquita, *Phocoena sinus*. PhD dissertation, University of California, Santa Cruz.

Silber, G. K. 1990b. Occurrence and distribution of the vaquita *Phocoena sinus* in the northern Gulf of California. Fishery Bulletin, 88:339-346.

Silber, G.K., and M. Brown. 1991. Death of a Bryde's whale: Killer whales attack in the Gulf of California. California Academy of Sciences, 44:16-17.

Silber, G. K., M.W. Newcomer, P.C. Silber, H.M. Pérez Cortez, and G.M. Ellis. 1994. Cetaceans of the northern Gulf of California: Distribution, occurrence and relative abundance. Marine Mammal Science, 10(3):283-298.

Silva-López, G. 1987. La situación actual de los monos araña (*Ateles geoffroyi*) y aullador (*Alouatta palliata*) en la Sierra de Santa Marta (Veracruz, México). Tesis de licenciatura, Universidad Veracruzana, Xalapa, Veracruz, México.

Silva-López, G., and D. Rumiz. 1995. Los primates de la reserva Río Bravo, Belice. La Ciencia y el Hombre, 20:49-64.

Silva-López, G., F. García, and E. Rodríguez. 1987. The present status of *Ateles* and *Alouatta* in non-extensive forest areas of the Sierra de Santa Marta, Veracruz, Mexico. Primate Conservation, 9:53-61.

Silva-López, G., J. Motta-Gill, and A.I. Sánchez-Hernández. 1995. The primates of Guatemala: Distribution and status. Final report of a project supported by the Wildlife Conservation Society.

Silva-López, G., J. Jiménez-Huerta, J. Benítez-Rodríguez, and M.R. Toledo-Cárdenas. 1993. Availability of resources to primates and humans in a forest fragment of Sierra de Santa Marta, Mexico. Neotropical Primates, 1:3-6.

Silva-Taboada, G. 1976. Historia y actualización taxonómica de algunas especies Antilles de murciélagos de los géneros: *Pteronotus, Brachyphylla, Lasiurus* y *Antrozous* (Mammalia: Chiroptera). Academia Científica Cuba, 153:1-24.

Silva-Taboada, G. 1979. Los Murciélagos de Cuba. Académia de Ciencias de Cuba, La Habana, Cuba.

Silver S.C., L. Ostro, L.K. Marsh, L. Maffei, A.J. Noss, M. Kelly, R.B. Wallace, H. Gómez, and G. Ayala. 2004. The use of camera traps for estimating jaguar *Panthera onca* abundance and density using capture/recapture analysis. Oryx, 38:148-154.

Siminski, P. 1990. The Mexican wolf recovery plan. Proceedings of the Arizona Wolf Symposium 90.

Simmons, N.B. 1996. A new species of Micronycteris (Chiroptera: Phyllostomidae) from northeastern Brazil, with comments on phylogenetic relationships. American Museum Novitates, 3158:1-34.

Simmons, N.B. 1998. 1. A reappraisal of interfamilial relationships of bats. Pp. 3-26, in: Bat Biology and Conservation (T.H. Kunz and P.A. Racey, eds.). Smithsonian Institution Press, Washington, DC.

Simmons, N.B. 2005. Order Chiroptera. Pp. 312-529, in: Mammal Species of the World: A Taxonomic and Geographic Reference (D.E. Wilson and D.M. Reeder, eds.). 3rd ed. Johns Hopkins University Press, Baltimore.

Simmons, N.B., and C.O. Handley Jr. 1998. A revisión of *Centronycteris* Gray (Chiroptera: Emballonuridae) with notes on natural history. American Museum Novitates, 3239:1-28.

Simmons, N.B., and R.S. Voss. 1998. The mammals of Paracou, French Guiana: A Neotropical lowland rainforest fauna. Part 1. Bats. Bulletin of the American Museum of Natural History, 237:1-219.

Simpson, G.G. 1964. Species density of North American recent mammals. Systematic Zoology, 13:57-73.

Simpson, G.G. 1980. Splendid Isolation: The Curious History of South American Mammals. Yale University Press, New Haven, CT.

Slijper, E.J. 1979. Whales. 2nd ed. Cornell University Press, Ithaca, NY.

Smith, A.T. 2008. Conservation of endangered lagomorphs. Pp. 297-316, in: Lagomorph Biology: Evolution, Ecology and Conservation (P.C. Alves, N. Ferrand, and K. Hackländer, eds.). Springer-Verlag, Berlin.

Smith, A.T., and J.M. Vrieze. 1979. Population structure of Everglades rodents: Responses to a patchy environment. Journal of Mammalogy, 60:778-794.

Smith, C.C. 1977. Feeding behavior and social organization in howling monkeys. Pp. 97-126, in: Primate Ecology (T.H. Clutton-Brock, ed.). Academic Press, London.

Smith, D.D., and J.K. Frenkel, 1984. *Besnoitia darlingi* (Apicomplexa, Sarcocystidae, Toxoplasmatinae): Transmission between opossums (*Didelphis marsupialis*) and cats (*Felis catus*). Journal of Protozoology, 31(4):584-587.

Smith, F.A. 1992. Evolution of body size among woodrats from Baja California, Mexico. Functional Ecology, 6:265-273.

Smith, F.A., B.T. Bestelmeyer, J. Biardi, and M. Strong. 1993. Anthropogenic extinction of the endemic woodrat *Neotoma bunkeri* Burt. Biodiversity Letters, 1:149-155.

Smith, J.D. 1970. The systematic status of the black howler monkey, *Alouatta pigra* Lawrence. Journal of Mammalogy, 51:358-369.

Smith, J.D. 1972. Systematics of the chiropteran family Mormoopidae. University of Kansas Publications, Museum of Natural History, 56:1-132.

Smith, J.D., and J.K. Jones Jr. 1967. Additional records of the Guatemalan vole, *Microtus guatemalensis* Merriam. The Southwestern Naturalist, 12:189-205.

Smith, J.L.D., and C. McDougal. 1991. The contribution of variance in lifetime reproduction to effective population size in tigers. Conservation Biology, 5:484-490.

Smith, J.L., D.C. McDougal, and M.E. Sunquist. 1987. Female land tenure system in tigers. Pp. 97-109, in: Tigers of the World (R.L. Tilson and U.S. Seal, eds.). Noyes Publications, Park Ridge, NJ.

Smith, M.F., and J.L. Patton. 1988. Subspecies of pocket gophers: Casual bases for geographic differentiation in *Thomomys bottae*. Systematic Zoology, 37:163-178.

Smith, M.H., W.V. Branan, R.L. Marchinton, P.E. Johns, and M.C. Wooten. 1986. Genetic and morphologic comparisons of red brocket, brown brocket and white-tailed deer. Journal of Mammalogy, 67:103-111.

Smith, S.A. 1990. Cytosystematic evidence against monophyly of the *Peromyscus boylii* species group (Rodentia: Cricetidae). Journal of Mammalogy, 71:654-667.

Smith, S.A., R.D. Bradley, and I. F. Greenbaum. 1986. Karyotypic conservatism in the *Peromyscus mexicanus* group. Journal of Mammalogy, 67:584-586.

Smith, S.A., I.F. Greenbaum, D. J. Schmidly, K.M. Davis, and T.W. Houseal. 1989. Additional notes on karyotypic variation in the *Peromyscus boylii* species group. Journal of Mammalogy, 70:603-608.

Smith, W.P. 1991. *Odocoileus virginianus*. Mammalian Species, 388:1-13.

Smolen, M.J., H.H. Genoways, and R.J. Baker. 1980. Demographic and reproductive parameters of the yellow-cheeked pocket gopher (*Pappogeomys castanops*). Journal of Mammalogy, 61:224-236.

Smythe, N. 1970. The adaptative value of the social organization of the coati (*Nasua narica*). Journal of Mammalogy, 51:818-820.

Smythe, N. 1978. The natural history of Central American agouti (*Dasyprocta punctata*). Smithsonian Contributions to Zoology, 257:1-52.

Smythe, N. 1983. *Dasyprocta punctata* and *Agouti paca*. Pp. 463-465, in: Costa Rican Natural History (D.H. Janzen, ed.). University of Chicago Press, Chicago.

Smythe, N. 1986. Competition and resource partitioning in the guild of Neotropical terrestrial frugivorous mammals. Annual Review of Ecology and Systematics, 17:169-188.

Smythe, N. 1991. Steps toward domesticating the paca (*Agouti cuniculus paca*) and prospects for the future. Pp. 202-216, in: Neotropical Wildlife Use and Conservation (J.G. Robinson and K.H. Redford, eds.). University of Chicago Press, Chicago.

Snow, J.L., J.K. Jones Jr., and W.D. Webster. 1970. *Centurio senex*. Mammalian Species, 138:1-3.

Solari, S., S.R. Hoofer, P.A. Larsen, A.D. Brown, R.J. Bull, J.A. Guerrero, J. Ortega, J.P. Carrera, R.D. Bradley, and R.J. Baker. 2011. Operational criteria for genetically defined species: Analysis of the diversification of the small fruit-eating bats, *Dermanura* (Phyllostomidae: Stenodermatinae). Acta Chiropterologica, 11:279-288.

Soler-Frost, A., R.A. Medellín, and G.N. Cameron. 2003. *Pappogeomys bulleri*. Mammalian Species, 717:1-3.

Solmsen, E. 1985. *Lonchorhina aurita* Tomes, 1863 (Phyllostominae, Phyllostomidae, Chiroptera) in West Lichen Ecuador. Zeitschrift Saugetierkunde, 50:329-337.

Soper, J.D. 1941. History, range, and home life of the northern bison. Ecological Monographs, 11:348-412.

Soriano, P.J. 2000. Functional structure of bat communities in tropical rainforests and Andean cloud forests. Ecotropicos, 13:1-20.

Sosa-Escalante, J.E., S. Hernández, and A. Segovia. 2003. *Chiroderma villosum* (Chiroptera: Phyllostomidae) en el Estado de Yucatán, México. Revista Mexicana de Mastozoología, 5:68-71.

Sosa, V.J. 1981. Contribución al conocimiento de la historia natural de la tuza *Pappogeomys tylorhinus tylorhinus* (Rodentia: Geomyidae) en una zona semiárida. Tesis de licenciatura, Universidad Nacional Autónoma de México, México.

Sosa, V.J., and V. Serrano. 1987. Distribución y reparto de recursos de dos especies de *Spermophilus* (Sciuridae) en la Reserva de Mapimí, Durango, México. III Simposio Latinoamericano de Mastozoología, Cancún, Quintana Roo.

Sosa, V., A. Hernandez, and A. Contreras. 1997. Gomez Farias Region (El Cielo Biosphere Reserve). In: Centres of Plant Diversity: A Guide and Strategy for Their Conservation. IUCN-World Wildlife Foundation.

Sosa, V., J.E. Hernández-S., D. Hernández-C., and A.A. Castro-L. 2008. Murciélagos. Capítulo 13. Pp. 181-192, in: Agroecosistemas Cafetaleros de Veracruz. Biodiversidad, Manejo y Conservación (R.H. Manson, V. Hernández Ortiz, S. Gallina, and K. Mehltreter, eds.). Instituto de Ecología, A.C. (INECOL), e Instituto Nacional de Ecología (INE-SEMARNAT). México D.F.

Sosa-Escalante, J., S. Hernández, A. Segovia, and V. Sánchez-Cordero. 1997. First record of the coyote, *Canis latrans* (Carnivora: Canidae), in the Yucatan Peninsula, Mexico. The Southwestern Naturalist, 42:494-495.

Sowls, L.K., 1974. Social behaviour of the collared peccary *Dicotyles tajacu* (L.). IUCN Publications (n.s.), 24:144-165.

Sowls, L.K., 1997. The Peccaries. University of Arizona Press, Tucson.

Spradling, T.A., S.V. Brant, M.S. Hafner, and C.J. Dickerson. 2004. DNA data support a rapid radiation of pocket gopher genera (Rodentia: Geomyidae), Journal of Mammalian Evolution, 11:105-12.

Spencer, S.G., P.C. Chouicar, and B.R. Chapman. 1988. Northward expansion of the southern yellow bat, *Lasiurus ega*, in Texas. The Southwestern Naturalist, 33:493.

Spencer, S.R., and G.N. Cameron. 1982. *Reithrodontomys fulvescens*. Mammalian Species, 174:1-7.

Spinola, R., and C. Vaughan. 1995. Dieta de la nutria Neotropical (*Lontra longicaudis*) en la estación biológica la Selva, Costa Rica. Vida Silvestre Neotropical, 4:125-132.

Stangl, F.B., Jr., and R.J. Baker. 1984a. Evolutionary relationships in *Peromyscus*: Congruence in chromosomal, genic, and classical data sets. Journal of Mammalogy, 65:643-654.

Stangl, F.B., Jr., and R.J. Baker. 1984b. A chromosomal subdivision in *Peromyscus leucopus:* Implications for the subspecies concept as applied to mammals. Pp. 139-145, in: Festschrift for Walter W. Dalquest in Honor of His Sixty-sixth Birthday (N. Horner, ed.). Department of Biology, Midwestern State University.

Starrett, A., and R.S. Casebeer. 1968. Records of bats from Costa Rica. Los Angeles Country Museum, Contributions in Science, 148:1-21.

Stehli, F.G., and S.D. Webb. 1985. The Great American Biotic Interchange. Plenum Press, New York.

Steindel, M., H.K. Toma, M.M.I. Ishida, S.M.F. Murta, and C.J.D.C. Pinto. 1995. Biological and isoenzymatic characterization of *Trypanosoma cruzi* strains isolated from sylvatic reservoirs and vectors from the state of Santa Catarina, southern Brazil. Acta Tropica, 60(3):167-177.

Stephen, C.F., C. Sánchez-Hernández, K.A. Shum Jr., W.R. Teska, and R.H. Baker. 1975. Algunas características poblacionales demográficas de pequeños mamíferos en dos hábitats Mexicanos. Anales del Instituto de Biología, UNAM, México, Serie Zoología, 46:101-124.

Stephenson, R.O., and D. James. 1982. Wolf movements and food habits in northwestern Alaska. Pp. 26-42, in: Wolves of the World: Perspectives of Behavior, Ecology and Conservation (F.H. Harrington and P.C. Paquet, eds.). Noyes Publications, Park Ridge, NJ.

Sterling, K.B. 1991. Two pioneering American mammalogists in Mexico: The field investigations of Edward William Nelson and Edward Alphonso Goldman, 1896-1906. Pp. 33-47, in: Latin American Mammalogy: History, Biodiversity, and Conservation (M.A. Mares and D. Schmidly, eds.). University of Oklahoma Press, Norman.

Stewart, B., and S. Leatherwood. 1985. Minke whale *Balaenoptera acutorostrata* Lacépède, 1804. Pp. 91-136, in: Handbook of Marine Mammals, vol. 3, The Sirenians and Baleen Whales (S.H. Ridgway and R. Harrison, eds.). Academic Press, New York.

Stewart, B.S., P.K. Yochem, R.L. Delong, and G.A. Antonelis Jr. 1987. Interactions between Guadalupe fur seals at San Nicholas and San Miguel Islands, California. Pp. 103-106, in: Status, Biology, and Ecology of Fur Seals (J.P. Croxall and R.L. Gentry, eds.). National Oceanic and Atmospheric Association, Technical Report, National Marine Fisheries Service, 51:1-212.

Stewart B., P. Yochem, H. Huber, R. De Long, R. Jameson, W. Sydeman, S. Allen, and B. Le Boeuf. 1994. History and present status of northern elephant seal population. Pp. 29-48, in: Elephant Seals (B.J. Le Boeuf and R. Laws, eds.). University of California Press, Berkeley.

Steyskal, G.C. 1972. The meaning of the term "sibling species." Systematic Zoology, 21:446.

Stirton, R.A. 1944. Tropical trapping. I. The water mouse *Rheomys*. Journal of Mammalogy, 25:337-343.

Stock, A.D. 1974. Chromosome evolution in the genus *Dipodomys* and its taxonomic and phylogenetic implications. Journal of Mammalogy, 55:505-526.

Stonehouse, B., and D. Gilmore. 1977. The Biology of Marsupials. University Park Press, Baltimore.
Stones, R.C., and C.L. Hayward. 1968. Natural history of the desert woodrat, *Neotoma lepida*. The American Midland Naturalist, 80:458-476.
Storer, T.J., and L.P. Tevis Jr. 1955. California Grizzly. University of California Press, Berkeley.
Storm, G.L. 1972. Daytime retreats and movements of skunks on farmland in Illinois. Journal of Wildlife Management, 36:31-45.
Stout, C.A., and D.W. Duszynski. 1983. Coccidia from kangaroo rats (*Dipodomys* spp.) in western United States, Baja California, and northern Mexico with descriptions of *Eimeria merriami* sp. n. and *Isospora* sp. Journal of Parasitology, 69:209-214.
Streubel, D.P. 1975. Behavioral features of sympatry of Spermophilus spilosoma and *Spermophilus tridecemlineatus* and some aspects of the life history of *S. spilosoma*. PhD dissertation, University of Northern Colorado, Greeley.
Streubel, D.P., and J.P. Fitzgerald, 1978. *Spermophilus spilosoma*. Mammalian Species, 101:1-4.
Stromberg, M.R., and M.S. Boyce. 1986. Systematics and conservation of the swift fox, *Vulpes velox*, in North America. Biological Conservation, 35:97-110.
Studholme, K.M., C.J. Phillips, and G.L. Forman 1986. Results of the Alcoa Fundation Suriname Expeditions. X. Patterns of cellular divergence and evolution in the gastric mucosa of two genera of phyllostomid bats, *Trachops* and *Chiroderma*. Annals of Carnegie Museum, 55:207-235.
Studier, E., B.C. Boyd, A.T. Feldman, R.W. Dapson, and D.E. Wilson. 1983. Renal function in the Neotropical bat, *Artibeus jamaicensis*. Comparative Biochemistry Physiology, 74:199-209.
Sudman, P.D., J.K. Wickliffe, P. Horner, M.J. Smolen, J.W. Bickham, and R.D. Bradley. 2006. Molecular systematics of pocket gophers of the genus *Geomys*. Journal of Mammalogy, 87:668-676.
Sullivan, J.M., and C.W. Kilpatrick. 1991. Biochemical systematics of the *Peromyscus aztecus* assemblage. Journal of Mammalogy, 72:681-696.
Sullivan, J., E. Arellano, and D.S. Rogers. 2000. Comparative phylogeography of Mesoamerican highland rodents: Concerted versus independent response to past climatic fluctuations. The American Naturalist, 155:755-768.
Sullivan, J.M., C.W. Kilpatrick, and P.D. Rennert. 1991. Biochemical systematics of the *Peromyscus boylii* species group. Journal of Mammalogy, 72:669-689.
Sullivan, J.P., S. Lavoué, and C.D. Hopkins. 2000. Molecular systematics of the African electric fishes (Mormyroidea: Teleostei) and a model for the evolution of their electric organs. Journal of Experimental Biology, 203:665-683.
Sullivan, J., J.A. Markert, and C.W. Kilpatrick. 1997. Phylogeography and molecular systematics of the *Peromyscus aztecus* species group (Rodentia: Muridae) inferred using parsimony and likelihood. Systematic Biology, 46:426-440.
Sullivan, R.M., and T.L. Best. 1997. Effects of environment on phenotypic variation and sexual dimorphism in *Dipodomys simulans* (Rodentia: Heteromyidae). Journal of Mammalogy, 78:798-810.
Sullivan, R.M., D.J. Hafner, and T.L. Yates. 1986. Genetics of a contact zone between three chromosomal forms of the grasshopper mouse (genus *Onychomys*): A reassessment. Journal of Mammalogy, 67:640-659.
Sumichrast, F. 1881. Enumeración de las especies de mamíferos, aves, reptiles, y batracios observados en la parte Central y Meridional de la República Mexicana. La Naturaleza, 5:199-214, 322-328.
Summerlin, C.T., and J.L. Wolfe. 1973. Social influences on trap responses of the cotton rat, *Sigmodon hispidus*. Ecology, 54:1156-1159.
Sunquist, M.E., and J.F. Eisenberg. 1993. Reproductive strategies of female Didelphis. Bulletin of the Florida Museum of Natural History, Biological Science, 36(4):109-140.
Sunquist, M.E., and G.G. Montgomery. 1973. Activity pattern of a translocated silky anteater (*Cyclopes didactylus*). Journal of Mammalogy, 54:782.
Sunquist, M., and F. Sunquist. 2002. Wild Cats of the World. University of Chicago Press, Chicago.
Sunquist, M.E., S.N. Austad, and F. Sunquist, 1987. Movement patterns and home range in the common opossum (*Didelphis marsupialis*). Journal of Mammalogy, 68(1):173-176.
Sunquist, M.E.; F. Sunquist, and D.E. Daneke. 1989. Ecological separation in a Venezuelan llanos carnivore community. Pp. 197-232, in: Advances in Neotropical Mammalogy (K.H. Redford and J.F. Eisenberg, eds.). Sandhill Crane Press, Gainesville, FL.
Sutton, D.F., and C.F. Nadler. 1969. Chromosomes of the North American chipmunks, genus *Eutamias*. Journal of Mammalogy, 50:524-535.
Svensen, G.E. 1982. Weasels. Pp. 613-628, in: Wild Mammals of North America: Biology, Management, Economics (J.A. Chapman and G.A. Feldhamer, eds.). Johns Hopkins University Press, Baltimore.
Svihla, R.O. 1930. Notes on the golden harvest mouse. Journal of Mammalogy, 11:53-55.
Svihla, R.O. 1931. Life history of the rice rat (*Oryzomys palustris texensis*). Journal of Mammalogy, 12:238-242.
Swanepoel, P., and H.H. Genoways. 1979. Morphometrics. Pp. 13-105, in: Biology of Bats of the New World Family Phyllostomatidae, Part III (R.J. Baker, J.K. Jones Jr., and D.C. Carter, eds.). Special Publications of the Museum, Texas Tech University, 16:1-441.
Swank, W.G., and J.G. Teer. 1989. Status of the jaguar--1987. Oryx, 23:14-21.
Swartz, S.L. 1986. Gray whale migratory, social and breeding behaviour. Pp. 207-229, in: Behaviour of Whales in Relation to Management (G.P. Donovan, ed.). Special Issue 8, Reports of the International Whaling Commission.
Sweanor, L.L., K.A., Logan, and M.G. Hornoker. 2000. Cougar dispersal patterns, metapopulation dynamics, and conservation. Conservation Biology, 14:798-808.
Swier V.J., R.D. Bradley, W. Rens, F.F. Elder, and R.J. Baker. 2009. Patterns of chromosomal evolution in *Sigmodon*, evidence from whole chromosome paints. Cytogenetic and Genome Research, 125:54-66.
Szteren, D., D. Aurioles, and L.R. Gerber. 2006. Population status and trends of the California sea lion (*Zalophus californianus californianus*) in the Gulf of California, Mexico. Pp. 369-403, in: Sea Lions of the World (A.W. Trites, S.P. Atkinson, D.P. DeMaster, L.W. Fritz, T.S. Gelatt, L.D. Rea, and K.M. Wynne, eds.). Alaska Sea Grant College Program, University of Alaska, Fairbanks.
Taddei, V.A., and V. Garutti. 1981. The southernmost record of the free-tailed bat, *Tadarida aurispinosa*. Journal of Mammalogy, 62:851-852.
Tanigoshi, L.K., and R.B. Loomis. 1974. The genus *Hyponeocula* (Acarina, Trombiculidae) of western North America. Melanderia, 17:1-27.
Tate, G.H.H. 1933. A systematic revision of the marsupial genus *Marmosa*, with a discussion of the adaptive radiation of the murine opossums (Marmosa). Bulletin of the American Museum of Natural History, 66:1-250.
Tavolga, M.C., and F.S. Essapian. 1957. The behavior of the bottlenosed dolphin (*Tursiops truncatus*): Mating, pregnancy, parturition and mother-infant behavior. Zoologica, 42:11-31.
Taylor, B.K. 1985. Functional anatomy of the forelimb in vermilinguas (anteaters). Pp. 163-171, in: The Evolution and Ecology of Armadillos, Sloths, and Vermilinguas (G.G. Montgomery, ed.). Smithsonian Institution Press, Washington, DC.
Taylor, W.P. 1935. Ecology and life history of the porcupine (*Erethizon epixanthum*) as related to the forests of Arizona and the southwestern United States. University of Arizona Bulletin, 6:1-177.
Taylor, P.W. 1954. Food habits and notes on life history of ring-tailed cat in Texas. Journal of Mammalogy, 35:55-63.
Taylor, W. 1956. The Deer of North America. Stackpole Books, Harrisburg, PA.
Tejedor, A. 2005. A new species of funnel-eared bat (Natalidae: *Natalus*) from Mexico. Journal of Mammalogy, 86:1109-1120.

Tejedor, A. 2006. The type locality of *Natalus stramineus* (Chiroptera: Natalidae): Implications for the taxonomy and biogeography of the genus Natalus. Acta Chiropterologica, 8:361-380.

Tejedor, A. 2011. Systematics of funnel-eared bats (Chiroptera: Natalidae). Bulletin of the American Museum of Natural History, 343:1-140.

Téllez-G., G., A. Mendoza, and G. Ceballos. 1997. Registros de mamíferos del oeste de México. Revista Mexicana de Mastozoología, 2:97-100.

Tello, Fr. A. 1652 [1968]. Crónica miscelánea de la sancta provincia de Xalisco. Gobierno del Estado de Jalisco, Universidad de Guadalajara, Instituto Jalisciense de Antropología e Historia, Instituto Nacional de Antropología e Historia. Guadalajara.

Terman, M.R. 1974. Behavioral interactions between Microtus and Sigmodon: A model for competitive exclusion. Journal of Mammalogy, 55:705-719.

Terrel, P.S. 1989. California ground squirrel trapping influenced by anal-gland odors. Journal of Mammalogy, 70:428-431.

Terry, C.J. 1981. Habitat differentiation among three species of Sorex and *Neurotrichus gibbsi* in Washington. The American Midland Naturalist, 106:119-125.

Tershy, B.R., D. Breese, and C.S. Strong. 1990. Abundance, seasonal distribution and population composition of Balaenopteriid whales in the Canal de ballenas, Gulf of California, México. Pp. 369-375, in: Individual Recognition of Cetaceans: Use of Photo-Identification and other Techniques to Estimate Population Parameters (P.S. Hammond, A. Mizroch, and G.P. Donovan, eds.). Special Issue 12, Reports of the International Whaling Commission.

Tershy, B.R., A. Acevedo-G, D. Breese, and C.S. Strong. 1993. Diet and feeding behavior of fin and Bryde's whales in the central Gulf of California, Mexico. Revista de Investigación Científica, 1:31-37.

Terwilliger, V.J. 1978. Natural history of Baird's tapir on Barro Colorado Island, Panama Canal Zone. Biotropica, 10:211-220.

Teska, W.R., E.N. Rybak, and R.H. Baker. 1981. Reproduction and development of the pygmy spotted skunk (*Spilogale pygmaea*). The American Midland Naturalist, 105:390-392.

Testa, J.W., D.F. Holleman, R.T. Bowyer, and J.B. Faro. 1994. Estimating populations of marine river otters in Prince William Sound, Alaska, using radiotracer implants. Journal of Mammalogy, 75:1021-1032.

Tewes, M.E. 1986. Ecological and behavioral correlates of ocelot spatial patterns. PhD dissertation, University of Idaho, Moscow.

Tewes, M.E., and D.J. Schmidly. 1987. The Neotropical felids: Jaguar, ocelot, margay, and jaguarundi. Pp. 697-712, in: Wild Furbearer Management and Conservation in North America (M. Novak, J.A. Baker, M.E. Obbard, and B. Malloch, eds.). Ministry of Natural Resources, Ontario, Canada.

Themotheo-Sobrinho, G.F. 1992. Ocurrencia de cetaceos no estado do ceara. Brasil. In: Memorias de la Quinta Reunión de Especialistas en Mamíferos Acuáticos de America del Sur. 28 septiembre–2 octubre, Buenos Aires, Argentina.

Theobald, D.P. 1983. Studies on the biology and habitat of the Arizona gray squirrel. Master's thesis, Arizona State University, Tempe.

Thomas, D. 1890. On a collection of mammals from central Veracruz, Mexico. Proceedings of the Zoology Society of London, 11:71-76.

Thomas, H.H., and T.L. Best. 1994. *Lepus insularis*. Mammalian Species, 465:1-3.

Thomas, J.W., and D.E. Toweill (eds.). 1982. Elk of North America: Ecology and Management. Stackpole Books, Harrisburg, PA.

Thomas, O. 1892. Note on Mexican examples of *Chylonycteris davyi* Gray. Annals Management Natural History (series 6), 10:410.

Thomas, O. 1893. On two new members of the genus *Heteromys* and two of *Neotoma*. Annals Management Natural History (series 6), 12:233-235.

Thomas, O. 1898a. On a badger from Lower California. Proceedings of the Zoology Society of London (for 1897):889.

Thomas, O. 1898b. On new mammals from western Mexico and Lower California. Annals Management Natural History (series 7), 1:40-46.

Thomas, O. 1903. On three new forms of *Peromyscus* obtained by Dr. Hans Gadow, F. R. S. and Mrs. Gadow in Mexico. Annals Management Natural History (series 7), 11:484-487.

Thomas, W.D. 1975. Observations on captive brockets *Mazama americana* and *Mazama gouazoubira*. International Zoological Yearbook, 15:77-79.

Thomas, W.D. 1992. Bats and old-growth forests: Are both vanishing? Bats, 10:4-9.

Thompson, J.E.S. 1994. Grandeza y decadencia de los mayas. 3rd ed. Fondo de Cultura Económica, México.

Thompson, P.O., L.T. Findley, and O. Vidal. 1992. 20-Hz pulses and other vocalizations of fin whales, *Balaenoptera physalus*, in the Gulf of California, Mexico. Journal of the Acoustic Society of America, 92:3051-3057.

Thompson-Olais, L.A. 1994. Sonoran Pronghorn Recovery Plan Revision (*Antilocapra americana*). U.S. Fish and Wildlife Service, Reg. 2.

Thornback, J., and M. Jenkins (eds.). 1982. The IUCN Mammal Red Data Book. Part 1. IUCN, Gland, Switzerland.

Thornback, J., and M. Jenkins. 1984. The IUCN Mammal Red Data Book. Part. I-IUCN. Gland, Switzerland.

Thorington, R.W., Jr., and R.S. Hoffmann. 2005. Family Sciuridae. Pp. 754-818, in: Mammal Species of the World: A Taxonomic and Geographic Reference (D.E. Wilson and D.M. Reeder, eds.). 3rd ed. Johns Hopkins University Press, Baltimore.

Tiemann-Boege, I., C.W. Kilpatrick, D.J. Schmidly, and R.D. Bradley. 2000. Molecular phylogenetics of *Peromyscus boylii* species group (Rodentia: Muridae) based on mitochondrial cytochrome b sequences. Molecular Phylogenetics and Evolution, 16:366-378.

Timm, R.M. 1984. Tent construction by *Vampyressa* in Costa Rica. Journal of Mammalogy, 65:166-167.

Timm, R.M. 1985. *Artibeus phaeotis*. Mammalian Species, 235:1-6.

Timm, R.M. 1987. Tent construction by bats of genera *Artibeus* and *Uroderma*. Pp. 187-212, in: Studies in Neotropical Mammalogy (B. D. Patterson and R.M. Timm, eds.). Fieldiana-Zoology, n.s. 39., Field Museum of Natural History, Chicago.

Timm, R.M., R.M. Salazar, A. Peterson, and A. Townsend. 1997. Historical distribution of the extinct tropical seal, *Monachus tropicalis* (Carnivora: Phocidae). Conservation Biology, 11(2):549-551.

Timm, R.M., D.E. Wilson, B.L. Clauson, R.K. LaVal, and C.S. Vaughan. 1989. Mammals of the La Selva-Braulio Carrillo Complex, Costa Rica. U.S. Fish and Wildlife Service. North American Fauna, 75:1-162.

Tinker, B. 1978. Mexican Wilderness and Wildlife. University of Texas Press, Austin.

Tipton, J.V., and E. Mendez. 1968. New species of fleas (*Sifonaptera*) from Cerro Potosi, Mexico with notes on ecology and host-parasite relationships. Pacific Insects, 10:177-214.

Toledo, V.M. 1988. La diversidad biológica de México. Ciencia y Desarrollo. México.

Toledo-Cárdenas, M.R. 1993. Locomoción, posturas y patrón de actividades del mono araña mexicano (Ateles geoffroyi vellerosus Kellogg y Goldman, 1944) en tres situaciones ambientales diferentes. Tesis de licenciatura, Universidad Veracruzana, Xalapa, Veracruz.

Tomich, Q.P. 1982. Ground squirrels *Spermophilus beecheyi* and allies. Pp. 192-208, in: Wild Mammals of North America: Biology, Management, Economics (J.A. Chapman and G.A. Feldhamer, eds.). John Hopkins University Press, Baltimore.

Tomilin, A.G. 1967. Mammals of the USSR and Adjacent Countries, vol. IX, Cetacea (V.G. Heptner, ed.). Nauk S.S.S.R., Mascú.

Torres, D.N. 1990. Colares plásticos en lobos finos Antarcticos. Otra evidencia de contaminación. Boletín Antárctico Chileno, 10:20-22.

Torres, G.A. 1991. Estudio demográfico del lobo fino de Guadalupe (*Arctocephalus towsendii*, Merrian, 1897) en la Isla Guadalupe, B.C., México. Tesis de licenciatura, Facultad de Ciencias, UNAM, México.

Torres, G.A., M.C. Esquivel, and G.G. Ceballos .1995. Diversidad y conservación de los mamíferos marinos de México. Revista Mexicana de Mastozoología, 1:22-43.

Torres-Ayala, J.M.. 1978. Comportamiento y datos ecológicos del vampiro *Desmodus rotundus murinus* Wagner 1840, en la cueva La "Chorrera," municipio de Linares, Nuevo León, México. Tesis Biólogo, F.C.B., U.A.N.L.

Tortato, F., and S. Althoff. 2007. Variações na coloração de iraras (*Eira barbara* Linnaeus, 1758--Carnivora, Mustelidae) da Reserva Biológica Estadual do Sassafrás, Santa Catarina, sul do Brasil. Biota Neotropica, 7:365-367.

Toweill, D.E., and J.E. Tabor. 1982. River otter: *Lutra canadensis*. Pp. 688-703, in: Wild Mammals of North America: Biology, Management, and Economics (J.A. Champman and G.A. Feldhamer, eds.). Johns Hopkins University Press, Baltimore.

Toweill, E.D., and G.J. Teer. 1977. Food habits on ringtails in the Edwards plateau region of Texas. Journal of Mammalogy, 58:660-663.

Townsend, C.H. 1912. Mammals collected by the "Albatross" expedition in Lower California in 1911, with descriptions of new species. Bulletin of the American Museum of Natural History, 31:117-130.

Trainer, C.E., M.J. Willis, G.P. Keiser Jr., and D.D. Sheely. 1983. Fawns' mortality and habitat use among pronghorn during spring and summer in southeastern Oregon, 1981-82. Department of Fish and Wildlife, Portland, OR.

Trajano, E. 1984. Ecología de populacoes de morcegos cavernícolas em uma regiao do Sudeste do Brasil. Revista Brasileira de Zoología. 2:255-320.

Trapido, H., and P.E. Crowe. 1942. Color abnormalities in three genera of northeastern cave bats. Journal of Mammalogy, 23:303-305.

Trapp, G.R. 1972. Some anatomical and behavorial adaptations of ringtails, *Bassariscus astutus*. Journal of Mammalogy, 53:549-557.

Trapp, G.R. 1978. Comparative behavorial ecology of the ringtail and gray fox in southwestern Utah. Carnivore, 1:1-32.

Traub, R., M. Rothschild, and J.F. Haddow. 1983. The Rothschild Collection of Fleas. The Ceroatophyllidae: Key to the Genera and Host Relationships with Notes on Their Evolution, Zoogeography and Medical Importance. Academic Press, London.

Travi, B.L., C. Jaramillo, J. Montoya, I. Segura, A. Zea, A. Goncalves, and I.D. Velez. 1994. *Didelphis marsupialis*, an important reservoir of *Trypanosoma* (*Schizotrypanum*) *cruzi* and *Leishmania* (*Leishmania*) *chagasi* in Colombia. American Journal of Tropical Medicine and Hygiene, 50(5):557-565.

Treviño, V.J. 1981. Datos ecológicos de la ardilla de tierra *Spermophilus spilosoma pallescens* Howell (1928), en el ejido Tokio Galeana, Nuevo León, México. Tesis profesional, Facultad de Ciencias Biológicas, Universidad Autónoma de Nuevo León, Monterrey, Nuevo León.

Treviño, J.C., and C. Jonkel. 1986. Do grizzly bears still live in Mexico? International Conference on Bear Research and Management, 6:11-13.

Treviño-Villareal, J. 1990. The annual cycle of the Mexican prairie dog (*Cynomys mexicanus*). University of Kansas Publications, Museum of Natural History, 139:1-27.

Treviño-Villarreal, J., and W.E. Grant. 1998. Geographic range of the endangered Mexican prairie dog (*Cynomys mexicanus*). Journal of Mammalogy, 79:1273-1287.

Treviño-Villarreal, J., I.M. Berk, A. Aguirre, and W.E. Grant. 1998. Survey for sylvatic plague in the Mexican prairie dog (*Cynomys mexicanus*). The Southwestern Naturalist, 43:147-154.

Trillmich, F. 1984. Natural history of the Galapagos fur seal (*Arctocephalus galapagoensis*, Heller). Pp. 215-223, in: Key Environments—Galapagos (R. Perry, ed.). Pergamon Press, Oxford.

Trolle, M., and M. Kery. 2003. Estimation of ocelot density in the Pantanal using capture-recapture analysis of camera-trapping data. Journal of Mammalogy, 84:607-614.

Trolle, M., and M. Kery. 2005. Camera-trap study of ocelot and other secretive mammals in the northern Pantanal. Mammalia, 69:405-412.

Tromba, F.G. 1954. Some parasites of the hoary bat, *Lasiurus cinereus* (Beauvois, 1796). Journal of Mammalogy, 34:253-254.

True, F.W. 1888. Description of *Geomys personatus* and *Dipodomys compactus*, two new species of rodents from Padre Island, Texas. Proceedings of the United States National Museum, 11:159-160.

Trujano-Álvarez, A.L, and S.T. Álvarez-Castañeda. 2007. Taxonomic revision of *Thomomys bottae* in the Baja California Sur lowlands. Journal of Mammalogy, 88:343-350.

Tschapka, M., E.B. Sperr, L.A. Caballero-Martínez, and R.A. Medellín. 2008. Diet and cranial morphology of *Musonycteris harrisoni*, a highly specialized nectar-feeding bat in western Mexico. Journal of Mammalogy, 89:924-932.

Tumlison, R. 1992. *Plecotus mexicanus*. Mammalian Species, 401:1-3.

Tumlison, R. 1993. Geographic variation in the lappet-eared bat, *Idionycteris phyllotis*, with descriptions of subspecies. Journal of Mammalogy, 74:412-421.

Turner, D.C. 1975. The Vampire Bat: A Field Study in Behavior and Ecology. Johns Hopkins University Press, Baltimore.

Turner, G.T., R.M. Hansen, V.H. Reid, H.P. Tietjen, and A.L. Ward. 1973. Pocket gophers and Colorado mountain rangeland. Colorado Agricultural Experimental Station Bulletin, 554:90.

Turner, J.C., and R.A. Weaver. 1980. Water. Pp. 100-112, in: The Desert Bighorn, Its Life History, Ecology, and Management (G. Monson and L. Sumner, eds.). University of Arizona Press, Tucson.

Tuttle, M.D. 1968. Feeding habits of *Artibeus jamaicensis*. Journal of Mammalogy, 49:787.

Tuttle, M.D. 1970. Distribution and zoogeography of Peruvian bats, with comments on natural history. University of Kansas Scientific Bulletin, 49:45-86.

Tuttle, M.D. 1976. Collecting techniques. Pp. 71-88, in: Biology of Bats of the New World Family Phyllostomatidae, Part I (R.J. Baker, J.K. Jones Jr., and D.C. Carter, eds.). Special Publications of the Museum, Texas Tech University, 10:1-218.

Tuttle, M.D. 1988. America's Neighborhood Bats. University of Texas Press, Austin.

Tuttle, M.D., and L.R. Heaney. 1974. Maternity habits of *Myotis leibii* in South Dakota. Bulletin of the Southern California Academy of Sciences, 73:80-83.

Tuttle, M.D., and M.J. Ryan. 1981. Bat predation and evolution of frog vocalizations in the Neotropics. Science, 214:677-678.

Tuttle, M.D., and D. Stevenson. 1982. Growth and survival of bats. Pp. 105-150, in: Ecology of Bats (T.H. Kunz, ed.). Plenum Press, New York.

Twente, J.W.J. 1955. Some aspects of habitat selection and other behavior of cavern-dwelling bats. Ecology, 36:706-732.

Tyndale-Biscoe, C.H., and R.B. Mackenzie. 1973. Reproduction in *Didelphis marsupialis* and *Didelphis albiventris* in Colombia. Journal of Mammalogy, 57(2):449-265.

Uhart, E.Q., and J.C. López-Vidal. 2008. Registro anómalo en la distribución del murciélago cara de viejo Centurio senex (Chiroptera: Mammalia). Revista Mexicana de Mastozoología, 12:176-179.

Underwood, H.T. 1986. Endohelmiths of three species of *Oryzomys* (Rodentia: Cricetidae) from San Luis Potosi, Mexico. The Southwestern Naturalist, 31:110-111.

Urbán, D. 1970. Raccoon populations, movement patterns, and predation on a managed waterfowl marsh. Journal of Wildlife Management, 34:372-382.

Urbán-R., J. 1983. Taxonomía y distribución de los géneros Tursiops, *Delphinus* y *Stenella* en las aguas adyacentes a Sinaloa y Nayarit, México (Cetacea: Delphinidae). Tesis profesional, Facultad de Ciencias, Universidad Nacional Autónoma de México, México.

Urbán-R., J. 1996. La población del rorcual común *Balaenoptera physalus* en el Golfo de California, México. Reporte a la CONABIO, Proyecto B040, México.

Urbán-R., J., and A. Aguayo L. 1987a. Cetáceos observados en la costa occidental de la Península de Baja California, México. septiembre 1981—enero 1986. Pp. 93-118, in: Memorias de la X Reunión Internacional sobre Mamíferos Marinos (Secretaría de Pesca, ed.), marzo 1985, La Paz, Baja California Sur, México.

Urbán-R., J., and A. Aguayo L. 1987b. Spatial and seasonal distribution of the humpback whale, *Megaptera novaeangliae*, in the Mexican Pacific. Marine Mammal Science, 3:333-344.

Urbán-R., J., and G.D. Aurioles. 1992. First record of the pygmy beaked whale *Mesoplodon peruvianus* in the North Pacific. Marine Mammal Science, 8:420-425.

Urbán-R., J., and S.R. Flores. 1996. A note on Bryde's whales (*Balaenoptera edeni*) in the Gulf of California, México. Reports of the International Whaling Commission, 46:453-457.

Urbán-R., J., and A. Jaramillo. 1992. Segundo varamiento de *Berardius bairdii* en la Bahía de la Paz, Baja California Sur. Revista de Investigación Cientifica, Número Especial SOMEMMA, Universidad Autónoma de Baja California Sur, 1:85-92.

Urbán-R., J., S. Ramírez-S., and J.C. Salinas-V. 1994. First record of bottlenose whales, *Hyperoodon* sp., in the Gulf of California. Marine Mammal Science, 10:471-473.

Urbán-R., J., A. Gómez-Gallardo U., M.A. Palmeros R., and G. Velazquez C. 1997. Los mamíferos de la Bahía de la Paz, B.C.S. Pp. 193-217, in: La Bahía de la Paz. Investigación y Conservación (J. Urbán-R. and M. Ramírez R., eds.). Universidad Autónoma de Baja California Sur-Pronatura, La Paz, Baja California Sur.

Urbán-R., J., A. Jaramillo L., M. Salinas Z., J. Jacobsen, K. Balcomb, P. Ladrón de Guevara P., and A. Aguayo L. 1994. Estimación de la abundancia de los rorcuales jorobados que habitan el Pacífico mexicano durante el período invernal. P. 40, in: Resúmenes XIX Reunión Internacional para el Estudio de los Mamíferos Marinos. Mayo, La Paz, Baja California Sur.

Urbán-R., J., C. Álvarez F., M. Salinas Z., J. Jacobsen, K.C. Balcomb III, A. Jaramillo L., P. Ladrón de Guevara P., and A. Aguayo L. 1999. Population size of humpback whale (*Megaptera novaeangliae*) in the Mexican Pacific. Fishery Bulletin, 97:1017-1024.

Urbán-R., J., A. Jaramillo L., A Aguayo L., P. Ladrón de Guevara P., M. Salinas Z., C. Álvarez F., L. Medrano G., J.K. Jacobsen, K.C. Balcomb, D.E. Claridge, J. Calambokidis, G.H. Steiger, J.M. Straley, O. von Ziegesar, J.M. Waite, S. Mizroch, M.E. Dahlheim, J.D. Darling, and C.S. Baker. 2000. Migratory destinations of humpback whales wintering in the Mexican Pacific. Journal of Cetacean Research and Management, 2:101-110.

Urbano-V. G., O. Sánchez-Herrera, G. Téllez-Girón, and R.A. Medellín. 1987. Additional records of Mexican mammals. The Southwestern Naturalist, 32:134-137.

Urdaneta-Morales, S., and I. Nironi. 1996. *Trypanosoma cruzi* in the anal glands of urban opossums. I. Isolation and experimental infections. Memorias do Instituto Oswaldo Cruz, 91(4):399-403.

Uribe-Alcocer, M., A. Laguarda-Figueras, and F. Rodríguez-Romero. 1977. The chromosomes of the population of *Microtus mexicanus* (Muridae, Rodentia) of central Mexico. Anales del Instituto de Biología Experimental, 48:57-63.

Uribe-Peña, Z., G. Gaviño De La Torre, and C. Sánchez-Hernández. 1981. Vertebrados del rancho "El Reparito" Municipio de Arteaga, Michoacán, México. Anales del Instituto de Biología, UNAM, México, Serie Zoología, 51:615-646.

U.S. Fish and Wildlife Service. 1991. Endangered and threatened wildlife and plants (those covered by the regulations for the U.S. Endangered Species Act), 50 CRF 17.11 17.12, July 15, 1991. U.S. Government Printing Office, Washington, DC.

Valdéz, M., and G. Ceballos. 1991. Historia natural, alimentación y reproducción de la ardilla terrestre (*Spermophilus mexicanus*) en una pradera intermontana. Acta Zoologica Mexicana (n.s.), 43:1-31.

Valdéz, M., and G. Ceballos. 1997. Conservation of endemic mammals of Mexico: The Perote ground squirrel (*Spermophilus perotensis*). Journal of Mammalogy, 78:74-82.

Valdés M., E. de La Cruz, E. Peters, and E. Pallares (eds.). 2006. El Berrendo en México, Acciones de Conservación. INE-SEMARNAT, México.

Valdéz, R., and R.K. LaVal. 1971. Records of bats from Honduras and Nicaragua. Journal of Mammalogy, 52:247-250.

Valdivia, J., P. Ramírez, H. Tovar, and F. Franco. 1981. Report of a cruise to mark and assess Bryde's whales of the "Peruvian stock," February 1980. Reports of the International Whaling Commission, 31:435-440.

Valencia, J.L. 1988. Estimaciones de las Poblaciones de *Trichechus manatus* en dos Regiones de Tabasco, México. IX Congreso Nacional de Zoología. Villahermosa, Tabasco Manuscrito.

Valenzuela, D. 1991. Estimación de la densidad y distribución de la población de venado cola blanca (*Odocoileus virginianus* Rafinesque 1832) en el Bosque La Primavera, Jalisco. Tesis de licenciatura, Escuela de Biología, Universidad Autónoma de Guadalajara, Guadalajara, Jalisco.

Valenzuela, D. 1998. Natural history of the white-nosed coati, *Nasua narica*, in the tropical dry forests of western México. Revista Mexicana de Mastozoología, 3:26-44.

Valenzuela, D. 1999. Efectos de la estacionalidad ambiental en la densidad, la conducta de agrupamiento y el tamaño de área de actividad de coatí (*Nasua narica*) en selvas tropicales caducifolias. Tesis doctoral, Instituto de Ecología, UNAM, México D.F.

Valenzuela, D., and G. Ceballos. 2000. Habitat selection, home range, and activity of the white-nosed coati, *Nasua narica*, in a Mexican tropical dry forest. Journal of Mammalogy, 81:810-819.

Valenzuela, D., G. Ceballos, and A. García. 2000. Mange epizootic in coatis at the Chamela-Cuixmala Biosphere Reserve, western México. Journal of Wildlife Diseases, 36:56-63.

Valtierra Azotla, M., and A. Garcia. 1998. Mating behavior of the Mexican mouse opossum (*Marmosa canescens*) in Cuixmala, Jalisco, Mexico. Revista Mexicana de Mastozoología, 3:146-147.

Van Den Bussche, R.A., J.L. Hudgeons, and R.J. Baker. 1998. Phylogenetic accuracy, stability, and congruence: Relationships within and among the New World bat genera *Artibeus, Dermanura,* and *Koopmania*. Pp. 59-71, in: Bat Biology and Conservation (T.H. Kunz and P.A. Racey, eds.). Smithsonian Institution Press, Washington, DC.

Van Den Bussche, R.A., R.J. Baker, H.A. Wichman, and M.J. Hamilton. 1993. Molecular phylogenetics of Stenodermatini bat genera: Congruence of data from nuclear and mitochondrial DNA. Molecular Biology and Evolution, 10:944-959.

Van Dyke, F.G., R.H. Brocke, H.G. Shaw, B.B. Ackerman, T.P. Hemker, and F.G. Lindzey. 1986. Reactions of mountain lions to logging and human activity. Journal of Wildlife Management, 50:95-102.

Van Gelder, R.G. 1959a. A taxonomic revision of the spotted skunks (genus *Spilogale*). American Museum of Natural History, 117:229-392.

Van Gelder, R.G. 1959b. Results of the Puritan-American Museum of Natural History expedition to western Mexico. 8. A new *Antrozous* (Mammalia, Vespertilionidae) from the Tres Marias Islands, Nayarit, Mexico. American Museum Novitates, 1973:1-14.

Van Gelder, R.G., and D.B. Wingate. 1961. The taxonomy and status of bats of Bermuda. American Museum Novitates, 2029:1-9.

Van Pijlen, I.A., B. Amos, and G.A. Dover. 1991. Multilocus DNA fingerprinting applied to population studies of the minke whale *Balaenoptera acutorostrata*. Reports of the International Whaling Commission (special issue 13), 245-254.

Van Zyll de Jong, C.G. 1972. A systematic review of the Nearctic and Neotropical river otters (genus *Lutra*, Mustelidae, Carnivora). Life Science Contributions, Research Ontario Museum, No. 80.

Van Zyll de Jong, C.G. 1984. Taxonomic relationships of Nearctic small-footed bats of the *Myotis leibii* group (Chiroptera: Vespertilionidae). Canadian Journal of Zoology, 62:2519-2526.

Van Zyll de Jong, C.G., and G.L. Kirkland Jr. 1989. A morphometric analysis of the *Sorex cinereus* group in central and eastern North America. Journal of Mammalogy, 70:110-122.

Vargas Contreras, J.A., and A. Hernandez Huerta. 2001. Distribucion altitudinal de la Mastofauna en la Reserva de la Biosfera "El Cielo," Tamaulipas, Mexico. Acta Zoologica Mexicaba (n.s.), 82:83-109.

Vargas-Contreras, J.A., J.R. Herrera-Herrera, and J. Enrique Escobedo-Cabrera. 2004. Noteworthy records of mammal from Campeche, México. Revista Mexicana de Mastozoología, 8:61-69.

Vargas-M., B., J. Ramírez-P., and G. Ceballos. 2008. Murciélagos del Estado de Puebla, México. Revista Mexicana de Mastozoología, 12:59-112.

Vargas-Sandoval, M., I. Bassols-Batalla, and O.J. Polaco. 1991. Un caso de mamíferos dispersores de ácaros en México. Anales de la Escuela Nacional de Ciencias Biologicas, 35:117-122.

Vaughan, T.A. 1954. A new subspecies of bat (*Myotis velifer*) from southeastern California and Arizona. University of Kansas Publications, Museum of Natural History, 7:507-512.

Vaughan, T.A. 1959. Functional morphology of three bats: *Eumops, Myotis, Macrotus*. University of Kansas Publications, Museum of Natural History, 12:1-153.

Vaughan, T.A. 1966. Morphology and flight characteristics of molossid bats. Journal of Mammalogy, 47:249-260.

Vaughan, T.A. 1978. Mammalogy. 2nd ed. W.B. Saunders Company, Philadelphia.

Vaughan, T.A., and T.J. O'Shea. 1976. Roosting ecology of the pallid bat, *Antrozous pallidus*. Journal of Mammalogy, 57:19-42.

Vázquez, H.L.B. 1997. Dieta y demografía de una comunidad de pequeños roedores en dos hábitats contrastantes. Tesis de licenciatura, División de Ciencias Biológicas y Ambientales, Universidad de Guadalajara, Guadalajara, Jalisco.

Vazquez, L.B., G.N. Cameron, and R.A. Medellín. 1999-2000. Hábitos alimentarios y biología poblacional de dos especies de roedores en el occidente de México. Revista Mexicana de Mastozoología, 4:5-21.

Vázquez, L.B., G.N. Cameron, and R.A. Medellín. 2001. *Peromyscus aztecus*. Mammalian Species, 649:1-4.

Vázquez-Domínguez, E., G. Ceballos, and J. Cruzado. 2004. Extirpation of an insular subspecies by a single introduced cat: The case of the endemic deer mouse *Peromyscus guardia* on Estanque Island, Mexico. Oryx, 38(3):347-350.

Vázquez-Yanes, C., A. Orozco, G. Francois, and L. Trejo. 1975. Observations on seed dispersal bats in a tropical humid region in Veracruz, Mexico. Biotropica, 7:73-76.

Veal, R., and W. Caire. 1979. *Peromyscus eremicus*. Mammalian Species, 118:1-6.

Vehrencamp, S.L., F.G. Stiles, and J.W. Bradbury. 1977. Observations on the foraging behavior and avian prey of the Neotropical carnivorous bat, Vampyrum spectrum. Journal of Mammalogy, 58:469-478.

Velarde, E., and R. Medellín. 1981. Predation on *Myotis vivesi*, the Baja California fishing bat, by the barn owl, Tyto alba. Abstracts of the 12th Annual North American Symposium on Bat Research, Cornell University, Ithaca, NY.

Velázquez, A. 1988. Especies y hábitats en peligro de extinción: el caso del conejo de los volcanes. Información Científica y Tecnológica, 10:45-49.

Velázquez, A. 1993. Landscape Ecology of Tláloc and Pelado Volcanoes, Mexico; with Special Reference to the Volcano Rabbit (*Romerolagus diazi*), Its Habitat, Ecology and Conservation. ITC Publication No. 16. Enschede, The Netherlands.

Velázquez, A., J.F. Mas, R. Mayorga-Saucedo, J.R. Díaz, C. Alcántara, R. Castro, T. Fernández, J.L. Palacio, G. Bocco, G. Gomez-Rodríguez, L. Luna-González, I. Trejo, J. López-García, M. Palma, A. Peralta, J. Prado-Molina, and F. González-Medrano. 2002. Estado actual y dinámica de los recursos forestales de México. Biodiversitas (CONABIO, México), 41:8-15.

Velázquez, A., F.J. Romero, and L. León. 1991. Estudios sobre la distribución y fragmentación del hábitat del zacatuche (*Romerolagus diazi* Ferrari-Pérez, 1893). Resumen 63. Pp. 37-38, in: I Congreso Nacional de Mastozoología, Asociación Mexicana de Mastozoología A.C., 7-9 noviembre 1991, Xalapa, Veracruz. México.

Velázquez, A., J.F. Mas, J.L. Palacio, J.R. Díaz, R. Mayorga, C. Alcántara, R. Castro, and T. Fernández. 2002. Análisis de cambio de uso del suelo. Informe técnico. Convenio INE-Instituto de Geografía, UNAM.

Velázquez, V., and S. Reyes. 1976. El Venado Bura en el Sur de la Península de Baja California, México. Boletín de Fauna No. 7. Secretaría de Agricultura y Ganadería.

Verts. B.J. 1967. The Biology of the Striped Skunk. University of Illinois Press, Urbana.

Verts, B.J., L.N. Carraway, and A. Kinlaw. 2001. *Spilogale gracilis*. Mammalian Species, 674:1-10.

Viale, D. 1984. The salt excretive function of the skin and changing coloration in Cetacea. Annuaire de l'Institut Océanographique de Paris, 60:87-93.

Vianna, J.A., R.K. Bonde, S. Caballero, J.P. Giraldo, R.P. Lima, A. Clark, M. Marmontel, B. Morales-Vela, M.J. de Sousa, L. Parr, M.A. Rodriguez-Lopez, A.A. Mignucci-Giannoni, J.A. Powell, and F.R. Santos. 2006. Phylogeography, phylogeny and hybridization in trichechid sirenians: Implications for manatee conservation. Molecular Ecology, 15:433-447.

Vickers, W.T. 1984. The faunal components of lowland South American hunting kills. Interciencia, 9:366-376.

Vickers, W.T. 1991. Hunting yields and game composition over ten years in an Amazon Indian territory. Pp. 53-81, in: Neotropical Wildlife Use and Conservation (J.G. Robinson and K.H. Redford, eds.). University of Chicago Press, Chicago.

Vidal, O. 1990. Population biology and exploitation of the vaquita *Phocoena sinus*. Report SC/42/SM24 to the International Whaling Commission, Amsterdam, The Netherlands.

Vidal, O. 1991. Catalog of Osteological Collections of Aquatic Mammals from Mexico. National Marine Fisheries Service, Seattle.

Vidal, O., and L.T. Findley. 1989. Aquatic mammals of Colombia: Diversity and distribution. In: 8th Biennial Conference on the Biology of Marine Mammals. California.

Vidal, O., L.T. Findley, and S. Leatherwood. 1993. Annotated checklist of the marine mammals of the Gulf of California. Proceedings of the San Diego Society of Natural History, 28:1-11.

Vila, U., and J. Castroviejo. 1990. Ecología del lobo en La Cabrera (León) y La Carballeda (Zamora). In: El lobo (*Canis lupus*) en España (J.C. Blanco, L. Cuesta, and S. Reig, eds.). ICONA.

Vilchez, R. 1978. La Pesca en la Crónica de los Siglos XVI–XVII y XVIII. Introducción, Recopilación y Comentarios. Departamento de Pesca, México.

Villa, C.B., and V.J. Sosa. 1984. Algunos aspectos reproductivos de la tuza *Pappogeomys tylorhinus tylorhinus* (Rodentia: Geomyidae) en el norte de la Ciudad de México. Anales del Instituto de Biología, UNAM, México, Serie Zoología, 54:199-205.

Villa-R., B. 1941. Una nueva rata de campo (*Tylomys gymnurus* sp. nov.). Anales del Instituto de Biología, UNAM, México, Serie Zoología, 12:763-766.

Villa-R., B. 1943. Algunos aspectos de la ecologia de *C. adocetus arceliae* Villa R. Anales del Instituto de Biologia, UNAM, México, Serie Zoología, 14:285-290.

Villa-R., B. 1948. Mamíferos del Soconusco, Chiapas. Anales del Instituto de Biología, UNAM, México, Serie Zoología, 19:485-528.

Villa-R., B. 1950. El murciélago blanco, género *Diclidurus*, por primera vez en México. Anales del Instituto de Biología, UNAM, México, Serie Zoología, 21:435-437.

Villa-R., B. 1953. Mamíferos silvestres del Valle de México. Anales del Instituto de Biología, UNAM, México, Serie Zoología, 23:269-492.

Villa-R., B. 1954. Distribución actual de los castores en México. Anales del Instituto de Biología, UNAM, México, Serie Zoología, 25:451-461.

Villa-R., B. 1955. El murciélago colorado de seminola (*Lasiurus borealis seminolus* Rhoads, 1895) en México. Anales del Instituto de Biología, UNAM, México, Serie Zoología, 26:237-238.

Villa-R., B. 1956. *Tadarida brasiliensis mexicana* (Sassure), el murciélago guaner, es una subespecie migratoria. Acta Zoologica Mexicana, 1:1-11.

Villa-R., B. 1959a. Mamíferos de caza. Pp. 123-148, in: Los Recursos Naturales del Sureste y su Aprovechamiento (E. Beltran, ed.). Instituto Mexicano de Recursos Naturales Renovables.

Villa-R., B. 1959b. *Pteronotus davyi fulvus*. El murciélago de espaldas desnudas en el Norte de Sonora, México. Anales del Instituto de Biología, UNAM, México, Serie Zoología, 29:375-378.

Villa-R., B. 1960. La Isla Socorro. Vertebrados terrestres. Monografías Instituto de Geofísica, 2:203-216.

Villa-R., B. 1961. Combate contra los coyotes y lobos en el Norte de México. Anales del Instituto de Biología, UNAM, México, Serie Zoología, 31:463-499.

Villa-R., B. 1963. *Thyroptera tricolor albiventer* (Tomes) el murciélago discóforo de la familia Thyropteridae, nueva para México, en el Sur del estado de Veracruz. Revista de la Sociedad Mexicana de Historia Natural, 24:45-48.

Villa-R., B. 1967. Los Murciélagos de México. Su Importancia en la Economía y la Salubridad. Su Clasificación Sistemática. Instituto de Biología, Universidad Nacional Autónoma de México.

Villa-R., B. 1971. Algunos mamíferos de la región de Charnela, Jalisco, México. Anales del Instituto de Biología, UNAM, México, Serie Zoología, 42:99-106.

Villa-R., B. 1978. Especies mexicanas silvestres en peligro de extinción. Anales del Instituto de Biología, UNAM, México, Serie Zoología, 49(1):303-320.

Villa-R., B. 1979. Algunas aves y la rata noruega *Rattus norvegicus* "versus" el murciélago insulano Pizonyx vivesi en las islas del Mar de Cortés, México. Anales del Instituto de Biología, UNAM, México, Serie Zoología, 50:729-736.

Villa-R., B., and F.A. Cervantes. 2003. Los Mamíferos de México. Grupo Editorial Iberoamérica e Instituto de Biología, UNAM, México.

Villa-R., B., and E.L. Cockrum. 1962. Migration in the guano bat *Tadarida brasiliensis mexicana* (Saussure). Journal of Mammalogy, 43:43-64.

Villa-R., B., and J. Ramírez Pulido. 1968. Diclidurus virgo Thomas, el murciélago blanco, en la costa de Nayarit. Anales del Instituto de Biología, UNAM, México, Serie Zoología, 1:155-158.

Villa-R., B., and M. Villa-C. 1971. Observaciones acerca de algunos murciélagos del norte de Argentina, especialmente de la biología del vampiro *Desmodus rotundus*. Anales del Instituto de Biología, UNAM, México, Serie Zoología, 42:107-148.

Villa-R., B., J.P. Gallo, and B. LeBoeuf, B. 1986. La foca monje *Monachus tropicalis* (Mammalia: pinnipedia) definitivamente extinguida en México. Anales del Instituto de Biología, UNAM, México. Serie Zoología, 2:573-588.

Villada, M. 1869-1870. Apuntes para la mamalogía mexicana. Memoria. La Naturaleza, 1:290-298.

Villalpando-R., J.A., and J. Arroyo-Cabrales. 1996. Una nueva localidad para *Rhogeessa mira* LaVal, 1973 (Chiroptera: Vespertilionidae) en la cuenca baja del Rio Balsas, Michoacán, México. Vertebrata Mexicana, 2:9-11.

Villarreal, J. 1986. Administración de un rancho cinegético de venado cola blanca (*Odocoileus virginianus texanus*) en el noreste de México. Pp. 139-201, in: I Simposio sobre el Venado en México, Universidad Nacional Autónoma de México, México.

Vorhies, C.T., and W.P. Taylor. 1922. Life history of the kangaroo rat *Dipodomys spectabilis spectabilis* Merriam. U.S. Department of Agricultural Bulletin, 1091:1-40.

Vorhies, C.T., and W.P. Taylor. 1940. Life history and ecology of the white-throated wood rat, *Neotoma albigula albigula* Hartley, in relation to grazing in Arizona. Technical Bulletin of the Agricultural Experiment Station, University of Arizona, 86:453-529.

Voss, R.S. 1988. Systematics and ecology of Ichthomyine rodents (Muroidea): Patterns of morphological evolution in a small adaptive radiation. Bulletin of the American Museum of Natural History, 188:259-493.

Voss, R.S., and S.A. Jansa. 2003. Phylogenetic studies on Didelphid marsupials. ii. Nonmolecular data and new irbp sequences: separate and combined analyses of didelphine relationships with denser taxon sampling. Bulletin of the American Museum of Natural History, 276:1-82.

Voss, R.S., and A.V. Linzey. 1981. Comparative gross morphology of male accessory glands among Neotropical Muridae (Mammalia: Rodentia) with comments on systematic implications. Ann Arbor, Museum of Zoology, University of Michigan, 159:1-41.

Wada, S., T. Kobayashi, and K. Numachi. 1991. Genetic variability and differentiation of mitochondrial DNA in minke whales. Reports of the International Whaling Commission (special issue 13), 203-215.

Wade, P.R. 1991. Absolute abundance estimates of cetaceans in the eastern tropical Pacific. Ninth Biennial Conference on the Biology of Marine Mammals. December. Chicago.

Wade, P.R., and T. Gerrodette. 1992. Estimates of dolphin abundance in the eastern tropical Pacific: Preliminary analysis of five years of data. Reports of the International Whaling Commission, 42:53-539.

Wade-Smith, J., and B.J. Verts. 1982. Mephitis mephitis. Mammalian Species, 173:1-7.

Waggoner, K.V. 1975. The effect of strip-mining and reclamation on small mammal communities. Master's thesis, Texas A&M University, College Station.

Wagner, J.A. 1840. Die Såugethiere in Abbildungen nach der Natur. Suppl. l., Abt. Die Affen und Flederthiere, München.

Wagner, J.W., J.L. Patton, A.L. Gardner, and R.J. Baker. 1974. Karyotype analysis of twenty-one species of molossid bats (Chiroptera: Molossidae). Canadian Journal of Genetics and Cytology, 16:165-176.

Wahlert, J. H. 1985. Skull morphology and relationships of geomyid rodents. American Museum Novitates, 2812:1-20.

Wakelyn, L.A. 1984. Analysis and comparison of existing and historic bighorn sheep ranges in Colorado. Master's thesis, Colorado State University, Fort Collins.

Walker, E.P. 1964. Mammals of the World. Johns Hopkins Press, Baltimore.

Walker, E.P. 1991. Mammals of the World. 2nd ed. Johns Hopkins University Press, Baltimore.

Walker, W.A. 1981. Geographical variation in morphology and biology of bottlenose dolphin (*Tursiops*) in the eastern North Pacific. National Oceanic and Atmospheric Administration Administrative Report No. LJ-81-03C. NMFS/Southwest Fisheries Center, La Jolla, CA.

Walker, W.A., S. Leatherwood, K.R. Coodeich, W.F. Perrin, and R.K. Strooud. 1986. Geographical variation and biology of the Pacific white-sided dolphin, *Lagenorhynchus obliquidens*, in the northeastern Pacific. Pp. 441-465, in: Research on Dolphins (M.M. Bryden and R.J. Harrison, eds.). Clarendon Press, Oxford.

Walkins, L.C. 1972. Nycticeius humeralis. Mammalian Species, 23:1-4.

Walton, D.W., and N.J. Siegel. 1966. The histology of the pararhinal glands of the pallid bat, *Antrozous pallidus*. Journal of Mammalogy, 47:357-360.

Walley, H.D., and W.L. Jarvis. 1972. Longevity record for *Pipistrellus subflavus*. Transactions of the Illinois State Academy of Science, 64:305.

Wallmo, O.C. 1981. Mule and black-tailed deer distribution and habitats. Pp. 366-386, in: Mule and Black-tailed Deer of North America (O.C. Wallmo, ed.). Wildlife Management Institute, University of Nebraska Press, Lincoln.

Wang, H.G., R.D. Owen, C. Sánchez-Hernández, and M. de L. Romero-Almaraz. 2003. Ecological characterization of bat species distributions in Michoacán, México, using a geographic information system. Global Ecology and Biogeography, 12:65-85.

Ward, H.L. 1887. Notes on the life-history of *Monachus tropicalis*, the West Indian seal. The American Naturalist, 21:257-264.

Warner, J.W. 1976. Chromosomal variation in the plains woodrat: Geographical distribution of three chromosomal morphs. Evolution, 30:593-598.

Warner, J.W., J.L. Patton, A.L. Gardner, and R.J. Baker. 1974. Karyotypic analysis of twenty-one species of molossid bats (Molossidae: Chiroptera). Canadian Journal of Genetic Cytology, 16:165-176.

Warner, R.M. 1982. *Myotis volans*. Mammalian Species, 191:1-3.

Warner, R.M. 1985. Interspecific and temporal dietary variation in an Arizona bat community. Journal of Mammalogy, 66:45-51.

Warner, R.M., and N.J. Czaplewski. 1981. Presence of *Myotis auriculus* (Vespetilionidae) in northern Arizona. The Southwestern Naturalist, 26:439-440.

Warner, R.M., and N.J. Czaplewski. 1984. *Myotis volans*. Mammalian Species, 224:1-4.

Waterhouse, G.R. 1838. Original description of *Lepus bachmani*. Proceedings of the Zoological Society of London, 103-105.

Watkins, L.C. 1977. *Euderma maculatum*. Mammalian Species, 77:1-4.

Watkins, L.C., J.K. Jones Jr. and H.H. Genoways. 1972. Bats of Jalisco, Mexico. Special Publications of the Museum, Texas Tech University, 1:1-44.

Watson, L. 1981. Sea Guide to the Whales of the World. E.P. Dutton, New York.

Watts, E.S., and V. Rico-Gray. 1987. Los primates de la Península de Yucatán, México: estudio preliminar sobre su distribución y estado de conservación. Biotica, 12:57-66.

Waver, R.H. 1965. Genus and species of shrew new for Utah. Journal of Mammalogy, 46:496.

Webb, R.G., and R.H. Baker. 1962. Terrestrial vertebrates of the Pueblo Nuevo area of southwestern Durango, Mexico. The American Midland Naturalist, 68:325-333.

Webb, R.G., and R.H. Baker. 1969. Vertebrados terrestres del suroeste de Oaxaca. Anales del Instituto de Biología, UNAM, México, Serie Zoología, 40:139-152.

Webb, R.G., and R.H. Baker. 1984. Terrestrial vertebrates of the Cerro Mohinora region, Chihuahua, Mexico. The Southwestern Naturalist, 29:243-246.

Webb, R.G., A. Martínez, and R.H. Baker. 1980. Algunos anfibios, reptiles y mamíferos del Mineral del Tigre, Nayarit. Anales del Instituto de Biología, UNAM, México, Serie Zoología, 51:699-702.

Weber, C., and L. de Roguin. 1983. Notes inèdites de Henri de Saussure sur les types de deux rongeurs du Mexique: *Microtus m. mexicanus* (Saussure) et *Reithrodontomys s. sumichrasti* (Saussure, 1861). Review Suisse Zoology, 90:747-750.

Weber, M. 1992. Valoración clínica del efecto de la Ivermectina, contra *Cephenemyia* spp. en venados cola blanca. Veterinaria México, 23:239-242.

Weber, M. 1993. Ganadería de ciervos: ¿alternativa de producción animal o amenaza a la conservación de la fauna nativa? Agrociencia, 3:99-112.

Weber, M. 2005. Ecology and conservation of sympatric tropical deer populations in the Greater Calakmul Region, Campeche, Mexico. Unpublished PhD dissertation, Durham University, Durham.

Weber, M., and C. Galindo-Leal. 1992. Distocia en venado cola blanca: Informe de un caso reincidente. Veterinaria México, 23:79-81.

Weber, M., and C. Galindo-Leal. 1994. History, needs and perspectives of deer research and conservation in Mexico. P. 40, in: Recent Developments in Deer Biology (J.A. Milne, ed.). Proceedings of the Third International Congress on the Biology of Deer. Edinburgh.

Weber, M., P. Rosas-Becerril., A. Morales-García, and C. Galindo-Leal. 1995. Biología reproductiva del venado cola blanca en Durango, México. Pp. 111-127, in: Ecología y Manejo del Venado Cola Blanca en México y Costa Rica (C. Vaughan and M. Rodríguez, eds.). Editorial de la Universidad Nacional, Heredia, Costa Rica.

Webster, W.D. 1983. Systematics and evolution of bats of the genus *Glossophaga*. PhD dissertation, Texas Tech University, Lubbock.

Webster, W.D. 1984. *Glossophaga leachii*. Mammalian Species, 226:1-3.

Webster, W.D., and J.K. Jones Jr. 1980. Taxonomic and nomenclatorial notes on bats of the genus *Glossophaga* in North America, with descriptions of a new species. Occasional Papers of the Museum, Texas Tech University, 71:1-12.

Webster, W.D., and J.K. Jones Jr. 1982a. *Artibeus aztecus*. Mammalian Species, 177:1-3.

Webster, W.D., and J.K. Jones Jr. 1982b. *Artibeus toltecus*. Mammalian Species, 178:1-3.

Webster, W.D., and J.K. Jones Jr. 1982c. *Reithrodontomys megalotis*. Mammalian Species, 167:1-5.

Webster, W.D., and J.K. Jones Jr. 1982d. A new subspecies of *Glossophaga comissarisi* (Chiroptera: Phyllostomidae) from western Mexico. Occasional Papers of the Museum, Texas Tech University, 76:1-6.

Webster, W.D., and J.K. Jones Jr. 1983. *Artibeus hirsutus* and *Artibeus inopinatus*. Mammalian Species, 199:1-3.

Webster, W.D., and J.K. Jones Jr. 1984. A new subspecies of *Glossophaga mexicana* (Chiroptera: Phyllostomidae) from southern Mexico. Occasional Papers of the Museum, Texas Tech University, 91:1-5.

Webster, W.D., and J.K. Jones Jr. 1985. *Glossophaga mexicana*. Mammalian Species, 245:1-2.

Webster, W.D., and J.K. Jones Jr. 1987. A new subspecies of *Glossophaga comissarisi* (Chiroptera: Phyllostomidae) from South America. Occasional Papers of the Museum, Texas Tech University, 109:1-6.

Webster, D.B., and M. Webster. 1971. Adaptive value of hearing and vision in kangaroo rat predator avoidance. Brain, Behaviour and Evolution, 4:310-322.

Webster, D., J.K. Jones Jr., and R.J. Baker. 1980. *Lasiurus intermedius*. Mammalian Species, 132:1-3.

Webster, W.D., L.W. Robbins, R.L. Robbins, and R.J. Baker. 1982. Comments on the status of *Musonycteris harrisoni* (Chiroptera: Phyllostomatidae). Occasional Papers of the Museum, Texas Tech University, 78:1-5.

Weigl, P.D., M.A. Steele, L.J. Sherman, J.C. Ha, and T.L. Sharpe. 1989. The ecology of the fox squirrel (*Sciurus niger*) in North Carolina: Implications for survival in the Southeast. Bulletin of Tall Timbers Research Station, 24:1-93.

Weir, B.J. 1974. Reproductive characteristics of hystricomorph rodents. Symposium of the Zoological Society, London, 34:265-301.

Weller, D.W., and A.J. Schiro. 1996. First account of a humpback whale (*Megaptera novaeangliae*) in Texas waters, with a reevaluation of historical records from the Gulf of Mexico. Marine Mammal Science, 12:133-137.

Wells, R.S., A.B. Irvine, and M.D. Scott. 1980. The social ecology of inshore odontocetes. Pp. 263-317, in: Cetacean Behavior: Mechanisms and Processes (L.M. Herman, ed.). John Wiley and Sons, Inc., New York.

Wells, R.S., B.G. Würsig, and K.S. Norris. 1981. A survey of the marine mammals of the upper Gulf of California, Mexico, with an assessment of the status of *Phocoena sinus*. National Technical Information Service, PB 81-168792.

Werbitsky, D., and C.W. Kilpatrick. 1987. Genetic variation and genetic differentiation among allopatric populations of *Megadontomys*. Journal of Mammalogy, 68:305-312.

Wetterer, A.L., M.V. Rockman, and N.B. Simmons. 2000. Phylogeny of phyllostomid bats (Mammalia: Chiroptera): Data from diverse morphological systems, sex chromosomes, and restriction sites. Bulletin of the American Museum of Natural History, 248:1-200.

Wettstein, P.J., M. Strausbauch, T. Lamb, J. States, R. Chakraborty, L. Jin, and R. Riblet. 1995. Phylogeny of 6 *Sciurus aberti* subspecies based on nucleotide sequences of cytochrome b. Molecular Phylogenetics and Evolution, 4:150-162.

Wetzel, R.M. 1975. The species of *Tamandua* Gray (Edentata, Myrmecophagidae). Proceedings of the Biological Society of Washington, 88:95-112.

Wetzel, R. M. 1977. The Chacoan peccary *Catagonus wagneri* (Rusconi). Bulletin of Carnegie Museum, Natural History. 3:1-36.

Wetzel, R.M. 1980. Revision of the naked-tailed armadillos, genus *Cabassous* McMurtrie. Annals of Carnegie Museum, 49:323-357.

Wetzel, R.M. 1982. Systematics, distribution, ecology and conservation of South American edentates. Pp. 345-375, in: Mammalian Biology in South America (M.A. Mares and H.H. Genoways, eds.). Special Publication Series, Pymatuning Laboratory of Ecology, University of Pittsburgh, Linesville.

Wetzel, R.M. 1985a. Taxonomy and distribution of armadillos, Dasypodidae. Pp. 23-46, in: The Evolution and Ecology of Armadillos, Sloths, and Vermilinguas (G.G. Montgomery, ed.). Smithsonian Institution Press, Washington, DC.

Wetzel, R.M. 1985b. The identification and distribution of recent Xenarthra (Mammalia, Edentata). Pp. 5-21, in: The Evolution and Ecology of Armadillos, Sloths, and Vermilinguas (G.G. Montgomery, ed.). Smithsonian Institution Press, Washington, DC.

Whitaker, J.O. 1973. External parasites of bats of Indiana. Proceedings of the Indiana Academy of Science, 83:469-472.

Whitaker, J.O. 1980. The Audubon Society, Field Guide to North American Mammals. Alfred A. Knopf, Inc., New York.

Whitaker, J.O., and W.A. Miller. 1974. Rabies in bats of Indiana, 1968-1972. Proceedings of the Indiana Academy of Science, 83:469-472.

Whitaker, J.O., Jr. 1972. Food habits of bats from Indiana. Canadian Journal of Zoology, 50:877-883.

Whitaker, J.O., Jr., 1974. *Cryptotis parva.* Mammalian Species, 43:1-8.

Whitaker, J.O., Jr. 1980. The Audubon Society Field Guide to North American Mammals. Alfred A. Knopf, Inc., New York.

Whitaker, J.O., Jr., and D.A. Easterla. 1975. Ectoparasites of bats from Big Bend National Park, Texas. The Southwestern Naturalist, 20:241-254.

Whitaker, J.O., Jr., and J.S. Findley. 1980. Food eaten by some bats from Costa Rica and Panama. Journal of Mammalogy, 61:540-544.

Whitaker, J.O., Jr., and R.E. Mumford. 1972a. Note on occurrence and reproduction of bats in Indiana. Proceedings of the Indiana Academy of Science, 81:376-383.

Whitaker, J.O., Jr., and R.E. Mumford. 1972b. Food and ectoparasites of Indiana shrews. Journal of Mammalogy, 53:329-335.

Whitaker, J.O., Jr., and N. Wilson, 1974. Host and distribution lists of mites (Acari), parasitic and phoretic, and in the hair of wild mammals of North America, north of Mexico. The American Midland Naturalist, 91:1-67.

Whitaker, J.O., Jr., J.I. Glendining, and W.I. Wren. 1991. Ectoparasites of *Sorex saussurei* from Michoacan, Mexico. The Southwestern Naturalist, 36:114-115.

Whitaker, J.O., Jr., W.J. Wrenn, and R.E. Lewis. 1993. Parasites. Pp. 386-478, in: Biology of the Heteromyidae (H.H. Genoways and J.H. Brown, eds.). Special Publication, American Society of Mammalogists.

Whitehead, H. 1990. Assessing sperm whale populations using natural markings: Recent progress. Pp. 377-382, in: Individual Recognition of Cetaceans: Use of Photo-Identification and Other Techniques to Estimate Population Parametres (P.S. Hammond, S.A. Mizroch, and G.P. Donovan, eds.). Special Issue 12, Reports of the International Whaling Commission.

Whitehead, H., and C. Glass. 1985. Orcas (killer whales) attack humpback whales. Journal of Mammalogy, 66:183-185.

Whitford, W.G. 1976. Temporal fluctuations in density and diversity of desert rodent populations. Journal of Mammalogy, 57:351-369.

Wiegert, R.G. 1972. Avian versus mammalian predation on a population of cotton rats. Journal of Wildlife Management, 36:1322-1327.

Wiegert, R.G., and J.C. Mayenschein. 1966. Distribution and trap response of a small wild population of cotton rats (*Sigmodon hispidus*). Journal of Mammalogy, 47:118-120.

Wilkins, K.T. 1986. *Reithrodontomys montanus.* Mammalian Species, 257:1-5.

Wilkins, K.T. 1989. *Tadarida brasiliensis.* Mammalian Species, 331:1-10.

Wilkins, K.T. 1991. *Lasiurus seminolus.* Mammalian Species, 280:1-5.

Wilkins, K.T., and C.D. Swearing. 1990. Factors affecting the historical distribution and modern geographic variation in the south Texas pocket gopher, *Geomys personatus.* The American Midland Naturalist, 124:57-72.

Wilkinson, G.S., and T.H. Fleming. 1996. Migration and evolution of lesser long nosed bats *Leptonycteris curasoae*, inferred from mitochondrial DNA. Molecular Ecology, 5:329-339.

Williams, C.F. 1986. Social organization of the bat, *Carollia perspicillata* (Chiroptera: Phyllostomidae). Ethology, 71:265-282.

Williams, D.F. 1978a. Karyological affinities of the species groups of silky pocket mice (Rodentia, Heteromyidae). Journal of Mammalogy, 59:599-612.

Williams, D.F. 1978b. Systematics and ecogeographic variation of the Apache pocket mouse (Rodentia: Heteromyidae). Bulletin of Carnegie Museum of Natural History, 10:1-57.

Williams, D.F., and J.S. Findley. 1979. Sexual size dimorphism in vespertilionid bats. The American Midland Naturalist, 102:113-126.

Williams, D.F., H.H. Genoways, and J.K. Braun. 1993a. Taxonomy and systematics. Pp. 38-190, in: Biology of the Heteromidae (H.H. Genoways and J.H. Brown, eds.). Special Publication 10, American Society of Mammalogists.

Williams, D.F., H.H. Genoways, and J.K. Braun. 1993b. Biogeography. Pp. 319-356, in: Biology of the Heteromyidae (H.H. Genoways and J.H. Brown, eds.). Special Publication 10, American Society of Mammalogists.

Williams, M.A.J., D.L. Dunkerley, P. De Dekker, A.P. Kershaw, and T. Stokes. 1993. Quaternary Environments. Arnold, London.

Williams, S.L. 1982a. *Geomys personatus.* Mammalian Species, 170:1-5.

Williams, S.L. 1982b. Phalli of recent genera and species of the family Geomyidae (Mammalia: Rodentia). Bulletin of Carnegie Museum of Natural History, 20:1-62.

Williams, S.L., and R.H. Baker 1974. *Geomys arenarius.* Mammalian Species, 36:1-6.

Williams, S.L., and H.H. Genoways. 1977. Morphometric variation in the tropical pocket gopher (*Geomys tropicalis*). Annals of Carnegie Museum of Natural History, 46:245-264.

Williams, S.L., J. Ramirez P., and R.J. Baker. 1985. *Peromyscus alstoni.* Mammalian Species, 242:1-4.

Williams, T.C., L.C. Ireland, and J.M. Williams. 1973. High altitude flights of the free-tailed bat, *Tadarida brasiliensis*, observed with radar. Journal of Mammalogy, 54:807-821.

Willig, M.R. 1983. Composition, microgeographic variation, and sexual dimorphism in Caatingas and Cerrado bat communities from northeast Brazil. Bulletin of Carnegie Museum of Natural History, 23:1-131.

Willig, M.R. 1985. Reproductive patterns of bats from Caatingas and Cerrado biomes in northeast Brazil. Journal of Mammalogy, 66:668-681.

Willis, K.B., M.R. Willig, and J.K. Jones Jr. 1990. *Vampyrodes caraccioli.* Mammalian Species, 359:1-4.

Willner, G.R., G.A. Feldhamer, E.E. Zucker, and J.A. Chapman. 1980. *Ondatra zibethicus.* Mammalian Species, 141:1-8.

Wilson, D.E. 1970. Opossum predation: *Didelphys* on *Philander.* Journal of Mammalogy, 51(2):386-387.

Wilson, D.E. 1971. Food habits of *Micronycteris hirusta* (Chiroptera: Phyllostomatidae). Mammalia, 35:107-110.

Wilson, D.E. 1973a. Bat faunas: A trophic comparison. Systematic Zoology, 22:14-29.

Wilson, D.E. 1973b. Reproduction in Neotropical bats. Periodicum Biologorum, 75:215-217.

Wilson, D.E. 1973c. The systematic status of *Perognathus merriami* Allen. Proceedings of the Biological Society of Washington, 86:175-192.

Wilson, D.E. 1974. Cranial variation in polar bears. Third International Conference on Bears: 447-453.
Wilson, D.E. 1979. Reproductive patterns. Pp. 317-378, in: Biology of Bats of the New World Family Phyllostomatidae, Part III (R.J. Baker, J.K. Jones Jr., and D.C. Carter, eds.). Special Publications of the Museum, Texas Tech University, 16:1-441.
Wilson, D.E. 1983a. Checklist of mammals. Pp. 443-447, in: Costa Rican Natural History (D.H. Janzen, ed.). University of Chicago Press, Chicago.
Wilson, D.E. 1983b. *Myotis nigricans* (murciélago pardo, black myotis). Pp. 477-478, in: Costa Rican Natural History (D.H. Janzen, ed.). University of Chicago Press, Chicago.
Wilson, D.E. 1985. New mammals records from Sinaloa: *Nyctinimops aurispinosa* and *Onychomys*. The Southwestern Naturalist, 30:323-324.
Wilson, D.E. 1991. Mammals of the Tres Marias Islands. Bulletin of the American Museum of Natural History, 206:214-250.
Wilson, D.E., and J.S. Findley. 1970. Reproductive cycle of a Neotropical insectivorous bat, *Myotis nigricans*. Nature, 225:1155.
Wilson, D.E., and J.S. Findley. 1971. Spermatogenesis in some Neotropical species of *Myotis*. Journal of Mammalogy, 52:420-426.
Wilson, D.E., and J.S. Findley. 1977. *Thyroptera tricolor*. Mammalian Species, 71:1-3.
Wilson, D.E., and R.K. LaVal. 1974. *Myotis nigricans*. Mammalian Species, 39:1-3.
Wilson, D.E., and D.M. Reeder (eds.). 2005. Mammal Species of the World: A Taxonomic and Geographic Reference. 2nd ed. Smithsonian Institution Press, Washington, DC.
Wilson, D.E., and S. Ruff (eds.). 1999. The Smithsonian Book of North American Mammals. Smithsonian Institution Press, Washington, DC.
Wilson, D.E., R.A. Medellín, D.V. Lanning, and H.T. Arita. 1985. Los murciélagos del noreste de México, con una lista de especies. Acta Zoologica Mexicana (n.s.), 8:1-26.
Wilson, E.O. 1993. The Diversity of Life. Belknap Press, Harvard University, Cambridge, MA.
Wilson, G.A., and K. Zittlau. 2004. Management strategies for minimizing the loss of genetic diversity in wood and plains bison populations at Elk Island National Park. Parks Canada Agency, Species at Risk.
Wimsatt, W.A. 1945. Notes on breeding behavior, pregnancy, and parturition in some vespertilionid bats of the eastern United States. Journal of Mammalogy, 26:23-33.
Winkelman, J.R. 1962a. Mammal records from Guerrero and Michoacan, Mexico. Journal of Mammalogy, 43:108-109.
Winkelman, J.R. 1962b. Additional records of *Mimon cozumelae*. Journal of Mammalogy, 43:112.
Winn, H.E., and L. Olla (eds.) 1979. Behavior of Marine Mammals, vol. 3, Cetaceans. Plenum Press, New York.
Winn, H.E., and N.E. Reichley. 1985. Humpback whale, *Megaptera novaeangliae* (Borowski, 1781). Pp. 241-273, in: Handbook of Marine Mammals (S.H. Ridgway and R. Harrison, eds.). Vol. 3. Academic Press, London.
Wolf, J.B.W., C. Harrod, S. Brunner, S. Salazar, F. Trillmich, and D. Tautz. 2008. Tracing early stages of species differentiation: Ecological, morphological and genetic divergence of Galápagos sea lion populations. BMC Evolutionary Biology, 8:150.
Wolfe, J.L. 1982. *Oryzomys palustris*. Mammalian Species, 176:1-5.
Wolfe, J.L. 1985. Population ecology of the rice rat (*Oryzomys palustris*) in a coastal marsh. Journal of Zoology (London), 205:235-244.
Woloszyn, D., and B.W. Woloszyn. 1982. Los Mamíferos de la Sierra de La Laguna Baja California Sur. Consejo Nacional de Ciencia y Tecnología, México.
Wood, A.E. 1935. Evolution and relationship of the heteromyid rodents with new forms from the tertiary of western North America. Annals Carnegie Museum, 35:73-262.
Wood, J.E. 1954. Food habits of furbearers of the upland post oak region in Texas. Journal of Mammalogy, 35:406-415.
Woodburne, M.O. 1968. The cranial myology and osteology of *Dicotyles tajacu*, the collared peccary, and its bearing on classification. California Academy of Sciences Bulletin, 7:48.
Woodgerd, W. 1964. Population dynamics of bighorn sheep on Wildhorse Island. Journal of Wildlife Management, 28:381-391.
Woodman, N. 2005. Evolution and biogeography of Mexican small-eared shrews of the *Cryptotis mexicana*-group (Insectivora: Soricidae). Pp. 513-524, in: Contribuciones Mastozoologicas en Homenaje a Bernardo Villa (V. Sanchez-Cordero and R. Medellín, eds.). R. A. Instituto de Biología e Instituto de Ecología, Universidad Nacional Autónoma de México, Mexico City.
Woodman, N., and D.A. Croft. 2005. Fossil shrews from Honduras and their significance for Late Glacial evolution in body size (Mammalia: Soricidae) of the genus *Cryptotis*. Fieldiana Geology (n.s.), 51:1-30.
Woodman N., and R.M. Timm. 1993. Intraspecific and interspecific variation in the *Cryptotis nigrescens* species complex of small-eared shrews (Insectivora: Soricidae), with the description of a new species from Colombia. Fieldiana Zoology (n.s.), 74:1-30.
Woodman, N., and R.M. Timm. 1999. Geographic variation and evolutionary relationships among broad-clawed shrews of the *Cryptotis goldmani*-group (Mammalia: Insectivora: Soricidae). Fieldiana Zoology (n.s.), 91:1-35.
Woodman, N., and R. M. Timm. 2000. Taxonomy and evolutionary relationships of Phillips' small-eared shrew, *Cryptotis phillipsii* (Schaldach, 1966), from Oaxaca, Mexico (Mammalia: Insectivora: Soricidae). Proceedings of the Biological Society of Washington, 113: 339-355.
Woods, C.A. 1973. *Erethizon dorsatum*. Mammalian Species, 29:1-6.
Woods, C.A. 1993. Suborder Hystricognathi. Pp. 771-805., in: Mammal Species of the World (D.E. Wilson and D.M. Reeder, eds.). Smithsonian Institute Press, Washington, DC.
Woodsworth, G.C., G.P. Bell, and M.B. Fenton. 1981. Observations on the echolocation, feeding behavior, and habitat use of Euderma maculatum in south-central British Columbia. Canadian Journal of Zoology, 59:1099-1102.
Worthington, D.A. 1970. The karyotype of the brush rabbit *Sylvilagus bachmani*. Mammalian Chromosome Newsletter, 2:21.
Wozencraft, C, 1993. Order Carnivora. Pp. 279-348, in: Mammal Species of the World (D.E. Wilson and D.M. Reeder, eds.). Smithsonian Institution Press, Washington, DC.
Wozencraft, W.C. 2005. Order Carnivora. Pp. 532-628, in: Mammal Species of the World: A Taxonomic and Geographic Reference (D.E. Wilson and D.A. Reeder, eds.). 3rd ed. Johns Hopkins University Press, Baltimore.
Wren, W.J., and R.B. Loomis. 1967. *Otrohinophila*, a new genus of chiggers (Acarina: Trombiculidae), from western North America. Acarologia, 9:152-178.
Wright, D. B. 1989. Phylogenetic relationships of *Catagonus wagneri*: Sister taxa from the Tertiary of North America. Pp. 281-308, in: Advances in Neotropical Mammalogy (K.H. Redford and J. F. Eisenberg, eds.). Sandhill Crane Press, Gainesville, FL.
Wright, R.L., and J.C. de Vas. 1986. Final report on Sonoran pronghorn status in Arizona. Arizona Fish and Game Department, Phoenix.
Wyrtki, K. 1967. Circulation and watermasses in the eastern equatorial Pacific Ocean. International Journal of Oceanology and Limnology, 1(2):117-147.
Ximénez, A., and P. André de C.F. 1992. Ocorrencia de golfinho de dentes rugosos *Steno bredanensis* (LESSSON, 1828) na lagoa da conceicao, ilha de Santa Catarina, Brasil. In: Memorias de la Quinta Reunión de Especialistas en Mamíferos Acuáticos de América del Sur, 28 Septiembre–2 Octubre 1992, Buenos Aires, Argentina.
Ximénez, F. 1722. Historia Natural del Reino de Guatemala. Sociedad Geografica y Historia de Guatemala, Publicaciones Especiales 14, Título VIII, José Pineda Ibarra.

Yates, T.L., and J. Salazar-Bravo. 2005. A revision of *Scapanus latimanus*, with the revalidation of a species of Mexican mole. Pp. 479-496, in: Contribuciones Mastozoológicas en homenaje a Bernardo Villa (V. Sánchez-Cordero and R.A. Medellín, eds.). Instituto de Biología e Instituto de Ecología, UNAM, México.

Yates, T.L., and D.J. Schmidly. 1975. Karyotype of the eastern mole (*Scalopus aquaticus*), with comments on the karyotype of the family Talpidae. Journal of Mammalogy, 56:902-905.

Yates, T.L., and D.J. Schmidly. 1977. Systematics of *Scalopus aquaticus* (Linnaeus, 1758) in Texas and adjacent states. Occasional Papers of the Museum, Texas Tech University, 45:1-36.

Yates, T.L., and D.J. Schmidly. 1978. *Scalopus aquaticus*. Mammalian Species, 105:1-4.

Yates, T.L., D.J. Schmidly, and K.L. Culbertson. 1976. Silver-haired bat in Mexico. Journal of Mammalogy, 57:205.

Yensen, E., and W.H. Clark. 1986. Records of desert shrews (*Notiosorex crawfordi*) from Baja California, Mexico. The Southwestern Naturalist, 31:529-530.

Yensen, E., and T. Tarifa. 2003. *Galictis vitatta*. Mammalian Species, 727:1-8.

Yensen, E., and M. Valdés-Alarcón. 1999. Family Sciuridae. Pp. 239-320, in: Mamíferos del Noroeste de México (S.T. Álvarez-Castañeda and J.L. Patton, eds.). CIBNOR, La Paz, México.

Yoakum, J. 1991. Literature Review of the Pronghorn: A Bibliography with Key Word and Reference Citation. U.S. Department of the Interior, Bureau of Land Management, Reno, NV.

Yoakum, J. 1994. Water Requirements for Pronghorn. 16th Proceedings of the Antelope Workshop, Lawrence, KS.

Yochem, P.K., and S. Leatherwood. 1985. Blue whale—*Balaenoptera musculus*. Pp. 193-240, in: Handbook of Marine Mammals, vol. 3, The Sirenians and Baleen Whales (S.H. Ridgway and R. Harrison, eds.). Academic Press, Orlando, FL.

Young, C.J., and J.K. Jones Jr. 1982. *Spermophilus mexicanus*. Mammalian Species, 164:1-4.

Young, C.J., and J.K. Jones Jr. 1983. *Peromyscus yucatanicus*. Mammalian Species, 196:1-3.

Young. C.J., and J.K. Jones Jr. 1984. *Reitrodontomys gracilis*. Mammalian Species, 218:1-3.

Young, P.J. 1979. Summer activity patterns of rock squirrels in Central Texas. Unpublished master's thesis, Texas Tech University, Lubbock.

Young, S.P., and E.A. Goldman. 1946. The Puma, Mysterious Cat. American Wildlife Institution, Washington, DC.

Young, S.P., and H. Jackson. 1978. The Clever Coyote. University of Nebraska Press, Lincoln.

Zakrzewski, R.J. 1985. The fossil record. Pp. 1-48, in: Biology of New World Microtus (R.M. Tamarin, ed.). Special Publication 8, American Society of Mammalogists.

Zárate B., E. 1993. Distribución del manatí (*Trichechus manatus*) en la porción sur de Quintana Roo, Mexico. Revista de Investigación Científica 1 (Número Especial SOMMEMA 1), UABCS, 1-11.

Zarza, H., G. Ceballos, and M.A. Steele. 2003. Marmosa canescens. Mammalian Species, 725:1-4.

Zarza, H., C. Chávez, and G. Ceballos. 2007. Uso de hábitat del jaguar a escala regional en un paisaje con actividades humanas en el sur de la Península de Yucatán. Pp. 101-110, in: Conservación y manejo del jaguar en México: estudios de caso y perspectivas (G. Ceballos, C. Chávez, R. List, and H. Zarza, eds.). CONABIO-UNAM-Alianza WWF-Telcel México.

Zarza, H., R. A. Medellín, and S. G. Pérez. 2003. First record of the Yucatan deer mouse, *Peromyscus yucatanicus* (Rodentia: Muridae) from Guatemala. The Southwestern Naturalist, 48:310-312.

Zavala, G. 1992. Estimación poblacional del Venado Cola Blanca (*Odocoileus virginianus*) en la Estación Científica Las Joyas, Reserva de la Biosfera Sierra de Manantlán, Jalisco. Tesis de licenciatura, Facultad de Ciencias Biológicas, Universidad de Guadalajara, Jalisco.

Zavala G., A. 1990. La población del lobo marino común *Zalophus californianus* (Lesson, 1828) en las Islas del Golfo de California, México. Tesis de licenciatura en biología, Facultad de Ciencias, Universidad Nacional Autónoma de México, México.

Zavala G., A. 1993. Biología poblacional del lobo marino de California, *Zalophus californianus californianus* (Lesson 1828), en la región de las Grandes Islas del Golfo de California, México. Tesis de maestría, Facultad de Ciencias, UNAM, México.

Zavala-González, A., and C. Esquivel. 1991. Observaciones y comentarios sobre la interacción demamíferos marinos con las pesquerías litorales en aguas mexicanas. Memorias XVI Reunión Internacional para el Estudio de los Mamíferos Marinos, Nuevo Vallarta, Nay., del 2 al 6 de abril de 1991.

Zavala-González, A., J. Urbán-Ramírez, and C. Esquivel-Macías. 1994. A note on artisanal fisheries interactions with small cetaceans in Mexico. Reports of the International Whaling Commission (special issue 15), 235-237.

Zeiner, D.C., W.F. Laudenslayer Jr., K.E. Mayer, and M. White. 1990. California's Wildlife, vol. III, Mammals. California Department of Fish and Game, Sacramento.

Zeng, Z., and J.H. Brown. 1987. Population ecology of a desert rodent: *Dipodomys merriami* in the Chihuahuan desert. Ecology, 68:1328-1340.

Zetek, J. 1930. The water opossum *Chironectes panamensis*. Journal of Mammalogy, 11:470-471.

Zezulak, D.S., and R.G. Schwab. 1979. A comparison of density, home range, and habitat utilization of bobcat populations at Lava Beds and Joshua Tree National Monuments, California. Pp. 74-79, in: Proceedings of the 1979 Bobcat Research Conference (P.C. Escherich and L. Blum, eds.). National Wildlife Federal Science and Technology Series 6.

Zezulak, D.S., and R.G. Schwab. 1980. Bobcat biology in a Mojave desert community. Federal Aid in Wildlife Restoration Project W-54-R-12, Job IV-4, California Department of Fish and Game, Sacramento.

Zimmerman, E.G. 1965. A comparison of habitat and food of two species of *Microtus*. Journal of Mammalogy, 46:605-612.

Zimmerman, E.G. 1970. Karyology, systematics and chromosomal evolution in the rodent genus Sigmodon. Publications of the Museum, Michigan State University, Biological Series, 4:385-454.

Zimmerman, E.G., and C.W. Kilpatrick. 1978. The genetics of speciation in the rodent genus *Peromyscus*. Evolution, 32:565-579.

Zimmerman, E.G., B.J. Hart, and C.W. Kilpatrick. 1975. Biochemical genetics of the *truei* and *boylii* groups of the genus *Peromyscus* (Rodentia). Comparative Biochemistry Physiology, 52B:541-545.

Index of Scientific Names